MODERN WELDING TECHNOLOGY

MODERN WELDING TECHNOLOGY

Fifth Edition

Howard B. Cary

Welding Engineer, Consultant
Past President
American Welding Society

Prentice
Hall

Upper Saddle River, New Jersey
Columbus, Ohio

Library of Congress Cataloging-in-Publication Data

Cary, Howard B.
    Modern welding technology / Howard B. Cary.—5th ed.
        p. cm.
    Includes bibliographical references and index.
    ISBN 0-13-030913-3 (alk. paper)
    1. Electric welding. 2. Welding. I. Title.
TS227 .C37 2002
671.512—dc21
                                            2001018557

Vice President and Editor in Chief: Stephen Helba
Executive Editor: Ed Francis
Production Editor: Christine M. Buckendahl
Production Coordination: Carlisle Publishers Services
Design Coordinator: Robin G. Chukes
Cover Designer: Galen Ludwick
Cover photo: FPG
Production Manager: Brian Fox
Marketing Manager: Jimmy Stephens

This book was set in Garamond Book by Carlisle Communications, Ltd., and was printed and bounded by R. R. Donnelley & Sons Company. The cover was printed by Phoenix Color Corp.

Prentice-Hall International (UK) Limited, *London*
Prentice-Hall of Australia Pty. Limited, *Sydney*
Prentice-Hall Canada, Inc., *Toronto*
Prentice-Hall Hispanoamericana, S.A., *Mexico*
Prentice-Hall of India Private Limited, *New Delhi*
Prentice-Hall of Japan, Inc., *Tokyo*
Prentice-Hall Singapore Pte. Ltd.
Editora Prentice-Hall do Brasil, Ltda., *Rio de Janeiro*

10 9 8 7 6 5 4 3 2 1
ISBN 0-13-030913-3

PREFACE

Welding continues to be the preferred method of permanently joining metal parts. As welding becomes more computer-driven, the technology becomes more complex. Worldwide, welding continues to grow, and that growth is dependent upon the growth of the steel and other metal industries. In the United States, a major change has been replacing the old-faithful stick welding, used for so many years, with wire welding.

The need to improve weld quality and reduce welding costs continues unabatedly. This is the highest priority because of improved materials and fabricating methods. Semi-automatic welding has largely replaced manual welding, and automatic and robotic welding are being widely accepted in the industry. Adaptive control is rapidly becoming more widely used. More powerful computer controls and more rugged sensors are becoming popular. All of this has helped take the human welder farther away from the arc and fumes, and helped clean up the welder's environment.

Welding power sources have experienced a revolution. The faithful motor generator welding machine is almost extinct. The buzz box transformer welding machine is extinct. These have been replaced by the new inverter power source, which offers many advantages. The inverter is smaller, lighter in weight, and very controllable; with new features it is becoming accepted for most applications.

Some welding processes have become more popular and others more refined. For example, the laser is more widely used, especially for cutting, and a new process, stir friction welding, is starting to be used to join aluminum for automotive and space applications.

Throughout the world many new alloys are being developed. Metals compete with plastics, composites, ceramics, and any material that will serve the need. The end result is the most economical material for a given application. Many new steels and alloys are being welded today,

including higher strength thermomechanically processed steels. Steels with lower carbon and lower impurity elements are available with high strengths based on the particular heat treatment. New steels for high-temperature applications have been developed. New grades of stainless steel that combat corrosion are appearing. New aluminums containing lithium and other elements are being utilized in the aircraft industry. Nonmetallic materials are advancing. Plastics have been greatly improved, and there are now composite beams available to build bridges. Ultimately, the most suitable material for the lowest price will be used for every application. The welding industry will determine the welding method.

Welding education and training are changing. Today there is less emphasis on skill training for stick welding, but more emphasis on technology training. We must be able to select the proper application of welding to increase productivity. A more thorough understanding is needed. That is the purpose of this book.

A major breakthrough has been accomplished by the joint American Welding Society (AWS) and the Welding Research Council program for providing the optimum way to make a quality weld. Standard welding procedures have been issued that show the preferred way to make a particular weld. This should greatly reduce welding costs since it saves the expense of duplicating qualifying procedures and allows the portability of welding credentials. It is a great step forward.

The American Welding Society continues to make welding-related occupations more professional. By standardizing the qualification and certification of personnel, public confidence in welding will increase. AWS has become the welding authority in the United States and is providing ways to educate welding inspectors, teachers, technicians, and engineers. This is done through increased training, testing, and certification of knowledge, based on proficiency testing.

The original concept of this book has been maintained, with emphasis on the arc welding processes and the use of steel for industrial and construction uses. The book still follows faithfully the standards, codes, and specifications provided by the AWS. It allows the reader to keep up-to-date as welding technical information and technology improvements advance. Truly, the industry is moving rapidly, and the welding is improved and more productive.

ACKNOWLEDGMENTS

The typing of the manuscript was done by two fine, efficient women who had to put up with more than they bargained for. Jeanne Sargent typed the original and Gayle Meyers the revisions. Many thanks for a job well done.

I would also like to thank the reviewers of this edition: Jeff Djonnerberg, Eastern Washington University; Angie Hill Price, Texas A&M University; and Gary Senff, Central Community College, Columbus Campus.

The illustrations were done by two gifted artists. Ernie Boller drew the originals, and I particularly enjoyed the fish in the underwater welding drawing. Russ Wogoman did the drawings in the revisions. Thanks for these fine drawings, which really help tell the story.

To make this book technically accurate, the official terminology of the American Welding Society is used. The book includes information from many AWS standards and codes. The society has graciously allowed the use of this information to help us all communicate welding information more accurately. My thanks to the society.

Hobart Brothers Company has given me permission to use its data, information, pictures, diagrams, and other material to help make this book as complete as possible. I wish specifically to thank Glenn Nally and William H. Hobart Jr.

I want to thank the many other people who furnished information and pictures. Many thanks to each. The list is long and I hope that I have not missed anyone.

Accra-Weld Controls
Advanced Manufacturing Engineering Technologies
AGA Gas, Inc.
Airflow Systems Inc.
Aluminum Association
American Iron and Steel Institute
American Petroleum Institute
American Society for Metals
American Society for Testing and Materials

American Society of Mechanical Engineers
Arc Air Co.
Arcmatic Integrated Systems
Association of Iron and Steel Engineers
Automated Production Concepts, Inc.

Battelle Columbus Laboratories
Berkeley Davis, Inc.
Berner, Susan
Bethlehem Steel Corp.
Bettermann Stud Welding
Boeing Aircraft
Boeing Petroleum Services Co.

Caterpillar Inc.
CBI Industries, Inc.
C. C. Peck Co.
Coastal CAD and Blueprint Inc.
Cincinnati Milacron
CRC Automatic

Dearman Div. of Cogsdille Tool Products Co.
Design Technologies & Mfg. Co.
DuPont—Aldyl Piping System
DuPont—Metal Cladding Section
Dual Draw Clean Air Work Station

Eagle Arc Metalizing Corp.
Edison Welding Institute
Engelhard Corp.
ESAB Automation, Inc.
Eutectic Corp.
Explosive Fabricators, Inc.

F. Bode and Son Ltd.
Frommelt Safety Products

General Electric
Gullco International

Heckendorn, Larry
Henning Hansen Inc.
H&M Pipe Beveling Machine Company, Inc.
Hypertherm, Inc.

InTech R&D
ITW Welding Products—McKay

Jefferson National Expansion NHS/National Park Service
Jet Line Engineering Inc.

KATBAK-Gullco Intl.
Keen Hinkel Inc.
Koike Aronson Inc.
Krall, Linda
Krautkramer Branson

Laramy Products Co.
Leybold Vacuum System, Inc.
Lincoln Electric Co.
L-Tec
Lumonics National Processing Corp.

Magnatech, The DSD Co.
Magnetrode Corp.
Maintenance Engineering Corp.
Manufacturing Technologies Inc.
McCreery Corp.
Microweld Products Co.
Miller Electric Manufacturing Co.
Mitsubishi Laser
Motoman Inc.

NASA
National Joint Steamfitter—Pipefitter Apprenticeship Committee
National Safety Council
Nederman Inc.
Newport News Shipbuilding

Oregon Graduate Institute

Panasonic Factory Automation
Pandjiris, Inc.
PHOENIX Products Company, Inc.
Pitt-Des Moines Steel Co.

Pow Con Inc.
Prestolite Electric Power
Preston-Easton Inc.

Ramstud (USA) Inc.

Sellstrom Manufacturing Co.
Servo Robot
Smith & Associates
Smith Welding Equipment, Division of Tescom Corporation
Stillwater Technologies
Stress Relief Engineering Company
Superior Flux Co.

TAFA, Inc.
Taylor Diving & Salvage Co., Inc.
Tec Torch Co.
Teledyne Precision—Cincinnati
Teledyne Readco
TEMPIL
Thermadyne Industries, Inc.
Thermosolda
3M Company—Industrial Specialties Division
Thompson Friction Welding Ltd.
Torsteknik
Trinity Marine Group, Trinity Industries Inc.
TRW Nelson Stud Welding

U.S. Navy

Vacuum/Atmospheres Co.
Victor Equipment Co.

Welding Design and Fabrication, Penton Publishing
The Welding Institute
Welding Services, Inc.
WeldLine Automation
Weldmatic, Inc.
Weld Mold Co.
Weld Tooling Corp.
Westinghouse Electric Corp., Industrial Equipment Div.

Yaskawa Electric America, Inc.

Contents

10

POWER SOURCES FOR ARC WELDING, 305

11

OTHER WELDING EQUIPMENT, 340

12

MECHANIZED, AUTOMATED, AND ROBOTIC ARC WELDING, 365

13

ELECTRODES AND FILLER METALS, 421

14

GASES USED IN WELDING, 440

15

METALS AND THEIR WELDABILITY, 458

16

WELDING STEELS, 495

17

WELDING NONFERROUS METALS, 516

18

WELDING SPECIAL AND DISSIMILAR METALS, 547

19

DESIGN FOR WELDING, 568

20

COST OF WELDING, 605

APPENDIES, 775

INDEX, 791

1

WELDING
BACKGROUND

1-1   THE IMPORTANCE OF WELDING

Welding is the only way of joining two or more pieces of metal to make them act as a single piece. It allows the production of a monolithic structure that is strong in all directions. It is used to join all the commercial metals and alloys and to join together metals of different types and strengths. It is vital to our economy. It is often said that more than 50% of the gross national product of the country is related to welding in one way or another. Welding is the most economical and efficient way to permanently join metals.

Welding began as a repair or maintenance tool and has become one of the most important manufacturing methods as well as the most important construction method. Almost everything made of metal is welded. Many products of the welding industry are shown in Figure 1–1. Construction equipment such as the bulldozer and dragline are all welded, as are railroad freight cars.

The famous trans-Alaskan pipeline is completely welded from end to end. A super-tanker, the largest moving weldment, would not be possible without welding. Elevated storage tanks for water supply systems are welded. Motorized and sailing yachts, hydrofoil boats, patio chairs, and even the tallest building in the United States, the Sears Tower in Chicago, are welded.

Today's automobile would be much more expensive if it were not for welding. The steel body and frame of today's car is robotic-spot welded, but arc welding is also used. Collision repair work is also done with welding. The space program exists because of welding, from the rocket engines to the shuttle's external fuel tank, which required a very special procedure for welding alloy aluminum. Long-span bridges, which usually have large girders, are all-welded assemblies and may be field-welded or field-bolted. Most large airplanes include much welding even though the all-welded plane is not yet here. The landing gear, the jet engines, the engine mounts, and much of the specialized equipment is welded, even though the body of the plane is riveted. Some military planes are largely welded. Finally, welded equipment makes possible the manufacture of miniature components for electronic equipment and telecommunication equipment.

Welding ranks high among the metalworking processes including machining, forging, forming, and casting. While it appears to be very simple, welding involves more sciences and variables than any other industrial

1

FIGURE 1–1 Welding is everywhere.

Automobiles

Airplanes

Coffeepots

Bridges

FIGURE 1–1 (*cont.*)

Bulldozers

Draglines

Furniture (finished and in the welding operation)

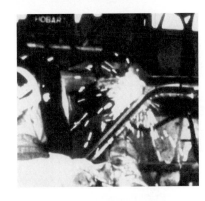

Collision Repairs

Jet Engines

FIGURE 1–1 (*cont.*)

In Space

Pipelines

Railroad Cars

Tank Farms

FIGURE 1–1 (*cont.*)

Sears Tower

LNG Tankers

Seagoing Hydrofoils

Water Storage Tankers

Yachts

process. It is only when it is understood that it becomes the most economical and efficient way to join metals. The purpose of this book is to provide information concerning the welding processes, the application of welding, and how welding is used. The book provides information concerning all of the popular processes, the properties of metal, how metals are welded, the design of weldments, and how welding is generally used to reduce production costs.

There are many different welding processes, many types of welds, and many ways to make a weld. The welder behind the hood making sparks is using a very popular welding process—arc welding. Some welding processes do not cause sparks, use electricity, or require added heat. Welding has become very complex and very technical. It requires knowledge to select the proper welding process for each application. The purpose of the book is to learn about welding so that you can utilize its many advantages. Some of these advantages are:

It is the lowest cost, permanent joining method
It affords lighter weight through better utilization of materials
It joins all commercial metals
It can be used anywhere
It provides design flexibility

It is also important to know the limitations of welding. These limitations are:

Procedures must be provided for all metals and applications
Manual welding depends on the human factor
Internal inspection is often required to assure quality

These limitations can be overcome by means of nondestructive evaluation, good supervision, qualified procedures, qualified personnel, and adopting mechanized welding methods.

It is important to remember that an intelligently designed, properly constructed weldment is the best and lowest cost solution for any metal product. Welded products will remain important to the economy as long as quality designed and produced products are available.

1-2 WELDING JOINS ALL METALS

All metals can be joined by one welding process or another. The metals that are easily weldable can be welded in thickness from the very thinnest to the thickest produced. The difficult-to-weld metals require special procedures and techniques that must be developed for specific applications.

Some metals may never be welded or joined. Later chapters will provide the properties of metals such as melting temperature, density, thermal conductivity, tensile strength, ductility, and so on. These properties provide information as to whether a metal can be welded. For example, mercury is a liquid at room temperature and cannot be welded, while sodium and potassium melt just below the temperature of boiling water and are of no use as a strength member and cannot be welded. In general, metals that have a low melting temperature or low strength would not be welded.

The physical and mechanical properties, availability, and price determine if a metal will be used in appli-

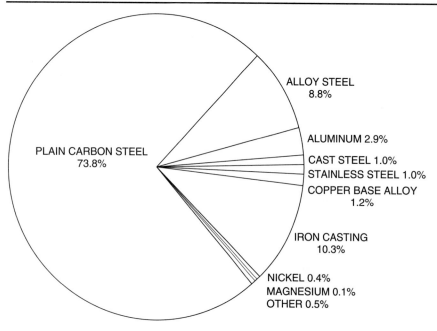

FIGURE 1–2 Metal production by types in the United States.

PLAIN CARBON STEEL 73.8%

ALLOY STEEL 8.8%

ALUMINUM 2.9%

CAST STEEL 1.0%
STAINLESS STEEL 1.0%
COPPER BASE ALLOY 1.2%

IRON CASTING 10.3%

NICKEL 0.4%
MAGNESIUM 0.1%
OTHER 0.5%

cations where welding is required. The more abundant and stronger metals are normally welded. Figure 1–2 shows the commercial metals according to their annual production in the United States. This may be somewhat misleading since many metals used in the United States are imported. However, the ratio of metals is important and it is believed that these ratios represent usage and are similar in most industrialized countries.

Plain carbon steel is by far the most widely used metal. This book provides much information for welding ordinary mild steel. Iron castings are the second largest type of metal produced, but most iron castings are used without welding. Sometimes, iron castings are joined or repaired, so it is necessary to know how different types of cast iron are welded. Alloy steels make up the next largest group, which includes many special types such as low-alloy high-strength steels, heat-treated steels, ultrahigh-strength steels, and many others. Each type of alloy steel involves different welding procedures.

Aluminum and aluminum alloys represent the next largest group and are continually finding wider application, especially when weight is a factor. The different aluminum alloys have different properties and require different procedures for welding.

Copper and copper-base alloys such as brasses and bronzes make up the next largest group. These metals are used when electrical conductivity, corrosion resistance, or heat conductivity is important.

Stainless steel and cast steels are tied for the next position. Cast steels are normally welded like rolled steels of the same composition. There are many different types of stainless steels having specific properties.

Nickel and nickel alloys represent the next largest group, a small percentage but very important since certain nickel alloys are the best metal to use for special service.

The magnesium group is the smallest, though important. Magnesium is the lightest structural metal and has many applications where welding is required.

The remaining metals represent small percentages but they are used when their particular characteristic is required and these applications often require that they be welded.

Steel is produced in many different forms. Figure 1–3 shows the production ratios of the various forms of steel. Sheet steel represents the greatest volume. This is not surprising considering that most of the steel used in automobiles, trucks, buses, home appliances, office machinery, and furniture is sheet steel. It is also used for many containers, cans, and so on.

Steel bars represent the second largest product form. Many bars are machined or cut and used, but a sizable portion of bars go with structural shapes, the fifth largest group, and are used in the construction industry.

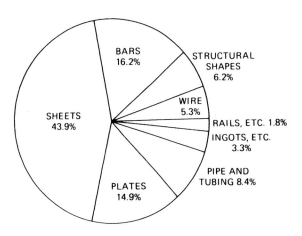

FIGURE 1–3 Steel production by types in the United States.

Steel plates represent the third largest group and are used to make tanks, boilers, machinery, ships, and other weldments.

Pipe and tubular products represent the fourth largest group. All large-diameter pipe and much of the smaller pipe is joined by welding.

The remaining groups—wires, rails, and others—may not involve much welding. However, rails are now being welded for the railroads.

Note the broad use of metal, especially the steels, and how welding relates to them. It is important that we know how to weld each metal in any form. Information about each metal is provided by specifications that are originated by engineering societies, technical groups, and trade associations.

A summary of metals is given in Figure 1–4. They are related to the common welding processes, and a rating system indicates how they are welded. All metals *can* be welded, but some are easier to weld than others. In other words, they possess **weldability**, defined as "the capacity of materials to be welded under the imposed fabrication conditions into a specific, suitably designed structure and to perform satisfactorily in the intended service." All metals cannot be joined by each welding process. Some welding processes were developed to join specific metals. Certain metals are known as "difficult to weld," which means that specific precautions and procedures are required.

The welding industry developed materials called filler metals to join the various metals and alloys. These materials fill the weld joint and provide joints as strong as the metals being joined. The term *filler metals* means "the metal to be added in making a welded, brazed, or soldered joint" and includes electrodes and welding rods. It is broadly interpreted to cover shielding methods, including fluxes and gases.

FIGURE 1–4 Summary of metals welded by various processes.

Base Metals Welded	Shielded Metal Arc	Gas Tungsten Arc	Plasma Arc	Submerged Arc	Gas Metal Arc	Flux-Cored Arc	Electroslag	Braze	Gas
Aluminums	C	A	A	No	A	No	Exp	B	B
Copper-base alloys									
Brasses	No	C	C	No	C	No	No	A	A
Bronzes	A	A	B	No	A	No	No	A	B
Copper	C	A	A	No	A	No	No	A	A
Copper nickel	B	A	A	No	A	No	No	A	A
Irons									
Cast, malleable, nodular iron	A	B	B	No	B	B	No	A	A
Wrought iron	A	B	B	A	A	A	No	A	A
Lead	No	B	B	No	No	No	No	No	A
Magnesium	No	A	B	No	A	No	No	No	No
Nickel-base alloys									
Inconel	A	A	A	No	A	No	No	A	B
Monel	A	A	A	C	A	No	No	A	A
Nickel	A	A	A	C	A	No	No	A	A
Nickel silver	No	C	C	No	C	No	No	A	B
Precious metals	No	A	A	No	Exp	No	No	A	B
Steels									
Alloy steel	A	A	A	B	A	A	A	A	A
Low-alloy steel	A	A	A	A	A	A	A	A	A
High- and medium-carbon steel	A	A	A	B	A	A	A	A	A
Low-carbon steel	A	A	A	A	A	A	A	A	A
Stainless steel	A	A	A	A	A	B	A	A	C
Tool steel	A	A	A	No	C	No	No	A	A
Titanium	No	A	A	Exp	A	No	No	No	No
Tungsten	No	B	A	No	No	No	No	No	No
Zinc	No	C	C	No	No	No	No	No	C

[a]Metal or process rating: A, recommended or easily weldable; B, acceptable but not best selection or weldable with precautions; C, possibly usable but not popular or restricted use or difficult to weld; No, not recommended or not weldable; Exp, experimental or research.

1-3 HISTORICAL DEVELOPMENT OF WELDING

Welding, one of the newer metalworking technologies, can trace its historic development back to ancient times. The earliest example comes from the Bronze Age. Small gold circular boxes were made by pressure-welding lap joints together. It is estimated that these boxes were made more than 2,000 years ago and are presently on exhibit at the National Museum in Dublin, Ireland. During the Iron Age, the Egyptians and people in the eastern Mediterranean area learned to weld pieces of iron together. Many tools were found that were made in approximately 1000 B.C. These items are on exhibit in the British Museum in London. Other examples of early welded art are displayed in the museums of Philadelphia and Toronto. Items of iron and bronze that exhibit intricate forging and forge welding operations have been found in the pyramids of Egypt.

During the Middle Ages the art of blacksmithing was developed and many items of iron that were welded by hammering were produced. One of the largest welds from this period was the Iron Pillar of Delhi in India, which was erected about the year A.D. 310. It was made from iron billets welded together. It is approximately 25 ft (7.6 m) tall with a diameter of 12 in (300 mm) at the top and 16 in. (400 mm) at the bottom. Its total weight is 12,000 lb (5.4 metric tons). Other pillars were erected in India at this same time and a few large weldments made by the Romans have been found in Europe and in England. Other welded works have been found in Scandinavia and in Germany. It was not until the nineteenth century that welding as we know it today was invented.

Edmund Davy of England is credited with the discovery of acetylene in 1836. The production of an arc between two carbon electrodes using a battery is credited to Sir Humphry Davy in 1800. In the mid-nineteenth century, the electric generator was invented and arc lighting became popular. The late nineteenth century provided a great number of discoveries. During this period gas welding and cutting were developed. Arc welding with the carbon arc and metal arc was developed, and resistance welding became a practical joining process. Auguste De Meritens, working in the Cabot Laboratory in France, used the heat of an arc for joining lead plates for storage batteries in the year 1881. It was his pupil, Nikolai N. Benardos, a Russian working in the French laboratory, who was granted a patent for welding. He, with a fellow Russian, Stanislaus Olszewski, secured a British patent in 1885 and an American patent in 1887.[1] The patents show an early electrode holder (Figure 1-5). This was the beginning of carbon arc welding. Benardos's efforts were restricted to carbon arc welding, although he was able to weld iron as well as lead. Carbon arc welding became popular during the late 1890s and early 1900s.

Apparently, Benardos was not successful with a metallic electrode, and in 1890, C. L. Coffin of Detroit was awarded the first U.S. patent for an arc welding process using a metal electrode.[2] This was the first record of the metal melted from the electrode carried across the arc to deposit filler metal in the joint to make a weld. At about the same time, N. G. Slavianoff, a Russian, presented the same idea of transferring metal across an arc, but to cast metal in a mold.[3]

In about 1900, Strohmenger introduced a coated metal electrode in Great Britain.[4] There was a thin coating of clay or lime, but it provided a more stable arc. Oscar Kjellberg of Sweden invented the covered or coated electrode during the period 1907 to 1914. Figure 1-6 shows his concept of welding with coated stick electrodes. They were produced by dipping short lengths of bare iron wire in thick mixtures of carbonates and silicates, and allowing the coating to dry.

Meanwhile, the resistance welding processes were developed, including spot welding, seam welding, projection welding, and flash butt welding. Elihu Thompson originated resistance welding. His patents began in 1885. He originated a company, the Thompson Electric Welding Company, and developed the different resistance welding processes between that time and 1900. Thermite welding was invented by a German named Goldschmidt in 1903 and was first used to weld railroad rails.

Gas welding and cutting were perfected in this period. The production of oxygen, and later the liquefying of air, along with the introduction in 1887 of a blowpipe or torch, helped the development of both welding and cutting. Before 1900 hydrogen and coal gas were used with oxygen. However, in about 1900 a torch suitable for use with low-pressure acetylene was developed.

World War I brought a tremendous demand for armament production, and welding was pressed into service. Many companies sprang up in America and in Europe to manufacture welding machines and electrodes to meet this requirement. The British built the first all-welded ship,

FIGURE 1–6 Kjellberg covered electrode.

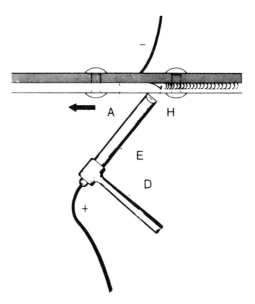

FIGURE 1–5 First electrode holder.

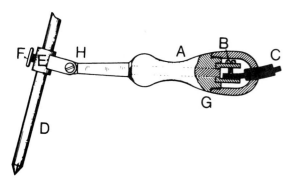

H.M.S. *Fulagar*, and the Dutch started to weld fuselages of fighter planes. The most famous incident was the repair work on German ships interned in New York harbor that were sabotaged. By means of arc welding, these ships were quickly repaired and put back into service to deliver material from the United States to Europe.

Immediately after the war, in 1919, 20 members of the Wartime Welding Committee of the Emergency Fleet Corporation, under the leadership of Comfort Avery Adams, founded the American Welding Society. It was founded as a nonprofit organization dedicated to the advancement of welding and allied processes.

In 1920, automatic welding was introduced. It utilized bare electrode wire operated on direct current and utilized arc voltage as the basis of regulating the feed rate. Automatic welding was invented by P. O. Nobel of the General Electric Company.[5] It was used to build up worn motor shafts and worn crane wheels. It was also used by the automobile industry to produce rear axle housings.

During the 1920s, various types of welding electrodes were developed. Mild steel with a carbon of 0.20% or less was used for welding practically all grades of steel. Higher-carbon electrodes and alloy steel electrodes were also developed. Copper alloy rods were developed for carbon arc welding and brazing.

There was considerable controversy during the 1920s about the advantage of the heavy-coated rods versus light-coated rods. Heavy-coated rods made by dipping were expensive. Coating applied by extrusion on the rod was less expensive. The heavy-coated electrodes, which were made by extruding, were developed by Langstroth and Wunder of A. O. Smith Company[6] and were used by the company in 1927. In 1929, Lincoln Electric Company produced extruded electrode rods that were sold to the public. By 1930, covered electrodes were widely used. Welding codes appeared requiring higher-quality weld metal, which increased the use of covered electrodes.

During the 1920s there was considerable research in shielding the arc and weld area by externally applied gases. The atmosphere of oxygen and nitrogen in contact with the molten weld metal caused brittle and sometimes porous welds. Research work was done utilizing gas-shielding techniques. Alexander and Langmuir did work in chambers using hydrogen as a welding atmosphere. They utilized two electrodes, starting with carbon but later changing to tungsten. The hydrogen was changed to atomic hydrogen in the arc. It was then blown out of the arc, forming an intensely hot flame of atomic hydrogen burning to the molecular form and liberating heat. This arc produced half again as much heat as an oxyacetylene flame. This became the atomic hydrogen welding process. Atomic hydrogen never became popular but was used during the 1930s and 1940s for special applications of welding and later on for welding of tool steels.[7]

H. M. Hobart and P. K. Devers were doing similar work but using atmospheres of argon and helium. In their patents applied for in 1926, arc welding utilizing gas supplied around the arc was a forerunner of the gas tungsten arc welding process.[8,9] They also showed welding with a concentric nozzle and with the electrode being fed as a wire through the nozzle. This was the forerunner of the gas metal arc welding process. These processes were developed much later.

Stud welding was developed in 1930 at the New York Navy Yard, specifically for attaching wood decking over a metal surface.[10] The process welded studs to the base metal by means of a special gun, which automatically controlled the arc. Fluxing elements on the end of the stud improved the properties of the weld. Stud welding became popular in the shipbuilding and construction industries.

The automatic process that became popular was the submerged arc welding process. This "under powder" or smothered arc welding process was developed by the National Tube Company for a pipe mill at McKeesport, Pennsylvania. It was designed to make the longitudinal seams in the pipe. This process was patented by Robinoff in 1930[11] and it was later sold to Linde Air Products Company, where it was renamed Unionmelt[12] welding. Submerged arc welding was used during the defense buildup in 1938 in shipyards and in ordnance factories. It is one of the most productive welding processes and remains popular today.

Gas tungsten arc welding had its beginnings from an idea by C. L. Coffin to weld in a nonoxidizing gas atmosphere, which he patented in 1890.[2] The concept was further refined in the late 1920s by Hobart, who used helium for shielding, and Devers, who used argon. The threat of World War II focused on the need to weld magnesium to build fighter planes. Engineers of the Northrup Aircraft Company, with Dow Chemical Company, developed a welding process for joining magnesium. The inert gas-shielded process developed by Hobart and Devers was ideal for welding magnesium and also for welding stainless steel and aluminum. It was perfected in 1941, patented by Meredith, and named Heliarc welding, since helium was initially used for shielding.[13] Figure 1–7 shows early torches developed by Meredith. It was later licensed to Linde Air Products, where the water-cooled torch was developed. The gas tungsten arc welding process has become one of the most important.

The gas-shielded metal arc welding (GMAW) process was successfully developed at Battelle Memorial Institute in 1948 under the sponsorship of the Air Reduction Company. This development utilized the gas-shielded arc similar to the gas tungsten arc, but replaced the tungsten electrode with a continuously fed electrode wire.[14] Figure 1–8 shows the first practical gas metal arc welding gun designed by the author. One of the basic changes that made the process more usable was the

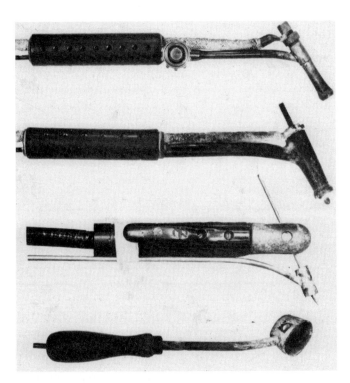

FIGURE 1–7 Meredith Heliarc torches.

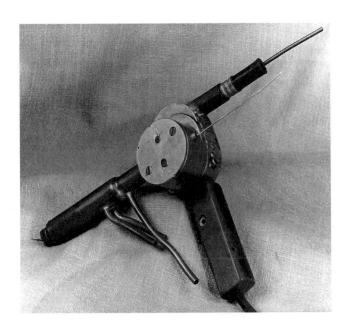

FIGURE 1–8 First gas metal arc semiautomatic gun.

small-diameter electrode wires and the constant-voltage power source. This principle had been patented earlier by H. E. Kennedy.[15] The initial introduction of GMAW was for welding nonferrous metals. The high deposition rate led users to try the process on steel. The cost of inert gas was relatively high and the cost savings were not immediately available.

In 1953, Lyubavskii and Novoshilov announced the use of welding with consumable electrodes in an atmosphere of CO_2 gas.[16] The CO_2 welding process immediately gained favor since it utilized equipment developed for inert gas metal arc welding but could now be used for economically welding steels. The CO_2 arc is a hot arc and the larger electrode wires required fairly high currents. The process became widely used with the introduction of smaller-diameter electrode wires and refined power supplies. This development was the short-circuit arc variation, which was known as Micro-wire, short arc, and dip transfer welding, all of which appeared late in 1958 and early in 1959. This variation allowed all-position welding on thin materials. It soon became the most popular of the gas metal arc welding process variations.[17]

Another variation was the use of inert gas with small amounts of oxygen, which provided the spray-type arc transfer. It became popular in the early 1960s. The latest variation is the use of pulsed current. The current is switched from a high to a low value at a rate of once or twice the line frequency. Now variable frequency is used. This process is becoming popular.

Soon after the introduction of CO_2 welding, a variation utilizing a special electrode wire was developed. This wire, described as an inside–outside electrode, was tubular in cross section, with the fluxing agents on the inside. The process was called Dualshield, which indicates that external shielding gas was utilized, as well as the gas produced by the flux in the core of the wire, for arc shielding. This process, invented by Bernard, was announced in 1954 but was patented in 1957, when it was reintroduced by the National Cylinder Gas Company.[18]

In 1959, an inside–outside electrode was produced that did not require external gas shielding. The absence of shielding gas gave the process popularity for noncritical work. This process was the self-shielding process named Innershield. Both the gas-shielded and self-shielding systems are widely used today and are growing in popularity.[19]

The electroslag welding process was announced by the Soviets at the Brussels World's Fair in Belgium in 1958. It had been used in the Soviet Union since 1951 but was based on work done in the United States by R. K. Hopkins, who was granted patents in 1940.[20] The Hopkins process was never used to a very great degree for joining. Figure 1–9 shows the Hopkins process for joining steel pieces. The process was perfected and equipment was developed at the Paton Institute Laboratory in Kiev, Ukraine, and also at the Welding Research Laboratory in Bratislava, Czechoslovakia. The first production use in the United States was at the Electromotive Division of General Motors Corporation in Chicago, where it was called the Electro-molding process. It was announced in December 1959 for the fabrication of welded diesel engine blocks.[21] The process and its variation, using a consumable guide tube, is used for welding thicker materials.

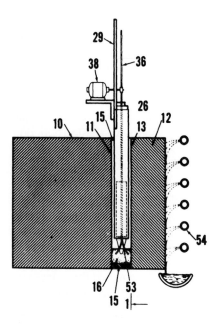

FIGURE 1–9   Hopkins electroslag welding.

Another vertical welding method, called Electrogas, was introduced in 1961 by the Arcos Corporation.[22] It utilized equipment developed for electroslag welding but employed a flux-cored electrode wire and an externally supplied gas shield. It is an open arc process since a slag bath is not involved. A newer development uses self-shielding electrode wires and a variation uses solid wire but with gas shielding. These methods allow the welding of thinner materials than can be welded with the electroslag process.

Plasma arc welding was invented by Gage in 1957. It is used for metal spraying and for cutting. It is used for spraying both wires and powders.

The electron beam welding process, which uses a focused beam of electrons as a heat source in a vacuum chamber, was developed in France.

Electron beam (EB) welding has gained widespread acceptance for welding. Its popularity is increasing due to recent developments in Japan for welding heavy-wall pressure vessels. In the United States the automotive and aircraft engine industries are major users of EB welding.

Friction welding, which uses rotational speed and upset pressure to provide friction heat, was developed in the Soviet Union. It is a specialized process and has applications only where a sufficient volume of similar parts are to be welded because of the initial expense of equipment and tooling. This process is also called inertia welding.

The newest welding process is laser welding. The laser was originally developed in 1951 and was used as a communications device. Because of the tremendous concentration of energy in a small space, it proved to be a powerful heat source. It has been used for cutting metals and nonmetals. The early problems involved short pulses

of energy; however, today continuous-pulse equipment is available. The laser is finding welding applications in automotive metalworking operations.

Many variations of these processes, which are not specifically processes themselves, will be developed and as the need arises they will be adapted to metalworking requirements.

1-4 THE WELDING INDUSTRY

Welding as an industry is poorly defined. It includes the welding equipment manufacturers, the welding research laboratories, the welding training schools and universities, the metal producers, as well as the many, many companies, in both industry and construction, that use welding.

The government developed a system to gather, tabulate, and analyze data related to all the different establishments throughout the country. An establishment is defined as an economic unit at a single location where business or industrial operations are performed. This refers, then, to those that use welding, are considered the welder-using industry, or produce welding equipment. All establishments used to be given Standard Industrial Classification (SIC) numbers. With the North American Free Trade Agreement (NAFTA), a cooperative system between Canada, Mexico, and the United States became necessary. A new system, known as the North American Industry Classification System (NAICS), began in 1997 and gave new code numbers to all establishments in those three countries. This is the most profound change for government statistical programs and will require years to be completed; unfortunately we are in the change period and data is not in complete synchronization.

NAICS numbers are assigned to every operation, and all establishments engaged in producing the same specific product or service are given the same code number. For example, three-digit numbers represent major groups and four-digit numbers represent subdivisions. Various government agencies periodically gather statistics from each establishment. This includes the value of products produced and the number and type of employees, including the number of welders and flame cutters. These numbers, then, can be used for ease of comparison when identifying and classifying establishments.

We are interested in the establishments that use welding. The Bureau of Labor Statistics provides the number of workers, including welders and flame cutters, employed at each location. An analysis of this data shows which industries and industry groups are the biggest users of welding. This is based on the assumption that the amount of welding performed at an establishment relates to the number of welders employed. This assumption can be misleading since there are many persons who weld but are not classified as welders. This, and the increasing trend toward welding automation, may have an effect on this data. This classification system provides a database which, when combined with the population data, allows us to determine which industries do the most arc welding. However, it doesn't include information about other welding processes such as electron beam welding, laser welding, and other specialties. Product value and employment data are gathered periodically by the government.

The major industrial groups that represent the welding industry use welding extensively but in different ways under different working conditions on different metals and according to different codes and specifications. Each group is briefly described.

Mining and Oil and Gas Extraction (NAICS Number 211-212)

This group includes mining, both deep shaft and open pit, for ores and coal, quarrying stone, sand, and gravel. This industry group also includes drilling and extraction of oil and gas. Welding codes are not normally employed.

Heavy Construction (NAICS Number 234)

This group of companies includes those that build tunnels, subways, dams, powerhouses, chemical plants, structural steel bridges, and buildings. Much of the work involves structural welding, which is covered by code. This group also includes piping contractors for cross-country and other pipelines, which are also governed by strict code.

Primary Metal Manufacturing (NAICS Number 331)

This group includes steel mills, iron and steel foundries, and smelting and refining plants. These companies produce steel plates, structural shapes, tubular products, sheet metal, and castings of all commercial metals. Most of the welding in this group is for maintenance repair on the facilities; however, a portion is reclamation welding of castings. Welding codes are usually not employed.

Fabricated Metal Products (NAICS Number 332)

This industry group includes manufacture of power boilers, pressure vessels, heat exchangers, tanks, refinery equipment, plate trusses, machine bases, sheet metal work, prefabricated metal buildings, and architectural and ornamental work. Most of the welding is governed by codes and standards.

Machinery Manufacturing (NAICS Number 333)

This industry group manufactures machinery: agricultural, construction, mining, and material handling equipment such as power shovels, bulldozers, cranes, metal working machines, press brakes, shears, stamping presses, and food processing, textile, woodworking, papermaking, printing, and office machinery. These products are normally built to American Welding Society (AWS) consensus codes and to company standards. Most of these products are made of steel utilizing plates, shapes, and castings.

Welding and Soldering Equipment (NAICS Number 333992)

This group includes manufacturers of welding power sources, gas welding equipment, resistance welding machines, welding robots, welding guns, tips, torches, soldering equipment, electrodes, and welding wire. The products must meet strict safety codes.

Electrical Equipment (NAICS Number 335)

This group includes companies that produce electrical generators, transformers, switch gear, electric motors, and household electrical appliances. Welding done by this group includes heavy massive welding and sheet metal welding. Welding codes are normally not employed.

Transportation—Motor Vehicles (NAICS Number 3361)

This group includes companies that manufacture automobiles, trucks, buses, and trailers. It also includes those companies that produce subassemblies or components for building automobiles. These products involve welding on a mass-production basis. It is a major user of automated welding, resistance welding, and welding robots. Some of this is governed by codes.

Transportation—Railroad Rolling Stock (NAICS Number 3365)

This includes manufacture of locomotives, railroad, street, and rapid transit cars, freight cars of all types, and track maintenance equipment. Welding on railroad rolling stock is strictly regulated by code.

Transportation—Ship and Boat Building (NAICS Number 3366)

Ship building and repair involve heavy plate and all position welding. This work is governed by codes or insurance regulations.

Repair and Maintenance (NAICS Number 811)

This group is involved entirely with maintenance and repair welding. A major portion of this is done on automobiles

(a)

(b)

(c)

FIGURE 1–10 St. Louis Arch.

NUMBER OF WELDERS AND CUTTERS EMPLOYED		
	1998 Employment	
Industry	Number	Percent Distribution
Total employment, all industries	367,708	100.00
Fabricated structural metal products	39,313	10.69
Motor vehicles and equipment	26,794	7.29
Construction and related machinery	24,637	6.70
All other repair shops and related services	19,313	5.25
Ship and boat building and repairing	15,755	4.28
Self-employed workers, primary job	15,281	4.16
Wholesale trade, other	14,965	4.07
All other transportation equipment	13,397	3.64
Personnel supply services	13,238	3.60
All other special trade contractors	12,933	3.52
Industrial machinery, nec	11,119	3.02
Heavy construction, except highway and street	9,448	2.57
Machinery, equipment, and supplies	7,915	2.15
Farm and garden machinery	7,393	2.01
Miscellaneous fabricated metal products	7,183	1.95
General industrial machinery	7,076	1.92
Special industry machinery	5,677	1.54
Refrigeration and service machinery	5,581	1.52
Nonresidential building construction	5,219	1.42
Oil and gas field services	5,140	1.40
Metal forgings and stampings	4,469	1.22
Blast furnaces and basic steel products	3,746	1.02
Self-employed workers, secondary job	3,745	1.02
Railroad transportation	3,728	1.01
Office and miscellaneous furniture and fixtures	3,622	0.99
Iron and steel foundries	3,154	0.86
Federal government	2,958	0.80
Metalworking machinery	2,884	0.78
Plumbing, heating, and air-conditioning	2,854	0.78
Manufactured products, nec	2,696	0.73
Household furniture	2,656	0.72
All other agriculture, forestry, and fishing, except secondary jobs	2,575	0.70
All others less than .7% distribution	61,244	16.67

FIGURE 1–11   Data from Bureau of Labor Statistics, National Industry-Occupation Employment Matrix, Occupation Reports.

in collision repair shops. However, much repair welding is done for the maintenance of industrial and construction equipment, electrical machinery, buildup surfacing, etc. Welding may be governed by codes applicable to the product.

Probably the smallest group, which is not defined, is involved with the production of welded sculptures. This group may be unlisted, but it does have an important impact on the public. The largest and most unusual welded sculpture is the 630-foot-tall stainless-steel St. Louis Arch (Figure 1–10). The arch is an inverted cate-

nary curve designed by Eero Saarinen and built by a Pittsburgh-Des Moines Steel Company for the Jefferson National Expansion Historical Association. This is the biggest, but there are many other welded sculptures and fountains throughout the country.

Since the new classification system and the data collection system are not in complete sync, we have different industry groups. The number of welders and cutters in each more narrowly defined group is shown by Figure 1–11. However, in order to more closely define the type of work performed, we now have an occupational

NUMBER OF WELDING MACHINE SETTERS, OPERATORS, AND TENDERS

Industry	1998 Employment	
	Number	Percent Distribution
Total employment, all industries	109,604	100.00
Fabricated structural metal products	16,617	15.16
Motor vehicles and equipment	12,813	11.69
Construction and related machinery	9,929	9.06
Metal forgings and stampings	9,111	8.31
Miscellaneous fabricated metal products	8,468	7.73
Industrial machinery, nec	6,713	6.12
Farm and garden machinery	6,142	5.60
Refrigeration and service machinery	3,690	3.37
General industrial machinery	3,329	3.04
Special industry machinery	3,144	2.87
All other transportation equipment	2,868	2.62
Office and miscellaneous furniture and fixtures	2,778	2.53
Electronic components and accessories	2,279	2.08
Blast furnaces and basic steel products	2,092	1.91
Cutlery, hand tools, and hardware	1,688	1.54
Household appliances	1,660	1.51
Metalworking machinery	1,453	1.33
Ship and boat building and repairing	1,413	1.29
Electric lighting and wiring equipment	1,310	1.19
Aircraft and parts	1,120	1.02
All others less than 1% distribution	10,987	10.03

FIGURE 1–11A   Data from Bureau of Labor Statistics, National Industry-Occupation Employment Matrix, Occupation Reports.

category called Welding Machine Setters, Operators and Tenders. This in effect indicates the trend to automatic, mechanized, and automated welding. The number of these employees in the more narrowly defined industry grouping is shown in Figure 1–11A. In time it is expected that the number of welders and cutters will decline and the number of welding machine setters, operators, and tenders will increase.

The group rankings of the different industry groups may change from year to year. This is based on the economy and the health of companies within each industry group.

1-5 THE FUTURE OF WELDING

The growth of the welding equipment and apparatus business shows that the use of welding is still increasing (see Figure 1–12, which compares the value of equipment shipped in 1993 to 1996 and projects for the year

1999). Arc welding equipment—which includes welding machines, power sources, and components—represents approximately half of the total welding equipment and is expected to grow at a 6% annual rate. Resistance welding

FIGURE 1–12   U.S. Department of Commerce: Bureau of the Census; International Trade Administration

WELDING APPARATUS TRENDS AND FORECASTS

(millions of dollars except as noted)

	1993	1996	1999	% increase
Value of shipments	3,101	3,496	4,261	6.3
Total employment (thousands)	20.3	19.7	—	—
Value of imports	535	751	1,118	18.4
Value of exports	655	850	1,322	8.0

equipment—including welding machines, transformers, controllers, and accessories—also has shown a great percentage of growth. Gas welding and cutting equipment is not expected to grow as fast. However, other welding equipment such as stud welding, laser beam welding, friction welding, electron beam welding, and ultrasonic welding will grow at a very high rate. Robot welding—welding equipment attached to robots—will grow at an extremely high rate, and the value of imports will grow at a high rate as well since most robots are manufactured abroad. Since the United States is the world's largest producer of welding and cutting apparatus, exports will remain strong. There are two groups of welding filler metal: stick electrodes, which will actually decrease in the future; and coiled and spooled welding wire, including solid as well as flux- and metal-cored electrodes, which will greatly increase. This is the trend of the future.

Welding use is expected to grow due to its economic advantage. Intelligently designed weldments will always be less expensive for similar applications than will products made by other manufacturing methods such as castings, forgings, riveted, and bolted assemblies. Welding is the only method that allows the designer to use the proper metal where it is required. Welding is the best method to protect and preserve metals by protecting their surface with special metal overlays. Corrosion and wear account for losses running into millions of dollars annually; weld surfacing is widely used to reduce costly abrasive and corrosive wear of metal parts.

The true impact of welding in the user industry should be measured in the amount of money saved by the use of welding over other metal fabrication processes. This information is not collected and therefore impossible to determine.

The future growth of welding depends largely on the future of each of the different welding-user industries previously described and on the materials that are used. The worldwide production of steel continues to increase at a steady rate. The amount of welding filler metals produced worldwide gains at a proportional rate. Both are expected to continue growing. However the production of plastics and composites is also increasing. This is expected since the world population is increasing and so is the standard of living. Concrete is becoming more advanced and stronger and is replacing structural steel in many medium-rise buildings. Composite beams are just now being used for short-span bridges. Plastics are increasingly used in the appliance industry for weight and cost savings, replacing many household items formerly made of metal.

The metal industry is producing many new materials and alloys that must be welded. These include the high-strength, low-alloy steels used in the construction equipment and transportation industries, and the new high-alloy, high-temperature steels used in the power-generating industries. The aluminum industry is producing higher-strength alloy aluminums, which may eventually be used in aircraft construction, bringing about the all-welded airplane. It will all depend on the competition for materials. Metals will be used where their strength and features are needed; welding will be used as new procedures and filler metals are developed.

Welding will continue to dominate the materials-joining industry because the productivity of welding is being increased. This is due to improved filler metals with higher deposition rates and the increased use of computer-controlled welding equipment and processes, which reduce overall labor costs.

Each process must be considered separately when forecasting the future of welding, since each has its own development and utilization and will have a different future. The arc welding processes will continue to dominate the welding industry. Shielded metal arc (stick) welding is rapidly losing ground to the wire and cored wire processes. Both the solid wire, the flux-cored, and the metal-cored wire processes are rapidly gaining because they can be used in robotic and automated welding processes. Gas tungsten arc (TIG) welding will continue to grow faster than the total welding market for three reasons: It will weld all metals; it is being used in high-quality work; and it can be used for welding newer, thin, specialty metals. It is increasingly being mechanized for more and more applications including miniature products. Plasma arc welding will also continue to grow but at a slower rate because the equipment is slightly more complex. However, plasma cutting will grow at a rapid rate because it is replacing fuel gas cutting in many applications.

Laser beam welding is growing quite rapidly due to its unique applications. Laser beam cutting is growing at a very fast rate because it is mechanized and is widely used in contour cutting equipment and robot cutting applications for metals and other materials. Resistance welding will continue its slow growth since it is becoming more refined and spot welds are becoming more reliable. Gas welding will decrease because it is expensive, but it will continue to be used for certain maintenance operations.

Robotic arc welding is becoming increasingly important in the automotive and high-volume, mass-production industries. Automated welding systems which involve computer controls with feedback are increasingly being used. These systems are used for both medium thicknesses and sheet metal applications, as well as on very thin material, or micro applications.

A new welding process, friction stir welding, is being used for the welding of aluminum for many applications. It will grow rapidly as more people become aware of its capabilities. The computer control of automated welding applications is greatly increasing its productivity. In addition, friction welding, magnetic propelled welding, diffusion welding, cold welding, and other new methods will continue to grow for special applications.

QUESTIONS

1–1. What is a monolithic structure?

1–2. How much of the gross national product of the United States is related to welding in one way or another?

1–3. Look around you. What products that are used in and around the home are welded?

1–4. What are some of the larger welded products?

1–5. What metal is not welded?

1–6. What metal is welded most often?

1–7. Are all metals welded with equal ease?

1–8. Steel is produced in many forms. What type is the most popular?

1–9. What is weldability?

1–10. What is a welding filler metal?

1–11. What ancient people used welding 2,000 years ago?

1–12. What important welding discovery was made by Sir Humphry Davy of England?

1–13. Where and when was arc welding invented?

1–14. Who invented metal arc welding?

1–15. Who invented the heavy-coated electrode and when?

1–16. What was automatic welding first used to weld?

1–17. What is the utilization factor of a covered electrode?

1–18. Why is this utilization factor so low?

1–19. Who was considered the inventor of gas tungsten arc welding? When was it put into practical use?

1–20. What arc welding process is the fastest growing?

REFERENCES

1. U.S. Patent 363,320, May 17, 1887, "Process of and Apparatus for Working Metals by the Direct Application of the Electric Current," N. N. Benardos and S. Olszewski, St. Petersburg, Russia.

2. U.S. Patent 419,032, Jan. 7, 1890, "Methods of Welding by Electricity," C. L. Coffin, Detroit, Mich.

3. U.S. Patent 577,329, Feb. 16, 1897, "Electrical Casting of Metals," N. Slavianoff, St. Petersburg, Russia.

4. U.S. Patent 1,041,525, Oct. 15, 1912, "Electric Deposition of Metal," A. P. Strohmenger, London.

5. U.S. Patent 1,508,711, Sept. 16, 1924, "Apparatus for Arc Welding," P. O. Nobel, Schenectady, N.Y.; also Patent 1,731,934, Oct. 15, 1929, same inventor, but filed Sept. 19, 1918.

6. U.S. Patent 1,643,274, Sept. 20, 1927, "Weld Rod for Arc Welding," C. B. Langstroth and G. G. Wunder, Milwaukee, Wis.

7. U.S. Patent 1,746,196, Feb. 4, 1930, "Method and Apparatus for Electric Arc Welding," I. Langmuir and P. P. Alexander, Schenectady, N.Y.

8. U.S. Patent 1,746,081, Feb. 4, 1930, "Arc Welding," H. M. Hobart, Niskayuna, N.Y.

9. U.S. Patent 1,746,191, Feb. 4, 1930, "Arc Welding," P. K. Devers, Lynn, Mass.

10. U.S. Patent 2,057,670, Oct. 20, 1936, "Apparatus for End-Welding," J. D. Creeca and S. S. Scott, St. Albans, N.Y.

11. U.S. Patent 1,782,316, Nov. 18, 1930, "Method of Welding," B. S. Robinoff, S. E. Paine, and W. E. Quillen, McKeesport, Pa.

12. U.S. Patent 2,043,960, June 9, 1936, "Electric Welding," L. T. Jones, H. E. Kennedy, and M. A. Rotermund, Berkeley, Calif.

13. U.S. Patent 2,274,631, Feb. 24, 1942, "Welding Torch," R. Meredith, Los Angeles, Calif.

14. U.S. Patent 2,504,868, April 18, 1950, "Electric Arc Welding," A. Muller, G. J. Gibson, and N. E. Anderson, assigned to Air Reduction Company.

15. U.S. Patents 2,532,410 and 2,532,411, Dec. 5, 1950, "Inert Monatomic Gas Shielded Metal Arc Welding Process" and "Constant Potential Submerged Metal-Arc Welding," H. E. Kennedy, Berkeley, Calif.

16. K. V. Lyubavskii and N. M. Novoshilov, "Welding with a Consumable Electrode in an Atmosphere of Protective Gases," *Avtogennoe Delo* 24 (1953).

17. U.S. Patent 2,886,696, May 12, 1959, "Electric Arc Welding," R. W. Tuthill and A. V. Welch, assigned to Air Reduction Company.

18. U.S. Patent 2,777,928, Jan. 15, 1957, "Arc Welding Method and Means," A. A. Bernard, Chicago, Ill.

19. U.S. Patent 2,909,650, Oct. 20, 1959, "Methods and Means for Bare Electrode Welding of Alloy Steels," G. G. Landis and D. M. Patton, South Euclid, Ohio.

20. U.S. Patents 2,191,471 and 2,191,482, Feb. 27, 1940, "Welding Method" and "Method for Manufacturing Composite Metal Articles," R. K. Hopkins, Staten Island, N.Y.

21. "Electromolding Technique," *Steel* (Dec. 14, 1959).

22. U.S. Patent 3,040,166, June 19, 1962, "Machine for Automatic Upward Electrical Welding of Vertical Joints," W. Krieweth and K. Dohm, assigned to Arcos of Belgium.

2 FUNDAMENTALS OF WELDING

OUTLINE

2-1 WELDING BASICS

To understand welding it is necessary to be familiar with the basic terms used by the industry. The American Welding Society (AWS) provides the majority of definitions; many of these are given in this chapter and throughout the book. The official AWS definitions will be used. There are, however, some slightly obscure definitions and slang terms. These and the AWS terms are presented in Section A-1 of the Appendix.

Welding is "a joining process that produces coalescence of materials by heating them to the welding temperature, with or without the application of pressure or by the application of pressure alone, and with or without the use of filler metal." It is used to make welds. A **weld** is "a localized coalescence of metals or nonmetals produced either by heating materials to the welding temperature, with or without the application of pressure, or by the application of pressure alone and with or without

the use of filler material. **Coalescence** means the growing together or growth into one body of the materials being welded. The word *coalescence* is not used in all the welding process definitions, since these definitions use the word *weld*.

A **weldment** is an assembly whose component parts are joined by welding. A weldment can be made of many or few metal parts. A weldment may contain metals of different compositions and the pieces may be in the form of rolled shapes, sheet, plate, pipe, forgings, or castings. To produce a usable structure or weldment there must be weld joints between the various pieces that make the weldment. The **joint** is "the junction of members or the edges of members that are to be joined or have been joined." There are five basic types of joints for bringing two members together. These joint types are also used by other skilled trades.

The five basic joints are shown in Figure 2-1:

* **Butt joint,** A: two parts in approximately the same plane
* **Corner joint,** B: two parts located approximately at right angles to each other
* **T joint,** C: parts at approximately right angles, in the form of a T
* **Lap joint,** D: between overlapping parts in parallel planes
* **Edge joint,** E: between the edges of two or more parallel parts

19

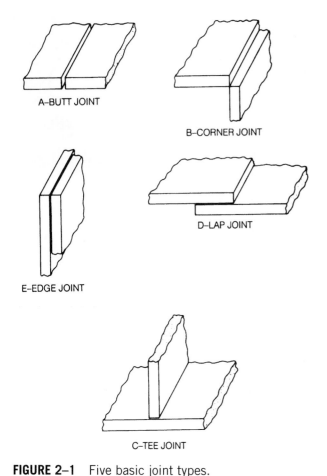

A–BUTT JOINT

B–CORNER JOINT

D–LAP JOINT

E–EDGE JOINT

C–TEE JOINT

FIGURE 2–1 Five basic joint types.

It is important to distinguish between the *joint* and the *weld*—each must be described to describe a **weld joint** completely. There are many different types of welds, and they are best described by their shape when shown in cross section. The most popular weld is the **fillet weld,** named after its cross-sectional shape. The second most popular is the **groove weld,** and there are seven basic types of groove welds. There are other types of welds: the **flange weld,** the **plug weld,** the **slot weld,** the **seam weld,** the **surfacing weld,** and the **backing weld.**

In making a weld, filler metal may or may not be used and heat with or without pressure is used, but the result is a continuity of solid metal. Joints are combined with welds to make weld joints. Welds and joints are described completely in Chapter 19.

Welding is often done on structures in the position in which they are found. In view of this, techniques have been developed to allow welding in any position. Certain welding processes have "all-position" capabilities, while others may be used in only one or two positions. There are four basic welding positions. Figure 2-2 shows the groove weld in the four positions:

* **Flat:** the welding position used to weld from the upper side of the joint; the weld axis is ap-

proximately horizontal. This is sometimes called downhand.

* **Horizontal:** position of welding in which the weld axis is approximately horizontal but the definition varies for groove and fillets.
* **Overhead:** position in which welding is performed from the underside of the joint.
* **Vertical:** position of welding in which the weld axis is approximately vertical.

There are approximately 50 different distinct welding processes. They are subdivided into seven groups. The arc welding group of processes is the most popular. There are nine distinct arc welding processes and numerous variations. In all of the arc welding processes, the heart of the arc welding system is the power source. This piece of equipment provides the electrical power to sustain the arc so that it can be used for making welds. There are many types, sizes, and variations. Some generate electricity from rotating energy sources and are called **welding generators.** Others take the power available from the lines and change it to power suitable for arc welding. These are known as **transformers** or **rectifier** or **inverter welding machines.** Both alternating and direct current can be used for some of the arc welding processes. The welding procedure and process will determine the type of power source required.

The most important part of the welding system is the welder or welding operator, the human element. The difference between welders and welding operators is a difference of the manipulative skills involved. The welder must have skill and ability to manipulate equipment to produce welds. The welding operator may monitor or operate an automatic welding machine. Robot welders may be supervised by human welders.

Another way of dividing or categorizing welding processes relates to whether filler metal is or is not used. If filler metal is not used, it is called an *autogenous weld.* **Filler metal** is "the metal or alloy to be added in making a welded, brazed, or soldered joint." It becomes the weld fillet or weld metal in a groove weld. In some welding processes, the filler metal is carried across the arc and deposited in the weld. In others, it is not carried across the arc but is melted by the heat of the arc and added to the molten puddle. If the weld metal passes through the arc, it is provided by an electrode. If it is melted by the heat of the arc and added to the pool, it is called a welding rod. Welding electrodes and welding rods have special composition requiring detailed specifications to describe them completely. Selection of filler metals is important; normally their properties should match the properties of the metal being welded. This metal, called the **base metal,** is defined as "the metal or alloy that is welded, brazed, soldered, or cut." This is the preferred term. In some countries the word *base material* or *parent* metal is used; for some processes the word *substrate* is used.

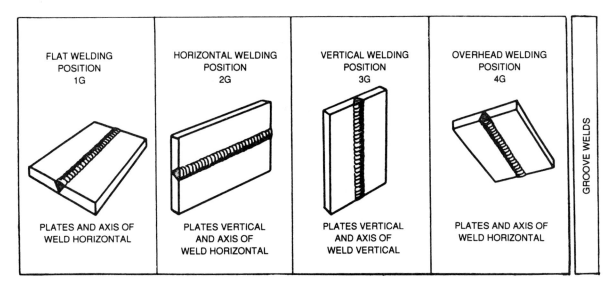

FIGURE 2–2 Welding positions for groove welds: plate.

The base metal often dictates the welding process that can be used.

To describe the making of a weld completely, it is necessary to specify the welding process and the method of applying the process. It is also necessary to describe the welding procedure, which is the detailed method and practices involved in the production of a weldment. This should include materials, joint design details, and method of welding, in order to describe how a particular weld or weldment is made. It is becoming more and more important to describe and document the entire welding procedure completely.

To ensure that the welds conform to demanding specifications, specialized inspection techniques are used. These include destructive and nondestructive testing methods. Nondestructive testing includes visual inspection, magnetic particle inspection, radiographic inspection, liquid penetrant inspection, and ultrasonic inspection. Welding quality control is required by most codes and is a necessary requirement for most manufactured products.

More terms and definitions will be presented in later chapters. It is important at the beginning to at least briefly define these terms so that you will better understand their meanings.

2-2 WELDING PROCESSES AND GROUPING

The American Welding Society (AWS) has made each welding and joining process definition as complete as possible so that it will suffice without reference to another definition. They define a *process* as "a grouping of basic opera- tional elements used in welding, joining, thermal cutting, or thermal spraying." The AWS master chart of welding and joining and allied processes is shown by Figure 2–3. All processes are broadly classed as welding and joining processes. Joining includes "any process used for connecting materials." This includes mechanical fastening and adhesive bonding, as well as welding and joining. Two processes are specifically designated as joining processes: brazing and soldering. AWS has grouped processes together according to the mode of energy transfer as a primary consideration. Capillary action distinguishes the joining processes from welding processes. The distinguishing feature of the welding process is the mode of energy transfer.

The AWS formulated process definitions from the operational instead of the metallurgical point of view. Thus the definitions proscribe the significant elements of operation instead of the significant metallurgical characteristics—"welding is a joining process that produces coalescence of materials by heating them to the welding temperature with or without the application of pressure or by the application of pressure alone and with or without the use of filler metal."

The society deliberately omitted the designation of *pressure* or *nonpressure* since the factor of pressure is an element of operation of the applicable welding process.

The designation *fusion welding* is not recognized as a grouping since fusion is involved with many of the processes. Other terms or factors—such as the type of current used in arc or resistance welding processes, whether electrodes are *consumable* or *nonconsumable* or *continuous* or *incremental,* or the method of application—are not shown in process groupings. These and other items characterize the methods by which the processes are performed. The AWS designation is used throughout this book.

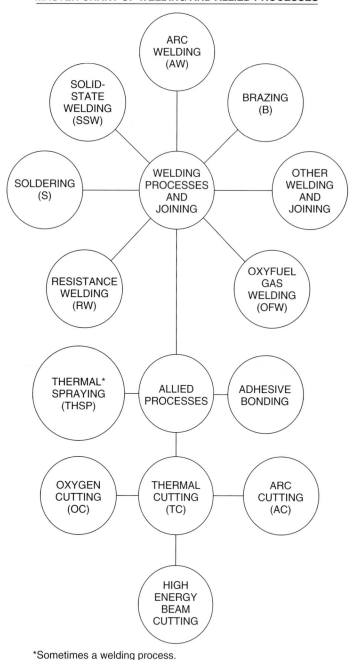

*Sometimes a welding process.

FIGURE 2–3 AWS master chart of welding and joining and allied processes.

Coalescence is defined as "the growing together or growth into one body of the materials being welded," and is applicable to all types of welding. The International Standard ISO 4063 is an alphabetical listing and definition of all welding processes.

Arc Welding

Arc welding (AW) is "a group of welding processes that produce coalescence of workpieces by heating them with an arc. The processes are used with or without the application of pressure and with or without filler metal." The arc welding processes are very common and consist of at least nine basic processes, some with several variations. There are two basic types of welding arcs. One uses a consumable electrode that is melted in the arc and the molten metal is carried across the arc gap. The other uses a nonconsumable electrode that does not melt in the arc and filler metal is added separately to the welding pool. These processes are described in detail in Chapters 5 and 6.

Oxyfuel Gas Welding

Oxyfuel gas welding (OFW) is "a group of welding processes that produces coalescence of workpieces by heating them with an oxyfuel gas flame. The processes are used with or without the application of pressure and with or without filler metal." There are four distinct processes within this group and in the case of two of them, *oxyacetylene welding* and *oxyhydrogen welding*, the classification is based on the fuel gas used. The heat of the flame is created by the chemical reaction or the burning of the gases. In the third process, *air acetylene welding,* air is used instead of oxygen, and in the fourth category, *pressure gas welding,* pressure is applied in addition to the heat from the burning of the gases. This welding process normally utilizes acetylene as the fuel gas. The oxygen thermal cutting processes have much in common with the welding processes.

Resistance Welding

Resistance welding (RW) is "a group of welding processes that produces coalescence of the faying surfaces with the heat obtained from resistance of the workpieces to the flow of the welding current in a circuit of which the workpieces are a part, and by the application of pressure." In general, the difference has to do with the design of the weld and the type of machine necessary to produce the weld. In almost all cases, the processes are applied automatically since the welding machines incorporate both electrical control and mechanical functions.

Other Welding and Joining Processes

This group of processes includes those which are not best defined under the other groupings. It consists of the following processes: *electron beam welding, laser beam welding, thermite welding, induction welding, percussion welding, electroslag welding,* and other miscellaneous welding processes.

Solid-State Welding

Solid-state welding (SSW) is "a group of welding processes that produces coalescence by the application of pressure without melting any of the joint components." The oldest of all welding processes, *forge welding,* belongs to this group. Others include *cold welding, diffusion welding, explosion welding, friction welding, hot pressure welding, roll welding, ultrasonic welding,* and *coextrusion welding.* These processes are all different and utilize different forms of energy for making welds.

Brazing

Brazing (B) is "a group of joining processes that produces coalescence of materials by heating them to the brazing temperature in the presence of filler metal having a liquidus above 450°C (840°F) and below the solidus of the base metal. The filler metal is distributed between the closely fitted faying surfaces of the joint by capillary action." A braze is a very special form of weld; the base metal is theoretically not melted. There are several variations within the brazing group. The source of heat differs among the processes. Braze welding relates to welding processes using brass or bronze filler metal, where the filler metal is not distributed by capillary action.

Soldering

Soldering (S) is "a group of joining processes that produces coalescence of materials by heating them to the soldering temperature and by using a filler metal having a liquidus not exceeding 450° C (840° F) and below the solidus of the base metals. The filler metal is distributed between the closely fitted faying surfaces of the joint by capillary action." There are a number of different methods identified by the way heat is applied.

2-3 METHODS OF APPLYING WELDING

There is more than one method of applying welding and some require manipulative skills. The title used for the individual doing the welding indicates the manipulative skill level involved. By definitions: The **welder** is "one who performs manual or semiautomatic welding." The **welding operator** is "one who operates adaptive control, automatic, mechanized, or robotic welding equipment."

These two definitions do not indicate the actual level of manipulative skill involved since they cover several methods of making welds. This tends to create confusion since a welder trained to do semiautomatic welding using one process may not be able to do manual welding with another process. This is not so important for the welding operator since the difference in skill for mechanized welding and automatic welding is not so great. The six methods of applying welding are shown in Figure 2–4. Each is defined as follows:

- **Manual welding (MA):** welding with the torch, gun, or electrode holder held and manipulated by hand.
- **Semiautomatic welding (SA):** manual welding with equipment that automatically controls one or more of the welding conditions.
- **Mechanized welding (ME):** welding with equipment that requires manual adjustment of the equipment controls in response to visual observation of the welding, with the torch, gun, or electrode holder by a mechanical device.
- **Automatic welding (AU):** welding with equipment that requires only occasional or no observation of the welding and no manual adjustment of the equipment controls.

Manual (MA)

Mechanized (ME)

Automatic (AU)

Semiautomatic (SA)

FIGURE 2–4 The methods of applying arc welding.

✴ **Robotic welding (RO):** welding that is performed and controlled by robotic equipment.

✴ **Adaptive control welding (AD):** welding with a process control system that determines changes in welding conditions automatically and directs the equipment to take appropriate action.

This concept of varying degree of control in the hands of the welder can be better understood when considering the normal activities involved to make an arc weld. These are defined and broken down into the following elements or functions:

1. *Starts,* maintains, and controls the arc
2. *Feeds* and directs the electrode into the arc (to control the placement of the weld deposit and fill the joint)
3. *Manipulates* the arc to control the molten metal weld pool
4. *Moves* the arc along the joint (travels) (to provide motion at proper speed to make the weld joint)
5. *Guides* the arc along the joint (to track the weld joint)
6. *Corrects* the arc to overcome deviations (to compensate for improper fitup)

Adaptive Control (AD)

Robotic (RO)

FIGURE 2–4 *(cont.)*

Closed loop means that real-time observations are made during welding and immediate corrections are made to compensate for deviations.

The person–machine relationship shown in Figure 2–5 shows that in manual welding the person has control over all of these functions and in automatic welding the same functions are completely controlled by the machine. The skill required is greatest when all functions are under the control of the person and diminishes as the functions are taken over by the machine.

Productivity is the amount of welding that can be performed by a welder or a welding machine in a day. This is determined by the operator factor, sometimes called duty cycle. Operator factor is the number of minutes per hour that the arc is actually making a weld. The different methods of application have different operating factors. Manual welding has the lowest operating factor, with semiautomatic welding approximately double this. Mechanized welding is higher with automatic or robotic welding with or without adaptive controls, the highest approaching 100%. The method to increase productivity and reduce welding costs is to move to the right of the application systems shown.

All of the arc welding processes can be analyzed with respect to the method of application. AWS also uses these methods of applying for brazing, soldering, and thermal cutting. The method of applying the resistance welding processes, solid-state welding processes, and most of the others is dictated by the process and the machine. Certain welding processes may be applied only as a manual process, while others are applied as semiautomatic, mechanized, or automatic. The method of application is extremely important when writing a procedure or assessing the economic capabilities of a process.

2-4 WELDING PROCEDURES

Welding is an accepted engineering technology which requires that the elements involved be identified in a standardized way. This is accomplished by writing a procedure which is simply a "manner of doing" or "the detailed elements [with prescribed values or range of values] of a process or method used to produce a specific result." The AWS definition for a **welding procedure** is "the detailed methods and practices involved in the production of a weldment."

A welding procedure is used to make a record of all of the elements, variables, and factors that are involved in producing a specific weld or weldment. Welding procedures should be written whenever it is necessary to:

✳ Comply with specifications and codes

✳ Maintain dimensions by controlling distortion

Method of Application Arc Welding Elements/Function	MA Manual (closed loop)	SA Semiautomatic (closed loop)	ME Mechanized (closed loop)	AU Automatic (open loop)	RO Robotic (open or closed loop)	AD Adaptive Control (closed loop)
Starts- maintains and controls the arc	Person	Machine	Machine	Machine	Machine	Machine
Feeds- and directs the electrode into the arc	Person	Machine	Machine	Machine	Machine	Machine
Manipulates- the arc to control the molten metal weld pool	Person	Person	Machine	Machine	Machine (robot) with or without sensor	Machine with sensor
Moves- the arc along joint (travel)	Person	Person	Machine	Machine via prearranged path	Machine (robot) with or without sensor	Machine with sensor
Guides- the arc along joint	Person	Person	Person	Machine via prearranged path	Machine (robot) with or without sensor	Machine with sensor
Corrects- the arc to overcome deviations	Person	Person	Person	Does not correct hence potential weld imperfection	Machine (robot) only with sensor	Machine with sensor

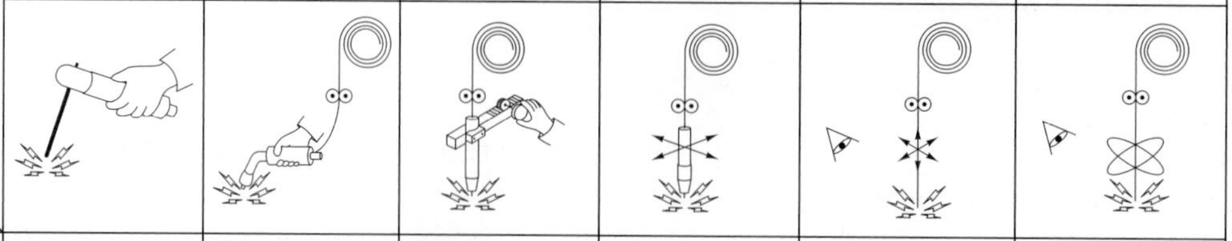

FIGURE 2–5 Person-machine relationship method of applying.

 ꙮ Reduce residual or locked-up stresses

 ꙮ Minimize detrimental metallurgical changes

 ꙮ Consistently build a weldment the same way

Welding procedures must be tested or qualified and they must be communicated to those who need to know. This includes the designer, the welding inspector, the welding supervisor, and most important, the welder.

When welding codes or high-quality work are involved, this becomes a **welding procedure specification,** known as a WPS, which is "a document providing the required welding variables for a specific application to assure repeatability by properly trained welders and welding operators." Different codes and specifications may have different requirements for a welding procedure, but in general a welding procedure consists of three parts as follows:

1. A detailed written explanation of how the weld is to be made
2. A drawing or sketch showing the weld joint design and the conditions for making each pass or bead
3. A record of the test results of the resulting weld

The variables involved in most specifications are considered to be essential variables. In some codes the term "nonessential variables" may also be used. Essential variables are those factors which must be recorded and if they are changed in any way, the procedure must be retested and requalified. Nonessential variables are usually of less importance and may be changed within prescribed limits and the procedure need not be requalified.

Essential variables usually include the following:

1. Welding process and its variation
2. Method of applying the process
3. Base metal type, specification, or composition
4. Base metal geometry, normally thickness
5. Base metal preheat or postheat
6. Welding position
7. Filler metal and other materials consumed in making the weld
8. Weld joint design
9. Electrical or operational parameters involved
10. Welding technique

Some specifications also include other variables, usually the following:

1. Travel speed
2. Travel progression (uphill or downhill)

3. Size of the electrode or filler wire
4. Use and type of weld backing

The procedure write-up must include each of these variables and describe in detail how it is to be done. The second portion of the welding procedure is the joint detail sketch and table or schedule of welding conditions.

Tests are performed to determine if the weld made to the WPS meets the standards described by the code or specification. If these tests meet the minimum requirements, the document becomes the welding **procedure qualification record** (PQR). This is a record of welding variables used to produce an acceptable test weldment, and the results of tests conducted on the weldment to qualify a welding procedure specification. The writing, testing, and qualifying procedures become quite involved and may be different for different specifications. This is covered in detail in Chapter 22.

In certain codes, welding procedures are prequalified. By using data provided in the code, individual qualified procedure specifications are not required for the standard joints on common base materials using specific arc welding processes.

The factors included in a procedure should be considered in approaching any new welding job. Using knowledge and experience, establish the optimum factors or variables in order to make the best and most economical weld on the material to be welded and in the position that must be welded.

Welding procedures take on added significance based on the quality requirements. When exact reproducibility and perfect quality are required, the procedures will become much more technical with added requirements, particularly in testing. Tests will become more complex to determine that the weld joint has the necessary properties to withstand the service for which the weld is designed.

Procedures are written to produce the highest-quality weld required for the service involved, but at the least possible cost and to provide weld consistency. It may be necessary to try different processes, different joint details, and so on, to arrive at the lowest-cost weld that will satisfy the service requirements of the weldment.

The contents of a welding procedure are brought out at this early stage to help the reader realize the importance of defining the factors involved in making a successful weld.

2-5 WELDING PHYSICS AND CHEMISTRY

Welding follows all the physical laws of nature. A good understanding of physics and chemistry will help you better understand how welds are made. Physics deals with energy and motion and is subdivided into such subjects as mechanics, sound, light, friction, magnetism, electricity, and heat. This section briefly describes some of these subjects.

The science of mechanics involves physical laws that relate to forces, motion, and direction. The term **force** is defined as a push or pull; specifically, a force is a tendency to produce a change in motion of the body upon which it acts. It is not necessary that the body be in motion; it is only necessary that there is a tendency to produce change. The first law of motion states that a body will remain at rest or in uniform motion if no force acts upon it. One of the forces that we live with daily is the pull of gravity, which acts on all objects on the earth's surface. Another of the laws states that for every action there is an equal and opposite action; that is, forces act in pairs. An opposite action force cannot exist before the action force takes place. If you push against the wall, the wall pushes back an equal or opposite force equal to your push; otherwise the wall would move. All forces and all opposite forces or reaction forces have direction and also have **magnitude,** which is the amount of force involved. It is possible to graphically represent different forces and magnitudes and direction by means of vector diagrams. Vectors indicate the direction, and the length of the vector indicates the magnitude. A thorough knowledge of this science is necessary in order to design welded structures. It is the basis for establishing the sizes of members and would be the basis for the size of welds to join members together.

The science of sound is important to welding. Sound is transmitted through most materials—metals, gases, liquids, and so on—but it will not pass through a vacuum. Sound is an alternating type of energy based on vibrations, which are regions of compaction and rarification. A compression wave and rarification wave are alternating pressures or vibrations which allow your eardrums to hear. The hearing of most people is sensitive between 20 and 20,000 vibrations per second. Sound has pitch, loudness, and quality. *Pitch* is defined as frequency; the higher the frequency, the higher the pitch. *Loudness* is subjective and is related to intensity, which is the basis of the energy in sound. *Quality* is the function of waveform based on the frequency and phase of combining vibrations. The use of sound in welding is in the higher-than-audible sound vibrations, above the normal hearing range for people. Ultrasonic vibrations are used to make welds and they are also used to detect voids in metals.

The frequency spectrum shown in Figure 2-6 is of interest when studying welding. This figure shows the spectrum from the lowest frequency and the longest wavelength to the highest frequency with the shortest wavelength, and covers both the sonic and the electromagnetic spectrum. Sonic frequencies are at the low end and, having a relatively low travel speed, must travel through some type of media. Electromagnetic radiation travels at the speed of light and will travel through a vacuum, and it is normally in the higher-frequency range.

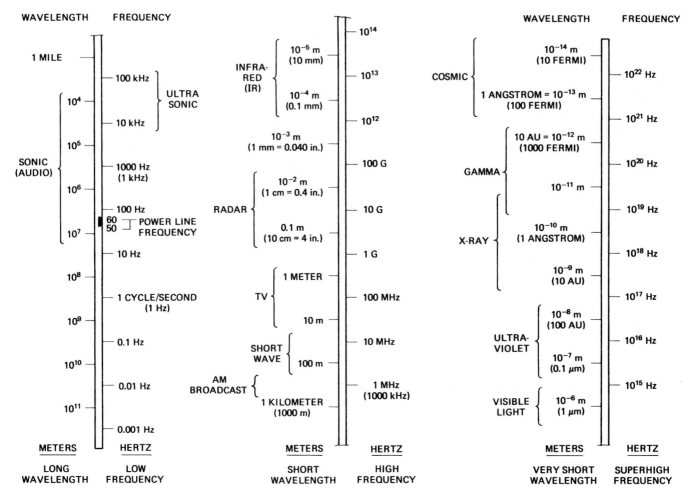

FIGURE 2-6 Frequency spectrum for sonic and electromagnetic radiation.

Sonic radiation includes the audio frequencies, which we normally hear, and the ultrasonic range frequencies, higher than the normal hearing range of people. The upper range for adults is normally from 15 to 17 kilohertz (kHz). Children have the ability to hear higher frequencies; dogs and some other animals can hear frequencies even higher. The musical scale ranges from below 20 Hz to over 4000 Hz, with middle Cat 261 Hz. The speed of travel of sound is the slowest through gases, which is approximately 200 meters per second (m/sec). It travels faster through liquids, ranging from approximately 1000 m/sec to almost 2000 m/sec, and has the highest speed through solids, ranging from 1000 to over 4000 m/sec in nonmetals and as high as 10,000 m/sec through metals. The rate is different for different metals. It is also different through liquids at different temperatures, traveling faster at lower temperatures. The pulsing rates used in welding and the frequency used in ultrasonic testing are in these ranges.

The electromagnetic frequency spectrum is much larger and includes radio broadcasting frequencies up through the infrared frequencies, visible light, x-rays, and gamma rays. Electromagnetic frequencies travel approxi-

mately 186,000 miles per second in air or vacuum, but change to a slower speed when going through liquids or transparent solids. The high-frequency current used for arc starting in gas tungsten arc welding is in the range of 2 megahertz. The frequency of power line alternating current is 60 Hz in North America and 50 Hz in many other parts of the world. It is important to keep these numbers in perspective since they vary over such a wide range. A study of Figure 2–6 will give you a better idea of the frequencies involved in different aspects of welding and weld testing.

The science of light also involves welding. The laser beam welding process utilized light energy at very high concentrations to create heat sufficient to cause melting, which can be used for welding or cutting. Light is a by-product of the arc welding processes. Light is given off by the arc and by heated electrodes and base metals. Light is a phenomenon that has never been completely explained to the satisfaction of scientists. Propagation of light is explained by alternating vibrations or the wave theory, while the energy transfer of light is explained by the particle theory. It is sufficient for us to know that light is transmitted through vacuums and gases, but not through all types of materials. Light is of interest to the welder from the point of view of protecting the body from the effects of the light rays. In the electromagnetic spectrum, light ranges from the lower frequencies of infrared through visible light up through the ultraviolet. The speed of light is independent of the light source, its intensity, or color. Light transmits energy, and the higher the frequency, the greater the energy; however, light radiates equally in all directions and its strength diminishes by the square of the distance.

Welding also involves the science of friction. Here we are interested in dynamic friction, better known as *sliding friction.* This is the force between two moving bodies, and if sufficient force is available, heat will be generated. This is the basis for the friction welding process.

The magnetic theory and its relationship to current flow are explained in the section on welding electricity. One aspect of magnetic fields can be detrimental. Welders call it arc blow, which is the deflection of an electric arc from its normal path due to magnetic forces. The leads from the welding machine to the electrode, and from the work back, carry a heavy current and create a magnetic field. The welding current flowing through the electrode and the base metal, providing it is ferromagnetic, also creates magnetic fields. The intensity of the magnetic field is directly proportional to the square of the current flowing. The distribution of a magnetic field in a welding circuit can become quite complex, particularly for nonuniform joint details and also when fixtures are employed. If the distribution of the magnetic field close to the arc is not uniform, it may cause the arc to deflect or attract toward the stronger portion, depending on polarity. Arc blow can create difficulties that affect weld quality. One advantage of alternating current is that the arc blow is minimized since the magnetic field is changing at line frequency and does not build up to as great a force.

Chemistry deals with the makeup of all matter. We are most interested in metals in connection with welding. By definition, matter is anything that occupies space and has mass or weight. Also by definition, elements are those particular kinds of matter that cannot be decomposed or broken down into simpler substances by ordinary means. Pure metals and pure gases are examples of elements. Compounds or mixtures can be broken down into their original elements. The elements are composed of atoms that are identical with each other atom of the same element but are different from atoms of other elements. The molecule is the smallest particle of the substance that has all the properties of that substance. It is a combination of two or more atoms of the same element or of different elements.

Many of the elements have similar properties; for example, some are inert gases, others are noble metals, others are active gases, and so on. This allows the elements to be classified and put into groups of families. This classification is called the *periodic table,* which is shown in all chemistry textbooks.

To better understand metals, we must first consider the atom and its structure. Scientists believe that each atom is composed of a very small compact nucleus surrounded by empty space in which one or more electrons revolve about the nucleus. It is believed that the nucleus of the atom is the most dense form of matter known. It is made of or contains two main types of particles known as protons and neutrons. These particles differ from each other in their charge, but they have about the same weight. The positive particle of matter is called a proton. All atoms have at least one proton in their nuclei. The number of protons in the nucleus is equal to the number of electrons outside the nucleus. Each proton has a charge of plus one; the charge on the nucleus is positive since it contains positively charged protons and no electrons. The neutron in the nucleus is a particle of matter which has a relative weight of 1 but it has no electrical charge. The third item is the electron and it is very light in comparison with the nucleus of the atom. Each electron has a charge of negative 1. Electrons of an atom are located in shells around the nucleus. These shells are more properly called energy levels because electrons in different shells have different amounts of energy. Electrons revolve around the nucleus in these shells in varying distances from the nucleus. Electrons are relatively far from the nucleus so that most of the atom consists of *empty space.* The difference between atoms of different elements is the result of differences in the number of protons and neutrons in the nucleus and the difference in the number and arrangement of the electrons surrounding the nucleus. Electrons in the outer shells have more energy than those in the inner shells. An electron can change from one shell or energy level to another: If it absorbs energy, it moves to an outer shell or to a higher energy level; if it gives off energy, it drops to a shell closer to the nucleus. Energy emitted when an electron drops from a higher to a lower energy level is in the form of electromagnetic radiation, light, or x-rays. The study of the makeup of the atoms is extremely technical and beyond the scope of this book. However, this brief explanation will help you better understand the makeup of metals.

Matter can exist in four states: solids, liquids, gases, and plasmas. Changes from one state to another are brought about by supplying energy in the form of heat. Water in the solid state is ice; by adding heat, the ice changes to water, which is its liquid state, and by adding additional heat, it will be converted to its gaseous state. The reverse can be done by removing heat energy. Then the gas (steam) turns to the liquid (water) and then to the solid (ice). Most substances can be changed from one physical state to another in the same manner. The temperature of these changes indicates the physical state the element will be in at normal room temperatures.

Each of the elements on the periodic table has its own name and symbol. Most of these elements will combine chemically to form compounds. There is the law of definite composition, which states that a chemical compound always contains the same elements with the same ratio of atoms of each. This is not true for mixtures. Most commercial metals are mixtures or alloys in that they are predominantly the element of the pure metal plus additions of other elements but not chemically combined as a compound. Gases also occur as elements, compounds, or mixtures. Air is a mixture of approximately 78% nitrogen and 21% oxygen, with small amounts of other elements. Carbon dioxide is a compound and always in the ratio of one atom of carbon and two atoms of oxygen. Argon is an element and, more important, an inert gas that will not combine chemically with any other element.

Several other chemical definitions relate to welding. One is known as burning or oxidation. This takes place when any substance combines with oxygen, usually at high temperatures. An example of this is the combining of acetylene with oxygen. This produces carbon dioxide plus water plus a large amount of heat. We use the heat produced by the burning of acetylene in the flame of the oxyacetylene torch to make welds. In all oxidation reactions, heat is given off. Oxidation can occur very slowly, as in the case of rusting. If iron is exposed to oxygen at high temperature, rapid oxidation or burning will occur with the liberation of more heat. Rapid oxidation or burning does not occur until the kindling temperature of the material is reached. In the case of a liquid this term is called the *flash point.* Oxidation is very important in welding operations since oxygen of the air is usually present as well as heat.

Another chemical definition is reduction, which is the process by which oxygen is taken from another material. The substance used to take the oxygen from the material is called a *reducing agent.* A reducing agent is anything that adds electrons to another material. Hydrogen is one of the most active reducing agents; however, in the case of the iron and oxygen reaction mentioned, the iron is the reducing agent. Whenever there is an oxidation reaction, there is also a reduction reaction. A common term in oxyacetylene welding is **reducing atmosphere** or **oxidizing atmosphere.** The flame can be adjusted to provide sufficient oxygen for complete combustion or an excess of acetylene and insufficient oxygen, resulting in incomplete combustion. Acetylene is high in hydrogen and carbon. The oxidizing flame would contribute excess oxygen; the reducing flame would contribute hydrogen and carbon.

The last two definitions are common in chemistry but less common in welding. The words are *acidic* and *basic* and refer to acid or base substances. A measure of basicity or acidity is by means of the pH scale. A pH of 7 is considered the neutral point. Pure water has a pH of 7 since it has the same number of hydrogen ions (H^+) as hydroxide ions (OH^-). Acids or acidic substances have an excess of hydrogen ions and have a pH value of less than 7. Bases or basic substances have an excess of hydroxide ions and have a pH value of more than 7. The terms are used in connection with nonmetallic slags used in welding and also used in steel-making. These slags are related to the coatings of electrodes and have acidic or basic characteristics when heated to steel-melting temperatures. Certain types of electrodes are called basic type because their coatings produce basic slags which react with impurities in the weld metal. Basic-type coatings are the low-hydrogen, sodium-, or potassium-containing types which produce basic slags; the basic slags remove appreciable amounts of undesirable phosphorus and sulfur from the molten weld metal. The cellulosic electrodes are sometimes called the acidic type since an excess amount of hydrogen is present in the arc atmosphere and in the slag.

 **QUESTIONS**

2-1. What is a weld?

2-2. What are the five basic joints?

2-3. What is the most popular weld type?

2-4. There are approximately how many welding processes?

2-5. How many arc welding processes are there?

2-6. How many welding positions are there?

2-7. Define the word *coalescence.* How does it apply to welding?

2-8. What is the difference between GTAW and GMAW?

2-9. Can all the arc welding processes be used in all positions?

2-10. Explain the difference between the welder and the welding operator.

2-11. Name the activities performed when making a manual arc weld.

2-12. Name the various methods of applying an arc weld.

2-13. Why is a written welding procedure required?

2-14. What method of application requires the highest level of manipulative skill?

2-15. What method of application requires the least manipulative skill?

2-16. What is the polarity of the electrode for reverse polarity welding?

2-17. Where does the deposited metal come from in a nonconsumable electrode welding arc?

2-18. Name the various branches of physics that relate to welding.

2-19. Pure metals and pure gases are elements or compounds?

2-20. What is the difference between a reducing atmosphere and an ordinary atmosphere?

WELDING PERSONNEL, TRAINING, AND CERTIFICATION

OUTLINE

3-1 THE WORK OF THE WELDER

Everything we use in our daily lives is welded or at least made by equipment that is welded. Welders work everywhere, from the small repair shop down the street to the largest factories. They help build space vehicles, coffeepots, buildings, oil drilling rigs, automobiles, and millions of other products. In construction welders are virtually rebuilding the world, extending subways, building bridges, and helping to improve the environment by building pollution control devices. The use of welding is practically unlimited. There is no lack of variety in the type of work that is done.

Welding is challenging and interesting. The actual progress and completion of a weld happens before the welder's eyes, providing a sense of accomplishment. It is exacting work. Every weld on a nuclear power reactor must be perfect. Welding is done under all kinds of conditions, both indoors and outdoors; in the construction trades that can mean working in all kinds of weather, at elevated heights, or even below the surface of the sea. In the construction industry welding is considered a tool of the trade.

Personnel who weld may be called boiler workers, plumber pipe-fitters, iron workers, sheet metal workers, and so on, even though the majority of their time is spent welding. This wasn't always the case; it became a legal question that was settled in 1917 during World War I. Then, welding was just becoming popular. In the railroad shops welding was used by blacksmiths, sheet metal workers, boilermakers, and others. Each craft union claimed jurisdiction over welders, which led to a nationwide strike. During the war, the U.S. government was running the railroads, and the strike caused a national emergency. To settle this issue, the U.S. Secretary of Labor ruled that welding was a tool of all of the trades and no one trade had jurisdiction over welding.[1] This still applies and welding is not an apprenticed trade. Today welding is included in apprenticeship programs of many trade unions. Often the welder must be a member of a labor organization of the craft for which welding is associated. The apprenticeship program can take several years. Sometimes when welders are urgently needed, apprenticeship requirements are canceled and the new welder can become a journeyman quickly.

Welding is demanding. Sometimes the welder must work in unusual and awkward positions. They may weld overhead, inside structures—or just about anywhere.

FIGURE 3–1 Welding on highly repetitive parts.

FIGURE 3–2 Welding on nuclear piping.

They weld on a variety of materials. Welding is a desirable career since welders receive high pay. Pay is similar to that of other skilled occupations and is sometimes based on the qualification of the welder. Qualification tests may be required and are of different levels of difficulty. Passing the more difficult test allows welders to work on the higher paying jobs. The hours of work are the same as for the others in the same factory or on the same project.

Welders are employed in industries that produce everything from appliances to ships. These welder-using industries were described in Chapter 1. A welder's job in different industries and companies and in different geographical locations varies tremendously. In manufacturing the variation of welding is very broad, but the work may be more repetitive unless it happens to be maintenance and repair welding. In some production shops, the welder may make the same type of weld on the same part day after day as shown in Figure 3-1. Welders in construction find each job totally different. A typical construction job is shown in Figure 3-2 on a nuclear piping assembly. Between these extremes are many variations.

There are some disagreeable parts to the welder's job. In some cases they must wear special clothing to protect them from the arc light and sparks. Usually they must wear a helmet to shield arc rays, which isolates them from their coworkers for a period of time. They are exposed to the heat generated by the arc and they may be exposed to fumes generated in the arc. In spite of this, it is very sat-

isfying to see the weld just made, the results of their skilled craftsmanship.

The government has provided occupational titles and job descriptions for many different jobs in the *Dictionary of Occupational Titles* (DOT), which includes many jobs related to welders, cutters, and related occupations. This is being replaced by the Standard Occupational Classification (SOC) system. It will provide an improved method to allow government agencies to gather and classify employees for statistical purposes. However, this two-volume handbook is not yet available.

Job descriptions are written for many occupations. An analysis of welders has been made to determine the work performed and the worker traits, including the following:

Work performed
1. Worker functions
2. Work fields
   a. Methods
   b. Machines, tools, and work aides
3. Materials, products, subject matter, and service

Worker traits
1. Training time
   a. General education development
   b. Special vocational programs

2. Aptitude
3. Temperament
4. Interest
5. Physical demands
6. Working environmental conditions

These are being updated and include such items as computer intelligence and communication skills.

The training time for general education development for welding ranged from a minimum of one day to a maximum of four years. Aptitude for welders indicated that spatial aptitude, form perfection, finger dexterity, and manual dexterity are the most significant for welders. Regarding temperament, welders have indicated a preference for activities dealing with things and objects and for activities that are carried on and related to process, machine, and technique.

The physical demands on welders are strength, climbing, balancing, stooping, kneeling, crouching or crawling, reaching, handling, fingering or feeling, talking, hearing, and seeing. Seeing is involved in all welding, and is important to welders because of the need to continually watch, observe, and see the weld as it is being made. Depth perception is especially important. The working environmental conditions show that welders work indoors primarily on factory jobs, and both indoors and outdoors for construction jobs. It shows that welders are exposed to certain hazards that are similar to those of other metalworking occupations in industry or construction.

Persons interested in becoming welders should take the trade aptitude test program called the General Aptitude Test Batteries. These are available at state employment services in cooperation with the Department of Labor. Scores for spatial aptitude, form perception, finger dexterity, and manual dexterity should be well above the minimum, although a strong desire to become a welder may overrule all aptitude tests. The welder, the flame cutter, and the welding operator represent the largest number of jobs in the welding field; however, there are others such as welder fitter, specialized welder, precision welder, and robot welder. Welders supervise arc welding robots since they are the best people qualified to program weld robots. Welding is a worthwhile and rewarding occupation that can be used as a stepping-stone for many related careers.

More data concerning the job of the welder is available from the American Welding Society and from the government.

3-2 THE JOB OUTLOOK FOR WELDERS

For many years the U.S. Department of Labor's *Occupational Outlook Handbook* has been a recognized source of career information. It is revised periodically and pro-

vides information on production occupations. It is available in your local library. The section entitled "Welders, Cutters, and Welding Machine Setters, Operators, and Tenders" describes the nature of the work, working conditions, employment, training, other qualifications and advancement, job outlook, earnings, and related occupations. This information is also available in the government's *Monthly Labor Review* which includes projections of employment for the next ten years. They state that there will be good employment opportunities for welders through the year 2006 in spite of the fact that much welding done in manufacturing is increasingly being automated. Employment for welders and cutters is expected to increase slowly while that of welding machine operators should remain unchanged through 2006. Even though production should increase, the number of welding machine operators is expected to remain level because of greater use of robots and other automated welding machines. Manual welders, especially those with a wide variety of skills, will increasingly be needed for sophisticated fabrication tasks and repair work that do not lend themselves to automation. Many job openings for welders will result from the need to replace experienced, retiring workers. The aging of the nation's infrastructure, which requires more products needing repair or replacement, will also provide employment opportunities.

Welders and cutters held about 367,000 jobs in 1998, a number that is expected to grow by 8.3% to a total of 398,000 in 2008. Welding machine setters, operators, and tenders held about 109,000 jobs in 1998, a number that is expected to grow 5.4% to a total of 115,000 in 2008.

In Chapter 1, "The Welding Industry" describes those industries that use welding to manufacture their products. There, Figures 1-11 and 1-12 provide a database of information showing the industries and the amount of welders in each classification; it is updated periodically. This database provides the foundation for Figure 3-3, which shows the numbers of welders and cutters employed in 1998 and then projects through 2008, and Figure 3-3, which does the same for welding machine setters, operators, and tenders. Note the differences in the two classifications.

3-3 WHERE WELDERS WORK

The breakdown of numbers of welders in various industries, as shown in Figures 3-3 and 3-4, is common to most industrialized countries; the ratios may change but the production requirements are similar. Developing countries, which also employ welders, are establishing welding jobs in different industries that vary as a country develops and manufactures more advanced products.

Many welders work in industries not listed. They work in ranching and farming, helping to build and

FIGURE 3–3 Number of welders and cutters employed

Industry	1998 Employment		Projected 2008 Employment	
	Number	Percent distribution	Number	Percent distribution
Total employment, all industries	367,708	100.00	398.362	100.00
Fabricated structural metal products	39,313	10.69	47,607	11.95
Motor vehicles and equipment	26,794	7.29	23,064	5.79
Construction and related machinery	24,637	6.70	31,540	7.92
All other repair shops and related services	19,313	5.25	17,274	4.34
Ship and boat building and repairing	15,755	4.28	16.872	4.24
Self-employed workers, primary job	15,281	4.16	16,163	4.06
Wholesale trade, other	14,965	4.07	15,342	3.85
All other transportation equipment	13,397	3.64	16.060	4.03
Personnel supply services	13,238	3.60	21.927	5.50
All other special trade contractors	12,933	3.52	14,860	3.73
Industrial machinery, nec	11,119	3.02	11,606	2.91
Heavy construction, except highway and street	9,448	2.57	9,809	2.46
Machinery, equipment, and supplies	7,915	2.15	8,327	2.09
Farm and garden machinery	7,393	2.01	6,845	1.72
Miscellaneous fabricated metal products	7,183	1.95	7,536	1.89
General industrial machinery	7,076	1.92	8,011	2.01
Special industry machinery	5,677	1.54	6,025	1.51
Refrigeration and service machinery	5,581	1.52	6,046	1.52
Nonresidential building construction	5,219	1.42	5,487	1.38
Oil and gas field services	5,140	1.40	5,552	1.39
Metal forgings and stampings	4,469	1.22	4,843	1.22
Blast furnaces and basic steel products	3,746	1.02	2,576	0.65
Self-employed workers, secondary job	3,745	1.02	3,701	0.93
Railroad transportation	3.728	1.01	2,897	0.73
Office and miscellaneous furniture and fixtures	3,622	0.99	4,046	1.02
Iron and steel foundries	3,154	0.86	3,233	0.81
Federal government	2,958	0.80	2,599	0.65
Metalworking machinery	2,884	0.78	2,712	0.68
Plumbing, heating, and air-conditioning	2,854	0.78	3,078	0.77
Manufactured products, nec	2,696	0.73	2,878	0.72
Household furniture	2,656	0.72	2,599	0.65
All other agriculture, forestry, and fishing; except secondary jobs	2,575	0.70	2,311	0.58
All others less than .7% distribution	61,244	16.67	64,936	16.30

Data from Bureau of Labor Statistics, National Industry-Occupation Employment Matrix, Occupation Report 2000.

maintain equipment; in the fruit and sugarcane industry; and in the printing and publishing industry. In fact, they work in any industry that uses metal equipment that is likely to break down and need repairing. Many welders work alone. They have their own rigs or own their own companies, and they work in all fields. This makes it an even more interesting occupation.

Welders work in different industries and locations in every state in the Union. Data collected by the Bureau of the Census shows the geographical location of welders and flame cutters. The largest number of welders is employed in the Great Lakes area: Pennsylvania ranks third with 6.7% of the welders, Ohio is fourth with 6.3%, Michigan is fifth with 5.9%, Illinois is sixth with 5.5%, New York is eighth with 3.7%, Indiana is ninth with 3.5%, and Wisconsin is tenth with 3.0%. Despite this, Texas ranks first with 9%, and California ranks second with 7.5% of the total. These rankings change from year to year, depending on the health of a specific industry and the business climate in each state.

The standard metropolitan statistic area ranking is as follows: Detroit, 1; Los Angeles/San Diego, 2; Chicago, 3; Houston, 4; Philadelphia, 5; Dallas/Fort Worth, 6; Pittsburgh, 7; St. Louis, 8; New York City, 9; and New Orleans, 10.

FIGURE 3-4 Number of welding machine setters, operators, and tenders

Industry	1998 Employment		Projected 2008 Employment	
	Number	Percent distribution	Number	Percent distribution
Total employment, all industries	109,604	100.00	115,507	100.00
Fabricated structural metal products	16,617	15.15	19.014	16.46
Motor vehicles and equipment	12,813	11.69	11,555	10.00
Construction and related machinery	9,939	9.06	10.988	9.51
Metal forgings and stampings	9,111	8.31	9,404	8.14
Miscellaneous fabricated metal products	8,468	7.73	9,307	8,06
Industrial machinery, nec	6,713	6.12	7,341	6.36
Farm and garden machinery	6,142	5.60	5,957	5.16
Refrigeration and service machinery	3,690	3.37	4,188	3.63
General industrial machinery	3,329	3.04	3,589	3.11
Special industry machinery	3,144	2.87	3,496	3.03
All other transportation equipment	2,868	2.62	3,165	2.74
Office and miscellaneous furniture and fixtures	2,778	2.53	3,251	2.81
Electronic components and accessories	2,279	2.08	2,935	2.54
Blast furnaces and basic steel products	2,092	1.91	1,674	1.45
Cutlery, hand tools, and hardware	1,688	1.54	1,551	1.34
Household appliances	1,660	1.51	1,174	1.02
Metalworking machinery	1,453	1.33	1,431	1.24
Ship and boat building and repairing	1,413	1.29	1,437	1.24
Electric lighting and wiring equipment	1,310	1.19	1,197	1.04
Aircraft and parts	1,120	1.02	1,505	1.30
All others less than 1% distribution	10,987	10.03	11,348	9.82

Data from Bureau of Labor Statistics, National Industry-Occupation Employment Matrix, Occupation Report 2000.

FIGURE 3-5 Welders and flame cutters employed, by industry groups

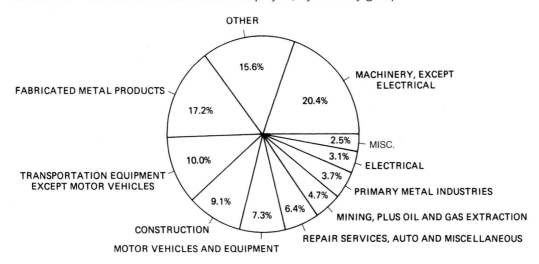

3-4 TRAINING PROGRAMS AND SCHOOLS

Welder training with emphasis on the manipulative skills is provided by a wide variety of schools. Welding is taught in high schools, vocational schools, technical schools, colleges and universities, apprentice training schools, company schools, and in the armed forces schools.

Most of these schools provide training programs based on a specific welding process. The depth of the program and the skills taught depends on the length of time provided. In general, these programs are as follows:

※ **Arc welding** (shielded metal arc welding) (SMAW), popularly called "stick" welding, involves up to 250 hours. This course should cover shielded metal arc welding in all positions, including pipe. It should teach the welding of thin and medium thickness steels using single and multiple pass procedures. It should provide training for a wide variety of electrode types and should provide training using different equipment. Related topics include welding safety, welding power sources, welding electrode selection and identification, weld testing, and inspection. The successful student should be able to pass qualification tests necessary for structural and most steel welding jobs.

※ **Gas welding,** which covers oxy-acetylene welding, brazing and flame cutting, can require up to 125 hours. This course would include the assembly of gas welding equipment, proper lighting and adjustment of torches, and welding of thin and medium thickness materials in all positions. It should include brazing and flame cutting of steel, both manually and with a machine. Related subjects include welding safety, the properties of gases, the combustion of fuel gases with oxygen, the selection of tips for different applications, the adjustment of the oxygen and fuel gas, and the selection of filler metals and fluxes.

※ **Gas tungsten arc welding** (GTAW), popularly called TIG, should require up to 60 hours. It should include welding, welding safety, assembly of equipment including the torch and adjustment, arc starting and technique, and welding nonferrous metals, particularly aluminum and stainless steel, in all positions. Related information should include equipment maintenance, the selection and its adjustment, the selection of shielding gas, and the selection of filler metals for different base metals.

※ **Gas-shielded metal arc welding** (GMAW), popularly called MIG, requires up to 50 hours. Flux-cored arc welding (FCAW) is very similar and should be included. This should include the assembly and adjustment of the equipment including the power source, the wire feeder, and the gun and cable assembly. It should also include the variations of solid wire and flux cored electrode wire and shielding gas requirements. Related topics include the principles of constant voltage power supplies and wire feeders, and the selection of filler metals and shielding gas for different base metals.

※ **Plasma arc welding** (PAW) or plasma cutting. Training requires a program of 25 class hours. This should include the explanation of the plasma process, the understanding of the plasma torch and its adjustment, the selection of the electrode and shielding gases as well as the technique involved in all positions, welding, and cutting of all metals.

A mastery of and experience with any one process would reduce the time required for learning the next welding process.

In general, SMAW and gas welding programs are being reduced or eliminated as these processes become less popular. GMAW, GTAW, and plasma cutting are becoming much more popular, with training programs available in most areas.

Many advanced programs are provided for welding structural shapes and pipe welding, both with uphill travel and downhill travel. Other programs are available for GTAW on particular metals and alloys such as titanium or other difficult-to-weld metals. Upon successful completion of these programs, students should have the necessary skill and ability to be employable for welding these materials. With enough experience, students can pass rigorous qualification tests required for welding under certain codes and specifications.

Qualification tests are required before a welder is employed for welding many types of products.

No matter what type of welding training program is provided, it should be presented in the following four steps.:

1. A lecture and discussion prior to each new phase of instruction. This provides background information that lasts from fifteen to thirty minutes and should include a question-and-answer period.

2. An explanation and demonstration of a new training phase. It may be repeated so that there are no questions about the technique.

3. Supervised individual practice after the discussion and demonstration. The trainee goes to an individual booth and practices the technique.

4. Periodic practical tests. The student is given a practical test and shall make the proscribed welds. Weld tests are very useful since they help students overcome the fear of qualification tests that are required many times throughout a welder's career.

Proper facilities and equipment are necessary to provide a successful training program. The well-planned welding shop must be designed for safety of the student. It should be of fireproof construction; exhaust ventilation, heating and air make-up should be provided. In warm climates, cooling should be provided. The weld shop must have adequate facilities: sewer, hot and cold water, compressed air, electricity. Electrical power requirements are considerably greater than for other shops, based on the amount of welding equipment provided. Welding shops should also have gas distribution systems for fuel gas, oxygen, and shielding gas. These systems must be designed by experts and must meet all applicable codes. Special precautions must be taken for fire protection: Eliminate all combustible materials in the shop and provide for fire extinguishers and other safety equipment. Local and state regulations must be followed.

The welding equipment must be representative of the type of machines used by local industry and should be of industrial quality. Limited-duty and light-duty-cycle machines should not be used.

Training aids including posters, movies, videocassettes, transparencies, slides, filmstrips, workmanship samples, cutaway equipment, and so on, should be provided. They should be selected with respect to the subject area being taught and should fit the objectives of the course. Training aids must use the correct terminology and safety practices and be technically accurate. Audiovisual equipment must be available. In addition, storage space for student projects, etc., should be provided.

The classroom area should be adequate for the size of the classes involved. This is true also for the demonstration areas.

Arrangements must be made for obtaining welding materials, gases, and electrodes. Arrangements can be made with local industry for base metal. The welding school shop must have heavy-duty furniture for each booth.

Welders: National Skill Standards

The American Welding Society (AWS), in cooperation with the U.S. Department of Education, has developed a *National Skill Standards* for the training and qualification of welders. This program will produce employable graduates who possess a certain level of competency as welders. AWS makes available curriculum guidelines and examination questions to be used and will recognize those institutions that adopt this program. Schools Excelling through National Skill Standards Education (SENSE) provides for three classes of welders: (1) entry-level welder (Level I), (2) advanced welder (Level II), and (3) expert welder (Level III).

The AWS provides guides that outline training programs for the three levels of welders. It also publishes specifications for qualifications and certification for each level of welder. The specifications at each level require passing performance of practical knowledge tests and worker qualification standards.

The written examination includes safety, welding, and cutting terms, theory fundamentals, non-destructive testing (NDT) inspection and testing, knowledge of drawings, welding and inspection symbols, codes, standards, welding procedures, and other topics. The performance tests are designed to show competency in practical skills. A candidate must read prints, make layouts, perform thermal cutting, fitup, assemble, and weld according to welding symbols, using specific welding procedures, and then inspect the work.

Schools may make application to the AWS to become a SENSE participating organization. SENSE schools will receive all necessary examination materials and a library of all referenced standards, and will become an AWS educational institution member.

Entry-Level Welders (Level I) An individual employed in this position is considered to possess a prerequisite amount of knowledge, attitude, skill, and habits required to perform routine, predictable, receptive, and proceduralized tasks involving motor skills and limited theoretical knowledge while working under close supervision.

- *Welding and cutting processes:* SMAW, GMAW, FCAW, and GTAW, plus OFC, PAC, and CAC-A (for all levels).
- *Base metals:* carbon steel, stainless steel, and aluminum.
- *Thickness and type:* 3/8-inch plate for SMAW, GMAW, and FCAW, and limited positions (depending on base metal) for GTAW.
- *Positions:* All positions using SMAW, GMAW, and FCAW, and limited positions (depending on base metal) for GTAW.
- *Qualifications:*
  - Workmanship samples: GMAW, FCAW, and GTAW
  - Performance qualifications: SMAW in 2G (horizontal) and 3G (vertical) positions

Advanced Welders (Level II) An individual employed in this position is considered to possess a prerequisite amount of knowledge, attitude, skill, and habits required to perform proceduralized tasks under general supervision, and complex tasks involving the use of theoretical knowledge and motor skills under close supervision.

- *Base metals:* Same as for entry level.
- *Thickness and type:* 3/8-inch plate-Schedule 80 pipe for SMAW, GMAW, and FCAW, and 11-18 gauge sheet and tubing for GTAW.
- *Positions:* All positions using SMAW, GMAW, FCAW, and GTAW.

- *Qualifications:*
  - Workmanship samples: SMAW, GMAW, FCAW, and GTAW
  - Performance qualification: SMAW in 6G (multiple, inclined) position

Expert Welders (Level III) An individual employed in this position is considered to possess a prerequisite amount of knowledge, attitude, skill, and habits required to perform tasks autonomously, including the selection and use of appropriate techniques and equipment, and to apply theoretical knowledge and motor skills with minimum supervision.

- *Base metals:* Same as entry level plus familiarization training in copper, nickel, magnesium, and/or titanium alloys.
- *Thickness and type:* Unlimited thickness range of pipe or tubing for SMAW, GMAW, FCAW, and GTAW.
- *Positions:* All positions using SMAW, GMAW, FCAW, and GTAW.
- *Qualifications:*
  - Workmanship samples: GMAW-P, FCAW, and GTAW in 6G (multiple, inclined) position
  - Performance qualification: SMAW in 6GR (restricted, multiple, inclined) position

For advanced and expert welders, performance qualification tests performed within the context of the AWS certified welder program may substitute for performance tests described in the respective guides. For level III expert welders, certification as an AWS certified welding inspector may substitute for some of the requirements.

Upon successful completion of the written examinations and performance tests, the candidate's name will be listed in the national registry for the entry level, advanced, or expert welder as applicable. AWS also provides document B5.13, the specification for the qualification of SENSE training facilities.

3-5 QUALIFYING AND CERTIFYING WELDING PERSONNEL

The American Welding Society is continuing its program to increase the professionalism of welders, welding inspectors, and other welding-related personnel. This is accomplished by providing technically accurate specifications for the different qualifications and facilities, and verifying that personnel do in fact have the knowledge and experience required. Two committees, the qualification committee and the certification committee, separately but collectively accomplish this mission.

The qualification committee writes qualification standards that specify the attributes of personnel fulfilling specific jobs. This includes describing the duties, responsibilities, knowledge, education, etc., of the job or position. These become consensus standards for the particular job or facility, which are then approved by the technical activities committee for technical accuracy. Since the American National Standards Institute (ANSI) approved the method by which AWS approves its technical standards, they then become American National Standards. These are reviewed and revised every five years. These standards provide guidelines for certification and educational programs and can be used by third-party organizations to qualify personnel.

Certification is the documentation of the satisfaction of specific qualification requirements. This gives employers or organizations such as AWS the opportunity to certify personnel.

Since welding, as a skill and technology, is widely employed throughout the world, the common standards will allow welding professionals to have the same experience, knowledge, and capabilities. Standards are available or are being prepared for the following welding personnel and welding facilities: welding inspectors, welding inspector specialists, welding educators, welding supervisors, welding engineers, welders and welding operators, and welding facilities, welding fabricators, and SENSE training facilities.

The certification committee administers an independent certification program, which acts as a third party with respect to the qualification standards. This group administers competency tests in the field. They have specific, strict rules for certification. They will write and provide examinations based on batteries of test questions. They will verify that they relate to actual job requirements. They will provide examinations at various locations and evaluate the results. Certification will demonstrate and verify that the individual has the capabilities appropriate to the subject. They will maintain necessary records and issue certificates under established rules.

Many of the different welding personnel are listed below with an outline of the status of the specifications and the requirements of each. If you have questions concerning certification or how to become certified, call AWS headquarters in Miami, Florida (see Appendix A-2).

Welding Inspector

This was the first AWS program for qualifying and certifying personnel. The program is more than 20 years old and has been extremely successful. It is accepted worldwide and AWS has given certification tests in many countries as well as in many locations in the United States. The original document QC1, "Standard for the Qualification and Certification of Welding Inspectors," has stood the test of time. It provides for the senior certified welding inspector (SCWI), the certified welding inspector (CWI), and the certified associate welding inspector (CAWI). However, it is being replaced by document B5.1 which is essentially the same. A new document, B5.2, "Specifications for the Training and Qualification of Welding Inspector Specialists and Welding Inspector Assistants," provides for the training and qualification of welding inspector specialists and assistants. The difference is that the welding inspector specialists and welding inspector assistants are certified by their employer and not by the American Welding Society.

A person desiring to be certified must make application to the Welding Society and provide a record of experience and training. The American Welding Society requires an eye examination and an intensive test related to a particular welding code or specification, with a hands-on examination that requires the use of inspection instruments. The applicant is required to inspect weld replicas and respond to specific questions. Upon approval of the application and successful completion of the tests, the person is certified as a CWI or as a CAWI, depending on test scores. The AWS maintains a roster of applicants who have passed the test.

Many states qualify inspectors for pressure vessel and pressure piping work. This is done by the State Industrial Commission in conjunction with the National Board of Boiler and Pressure Vessel Inspectors and casualty insurance companies. The national board provides training and testing to determine that inspectors have the necessary knowledge and ability. There are government inspectors for the U.S. Department of Defense; various cities have inspectors for structural welding. Most state departments of transportation have welding inspectors. It is impossible to include all the different types of inspectors and the tests and qualifications that are required. Contact the organization of interest to determine their requirements.

Welding Engineer

Welding engineers may receive a degree from a number of colleges and universities in the United States, and in Europe and Japan. Welding engineering degree programs can be accredited by the Accreditation Board for Engineering and Technology (ABET) of New York City. The welding engineering baccalaureate degree program at Ohio State University is the only ABET-accredited welding engineering program in the country. Welding engineer associate degree programs are also accredited. Master's and doctorate degrees in welding engineering are also available and in engineering disciplines closely associated with welding in the metallurgical, mechanical, or structural engineering departments are also available at several universities.

State registration of welding engineers is desirable. If the welding engineer is involved with public works,

registration as a professional engineer under state law is necessary. The state of Ohio offers welding engineering professional registration and will cooperate with other states to provide welding engineering registration.

AWS provides a document B5.16, "Specifications for the Qualification of Welding Engineers." It requires personnel to have a specific combination of education and experience to be eligible to take the examination, which consists of four parts. Individuals passing an international or European welding engineer exam, and with one year of experience, are eligible to take the welding engineering examination. Continued education is required to maintain qualification. Qualification to this standard does not imply the status of a state-registered professional engineer (PE). However, a state-registered professional welding engineer is qualified in accordance with this specification.

Welding Educator

AWS provides document B5.5 entitled, "Specifications for the Qualification of Welding Educators." There are three levels. The senior welding educator (SWE), the welding educator (WE), and the associate welding educator (AWE). All welding educators must demonstrate competence of welding skills of the different arc welding and cutting processes. They must be able to explain welding codes, drawings, and specifications, and to explain welding base metals and materials. They must evaluate instructional plans and welding students. They must be capable of visually inspecting welds and preparing reports. The level of welding educator depends on the level of education and experience. All welding educators must successfully pass both a closed book examination and the CWI practical examination to demonstrate knowledge and skills. You may qualify for the certified welding educator if you are a qualified welding inspector, if you have specific teaching responsibilities, and if you hold a welder's certificate. Certification requirements are covered by document QC5, entitled "Standard for AWS Certification of Welding Educators."

Welding Technician

Welding technicians, most of whom are highly skilled welders, have received specialized training by taking courses to supplement their knowledge of welding. The training, available through special courses at local schools and colleges, technical society programs, correspondence courses, and company-sponsored seminars, covers subjects that are associated with welding. Many technicians are graduates of associate degree programs from technical schools or junior colleges. An associate degree is obtained with two years of post–high school study. On the job they perform semiprofessional engineering functions, normally with supervision by an engineer. The engineering technologist must master the language of engineering and specialized subject matter. The technician must be able to apply theory and use capabilities of skilled craftsmen to achieve practical results. The welding technician must know, understand, and be able to operate processes, procedures, and equipment of the welding industry and must verify the existence of a predetermined quality level. Upon graduation, the employment opportunities for welding technicians are almost unlimited.

Welding Supervisor

Welding supervisors coordinate and supervise the activity of welders. The supervisor must have welding experience but need not be the most skilled welder. The supervisor must have knowledge of the welding processes being used and must have a background of technical information about welding. The supervisor must also have training and experience in supervision and management. AWS provides document B5.9, "Specifications for the Qualifications of Welding Supervisors." AWS will also provide a welding certification program for welding supervisors.

Robotic Arc Welding Support Personnel

Due to the rapid increase in the use of arc welding robots, there is a need for personnel knowledgeable in both arc welding and robotic technology. The AWS has responded by establishing a "Specification for the Qualification of Robotic Arc Welding Personnel," AWS D16.4:2000, and the

"Standard for AWS Certification of Robotic Arc Welding Support Personnel" (CRAWSP), AWS QC 19:2000.

The society will certify personnel, but the employer is responsible for establishing capabilities, providing training, establishing ability to perform duties involved, and other required qualifications. An eye examination, a written examination and a performance examination are all required.

There are two levels of AWS certification: Level A is titled "Certified Robotic Arc Welding Operator," and Level B is titled "Certified Robotic Arc Welding Technician/Inspector." Contact AWS for more details (see Appendix A-2).

Welders

AWS has several programs for qualifying and certifying welders including QC7, "Specifications for AWS Certified Welders," and several supplements. This is a national program where welders can take proscribed tests at qualified welder test facilities. Upon successful completion of the tests, they will be placed on the AWS national registry of welders. This makes possible the transfer of welder qualifications from employer to employer around the country without retesting.

Welder qualification is an extremely complex situation. More complete information concerning the requirements for the qualification and certification of welders based on specific codes is covered in Chapter 22.

 QUESTIONS

3–1. Where are plants doing welding generally located?

3–2. What products do welders help make?

3–3. What tests can be taken to determine if a person should become a welder?

3–4. Name some of the future jobs open to welders who take extra training.

3–5. Where do welders learn to do construction welding?

3–6. In construction, what job titles may a welder have?

3–7. How does welder's pay compare with others?

3–8. What are the physical demands on a welder?

3–9. Is seeing important for a welder?

3–10. Are manipulator skills important for a welder?

3–11. Who normally programs arc welding robots?

3–12. What type of school must one attend to learn to weld?

3–13. What are the four steps of a welding training program?

3–14. What specifications involve welder qualifications?

3–15. What does a welding inspector do?

3–16. Do welding sales representatives need to know how to weld?

3–17. What is a CWI?

3–18. What is a CAWI? What is the difference between a CWI and a CAWI?

3–19. What is a CWE? How do you become a CWE?

3–20. How can you advance to higher-paying welding-related jobs?

REFERENCES

1. Statistical Policy Division, Office of Management and Budget, "Standard Industrial Classification Manual," U.S. Government Printing Office, Washington, D.C.

2. "Where Are They Now," *Welding Engineer,* July 1973, p. 31.

3. U.S. Department of Labor, "Dictionary of Occupational Titles," 4th Edition. U.S. Government Printing Office, Washington, D.C., 1977, revised 1991.

4

SAFETY AND HEALTH OF WELDERS

OUTLINE

4-1 PERSONNEL PROTECTION AND SAFETY RULES

Your safety and health is extremely important. All workers engaged in production and construction are continually exposed to potential hazards. There are a number of safety and health problems associated with welding. When correct precautionary measures are followed, welding is a safe occupation. Health officials state that welding, as an occupation, is no more hazardous or injurious to the health than other metalworking occupations.

Governments have become increasingly active concerning the safety and health of workers and have enacted laws prescribing safety regulations and the publi-cation of safety warnings to ensure the safety of workers. In the United States, the provisions of the Occupational Safety and Health Act (OSHA)[1] are the law. It makes many national consensus standards enforceable. The most important is the American National Standard "Safety in Welding and Cutting,"[2] which states that welding and cutting operations pose potential hazards from fumes, gases, electric shock, heat radiation, and sometimes noise. All personnel must be warned against these hazards where applicable by the use of adequate precautionary labeling. The precautionary label for welding processes and equipment is shown in Figure 4-1. There are other hazards that apply to all metalworking occupations. These are accidents resulting from falling, from being hit by moving objects, from working around moving machinery, from exposure to hot metal, and so on. Normal precautions are required with regard to these hazards as well.

The hazards that relate to welding are:

* Electrical shock
* Arc radiation
* Air contamination
* Fire and explosion
* Compressed gases
* Welding cleaning
* Other hazards related to specific processes or occupations

Welders work under a variety of conditions: outdoors, indoors in open areas, in confined spaces, high above the

41

WARNING: PROTECT yourself and others. Read and understand this label.

FUMES AND GASES can be dangerous to your health.
ARC RAYS can injure eyes and burn skin.
ELECTRIC SHOCK can KILL.

- Before use, read and understand the manufacturer's instructions, Material Safety Data Sheets (MSDSs), and your employer's safety practices.
- Keep your head out of the fumes.
- Use enough ventilation, exhaust at the arc, or both, to keep fumes and gases from your breathing zone and the general area.
- Wear correct eye, ear, and body protection.
- Do not touch live electrical parts.
- See American National Standard Z49.1, *Safety in Welding and Cutting*, published by the American Welding Society, 550 N.W. LeJeune Rd., P.O. BOX 351040, Miami, Florida 33135; OSHA Safety and Health Standards, 29 CFR 1910, available from U.S. Government Printing Office, Washington, DC 20402.

DO NOT REMOVE THIS LABEL

FIGURE 4–1 Warning label for arc processes and equipment. (From Ref. 2.)

ground, and under water. They utilize a large number of welding and cutting processes. Most have exposure to fumes, gases, radiation, and heat. Welders are exposed to a number of factors simultaneously. The use of specific welding processes or welding on particular metals presents potential health risks. Additional information is available in the American Welding Society publication entitled "Effects of Welding on Health" published from 1979 to 1987.[3]

Welding Workplace Safety

The welding shop management and supervisors are responsible for providing training for workers in the safe conduct of their day-to-day activities. Employees must be informed and trained so that they are able to detect when hazards are present and protect themselves from them.

The welders and other employees have an obligation to learn and use safe practices and to obey safety rules and regulations and are expected to work in a safe manner. They are responsible for the safe use of equipment and materials. It is the responsibility of supervisors to enforce safety rules and regulations.

Good housekeeping practices should always be employed in the welding shop. Adequate safety devices should be provided, such as proper fire extinguishers, life-saving and support equipment, first-aid kits, and so on, plus the training of personnel to utilize this equipment properly. Only approved equipment should be used, and it must be properly installed and maintained in good working order.

Heat Exposure

Welders are sometimes required to weld on, or inside, preheated weldments. The preheat temperatures required for welding special materials can be quite high and the welder must be protected from coming into contact with hot metal. Workers should be supplied with sufficient cool air to avoid breathing excessively hot air. Special precautions must be taken and special procedures must be adopted to protect the welder from the heat. Protective clothing should be worn, which helps insulate the welder from excessive heat. Consultation with safety experts and just plain common sense are required in these situations.

Protective Clothing

Welders should wear work or shop clothes without openings or gaps to prevent the arc rays from contacting the skin. If the arc rays contact the skin for a period of time, painful "sunburns" or "arc burns" will result. People working close to arc welding should also wear protective clothing.

For light-duty welding, normally 200 A or lower, the level of protection can be reduced. Figure 4–2 shows a welder dressed for light-duty work. Woolen clothing is much more satisfactory than cotton since it will not disintegrate from arc radiation or catch fire as quickly. Cloth gloves can be used for light-duty work. For heavy-duty work, more thorough protective clothing is required. Figure 4–3 shows a welder dressed for heavy-duty welding work, wearing leather gauntlet gloves, a leather jacket, leather apron, and spats, which also protect against sparks and molten metal. When welding in the vertical and overhead position, this type of clothing is required. In all cases a headcap should be used. Flame-retardant clothing should be worn. Clothing should always be kept dry, and this applies to gloves as well. High-top shoes with safety toes are recommended. The leather clothes should be of the chrome-tanned type. Leather gloves should not

FIGURE 4–2 Welder dressed for light-duty welding. Most also wear cap and safety glasses.

FIGURE 4–3 Welder dressed for heavy-duty welding. Most also wear cap and safety glasses.

be used to pick up hot items since this will cause the leather to become stiff and crack. Protective clothing must be kept in good repair. Hard hats should be checked occasionally. Gloves should be clean and not oily. Welding helmets should be checked for cracks, and filter glasses should be replaced if damaged.

Signs should be posted in the welding department pointing out precautions that must be taken by employees and visitors in the welding shop. These signs should be in agreement with ANSI Standard "Specifications for Accident Prevention Signs."[6] The welding department should also post signs warning people with pacemakers that they should not enter or should take special precautions.

Safety Rules

The following sets of 20 rules, "Safety Precautions for Arc Welding" and "Safety Precautions for Oxyacetylene Welding and Cutting," should be posted in the welding shop.

Safety Precautions for Arc Welding

1. Make sure that your arc welding equipment is installed properly and grounded and is in good working condition.

2. Always wear protective clothing suitable for the welding to be done.

3. Always wear proper eye protection when welding, cutting, or grinding. Do not look at the arc without proper eye protection.

4. Avoid breathing the air in the fume plume directly above the arc.

5. Keep your work area clean and free of hazards. Make sure that no flammable, volatile, or explosive materials are in or near the work area.

6. Handle all compressed gas cylinders with extreme care. Keep caps on when not in use.

7. Make sure that compressed gas cylinders are secured to the wall or to other structural supports.

8. When compressed gas cylinders are empty, close the valve and mark the cylinder "empty."

9. Do not weld in a confined space without special precautions.

10. Do not weld on containers that have held combustibles without taking special precautions.

11. Do not weld on sealed containers or compartments without providing vents and taking special precautions.

12. Use mechanical exhaust at the point of welding when welding lead, cadmium, chromium, manganese, brass, bronze, zinc, or galvanized steel, and when welding in a confined space.

13. When it is necessary to weld in a damp or wet area, wear rubber boots and stand on a dry insulated platform.

14. Do not use cables with frayed, cracked, or bare spots in the insulation.

15. When the electrode holder is not in use, hang it on the brackets provided. Never let it touch a compressed gas cylinder.

16. Dispose of electrode stubs in proper containers since stubs on the floor are a safety hazard.

17. Shield others from the light rays produced by your welding arc.

18. Do not weld near degreasing operations.

19. When working above ground, make sure that the scaffold, ladder, or work surface is solid.

20. When welding in high places, use a safety belt or lifeline.

Safety Precautions for Oxyacetylene Welding and Cutting

1. Make sure that all gas apparatus shows UL or FM approval, is installed properly, and is in good working condition. Make sure that all connections are tight before lighting the torch. Do not use a flame to inspect for tight joints. Use soap solution to detect leaks.

2. Always wear protective clothing suitable for welding or flame cutting.

3. Keep work area clean and free of hazardous materials. When flame cutting, sparks can travel 30 to 40 feet (10 to 15 meters). Do not allow flame cut sparks to hit hoses, regulators, or cylinders.

4. Handle all compressed gas cylinders with extreme care. Keep cylinder caps on when not in use.

5. Make sure that all compressed gas cylinders are secured to the wall or to other structural supports. Keep acetylene cylinders in the vertical position.

6. Store compressed gas cylinders in a safe place with good ventilation. Acetylene cylinders and oxygen cylinders should be kept apart.

7. When compressed gas cylinders or fuel gas cylinders are empty, close the valve and mark the cylinder "empty."

8. Use oxygen and acetylene or other fuel gases with the appropriate torches and only for the purpose intended.

9. Avoid breathing the air in the fume plume directly above the flame.

10. Never use acetylene at a pressure in excess of 15 psi (103.4 kPa). Higher pressure can cause an explosion.

11. Never use oil, grease, or any material on any apparatus or threaded fittings in the oxyacetylene or oxyfuel system. Oil and grease in contact with oxygen may cause spontaneous combustion.

12. Do not weld or flame cut in a confined space without taking special precautions.

13. When assembling apparatus, crack the gas cylinder valve before attaching regulators (*cracking* involves opening the valve on a cylinder slightly, then closing). This blows out any accumulated foreign material. Make sure that all threaded fittings are clean and tight.

14. Always use this correct sequence and technique for lighting a torch.
    (a) Open acetylene cylinder valve.
    (b) Open acetylene torch valve 1/4 turn.
    (c) Screw in acetylene regulator adjusting valve handle to working pressure.
    (d) Turn off acetylene torch valve (you will have purged the acetylene line).
    (e) Slowly open oxygen cylinder valve all the way.
    (f) Open oxygen torch valve 1/4 turn.
    (g) Screw in oxygen regulator screw to working pressure.
    (h) Turn off oxygen torch valve (you will have purged the oxygen line).
    (i) Open acetylene torch valve 1/4 turn and light with lighter (use friction-type lighter or special provided lighting device only).
    (j) Open oxygen torch valve 1/4 turn.
    (k) Adjust to neutral flame.

15. Always use this correct sequence and technique of shutting off a torch.
    (a) Close acetylene torch valve first, then close oxygen torch valve.
    (b) Close cylinder valves—the acetylene valve first, then the oxygen valve.
    (c) Open torch acetylene and oxygen valves (to release pressure in the regulator and hose).
    (d) Back off regulator adjusting valve handle until no spring tension is felt.
    (e) Close torch valves.

    Note: Different torch manufacturers recommend different shutdown procedures for the torch acetylene and oxygen valves. Follow the procedure recommended for the torch in use. If the oxygen valve is closed first, the yellow, sooty acetylene flame enlarges appreciably and could burn the welder. The carbon soot will deposit in the area. If the acetylene valve is closed first, there will be a loud "bang," which may distract nearby welders. In either case the other valve should be closed quickly.

16. Use mechanical exhaust when welding or cutting lead, cadmium, chromium, manganese, brass, bronze, zinc, or galvanized steel.

17. If you must weld or flame cut with combustible or volatile materials present, take extra precautions, make out hot work permit, provide for a lookout, and so on.

18. Do not weld or flame cut on containers that have held combustibles without taking special precautions.

19. Do not weld or flame cut into sealed container or compartment without providing vents and taking special precautions.

20. Do not weld or cut in a confined space without taking special precautions.

The "Safety in Welding and Cutting" standard also provides a warning label for oxyfuel gas processes (Figure 4-4).

If the hazards mentioned in this chapter are handled properly, the welder is as safe as any other industrial worker. There must be continual vigilance over safety conditions and safety hazards. Safety meetings should be held regularly. The safety rules should be reissued annually and they must be completely understood and enforced.

Material Safety Data Sheets

OSHA requires that employers must have a comprehensive hazard communication program to inform employees about hazardous substances used in the workplace. The employer must maintain continuous training concerning such materials and safety in general. Provisions to safeguard employees are included in Material Safety Data Sheets as prescribed by the Hazard Communication Standard of the U.S. Department of Labor.[4] Information must be provided for all substances taken into the workplace, except foods, drugs, cosmetics, or tobacco products used for personal consumption. More than 600 substances are covered. The use of these data sheets in all manufacturing workplaces has been mandated since 1985. Employees must be taught how to read and interpret information on labels and material data sheets.

Each data sheet for welding products includes information about every hazardous component comprising 1% or more of the contents, and for every potential carcinogen (cancer inciting or producing) comprising 0.1% or more. The components are included in the listing by the American Conference of Governmental Industrial Hygienists, with threshold limit values.[5] Threshold limit values are being reduced when some Material Safety Data Sheets are revised.

Material Safety Data Sheets should be obtained from suppliers of every product used in the welding shop. This includes welding electrodes, coated stick, cored wire, solid wire, tungsten and carbon nonconsumable electrodes. It also includes base metals, steel, stainless steel, aluminum, nickel, alloys, titanium, copper, and so forth. All of the gases used in the shop, either for shielding or for fuel gas, are also included, as well as all types of fluxes— brazing fluxes, submerged arc fluxes, metal powders for cutting, and also items for cleaning, lubricating, etc. It is all inclusive. These safety data sheets should be kept on file in the personnel or welding department. The training program regarding Material Safety Data Sheets must cover not only welders, but others working in the welding area, including service personnel, maintenance personnel, and others who regularly visit the welding shop.

A typical Material Safety Data Sheet for a flux-cored arc welding electrode is shown by Figures 4-5 and 4-6.

FIGURE 4-4 Warning label for oxyfuel gas processes. (From Ref. 2.)

WARNING: PROTECT yourself and others. Read and understand this label.

FUMES AND GASES can be dangerous to your health.
HEAT RAYS (INFRARED RADIATION from flame or hot metal) can injure eyes.

- Before use, read and understand the manufacturer's instructions, Material Safety Data Sheets (MSDSs), and your employer's safety practices.
- Keep your head out of the fumes.
- Use enough ventilation, exhaust at the flame, or both, to keep fumes and gases from your breathing zone and the general area.
- Wear correct eye, ear, and body protection.
- See American National Standard Z49.1, *Safety in Welding and Cutting*, published by the American Welding Society, 550 N.W. LeJeune Rd., P.O. Box 351040, Miami, Florida 33135; OSHA Safety and Health Standards, 29 CFR 1910, available from the U.S. Government Printing Office, Washington, DC 20402.

DO NOT REMOVE THIS LABEL.

Hazards in the filler metal or flux itself.

Chemical Abstracts Service No. (safety information, immediately available by telephone to physicians and paramedics).

PEL = permissible exposure limit, mg/m³.

American Conference of Governmental Industrial Hygienists TLV (mg/m³), time-weighted average.

Hazards created by the welding arc or torch.

Typical fumes.

MSDS NO: FCAW

REVISED: 8-85

MATERIAL SAFETY DATA SHEET
For U.S. Manufactured Welding Consumables and Related Products
May be used to comply with OSHA's Hazard Communication Standard, 29 CFR 1910. 1200.
Standard must be consulted for specific requirements.

SECTION 1 — IDENTIFICATION

Manufacturer/Supplier Name: COMPANY ABC	Telephone No: 1-825-600-0000
Address: MAIN STREET, US A	Emergency No: 1-825-600-0000
Trade Name FCAW E 70 T−1	Classification: AWS A5.20

Product Type: MILD STEEL FLUX CORED ARC WELDING (FCAW)

SECTION 2 — HAZARDOUS MATERIALS

IMPORTANT

This section covers the materials from which this product is manufactured. The fumes and gases produced during welding with normal use of this product are covered by Section 5.

The term "hazardous" in "Hazardous Materials" should be interpreted as a term required and defined in OSHA Hazard Communication Standard (29 CFR Part 1910.1200).

Ingredient	CAS No.	Exposure Limit (mg/m³)	
		OSHA PEL	ACGIH TLV
IRON	7439-89-6	5	Not Reported
MANGANESE	7439-96-5	5 CL*	1 CL* (Fume)
TITANIUM OXIDE	13463-67-7	15	10, 20 STEL**
MAGNESIUM OXIDE	1309-48-4	15	10
SILICON	7440-21-3	Nothing Found	10, 20 STEL**
FLUORSPAR	7789-75-5	2.5 (as F)	2.5 (as F)

*CL — Ceiling Limit   **STEL — Short Term Exposure Limit

SECTION 3 — PHYSICAL/CHEMICAL CHARACTERISTICS

Not Applicable

SECTION 4 — FIRE AND EXPLOSION HAZARD DATA

Non Flammable: Welding arc and sparks can ignite combustibles. See Z49.1 referenced in Section 7.

SECTION 5 — REACTIVITY DATA

Hazardous Decomposition Products

Welding fumes and gases cannot be classified simply. The composition and quantity of both are dependent upon the metal being welded, the process, procedures, and electrodes used. Other conditions which also influence the composition and quantity of the fumes and gases to which workers may be exposed include: coatings on the metal being welded (such as paint, plating, or galvanizing), the number of welders and the volume of the work area, the quality and amount of ventilation, the position of the welder's head with respect to the fume plume, as well as the presence of contaminants in the atmosphere (such as chlorinated hydrocarbon vapors from cleaning and degreasing activities).

When the electrode is consumed, the fume and gas decomposition products generated are different in percent and form from the ingredients listed in Section 2. Decomposition products of normal operation include those originating from the volatilization, reaction, or oxidation of the materials shown in Section 2, plus those from the base metal and coating, etc., as noted above.

It is understood, however, that the elements and or oxides to be mentioned are virtually always present as complex oxides and not as metals. [Characterization of Arc Welding Fume: American Welding Society]. The elements or oxides listed below correspond to the ACGIH categories located in [TLV Threshold Limit Values for Chemical Substances and Physical Agents in the Workroom Environment].

Reasonably expected constituents of the fume would include: complex oxides of iron, manganese, silicon, titanium and magnesium. Fluorides would also be present.

Substance	CAS No.	Exposure Limit (mg/m³)	
		OSHA PEL	ACGIH TLV
IRON OXIDE	1309-38-2	5	10 (as Fe₂O₃)
MANGANESE	7439-96-5	5 CL*	1 CL* (Fume)
SILICON OXIDE	7631-86-9	5	3
TITANIUM OXIDE	13463-67-7	15	10, 20 STEL**
MAGNESIUM OXIDE	1309-48-4	15	10
FLUORIDES		2.5 (as F)	2.5 (as F)

*CL — Ceiling Limit   **STEL — Short Term Exposure Limit

Page 1 of 2

FIGURE 4–5   Page 1 of Material Safety Data Sheet of flux-cored electrode.

Gases. Gaseous reaction products may include carbon monoxide and carbon dioxide. Ozone and nitrogen oxides may be formed by the radiation from the arc.

How to sample actual fumes. One recommended way to determine the composition and quantity of fumes and gases to which workers are exposed is to take an air sample inside the welder's helmet if worn or in the worker's breathing zone. [See ANSI/AWS F1.1, available from the "American Welding Society," P. O. Box 351040, Miami, FL 33135. Also, from AWS is F1.3 "Evaluating Contaminants in the Welding Environment — A Sampling Strategy Guide," which gives additional advice on sampling]. At a minimum, materials listed in this section should be analyzed.

SECTION 6 — HEALTH HAZARD DATA

Threshold Limit Value:

The ACGIH recommended general limit for Welding Fume NOC (Not Otherwise Classified) is 5 mg/m³. ACGIH-1985 or latest date) preface states "The TLV-TWA should be used as guides in the control of health hazards and should not be used as fine lines between safe and dangerous concentrations." See Section 5 for specific fume constituents which may modify this TLV.

Effects of Overexposure

Electric arc welding may create one or more of the following health hazards:

Note effects of overexposure. FUMES AND GASES can be dangerous to your health.
SHORT - TERM (ACUTE) OVEREXPOSURE to welding fumes may result in discomfort such as dizziness, nausea, or dryness or irritation of nose, throat or eyes.
LONG-TERM (CHRONIC) OVEREXPOSURE may lead to siderosis (iron deposits in lungs) and is believed by some investigators to affect pulmonary functions.
ARC RAYS can injure eyes and burn skin.

Apply first aid. ELECTRIC SHOCK can kill.
See Section 7.

Emergency and First Aid Procedures

Beware! (Carcinogenic means it may produce cancer.) Call for medical aid. Employ first aid techniques recommended by the American Red Cross.
Eyes & Skin: If irritation or flash burns develop after exposure, consult a physician.

Carcinogenicity

These products do not contain ingredients that are defined as carcinogenic per 29CFR 1910.1200 - Hazard Communication Standard.

SECTION 7 — PRECAUTIONS FOR SAFE HANDLING AND USE/APPLICABLE CONTROL MEASURES

Read and understand the manufacturer's instructions and the precautionary label on the product. (See American National Standard Z49.1. Safety in Welding and Cutting published by the American Welding Society, P. O. Box 351040, Miami, FL 33135 and OSHA Publication 2206 (29CFR1910), U.S. Government Printing Office, Washington, D.C. 20402. For more detail on many of the following:)

Ventilate weld area. VENTILATION: Use enough ventilation, local exhaust at the arc, or both, to keep the fumes and gases below TLV's in the worker's breathing zone and the general area. Train the welder to keep his head out of the fumes.

Use respirator when necessary. RESPIRATORY PROTECTION: Use NIOSH approved or equivalent fume respirator or air supplied respirator when welding in confined space or where local exhaust or ventilation does not keep exposure below TLV.

Wear helmet, filter lens. EYE PROTECTION: Wear helmet or use face shield with filter lens. As a rule of thumb begin with Shade Number 14. Adjust if needed by selecting the next lighter and/or darker shade number. Provide protective screens and flash goggles, if necessary, to shield others.

Protect from radiation, sparks, electric shock, hot metal, sharp edges, pinch points, falls. PROTECTIVE CLOTHING: Wear hand, head, and body protection which help to prevent injury from radiation, sparks, and electrical shock. See ANSI Z49.1. At a minimum this includes welder's gloves and a protective face shield, and may include arm protectors, aprons, hats, shoulder protection, as well as dark substantial clothing. Train the welder not to touch live electrical parts and to insulate himself from work and ground.

PROCEDURE FOR CLEANUP OF SPILLS OR LEAKS: Not applicable

WASTE DISPOSAL: Prevent waste from contaminating surrounding environment. Discard any product, residue, disposable container or liner in an environmentally acceptable manner, in full compliance with federal, state and local regulations.

Never exceed permissable exposure limits. SPECIAL PRECAUTIONS: IMPORTANT: Maintain exposure below the PEL/TLV. Use industrial hygiene monitoring to ensure that your use of this material does not create exposures which exceed PEL/TLV. Always use exhaust ventilation. Refer to the following sources for important additional information.

ANSI Z49.1 The American Welding Society, P. O. Box 351040, Miami, FL 33135 — OSHA (29CFR1910) U.S. Dept. of Labor, Washington, D.C. 20210.

Protect yourself. The manufacturer disclaims any responsibility. Manufacturer believes these data to be accurate and to reflect qualified expert opinion regarding current research. However, Manufacturer cannot make any express or implied warranty as to this information.

Page 2 of 2

FIGURE 4–6 Page 2 of Material Safety Data Sheet of flux-cored electrode.

Particular points of interest are highlighted to provide more data for intelligent interpretation of this information. The OSHA Hazard Communications Standard includes Appendix A, which requires employers to report any adverse health effects for which there is scientific evidence. Appendix B provides guidance in recognizing hazards. The Hazardous Communication Program and Welding Safety Training Programs must be ongoing.

4-2 ELECTRICAL SHOCK HAZARD

The shock hazard is associated with all electrical equipment. This includes extension lights, electric hand tools, and all types of electrically powered machinery. Ordinary household voltage (115 V) is higher than the output voltage of a conventional arc welding machine.

Use only welding machines that meet recognized national standards. Most industrial welding machines meet the National Electrical Manufacturers Association (NEMA) standards for electric welding apparatus.[7] This is mentioned in the manufacturer's literature and is shown on the nameplate of the welding machine. In Canada approval by the Canadian Standards Association is required for certain types of welding machines and this is also indicated on the nameplate.[8] In certain parts of the United States, and for certain applications, Underwriters'

Laboratories approval is required for transformer-type welding power sources.[9] The NEMA specification provides classes of welding machines, duty cycle requirements, and no-load voltage maximum requirements.

The new International Standard IEC 974-1, "Arc Welding Equipment, Part 1: Welding Power Sources,"[10] is finding more and more acceptance in many industrial countries. This specification provides classes of welding machines, duty cycle requirements, no-load voltage maximum requirements, and many other details related to construction. OSHA has additional requirements to improve the safety of welding machines. They have also made ventilating holes smaller so that the welder cannot come in contact with high voltage inside the case. They have changed the case of the welding machine so that "tools" are required to open the case where high voltage is exposed.

The new regulations also cover the location and insulation of the welding machine terminals. These are to be located or insulated so that the terminals are protected. This takes several forms. In some cases the terminals are recessed on the front of the power source. In other cases they are recessed but also covered by a protection plate. For semiautomatic machines the power cable is inserted in the front of the welding machine as well as the work lead. In this way the terminals are protected from any type of accident. These different arrangements for terminal protection are shown in Figure 4-7.

FIGURE 4–7 Terminal protection with cover plate open and closed.

Only insulated-type welding electrode holders should be used for shielded metal arc welding. Semiautomatic welding guns for continuous wire processes should utilize low-voltage control switches so that high voltage is not brought into the hands of the welder. In fully automatic equipment, higher voltages are permitted but are inaccessible to the operator during normal operation.

Installation of Welding Machines

All electric arc welding machines must be installed in accordance with the National Electrical Code®[11] and all local codes. Installation instructions are included in the manufacturer's manual that accompanies the welding machine. The manual also gives the size of the power cable that should be used to connect the machines to the main line. Motor generator welding machines feature complete separation of the primary power and the welding circuit since the generator is mechanically connected to the electric motor. However, the metal frames and cases of motor generators must be grounded to earth since the high voltage from the main lines does come into the case. In transformer, rectifier, and inverter machines, the primary and secondary transformer windings are electrically isolated from each other by insulation. This insulation may become defective in time if proper maintenance practices are not observed. The metal frame and cases of transformers, rectifier, and inverter machines must be grounded to earth. The work terminal of the welding machine should not be grounded to earth. Disconnect switches should be employed with all power sources so that they can be disconnected from the main lines for maintenance.

It is extremely important when several welding machines are working on the same weldment that the phases of a three-phase power line be accurately identified. This will ensure that the machines will be on the same phase and *in phase* with one another. It is easy to check this by connecting the work leads together and measuring the voltage between the electrode holders of the different machines. This voltage should be practically zero. If it is double the normal open-circuit voltage, it means that either the primary or secondary connections are reversed. If the voltage is approximately one and one-half times the normal open-circuit voltage, it means that the machines are connected to different phases of the three-phase power line. Corrections must be made before welding begins.

When large weldments, such as ships, buildings, or structural parts, are involved, it is normal to have the work terminal of many welding machines connected to them. It is extremely important that the machines be connected to the proper phase and have the same polarity. This can be checked by measuring the voltage between the electrode holders of the different machines mentioned above. The situation can also occur with respect to direct-current power sources when they are connected to a common weldment. If one machine is connected for straight polarity and one for reverse polarity, the voltage between the electrode holders will be double the normal open-circuit voltage. Precautions should be taken to see that all machines are of the same polarity when connected to a common weldment. Simultaneous welding with ac and dc welding machines must not be permitted on the same weldment.

The welding electrode holders must be connected to machines with flexible cables designed for welding application. There must be no splices in the electrode cable within 10 ft (3 m) of the electrode holder. Splices, if used in work or electrode leads, must be insulated.

Finally, it is important to locate welding machines where they have adequate ventilation and that ventilating ports be located where they cannot be obstructed.

Use of Welding Machines

Electrode leads and work leads should not be coiled around the welding machines, nor should they ever be coiled around the welder. Electrode holders should not be hung where they can accidentally come in contact with the other side of the circuit. Electrodes should be removed from holders whenever they are not in use. It is absolutely essential that power cables or primary power coming to a welding machine not be intermixed or come in contact in any way with the welding cables. The welding machine must be kept dry, and if it should become wet, it should be dried properly by competent electrical maintenance personnel. In addition, the work area must be kept dry. Welders should never work in water or damp areas since this reduces the resistance to the welder and increases potential electrical hazard.

Welders should not make repairs on welding machines or associated equipment. Welders should be instructed not to use tools to open cases of welding machines. They should be instructed not to perform maintenance on electrode holders, welding cables, welding guns, wire feeders, and so on. Instead, they should be advised to notify their supervisors of maintenance problems or potential hazards so that qualified maintenance personnel can make needed repairs.

Maintenance of Welding Machines

Welding machines and auxiliary equipment must be inspected periodically and maintained by competent electricians. During maintenance the equipment must be disconnected from power lines so that there is no possibility of anyone coming in contact with the high input voltage. Maintenance records should be kept on welding power supplies to comply with OSHA regulations. Supervisors and maintenance personnel should make routine inspections of welding cables and electrode holders, guns, and work clamps. Welders should report defective equipment or problems to their supervisors. Electrode holders with worn or missing insulators, and worn and frayed cables, should be repaired or replaced. Wire feeding semiautomatic equipment and specialty equipment, designed for gas tungsten arc welding, normally utilize power contractors. This means that the electrode wire or torch is electrically "cold" except while welding. The trigger on the welding gun or foot switch or programmer closes the contractors that energize the welding circuit. Arc voltage is normally nonhazardous.

4·3 ARC RADIATION HAZARD

The electric arc is a powerful source of light: visible, ultraviolet, and infrared. It is necessary that welders and others close to the welding arc wear suitable protection from the arc radiation. The brightness and exact spectrum of a welding arc depend on the welding process, the metals in the arc, the arc atmosphere, the length of the arc, and the welding current. The higher the current and arc voltage, the more intense the light from the arc. Like all radiation, arc light radiation decreases with the square of the distance. Those processes that produce smoke surrounding the arc have a less bright arc since the smoke acts as a filter. The spectrum of the welding arc is similar to that of the sun. Exposure of the skin and eyes to the arc is the same as exposure to the sun. If they are using a thoriated tungsten electrode for the gas tungsten arc welding process, radiation is minute.

Heat is radiated from the arc in the form of infrared radiation. The infrared radiation is harmless, provided that the proper eye protection and clothing are worn. To minimize light radiation, screens should be placed around the welding area so that people working nearby are shielded from the arc. Welders should attempt to screen nearby people from their arc. Screens and surrounding areas, especially welding booths, should be painted with flat finish paints that absorb ultraviolet radiation yet do not create high contrast between the bright and dark areas. The flat paint finish should have a low reflectivity to ultraviolet radiation. Light pastel colors of a zinc oxide or a titanium dioxide paint are recommended. Black paint or glossy finish paint should not be used.

Eye Protection

Welders must wear protective welding helmets with special filter plates or filter glasses. The welding helmets should be in good repair since openings or cracks can allow arc light to get through and create discomfort. The curved front welding helmets are preferred over straight front because they reduce the amount of welding fumes that come to the welder's breathing zone. Figure 4–8 shows welding helmets. Fiberglass is recommended for its light weight, but the newer nylon helmets are lighter.

FIGURE 4–8 Welding helmets with large and standard size filter lenses.

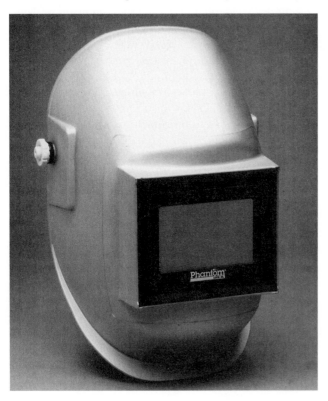

Welding helmets can be attached to safety hard hats for industrial and construction work. Welding helmets have lens holders for inserting the cover glass and filter glass or plate. The standard-size filter plate is $2 \times 4\frac{1}{4}$ in. (50 × 108 mm). In some helmets, the lens holders will open or flip upward. Helmets that accommodate larger-size filter lenses are also available. The larger filter glasses are $4\frac{1}{2} \times 5\frac{1}{4}$ in. (115 × 133 mm) and are more expensive. The filter glasses or plates come in various optical densities to filter out a portion of the arc rays. The shade of the filter glass used is based on the welding current. Figure 4-9 shows the proper filter shades. A cover plate should be placed on the outside of the filter glass to protect it from weld splatter. Plastic or glass plates are used. Some welders also use magnifier lenses behind the filter plate to provide clearer vision. The filter glass must be tempered so that it will not break if hit by flying objects. Filter glasses must be marked showing the manufacturer, the shade number, and the letter *H*, indicating that it has been treated for impact resistance.

FIGURE 4-9 Eye protection: guide for filter shade number. (From Ref. 2.)

Welding or Cutting Operation	Electrode Size (mm) or Metal Thickness	Welding Current (A)	Minimum Protective Shade	Suggested Shade Number[a] (Comfort)
Shielded metal arc welding	Less than 3 (2.5)	Less than 60	7	—
	3–5 (2.5–4)	60–160	8	10
	5–8 (4–6.4)	160–250	10	12
	More than 8 (6.4)	250–550	11	14
Gas metal arc welding and flux-cored arc welding		Less than 60	7	—
		60–160	10	11
		160–250	10	12
		250–500	10	14
Gas tungsten arc welding		Less than 50	8	10
		50–150	8	12
		150–500	10	14
Air carbon	(Light)	Less than 500	10	12
Arc cutting	(Heavy)	500–1000	11	14
Plasma arc welding		Less than 20	6	6 to 8
		20–100	8	10
		100–400	10	12
		400–800	11	14
Plasma arc cutting	(Light)[b]	Less than 300	8	9
	(Medium)[b]	300–400	9	12
	(Heavy)[b]	400–800	10	14
Torch brazing	—	—	—	3 or 4
Torch soldering	—	—	—	2
Carbon arc welding	—	—	—	14

	Plate Thickness			
	in.	mm		
Gas welding				
Light	Under $\frac{1}{8}$	Under 3.2		4 or 5
Medium	$\frac{1}{8}$ to $\frac{1}{2}$	3.2 to 12.7		5 or 6
Heavy	Over $\frac{1}{2}$	Over 12.7		6 or 8
Oxygen cutting				
Light	Under 1	Under 25		3 or 4
Medium	1 to 6	25 to 150		4 or 5
Heavy	Over 6	Over 150		5 or 6

[a]As a rule of thumb, start with a shade that is too dark to see the weld zone. Then go to a lighter shade that gives sufficient view of the weld zone without going below the minimum. In oxyfuel gas welding or cutting where the torch produces a high yellow light, it is desirable to use a filter lens that absorbs the yellow or sodium line in the visible light of the (spectrum) operation.
[b]These values apply where the actual arc is clearly seen. Experience has shown that lighter filters may be used when the arc is hidden by the workpiece.

Several new types of filter lens for welding helmets have been introduced recently. One type of filter glass utilizes a thin layer of liquid crystals sandwiched between two pieces of clear glass. The liquid crystals employed have special properties, so that when an electrical signal is placed across them they will change their ability to transmit light. When electrically changed the liquid crystals produce a screen with the same approximate density as the welding filter glass. A photosensor on the helmet is triggered by the light from the arc. Within a hundredth of a second, this signal is transmitted through the liquid crystals, which change the density of the filter glass. Another type of filter becomes darker when exposed to the bright light of the arc. These filters are becoming more popular since they eliminate the need for opening and closing or repositioning the welding helmet. These new-style filter lenses have not yet been included in the safety standards; however, testing is under way. In addition to helmets using automatic darkening welding lenses, there are helmets that have mechanical motion of the filter glass. The glass flips up and down outside the helmet and in other cases works within the helmet. The idea is to improve the comfort of the welder and still provide the necessary eye protection. They are considerably more expensive than the standard helmet.

Safety glasses should be worn underneath the welding helmet. These are required since the helmet is usually lifted when slag is chipped or welds are ground. Tinted safety glasses with side shields are recommended. People working around welders should also wear tinted safety glasses with side shields. Safety glasses should meet all the requirements of the eye and face protection standard.[12]

Contact Lenses

The wearing of contact lenses by welders is the subject of erroneous and recurring rumors. Various authorities, including the National Society to Prevent Blindness, the Contact Lens Association of Ophthalmologists, and others, state that the normal eye protection required by OSHA for welding, brazing, and soldering is the same with or without contacts. The American Optometric Association adopted a policy statement saying that contact lenses may be worn in hazardous environments with appropriate normal safety eye wear. Contact lenses themselves do not provide eye protection in the industrial sense. As a general rule, if an employee habitually wears contact lenses, the welder should be allowed to wear them in addition to normal safety equipment. It was further noted that the heat from the welding arc or flash is not intense enough to affect the durable plastic from which contact lenses are made. Welders or anyone who may be exposed to a welding flash or arc should wear appropriate safety goggles over their contact lenses. Eye experts unanimously agree that it is impossible for an electric arc to weld contact lenses to the eye. The American Optometric Association says that reports of this hazard are based on rumor and have been thoroughly discredited. Both OSHA and the U.S. Food and Drug Administration stated that the reports of this accident were false and there is no such danger.

On occasion, welders and others will have their eyes exposed to the arc for a short period. This will result in what is known as *arc burn, arc flash,* or *welding flash* and is technically called *photokeratitis*. It is very similar to a sunburn of the eye. For a period of approximately 24 hours the welder will have the painful sensation of sand in the eyes. The condition is normally of temporary duration and should not last over 48 hours. The welder who receives an arc flash may not be aware of it at the time. The first indication of an arc burn may occur 6 to 12 hours later in the middle of the night. Temporary relief can be obtained by using eyedrops and eyewashes. If the painful sensation lasts beyond one day, a doctor should be consulted for treatment.

Transparent Welding Curtains

Transparent welding curtains made of polyvinyl chloride plastic film are sometimes used for screening welding operations (Figure 4–10). The material is about 0.012 in. thick (0.3 mm), relatively tough, available in large sheets, and comes in blue, green, gray, and yellow. Tests have been performed by the National Institute of Occupational Safety and Health,[13] which concluded that these curtains provide protection in the ultraviolet range. The gray color provides the most protection, with yellow providing the least; all They meet OSHA requirements. The age of curtains may have an effect. The material is flame resistant. In no case can this curtain material be substituted for filter glass in helmets. It is intended to protect nearby workers from arc flash and improve communication with welders. In this application it is an improvement over opaque curtains or shields. These curtains are also available as strips for accessibility and can be placed on rollers.

Other Factors

Welding operations should be isolated from metal-degreasing or solvent-cleaning operations. Chemical-degreasing tanks may use trichloroethylene or other chlorinated hydrocarbons which will decompose to phosgene gas when exposed to arc (ultraviolet) radiation. Phosgene can build up to dangerous concentrations that are harmful. Fortunately, the odor of phosgene gas is quickly recognized (it smells like new-mown hay), and if it is detected, the area should be evacuated and ventilated. Degreasing operations should be at least 200 ft away from welding operations. If this is not possible, adequate ventilation is required. Care should be taken when

FIGURE 4–10 Welding station using transparent welding curtains.

welding parts that have been cleaned with these solvents. The surface must be thoroughly dry before welding.

Warning signs should be posted in welding departments advising visitors not to look at the arc, since arc flash will injure eyes.

4-4 AIR CONTAMINATION HAZARD

Arc welding and flame cutting produce air contamination. This is identified as smoke rising above the welding or flame cutting operation. The smoke or plume appears similar to smoke rising from a wood fire. Normal ventilation practice reduces the hazards of smoke from either welding or an open fire. The welding fumes contain two types of air contamination: particulate matter and gases.

The welding industry sponsored research to investigate the welding atmosphere and to recommend precautions to avoid potential hazards. This includes a series of reports entitled "Effects of Welding on Health," mentioned previously,[13] starting in 1979. The American Welding Society's study entitled "The Welding Environment"[14] and several foreign studies indicate that there is no significant health difference between welders and nonwelders when the welding process is carried out with adequate ventilation.

A warning label introduced in 1967 states: "Caution: Welding may produce fumes and gases hazardous to health. Avoid breathing these fumes and gases. Use adequate ventilation. See American National Standard Z49.1 'Safety in Welding and Cutting' published by the American Welding Society."[2]

This label has been revised to be more encompassing and is shown in Figure 4-1. A similar warning label for oxyfuel gas processes is shown in Figure 4-4. The purpose of these labels is to remind welders and companies employing welders of the potential hazard, so that adequate steps are taken to protect personnel from concentrations that might be harmful. The potential harm from fumes and gases depend on:

❋ The chemical composition of the particulate matter
❋ The concentration at the welder's breathing zone
❋ The length of time of exposure to these fumes and gases

Particulate Matter

Particulate matter is extremely small solids suspended in the air. Smoke is an example of particulate matter. Particulate matter includes common house dust, powders, pollen, smog, fly ash, grinding dust, and so on. These range in size from less than 0.1 micron (μm) to over 100 μm. The smaller-diameter particulates can only be seen with a microscope, while the larger ones can be seen with the human eye. In welding the type of particulate matter relates to the welding process, the type of electrode or filler metal, the welding current employed, and the welding location, atmospheric conditions, wind, and so on. It also depends on the composition of the base metal being welded and on any coating on the base metal near the arc. All welding smoke is not the same, and the concentration can vary over a wide range.

Many investigations and tests have been made to determine the composition of fumes generated. This is presented in the AWS publication "The Welding Environment" mentioned previously,[14] and was based on using different welding and allied processes. Many data are presented in the document "Fumes and Gases in the Welding Environment."[15] Research to determine fumes generated by arc welding is given in the document "Characterization of Arc Welding Fumes."[16] In general, welding with mild steel electrodes on clean steel produces fumes containing a high proportion of iron oxide and small amounts of calcium oxide, titanium oxide, and amorphous silica. The fumes produced when welding with low-hydrogen-type electrodes contain the same oxides and fluorides. When welding with stainless steel electrodes, the iron oxide is lower but there are now oxides of chromium and nickel as well as fluorides. Electrode manufacturers supply Material Safety Data Sheets (MSDSs) in each container of filler metals, which show the composition of the coating on electrodes, fluxes, or flux cores. Data sheets may also include the composition of particulate matter produced as these electrodes are consumed in the arc. Due to the high temperature of the arc, the composition of the particulate matter is different from that of the coating.

The flux-cored arc welding process seems to produce the most particulate matter, or smoke. However, compared to the amount of weld metal deposited, the particulate matter of SMAW and FCAW is very similar. The gas metal arc welding process produces less particulate matter, and the submerged arc process produces a very small amount of particulate matter, as do the gas tungsten and plasma arc welding processes. Recent research on the GMAW pulsed-spray transfer mode of operation indicates that careful control of the metal transfer may reduce the fume emission. This is based on using a power source that does not overheat the metal droplets and reduces vaporization during transfer. This mode reduces the fume produced in the arc and helps meet lower emission standards.

The base material is another source of particulate matter. When melted by an arc, the base metal may volatilize and produce airborne contaminants. Chromium and nickel compounds are found in the fume when stainless steels are arc welded. The American Welding Society has developed a standardized method for measuring and determining the particulate matter produced by different welding processes. This method is outlined by the AWS document "Method for Sampling Airborne Particulates Generated by Welding and Allied Processes."[17] By using this technique, measurements can be made to determine contamination.

Certain metals should not be welded without the use of mechanical exhaust systems because the vaporized metals are potentially hazardous. The metals that create hazardous airborne contaminations are beryllium, brass, bronze, cadmium, chromium, cobalt, copper, lead, manganese, nickel, vanadium, and zinc. Arc welding should not be done on any of these metals unless mechanical ventilation is employed or unless the welder is protected in some manner.

Certain of the metals mentioned above may be used as a coating on steel. The common steel coatings are cadmium, zinc, and lead as well as chromium, nickel, and copper. Mechanical ventilation must be employed when welding on these coating materials. In addition, some brazing filler metals may contain cadmium, and protection should be provided. This is described in Figure 4-11.

Airborne contaminants are produced when welding or flame cutting on coated materials. Base metal coated with any of the metals listed above must be treated with caution, and mechanical ventilation must be provided. Other coatings, such as paint, varnish, plastic, and oil, can also generate contamination. The coatings must be removed from the welding area or mechanical ventilation must be provided. A serious problem can be encountered when old steel work is flame cut or welded. Often, older structural steel may be covered with many coats of lead-bearing paint. The heat of the arc or flame will cause the coating to volatilize and produce smoke containing lead. New pipe is often coated with a protective material. This must be removed from the arc area. In any case, adequate ventilation or protection for the welder must be employed.

FIGURE 4-11 Warning label for brazing filler metals containing cadmium. (From Ref. 2.)

DANGER: CONTAINS CADMIUM. Protect yourself and others. Read and understand this label.

FUMES ARE POISONOUS AND CAN KILL.

- Before use, read, understand and follow the manufacturer's instructions, Material Safety Data Sheets (MSDSs), and your employer's safety practices.
- Do not breathe fumes. Even brief exposure to high concentrations should be avoided.
- Use only enough ventilation, exhaust at the work, or both, to keep fumes from your breathing zone and the general area. If this cannot be done, use air supplied respirators.
- Keep children away when using.
- See American National Standard Z49.1, *Safety in Welding and Cutting*, published by the American Welding Society, 550 N.W. LeJeune Rd., P.O. Box 351040, Miami, Florida 33135; OSHA Safety and Health Standards, 29 CFR 1910, available from U.S. Government Printing Office, Washington, DC 20402.

If chest pain, shortness of breath, cough, or fever develop after use, obtain medical help immediately.

DO NOT REMOVE THIS LABEL.

Gases

Gases are produced or may be involved in many of the welding processes in oxygen flame cutting and allied processes. Gases are produced as products of combustion with the fuel gas. Gas is produced when steel is melted in the arc. Gas is produced by some of the constituents of the coating on the shielded metal arc welding electrode or the material contained in the core of a flux-cored electrode wire. These coating and contained materials are designed as a part of the consumable filler metal to produce gases to help shield the arc area from the atmosphere. Packages of filler metals carry a warning label which is the same as Figure 4-1.

Fluxes used for gas welding and brazing, and for submerged arc welding and electroslag welding, will also produce gases when they are heated. Brazing and gas welding fluxes sometimes contain fluoride, and heating or melting produces small amounts of fluorine in the atmosphere. Packages containing these types of fluxes are labeled as shown in Figure 4-12. These products produce potentially harmful gases, and adequate ventilation should be employed. Carbon dioxide is the most common gas produced by the disintegration of electrode coatings or materials in flux-cored electrode wires. The CO_2 is used to help protect the arc area from the atmosphere. There is a possibility of carbon monoxide gas being produced in the arc. Carbon monoxide, however, readily recombines with available oxygen in the heated atmosphere to produce CO_2 gas. Carbon monoxide is rarely found beyond a short distance away from the arc.

Ultraviolet rays from the arc, particularly the high-intensity gas tungsten arc, react with the oxygen in the atmosphere to produce ozone. Ozone is an active form of oxygen which has a sweet smell. It is sometimes evident after a lightning strike or in the generating room of a powerhouse. It is relatively unstable and quickly recombines to oxygen. Exposure to ozone will cause a burning sensation in the throat, coughing or chest pains, or wheezing in the chest during breathing. Ventilation should be used so that ozone concentration will be below the threshold limit values.

The gas-shielded welding processes utilize various gases to shield or protect the arc area from the atmosphere. Inert gases are used for gas tungsten arc welding and for plasma arc welding, but active gases or mixtures of active and inert gases are used for gas metal arc and flux-cored arc welding. Adequate ventilation is required to remove these gases from the welder's breathing zone.

Confined or Enclosed Areas

All welding, flame cutting, and associated operations carried out in confined or restricted spaces must be adequately ventilated to prevent the accumulation of toxic materials, combustible gases, or oxygen deficiency.

An enclosed area, also called a confined space, is a relatively small or restricted space such as a tank, vat, pressure vessel, boiler, compartment, small room, or any enclosure that may have poor ventilation. Enclosed areas, which also include tunnels, pose problems not only for

---

FIGURE 4-12 Warning label for fluxes that contain fluorides. (From Ref. 2.)

WARNING: CONTAINS FLUORIDES. Protect yourself and others. Read and understand this label.

FUMES AND GASES CAN BE DANGEROUS TO YOUR HEALTH. BURNS EYES AND SKIN ON CONTACT. CAN BE FATAL IF SWALLOWED.

- Before use, read, understand and follow the manufacturer's instructions, Material Safety Data Sheets (MSDSs), and your employer's safety practices.
- Keep your head out of the fume.
- Use enough ventilation, exhaust at the work, or both, to keep fumes and gases from your breathing zone and the general area.
- Avoid contact of flux with eyes and skin.
- Do not take internally.
- Keep out of reach of children.
- See American National Standard Z49.1, *Safety in Welding and Cutting,* published by the American Welding Society, 550 N.W. Lejeune Rd., P.O. Box 351040, Miami, Florida 33135; OSHA Safety and Health Standards, 29 CFR 1910, available from U.S. Government Printing Office, Washington, DC 20402.

First Aid: If contact in eyes, flush immediately with clean water for at least 15 minutes. If swallowed, induce vomiting. Never give anything by mouth to an unconscious person. Call a physician.

DO NOT REMOVE THIS LABEL

welders but for anyone working inside them. The potential hazards range from deficiency of oxygen, too much oxygen, poisonous gases, and flammable or explosive gases, to the accumulation of dense smoke or particulate matter. Welding, flame cutting, or allied processes should never be started without taking special precautions. Welding or cutting apparatus should never be taken into the enclosed area.

Everyone knows the risk of remaining in a closed garage with an automobile engine running. This can also be a potential problem with an engine-driven welding machine. The exhaust gas given off by the engine should always be channeled to the outside. In enclosed areas, even large rooms, an engine-driven welding machine, if not exhausted to the outside, can produce a buildup of carbon monoxide and carbon dioxide gas hazardous for people working within the room.

The same problem can occur when preheating weldments using the combustion of fuel gases, coal, or charcoal for heat. The burning of these fuels will produce carbon monoxide and carbon dioxide, which must be exhausted to the outside. Serious harm can happen to the workers in an enclosed area.

A "lookout," or watcher or attendant, must be assigned to watch the welders and other workers continually and to have occasional voice contact with those in the enclosed area. One lookout should be assigned to a team of welders working in a specific enclosed area. In hazardous cases, lifelines with harnesses should be employed. Lifelines should be attached so that workers can be removed through manholes with ease.

Prior to entering enclosed areas, special precautions should be taken to determine the atmosphere within the enclosed area. Explosive concentrations of gases sometimes build up in an enclosed area. This can occur if an acetylene torch is left inside a compartment, if products of decomposition are enclosed, or from a fuel gas leak into the compartment. The atmosphere within the enclosed area must be tested prior to entering the area. Portable explosimeters are available for sampling the atmosphere to determine if any explosive mixture is present.

Another problem relating to confined or enclosed areas involves oxygen-enriched atmospheres. Such atmospheres can result from the oxy flame cutting torch being left in the compartment and a leak of the oxygen line. Normally, the atmosphere contains approximately 21% oxygen. If the oxygen were to increase by 5% or more, the enriched atmosphere would support rapid combustion or even an explosive mixture. Striking an arc or lighting a flame could be extremely hazardous. Clothes, oily cloth, and other combustible items would burn rapidly and create a hazardous condition. Oxygen from a compressed gas cylinder should never be used to help ventilate an enclosed compartment. It should never be used in place of compressed air. Portable instruments indicating oxygen concentration are available and should be used to sample the atmosphere before entering an enclosed compartment.

Oxygen deficiency can be another hazard for workers in an enclosed area. When using the gas-shielded metal arc process, the two most popular shielding gases are both heavier than air. Both argon and carbon dioxide weigh approximately 1 1/2 times the weight of air and will displace it. The used shielding gas will in time displace the air so that the atmosphere at the welder's breathing zone will become rich in the shielding gas atmosphere and there will be a deficiency of oxygen. If the oxygen content in the breathing zone is reduced by 5% or more, serious damage can be done to the worker. The atmosphere in an enclosed area must be monitored with a portable oxygen indicator.

Mechanical ventilation must be used for ventilating enclosed areas. Both air exhaust systems and fresh-air supply systems should be employed. When welding, cutting, or allied processes are used in any area that cannot be adequately mechanically ventilated, positive-pressure, self-contained breathing apparatus or air-line respirators must be used.

If you have questions concerning monitoring atmospheres or monitoring instruments, or special breathing apparatus, contact your company's safety department or your local fire department or state industrial commission representative.

Ventilation

Adequate ventilation must be provided for all welding, cutting, brazing, and related operations. Adequate ventilation means sufficient ventilation so that hazardous concentrations of airborne contaminants are below the allowable levels specified by OSHA or the American Conference of Governmental Industrial Hygienists (ACGIH). Adequate ventilation depends on the following:

1. Volume and configuration of space where welding occurs
2. Number and type of operations generating contaminants
3. Allowable levels of specific toxic or flammable contaminants being generated
4. Natural airflow and general atmospheric conditions where work is being done
5. Location of welders and other persons' breathing zones in relation to the contamination, contaminants, or sources

Adequate ventilation for welding can be obtained in three different ways.

1. Natural ventilation
2. General mechanical ventilation
3. Local exhaust ventilation

Natural ventilation occurs when the welding is done out of doors. Natural ventilation occurs indoors if the welding shop is sufficiently large, with a space of 10,000 ft³ (284 m³) per welder; if there is a ceiling height of more than 16 ft (5 m) and the welding space does not contain partitions, balconies, or other structural barriers that obstruct ventilation; and finally, if the welding is not done in a confined area. Natural ventilation must be supplemented when welding on hazardous materials.

General mechanical ventilation utilizing roof exhaust fans, wall exhaust fans, or similar large-area air movers must be used if the space per welder is less than 10,000 ft³ (284 m³), or if the ceiling height is less than 16 ft (5 m) or the shop includes partitions, balconies, or other structural barriers that obstruct cross ventilation. General mechanical ventilation is recommended to maintain a low level of airborne contaminants and to prevent the accumulation of explosive gas mixtures. General mechanical ventilation is used for individual welding booths (Figure 4–13). If general mechanical ventilation is not sufficient to maintain the general background level of airborne con-

taminants below the limits recommended, local exhaust ventilation or local forced ventilation is required.

There are basically two types of **local exhaust ventilation** systems. One is a low-volume, high-velocity fume exhaust system and the other is a high-volume, low-velocity fume exhaust system. They are different. They both use a fixed or movable suction pickup device to capture contaminants to keep the level of pollutants below the legal requirements. The high-volume, low-velocity system moves much more air and uses a relatively low vacuum, of 10 to 15 in. H₂O. This system extracts fume from near the arc to up to 1 ft from the arc. The amount of air to be moved is related to the distance between the arc and the collection hood. For example, if the hood is 6 in. (150 mm) from the arc, 250 ft³/min is required. At 9 in. (225 mm), 400 ft³/min is required and at 12 in. (300 mm) 1000 ft³/min is required. The hood is normally attached to a freely movable ventilating duct and is placed close to the welding operation, as shown in Figure 4–14. These systems should move 800 to 1000 ft³ of air per minute. An example of this type of system using a self-contained portable exhaust unit is shown in Figure 4–15. Another type of local exhaust system uses a downdraft welding worktable. These tables require 100 ft³/min (47 m³/mm) of volume per square meter of surface work area. A downdraft table is shown in Figure 4–16. They are very popular in Europe. Fixed enclosure around the welding area with

---

FIGURE 4–13 Welding booth with mechanical ventilation in ceiling and wall.

FIGURE 4–14 Local exhaust ventilation using movable hood.

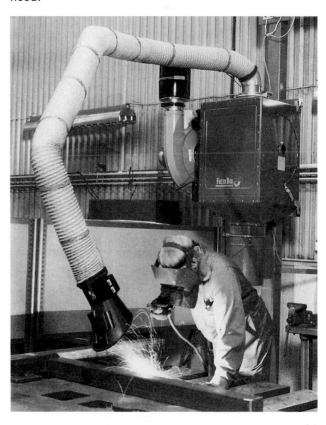

FIGURE 4–15 Local exhaust ventilation using portable exhaust system.

FIGURE 4–16 Downdraft welding table.

a top and at least two sides can also be used with positive airflow.

The newer low volume–high velocity system has become popular for GMAW and FCAW applied semiautomatically or with a welding robot. This system uses a high vacuum of 50 to 75 in. H_2O. An example is based in collecting the fume as close as possible to the arc. This system is illustrated in Figure 4-17, which shows an exhaust nozzle on a welding gun and a comparison with and without the exhaust system in operation. Special vacuum pickups for the semiautomatic guns and for manual SMAW are available. This system is economical since much less air is exhausted. Its use reduces the size of the air makeup unit to replace heated or cooled air in the shop.

The exhausted air should be filtered before it is discharged into the atmosphere or returned to the welding shop. It is important to keep the low volume–high velocity and high volume–low velocity central collection systems separated. They are designed for different air volume, velocity, and pressure ratings.

The collection system can be a central shop-wide system using 6- to 8-in.-diameter ducts and a single high-

FIGURE 4–17 Local exhaust ventilation comparison with and without exhaust.

volume fan. A central system can be extended up to 1000 ft and can accommodate up to 25 welding stations. The other type of collection system uses individual collectors for each station. These can be self-contained portable or wall or ceiling units mounted with built-in filters. These units can be noisy.

In all cases the fume collected must be filtered to remove the particulate matter. Filters, which are designed for individual systems or central systems, will not separate gaseous matter. There are three types of filters: (1) fabric collectors, known as barrier filters; (2) electronic air cleaners or electrostatic precipitators; and (3) cartridge collectors. Particulate collection depends on the filter efficiency ratio and the source capture efficiency, which must be studied and selected with care.

Compact gas measuring and warning instruments are sometimes needed. These are used to warn people of the presence of hazardous gases in dangerous concentrations. Many of these instruments provide visual and audible alarms. Instruments are available for measuring CO, H_2S, and oxygen, and other potentially hazardous gases. Consult with a safety expert regarding the need of such devices.

Local forced ventilation is a local air-moving system such as a fan placed so that it moves the air at right angles to the welder across the welder's face. It should produce a velocity of approximately 100 ft/min (30 m/min) and be maintained for a distance of approximately 2 ft (0.6 m) directly across the work area. Air velocity is relatively easy to measure using a velometer or airflow meter; thus it is easy to check the efficiency of local forced ventilation.

For serious ventilation situations, welders should use faceplate respirators. There are different kinds. Some include filter cartridges, others utilize air filters or air supply filters, and others use external air supply. This is a very complex subject, and decisions should only be made by on-site safety specialists or industrial hygienists. Furthermore, faceplate respirators should not be used in special atmospheres, nor should they be used in atmospheres of less than 19.5% oxygen. Faceplate respirators can be combined with welders' helmets for practicability. All units should be approved by NIOSH.

There is one foolproof method to determine if proper ventilation is being provided. This is done by collecting samples of the atmosphere at the welder's breathing zone under the helmet. A special pickup device mounted inside the helmet is usually used. The atmosphere samples are collected by specialized instruments for a specific period. The samples are then chemically analyzed in calibrated instruments which determine the value of all elements that are found in the welder's breathing zone.

The AWS has a new standard F3.2-2000, called "Ventilation Guide for Weld Fume." This is also an American National Standards Institute standard. This guide outlines recommended principles of ventilation systems for facilities with welding and allied processes. It is intended to help select and design proper ventilation systems. The primary objective is to enhance the health and safety of the industrial environment. Another objective is to conserve energy. It does not provide information on respiratory protection devices or specific precautions for confined spaces. It does, however, include references to all the major documents that are involved. It is highly recommended that this document be referred to with respect to any welding shop ventilation situation.

4-5 FIRE AND EXPLOSION HAZARD

A large number of the fires in industrial plants are caused by cutting and welding with portable equipment in areas not specifically designated or approved for such work. The three elements of the fire triangle—fuel, heat, and oxygen—are present in most welding operations. The heat is from the torch flame, the arc, or hot metal. The fuel is from the fuel gas employed or from combustibles in the welding area. The oxygen is present in the air but may be enriched by oxygen used with the fuel gas. Many industrial fires have been caused by sparks, which are globules of oxidized molten metal that can travel up to 40 ft (13 m). Sparks may also fall through cracks, pipe holes, or other small openings in floors and partitions and start fires in other areas, which may temporarily go unnoticed.

Hot pieces of metal may come in contact with combustible materials and start fire. Fires and explosions have also been caused when this heat is transmitted through walls of containers to flammable atmospheres or to combustibles within containers. Anything that is combustible or flammable is susceptible to ignition by cutting and welding. Welding or cutting on metal that is in contact with urethane foam insulation is prohibited. All insulating organic foams, whether or not indicated to be fire retardant, should be considered combustible and handled accordingly.

Cutting and welding fires can be prevented by eliminating all combustibles from the welding area. Welding arcs or oxyfuel gas flames rarely cause fires when used in the workshop area designed for welding and cutting. Fire and explosion hazards should be considered from two points of view: welding in designated workshop areas and welding with portable equipment in all other areas.

Work Area

A safe workplace must be provided for welding and cutting operations. Floors, walls, ceilings, and so on, must be constructed of noncombustible materials. The work area must be kept clean and free of combustible and flammable materials. All fuel gas lines, manifolds, branches, and so on, must be installed in accordance with specifications and codes.

Fire and Extinguishers

In every situation where welding is done, in the welding shop and, with portable equipment, in all other areas, fire extinguishers should be available. The classic fire triangle is complete in a welding situation; hence the potential is great and precautions should be provided. Figure 4–18 shows the recommended fire extinguishers for class A, class B, class C, and class D fires. Depending on the work area, the appropriate extinguishers should be available at the work site.

Fuel Gases

There are many different fuel gases used for welding and flame cutting. The most familiar is acetylene, but propane, natural gas, methylacetylene–propadiene stabilized, and others are also used (see Chapter 14 for details).

Acetylene is sometimes produced on the premises by an acetylene generator. Portable acetylene generators are sometimes used in the field. An acetylene generator uses carbide and water to produce acetylene, which is then piped through the plant to the welding and cutting departments. Acetylene generators must be properly installed and maintained. They should be operated only by trained and qualified personnel. Observe the standards for safety for acetylene generators issued by Underwriters' Laboratory (UL297) (UL408) (UL409).[18] Carbide must be stored away from the acetylene generator and in a dry place not exposed to moisture or water.

Acetylene cylinders and other fuel gas cylinders should be stored in a specified well-ventilated area or outdoors away from oxygen and in the vertical position. All cylinders in storage should have their caps on, and both filled and empty cylinders should have their valves closed. In a fire situation special precautions should be taken for acetylene cylinders.[19] All acetylene cylinders are equipped with safety relief devices filled with a low-melting-point metal. This fusible metal melts at about the boiling point of water (212°F or 100°F). If fire occurs on or near an acetylene cylinder, the fuse plug will melt. The escaping acetylene may be ignited and will burn with a roaring sound. Evacuate all people from the area immediately. It is difficult to put out such a fire. The best action is to pour water on the cylinder to keep it cool and to keep all other acetylene cylinders in the area cool. Attempt to remove the burning cylinder from close proximity to other acetylene cylinders, from flammable or hazardous materials, or from combustible buildings. It is best to allow the gas to burn rather than to allow acetylene to escape, mix with air, and possibly explode.

If the fire on a cylinder is a small flame around the hose connection, the valve stem, or the fuse plug, try to put it out as quickly as possible. A wet glove, wet heavy cloth, or mud slapped on the flame will frequently extinguish it. Thoroughly wetting the gloves and clothing will help protect the person approaching the cylinder. Avoid getting in line with the fuse plug, which might melt at any time.

FIGURE 4-18 Fire extinguisher data.

KNOW YOUR FIRE EXTINGUISHERS

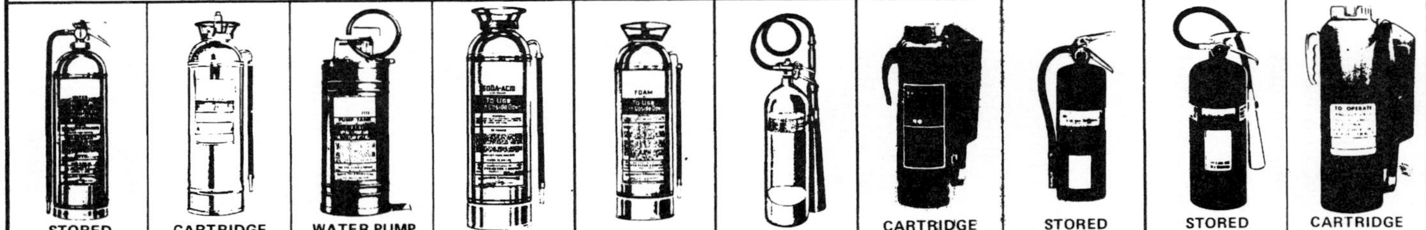

	WATER TYPE				FOAM	CARBON DIOXIDE	DRY CHEMICAL			
							SODIUM OR POTASSIUM BICARBONATE		MULTI-PURPOSE ABC	
	STORED PRESSURE	CARTRIDGE OPERATED	WATER PUMP TANK	SODA ACID	FOAM	CO2	CARTRIDGE OPERATED	STORED PRESSURE	STORED PRESSURE	CARTRIDGE OPERATED
CLASS A FIRES — ORDINARY COMBUSTIBLES — WOOD, PAPER, TRASH HAVING GLOWING EMBERS	YES	YES	YES	YES	YES	NO (BUT WILL CONTROL SMALL SURFACE FIRES)	NO (BUT WILL CONTROL SMALL SURFACE FIRES)	NO (BUT WILL CONTROL SMALL SURFACE FIRES)	YES	YES
CLASS B FIRES — FLAMMABLE LIQUIDS — FLAMMABLE LIQUIDS, GASOLINE, OIL, PAINTS, GREASE, ETC.	NO	NO	NO	NO	YES	YES	YES	YES	YES	YES
CLASS C FIRES — ELECTRICAL — ELECTRICAL EQUIPMENT	NO	NO	NO	NO	NO	YES	YES	YES	YES	YES
CLASS D FIRES — COMBUSTIBLE — COMBUSTIBLE METALS	SPECIAL EXTINGUISHING AGENTS APPROVED BY RECOGNIZED TESTING LABORATORIES									
METHOD OF OPERATION	PULL PIN- SQUEEZE HANDLE	TURN UPSIDE DOWN AND BUMP	PUMP HANDLE	TURN UPSIDE DOWN	TURN UPSIDE DOWN	PULL PIN- SQUEEZE LEVER	RUPTURE CARTRIDGE- SQUEEZE LEVER	PULL PIN- SQUEEZE HANDLE	PULL PIN- SQUEEZE HANDLE	RUPTURE CARTRIDGE- SQUEEZE LEVER
RANGE	30'-40'	30'-40'	30'-40'	30'-40'	30'-40'	3'-8'	5'-20'	5'-20'	5'-20'	5'-20'
MAINTENANCE	CHECK AIR PRESSURE GAUGE MONTHLY	WEIGH GAS CARTRIDGE ADD WATER IF REQUIRED ANNUALLY	DISCHARGE AND FILL WITH WATER ANNUALLY	DISCHARGE ANNUALLY -RECHARGE	DISCHARGE ANNUALLY -RECHARGE	WEIGH SEMI- ANNUALLY	WEIGH GAS CARTRIDGE- CHECK CONDITION OF DRY CHEMICAL ANNUALLY	CHECK PRESSURE GAUGE AND CONDITION OF DRY CHEMICAL ANNUALLY	CHECK PRESSURE GAUGE AND CONDITION OF DRY CHEMICAL ANNUALLY	WEIGH GAS CARTRIDGE- CHECK CONDITION OF DRY CHEMICAL ANNUALLY

the industrial commission of ohio
division of safety and hygiene
in cooperation with
ohio association
of broadcasters

Apparatus

Gas welding and cutting apparatus must show the approval of an independent testing laboratory. When ordering gas welding or cutting apparatus, specify that it must carry the Underwriters' Laboratories (UL) or Factory Mutual Engineering Corporation (FM) seal of approval.

Gas apparatus must be properly maintained and repaired by qualified people. All too often apparatus is allowed to deteriorate before maintenance is performed. Pressure gauges, welding regulators, welding torches, welding tips, and so on, should all be carefully inspected periodically and maintained. Oil or grease should never be used on any gas welding or cutting apparatus.

Only approved gas hoses should be used with oxy-fuel gas equipment. Single lines, double vulcanized, or double or multiple stranded lines are available. The double vulcanized or twin hose is preferred. The size of hose should be matched to the connectors, the regulators, and torches. In the United States the color green is used for oxygen, red for the acetylene or fuel gas, and black for inert gas and compressed air. The international standard calls for blue for oxygen and orange for the fuel gas. The connections on hoses are right-handed for inert gases and oxygen, and left-handed for fuel gases. The nuts on fuel gas hoses are identified by a groove machined in the center of the nuts. Hoses should be inspected periodically for burns, worn places, or leaks at the connections. They must be kept in good repair and should be no longer than necessary.

Hot Work Permits

Welding permits, or, as they are sometimes called, "hot work permits," are often required. These permits must be used when welding or flame cutting is done on items that involve hazards. Special precautions such as the making out of the welding permit, the stationing of a fire watcher or lookout with proper fire control equipment, and the signing on and off of the welding operation are prescribed. Other types of operations require hot work permits. These are outlined in the instructions given by the National Safety Council for "Hot Work Permits."[20] Their sample permit form is shown in Figure 4–19. Instructions and precautions against fire appear on the reverse side. Your casualty insurance company may have similar tags. The National Fire Protection Association in its specifications for cutting and welding processes[21] also recommends the use of a hot work permit. A welding permit or hot work permit should be used when portable welding or flame cutting equipment is used for maintenance welding in plants where combustible materials are present, on ships, and on any other type of potentially hazardous operations. When using portable equipment, these are invaluable and can avoid headlines stating that the fire was caused by "a welder's torch."

Welding on Containers

Any container or hollow body, such as a can, a tank, a hollow compartment in a weldment, or a hollow area in a casting, even though it may contain only air, must be given special attention before welding. The heat from welding will raise the temperature of the enclosed air or gas to a possible dangerously high pressure so that the part or container may explode. Always vent confined air before welding or cutting on a hollow area. Hollow areas may contain oxygen-enriched air or fuel gases, either of which is extremely dangerous when heated or exposed to an arc or flame.

Explosions and fires may result if welding or cutting is done on empty containers that are not entirely free of combustible solids, liquids, vapors, dust, and gases. Containers can be made safe for welding or cutting by following prescribed steps. Refer to the welding society's "Recommended Safe Practices for the Preparation for Welding and Cutting of Containers and Piping That Have Held Hazardous Substances."[22] No container should be considered clean or safe until proved so by tests. Cleaning the container, which is normally made of metal, is necessary in all cases before welding or cutting. Cleaning should be done outdoors; if this is impractical, the inside work area should be well ventilated so that flammable vapors will be quickly carried away. Drain all material from the container and remove sludge, sediment, and the like. Dispose of residue before starting to weld or cut. Identify the material that was in the container and match the cleaning method to the material it contained. If the material is water soluble, the container can be cleaned with water. If the material is not readily soluble in water, the container should be cleaned by a hot chemical solution or by steam. Mix the chemicals for cleaning in hot water and pour into the container, fill the container completely with water, introduce live steam to heat, and agitate the solution. If hot water and steam are not available, the cold water method can be used; however, it is less effective, and agitation should be handled by means of compressed air. Another way of cleaning the container is to fill the container 25% full with cleaning solution and clean thoroughly. Following this, introduce low-pressure steam into the tank, allowing it to vent through openings. Continue to flow steam through the tank for several hours. None of these cleaning methods is perfect, and after cleaning, the tank should be inspected to determine that it is thoroughly clean. If it is not, continue the cleaning operation.

After the container is cleaned, close all openings and after 15 minutes test a sample of the gas inside the container. Use a combustible gas indicator instrument. If the concentration of flammable vapors in the sample is not below the limit of flammability, repeat the cleaning operation. When it is determined that the gas or air inside the container is safe, the container should be so marked, signed, and dated. Even after tanks have been made safe

PERMIT NO. _____

For electric and acetylene burning and welding with portable equipment in all locations outside of shop.

Date _____ _____

Time Started _____ Finished _____

Building _____

Dept. _____ _____ Floor _____

Location on Floor _____

Nature of Job _____

_____

Operator _____

Clock No. _____

All precautions have been taken to avoid any possible fire hazard, and permission is given for this work.

Signed _____
                                                    Foreman

Signed _____
                                        Safety supervisor or
                                        plant superintendent

_____

PERMIT NO. 10534

Date _____ _____

Bldg _____ Floor _____

Nature of Job _____

Operator _____

_____

INSTRUCTIONS TO OPERATORS

This permit is good only for the location and time shown. Return the permit when work is completed.

PRECAUTIONS AGAINST FIRE

1. Permits should be signed by the foreman of the welder or cutter and by the safety supervisor or plant superintendent.

2. Obtain a written permit before using portable cutting or welding equipment anywhere in the plant except in permanent safe-guarded locations.

3. Make sure sprinkler system is in service.

4. Before starting, sweep floor clean, wet down wooden floors, or cover them with sheet metal or equivalent. In outside work, don't let sparks enter doors or windows.

5. Move combustible material 25 feet away. Cover what can't be moved with asbestos curtain or sheet metal, carefully and completely.

6. Obtain standby fire extinguishers and locate at work site. Instruct helper or fire watcher to extinguish small fires.

7. After completion, watch scene of work a half hour for smoldering fires, and inspect adjoining rooms and floors above and below.

8. Don't use the equipment near flammable liquids, or on closed tanks which have held flammable liquids or other combustibles. Remove inside deposits before working on ducts.

9. Keep cutting and welding equipment in good condition. Carefully follow manufacturer's instructions for its use and maintenance.

FIGURE 4–19 Hot work permit. (From Ref. 20.)

they should be filled with water, as an added precaution before welding or cutting. Place the container so that it can be kept filled with water to within a few inches of the point where welding and cutting are to be done. Make sure that the space above the water level is vented so that the heated air can escape (Figure 4-20).

As an alternative to the water-filling method, fill the container with an inert gas. Flammable gases and vapors will be rendered nonflammable and nonexplosive if mixed with a sufficient amount of inert gas. Nitrogen or carbon dioxide is normally used. The concentration of flammable gases and vapors must be checked by testing as mentioned previously. The inert gas concentration must be maintained during the entire welding and cutting operation. Hot work or welding permits should be utilized for all welding or cutting operations on containers that have held combustibles.

FIGURE 4–20 Safe way to weld containers that hold combustibles.

Hot Tapping

In pipe lining, welders sometimes do hot tapping. This is the welding of a special fitting to a line carrying a combustible liquid or gas, then cutting a hole in the pipe after the fitting has been welded to it. This must be done by experienced people using special equipment with proper precautions. Before attempting such work, refer to the American Petroleum Institute (API) publication "Welding or Hot Tapping on Equipment Containing Flammables."[23]

4-6 COMPRESSED GASES HAZARD

All compressed gas cylinders are potential hazards. The major hazard is the possibility of sudden release of the gas by removal or breaking off of the valve. Escaping gas that is under high pressure will cause the cylinder to act as a rocket, smashing into people and property. Escaping fuel gas can also be a fire or explosion hazard. See Chapter 14 for more information about gas apparatus.

Treatment of Gas Cylinders

Gases used for welding—fuel gases, oxygen, or shielding gases—are normally delivered in cylinders which are manufactured and maintained by the gas supplier in accordance with the regulations of the U.S. Department of Transportation (DOT). In Canada the Board of Transport Commissioners for Canada has this responsibility. In most countries there are laws and regulations concerning manufacturing, maintaining, and periodic inspection of portable cylinders for the storage and shipment of compressed gases. All compressed gas cylinders must be legibly marked to identify the gases contained by either the chemical or the trade name of the gas. There is no international uniform color coding for identification purposes; however, some countries have standardized color marking systems. There is a uniform standard for connection threads in North America in accordance with the American–Canadian Standard "Compressed Gas Cylinder Valve Outlet and Inlet Connections."[24]

In North America the authorities require that a cylinder be condemned when it leaks or when internal or external corrosion, denting, bulging, or evidence of rough usage exists to the extent that the cylinder is likely to be appreciably weakened. Always inspect cylinders for suspicious areas and report this or any damage done to a cylinder to your gas supplier.

Cylinder Storage

Oxygen cylinders should be stored separately from fuel gas cylinders and separate from combustible materials. Store cylinders in cool, well-ventilated areas. The temperature of the cylinder should never be allowed to exceed 130 °F (54 °C). Cylinders should be stored vertically and secured to prevent falling. The valve protection caps must be in place. When cylinders are empty they should be marked empty and the valves must be closed to prohibit contamination from entering. When the gas cylinders are in use, a regulator is attached and the cylinder should be secured by means of chains or clamps. Cylinders for portable apparatus should be securely mounted in specially designed cylinder trucks. Cylinders should be handled with respect. They should not be dropped or struck. They should never be used as rollers. Hammers or wrenches should not be used to open cylinder valves that are fitted with hand wheels. They should never be moved by electromagnetic cranes. They should never be in an electric circuit so that the welding current could pass through them. An arc strike on a cylinder will damage the cylinder, causing possible fracture and requiring the cylinder to be condemned and removed from service.

Oxygen

Oxygen is one of the most common gases carried in portable high-pressure cylinders. It should always be called "oxygen," never "air." Combustibles should be kept away from oxygen, including the cylinder, valves, regulators, and hose apparatus. Oxygen cylinders or oxygen apparatus should not be handled with oily hands or oily gloves. Oxygen does not burn but will support and accelerate combustion of oil and grease and other hydrocarbon materials, causing them to burn with great intensity. Oil or grease in the presence of oxygen may ignite spontaneously and burn violently or explode. Oxygen should never be used in any air tools or for any of the purposes where compressed air is normally used. Escaping oxygen can enrich the work area, especially enclosed areas, and be a fire or explosion hazard.

Fuel Gases

Many different fuel gases are used in welding shops. Their properties are given in Chapter 14. All are compounds of carbon and hydrogen, and all are potentially hazardous. They must be treated with respect. See the information given under the heading "Fuel Gases" in Section 14-2.

When welding or cutting with oxygen and fuel gases, the welder should be alert to backfires and flashbacks. A *backfire* is an explosion in the torch head, usually accompanied by a loud popping sound associated with a momentary extinguishment and reignition of the flame at the torch tip. It is caused by obstructing the gas flow, which may occur from having the torch positioned too close to the work, by an overheated or damaged tip, by loose connections, or by incorrect gas pressure. If it occurs, the equipment should immediately be shut down and corrective action taken.

A *flashback* is the burning of the flame in the tip or torch, or even in the hose, when an explosive mixture is present. It is usually accompanied by a hissing or squealing sound and has the characteristic smoky or sharp-pointed flame. It can be caused by improper pressure, damaged or loose tips, damaged seats, kinked hose, and so on. Both backfires and flashbacks are dangerous and should be avoided by the use of flashback arresters. Torch flashback arresters consist of a check valve. Regulator flashback arresters include a check valve plus a sintered metal filter that allows the passage of gas but not of the flame. Both torch and regulator-mounted flashback arresters should be used, and they should be used on the oxygen line as well as on the fuel gas line. Flashback arresters are discussed in Section 7-1.

Shielding Gases

Shielding gases are either inert or active. True inert gases are argon and helium and are stored in high-pressure cylinders. Nitrogen, considered inert at low temperatures, is also stored in high-pressure cylinders. These cylinders must be treated with the same precautions as those used with oxygen cylinders. The active gas normally used for weld shielding is carbon dioxide (CO_2). It is stored as a liquid but gasifies upon release. More information is given in Chapter 14.

Gas Cylinder Adapters

Adapters are connectors that convert one type of valve outlet to another to allow connections to be made to devices such as regulators with different connection threads. This is done to allow a regulator or other component originally designed for a particular gas to be used with another gas. Gas fueling operations often use adapters to allow connections to be made to a manifold that handles a number of different gases. These are for the mixing of gases for special applications. When used judicially by well-trained personnel who understand the potential hazards of bringing incompatible gases together, adapters can serve a useful purpose. On the other hand, the indiscriminate use of adapters can lead to drastic consequences. Adapters must not be used that connect a high-pressure source to a low-pressure tank or cylinder. It is wise to review any adapters employed and control or limit those and prohibit the making of special adapters without thorough investigation of their use and potential hazards.

4-7 WELD CLEANING AND OTHER HAZARDS

The slag that often covers the deposited weld metal must be removed. Welds are often chipped and ground. Hand and power tools are employed, and the materials removed are propelled through the air to become potential hazards. Safety glasses with side shields should be worn under the welding helmet.

Radioactive Hot Areas

Welders may be required to work in radioactive "hot" areas. This is due to repair and maintenance operations necessary in nuclear power plants. In such cases, extra-special care and precautions must be observed to determine the radiation levels, time of exposure, radiation protection, and all other factors involved. The exposure time may be extremely short, and welders may be used to set up automated welding equipment and then leave the hot area to operate the devices remotely. Only qualified personnel with knowledge of working in and around radioactive areas should be permitted to make judgments of this type.

Noise

Weld chipping and weld peening produce excessive noise and should be controlled. Excessive noise can damage hearing and cause other injury. Noise exposure can cause either temporary or permanent hearing loss. The requirements of OSHA regulation prescribe allowable noise exposure levels. Carbon arc air gouging at high currents produces large amounts of noise, and plasma arc cutting with high current also creates excessive noise. Ear protection is required, for both.

Noise measurement instruments are available and should be used to check noise in the work area so that precautionary measures can be taken. Normal arc welding operations do not exceed noise-level requirements as specified by OSHA. In combination with other noise-producing industrial machinery, noise levels may be excessive. Noise levels can be measured and monitored by means of specialized instruments. The AWS "Method of Sound Level Measurement of Manual Arc Welding and Cutting Processes"[25] should be consulted. It is necessary that trained personnel be used to measure noise. You can request help from your company's safety department or from the state industrial commission representatives. Noise levels are reduced fairly rapidly as the worker moves farther away from the source of the noise. Sound attenuation equipment can be installed to reduce excessive noise.

Other Hazards

Falling items create hazards. Hard hats should be worn in connection with welding helmets on construction sites and in some plants. Other hazards, such as falling from high places, working with heavy objects, and working around heated metals, are similar to the hazards encountered by all employees in steel plants, forging shops, structural shops, and so on. Welding electrode stubs, the unused ends gripped in the holder, are usually short and round and act as a roller when stepped on at the wrong angle. Electrode stubs should be placed in containers and not thrown or allowed to remain on the floor or working surface.

4-8 SAFETY FOR SPECIFIC WELDING PROCESSES AND OCCUPATIONS

The previous sections dealt with arc welding and oxyfuel gas welding, cutting, and torch brazing. The other welding and allied processes can also be hazardous if safety precautions are ignored. The potential hazards mentioned previously apply to most welding and allied processes since electricity, compressed gases, flames, heated metals, or fumes are usually involved. Specific process applications or welding occupations involve other particular hazards. Each section that relates to a particular process includes specific safety information. The following is an overview of these safety situations.

Welders may be required to work inside heated tanks or containers. They should be supplied with cool air, preferably from a portable air conditioner, while they are inside a heated chamber. To avoid heat stress, they should use a body cooler or body core cooling system. The welder wears a specially designed vest that helps reduce the body temperature while in the preheated weldment. This vest is made of thin, nontoxic, nonflammable material that contains pouches. The numerous pouches are designed to hold special polymer gel thermal strips. This material is cooled prior to inserting in the pouches and will provide the welder up to four hours of comfort. The vests are lightweight, fit comfortably no matter the position the welder is in, and can be taken off and put on easily.

Underwater welding is a very dangerous welding occupation. Underwater work of any type is dangerous, at any working depth. "Welding in the dry," underwater, is welding in an atmosphere which is under pressure that is greater than sea-level atmosphere pressure. Higher operating pressures create special hazards. The hazards of underwater "welding in the wet," in contact with the water or in a habitat, are very complex and are mentioned only briefly here. More complete information concerning all aspects of underwater welding is provided in Section 26-5.

Robotic and automated welding are becoming more popular. Robot welding combines the potential hazards of welding with the hazards of moving machinery. Robots operate outside their machine base area. They involve unanticipated motion, start unexpectedly, and operate at relatively high rates of speed. Robots are normally safe since operators work outside the operating envelope of the robot. However, when programming robots or maintaining equipment, or troubleshooting welding problems, people work in close proximity to the robot's welding torch and are thus exposed to potential hazards. More information on robotic arc welding is presented in Section 12-7.

Automated brazing and soldering involves motion equipment and the associated hazards. However, fluxes and filler metals employed may give off noxious fumes when heated. Adequate mechanical ventilation should be provided for all automated brazing and soldering operations to remove toxic gases. In addition, large quantities of liquid-heated flux or filler material metal create hazards. Guards on motion devices must be properly designed and always in place.

Resistance welding operations involve some potential hazards. These are largely involved with motion, since it is present with resistance welding equipment. Dual palm buttons are normally used to provide operator safety. Operators should wear face shields, spectacles, or goggles to protect the face and eyes from flying sparks that may be ejected from the weld area.

Air arc cutting and gouging and plasma arc cutting at high currents create noise of a level that may be harmful. Ear protection should be worn.

Electron beam welding is an automated process, but the motion is normally enclosed. In most cases a vacuum is involved with the welding chamber and normal precautions are required. In high-voltage electron beam systems, x-rays are generated as the electron beam strikes the workpiece. Adequate shielding must be provided to protect the operator from x-rays.

Thermal spraying involves potential hazards in addition to those involved with arc welding and oxyfuel gas welding. These involve the use of powders or wires which are atomized and sprayed on the workpiece. Large amounts of particulate matter are produced, which can create problems. Additional information is provided in Section 9-3.

Laser welding is usually an automated operation. Lasers are used not only for welding but also for cutting and surface metal treatment. The equipment must definitely be installed in accordance with the manufacturer's recommendations. Certain classes of lasers generate radiation, which can produce eye damage. This also relates to reflected laser light. Safety precautions require the use of special glasses and other protective materials.

Continued attention to safe practice is required for all welding, cutting, and allied processes. Common sense and the adoption of practices recommended in this book will help provide a safe workplace.

 QUESTIONS

4–1. Why wear safety glasses?

4–2. What is the advantage of a curved-front helmet?

4–3. Should oxygen or fuel gas be turned on first when lighting a flame cutting torch?

4–4. Is welding more dangerous or less dangerous than other metalworking jobs?

4–5. Why shouldn't a welder open up the case of an arc welding machine?

4–6. Which is more dangerous, welding open-circuit voltage or the voltage in an electric drill?

4–7. Why is eye and skin protection required against an electric arc?

4–8. Name the metals that create hazards when welded.

4–9. What is an oxygen-deficient atmosphere? How is it related to welding?

4–10. Why is it dangerous to weld on a closed tank or can?

4–11. Give two reasons for checking an enclosed area before welding in it.

4–12. What is the purpose of a hot work permit?

4–13. What should you do if an acetylene cylinder catches on fire?

4–14. Why is maintenance welding more dangerous than production welding?

4–15. How do you make a tank that has held a combustible safe for welding?

4–16. Why should oil and grease be kept away from pure oxygen?

4–17. What is the purpose of warning labels?

4–18. Why should welders keep out of the plume?

4–19. What causes most fires during maintenance work?

4–20. What is dangerous about welding in a wet situation?

REFERENCES

1. "Occupational Safety and Health Standards," Code of Federal Regulations Title 29, Labor, Part 1910, Subpart Q, U.S. Government Printing Office, Washington, D.C.; *Federal Register,* April 1990.

2. "Safety in Welding, Cutting, and Allied Processes." ANSI/ASCZ49.1, American Welding Society, Miami, Fla.; and "Code for Safety in Welding and Cutting," CSA Standard W117.2, Canadian Standards Association, Rexdale, Ontario, Canada.

3. "Effects of Welding on Health," Part I, 1979; Part II, 1981; Part III, 1983; Part IV, 1985; Part V, 1987; American Welding Society, Miami, Fla.

4. "Inspection Procedures for the Hazard Communication Standard," 29 CFR 1910, 1200 2nd revision, U.S. Department of Labor, Washington, D.C., May 16, 1986.

5. "Threshold Limit Values and Biological Exposure Indices for 1986–1987," American Conference of Governmental Industrial Hygienists, Cincinnati, Ohio.

6. "Specifications for Accident Prevention Signs," ANSI Z535, American National Standards Institute, New York.

7. "Electric Arc Welding Apparatus," EW-1, ANSI C87.1, National Electrical Manufacturers Association, Washington, D.C.

8. "Construction and Test of Arc-Welding Equipment, Transformer Type," CSA C22.2, No. 60, Canadian Standards Association, Rexdale, Ontario, Canada.

9. "Safety Standards for Transformer Type Welding Machines," ANSI C33.2, Underwriters' Laboratories, Chicago.

10. "Arc Welding Equipment, Part 1: Welding Power Sources," 1EC 974-1. International Electrotechnical Commission; Geneva, Switzerland.

11. "National Electrical Code," No. 70, ANSI C11-1971, National Fire Protection Association, Boston.

12. "Practice for Occupational and Educational Eye and Face Protection," ANSI Z87.1, American National Standards Institute, New York.

13. D. H. Sliney, C. E. Moss, C. G. Miller, and J. B. Stephens, "Transparent Welding Curtains," *Welding Journal* (May 1982).

14. "The Welding Environment," American Welding Society, Miami, Fla.

15. "Fumes and Gases in the Welding Environment," American Welding Society, Miami, Fla.

16. "Characterization of Arc Welding Fume," American Welding Society, Miami, Fla.

17. "Method for Sampling Airborne Particulates Generated by Welding and Allied Processes," AWS F1.1, American Welding Society, Miami, Fla.

18. Standards for Safety: UL297, "Acetylene Generators Portable, Medium Pressure"; UL408, "Standard for Acetylene Generators Stationary, Medium Pressure"; UL-409, "Standard for Acetylene Generators Stationary, Low Pressure"; Underwriters' Laboratories, Chicago.

19. "Handling Acetylene Cylinders in Fire Situations," Safety Bulletin SB-4, Compressed Gas Association, New York.

20. "Hot Work Permits (Flame or Sparks)," Data Sheet 522, Revision A, National Safety Council, Chicago.

21. "Cutting and Welding Practices," No. 51B, National Fire Protection Association, Boston.

22. "Recommended Safe Practices for the Preparation for Welding and Cutting of Containers and Piping That Have Held Hazardous Substances," AWS F4.1, American Welding Society, Miami, Fla.

23. "Welding or Hot Tapping on Equipment Containing Flammables," Petroleum Safety Data Sheet 2201, American Petroleum Institute, New York.

24. "Compressed Gas Cylinder Valve Outlet and Inlet Connections," American Standard B57.1 and Canadian Standard CSA B96.

25. "Method of Sound Level Measurement of Manual Arc Welding and Cutting Processes," AWS F6.1, American Welding Society, Miami, Fla.

ARC WELDING WITH A NONCONSUMABLE ELECTRODE

OUTLINE

5-1 THE NONCONSUMABLE WELDING ARC

There are two basic types of welding arcs: One uses a nonconsumable electrode and the other a consumable electrode. The nonconsumable electrode does not melt in the arc, and filler metal is not carried across the arc gap. The welding processes that use a nonconsumable electrode arc are: gas tungsten arc welding, plasma arc welding, and carbon arc welding. The main function of the arc is to produce heat. At the same time it produces a bright light, noise, and ionic bombardment that removes the oxide surface of the base metal.

A welding arc is a sustained high-current, low-voltage electrical discharge through a high conducting plasma that produces sufficient thermal energy which is useful for joining metals by fusion.[1] The welding arc is a steady-state condition maintained at the gap between the end of an electrode and a workpiece that carries current. Welding arcs range from as low as 1 A to as high as 3000 A and a voltage as low as 10 V to over 40 V. The welding arc has a point-to-plane geometric configuration, the point being the arcing end of the electrode and the plane the arcing area of the molten pool. A nonconsumable electrode arc is shown in Figure 5-1. Whether the electrode is positive or negative, the arc is restricted at the electrode and spread out toward the workpiece.

The length of the arc gap is proportional to the voltage across the arc, if other conditions remain the same. If the arc length is increased beyond a certain point, the arc will go out. The arc length for welding is the dimension equal to the electrode diameter, up to about four times the electrode diameter. There is a certain current necessary to sustain an arc of different lengths. If a higher current is used, a longer arc can be maintained.

The arc column is normally round in cross section and is made of two concentric zones: an inner core or plasma and an outer flame. The plasma carries most of the current and has the highest temperature. The outer flame of the arc is much cooler and tends to keep the plasma in the center. The temperature and the diameter of the central plasma depend on the amount of current passing through the arc, the shielding atmosphere, electrode size, and type. The relationship between current and arc voltage is not a straight line. The curve of a nonconsumable arc (Figure 5-2) takes a nonlinear form.[2] In general, the arc voltage increases slightly as the current increases. The voltage is higher for longer arcs and for arcs in a helium

FIGURE 5–1 Nonconsumable electrode arc.

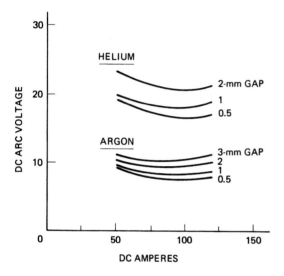

FIGURE 5–2 Arc characteristic volt-ampere curve in argon and helium. (From Ref. 2.)

thermal ionization of some of the atoms of the shielding gas. The positively charged gaseous atoms are attracted to the negative electrode, where their kinetic (motion) energy is converted to heat. This heat keeps the tungsten electrode hot enough for electron emission. Emission of electrons from the surface of the tungsten cathode is known as **thermionic emission.** Positive ions also cross the arc. They travel from the positive pole, the work, to the negative pole, the electrode. Positive ions are much heavier than the electrons but help carry the current flow of the relatively low-voltage welding arc. The largest portion of the current flow, approximately 99%, is via electron flow rather than the flow of positive ions. The continuous feeding of electrons into the welding circuit from the power source accounts for the continuing balance between electrons and ions in the arc. The electrons colliding with the work create the intense localized heat, which provides melting of the base metal.

In the dc tungsten-to-base metal arc in an inert gas atmosphere, the maximum heat occurs at the positive pole (anode).[3] When the electrode is positive (anode) and the work is negative (cathode) (Figure 5–3b), the electrons flow from the work to the electrode, where they create intense heat. The electrode tends to overheat, so a larger electrode with more heat-absorbing capacity is used for DCEP than for DCEN for the same welding current. In addition, since less heat is generated at the work, the penetration is not so great. One result of DCEP welding is the *cleaning effect* on the base metal adjacent to the arc area. This appears as an etched surface and is known as *cathodic etching;* it results from positive ion bombardment. This bombardment also occurs during the reverse-polarity half-cycle when using alternating current.

The arc length or gap between the electrode and the work can be divided into three regions: a central region, a region adjacent to the electrode, and a region adjacent to the work. At the end regions the cooling effects of the electrode and the work cause a rapid drop in potential. These two regions are known as the anode and cathode drop, according to the direction of current flow. The length of the central region or arc column represents 99% of the arc length and is linear with respect to arc voltage. Figure 5–4 shows the distribution of heat in the arc, which varies in these three regions. In the central region, a circular magnetic field surrounds the arc. This field, produced by the current flow, tends to constrict the plasma and is known as the *magnetic pinch effect.* The constriction causes high pressures in the arc plasma and extremely high velocities, and this in turn produces a plasma jet. The speed of the plasma jet approaches sonic speed.

The cathode drop is the electrical connection between the arc column and the negative pole (cathode). There is a relatively large temperature and potential drop at this point. This is the point at which the electrons are emitted by the cathode and given to the arc column. The stability of an arc depends on the smoothness of the flow

atmosphere. The conductivity of the arc increases faster than simply proportionally to current.

The arc occurs when electrons are emitted from the surface of the negative pole (cathode) and flow across a region of hot electrically charged plasma to the positive pole (anode), where they are absorbed.

Arc heat can best be explained by considering the dc tungsten electrode arc in the inert-gas atmosphere (Figure 5–3). In Figure 3a the tungsten arc is connected for direct-current electrode negative (DCEN). When the arc is started, the electrode becomes hot and emits electrons. The emitted electrons are attracted to the positive pole, travel through the arc gap, and raise the temperature of the shielding gas atoms by colliding with them. The collisions of electrons with atoms and molecules produce

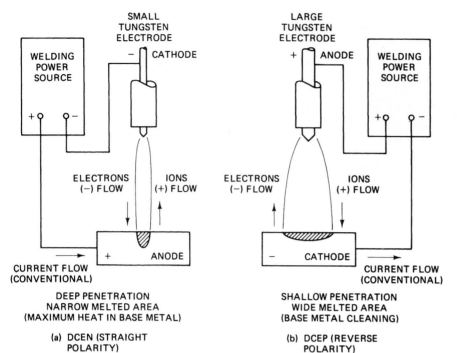

FIGURE 5–3 Nonconsumable arc showing polarity and heat.

of electrons at this point. Tungsten and carbon provide thermionic emissions since both are good emitters of electrons. They have high melting temperatures, are practically nonconsumable, and are therefore used for welding electrodes. Since tungsten has the highest melting point of any metal, it is preferred.

The anode drop occurs at the other end of the arc and is the electrical connection between the positive pole (anode) and the arc column. The temperature changes from that of the arc column to that of the anode, which is considerably lower. The reduction in temperature occurs because there are fewer ions in this region. The heat liberated at the anode and at the cathode is greater than that from the arc column. Figure 5–5 shows the approximate temperature in the arc. Arc temperature relates to the location within the arc and to the current in the arc.

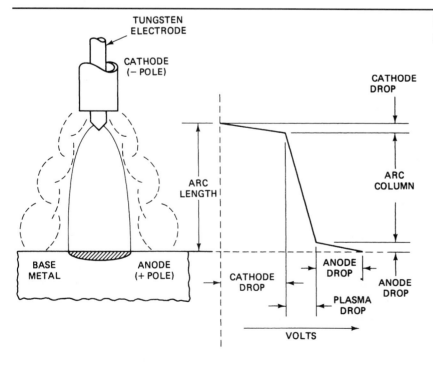

FIGURE 5–4 Arc region versus voltage and heat. (From Ref. 4.)

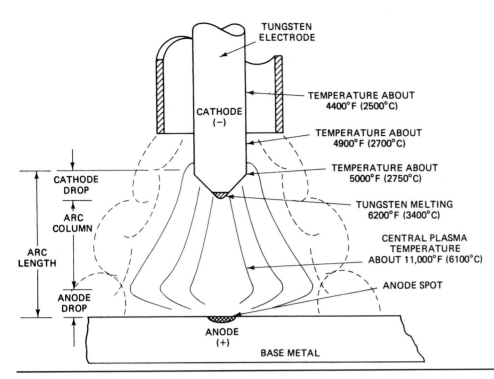

TUNGSTEN
ELECTRODE

CATHODE
(−)

TEMPERATURE ABOUT
4400°F (2500°C)

TEMPERATURE ABOUT
4900°F (2700°C)

TEMPERATURE ABOUT
5000°F (2750°C)

TUNGSTEN MELTING
6200°F (3400°C)

CENTRAL PLASMA
TEMPERATURE
ABOUT 11,000°F (6100°C)

ANODE SPOT

CATHODE
DROP

ARC
COLUMN

ARC
LENGTH

ANODE
DROP

ANODE
(+)

BASE METAL

FIGURE 5–5 Temperature of a 100-A dc tungsten arc in argon.

In the carbon arc, a stable dc arc is obtained when the carbon is negative. In this condition about one-third of the heat occurs at the negative pole (cathode), the electrode, and about two-thirds of the heat occurs at the positive pole (anode), the workpiece.

Researchers theorized that if the arc column of the tungsten arc could be reduced in cross section, its temperature would increase. If the cross-sectional area of the conducting column is reduced by constricting it, the only way that the same current would still pass is for its conductivity and its temperature to increase. Constriction occurs in a plasma arc torch by making the arc pass through a small hole in a water-cooled copper nozzle. It is a characteristic of the arc that the more it is cooled, the hotter it gets; however, it requires a higher voltage. By flowing additional gas through the small hole, the arc is further constricted and a high-velocity, high-temperature gas jet or plasma emerges. This plasma is used for welding, cutting, and for metal spraying. Its temperature is higher than that of the unrestricted arc.

The thermal energy generated in the arc is the product of welding current and arc voltage. The heat raises the temperature of the base metal, causing melting and resulting in a molten pool. The heat of the arc is distributed through radiation, convection, and conduction to the base metal. Figure 5–6 shows the frozen molten weld pool, known as the crater of an aluminum bead on a plate weld.

Penetration of the arc depends on several factors, including the polarity of the arc and the composition of the shielding gas. In addition, penetration depends on the mass of the base metal and its composition. The preheat temperature also affects penetration. The composition determines base metal thermal conductivity and melting temperature. As the welding current increases, the depth of penetration increases. The current-carrying capacity of the electrode determines the heat in the arc area, which depends on the type of electrode, pure or alloyed, and the tip configuration. As travel speed increases, the depth of penetration decreases. All of these factors interrelate.

The success of the operation depends on the control of the molten weld metal. This depends greatly on the welding position. Control depends on the skill of the welder for manual application and the complexity of the control system for machine welding. The gas composition

FIGURE 5–6 Aluminum bead on plate weld showing the crater.

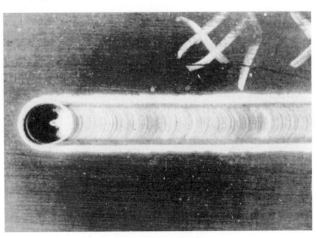

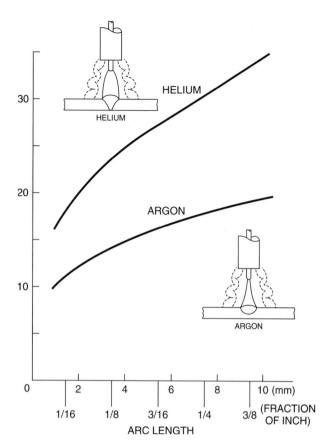

FIGURE 5–7 Arc voltage—arc length: shielding gas.

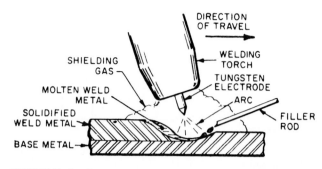

FIGURE 5–8 Process diagram for GTAW.

of the area surrounding the arc has an influence on the characteristics of the arc. Figure 5–7 is a comparison of a tungsten arc in argon and in helium. The voltage of a helium-shielded arc is higher than that of an argon-shielded arc for the same length carrying the same current. This is due to the higher ionization potential for helium, which is 24.5 V. The ionization potential for argon is 15.7 V. The ionization potential is the voltage necessary to remove an electron from a gas atom, making it an ion or charged atom. The arcs shown are a 300-A arc (DCEN) in argon and helium. The helium arc column is larger and has deeper penetration. This is why the arc shielded with helium has more power (heat) and can do more work. The helium-shielded arc column is larger, produces more penetration, can use a higher travel speed, and can weld heavier base metals.

This brief explanation of the nonconsumable electrode arc just touches on more common knowledge. The tungsten arc has been investigated by many scientists, yet many mysteries remain unanswered, which indicates why the welding arc is such a complicated industrial tool.

5-2 GAS TUNGSTEN ARC WELDING

Gas tungsten arc welding (GTAW) "is an arc welding process that uses an arc between a tungsten electrode (nonconsumable) and the weld pool (Figure 5–8). The

process is used with shielding gas and without the application of pressure. Filler metal may or may not be used." This process was developed in the late 1930s as heliarc or TIG welding, and was used to weld nonferrous metals, particularly magnesium and aluminum, and to join hard-to-weld metals. TIG stands for *tungsten inert gas welding,* and in Europe it is called WIG welding, using *wolfram,* the German word for tungsten.

Principles of Operation

The gas tungsten arc welding process, shown in Figure 5–9, utilizes the heat of an arc between a nonconsumable tungsten electrode and the base metal. The welder's view of the gas tungsten arc is shown in Figure 5–10. The arc develops intense heat, approximately 11,000 °F (6100 °C), which melts the surface of the base metal to form a molten pool. Filler metal is not added when thinner materials, edge joints and flange joints are welded. This is known as *autogenous welding.* For thicker materials an externally fed or cold filler rod is generally used. The filler metal is not transferred across the arc but melted by it. The arc area is protected from the atmosphere by the inert shielding gas, which flows from the nozzle of the torch. The shielding gas displaces the air, so

FIGURE 5–9 Gas tungsten arc welding process.

FIGURE 5–10 Welder's view of gas tungsten arc.

that the oxygen and the nitrogen of the air do not come in contact with the molten metal or the hot tungsten electrode. As the molten metal cools, coalescence occurs and the parts are joined. There is little or no spatter and little or no smoke. The resulting weld is smooth and uniform and requires minimum finishing.[5]

Advantages and Major Uses

The outstanding features of the gas tungsten arc welding process are:

1. It will make high-quality welds in almost all metals and alloys.
2. Very little, if any, postweld cleaning is required.
3. The arc and weld pool are clearly visible to the welder.
4. There is no filler metal carried across the arc, so there is little or no spatter.
5. Welding can be performed in all positions.
6. There is no slag produced that might be trapped in the weld.

This process allows the welder extreme control for precision work. Heat can be controlled very closely and the arc can be accurately directed. GTAW is used in many welding manufacturing operations, primarily on thinner materials. It is very useful for maintenance and repair work and for welding unusual metals. Gas tungsten arc welding is widely used for joining thin wall tubing and for making root passes in pipe joints. The gas tungsten arc welds are usually of extremely high quality.

The manual method of applying is used for the greatest majority of work. However, both mechanized and automatic methods are increasingly used. Torches equipped with filler metal wire feed systems are available for semiautomatic welding, but they have limited application.

The gas tungsten arc welding process is an all-position welding process shown by Figure 5–11. Welding in other-than-flat positions depends on the base metal, the welding current, and the skill of the welder. This process was originally developed for the hard-to-weld metals. It can be used to weld more different kinds of metals than any other arc welding process (Figure 5–12).

This process can weld extremely thin metals normally by the automatic method and without the addition of filler metal. Above 0.125 in. (3.2 mm), a joint preparation is usually required; however, this depends on the base metal type and welding position. Also, above this thickness, multi-pass technique is usually required (Figure 5–13).

Joint Design

The joint designs used for gas tungsten arc welding are essentially the same as those used for shielded metal arc and gas welding. Some changes are made, but these are usually involved with different metals or for welding pipe in the fixed position. Joint detail variations for different

FIGURE 5–11 Welding position capabilities of GTAW.

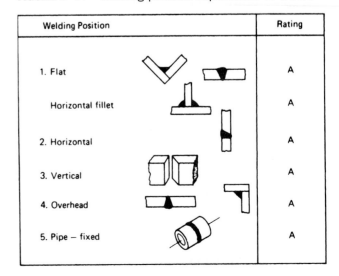

Welding Position		Rating
1. Flat		A
Horizontal fillet		A
2. Horizontal		A
3. Vertical		A
4. Overhead		A
5. Pipe – fixed		A

FIGURE 5–12 Metals weldable by GTAW.

Base Metal	Weldability
Aluminum	Weldable
Bronze	Weldable
Copper	Weldable
Copper nickel	Weldable
Cast iron, malleable, nodular	Possible but not popular
Wrought iron	Possible but not popular
Lead	Possible
Magnesium	Weldable
Inconel	Weldable
Nickel	Weldable
Monel	Weldable
Precious metals	Weldable
Low-carbon steel	Weldable
Low-alloy steel	Weldable
High- and medium-carbon steel	Weldable
Alloy steel	Weldable
Stainless steel	Weldable
Tool steel	Weldable
Titanium	Weldable
Tungsten	Possible

metals are covered in Chapter 15, and special joints for pipe welding are covered in Chapter 25.

Welding Circuit and Current

Welding circuit for gas tungsten arc welding is shown in Figure 5–14. This circuit diagram shows several optional items. One is the "cold" filler rod, the second is the foot pedal which can be used to regulate the current while welding, and the third is cooling water used for the welding torch, recommended when welding at high current. Constant current (CC) is used, and it may be alternating or direct current. Direct current can be used with either polarity, depending on the job requirements.

When superimposed high frequency is used with AC gas tungsten arc welding, certain precautions are required. These are necessary since welding power sources equipped with high-frequency spark gap oscillators inherently radiate power at frequencies that may interfere with radio communications and television transmission. In view of this, their operation in the United States is subject to control by the Federal Communication Commission (FCC). Most countries have similar regulations.

Welding machines containing high-frequency stabilizers or separate high-frequency stabilizers must be installed with special attention to provide earth grounding and special shielding. Manufacturers provide special installation instructions that limit high-frequency radiation. These instructions require that all metal conductors in the area of the machine must be earth grounded. If these instructions are followed, the user can post a certificate stating that the high-frequency stabilizer may reasonably be expected to meet FCC regulations.

Equipment Required

The main component of the GTAW system is the welding power source. The constant-current (CC) power source is used for gas tungsten arc welding. Conventional welding machines used for shielded metal arc welding can be employed for GTAW. Conventional AC welding machines must be derated 25% due to rectification in the arc when welding aluminum. It is recommended that machines designed for GTAW be used since they include special features such as high-frequency stabilization and gas and water valves. Some include remote control and programmers. A typical GTAW welding machine operates with a range of 3 to 200 A or 5 to 300 A, with a range of 10 to 35 V at a 60% duty cycle. The newer inverter type of power sources offer additional features, such as complex programmers, pulsing, variable polarity, and others. This is covered more completely in Chapter 10.

The torches used for gas tungsten arc welding are designed and used only for GTAW. There are four basic types: (1) torches for automatic welding, (2) torches for manual welding, (3) air-cooled torches for lower current, and (4) water-cooled torches for high-current welding. Gas tungsten arc welding torches have cables attached that connect to the power source to provide the power, shielding gas, and water cooling when used. The cables

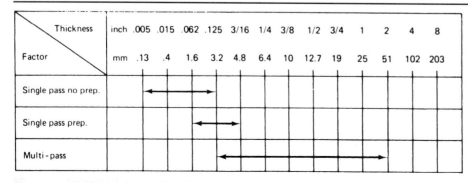

FIGURE 5–13 Base metal thickness range for GTAW.

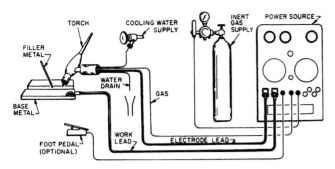

FIGURE 5–14 Circuit diagram for GTAW.

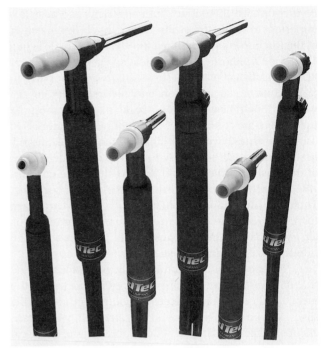

(a)

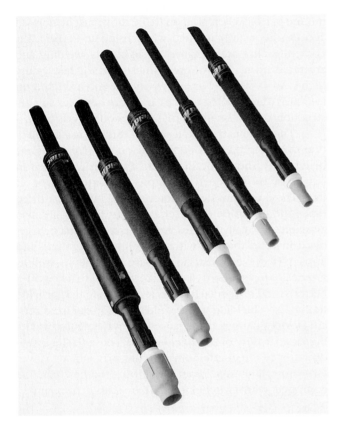

FIGURE 5–15 GTAW torches for automatic welding.

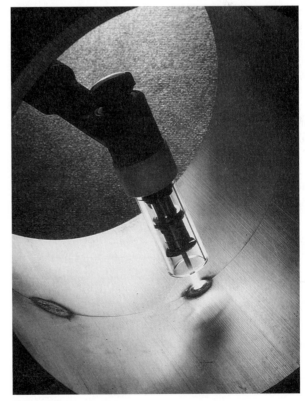

(b)

FIGURE 5–16 Manual GTAW torches.

are $12\frac{1}{2}$ or 25 ft long and may be of one- or three-piece construction. They normally contain a cable block adapter to attach to the power source. The internal design and construction of automatic and manual torches are very similar. The main difference is addition of a handle for manual torches. Automatic torches, sometimes called straight-line or pencil torches, usually include a rack. They are mounted in a bracket that includes an adjusting knob and pinion that engages the rack for adjustment. Typical torches for automatic welding are shown in Figure 5–15.

Different types and sizes of manual torches are shown in Figure 5–16. Different torch sizes allow the welder to select the size to match the job. All torches, whether air- or water-cooled, are of rugged construction

designed to hold the tungsten electrode firmly. This is necessary to transmit the welding current to the electrode, and to transmit heat from the electrode to the torch for cooling. The metal torch parts should be hard alloys of copper or brass to conduct both electricity and heat. It is important to locate the electrode accurately in the center of the torch. Ports surrounding the electrode are designed to provide a uniform flow of shielding gas to the arc area. Plastic insulating material surrounds the metal parts to provide electrical insulation and for the safety of the welder. Tough heat-resistant plastic materials must be used. There are at least three different plastic materials in use. Some types have better high-frequency insulating values and others are less likely to chip. The outside surface of the plastic handles and other parts should be comfortable and provide textured surfaces for ease of handling. There are no national standards for torches; however, there is agreement among manufacturers to provide collets with specific inside diameters to accommodate the standard sizes of tungsten electrodes.

The electrical conductor and the hoses that provide shielding gas and water must be securely attached to the metal head of the torch. Electrical conductors are usually silver brazed and the hoses are attached with small clamps. In the case of water-cooled torches, the electrical conductors are usually inside the hose carrying the cooling water to and from the torch. This allows the use of more flexible cable with smaller conductors and lighter weight. The manual torch shown in cross-sectional view in Figure 5–17 has a plastic back cap surrounding the tungsten electrode. Back caps come in different lengths. This type of torch allows the use of longer tungsten electrodes and provides gas tightness and electrical insulation. Torches without the back cap must be loaded from the front and utilize shorter lengths of tungsten electrode.

The head angle can be varied on some types of manual torches. This angle is measured by the angle between the centerline of the handle and the centerline of the tungsten electrode from the arcing end. The head angle is a matter of welder preference. Some torches have a flexible neck that allows the torch to be adjusted to different positions with respect to the handle. A rigid neck torch is more durable, however.

Gas tungsten arc welding torches of various sizes and capacities are described in Figure 5–19. Torches are generally rated according to their current carrying capacity using direct current with the electrode negative (DCEN) at normally 100% duty cycle based on a 10-minute period. When used on DCEP they must be derated and can be slightly uprated when using ac current. Air-cooled torches are designed for light-duty welding and are rated up to 200 A. Torches designed for heavy-duty high-current welding are water-cooled and are rated up to 600 A. Water-cooled torches are more complex in construction, due to small internal water-cooling passages.

The various types and sizes of torches use a variety of nozzles, or cup sizes, depending on the size of the arc being used. The inside diameter of a nozzle is given in inches or by numbers that represent sixteenths of an inch. The nozzle inside diameter should be at least three times the electrode diameter. Different lengths are also available. Normally, nozzles are made of alumina or a ceramic material that will withstand high temperatures and impact. However, metal nozzles are available as well as nozzles of fused quartz, a glasslike material that provides better visibility. Special nozzles are available that provide trailing gas shielding when welding easily oxidized metals. Devices known as gas lenses are available for providing laminar flow of the shielding gas. These fit the inside diameter of the nozzle. Some torches may include valves for controlling shielding gas flow. Torches can be equipped with switches to start or stop a program or

FIGURE 5–17 Cross-sectional view of water-cooled torch.

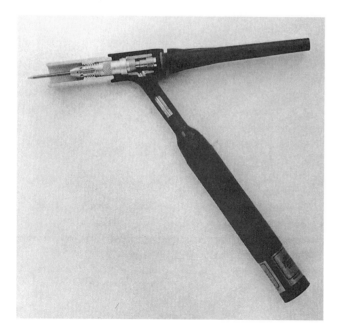

FIGURE 5–18 Cold wire spool gun

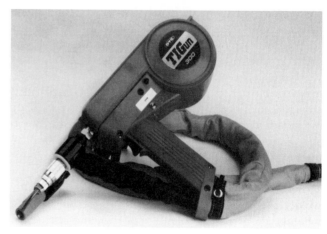

with small rheostats to control the welding current. Spare parts, for repairs, are available from most manufacturers.

Normally, when filler metal is added to the weld, it is added manually in the arc area. This gives the welder control over the size and the shape of the weld and the number of layers that are employed. In some cases, however, the filler metal is added to the arc area by a wire feeder that feeds the wire in at a predetermined rate. The wire feeder may be turned off and on. The wire feed must be adjusted to melt at the proper rate based on welding current, joint type, etc. This can be automated by controlling starting and stopping. The spool gun, shown in Figure 5-18, provides cold wire feed with an adjustable wire feed rate and guide to direct the filler metal into the arc area at the proper location. These devices ensure higher productivity with the normal high quality of a gas tungsten arc weld.

There is one other special type of GTAW torch or gun, the TIG spot gun, which is designed more like a pistol. It is used for spot welding thin sheets when the arc is accessible to one side only. This process is covered more completely in Chapter 26.

Materials Used

Materials used in gas tungsten arc welding are the filler metal, the shielding gas, and, to a lesser degree, the tungsten electrode. Filler metal is not used when welding extremely thin metals; however, for most applications filler metal is added. The size of the filler metal rod depends on the thickness of the base metal, which usually dictates the welding current. Filler metal is normally added to the pool manually, but automatic feed can be used.

The AWS provides specifications for filler metal for GTAW, the most popular being "Specifications for Carbon Steel Filler Metals for Gas Shielded Arc Welding," which covers filler metals for GTAW, PAW, and GMAW. The prefix for welding rods is the letter *R,* which indicates that the rod is not part of the electrical circuit. The specifications for all filler wires are given in Figure 13-1. GTAW uses straight lengths of weld rod for manual application. Automatic application may use filler metal from coils or spools. AWS specifications cover filler metals for most weldable metals. For more information, see the chapter for the metal to be welded.

The electrode material for GTAW is tungsten or tungsten alloys, since tungsten has the highest melting point of any metal, at 6170 °F (3410 °C). Nine classes of tungsten electrodes have been standardized by AWS in their specification for tungsten and tungsten alloy electrodes. These electrodes are listed in Figure 5-20, which shows the AWS classification, the approximate composition, and the color code for the tip end. Tungsten electrodes are available as either clean finish or ground finish. *Clean finish* has been chemically cleaned after drawing or swedging. *Ground finish* means that the electrode has been centerless ground to a uniform size and has a bright polished surface. The ground finish provides an extremely smooth and perfectly round electrode that is better able to conduct heat to the collet of the torch. Tungsten electrodes are available with diameters ranging from 0.20 in. (0.5 mm) to $\frac{1}{4}$ in. (6.4 mm). They are available in lengths of 3 in. (75 mm) to 24 in. (610 mm).

The AWS classification system uses the letter *E* at the beginning, which stands for "electrode." The letter *W* indicates that the electrode is primarily tungsten (wolfram). The letter *P* indicates that the electrode is essentially pure tungsten. *Ce, La, Th,* and *Zr* indicate that the electrode is alloyed with oxides of cerium, lanthanum, thorium, or zirconium, respectively. The number at the end of some classifications indicates a different composition level within a specific group. The letter *G* indicates

FIGURE 5-19 Size and capacity of GTAW torches.

Current Capacity 100% Duty Cycle	AC	DCEN	Head Angle (deg)	Cooling Method	Nozzle Inside Diameter (in.)	Tungsten Diameter in.	mm	Tungsten Length in.	mm	Torch Weight oz	g
50 at 60%		×	135	Air	$\frac{1}{4}$	0.040-$\frac{3}{32}$	1.0-2.4	1	25.4	3	85
75	×	×	120	Air	$\frac{5}{16}$	0.020-$\frac{1}{16}$	0.5-1.6	3-7	76.2-177.8	4	113.4
100	×	×	120 or 135	Air	$\frac{5}{16}$	0.020-$\frac{5}{32}$	0.5-4.0	2-3-7	50.8-76.2-177.8	4	113.4
150	×	×	120	Air	$\frac{5}{16}$	0.020-$\frac{5}{32}$	0.5-4.0	3-7	76.2-177.8	5	141.75
200	×	×	120 or 135	Water	$\frac{5}{16}$	0.020-$\frac{3}{32}$	0.5-2.4	2-3-7	50.8-76.2-177.8	5	141.75
250	×	×	120 or 135	Water	$\frac{3}{8}$	0.020-$\frac{1}{8}$	0.5-3.2	2-3-7	50.8-76.2-177.8	5	141.75
300	×	×	120	Water	$\frac{3}{8}$	0.020-$\frac{5}{32}$	0.5-4.0	3-7	76.2-177.8	6	170
350	×	×	120	Water	$\frac{11}{16}$	0.010-$\frac{5}{32}$	0.3-4.0	3-6	76.2-152.4	6	170
500	×	×	120	Water	$\frac{11}{16}$	0.010-$\frac{3}{16}$	0.3-4.8	3-7	76.2-177.8	6	170
650	×	×	135	Water	$\frac{5}{8}$	$\frac{1}{8}$-$\frac{5}{16}$	3.2-7.9	2-$\frac{1}{2}$	63.5	7	198.45

AWS Classification	Approximate Composition	Color Code
EWP	Pure Tungsten	Green
EW Ce-2	97.3% Tungsten, 2% Cerium Oxide	Orange
EW La-1	98.3% Tungsten, 1% Lanthanum Oxide	Black
EW La-1.5	97.8% Tungsten, 1.5% Lanthanum Oxide	Gold
EW La-2	97.3% Tungsten, 2% Lanthanum Oxide	Blue
EW Th-1	98.3% Tungsten, 1% Thorium Oxide	Yellow
EW Th-2	97.3% Tungsten, 2% Thorium Oxide	Red
EW Zr-1	99.1% Tungsten, 0.25% Zirconium Oxide	Brown
EW G	94.5% Tungsten, remainder not specified	Grey

FIGURE 5–20 Type, composition, and color code of tungsten electrodes.

that the electrode is a general classification and may not specify alloying elements.

The AWS electrode classifications are shown in Figure 5–20. The EWP class is pure tungsten. They are the least expensive and should be used for general-purpose work on different metals. The EWCe-2 contains cerium oxide, referred to as ceria. Ceria increases the ease of starting, improves arc stability, and reduces the rate of burn-off (erosion).

The classification EWLa-1 contains approximately 1% lanthanum oxide, referred to as lanthana. The advantages and operating characteristics are very similar to those of the EWCe-2 electrode. The classifications EWTh-1 and EWTh-2 contain approximately 1% or 2% thorium oxide, referred to as thoria. Thoria is a very low-level radioactive material. These electrodes are designed for dc applications. They provide for easier arc starting, provide a more stable arc, and can be operated at slightly higher temperatures. The 2% has better starting and is more stable, and has higher current-carrying capacity. Precautions should be exercised concerning electrode grinding dust when sharpening the electrode. The new classifications

EW La-1.5 and EW La-2 were created as replacements for the radioactive compositions.

The classification EWZr-1 contains approximately $\frac{1}{2}$ of 1% zirconium oxide, referred to as zirconia. The addition of zirconium makes the tungsten alloy better able to emit electrons, provides for increased current-carrying capacity, and gives a more stable arc with better arc starting. It also provides longer electrode life. The classification EWG provides for a tungsten electrode containing a specified addition of an unspecified rare earth oxide or combination. The addition must be specified by the electrode manufacturer.

Figure 5–21 gives the continuous welding current range for each type of electrode related to the type of current and the electrode size. When installing a new tungsten electrode in a torch, the color tip should be at the back end of the torch so that it is not destroyed by the arc. The collet must be the proper size for the electrode, and the entire assembly must be tight so that the heat of the arc is transmitted to the torch body and carried away. The electrode extension from the collet should be kept to a minimum.

Tungsten Electrode Diameter		DCEN EWX-X	DCEP EWX-X	AC Unbalanced Wave		AC Balanced Wave	
in.	mm			EWP	EWX-X	EWP	EWX-X
0.010	0.30	Up to 15	NA[a]	Up to 15	Up to 15	Up to 15	Up to 15
0.020	0.50	5–20	NA	5–15	5–20	10–20	5–20
0.040	1.00	15–80	NA	10–60	15–80	20–30	20–60
0.060	1.60	70–150	10–20	50–100	70–150	30–80	60–120
0.093	2.40	150–250	15–30	100–160	140–235	60–130	100–180
0.125	3.20	250–400	25–40	150–200	225–325	100–180	160–250
0.156	4.00	400–500	40–55	200–275	300–400	160–240	200–320
0.187	5.00	500–750	55–80	250–350	400–500	190–300	290–390
0.250	6.40	750–1000	80–125	325–450	500–630	250–400	340–525

Note: All are values based on the use of argon gas.
[a]NA, not applicable.

FIGURE 5–21 Current ranges for tungsten electrodes.

The collet should be kept clean, as should the nozzle, so as not to result in poor gas flow. The electrode size must be related to the type of current, type of work, and the amount of current that will be used. The amount of welding current required is found in the welding procedure schedules for welding a particular metal. The data provided in these tables constitute the starting point; for example, if a welding current of 100 A is required and alternating current is used, it would indicate that a $\frac{3}{32}$-in. (2.4-mm) pure tungsten or a $\frac{1}{8}$-in. (3.2-mm) pure tungsten electrode could be used, or if an alloyed electrode were to be used, the $\frac{1}{16}$-in. (1.6-mm) or the $\frac{3}{32}$-in. (2.4-mm) electrode could be used. If the procedure schedule calls for dc electrode negative (DCEN) or dc electrode positive (DCEP), different electrode sizes would be needed. Welder preference and experience enter into the size selection. If the welder is using pure tungsten electrode and it tends to become overheated or appears to have a wet surface, the current is too high for the size of the electrode. When the tungsten has this wet surface appearance, it becomes more susceptible to picking up contamination from the base metal. A larger electrode size of the same type should be selected or an alloyed-type electrode of the same size could be used. Too much current or an electrode too small will cause excessive tungsten erosion. Tungsten particles may become deposited in the weld metal.

If the current is too low, or the tungsten electrode is too large in diameter, the arc will wander erratically over the end of the electrode. Grinding the electrode to a point will reduce this problem. It will also help to direct the arc. Choose the size of electrode that will be working as close to its maximum current-carrying capacity as possible. The electrode should remain shiny after use and should never be allowed to touch the molten metal. If this happens, it will become contaminated and must be reprepared. The electrode should show a balled end and the balled end should not exceed $1\frac{1}{2}$ times the diameter of the electrode (Figure 5–22). The angle of pointing the electrode should be related to the welding current and the thickness of the metal being welded. It usually ranges from 30° to 120°; 60° is the most common angle.

The **shielding gas** used for gas tungsten arc welding must be an inert gas. Only argon and helium are used since the other inert gases are much too expensive. Gas selection is based on the metal to be welded. It is necessary to consult the procedure schedule. These tables show the gas recommended and the gas flow rate. More information concerning shielding gas is given in Chapter 14. Argon is more commonly used. It is readily available and is heavier than helium and slightly heavier than air, which provides for a more efficient arc shielding at lower flow rates. Argon is better for arc starting and operates at a lower arc voltage. Helium is much lighter than argon or air and tends to float away from the weld zone, and higher flow rates are required. Helium provides a higher voltage and more heat in the arc. It is also possible to weld at a higher speed with helium than with argon. In some cases helium and argon are mixed for the optimum shielding gas for a particular metal or weld schedule.

Deposition Rates

The gas tungsten arc is used as a source of heat to melt base metal and filler metal to provide a manageable weld pool. It is similar to oxyacetylene and carbon arc welding. The gas tungsten arc welding process is not a high-production or high-deposition-rate welding process. The graph shown in Figure 5–23 is a deposition-rate curve relating to welding current. This deposition rate is relatively low, which is one reason why gas tungsten arc welding is not used for welding heavy material.

Quality of Welds

The quality of gas tungsten arc welds ranks higher than the quality of any of the arc welding processes. This high level of quality is obtained when all necessary precautions are

---

FIGURE 5–22 Tungsten electrode arc end condition.

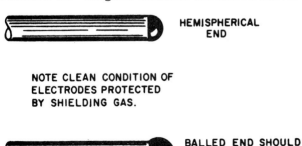

HEMISPHERICAL END

NOTE CLEAN CONDITION OF ELECTRODES PROTECTED BY SHIELDING GAS.

BALLED END SHOULD NOT EXCEED 1-1/2 TIMES DIAMETER OF ROD.

FIGURE 5–23 Deposition rate.

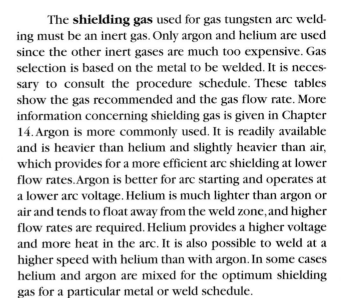

taken. Since much of the work done by the gas tungsten arc welding process is on nonferrous metals, it is absolutely essential that cleanliness be considered every step of the way. The work area should be extremely clean. The work tables and fixtures must be clean. The gloves used by welders should be clean. Filler metal should be clean, the gas must be welding grade, and the apparatus must be in excellent condition. If these conditions are followed and if the welder has sufficient skill, high-quality welds will result.

Heat input and welder technique has much to do with weld quality. Figure 5–24 shows these factors when welding on aluminum. When the heat input is too low, which can occur from too low welding current or welding speed too fast, the high small bead is evident and penetration is minor. When the welding current is too low, the bead is too high, there is poor penetration, and the possibility of overlapping at the edges. When the welding speed is too fast, the bead is too small and the penetration is minimal. When the heat input is too great, which can occur from too high a welding current or too low welding speed, the bead becomes extremely large, usually wide and flat. There is too much penetration and there may be spatter. When the torch is too far from the work, a long arc occurs, the efficiency of the gas shielding is reduced and poor weld appearance will result, especially in welding aluminum.

Weld Metal Porosity Porosity is usually caused by oily, wet, dirty base metal, insufficient inert gas coverage, or dirt and heavy oxide coating on the filler rod. In groove welds there should be backing or purging gas and cleaning of the underside of the groove adjacent to the weld. In the case of aluminum, a stainless steel wire brush or chemical cleaning should be used. Inefficient gas shielding can be caused by side drafts of air which disturb the shielding gas envelope. It may also be caused by leaks in the gas system or it can be caused by impure shielding gas. Equipment should be checked frequently to make sure that the gas system is tight. It is also important that no moisture from cooling water gets inside the gas supply hose.

Dirty welds, particularly on aluminum, can result from a problem in the shielding gas supply. There can be a leak in the hose connection, poor-quality shielding gas, an oversized gas nozzle, insufficient gas flow, or anything else that contributes to poor shielding of the arc area—the nozzle being held too far from the work.

Poor penetration is primarily a heat input problem and is related to travel speed and welding current to the base metal thickness and conductivity and joint type. Too much amperage will make the bead too flat and rough and may cause cracking. Insufficient amperage will produce an uneven high crown bead. If the speed is too fast,

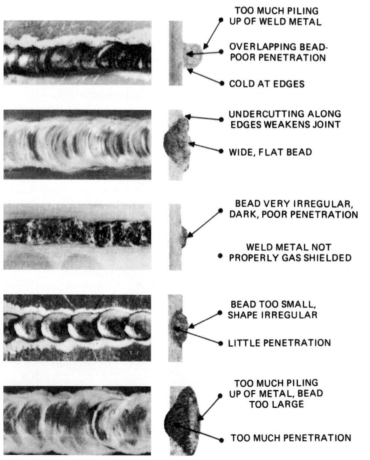

FIGURE 5–24 Quality factors of GTAW welding of aluminum.

TOO MUCH PILING UP OF WELD METAL

OVERLAPPING BEAD-POOR PENETRATION

COLD AT EDGES

UNDERCUTTING ALONG EDGES WEAKENS JOINT

WIDE, FLAT BEAD

BEAD VERY IRREGULAR, DARK, POOR PENETRATION

WELD METAL NOT PROPERLY GAS SHIELDED

BEAD TOO SMALL, SHAPE IRREGULAR

LITTLE PENETRATION

TOO MUCH PILING UP OF METAL, BEAD TOO LARGE

TOO MUCH PENETRATION

the bead will have a high crown, be rough, and have insufficient penetration. The welding speed may need to be varied or changed, particularly on small weldments where heat buildup occurs. When the base metal is cold, a lower speed is required; as the workpiece absorbs heat and rises in temperature, the speed should be increased.

A common quality problem is tungsten inclusions in the weld deposit, which can be detected by radiography. This is sometimes called *tungsten spitting* and is based on using too much current for the size or type of the tungsten electrode. The tungsten type may have to be changed or a larger tungsten electrode employed.

Another serious problem is an unstable arc, which is normally a result of a contaminated or dirty electrode. The electrode will become oxidized if the inert gas is not continually surrounding it while it is hot. This is the reason for post-flow controls on most gas tungsten arc welding machines. The gas flow should continue after the welding is stopped to keep the tungsten surrounded by inert gas while it is still at an elevated temperature. Another problem can be that the tungsten electrode protrudes too far beyond the end of the gas nozzle. Normally it should not protrude more than $\frac{1}{16}$ in. (1.6 mm). Postflow should be adjusted for a longer period of time if the tungsten becomes dirty. The other problem with tungsten electrodes is the contamination that occurs if the electrode is allowed to touch the molten metal. This will immediately cause the electrode and the arc to become unstable. Welding should be stopped and the electrode redressed.

Filler metal with an excessive oxide coating can also create dirty welds. Moisture will collect in this heavy oxide coating. The filler rod should be cleaned with sandpaper.

Welding on dirty, oil-impregnated material or attempting to repair cracks in machinery parts requires the removal of defective material, thorough cleaning, and preheating to help eliminate any absorbed oil, grease, moisture, and so on.

Water leaks in the torch can usually be detected by the coloring of the weld surface. Condensation can occur on the inside of gas hoses, and water vapor in the arc will cause the tungsten to become contaminated.

In summary, good-quality welds require that all conditions be correct, that materials used be of the correct specification and cleanliness, that the apparatus be in good working order, and that the proper welding technique be employed.

Weld Schedules

For gas tungsten arc welding it is important to use welding currents based on the type of metal and the weld joint detail. Welding schedule tables are provided in the chapter for the metal being welded. Once the welding current level is determined, the type of welding current to be used and the type of tungsten electrode recommended, it is then possible to establish tungsten electrode size.

The welding procedure tables also provide weld travel speed, which must be used in determining heat input. Under mechanized conditions, the weld speeds can normally be increased over manual application. The feed rate for filler rod is not given since this is a matter of technique for manual welding.

The weld schedules are related to the weld pool that must be carried by the welder. This relates to welding position, type of metal, weld joint detail, and so on. More experienced welders can carry larger molten pools and make welds at a higher speed.

Welding Variables

Gas tungsten arc welding involves a number of variables. Each variable has a specific effect on the weld, and there is an interrelationship among variables that affects the weld. The preselected variables include tungsten type, tungsten size, nozzle size, and gas type. These are part of a welding procedure. The primary variables are welding current, arc voltage, travel speed, pulsing when used, and upslope and downslope, when used, for programmed welding. The secondary variables include rod feed speed, torch angles, and possibly tungsten angles. There are other factors that affect the weld quality, and these include clamping, fixturing, heat sinks, heat buildup, backing, purging gas, and high frequency for arc starting.

The factors that are of interest in a weld are penetration in the base metal, bead width, and weld reinforcement or height. It is assumed that weld surface appearance is acceptable and weld metal deposit is of the required quality. These factors are all influenced by heat input. However, it is important to consider the conductivity of the metal and the heat-sink effect of any fixturing that might be employed.

A major reason for developing pulsed welding was to provide deep penetration from the high current, while reducing the total heat input to avoid too much molten weld metal. Programming is also useful, particularly, on small welds, where heat buildup occurs.

Safety Considerations

The safety factors are very similar to those involved with the other arc welding processes. The gas tungsten arc seems brighter at the same current than the arc of shielded metal arc welding. This is because the smoke is not present. The brightness of the arc tends to cause air to break down and form ozone. Adequate ventilation should be provided. The bright arc rays cause fumes from hydrochlorinated cleaning materials or degreasing agents to break down and form phosgene gas. Cleaning operations using these materials should be shielded from the arc rays of the gas tungsten arc.

The final hazard is the possibility of displacing the air when welding in enclosed areas such as tanks. Ventilation and other precautions for welding in enclosed areas should be followed.

Limitations of the Process

The major limitation of gas tungsten arc welding is its low productivity. The power source and the torch are more expensive. The justification for this is the ability of the process to weld so many metals in thicknesses and positions not possible by shielded metal arc welding.

Variations of the Process

The more popular variations are:

* ✹ Pulsed-current GTAW
* ✹ Manual programmed GTAW
* ✹ Hot-wire GTAW (automatic)
* ✹ Dabber welding
* ✹ Increased penetration GTAW

The pulsed-current mode of welding is a way to control heat input. It offers a number of advantages over conventional or steady-current welding as follows:

1. Control of molten pool: size and fluidity (especially out of position)
2. Increased penetration
3. Oscillation travel and dwell control
4. Travel speed control
5. Better consistent quality

The pulsed-current mode provides a system in which the welding current continuously changes between two levels as shown in Figure 5–25. During the periods of high-pulsed current, heating and fusion take place; during the low-pulse current periods, cooling and solidification take place. It is as if the foot rheostat were moved up and down to increase and decrease the welding current on a regular basis. Newer GTAW power sources automatically switch to high current then to low current and will hold each value for a specific time. The pulsed gas tungsten arc can make a weld seam of overlapping arc spot welds. Each arc spot type weld is produced during the high-current pulse time. The current then decreases to the "low pulse current" or background current, which allows the weld to partially cool and solidify while maintaining a low-current consistent arc. The torch is then moved to the next point along the weld joint and held motionless during the next high-current pulse.

Four factors must be controlled to weld with pulsed-current: *high-pulse current* or *pulse current, low-pulse current* or *background current, high-pulse time,* and *low-pulse time.*

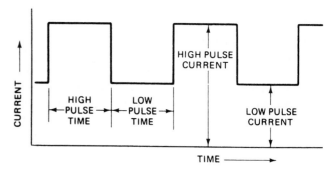

FIGURE 5–25 Pulsed-current current–time relationship.

Distortion and warpage are reduced with pulsed-current GTA welding on thinner material, due to the lower heat input. Misalignment of joints and the welding of light to heavy sections are made easier with pulsed-current welding. Pulsed-current welding can be done manually or automatically with or without filler wire.

Pulsing of GTAW has certain advantages. As automatic pulsing gained popularity, it was used for specific hard-to-weld joints, such as thick to thin variable root face, different metals, and so on. Low-frequency pulsing increased the penetration due to the higher peak current and seemed to reduce the width of the molten weld pool. Pulsing at high frequency, in the audible range, greatly increased penetration due to the high current pulse, and the width of the molten weld metal was decreased further. This increased welding speed since a smaller mass of molten metal was produced. Pulsing also improved the weld quality. Pulsing is used with tube heads.

The programming of weld current is often used in automatic welding and can be used for manual application. Programmers are used to make the welding current rise or fall at specific rates to specific values. A finger switch mounted on the torch shown in Figure 5–26 will start the preselected program. The torch switch can be used to stop the program or make it repeat. This is known as manual programmed GTAW and is popular for welding tubing and root pass welding of pipe.

The *hot-wire TIG welding* variation uses electrical power on the filler metal. The filler rod that is fed into the weld puddle is "electrically" hot. The electrical hot wire carries a low-voltage current that preheats the filler rod. It enters the weld pool at an elevated temperature and melts quicker, thus increasing the deposition rate. One of the major applications for the hot-wire GTAW variation is for weld surfacing. It is used in automatic welding since the hot wire must always be in contact with the molten pool in order to conduct the preheating current (Figure 5–27).

Dabber Welding

The Dabber welding variation was developed for the precise placement of weld metal on thin edges. It is being uti-

FIGURE 5–26   Repairing an aluminum assembly.

FIGURE 5–28   Making Dabber weld.

FIGURE 5–27   Hot-wire feed entering weld pool.

lized for rebuilding knife-edge seals for jet engines. Dabber welding operation is shown in Figure 5-28, and a cross section of welded seal buildup is shown in Figure 5-29a. This variation uses a coordinated motion of the cold filler wire end and the welding torch, which varies the arc length. This motion duplicates the motions of a manual welder but is performed automatically. The cold filler wire is fed continuously (Figure 5-29b); however, it is fed into, and removed from, the arc by a dabbing or oscillating motion. At the same time, the arc is lengthened and shortened together with the feeding of the cold wire.

The dabbing stroke length is sufficient to pull the end of the heated filler wire from which a droplet has just been detached. At the same time, the torch moves toward the work and the arc is shortened. This simulates the two-handed action of a human welder. It was accomplished by moving the wire guide in and out and the torch up and down. The wire approaches at a very shallow angle. This coordinated motion variation can also be used with pulsed current. It is used to weld many special alloys, such as titanium, high-nickel alloys, and tool steels. In addition to being used for rebuilding jet engine seals, it can be used to provide buildup on jet engine blades, saw blades, valve seats, milling cutters, drill bits, mower blades, and other devices.

Increased Penetration GTAW

At the Paton Welding Institute in Kiev, Ukraine, it was discovered that flux painted on the surface of austenitic stainless steel greatly increased the penetrating capability of gas tungsten arc welds. The mechanism for increasing the penetration of the process has not been completely explained, however it seems to be related to arc constriction. The arc constriction due to the flux increases the current density at the arc root and thus increases the arc force, which produces the substantial increase of penetration of the weld. The use of this surface-applied

(a)

(b)

FIGURE 5–29 (a) Cross section of welded seal buildup; (b) Filler wire feeding for semiautomatic welding.

flux increases the penetration depth by at least two times the depth produced with conventional gas tungsten arc welding. Use of this flux reduces the need for edge preparation and increases productivity due to the reduction in the number of weld passes required to make a joint. The need to provide groove preparation is not required on thin materials. In some cases, the square groove joint, when welding from both sides, eliminates groove joint preparations requiring several passes. The flux is a mixture of inorganic powders which are suspended in a volatile liquid media such as acetone. Immediately prior to the welding, a 0.005-inch-thick layer of the mixture is applied to the top surface of the base metal, where the weld is to be made. The flux is not intended to provide shielding. Sound welds result, which pass radiographic inspection. The flux does not significantly alter the chemical composition of the weld metal. The ultimate strength and ductility of the welds exceed the minimum specified for the base metal.

Industrial Use and Typical Applications

The aircraft industry is one of the principal users of gas tungsten arc welding. Space vehicles are fabricated by gas tungsten arc welding. This includes the shells, structures, the various tanks that are required, and the thousands of feet of tubing involved in rocket engines.

Small-diameter thin-wall tubing is welded by this process. Tubes are also welded to tube sheets for heat exchangers with programmed GTAW. Another important use is the making of root-pass welds in piping—large and small diameter for the process and power industries where high-quality welding is required. Virtually every industry uses GTAW for welding thin materials, especially the nonferrous metals.

The repair and maintenance industry is a major user. GTAW is used for repairing tools and dies, for repairing aluminum and magnesium parts, and for repairing highly critical items. Figure 5–26 shows an example of repairing an aluminum assembly.

5-3 PLASMA ARC WELDING

Plasma arc welding (PAW) is "an arc welding process that uses a constricted arc between a nonconsumable electrode and the weld pool (transferred arc) or between the electrode and the constricting nozzle (nontransferred arc). Shielding is obtained from the ionized gas issuing from the torch, which may be supplemented by an auxiliary source of shielding gas. The process is used without the application of pressure." Filler metals may or may not be used. The transferred plasma arc process is shown in Figure 5–30.

FIGURE 5–30 Welder's view of plasma welding arc.

Plasma arc welding was invented by Robert Gage of Buffalo, New York, in 1957 (US Patent 2,806,124). The plasma arc is also used for metal cutting and for metal spraying. As a cutting process, it is very popular for nonferrous metals and competes with oxyfuel gas cutting of heavy-plate steel. For metal spraying, it is used both with wires and powders.

Principles of Operation

The plasma arc welding process (Figure 5–31) is compared to the GTAW process because of the many similarities. If an electric arc between a tungsten electrode and the work is constricted or reduced in cross-sectional area, its temperature increases since it carries the same amount of current. This constricted arc is called a plasma, and plasma is the fourth state of matter.[6] There are two modes of operation, the nontransferred arc and the transferred arc. In **nontransferred arc mode** the current flow is from the electrode inside the torch to the nozzle containing the orifice and back to the power supply. The nontransferred mode is normally used for plasma spraying or for generating heat in nonmetals. In the **transferred arc mode** the current is transferred from the tungsten electrode inside the welding torch through the orifice to the workpiece and back to the power supply. The difference between these two modes of operation is shown in Figure 5–32. The transferred arc mode is most used for welding except for very low-current applications.

The plasma is generated by constricting the electric arc passing through the orifice of the nozzle and the hot ionized gases that are forced through this opening. The plasma has a stiff columnar form and is fairly parallel sided so that it does not flare out in the same manner as the gas tungsten arc. When directed toward the work, the high-temperature stiff plasma arc shown in Figure 5–33 will melt the base metal surface and the filler metal that may be added to make the weld. In this way, the plasma

acts as an extremely high temperature heat source to form a molten weld pool in the same manner as the gas tungsten arc. The higher-temperature plasma causes this to happen faster. When the plasma is used in this way, it is known as the *melt-in mode* of operation. A high-temperature, high-velocity plasma jet provides an increased heat transfer rate over gas tungsten arc welding at the same current. This results in faster welding speeds and deeper weld penetration.

FIGURE 5–32 Modes of operation of PAW.

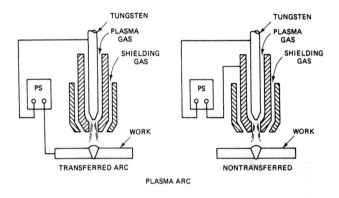

PLASMA ARC

FIGURE 5–33 High-temperature stiff plasma arc.

FIGURE 5–31 Process diagram for PAW.

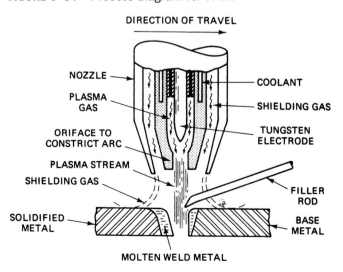

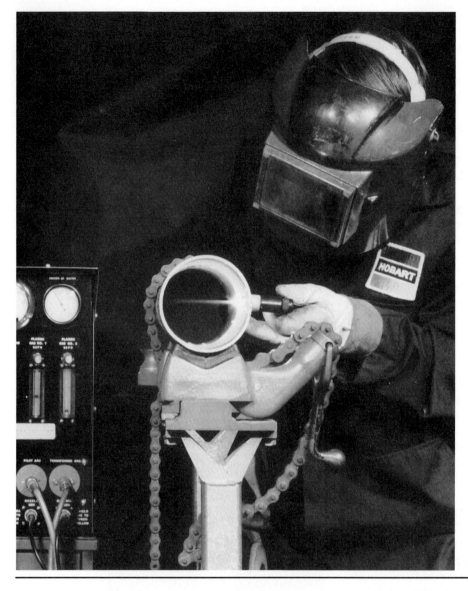

FIGURE 5–34 Plasma pipe welding.

The other method of welding with plasma is known as *keyhole welding*. In this method the plasma jet penetrates through the workpiece and forms a hole known as a keyhole. Surface tension forces the molten base metal to flow around the keyhole to form the weld. The keyhole method can be used only for joints where the plasma can pass through the joint. It is used for base metals $\frac{1}{16}$ in. (1.6 mm) to $\frac{1}{2}$ in. (12 mm) in thickness and is affected by the base metal composition and the welding gases. The keyhole method provides for full-penetration single-pass welding, which may be applied either manually or automatically in all positions. Keyhole welds have been made in aluminum $\frac{3}{4}$ in. (19 mm) thick.

Advantages and Major Uses

When compared with gas tungsten arc welding, plasma arc welding has several advantages. PAW has a higher energy concentration. Its higher temperature, its con-

stricted cross-sectional area, and the velocity of the plasma jet create a higher heat content. In addition, the stiff columnar plasma does not flare like the gas tungsten arc. These factors provide the following advantages:

1. The torch-to-work distance is less critical than for gas tungsten arc due to the columnar form of the plasma. This is important for manual operation.

2. High temperature and high heat concentration of the plasma allow for the keyhole effect, which provides complete-penetration single-pass welding of many joints. The heat-affected zone is smaller and the shape of the weld is more desirable. The weld tends to have parallel sides, which reduces angular distortion.

3. The higher heat concentration and the plasma jet allow for higher travel speeds. The plasma arc is thus more stable and is not as easily deflected to the closest point of base metal. Greater variation in joint

FIGURE 5–35 Metals weldable by the plasma arc process.

Base Metal	Weldability
Aluminum	Weldable
Bronze	Possible but not popular
Copper	Weldable
Copper nickel	Weldable
Cast iron, malleable, nodular	Possible but not popular
Wrought iron	Possible but not popular
Lead	Possible but not popular
Magnesium	Possible but not popular
Inconel	Weldable
Nickel	Weldable
Monel	Weldable
Precious metals	Weldable
Low-carbon steel	Weldable
Low-alloy steel	Weldable
High- and medium-carbon steel	Weldable
Alloy steel	Weldable
Stainless steel	Weldable
Tool steel	Weldable
Titanium	Weldable
Tungsten	Weldable

alignment is possible. The plasma weld has deeper penetration and produces a narrower weld. Faster travel speeds result in higher production rates.

The plasma arc welding process is used for manufacturing tubing, making small welds on instruments and components made of thin metal, making root-pass welds on pipe, and making butt joints of thin-wall tubing. It also is used to do work similar to that done by electron beam welding in the open, with much lower equipment cost.

Plasma arc welding is normally applied as a manual process, but it is also used in automatic and mechanized applications. Figure 5–34 shows plasma pipe welding.

The plasma arc welding process is able to join practically all of the commercially available metals (Figure 5–35). It will join all the metals that the gas tungsten arc welding process will weld.

Regarding the range of thickness welded by plasma (Figure 5–36), consider the keyhole mode of operation, which can only be used where the plasma jet can penetrate the joint. In this mode the process can be used for welding material from $\frac{1}{16}$ in. (1.6 mm) through $\frac{3}{4}$ in. (19 mm). Thickness ranges vary with different metals. The melt-in mode is used to weld material as thin as 0.002 in. (0.05 mm) up through $\frac{1}{8}$ in. (3.2 mm). Using multi-pass techniques, it can weld up to an unlimited thickness. Note that filler rod is used for making welds in thicker material.

Joint Design

Joint design is based on the thicknesses of the metal to be welded and by the mode of operation. For the keyhole mode, the joint design is restricted to full penetration. The preferred joint design is the square groove, with a minimum root opening. For root-pass work, on heavy-wall pipe, the U-groove design is used. The root face should be $\frac{1}{8}$ in. (3.2 mm) to allow for full keyhole penetration.

For the melt-in method for welding thin gauge, 0.020 in. (0.5 mm) to 0.100 in. (2.5 mm) metals, the square groove weld should be utilized. For welding foil thickness, 0.005 in. (0.13 mm) to 0.020 in. (0.5 mm), the edge flange joint should be used. The flanges are melted to provide filler metal for making the weld.

When using the melt-in mode for thick materials, the same general joint detail as used for shielded metal arc or gas tungsten arc can be employed. It can be used for fillets, flange welds, all types of groove welds, and so on. It can also be used for lap joints using arc spot welds and arc seam welds. Figure 5–37 shows various joint designs that can be welded.

Welding Circuit and Current

The welding circuit for plasma arc welding is more complex than for gas tungsten arc welding because an extra component is required: the control circuit necessary for starting and stopping the plasma arc. The same power source is normally used. There are two gas systems, one to supply the plasma gas and the second for the shielding gas. The welding circuit for plasma arc welding is shown in Figure 5–38.

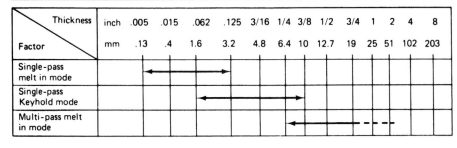

Thickness	inch	.005	.015	.062	.125	3/16	1/4	3/8	1/2	3/4	1	2	4	8
Factor	mm	.13	.4	1.6	3.2	4.8	6.4	10	12.7	19	25	51	102	203
Single-pass melt in mode														
Single-pass Keyhold mode														
Multi-pass melt in mode														

FIGURE 5–36 Base metal thickness range for PAW.

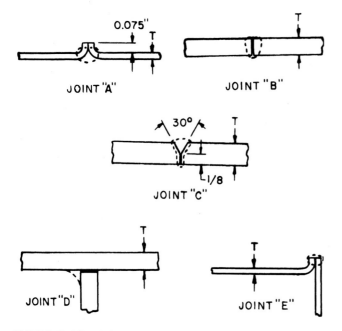

FIGURE 5–37 Joint designs for plasma arc.

Direct-current constant-current (CC) power is used. Alternating current is used for a few applications.

Equipment Required

Power Source A CC drooping characteristic power source supplying dc welding current is recommended; however, an ac/dc power source can be used. The power source should have an open-circuit voltage of 80 V and should have a duty cycle of 60%. It should have a contactor and provisions for remote control current adjustment. For welding very thin metals, it should have a minimum

amperage of 1 A. A maximum of 500 A is adequate for most plasma welding applications.

The *welding torch* for plasma arc welding is similar in appearance to a gas tungsten arc torch, but it is more complex. Figure 5–39 shows typical plasma torches. Plasma torches are water cooled, even for the lowest-current range torch. This is because the arc is contained inside a chamber in the torch where it generates heat. If water flow is interrupted briefly, the head assembly may melt. A cross section of a plasma arc torch head is shown in Figure 5–40. During the nontransferred period the arc will be struck between the orifice and the tungsten electrode. Manual plasma arc torches are made in various sizes, starting with 100 A up through 300 A. Automatic torches for machine operation of the same basic ratings are available. Cable assemblies come with the torches.

The torch utilizes the 2% thoriated *tungsten electrode*. Since the tungsten electrode is located inside the torch, it is almost impossible to contaminate it with base metal. Tungsten electrodes are specified by the AWS.

A **control circuit** is required. The plasma torch connects to the control console or to the power source. The control console includes a power source for the pilot arc, a timer for transferring from the pilot arc to transferred arc, and water and gas valves and separate flowmeters for the plasma gas and the shielding gas. Usually, the console is connected to the power source and may operate the contactor. The control console will also contain a high-frequency arc starting unit, torch protection circuit, and an ammeter. The high frequency is used to initiate the pilot arc. Torch protective devices include water and plasma gas pressure switches, which interlock with the contactor. A **wire feeder** may be used for mechanized or automatic welding and must be the constant speed type. The wire feeder must have a speed adjustment covering

FIGURE 5–38 Circuit diagram for PAW.

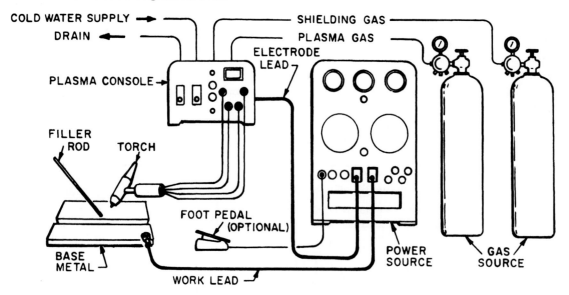

FIGURE 5–39 Typical manual plasma arc torches.

the range from 10 in. (254 mm) per minute to 125 in. (3.18 m) per minute feed speed.

Materials Used

Filler metal is used except when welding the thinnest metal. The composition of the filler metal should match the base metal. The size of the filler metal rod depends on the thickness of the base metal and the welding current. The filler metal is usually added to the pool manually but can be added automatically. The specifications are the same as for GTAW.

Plasma and Shielding Gas An inert gas, either argon, helium, or a mixture, is used for shielding the weld area from the atmosphere. Argon is more commonly used because it is heavier and provides better shielding at lower flow rates.

For flat and vertical welding, a shielding gas flow of 15 to 30 ft^3/hr (7 to 14 liters/min) is sufficient. Overhead position welding requires a higher flow rate. Argon is usually used for the plasma gas at a flow rate of 1 ft^3/hr (0.5 liters/min) up to 5 ft^3/hr (2.4 liters/min) for welding, depending on torch size and the application. Active gases are not recommended for the plasma gas.

Quality, Deposition Rates, and Variables

The quality of the plasma arc welds is extremely high. The skill of the welder is a major factor with respect to weld quality. A welder will find the plasma arc welding process easier to use than the gas tungsten arc, which helps ensure weld quality.

Deposition rates for plasma arc welding are higher than for gas tungsten arc welding (Figure 5–41). Weld schedules for the plasma arc process are shown in Figure 5–42.

The quality factors for plasma arc welding are shown in Figure 5–43. The variables for plasma arc are similar to the other arc welding processes, with two exceptions: the plasma gas flow and the orifice diameter in the nozzle. The major variables exert considerable control in the process. The minor variables are generally fixed at optimum conditions for the given application. All variables should appear in the welding procedure. Variables such as the angle of the tungsten electrode, the setback of the electrode, and the electrode type are considered fixed for the application. The stand-off or torch-to-work distance is less sensitive with plasma, but the torch angle

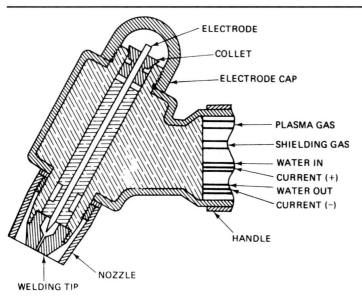

FIGURE 5–40 Cross section of plasma arc torch head.

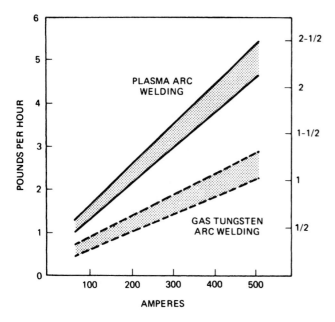

FIGURE 5–41 Deposition rates for PAW.

when welding parts of unequal thicknesses is more important than with gas tungsten arc.

Tips for Using the Process

The most important tip is to properly maintain the welding torch. The tungsten electrode must be precisely centered and located with respect to the orifice in the nozzle. The pilot arc current must be kept low, just high enough to maintain a stable pilot arc. When welding extremely thin materials in the foil range, the pilot arc may be all that is necessary.

When filler metal is used, it is added in the same manner as gas tungsten arc welding. However, with the torch-to-work distance a little greater, there is more freedom for adding filler metal. Equipment must be properly adjusted so that the shielding gas and plasma gas are in the right proportions. Proper gases must also be used. Plasma gas flow also has an important effect. The safety considerations for plasma arc welding are the same as for gas tungsten arc welding.

Limitations of the Process

The major limitations of the process have to do more with the equipment and apparatus. The torch is more delicate and complex and must be water cooled. The tip of the tungsten and the alignment of the orifice must be maintained within very close limits. The current level of the torch cannot be exceeded without damaging the tip. The water-cooling passages in the torch are very small, and water filters and deionized water are recommended. The control console adds another piece of equipment, which makes the system more expensive.

FIGURE 5–42 Weld procedure schedules for manual PAW.

Material	Material Thickness (in.)	Type of Weld	Orifice Diameter (in.)	Filler Diameter (in.)	Shield Gas at 20 ft³/hr	Plasma Gas Flow (ft³/hr argon)	Weld Current (A)	Number of Passes	Travel Speed (in./min)
Stainless	0.008	Edge butt	0.093	—	A	0.5	12 DCEN	1	7
steel[a]	0.008	Edge butt	0.093	—	A-5H₂	0.5	10 DCEN	1	13
	0.020	Square groove	0.046	—	A-5H₂	0.5	12 DCEN	1	21
	0.030	Square groove	0.046	—	A-5H₂	0.5	34 DCEN	1	17
	0.062	Square groove	0.081	—	A-5H₂	0.7	65 DCEN	1	14
	0.093	Square groove	0.081	—	A	2.0	85 DCEN	1	12
	0.093	Square groove	0.081	—	A-5H₂	2.0	85 DCEN	1	16
	0.125	Square groove	0.081	—	A	2.5	100 DCEN	1	10
	0.125	Square groove	0.081	—	A-5H₂	2.5	100 DCEN	1	16
	0.187	Square groove	0.081	—	A-5H₂	3.5	100 DCEN	1	7
	0.250	V-groove	0.081	—	A-5H₂	3.0	100 DCEN	First	5
	0.250	V-groove	0.081	$\frac{3}{32}$	A-5H₂	1.4	100 DCEN	Second	2
Mild steel	0.030	Square groove	0.081	—	A	0.5	45 DCEN	1	26
	0.080	Square groove	0.081	—	A	1.0	55 DCEN	1	17
Copper[a]	0.016	Edge butt	0.093	—	He	0.5	18 DCEN	1	24
Aluminum	0.036	Square groove	0.081	$\frac{1}{16}$	He	0.05	47 DCEP	1	24
	0.050	Edge joint	0.081	—	He	0.5	48 DCEP	1	22
	0.090	Fillet	0.081	$\frac{3}{32}$	He	1.4	34 DCEP	1	4

[a]Backing gas 5 to 10 CFH argon.

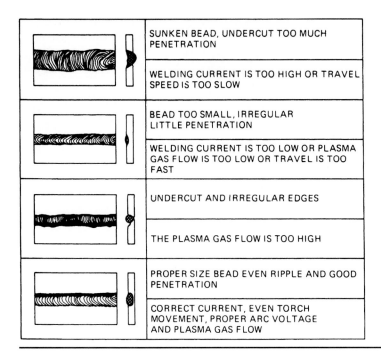

		SUNKEN BEAD, UNDERCUT TOO MUCH PENETRATION
		WELDING CURRENT IS TOO HIGH OR TRAVEL SPEED IS TOO SLOW
		BEAD TOO SMALL, IRREGULAR LITTLE PENETRATION
		WELDING CURRENT IS TOO LOW OR PLASMA GAS FLOW IS TOO LOW OR TRAVEL IS TOO FAST
		UNDERCUT AND IRREGULAR EDGES
		THE PLASMA GAS FLOW IS TOO HIGH
		PROPER SIZE BEAD EVEN RIPPLE AND GOOD PENETRATION
		CORRECT CURRENT, EVEN TORCH MOVEMENT, PROPER ARC VOLTAGE AND PLASMA GAS FLOW

FIGURE 5–43 Quality and common faults of PAW.

Variations of the Process

The variations of plasma arc welding are essentially the same as for GTAW:

- Pulsed-current PAW
- Manually programmed PAW
- Hot-wire PAW
- Micro plasma low-current precision PAW
- Variable-polarity plasma arc (VPPA)
- Plasma transferred arc

The welding current may be pulsed to gain the same advantages as pulsing provides for GTAW. A high current pulse is used for maximum penetration, and the low current allows for solidification. This gives a more easily controlled pool for difficult work. The same control systems used for GTAW can be used for PAW.

Programmed welding is used for plasma arc welding. A programmable power source is used and offers advantages for different types of work. The complexity of the program depends on the needs of the specific application. In addition to programming the welding current, it is sometimes necessary to program the plasma gas flow. This is important when closing a keyhole, which is required when making a root pass for pipe welding.

Filler metal is fed into the plasma arc weld pool in the same manner as for GTAW. The hot-wire procedure can be used. It is used for roll welding pipe and for surfacing.

Micro plasma welding using 1 to 10 A of current is used for precision welding of extremely thin material using precision motion devices and accurate fixturing. These are automatic applications and are widely used in the aircraft, jet engine, precision instrument industry, and so on.

One of the latest variations is variable-polarity plasma arc welding, known as VPPA. This variation was developed by the aerospace industry for welding thicker sections of alloy aluminum, specifically for the external fuel tank of the space shuttle. Research indicated that the shorter time period of positive polarity would provide adequate cleaning and allow more negative polarity for making the weld. It was found that 2 to 5 milliseconds (msec) of positive polarity should be used with 15 to 20 msec of negative polarity. With a special power source, which is actually two power sources connected together by an electronic switch, the negative and positive half-cycles can each be varied with respect to current and time. This waveform provides sufficient cleaning, yet provides maximum heat in the base metal. This waveform is coupled to the plasma torch so that a keyhole could be carried from the start to the finish of the weld. Figure 5–44 shows the variable-polarity weld being made on the external fuel tank. Figure 5–45 shows the difference in angular warpage between a gas tungsten arc-welded plate and a VPPA welded plate. The VPPA system is now used to make all joints in the external fuel tank for the U.S. Space Shuttle program.

Plasma Transferred Arc The plasma transferred arc (PTA) process variation is often referred to as a thermal spray process. The equipment is similar to that used for plasma thermal spray; however, fusion occurs between the deposit and the base metal. In the PTA variation, shown in Figure 5–46, the deposited metal is applied to greater thicknesses than that of a thermal spray coating. There is no slag to be removed, and the complete deposit is smooth and uniform. The deposit is more localized, 100% dense, and is metallurgically welded to the work-

FIGURE 5–44 Variable-polarity weld being made.

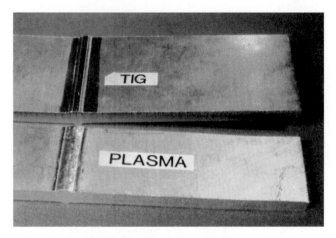

FIGURE 5–45 Angular warpage GTAW versus VPPA.

piece. Dilution levels can be controlled to give consistently low values with single-pass deposits. The filler metal is usually in the form of powder because the alloys usually involved, which produce high-hardness deposits, cannot be drawn into wire. A common application is the

abrasion resistance overlay to wear parts. Figure 5–47 shows the surfacing of wear edges of a material-moving auger. Overlays can be created in the exact shape and thickness required, from as thin as 0.050 in. (1.25 mm) to as thick as $\frac{3}{16}$ in. (4.75 mm) in a single pass. Another typical application is the overlay of seating surfaces of valves, shown in Figure 5–48. This surface must be ground for a precise fit.

FIGURE 5–46 Plasma transferred arc system.

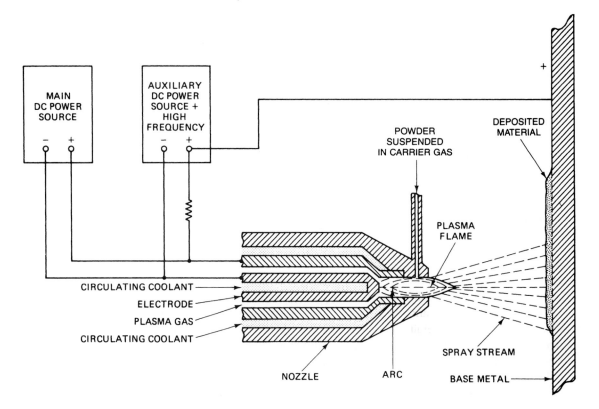

FIGURE 5–47   Surfacing wear portion of an auger.

FIGURE 5–48   Seating surface of valves overlayed by PTA-welding.

5-4 CARBON ARC WELDING

Carbon arc welding (CAW) is "an arc welding process that uses an arc between a carbon electrode and the weld pool. The process is used with or without shielding gas and without the application of pressure. Filler metal may or may not be used." The carbon arc welding process is the oldest of all the arc welding processes and is considered to be the beginning of arc welding. The single-electrode ver-

FIGURE 5–49   Carbon arc welding, single electrode.

sion is shown in Figure 5–49. Carbon arc welding is not popular today and is included here for historical purposes.

Principles of Operation

There are two variations. One uses a single electrode with an arc between it and the base metal, and the other uses two electrodes with an arc between them. In carbon arc welding, the heat of the arc between the carbon electrode and the work melts the base metal and a filler rod when used. As the molten metal solidifies, a weld is produced. The carbon graphite electrode, considered to be nonconsumable, erodes away fairly rapidly and as it disintegrates produces a shielding atmosphere of carbon monoxide and carbon dioxide gas. These gases displace the atmosphere and prohibit the oxygen and nitrogen from coming into contact with the molten metal. Filler metal when employed is normally the same composition as the base metal.

Advantages and Major Uses

The single-electrode carbon arc welding process is used for welding copper since it can be used at high currents to develop the high heat usually required. It is also used for making bronze repairs on cast iron parts. When welding thinner materials, the process is used for making autogenous welds or welds without adding filler metal. Carbon arc welding is also used for joining galvanized steel. In this case the bronze filler rod is added by placing it between the arc and the base metal.

The carbon arc welding process can be applied several ways, as is shown in Figure 5-50.

The manual carbon arc process is an all-position welding process. It is used as a heat source to generate the weld pool which can be carried in any position. Figure 5-51 shows the process diagram.

Weldable Metals

Mild and low-carbon steels are most widely welded with the carbon arc process, followed by copper. The carbon arc has been used for welding other nonferrous metals. The greatest use of the carbon arc is for brazing and to deposit wear-resistant surfaces. The carbon arc is used for repairing iron castings. The filler metal can be cast iron or bronze. The base metal thickness range and the joint design used are very similar to those of shielded metal arc welding.

Welding Circuit and Current

The welding circuit for carbon arc welding is the same as for shielded metal arc welding. The power source is the conventional or constant current type with drooping volt-ampere characteristics. A 60% duty cycle power source is utilized. The power source should have a voltage rating of 50 V since this voltage is required when welding copper with the carbon arc.

Carbon Arc Welding Electrode Holders

An electrode holder for carbon arc welding is shown in Figure 5-52. It comes in four sizes, based on the welding current capacity, which relates to the carbon electrode size. These electrode holders are not insulated, and the carbon is usually gripped by collet action or by a setscrew.

Single-electrode carbon arc welding is always used with direct-current electrode negative DCEN (straight polarity). In the carbon-to-steel arc the positive pole (anode) is the pole of maximum heat. If the electrode were positive, it would erode very rapidly and would cause black carbon smoke and excess carbon that could be absorbed by the weld metal. Alternating current is not recommended for single-electrode carbon arc welding. The electrode should be adjusted often to compensate for the erosion of the carbon.

Carbon Electrode

There are two types of electrodes. One is made of pure graphite and the other of baked carbon. The pure graphite electrode does not erode away as quickly as the carbon electrode. It is more expensive and more fragile. Electrodes are available in diameters ranging from $\frac{5}{32}$ in. (4 mm) through $\frac{3}{4}$ in. (19 mm) in diameter, with a length of either 12 or 17 in. (310 or 460 mm). AWS does not have a specification for carbon electrodes.[7]

Welding Schedules

The welding schedule for carbon arc welding galvanized iron using silicon bronze filler metal is given in Figure 5-53. A short arc must be used to avoid damaging the gal-

FIGURE 5-52 Carbon arc single-electrode holder.

FIGURE 5-50 Methods of applying CAW.

Method of Applying	Rating
Manual (MA)	Most popular
Semiautomatic (SA)	Note used
Mechanized (ME)	Not popular
Automatic (AU)	Not popular

FIGURE 5-51 Process diagram for CAW.

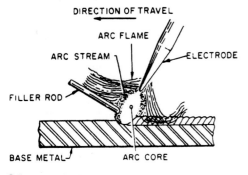

DIRECTION OF TRAVEL
ARC FLAME
ARC STREAM
ELECTRODE
FILLER ROD
BASE METAL
ARC CORE

vanizing. The arc must be directed on the filler wire, which will melt and flow onto the joint.

For welding copper, use a high arc voltage and follow the schedule given in Figure 5–54. Figure 5–55 shows the welding current to be used for each size of the two types of carbon electrodes.

Twin Carbon Arc Welding

Twin carbon arc welding (CAW-T) is a variation that produces coalescence of metals by heating them with an electric arc between two carbon electrodes. No shielding

Material Thickness			Electrode Size		Filler Rod Size		Welding Current (A dc)	Arc Voltage (Electrode Negative)
Gauge	in.	mm	in.	mm	in.	mm		
24	0.024	0.6	$\frac{3}{15}$	5	$\frac{3}{32}$	2.4	25–30	13–15
22	0.020	0.7	$\frac{3}{15}$	5	$\frac{3}{32}$	2.4	25–30	13–15
20	0.036	0.9	$\frac{3}{15}$	5	$\frac{3}{32}$	2.4	30–35	14–16
18	0.048	1.2	$\frac{1}{4}$	6.4	$\frac{1}{8}$	3.2	30–35	14–16
16	0.060	1.5	$\frac{1}{4}$	6.4	$\frac{1}{8}$	3.2	30–35	14–16
14	0.075	1.9	$\frac{1}{4}$	6.4	$\frac{1}{8}$	3.2	30–35	14–16
12	0.105	2.7	$\frac{1}{4}$	6.4	$\frac{1}{8}$	3.2	35–40	15–17

FIGURE 5–53 Welding procedure schedule: galvanized steel.

Thickness of Copper			Diameter of Electrode and Filler Rod				Welding Current (A dc)	Voltage (Electrode Negative)
Decimal Inches	Fraction Inches on U.S. Gauge		Electrode Carbon		Filler Rod Size			
	in.	mm	in.	mm	in.	mm		
0.05	18				$\frac{3}{32}$		80	
			$\frac{3}{16}$	4.8				
0.0563	17				$\frac{3}{32}$	2.4	90	3.5
0.0625	$\frac{1}{16}$	1.6	$\frac{3}{16}$	4.8	$\frac{1}{8}$		90	
0.07	15				$\frac{1}{8}$	3.2	100	40
0.078	$\frac{5}{64}$	1.9			$\frac{5}{32}$		120	
0.094	$\frac{3}{32}$	2.4	$\frac{1}{4}$	6.4	$\frac{5}{32}$		135	
0.109	$\frac{7}{64}$	2.8			$\frac{5}{32}$	11.9	140	40
0.125	$\frac{1}{8}$	3.2			$\frac{3}{16}$		150	
0.141	$\frac{9}{64}$	3.6			$\frac{3}{16}$		160	
0.156	$\frac{5}{32}$	3.9	$\frac{1}{4}$	6.4	$\frac{3}{16}$		165	
0.172	$\frac{11}{64}$	4.4			$\frac{3}{16}$		170	
0.1875	$\frac{3}{16}$	4.8			$\frac{3}{16}$	4.8	185	45
0.203	$\frac{13}{64}$	5.2			$\frac{1}{4}$		200	
0.219	$\frac{7}{32}$	5.6			$\frac{1}{4}$		200	
0.234	$\frac{15}{64}$	5.9	$\frac{5}{16}$	7.9	$\frac{1}{4}$		205	
0.25	$\frac{1}{4}$	6.4			$\frac{1}{4}$		215	
0.266	$\frac{17}{64}$	6.7			$\frac{1}{4}$	6.4	225	45
0.281	$\frac{9}{32}$	7.1			$\frac{5}{16}$		250	
0.3125	$\frac{5}{16}$	7.9			$\frac{5}{16}$		250	
0.344	$\frac{11}{32}$	8.7	$\frac{5}{16}$	7.9	$\frac{5}{16}$		255	
0.375	$\frac{3}{8}$	9.5			$\frac{5}{16}$		270	
0.406	$\frac{13}{32}$	10.3			$\frac{5}{16}$	7.9	290	50
0.4375	$\frac{7}{16}$	11.1			$\frac{3}{8}$		300	
0.4688	$\frac{15}{32}$	11.9	$\frac{3}{8}$	9.5	$\frac{3}{8}$		310	
0.5	$\frac{1}{2}$	12.7			$\frac{3}{8}$	9.5	325	50

FIGURE 5–54 Welding procedure schedule: copper.

Electrode Diameter		Welding Current (amps DCEN)	
in.	mm	Carbon Electrodes	Graphite Electrodes
$\frac{1}{8}$	3.2	15–30	15–35
$\frac{3}{16}$	4.8	25–55	25–60
$\frac{1}{4}$	6.4	50–85	50–90
$\frac{5}{16}$	7.9	75–115	80–125
$\frac{3}{8}$	9.5	100–165	110–165
$\frac{7}{16}$	11.1	125–184	140–210
$\frac{1}{2}$	12.7	150–225	170–260
$\frac{5}{8}$	15.9	200–310	230–370
$\frac{3}{4}$	19.0	250–400	290–490
$\frac{7}{8}$	22.2	300–500	400–750

FIGURE 5–55 Welding current for carbon electrode types and size.

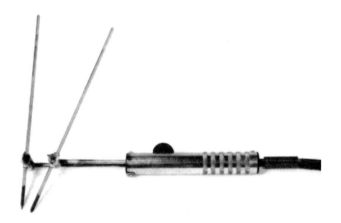

FIGURE 5–56 Twin carbon electrode holder.

FIGURE 5–57 Welding with twin carbons.

Carbon Electrode Diameter		Welding Current (A ac)	Arc Voltage (V)	Base Metal Thickness	
in.	mm			in.	mm
$\frac{1}{4}$	6.4	55	35–40	$\frac{1}{16}$	1.6
$\frac{5}{16}$	7.9	75	35–40	$\frac{1}{8}$	3.2
$\frac{3}{8}$	9.5	95	35–40	$\frac{1}{4}$	6.4
$\frac{3}{8}$	9.5	120	35–40	Over $\frac{1}{4}$	Over 6.4

FIGURE 5–58 Welding current for carbon electrodes (twin torch).

is used. Pressure and filler metals may or may not be used. The twin carbon arc can also be used for brazing.

The electrode holder (Figure 5-56) is used for twin electrodes. It comes in only one size but will accommodate several sizes of electrodes. It is not insulated. The twin carbon electrode holder is designed so that one electrode is movable and can be touched against the other to initiate the arc. The carbon electrodes are held in the holder by means of setscrews and are adjusted so that they protrude equally from the clamping jaws. When the two carbon electrodes are brought together, the arc is struck and established between them. The angle of the electrodes provides an arc that forms in front of the apex angle and fans out as a soft source of concentrated heat or arc flame. It is softer than that of the single carbon arc. The temperature of this arc flame is between 8000 and 9000 °F (4426 and 4982 °C).

Alternating current is used for the twin carbon welding arc. With alternating current the electrodes will burn off or disintegrate at equal rates. Direct-current power can be used, but when it is, the electrode connected to the positive terminal should be one size larger than the electrode connected to the negative terminal. This will ensure an even burning of the carbon electrodes since the positive electrode disintegrates at the higher rate. The arc gap or spacing between the two electrodes is adjustable during welding and must be adjusted more or less continuously to provide the fan-shaped arc shown by Figure 5-57.

The twin carbon arc is a very useful source of heat that can be used for many applications in addition to welding, brazing, and soldering. It can be used as a heat source to bend or form metal. The welding current settings or schedules for different sizes of electrodes are shown in Figure 5-58.

The twin carbon electrode method is used by the hobbyist and for maintenance work, in the small shop, and on the farm. It is used with the low-duty cycle single-phase limited-input ac transformer welding machines. It can be used in any position and on any materials where the heat is required. It is relatively slow and for this reason does not have too much use as an industrial welding process.

5-5 STUD WELDING

Arc stud welding (SW) is "an arc welding process that uses an arc between a metal stud, or similar part, and the other workpiece. The process is used without filler metal, with or without shielding gas or flux, with or without partial shielding from a ceramic ferrule surrounding the stud, and with the application of pressure after the faying surfaces are sufficiently heated." This is a special application process developed in the mid-1930s. A stud weld being made is shown in Figure 5-59. There are several variations of the process.

Principles of Operation

It is questionable if arc stud welding is a true arc welding process. It has a very specialized field of application and is not a metal joining process in the same manner as the others discussed previously. It welds prepared studs at the end to the base metal. The process is a combination of arc welding and forge welding. It is based on two steps. First, electrical contact between the end of the stud and base metal occurs and an arc is established. The heat of the arc melts the surface of the end of the stud and the work surface. As soon as the entire cross section of the stud and an area of equal size on the base metal are melted, the stud is forced against the base metal. The molten end of the stud joins with the molten pool on the work surface, and as the metal solidifies the weld is produced. Partial shielding is normally accomplished by means of a ceramic ferrule that surrounds the arc area and by fluxing ingredients sometimes placed on the arcing end of the stud.

The making of an arc stud weld is shown in Figure 5-60. The stud gun (step A) holds the stud in contact with

FIGURE 5-59 Stud welding.

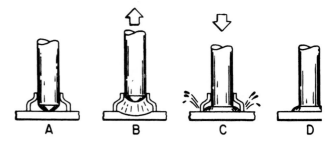

FIGURE 5-60 Stud welding process.

the workpiece until the welder depresses the gun trigger switch. This causes welding current to flow from the power source through the stud (which acts as an electrode) to the work surface. The welding current flow actuates a solenoid within the stud gun, which draws the stud away from the work surface (step B) and establishes the arc. The arc time duration is controlled by a timer in the control unit. At the appropriate time the welding current is shut off, the gun solenoid releases its pull on the stud, and the spring-loaded action plunges the stud into the molten pool of the workpiece (step C). The molten metal solidifies and produces the weld, plus a small reinforcing fillet. After solidification the gun is released from the stud and the ceramic ferrule is broken off, revealing the weld (step D).

Advantages and Method of Application

The arc stud welding process is unique and is a special application process. It offers tremendous cost savings when compared to drilling and tapping for studs, or to manually welding studs to base metal. Stud welding does not destroy the water tightness or weaken the base metal in the way that drilled and tapped holes do.

The semiautomatic method of applying is the most common for construction work and for ship work. Automatic feed and automatic location are becoming increasingly popular. Robots are also being used. The various methods of applying are listed in Figure 5-61. This process is practically an all-position process (Figure 5-62), however, vertical and overhead positions are difficult.

This process is widely used for welding studs to mild steels, low-alloy steels, and some of the austenitic

FIGURE 5-61 Method of applying SW.

Method of Application	Rating
Manual (MA)	No
Semiautomatic (SA)	A
Mechanized (ME)	B
Automatic (AU)	B
Robot	B

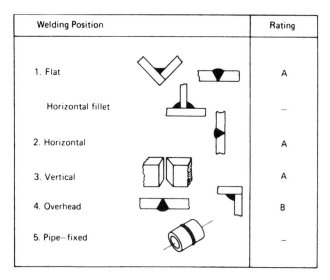

Welding Position		Rating
1. Flat		A
Horizontal fillet		—
2. Horizontal		A
3. Vertical		A
4. Overhead		B
5. Pipe—fixed		—

FIGURE 5–62 Welding position capabilities for SW.

Stud Base Diameter			Base metal Thickness		
Fraction	in.	mm	in.	Gauge	mm
$\frac{3}{16}$	0.187	4.7	0.059	16	1.5
$\frac{1}{4}$	0.250	6.3	0.075	14	1.9
$\frac{5}{16}$	0.312	7.9	0.104	12	2.6
$\frac{3}{8}$	0.375	9.5	0.117	11	2.9
$\frac{1}{2}$	0.500	12.7	0.164	—	4.1
$\frac{5}{8}$	0.625	15.8	0.209	—	5.3
$\frac{3}{4}$	0.750	19.0	0.250	$\frac{1}{4}$ in.	6.3
$\frac{7}{8}$	0.875	22.2	0.312	$\frac{5}{16}$ in.	7.9
1	1.000	25.4	0.375	$\frac{3}{8}$ in.	9.5

FIGURE 5–64 Minimum recommended base metal thickness (steel) for SW.

stainless steels. Some of the process variations can be used on nonferrous metals. Figure 5-63 shows the base metals that can be welded. The stud should have the same analysis as the base metal.

The minimum recommended plate thickness to permit efficient welding without burn-through or excessive distortion is shown in Figure 5-64. As a general rule, the minimum thickness of the plate or base metal is 20% of the stud base diameter. To develop full strength of the stud, the plate thickness should be not less than 33% of the stud base diameter.

Joint Design

Stud welding is a special application process. The joint would be considered a T. The weld would be a square groove with a small reinforcing fillet all around. A variety of studs is shown in Figure 5-65.

FIGURE 5–65 Variety of studs available.

FIGURE 5–63 Metals weldable by SW and variations.

		Stud Welding Variations	
Base Metal	Conventional Arc Study	Contact Capacitor Discharge	Drawn Arc Capacitor Discharge
Aluminum	No	Weldable	Weldable
Brass-bronze (lead free)	No	Weldable	Weldable
Copper (lead free)	No	Weldable	Weldable
Low-carbon steel	Weldable	Weldable	Weldable
Low-alloy steel	Weldable	Weldable	Weldable
Medium-carbon	Limited	Weldable	Weldable
Stainless steel	Weldable	Weldable	Weldable
Zinc	No	Weldable	Weldable (limited)
Dissimilar metals	Limited	Weldable	Weldable (limited)

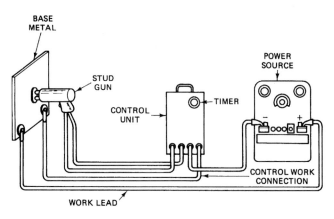

FIGURE 5-66 Circuit diagram for stud welding.

FIGURE 5-67 Apparatus for stud welding.

Equipment Required

Figure 5-66 is the circuit diagram for arc stud welding. It shows the welding power source, the stud gun, and special control unit. Figure 5-67 shows typical apparatus.

Direct current is preferred for arc stud welding, and the stud gun (electrode) is connected to the negative terminal (DCEN) or straight polarity. The workpiece is attached to the positive pole. Ac is not recommended for stud arc welding, and direct-current constant voltage is usable but not recommended.

The *power source* for arc stud welding is normally a direct-current constant-current welding machine. The size of the welding machine is based on the size of the stud to be welded (Figure 5-68). Welding machines can be paralleled to provide sufficient current. Stud welding has a short arc period, rarely lasting over 1 second. Overcapacity currents are drawn from the machine for a very short period. The welding machine must have sufficient overload capacity. See Figure 5-68 for the recommended welding power sources and cable for welding different sizes of studs.

The *stud welding gun* is designed only for stud welding. The gun resembles a pistol with a trigger switch for starting the weld cycle. A popular stud gun is shown in Figure 5-69. The gun contains the solenoid lifting mechanism as well as the spring for plunging the stud into the molten weld pool. The welding current must flow through the gun. The gun must also contain adjustments to establish the arc length and to provide for accurate plunge dimensions. The gun is equipped with a standoff device, to hold it in proper relationship to the work. In addition, the gun allows the use of different types of collets or chucks for each size and type of stud. The stand-off device or arc shield holder must be adjustable to accommodate different types of chucks.

FIGURE 5-68 Power source requirements for different sizes of studs.

Stud (in.)	Diameter (mm)	Amperes Required	Weld Cable		Power Source		Alternative Power Source	
			Size	Number Required	Number	Rated Size (A)	Number	Rated Size (A)
$\frac{3}{16}$	4.76	300	2/0	1	1	300		
$\frac{1}{4}$	6.35	400	2/0	1	1	400		
$\frac{5}{16}$	7.94	500	2/0	1	1	400		
$\frac{3}{8}$	9.53	600	2/0	1	1	600	2	300
$\frac{7}{16}$	11.11	600	2/0	1	1	600	2	300
$\frac{1}{2}$	12.70	700	2/0	1	1	600	2	400
$\frac{5}{8}$	15.87	1150	4/0	1	1	1000	2	600
$\frac{3}{4}$	19.05	1600	4/0	2	2	1000	2	600
$\frac{7}{8}$	22.23	1800	4/0	2	2	1000		
1	25.40	2000	4/0	2	2	1000		

Note: Generator welders have greater overload capacity than rectifiers.

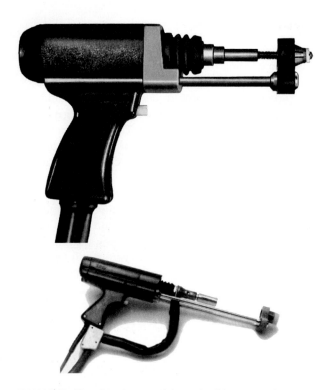

FIGURE 5–69 Stud gun with and without stud.

The weld timer controller establishes the arc time required for proper arcing. The control unit contains a contactor capable of breaking the welding current at the end of the arcing period. It is calibrated in electrical cycles: $\frac{1}{60}$ of a second for 60-Hz power. The control units are contained within the cabinet of the power source. However, weld timer controllers may be separate. Control units for stud welding are designed specifically for stud welding guns and power sources. Stud guns of one manufacturer should not be used with control timers or studs of another manufacturer. The control cable is between the stud gun and the timer only.

Materials Used

Electrode and filler metals are combined in the stud. Studs are made in many sizes and shapes. A variety is shown in Figure 5-65. Normally, studs are made of low-carbon steel. Stud types can be threaded fasteners, internally threaded fasteners, flat fasteners with rectangular cross sections, header pins, eye bolts, slotted pins, keys, and so on. Studs are also made of different metals. The stud gun manufacturers and stud manufacturers offer catalogs with engineering data pertaining to the exact design of the studs they manufacture. Studs up to 1 in. (25 mm) in diameter can be welded. The round stud is the most common. Square and rectangular shaped studs are available.

Most studs include a method of fluxing. Flux acts as a deoxidizer and arc stabilizer. The most common method

is to utilize a solid flux insert in the arcing end of the stud. A short portion of the stud is melted off during the arcing period so that the finished length is less than the original length of the stud. The amount of burn-off material depends on the diameter of the stud and on the application.

A ceramic *ferrule* must be used for each stud. This is sometimes called an *arc shield*. This ferrule is placed over the stud and held in position by a holder or grip on the gun. The ceramic ferrule performs the following functions:

1. Concentrates the heat of the arc in the weld area
2. Reduces oxidation of the molten metal during welding by restricting contact with the atmosphere
3. Confines the molten metal
4. Partially shields the arc from the welder

The inner surface shape of the ferrule is the same shape as the stud being used, usually cylindrical. It has a serrated shape at the base to form vents for escaping shielding gas. Internally it is shaped to help mold the molten metal around the base of the stud to form a small fillet. Specially designed ferrules may be obtained for specific applications. The ferrule is broken off the weld at the completion of the weld and discarded. It can be used only once.

The weld quality of properly made stud welds is excellent and usually exceeds the strength of the stud. Properly made welds depend on the weld schedule, the size of the power source, the gun chuck and hold down, the use of correct size of cables, and a good work connection.

Mechanical tests may be made by shearing the stud from the work or by pulling it from the work. The AWS Structural Code and Navy welding code specifies how these tests are to be made.[8]

The welding conditions for stud welding are given in Figure 5-70.

FIGURE 5–70 Stud arc welding conditions.

Stud Diameter (in.)	Amps DCEN	Welding Volts	Time Cycle[a]	Lift (in.)	Plunge (in.)
$\frac{3}{16}$	300	30	7	$\frac{1}{16}$	$\frac{1}{8}$
$\frac{1}{4}$	400	30	10	$\frac{1}{16}$	$\frac{1}{8}$
$\frac{5}{16}$	500	30	15	$\frac{1}{16}$	$\frac{1}{8}$
$\frac{3}{8}$	600	28	20	$\frac{1}{16}$	$\frac{1}{8}$
$\frac{7}{16}$	700	28	25	$\frac{1}{16}$	$\frac{1}{8}$
$\frac{1}{2}$	900	28	30	$\frac{3}{32}$	$\frac{5}{32}$
$\frac{5}{8}$	1150	28	40	$\frac{3}{32}$	$\frac{5}{32}$
$\frac{3}{4}$	1600	26	50	$\frac{1}{8}$	$\frac{3}{16}$
$\frac{7}{8}$	1800	24	60	$\frac{1}{8}$	$\frac{3}{16}$
1	2000	24	70	$\frac{1}{8}$	$\frac{3}{16}$

[a]Based on 60 Hz.

Welding Variables and Tips for Using

Stud welding is a relatively simple and foolproof welding process. The apparatus must be properly set up and adjusted for the size and type of studs to be welded. The settings and control schedules must be in agreement with the schedules based on stud sizes. Further, the stud gun must be equipped with the proper size stud-holding devices.

Stud welding does not require manipulative skill necessary for the other arc welding processes. The most popular method of applying is semiautomatic, where the welder holds the stud gun during the welding operation. When the trigger is pressed, the entire weld cycle is automatic, based on the settings of the control unit. The gun should not be moved during the welding operation. The arc will be noticed, and plunging action will be heard at the completion of the welding cycle. After the plunging operation, the gun should be held steady for at least a half-second before withdrawing it from the welding stud.

A variety of problems might occur in making a stud weld. If the current is too low, a full-strength weld will not result. If it is too high, there will be too much metal discharged from the weld. The time cycle must be set correctly. If the plunge is too short, a full weld will not result. If the ferrule is not in place properly, the weld will be off-side. If the gun is not perpendicular to the work, the stud will be welded at an angle. Other problems can result if there is dirt, paint, or foreign material on the workpiece. Another problem may result if the welding leads from the power source to the work and to the stud gun or control are not tight. There is also a problem if these cables are longer than recommended.

Every time a new job is started or a new setup is made, the welding procedure should be verified. This is done by bending a stud with hammer blows to see that the weld procedure produces a quality weld.

Safety Considerations

The ceramic ferrule, or arc shield, shields the arc from the welder and eliminates the need for the normal welding hood. The welder should wear safety glasses or flash goggles with a tinted shade. For overhead and vertical position welding above the welder's head, protective clothing is required. Gloves are recommended for all arc stud welding.

Limitations of the Process

The arc stud process is limited primarily to the mild and low-alloy steels. High-carbon and high-alloy steels should not be stud welded unless a heat-treating operation is performed. Stud welding can be performed on the austenitic types of stainless steels only. It is used to weld aluminum with gas shielding.

Variations of the Process

There are three variations of stud welding:

1. The capacitor discharge stud welding method
2. The drawn arc capacitor discharge stud welding method
3. Stud friction welding

In the contact capacitor discharge stud welding variation, the energy for making the weld is stored at a low voltage in high-capacity capacitors. This method is called the *stored energy system* or *percussive stud welding*. The stud is slightly different since it has a small tip on the end of the stud. In operation this small tip is brought into contact with the base material and then pressure is applied by the gun. The stored energy is discharged through the small tip or projection at the base of the stud, and it presents a high resistance to the electrical energy and rapidly disintegrates. This creates an arc, which heats the surface of the stud and base metal. During this arcing period, the stud is plunged to the base metal by means of a spring or air pressure. The weld is completed immediately following the high-intensity arc, in a short period of 3 to 4 msec. It is done so quickly that the heat effect on the parts is minimal. Figure 5–71 shows the sequence of operations. For welding mild steel, neither flux nor shielding is used. However, only smaller studs of $\frac{1}{4}$ in. (6.4 mm) maximum diameter are welded with this method. The power source, stud gun, and controller are designed especially for this stud welding variation.

In the drawn arc capacitor discharge stud welding variation, arc initiation is obtained in the same manner as arc stud welding. The sequence is shown in Figure 5–72. The stud is placed against the work, then it lifts from the work to draw an arc. Studs up to $\frac{1}{4}$ in. (6.4 mm) diameter are used. The arc time varies from 6 to 15 milliseconds, and then the stud is plunged into the molten pool and the weld is completed. Flux is not required, but shielding gas

---

FIGURE 5–71 Sequence of operations: contact capacitor discharge.

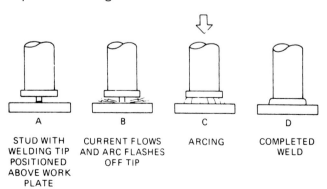

| A | B | C | D |
| STUD WITH WELDING TIP POSITIONED ABOVE WORK PLATE | CURRENT FLOWS AND ARC FLASHES OFF TIP | ARCING | COMPLETED WELD |

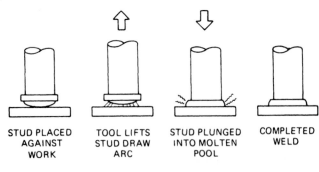

STUD PLACED AGAINST WORK TOOL LIFTS STUD DRAW ARC STUD PLUNGED INTO MOLTEN POOL COMPLETED WELD

FIGURE 5–72 Sequence of operations: drawn arc capacitor discharge.

FIGURE 5–73 Making stud welds on bridge girder.

may be used for welding such metals as aluminum. A special gun, control, and power source are required for this variation of stud welding. This method including power source and associated equipment is very similar to that used for contact capacitor discharge stud welding.

Industrial Use and Typical Applications

The construction industry is a major user of stud welding for attaching shear connectors (Figure 5–73), conduits, piping, electrical switch boxes, and so on, to metal work. The shipbuilding industry uses stud welding for attaching wood decking to metal decking, also for attaching insulation to the interior steel portions of ships. Machinery manufacturers use studs for the attachment of inspection cover plates. The automotive industry uses stud welding

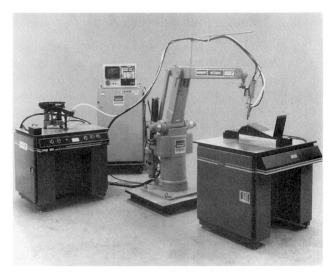

FIGURE 5–74 Robotic stud welding.

for frames and for attaching trim to auto bodies. These installations utilize an automatic stud feeding mechanism. Figure 5–74 shows a robotic stud welding machine. This machine has a moving table with a fixture for holding small parts. It is controlled by a microprocessor.

5-6 OTHER NONCONSUMABLE ARC WELDING PROCESSES

Atomic hydrogen welding (AHW) is an arc welding process that uses an arc between two tungsten electrodes in a shielding atmosphere of hydrogen and without the application of pressure. Filler metal may or may not be used. Atomic hydrogen welding is no longer of industrial significance.

This process was invented by Irving Langmuir of General Electric Company in the mid-1920s. Figure 5–75 is a diagram of the atomic hydrogen process. The arc is between two tungsten electrodes, and the hydrogen passes from the electrode holder through the arc. The arc stream assumes a fan shape and is characterized by a sharp singing sound. The arc area is usually $\frac{3}{8}$ to $\frac{3}{4}$ in. (9 to 20 mm) in diameter. As the hydrogen passes through the arc, molecules are separated into atoms and this gives the process its name, *atomic hydrogen*. As the gas in its atomic state leaves the arc, it recombines to molecular form, giving up its heat of disassociation to produce the extremely high welding temperature flame. The arc is independent of the work and can be moved closer to or farther from the work for heat control.

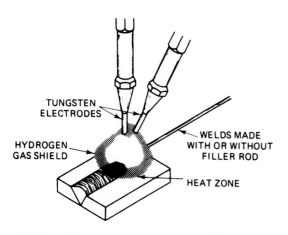

FIGURE 5–75 Atomic hydrogen welding process.

Hydrogen is a powerful reducing agent, and in the presence of molten metal it tends to reduce any gas-forming material and produce a sound porosity-free weld. In addition, the hydrogen prevents contamination of the arc and weld puddle by excluding atmospheric oxygen and nitrogen from the weld area. Years ago the process was popular for welding hard-to-weld metals such as nickel-base alloys, molybdenum, high-alloy steels, and steels for making tools and dies. Today, it is no longer used for production.

The arc is visible to the welder, and welding is done in the flat and horizontal positions. Safety precautions must be followed since the open-circuit voltage of the power source approaches 300 V.

The equipment needed for atomic hydrogen welding is shown in Figure 5–76. The power source has a high open-circuit voltage, and for this reason it contains a contactor operated by a foot switch. It also contains a solenoid valve for controlling the flow of the hydrogen gas. The atomic hydrogen welding torch is shown in Figure 5–77.

Magnetic Rotating Arc Welding

The magnetic rotating arc process is also called magnetically impelled arc butt (MIAB) welding. This process was developed in Europe in the mid-1970s and is now widely used for high-volume-production welded parts. It is "an arc welding process in which an arc is created between the butted ends of tubes and propelled around the weld joint by a magnetic field, followed by an upsetting operation." It combines arc welding with gas shielding and forge welding and is applied automatically. It is used to weld mild and low alloy steels and for welding relatively thin parts that are cylindrical in shape or nearly so. If the part is not cylindrical, it must at least be similar in geometry and have a continuous adjacent joint face. It is also called "Magnetarc Welding." (Figure 5–78).

The sequence of operations for making a magnetic rotating arc welding is shown in Figure 5–79. The parts are first clamped in the machine and this is the setup. They are then moved together linearly until they touch. The current for the arc is initiated by a high-frequency discharge and the arc starts. Immediately the parts are separated to establish the arc. A magnetic coil

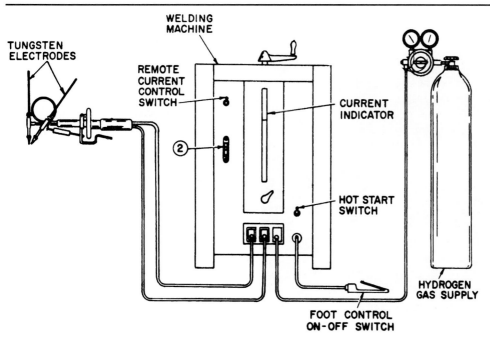

FIGURE 5–76 Atomic hydrogen: circuit diagram.

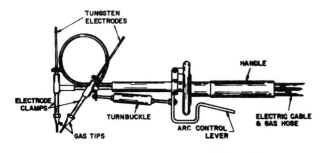

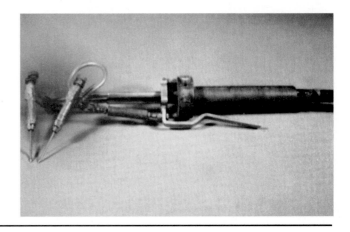

FIGURE 5–77 Atomic hydrogen torch.

surrounds the assembly and causes the arc to rotate around the periphery of the mating parts. The arc area is usually shielded by CO_2 gas to prohibit atmospheric contamination. This heats them at the surface until they are red hot and plastic. This arcing period takes from $\frac{1}{2}$ to 2 seconds, depending on the thickness of the metal being heated. At the proper time the parts are pushed together. This extinguishes the arc and completes the forge weld. This sequence, the time period for each activity, and the forge pressure are programmed into the computer-controller. The total time cycle is very short. The welding machine can be quickly reprogrammed and changed for a different part. Figure 5–80 shows examples of welded assemblies of many types.

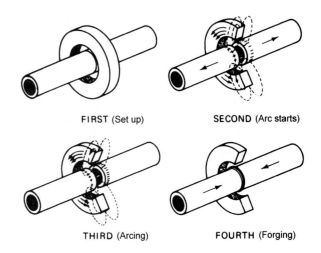

FIRST (Set up)   SECOND (Arc starts)

THIRD (Arcing)   FOURTH (Forging)

FIGURE 5–79 Sequence of operations

FIGURE 5–78 Magnetic rotating arc welding.

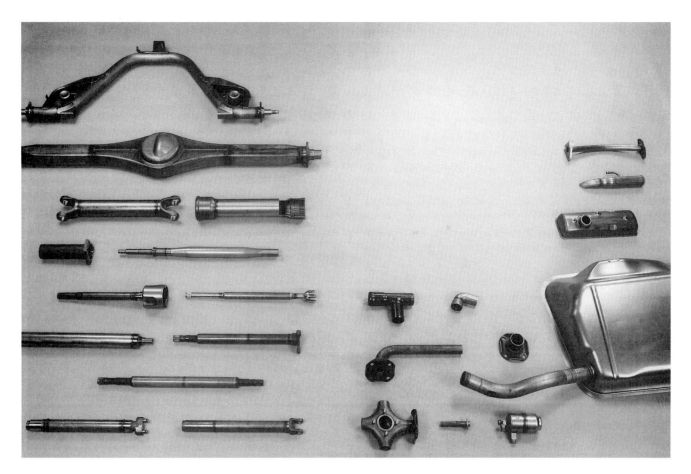

FIGURE 5–80 Examples of magnetic weldments.

FIGURE 5–81 MIAB welded shock absorber.

The MIAB welding process has several advantages. It is completely automatic, very fast, and uses less power than would be required by resistance welding. It can be used for welding thin material. The parts can be cylindrical or not. It produces high-quality, repeatable welds with a minimum of distortion. The rotating arc process does not require as much material to be consumed as does flash welding. This process is widely used by the automotive industry and will find increasing use in the mass production industries. Currently, it is being used to join the cap on the end of a tube to manufacture shock absorbers (Figure 5–81).

QUESTIONS

5–1. How does a nonconsumable arc allow you to make a weld?

5–2. What is the polarity of maximum heat for tungsten arc welding?

5–3. What shielding gas provides a higher arc voltage?

5–4. What polarity provides for cathodic cleaning of the workpiece?

5–5. Where is the maximum voltage drop in the nonconsumable arc?

5–6. What is the approximate temperature of the GTAW arc?

5–7. Why is tungsten used in the nonconsumable welding arc?

5–8. Should CO_2 be used with a tungsten arc?

5–9. In GTAW the filler metal is added by a separate rod. How does this minimize spatter?

5–10. What is the major advantage of gas tungsten arc welding?

5–11. What are the disadvantages of gas tungsten arc welding?

5–12. Why is alternating current used for welding aluminum?

5–13. What determines the size of the tungsten electrode?

5–14. Why is stand-off distance less critical for PAW than GTAW?

5–15. Explain the difference between keyhole welding and melt-in mode.

5–16. Must the plasma arc torch always be water cooled? Why?

5–17. Why is carbon arc welding becoming less important industrially?

5–18. True or false: Arc stud welding is a combination of arc and forge welding.

5–19. Explain the difference between arc stud and the discharge variations.

5–20. What is MIAB welding? Where is it used?

REFERENCES

1. C. E. Jackson, "The Science of Arc Welding," *Welding Journal,* Research Supplement (April 1960): 1295 and (June 1960): 2255.

2. J. F. Lancaster, *The Physics of Welding* International Institute of Welding, Pergamon Press, Elmsford, N.Y., 1984.

3. E. F. Gibbs, "A Fundamental Study of the Tungsten Arc," *Metal Progress* (July 1960): 84.

4. D. R. Milner, G. R. Salter, and J. B. Wilkinson, "Arc Characteristics and Their Significance in Welding," *British Welding Journal* (February 1960): 73.

5. "Recommended Practices for Gas Tungsten Arc Welding," C5.5, American Welding Society, Miami, Fla.

6. "Recommended Practices for Plasma Arc Welding," AWS C5.1, American Welding Society, Miami, Fla.

7. "Electrode, Cutting and Welding Carbon-Graphite Uncoated and Copper Coated," Military Specification MIL-E-17777D, U.S. Department of Defense, Washington, D.C.

8. "Recommended Practices for Stud Welding," AWS C5.4, American Welding Society, Miami, Fla.

6

Arc Welding with a Consumable Electrode

OUTLINE

6-1 THE CONSUMABLE WELDING ARC

In the consumable electrode welding arc, which is the second basic type of welding arc, the electrode is melted and the molten metal is carried across the arc gap. A uniform arc length is maintained between the melting end of the electrode and the weld pool by feeding the electrode into the arc as fast as it melts.

The arc welding processes that use a consumable electrode are: shielded metal arc welding, gas metal arc welding, flux cored arc welding, electrogas welding, and submerged arc welding.

The main function of the arc is to produce heat. It also produces a bright light, noise, and ionic bombardment, which removes the surface of the base metal. The consumable electrode welding arc, also known as a metallic arc, is a sustained high-current low-voltage electrical discharge through a highly conductive plasma that produces sufficient thermal energy which is useful for joining metals by fusion.

The consumable electrode welding arc (Figure 6-1) is a steady-state condition maintained at the gap between the tip of the melting electrode and the molten pool of the workpiece. The electrode is continuously fed into the arc and is melted by the heat of the arc. The molten metal of the electrode transfers across the arc gap to the workpiece, where it is deposited and upon solidification becomes the deposited weld metal. This is a very complex operation that is not completely understood.

The consumable electrode welding arc is a column of electrically and thermally excited gas atoms and ionized metal vapors from the electrode material known as a plasma. This plasma conducts current ranging from a few amperes to hundreds of amperes of either alternating current or direct current of either polarity. It has a voltage or potential drop of 10 to 50 V. The plasma operates at a very high temperature, approximately 6000° C (10,000° F). A consumable electrode welding arc has a point-to-plane geometric configuration. Details of the arc region are shown in Figure 6-2.

FIGURE 6–1 Consumable electrode welding arc.

In the metallic arc the high-temperature plasma causes the gas atoms in the arc to break down into positive ions and negative electrons. Electrons (negative) move from the cathode (negative) to the anode (posi-

tive). The ions (positive) move from the anode (positive) to the cathode (negative). In direct-current reverse-polarity (DCEP) gas metal arc welding, the common welding polarity, the electrons move from the workpiece to the electrode, and the positive ions move from the electrode to the workpiece. The largest portion of current flow is carried by the electrons. Conventional current flows from the electrode to the workpiece.

The potential gradient along the axis of the arc is not uniform. The three voltage regions, known as the *anode drop*, the arc column or *plasma drop*, and the *cathode drop*, are shown in Figure 6–2. The anode and cathode drops are extremely short in length but represent the largest gradient of the voltage potential. The total voltage potential in the arc is the sum of these three potential drop regions.[1] The theoretical arc heat energy available is governed by the welding current and the voltage drops of these three regions. In the total arc region there is another potential drop, known as the electrode *resistance drop*. This is the resistance to current flow through the electrode extension or stickout. This represents a fairly large drop based on the welding current, the diameter of the electrode, and the electrode composition. It is calculated as: electrode resistance drop $E = I^2 \times R$. The welding current is I and the resistance of the electrode wire is R for the length of the extension. The heating of the electrode extension has a great effect on burn-off rates.

The relationship between welding current and arc voltage is not a straight line. The curve shown in Figure 6–3

FIGURE 6–2 Arc region of the consumable electrode arc.

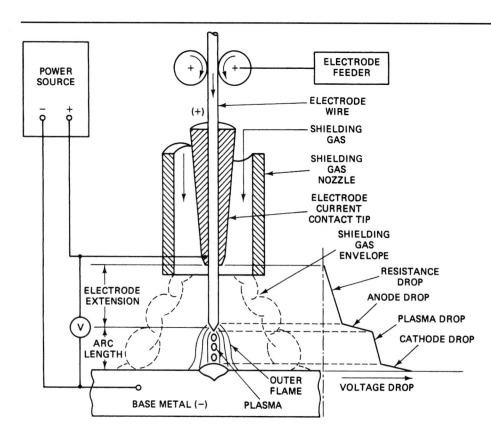

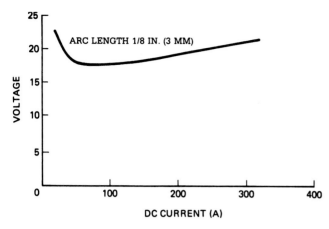

FIGURE 6–3 Arc voltage versus welding current of the metal arc in argon. (From Ref. 2.)

is nonlinear and in the low-current area has a negative slope. The major part of the curve shows that the arc voltage does increase with welding current, other conditions remaining the same. The welding current can be varied over a wide range of values from at least 20 A to over 500 A direct current.

The length of the arc, or the arc gap, affects the arc voltage. A short arc, which is approximately equal to one diameter of the electrode wire, has the lowest voltage. The medium-length arc is in the medium-voltage range. The long arc is equal to about five times the diameter of the electrode and has the highest voltage. This is shown

in Figure 6–4, which shows an aluminum arc in helium. The long arc becomes uncontrollable and it will not deposit metal. If the gap becomes too long, the arc will go out; however, higher currents will sustain a longer arc length. If the gap is too short, the arc will short out. With covered electrodes a higher current allows a shorter arc.

The arc atmosphere affects arc voltage. Figure 6–5 shows the effect of argon and helium gas on the arc length and arc voltage when welding on aluminum.[3] This relationship would apply to different metals. This is shown in another way in Figure 6–6, which shows the arc volts and welding current for argon, argon + 5% oxygen, and helium when welding steel.[4] With a higher arc voltage based on the shielding atmosphere, a hotter arc will result. This increases the thermal energy in the arc, which will slightly increase the melt-off rate of the electrode. This is a minor factor in determining burn-off rates.

Good-quality welding and high-productivity welding depend on two major factors, the penetration of the weld into the base metal and the melt-off rate of the electrode. Figure 6–7 shows the polarity and heat relationship for gas metal arc and for shielded metal arc welding. The polarity of maximum heat in a metallic welding arc depends on the arc atmosphere. In shielded metal arc (covered electrode) welding, the arc atmosphere depends on the composition of the coating on the electrode. The maximum heat normally occurs at the negative pole (cathode). When straight-polarity welding (DCEN) with an E6012 electrode, the electrode is the negative pole and the melt-off rate is high, but the penetration of the base

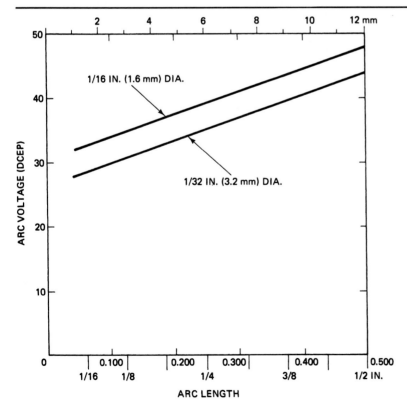

FIGURE 6–4 Arc length–arc voltage relationship. (From Ref. 3.)

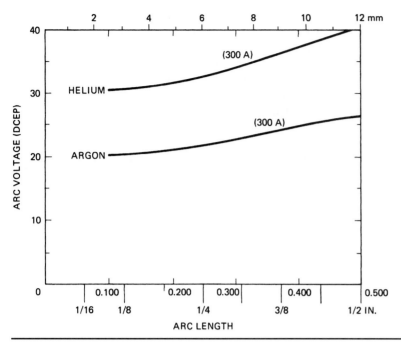

FIGURE 6–5 Arc length–arc voltage relationship in a variety of atmospheres. (From Ref. 3.)

metal is low. When reverse-polarity welding (DCEP) with an E6010 electrode, the maximum heat still occurs at the cathode (negative pole), but this is now the base metal, where deep penetration occurs. With a bare steel electrode on steel in air, the polarity of maximum heat is the anode (positive pole). This is why in the early days bare electrodes were operated on straight polarity (DCEN) so that adequate penetration would result. When coated electrodes are operated on alternating current (AC), the same amount of heat is produced at each polarity of the arc.

The same characteristics are exhibited by the gas metal arc welding process. Direct-current electrode negative is normally used only with an emissive electrode wire, which is not very popular in the United States. This

is because the emissive coated electrode wire has a relatively short storage life.

The relationship between melt-off rates and polarity for metallic arc welding is shown in Figure 6–8. The melt-off rate with reverse polarity (DCEP) is lower than when straight polarity is used. Submerged arc welding is shown in part a[5] and gas metal arc welding is shown in part b.[6] This shows a $\frac{3}{32}$-in. (3.2-mm)-diameter steel electrode wire and approximately $\frac{1}{4}$-in. (6.3-mm) arc length. The higher melting rates occurring when the electrode is negative (straight polarity) can be up to 50% faster. DC straight or reverse polarity or ac welding for submerged arc is very common. Straight polarity (DCEN) with gas metal arc welding is not common but is used with flux

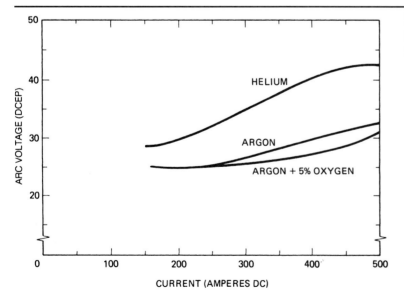

FIGURE 6–6 Arc voltage–welding current relationship in a variety of atmospheres. (From Ref. 4.)

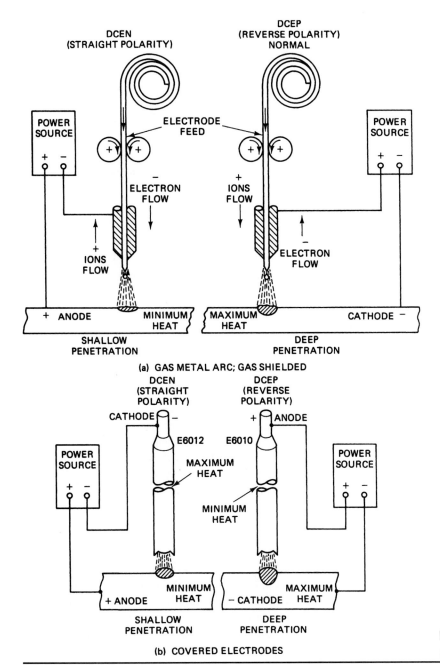

DCEN
(STRAIGHT POLARITY)

DCEP
(REVERSE POLARITY)
NORMAL

POWER
SOURCE
+ −

ELECTRODE
FEED

POWER
SOURCE
+ −

−
ELECTRON
FLOW

+
IONS
FLOW

+
IONS
FLOW

−
ELECTRON
FLOW

+ ANODE   MINIMUM
HEAT

MAXIMUM
HEAT   CATHODE −

SHALLOW
PENETRATION

DEEP
PENETRATION

(a) GAS METAL ARC; GAS SHIELDED

DCEN
(STRAIGHT
POLARITY)

DCEP
(REVERSE
POLARITY)

CATHODE −

+ ANODE

E6012   E6010

POWER
SOURCE
+ −

MAXIMUM
HEAT

MINIMUM
HEAT

POWER
SOURCE
+ −

+ ANODE   MINIMUM
HEAT

− CATHODE   MAXIMUM
HEAT

SHALLOW
PENETRATION

DEEP
PENETRATION

(b) COVERED ELECTRODES

FIGURE 6–7 Metallic arc showing polarity and heat.

cored arc welding. Alternating-current (AC) welding is impossible using a sinusoidal waveform with GMAW or FCAW.

Welders soon learn that when using high current the electrode is melted off rapidly, and when using low current the electrode melts off slowly. This relates to the thermal energy or power in the arc and arc area. This basic relationship, shown in Figure 6-9, applies to all the consumable electrode welding processes. The thermal energy produced in the arc is the product of the welding current and the arc voltage. It is a measure of work that can be performed. The thermal energy is used to melt the base metal to provide weld penetration and to melt the

welding electrode, and in addition, it heats the atmosphere, the welding gun, the gas nozzle, and the contact tip. The heat required to melt the electrode is a physical relationship between the current and the weight of metal melted. This is known as melt-off rate or burn-off rate, which is the weight of metal melted per unit of time. The deposition rate is the weight of metal deposited per unit of time and takes into consideration spatter and slag losses.

There are a number of factors that affect the melt-off rate. Most important is the melting point of the material. For example, aluminum, with a low melting temperature, will have a high linear burn-off rate. However, the

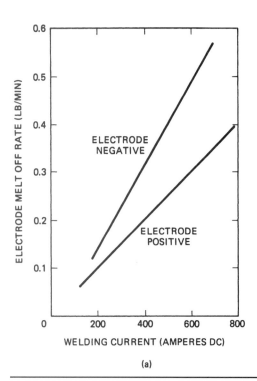

(a)

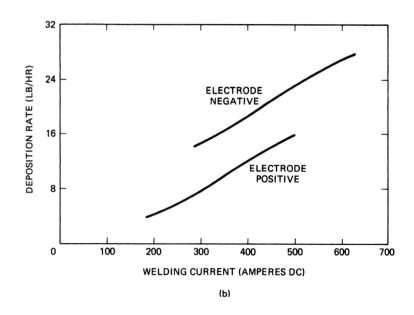

(b)

FIGURE 6–8 Melt-off and deposition rates versus polarity for steel. (From Refs. 5 and 6.)

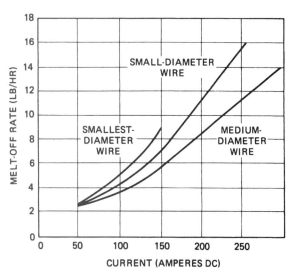

FIGURE 6–9 Melt-off rates versus welding current (DCEP).

weight of material depends on its density; hence a high linear burn-off rate (wire feed speed) does not necessarily mean a high deposition rate based on pounds per hour. Steel has a higher melting temperature but has a greater density. Melt-off rates for different metals and electrode sizes are shown in Figure 6–10.

The size of the electrode wire also has an effect. This is based on the current density, which is the welding current divided by the cross-sectional area of the electrode wire. Figure 6–11 shows the current density versus

sizes of electrodes. This seems to be contradictory to an earlier statement, but is because of the heating in the electrode extension. Current density also has an effect on depth of penetration. The smaller the electrode diameter, the higher the current density at the same welding current. At higher current densities, the melting rate increases for the same current, which explains the divergence of the curves in the melt-off rate versus welding current curve.

The other factor that has a major contribution to melt-off rate is the effect of the electrode extension. This is the resistance heating of the electrode extension. This heating can be called self-preheating and is a technique used to increase deposition rates. The effect of electrode extension and welding current on the melt-off rate is shown in Figure 6–12. This shows that melt-off rate is dependent on both factors, which are interrelated.

The heat generated in this extension can be quite high and will cause the electrode wire to lose its stiffness between the contact tip and the arc. Special nozzle adapters or extended pickup tips are often used. These are electrode wire guides with insulation so that the current is introduced at the pickup tip but the extension keeps the electrode pointed in the proper direction. Too much preheating will reduce the penetration of the arc. Electrode extension is very useful with the flux-cored arc welding process. It preheats the electrode and drives off hydrogen that might be present due to moisture of the ingredients or drawing compounds. Special nozzles, as mentioned above, are quite popular for dc electrode negative (straight polarity) welding with self-shielding flux cored electrode wires.

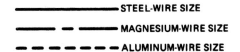

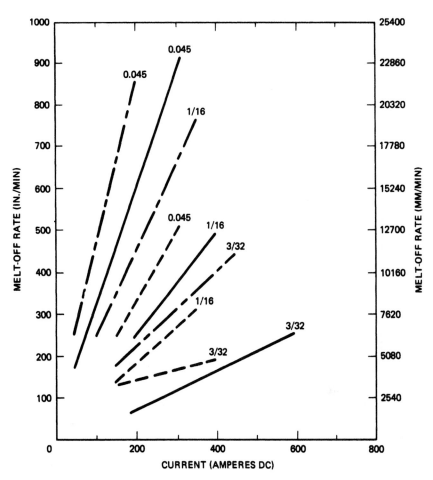

FIGURE 6–10 Melt-off rate for various metals and electrode sizes.

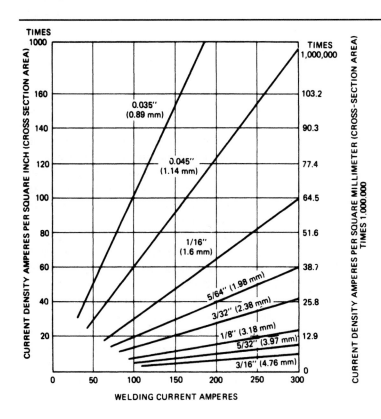

FIGURE 6–11 Current density for a variety of electrode diameters.

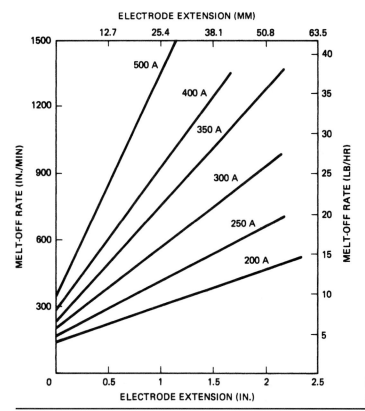

FIGURE 6–12 Electrode extension related to current and melt-off rate. (From Ref. 7.)

In all the consumable electrode arc welding processes, a stable arc is required for successful operation. A stable sustained metallic arc is obtained only when the melting rate of the electrode is equal to the feed rate of the electrode into the arc. This applies whether feeding is done manually, as with coated electrodes, or mechanically with the other consumable electrode arc welding processes. The effect of other variables, such as electrode angle, travel speed, and work position, are discussed later in the chapter.

6-2 METAL TRANSFER ACROSS THE ARC

The forces that cause metal to transfer across the arc are similar for all the consumable electrode arc welding processes. The metal being transferred ranges from small droplets, smaller than the diameter of the electrode, to droplets much larger in diameter than the electrode. The mechanism of transferring liquid metal across the arc gap is controlled by surface tension, the plasma jet, gravity, and electromagnetic force, which provides the pinch effect.[8] It is a combination of these forces that acts on the molten droplet and determines the transfer mode.

Surface tension of a liquid causes the surface of the liquid to contract to the smallest possible area. Surface tension tends to hold the liquid drops on the end of a melting electrode without regard to welding position.

This force works against the transfer of metal across the arc. It also helps keep molten metal in the weld pool when welding in the overhead or vertical positions.

The arc contains a plasma jet that flows along the center of the arc column between the electrode and base metal. Molten metal drops in flight are accelerated toward the workpiece by the plasma jet. Under some conditions the plasma jet may interfere with the transfer of metal across the arc gap.

Earth's gravity tends to detach the liquid drop when the electrode is pointed downward and is a restraining force when the electrode is pointing upward. Earth's gravity has a noticeable effect only at low welding currents. The difference between the mass of the molten metal droplet and the mass of the workpiece has a gravitational effect, which tends to pull the droplet to the larger mass, the workpiece. An arc between two electrodes will not deposit metal on either one.

Electromagnetic force creates the pinch effect force, which helps transfer metal across the arc. When the welding current flows through the electrode, a magnetic field is set up around it. Electromagnetic force acts on the liquid metal drop when it is about to detach from the electrode. As the metal melts, the cross-sectional area of the electrode changes at the molten tip. The electromagnetic force acts to detach a molten drop at the tip of the electrode. When the molten drop is larger in diameter than the electrode, the magnetic force tends to detach the drop. When there is a constriction, or necking down, which occurs when the molten drop is about to detach, the magnetic force acts away from the point of constriction in both directions and the drop, which has started to separate, will be given a push, which increases the rate of separation. This is known as the *pinch force* (Figure 6–13). Pinch force is proportional to the square of the current. Figure 6–14 is a series of high-speed movie photographs of the welding arc. Part (a) shows the start of the constriction, part (b) shows the droplet just before separation, and parts (c) and (d) show the drop in free flight across the arc gap. The rate of change of welding current can regulate the strength of the pinch effect. This is determined by slope of output current of the machine, but more dramatically by pulsing the current which controls the detachment of liquid drops from the end of the electrode.

Magnetic forces also set up a pressure within the liquid drop. The maximum pressure is radial to the axis of the electrode and at high currents causes the drop to become elongated. It gives the drop stiffness and causes it to project in line with the electrode, regardless of the welding position.

The mode of metal transfer across the arc is related to the welding process; the metal involved; the arc atmosphere; the size, type, and polarity of the electrode; the characteristics of the power source; the welding position; and the welding current, current density, and heat input. In gas metal arc welding, reverse polarity (DCEP) is normally employed. For straight polarity (DCEN), an emissive coating is often placed on the electrode surface.

Metal transfer can be defined as a *free-flight* transfer mode, which includes spray and globular transfer, or as a *contact atmosphere mode,* which includes short-circuit transfer. The most common way to classify metal transfer is according to size and frequency and characteristics of the metal drops being transferred. There are four major types of metal transfer:

1. Spray transfer
2. Globular transfer
3. Short-circuiting transfer
4. Pulsed-spray metal transfer

These types of metal transfer are well defined. There is an intermediate form of transfer in the transition zone between two modes where both types of transfer may occur simultaneously. The stability of the welding arc and the metallurgical changes in the weld metal are dependent on the metal transfer mode. Welding procedures are grouped according to the metal transfer mode.

Spray Transfer

The transition between the globular mode and the spray transfer mode occurs in the mid-200-A range on carbon steel. This is based on the size of the electrode and the current density. This transition range of a mild steel electrode in an atmosphere of 95% argon + 5% oxygen is shown in Figure 6–15. In the range below the transition or full spray mode, the drops of molten metal are approximately the same size as the electrode wire. In this transition mode, metal transfer is not as smooth and there is more spatter. The frequency of drop detachment is regular and at a higher frequency than globular transfer.[9]

Spray transfer, sometimes called axial spray, is a very smooth mode of transfer of molten metal droplets from the end of the electrode to the molten weld pool. It was the original type of metal transfer used when gas metal

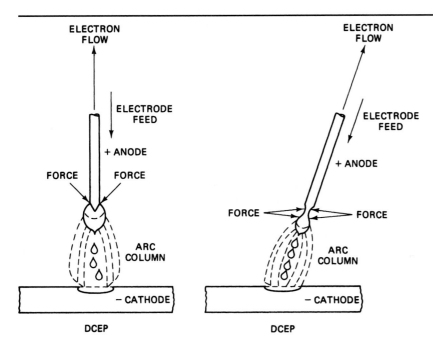

FIGURE 6–13 Electromagnetic force on drop about to transfer.

ARC WELDING WITH A CONSUMABLE ELECTRODE

ARC WELDING WITH A CONSUMABLE ELECTRODE   **115**

(a)

(b)

(c)

(d)

FIGURE 6–14 High-speed photographs of metal transfer.

arc welding was initially developed. Spray transfer is shown graphically in Figure 6-16. Spray transfer occurs in an inert gas atmosphere, usually with a minimum of 80% argon shielding gas. The droplets crossing the arc are smaller in diameter than the electrode. The tip of the electrode is pointed. It occurs at a relatively high current density and there is a minimum current level for each electrode size. As the current increases, the drop size decreases and the frequency of drops increases. The drops have an axial flow, which means that they follow the centerline of the electrode and travel directly to the weld pool. There is no short circuiting in spray transfer. The electromagnetic forces are the dominant forces due to the high current density. The pinch effect on the molten tip of the electrode physically limits the size of the molten metal droplet that can form. Therefore, only small droplets are formed, which transfer across the arc at relatively high frequency. With spray transfer the deposition rate and efficiency is relatively high. The arc is very smooth, stable, and stiff and the weld bead has a nice appearance and a good wash into the sides. Spray transfer is shown by a high-speed photograph in Figure 6-17.

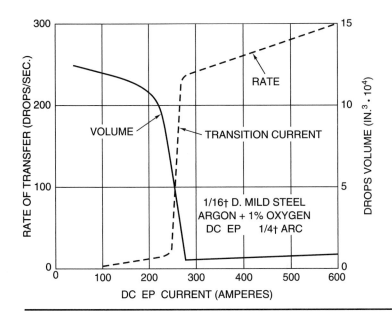

FIGURE 6–15 Transition current related to drop size and frequency.

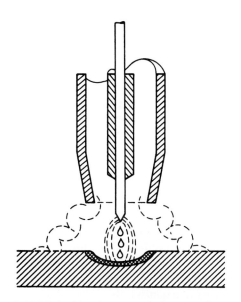

FIGURE 6–16 Spray transfer metal transfer mode.

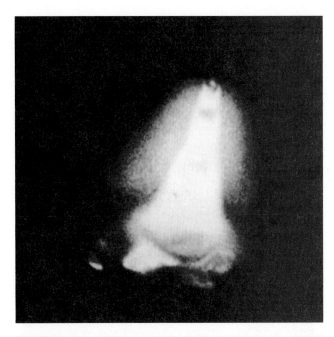

FIGURE 6–17 High-speed photograph of spray transfer mode.

In spray transfer a large amount of heat is involved, which creates a large weld pool with good penetration that can be difficult to control. Thus normal spray transfer is limited to the flat and horizontal positions and is not used to weld thin materials. As current is increased beyond axial spray transfer range, the line of metal drops begins to rotate rapidly about the axis of the electrode, still leaving from the tip of the electrodes. As current is increased further, the diameter of rotation increases and spatter increases. This is known as rotational spray transfer. The solution to this is to use a larger-diameter electrode or decrease the current.

Globular Transfer

Below the transition level, the metal transfer mode is called *globular transfer* and is shown graphically in Figure 6–18. This type of transfer usually occurs when CO_2 shielding gas is used. It was a type of metal transfer originally encountered with the development of CO_2 gas-shielded welding. A high-speed photo of globular metal transfer is shown in Figure 6–19.

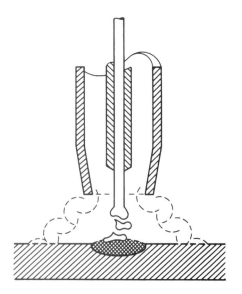

FIGURE 6–18 Globular transfer metal transfer mode.

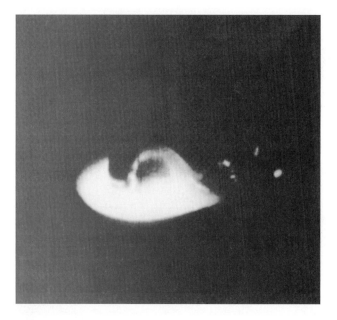

FIGURE 6–19 High-speed photograph of globular transfer mode.

Variations of globular transfer include one known as *drop transfer,* where gravity is the dominant force, and one known as the *repelled transfer,* where forces due to the plasma jet occur even though gravity force is a factor.

Arcs in a CO_2 atmosphere are longer than those in an argon atmosphere, hence the voltage is higher. The *cathode jet* is involved and the electrode is positive (anode). The cathode jet originates from the workpiece (cathode) and actually supports the molten drop of metal on the tip of the electrode. The molten globule can grow in size until its diameter reaches $1\frac{1}{2}$ to 3 times the diameter of the electrode, due to this repelling force. When the

globule grows on the tip of the electrode, it takes on unusual shapes and moves around on the tip of the electrode. It separates from the electrode and is transferred across the arc by electromagnetic and gravity forces. The globule transfers across the arc in an irregular path. It changes its irregular shape during flight and sometimes has a rotating motion. The irregular shape, motion, and flight direction sometimes cause the globule to reconnect with the electrode and touch the work as well. This causes a short circuit, which momentarily extinguishes the arc. In the globular transfer mode, the drop splashes into the weld pool and produces much more spatter than spray transfer. The spatter comes from the violent molten pool as well as from the metal transferring across the arc. The frequency of globular detachment and flight across the arc is random but of relatively low frequency. It takes place at a relatively low current density. The resulting welding deposit is not as smooth as that produced by spray transfer. It is used primarily in the flat and horizontal positions for welding steel. A variation of globular transfer is known as the *buried arc.* This is obtained by increasing welding current and adjusting the welding voltage so that the tip of the electrode is below the surface of the molten weld pool. The arc is within a cavity generated by the force of the arc. It occurs in a CO_2 arc atmosphere and provides very deep penetration. Spatter is reduced. It is used only in the flat position and is useful when making arc spot welds in heavy steel sections.

Short-Circuiting Transfer

Out-of-position welding and welding on thin materials were extremely difficult, if not impossible, using spray or globular metal transfer. The *short-circuiting mode of transfer* was introduced in the late 1950s for gas metal arc welding of thin metal and for out-of-position welding.[10] It is a low-energy mode of metal transfer. The short-circuiting transfer mechanism, also called *short arc* and *dip transfer,* is shown in Figure 6–20. The mechanism of short-circuiting metal transfer is as follows: The molten tip of the electrode is supported by the cathode jet and may grow to $1\frac{1}{2}$ times the diameter of the electrode. The electrode is feeding at such a high relative rate of speed that the molten tip will periodically come in contact with the molten weld pool. This is a short circuit that creates a bridge across the gap between the electrode and the molten pool, and the arc is extinguished. Molten metal is transferred from the electrode by the surface tension of the weld pool, which draws the molten metal of the electrode tip into the molten weld pool. The electrode will then separate from the weld pool and reestablish an arc (Figure 6–21). Occasionally when the electrode contacts the weld pool, the electrode will act like a fuse and literally explode, due to the high current density. The explosion reestablishes the arc. These conditions continue at a random frequency. These arc outages occur so rapidly

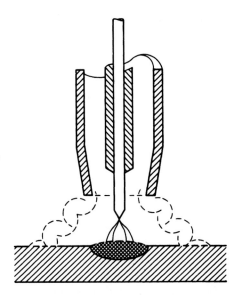

FIGURE 6–20 Short-circuiting transfer mode.

that they are not noticed by the human eye. They are, however, caught by high-speed photographs (Figure 6–22).

The short-circuiting mode of metal transfer allows all-position welding and the welding of thin materials. It is obtained with specific welding parameters normally limited to a maximum of 200 A DCEP on a 0.035-in. (0.9-mm) steel electrode and CO_2 or 75% argon–25% CO_2 shielding gas. It uses a constant-voltage (CV) power source with the correct impedance, which provides the proper rate of increase of current during the short circuit to maintain a stable arc. The short-circuiting mode of metal transfer will sometimes cause cold lap defects in the weld and may create undercutting if proper technique is not employed. The short-circuiting mode is normally used with CO_2-rich shielding atmospheres and is used basically on ferrous metals. It cannot be used on nonferrous metals.

There is a small amount of spatter involved; however, the weld pool is relatively small and easily controlled. Base metal penetration can be controlled by technique. This mode of metal transfer has the ability to bridge gaps between piece parts that are wider than the thickness of the metal.

Pulsed-Spray Metal Transfer

The spray transfer mode described previously has many advantages, including smooth bead appearance and minimum spatter. It has several disadvantages, which prohibit it from being used for some applications. The transition current for spray transfer is relatively high, which creates a large molten metal weld pool and deep penetration. Spray transfer could not be used when welding on thin materials, and the large weld pool could not be controlled when welding in the vertical or overhead position. In the late 1960s, J. C. Needham of Great Britain determined that a high short peak or pulse of current would transfer one drop of metal across the arc. This was named the *pulsed-spray metal transfer mode*. This technique produces droplets of approximately the same or smaller size than

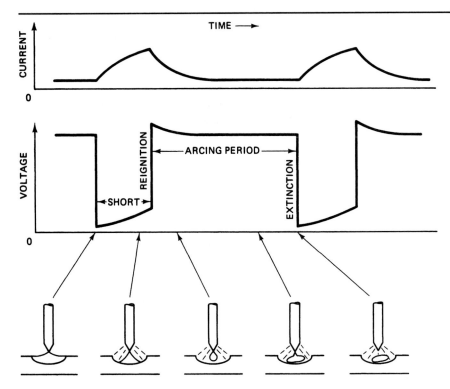

FIGURE 6–21 Short-circuiting transfer mechanism.

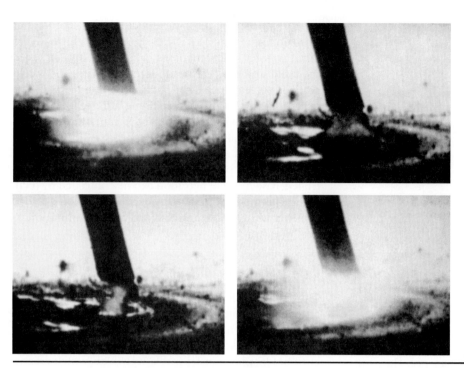

FIGURE 6–22 High-speed photographs of the short-circuiting arc.

the electrode diameter. This mode of metal transfer is shown in Figure 6–23. The mechanism of pulsed-spray metal transfer is based on a special pulsed waveform of the welding current, shown in Figure 6–24. The current output is pulsed at high speed from a low to a high current peak, known as *peak current* (I_p), which is above the transition current, shown by Figure 6–25. The time period for the peak current is known as *peak time* (t_p), sometimes called *pulsed width*. The current level the remaining time is the background current (I_b), known as *low-level current*. The background current is sufficient to maintain the arc. The pulsing waveform continues at a consistent manner at a frequency range of 30 to about 400 pulses per second. The pulsed-spray mode allows the use of larger-diameter electrode wire. It allows welding of thin materials in all positions. It can be used to weld most metals. It uses at least 85% to 90% argon-rich shielding gas with mixtures of helium, hydrogen, oxygen, or CO_2. It allows the use of from 5% to 15% CO_2 in argon when welding mild steel. It is recommended for high-quality precision welding for semiautomatic application or mechanization or when robot welding is used.

The theoretical square waveform, shown in Figure 6–24, is not precisely the waveform that occurs in practice. The normal waveform is more rounded and the vertical lines are not precisely perpendicular. The proper values for peak current, background current, and width of peak current, and the distance between peak current pulses, strictly control the metal transfer across the arc. These values must be set properly when using different atmospheres and electrodes of various materials and sizes. The pulsed-spray metal transfer mode gives spray

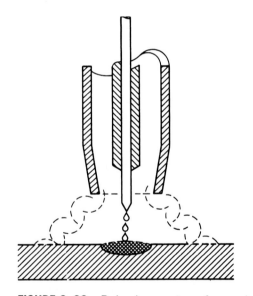

FIGURE 6–23 Pulsed-spray transfer mode.

transfer characteristics at a much lower average current. This reduces penetration and the size of the molten weld pool and allows puddle control.

The original research used pulses at line frequency (i.e., 50 or 60 pulses per second), or at double line frequency (100 or 120 pulses per second). The background current had to be balanced with the peak current and the electrode feed rate in order to obtain a stable arc. Power sources designed to provide this type of pulsing were extremely difficult to adjust and never became popular.

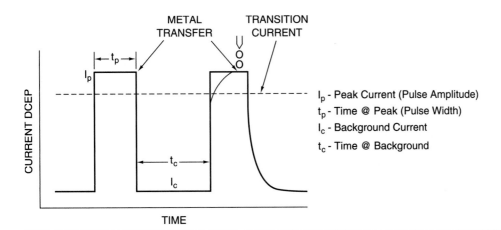

FIGURE 6–24 Power source output waveform.

I_p - Peak Current (Pulse Amplitude)
t_p - Time @ Peak (Pulse Width)
I_c - Background Current
t_c - Time @ Background

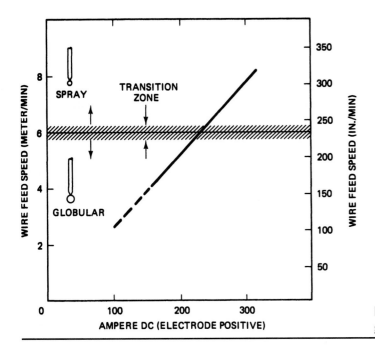

FIGURE 6–25 Transition zone between spray and globular modes. (From Ref. 9.)

To obtain a small controllable weld pool and excellent weld appearance required more developmental work. The development of the inverter power source controlled by a microprocessor provides a precisely regulated pulsing system. This research provided a better understanding of the variables involved in pulsed-spray welding frequencies, pulse shapes, pulse amplitude, pulse width, and so on. Pulsed-spray metal transfer is now a practical and increasingly popular method of gas metal arc welding.

Peak Current (I_p) Also known as *pulse amplitude*. This is above the transition current and takes the arc into the spray transfer mode. It controls the metal droplet size and provides shorter arc lengths with higher peak currents. If the arc is too long, droplets become too large and may cause droplets to be ejected in front of the weld pool or continuous spraying of fine spatter in all directions. If the

arc length is too short, it takes more than one pulse to transfer the droplets, causing a more globular type of transfer.

Peak Time (t_p) Also known as *pulse width*. This controls the metal droplet size and high peak current with short peak times, and increases penetration. If the time period is too long, droplet size becomes too large or more than one drop will transfer per pulse. This may also cause droplets to be ejected in front of the weld pool.

Background Current (t_b) Also known as *low current pulse*. This maintains the arc and keeps the electrode wire tip hot. It does not have sufficient energy to cause the metal to melt off the end of the electrode and transfer across the arc. If it is too low, the arc will tend to short circuit; that is, the electrode wire will touch the work. If it is too high, metal may transfer during background current, causing a globular type of transfer. It must be carefully coordinated

ARC WELDING WITH A CONSUMABLE ELECTRODE **121**

with the wire feed speed. Wire feed speed changes with average welding current. If the wire feed speed is too high, the arc goes into short-circuiting transfer mode, and if it is too low, the arc length becomes too long. Changes must be coordinated with the electrical stickout to maintain a consistent arc voltage and arc length.

Pulse Frequency   This must be adjusted to provide one drop per pulse. This is controlled by the peak current time and the background current time. Pulsing frequency is defined as $1/t_p - t_b$ in hertz. The applicable amperage of peak and background current adjusted to the wire feed speed maintains the requirement of one drop per pulse and constant droplet size.

As advanced pulsing techniques were developed, they become known as *synergic welding. Synergic* comes from a Greek word meaning "working together." This type of welding requires that the power source and wire feeder work together to provide the proper welding procedure. The power source, and in some cases the wire feeder, are controlled by a microprocessor with appropriate software. The peak pulse current, the pulse width, and the frequency of pulsing and waveshape of the pulse are coordinated and controlled by a single knob which indicates average current. Software programs provide a welding procedure based on specific electrode metal and size, and shielding gas. These can be dialed in and/or modified as required.

There are several concepts of synergic welding with matched inverter power sources and wire feeders. In one case, the frequency of pulsing remains the same with a maximum of peak pulse current, and the background current varies. In the other case, the pulsing frequency will vary but the background current remains the same. The wire feeder must be coordinated with the output of the power source. Some power sources have a subprogram that provides the pulsed waveform for each pulse. This waveform can be changed for different applications based on the filler metal type, size, and shielding gas atmosphere. The advantage of this type of system is to provide a very controllable weld pool with a simple-to-control power source and wire feeder. More information on pulsed-spray welding is given in Chapter 10.

In the original pulsed-spray system, the frequency of peak current pulses was at utility line frequency or double the utility line frequency: normally at 50 or 60 pulses per second, or 100 or 120 pulses per second. The pulse current duration was fixed and related to the line frequency. The background current had to be balanced with the peak current and the electrode feed speed in order to obtain good arc stability, which in turn provided the proper welding power and proper bead shape. The machines that pulsed at line or double line frequency never became popular.

Metal Transfer: Other Processes

The metal transfer for flux cored arc welding electrodes is different than for solid electrodes. Metal transfer is different when the core is filled with minerals for atmosphere control and for fluxing than when metal and ferrous alloys are used. The sheath or outer metal part of the electrode carries the welding current. Arcing occurs at the metallic sheath, which starts melting. The core material melts from the heat generated by the arc. The electrode extension portion is preheated, which causes smoother metal transfer. Flux also transfers across the arc and sometimes bridges the gap. The type of transfer does not relate to the welding procedure, which is largely based on electrode diameter, current, and voltage.

The metal transfer across the arc with submerged arc welding is hidden from the human eye. It has been studied, however, and is very similar to the transfer in gas metal arc welding. The metal transfer mode is of less importance for submerged arc welding.

The metal transfer across the arc with covered electrodes is somewhat different since several other factors are involved. With a larger-diameter electrode, the arc does not constantly cover the entire cross section or end of the core wire. The slag formed from the melting of the electrode coating covers a portion of the tip of the core wire. This slag and the atmosphere created by the burning of the coating affects the metal transfer. The drop size can be rather large up to the diameter of the core wire, or small, much smaller, than the core wire diameter. With heavy-coated lime base (EXX15) electrodes, the metal transfer has fairly large, coarse drops with short circuits. With the low hydrogen iron powder type (EXX18) and the cellulose (EXX10 and 11) electrodes, relatively large drop transfer occurs. With rutile-type (EXX12) electrodes, relatively fine drop transfer occurs. The droplets are covered with flux during the flight across the arc.

The Weld Pool

The *weld pool* is the localized volume of molten metal in a weld prior to its solidification as weld metal and is shown in Figure 6–26. This shows the depth of penetration D, the width W, and the length L. It is also called the *weld puddle* and is present when using a consumable electrode arc welding process. It is similar but not exactly the same when using shielded metal, gas metal, or flux cored arc welding. Welders learning manual or semiautomatic arc welding must first learn how to control the puddle or molten weld metal. In machine or automated welding, control circuit sensors and motion systems are designed to control the molten weld metal. All of the variables required to develop a welding procedure, including the mode of metal transfer, must be specified with acceptable values in order to produce a controllable weld

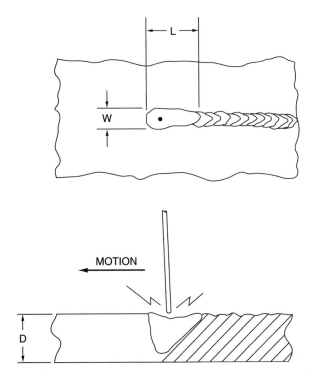

FIGURE 6–26 Weld pool for consumable electrode.

pool. The weld pool is very complicated, and there are many aspects that must be considered from different points of view.

Controlling the weld pool largely involves the power source and wire feeder adjustments and arc manipulation, particularly in vertical position or overhead position welding. If the weld pool is too large, gravity will cause the molten metal to run out and damage the weld. If the weld pool is too deep, it will cause burn-through on thinner materials. On the other hand, if the weld pool is not large enough, a weld may not be made. For thin materials welded at high speed, the volume of molten metal is very small and it will freeze almost immediately and produce a high-quality weld. The power source dynamic response will also affect weld pool stability.

The fact that the weld pool is constantly moving is another complicating condition. The heat input from arc energy must be great enough to melt the base material under the arc. This depends on the total energy in the weld, which can be measured. The most common way is to calculate the energy per unit of length on a per minute basis. The following formula is used:

$$H \text{ (joules per linear measurement per minute)} = \frac{E \text{ (volts)} \times I \text{ (amperes)} \times 60}{S \text{ (speed in the linear measure per minute)}}$$

This can be calculated in joules per inch or joules per metric length. All of the heat generated in the arc does not enter the base metal. Some energy is lost by radiation, and some is used to melt the electrode. It varies from as low as 20% to as high as 75%, based on the welding process and other conditions.

The amount of metal melted in the weld pool depends on many factors, including the temperature of the arc; the amount of heat transferred to the base metal; the melting temperature of the base metal; the mass of base metal, particularly its thickness; the conductivity of the base metal; and the initial temperature of the base metal if preheat is used. The heat input involves the size and polarity of the electrode, the arc atmosphere, the process and the welding current being used, the arc length, and the travel speed. Success in controlling the molten weld pool requires an understanding of the interrelationship of all of these variables. These variables also affect the cooling rate or solidification rate of the weld pool.

Arc welding creates metallurgical factors that are affected by the rapid rate of heating and cooling. The rate of cooling has an influence on the metallurgical properties of the weld and the heat-affected zone, particularly on higher-carbon or higher-alloy steels. Another factor relates to the alloying in the weld pool, especially if the analysis of the welding rod is different from the base metal. These factors and their relationship to the weld pool will be covered more thoroughly later in the book.

The molten weld pool can be used as an indicator for proper adjustment of variables and arc manipulation by the welder. In automatic welding, sensors can be used to view the molten weld pool to control the variables during welding. The depth and width of the molten pool are extremely important for a high-quality weld.

The molten weld pool also indicates the possibility of specific defects. Ever-increasing speed requirements to make the welds more quickly can reveal the tendency for undercutting and *humping*. Humping consists of a regular series of swellings in the weld bead. These potential defects occur at speeds around or exceeding 50 in./min. In undercutting, the width of the groove gouged out by the arc depends on the arc energy and particularly, the arc voltage. Undercutting occurs when the weld metal solidifies too rapidly before the groove is completely filled. Solidification starts at the edge of the molten metal before it has spread out to the edges of the groove. This is based on speed and influenced by the fluidity of the molten metal and the wettability of the molten steel. Wettability depends on the relationship of the surface tension forces involved. The surface tension of the metal oxides are appreciably lower than the surface tension of the pure metal. Humping occurred at the higher speeds and is related to the angle of the electrode and the amount of oxygen available either through shielding gas or coatings on the base metal. More information concerning these potential defects is found in Reference 11.

6-3 SHIELDED METAL ARC WELDING

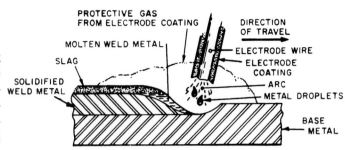

FIGURE 6–28 Shielded metal arc welding process diagram.

Shielded metal arc welding (SMAW) is "an arc welding process with an arc between a covered electrode and the weld pool. The process is used with shielding from the decomposition of the electrode covering, without the application of pressure, and with filler metal from the electrode." This was the development that quickly followed the carbon arc process. Shielded metal arc welding was an outgrowth of bare metal arc welding, which used a bare or lightly coated electrode, which is an obsolete welding process. SMAW is also known as *stick electrode welding.* Figure 6-27 shows this popular process.

Principles of Operation

The shielded metal arc welding process, shown in Figure 6-28, consists of an arc between a covered electrode and the base metal. A welder's view of SMAW is shown in Figure 6-29. The arc is initiated by touching the electrode momentarily to the workpiece. The heat of the arc melts the surface of the base metal to form a molten pool. The metal melted from the electrode is transferred across the arc into the molten pool. When it solidifies it becomes the deposited weld metal. The molten pool, sometimes called the weld puddle, must be properly controlled for successful application of the SMAW process. The size of the weld pool and the depth of penetration determine the mass of molten metal under the control of the welder. If

FIGURE 6–27 Shielded metal arc welding.

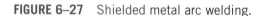

FIGURE 6–29 Welder's view of SMAW arc.

the current is too high, the depth of penetration will be excessive and the volume of molten weld metal will become uncontrollable. A higher speed of travel reduces the size of the molten weld pool. When welds are not made in the flat position, the molten metal may run out of the pool and create problems. Adjusting the welding variables and manipulating the arc will allow the welder to control the molten metal pool properly. The weld metal deposit is covered by a slag from the electrode covering. The arc in the immediate arc area is enveloped by an atmosphere of protective gas produced by the disintegration of the electrode coating. Most of the electrode core is transferred across the arc; however, small particles escape from the weld area as spatter.

Advantages and Major Uses

The shielded metal arc welding process is one of the most popular arc welding processes. It has maximum flexibility and can weld many metals in all positions from near minimum to maximum thickness. The investment for equipment is relatively small. It is used in manufacturing and in field work for construction and maintenance.

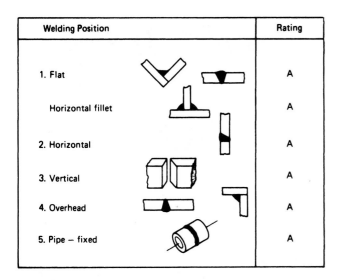

Welding Position		Rating
1. Flat		A
Horizontal fillet		A
2. Horizontal		A
3. Vertical		A
4. Overhead		A
5. Pipe – fixed		A

FIGURE 6–30 Welding position capabilities for SMAW.

Base Metal	Weldability
Aluminum	Possible but not popular
Bronze	Weldable
Copper	Possible but not popular
Copper nickel	Acceptable
Cast and malleable iron	Weldable
Wrought iron	Weldable
Inconel	Weldable
Nickel	Weldable
Monel	Weldable
Low-carbon steel	Weldable
Low-alloy steel	Weldable
High- and medium-carbon steel	Weldable
Alloy steel	Weldable
Stainless steel	Weldable

FIGURE 6–31 Metal weldable by SMAW.

Method of Application

The normal method of applying SMAW is the manual (MA) method. This is the most common method and represents 99% of all use of the process. Semiautomatic (SA) and mechanized (ME) methods are not used. The automatic (AU) method is used and is called gravity welding but has very limited applications and is no longer popular.

Welding Position Capabilities

This process has all-position capabilities (Figure 6–30). Welding in the horizontal, vertical, and overhead positions depends on the type and size of the electrode, the welding current, and the skill of the welder.

Weldable Metals

This process can be used to weld steels and some of the nonferrous metals. Its major use is for joining steels, including low-carbon or mild steels, low-alloy steels, high-strength steels, quenched and tempered steels, high-alloy

steels, stainless steels, and corrosion-resistant steels, and for welding cast iron and malleable irons. It is used for welding nickel and nickel alloys and to a lesser degree for welding copper and some copper alloys. It can be, but rarely is, used for welding aluminum. It is not used for welding magnesium, the precious metals, or the refractory metals. Figure 6–31 shows the weldable base metals. Shielded metal arc welding is also used for surfacing.

Base Metal Thickness Range

The range of thickness of base metal normally welded is shown in Figure 6–32. The minimum thickness that can be welded is largely dependent on the skill of the welder. Steel of $\frac{1}{16}$ in. (1.6 mm) can be welded by a skilled welder. Steel up to $\frac{1}{4}$ in. (6.4 mm) can be welded without a groove if sufficient root opening is provided. Thicker material requires joint preparation and multiple passes. The largest fillet weld that can normally be made in one pass in the horizontal position is $\frac{5}{16}$ in. (8 mm). In the vertical position larger fillets can be made; however, quality deteriorates if fillets are made over $\frac{3}{8}$ in. (10 mm) in a single pass. Maxi-

FIGURE 6–32 Base metal thickness range for SMAW.

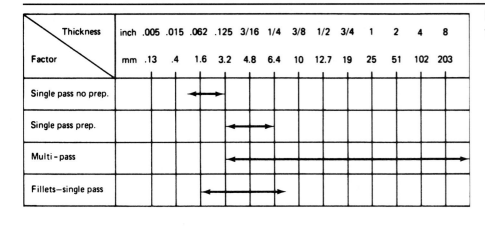

Thickness	inch	.005	.015	.062	.125	3/16	1/4	3/8	1/2	3/4	1	2	4	8
Factor	mm	.13	.4	1.6	3.2	4.8	6.4	10	12.7	19	25	51	102	203
Single pass no prep.				←→										
Single pass prep.					←——→									
Multi - pass					←——————————————→									
Fillets—single pass				←————→										

mum thickness is practically unlimited but requires multiple-pass technique.

Joint Design

When welding materials thicker than $\frac{1}{8}$ in. (3.2 mm), space must be made available to deposit the weld metal. Various weld groove designs are used but the fillet is the most common weld made. Complete information on joint design is given in Chapter 19.

Welding Circuit

Figure 6–33 shows the circuit diagram for shielded metal arc welding. It shows the welding cables used to conduct the welding current from the power source to the arc. The electrode lead forms one side of the circuit and the work lead is the other side of the circuit. They are attached to the terminals of the welding machine.

Welding can be accomplished with either alternating current (ac) or direct-current electrode negative (DCEN), straight polarity, or electrode positive (DCEP) reverse polarity.

Equipment Required

The welding machine or power source is the heart of the shielded metal arc welding system. Its primary purpose is to provide electric power of the proper current and voltage to maintain a controllable and stable welding arc.

The output characteristics of the power source must be of the constant-current (CC) type. The normal current range is from 25 to 500 A using conventional-size electrodes. The arc voltage varies from 15 to 35 V. Power sources for welding are covered in Chapter 10.

The next important piece of apparatus is the electrode holder, which is held by the welder. The holder firmly grips the electrode and transmits the welding current to it. Electrode holders, shown in Figure 6–34, come in two basic designs, the pincher type and the collet or

FIGURE 6–34 SMAW electrode holders.

twist type. Each style has its proponents and the selection is largely personal preference. Electrode holders are designated for their current-carrying capacity. The basis for selection is shown in Figure 6–35, which gives the current rating, duty cycle, maximum electrode size, and the cable size that is to be used and a nominal weight range. There are no standards or specifications for electrode holders in the United States; however, several European countries have them. The holders range from 8 to 14 in. long based on the rating. The weights shown are without the cable. All electrode holders should be fully insulated. It is desirable to select the lightest-weight holder that will accommodate the required electrode size to be used.

Electrode holders—specifically, the insulators—deteriorate rapidly since they are so close to the arc and exposed to high heat. It is extremely important to maintain electrode holders so that they retain their current-carrying efficiency and their insulating qualities. Manufacturers supply spare parts so that the holders can be rebuilt and maintained for peak performance.

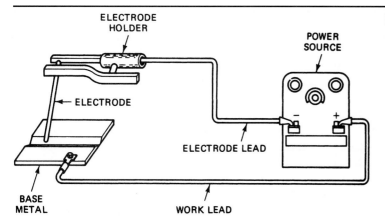

FIGURE 6–33 Circuit diagram for SMAW.

Electrode Holder Classification	Rating		Maximum Electrode Size (in.)	Maximum Cable Size (AWG)	Nominal Weight (oz.)
	Maximum Current (A)	Duty Cycle (%)			
Small	100	50	$\frac{1}{8}$	1	10–12
	200	50	$\frac{5}{32}$	1/0	10–14
Medium	300	60	$\frac{7}{32}$	2/0	12–20
Large	400	60	$\frac{1}{4}$	3/0	16–26
Extra large	500	75	$\frac{5}{16}$	4/0	22–30
Super extra large	600	75	$\frac{3}{8}$	4/0	28–36

FIGURE 6–35 Size and capacity of electrode holders.

Covered Electrodes

The covered electrode is the only item of material normally required. The selection of the covered electrode for specific work is based on the electrode usability and the composition and properties of the deposited weld metal. In order to properly select an electrode, it is necessary to understand the function of the coating, the basis of specifying, the usability factors, and the deposited weld metal properties.

The coating on the electrode provides (1) gas from the decomposition of certain ingredients of the coating to shield the arc from the atmosphere, (2) the deoxidizers for scavenging and purifying the deposited weld metal, (3) slag formers to protect the deposited weld metal with a slag from atmospheric oxidation, (4) ionizing elements to make the arc more stable and to operate with alternating current, (5) alloying elements to provide special characteristics to the deposited weld metal, and (6) iron powder to improve productivity of the electrode. Before the late 1920s, bare or lightly coated electrodes were used. They were more difficult to use and did not produce high-quality weld metal, so they are no longer used.

Welding electrodes have been divided into categories based on composition. More information about electrodes—manufacturing, storage, and so on—is given in Chapter 13.

The American Welding Society has established a system for identifying and specifying the different types of electrodes[12,13] (Figure 6–36). The mild steel and low-alloy steel-covered electrodes are prefixed by the letter *E*, followed by a four- or five-digit number. The prefix *E* means *electrode*.

The first two (or three) digits indicate the tensile strength in thousand pounds per square inch (KSI) of the deposited weld metal. The third (or fourth) digit indicates three things: the position in which the electrodes are useable, the type of covering, and the welding current type. The third digit indicates positional usability with No. 1— meaning all positions, flat, horizontal, vertical, and over-

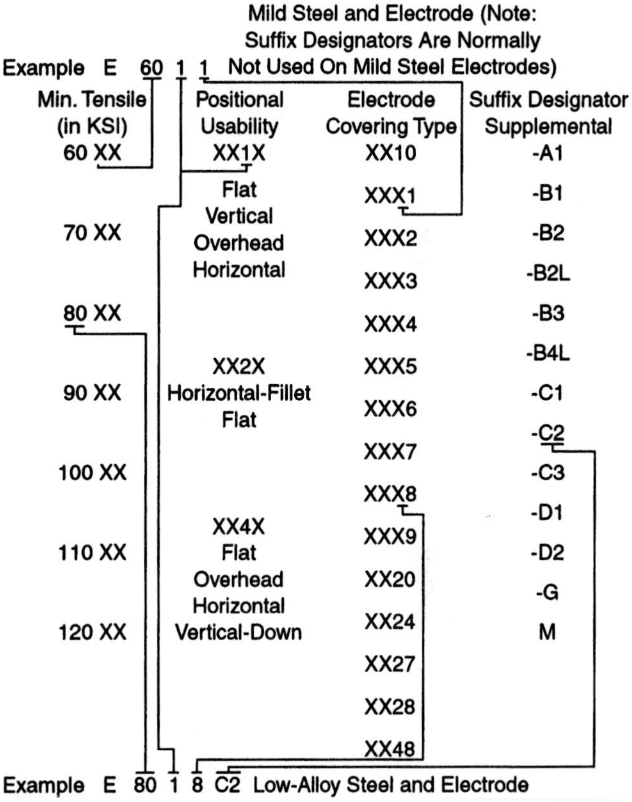

FIGURE 6–36 AWS system for identifying covered electrodes.

head—and No. 2—meaning horizontal fillet and flat. There is no No. 3 positional usability. The No. 4 means all positions and vertical down. The fourth (or fifth) digit is an indication of the electrode covering type. This also relates to the type of welding current that can be employed. The exact meaning of each code number is given in Figure 6–37. Note that when the fourth (or fifth) digit is zero, the type of coating and the current to use is determined by the third digit. For example, E6010 indicates

FIGURE 6–37 Details of the classification system for mild and low-alloy steel electrodes.

First Two or Three Digits Indicates Tensile Strength and Other Mechanical Properties					
AWS Classification[a]	Minimum Tensile Strength		Minimum Yield Strength		Minimum Elongation (%)
	ksi	MPa	ksi	MPa	
E60XX					17
E70XX	70	450	57	390	22, 25
E80XX	80	550	67	460	16, 19, 24
E90XX	90	620	77	530	14, 17, 24
E100XX	100	690	87	600	13, 16, 20
E110XX	110	760	97	670	15, 20
E120XX	120	830	107	740	14, 18

[a]110XX and 120XX refer to the low-hydrogen type of coating only.

Third (or) Fourth Digit Indicates the Welding Position Usability				
Classification	Flat (F)	Horizontal (H)	Vertical (V)	Overhead (OH)
EXX1X	Yes	Yes	Yes	Yes
EXX2X	Yes	Fillet	No	No
EXX4X	Yes	Yes	Down	Yes

Electrode Covering Type and Current Used						
Classification						
AWS	ASME	Current	Welding—Type of Covering	Position	Arc Type	Approximate Iron Powder[a](%)
6010	F-3	DCEP	Cellulose-sodium	All	Digging	0–10
6011	F-3	Ac and DCEP	Cellulose-potassium	All	Digging	0
6012	F-2	Ac and DCEN	Rutile-sodium	All	Medium	0–10
6013	F-2	Ac and dc	Rutile-potassium	All	Light	0–10
6019	F-2	Ac and dc	Iron oxide rutile-potassium	All	Medium	0–10
6020	F-1	Ac and dc	High iron oxide	F, Hf	Medium	0
6022	F-1	Ac and dc	High iron oxide	F, H	Hi speed	0–10
6027	F-1	Ac and dc	Iron oxide-iron powder	F, Hf	Medium	50
7014	F-2	Ac and dc	Rutile-iron powder	All	Light	25–40
7015	F-4	DECP	Low hydrogen-sodium	All	Medium	0
7016	F-4	Ac and DCEP	Low hydrogen-potassium	All	Medium	0
7018	F-4	Ac and DCEP	Low hydrogen-potassium-iron powder	All	Medium	25–40
7018M	F-4	DCEP	Low hydrogen-iron powder	All	Medium	10–25
7024	F-1	Ac and dc	Rutile-iron powder	F, Hf	Light	50
7027	F-1	Ac and dc	Iron oxide-iron powder	F, Hf	Medium	50
7028	F-1	Ac and DCEP	Low hydrogen-potassium-iron powder	F, Hf	Medium	50
7048	F-4	Ac and DCEP	Low hydrogen-potassium-iron powder	All	Medium	25–40

[a]Iron powder percentage based on weight of the covering.

cellulose sodium coating and operates on DC electrode positive, while E6020 indicates a high iron oxide coating that operates on AC and DC, while an E7018 has an iron powder low hydrogen coating and operates on DC electrode positive or on AC. To identify electrodes, each is type marked or printed with the identifying number shown by Figure 6-38.

The fifth (or sixth) digit, which follows a dash, is known as a suffix designator supplemental number. It is usually a letter followed by a number. This designates the chemical composition of the undiluted weld metal produced by the electrode. It is used for low-alloy steel electrodes. The meaning of exact suffix designator supplemental numbers is given in Figure 16-1.

FIGURE 6–38 Printed type marking on electrodes.

Another, or optional, supplemental designation may be employed for low-hydrogen electrodes. This is shown by the letter *H*, followed by a number, usually 4, 8, or 16. This indicates the average value of H_2, not to exceed 100 grams of deposited metal. The number is then followed by the letter *R*, which indicates that the electrode meets the requirements of the absorbed moisture test for all low-hydrogen electrodes.

The mechanical properties of the deposited weld metal must equal or exceed those of the base metal. Weld metal must also have approximately the same composition and physical properties.

The base metal must be identified so that its mechanical properties and composition are known. If the base metal is mild steel, select any E60XX electrode because its deposited metal will overmatch mechanical properties.

The following will help in selecting the proper electrodes:

⁕ E6010 is an all-position electrode for deep penetrating digging arc. It produces good mechanical properties and is widely used for structural welding and pipe welding. It is operated DCEP.

⁕ E6011 is essentially the same as an E6010 but is designed for AC welding.

⁕ E6012 is an all-position electrode producing medium penetration but high productivity. It is designed for DCEN or AC. It has relatively poor impact properties.

⁕ E6013 is an all-position electrode for sheet metal welding. It has medium penetration and is used on AC or DCEN.

⁕ E6014 is similar to the E6013 but with iron powder additions to improve productivity. It is used for a drag technique of welding.

⁕ E6015 was the initial low-hydrogen electrode designed for DCEP and has medium penetration but excellent impact properties.

⁕ E6016 replaced the E6015 and is used for ac as well as dc.

⁕ E6018 is a low-hydrogen electrode that has largely replaced both the E6015 and E6016 electrodes. It has increased productivity due to the addition of iron powder. It has excellent impact properties and is used to weld many low-alloy and "hard to weld" steels.

⁕ E6024 is similar to the E6012 but has higher productivity due to the addition of iron oxide. It is restricted to the flat or horizontal fillet position.

⁕ E6028 is similar to the E6018 but with a much higher deposition rate due to an increased amount of iron powder. It is used in the flat position and for horizontal fillet.

The abridged specifications for carbon steel-covered arc welding electrodes are shown in Figure 6–39. This provides the AWS classification, the radiographic standard, the mechanical properties, and also the minimum V-notch impact requirements for the E60XX and E70XX class of electrodes. The radiographic standard is related to the allowable porosity limits as shown by the specification. The mechanical properties are the minimum tensile strength in thousand pounds per square inch or MPa, the minimum yield point in thousand pounds per square inch, and the minimum elongation in 2 in. in percent. The V-notch impact requirements are a minimum at two temperatures, either 0° F (−18° C) or −20° F (−29° C). There are other requirements in this specification, and for this reason the specification must be referred to for details.

When welding any material other than low-carbon mild steel, special covered electrodes should be used. The welding society provides specifications for low-alloy steels and special composition steels. These are covered in "Specifications for Low-Alloy Steel-Covered Arc Welding Electrodes."[13] The electrode classification numbers include suffix digits, which consist of a letter followed by a single number and sometimes followed by another letter. These describe the analysis of the deposit weld metal and help make the selection of the electrode to match the composition of low-alloy base metal. These suffixes are

AWS Classification	Minimum Tensile Strength		Minimum Yield Strength at 0.2% Offset		Elongation (%)	Radiographic Standard AWS Grade	Minimum V-Notch Impact[a] (ft-lb)
	ksi	MPa	ksi	MPa			
E6010	62	430	50	340	22	2	20 at −20° F
E6011	62	430	50	340	22	2	20 at −20° F
E6012	67	460	55	380	17	Not required	Not required
E6013	67	460	55	380	17	2	Not required
E6020	62	430	50	340	20	1	Not required
E6022	67	460	Not required		Not required	Not required	Not required
E6027	62	430	50	340	20	2	20 at −20° F
E7014	72	500	60	420	17	2	Not required
E7015	72	500	60	420	22	1	20 at −20° F
E7016	72	500	60	420	22	1	20 at −20° F
E7018	72	500	60	420	22	1	20 at −20° F
E7024	72	500	60	420	17	2	Not required
E7027	72	500	60	420	22	2	20 at −20° F
E7028	72	500	60	420	22	2	20 at −20° F
E7048	72	500	60	420	22	1	20 at −20° F

[a] 20 ft-lb at −20° F = 27 joules at −29° C and 20 ft-lb at 0°F = 27 joules at −18°C.

FIGURE 6–39 Abridged specifications for mild steel-covered electrodes.

explained in Figure 16–1, which includes much data for welding the alloy steels.

Welding Position Electrodes are designed to be used in specific positions. The third (or fourth) digit of the electrode classification indicates the welding position that can be used. Match the electrode to the welding position to be encountered.

Welding Current Some electrodes are designed to operate best with direct current (dc), and others operate best with alternating current (ac). Some will operate on either. The last two digits together indicate welding current usability. Select the electrode to match the type of power source to be used.

Joint Design and Fitup Welding electrodes are designed with a digging, medium, or soft arc for deep, medium, or light penetration. The last two digits of the classification taken together also indicate this factor. Deep penetrating electrodes with a digging arc should be used when edges are not beveled or fitup is tight, but light penetrating electrodes with a soft arc are required when welding on thin material or when root openings are too wide.

Service Conditions or Specifications For weldments subject to severe service conditions, such as low temperature, high temperature, or shock loading, select the electrode that matches base metal composition, ductility, and impact resistance properties. The low-hydrogen types should be used.

Production Efficiency and Job Conditions Some electrodes are designed for high deposition rates but may be used only under specific position requirements. If they can be used, select the high-iron-powder types: EXX24, 27, 28, or 48. Other conditions may be present, which will require experimentation to determine the most efficient electrode.

For the usability of covered electrodes, the mild steel electrodes are classified into four general groups.

- ☀ F-1, high-deposition group: iron powder types
- ☀ F-2, mild penetration group: rutile (titania) types
- ☀ F-3, deep penetration group: high-cellulose types
- ☀ F-4, low-hydrogen group: lime types

Figure 6–40 is a guide to selecting the covered electrode for specific welding jobs based on the welding position, metal thickness, and weld type.

Electrodes in the same grouping operate and are run the same way. These *F* numbers correspond to the classification system used in Section IX of the ASME Pressure Vessel Boiler Code.

Deposition Rates

The melting rate of the electrode is related to the welding current. A portion of the arc energy is used to melt the surface of the base metal and a portion to melt the electrode. The electrode coating also affects deposition rates. The iron oxide types and iron powder types have higher deposition rates.

Welding Position / Weld Type	Fillet Welds — Inside or Outside				Groove Welds					
					Square		Vee (Open Root)		U	
Material Thickness →	Very Thin	Thin	Medium	Thick	Very Thin	Thin	Medium	Thick	Medium	Thick
1 FLAT	F-2	F-2	F-1	F-1 F-4	F-2	F-3	F-3	F-3 F-4	F-4	F-4
1A HORIZ FILLET	F-2	F-3	F-1	F-1 F-4	—	—	—	—	—	—
2 HORIZ	F-2	F-3	F-3 F-4	F-3 F-4	F-2	F-2 or F-3	F-3 and F-4	F-3 F-4	F-4	F-4
3 VERT UP	F-2	F-3	F-4	F-4	F-2	F-2 or F-3	F-3 and F-4	F-3 and F-4	F-4	F-4
3A VERT DOWN	F-2	F-3	—	—	F-2	F-2 or F-3	F-3	F-3	F-3	F-3
4 OVER-HEAD	F-2	F-3	F-3 F-4	F-3 F-4	F-2	F-2 or F-3	F-3 and F-4	F-3 and F-4	F-4	F-4
5 PIPE FIXED DOWNHILL	—	—	—	—	F-2	F-2	F-3	F-3	F-3	F-3
5A PIPE FIXED UPHILL	—	—	—	—	F-3	F-3	F-3 and F-4	F-3 and F-4	F-4	F-4

Material Thickness: Very thin = .005″ to .063″ (.125-1.6 mm),
Thin = $\frac{1}{16}$″ – $\frac{1}{4}$″ (1.6-6.3 mm), Medium = $\frac{1}{4}$″ – $\frac{3}{4}$″ (6.3-19 mm),
Thickness = $\frac{3}{4}$″-up (19 mm-up)

FIGURE 6–40 Usability rating guide for selecting mild- and low-alloy electrodes.

The melting rate to current is a fairly direct relationship (Figure 6-41). With higher current, the current density in the electrode increases and this increases the melting rate, which increases the deposition rate. Electrode size is determined by the job, the welding position, the joint detail, and the skill of the welder.

Quality of Welds

The quality of the weld depends on the design of the joint, the electrode, the technique, and the skill of the welder. If joint details are varied greatly from established design details, lower quality can result. The fitup of joints must match the design. Some electrodes deposit higher-quality weld metal than others, based on their specifications. See Figure 6-42 for weld appearance and variables.

Weld Schedules

Weld schedules are tables of operating parameters that will provide high-quality welds under normal conditions. Strict welding schedules are not as important for manual

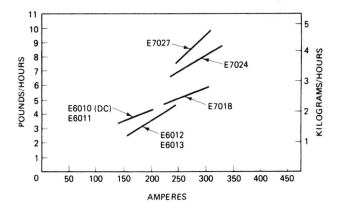

FIGURE 6–41 Deposition rates for various electrodes.

shielded metal arc welding as for semiautomatic and automatic welding for several reasons. First, in manual welding the welder controls welding conditions more by the manipulation of the arc than in any of the other arc welding processes. The welder directly controls the arc voltage and travel speed and indirectly the welding current.

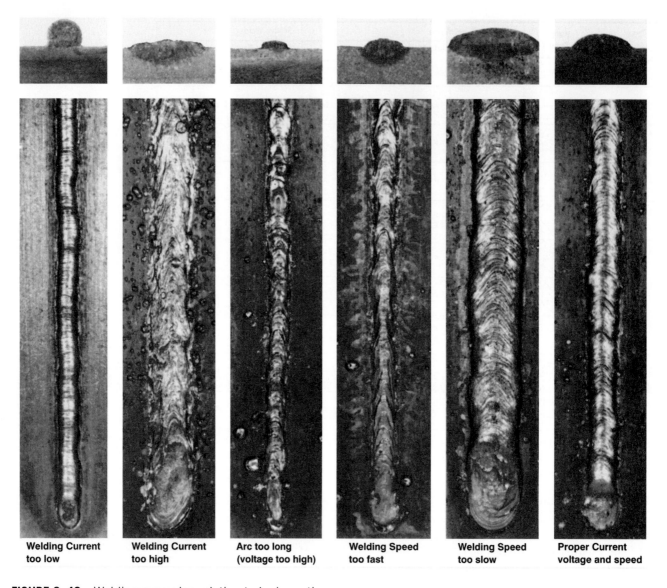

Welding Current too low **Welding Current too high** **Arc too long (voltage too high)** **Welding Speed too fast** **Welding Speed too slow** **Proper Current voltage and speed**

FIGURE 6–42 Welding examples relating to bad practices.

Second, in shielded metal arc welding, meter readings are rarely used for the duplication of jobs. It is felt that recommended welding conditions for the different types of electrodes are sufficient for most operations (Figure 6-43). However, when more complete information is needed, see the data provided in Figure 6-44. The settings given in these tables are not necessarily the only welding settings that can be used under every condition. For high-production work, the current settings could be increased considerably over those shown. Such factors as weld appearance, welder skill, and quality level will allow variations from the settings. As the requirements of a new application become better known, the settings can be adjusted to obtain optimum welding conditions. Trials are required prior to the establishment of a firm procedure for specific applications.

These tables are based on welding low-carbon mild steels under normal conditions and show the suggested electrode types for different weld types. Other electrode types and joints may be used. They are made in a consistent manner and are based on the type of welds and the position of welding, and are made by a welder of normal skill.

Tips for Using the Process

There is a definite relationship between the welding current, the size of the welding electrodes, and the welding position. These must be selected so that the welder has the molten weld pool under complete control at all times. If the pool becomes too large, it becomes unmanageable and molten metal may run out of the pool, particularly in out-of-position welding.

Type of Electrode	Size		Direct Current		Alternating Current	
	mm	in.	A	V	A	V
E–6010 F–3	2.38	$\frac{3}{32} \times 14$	40–80	24–26	—	—
All positions	3.18	$\frac{1}{8} \times 14$	70–130	24–26	—	—
Deep penetration	3.97	$\frac{5}{32} \times 14$	110–165	21–23	—	—
DCEP	4.76	$\frac{3}{16} \times 14$	140–225	20–26	—	—
	5.56	$\frac{7}{32} \times 18$	160–300	20–32	—	—
	6.35	$\frac{1}{4} \times 18$	200–400	24–34	—	—
E–6011 F–3	2.38	$\frac{3}{32} \times 14$	50–70	21–26	50–70	25–29
All positions	3.18	$\frac{1}{8} \times 14$	75–130	20–30	75–130	25–28
Deep penetration	3.97	$\frac{5}{32} \times 14$	120–160	16–26	120–160	24–29
AC or DCEP	4.76	$\frac{3}{16} \times 14$	150–190	20–28	150–190	24–33
	5.56	$\frac{7}{32} \times 18$	180–250	22–28	180–250	25–35
	6.35	$\frac{1}{4} \times 18$	200–300	22–28	200–300	30–40
E–6012 F–2	2.38	$\frac{3}{32} \times 14$	50–90	16–26	80–120	17–23
All positions	3.18	$\frac{1}{8} \times 14$	76–135	16–22	80–120	17–23
Medium penetration	3.97	$\frac{5}{32} \times 14$	120–205	16–24	120–190	17–25
AC or DCEN	4.76	$\frac{3}{16} \times 14$	140–255	17–26	140–240	17–28
	5.56	$\frac{7}{32} \times 18$	220–335	12–28	210–330	18–28
	6.35	$\frac{1}{4} \times 18$	200–400	16–25	220–350	18–28
E–6013 F–2	2.38	$\frac{3}{32} \times 14$	50–100	21–24	50–80	20–27
All positions	3.18	$\frac{1}{8} \times 14$	80–140	18–20	80–120	22–24
Light penetration	3.97	$\frac{5}{32} \times 14$	120–190	20–25	120–170	22–26
AC or DC	4.76	$\frac{3}{16} \times 14$	160–220	25–28	190–220	20–22
	5.56	$\frac{7}{32} \times 18$	240–270	25–29	200–240	20–22
	6.35	$\frac{1}{4} \times 18$	270–350	18–21	270–350	20–22
	7.94	$\frac{5}{16} \times 18$	320–420	23–26	320–420	24–30
E–6020 F–1	2.38	$\frac{1}{8} \times 14$	120–145	20–22	120–145	20–22
Flat and HF positions	3.18	$\frac{5}{32} \times 14$	150–175	24–26	150–175	24–26
Medium penetration	3.97	$\frac{3}{16} \times 18$	210–240	23–27	210–240	23–27
High deposit rate	4.76	$\frac{7}{32} \times 18$	240–275	24–28	240–275	24–28
AC or DC	5.56	$\frac{1}{4} \times 18$	290–320	29–30	290–320	29–30
E–7010			Same as E–6010 above			
E–7014 F–2	2.38	$\frac{3}{32} \times 14$	70–90	20–24	70–90	20–24
All positions	3.18	$\frac{1}{8} \times 14$	120–145	14–17	120–45	14–17
Light penetration	3.97	$\frac{5}{32} \times 14$	140–250	17–19	140–210	17–19
AC or DC	4.76	$\frac{3}{16} \times 18$	180–280	20–24	180–280	20–24
	5.56	$\frac{7}{32} \times 18$	250–375	28–35	250–375	28–35
	6.35	$\frac{1}{4} \times 18$	300–420	26–31	300–420	26–31
	7.94	$\frac{5}{16} \times 18$	375–500	28–33	375–500	28–33
E–7016 F–4	2.38	$\frac{3}{32} \times 14$	70–100	17–21	70–100	20–25
All positions	3.18	$\frac{1}{8} \times 14$	80–130	17–22	40–130	20–22
Medium penetration	3.97	$\frac{5}{32} \times 14$	120–170	18–19	120–170	18–20
AC or DCEP	4.76	$\frac{3}{16} \times 14$–18	170–250	17–22	170–250	17–23
	5.56	$\frac{7}{32} \times 18$	250–325	18–24	180–260	18–24
	6.35	$\frac{1}{4} \times 18$	300–350	21–25	200–275	19–24
	7.94	$\frac{5}{16} \times 18$	235–375	22–26	220–300	20–25

FIGURE 6–43   Recommended welding conditions for covered electrodes.

Type of Electrode	Size		Direct Current		Alternating Current	
	mm	in.	A	V	A	V
E–7018 F–4	2.38	$\frac{3}{32} \times 14$	80–110	20–22	80–110	24–26
All positions	3.18	$\frac{1}{8} \times 14$	90–150	20–21	90–150	19–23
Medium penetration	3.97	$\frac{5}{32} \times 14$	110–230	20–22	110–230	20–25
AC or DCEP	4.76	$\frac{3}{16} \times 14$	150–300	20–22	150–300	21–28
	5.56	$\frac{7}{32} \times 18$	250–350	20–24	250–350	20–29
	6.35	$\frac{1}{4} \times 18$	300–400	20–24	300–400	24–30
	7.94	$\frac{5}{16} \times 18$	325–375	21–25	320–450	25–30
E–7024 F–1	2.38	$\frac{3}{32} \times 12$	100–140	27–32	100–140	27–32
Flat and HF positions	3.18	$\frac{1}{8} \times 14$	130–180	27–30	130–180	27–30
Light penetration	3.97	$\frac{5}{32} \times 14$	180–240	30–33	180–240	30–33
High deposition rate	4.76	$\frac{3}{16} \times 14$	200–280	22–28	200–280	22–28
AC or DC	5.56	$\frac{7}{32} \times 18$	250–375	29–32	250–375	29–32
	6.35	$\frac{1}{4} \times 18$	300–420	28–34	300–420	28–34
	7.94	$\frac{5}{16} \times 18$	425–500	29–35	425–500	29–35
E–7028 F–1	2.38	$\frac{1}{8} \times 14$	—	—	—	—
Flat and HF positions	3.18	$\frac{5}{32} \times 14$	—	—	—	—
Medium penetration	3.97	$\frac{3}{16} \times 14$	240–300	31–33	260–320	30–33
High deposition rate	4.76	$\frac{7}{32} \times 18$	300–400	36–40	320–400	30–35
AC or DCEP	5.56	$\frac{1}{4} \times 18$	350–450	37–41	370–470	31–37
	7.94	$\frac{5}{16} \times 18$	450–550	38–42	400–500	32–38
E-8018			Same as E-7018 above			

Note: Voltage is based on a normal arc length measured at the arc.

FIGURE 6–43 *(cont.)*

The welder should maintain the steady frying and crackling sound that comes with correct procedures. The shape of the molten pool and the movement of the metal at the rear of the pool serve as a guide in checking weld quality. The ripples produced on the bead should be uniform, and the bead should be smooth with no overlap or undercut. The following seven factors are essential for maintaining high-quality welding.

1. *Correct electrode type.* It is important to select the proper electrode for each job.
2. *Correct electrode size.* Electrode size choice involves the type of electrode, welding position, joint preparation, weld size, welding current, the thickness of the base metal, and the skill of the welder.
3. *Correct current.* If the current is too high, the electrode melts too fast and the molten pool is large, irregular, and hard to control. If the current is too low, there is not enough heat to melt the base metal and the molten pool will be too small and will pile up and be irregular (Figure 6–45a).
4. *Correct arc length.* If the arc is too long, the metal melts off the electrode in large globules that wobble from side to side, giving a wide, spattered, and irregular bead with poor fusion to the base metal. It may also result in porosity, especially with low-hydrogen electrodes. If the arc is too short, there is not sufficient heat in the arc at the start to melt the base metal sufficiently and the electrode may stick to the work.
5. *Correct travel speed.* When the speed is too fast, the weld pool freezes too quickly. Impurities and gases are not allowed to be released. The bead is narrow and the ripples pointed. When the speed is too slow, the metal piles up and the bead is too high and wide with a rather straight ripple (Figure 6–45b). The factors: correct current, correct arc length (or voltage), and correct travel speed all relate to heat input.
6. *Correct electrode angle.* The electrode angle is important, particularly in fillet welding and in deep groove welding. When making a fillet, the electrode should be held so that it bisects the angle between the plates and is perpendicular to the line of the weld. When undercut occurs in the vertical member, lower the angle and direct the arc toward the vertical member.
7. *Correct manipulation pattern.* Different manipulation patterns are used for different types of electrodes, different weld designs, and different welding positions. Knowledge of the different patterns is learned in a good welding training program.

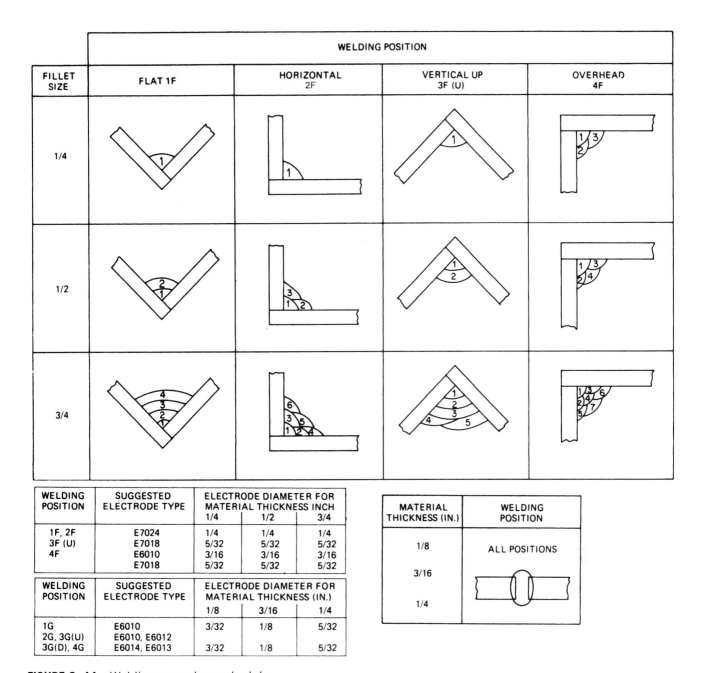

FIGURE 6–44 Welding procedure schedule.

Breaking the Arc

If the weld is to be continued, the crater should remain, and the arc should be quickly reestablished. If it is the end of the weld, the arc should not be broken until travel has stopped momentarily to allow the crater to fill.

When weaving, the width of the weave and the pause at the end of the weave and other movements are important. The welder must pause at each end of the weave to allow for complete fusion into the side. The welder should quickly move across the center of the weld since heating is more concentrated in the center than at the edges. The width of the weave for low-hy-drogen electrodes should not exceed $2\frac{1}{2}$ times the core wire diameter. For other electrode types, this can be doubled.

Safety Considerations

The safety factors and potential hazards involved with shielded metal arc welding were discussed in detail in Chapter 4. This information is so important and needs to be repeated. Therefore it is summarized and shown in Figure 6–46, which shows the warning label for arc welding processes and equipment and is on every container of filler metal.

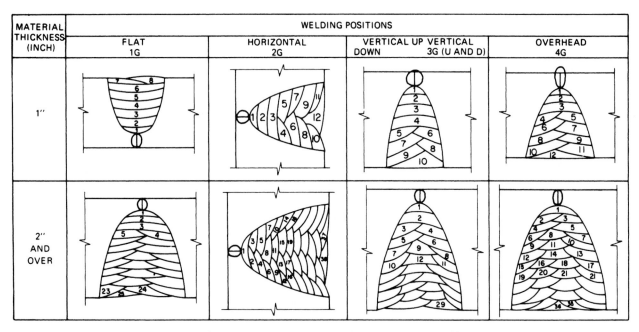

WELDING POSITION	SUGGESTED ELECTRODE TYPE	ELECTRODE DIAMETER FOR MATERIAL THICKNESS (IN.)		
		3/8	1/2	5/8
1G	E6010	3/16	3/16	3/16
2G	E6010	3/16	3/16	3/16
	E7018	5/32	5/32	5/32
3G (U)	E6010	5/32	5/32	5/32
3G (D), 4G	E7018	5/32	5/32	5/32

WELDING POSITION	SUGGESTED ELECTRODE TYPE	ELECTRODE DIAMETER FOR MATERIAL THICKNESS (IN.)	
		1	2
1G	E7018	1/4	1/4
2G	E7018	5/32	5/32
3G (U)	E6010	3/16	3/16
3G (D), 4G	E7018	5/32	5/32

FIGURE 6–44 (cont.)

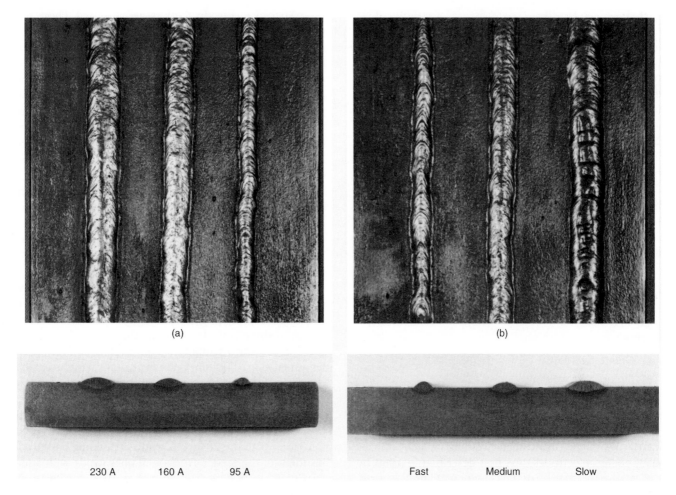

| 230 A | 160 A | 95 A | | Fast | Medium | Slow |

FIGURE 6–45 Welding variable travel speed and current using E7024 covered electrode: (a) constant travel speed, changing current; (b) all beads 160A, changing travel speed.

FIGURE 6–46 Warning label for arc welding processes and equipment.

WARNING: PROTECT yourself and others. Read and understand this label.

FUMES AND GASES can be dangerous to your health.
ARC RAYS can injure eyes and burn skin.
ELECTRIC SHOCK can KILL.

- Before use, read and understand the manufacturer's instructions, Material Safety Data Sheets (MSDSs), and your employer's safety practices.
- Keep your head out of the fumes.
- Use enough ventilation, exhaust at the arc, or both, to keep fumes and gases from your breathing zone and the general area.
- Wear correct eye, ear, and body protection.
- Do not touch live electrical parts.
- See American National Standard Z49.1, *Safety in Welding and Cutting,* published by the American Welding Society, 550 N.W. LeJeune Rd., P.O. Box 351040, Miami, Florida 33135; OSHA Safety and Health Standards, 29 CFR 1910, available from U.S. Government Printing Office, Washington, D.C. 20402.

DO NOT REMOVE THIS LABEL

Limitations of the Process

The major limitation of the shielded metal arc process is the *built-in break*. Whenever an electrode is consumed to within 2 in. (50 mm) of its original length, the welder must stop. Welding cannot continue because the bare portion of the electrode in the electrode holder should not be used. The welder must stop, chip slag, remove the electrode stub, and place a new electrode in the holder. This occurs many times during the workday and is controlled by the size and length of the electrode. This prohibits the welder from attaining an operator factor or duty cycle much greater than 25%.

Another limitation is the filler metal utilization. The electrode stub loss and the coating loss allows for a total utilization of covered electrode of approximately 65%.

Variations of the Process

There are four variations of the shielded metal arc welding process:

✳ Gravity welding
✳ Firecracker welding
✳ Massive electrode welding
✳ Arc spot welding

Gravity Welding Gravity-feed welding, which utilizes heavy-coated electrodes, was first described in 1938 by K. K. Madsen of Denmark.[14] Gravity welding is considered an automatic method of applying the shielded metal arc welding process. It utilizes a low-cost mechanism that includes an electrode holder attached to a bracket that slides down an inclined bar arranged along the line of the weld. Heavy-coated electrodes are maintained in contact with the workpiece by the weight of the electrode holder bracket and the electrode. Once the process is started, it continues automatically until the electrode is consumed. When the electrode has burned to a short stub, the bracket and electrode holder are automatically kicked up to break the arc.

This welding method received considerable publicity in the early 1960s, based on work done in Japanese shipyards. Credit should be given to the Japanese shipbuilders for perfecting and utilizing the gravity welding system on a large scale.

The gravity welding system provides welding economies over manual welding since the operator can use a number of the gravity feeders simultaneously. This increases the productivity, reduces welder fatigue, minimizes operator training, and provides substantial savings in welding labor cost. Initially, electrodes of standard length 18 in. (550 mm) were used; however, the Japanese produced extra-long electrodes to make the process more productive. The most common length is now 28 in. (800 mm). This reduces the stub end loss ratio.

Special electrode feeders are required. There are two types of feeders. One is the tripod or gravity feeder, which has an electrode holder mounted on a bracket or carriage that slides down an incline (Figure 6–47). It is used to produce horizontal fillet welds. The second type is a spring-loaded holder, used for making flat-position groove welds, as well as horizontal fillet welds (Figure 6–48). The tripod type of feeder is the most popular since it provides a more uniform weld throughout its entire length. The spring-loaded feeder has the advantage of being usable in less accessible spaces.

The power source for gravity welding is a conventional constant-current welding machine. Gravity welding can attain a 90% duty cycle and currents of up to 400 A can be used. The conventional 60% duty cycle welding power source must be derated to allow for the 90% duty cycle. Both ac an dc power sources are used.

Heavy-coated welding electrodes are used for gravity welding. The E6027 and E7024 types are most often used; however, the E7028 type can also be used. The most common size electrodes used are the $\frac{7}{32}$-in. (5.6-mm) and the $\frac{1}{4}$-in. (6.4-mm) diameters. The most commonly used length is 28 in. (800 mm); however, other lengths can be used. The size and length of fillet welds produced by these electrodes can be varied with the gravity feeders based on changing the angle of the inclined track. It is possible to obtain welds from 20 to 40 in. (500 to 1000 mm) long with the 28-in. electrode, depending on the adjustment of the feeder. Horizontal fillet welds with a leg length of $\frac{7}{32}$ in. (5.6 mm) to $\frac{3}{8}$ in. (9.5 mm) can be obtained. The properties of the weld metal deposited when using the gravity feed system are the same as when electrodes are used manually.

There are two variations of the tripod gravity feeder. One is preset and allows no adjustment of the incline of the track, whereas the other type allows the tripod legs to be adjusted. The preset feeder will produce welds in lengths that vary from approximately 27 to 31 in. (675 to 775 mm) long, using the 28-in. (800-mm) electrode. Figure 6–49 shows the welding procedure when using the preset type of gravity feeder. With two different sizes of electrodes available, two different sizes of fillet welds can be obtained. The electrode type and size and the welding current will determine the specific weld fillet size and weld length. Two different levels of current are used. The higher current will produce the larger fillet size.

The second variation of the tripod gravity feeder provides for leg-length adjustments. This provides a different angle of the inclined track. With this type of feeder, a variety of fillet sizes can be produced with the same electrode size operating at the same welding current. The fillet weld size is determined by the electrode diameter and by the weld-length setting used. The weld length may be varied depending on angle adjustments, which are established between the axis of the weld, the centerline of the electrode, and the inclined track of the gravity feeder.

FIGURE 6–47 Gravity feeders used for shipbuilding.

FIGURE 6–48 Spring-loaded feeder.

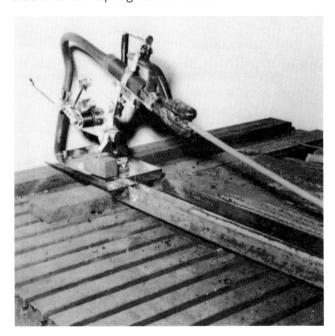

The actual range of weld sizes obtainable will depend on the specific feeder being used.

The spring-loaded feeder produces welds 24 to 27 in. (600 to 675 mm) long using a 28-in. (800-mm)-long electrode. A normal stub length of 2 in. (50 mm) results. The weld size is established by the electrode diameter. Figure 6–49 shows the fillet weld size and is related to the electrode size and welding current.

When using the spring-loaded feeders, there is a small size variation of the fillet from the beginning of the weld to the end of the weld. There is also a change in penetration from start to end. This is due to the change in the approach angle that occurs during the melting of the electrode.

When direct current is used, the electrode is negative (straight polarity). When alternating current is used, the change in bead smoothness and spatter level is less dependent on welding direction related to the work lead connection.

The economic advantage depends on one welding operator utilizing two or more feeders simultaneously. It is the labor cost that is largely involved in using more than

Feeder	Fillet Size (in.)	Electrode	Electrode Size (in.)	Weld Length (in.)	AC Weld Current
Preset	$\frac{9}{32}$	E6026	$\frac{7}{32}$	27	260
gravity	$\frac{11}{32}$	E6027	$\frac{1}{4}$	27	290
type	$\frac{1}{4}$	E6027	$\frac{7}{32}$	31	260
(fixed	$\frac{5}{16}$	E6027	$\frac{1}{4}$	31	290
legs)	$\frac{9}{32}$	E7024	$\frac{7}{32}$	27	290
	$\frac{11}{32}$	E7024	$\frac{1}{4}$	27	350
	$\frac{1}{4}$	E7024	$\frac{7}{32}$	31	290
	$\frac{5}{16}$	E7024	$\frac{1}{4}$	31	350
	$\frac{1}{4}$	E6027	$\frac{7}{32}$	27	240
Spring-	$\frac{5}{16}$	E6027	$\frac{1}{4}$	27	270
loaded	$\frac{1}{4}$	E7024	$\frac{7}{32}$	27	270
type	$\frac{5}{16}$	E7024	$\frac{1}{4}$	27	300

FIGURE 6–49 Weld procedures for gravity welding.

one feeder. When using long electrodes in gravity feeders, the welding current is less than normally used for manual welding with standard length electrodes. It is necessary to use at least two gravity feeders to obtain an economic advantage. Additional cost reduction is possible by using three or four automatic feeders. The most economical operation is to have a sufficient number of feeders in a small area so that the operator can move from one to another and reload the holder and reestablish the arc of one feeder while all of the other feeders are welding. Figure 6–50 shows a comparison of the deposition rate based on pounds per hour when using one electrode manually versus two, then three, four, or five gravity feeders. Gravity welding is becoming less popular due to more widespread use of semiautomatic welding using flux cored electrodes.

Firecracker Welding Firecracker welding is a variation that uses a length of covered electrode placed along the joint in contact with the workpiece. During welding the stationary electrode is consumed as the arc travels the length of the electrode. This method was developed in the late 1930s by George Hafergut of the Elin Company of Austria and is known as the Elin-Hafergut welding

method (U.S. Patent 2,269,369). It is considered an automatic method of application since human involvement is not required after the arc is initiated. It can be used for making square groove butt welds in materials from 0.030 in. (0.8 mm) to 0.120 in. (3 mm), for making full fillet lap welds in materials of similar thicknesses, and for making fillet welds from $\frac{3}{16}$ in. (5 mm) and heavier. AWS types E6024 and E7028 can be used.

To make a firecracker fillet weld, the work is positioned so the weld is in the flat position. The welding electrode is placed in the joint and a retaining bar is placed over it. The arc is started by shorting the end of the electrode to the work. The arc length depends on the thickness of the coating. As the arc travels along the electrode, the electrode melts and makes a deposit on the material immediately underneath it. Once the arc is started, it proceeds to completion automatically. Figure 6–51 shows the method being used for making butt welds and fillet welds. Electrodes up to 39 in. long (1000 mm) and with a 5-, 6-, 7-, or 8-mm core diameter have been used. Both alternating current and direct current are used, and there seems to be a preference for alternating current because of arc blow with direct current. The quality of the weld metal is equal to that produced by manual welding.

FIGURE 6–50 Comparison of deposition rates for gravity welding.

Method of Application to Make $\frac{5}{16}$ in. Fillet	Deposition Rate, E6027 (lb/hr)
Manual, one arc	10
Gravity, two arcs	17
Gravity, three arcs	26
Gravity, four arcs	34
Gravity, five arcs	43

FIGURE 6–51 Firecracker welding.

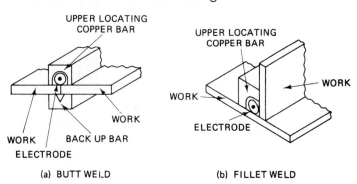

(a) BUTT WELD   (b) FILLET WELD

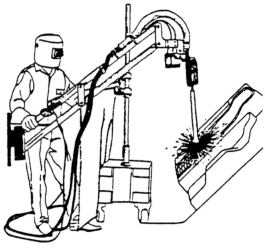

FIGURE 6–52 Using massive electrodes.

One operator can make several firecracker welds simultaneously. This method is little used today.

Massive Electrode Welding Massive electrode welding is sometimes called progressive manual casting. It is a variation that employs extremely large diameter and long electrodes. These electrodes are made for repairing forging dies, steel castings, or any large component and for the construction of large fabrications. The electrodes are so large and heavy that they require a manipulator to hold them and feed them into the weld pool. Figure 6-52 shows these large electrodes being used to repair a forging die. Up to 60 lb per hour of weld metal can be deposited with this welding variation.

The electrodes are mainly nickel–chromium–molybdenum alloys with hardnesses ranging from Rockwell C15 to 45, depending on the grade. Extremely high tensile strengths are achieved with proper preheat and postheat procedures. A 1500-A dc constant-current power source is necessary to weld with these large electrodes. Conventional power sources can be paralleled to obtain this welding current. Dc electrode positive (reverse polarity) is used. Figure 6-53 shows the deposition rate for these large electrodes. This method of welding is an excellent choice for large repairs and fabrication.

Arc Spot Welding Shielded metal arc welding can be used to make arc spot welds. Special spring-loaded feeders are used with small-diameter electrodes for arc spot welding thin sheet metal. This method is used by automotive body repair shops. Arc spot welding is discussed in detail in Chapter 26.

Industrial Use and Typical Applications

Typical applications of the shielded metal arc welding process are as varied and widespread as arc welding itself. Shielded metal arc welding will probably always be

FIGURE 6–53 Deposition rate chart for massive electrodes.

Electrode Size (×) Diameter Length (in.)	Current Range DCEP (A)	Approx. Pool Size (in^2)	Weld Metal Deposition Rate (lb/hr)
$\frac{5}{16}$ × 24	300–500	2–6	10
$\frac{3}{8}$ × 24	400–600	6–10	25
$\frac{1}{2}$ × 30	600–950	10–20	25
$\frac{5}{8}$ × 30	600–1,500	20–36	45
$\frac{3}{4}$ × 36	1,200–2,100	36–60	60

the mainstay for maintenance and repair welding; because welding is required at remote locations, the jobs are relatively small and each and every one is different. Shielded metal arc welding will also remain popular in small production shops where limited capital is available and where the amount of welding is relatively minor compared to other manufacturing operations.

6-4 GAS METAL ARC WELDING

Gas metal arc welding (GMAW) is "an arc welding process that uses an arc between a continuous filler metal electrode and the weld pool. The process is used with shielding from an externally supplied gas and without the application of pressure." It was developed in the late 1940s for welding aluminum and has become very popular. This process, also called *metal inert gas* (MIG) *welding*, is shown in Figure 6-54. The welder's view of the arc is shown in Figure 6-55. There are many variations depending on the type of shielding gas, the type of metal transfer, the type of metal welded, and so on. It has been given many names: for example, MIG welding, CO_2 welding, fine wire welding, spray arc welding, pulse arc welding, dip transfer welding, short-circuit arc welding, and various trade names.

Principles of Operation

The gas metal arc welding process, shown in Figure 6-56, utilizes the heat of an arc between a continuously fed consumable electrode and the work to be welded. The

FIGURE 6–54 Gas metal arc welding.

FIGURE 6–55 Welder's view of gas metal arc welding.

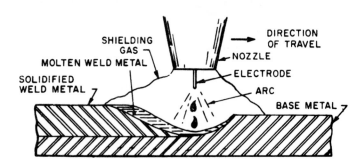

FIGURE 6–56 Process diagram for GMAW.

heat of the arc melts the surface of the base metal and the end of the electrode. The metal melted off the electrode is transferred across the arc to the molten pool. The molten weld metal, sometimes called the weld puddle, must be properly controlled to provide a high-quality weld. The depth of penetration is controlled by many factors, but the primary one is the welding current. If the depth of penetration is too great, the arc will burn through thinner material and reduce the quality of a weld. The width of the molten pool is also based on many factors, but the primary one is the travel speed. If the molten pool is too large, particularly when welding other than in the flat position, the molten metal will run out and create a welding problem. Many factors, including electrode size and the mode of metal transfer, relate to the weld pool size.

Shielding of the molten pool, the arc, and the surrounding area is provided by an envelope of gas fed through the nozzle. This shielding gas, which may be an inert gas, an active gas, or a mixture, surrounds the arc area to protect it from contamination from the atmosphere. The electrode is fed into the arc automatically, usually from a coil of wire. The arc is maintained automatically and travel and guidance can be handled manually or by machine.

The metal being welded dictates the composition of the electrode and the shielding gas. The shielding gas and the type and size of the electrode affect the mode of metal transfer. The metal transfer mode is one way of identifying the variation of the process.

Advantages and Major Uses

The GMAW process has become one of the most popular arc welding processes. There are 4 major variations of the process based on the mode of metal transfer. The early development of GMAW was for welding aluminum using inert gas for shielding. This utilized the spray mode of transfer, with argon shielding and a relatively large diameter electrode. This produced a smooth weld bead with a relatively small amount of spatter but utilized a large, sometimes uncontrollable molten weld pool.

For welding steels, inert gases were too expensive and an active gas, CO_2, was selected. This was based on the analysis of gases produced by disintegration of the coating of covered steel electrodes. CO_2 welding was adopted for welding mild steels in the flat position using relatively large [1/16 in. (1.6 mm)] electrode wires. The metal transfer was globular and the spatter was greater than desired. This did not become too popular with welders because of the high heat and high travel speed.

Efforts to refine this variation led to an all-position variation still using CO_2 gas shielding but with lower currents and smaller-diameter electrodes [0.35 in. (0.9 mm) and 0.45 in. (1.1 mm)]. This variation provides a short-circuiting mode of metal transfer and is called short-circuiting or dip transfer welding. Improvements were made by using shielding gas mixtures of argon and CO_2. This provided a smooth, nice-appearing weld surface that could be used to weld thin materials in all positions. The short-circuiting mode of welding should not be used on structural applications.

The spray transfer mode used originally was improved for welding steels. Adding up to 5% oxygen to argon produced a spray-type metal transfer and produced welds with an extremely smooth surface with minimum spatter. However, the weld pool was still large and difficult to control in out-of-flat-position welding. This led to development of the pulsed-spray mode, which utilizes pulses of current to produce small drops of metal across the arc. This variation depends on specially designed power sources. It provides a controlled weld pool for welding thin materials in any position. It produces a smooth weld and minimum spatter and has become very popular.

The major advantages of gas metal arc welding are:

※ High operator factor
※ High deposition rates
※ High utilization of filler metal
※ Elimination of slag and flux removal
※ Reduction in smoke and fumes
※ Lower skill level in a semiautomatic method of application than that required for manual shielded metal arc welding
※ Automation possible
※ Extremely versatile, with wide and broad application ability

Methods of Application and Position Capabilities

The most popular method of applying is the semiautomatic method, where the welder provides manual travel and guidance of a welding gun. Second is the fully automatic method, where the welding operation is automated. This process cannot be applied manually.

The gas metal arc welding process is an all-position process. However, each of the variations has its own position capabilities, depending on electrode size and metal transfer. The CO_2 welding variation, utilizing large electrode wires, is used primarily in the flat and horizontal fillet position. The spray arc variation is normally used in the flat and horizontal position. It can be used in the vertical and overhead positions if smaller electrodes are employed. The short-circuiting and pulsed variations can be used in all positions.

Weldable Metals and Thickness Range

The gas metal arc welding process can be used to weld most metals. Carbon dioxide welding is restricted to steels. Electrodes are matched to the base metals (Figure 6–57). The process can also be used for surfacing and for buildup using special metals for bearing surfaces, corrosion-resistant surfaces, and so on.

Metal thickness from 0.005 in. (0.13 mm) upward can be welded. The short-cutting variation and the pulsed arc variation are used for welding the thinner materials in all positions. Heavier thicknesses can be welded with large-wire CO_2 variation. Weld grooves and multiple-pass technique will allow welding of practically unlimited thickness (Figure 6–58). The extreme versatility of the process allows welding of the thinnest up to the thickest.

Joint Design

The gas metal arc welding process can utilize the same joint design details used for the shielded metal arc welding process. These joint details are given in Chapter 19. For maximum economy and efficiency, groove welds should be modified. The diameter of the electrodes employed by gas metal arc welding are smaller than those employed for shielded metal arc welding. Because of this,

Base Metal	Short-Circuiting Arc	Spray Arc	Globular (CO_2)	Pulsed Spray
Aluminum	No	Yes	No	Yes
Bronze	No	Yes	No	Yes
Copper	No	Yes	No	Yes
Copper–nickel	No	Yes	No	Yes
Cast iron	Yes	No	No	Yes
Magnesium	No	Yes	No	Yes
Inconel	No	Yes	No	Yes
Nickel	No	Yes	No	Yes
Monel	No	Yes	No	Yes
Low-carbon steel	Yes	Yes	Yes	Yes
Low-alloy steel	Yes	Yes	Yes	Yes
Medium-carbon steel	Yes	Yes	Yes	Yes
Stainless steel	Yes	Yes	No	Yes
Titanium	No	Yes	No	Yes

FIGURE 6–57 Metals weldable by GMAW.

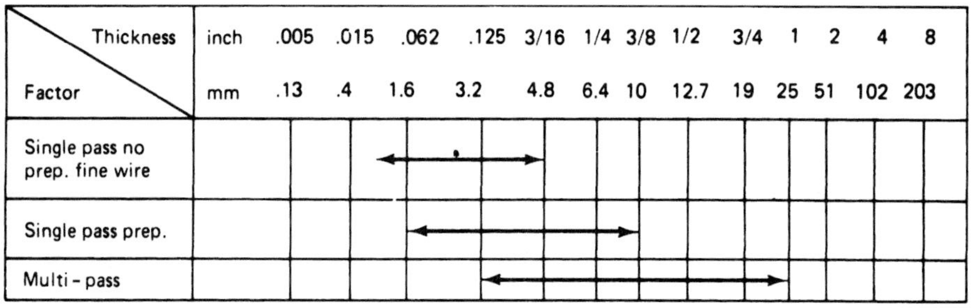

Thickness	inch	.005	.015	.062	.125	3/16	1/4	3/8	1/2	3/4	1	2	4	8
Factor	mm	.13	.4	1.6	3.2	4.8	6.4	10	12.7	19	25	51	102	203
Single pass no prep. fine wire				←	•	→								
Single pass prep.					←		→							
Multi-pass					←						→			

FIGURE 6–58 Base metal thickness range for GMAW.

the groove angles can be reduced (Figure 6–59). Reducing groove angles will still allow the electrode to be directed to the root of the weld joint so that complete penetration will occur.

The different variations require special attention concerning weld design. The CO_2 variation provides extremely deep penetrating qualities, and in designing fillet welds, the size of the fillet can be reduced at least one size when converting from shielded metal arc welding to CO_2 welding.

The variation using inert gas on nonferrous metals can use the standard joint details recommended for shielded metal arc welding, except that the groove angle should be reduced. The joint designs used for pipe welding with shielded metal arc welding or gas welding are normally used for gas metal arc welding.

Welding Circuit and Current

The welding circuit employed for gas metal arc welding (Figure 6–60) uses a wire feeder system that controls the electrode wire feed and welding arc, as well as the flow of shielding gas and cooling water. The power supply is normally the constant-voltage (CV) type. A gun or torch is used for directing the electrode and shielding gas to the arc area. A travel system is required for mechanical welding.

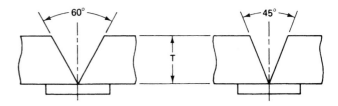

FIGURE 6–59 Weld joint design difference for GMAW.

The gas metal arc welding process uses direct current. Alternating current has not been successfully used. Direct current is normally used with the electrode positive DCEP (reverse polarity). Direct-current electrode negative DCEN (straight polarity) can be used with special emissive-coated electrode wires, which provide for better electron emissions. DCEN is rarely used because the emissive-coated electrodes are not popular.

The shorting arc variation became popular when the CV system of welding power was introduced. The CV system reduced the complexity of the wire feed control circuits and eliminated electrode burnback to the contact tip or stubbing to the work. It also provided positive arc starting.

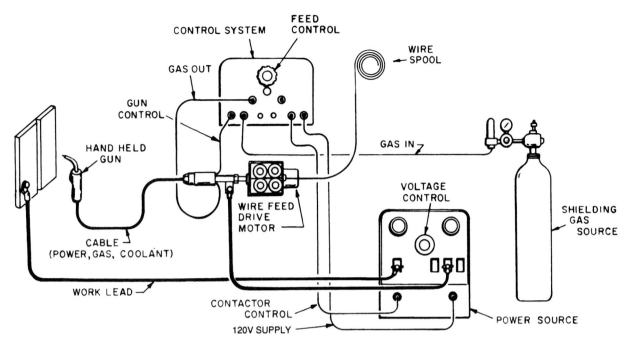

FIGURE 6–60 Circuit diagram for GMAW.

The pulsed-current variation requires a special power source which changes from a lower to a higher current at a programmed frequency. The welding current varies from as low as 20 A at a voltage of 18 V to as high as 750 A at an arc voltage of 50 V. This broad range of current and voltage encompasses all the variations.

Equipment Required

The equipment required for a GMAW system (Figure 6–60) consists of (1) the power source, (2) the electrode wire feeder and control system, (3) the welding gun and cable assembly for semiautomatic welding or the welding torch for automatic welding, (4) the gas and water control system for the shielding gas and cooling water when used, and (5) travel mechanism and guidance for automatic welding.

The power source can be a rectifier, an inverter, or, for field use, a generator welding machine. For the short-circuiting arc variation a 200-A machine is normally used. For CO_2 welding and spray transfer arc welding, higher current power sources, up to 500 A, are used. For pulsed-spray welding, special power sources with complex controls must be used. These can be rectifier or inverter machines. The volt-ampere characteristic curve of the machine is different for the different process variations. The machine must be designed or tuned to the necessary welding procedure requirements.

Normally, the constant-speed wire feeder is used with a constant-voltage power source since it provides a self-regulated arc. When using the pulsed-spray mode of transfer, a constant-speed wire feeder is normally used. However, for special variations of the pulsed-spray mode,

controllable wire feeders are matched to specially designed power sources. The wire feeder must match the power source for these applications.

A welding gun, or welding torch, is used to carry the electrode, the welding current, and the shielding gas to the arc. For the short-circuiting variation, air-cooled welding guns are normally used. When larger-diameter electrodes are used with CO_2 shielded gas, air-cooled guns are also used, since CO_2 is a cooling medium for the gun. When inert gas or argon oxygen mixtures are used for spray or pulsed-spray welding, the gun must be water cooled if high current is employed. Information about semiautomatic guns and automatic torches is given in Section 11–1.

Electrodes and Shielding Gas

Two materials are used for the gas metal arc welding process: the electrode and the shielding gas. They must be carefully selected with regard to the base metal to be welded and the process variation to be employed. The electrode wire is related to the strength requirements of the deposited weld metal, as well as to the composition. The following factors govern the selection of the electrode.

1. *Metal to be welded.* The composition and mechanical properties of the base metal are of primary importance.
2. *Thickness and joint design.* Thicker sections and complex joint designs require filler metals that provide high weld metal ductility.

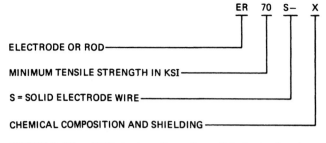

ELECTRODE OR ROD ——————————————

MINIMUM TENSILE STRENGTH IN KSI ——————

S = SOLID ELECTRODE WIRE ——————————

CHEMICAL COMPOSITION AND SHIELDING ———

FIGURE 6–61 AWS designations for solid electrode wire.

3. *Surface conditions.* The surface of the base metal to be welded, whether it is scaly, rusty, or such, has an important effect on the electrode wire to be used.

4. *Specifications or service conditions.* Specifications may dictate the electrode to be used. If specifications are not involved, consider the service requirements that the weldment will encounter.

The welding society has provided specifications for electrodes for use with the gas metal arc welding process. The AWS classification system for carbon steel electrodes[15] is shown in Figure 6-61. The first digit, *E*, indicates that it is an electrode. The next two digits stand for the minimum tensile strength in ksi. For carbon steel electrodes, this number is 70. For low-alloy steel electrodes,

the number can be 80 or higher. The next digit, which is the letter *S*, indicates a solid electrode wire. The last digit (*X*) follows a dash and indicates the shielding gas or chemical composition. These data are summarized with mechanical properties in Figure 6-62. The chemical composition of the electrodes are given in Figure 6-63. The chemical composition and shielding gases used for low-alloy electrodes is more specific and is given by the AWS "Specification for Low Alloy Steel Filler Metals for Gas Shielded Arc Welding."[16] More information for selecting the proper electrode to match a particular base metal is covered in detail in the chapter on the specific base metal. The size of the electrode depends on the welding position and the variation of the process.

The basis for selecting the shielding gas involves the electrode, the base metal, the welding position, the variation of the process, and the desired weld quality. The recommended shielding gases for different metals and process variations are covered in the chapter for the particular metal being welded.

To establish a basis for selection of process variation, it is necessary to know the capabilities and normal applications for each of the process variations. Figure 6-64 shows the variations, the type of metal transfer for steel, the welding position capabilities, and the recommended welding shielding gas.

FIGURE 6–62 Summary of carbon steel electrodes for GMAW.

AWS Classification	Welding Conditions		Strength Requirements (as Welded)			
	Current Electrode Polarity	External Gas Shield	All Tensile (min. psi)	Weld Yield (min. psi)	Percent Metal Elongation (min. 2 in.)	Impact Test Charpy V
E70S–2	DCEP	CO_2	72,000	60,000	22	20 at −20° F
E70S–3	DCEP	CO_2	72,000	60,000	22	20 at −20° F
E70S–4	DCEP	CO_2	72,000	60,000	22	Not required
E70S–5	DCEP	CO_2	72,000	60,000	22	Not required
E70S–6	DCEP	CO_2	72,000	60,000	22	20 at −20° F
E70S–7	DCEP	CO_2	72,000	60,000	22	20 at −20° F
E70S–G	Not specified	Not specified	72,000	60,000	22	Not required

Note: P, 0.025 max.; S, 0.035 max. Shielding gas may be argon–CO_2 or argon–O_2 mixture.

FIGURE 6–63 Composition of GMAW electrodes.

AWS Classification	Composition (%)					
	C	Mn	Si	Other		
E70S–2	0.6		0.40–0.70	Ti—0.05–0.15	Zi—0.02–0.12	Al—0.05–0.15
E70S–3	0.06–0.15	0.90–1.40	0.45–0.70			
E70S–4	0.07–0.15		0.65–0.85			
E70S–5	0.07–0.19		0.30–0.60	A1—0.50–0.90		
E70S–6	0.07–0.15	1.40–1.85	0.80–1.15			
E70S–7	0.07–0.15	1.50–2.00	0.50–0.80			
E70S–G	No chemical requirements					

Metal Transfer	Globular	Short-Circuiting	Spray	Pulsed-Spray
Shielding gas	CO_2	CO_2 or CO_2 + argon (C–25)	Argon + oxygen and others	Argon + oxygen and others
Metals to be welded	Low-carbon and medium-carbon steel, low-alloy high-strength steels	Low-carbon and medium-carbon steels, low-alloy high-strength steels, some stainless steels	Low-carbon and medium-carbon steels, low-alloy high-strength steels	All steels, aluminum and many alloys
Metal thickness	10 gauge (0.140 in.); up to $\frac{1}{2}$ in. without bevel preparation	20 gauge (0.038 in.), up to $\frac{1}{4}$ in.; economical in heavier metals for vertical and overhead welding	$\frac{1}{4}$ to $\frac{1}{2}$ in. with no preparation; maximum thickness practically unlimited	Thin to unlimited thickness
Welding positions	Flat and horizontal	All positions (also pipe welding)	Flat and horizontal with small electrode wire all positions	All positions
Major advantages	Low-cost gas, high travel speed, deep penetration, high deposition	Thin material, will bridge gaps, minimum cleanup	Smooth surface, deep penetration, high travel speed	Uses larger electrode
Limitations	Spatter removal sometimes required, high heat	Uneconomical in heavy thickness-except out of position	Position, minimum thickness	Special power source
Appearance of weld	Relatively smooth, some spatter	Smooth surface, minor spatter	Smooth surface, minimum spatter	Smooth surface, minimum spatter
Travel speeds	Up to 250 in./min	Max. 50 in./min	Up to 150 in./min	Up to 200 in./min
Range of electrode wire sizes (in.)	Diameter: 0.045, $\frac{1}{16}$, $\frac{5}{64}$, $\frac{3}{32}$	Diameter: 0.030, 0.035, 0.045	Diameter: 0.035, 0.045, $\frac{1}{16}$, $\frac{3}{32}$	Diameter: $\frac{1}{16}$, $\frac{5}{64}$, $\frac{3}{32}$, $\frac{1}{8}$

FIGURE 6–64 Variations of the GMAW process.

Deposition Rates and Quality of Welds

Each of the variations has a considerable range of deposition rates based on the weld procedure employed. Figure 6–65 shows the relationship of deposition rates for the steady current and different electrode sizes used. This chart is based on carbon steel base metals and electrodes. For welding nonferrous metals, deposition rates vary considerably due to the density of the metals.

The deposition rates of gas metal arc welding are higher for the same welding currents than are obtained with shielded metal arc welding. These higher rates occur because there is no electrode coating that must be melted. The current density on the small-diameter electrode wires is much higher than with covered electrodes, which contributes to the higher deposition rates for the same welding current. The tip-to-work distance affects deposition rates, and as the distance is increased, the preheating of the electrode wire contributes to higher deposition rates. Excessive stickout cannot be used without losing the ability to point the electrode wire in the joint accurately.

GMAW is a no-hydrogen or low-hydrogen welding process. There is no hydrogen in the shielding gas

FIGURE 6–65 Deposition rates for steady current and different steel electrode sizes.

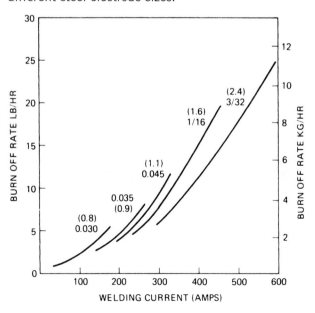

atmosphere or in any component involved in making the weld. This is based on the use of welding-grade shielding gas and clean, dry electrodes on clean, dry surfaces.

The quality of welds made with the GMAW process can be extremely high. The quality of the weld deposited depends on the electrode wire, its cleanliness, the cleanliness of the joint, the welding procedure, the welding position, and the skill of the welder.

Certain factors may detract from the quality of the deposited weld metal. One is the possibility of reduced efficiency of the gas shielding envelope. Breezes in the weld area may blow the shielding gas envelope away and allow the atmosphere to come in contact with the molten pool. The loss of shielding gas is normally noticed by the welder. This will cause a dirty-appearing weld or

will create an unstable welding condition or porosity. Another factor is impure gas that may contain water vapor, oil, or other impurities. Welding with electrode wire that is dirty, oily, or greasy will contribute to inferior weld deposits. Welding on dirty surfaces—wet, oily, or otherwise—will reduce the quality of the weld metal. Figure 6–66 shows weld surfaces related to welding problems.

Weld Schedules

The welding schedule for short-circuiting transfer or with fine-wire welding is shown in Figure 6–67. (Short-circuiting transfer procedures should not be used for structural application or where extra high quality welds are required.) This schedule can be utilized on all types of joint

---

FIGURE 6–66 Quality factors for GMAW, short-circuiting transfer mode.

INCOMPLETE FUSION

1. Current too high — Voltage too low
2. Holding electrode wire too far back from leading edge of puddle
3. Too short a pause at edge of bead

EXCESSIVE PENETRATION

1. Root opening too wide
2. Travel speed too slow
3. Current too high
4. Tip-to-work distance too close

INSUFFICIENT PENETRATION

1. Current too low
2. Tip-to-work distance too great
3. Travel speed too fast
4. Improper welding technique

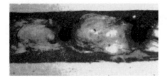

MELT-THRU

1. Current too high
2. Travel speed too slow
3. Root opening too wide
4. Root face too small

SURFACE POROSITY

1. Lack of gas coverage
2. Joint not clean
3. Strong air drafts
4. Damp gas-contaminated gas

POOR APPEARANCE

1. Current too high
2. Faulty joint preparation
3. Improper welding technique

WHISKERS

1. Travel speed too fast: electrode too close to leading edge of puddle
2. Current too high
3. Tip-to-work distance too close

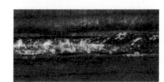

UNEQUAL PENETRATION

1. Improper work angle
2. Improper welding technique

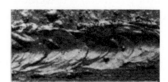

CONCAVE ROOT SURFACE

1. Voltage too high
2. Travel speed too slow
3. Root opening too wide

details and is relatively narrow in range based on electrode size, welding current, and voltage. It enables all-position welding of carbon steels on the metal thicknesses shown. In the flat position, this variation becomes less economical. For vertical and overhead welding, it is more productive than shielded metal arc welding. Welding techniques must be correct to obtain high-quality joints.

The welding schedules for globular transfer CO_2 welding are shown in Figure 6–68. Welding currents are much higher, and deposition rates and productivity are greatly

FIGURE 6–67 Short-circuiting transfer schedules.

Material Thickness[a]			Electrode Diameter		Welding Current (A dc)	Arc Voltage (Electrode Positive)	Wire Feed (in./min)	Travel Speed (in./min)	Shielding Gas Flow[b] (ft³/hr)
Fraction	in.	mm	in.	mm					
26 gauge	0.020	0.5	0.030	0.8	25–40	15–17	75–90	10–20	15–20
24 gauge	0.025	0.6	0.030	0.8	30–50	15–17	85–100	12–20	15–20
22 gauge	0.031	0.8	0.030	0.8	40–60	15–17	90–130	18–22	15–20
20 gauge	0.037	0.9	0.035	0.9	55–85	15–17	70–120	35–40	15–20
18 gauge	0.050	1.3	0.035	0.9	70–100	16–19	100–160	35–40	15–20
$\frac{1}{16}$ in.	0.063	1.6	0.035	0.9	80–110	17–20	120–180	30–35	20–25
$\frac{5}{64}$ in.	0.078	2.0	0.035	0.9	100–130	18–20	160–220	25–30	20–25
$\frac{1}{8}$ in.	0.125	3.2	0.035	0.9	120–160	19–22	210–290	20–25	20–25
$\frac{1}{8}$ in.	0.125	3.2	0.045	1.1	180–200	20–24	210–240	27–32	20–25
$\frac{3}{16}$ in.	0.187	4.7	0.035	0.9	140–160	19–22	210–290	14–19	20–25
$\frac{3}{16}$ in.	0.187	4.7	0.045	1.1	180–205	20–24	210–245	18–22	20–25
$\frac{1}{4}$ in.	0.250	6.4	0.035	0.9	140–160	19–22	240–290	11–15	20–25
$\frac{1}{4}$ in.	0.250	6.4	0.045	1.1	180–225	20–24	210–260	12–18	20–25

Note: Single-pass flat and horizontal fillet positions. Reduce current 10% to 15% for vertical and overhead welding.

[a]For fillet and groove welds. For fillet welds, size equals metal thickness. For square-groove welds, the root opening should equal one-half the metal thickness.

[b]Shielding gas is CO_2 or mixture of 75% argon + 25% CO_2.

Wire feed speed is approximate.

FIGURE 6–68 Globular transfer (CO_2) schedules.

Material Thickness			Type of Weld[a]	Electrode Diameter		Welding Current (A dc)	Arc Voltage (Electrode Positive)	Wire Feed (in./min)	Travel Speed (in./min)	CO_2 Gas Flow (ft³/hr)
Gauge	in.	mm		in	mm					
18	0.050	1.3	Fillet	0.045	1.1	280	26	350	190	20–25
			Square groove	0.045	1.1	270	25	340	180	20–25
16	0.063	1.6	Fillet	0.045	1.1	325	26	360	150	30–35
			Square groove	0.045	1.1	300	28	350	140	30–35
14	0.078	2.0	Fillet	0.045	1.1	325	27	360	130	30–35
			Square groove	0.045	1.1	325	29	360	110	30–35
			Square groove	0.045	1.1	330	29	350	105	30–35
11	0.125	3.2	Fillet	$\frac{1}{16}$	1.6	380	28	210	85	30–35
			Square groove	0.045	1.1	350	29	380	100	30–35
$\frac{3}{16}$ in.	0.188	4.8	Fillet	$\frac{1}{16}$	1.6	425	31	260	75	30–35
			Square groove	$\frac{1}{16}$	1.6	425	30	320	75	30–35
			Square groove	$\frac{1}{16}$	1.6	375	31	260	70	30–35
$\frac{1}{4}$ in.	0.250	6.4	Fillet	$\frac{5}{64}$	2.0	500	32	185	40	30–35
			Square groove	$\frac{1}{16}$	1.6	475	32	340	55	30–35
$\frac{3}{8}$ in.	0.375	9.5	Fillet	$\frac{3}{32}$	2.4	550	34	200	25	30–35
			Square groove	$\frac{3}{32}$	2.4	575	34	160	40	30–35
$\frac{1}{2}$ in.	0.500	12.7	Fillet	$\frac{3}{32}$	2.4	625	36	160	23	30–35
			Square groove	$\frac{3}{32}$	2.4	625	35	200	33	30–35

[a]For mild-carbon and low-alloy steels on square-groove welds, backing is required.

Material Thickness[a]		Type of Weld	Number of Passes	Electrode Diameter		Welding Current (A dc)	Arc Voltage (Electrode Positive)	Wire Feed (in./min)	Travel Speed (in./min)	Shielding Gas[b] Flow (ft³/hr)
in.	mm			in.	mm					
$\frac{1}{16}$	1.6	All	1	0.045	1.1	250	22	400	35	35–40
$\frac{1}{8}$	3.2	Fillet or square groove	1	$\frac{1}{16}$	1.6	300	24	165	35	40–50
$\frac{3}{16}$	4.8	Fillet or square groove	1	$\frac{1}{16}$	1.6	350	25	230	32	40–50
						325	24	210		
$\frac{1}{4}$	6.4	V-groove	2	$\frac{1}{16}$	1.6	375	25	260	30	40–50
						400	26	100		
$\frac{1}{4}$	6.4	V-groove	2	$\frac{3}{32}$	2.4	450	29	120	35	40–50
$\frac{1}{4}$	6.4	Fillet	1	$\frac{1}{16}$	1.6	350	25	230	32	40–50
$\frac{1}{4}$	6.4	Fillet	1	$\frac{3}{32}$	2.4	400	26	100	32	40–50
						325	24	210		
$\frac{3}{8}$	9.5	V-groove	2	$\frac{1}{16}$	1.6	375	25	260	24	40–50
						400	26	100		
$\frac{3}{8}$	9.5	V-groove	2	$\frac{3}{32}$	2.4	450	29	120	28	40–50
$\frac{3}{8}$	9.5	Fillet	2	$\frac{1}{16}$	1.6	350	25	230	20	40–50
$\frac{3}{8}$	9.5	Fillet	1	$\frac{3}{32}$	2.4	425	27	110	20	40–50
						325	24	210		
$\frac{1}{2}$	12.7	V-groove	3	$\frac{1}{16}$	1.6	375	26	260	24	40–50
						375	26	250		
						400	26	100		
						450	29	120		
$\frac{1}{2}$	12.7	V-groove	3	$\frac{3}{32}$	2.4	425	27	110	30	40–50
$\frac{1}{2}$	12.7	Fillet	3	$\frac{1}{16}$	1.6	350	25	230	24	40–50
$\frac{1}{2}$	12.7	Fillet	3	$\frac{3}{32}$	2.4	425	27	105	26	40–50
								110		
						325	24	210		
						375	26	260		
$\frac{3}{4}$	19.1	Double V-groove	4	$\frac{1}{16}$	1.6	350	25	230	24	40–50
						400	26	100		
						450	29	120		
$\frac{3}{4}$	19.1	Double V-groove	4	$\frac{3}{32}$	2.4	425	27	110	24	40–50
$\frac{3}{4}$	19.1	Fillet	5	$\frac{1}{16}$	1.6	350	25	230	24	40–50
$\frac{3}{4}$	19.1	Fillet	4	$\frac{3}{32}$	2.4	425	27	110	26	40–50
1	24.1	Fillet	7	$\frac{1}{16}$	1.6	350	25	230	24	40–50
1	24.1	Fillet	6	$\frac{3}{32}$	2.4	425	27	110	26	40–50

Note: Use only in flat and horizontal fillet positions.

[a]For fillet welds, material thickness indicates fillet weld size.

[b]Shielding gas is argon plus 1% to 5% oxygen.

Note: Wire feed speed is approximate.

FIGURE 6–69 Spray arc transfer schedules.

increased. It is normally used in the flat and horizontal position. The basic difference between the two welding procedure schedules is the position capabilities. Note the extremely high currents that can be used on carbon steels.

For spray transfer, high currents can be employed and these data are shown by the welding procedure schedules in Figure 6–69. This is a highly productive variation for welding mild and low-alloy steels. Spray transfer normally uses larger size electrode wires which are less expensive. Welding procedure schedules for nonferrous metals will be found in the chapter for each metal.

The pulsed-spray mode has two variations: fixed-frequency pulsed-current welding and variable-frequency (synergic) pulsed-current welding. Both variations utilize special power sources. Synergic is becoming the more popular. With synergic equipment the pulsed variables are pro-

Material Thickness			Electrode Diameter		Average Current (A)	Peak Current (A)	Background Current (A)	Arc Voltage (Electrode positive)
Gauge	in.	mm	in.	mm				
22	0.031	0.8	0.035	0.9	50	150	20	16
20	0.037	0.9	0.035	0.9	60	160	20	17
18	0.050	1.3	0.035	0.9	70	180	20	18
16	0.063	1.6	0.045	1.2	80	200	25	19
14	0.078	2.0	0.045	1.2	90	250	35	21
11	0.125	3.2	0.045	1.2	120	250	150	22
$\frac{3}{16}$ in.	0.188	4.8	0.045	1.2	150	250	200	23
$\frac{1}{4}$ in.	0.250	6.4	0.052	1.3	120	275	90	24
$\frac{3}{8}$ in.	0.375	9.5	0.052	1.3	200	350	150	26

Note: For square groove or fillet, use a root opening of one-half the material thickness. Fillet equal to thickness. For mild-carbon and low-alloy steels, shielding gas 95% argon + 5% oxygen.

Material Thickness			Wire Feed Speed		Travel Speed		Shielding Gas Flow	
Gauge	in.	mm	in./min	mm/min	in./min	mm/min	ft³/min	liters/min
22	0.031	0.8	75	1,900	30	760	20	9
20	0.037	0.9	90	2,300	30	760	20	9
18	0.050	1.3	115	2,900	30	760	20	9
16	0.063	1.6	80	2,000	20	500	25	12
14	0.078	2.0	120	3,000	20	500	25	12
11	0.125	3.2	200	5,000	15	375	25	12
$\frac{3}{16}$ in.	0.188	4.8	240	6,000	10	250	25	12
$\frac{1}{4}$ in.	0.250	6.4	215	5,500	9	225	25	12
$\frac{3}{8}$ in.	0.375	9.5	300	7,500	8	200	25	12

FIGURE 6–70 Pulsed-spray transfer schedules, variable-frequency (synergic) variation.

grammed in the welding machine. This provides the ratio of peak to background of the pulse and the time duration of the pulse. The specific program relates to the type of metal being welded, the electrode diameter, and the shielding gas composition. The average current is adjusted by the welder; this, in turn, changes the pulsing frequency or the background current depending on the design of the machine. The manufacturer's data must be used for each welding situation. Approximate welding procedure schedules for synergic pulsed spray welding are shown in Figure 6–70.

Tips for Using the Process

Semiautomatic welding using short-circuiting metal transfer is easy to use. Experienced shielded metal arc welders or people with no welding experience can learn this process variation in a relatively short time. Production welding can be learned in a few days, whereas pipe welding may require 80 to 120 hours of training.[17]

It is important to use the correct welding technique when welding semiautomatically. The electrode wire should be directed to the leading edge of the pool for optimum results. For out-of-position welding, the pool should remain small for best control.

The gun tip-to-work distance known as stickout must be closely controlled. If the stickout becomes too long, the electrode will become overheated and will minimize penetration. Also, when the gun nozzle is too far from the arc, the shielding gas efficiency is reduced. Normal nozzle-to-work distance should be approximately 1 to $1\frac{1}{2}$ times the inside diameter of the gas nozzle.

Another important factor is the angle the gun nozzle makes with the work. Two angles are involved. One is known as the *travel angle,* the other is the *work angle.* The work angle is normally half the included angle between the plates forming the joint. When making fillet welds, the gun should be at a 45° angle but directed slightly toward the horizontal plate by one electrode wire diameter from the bisecting angle.

The travel angle can be a *drag angle* or a *push angle.* The push angle pointing forward is used when pure inert gases are employed. The drag angle pointing backward is used when CO_2 is used—with short circuiting or globular transfer.

The welding equipment must be in good operating condition. The drive rolls and contact tip must be proper

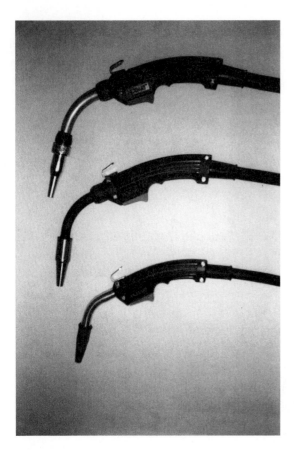

FIGURE 6–71 Gun nozzle extensions.

for the electrode size being used. The conduit tube in the gun cable assembly must be kept clean and any centering guides, lineup rolls, and so on, must be properly aligned. The nozzle of the gun must be kept clean and all portions of the gas supply system must be tight and operating

properly. Finally, the work cable must be tightly connected to the work for trouble-free operation.

The welding parameters must be set in accordance with welding procedure schedules. The correct gas flow rates must be employed for optimum results. The welding polarity must be correct. For almost all gas metal arc welding, dc electrode positive is employed.

Safety Considerations

Safety factors and potential hazards involved with the GMAW process are covered in detail in Chapter 4. In general, GMAW is a less hazardous process than manual shielded metal arc welding.

Limitations of the Process

One problem has been the inability to reach inaccessible welding areas with the available guns. GMAW guns are not as flexible as the covered electrode used for shielded metal arc welding. However, extensions can be placed on welding guns to reach relatively inaccessible areas. A variety of gun nozzle extensions is shown in Figure 6–71.

There is a problem feeding small-diameter soft electrode wire. The solution is the use of a spool gun shown in Figure 6–72. Spool guns usually employ constant-speed feed motors and carry the electrode supply on the gun. Electrode wire is fed only a few inches and will accommodate small-diameter aluminum wire and other non-ferrous soft-type electrode wires. In some cases, the feed rate can be adjusted at the gun and the solenoid valve controlling the gas flow is in the gun. Even so, they are lightweight and easily maneuvered. It is the best solution for feeding thin-diameter aluminum electrode wires for any application.

FIGURE 6–72 Spool Gun.

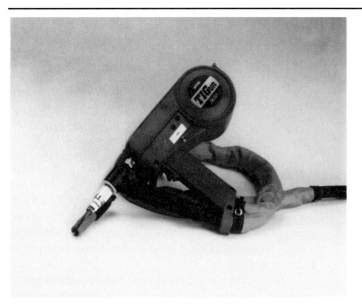

An objection is the problem of wind and drafts affecting the efficiency of the gas shielding envelope around the arc area. It can be overcome by establishing windbreaks or shielding the welding area from direct exposure to fans, open doors, or the wind. With a little experience, welders are able to use their bodies to shield the arc area from drafts and breezes.

Industrial Use and Typical Applications

A typical application is the welding of aluminum bus bars in the electrical industry. The most aggressive user has been the sheet metal industry. Many submerged arc welding applications have been changed to GMAW since it is better for automatic fixturing and avoids the problem of abrasive flux in fixtures. A typical application is the semiautomatic welding of a sheet metal assembly (Figure 6-73).

Pipe welding also utilizes the GMAW process. A typical application is shown in Figure 6-74. There are many, many applications for GMAW.

FIGURE 6–73 Welding sheet metal assembly.

FIGURE 6–74 Welding pipe.

6-5 FLUX-CORED ARC WELDING

Flux-cored arc welding (FCAW) is "an arc welding process that uses an arc between a continuous filler metal electrode and the weld pool. The process is used with shielding gas from a flux contained within the tubular electrode with or without additional shielding from an externally supplied gas, and without the application of pressure." This is a variation of GMAW and is based on the configuration of the electrode. This process is shown in Figure 6-75. The welder's view is shown in Figure 6-76.

FIGURE 6–75 Flux cored arc welding.

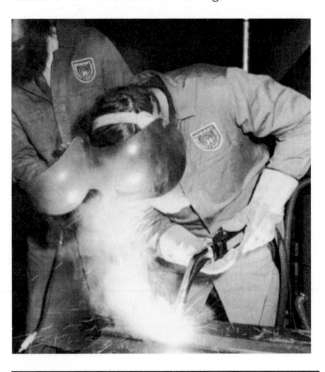

FIGURE 6–76 Welder's view of flux cored arc using CO_2 shielding gas.

There are two variations of process. One uses externally supplied shielding gas, and the second relies entirely on shielding gas generated from the disintegration of the flux within the electrode. Of the variation using externally supplied shielding gas, there are two distinct types of electrodes. One is the original flux-cored type electrode, where the materials contained within the tubular electrode are primarily fluxing agents. Another type, known as cored wire welding electrodes, contains alloy elements as well as powdered iron to enhance productivity with a minimum amount of fluxing material. Both flux-cored and metal-cored electrode wires require external gas shielding. This is more logically called cored wire welding and is almost identical to gas metal arc welding, except for the electrode. In some countries, cored wire welding, either flux-cored or metal-cored, is considered a variation of gas metal arc welding.

Principle of Operation

The FCAW process (Figure 6–77) utilizes the heat of an arc between a continuously fed consumable cored electrode and the work. The heat of the arc melts the surface of the base metal and the end of the electrode. The metal melted off the electrode is transferred across the arc to the workpiece, where it becomes the deposited weld metal. Shielding is obtained from the disintegration of ingredients contained within the flux-cored electrode. Additional shielding is obtained from an envelope of gas supplied through a nozzle to the arc area. Ingredients within the electrode produce gas for shielding and also provide deoxidizers, ionizers, purifying agents, and in some cases alloying elements. These ingredients form a glasslike slag, which is lighter in weight than the deposited weld metal, and floats on the surface of the weld as a protective cover. The electrode is fed into the arc automatically, from a coil. The arc is maintained automatically, and travel can be manual or by machine.

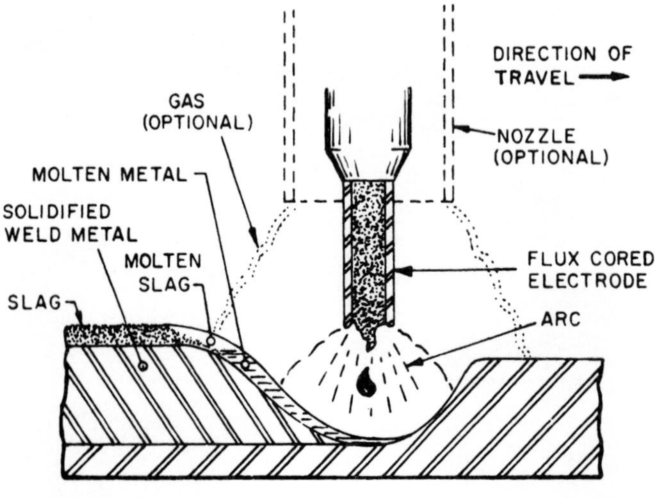

Advantages and Major Uses

The FCAW process introduced in early 1950 is an outgrowth of the GMAW process. FCAW has many advantages over the manual SMAW process. It also provides advantages over submerged arc welding and the GMAW processes. Simply stated, the FCAW process provides high-quality weld metal at lower cost with less effort on the part of the welder than SMAW. It is more forgiving than GMAW and is more flexible and adaptable than submerged arc welding. These advantages are as follows:

- High-quality weld metal deposit
- Excellent weld appearance: smooth, uniform welds
- Excellent contour of horizontal fillet welds
- Welds a variety of steels over a wide thickness range
- High operating factor: easily mechanized
- High deposition rate: high-current density
- Relatively high electrode metal utilization
- Relatively high travel speeds
- Economical engineering joint designs
- Gasless variation can be used outdoors
- Visible arc: easy to use
- Less precleaning required than for gas metal arc welding
- Reduced distortion over shielded metal arc welding

This process is becoming increasingly popular. It is widely used on medium thickness steel fabricating work, where the fine wire GMAW process would not apply and where the fitup is such that submerged arc welding would be unsuitable.

Methods of Application and Position Capabilities

The most popular method of applying FCAW is by the semiautomatic method. Second is the fully automatic method. The process can also be applied by mechanized methods, but it cannot be applied manually. The FCAW process is an all-position welding process depending on electrode size.

Weldable Metals and Thickness Range

The FCAW process is used to weld low- and medium-carbon steels, low-alloy high-strength steels, quenched and tempered steels, certain stainless steels, and cast iron. The process is also used for surfacing and for buildup. The metals welded by the process are shown in Figure 6–78.

The metal thickness range for the two variations—shelf shielding and using external gas shielding—is different. With a gas-shielded atmosphere, weld penetration

FIGURE 6–77 Process diagram for FCAW.

Base Metal	Weldability
Cast-iron	Using special electrode
Low-carbon steel	Weldable
Low-alloy steel	Weldable
High- and medium-carbon steel	Weldable
Alloy steel	Weldable
Stainless steel—selected	Weldable
Nickel alloys	Limited types

FIGURE 6–78 Metals weldable by FCAW.

is considerably deeper and metal thicknesses from $\frac{1}{16}$ in. (1.6 mm) to $\frac{1}{2}$ in. (13 mm) can be welded with no edge preparation. When external gas is not used, the maximum penetration is only $\frac{1}{2}$ or $\frac{1}{4}$ in. (6 mm). With edge preparation, welds can be made with a single pass on material from $\frac{1}{4}$ in. (6 mm) through $\frac{3}{4}$ (19 mm), with either variation. With a multi-pass technique and joint preparation, the maximum thickness is practically unlimited (Figure 6–79). Horizontal fillets can be made up to $\frac{3}{8}$ in. (9.5 mm) in a single pass, and in the flat position fillet welds can be made up to $\frac{3}{4}$ in. (19 mm).

Joint Design

The FCAW process can utilize the same joint design details used by the SMAW process. For maximum utilization and efficiency, different joint details are suggested.

For groove welds, the square-groove design can be used up to $\frac{1}{2}$ in. (13 mm) thickness. Beyond this thickness, bevels are required; however, the included angle of bevel-groove welds can be reduced 35% to 50% over that normally used for SMAW. This is because the smaller electrode wire can get deeper into the joint. Open roots can be used; however, a root face is normally required to avoid burning through. In many structural applications the weld is made with a tight root opening, and the back side is gouged and rewelded. When welding fillet welds using the gas-shielded version, the fillet size can be smaller yet will have the same strength as shielded metal arc welds (Figure 6–80). The self-shielding electrode wire does not

have the deep penetrating qualities of the gas-shielded deposit; therefore, the fillet size cannot be reduced when using the gasless variation.

Welding Circuit and Current

The welding circuit employed for FCAW is identical to that used by the GMAW process (Figure 6–81). In the case of self-shielding electrode wires, the gas system is eliminated.

The FCAW process normally uses direct current with the electrode positive (DCEP). Some electrodes operate with the electrode negative (DCEN). Direct current with constant-voltage power is used.

Ac FCAW is used in some situations with specially formulated flux-cored electrodes. When ac electrodes are used, a drooping characteristic (CC) type of power source and voltage-sensing feeders are employed. The welding current for FCAW can vary from as low as 50 A to as high as 750 A depending on the electrode size.

Equipment Required

The equipment required for FCAW is shown in Figure 6–81. These components, when using the externally gas-shielded version, are identical to the GMAW process. The only difference is that higher current power sources and larger welding guns or torches are used.

When the gasless version is used, the entire shielding gas supply system is eliminated. This eliminates the gas cylinders, the regulator and flow meter, the hoses, the solenoid valve, and the nozzle on the welding gun. Since the nozzle is removed from the welding guns, the guns can be designed with different tip configurations. The end of the welding gun is smaller, and visibility is improved. However, in view of the amount of smoke produced by FCAW, it is becoming increasingly necessary to include smoke suction nozzles surrounding the gun nozzle to reduce smoke and fumes. Guns for self-shielding electrodes normally use special wire guides that include *electrical stickout*. This means that the current is introduced to the electrode before the end of the tip. This preheats the electrode wire and makes it more productive.

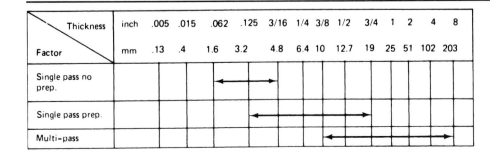

FIGURE 6–79 Base metal thickness range for FCAW.

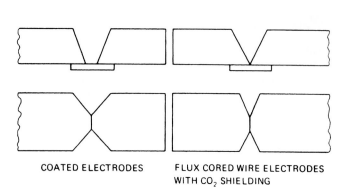

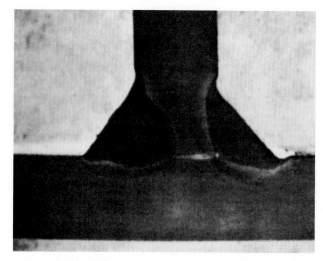

COATED ELECTRODES   FLUX CORED WIRE ELECTRODES
WITH CO_2 SHIELDING

FILLET WELDS OF EQUAL STRENGTH

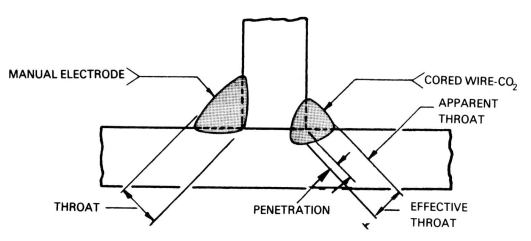

MANUAL ELECTRODE

CORED WIRE-CO_2

APPARENT THROAT

THROAT

PENETRATION

EFFECTIVE THROAT

FIGURE 6–80   Welding joint design details for FCAW.

FIGURE 6–81   Circuit diagram for FCAW.

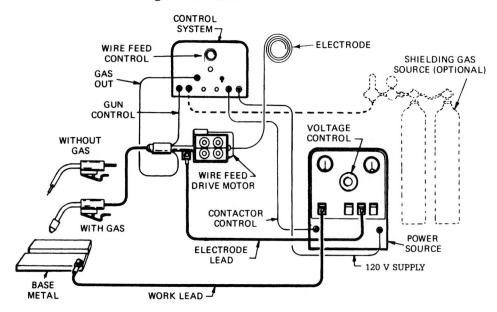

CONTROL SYSTEM

WIRE FEED CONTROL

ELECTRODE

SHIELDING GAS SOURCE (OPTIONAL)

GAS OUT

GUN CONTROL

WITHOUT GAS

VOLTAGE CONTROL

WIRE FEED DRIVE MOTOR

WITH GAS

CONTACTOR CONTROL

POWER SOURCE

ELECTRODE LEAD

120 V SUPPLY

BASE METAL

WORK LEAD

Electrodes

The flux cored electrode wires have been called inside-outside electrodes since the fluxing and alloying compounds are on the inside rather than on the outside, as with a covered electrode. The flux cored electrode consists of a metal sheath surrounding the core of chemicals, which perform the same function as the coating on a covered electrode. The design of the gas-shielded wires and the self-shielding wires are different. The core materials of the self-shielding electrode wires include additional gas-forming chemicals. These are necessary to prohibit the oxygen and nitrogen of the air from contacting the metal transferring across the arc and the molten weld pool. Self-shielding electrodes also include extra deoxidizing and denitriding elements. They are usually more voltage sensitive and require electrical stickout for smooth operation.

The properties of the weld metal deposited by the gas-shielded wires are extremely good, while those deposited by self-shielding wires are somewhat less. This is borne out by the specification requirements. Caution should be exercised when using the self-shielded wires because of the extra amounts of deoxidizer chemicals included in the core. These elements may build up in multipass welds, lower the ductility, and reduce the impact values. Some codes prohibit the use of self-shielding wires on high-yield-strength steels or on dynamically loaded structures.

Flux cored electrode wires are designed for use on carbon steels and low-alloy steels. In addition, they are used for welding cast iron. Certain flux cored wires are available for welding stainless steel, and some are designed for surfacing applications. Flux cored wires are sometimes used with the submerged arc welding process. Metal-cored wires are tubular electrode wires that contain metal powders and a minimum amount of chemicals for flux, deoxidizers, and gas formers. This represents less than 5% of the core materials. Metal-cored wires have high deposition efficiencies, come in different classes, and must be used with external gas shielding.

The system for identifying flux cored electrodes is complicated. The most common system is shown in Figure 6–82, which shows the numbering system for electrodes for welding carbon steels. The numbering system for electrodes for low-alloy steels, corrosion-resisting steels, surfacing applications, and for welding cast iron are all slightly different. For carbon steel electrodes, the E indicates an electrode, and this is common for all specifications. The next digit stands for the minimum tensile strength, as welded, in 10 ksi, usually a 6 or 7. The next digit stands for welding position; 0 indicates flat or horizontal position welding, and 1 indicates all position welding. The next digit, a T, indicates a tubular or flux cored electrode. The next digit following a dash designates the external shielding medium and welding power to be em-

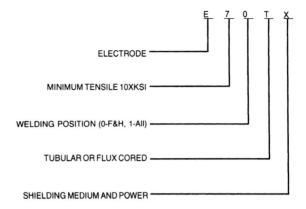

FIGURE 6–82 AWS designation for tubular electrode wire for carbon steel per AWS A5.20. (From Ref. 18.)

ployed. See Figure 6–83 for details concerning shielding gas, polarity, etc.

In the case of low-alloy electrodes,[19] the first digit following the T indicates usability and performance capabilities. This is followed by a dash and other digits that designate the chemical composition of the deposited weld metal. The specification for low-alloy-steel electrodes must be consulted for exact identification details for FCAW.

In the case of stainless steel electrodes, the three digits following the E indicate the AISI designation of the stainless steel. The digit following the T indicates the shielding medium to be employed. In the case of electrodes for cast iron, the information is given in the section for cast iron welding.

The flux cored electrode wires are considered to be low hydrogen since materials used in the core do not contain hydrogen. However, certain of these materials are hygroscopic and tend to absorb moisture when exposed to a high-humidity atmosphere. Electrode wires are therefore packaged in waterproof containers to prevent moisture pickup. It is recommended that the flux cored electrode wires be stored in a dry room. Chapter 13 gives more details, including the method of manufacturing flux-cored electrode wires.

Metal Core Electrodes

The welding industry has developed special high-deposition-rate cored electrode wires. These electrode wires contain little or no fluxing agent in the core. They have extremely high deposition efficiencies of 95% and greater, which are on the order of those of solid electrode wires. The deposition rates can be in excess of 20 lb/hr. They require shielding gas, but they operate with an extremely smooth arc. They have low spatter levels, low smoke levels, and a minimum slag coverage. They are produced in all common wire diameters but also in extremely small diameters. They are classified as composite

AWS Classification	Position	Shielding	Current and Polarity	Single-Pass/ Multi-pass
E70T–1	H and F	CO_2	DCEP	Multi
E70T–1M	H and F	75–80% Ar-CO_2	DCEP	Multi
E71T–1	H, F, VU, OH	CO_2	DCEP	Multi
E71T1–1M	H, F, VU, OH	75–80% AR–CO_2	DCEP	Multi
E70T–2	H and F	CO_2	DCEP	Single
E70T–2M	H and F	75–80% Ar–CO_2	DCEP	Single
E71T–2	H, F, VU, OH	CO_2	DCEP	Single
E71T–2M	H, F, VU, OH	75–80% Ar–CO_2	DCEP	Single
E70T–3	H and F	Self	DCEP	Single
E70T–4	H and F	Self	DCEP	Multi
E70T–5	H and F	CO_2	DCEP	Multi
E70T–5M	H and F	75–80% Ar–CO_2	DCEP	Multi
E71T–5	H, F, VU, OH	CO_2	DCEP or DCEN	Multi
E71T–5M	H, V, VU, OH	75–80% Ar–CO_2	DCEP or DCEN	Multi
E70T–6	H and F	Self	DCEP	Multi
E70T–7	H and F	Self	DCEN	Multi
E71T–7	H, F, VU, OH	Self	DCEN	Multi
E70T–8	H and F	Self	DCEN	Multi
E71T–8	H, F, VU, OH	Self	DCEN	Multi
E70T–9	H and F	CO_2	DCEP	Multi
E70T–9M	H and F	75–80% Ar–CO_2	DCEP	Multi
E71T–9	H, F, VU, OH	CO_2	DCEP	Multi
E71T–9M	H, F, VU, OH	75–80% Ar–CO_2	DCEP	Multi
E70T–10	H and F	Self	DCEN	Single
E70T–11	H and F	Self	DCEN	Multi
E71T–11	H, F, VD, OH	Self	DCEN	Multi
E70T–12	H and F	CO_2	DCEP	Multi
E70T–12M	H and F	75–80% Ar–CO_2	DCEP	Multi
E71T–12	H, F, VU, OH	CO_2	DCEP	Multi
E71T–12M	H, F, VU, OH	75–80% Ar–CO_2	DCEP	Multi
E61T–13	H, F, VD, OH	Self	DCEN	Single
E71T–13	H, F, VD, OH	Self	DCEN	Single
E71T–14	H, F, VD, OH	Self	DCEN	Single
EX0T–G	H and F	Note	Note	Multi
EX1T–G	H, F, V, OH	Note	Note	Multi
EX01–GS	H and F	Note	Note	Single
EX1T–GS	H, F, V, OH	Note	Note	Single

Note: New or proprietary wire, properties as specified by the supplier.

FIGURE 6–83   Summary of flux-cored electrodes.

metal-cored electrodes under AWS A5.18, not as tubular electrode wires under AWS A5.20. Although they are not classified as tubular wires, they meet most of the requirements for E70T-1 and E71T-1 applications. Some are designed for single- and multi-pass welding and, depending on the size, can be used in all positions. The shielding gas composition is based on the manufacturer's recommendations.

A summary of the carbon steel flux cored electrode is given in Figure 6-83. The following is a summary of flux-cored electrode classifications and their application.

EXXT-I and EXXT-IM Classifications   Electrodes of these classifications are used for single- and multiple-pass welding uses DCEP polarity. The larger diameters [$\frac{5}{64}$ in. (2.0 mm) and larger] are used for welding in the flat and horizontal fillet positions. The smaller diameters [$\frac{1}{16}$ in. (1.6 mm) and smaller] are used for all position welding. Electrodes of the EXXT-1 group are used with CO_2 shielding gas. Electrodes of the EXXT-1M group are classified with 75% to 80% argon, balance CO_2 shielding gas. The argon–CO_2 mixture will increase the manganese and silicon content of the weld metal, which will increase the yield and tensile strength and may affect impact properties. These electrodes have a spray transfer with low spatter loss and produce a flat to slightly convex bead contour.

EXXT-2 and EXXT-2M Classifications Electrodes of these classifications are very similar to EXXT-1 and EXXT-1M with higher manganese and silicon. They are used primarily for single-pass welding in flat and horizontal fillet position welding. The higher levels of deoxidizers allow single-pass welding of heavily oxidized or rim steels. They are used with the same external shielding gas as the EXXT-1 and EXXT-1M classification. These electrodes give good mechanical properties in single-pass welds. The arc characteristics and deposition rates are similar to the previous classification.

EXXT-3 Classification Electrodes of this classification are self-shielded using DCEP and have a spray-type transfer. The electrodes are used for single-pass welds in the flat, horizontal, and vertical (down) position on sheet metal. They are not recommended for medium and heavy plate welding. The slag system is designed for high-speed welding.

EXXT-4 Classification These electrodes are self-shielded and operate on DCEP. They have a globular-type metal transfer. They have high deposition rates and are designed for low penetration and can be used with poor fitup and for single- and multiple-pass welding.

EXXT-5 and EXXT-5M Classifications These classifications are designed for use with CO_2 shielding gas; however, the T-5M classification is designed for use with 75% to 80% argon, balance CO_2 shielding gas. These electrodes are used primarily for single- and multiple-pass welding in the flat and horizontal fillet position. Electrodes have globular transfer and produce slightly convex bead contour and a thin slag covering. The smaller-diameter electrodes can be used in all positions; however, welder appeal of these electrodes are not as good as the T-4 classification.

EXXT-6 Classification These electrodes are self-shielded, operate on DCEP, and have a spray metal transfer. The slag system is designed to give good low-temperature impact properties and good penetration with excellent slag removal. These electrodes are used for single- and multiple-pass welding in flat and horizontal positions.

EXXT-7 Classification These electrodes are self-shielded, operate on DCEN, and have a small droplet-to-spray-type transfer. The slag system allows for high deposition rates and can be used in the horizontal and flat positions. Smaller sizes can be used in all positions. These electrodes are used for low-sulfur weld deposits and have good resistance to cracking.

EXXT-8 Classification These electrodes are self-shielding, operate on DCEN, and have a small droplet-to-spray transfer. The smaller-diameter electrodes are suitable for welding in all positions and the weld metal has good low-temperature properties. They can be used for single- and multiple-pass welds.

EXXT-9 and EXXT-9M Classifications Electrodes of the T-9 group are used with CO_2 shielding gas and those of the T-9M group are to be used with 75% to 80% argon, balance CO_2 shielding gas. They are used for all position applications. The addition of argon will result in higher manganese and silicon deposits, which will have an effect on the weld metal properties. They are designed for single- or multiple-pass welding. They have high deposition rates and fine globular transfer with low spatter loss. The beads are flat to slightly convex and a moderate volume of slag completely covers the weld bead. The weld joint should be clean and free of oil and dirt, excessive oxides, and scale in order to produce radiographic quality welds.

EXXT-10 Classification These electrodes are self-shielded, operate on DCEN, and have a small droplet metal transfer. These electrodes are used for single-pass welds at high travel speeds on metal material of any thickness in the flat, horizontal, and vertical positions, depending on size.

EXXT-11 Classification These electrodes are self-shielded, operate on DCEN, and have a smooth spray transfer. They are general-purpose electrodes for single- and multiple-pass welding in all positions, depending on size. They are not generally used on heavier thicknesses.

EXXT-12 and EXXT-12M Classifications These electrodes are essentially the same as EXXT-1 and EXXT-1M electrodes, which have been modified to improve impact toughness and to meet the lower manganese requirements of the ASME A-1 analysis group. They have a decreased tensile strength and hardness and must be matched to the base material.

EXXT-13 Classification These are self-shielded electrodes that operate on DCEN. These electrodes can be used in all positions and for the root pass on circumferential pipe girth welds. The electrode can be used on all low-carbon steel pipe and all thicknesses, but are only recommended for the root pass. They generally are not recommended for multiple-pass welding.

EXXT-14 Classification These electrodes are self-shielded and operate on DCEN. They are designed for high-speed welds. They are used for welding sheet metal and can be used on all positions, depending on size. They are specifically designed for galvanized, aluminized, and other coated steels. They are not recommended for T or lap joints, butt, edge, or corner joints in heavier material.

EXXT-G Classification These electrodes are for multiple-pass welding that is not covered by any of the classifications above. The requirements for this classification are not specified except for chemical requirements to assure a carbon steel deposit and tensile strength. Consult the supplier for details.

EXXT-GS Classification These electrodes are for single-pass welding and are not covered by any of the classifications above. The tensile strength is specified; the other requirements are not. Consult with the supplier for detailed information.

Shielding Gas

The other material used with gas-shielded flux-cored arc welding is the shielding gas. Carbon dioxide is used most often. However, the CO_2–argon mixture (75% argon plus 25% CO_2) and the argon–oxygen mixtures are sometimes used. These gas mixtures are used for out-of-position welding, for welding piping, or when an extra smooth weld is required. It is important to know if an electrode can be used with a gas mixture since the majority of the electrodes are designed for use with CO_2 shielding. Caution should be exercised to determine if this will be detrimental to the weldment. Certain of the deoxidizers included in the electrode wire, such as silicon, manganese, and possibly aluminum, will carry across the arc rather than be oxidized in the arc. These elements may build up in multiple-pass welds to a level that could be unacceptable. Shielding gas of 98% argon plus 2% oxygen is also used for some electrodes.

Deposition Rates and Quality of Welds

The deposition rates for flux-cored electrodes are shown in Figure 6–84. These curves show deposition rates when welding with mild and low-alloy steels using direct-current electrode positive (DCEP). Deposition rates of the smaller size flux-cored wires exceed that of the covered electrodes. The metal utilization of the flux-cored electrode is higher. Flux-cored electrodes have a much broader current range than covered electrodes.

The quality of the deposited weld metal produced by the flux-cored welding process is high. The deposited weld metal will match or exceed the properties shown for the electrode used. This assures the proper matching of base metal, flux-cored electrode type and shielding gas. Quality depends on the efficiency of the gas shielding envelope, on the joint detail, on the cleanliness of the joint, and on the skill of the welder.

The mechanical properties of weld and metal deposited by the self-shielding electrode wires are slightly lower than that produced by electrodes that utilize external gas shielding. This is seen by reviewing the properties of the deposited metal of both types of electrodes. The self-shielding and the gas-shielded electrodes both produce x-ray quality welds.

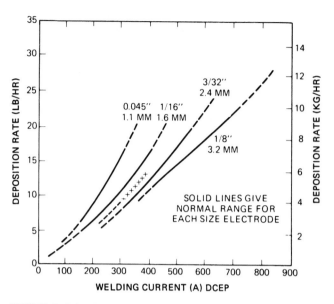

FIGURE 6–84 Deposition rates of steel flux-cored electrodes.

Weld Schedules

The welding procedure schedules for FCAW are given in two ways. The first is shown in Figure 6–85, which provides information for using electrode-positive electrodes and electrode-negative electrodes. The welding parameters are given for each size of electrode for welding in different positions. In this or any welding procedure schedule, the ranges can be expanded. Higher currents can be used when automatic travel is used. The voltage range can also be expanded and will increase when a longer tip-to-work or stickout distance is used. Normally, for the 0.45-in. (1.1-mm) size, the stickout is $\frac{1}{2}$ to $\frac{5}{8}$ in. For the $\frac{1}{16}$-in. (1.6-mm) size, the stickout is increased to $\frac{3}{4}$ in., and for the $\frac{5}{64}$-in. (2.0-mm) size and above, the stickout is increased to $\frac{3}{4}$ to 1 in. CO_2 shielding gas is used with the electrode-positive electrode in the range 35 ft³/hr (16.5 liters/min). Wire feed speed is given, which provides a deposition rate.

The second set of tables for welding procedure schedules is shown in Figure 6–86. In this series, specific weld details are shown and weld schedules are given for each detail. This is based on generalized conditions using manual travel on carbon steel and can be altered for specific situations. These tables are starting points and should be verified by qualification tests or production runs. The weld cross-sectional area for FCAW can be reduced over that utilized for coated electrodes, and for this reason joint details shown on these charts can be modified to reduce the included angle.

When utilizing the self-shielding type wires that operate with direct-current electrode negative (DCEN) cur-

Welding Data: For CO_2 Gas-Shielded Electrode with Reverse Polarity (Electrode Positive)									
Diameter					Approx. Wire Feed Speed		Deposition Rate		
in.	mm	Weld Position	Amperage DCEP	Voltage	in./min	mm/min	lb/hr	kg/hr	Stickout $\pm\frac{1}{4}$ in.
0.035	0.9	Flat–horiz[a]	100–250	23–29	245–790	6,223–20,066	3.3–10.2	1.49–4.62	
0.035	0.9	Vertical[a]	125–225	23–28	245–640	6,223–16,256	3.3–8.5	1.49–3.85	
0.035	0.9	Overhead[a]	125–225	23–28	245–640	6,223–16,256	3.3–8.5	1.49–3.85	
0.045	1.1	Flat–horiz[a]	100–300	24–33	220–730	5,588–18,542	4.6–15.4	2.08–6.98	
0.045	1.1	Vertical[a]	150–220	24–27	220–430	5,588–10,922	4.6–8.4	2.08–3.81	
0.045	1.1	Overhead[a]	150–250	24–29	220–530	5,588–13,462	4.5–10.3	2.04–4.67	
0.052	1.3	Flat–horiz[a]	150–350	23–35	160–585	4,064–14,859	4.2–15.0	1.90–6.80	
0.052	1.3	Vertical[a]	150–250	24–28	160–320	4,064–8,128	4.2–8.4	1.90–3.81	
0.052	1.3	Overhead[a]	150–250	24–28	160–320	4,064–8,128	4.2–8.4	1.90–3.81	
$\frac{1}{16}$	1.6	Flat–horiz[b]	200	25	138	3,505	4.7	2.13	$\frac{3}{4}$
$\frac{1}{16}$	1.6	Flat–horiz[b]	250	26	177	4,495	6.0	2.72	$\frac{3}{4}$
$\frac{1}{16}$	1.6	Flat–horiz[b]	300	27	230	5,842	8.4	3.81	$\frac{3}{4}$
$\frac{1}{16}$	1.6	Flat[b]	350	28	280	7,112	10.9	4.94	$\frac{3}{4}$
$\frac{1}{16}$	1.6	Flat[b]	375	29	311	7,899	11.6	5.26	$\frac{3}{4}$
$\frac{5}{64}$	2.0	Flat–horiz[b]	250	26	119	3,040	6.6	2.99	1
$\frac{5}{64}$	2.0	Flat–horiz[b]	300	29	145	3,683	8.4	3.81	1
$\frac{5}{64}$	2.0	Flat–horiz[b]	350	31	181	4,597	10.2	4.63	1
$\frac{5}{64}$	2.0	Flat[b]	400	33	226	5,740	12.1	5.49	1
$\frac{3}{32}$	2.4	Flat–horiz[b]	350	26	120	3,048	9.2	4.17	1
$\frac{3}{32}$	2.4	Flat–horiz[b]	400	29	142	3,606	11.5	5.22	1
$\frac{3}{32}$	2.4	Flat[b]	450	32	174	4,419	13.7	6.21	1
$\frac{3}{32}$	2.4	Flat[b]	500	34	201	5,105	15.2	6.89	1
$\frac{3}{32}$	2.4	Flat[b]	550	36	234	5,943	18.1	8.21	1
$\frac{7}{64}$	2.8	Flat[b]	500	30	125	3,175	13.4	6.08	1
$\frac{7}{64}$	2.8	Flat[b]	550	32	145	3,683	15.5	7.03	1
$\frac{7}{64}$	2.8	Flat[b]	600	34	176	4,470	18.5	8.39	1
$\frac{7}{64}$	2.8	Flat[b]	650	36	196	4,978	20.6	9.34	1
$\frac{7}{64}$	2.8	Flat[b]	700	36	221	5,613	23.6	10.70	1
$\frac{1}{8}$	3.2	Flat[b]	600	32	120	3,048	17.8	8.07	1
$\frac{1}{8}$	3.2	Flat[b]	650	34	130	3,302	19.7	8.93	1
$\frac{1}{8}$	3.2	Flat[b]	700	36	143	3,632	21.4	9.70	1
$\frac{1}{8}$	3.2	Flat[b]	750	38	155	3,937	22.0	9.97	1
$\frac{1}{8}$	3.2	Flat[b]	800	38	166	4,216	24.6	10.88	1

[a]Use 80% argon, 20% CO_2 shielding gas.

[b]Use CO_2 shielding gas at 35 to 40 ft^3/hr (16 to 19 liters/min).

FIGURE 6–85 Welding procedure schedule for FCAW.

rent, levels are reduced approximately 20%. Electrical stickout is required for most self-shielding electrodes. The amount varies by electrode type.

Welding Variables

The welding variables involved with FCAW are essentially the same as those associated with gas metal arc welding. FCAW does have an extremely wide range of welding current and voltage. These are quite different for the electrodes that operate electrode positive and those that operate electrode negative. This information is summarized in Figure 6–87, which shows the operating range for both types.

Tips for Using the Process

These tips are essentially the same as those given for gas metal arc welding. There is only one major difference: The slag covering on the weld deposit allows the possibility of slag entrapment with FCAW. This requires manipulation of the welding arc in the same manner as used with

Diameter		Weld Position	Current Amp	Voltage DCEN	Approx. Wire Feed Speed		Deposition Rate	
in.	mm				in./min	mm/min	lb/hr	kg/hr
0.030	0.8	Flat & Horiz	25–200	14–18	55–565	1,397–14,351	.3–5.4	.13–2.45
0.030	0.8	Vert & Ovhd	25–150	14–17	80–300	2,032–7,620	.3–4.1	.13–1.86
0.035	0.9	Flat & Horiz	50–225	13–20	65–465	1,651–11,811	.6–6.3	.27–2.85
0.035	0.9	Vert. & Ovhd	75–175	13–19	52–300	1,321–7,620	.6–4.2	.27–1.91
0.045	1.1	Flat & Horiz	75–250	15–19	65–290	1,651–7,366	1.1–6.0	.49–2.72
0.045	1.1	Vert & Ovhd	100–200	16–18	80–190	2,032–4,826	1.5–3.8	.68–1.72
0.052	1.3	Flat & Horiz	125–300	16–20	60–191	1,524–4,851	1.1–4.2	.49–1.91
0.052	1.3	Vert & Ovhd	125–250	15–18	64–158	1,625–4,013	1.5–4.7	.68–2.13
$\frac{1}{16}$	1.6	Flat & Horiz	150–325	16–20	61–185	1,549–4,699	1.2–7.5	.54–3.40
$\frac{1}{16}$	1.6	Vert & Ovhd	150–250	16–18	61–110	1,549–2,794	1.2–4.8	.54–2.17
0.068	1.7	Flat & Horiz	175–325	15–18	53–133	1,346–3,378	1.5–7.6	.68–3.44
$\frac{5}{64}$	2.0	Flat & Horiz	200–400	16–23	30–104	762–2,642	3.0–7.8	1.36–3.54
$\frac{3}{32}$	2.4	Flat & Horiz	300–450	18–24	70–80	1,778–2,032	6.1–7.2	2.76–3.26

Note: No external shielding gas is used.

Stickout is $\frac{1}{2}$ in. for 0.30 thru 0.45 min dia and $\frac{3}{4}$ in. for 0.52 and larger.

FIGURE 6–85 (*cont.*)

shielded metal arc welding to avoid flux entrapment. The electrode should be directed toward the leading edge of the weld puddle and the tip-to-work or stickout length should be kept uniform. Special guns are available that incorporate electrical stickout, and these increase deposition rates. However, penetration is reduced and therefore the proper current and voltage should be employed to ensure root penetration.

Safety Considerations

Safety considerations for FCAW are the same as those for the other arc welding processes, and this was completely covered in Chapter 4. One factor with FCAW is the amount of smoke and fumes produced. This process produces more smoke than SMAW with covered electrodes; however, much more weld metal is being deposited per hour with this process. Proper positioning of the welder's head and the use of curved front welding hoods will greatly reduce the smoke that will reach the breathing zone. For more efficient collection of smoke, exhaust welding guns are recommended.

Limitation of the Process

The following are some of the limitations to this process:

  ❊ FCAW is used only to weld ferrous metals, primarily steels.
  ❊ The process normally produces a slag covering that must be removed.

  ❊ Cored electrode wire is more expensive on a weight basis than are solid electrode wires.

Industrial Use and Typical Applications

The FCAW process is replacing shielded metal arc for many applications, replacing GMAW, primarily the CO_2 version, and replacing submerged arc welding for thinner metal. The construction equipment industry has used the process to the greatest degree. Figure 6–88 shows the application of the process on a loader bucket.

The industrial equipment industry that produces machine tool bases, press frames, and so on, also is a large user of FCAW. A typical application on a press frame is shown in Figure 6–89. This type of work involves relatively heavy plate fabrication, usually carbon or low-alloy steel.

The tank and vessel industry also utilizes FCAW. The process meets the requirements of ASME for pressure vessel work. It is used on many applications that conform to the code. Figure 6–90 shows a typical vessel being welded.

The structural steel industry has switched to FCAW, for both in-plant fabrication and erection work. For the erection of structural steel, it is particularly useful for column splices, beam splices, and beam-to-column connections. Figure 6–91 shows a typical heavy box column being welded with the FCAW process.

The piping industry uses FCAW with the small-diameter electrode wires that provide all-position capabilities. The FCAW process is gaining in popularity because

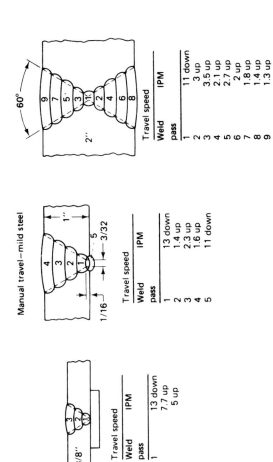

Travel speed

Weld pass	IPM
1	11 down
2	3 up
3	3.5 up
4	2.1 up
5	2.7 up
6	2 up
7	1.8 up
8	1.4 up
9	1.3 up

Manual travel—mild steel

Travel speed

Weld pass	IPM
1	13 down
2	1.4 up
3	2.3 up
4	1.6 up
5	11 down

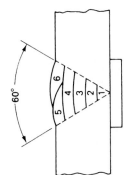

Travel speed

Weld pass	IPM
1	13 down
	7.7 up
	5 up

Material Thickness T		Type of Joint	Number of Passes	Root Opening		Electrode Diameter		Volts EP	Amps DC
in.	mm			in.	mm	in.	mm		
3/8	9.5	60° single vee	3	0	0	.045	1.1	22	180
1	25.4	60° single vee	5	3/32	2.4	.045	1.1	22	180
2	50.8	60° single vee	9	1/16	1.6	.045	1.1	22	180

(b)

Mild steel with backup

Material Thickness T		Type of Joint	Number of Passes	Root Opening R		Electrode Diameter		Welding Power Volts EP	Amps DC	Travel Speed per pass, ipm
in.	mm			in.	mm	in.	mm			
1/8	3.2	square	1	1/32	0.8	3/32	2.4	24-26	325	56
1/4	6.4	60° v	1	0	0	3/32	2.4	25-27	375	41
1/2	1.27	60° v	1	0	0	1/8	3.2	27-30	550	14
3/4	19.0	60° v	3	0	0	1/8	3.2	27-30	550	18
1	25.4	60° v	6	0	0	1/8	3.2	27-30	550	11

(a)

Flat

Horizontal

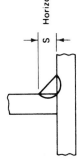

Weld Size (S)		Material Thickness T		Number of Passes	Electrode Diameter		Welding Power Volts EP	Amps DC	Travel Speed IMP (per pass)
in.	mm	in.	mm		in.	mm			
1/8	3.2	1/8	3.2	1	3/32	2.4	24-26	300-350	44-60
1/4	6.4	1/4	6.4	1	3/32	2.4	24-26	350-400	22-24
1/4	6.4	1/4	6.4	1	1/8	3.2	25-27	450-500	26-30
3/8	9.5	3/8	9.5	1	3/32	2.4	26-30	375-500	13-17
3/8	9.5	3/8	9.5	1	1/8	3.2	28-31	500-575	16-20
5/8	15.9	5/8	15.9	3	3/32	2.4	26-31	450-475	12-14
5/8	15.9	5/8	15.9	3	1/8	3.2	27-30	450-500	12-14

(c)

FIGURE 6–86 Welding procedure schedule: joint details.

Welding Range for E70T–1 with CO_2 Shielding (DCEP)									
		Minimum				Maximum			
Diameter				Wire Feed Speed				Wire Feed Speed	
in.	mm	Amperes	Volts	in./min	mm/min	Amperes	Volts	in./min	mm/min
0.045	1.2	120	21	168	4,267	300	30	625	15,875
$\frac{1}{16}$	1.6	150	24	100	2,540	425	31	400	10,160
$\frac{5}{64}$	2.0	200	26	95	2,413	450	33	270	6,858
$\frac{3}{32}$	2.4	300	26	95	2,413	600	36	255	5,477
$\frac{7}{64}$	2.8	450	30	110	2,794	750	38	237	6,019
$\frac{1}{8}$	3.2	550	32	98	2,489	850	39	175	4,445

Welding Range for E71T–11 Self-Shielding (DCEN)									
		Minimum				Maximum			
Diameter				Wire Feed Speed				Wire Feed Speed	
in.	mm	Amperes	Volts	in./min	mm/min	Amperes	Volts	in./min	mm/min
0.045	1.1	95	13	65	1,651	180	18.5	200	5,080
$\frac{1}{16}$	1.6	100	15	47	1,193	300	22	189	4,800
0.068	1.7	125	17	49	1,245	300	23	184	4,673
$\frac{5}{64}$	2.0	150	18	47	1,193	300	22.5	124	3,149
$\frac{3}{32}$	2.4	200	17	40	1,016	350	22	93	2,410

FIGURE 6–87 Welding current range for flux-cored electrodes.

of its proven economies. FCAW provides high deposition rates. It provides a higher operator factor or duty cycle. The electrode wires have a higher rate of utilization, and more economical weld joint details can be employed. This results in lower-cost weldment, which is the goal for weldment fabricators.

FIGURE 6–89 Welding press frame.

FIGURE 6–88 Welding loader bucket.

FIGURE 6–90 Welding pressure vessel.

FIGURE 6–92 Submerged arc welding.

FIGURE 6–91 Welding box column splice.

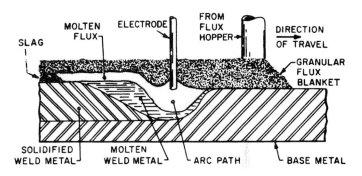

FIGURE 6–93 Process diagram for SAW.

6-6 SUBMERGED ARC WELDING

Submerged arc welding (SAW) is "an arc welding process that uses an arc or arcs between a bare metal electrode or electrodes and the weld pool. The arc and molten metal are shielded by a blanket of granular flux on the workpieces. The process is used without pressure and with filler metal from the electrode and sometimes from a supplemental source (welding rod, flux, or metal granules)." This normally applied automatic welding process is shown in Figure 6-92. This process, also known as *under*

powder welding or *smothered arc welding,* was developed in the late 1920s and introduced in the early 1930s.

Principles of Operation

The submerged arc welding process shown in Figure 6-93 utilizes the heat of an arc between a continuously fed electrode and the work. The heat of the arc melts the surface of the base metal and the end of the electrode. The metal melted off the electrode is transferred through the arc to the workpiece, where it becomes the deposited weld metal. Shielding is obtained from a blanket of granular flux, which is laid directly over the weld area. The flux close to the arc melts and intermixes with the molten weld metal and helps purify and fortify it. The flux forms a glasslike slag that is lighter in weight than the deposited weld metal and floats on the surface as a protective cover. The weld is submerged under this layer of flux

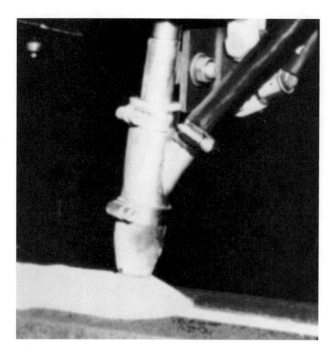

FIGURE 6–94 Welder's view of submerged arc welding.

and slag—hence the name *submerged arc welding.* The welder's view is shown in Figure 6-94. The flux and slag normally cover the arc so that it is not visible. The unmelted portion of the flux can be reused. The electrode is fed into the arc automatically from a coil. The arc is maintained automatically and travel can be manual or by machine. The arc is initiated by a fuse-type start or by a reversing feed system. The metal transfer mode is less important in submerged arc welding.

Advantages and Major Uses

The submerged arc welding process is one of the older automatic processes and was originally used to make the longitudinal seam in large pipe. It was developed to provide high-quality deposited weld metal by shielding the arc and the molten metal from the contaminating effects of the air. The major advantages of the process are:

✻ High-quality weld metal
✻ Extremely high deposition rate and speed
✻ Smooth, uniform finished weld with no spatter
✻ Little or no smoke
✻ No arc flash, thus minimal need for protective clothing
✻ High utilization of electrode wire
✻ Easily automated for high operator factor
✻ Manipulative skills not involved

The submerged arc process is widely used in heavy steel plate fabrication work. This includes the welding of

structural shapes and the longitudinal seam of larger diameter pipe, the manufacture of machine components for all types of heavy industry, and the manufacture of vessels and tanks for pressure and storage use. It is widely used in the shipbuilding industry for splicing and fabricating subassemblies, and by many other industries where steels are used in medium to heavy thickness. It is also used for surfacing and buildup work, maintenance, and repair.

Methods of Application and Position Capabilities

The most popular method of applying is the mechanized method where the operator monitors the welding operation. Second in popularity is the automatic method, where welding is a push-button operation. The process can be applied semiautomatically; however, this method of application is not too popular. The process cannot be applied manually because it is impossible for a welder to control an arc that is not visible.

The submerged arc welding process is a limited-position welding process. Welding can be done in the flat position and in the horizontal fillet position with ease. The welding positions are limited because of the large pool of molten metal which is very fluid. The slag is also very fluid and will tend to run out of the joint. Under special controlled procedures it is possible to weld in the horizontal position, sometimes called 3-o'clock welding. This requires special devices to hold the flux up so that the molten slag and weld metal cannot run away. The process is not used in the vertical or overhead position.

Weldable Metals and Thickness Range

Submerged arc welding is used to weld low- and medium-carbon steels, low-alloy high-strength steels, quenched and tempered steels, and many stainless steels. Experimentally it has been used to weld certain copper alloys, nickel alloys, and even uranium. This information is summarized in Figure 6-95. Submerged arc welding is also used for hard surfacing and overlay operations.

Metal thickness from $\frac{1}{16}$ in. (1.6 mm) to $\frac{1}{2}$ in. (12 mm) can be welded with no edge preparation. With edge preparation, welds can be made with a single pass on ma-

FIGURE 6–95 Metals weldable by SAW.

Base Metal	Weldability
Wrought iron	Weldable
Low-carbon steel	Weldable
Low-alloy steel	Weldable
High- and medium-carbon steel	Possible but not popular
Alloy steel	Possible but not popular
Stainless steel	Weldable

terial from $\frac{1}{4}$ in. (6.4 mm) through 1 in. (25 mm). When multi-pass technique is used, the maximum thickness is practically unlimited (Figure 6-96). Horizontal fillet welds can be made up to $\frac{3}{8}$ in. (9.5 mm) in a single pass, and in the flat position fillet welds can be made up to 1 in. (25 mm) size.

Joint Design

The submerged arc welding process can utilize the same joint design details as the shielded metal arc welding process. These joint details are given in Chapter 19. However, for maximum utilization and efficiency of submerged arc welding, different joint details are suggested.

For groove welds, the square groove design can be used up to a $\frac{5}{8}$-in. (16-mm) thickness. Beyond this thickness bevels are required. Open roots are used but backing bars are necessary since the molten metal will run through the joint. When welding thicker metal, if a sufficiently large root face is used, the backing bar may be eliminated. However, to assure full penetration when welding from one side, backing bars are recommended. Where both sides are accessible, the backing weld can be made, which will fuse into the original weld to provide full penetration. Recommended submerged arc joint designs are shown in Figure 6-97.

Welding Circuit and Current

The welding circuit employed for single-electrode submerged arc welding is shown in Figure 6-98. The submerged arc welding process uses either direct or alternating current for welding power. Direct current is used for most applications that employ a single arc. Both direct-current electrode positive (DCEP) and electrode negative (DCEN) are used, as is alternating current.

The constant-voltage type of dc power is more popular for submerged arc welding with $\frac{1}{8}$ in. (3.2 mm) and smaller-diameter electrode wires. The constant-current power system is normally used for welding with $\frac{5}{32}$-in. (4-mm) and large-diameter electrode wires. The control circuit for CC power is more complex since it duplicates the actions of the welder to retain a specific arc length. The wire feed system must sense the voltage across the arc and feed the electrode wire into the arc to maintain this voltage. As conditions change, the wire feed must slow down or speed up to maintain the prefixed voltage across the arc. This adds complexity to the control system, and the system cannot react instantaneously. Arc starting is more complicated with the constant current system since it requires the use of a reversing system to strike the arc, retract, and then maintain the preset arc voltage.

For ac welding the constant-current power is always used. When multiple-electrode-wire systems are used with both ac and dc arcs, the constant-current power system is utilized. The constant-voltage system, however, can be applied when two wires are fed into the arc supplied by a single power source. Welding current for submerged arc welding can vary from as low as 50 A to as high as 2,000 A. Most submerged arc welding is done in the range 200 to 1,200 A.

Equipment Required

The equipment components required for submerged arc welding (Figure 6-98) consist of (1) welding machine or power source, (2) the wire feeder and control system, (3) the welding torch for automatic welding, or the welding gun and cable assembly for semiautomatic welding, (4) the flux hopper and feeding mechanism and usually a flux recovery system, and (5) travel mechanism for automatic welding.

The power source for submerged arc welding must be rated for a 100% duty cycle, since the submerged arc welding operations are continuous and the length of time for making a weld may exceed 10 minutes. If a 60% duty cycle power source is used, it must be derated according to the duty cycle curve for 100% operation.

When constant current is used, either ac or dc, the voltage-sensing electrode wire feeder system must be used. The CV system is only used with direct current.

Both generator and transformer-rectifier power sources are used, but the rectifier machines are more popular. Welding machines for submerged arc welding range in size from 300 to 1500 A. They may be connected in parallel to provide extra power for high-current applications. Direct-current power is used for semiautomatic applications, but alternating current power is used primarily with the mechanized or the automatic method.

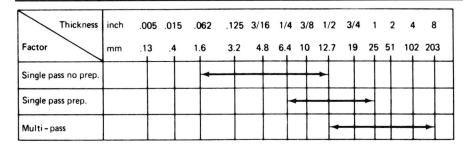

Thickness / Factor	inch	.005	.015	.062	.125	3/16	1/4	3/8	1/2	3/4	1	2	4	8
	mm	.13	.4	1.6	3.2	4.8	6.4	10	12.7	19	25	51	102	203
Single pass no prep.				←				→						
Single pass prep.								←		→				
Multi - pass									←			→		

FIGURE 6-96 Base metal thickness range for SAW.

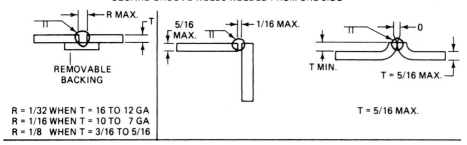

SQUARE-GROOVE WELDS WELDED FROM ONE SIDE

R = 1/32 WHEN T = 16 TO 12 GA
R = 1/16 WHEN T = 10 TO 7 GA
R = 1/8 WHEN T = 3/16 TO 5/16

T = 5/16 MAX.

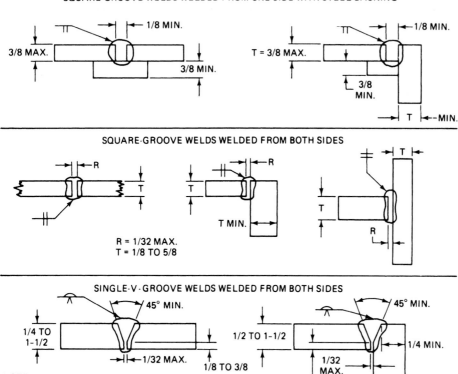

SQUARE-GROOVE WELDS WELDED FROM ONE SIDE WITH STEEL BACKING

SQUARE-GROOVE WELDS WELDED FROM BOTH SIDES

R = 1/32 MAX.
T = 1/8 TO 5/8

SINGLE-V-GROOVE WELDS WELDED FROM BOTH SIDES

NOTE:
TO OBTAIN FULL PENETRATION WELD FROM
ONE SIDE USE THE SMALL ROOT FACE
DIMENSION AND REMOVABLE BACKING.

FIGURE 6–97 Weld joint designs for SAW.

Multiple-electrode systems require specialized types of circuits, especially when ac is employed.

For semiautomatic application, a welding gun and cable assembly are used to carry the electrode and current and to provide the flux at the arc. An application of semiautomatic submerged arc welding is shown in Figure 6–99. A small flux hopper is attached to the end of the cable assembly and the electrode wire is fed through the bottom of this flux hopper through a current pickup tip to the arc. The flux is fed from the hopper to the welding area by means of gravity. The amount of flux fed depends on the height the gun is held above the work. The hopper gun may include a start switch to initiate the weld, or it may utilize a "hot" electrode so that when the electrode is touched to the work, feeding will begin automatically. Pressure flux feed systems are also used.

For automatic welding, the torch is normally attached to the wire feed motor and the flux hopper is at-tached to the torch. The flux hopper may have a magnetically operated valve that can be opened or closed by the control system.

The final piece of equipment sometimes used is a travel carriage, which can be a simple tractor or a complex moving specialized fixture. A typical travel carriage is shown in Figure 6–100. A flux recovery unit can be provided to collect the unused submerged arc flux and return it to the supply hopper. Submerged arc welding systems can become more complex by adding such devices as seam followers, weavers, or work movers.

Electrode and Flux

Two materials are used in submerged arc welding, the welding flux and the consumable electrode. The American Welding Society has two specifications[20,21] that provide a classification system for both the flux and the electrode.

SINGLE-GROOVE WELDS WELDED FROM ONE SIDE WITH STEEL BACKING

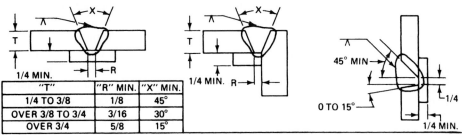

"T"	"R" MIN.	"X" MIN.
1/4 TO 3/8	1/8	45°
OVER 3/8 TO 3/4	3/16	30°
OVER 3/4	5/8	15°

DOUBLE-V-GROOVE WELDS WELDED FROM BOTH SIDES

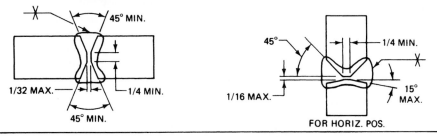

DOUBLE-BEVEL-GROOVE WELDS WELDED FROM BOTH SIDES

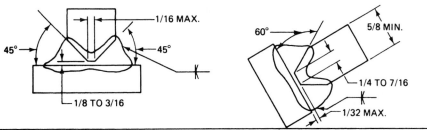

SINGLE-U-GROOVE WELDS WELDED FROM ONE OR BOTH SIDES

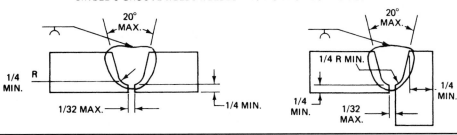

FIGURE 6–97 (*cont.*)

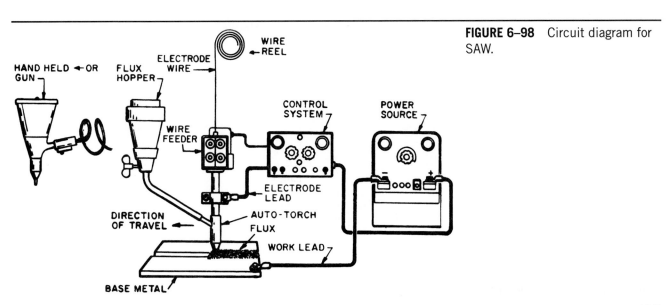

FIGURE 6–98 Circuit diagram for SAW.

FIGURE 6–99 Semiautomatic submerged arc welding.

FIGURE 6–100 Submerged arc head on travel carriage.

The flux is specified by the letter *F* followed by a two- or three-digit number that indicates the minimum tensile strength in increments of 10,000 psi. This is followed by a letter that indicates the condition of heat treatment for testing the welds. *A* stands for "as welded," and *P* stands for "postweld heat treated." This is followed by a one- or two-digit number, which indicates the minimum temperature in Fahrenheit of impact tests to provide 20 ft-lb of energy absorption (or the minimum temperature in Celsius of an impact test to provide 27 joules of energy absorption). There are eight classifications for impact strength. The classification for the flux is summarized in Figure 6–101.

The electrode is specified by the letter *E* followed by three digits. Note, however, that the letter *E* can be followed by the letter *C* if the electrode is of composite construction. Omission of a *C* indicates a solid electrode. The next digit is to designate the manganese content. This is followed by a one- or two-digit number used to indicate nominal carbon content in hundredths of a percent carbon. These digits are sometimes followed by a letter *K* which indicates that the electrode steel was silicon killed. If the steel is of another type, a *K* will not appear. This is sometimes followed by two digits, which indicate the alloys that are present. Figure 6–102 shows the electrode classification system for carbon steels.[20] This does not, however, cover the alloy steels. For complete information on the alloy steels, refer to the AWS specification.[21] The composition requirements for submerged arc carbon steel electrodes are shown in Figure 6–103.

An example of the flux–electrode classification system is as follows:

> *F7A6-EM12K:* indicates a flux–electrode combination that will produce weld metal that in the as-welded condition will have a tensile strength of not less than 70,000 psi and a Charpy V-notch impact strength of at least 20 ft-lb at −60° F when deposited with

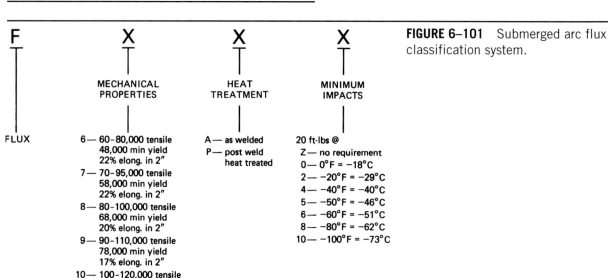

FIGURE 6–101 Submerged arc flux classification system.

F	X	X	X
	MECHANICAL PROPERTIES	HEAT TREATMENT	MINIMUM IMPACTS

FLUX

6 — 60–80,000 tensile
48,000 min yield
22% elong. in 2″

7 — 70–95,000 tensile
58,000 min yield
22% elong. in 2″

8 — 80–100,000 tensile
68,000 min yield
20% elong. in 2″

9 — 90–110,000 tensile
78,000 min yield
17% elong. in 2″

10 — 100–120,000 tensile
88,000 min yield
16% elong. in 2″

A — as welded
P — post weld
heat treated

20 ft-lbs @
Z — no requirement
0 — 0°F = −18°C
2 — −20°F = −29°C
4 — −40°F = −40°C
5 — −50°F = −46°C
6 — −60°F = −51°C
8 — −80°F = −62°C
10 — −100°F = −73°C

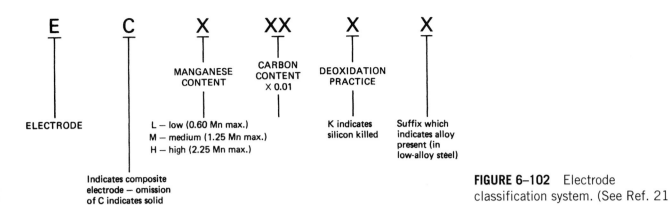

FIGURE 6–102 Electrode classification system. (See Ref. 21 for low-alloy steels.)

AWS Classification	Chemical Composition (Wt %)[a, b]					
	C	Mn	Si	S	P	Cu[c]
Low-manganese steel electrodes						
ELB	0.10	0.25/0.60	0.07	0.035	0.035	0.35
EL8K	0.10	0.25/0.60	0.10/0.25	0.035	0.035	0.35
EL12	0.05/0.15	0.25/0.60	0.07	0.035	0.035	0.35
Medium-manganese steel electrodes						
EM12	0.06/0/15	0.80/1.25	0.10	0.035	0.035	0.35
EM12K	0.05/0.15	0.80/1.25	0.10/0.35	0.035	0.035	0.35
EM13K	0.07/0.19	0.90/1.40	0.35/0.75	0.035	0.035	0.35
EM15K	0.10/0.20	0.80/1.25	0.10/0.35	0.035	0.035	0.35
High-manganese steel electrodes						
EH14	0.10/0.20	1.70/2.20	0.10	0.035	0.035	0.35

[a]Single values are maximums.

[b]Electrodes are to be analyzed for those elements for which specific values are shown. Elements other than those shown, which are intentionally added (except iron), are also to be reported. The total of these latter elements and all other elements that are not intentionally added must not exceed 0.50%.

[c]The copper limit includes any copper coating that may be applied to the electrode.

FIGURE 6–103 Electrode composition. (See Ref. 21 for low-alloy steels.)

an EM12K electrode under standard conditions called for in the AWS specification.

The flux shields the arc and molten weld metal from atmospheric oxygen and nitrogen. The flux contains deoxidizers and scavengers, which help remove impurities from the weld metal. Flux also provides a means for introducing alloys into the weld metal. Alloys and deoxidizers may also be introduced from the welding electrode.

As the molten flux cools, it forms a glassy slag covering, which protects the surface of the weld. The nonmelted portion of the flux does not change its form; its properties are not affected, so this unmelted flux can be recovered and reused. The flux that melts and forms the slag covering must be removed from the weld. This is easily done after the weld cools and, in many cases, will peel for removal without special effort. In a groove weld the solidified slag may have to be removed by a chipping hammer. The fused flux that is removed must be discarded since the alloying elements and deoxidizers are exhausted during the melting phase.

Selection of Flux Wire Combination

In submerged arc welding it is necessary to select an electrode and flux combination to match the base metal composition and properties. Fluxes of different manufacturers are not interchangeable without making tests. Fluxes may be neutral or active. Neutral fluxes will not produce any significant changes in weld metal chemistry. They are normally used for multi-pass welding. Active fluxes contain small amounts of manganese and/or silicon used to reduce porosity and weld cracking. They are normally used for single-pass applications. The (1) neutral, (2) active, (3) alloy is alloy fluxes, which when used with plain carbon steel electrodes produce alloy weld deposits. This is done to match particular base metals or, with additional alloys, is used for hardfacing applications.

Variations in arc voltage change flux consumption. Higher arc voltage (long arc length) increases the amount of flux melted or consumed. This can cause more alloy to be deposited; hence it is important to follow the manufacturer's recommended voltages when using a particular flux.

In general, the flux is selected based on the mechanical properties required of the weld deposit. The electrode would be selected in conjunction with the flux to deliver these mechanical properties. Manufacturers usually list fluxes with several combinations of electrodes for welding different steels. The manufacturer's recommendations should be followed with respect to single- or multiple-pass type of application related to the base metal properties. If weld requirements are critical, tests should be made to qualify the procedure that will produce the weld properties desired.

Deposition Rates and Quality of Welds

The deposition rates of the submerged arc welding process are higher than any other arc welding process. Deposition rates for single electrodes are shown in Figure 6-104. There are four factors that control the deposition rate of submerged arc welding: polarity, long stickout, additives in the flux, and additional electrodes. The deposition rate is the highest for direct current electrode negative DCEN. The deposition rate for alternating current is between DCEP and DCEN. The polarity of maximum heat is the negative pole.

The deposition rate can be increased by extending the stickout. This is the distance from the point where current is introduced into the electrode to the arc. It is also called I^2R welding. Normally, the distance between the contact tip and the work is 1 to $1\frac{1}{2}$ in. (25 to 38 mm). If the stickout is increased, it will cause preheating of the electrode wire, which will greatly increase the deposition rate. As stickout is increased, the penetration into the base metal decreases. This factor must be given serious consideration because in some situations the penetration is required. The relationship between stickout and deposition rate is shown in Figure 6-105. The deposition rates can be increased by metal additives in the submerged arc flux. Additional electrodes can be used to increase the overall deposition rate.

The quality of the weld metal deposited by the submerged arc welding process is high. The weld metal strength and ductility exceeds that of the mild steel or low-alloy base material when the correct combination of electrode wire and submerged arc flux is used. In general, the weld bead size per pass is much greater with submerged arc welding than with any of the other arc welding processes. The heat input is higher and cooling rates are slower, and, for this reason, gases are allowed more

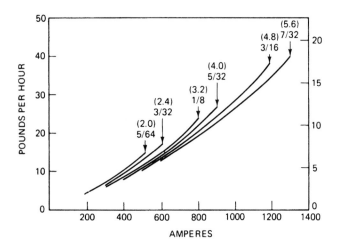

FIGURE 6-104 Welding deposition rates for SAW.

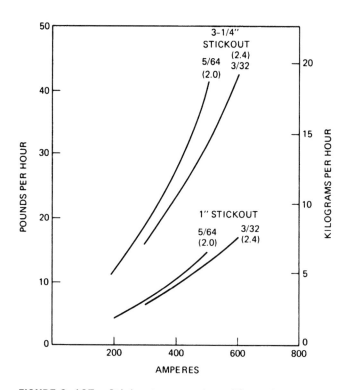

FIGURE 6-105 Stickout versus deposition rate.

time to escape. Uniformity and consistency are advantages of this process when applied automatically.

Several problems may occur when using the semiautomatic application method. The electrode wire may be curved when it leaves the nozzle of the welding gun. This curvature can cause the arc to be struck in a location not expected by the welder. When welding in deep grooves, the curvature may cause the arc to be against one side of the weld joint rather than at the root. This will cause incomplete root fusion, and flux will be trapped at the root of the weld. Another problem with semiautomatic welding is the problem of completely filling the weld groove or maintaining exact size, since the weld is hidden and cannot be observed while it is being made. This requires making an extra pass or in some cases too much weld is deposited. Variations in root opening affect the travel speed, and if travel speed is uniform the weld may be under- or overfilled in different areas. High operator skill and experience will overcome this problem.

There is another quality problem associated with extremely large single-pass weld deposits. When these large welds solidify, the impurities in the melted base metal and in the weld metal all collect at the last point to freeze, which is the centerline of the weld. If there is sufficient restraint and enough impurities are collected at this point, centerline cracking may occur. This can happen when making large single-pass flat fillet welds if the base metal plates are 45° from flat. A simple solution is to avoid placing the parts at a true 45° angle. It should be varied approximately 10° so that the root of the joint is not in line with the centerline of the fillet weld. Another solution is to make multiple passes rather than attempting to make a large weld in a single pass.

Excessively hard weld deposits contribute to cracking of the weld during fabrication or during service. A maximum hardness level of 225 Brinell is recommended. The reason for the hard weld in carbon and low-alloy steels is too rapid cooling, inadequate postweld treatment, or excessive alloy pickup in the weld metal. Excessive alloy pickup is due to selecting an electrode that has too much alloy, or a flux that introduces too much alloy into the weld, or the use of excessively high welding voltages.

In automatic and mechanized welding, defects may occur at the start or at the end of the weld. The best solution is to use runout tabs so that starts and stops will be on the tabs rather than on the product.

Weld Schedules

The submerged arc welding process applied fully automatically should be done in accordance with welding procedure schedules. Figure 6–106 shows some suggested welding schedules using a single electrode on mild and low-alloy steels. These tables can be used for welding other ferrous materials, but were developed for mild steel. All of the welds made by this procedure should pass qualification tests. If the schedules are varied more than 10%, qualification tests should be performed to determine the weld quality.

Welding Variables

The welding variables for submerged arc welding are similar to the other arc welding processes, with several exceptions. The electrode size is related to the weld joint size and the current recommended for the particular joint. This must also be considered in determining the number of passes or beads for a particular joint. Welds for the same joint dimension can be made with many or few passes; this depends on the weld metal metallurgy desired. Multiple passes are more expensive but usually deposit higher-quality weld metal. The polarity is established initially and is based on whether maximum penetration or maximum deposition rate is required.

The major variables that affect the weld involve heat input and include the welding current, arc voltage, and travel speed. Welding current is the most important. For single-pass welds, the current should be sufficient for the desired penetration without burn-through. The higher the current, the deeper the penetration. In multipass work, the current should be suitable to produce the size of the weld expected in each pass.

The arc voltage is varied within narrow limits. It has an influence on the bead width and shape. Higher voltages will cause the bead to be wider and flatter. Extremely high arc voltage can cause cracking. This is because an abnormal amount of flux is melted and excess deoxidizers may be transferred to the weld deposit, lowering its ductility. Higher arc voltage also increases the amount of flux consumed. The low arc voltage produces a stiffer arc that improves penetration, particularly in the bottom of deep grooves. If the voltage is too low, a very narrow bead will result. It will have a high crown, and the slag will be difficult to remove.

Travel speed has an influence on both bead width and on penetration. Faster travel speeds produce narrower beads that have less penetration. This can be an advantage for sheet metal welding, where small beads and minimum penetration are required. If speeds are too fast, however, there is a tendency for undercut and porosity, since the weld freezes quicker. If the travel speed is too slow, the electrode stays in the weld pool too long, which will create poor bead shape and may cause excessive spatter and flash through the layer of flux.

The secondary variables include the angle of the electrode to the work, the angle of the work itself, the thickness of the flux layer, and, most important, the distance between the current pickup tip and the arc also called electrode stickout.

The depth of the flux layer must be controlled. If it is too thin, there will be too much arcing through the flux or arc flash. This also may cause porosity. If the flux depth is too heavy, the weld may be narrow and humped. On the subject of flux, too many fines (small particles of flux) in the flux can cause surface pitting since the gases generated in the weld may not escape. These are sometimes called *pock marks* on the bead surface.

Tips for Using the Process

One of the major applications for submerged arc welding is on circular welds where the parts are rotated under a fixed head. These welds can be made on the inside or outside diameter. Submerged arc welding produces a large

| Material Thickness | | | Type of Weld | Electrode Dia. (in.) | Arc Welding Current (A dc) | Voltage (Electrode Positive) | Wire Feed (in./min) | Travel Speed (in./min) |
Gauge	in.	mm						
18 and thinner			Square groove	$\frac{1}{16}$	200	20 to 22	85	100–140
16	0.063	1.6	a Square groove	$\frac{3}{32}$	300	22	68	100–140
			b Square groove	$\frac{1}{8}$	425	26	53	95–120
14	0.078	2	a Square groove	$\frac{3}{32}$	375	23	85	100–140
			b Square groove	$\frac{1}{8}$	500	27	65	75–85
12	0.109	2.8	a Square groove	$\frac{1}{8}$	400	23	51	70–90
			b Square groove	$\frac{1}{8}$	550	27	65	50–60
			d Fillet	$\frac{1}{8}$	400	25	51	40–60
10	0.140	3.5	a Square groove	$\frac{1}{8}$	425	26	53	50–80
			b Square groove	$\frac{5}{32}$	650	27	55	40–45
$\frac{3}{16}$ in.	0.188	4.8	a Square groove	$\frac{5}{32}$	600	26	50	40–75
			b Square groove	$\frac{3}{16}$	875	31	55	35–40
			d Fillet	$\frac{1}{8}$	525	26	67	35–40
$\frac{1}{4}$ in.	0.250	6.3	a Square groove	$\frac{3}{16}$	800	28	50	30–35
			b Square groove	$\frac{3}{16}$	875	31	56	22–25
			d Fillet	$\frac{5}{32}$	650	28	56	30–35
			e V-groove	$\frac{3}{16}$	750	30	47	25–40
$\frac{3}{8}$ in.	0.375	9.5	b Square groove	$\frac{3}{16}$	950	32	61	20–25
			f Square groove	$\frac{3}{16}$	First pass 500	32	27	30
					Second pass 750	33	47	30
			e V-groove	$\frac{3}{16}$	900	33	57	23–25
			d Fillet	$\frac{3}{16}$	950	31	61	30–35
$\frac{1}{2}$ in.	0.500	12.6	c V-groove	$\frac{3}{16}$	975	33	63	12–17
			f Square groove	$\frac{3}{16}$	First pass 650	34	40	25
					Second pass 850	35	54	23–27
			e V-groove	$\frac{3}{16}$	950	35	61	18–20
			d Fillet	$\frac{3}{16}$	950	33	61	14–17
$\frac{3}{4}$ in.	0.75	19	c V-groove	$\frac{7}{32}$	1,000	35	49	68
			f Square groove	$\frac{3}{16}$	First pass 925	37	59	12
					Second pass 1,000	40	65	11
			e V-groove	$\frac{7}{32}$	950	36	46	10–12
			d Fillet	$\frac{7}{32}$	1,000	35	49	6–8
			g V-groove	$\frac{7}{32}$	First pass	34	46	15
					Second pass 750	34	25	22
			h Double V-groove	$\frac{3}{16}$	First pass 700	35	42	20–22
					Second pass 1,000	36	65	14–16
1 in.	1.000	25.4	g V-groove	$\frac{7}{32}$	First pass 1,150	36	58	11
					Second pass 850	36	40	20
			h Double V-groove	$\frac{7}{32}$	First pass 900	36	42	13–15
					Second pass 1,075	36	52	12–14
$1\frac{1}{4}$ in.	1.25	32	h Double V-groove	$\frac{7}{32}$	First pass 1,000	36	50	13
					Second pass 1,125	37	56	8
$1\frac{1}{2}$ in.	1.50	38	h Double V-groove	$\frac{7}{32}$	First pass 1,050	36	51	9
					Second pass 1,125	37	56	7

Note: Wire feed speed is approximate.

FIGURE 6–106 Welding procedure schedules for SAW.

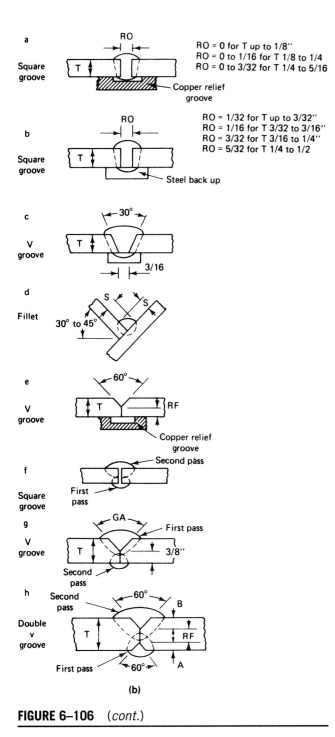

a Square groove	RO = 0 for T up to 1/8" RO = 0 to 1/16 for T 1/8 to 1/4 RO = 0 to 3/32 for T 1/4 to 5/16
b Square groove	RO = 1/32 for T up to 3/32" RO = 1/16 for T 3/32 to 3/16" RO = 3/32 for T 3/16 to 1/4" RO = 5/32 for T 1/4 to 1/2

FIGURE 6–106 *(cont.)*

molten weld pool and molten slag which tends to run. This dictates that on outside diameters the electrode should be positioned ahead of the extreme top, or 12-o'clock position, so that the weld metal will begin to solidify before it starts the downside slope. This becomes more of a problem as the diameter of the part being welded is smaller. Improper electrode position will increase the possibility of slag entrapment or a poor weld surface. The angle of the electrode should also be changed and pointed in the direction of travel of the rotating part.

When the welding is done on the inside circumference, the electrode should be angled so that it is ahead of bottom center, or the 6-o'clock position. Figure 6–107 illustrates these two conditions.

Sometimes the work being welded is sloped downhill or uphill to provide different types of weld bead contours. If the work is sloped downhill, the bead will have less penetration and will be wider. If the weld is sloped uphill, the bead will have deeper penetration and will be narrower. These are based on all other factors remaining the same (Figure 6–108).

The weld will be different depending on the angle of the electrode with respect to the work when the work is level. This is the travel angle, which can be a drag or push angle, and has a definite effect on the bead contour and weld metal penetration. Figure 6–109 shows the relationship.

FIGURE 6–107 Welding on rotating circular parts.

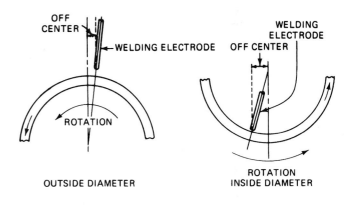

FIGURE 6–108 Angle of slope of work versus weld.

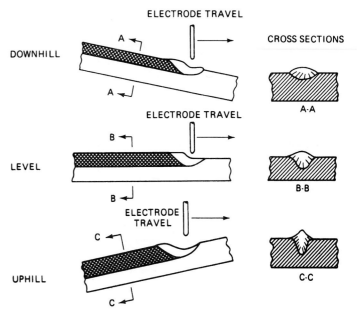

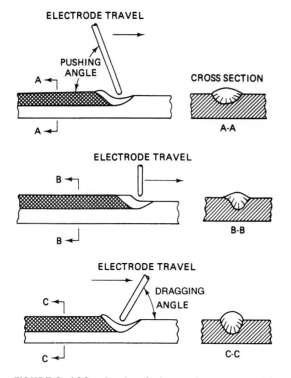

FIGURE 6–109 Angle of electrode versus weld.

FIGURE 6–110 Two-electrode wire systems.

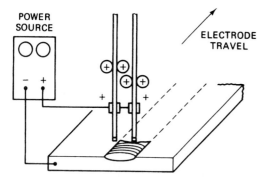

TRANSVERSE ELECTRODE POSITION

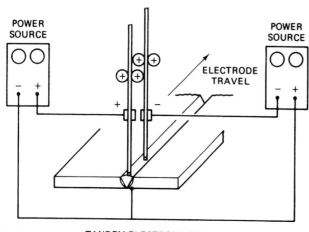

TANDEM ELECTRODE POSITION

One-side welding with complete root penetration can be obtained with submerged arc welding. When the weld joint is designed with a tight root opening and a fairly large root face, high current and electrode positive should be used. If the joint is designed with a root opening and a minimum root face, it will be necessary to use a backing bar since there will be nothing to support the molten weld metal. The molten flux is very fluid and will run through narrow openings. If this happens, the weld metal will follow and the weld will burn through the joint. Backing bars are needed whenever there is a root opening and a minimum root face.

Copper backing bars are useful when welding thin steel. Without backing bars, the weld would tend to melt through and the weld metal would fall away from the joint. The backing bar holds the weld metal in place until it solidifies. The copper backing bars may be water cooled to avoid the possibility of melting and copper pickup in the weld metal. For thicker materials, the backing may be submerged arc flux or other specialized type flux. More details of one-side welding are given in Chapter 26.

Safety Considerations

Safety precautions for submerged arc welding are somewhat fewer and less exacting than for other arc welding processes because of the nature of the submerged arc process and because most submerged arc welding is applied automatically. See Chapter 4 for details.

The welding arc is normally not visible in the submerged arc welding process. Only small amounts of sparks or flash are produced; therefore, it is not necessary to wear a welding face helmet. It is necessary to wear tinted safety glasses.

Limitations of the Process

A major limitation of submerged arc welding is its limited welding positions capability. The other limitation is that it is used primarily to weld steels.

The high-heat input, slow-cooling cycle can be a problem when welding quenched and tempered steels. The heat input limitation of the steel in question must be strictly adhered to when using submerged arc welding. This may require the making of multi-pass welds where a single pass weld would be acceptable in mild steel. In some cases, the economics may be reduced to the point where flux-cored arc welding or some other process should be considered.

In semiautomatic submerged arc welding, the inability to see the arc and puddle can be a disadvantage in reaching the root of a groove weld and properly filling or sizing.

Variations of the Process

There are a large number of variations to the process that give it additional capabilities. Some of the more popular variations are:

1. Two-wire systems having the same power source
2. Two-wire systems having a separate power source
3. Three-wire systems having a separate power source
4. Strip electrode for surfacing
5. Iron powder additions to the flux
6. Long stickout welding (mentioned previously)
7. Electrically "cold" filler wire

The multiwire systems offer advantages since deposition rates and travel speeds can be improved by using more electrodes. Figure 6–110 shows the two methods of utilizing two electrodes, one with a single-power source and one with two-power sources. When a single-power source is used, the same drive rolls are used for feeding both electrodes in the weld. When two power sources are used, individual wire feeders must be used to provide electrical insulation between the two electrodes. With two electrodes and separate power, it is possible to utilize different polarities on the two electrodes or to utilize alternating current on one and direct current on the other. The electrodes can be placed side by side, in what is called *transverse electrode position,* or they can be placed one in front of the other in the *tandem electrode position.*

The two-wire tandem electrode position with individual power sources is used where extreme penetration is required. The leading electrode is positive, with the trailing electrode negative. The first electrode creates a digging action and the second electrode will fill the weld joint. When two dc arcs are in close proximity, there is a tendency for arc interference between them. In some cases, the second electrode is connected to alternating current to avoid the interaction of the arc. The three-wire tandem system normally uses ac power on all three electrodes connected to three-phase power systems. The three-wire systems are used for making high-speed longitudinal seams for large-diameter pipe and for fabricated beams. Extremely high currents can be used, with correspondingly high travel speeds and deposition rates.

The strip welding system is used to overlay mild and alloy steels, usually with stainless steel. A wide bead is produced that has a uniform and minimum penetration. This process variation is shown in Figure 6–111. It is used for overlaying the inside of vessels to provide the corrosion resistance of stainless steel while utilizing the strength and economy of the low-alloy steels for the wall thickness. A strip electrode feeder is required and special flux is normally used. When the width of the strip is over 2 in. (50 mm), a magnetic arc oscillating device is employed to provide for even burn-off of the strip and uniform penetration.

Another way of increasing the deposition rate of submerged arc welding is to add iron base ingredients to the joint under the flux. This is sometimes called "bulk" welding. The iron in this material will melt in the heat of the arc and will become part of the deposited weld metal.

This greatly increases deposition rates without decreasing weld metal properties. Metal additives can also be used for special surfacing applications. This variation can be used with single-wire or multiwire installations. Figure 6–112 shows the increased deposition rates attainable.

Another variation is the use of an electrically cold filler wire fed into the arc area. The cold filler rod can be solid or flux cored to add special alloys to the weld metal. By regulating the addition of the proper material, the properties of the deposited weld metal can be improved. It is possible to utilize a flux-cored electrode for one of the multiple electrodes to introduce special alloys into the weld metal deposit. Each of these variations requires special engineering to ensure that the proper material is added to provide the desired deposit properties.

Industrial Use and Typical Applications

The submerged arc welding process is widely used to manufacture most heavy steel products. These include pressure vessels, boilers, tanks, nuclear reactors, chemical vessels, and so on. Another use is in the fabrication of trusses and beams. It is used for welding flanges to the web (Figure 6–113). The heavy equipment industry is a major user of submerged arc welding. An unusual application is the simultaneous welding of two joints on a triangular-shaped boom of an excavator (Figure 6–114). Another application is the welding together of forged halves to make tractor rollers (Figure 6–115). Shipbuilding is another major user of submerged arc welding, using both mechanized and automatic methods. Fully automatic submerged arc machines splice plates together and weld stiffeners to plates. Much of the machine welding is done using welding heads mounted on small carriages (Figure 6–116). The overlay of surfaces for corrosion resistance and for wear resistance is also a major use for submerged arc welding.

6-7 ELECTROSLAG WELDING

Electroslag welding (ESW) is "a welding process that produces coalescence of metals with molten slag that melts the filler metal and the surfaces of the workpieces. The weld pool is shielded by the slag, which moves along the full cross section of the joint as welding progresses. The process is initiated by an arc that heats the slag. The arc is then extinguished by the conductive slag, which is kept molten by its resistance to electric current passing between the electrode and the workpieces." Normally, molding shoes are used to confine the molten weld metal and slag; pressure is not used. This process was invented in the early 1930s in the United States, but became popular when equipment was designed for its use in the Soviet Union in the late 1940s.

Electroslag welding is normally used to make welds in the vertical position and on steels. It is applied as a

FIGURE 6–111 Strip electrode surfacing for SAW.

mechanized welding method. There are two major variations: the consumable guide system, which is shown in Figures 6–117 and 6–118, and the upward-moving system. The upward-moving system has not been used appreciably in the United States.

In the mid-1970s, electroslag welding became a well-established fabricating process for joining thick components for bridges, buildings, ships, pressure vessels, and more. Electroslag welding was utilized for splicing flanges, beams, and cover plates of heavy steel girders and rolled beams. However, certain problems began to surface in terms of weld imperfections and insufficient properties. During the mid-1970s, research in the field turned up problems of electroslag welded bridge beams.

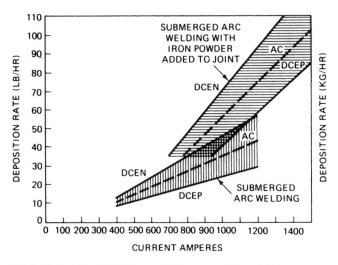

FIGURE 6–112   Welding with iron powder additives.

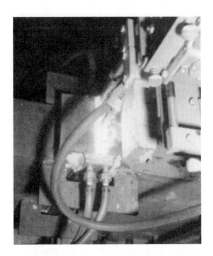

FIGURE 6–113   Welding a bridge girder.

FIGURE 6–114   Welding triangular excavator boom—two heads.

FIGURE 6–115   Welding tractor roller forgings.

FIGURE 6–116   Welding head on carriage in shipyard.

In view of this, the Federal Highway Administration placed a moratorium on the use of electroslag welding (February 1977) for weldments on primary structural tension members on bridges. This led to research at the Oregon Graduate Institute of Science and Technology to improve fracture toughness and fatigue characteristics of the ESW.[22] This research led to modified procedures for consumable guide electroslag welding. An electroslag welding process utilizing the upward-moving head version has been widely used in Europe but not in the United States.

FIGURE 6–117 Electroslag welding.

FIGURE 6–118 Process diagram for ESW consumable guide.

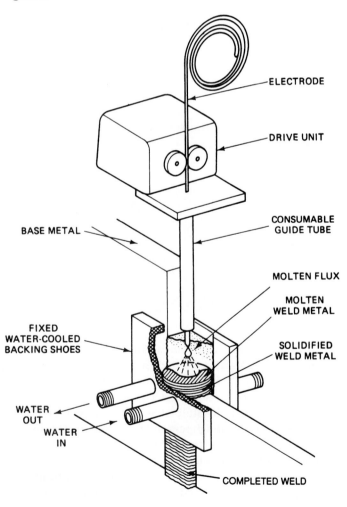

ELECTRODE

DRIVE UNIT

BASE METAL

CONSUMABLE
GUIDE TUBE

MOLTEN FLUX

MOLTEN
WELD METAL

FIXED
WATER-COOLED
BACKING SHOES

SOLIDIFIED
WELD METAL

WATER
OUT

WATER
IN

COMPLETED WELD

Principles of Operation

Electroslag welding is not an arc welding process. It is included here since it utilizes the same basic equipment as the other consumable electrode welding processes described in this chapter.

The electroslag welding process consumable guide variation is shown in Figure 6–118. Electroslag welding is done in the vertical position using molding shoes, usually water cooled, in contact with the joint to contain the molten flux and weld metal. The electrode is fed through a guide tube to the bottom of the joint. The guide tube carries the welding current and transmits it to the electrode. The guide tube is normally a heavy wall tube. At the start of the weld, granulated flux is placed in the bottom of the cavity. The electrode is fed to the bottom of the joint and for a brief period will create an arc. In a very short time the granulated flux will melt from the heat of the arc and produce a pool of molten flux. The flux is electrically conductive and the welding current will pass from the electrode through the molten flux to the base metal. The passage of current through the conductive flux causes it to become very hot, and it reaches a temperature in excess of the melting temperature of the base metal. The high-temperature flux causes melting of the edges of the joint as well as melting of the electrode and the end of the guide tube. The melted base metal, electrode, and guide tube are heavier than the flux and collect at the bottom of the cavity to form the molten weld metal. As the molten weld metal slowly solidifies from the bottom, it joins the parts to be welded. Shielding of the molten metal from atmospheric contamination is provided by the pool of molten flux. Surface contour of the weld is determined by the contour of the backing shoes.

The consumable guide variation of electroslag welding normally uses fixed backing shoes. The welding head does not move vertically and is normally mounted on the work at the top of the weld joint. Multiple electrodes and guides may be employed for welds of larger cross section. It is also possible to oscillate the electrode and guide tube across the width of the joint.

The surface of the solidified weld metal is covered with an easily removed thin layer of slag. The slag loss must be compensated for by adding flux during the welding operation. A starting tab is necessary to build up the proper depth of the flux so that the molten pool is formed at the bottom of the joint. Runoff tabs are required at the top of the joint so that the molten flux will rise above the top of the joint. Both starting and runoff tabs are removed from the ends of the joint after the weld is completed.

Advantages and Major Uses

The electroslag welding process is a very productive welding process. Some of its advantages are:

1. *Extremely high metal deposition rates.* Electroslag has a deposition rate of 35 to 45 lb/hr per electrode.

2. *Ability to weld thick materials in one pass.* There is only one setup and no interpass cleaning since there is only one pass.

3. *High-quality weld deposit.* Weld metal stays molten longer, allowing gases to escape.

4. *Minimized joint preparation and fitup requirements.* Mill edges and square flame-cut edge are normally employed.

5. *Mechanized process.* Once started, the process continues to completion. There is little operator fatigue since manipulative skill is not involved.

6. *Minimized materials handling.* The equipment may be moved to the work rather than the work moved to the equipment.

7. *High filler metal utilization.* All the welding electrode is melted into the joint. In addition, the amount of flux consumed is small.

8. *Minimum distortion.* There is no angular distortion in the horizontal plane. There is minimum distortion (shrinkage) in the vertical plane.

9. *Minimal time.* It is the fastest welding process for large, thick material.

10. *No weld spatter and minimal metal finishing* of the weld.

11. *No arc flash.* A welding helmet is not required. Tinted safety glasses are required.

Methods of Application and Position Capabilities

The consumable guide version of electroslag welding is applied as a machine operation. Once the process is started, it should be continued until the weld joint is completed. The apparatus should be monitored by the welding operator, although little is done in guiding or directing it. Flux is added manually periodically and the welding operator must monitor the depth of the molten flux pool.

Weldable Metals and Thickness Range

The metals welded by the consumable guide electroslag process are low-carbon steels, low-alloy high-strength steels, medium-carbon steels, and certain stainless steels. Quenched and tempered steels can be electroslag welded; however, a postheat treatment is necessary to compensate for the softened heat-affected zone.

Under normal conditions the minimum-thickness metal welded with the consumable guide method is $\frac{3}{4}$ in. (20 mm). Maximum thickness that has been successfully welded with electroslag is 36 in. (950 mm). To weld this thickness, six individual guide tubes and electrodes were used.

A single electrode is used on materials ranging from 1 to 3 in. (25 to 75 mm) thick. From 2 to 5 in. (50 to 125 mm) thick, the electrode and guide tube are oscillated in the joint. From 5 to 12 in. (125 to 320 mm) thick, two electrodes and guide tubes are used and are oscillated in the joint. If oscillation is not employed, additional guide tubes and electrodes are required. This necessitates additional power sources and wire feed systems, and, therefore, oscillation is preferred where it can be used.

The height of the joint has a definite relationship and must be considered. The process can be used for joints as short as 4 in. (100 mm) and as long or high as 20 ft (6.5 m). It is difficult to oscillate extremely long guide tubes since they become heated and flexible. When two guide tubes are used and properly secured together, it is possible for oscillation; however, as the number of tubes increases, the height of the joint must be decreased. The relationship of joint thickness and joint length or height is shown in Figure 6–119.

Joint Design

In electroslag welding, there is just one basic weld, the square-groove weld (Figure 6–120). The square-groove weld can be used to produce butt joints, T-joints, corner joints, and even lap and edge joints. The square-groove butt configuration is used for the transition joint, where two thicknesses of plate are joined with a smooth contour from one thickness to the other. The transition can be in the weld metal. Bead or overlay welds can also be made with electroslag. The different weld joints made with the process are shown in Figure 6–121.

In a square-groove weld, there are only two dimensions: the thickness of the parts being joined, *T,* and the root opening between the parts, *RO.* It is desirable to

Plate Thickness		Root Opening		Joint Height		Number of Electrodes	Oscillation	Welding Voltage (Electrode Positive)	Total Current (A dc)	Vertical Speed	
in.	mm	in.	mm	ft	m					in./min	mm/min
$\frac{3}{4}$	19.0	1	25.4	20	6	1	No	35	500	1.40	36.0
1	25.4	1	25.4	20	6	1	No	38	600	1.20	30.0
2	50.8	1	25.4	20	6	1	No	39	700	1.00	25.0
3	76.2	1	25.4	20	6	1	No	52	700	0.80	20.3
2	50.8	$1\frac{1}{4}$	31.8	5	1.5	1	Yes	39	700	0.76	19.3
3	76.2	$1\frac{1}{4}$	31.8	5	1.5	1	Yes	40	750	0.64	16.3
4	101.6	$1\frac{1}{4}$	31.8	5	1.5	1	Yes	41	750	0.52	13.2
5	127.0	$1\frac{1}{4}$	31.8	5	1.5	1	Yes	46	750	0.40	10.2
3	76.2	1	25.4	20	6	2	No	40	850	0.50	12.7
4	101.6	1	25.4	20	6	2	No	41	850	0.44	11.2
5	127.0	1	25.4	20	6	2	No	46	850	0.38	9.7
5	127.0	$1\frac{1}{4}$	31.8	10	3	2	Yes	41	1,500	0.80	20.3
6	127.0	$1\frac{1}{4}$	31.8	10	3	2	Yes	42	1,500	0.72	18.2
8	203.2	$1\frac{1}{4}$	31.8	10	3	2	Yes	45	1,500	0.54	13.7
10	254.0	$1\frac{1}{4}$	31.8	10	3	2	Yes	48	1,500	0.47	11.9
12	304.8	$1\frac{1}{4}$	31.8	10	3	2	Yes	51	1,500	0.36	9.1
12–18	304.8–457.2	$1\frac{1}{2}$	38.1	6	1.8	3	Yes	55	1,800	0.18	4.6
18–24	457.2–609.6	$1\frac{1}{2}$	38.1	5	1.5	4	Yes	55	2,400	0.18	4.6
24–30	609.6–762.0	$1\frac{1}{2}$	38.1	4	1.2	5	Yes	55	3,000	0.18	4.6
30–3	762.0–914.4	$1\frac{1}{2}$	38.1	3	1	6	Yes	55	3,600	0.18	4.6

FIGURE 6–119 Base metal thickness and height that can be welded.

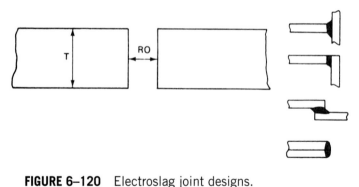

FIGURE 6–120 Electroslag joint designs.

have the root opening as small as possible to use a minimum amount of weld metal. A limiting factor is the size of the consumable guide tube and the insulators that are required to keep it from touching the sides of the joint. The root opening must be large enough to provide sufficient volume of the molten flux to ensure stable welding conditions.

The water-cooled backing shoes are designed to accommodate the different types of joints. Shoes are available for the square-groove welding with reinforcing used for butt joints and for other joints where the surfaces of the plates to be joined are flush. For square-groove welds involving corner or T-joints, fillet-type shoes are used.

Welding Circuit and Current

The welding circuit used is shown in Figure 6–122. For the consumable guide system, dc welding power is employed. The electrode is positive (DCEP). The constant voltage (CV) system with the constant-adjustable-speed wire feeder is used.

In the normal electroslag welding process, alternating current is often used, especially for three-wire systems. In these cases, the constant-feed electrode wire drive motor is used and the characteristics of the power source are close to flat. The electrical characteristics of the conductive molten flux are similar to those of a high-current welding arc.

Welding current per electrode wire may range from as low as 400 A to as high as 800 A. The weld voltage will range from 25 to 55 V. The high voltage is extremely important, especially when using long guide tubes.

Equipment Required

The equipment required for the consumable guide electroslag welding process is also shown in Figure 6–122. The systems become more complex as additional electrodes are added. The use of oscillation provides greater latitude of the consumable guide method. All the electrode wires are mounted on one oscillating assembly, so only one oscillating device and control are required.

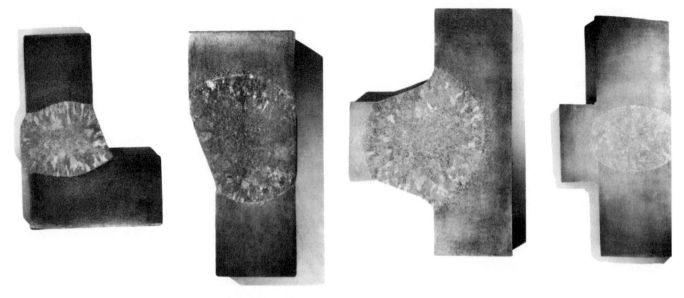

FIGURE 6–121   Different weld joints made with electroslag welding.

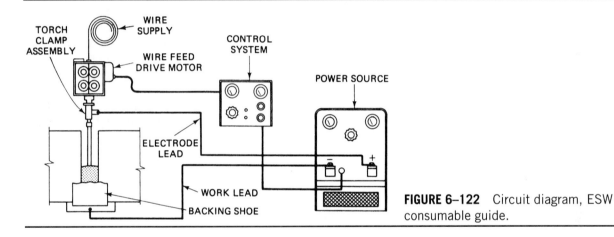

FIGURE 6–122   Circuit diagram, ESW consumable guide.

The power source used for the consumable guide electroslag welding process should be a direct-current welding machine of the CV type. It must be rated at a 100% duty cycle since some electroslag welds take hours to complete. The power source should have high voltage ratings since starting voltages as high as 55 V are sometimes required. Transformer–rectifier machines are best suited for electroslag welding. Primary contactors and provisions for remote control, including voltage adjustments, should be included.

When water-cooled backing shoes are used, a system for water circulation and heat removal is required. When running water is available and when it can be easily disposed of, this is the simplest solution. However, water circulating systems, which include heat exchangers, can be used.

Materials Used

Three materials are routinely used in making consumable guide electroslag welds: the flux, electrode wire, and the guide tube. These are specified by the AWS "Specifications

for Consumables Used for Electroslag Welding of Carbon and High Strength Low Alloy Steels."[23] Other materials used, including runoff tabs and the starting sump, are reusable and must be the same thickness and composition as the base metal. Insulating material is used for certain applications. Insulators are sometimes required around the bare guide tube to avoid short-circuiting the system if the guide tube comes in contact with the retaining shoes or the face of the weld joint. Other reusable items are the strong backs used to hold the retaining shoes against the weld joint. Wedges are used to hold the retaining shoes in place. The strong backs and wedges are reused many times. When more than one electrode is used, a steel wool ball is placed at the bottom of the joint under the electrode wire to aid arc initiation. Steel wool also can be used for single wire applications, although it is not normally required.

When the work surface is irregular, it is necessary to install a puttylike material to seal the cracks between the shoes and the work. Commercial materials such as furnace sealing compound can be used.

The functions for an electroslag flux are:

1. Providing heat to melt the electrode and base metal
2. Conducting the welding current
3. Protecting the molten weld metal from the atmosphere
4. Purifying or scavenging the deposited weld metal
5. Providing stable operation

There are two types of granular fluxes normally used for electroslag with the consumable guide tube. One is a starting flux, and the other is a running flux. The starting flux is designed to bring the electroslag process into quick stabilization. It melts quickly and wets the bottom of the sump to facilitate starting. The running flux is designed to provide the proper balance for correct electrical conductivity, correct bath temperature and viscosity, and the proper chemical analysis. Running flux will operate over a wide range of conditions. Only a relatively small amount of electroslag flux is used: Approximately $\frac{1}{4}$ lb (100 g) of flux is used per vertical foot (320 mm) of the joint or height.

The *electrode* for consumable guide electroslag welding supplies over 80% of the deposited weld metal. The guide tube supplies the remainder. The electrode wire must match the base metal. Since an electroslag weld deposit is similar to a casting, it is essential that the properties of this *as-cast* metal should overmatch the mechanical properties of the parts being joined. It is important to consider the dilution factor provided by the base metal. In a consumable guide weld, the dilution runs from a low of 25% to a high of 50% base metal. The amount of dilution of base metal depends on the welding conditions.

The flux adds no alloys and has little effect on the weld deposit in relationship to the analysis of the wire. Electrode wires designed for gas metal arc welding and submerged arc welding are employed for electroslag welding. The $\frac{3}{32}$-in. (2.4-mm) electrode size is the most common. It is the most easily used to feed through a guide tube and produces the highest deposition rate.

The *consumable guide tube* melts just above the surface of the molten slag bath. A guide tube must be used whenever the length of the weld is 6 in. (160 mm) or over, assuming that the head is stationary.

When a bare guide tube is used and if the weld is over 12 in. long (304 mm), insulators should be placed on the tube to avoid the guide tube coming in contact with the sidewall or face of the joint or the retaining shoes. Coated guide tubes are also available and the coating is an effective insulator, particularly when working in tight joints.

There are several variations of the consumable guide tube system. In some cases bars are tack-welded to the guide tube, or tubes are tacked on edges of bars. These bars contribute metal to the weld deposit.

Deposition Rates and Quality of Welds

Deposition rates of the electroslag welding process are among the highest. Figure 6–123 shows the deposition rate versus welding current of the $\frac{3}{32}$-in. (2.4-mm) electrode wire and of the $\frac{1}{8}$-in. (3.2-mm) electrode wire.

The electroslag welding process produces a high-quality weld metal deposit. The high quality of electroslag weld metal is the result of progressive solidification, which begins at the bottom of the joint or cavity. There is always molten metal above the solidifying weld metal, and the impurities, which are lighter, rise above the deposited metal and collect only at the very top of the weld in the area that is normally discarded.

Electroslag welding is a low-hydrogen welding process; hydrogen is not present in any of the materials involved in making the weld. Because of the slow cooling rate, any impurities that are in the base metal and are melted during the welding process have time to escape. The cooling rate of the electroslag weld is much slower than the cooling rate of welds made by other arc welding processes. The slow cooling rate allows large grain growth in the weld metal and also in the heat-affected zone of the base metal. The slow cooling rate minimizes the risk of cracking and reduces the hardness in the heat-affected zone sometimes found in conventional arc welds.

Weld metal produced by electroslag welding will qualify under the most strict codes and specifications. The ductility of the weld metal is relatively high, in the

FIGURE 6–123 Deposition rate for ESW.

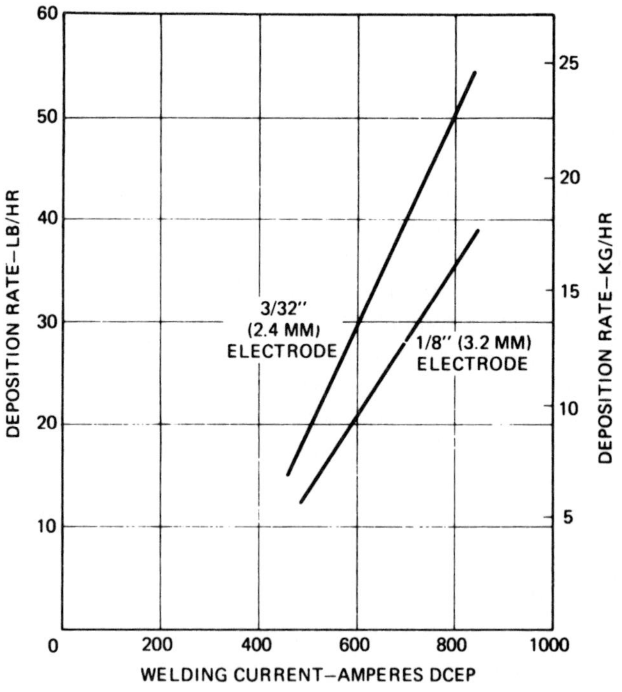

range 25% to 30%. The impact requirements for electroslag welds will meet those required by the AWS structural welding code. V-notch Charpy impact specimens producing 5 to 30 ft-lb at 0° F are normal and expected.

Weld Schedules

Welding procedure schedules for electroslag welding with the consumable guide method are provided by the equipment manufacturer. Welding procedure schedules are based on welding low-carbon steel under normal conditions using water-cooled copper shoes and the $\frac{5}{8}$-in. (16-mm) outside-diameter guide tube with a $\frac{1}{8}$-in. (3.2-mm) inside diameter unless otherwise specified. The electrode diameter is $\frac{3}{32}$ in. (2.4 mm), and proprietary starting and running fluxes are used. Oscillation speed is based on the number of seconds per cycle, which is shown as a rate of speed. There is a dwell time at each end of oscillation, which is normally 4 seconds.

Welding Variables

Electroslag welding differs from arc welding processes in that the base metal melting results from localized heat generated in the molten slag pool instead of from an arc. The heating involved in electroslag welding is concentrated in a volume of molten flux, which is the product of the metal thickness by the root opening and by the slag pool depth.

In the arc welding processes, the localized heating is confined to the much smaller area of an arc and pool, but the arc is at a much higher temperature. The operation of electroslag welding is thus different from the familiar arc processes.

In electroslag welding the metal surface to be melted (joint sidewalls) is parallel to the axis of the electrode. Thus increasing welding current does not increase the depth of penetration of the sidewalls of the base metal. The higher the welding current, the higher the deposition rate.

With all arc welding processes, an increase in arc voltage causes the weld bead to widen. In electroslag welding, the same thing is true, but now this widening causes an increase in the depth of penetration into the sidewall. The increased voltage raises the slag bath temperature and causes more of the base metal or sidewall to melt. Increase voltage to increase depth of fusion. Excessively high voltage will cause undercutting. Too low a voltage may result in arcing between electrode wire and the molten weld metal at the bottom of the flux pool. The operator must be continually alert to make adjustments as required during the welding operation. The operator must have a good operating knowledge of electroslag welding because of the different effects of changing the various parameters.

The depth of the molten slag pool should be checked if possible. When the pool is accessible to the operator, a dipstick can be used to determine its depth. Experience will quickly show that when the pool is quiet and the process is running without sparking or sputtering, the pool depth is correct. If the pool depth becomes too shallow, sparks will emit from the surface and can be seen by the operator. Additional flux should be added to the pool.

If the backing shoes leak and water gets into the weld cavity, the operation must be stopped. This can create a safety hazard and will create gross porosity in the weld metal. With respect to water-cooled shoes, the operator must ensure that the water flow is uninterrupted during the entire welding operation.

Safety Considerations

The major safety factor is the presence of a large mass of molten slag and molten weld metal. The high welding current creates a large mass of metal that must be contained within the weld cavity. If the backing shoes should fail and allow the molten metal to escape, it is best to evacuate the area, turn off the equipment, and wait for the metal to solidify. Obviously, the surface under the welding operation should be noncombustible. The work being welded must be securely braced to eliminate the possibility of it falling.

Limitations of the Process

The major limitation is the welding position limitation. It can be used only when the axis of the weld is vertical. A tilt of up to 15° is permitted, but beyond this the process may not function correctly. The second limitation is that the process is used only on steels.

OGI Research and Improvement

At the Oregon Graduate Institute of Science and Technology (OGI), researchers worked to improve the metallurgical quality of electroslag welds by modifying the welding variables. It was felt that if the procedure could be made more technical, it would be mandatory in the field to check and monitor more closely the variables involved. The procedures developed at OGI allowed for faster cooling rates and an improved weld metal chemistry due to use of a cored electrode wire. In general, the guide tube design was altered, the root opening was reduced, automatic flux feeding was introduced, and a new electrode wire was developed that included alloying elements. This modified procedure produced welds that provided weld joints that consistently met the federal highway requirements. In view of this, the Federal Highway Administration started to promote the OGI procedure for welding tension members of bridges. The improved procedure is now being used on a limited basis, and in time it is expected that use of the improved procedure will again allow electroslag welding to be widely used.

Variations of the Process

Electroslag Cladding   This is a variation that deposits surfacing materials on base metals. It is very similar to strip cladding with the submerged arc welding process except that the heat required to melt the surface of the base metal, the strip, and the flux is generated by resistance heating from the current flow to the strip and through a shallow layer of electroconductive slag (Figure 6–124). Electroslag cladding has become popular because it provides for high deposition rates and low dilution. In addition, it can be used with the same equipment as that used for submerged arc strip cladding. Magnetic oscillation of the arc is recommended for best results. The electroslag process will deposit approximately two times as much metal per hour as the submerged arc method. Dilution is controllable and is usually less than with submerged arc. It will range from 10% to 20%. It is possible to clad ferritic, martensitic, and austenitic stainless steels, nickel-base alloys, and some hard-surfacing materials. Strip width is approximately 2 in. (50 mm) to $2\frac{1}{2}$ in. (61 mm). The major use for electroslag cladding is the deposition of austenitic stainless steel on carbon steel for tube sheets and other corrosive-resistant application. Another use is the deposition of corrosion-resistant coatings on large shafts.

Industrial Use and Typical Applications

The major user of conventional consumable guide tube electroslag welding has been the heavy plate fabrication industry, which includes manufacturers of frames, bases, metalworking machinery, and so on. A frequent use of the process is the splicing of rolled steel plates to obtain a larger piece for a specific application. A typical application is the splicing of an 8-in. (203-mm)-thick plate, 12 ft (4 m) wide, to make a press frame (Figure 6–125). The weld joint was set up in the vertical position, strong backs were attached, and the two-wire feeder was placed at the top of the joint. Four water-cooled backing shoes were used, two on either side, which were moved progressively from bottom to top until the weld was completed. Previously, with submerged arc welding and turning the plate over after every few passes, the splicing operation required 80 hours. With the consumable guide electroslag process, the weld was completed in slightly over 4 hours.

Another machinery application is shown in Figure 6–126. This is the end frame for a steel mill roll. Five electrode wires are being used to make this massive weld. The top side will be positioned and then welded the same way.

FIGURE 6–124   Electroslag cladding.

FIGURE 6–125 Splicing 8-in.-thick 12-ft-wide plates.

FIGURE 6–126 Large end frame for steel mill rolls electroslag welded.

Another user of electroslag is the structural steel industry, for making subassemblies for steel buildings. It has also been used for field erection on the building site. A common application is the welding of continuity plates inside box columns. The continuity plate carries the load from one side of the column to the other side at the point of beam-to-column connections. Continuity plates must be welded with complete penetration welds to the two sides of the box column. Figure 6–127 shows an auto-

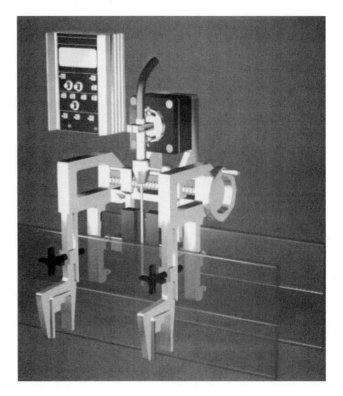

FIGURE 6–127 Automatic machine for box column continuity plate.

matic fixture, which provides for two welding operations simultaneously.

The electrical machinery industry also utilizes electroslag welding. Electric motor housings are rolled from a single plate, and a single weld is made to join the two abutting edges. In other cases, the material is heavier or the housing might be square. Figure 6–128 shows two sides of a motor housing being welded simultaneously. Electroslag reduces total cycle time because the housing was previously welded with submerged arc one joint at a time.

There are many, many other applications of the consumable guide version, both in the shop and in the field.

6-8 ELECTROGAS WELDING

Electrogas welding (EGW) is "an arc welding process that uses an arc between a continuous filler metal electrode and the weld pool, employing vertical position welding with backing to confine the molten weld metal. The process is used with or without an externally supplied shielding gas and without the application of pressure." It is a limited-position arc welding process and is a single-pass process that produces square-groove welds for butt and tee joints.

There are two basic variations: One utilizes the upward-moving-head variation, the other variation uses the

FIGURE 6–128 Motor housing: two welds made simultaneously.

consumable guide tube system described earlier for use with electroslag welding. In addition, both of these variations have two variations. One uses the solid consumable electrode wire and externally supplied shielding gas, normally CO_2. The second utilizes a flux-cored electrode wire and does not use an external shielding gas since shielding gases are produced by the flux-cored electrode as it is consumed in the arc. Figure 6–129 shows the movable-head variation.

Principles of Operation

The primary difference between electrogas welding and electroslag welding is that the arc is continuous in the electrogas process. Electrogas welding utilizes an arc between a continuously fed consumable electrode wire and the molten weld pool. The heat of the arc melts the surface of the base metal and the end of the electrode. The metal melted from the electrode, the metal melted from the surface of the abutting joints, and metal melted from the consumable guide tube, when used, collect at the bottom of the cavity formed between the parts and the molding shoes. This molten weld solidifies from the bottom of the joint and joins the part to be welded. Shielding of the molten metal from the atmosphere is provided by shield-

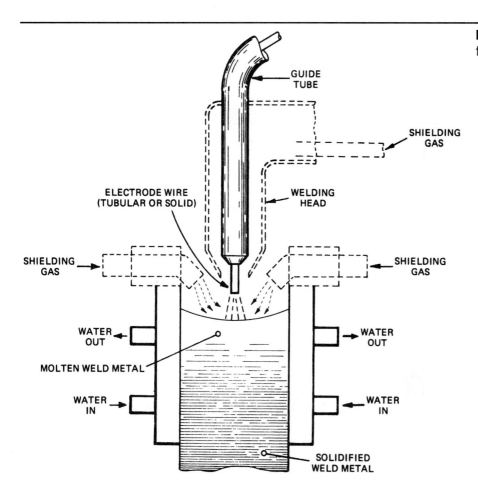

FIGURE 6–129 Process diagram for EGW: movable-head variation.

ing gas from the external source or from the disintegration of the ingredients in the cored electrode wire.

In the moving-head version, the electrode is fed to the bottom of the joint by means of the wire feeder guide tube or contact tip. This assembly will travel vertically along the joint to maintain the normal arc length between the electrode and the molten weld metal. In some cases one backing shoe is stationary and can be made of steel, thus becoming a part of the joint. Or the molding shoe can be made of copper and does not become a part of the joint, but moves upward. On the side with the wire feeding mechanism, the moving shoe is normally employed, which rises with the wire feeder assembly to maintain the weld metal within the cavity. Normally, only one electrode wire is used for making a weld. In the consumable guide variation, external shielding gas is not used, but the consumable electrode wire must be of the flux-filled type. Both shielding gas and flux are fed into the arc. The flux contained within the electrode wire turns into slag, which covers the weld deposit. Since the correct amount of flux cannot be provided based on the plate thickness and the root opening, a surplus of flux will be generated. The flux must not accumulate and become too deep, which would extinguish the arc. If so, the operation would become an electroslag weld. The excess molten flux leaks through the slots in the water-cooled retaining shoes to avoid flux buildup. Figure 6–130 shows the setup using the consumable guide tube variation prior to welding and after welding. A starting tab and runoff tabs are required. Both starting and runoff tabs are removed from the ends of the weld joint after the weld is completed.

Methods of Application and Position Capabilities

The electrogas welding process is continuous. Once the process has started, it should be continued until the weld joint is completed. The welding operation should be monitored by the operator. The principal purpose is to provide guidance or ensure that the electrode and arc is centered in the joint. It is important to maintain shielding gas flow during the entire welding operation. The arc voltage is utilized to provide control of the vertical motion of the apparatus. The motion is controlled so that the arc length will remain constant.

FIGURE 6–130 EGW prior to and after welding.

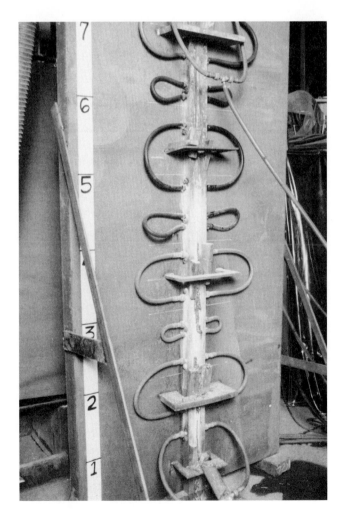

FIGURE 6–130 *(cont.)*

The electrogas welding process is a limited position process. It can be used only when the axis of the weld joint is vertical or varies from the vertical by not over 15°.

Weldable Metals and Thickness Range

The metals welded by the electrogas process are low-carbon steels, low-alloy high-strength steels, medium-carbon steels, and certain stainless steels. The process can also be used for welding quenched and tempered steels providing that the correct heat input is maintained for the type of steel being welded.

Under normal conditions the minimum thickness of metal welded with electrogas is $\frac{3}{8}$ in. (10 mm). The maximum thickness utilizing one electrode is $\frac{3}{4}$ in. (20 mm). Additional is welded by using additional consumable fillers.

The height (or length) of the joint is practically unlimited. The process can be used for joints as short as 4 in. (100 mm) and as long or high as 50 ft (18 m). The only limitation is the weight of the elevating mechanism for moving the weld head vertically.

Joint Design

Fillet welds and groove welds can be produced by the electrogas process. For making fillet welds, a single backing shoe is required. This shoe fits on the face of the fillet and provides the fillet size. For groove welds, the square-groove design can be used up to the maximum possible with one electrode (usually $\frac{3}{4}$ in.).

Welding Circuit and Current

The welding circuit used for the electrogas welding process is essentially the same as for the other continuous or consumable electrode processes. The block diagram for electrogas welding is shown in Figure 6–131. Direct-current welding power is employed and the electrode is positive (DCEP). The constant-voltage system with the constant/adjustable speed wire feeder is used. The welding current may range from as low as 100 A to as high as 800 A. The welding voltage will range from 30 to 50 V. The welding voltage is used to control the vertical travel speed of the welding head. The welding head apparatus normally includes the moving shoe required on the electrode side of the weld joint.

Equipment Required

The equipment required for the electrogas welding process is shown in Figure 6–131. The system normally utilizes one electrode. The welding head assembly is normally mounted on a carriage, which is elevated as the weld progresses. For shipbuilding, the entire apparatus, which may also carry the welding operator, will move from the bottom of the joint to the top. This is done with a precision elevating system controlled by the welding arc voltage. The control for the entire operation is mounted with the welding head and available to the welding operator. This enables the operator to start the weld and have it run continuously until the joint is completed.

The power source used for electrogas welding should be a direct-current machine of the constant-voltage (CV) type. It must be rated at 100% duty cycle since some electrogas welds take over an hour to complete.

The wire drive feed motor and control system is the same as that used for other consumable electrode wire processes. Normally, the wire feed motor is mounted adjacent to the weld joint, with a contact tube delivery system bringing the electrode into the center of the joint and pointed downward within the cavity.

The shielding gas delivery system, when used, must provide efficient shielding of the molten metal to avoid atmospheric contamination.

The moving backing shoes are normally water cooled and designed for the specific joint design. They provide a water flow channel and are made of copper to

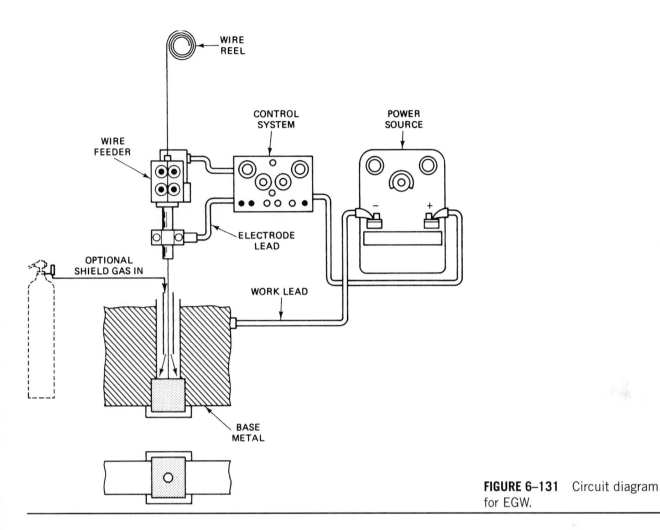

WIRE REEL

CONTROL SYSTEM

POWER SOURCE

WIRE FEEDER

ELECTRODE LEAD

OPTIONAL SHIELD GAS IN

WORK LEAD

BASE METAL

FIGURE 6–131 Circuit diagram for EGW.

avoid melting. The water circulation should be of such volume as to avoid any surface melting of the shoes. A water circulator that includes a heat exchanger is normally used. More details concerning electrogas welding are given in the AWS "Recommended Practices for Electrogas Welding."[24]

Electrode Wire

The electrode wire must be matched to the material being welded and can be specified according to the AWS "Specifications for Consumables Used for Electrogas Welding of Carbon Steels and High Strength Low Alloy Steels."[25] This covers the solid wires and the flux-cored wires. The shielding gas, which is normally CO_2, would be specified as welding grade.

Deposition Rates and Quality of Welds

The deposition rate of electrogas welding is relatively high. Flux-cored wire deposition rates vary with wire types and manufacturers since the ratio of fill to metal varies.

Electrogas welding is considered a low-hydrogen type of welding process since hydrogen is not present in any of the materials involved in making a weld. Electrogas welds possess properties and characteristics that surpass welds made with shielded metal arc welding. The higher-than-normal heat input of electrogas welds reduces the cooling rate, which helps reduce impurities. This allows larger grain growth of the weld metal and also in the heat-affected zone of the base metal. This lower cooling rate minimizes the risk of cracking and reduces the high hardness zones in the weld and heat-affected zone sometimes found with shielded metal arc welding. The hardness of the weld is normally uniform across the weld's cross section and is very similar to the unaffected base metal.

Weld metal produced by electrogas welding will qualify under most codes and specifications. The ductility of the weld metal of electrogas weld is relatively high, in the range of 25% elongation. Impact requirements for electrogas welds will meet those required by the AWS Structural Welding Code. V-notch Charpy impact specimens producing 5 to 30 ft-lb at 0° F (19 to 60 joules at −10 to −34° C) are normal and expected.

Weld Schedules

Welding procedure schedules for electrogas welding may not necessarily be the only conditions that can be used. It is possible that welding parameters can be adjusted to obtain optimum results; however, qualification tests should be made before utilizing published welding procedure schedules, especially when welding critical jobs.

Tips for Using the Process

Even though the electrogas process is a machine process, the operator must be continually alert and make adjustments during the welding operation. The operator must have a good operating knowledge of electrogas welding because of the different effect of changing various parameters.

If the electrode wire is not properly centered, the penetration on the opposite sides of the weld joint will be nonuniform. The electrode should be centered between the backing shoes. However, if one shoe is steel rather than copper, the electrode should be directed to the side of the joint with the copper shoe.

If the backing shoe does not fit tightly along the joint, the molten weld metal may run out of the cavity. If this happens, steps must be taken immediately to stop the leak. This is done by using a puttylike sealing preparation made of clay. Any leaks should be immediately sealed off to avoid loss of the weld.

Sufficient wire should be available on the machine prior to starting the weld. Once the weld is started, it should run continuously until it is finished. If the operation stops for any reason, the machine should immediately be turned off, correction made, and the weld restarted. At the point of stopping and restarting, there is normally an unfused area that must be gouged out and rewelded with an arc welding process capable of welding in the vertical position.

With respect to water-cooled shoes, the operator must ensure that water flow is uninterrupted during the entire welding operation.

Safety Considerations

The safety factors involved with electrogas welding are much the same as for the other continuous wire arc welding processes. A welding helmet and shield should be worn because the arc is continuous from start to finish. A safety factor involved is the presence of larger-than-normal amounts of molten weld metal. If this metal escapes, it creates a safety hazard and a fire hazard. The work being welded must be securely braced to eliminate the possibility of it falling. In addition, since vertical heights are involved, the equipment should be safety-related to protect personnel from falling.

FIGURE 6–132 Electrogas welding used in shipbuilding.

Limitations of the Process

The major limitation is the welding position limitation. The process should not be used if the joint is at an angle in excess of 15° from vertical. The other limitation is that only steel can be welded. For more information, see the AWS "Recommended Practices for Electrogas Welding."[24]

Industrial Use and Typical Applications

The major use of electrogas welding has been in the field erection of storage tanks and for splicing. Another user is the shipbuilding industry, for joining shell plates (Figure 6–132).

6-9 OTHER CONSUMABLE ELECTRODE WELDING PROCESSES

Automatic welding began in the early 1920s, but the covered electrode, which produced higher-quality weld metal, replaced bare electrode automatic welding applications. Automatic welding was improved by using lightly coated wires and later by using knurled wires, which held more of the stabilizing and deoxidizing coating materials. The quality was not as good as that produced by covered electrodes.

Many efforts were made to produce an all-position automatic welding process that would produce high-quality weld metal. The coating on the *covered wire* created two problems. The coating was fragile and brittle and could not be bent into coils without cracking and falling off. The coatings were insulators and the welding current could not be introduced into the metal core wire in an efficient manner. Several developments deserve brief consideration.

One early attempt was made to extrude coatings on large-diameter electrode wire and then coil the wire into extremely large diameter coils. At the welding head a cutter removed the coating from one small area of the wire. The welding current was conducted to the core wire through the slot in the coating by multiple pickup shoes. This process had limited success and has been discontinued.

Another variation, known as the Una-Matic process, used *impregnated tape*. Shielding is obtained from decomposition of an impregnated tape wrapped around the bare electrode as it is fed into the arc. The tape, similar to insect screen wire, was woven in narrow widths, impregnated with stabilizing and deoxidizing chemicals, and coiled. This tape was wrapped around the bare electrode wire below the current pickup jaws and above the arc. This process had limited success and has been discontinued.

Another system utilized the magnetic field surrounding the electrode wire to carry flux into the arc. The bare electrode wire was fed into the arc, surrounded by carbon dioxide gas that carried powdered magnetic flux. The *magnetic flux* is attracted to the electrode wire by the magnetic field and covered the electrode wire as it entered the arc. The flux performed the normal function. The carbon dioxide gas shielded the molten weld metal from the atmosphere. It never became popular, due largely to the problems of feeding the flux.

A system developed in England, called Fusarc, utilizes a composite electrode of special design. The large diameter center core wire is wrapped with small wires spiraled in both directions to create cavities. This assembly is run through an extrusion press, and flux is forced into the cavities to provide protection in the same manner as the covering on a conventional coated electrode. It produced quality welds when applied automatically. The core electrode wire is fairly large and too stiff to be manipulated for semiautomatic welding. Fusarc has been discontinued and is no longer in use.

Hidden arc welding (HAW) is an automatic arc welding process recently developed by the Moscow Energy Institute in Russia. It uses a square-groove weld design and is usually used in the vertical position. The welding electrode is cut of approximately 1/4-in. thick material the exact size of the joint to be welded. The material is coated with a welding flux, or a flux containing insulation, and is placed on both sides of the electrode. The electrode assembly is clamped between the parts to be welded. The electrode is connected to one pole of the welding machine and the material to the other pole. The arc is initiated at the bottom of the joint by momentarily shorting the electrode to the work. The entire periphery of the joint is encased by copper bars. The arc, once struck, automatically oscillates from one side of the joint to the other and gradually travels from the bottom to the top. No operator assistance or attention is required. When the weld is completed, the arc is extinguished. The copper bars are removed, and the resulting weld is complete. No further work is required. The deposit weld metal is equal to, or exceeds, the quality of the parts being welded. So far, the HAW process has limited applications.

6-10 ARC WELDING VARIABLES

During the manual welding operation, the welder has control over factors that affect the weld. For example, the welder can increase or decrease the speed of travel along the weld joint. The welder can increase the length of the arc, which increases the voltage, or decrease the length of the arc, which decreases the voltage. In this way the welder is also changing the welding current. The welder can also change the angle of the electrode or the torch to either push or drag, and these changes can be made while welding. When all the variables are in proper balance, the welder will have control over the molten metal and will deposit high-quality weld metal. This section will explain how these welding variables interrelate and how some of them are more easily changed and are useful for control.

The effect of changing these variables and the resulting change is essentially the same for all of the arc welding processes when the weld metal crosses the arc. All welds shown are on steel, however; the same factors apply to other metals.

Welding variables can be divided into three classifications: primary adjustable variables, secondary adjustable variables, and preselected or distinct level variables. The *primary* adjustable variables are those most commonly used to change the characteristics of the weld. These are: travel speed, arc voltage, and welding current. They can be easily measured and continually adjusted over a wide range. These primary variables control the formation of the weld by influencing the depth of penetration, the bead width, and the bead height (or reinforcement). They also affect deposition rate, arc stability, spatter level, and so on. Specific values are assigned to these variables. They are included in welding procedure schedules and can be duplicated time after time.

The *secondary* adjustable variables can also be changed continuously over a wide range. The secondary adjustable variables do not directly affect bead formation; instead, they cause a change in a primary variable, which in turn causes the change in weld formation. Secondary adjustable variables are more difficult to measure and accurately control. They are assigned values and are usually included in welding procedure schedules. They include tip-to-work distance (stickout) and electrode or nozzle angle.

The third class of variables is known as *distinct level* variables, since they cannot be changed in a continuous fashion, but instead normally in increments or in specific steps. Distinct level variables must be preselected and are fixed during a particular weld. They have considerable influence on the weld formation. Distinct level variables that must be preselected are included in welding procedure schedules. They include the electrode size, the electrode type, welding current type and polarity, shielding gas composition, and flux type. These variables are selected with regard to the type and thickness of the material, joint design, welding position, deposition rate, and appearance.

Primary Adjustable Variables

The primary adjustable variables are welding current, arc voltage, and travel speed. To explain the effect of these variables, bead on plate welds are shown with the three characteristics involved: weld penetration, the weld bead width, and the weld reinforcement. Each variable has a distinct effect on the three characteristics.

When making welds to establish a welding procedure or in reviewing welds that did not meet requirements, judgment is based on these three weld characteristics. Weld penetration is the most important, and it is affected by all three of the variables. Penetration is also affected by the secondary adjustable variable and by the preselected variables.

In analyzing the weld if it is decided that penetration must be increased, one of the preselected variables may be changed. For example, if the maximum welding current for a particular electrode size is being used, it would be necessary to change to a larger electrode size to further increase the welding current. This same rationale may have to be followed to change the bead width or weld reinforcement if the limit of a primary variable is reached without obtaining the desired results.

Three sets of curves show the effect of the three primary variables on the three weld characteristics. Figure 6-133 shows the effects of the three variables on weld penetration. Penetration is the distance that the fusion zone extends below the original surface of the parts being welded. Joint design is also a factor that must be considered. This curve is based on flux-cored arc welding, but would apply to GMAW, submerged arc welding, and, to a fairly large degree, to SMAW. The values may change, but the relationships are similar. To explain this, Figure 6-134 shows bead appearance and a cross section of welds made with the FCAW process. Welding conditions were $\frac{3}{32}$ in. (2.4 mm) electrode, 29 V, electrode positive, and travel speed 20 in./min (510 mm/min). The depth of penetration increases as the current level increases. The welding current and weld penetration relationship is almost a straight line and is the most effective in controlling this weld characteristic. It should be considered first when a change of penetration is required.

The relationship between travel speed and weld penetration also is a relatively straight-line relationship.

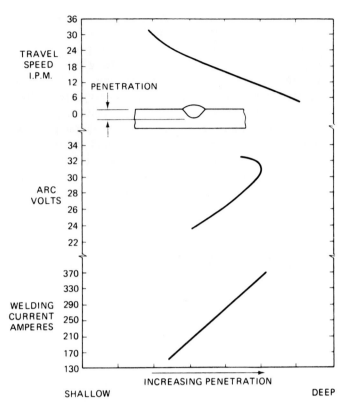

FIGURE 6-133 Weld penetration related to primary variable.

Penetration is increased as travel speed is decreased. Travel speed should not be used as the major control since, for economical reasons, it is desired to weld at a maximum speed possible.

The relationship of penetration and arc voltage is not a straight-line relationship. The curve shows that there is an optimum arc voltage where penetration is maximum. Raising or lowering arc voltage from this point reduces penetration. Thus a long arc or a short arc will decrease penetration. For a given welding current, there is a certain voltage that will provide the smoothest welding arc. It is for this reason that arc voltage is not recommended as a control for penetration.

The weld bead width relationship to the primary variables is shown in Figure 6-135. Bead width is an important characteristic of a weld, particularly when using automatic equipment to fill a weld groove. The arc voltage variable, or arc length, is a straight-line relationship with weld bead width. As the arc voltage is increased, bead width increases. This can be explained by considering the welding arc. The welding arc has a point-to-plane relationship and is conical in shape with the point of the cone at the end of the electrode and the wide portion at the surface of the weld. This is shown in Figure 6-136 and explains the relationship between the longer arc with higher voltage and the bead width. This shows the arc voltage at different arc lengths and how the arc spreads out and

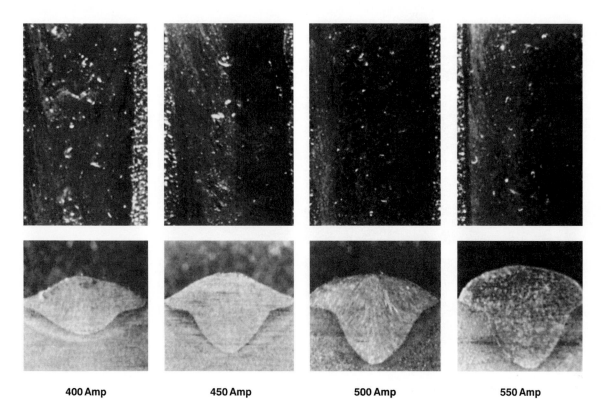

| 400 Amp | 450 Amp | 500 Amp | 550 Amp |

FIGURE 6–134   Weld penetration versus welding current.

FIGURE 6–135   Weld bead width related to primary variable.

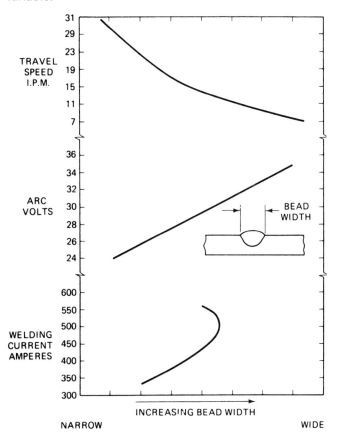

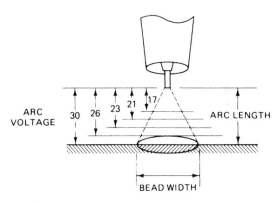

FIGURE 6–136   Weld bead width related to arc voltage.

makes a wider bead. This relationship is also shown in Figure 6–137, which shows the weld surface appearance and cross section of flux-cored arc welds made at different arc voltages. Welding conditions: electrode size and travel speed are the same as previously; but the current is maintained at 450 A. Since increasing the arc voltage makes the bead wider, the reinforcement is reduced because the same volume of weld metal is involved. Conversely, reducing the arc voltage makes the bead narrower and increases the height of the reinforcement.

Travel speed is the second choice for changing bead width, since it has a relatively straight-line relationship. The welding current is not a straight-line relationship, as is

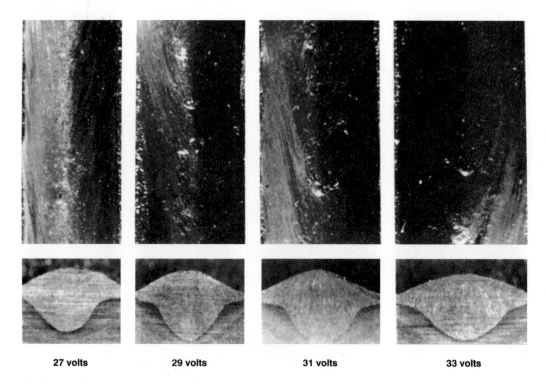

| 27 volts | 29 volts | 31 volts | 33 volts |

FIGURE 6–137 Arc length–weld bead width relationship.

shown by the curve. Therefore, welding current is not used as a control for weld bead width.

The weld bead reinforcement or height related to the three primary variables is shown by the curve of Figure 6-138. Weld reinforcement is important in automatic welding for filling a groove weld with the proper amount of metal using the desired number of passes. The weld height is most effectively controlled by arc voltage because of the relatively straight-line relationship. It should be the first choice for changing weld reinforcement.

The welding current to bead height is a relatively straight-line relationship. This is based on the mass or amount of weld metal deposited. The travel speed relationship to the weld characteristics is shown in Figure 6-139, which shows the weld surface appearance and cross section of flux-cored arc welds made at different speeds. Welding conditions: electrode size is the same as for the previous; welding current is 450 A, voltage is 29 V. At the lower travel speeds the weld is large in mass, whereas at the high travel speed it is smaller in mass. This relationship is very easily determined by relating the cross-sectional area of the welding electrode times the wire feed speed to the cross-sectional area of the weld times the travel speed. As more electrode is fed into the arc, based on higher welding current, a greater mass of metal is deposited. However, as the speed of travel is increased, this mass of metal will be spread out over a longer length. The relationships shown relate penetration, bead width, and reinforcement to welding current, arc voltage, and travel speed. Notice the interaction that

FIGURE 6–138 Weld bead reinforcement related to primary variable.

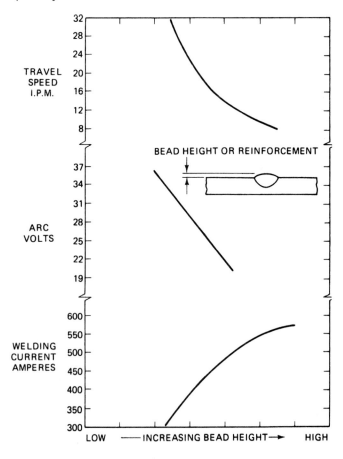

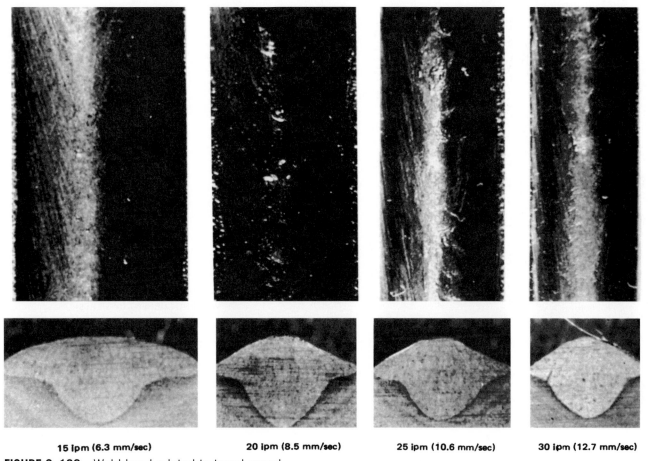

| 15 ipm (6.3 mm/sec) | 20 ipm (8.5 mm/sec) | 25 ipm (10.6 mm/sec) | 30 ipm (12.7 mm/sec) |

FIGURE 6–139 Weld bead related to travel speed.

occurs. These relationships can only be varied within limits, since there is a fixed relationship between arc voltage and welding current within the stable operating range. This relationship changes for different processes, shielding gas atmospheres, and electrode sizes. The relationship is shown in Figure 6–140 for FCAW.

All these relationships are relative. Different values would be used for different processes. The shape of the curves and the changes in weld bead characteristics would be basically the same.

Secondary Adjustable Variables

The secondary adjustable variables include the stickout and torch or electrode angle. These variables change the weld characteristics because they influence one of the primary variables. When using the CV welding system, welding current is controlled by the electrode wire feed speed. Therefore, penetration is directly influenced by wire feed speed when all other conditions are the same. Since welding current can be easily measured, penetration is normally related to it rather than to wire feed speed. The wire feed speed–current relationship can be changed by changing polarity, shielding media, electrode wire size, and the stickout. Stickout is shown in Figure

FIGURE 6–140 Welding voltage–current relationship.

(graph: ARC VOLTAGE (VOLTS) vs AMPERES DIRECT CURRENT (ELECTRODE POSITIVE))

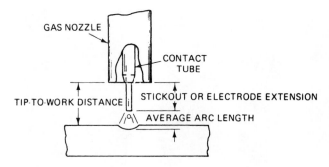

FIGURE 6–141 Stickout or electrode extension.

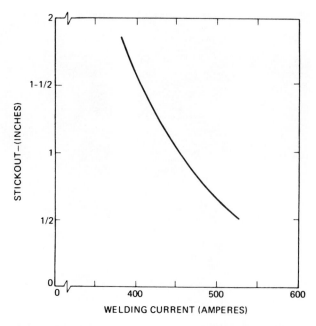

FIGURE 6–143 Stickout versus welding current.

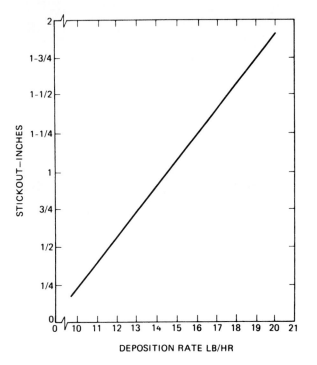

FIGURE 6–142 Stickout versus deposition rate.

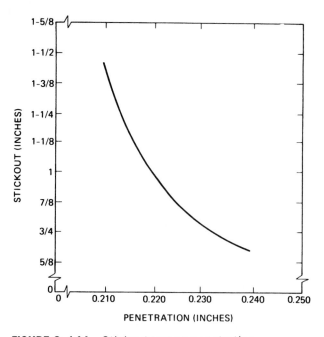

FIGURE 6–144 Stickout versus penetration.

6-141. The electrode wire extending from the current pickup tip to the arc is heated by the current carried by the wire. The heat generated in this portion of the electrode wire is equal to I^2R, which is the current squared times the resistance of the electrode wire. This preheats the electrode wire so that when it enters the arc it is at an elevated temperature, which increases the melt-off rate. Increasing stickout increases deposition rate *only if the wire feed speed is increased* sufficiently to maintain the current at a constant value (Figure 6-142).

The relationship between stickout and welding current is shown in Figure 6-143. Increasing the stickout will reduce the welding current in the arc by almost 100 A when the wire feed speed rate is not changed. This reduces penetration. In semiautomatic welding, the stickout is adjusted by the welder and is an excellent means of compensating for joint variations without stopping the

weld. Stickout exerts an influence on penetration through its effect on welding current (Figure 6-144). Stickout is thus a control during the welding operation.

Stickout influences the welding current. Increasing the electrode extension increases the resistance in the circuit and, with the voltage constant, the current will be reduced in accordance with Ohm's law. The voltage between the current pickup tip and the work is the sum of the voltage across the arc and the voltage drop in the electrode extension. As the electrode extension or stickout in-

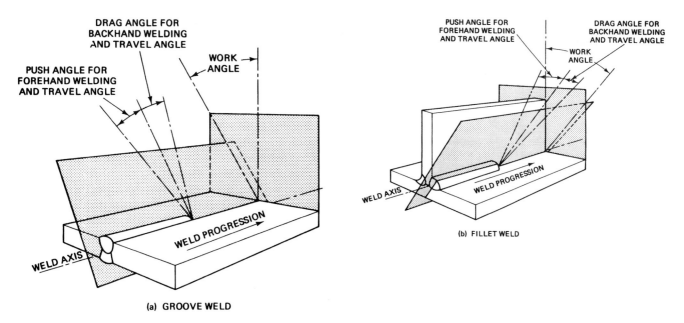

FIGURE 6-145 Travel and work angles: groove and fillet welds.

creases, the circuit resistance increases and the welding current stays constant; therefore, more voltage occurs across the extension and thus less voltage occurs across the arc. The decrease of both voltage and current will reduce the penetration of the arc. There is a limit to the stickout that can be used.

Conversely, as the stickout (electrode extension) decreases, the preheating effect is reduced and the welding power source furnishes more current. This increase in welding current provides a proportionate increase in penetration.

Another secondary adjustable variable is the electrode or nozzle travel angle, which has an appreciable effect on penetration. Two angles are required to define the position of an electrode or welding gun nozzle: (1) the travel angle, and (2) the work angle.

The work angle is the angle, less than 90°, between a line perpendicular to the major workpiece surface and a plane determined by the electrode or centerline of the welding gun and the weld axis. In a T-joint or a corner joint, the line is perpendicular to the nonbutting member (Figure 6-145).

The travel angle is the angle, less than 90°, between the electrode centerline, or centerline of the welding gun, and the line perpendicular to the weld axis, in a plane determined by the electrode axis and the weld axis. This is also shown in Figure 6-145.

The travel angle is described further as being either a drag angle or a push angle. The drag angle is the travel angle when the electrode is pointing in a direction opposite to the progression of welding (points backward). The push angle is the travel angle when the electrode is pointing in the direction of weld progression (points forward). The push angle is also known as forehand welding, and the drag angle is also known as backhand welding.

In pipe welding, the work angle is the angle, less than 90°, between a line that is perpendicular to the pipe surface at the point of intersection of the weld axis and the centerline of the electrode or welding gun, in a plane determined by the centerline of the electrode and a line tangent to the pipe at this same point (Figure 6-146). The travel angle for a pipe weld is the angle, less than 90°, in the electrode centerline or the torch centerline and a line perpendicular to the weld axis at its point of intersection with the electrode centerline, in a plane determined by the electrode centerline and a line tangent to the pipe surface at the same point. This is also shown.

It is found that maximum penetration is obtained when a drag angle of 15 to 20° is used. If the gun travel angle is changed from this optimum condition, penetration decreases. From a drag angle of 15° to a push angle of 30°, the relationship between penetration and travel angle is almost a straight line. Therefore, good control of penetration can be obtained in this range. A drag angle greater than 25° cannot be used. The gun travel angle variable can also be used to change bead height and width, since the gun travel angle does affect bead contour. A drag travel angle tends to produce a high, narrow bead. As the drag angle is reduced, the bead height decreases and the width increases. This relationship is shown by Figure 6-147. The push travel angle is used for high travel speeds. These angles vary slightly with different processes and procedures.

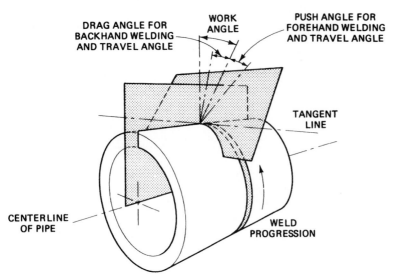

DRAG ANGLE FOR
BACKHAND WELDING
AND TRAVEL ANGLE

WORK
ANGLE

PUSH ANGLE FOR
FOREHAND WELDING
AND TRAVEL ANGLE

TANGENT
LINE

CENTERLINE
OF PIPE

WELD
PROGRESSION

FIGURE 6–146 Travel and work angles: pipe welding.

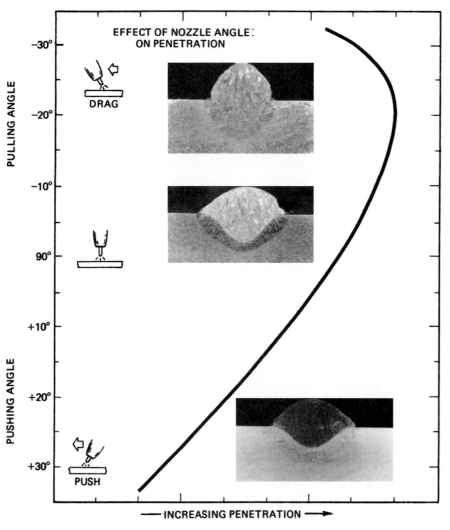

FIGURE 6–147 Travel angle versus penetration.

EFFECT OF NOZZLE ANGLE
ON PENETRATION

−30°

−20°

PULLING ANGLE

DRAG

−10°

90°

+10°

PUSHING ANGLE

+20°

+30°

PUSH

← INCREASING PENETRATION →

Distinct Level Variables

The most important distinct level variable is the selection of the welding process. Once this is done, the next important variable can be the polarity of welding. In general, direct-current electrode positive DCEP produces greater penetration than electrode negative DCEN. Alternating current produces penetration between that produced by electrode positive and electrode negative. The polarity of the electrode, and its influence on penetration, is shown by the three curves in Figure 6-148. This is also shown by Figure 6-149, which was made using the submerged arc welding process.

The other variable has to do with electrode size. Smaller electrodes tend to produce deeper penetration. This is related to the geometry of the arc and the point-to-plane relationship. The higher the current density on the electrode wire, the deeper the penetration. Larger-wire electrodes produce wider beads and less penetration (Figure 6-150).

In GMAW and FCAW, the use of CO_2 gas shielding provides deeper penetration. It is the characteristic of carbon dioxide to provide deep penetration. The shielding gas relationship to penetration is shown in Figure 6-151, made using a self-shielding flux-cored electrode wire.

FIGURE 6-148 Welding polarity versus penetration: curve.

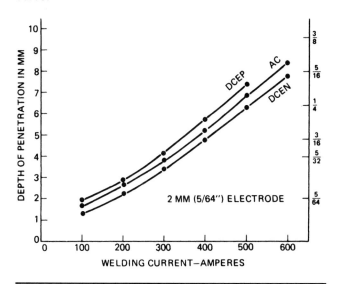

FIGURE 6-150 Electrode size versus penetration.

3/16 in.   5/32 in.   1/8 in.

FIGURE 6-149 Welding polarity versus penetration: cross section (SAW).

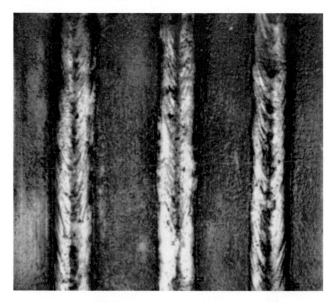

DC EH   DI EP   AC

FIGURE 6-151 With and without shielding gas (FCAW).

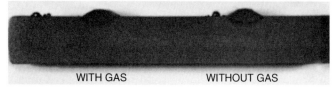

WITH GAS   WITHOUT GAS

Also, in GMAW the type of shielding gas affects the weld bead shape and penetration pattern. Argon has a characteristic deep center or pointed penetration, while CO_2 provides a wider pattern. The CO_2-argon mixture is between these. The cross-sectional view and weld surface appearances are shown in Figure 6-152.

By fully understanding the relationship of the variables and their effect on weld characteristics, it is possible to establish a welding procedure to provide the exact type, shape, size of weld, and welding production rate that are required. A summary of the welding variables that can be changed to change the characteristics of the weld is given in Figure 6-153.

6-11 ARC WELDING PROCESS SELECTION

The criteria for selecting an arc welding process are extremely complex. In view of this complexity, it is well to establish a basis for choosing the welding process. The factors that must be considered are as follows:

1. The ability to join the metals involved
2. The quality or reliability of the resulting joint
3. The capability of the process to join the metals in the thickness and position required

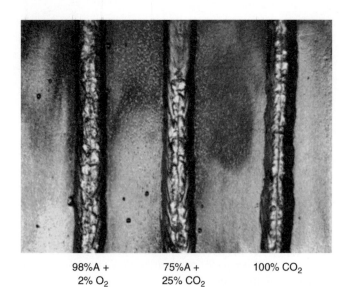

| 98%A + 2% O_2 | 75%A + 25% CO_2 | 100% CO_2 |

FIGURE 6–152 Shielding gas versus weld bead shape.

FIGURE 6–153 Recommended variable adjustments for consumable electrode arc welding.

Change Required	Welding Variable	Arc Voltage	Welding Current (Feed Speed)	Travel Speed	Travel Angle	Stickout or Tip-to-Work Distance	Electrode Size
Deeper penetration			[1]Increase		[3]Drag max. 25°	[2]Decrease	[5]Smaller[a]
Shallower penetration			[1]Decrease		[3]Push	[2]Increase	[5]Larger[a]
Bead height	Larger bead		[1]Increase	[2]Decrease		[3]Increase[a]	
and	Smaller bead		[1]Decrease	[2]Increase		[3]Decrease[a]	
Bead width	Higher narrower bead	[1]Decrease			[2]Drag trailing	[3]Increase	
	Flatter wider bead	[1]Increase			[2]90°	[3]Decrease	
Faster deposition rate			[1]Increase			[2]Increase[a]	[3]Smaller
Slower deposition rate			[1]Decrease			[2]Decrease[a]	[3]Larger

Key: 1, First choice; 2, second choice; 3, third choice; 4, fourth choice; 5, fifth choice.
[a]It is assumed that the wire feed speed is readjusted to hold welding current constant.

4. The most economical way of joining metals
5. The availability of the necessary equipment
6. The familiarity of the personnel involved in making the joint
7. Other factors, such as the engineering capabilities to design, the user reaction to the method, and so on

The ability to join the given metal must be the very first consideration. In many cases the metal can be welded with a number of welding processes; then the selection depends on other factors. The quality of the joint produced by the processes is the second basis for process determination. The designer must be aware of the quality requirement of the product, which involves the service requirements, specifications, codes, and environmental exposure that can be expected. The metals to be joined must be selected on the same basis. It is then necessary to determine the joining process that will provide a quality joint.

The third factor is the thickness of the metals to be joined and the position of the joints to be welded. Some processes have all-position capabilities, while others are limited to a few welding positions. This information is summarized in Figure 6-154. The position capability may not be important since many assemblies can be positioned to place them in the position for most advantageous welding. The position cannot be altered in some cases: for example, in the field, erection of large products and in the repair welding of products that cannot be moved.

These factors will narrow the choice of welding processes. After analyzing these three, there still is the requirement to establish the optimum and most economical welding process. The welding cost factor should be used. The two major components in welding cost are the cost of labor to apply the welds and the cost of the materials used.

The cost of labor continually increases. It is important to select processes that are most productive. The productivity is related to its deposition rates. Deposition rate data for each of the processes are presented in the process section. A summary of deposition rates based on 100% duty cycle is shown in Figure 6-155. These data are an indication but are not the entire story. Some processes provide high deposition rates but require more weld metal to complete the weld joint. Joint design and the amount of metal required to make the weld joint enter into this. Process productivity relates to labor cost, since each process may be applied in more than one way. The method of applying the continuous electrode wire processes is summarized in Figure 6-156. Each method of applying has a specific operator factor on duty cycle based upon the amount of time that the process is in actual operation depositing metal, versus the time available.

The other factor has to do with cost of materials. It is important to recognize that all filler metals are not utilized to the same degree. The filler metal with the lowest utilization is the covered electrode. Only about 65% of the weight of the electrodes purchased becomes deposited weld metal in the product. In gas metal arc welding and electroslag welding the amount of purchased weld metal deposited in the joint is 95%, while the amount of flux-cored electrode wire deposited in the joint approaches 85%.

The availability of equipment also has a bearing on the process selection. Products similar to those normally produced will utilize the same equipment. If the new item is sufficiently different from existing products, its production may require different equipment. The cost and the availability of the equipment are important factors. Tooling for high-volume production or specialized precision welding must also be considered.

It is important to be objective and question whether or not the availability of equipment overweighs new equipment that might be more productive. Changing to a more productive welding process can make available equipment obsolete. This may be justified by a quick repayment of the cost of the new equipment. This is very important when considering robot welding.

The familiarity of the personnel can have an important bearing on process selection. It may be necessary to provide a training program to teach existing personnel new skills. An objective view should be taken to compare the availability of skill of the existing workforce versus

Welding Position		Welding Process Rating			
		SAW	GMAW	FCAW	ESW
1. Flat		A	A	A	No
Horizontal fillet		A	A	A	No
2. Horizontal		C	A	A	No
3. Vertical		No	A	A	A
4. Overhead		No	A	A	No
5. Pipe-fixed		No	A	A	No

FIGURE 6-154 Summary of welding positions for consumable electrodes.

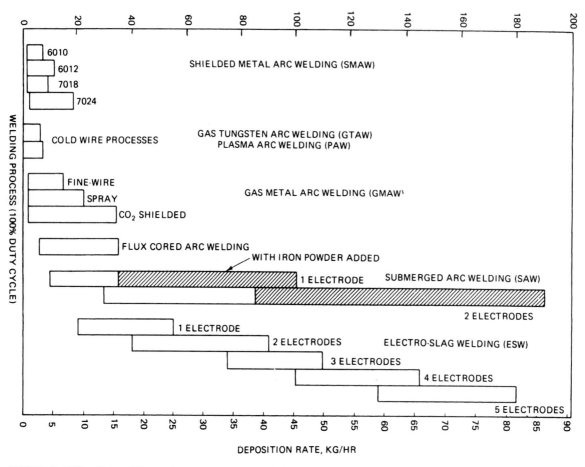

FIGURE 6–155 Deposition rate summary by welding processes.

Method of Applying	Welding Process			
	SAW	GMAW	FCAW	ESW
Manual (MA)	Not used	Not used	Not used	Not used
Semiautomatic (SA)	Not popular	Most popular	Most popular	Not used
Mechanized (ME)	Most popular	Most popular	Most popular	Most popular
Adaptive control (AD)	Most popular	Most popular	Most popular	Not used
Robotic (RO)	Not popular	Most popular	Most popular	Not used

FIGURE 6–156 Summary of methods of application.

what would be required if the process were changed. It is necessary to consider inspection, material preparation, and supervisory and administrative personnel, as well.

There are other factors that must be considered. One has to do with the ability of engineering and design personnel to adapt to a new process. There is reluctance to switch from a well-known process to a new unknown one. Another factor has to do with the user of the end product. Users are familiar and apparently satisfied with the present method of production. Change can upset relationships and cause users to discontinue the product

made by a new process. When suddenly confronted with a welded assembly, they may have questions concerning the new part and its ability to withstand rough service. There is also the reluctance to change versus the necessity to improve products and reduce costs.

When all these factors are considered, a manageable plan evolves. It will enable the metalworking executive to chart a course with the known factors and to arrive at the preferred welding process. Alternative selections may result and practical production tests should be made to obtain the best answer.

QUESTIONS

6–1. Is CO_2 shielding gas used for GMAW welding of aluminum?

6–2. What is the main purpose of the coating on stick electrodes?

6–3. What are three types of metal transfer in an arc welding process?

6–4. What is the main purpose of submerged arc flux? Other purposes?

6–5. Why isn't submerged arc welding an all-position welding process?

6–6. If ac is used for submerged arc, is it CC or CV?

6–7. Is a welding helmet required for submerged arc welding?

6–8. For GMAW, what changes are required in weld joint design?

6–9. Describe the electrogas process of welding. For what position is it used?

6–10. When is cooling water recommended for the GMAW torch or gun?

6–11. What is the major difference between FCAW and GMAW?

6–12. What equipment changes are necessary to change processes?

6–13. What are the three classifications of welding variables?

6–14. What adjustments change weld penetration? Explain.

6–15. What adjustments change weld bead width? Explain.

6–16. What adjustments change weld bead reinforcement? Explain.

6–17. What is stickout, and what effect does it have?

6–18. Define *work angle* and *travel angle*.

6–19. What makes electroslag welding different from submerged arc welding?

6–20. What welding groove design is normally used for electroslag welding?

REFERENCES

1. K. E. Dorschu, "Control of Cooling Rates in Steel Weld Metal," *Welding Journal,* Research Supplement (February 1968): 495.

2. L. N. Gourd, (*Principles of Welding Technology,* London: Edward Arnold, 1980.

3. J. F. Lancaster, *The Physics of Welding,* International Institute of Welding, Elmsford, N.Y.: Pergamon Press, 1984.

4. C. E. Jackson, "The Science of Arc Welding, Part II," *Welding Journal,* Research Supplement (May 1960): 177s.

5. C. E. Jackson, "Control of Welding Performance by Selected Technique Parameters," Document X11K85-77, International Institute of Welding.

6. H. T. Herbst and T. McElrath, Jr., "Sigma Welding of Carbon Steels," *Welding Journal* (December 1951): 1084.

7. A. Lesnewich, "Control of Melting Rate and Metal Transfer in Gas Shielded Metal-Arc Welding, Part I," *Welding Journal* (August 1958): 343s.

8. J. C. Needham, C. J. Cooksey, and D. R. Milner, "Metal Transfer in Inert-Gas Shielded-Arc Welding," *British Welding Journal* (February 1960): 101.

9. A. Lesnewich, "Control of Melting Rate and Metal Transfer in Gas-Shielded Metal-Arc Welding, Part II," *Control of Metal Transfer Welding,* Research Supplement (September 1958): p. 418s.

10. A. A. Smith, "Characteristics of the Short Circuiting CO_2 Shielded Arc," (proceedings of a symposium on Physics of the Welding Arc, The Welding Institute, London, Oct. 29–Nov. 2, 1962): 75–91.

11. B. J. Bradstreet, "Effect of Surface Tension and Metal Flow on Weld Bead Formation," *Welding Journal* (July 1968): 314s.

12. "Specifications for Mild Steel-Covered Arc Welding Electrodes," AWS A5.1, American Welding Society, Miami, Fla.

13. "Specifications for Low-Alloy Steel-Covered Arc Welding Electrodes," AWS A5.5, American Welding Society, Miami, Fla.

14. H. B. Cary, "Gravity Welding—A Variation of Shielded Metal-Arc Welding," *Welding Journal* (November 1979): 36.

15. "Specification for Carbon Steel Filler Metals for Gas Shielded Arc Welding," AWS A5.18, American Welding Society, Miami, Fla.

16. "Specification for Low Alloy Steel Filler Metals for Gas Shielded Arc Welding," AWS A5.28, American Welding Society, Miami, Fla.

17. "Recommended Practices for Gas Metal Arc Welding," AWS C5.6, American Welding Society, Miami, Fla.

18. "Specification for Carbon Steel Electrodes for Flux Cored Arc Welding," AWS A5.20, American Welding Society, Miami, Fla.

19. "Specification for Low Alloy Steel Flux Cored Welding Electrodes," AWS A5.29, American Welding Society, Miami, Fla.

20. "Specification for Carbon Steel Electrodes and Fluxes for Submerged Arc Welding," AWS A5.17, American Welding Society, Miami, Fla.

21. "Specifications for Bare Low-Alloy Steel Electrodes and Fluxes for Submerged Arc Welding," AWS A5.23, American Welding Society, Miami, Fla.

22. "Improved Fracture Toughness and Fatigue Characteristics of Electroslag Welds," FHWA/RD-87/026, Federal Highway Administration, Washington, D.C., Oct. 1986.

23. "Specification for Consumables Used for Electroslag Welding of Carbon and High Strength Low Alloy Steels," AWS A5.25, American Welding Society, Miami, Fla.

24. "Recommended Practices for Electrogas Welding," AWS C5.7, American Welding Society, Miami, Fla.

25. "Specifications for Consumables Used for Electrogas Welding of Carbon Steels and High Strength Low Alloy Steels," AWS A5.26, American Welding Society, Miami, Fla.

GAS WELDING, BRAZING, SOLDERING, AND SOLID-STATE WELDING

OUTLINE

7-1 OXYFUEL GAS WELDING

Oxyfuel gas welding (OFW) is a group of welding processes that produce coalescence of workpieces by heating them with an oxyfuel gas flame. The processes are used with or without the application of pressure, and with or without filler metal. There are three major processes within this group: oxyacetylene welding, oxyhydrogen welding, and pressure gas welding. There is another process, but of minor industrial significance, known as airacetylene welding.

The most popular process in this group is oxyacetylene welding, which is an oxyfuel gas welding process that uses acetylene as the fuel gas. Oxyhydrogen welding uses hydrogen as the fuel gas. It is not popular. The third major process is pressure gas welding (PGW), which is an oxyfuel gas welding process that produces a weld simultaneously over the entire faying surfaces. The process is used with the application of pressure and without filler metal.

Oxyacetylene Welding

The **oxyacetylene welding** (OAW) process (Figures 7–1 and 7–2) consist of high-temperature flame produced by the combustion of acetylene with oxygen and directed by a torch. The intense heat of the flame 6300° F (3482° C) melts the surface of the base metal to form a molten pool. Filler metal is added to fill gaps or grooves. As the flame moves along the joint, the melted base metal and filler metal solidify to produce the weld.

The temperature of the oxyacetylene flame is not uniform throughout its length and the combustion is also different in different parts of the flame. Figure 7–3 shows the temperature in different portions of the flame. The temperature is the highest just beyond the end of the inner cone and decreases gradually toward the end of the flame. The chemical reaction for a 1:1 ratio of acetylene and oxygen plus air is as follows:

$$C_2H_2 + O_2 = 2CO + H_2 + heat$$

This is the primary reaction; however, both carbon monoxide and hydrogen are combustible and will react with oxygen from the air:

$$2CO + H_2 + 1.5O_2 = 2CO_2 + H_2O + heat$$

This is the secondary reaction, which produces carbon dioxide, heat, and water.

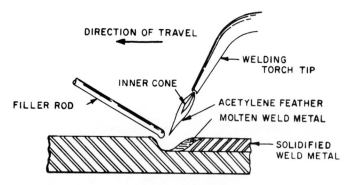

FIGURE 7-1 Oxyacetylene welding process.

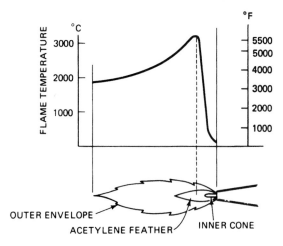

FIGURE 7-2 Using the oxyacetylene welding process.

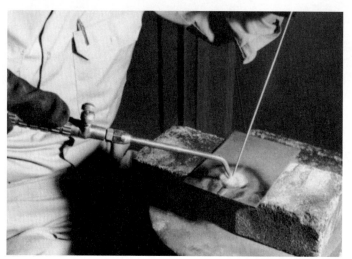

FIGURE 7-3 Temperature of the oxyacetylene flame.

There are three basic flame types: neutral (or balanced), excess acetylene (carborizing), and excess oxygen (oxidizing) (Figure 7-4). The neutral flame has a 1:1 ratio of acetylene and oxygen. It obtains additional oxygen from the air and provides complete combustion. It is

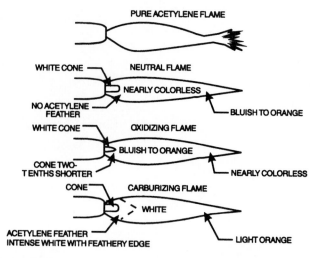

FIGURE 7-4 Four types of oxyacetylene flame.

generally preferred for welding. The neutral flame has a clear, well-defined, luminous cone, indicating that combustion is complete. The carborizing flame has excess acetylene. This is indicated in the flame when the inner cone has a feathery edge extending beyond it. This white feather is called the acetylene feather. If the acetylene feather is twice as long as the inner cone, it is known as a 2X flame, which is a way of expressing the amount of excess acetylene. The carborizing flame may add carbon to the weld metal. The oxidizing flame, which has an excess of oxygen, has a shorter envelope and a small pointed white cone. The reduction in length of the inner core is a measure of excess oxygen. This flame tends to oxidize the weld metal and is used only for welding specific metals. Most welding procedures use the neutral flame. The welder soon learns proper flame adjustment.

Advantages and Major Uses

The oxyacetylene welding process has many advantages. The equipment is very portable, is relatively inexpensive, and can be used in all welding positions. The equipment is also versatile: It can be used for welding, brazing, soldering, and, with proper attachments, flame cutting. The equipment can be used as a source of heat for bending, forming, straightening, hardening, and so on. Another advantage is that the molten pool is visible to the welder.

The oxyacetylene welding process is normally used as a manual process. It can be mechanized, but this is not common. It is rarely used for semiautomatic applications. Oxyacetylene welding is used to weld most of the common metals (Figure 7-5).

When welding any metal, the appropriate filler material must be selected. The filler metal must match the composition of the base metal to be welded and normally contains deoxidizers to aid in producing sound welds. Flux is also required for welding certain materials.

Base Metal	Filler Metal Type	Flame Type	Flux Type
Aluminum	Match base metal	Slightly reducing	Al. flux
Brass	Navy brass	Slightly oxidizing	Borax flux
Bronze	Copper-tin	Slightly oxidizing	Borax flux
Copper	Copper	Neutral	None
Copper-nickel	Copper-nickel	Reducing	None
Inconel	Match base metal	Slightly reducing	Fluoride flux
Iron, cast	Cast iron	Neutral	Borax flux
Iron, wrought	Steel	Neutral	None
Lead	Lead	Slightly reducing	None
Monel	Match base metal	Slightly reducing	Monel flux
Nickel	Nickel	Slightly reducing	None
Nickel-silver	Nickel-silver	Reducing	None
Steel, high carbon	Steel	Reducing	None
Steel, low alloy	Steel	Slightly reducing	None
Steel, low carbon	Steel	Neutral	None
Steel, medium carbon	Steel	Slightly reducing	None
Steel, stainless	Match base metal	Slightly reducing	SS flux

FIGURE 7–5 Base metals weldable by the oxyacetylene process.

The oxyacetylene welding process is normally used for welding thinner materials up to $\frac{1}{4}$ in. (6.4 mm) thick. It can be used for welding heavier material but is rarely used for thick metals. Its major industrial applications are in the field of maintenance and repair, and in welding small-diameter pipe.

Welding Apparatus

The apparatus and equipment employed for oxyacetylene welding are shown in Figure 7-6. This diagram shows the (1) welding torch and tips, (2) oxygen and acetylene hose, (3) oxygen and acetylene regulators, (4) oxygen cylinder, and (5) acetylene cylinder. A spark lighter is normally used. The welding torch, sometimes called a *blowpipe,* is the major piece of equipment for this process. It performs the function of mixing the fuel gas with oxygen and provides the required flame, which is directed as desired. The torch consists of a handle or body, which contains the hose connections for the oxygen and the fuel gas. It also contains an oxygen and acetylene valve for regulating gas flow and a mixing chamber. Various sized tips can be attached. The torch must be well constructed and rugged, since it is in contact with the high-temperature flame.

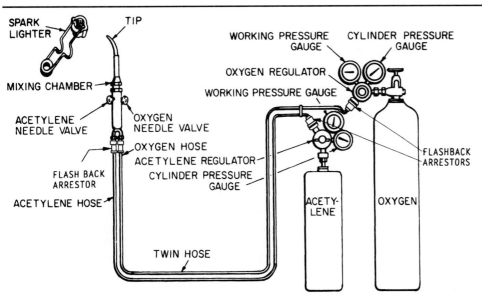

FIGURE 7–6 Apparatus required for welding with OAW.

FIGURE 7–7 Medium-pressure oxyacetylene torch.

There are two basic types of torches, the medium-pressure torch, which is most popular, and the low-pressure or injector type. When using the medium-pressure torch, both oxygen and acetylene are supplied at approximately the same pressure, which may vary from 1 to 10 psi, depending on the size of the tip being used. The two gases are mixed together in the mixing chamber in the torch handle. Some torches have the mixing chamber in the tip (Figure 7–7).

The injector type of torch uses acetylene at pressures less than 1 psi and are designed so that the oxygen at a higher pressure draws the acetylene into the mixing chamber. Any change in oxygen flow will produce a relative change in acetylene flow so that the proportion of the two gases remains constant. This type of torch is not popular.

The valves on the body of the torch control the amount of oxygen and acetylene or fuel gas that flows to the mixing chamber, where they are combined. Different welding tips are available so that the same torch handle can be used for a wide variety of operations.

Welding tips are available in a variety of sizes, which are determined by the drill size of the orifice or hole in the flame end of the tip. The larger tip will have a larger orifice, which will produce a larger flame and use more gas. The larger flame supplies greater amounts of heat. The proper tip size must be selected for welding different metals and metal thicknesses. Welding procedure schedules indicate the tip size and gas pressures to be employed. This will determine the volume of gases used. Tips must fit properly and tight to the torch. They must be kept clean for proper operation. Some welding torches are designed so that they can be converted to an oxyacetylene flame cutting torch by replacing the welding tip with a cutting attachment.

The connections for oxygen and acetylene or fuel gas hoses to the torch handle have special threads so that they cannot be incorrectly attached. The oxygen fittings have right-hand threads and the acetylene or fuel gas fittings have left-hand threads. There is a special sequence for opening and closing valves and lighting the torch, which must be followed for safe operation.

Gas pressure regulators are required for both the oxygen and acetylene. Regulators reduce the pressure of the gas in the cylinder or supply system to the pressure

used in the torch. The pressure in an oxygen cylinder can be as high as 2200 psi (15.2 MPa) and this must be reduced to a working pressure of from 1 to 25 psi (6.9 to 172 kPa). The pressure of acetylene in an acetylene cylinder can be as high as 250 psi (1.7 MPa) and this must be reduced to a working pressure of from 1 to 12 psi (6.9 to 82 kPa). When gases are piped to the workstations from a central supply, the pressure is lower but regulators are still required. A gas pressure regulator will automatically deliver a constant volume of gas to the torch at the adjusted working pressure. The regulators for oxygen, for acetylene, and for liquid petroleum fuel gases are of different construction. They must be used only for the gas for which they are designed.

There are two types of regulators, the single-stage regulator and the two-stage regulator. The *single-stage regulator* reduces the cylinder pressure of the gas to a working pressure in one step. Single-stage regulators must be readjusted from time to time to maintain the required working pressure. The gas pressure in the cylinder decreases gradually as gas is withdrawn. Single-stage regulators are less expensive than two-stage regulators and are more popular. Figure 7–8 shows a cutaway view of a single-stage regulator.

FIGURE 7–8 Cutaway view of a single-stage regulator.

The *two-stage regulator* makes the reduction of pressure in two steps. The first step reduces the cylinder pressure to an intermediate pressure. The second step reduces this intermediate pressure to the desired working pressure. A two-stage regulator is simply two single-stage regulators in the same case. The two-stage regulator provides more accurate regulation and eliminates the need to readjust the regulator as the pressure in the supply tank is reduced. Regulators have two pressure gauges: One shows the pressure of the gas inside the cylinder, the other shows the working pressure that is being supplied to the torch. Figure 7–9 shows a cutaway view of a two-stage gas regulator.

The operation of a regulator for controlling gas pressure and providing for uniform gas flow is rather straightforward. The regulator consists of a flexible diaphragm, which controls a needle valve between the high-pressure zone and the working pressure zone; a compression spring; and an adjusting screw, which compensates for the pressure of the gas against the diaphragm. The needle valve is on the side of the diaphragm exposed to high gas pressure, while the compression spring and adjusting screw are on the opposite side in a zone vented to the atmosphere. The spring is compressed by the adjusting knob on the outside of the regulator. In the closed position, the diaphragm is flat and the needle valve closes the orifice to the high pressure zone. In this condition the compression spring is not loaded and the adjusting knob is backed off. As the knob is turned clockwise, it compresses the spring, which in turn presses against the diaphragm to open the needle valve into the high-pressure zone. As the valve opens, high-pressure gas

enters the chamber and tends to push the diaphragm against the compression spring. This will tend to close the needle valve. When the spring is further compressed, it will open the needle valve further, allowing more gas through the orifice and into the pressure zone. By balancing the compression spring against the pressure of the gas, the needle valve is kept at the right opening to allow the correct flow of gas through the orifice. The compression spring balanced against the gas pressure keeps the valve at the proper opening to allow the required flow and pressure on the low-pressure side of the regulator. Figure 7–10 shows a diagram of a gas regulator. Single-stage regulators are used for plant piping systems since the pressure in the pipe system is much lower than in the cylinder. Torches, regulators, and other gas apparatuses must be approved by one of the approving agencies. They must also be properly handled and maintained.

Flashback Arresters

A backfire is an explosion in the welding or cutting torch head. A flashback is combustion in the torch. Both are dangerous and highly undesirable. Backfires and flashbacks can be avoided by the use of flashback arresters. There are two types, as shown in Figure 7–11. The torch-mounted

FIGURE 7–9 Cutaway view of a two-stage regulator.

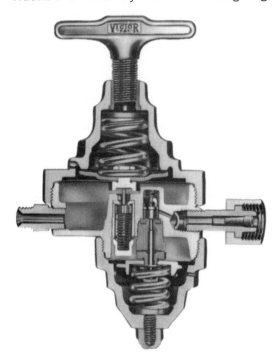

FIGURE 7–10 Gas regulator operation.

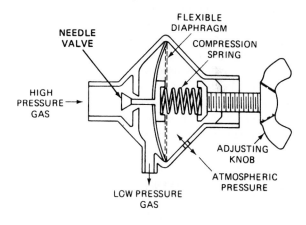

FIGURE 7–11 Flashback arresters.

flashback arrester is primarily a check valve that prohibits reverse gas flow from the torch. The regulator-mounted flashback arrester is a combination check valve and porous sintered metal barrier. The gas flows through it, but the flame is unable to penetrate it. Both torch-mounted and regulator-mounted flashback arresters should be employed for maximum safety. They should be installed in both the fuel gas and oxygen lines. They should be checked periodically for proper operation.

Gas Hose

Gas hose, used between pieces of apparatus, is described by standard ANSI/1P-7, "Specifications for Rubber Welding Hose."[1] Hose is available as single or double vulcanized lines. The double lines consist of two hoses, one for oxygen and one for fuel gas, connected continuously along the sides so that the two hoses are an integral unit. This is for ease in handling. The maximum working pressure is 150 psi (1.03 MPa). Hose is available in three strengths, with the standard duty "type S" the most popular. Welding hose is specified by its inside diameter. The nominal inside diameter most popular is the $\frac{1}{4}$ in. (6.4 mm) size. The hose fittings have the same threaded connections as connections on the torch and regulator (i.e., oxygen has right-hand threads and fuel gas has left-hand threads). In addition, the acetylene hose connection has a groove around the outside to distinguish it from oxygen hose connections. There is no international color code for gas hose. In North America, green is used for oxygen hose and red is used for acetylene or fuel gas hose. In Europe, blue is used for oxygen hose and orange is used for acetylene or fuel gas hose. Black is used for oxygen hose in some parts of the world. Hose should never be used for oxygen gas if it was used previously for a fuel gas. Hose and connections should be kept in good repair, and if leaks occur, repair them immediately or replace the hose.

A torch must be lighted by a spark lighter consisting of a flint on a lever so it can be moved across a piece of roughened steel. Matches or cigarette lighters should not be used to light an oxyacetylene torch since that brings the hand too close to the flame. A convenient accessory is a gas saver or economizer, which includes a bracket for hanging the torch. When the torch is hung on the bracket, it closes the valves to stop the flow of oxygen and acetylene. The device has a pilot light so that when the torch is lifted from the bracket near the pilot light, the torch can be lighted. This apparatus is shown in Figure 7-12.

Gas Supply

The oxygen required for oxyacetylene welding can be supplied in several ways. The more common way is the use of high-pressure oxygen cylinders at the welding station. Another way is to manifold oxygen cylinders and run the oxygen through a piping system to the welding stations. This is common where there are a large number

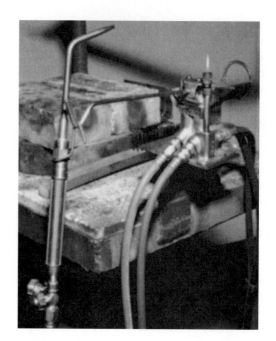

FIGURE 7-12 Gas valve and pilot light.

of welding or cutting stations that use oxygen. When a piping distribution system is used, it can be supplied by liquid oxygen if a large amount of oxygen is used.

Acetylene or fuel gases are often supplied in cylinders taken to the welding station. However, they can be piped throughout the plant in the same manner as oxygen. The acetylene may be supplied to the piping system by manifolding cylinders or by an acetylene generator. The acetylene generator produces acetylene at the plant site by the reaction of carbide and water. In all cases, the installation and operation of piping systems must be in accordance with strict specifications and safety requirements.

Welding Rod

The American Welding Society provides specification AWS A5.2 covering welding rods or filler metal used for the oxyacetylene or oxyfuel gas welding process. There are four grades shown by Figure 7-13. The specification number has the prefix letter *R*. This is followed by two or three digits, 45, 60, 65, or 100, which designate the approximate

FIGURE 7-13 Gas welding rods.

AWS Classification	Minimum Tensile Strength		Elongation in 1 in. (25 mm) (min. %)
	ksi	MPa	
R45	—	—	—
R60	60	415	20
R65	65	462	16
R100	100	690	14

tensile strength in ksi (1,000 psi). The chemical composition requirements are given in the specification.

In the case of nonferrous filler metals, the prefix *R* is followed by the chemical symbol of the principle constituent metal in the wire. The initials for one or two elements will follow. If there is more than one alloy containing the same elements, a suffix letter or number may be added.

Figure 7-5 shows the base metals weldable by the oxyacetylene welding process. This table also shows the type of filler metal required, the flame type, and the type of flux.

Welding flux is required to maintain cleanliness of the base metal, at the welding area, and to help remove the oxide film on the surface of the metal. The welding area should be cleaned. Flux melts at about the melting point of the base metal and helps protect the molten metal from the atmosphere. The molten flux combines with base metal oxides and removes them. There is no national standard for gas welding fluxes. They are categorized according to the base ingredient in the flux or the

base metal for which they are to be used. Fluxes are usually in powder form. These fluxes are often applied by sticking the hot filler metal rod in the flux. Sufficient flux will adhere to the rod to provide proper fluxing action as the filler rod is melted in the flame. Other types of fluxes are of a paste consistency, which are usually painted on the filler rod or on the work to be welded. Welding rods with a covering of flux are also available. Fluxes are available from welding supply companies and should be used in accordance with the directions accompanying them.

Quality of Welds

The quality of a weld made with the oxyacetylene process can equal the quality of the base metal being welded. This is based on the use of the proper filler metal, the proper flux, and the skill of the welder. The procedure will show the proper tip size, torch adjustment for the proper type of flame, and the travel speed. Figure 7-14 shows a good weld and common welding mistakes.

A GOOD WELD

 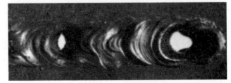

FIGURE 7-14 Quality of oxyacetylene welds.

GOOD, EVEN PENETRATION   EVEN EDGES   SMOOTH, EVEN RIPPLE

COMMON WELDING MISTAKES

GAS PRESSURE TOO HIGH   TOO MUCH OXYGEN

TRAVEL SPEED TOO FAST   IRREGULAR TRAVEL SPEED

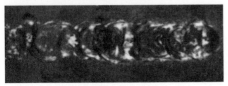 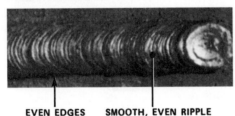

NOT ENOUGH HEAT   TOO MUCH HEAT

Welding Schedules

The oxyacetylene welding process is rarely used for joining heavy thicknesses. Figure 7–15 is a schedule that can be used for welding material, ranging from the thinnest up to the heaviest. The tip size is given by showing the orifice size and the equivalent drill size, since manufacturers utilize different numbering systems for their tips. Each manufacturer relates tip size number to either the drill size or the orifice size. The length of the inner cone is shown, as well as the recommended oxygen and acetylene pressure. The diameter of the filler rod is also shown. This schedule can be used for all-position welding. The major requirement for out-of-position welding is the control of the weld pool, which relates to the skill of the welder.

This schedule is based on welding clean mild steel using a neutral flame and not using a flux. Information concerning the welding of the different metals is provided in the metal chapters.

Safety Considerations

The oxyacetylene process is a safe welding process, provided that proper precautions are taken. Normal precautions involve the respect for open flames, for compressed gases, for combustible gases, and for hot metal. Eye and skin protection is different since the flame is not nearly as bright as an arc. Goggles with the appropriate colored lenses are used, and headshields are normally not required. Installation of the equipment and apparatus is extremely important, and if piping and manifold systems are used, they must be installed with strict compliance to codes. Torches should always be properly stored when not in use. Hoses and torches should be bled so that gas pressure does not remain in the torch or hoses. Oil or grease should never be used on gas apparatus. Review Chapter 4 for complete safety information.

Limitations of the Process

The equipment for oxyacetylene welding is fairly inexpensive. It is one of the slowest processes due to the heat transfer and temperature involved. For this reason oxyacetylene welding is rarely used for manufacturing operations. The most popular uses for oxyacetylene are torch brazing and oxygen flame cutting.

Variations of the Process

The main variation is gas pressure welding. In this process the entire area of abutting surfaces is heated with gas flames. When the heating is completed, the flames are removed and pressure is applied to achieve the weld. This process has been used for joining tubular members such as pipe. It has also been used for joining railroad rails and other parts. It is not of major industrial significance today.

The other variation is the use of hydrogen instead of acetylene. If hydrogen is used, the apparatus must be proper for hydrogen and equipment designed for acetylene cannot be used. Oxyhydrogen welding is not popular and details are not presented.

7-2 BRAZING

Brazing (B) is a group of joining processes that produces coalescence of materials by heating them to the brazing temperature in the presence of a filler metal having a liquidus above 840° F (450° C) and below the solidus of the base metal. The filler metal is distributed between the closely fitted faying surfaces of the joint by capillary action.

FIGURE 7–15 Schedule of oxyacetylene welding of mild steel.

Material Thickness Range (in.)	Filler Rod Diameter (in.)	Tip Drill Size	Orifice Size (in.)	Cone Length of Flame (in.)	Approx. Gas Pressure (psi)		Approx. Gas Consumption (ft³/h4)	
					Acetylene	Oxygen	Acetylene	Oxygen
22–16 gauge	$\frac{1}{16}$	69	0.029	$\frac{3}{16}$	1	1	2	2
$\frac{1}{16}$–$\frac{1}{8}$	$\frac{3}{32}$	64	0.036	$\frac{1}{4}$	2	2	4	4
$\frac{1}{8}$–$\frac{3}{16}$	$\frac{1}{8}$	57	0.043	$\frac{5}{16}$	3	3	10	10
$\frac{3}{16}$–$\frac{5}{16}$	$\frac{1}{8}$	55	0.052	$\frac{3}{8}$	4	4	20	20
$\frac{5}{16}$–$\frac{7}{16}$	$\frac{5}{32}$	52	0.064	$\frac{7}{16}$	5	5	45	45
$\frac{7}{16}$–$\frac{1}{2}$	$\frac{3}{16}$	49	0.073	$\frac{1}{2}$	6	6	60	60
$\frac{1}{2}$–$\frac{3}{4}$	$\frac{3}{16}$	45	0.082	$\frac{1}{2}$	7	7	70	70
$\frac{3}{4}$–1	$\frac{1}{4}$	42	0.094	$\frac{9}{16}$	8	8	80	80
Over 1 inch	$\frac{1}{4}$	36	0.107	$\frac{5}{8}$	9	9	90	90
Heavy duty	$\frac{1}{4}$	28	0.140	$\frac{3}{4}$	10	10	100	100

Note: Based on use of neutral flame. For welding clean steel, flux is not normally used. There is no standardized tip size for gas torches—this table gives data based on tip orifice size in drill size and inch diameter. There are no metric equivalents.

The *solidus* is the highest temperature at which a metal or an alloy is completely solid, that is, the temperature at which melting starts. The *liquidus* is the lowest temperature at which a metal or an alloy is completely liquid, that is, the temperature at which freezing starts. The solidus and liquidus for a particular metal or alloy are definite.

There are many different ways of brazing, and they all pertain to the method of applying heat.

Dip Brazing

There are two methods of **dip brazing** (DB): molten chemical bath dip brazing and molten metal bath dip brazing. In both cases brazing is accomplished by immersing clean and assembled parts into a molten bath. Dip brazing is used for brazing small parts. The assembly should be self-jigging so the parts will maintain the proper relationship until the brazing filler metal has completely solidified. By using a high-quality furnace and controller, close temperature control is obtained with the dip brazing method.

The molten material is contained in a pot-type furnace, which is heated by oil, gas, or electricity, or by means of electrical resistance units placed in the bath. Normally, the parts to be brazed are first preheated in an air-circulating furnace. When the parts have reached the preheat temperature, they are immersed in the molten bath.

The difference between the two methods of dip brazing is the molten material in the pot. In the molten chemical bath, the bath is called a flux bath. When a chemical bath is used, the filler metal must be preplaced in the joints that are to be brazed. Fluxing of the assembly is not required.

The other method of dip brazing utilizes molten metal. The molten brazing material will flow into the joints to be brazed by capillary action. The parts must be clean and fluxed prior to dip brazing. A flux cover should be maintained over the surface of the molten metal bath.

Dip brazed parts normally distort less than torch brazed parts because of the uniform heating. It is suited for moderate- to high-production runs because the tooling is relatively complex. The method of applying the process may be manual or automatic. The process is suited for brazing small to medium parts with multiple or hidden joints. It can be used for all the metals that can be brazed and is particularly suited for aluminum and other alloys that have melting points very close to the brazing temperature. The brazing operation can also perform certain of the heat-treating operations on aluminum.

Furnace Brazing

Furnace brazing (FB) is accomplished by placing cleaned parts in a furnace. The parts should be self-jigging and assembled, with filler materials preplaced near or in the joint. The preplaced brazing filler material may be in the form of wire, foil, filings, slugs, powder, paste, tape, and so on. The furnaces are usually heated by electrical resistance. Other types of fuel can be used but only for muffle-type furnaces. Automatic temperature controllers are required so that they can be programmed for the brazing temperatures and for cooling. The batch-type furnace is used for medium-production work. A continuous-conveyor furnace is used for high-volume work. When continuous furnaces are used, several temperature zones may be employed, which provide the proper preheat, brazing, and cooling temperatures. In either type, specialized holding fixtures are required.

Flux is employed except when an atmosphere is specifically introduced in the furnace to perform this function. Flux should not be used where postbraze cleaning is made difficult by the complexity of the design of the brazed parts. Furnace brazing is often done without the use of flux but by the use of special atmospheres in the brazing furnace. Flux is not necessary if the brazing is done in a reducing-gas atmosphere, such as hydrogen or other special gases. Inert gases—argon or helium—are sometimes employed to obtain special properties. Furnace brazing can also be performed in a vacuum, which prevents oxidation and may eliminate the need for flux. Vacuum brazing is widely used in the aerospace and nuclear fields, where reactive metals are being joined and where entrapped fluxes would not be acceptable. In vacuum brazing, the vacuum is maintained by continuous pumping, which will remove volatile constituents liberated during the brazing operation. There are some base metals and filler metals which cannot be brazed in a vacuum since low-boiling-point or high-vapor-pressure constituents would be volatized and lost. The vacuum is a relatively economical method and is an accurately controlled atmosphere. It provides for surface cleanliness and allows the flow of filler metals without the use of fluxes.

It is important to select the correct atmosphere based on the type of base metals and filler metals being employed. For example, copper brazing of steels is normally done in a reducing atmosphere of high-purity hydrogen. For more information, see specification for furnace brazing ANSI AWS C3.6.

The distortion of furnace brazed assemblies is less than torch brazed parts. Furnace brazing is suitable for joining thin sections to thick sections. It can be used for brazing parts of all sizes having multiple joints and hidden joints. Most metals are completely annealed as a result of furnace brazing. In some cases the heat-treating operation can be done in conjunction with the brazing cycle, provided that the two programs are compatible.

Induction Brazing

The heat for induction brazing (IB) is obtained from the resistance of the work to an electrical current induced in the parts to be brazed. The parts or joints are placed in an alternating-current field but do not become a part of the electrical circuit. High-cycle alternating current ranging from 5,000 to 5,000,000 Hz can be used. In general, solid-state oscillator units operating in the range 200,000 to 5,000,000 Hz are used. The output is fed into a copper tubing work coil, usually water cooled, designed specifically to fit the shape of the parts to be brazed. They do not touch the parts but are coupled to them by the electrical field. The design of work coils and how they are coupled to the workpieces is quite complex. The frequency of the power source determines the type of heat that will be induced in the part. High-frequency power sources produce skin heating in the parts. Low-frequency current results in deeper heating and is used for brazing heavier sections. Heating of the part usually occurs within 10 to 60 seconds. Sufficient time must be provided for the filler metal to flow through the entire joint and to form good fillets at the interface.

Induction brazing is ideally suited for high-volume manufactured parts. Mechanized systems for moving the parts to and from the coil are quite common. The filler metal is normally preplaced in the joint, and the brazing operation can be done in air, in an inert-gas atmosphere, or in a vacuum—which dictates whether or not brazing fluxes are used. The major advantage of induction brazing is the rapid heating rates that make it suitable for brazing with filler metal alloys that tend to vaporize or segregate. A disadvantage of induction brazing is that the heat may not be uniform. Thin sections tend to heat up quicker than heavy sections, and thin sections may tend to overheat. Field or shading coils are sometimes used to reduce the problem of overheating. Induction brazing is applied as an automatic process.

Infrared Brazing

In **infrared brazing** (IRB) the heat is obtained from infrared heat or a *black* heat source below the red rays in the spectrum. There is some visible light involved but the principal heating is done by the invisible radiation. Heat sources or lamps capable of delivering up to 5,000 W of radiant energy are used. The lamps do not necessarily need to follow the contour of the parts being brazed, even though the heat input varies by the square of the distance from the source. Radiation concentrating reflectors are often used. Sources other than electrical lamps that supply infrared radiation can be used. Parts to be brazed are positioned so that the radiant energy will impinge on the joint.

With infrared brazing, as in furnace brazing and induction brazing, the parts can be contained in air, in an inert atmosphere, or in a vacuum. The same comments concerning fluxes and atmospheres apply. Infrared brazing is not as fast as induction brazing; however, the equipment is much less expensive. Infrared brazing is designed for automatic application and is not applied manually.

Normally, the parts to be brazed are self-jigging, and the filler material is preplaced in or near the joint.

Resistance Brazing

In **resistance brazing** (RB) the heat is obtained from the resistance to the flow of an electrical current through the parts being brazed. The parts become a part of the electrical circuit. Electrodes may be copper alloys or carbon-graphite material. Resistance-welding machines can be used for supplying the electric current to the parts. When resistance-welding equipment is used, it is used at a lower power input than when it is used for resistance welding. Specially designed machines for resistance brazing are also used. Alternating current is normally employed. Direct current may be used but is not as common. The parts to be brazed are held between the two electrodes while the correct pressure and electrical current are applied. The pressure is maintained until the filler metal has solidified. High-amperage current at low voltage is used. The heat is generated at the brazed joint interface and at the electrode to the part interface, and depends on the resistance to the current flow at these locations.

Resistance brazing is normally limited to applications where the brazing filler metal is preplaced; however, face feeding of the filler metal into the joint may be used. Resistance brazing is normally used for low-volume production where heating is localized at the area to be brazed.

The flux used for resistance brazing must be given specific attention since the conductivity of the flux is important. Normally, brazing fluxes are insulators when cool and dry. When they become molten from the heat of the brazing operation, they may become conductive. Fluxes are normally employed, except when an atmosphere is utilized to perform the same function.

Torch Brazing

Torch brazing (TB) is done by heating the parts to be brazed with the flame of a gas torch or torches. The temperature and the amount of heat required determine the gases used. The flame can be supplied by any fuel gas that may be burned with air, or oxygen. For manual torch brazing, normally a single torch and tip are used, but for automatic torch brazing, multiple tips may be required. Manual torch brazing is probably the most widely used brazing method and is shown in Figure 7-16. Torch brazing is very useful on assemblies that involve heating sections of different mass. The flame can be directed to the heavier part to produce uniform heating. Manual brazing is particularly useful for repair work. Automatic torch brazing is used in manufacturing operations where the rate and volume of production warrants the expense. An automatic brazing operation for an electrical part is shown in Figure 7-17. Normally, the work will move under or between multiple torches, as shown in this application.

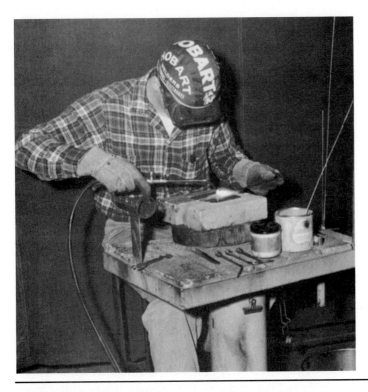

FIGURE 7–16   Torch brazing manually applied.

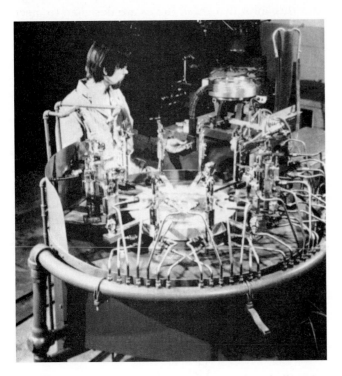

FIGURE 7–17   Automatic torch brazing electrical parts.

For torch brazing the atmosphere is the product of the combustion of the flame. The neutral or reducing flame is normally used. A slightly oxidizing flame may be used for certain materials. The brazing filler metal may be preplaced in or at the joint, or it may be face-fed manually.

The parts must be assembled and self-jigging or held in place by mechanical means during the heating and cooling cycle. The parts must be cleaned, and flux is added to assist the capillary flow of the filler material. Position is not too important; however, gravity can be helpful in assisting capillary flow in complex joints.

Torch brazing can be used for a variety of materials. It can be used when parts of unequal mass are being brazed; however, because of the poor temperature control, metals that have a melting point close to the brazing temperature are not normally torch brazed. Torch-brazed parts tend to have warpage or distortion when joining parts of different thicknesses. Torch brazing is used when the part to be brazed is too large, is an unusual shape, or cannot be heated by the other methods. Brazers can be certified by AWS or qualified according to ASME Section IX.

Other Brazing Methods

Other methods of brazing involve different methods of supplying the heat. Exothermic brazing, which utilizes the heat of specific chemical reactions, can be used. An exothermic chemical reaction is any reaction between two or more chemicals in which heat is given off due to the free energy of the reaction. The solid-state or nearly solid-state metal oxide reactions are used. This method is not employed widely.

Other sources of heat can be any of the arc processes, where the arc is used as a source of heat. This includes the carbon arc, the twin-carbon arc, the gas tungsten arc, and the plasma arc processes. The laser beam and the electron beam can also be used. The heat input–heat loss relationship must be carefully controlled to avoid melting the parts to be brazed.

GAS WELDING, BRAZING, SOLDERING, AND SOLID-STATE WELDING   **217**

Almost all of the common metals can be brazed. The different metals and their **brazeability**—that is, the capacity of a metal to be brazed under the fabrication conditions—are listed in Figure 7-18.

The thicknesses of the parts to be joined are dependent on the ability of providing sufficient heat to the heavier sections without excessive heat to the thinner sections. This can be a problem with some of the brazing methods and is a major consideration in selecting the heating method.

Filler Metals

Filler materials for brazing are covered by Specifications for Filler Metals for Brazing AWS A5.8. They are classified according to analysis: Aluminum–silicon, copper, copper–zinc, copper–phosphorus, nickel–gold, heat-resisting materials, magnesium, and silver are the basic groupings. The composition of the different filler metals, as well as the operating range and recommended uses, are given in Figure 7-19. Filler metal selection is based on the metal being brazed.

Certain brazing filler metals contain cadmium in significant amounts. When these are used, adequate ventilation is required. Filler metals are available in many forms; the most common is the wire or rod. Filler metal is also available as thin sheet, powder, paste, or as a clad surface of the part to be brazed.

The placement of the filler metal affects the quality of the joint. For normal lap joints the filler metal should be supplied at only one end and allowed to flow completely through the joint by capillary action. If the filler metal is supplied at both ends, gas will be trapped in the joint and will create voids, which will drastically reduce the effective area of the braze. Another advantage of supplying filler metal at only one end is for quality control. It will be apparent that the brazed joint is complete if the filler material creates a fillet at the other end. Filler metal cannot be made to flow by means of capillary action into a blind joint. Gas will be trapped and will not allow complete flow of filler metal throughout the faying surfaces. In such cases venting must be provided. This applies also to small tanks or vessels. The gas in these containers will expand as a result of the heat and will prevent filler metals from penetrating the abutting surfaces.

Brazing Fluxes

The correct brazing flux must be used to ensure a successful joint. The American Welding Society provides a specification for fluxes for brazing and braze welding, AWS A5.31. This specification has 15 classifications of fluxes, which satisfy most brazing requirements. Figure 7-20 is a summary of these fluxes, showing the recommended use, types of filler metals, temperature range, and flux form that these fluxes are designed to meet. The letters *FB* at the beginning of the classification stand for "flux for brazing or braze welding." The third character is a number that stands for a group of base metals. The fourth character, a letter, designates a change in form and composition within the broader base metal classification.

FIGURE 7–18 Brazeability of base metals using torch brazing.

Base Metal	Filler Metal Type	Torch Brazing Flame Type	AWS Flux Classification
Aluminum	Aluminum–silicon	Slightly reducing	FB1-A
Brass	Silver alloy	Slightly reducing	F33-A or C
Bronze	Copper–zinc	Slightly reducing	FB3-C
Copper	Copper–zinc	Slightly reducing	FB3-C
Inconel	Silver alloy	Slightly reducing	FB3-A or C
Iron, case, nodular	Silver alloy	Neutral/slightly oxidizing	FB3-A or C
Iron, wrought	Copper–zinc	Slightly reducing	FB3-C
Monel	Silver alloy	Slightly reducing	FB3-A or C
Nickel	Silver alloy	Slightly reducing	FB3-A or C
Nickel–copper	Copper	Slightly reducing	FB3-C
Nickel–silver	Silver alloy	Slightly reducing	FB3-A or C
Precious metals	Variable	Variable	Variable
Steel, high carbon	Copper–zinc	Slightly reducing	FB3-C
Steel, low alloy	Copper–zinc	Slightly reducing	FB3-C
Steel, low carbon	Copper–zinc	Slightly reducing	FB3-C
Steel, medium carbon	Copper–zinc	Slightly reducing	FB3-C
Steel, stainless	Silver alloy	Slightly reducing/neutral	FB4-A

Note: In many cases, different filler metals may be used. The easiest to use filler metal is shown above. Color matching has been ignored.

AWS Classification	Brazing Temperature Range		Approximate Composition (%)
	°F	°C	
Silver alloys			
BAg-1	1145–1400	618–760	45 Ag, 15 Cu, 16 Zn, 24 Cd
BAg-1a	1175–1400	635–760	50 Ag, 15 Cu, 16 Zn, 18 Cd
BAg-2	1295–1550	702–843	35 Ag, 26 Cu, 21 Zn, 18 Cd
BAg-2a	1310–1550	710–843	30 Ag, 27 Cu, 23 Zn, 20 Cd
BAg-3	1270–1500	688–816	50 Ag, 15 Cu, 15 Zn, 16 Cd, 3 Ni
BAg-4	1435–1650	779–899	40 Ag, 30 Cu, 28 Zn, 2 Ni
BAg-5	1370–1550	743–843	45 Ag, 30 Cu, 25 Cd
BAg-6	1425–1600	774–871	50 Ag, 34 Cu, 16 Zn
BAg-7	1205–1400	652–760	56 Ag, 22 Cu, 17 Zn, 5 Sn
BAg-8	1435–1650	779–899	72 Ag, 28 Cu
BAg-8a	1410–1600	766–871	72 Ag, 27 Cu, 0.4 Li
BAg-9	1325–1550	718–843	65 Ag, 20 Cu, 15 Zn
BAg-10	1360–1550	738–843	70 Ag, 20 Cu, 10 Zn
BAg-13	1575–1775	857–968	54 Ag, 40 Cu, 5 Zn, 1 Ni
BAg-13a	1600–1800	871–982	56 Ag, 42 Cu, 2 Cd
BAg-18	1325–1550	718–843	60 Ag, 30 Cu, 10 Sn
BAg-19	1610–1800	877–982	92.5 Ag, 7.25 Cu, 0.25 Li
BAg-20	1410–1600	766–871	30 Ag, 38 Cu, 32 Zn
BAg-21	1475–1650	802–899	63 Ag, 28.6 Cu, 2.5 Ni, 6 Sn
BAg-22	1290–1525	699–830	49 Ag, 16 Cu, 23 Zn, 5 Ni, 7 Mn
BAg-23	1780–1900	970–1038	85 Ag, 15 Mn
BAg-24	1306–1550	750–843	50 Ag, 20 Cu, 28 Zn, 3 Ni
BAg-26	1474–1600	800–870	25 Ag, 38 Cu, 33 Zn, 2 Ni, 2 Li
BAg-27	1373–1575	745–860	25 Ag, 35 Cu, 26.5 Zn, 13.5 Ni
BAg-28	1310–1550	710–843	40 Ag, 30 Cu, 28 Zn, 2 Sn
BAg-33	1260–1400	681–760	25 Ag, 30 Cu, 27.5 Zn, 17 Cd
BAg-34	1330–1550	721–843	38 Ag, 32 Cu, 28 Zn, 2 Sn
Gold Alloys			
BAu-1	1860–2000	1016–1093	37.5 Au, 62.5 Cu
BAu-2	1635–1850	891–1010	80 Au, 20 Cu
BAu-3	1885–1995	1029–2091	35 Au, 62 Cu, 3 Ni
BAu-4	1740–1840	949–1004	82 Au, 18 Ni
BAu-5	2130–2250	1166–1232	30 Au, 34 Pd, 36 Ni
BAu-6	1915–2050	1046–1121	69 Au, 8 Pd, 22 Ni
Aluminum and magnesium alloys			
BAlSi-2	1110–1150	599–621	7 Si, 1 Fe, 91.5 Al
BAlSi-3	1060–1120	571–604	10 Si, 4 Cu, 84.5 Al
BAlSi-4	1080–1120	582–604	12 Si, 87 Al
BAlSi-5	1090–1120	588–604	10 Si, 88.5 Al
BAlSi-7	1090–1120	588–604	10 Si, 1.5 Mg, 87.5 Al
BAlSi-9	1080–1120	582–604	12 Si, 87 Al
BAlSi-11	1090–1120	588–604	10 Si, 1.5 Mg, 87 Al
BMg-1	1120–1160	604–627	88 Mg, 2 Zn, 1 Mn, 9 Al
Copper, copper–zinc, and copper–phosphorus alloys			
BCu-1	2000–2100	1093–1149	99.9 Cu
BCu-1a	2000–2100	1093–1149	99 Cu
BCu-2	2000–2100	1093–1149	86.5 Cu
RBCuZn-A	1670–1750	910–954	59 Cu, 41 Zn, 1 Sn
RBCuZn-C	1670–1750	910–954	58 Cu, 39.5 Zn, 1 Sn, 1 Fe, 0.5 Mn
RBCuZn-D	1720–1800	938–982	48 Cu, 42 Zn, 10 Ni
BCuP-1	1450–1700	788–927	95 Cu, 5 P
BCuP-2	1350–1550	732–843	93 Cu, 7 P
BCuP-3	1325–1500	718–816	89 Cu, 5 Ag, 6 P
BCuP-4	1275–1450	691–788	87 Cu, 6 Ag, 7P
BCuP-5	1300–1500	704–816	80 Cu, 15 Ag, 5P
BCuP-6	1350–1500	732–816	91 Cu, 2 Ag, 7 P
BCuP-7	1300–1500	704–816	88 Cu, 5 Ag, 7 P

FIGURE 7–19 Filler metals for brazing (AWS A5.8).

AWS Classification	Brazing Temperature Range		Approximate Composition (%)
	°F	°C	
Nickel and cobalt alloys			
BNi-1	1950–2200	1066–1204	73.5 Ni, 14 Cr, 3 B, 4.5 Si, 5 Fe
BNi-1a	1970–2200	1077–1204	73 Ni, 14 Cr, 3 B, 5 Si, 5 Fe
BNi-2	1850–2150	1010–1177	81.5 Ni, 7 Cr, 3 B, 5 Si, 3.5 Fe
BNi-3	1850–2150	1010–1177	92 Ni, 3 B, 4.5 Si, 0.5 Fe
BNi-4	1850–2150	1010–1177	92.5 Ni, 2 B, 4 Si, 1.5 Fe
BNi-5	2100–2200	1149–1204	72 Ni, 18 Cr, 10 Si
BNi-6	1700–2000	927–1093	89 Cr, 11 P
BNi-7	1700–2000	927–1093	76 Ni, 14 Cr, 10 P
BNi-8	1850–2000	1010–1093	69.5 Ni, 7 Si, 18 Mn, 5.5 Cu
BNi-9	1950–2200	1066–1204	80 Ni, 15 Cr, 3.5 B, 1.5 Fe
BNi-10	2100–2200	1149–1204	63 Ni, 12 Cr, 2.5 B, 3.5 Si, 3 Fe, 16 W
BNi-11	2100–2200	1149–1204	68.5 Ni, 10 Cr, 3 B, 3 Si, 3.5 Fe, 12 W
BCo-1	2100–2250	1149–1232	16 Ni, 19 Cr, 1 Fe, 4 W, 60 Co

FIGURE 7–19 (*cont.*)

Many of the brazing fluxes contain fluorides. The package will show a warning label stating that the brazer must protect himself and others and must read and understand the label. This label further states, "Fumes and gases can be dangerous to your health. Burns eyes and skin on contact. Can be fatal if swallowed. Use sufficient ventilation to keep fumes and gases from your breathing zone and the general area." The "AWS Brazing Manual" 4th edition provides specific safety instruction, as well as the safety and welding and cutting document.

The appendix of the specification gives a description and intended use for each brazing flux classification. The form available for each classification is a clue to its use. Flux in powder form is usually used for furnace braz-

FIGURE 7–20 Fluxes for brazing.

AWS Flux Classification	Base Metal Common Name	AWS Filler Metal	Normal Temperature Range		Form Available
			°F	°C	
FB1-A	Aluminum	BAlSi	1080–1140	580–615	Powder
FB1-B	Aluminum	BAlSi	1040–1140	560–615	Powder
FB1-C	Aluminum	BAlSi	1000–1040	540–615	Powder
FB2-A	Magnesium	BMg	900–1150	480–620	Powder
FB3-A	Carbon steel	BAg and BCuP	1050–1600	565–870	Paste
FB3-C	Stainless steel	BAg and BCuP	1060–1700	565–925	Paste
FB3-D	Stainless steel	BAg, BCu, BNi, BAu, and BCuZn	1400–2200	760–1205	Paste
FB3-E	Stainless steel	BAg and BCuP	1050–1600	565–870	Liquid
FB3-F	Carbon steel	BAg and BCuP	1200–1600	650–870	Powder
FB3-G	Carbon steel	BAg and BCuP	1050–1600	565–870	Slurry
FB3-H	Carbon steel	BAg	1050–1700	565–925	Slurry
FB3-I	Stainless steel	BAg, BCu, BNi, BAu, and BCuZn	1400–2200	760–1205	Slurry
FB3-J	Stainless steel	BAg, BCu, BNi, BAu, and BCuZn	1400–2200	760–1205	Powder
FB3-K	Carbon steel	BAg and BCuZn	1400–2200	760–1205	Liquid
FB4-A	Aluminum bronze	BAg and BCuP	1100–1600	595–870	Paste

ing and dip brazing. In many cases flux is made into a slurry by the addition of water or alcohol. Paste flux is usually used for torch brazing but can be used, for example, for induction brazing. Flux in liquid form is used for torch brazing of jewelry. FB3-K is used in torch brazing, with the fuel gas being passed through the container of liquid flux and carried to the workplace.

Placement of the flux affects the quality of the brazed joint. Paste flux is the most common form and is usually spread over the surfaces to be joined. It is also painted on the preplaced brazing filler materials. Brazing fluxes can be sprayed for high-volume production. In addition, liquid flux can be introduced into the fuel gas and supplied to the flame for torch brazing at the point where it is needed. Flux in the flame may not be satisfactory for large, deep, or complex joints. In such cases preplaced paste flux may also be required.

Joint Designs

When designing a joint for brazing, the following six factors must be considered:

1. The type of joint required.
2. The clearance between the parts
3. The surface finish of the faying surfaces
4. Placement of the filler metal
5. The placement of the flux when used
6. The possibility of gas entrapment

Brazed joints fall into two general types, *butt joints* and *lap joints*. Butt joints are subjected to tensile or compressive loads, and lap joints are normally subjected to shear loading. There are many variations of these two types. Butt joints provide a limited area for brazing. The strength of the filler material is usually less than the strength of the base metal. A butt joint will not provide 100% joint efficiency. If the joint is scarfed to form a bevel, additional area will be provided which will increase the strength of the brazed joint. The bevel area should provide at least three times the area that is obtained with a simple square butt joint. Unfortunately, scarf joints are more difficult to hold in alignment than the square butt or lap joints.

Lap joints are more widely used since they can be designed to provide sufficient brazed area so that the joint is as strong as the base metal. Unfortunately, lap joints tend to be unbalanced joints and this produces stress concentrations that adversely affect the joint strength. Every effort should be made to provide a balanced lap joint to properly carry the load. Figure 7-21 shows the different brazed joints and the joint detail.

The clearance between the parts being joined is important. If the joint clearance is too small, it will not allow capillary action to cause the filler metal to flow uniformly

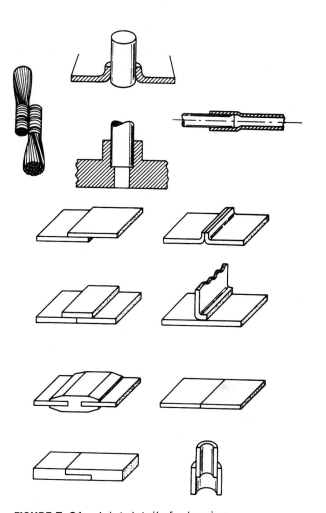

FIGURE 7–21 Joint details for brazing.

throughout the entire joint. If the clearance is too great, filler metal may now flow throughout the joint, and a low strength joint will result. The brazing filler metal also has an influence on the clearance. Another factor is the length or area of the joint. For smaller areas, a smaller joint clearance can be used. In general, when using an atmosphere system, smaller joint clearances can be used. Where fluxes are required, the clearances are normally larger. Clearances range from 0.001 to 0.025 in (0.0225 to 0.635 mm) for clearance when fluxes are involved. The recommended clearances of different groups of brazing filler metals are shown in Figure 7-22.

It is important to compensate for unequal expansion and contraction of a joint design. This can occur when brazing dissimilar metals and when the difference of thermal expansion would create tensile loads on the filler metal during cooling.

The surface finish of the faying surfaces should be between 30 and 80 microinches for best joint strength. The filler metal may not wet the surfaces completely if they are too smooth. Furthermore, the filler metal will not distribute itself throughout the complete joint by capillary action if they are too smooth. If the surfaces are too

AWS Classification	Joint Clearance in.	Joint Clearance mm	Brazing Conditions
BAlSi group	0.002–0.008 0.008–0.010	0.051–0.203 0.203–0.254	For length of lap less than $\frac{1}{4}$ in. (6.4 mm) For length of lap greater than $\frac{1}{4}$ in. (6.4 mm)
BCuP group	0.001–0.005	0.025–0.127	No flux or mineral brazing fluxes
BAg group	0.002–0.005 0.000–0.002	0.051–0.127 0.000–0.051	Mineral brazing fluxes Gas-atmosphere brazing fluxes
BAu group	0.002–0.005 0.000–0.002	0.051–0.127 0.000–0.051	Mineral brazing fluxes Gas-atmosphere brazing fluxes
BCu group	0.000–0.002	0.000–0.051	Gas-atmosphere brazing fluxes
BCuZn group	0.002–0.005	0.051–0.127	Mineral brazing fluxes
BMg	0.004–0.010	0.102–0.254	Mineral brazing fluxes
BNi group	0.002–0.005 0.000–0.002	0.051–0.127 0.000–0.051	General applications flux or atmosphere Free-flowing types, atmosphere brazing

FIGURE 7–22 Recommended joint clearance for brazing filler metals.

rough, only the high points may be properly brazed. With very rough surfaces the clearance will be too great to provide optimum strength of the brazed joint.

Joint Cleanliness

It is important to have extremely clean surfaces for the brazed joint. Mechanical surface preparations such as grinding, sandblasting, wire brushing, filing, and machining can be used. However, in every case care must be taken to make sure that the surface is clean. For example, grit should not become embedded in the surface. Wire brushing can result in the folding in of oxides and burnishing of the surface. Chemical cleaning can be used to remove dirt and oils. Solvents, alkaline baths, acid baths, salt bath pickling, and ultrasonic cleaning have all been used successfully. When the surfaces have been cleaned, flux is used to protect the surface from oxidation or from other undesirable chemical action during the heating and brazing operation. Fluxes are not designed to clean joints. They are designed to keep cleaned joints clean during the brazing operation. They will combine with, dissolve, or inhibit the formation of chemical compounds that might interfere with the quality of the brazed joint.

Braze Quality

Close adherence to the design factors, filler metal selection, flux selection, and cleanliness will insure quality brazed joints. When the joint does not exhibit the quality required, investigate using the following troubleshooting hints:

1. The brazing filler metal does not wet the surface and balls up instead of flowing into the joint.
   (a) Increase the amount of flux used.
   (b) Roughen the surface slightly, especially the surface of cold-drawn or cold-rolled stock.
   (c) Acid pickle parts to remove surface oxides.
   (d) Change work position so that gravity will help the filler metal fill the joint.

2. The brazing alloy does not flow through the joint even though it melts and forms a fillet.
   (a) Allow more time for heating.
   (b) Heat to a higher temperature.
   (c) Determine the clearance in the joint and, if required, rework it to be looser or tighter.
   (d) Apply flux to both the base metals and brazing filler metal.
   (e) Do a more thorough cleaning job before assembly.

3. The assembled joint was tight and it opens up during brazing.
   (a) The clearance was too small and a load was introduced into one part, which causes the opening.
   (b) The parts have unequal coefficients of expansion due to dissimilar metals.
   (c) Unsupported section might cause improper clearances due to sag from heating.

4. The brazing filler metal melts but does not flow.
   (a) Coat the filler metal with flux before using and apply flux generously to the base metal.
   (b) Mechanically or chemically clean the filler metal if there are surface oxides present.

5. The brazing filler metal flows away from the joint instead of into the joint.
   (a) Provide a reservoir in the joint into which the brazed filler metal can flow.
   (b) Reposition the assembly so that gravity will help the filler metal flow into the joint.
   (c) Remove burrs, edges, or other obstacles over which the brazing alloy might not flow.

Above all, make sure that the filler metal alloy is compatible with the base metal and that the proper temperatures and fluxes are employed. To determine the strength of a brazed joint, the standard method should be used. The AWS standard AWS C3.2 outlines the procedure to be used for making tests that are comparable to others.

For certain work the brazer, or one who performs a manual or semiautomatic brazing operation, must be qualified. Qualification is in accordance with Section IX of the "ASME Boiler and Pressure Vessel Code." Part C pertains to brazing ferrous and nonferrous materials. This specification must be read carefully. It introduces new uses for positions in flat flow, vertical down flow, vertical up flow, horizontal flow, and special positions. The ANSI/AWS "Standard for Brazing Procedure and Performance Qualification," B2.2, is similar and may be used.

Disadvantages and Uses

There is one disadvantage to brazing. It is the possibility of lack of *color match* of the parts being brazed and the brazing filler material.

Brazing is widely used throughout industry, and applications are so numerous that it is impossible to list them. Three major industries using brazing are the electrical industry, the utensil-manufacturing industry, and the maintenance and repair industry.

7-3 SOLDERING

Soldering (S) is "a group of joining processes that produce coalescence of material by heating them to the soldering temperature and by using a filler metal having a liquidus not exceeding 840° F (450° C) and below the solidus of the base metals. The filler metal is distributed between closely fitted faying surfaces of the joint by capillary action or by wetting the surfaces of the workpieces."

Solder is a filler metal used in soldering that has a liquidus not exceeding 840° F (450° C). The solder is normally a nonferrous alloy. The temperature of 840° F (450° C) is what differentiates soldering from brazing. Most of the factors involved with brazing apply to soldering. In fact, slang use of the terms *soft solder* and *hard solder* or *silver solder* attempts to differentiate between soldering and brazing.

There are at least eight popular soldering methods in wide use. These can be classified into three general groups. One group relates to the means of applying heat and two groups are related to the means of applying the solder. There are several older methods of little industrial use today. The soldering methods are described below.

Dip Soldering

Dip solder (DS) is a soldering process using the heat furnished by a molten metal bath that provides the solder filler metal. The solder may be kept molten by any source of heat. Solder dipping machines are automated and are programmed to preclean, flux, preheat, and insert into the solder.

Furnace Soldering

Furnace soldering (FS) is a soldering process in which the parts to be joined are placed in a furnace and heated to the soldering temperature. In furnace soldering the parts must be assembled and fixed in their proper position. The solder must be preplaced in the joint. The furnace can be fired by any suitable fuel.

Induction Soldering

Induction soldering (IS) is a soldering process in which the heat required is obtained from the resistance of the workpieces to an induced electric current. This is similar to induction brazing.

Infrared Soldering

Infrared soldering (IRS) is a soldering process in which the heat required is furnished by infrared radiation. This is similar to infrared brazing.

Iron Soldering

Iron soldering (INS) is a soldering process in which the heat required is obtained from a soldering iron. The part of the soldering iron that is heated and transfers the heat and the solder to the joint, called a *bit*, is made of copper. The bit of soldering iron may be heated different ways. It can be electrically heated by internal resistance coil—hence the electric soldering iron—it can be heated in a flame, or it can be heated in a furnace. There is also the soldering gun, which utilizes resistance heating to heat the bit. The bit is the high-resistance part of the electrical circuit. Soldering guns are very popular and widely used for electronic assembly work. The solder is applied manually.

Resistance Soldering

Resistance soldering (RS) is a soldering process that uses heat from the resistance to electric current flow in a circuit of which the workpieces are a part. This is slightly different from resistance brazing and is usually done with a handheld tool using carbon blocks and introducing a low voltage and relatively high current to the part to be soldered. This is a very common method of manufacturing electrical machinery involving soldered joints. Figure 7–23 shows this soldering method in use, attaching lugs to welding cables. The solder is applied manually. It is also used for soldering copper plumbing fittings.

FIGURE 7–23 Resistance soldering.

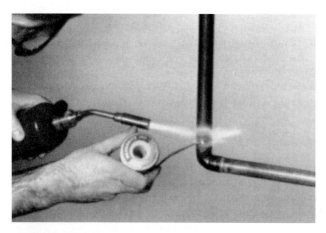

FIGURE 7–24 Torch soldering.

Torch Soldering

Torch soldering (TS) is very similar to torch brazing except that lower temperatures are involved and air is used rather than oxygen. Small cylinders of propane are available for this use. The cylinder becomes the handle when a torch head is attached to it. The solder is applied manually. Torch soldering is widely used in the plumbing trade for soldering copper tubing to copper fittings. It is shown in Figure 7-24.

Wave Soldering

Wave soldering (WS) is an automatic soldering process where workpieces are passed through a wave of molten solder. This method is used in the production of printed circuit boards. The circuit boards are assembled with electronic components. The pigtails stick through the circuit board and are crimped over the printed metal circuit on the underside of the board. The boards are then placed over the tank holding the molten solder, and the wave of solder touches the metal circuit and joints it to the pigtails of the electronic component with a soldered joint. This is completely automatic and produces high-quality soldered joints. It is widely used in the electronics industry. Figure 7-25 shows a dual-wave soldering apparatus.

Other Soldering Methods

There are several other methods of soldering. One is *ultrasonic soldering*. It is a method in which high-frequency sonic energy is transmitted through molten solder to remove undesirable surface films and promote wetting of the base metal. Flux is not normally used. In this method the ultrasonic vibrations are transmitted to the soldering iron and then to the solder and to the work. A source of ultrasonic energy is required as well as a specialized soldering iron. This method is used for soldering aluminum.

Wipe soldering is a method of producing a joint with the heat applied by the molted solder poured onto the joint. The solder is manipulated with a handheld cloth or paddle so as to obtain the required size and contour. The filler metal is also distributed into the joint by capillary action. Wipe soldering was formerly used by the plumbing industry and is of minor industrial importance.

Sweat soldering is a method in which two or more parts are precoated with solder, assembled into a joint, and reheated without the use of additional solder. This is used in the electrical industry for joining wires to conductors.

Soldering Mechanism

The mechanism for joining by soldering involves three closely related factors: (1) wetting, (2) alloying, and (3) capillary action. Wetting is the bonding or spreading of a liquid filler metal or flux on a solid base metal. For soldering it is more specific. When molted solder leaves a continuous permanent film on the surface of the base metal, it is said to have wet that surface. Wetting occurs when there is a stronger attraction between certain atoms of the solder and the base metal than between the atoms of the solder. Wetting is essentially a chemical reaction. It occurs when one or more elements of the solder react with the base metal being soldered to form a compound.

The ability of a solder to alloy with the base metal is related to its ability to wet the surface. Heat is applied to facilitate wetting. Alloying is related to the cleanliness of the base metal. The base metal must be oxide free, and this is accomplished by cleaning and using a flux. For alloying to occur at the interface, there must be intimate contact between the solder and the base metal. The temperature of wetting may not correspond with the liquidus temperature of the solder alloy.

FIGURE 7–25 Dual-wave soldering machine.

The fluidity of the molten solder must be such that it can flow into narrow spaces by capillary action. Tests are available to determine this quality. The fluidity of the molten solder is the property that influences the spreading of the solder over the base metal surface. The flowability or spread of solder can also be determined by tests.

Actual application of the solder involves two steps: wetting of the base metal surface with solder and filling of the gap between the wetted surfaces with the solder. These two steps are normally carried out together, depending on conditions and application; however, for "difficult-to-solder" metals it is desirable to wet the surface of the base metal with solder prior to making the joint. This is called tinning or precoating.

The strength of a soldered joint depends on the design of the joint and its clearance. Joints stressed in tension are not successful. The normal joint is the lap joint, where sufficient overlap occurs to provide enough area for required strength. The clearance between the parts being joined must be held to close limits. The maximum strength of the joint will provide a joint considerably stronger than the strength of the solder alloy. A clearance of from 0.003 to 0.005 in. (0.07 to 0.12 mm) is recommended for most applications. If clearance between the joint is excessive, the strength of the joint drops to the strength level of the soldering alloy. The joint design is the same as used for brazing.

Soldering Flux

The flux helps remove oxides, but the flux must be designed so that it can be removed after the joint is soldered. It should be fluid at a lower temperature than the liquidus of the solder. It should have a lower specific gravity than the solder so that the solder will displace it in the joint. It should promote wetting of the surface by the solder. Stronger fluxes are the acid fluxes or the inorganic type. Intermediate fluxes are less severe. The organic fluxes, such as resin, are the mildest and are often referred to as pure water, white resin, or nonactivated resin. Resin contains an abietic acid, which is very mild, has a long melting point, remains effective to the highest-melting-point solder, and is normally used. Flux should be applied to the base metal to protect it from oxidation.

Solder is available with the flux contained in its core. The amount of flux in the core ranges from about 0.5% to over 3%, 2.2% being most common. Resin-core solder and acid-core solders are available. The resin-core type is used for electrical work; the acid-core type is used for sheet metal work.

Solder

There is one ASTM specification, ASTM B-32, that covers all the different types of solder. Figure 7–26 shows the approximate composition and melting range for the various ASTM alloy grades. Solders are classified according to whether or not they contain lead. Solders containing lead should not be used for drinking water systems or for food equipment.[2]

The most common general-purpose solder (for non-water supply use) is the 50% lead–50% tin composition. Solder selection is based on its ability to wet the surface of the metals being joined. The grade containing the least amount of tin that provides suitable flowing and wetting action should be used. The appendix of the

ASTM Alloy Grade	Approximate Composition (%)					Melting Range			
	Tin Sn	Lead Pb	Antimony Sb	Silver Ag	Other	Solidus °F	Solidus °C	Liquidus °F	Liquidus °C
Solder alloys containing less than 0.2% lead									
Sn96	96.2	0.10	0.12	3.6		430	221	430	221
Sn95	95.2	0.10	0.12	4.6		430	221	473	245
Sn94	94.2	0.10	0.12	5.6		430	221	536	280
Sb5	94.0	0.20	5.0			450	233	464	240
E	95.4	0.10	0.05	0.5	4.0 Cu	440	225	660	349
HA	90.9	0.10	2.5	2.5	1.5 Cu, 3.5 Zn	420	216	440	227
HB	89.7	0.10	5.0	0.20	3.0 Cu 1.5 Ni	460	238	660	349
Solder alloys containing lead									
Sn70	70.5	29.0	0.5			361	183	377	193
Sn63	63.0	36.5	0.5			361	183	361	183
Sn62	63.0	34.5	0.5	2.0		354	179	372	189
Sn60	60.5	39.0	0.5			361	183	374	190
Sn50	50.5	49.0	0.5			361	183	421	216
Sn45	45.5	54.0	0.5			361	183	441	227
Sn40A	40.5	59.0	0.5			361	183	460	238
Sn40B	40.5	57.5	2.0			365	185	448	231
Sn35A	35.5	64.0	0.5			361	183	447	247
Sn35B	35.5	62.7	1.8			365	185	470	243
Sn30A	30.5	69.0	0.5			361	183	491	255
Sn30B	30.5	67.9	1.6			365	185	482	250
Sn25A	25.5	74.0	0.5			361	183	511	266
Sn25B	25.5	74.2	1.3			365	185	504	263
Sn20A	20.5	79.0	0.5			361	183	531	277
Sn20B	20.5	78.5	1.0			363	184	517	270
Sn15	15.5	84.0	0.5			437	225	554	290
Sn10A	10.0	89.5	0.5			514	268	576	302
Sn10B	10.0	87.8	0.2	2.0		514	268	570	299
Sn5	5.0	94.5	0.5			586	308	594	312
Sn2	2.0	97.5	0.5			601	316	611	322
Ag1.5	1.0	97.1	0.4	1.5		588	309	588	309
Ag2.5	0.25	96.85	0.4	2.5		580	304	580	304
Ag2.5	0.25	93.85	0.4	5.5		580	304	716	380

FIGURE 7–26 Solders (ASTM B-32).

ASTM specification provides much information on solder selection based on the materials being joined.

Solder Procedures

The joint must be properly cleaned. It must be free of all oil, grease, dirt, oxides, and so on. This is best done by mechanical or chemical cleaning. Solder will not wet a dirty surface or a surface covered with oxides. Cleaning can be accomplished by brushing, filing, machining, sanding, and by the use of chemicals.

Heat is applied to the joint by many different mechanisms and these were described by each method. After the metal surfaces have been wetted and the space between them has been filled with solder, the joint is cooled to room temperature. Self-jigging joints are often employed, or staking, bending, or other temporary assembly methods are used. Cooling is accomplished by removing the heat source and/or utilizing an air blast. After the solder joint is cooled, post-cleaning is necessary. Certain fluxes are considered noncorrosive and may not need to be removed unless they affect the appearance or later processing. Fluxes identified as corrosive, such as the acid types, must be removed. They should be neutralized and removed to provide for a successful joint to provide useful service.

Soldering is a very widely used metals-joining method. More information on this subject can be found in the "AWS Soldering Manual," 2nd edition.

7-4 THERMITE WELDING

Thermite welding (TW) is "a welding process that produces coalescence of metals by heating them with superheated liquid metal from a chemical reaction between a metal oxide and aluminum with or without the application of pressure. Filler metal is obtained from the liquid metal." This is one of the older welding processes but is still used for specific applications. It was invented by H. Goldberg of Essen, Germany in 1903 (US Patent 729,573, 1903, 735,244 1903 and 875,345 1907).

The heat for welding is obtained from an exothermic reaction between iron oxide and aluminum. This reaction is shown by the following formula:

$$8Al + 3Fe_3O_4 = 9Fe + 4Al_2O_3 + heat$$

The temperature from this reaction is approximately 4500° F (2500° C). The superheated steel is contained in a crucible located immediately above the weld joint. The exothermic reaction requires 20 to 30 seconds no matter how much of the chemicals are involved. The superheated steel runs into a mold that is built around the parts to be welded. Since it is almost twice as hot as the melting temperature of the base metal, melting occurs at the edges of the joint and alloys with the molten steel from the crucible. Normal heat losses cause the mass of molten metal to solidify, coalescence occurs, and the weld is completed. The thermite welding process is applied in the automatic mode. Once the reaction is started, it continues until it goes to completion. Welding utilizes gravity, which causes the molten metal to fill the cavity between the parts being welded. Making a thermite weld is shown in Figure 7-27.

The thermite material is a mechanical mixture of metallic aluminum and processed iron oxide. This mixture may also include various elements for alloying the weld metal. Thermite mixtures can be designed to produce specific weld metal deposits. The normal analysis of thermite employed to weld mild and medium carbon steel is as follows:

✳	Carbon	0.20–0.30
✳	Manganese	0.50–0.60
✳	Silicon	0.25–0.50
✳	Aluminum	0.07–0.18
✳	Iron	balance

The mechanical properties of normal thermite are approximately the same as mild steel.

Thermite powder will not ignite until it is brought to a temperature of 2400° F (1300° C). It is started by using a special ignition powder. During the exothermic reaction, the molten steel will go to the bottom of the crucible. The aluminum oxide will float to the top as a slag, which protects the molten steel from the atmosphere. The molten metal is tapped by a tapping pin at the bottom of the crucible. The superheated molten metal immediately flows into the mold through the pouring gate into the cavity making the weld. Venting must be provided to ensure that the cavity is completely filled.

The parts to be welded must be prepared with a square-groove joint. The root opening between the parts is related to the cross-sectional area of the weld. The root opening should range from $\frac{3}{4}$ to $1\frac{1}{2}$ in. (19 to 37 mm) for joining railroad rails. For larger sections the root openings should be greater. Parts to be welded are properly aligned and braced. A mold is then made around the joint. The lost-wax technique is sometimes employed for making

FIGURE 7–27 Steps in making a thermite weld.

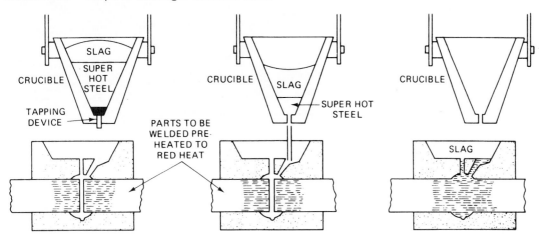

molds of unusual shapes. The lost-wax method involves filling the weld joint or cavity with wax, then making the mold around this assembly. A riser must be provided as well as a pouring gate. A heating gate at the lower portion of the joint is provided for preheating, which is usually required. This will melt the wax, which will run out of the mold and provide the cavity for the molten weld metal. After heating is completed, the heating gate is sealed with a plug, as shown in Figure 7-28.

The amount of thermite is calculated to provide sufficient metal to produce the weld. The amount of steel produced by the reaction is approximately one-half the original quantity of thermite material by weight and one-third by volume.

After the weld has cooled, the mold is broken away and discarded, and the gates and risers are removed by oxyacetylene flame cutting. The surface of the completed weld is usually sufficiently smooth and contoured so that it does not require additional metal finishing.

The deposited weld metal is homogeneous and quality is relatively high. Distortion is minimized since the weld is accomplished in one pass and since cooling is uniform across the entire weld cross section. There is normally shrinkage across the joint, but no angular distortion. Welds can be made with the parts to be joined in almost any position as long as the cavity has vertical sides.

Thermite welding has been used for many special applications, such as welding stern frames for Liberty ships during World War II. These frames were so large they could not be cast in one piece and were cast in four sections, which were joined together by four thermite welds. Thermite welds have also been used to weld large, thick I-beams and railroad and craneway rails. They have also been popular for welding reinforcing bars. Special mixtures of thermite are required for welding alloy steels of this type.

When the thermite process is used for joining rails and reinforcing bars, standardized semipermanent molds are used. These molds are made in halves, which are clamped around the rail or bar. The molds, which can be reused, are available for various sizes of rails and reinforcing bars. Figure 7-29 shows the welding of railroad rails.

Thermite welds are relatively inexpensive since little or no equipment is required. The primary cost is the cost of the thermite material, which becomes the deposited weld metal. Thermite reinforcing bar welds meet the requirements of the Concrete Institute.

Thermite welds can also be used for welding nonferrous materials. The most popular uses of nonferrous thermite welding is the joining of copper and aluminum conductors for the electrical industry. In these cases the exothermic reaction is a reduction of copper oxide by aluminum, which produces molten superheated copper. In welding copper and aluminum cables, the molds are made of graphite and can be used over and over. When welding nonferrous materials, the parts to be joined must be extremely clean and flux is normally applied to the joint prior to welding. Special kits are available that provide the molds for different sizes of cable and provide the premixed thermite material. This material also includes enough of the igniting material so that the exothermic reaction is started by means of a special lighter. An example of a thermite weld made joining heavy copper cables is shown in Figure 7-30.

FIGURE 7-29   Making thermite rail weld.

FIGURE 7-28   Details of the crucible and mold.

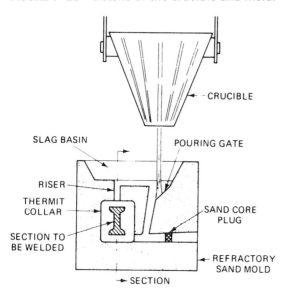

FIGURE 7-30   Thermite weld of copper cables.

7-5 SOLID-STATE WELDING

Solid-state welding (SSW) is a group of welding processes that produces coalescence by the application of pressure without melting any of the joint components. Sometimes erroneously called solid-state bonding processes, this group of welding processes includes cold welding, diffusion welding, explosion welding, forge welding, friction welding, hot pressure welding, roll welding, and ultrasonic welding. In all these processes, time, temperature, and pressure individually or in combination produce coalescence of the base metal without significant melting of the base metals.

Solid-state welding includes some of the oldest welding processes and some of the newest. Some of the processes offer advantages since the base metal does not melt and form a nugget. The metals being joined retain their original properties without the heat-affected zone problems involved when there is base metal melting. When dissimilar metals are joined, their thermal expansion and conductivity are of much less importance with solid-state welding than with arc welding.

Time, temperature, and pressure are involved; however, in some processes the time element is extremely short, up to a few seconds. In other cases, the time is extended to several hours. As temperature increases, time is usually reduced.

Cold Welding

Cold welding (CW) is a solid-state welding process in which pressure is used to produce a weld at room temperature with substantial deformation of the weld. Welding is accomplished by using high pressures on clean interfacing materials. Sufficiently high pressure can be obtained with simple hand tools when extremely thin materials are being joined. When cold welding heavier sections, a press is required to exert sufficient pressure to make a successful weld. Indentations are made in the parts being cold welded. The process is readily adaptable to joining ductile metals. Aluminum and copper are readily cold welded. Aluminum and copper can be joined together by cold welding; this is shown in Figures 7–31 and 7–32.

Diffusion Welding

Diffusion welding (DFW) is a solid-state welding process that produces a weld by the application of pressure at elevated temperature, with no macroscopic deformation or relative motion of the workpieces. A solid filler metal may be inserted between the faying surfaces.

The process is used for joining refractory metals at temperatures that do not affect their metallurgical properties. Heating is usually accomplished by induction, re-

FIGURE 7–31 Copper welded to aluminum.

sistance, or furnace. Atmosphere and vacuum furnaces are used, and for refractory metals a protective inert atmosphere is desirable. Successful welds have been made on refractory metals at temperatures slightly over half the normal melting temperature of the metal. To accomplish this type of joining, extremely close tolerance joint preparation is required and a vacuum or inert atmosphere is used. The process is used quite extensively for joining dissimilar metals. The process is considered diffusion brazing when a layer of filler material is placed between the faying surfaces of the parts being joined. These processes are used primarily by the aircraft and aerospace industries.

Explosion Welding

Explosion welding (EXW) is a solid-state welding process that produces a weld by high-velocity impact of the workpieces as the result of controlled detonation. The explosion welding process is shown in Figure 7–33. Explosive welding was developed in the mid-1940s and the first patent was granted in 1957.[3] Even though heat is not applied in making an explosion weld, it appears that the metal at the interface is molten during welding. This heat comes from the shock wave associated with impact and from the energy expended in collision. Heat is also released by plastic deformation associated with jetting and ripple formation at the interface between the parts being welded. Plastic interaction between the metal surfaces is especially pronounced when surface jetting occurs. It is necessary to allow the metal to flow plastically in order to provide a good-quality weld. The interface or weld of explosion-welded parts is shown in Figure 7–34. Explosion welding creates a strong weld between almost all metals. It has been used to weld dissimilar metals that are not weldable by the arc processes. The weld apparently does not disturb the effects of cold work or other forms

FIGURE 7–34 Interface of explosion weld.

FIGURE 7–32 Making a cold weld.

FIGURE 7–33 Explosion welding.

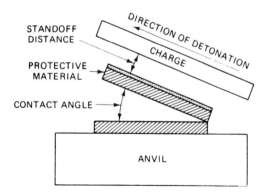

FIGURE 7–35 Condenser sheet of $\frac{1}{4}$-in. titanium clad on $1\frac{1}{4}$-in. steel.

of mechanical or thermal treatment. The process is self-contained, it is portable, and welding can be achieved quickly over large areas. The strength of the weld joint is equal to or greater than the strength of the weaker of the two metals joined.

Explosion welding has not become widely used except in limited fields. One of the most widely used applications of explosion welding has been in the cladding of steel with thinner metals as shown in Figure 7–35. The photomicrograph shown in Figure 7–34 is a cross section of a weld between dissimilar metals. A popular application for explosion welding is the joining of tube-to-tube sheets for the manufacture of heat exchangers. Another application is the joining of pipes in a socket joint.

Forge Welding

Forge welding (FOW) is a solid-state welding process that produces a weld by heating the workpieces to welding temperature and applying blows sufficient to cause permanent deformation at the faying surfaces. This is one of the older welding processes and at one time was called hammer welding. Forge welds made by blacksmiths were made by heating the parts to be joined to a red heat considerably below the molten temperature. Normal practice was to apply flux to the interface. The blacksmith, by skillful use of a hammer and an anvil, was able to create pressure at the faying surfaces sufficient to cause a weld. This process is of minor industrial significance today; however, it is often demonstrated as an old craft.

Friction Welding

Friction welding (FRW) is a solid-state welding process that produces a weld under compressive force contact of workpieces rotating or moving relative to one another to produce heat and plastically displace material from the faying surfaces. This process, shown in Figure 7–36, usually involves the rotating of one part against another to generate frictional heat at the junction.

When a suitable high temperature has been reached, rotational motion ceases, additional pressure is

---

FIGURE 7–36 Friction welding process.

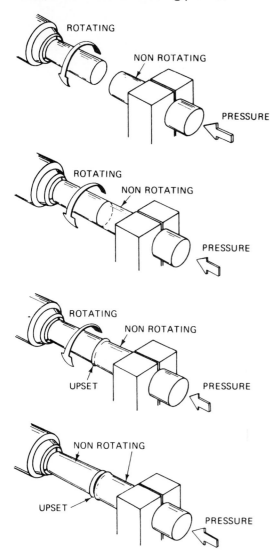

applied and coalescence occurs. Friction welding was developed in the Soviet Union in 1957. The results of research were published by V. I. Vill.[4] It was not until 1960 that the process was used in the United States.

There are two variations. In the original equipment, known as direct drive, one part is held stationary and the other part is rotated by a motor which maintains an essentially constant rotational speed. The two parts are brought in contact under pressure for a specified period of time with a specific pressure. Rotating power is disengaged from the rotating piece and the pressure is increased. When the rotating piece stops, the weld is completed. This process can be accurately controlled when speed, pressure, and time are closely regulated.

The other variation is called *inertia friction welding.* Here a flywheel is revolved by a motor until a preset speed is reached. It rotates one of the pieces to be welded. The motor is disengaged from the flywheel and the other part is brought in contact under pressure with the rotating piece. During the predetermined time during which the rotational speed of the part is reduced, the flywheel is brought to an immediate stop and additional pressure is provided to complete the weld.

Among the advantages of friction welding is the ability to produce high-quality welds in a short cycle time. No filler metal is required and flux is not used. The process is capable of welding most of the common metals. It can also be used to join many combinations of dissimilar metals. Friction welding requires relatively expensive apparatus similar to a machine tool. Figure 7–37 shows a friction welding machine.

There are three important factors involved in making a friction weld:

1. *Rotational speed.* This is related to the material to be welded and the diameter of the weld at the interface.
2. *Pressure between the two parts to be welded.* Pressure changes during the weld sequence. At the start it is very low, but it is increased to create the frictional heat. When the rotation is stopped, pressure is rapidly increased.
3. *Welding time.* Time is related to the shape and the type of metal and the surface area. It is normally a matter of a few seconds. The actual operation of the machine is automatic and is controlled by a sequence controller which can be set according to the weld schedule established for the parts to be joined.

Normally, for friction welding one of the parts to be welded is round in cross section; however, this is not an absolute necessity. Visual inspection of weld quality can be based on the flash, which occurs around the outside perimeter of the weld. This flash will extend beyond the outside diameter of the parts and will curl around back

FIGURE 7–37 Friction welding machine.

toward the part. If the flash sticks out relatively straight from the joint, it is an indication that the time was too short, the pressure was too low, or the speed was too high. These joints may crack. If the flash curls too far back on the outside diameter, it is an indication that the time was too long and the pressure was too high. Between these extremes is the correct flash shape. The flash is normally removed after welding. Figure 7-38 shows a circular friction welded part. Note both the inside flash and the outside flash. Provisions were made in this part so that the inner flash would not extend to inside the chamber where it might interfere with the function of the part. Figure 7-39 shows a variety of parts made by friction welding.

One special application of friction welding deserves attention. This is a stud weld made by friction welding. In *stud friction welding* the stud is the rotating workpiece under pressure in contact with the stationary workpiece, as shown in Figure 7-40. When the interface reaches the welding and forging temperature, rotation is stopped automatically and pressure is maintained for a short period of time. Weld time will vary, depending on the materials being welded and the stud diameter. The stud friction welding tool provides the rotation motion and is either hydraulic or pneumatic powered. A holding

tool is utilized that forces the stud friction welding head against the stationary workpiece. All of the factors relating to friction welding relate also to stud friction welding.

Stud friction welding can be used to weld underwater in the presence of a foam that keeps water from coming in contact with the stud material. As the weld progresses, the hot metal burns into the foam, causing a chamber to form that fills with an inert gas. The weld properties of underwater welding match those made in air. The rotating motion of the friction welding machine can be either hydraulic or air powered. Figure 7-41 is a cross-sectional view of a stud friction weld.

More information on friction welding is available in the AWS publication "Recommended Practices for Friction Welding" (AWS C6.1).

Friction Stir Welding

Friction stir welding (FSW) is "a variation of friction welding that produces a weld between two abutting workpieces by the friction heating and plastic material displacement caused by a high-speed rotating tool that traverses along the weld joint." This new friction welding method was developed by TWI in Great Britain in 1991. This method uses a nonconsumable rotating tool to fric-

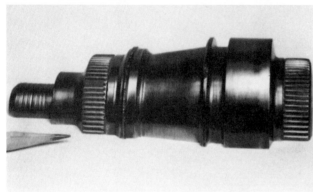

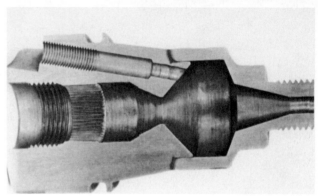

FIGURE 7–38   Part friction welded.

tion weld metals. It is shown by Figure 7–42. It is unlike conventional friction welding in that the abutting parts do not move relative to each other. It relies upon friction heat generation and subsequent flow of plasticized material when a rotating tool pin or probe is plunged into the abutting edges of the parts to be joined and then traverses along the joint. It was first used to thin aluminum sheets up to 13 mm ($\frac{1}{2}$ inch). Today the technology has been improved to the extent that it is now possible to weld 50-mm (2-in.)-thick aluminum in one pass. So far the process has been used to weld aluminum, titanium, lead, magnesium, zinc, copper, and some types of steel. The temperature attained is less than the melting point of the metal. It joins metal in the solid phase. It is autogenous (no filler metal), does not need gas shielding, and requires little surface preparation prior to welding. The parts to be joined are clamped to a backing plate. The tool, which rotates at about 1500 rpm, is pushed along the joint between the pieces with great force. The rotational frictional heat causes the parts to become plastic. As the tool moves along the joint, the plasticized material fills the space behind it and forms a weld. This process continues as long as the rotating tool is moved through the joint. The speed of welding depends on the thickness, the alloy, and

the tool rotational speed and force. Square-groove welds are routinely welded. It is claimed that lap joints, corner joints, and T-joints can also be made (see Figure 7–43). The welding speed is approximately 7 in./min (180 mm/min). The weld joint is equal in strength to the base material.

The welding variables depend on the particular alloy and its thickness. The advantage of the process is that welds can be made with a single pass and that shielding gases and consumables are not required. Joint preparation is a square-groove weld. A typical application is shown by cross section shown in Figure 7–44. The disadvantage is that only highly ductile material can be welded. Equipment required for making friction stir welds is expensive. It is expected that many additional applications will be found for friction stir welding in the near future.

Hot Pressure Welding

Hot pressure welding (HPW) is a solid-state welding process that produces coalescence of metals with heat and application of pressure sufficient to produce macrodeformation of the base metal. Vacuum or shielding gas may be used.

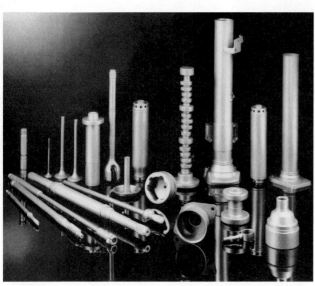

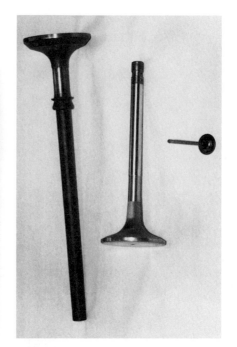

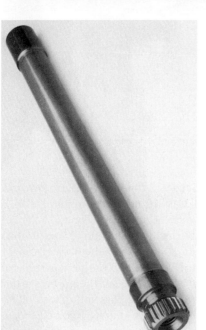

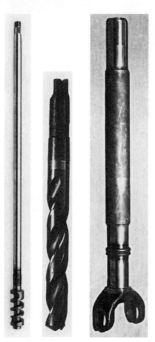

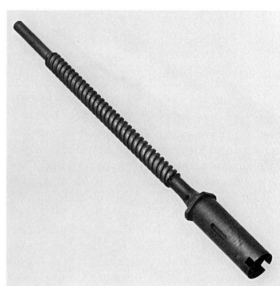

FIGURE 7–39 Parts assembled by friction welding.

In this process, coalescence occurs at the interface between the parts because of pressure and heat, which is accompanied by noticeable deformation. The deformation of the surface cracks the surface oxide film and increases the areas of clean metal. Welding the abutting parts is accomplished by diffusion across the interface so that coalescence of the faying surface occurs. This type of operation is carried on in closed chambers when vacuum or a shielding gas may be used. It is used in the aerospace industry. A variation is the hot isostatic pressure welding method. In this case, the pressure is applied by means of a hot inert gas in a pressure vessel.

Roll Welding

Roll welding (ROW) is a solid-state welding process that produces a weld by the application of heat and sufficient pressure with rolls to cause deformation at the faying surfaces. This process is similar to forge welding except that pressure is applied by means of rolls rather than by ham-

FIGURE 7–40 Friction welding stud to plate.

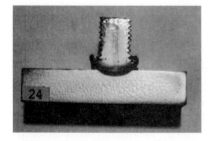

FIGURE 7–41 Cross-sectional view of friction welded stud.

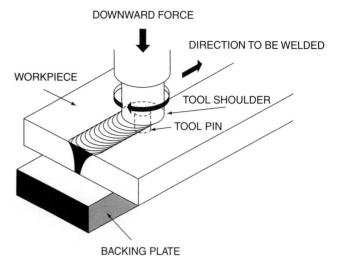

FIGURE 7–42 Stir friction welding method.

mer blows. Coalescence occurs at the interface between the two parts by means of diffusion at the faying surfaces.

One of the major uses of this process is the cladding of mild or low-alloy steel with a high-alloy material such as stainless steel. It is also used for making bimetallic materials for the instrument industry. It is used to produce the *sandwich* coins used in the United States. Figure 7–45 is a photomicrograph of the weld of a coin made in this manner. This is a weld between copper and a nickel–copper alloy.

Ultrasonic Welding

Ultrasonic welding (USW) is a solid-state welding process that produces a weld by the local application of high-frequency vibratory energy as the workpieces are held together under pressure. Welding occurs when the ultrasonic tip or electrode, the energy-coupling device, is clamped against the workpieces and is made to oscillate in a plane parallel to the weld interface. The combined clamping pressure and oscillating forces introduce dynamic stresses in the base metal. This produces minute

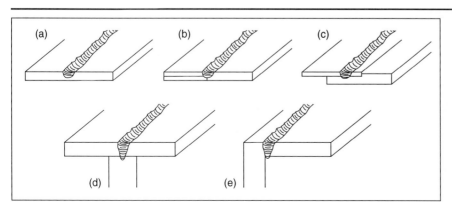

FIGURE 7–43 Joint configurations welded with stir friction welding:
(a) square groove butt;
(b) combined butt and lap;
(c) single lap; (d) two-piece T butt;
(e) edge butt.

GAS WELDING, BRAZING, SOLDERING, AND SOLID-STATE WELDING   **235**

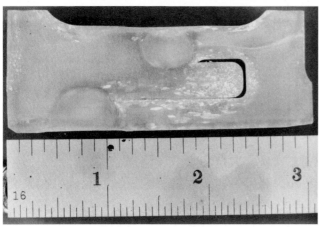

FIGURE 7–44 Cross section at stir friction weld.

FIGURE 7–45 Interface of roll weld.

deformations that create a moderate temperature rise in the base metal at the weld zone. This, coupled with the clamping pressure, provides for coalescence across the interface to produce the weld. Ultrasonic energy will aid in cleaning the weld area by breaking up oxide films and causing them to be carried away. The vibratory energy that produces the minute deformation comes from a transducer that converts high-frequency alternating electrical energy into mechanical energy. The transducer is coupled to the work by various types of tooling, ranging from tips similar to resistance welding tips to resistance roll welding electrode wheels. The normal weld is the lap joint weld. The temperature at the weld is not raised to the melting point and therefore has no nugget similar to resistance welding. Weld strength is equal to the strength of the base metal. Most ductile metals can be welded together, and there are many combinations of dissimilar metals that can be welded. The process is restricted to relatively thin materials normally in the foil or extremely thin gauge thicknesses.[5]

This process is used extensively in the electronics, aerospace, and instrument industries. It is also used for

FIGURE 7–46 Ultrasonic welding.

producing packages and containers and for sealing them. Figure 7–46 shows an ultrasonic weld being made by a continuous seam welder. In this picture a rotating electrode is the active welding tip that delivers ultrasonic energy to the work. The process can also be used for joining plastics and is finding wider use in this field than in joining metals.

7-6 MISCELLANEOUS WELDING PROCESSES

Percussion Welding

Percussion welding (PEW) is a "welding process that produces coalescence with an arc resulting from a rapid discharge of electrical energy. Pressure is applied percussively during or immediately following the electrical discharge." This process is quite similar to resistance flash welding and upset welding, but is limited to parts of the same geometry and cross section. It is more complex than the other two processes, in that heat is obtained from an arc produced at the abutting surfaces by the very rapid discharge of stored electrical energy across a rapidly decreasing air gap. This is immediately followed by application of pressure, providing an impact that brings the two parts together in a progressive percussive man-

ner. The advantage of the process is that there is an extremely shallow depth of heating and the time cycle is very short. It is used only for parts with fairly small cross-sectional areas. It can be used for welding a large number of dissimilar metals. It is used for very specialized applications, and the process is entirely automatic.

Plasma MIG Welding

A welding process developed by the Philips company in Holland in 1971 is named **plasma-MIG welding.**[6] It is a combination of gas metal arc welding and plasma arc welding. A plasma torch is modified by placing the tungsten electrode slightly to one side of the center of the nozzle orifice. A contact tip is inserted in the torch on the centerline, and a continuous electrode wire is fed through the nozzle orifice along with the plasma gas. The gas metal arc consumable electrode and the arc are contained within the plasma column. Separate power supplies are used—one for the plasma and one for the electrode wire. Under certain conditions the gas metal arc rotates within the plasma. Extremely high temperatures are attained, resulting in high deposition rates.

The torch is larger than a standard plasma torch and more complex. It is normal to move the work under a fixed torch. One major application for the process is cladding of steel with dissimilar material such as stainless steel. This process has not attained sufficient industrial use.

Ion Beam Welding

Ion beam welding has not yet been developed to practical use. It is similar to electron beam welding, and heat is created by the bombardment of ions on the surface to be welded. The reason for interest in ion beam welding is that it has different characteristics from electron beam welding. Ion bombardment, which is used in gas tungsten arc welding for the cleaning action, becomes part of this welding process. The ion beam is less sensitive to external magnetic fields than the electron beam and there is no x-radiation. To make this process practical, it is necessary to have a source of ions, to accelerate the beam of ions to the workpiece, and to be able to focus the beam on the weld area. Efforts to perfect this process are progressing, and in the future ion beam welding may be practical.

Solar Energy Welding

All the energy beam welding processes use highly concentrated sources of energy to heat metals so that a weld can be made. The sun is a source of heat. For many years solar furnaces have been used to melt metals. This is accomplished by focusing the sun's rays with a parabolic mirror to a crucible. The major part of **solar energy welding** is a *heliostat*, a device utilizing mirrors, moved by clockwork, to direct the sun's rays to a fixed point. The rays are directed into a parabolic mirror, which deflects them to form a small focal spot. The focal spot is then directed to the work and concentrated to the actual welding point. Welding is accomplished by moving the workpiece with respect to the stationary focal spot. Fixturing and travel mechanisms are similar to those used for automatic welding. To prevent the atmosphere from coming in contact with the molten base metal, the weld area is shielded by an inert gas. Welds have been made successfully on material from 0.039 to 0.118 in. (1 to 3 mm) thick in a single pass. Filler metal was not utilized. Travel speeds up to 2 in. (50 mm) per minute have been accomplished. Welds have been made in stainless steels with qualities equal to the base metal. So far this process is a laboratory curiosity.

Spot-Adhesive Welding

Spot-adhesive welding combines spot welding with adhesives to produce joints that are stronger, more durable, and more resistant to fatigue than joints produced by either method alone. This process has been called *weld bonding* and has been used on aircraft structures.

It is practiced by the application of an adhesive to the faying surfaces of the weld joint. Resistance spot welds are made through the lap joints. The joint is then heated in an oven to cure the adhesive. Another variation is accomplished by first making the spot weld joint and then applying the adhesive at the edge of the joint. By means of capillary action, the adhesive flows between the faying surfaces and around the resistance spot-welded nuggets. After the adhesive dries, the assemblies are placed into an oven for curing. The process has been used on aluminum aircraft subassemblies.

Ultra Pulse Welding

Ultra pulse welding is a resistance welding process that uses an extremely short weld time. The principle of operation is a capacitor that is charged from the power lines and discharged through a transformer and resistance-type electrodes to the work. The charging time is relatively long compared to the discharge or welding time, which is in the order of a few milliseconds. The advantage of this system is that it reduces power line draw and makes welds in an extremely short time. Heat does not build up in the workpiece. It is used for welding extremely small and thin parts. The process is used for producing electronic components and instruments.

Other Welding Processes

There are other welding processes that have been used but today are of little industrial significance. Undoubtedly, there are other welding processes that will be developed in the future. Some of the currently used processes will fall into disuse. These continuing changes will ultimately help produce weldments at lower costs.

 # QUESTIONS

7–1. Where is the hottest part of the oxyacetylene flame?

7–2. What is produced by the secondary reaction besides heat and CO_2?

7–3. What are the three types of oxyacetylene flames? How is each produced?

7–4. Why shouldn't you use a cigarette lighter to light the welding torch?

7–5. Explain the operation of a gas pressure regulator. What is the difference between a single- and a two-stage regulator?

7–6. Why shouldn't oil or grease be used on an oxygen apparatus?

7–7. Is flux required for making oxyacetylene welds on clean mild steel? Why?

7–8. Does brazing require the use of a filler metal that has a higher or lower melting temperature than the base metal?

7–9. Is it possible to obtain a brazed joint stronger than the filler metal?

7–10. What is preplaced filler metal?

7–11. Explain how capillary action is needed for brazing and soldering.

7–12. What are the hazards of certain brazing filler metal and certain brazing fluxes? What precautions are required?

7–13. What is the difference between soldering and brazing?

7–14. Why is flux required for soldering?

7–15. Explain how gravity is used in making a thermite weld.

7–16. Is it possible to thermite weld copper or aluminum?

7–17. Can cold welding be applied manually?

7–18. The clad coins of the United States are sometimes made by what welding process?

7–19. Explain the difference between friction welding and cold welding.

7–20. Can ultrasonic welding be applied to plastics as well as to thin metals?

REFERENCES

1. "Specifications for Rubber Welding Hose," ANSI IP-7, Rubber Manufacturers Association, Washington, D.C., and Compressed Gas Association, Arlington, Va.

2. Safe Drinking Water Act, Amendments of 1986, Public Law 99-339, June 19, 1986, 99th Congress, Section 109, "Lead Free Drinking Water."

3. "Explosive Welding," National Seminar, September 1975, London, Welding Institute, Cambridge, England.

4. V. I. Vill, "Friction Welding of Metals," February 1962 (translated from Russian), American Welding Society, Miami, Fla.

5. Arthur L. Phillips, ed., "Ultra Sonic Welding," American Welding Society, Miami, Fla.

6. W. G. Essers, G. Jelmarini, and G. W. Tichelaar, "Arc Characteristics and Metal Transfers with Plasma-MIG Welding," *Metal Construction and British Welding Journal* (Dec. 1972).

8

RESISTANCE, ELECTRON BEAM, AND LASER BEAM WELDING AND CUTTING

OUTLINE

8-1 RESISTANCE WELDING

Resistance welding (RW) is "a group of welding processes that produces coalescence of the faying surfaces with the heat obtained from resistance of the workpieces to the flow of the welding current in a circuit of which the workpieces are a part, and by the application of pressure." There are at least 10 different resistance welding processes and many variations. They are as follows:

* Flash welding (FW)
* Percussion welding (PEW)
* Projection welding (PW)
* Resistance seam welding (RSEW)
   * High frequency (RSEW-HF)
   * Induction (RSEW-1)
* Resistance spot welding (RSW)
* Upset welding (UW)
   * High frequency (UW-HF)
   * Induction (UW-1)

In addition, there is resistance brazing, which uses a filler metal.

Spot welding was invented by Elihu Thomson of Massachusetts in 1877. In the early 1880s, it was being used commercially and was called *incandescent welding.* Thomson went on to develop other resistance welding processes.

The resistance welding processes share a common definition, but many of them are considerably different. The more important processes and variations will be explained.

Principles of Operation

The resistance welding processes differ from arc welding in that pressure is used but filler metal or fluxes are not. Four factors are involved in making a resistance weld. They are (1) the amount of current that passes through the work, (2) the pressure that the electrodes transfer to the work, (3) the time the current flows through the work, and (4) the area of the electrode tip in contact with the work. Heat is generated by the passage of electrical current through a resistance circuit. The maximum amount of heat is generated at the point of maximum resistance, which is at the surface between the parts being joined. The high current, up to 100,000 A at low voltage, generates sufficient heat at this resistance point so that the metal reaches a molten state. The force applied before, during, and after the current flow forges the heated parts together so that coalescence will occur. Pressure is required throughout the entire welding cycle to assure a continuous electrical circuit. The amount of current employed and the time period are related to the heat input required to overcome heat losses and raise the temperature of the metal to the welding temperature.

The concept of resistance welding is most easily understood by relating it to spot welding. Spot welding is shown in Figure 8–1. High current at a low voltage flows through the circuit in accordance with Ohm's law:

$$I = \frac{E}{R} \quad \text{or} \quad R = \frac{E}{I} \quad \text{or} \quad E = I \times R$$

where I = current in amperes
E = voltage in volts
R = resistance of the materials in ohms

FIGURE 8–1 Resistance spot welding.

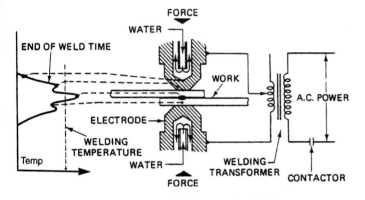

The total energy is expressed by the formula heat energy H equals $I \times E \times T$, in which T is the time in seconds during which current flows in the circuit. Combining these two equations gives:

$$H \text{ (heat energy)} = I^2 \times R \times T$$

For practical reasons a factor that relates to heat loss should be included; therefore, the actual resistance welding formula is

$$H \text{ (heat energy)} = I^2 \times R \times T \times K$$

where I = current in amperes
R = resistance of the work in ohms
T = time of current flow in seconds
K = heat losses through radiation and conduction
H = heat energy in watt seconds

Welding heat is proportional to the square of the welding current. If the current is doubled, the heat generated is quadrupled. Welding heat is proportional to the total time of current flow. If current is doubled, the time can be reduced, which is recommended. The welding heat generated is directly proportional to the resistance, which is related to the material being welded, the contact area, and the pressure applied. Resistance is also related to coating on the metal. There must be sufficient time to melt the coating or otherwise eliminate it. Resistance is also affected by dirt, paint, oil, or other materials on the surface. Mechanical pressure, which forces the parts together, helps contain the molten metal until it solidifies.

Heat is also generated at the contact between the welding electrodes and the work. This amount of heat generated is lower since the resistance between high-conductivity electrode material and the work is less than that between the two workpieces. In most applications the electrodes are water cooled to minimize the heat generated.

Resistance welds are made very quickly; however, each process has its own time cycle. Resistance welding operations are automatic. Good-quality welds made on press or rocker arm machines do not depend on welding operator skill but on the proper setup and adjustment of the equipment and adherence to weld schedules. Operator skill is important when using portable gun machines. The position of making resistance welds is not a factor, particularly when welding thinner materials.

Resistance welding is widely used by mass-production industries, where production runs and consistent conditions are maintained. Welding is performed by operators who normally load and unload the welding machine and push the switch to initiate the weld operation. The automotive industry is the major user, followed by the appliance industry. It is used by many industries manufacturing a variety of products made of thinner gauge metals and for manufacturing pipe, tubing, and smaller structural sections. Resistance welding has the advantage of producing a high volume of work at high speeds that are reproducible at high quality.

Weldable Metals

Metals that are weldable, the thicknesses that can be welded, and joint design are related to specific resistance welding processes. Most of the common metals can be welded by many of the resistance welding processes (Figure 8–2). However, difficulties may be encountered when welding certain metals in heavier thicknesses. Some metals require heat treatment after welding for satisfactory mechanical properties. Weldability of a metal is controlled by three factors: (1) resistivity, (2) thermal conductivity, and (3) melting temperature. Metals with a high resistance to current flow and with a low thermal conductivity and a relatively low melting temperature are easily weldable. Ferrous metals all fall into this category. Metals that have a lower resistivity but a higher thermal conductivity will be more difficult to weld. This includes the light metals—aluminum and magnesium. The precious metals are difficult to weld because of their high thermal conductivity. The refractory metals, which have extremely high melting points, are difficult to weld.

These three properties can be combined into a formula that will provide an indication of the ease of welding a metal. This formula is

$$W = \frac{R}{FKt} \times 100$$

where W = weldability
R = resistivity
F = melting temperature of the metal in °C
Kt = relative thermal conductivity with copper equal to 1.00

FIGURE 8–2 Metals weldable by the spot welding process.

Metal	Weldability	Weldability Rating
Aluminum	Weldable	0.75–2+
Magnesium	Weldable	1.80
Inconel	Weldable	2+
Nickel	Weldable	2.15
Brass and bronze	Variable	0.5–10+
Monel	Weldable	2+
Precious metals	Variable	0.16–3.0
Low-carbon steel	Weldable	10+
Low-alloy steel	Weldable	10+
High- and medium-carbon steel	Possible	10+
Stainless steel	Weldable	35+
Titanium	Weldable	50+

If W (weldability) is below 0.25, it is a poor rating. If W is between 0.25 and 0.75, weldability is fair. Between 0.75 and 2.0 weldability is good, and above 2.0 weldability is excellent. In this formula mild steel would have a weldability rating of over 10. Aluminum has a weldability factor of from 0.75 to 2, depending on the alloy, considered to be a good weldability rating. Copper and most brasses have a low weldability factor and are known to be very difficult to weld.[1] This applies primarily to spot welding, but would be an indication for the other resistance welding processes where arcing does not take place.

There are more coated metals being spot welded. This includes zinc-coated, tin-coated (tern), aluminum-coated, painted material, and plastic-coated materials. These add to the complication of making spot welds, and in general require more sophisticated control systems. In addition, the electrode tips deteriorate much more quickly when welding coated sheet metal. Special procedures and techniques have been developed for coated steels.

Resistance Spot Welding

Resistance spot welding (RSW) is "a resistance welding process that produces a weld at the faying surfaces of a joint by the heat obtained from resistance to the flow of welding current through the workpieces from electrodes that serve to concentrate the welding current and pressure at the weld area." This was shown in Figure 8–1. The size and shape of the individually formed spot welds are determined by the size and contour of the electrodes. Spot welding is the most popular resistance welding process and is described in more detail. The basic concepts pertaining to equipment, controls, electrodes, pressure application, and mechanization are generally true of the other processes.

A spot welding system needs at least the following components:

❈ Welding transformer for supplying power
❈ A means of applying pressure
❈ A controller/contactor
❈ Electrode tips for conducting welding current to the work

Welding machines include all of these functions. They are available from the very smallest to extremely large complex machines.

Spot Welding Machines

Spot welding machines are available in two categories: single-point or single-spot machines and multiple-point machines. Single-point spot welding machines can be relatively simple. The simplest is manually operated, rated at 2 kVA with a short circuit current of 6,000 A and capable of welding 20-gauge and thinner carbon steel. A light-duty hand-operated machine is shown in Figure 8–3. Machines

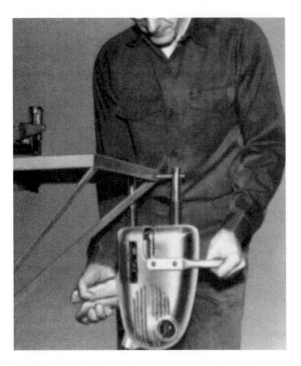

FIGURE 8–3 Manually operated spot welding gun.

FIGURE 8–4 Rocker arm spot welding machine.

FIGURE 8–5 Press-type spot welding machine.

of this type are used for maintenance, automobile body repair, and light-duty operations.

The more popular machines are stationary single-point spot welding machines of either the horn "rocker arm" type or the press type. The horn-type machines have a pivoted or rocking upper electrode arm, which is actuated by either the operator's physical power or by air or hydraulic power. They are used for a wide variety of work, but are restricted to 50 kVA and are used for thinner gauge. Figure 8–4 shows this type of machine.

For heavier requirements, press-type machines are used. This type of machine, shown in Figure 8–5, is normally rated at 50 kVA and up. In the press-type machine the upper electrode moves in a slide. The pressure and motion are provided on the upper electrode by hydraulic or pneumatic pressure, or are motor operated. Press resistance welding machines are used for welding medium-gauge up to the heaviest-gauge materials.

Both the press and rocker arm machines include the welding transformer. The transformer must be closely coupled to the upper and lower electrodes. The control circuit is usually in a separate enclosure mounted on the side of the machine. For all but the smallest machines, water cooling is used to cool the electrodes. The Resistance Welders Manufacturers Association has standardized and classified spot welders.[2] This information is shown in Figure 8–6, which gives the size, the kVA rating, and throat depth. A resistance welding machine rated according to RWMA standards would have a 50% duty cycle. Thus a 30-kVA RWMA machine provides 30 kVA for 30 seconds of

Type of Welding Machine	Size RWMA	Rating (kVA)	Electrode Cooling	Nominal Throat Depth (in.)
Rocker-arm spot welding machines	000	5	Air	8, 12, 16
	00	7.5	Air	8, 12, 16
	0	10	Air	8, 12, 16
	1	15	Air or water	12, 18, 24
			Water	12, 18, 24, 30, 36
	2	30	Water	12, 18, 24, 30, 36
	3	50		
Press-type and project welding machines	000	5	Water	6, 8
	00	20	Water	6, 8
	0	30		
		50	Water	6, 8
		30		
	1	50		
		75	Water	12, 18, 24, 30, 36
		100		
	2	150	Water	12, 18, 24, 30, 36
		150		
	3	200	Water	12, 18, 24, 30, 36
		300		
	4	400		
		500	Water	12, 18, 30

FIGURE 8–6 RWMA standard spot and projection welding machines. (From Ref. 2.)

every minute, operating continuously without overheating. Machines not rated to RWMA standards may be built to a lower duty cycle, rated as low as 10% to 30%, and will overheat at higher duty cycles unless used at reduced power. The RWMA resistance welding handbook also provides the size of welding machines required to weld different metals and metal thicknesses.

When the work is too bulky to take to the welding machine, a portable spot welding machine can be used. The portable machine is moved from one welding location or fixture to another, and a trigger on the gun actuates the welding cycle. Portable units are normally operated by air pressure. There are three types of portable welding guns. In one case, the welding transformer is separated from the welding gun; the welding gun has its own pressure mechanism (Figure 8–7a). The portability of this type of machine is limited by heavy cables connected between the gun and the transformer. The cables are usually coaxial, to avoid movement due to magnetic forces. In the smaller machines, the transformer is included as a part of the gun (Figure 8–7b). The guns are manually manipulated by the operator. The weight of the gun is handled by a balancing mechanism, but the location is established by the operator (Figure 8-9).

One of the early applications of robotic spot welding was the manipulation of the welding gun by a robot; nowadays, spot welding guns manipulated by robots are replacing manually manipulated spot welding guns. This type of application is becoming more widely used in the automotive industries.

Robot-mounted guns are also replacing fixed location rocker arm- and press-type spot welding machines. It depends on which is the easiest to move: the work or the machine. Figures 8–8A and B show two different types of portable guns mounted on robots.

A typical spot weld robot application is shown in Figure 8–10A. This shows an automobile seat frame assembly on a rotating positioner with two robots making the welds. Figure 8–10B is a close-up showing the weld detail. The spot welding of almost all automobile bodies produced today is done on robotic lines. Each spot welding gun will move around and make many welds. The robot always makes the spot welds in the same sequence and produces higher-quality welds and more accurate bodies than do individual operators.

A new type of resistance welding gun is being utilized by robots. This is known as a servo-driven spot weld gun, or servo gun. These guns are controlled by the robot controller and provide motion by means of a servo motor rather than by an air cylinder. This provides better control and results in higher quality welds. This is probably due to improved weld force repeatability and a reduction in the amount of noise produced. Shorter weld cycles are available, and less weld expulsion occurs. It is actually a faster, more quality weld that is produced. A servo gun is shown by Figure 8–7(c).

(a)

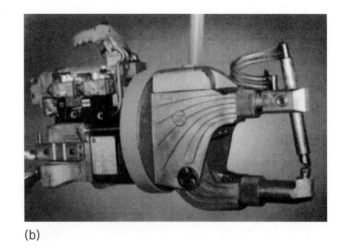

(b)

(c)

FIGURE 8–7 Spot welding guns.

FIGURE 8–8A Portable spot welding gun mounted on robot.

FIGURE 8–8B Portable spot welding gun on robot.

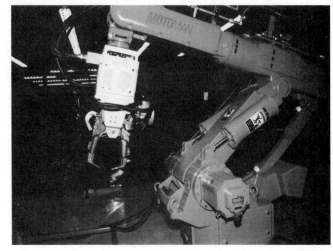

FIGURE 8–9 Manual portable spot welding gun in use.

FIGURE 8–10B Close-up showing weld detail.

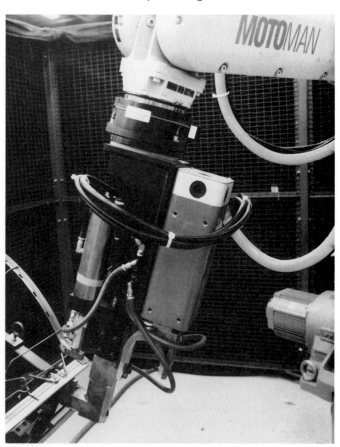

For high-volume production work such as subassemblies in the automotive industry, multiple-point spot sheet welding machines are used. These are generally in the form of a press in which individual guns carrying electrode tips are mounted. Welds are made in sequential order so that all electrodes are not carrying current at the same time. Figure 8–11 illustrates two typical multiple-point spot welding machines. They must be

FIGURE 8–10A Auto seat frame on rotating positioner.

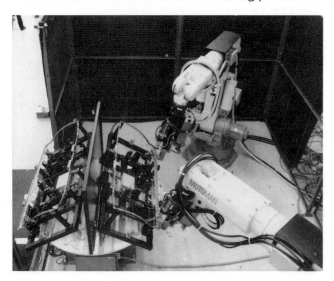

FIGURE 8–11   Multiple-point spot welding machines.

designed for a particular product and are considered dedicated machines. However, in press welders the part that is changed for model changeover is the platen that carries the welding guns. Each individual gun has its own piston, so that it can be moved, pressure applied, and retracted independently.

The working part of the resistance welding machine is the electrode. The electrode is the means for conducting welding current to the work and for providing the force necessary to make welds and for dissipating some of the heat generated. Resistance welding electrodes, the electrode holders, and the electrode material specifications are standardized by the Resistance Welders Manufacturers Association.[3] This standard separates electrode composition into two basic groups: group A, copper-base alloys, with five classes; and group B, refractory metal compositions, also with five classes. These various groups and classes identify the analysis of the electrode alloy, the electrode hardness, strength and conductivity. RWMA recommends different electrode materials for welding different metals.

RWMA provides a method that identifies standard straight tips by a five-digit code (Figure 8-13). The taper relates to the taper on the end of the electrode and in the electrode holder. This fit must be watertight and provides an area to transfer the welding current from the holder to the electrode. Normally, electrodes are straight, but bent tips and double-bent tips are available. Many special electrodes are made particularly for gun welders. The straight-type electrode holders are also standardized, some with

an ejector tube and some without. There are also offset holders and other features. Various companies supply electrodes and electrode holders to RWMA standards and to special requirements.

Welding Controllers

As spot weld quality and consistency requirements increase, better controllers become available. Originally the controller-contactor was a simple on/off timer based on the time of the weld. The new controllers include computers and can accommodate complex programs such as the ones shown in Figure 8-12. These include pulsing currents, delay time, upslope time, heat time, downslope, and temper time. Newer controllers contain adaptive controls, which read signals produced at the weld tips, to vary the welding schedule. This includes thermal reaction while the weld is being made. Controllers also provide monitoring and data presentation that can be remotely transmitted. They also include limit controls, which will shut down the operation or provide warnings when the process is out of control. New sophisticated controllers for seam welding also include travel speed mechanisms and feedback data.

Joint Types

The joint type used most often for spot welding is the lap joint. The joint overlap has a minimum requirement, based on the nugget size, which is related to the electrode

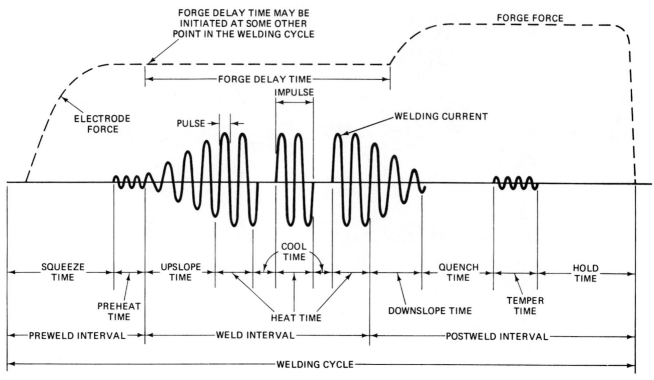

FIGURE 8–12 Resistance welding program.

FIGURE 8–13 RWMA alloy class and standard straight tips description.

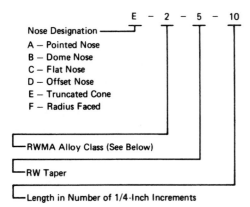

Group	Class	Conductivity (%)	Hardness Rockwell	Tensile Compressive (psi)
A	1	80	65B	60K
	2	75	75B	65K
	3	45	90B	100K
	4	20	33C	140K
	5	10 to 15	65 to 85B	65 to 75K
B	10	35	72B	135K
	11	28	94B	160K
	12	27	98B	170K
	13	30	69B	200K
	14	30	85B	200K

size. The distance from the centerline of the nugget to the edge of the sheet, known as the *edge distance,* should be at least $1\frac{1}{2}$ times the nugget diameter. The separation between the sheets being welded should not exceed 10% of the thinnest sheet. Design parameters for spot welding are given in the resistance welding handbook.

Butt joints can also be accomplished with resistance welding processes. This is the only joint for flash butt and upset welding. However, high-frequency welding can accomplish butt welds and T-welds. The T-weld is used for making small beams in high-frequency resistance welding mills. Another joint, known as the lip joint, is a flanged joint. The flange width should be sufficient to allow spot welding.

The spacing of spot welds and the spacing of roll seam welds are important. If the nuggets overlap, the joint will be watertight. If they do not, water can escape between the nuggets. The resistance welding handbook also provides the size of welding machines required to weld different metals and metal thicknesses.

Spot Weld Quality

Each resistance spot weld nugget is expected to be perfect. The test used has been to pull the parts apart. If the button pulled out of the base metal, it is a quality spot weld. If it fails in any other manner, it is a substandard weld. Many specifications require destructive tests to be made after a prescribed number of welds; in the mean-

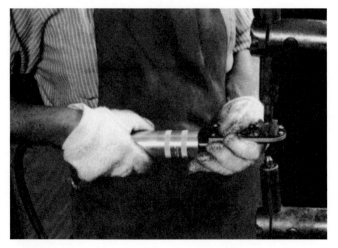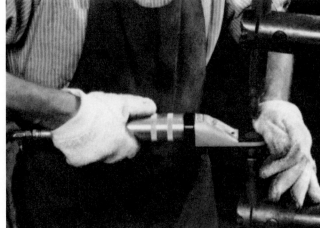

FIGURE 8–14   Redressing spot welding electrodes.

time, monitor the input power, the welding current time and pressure. To ensure a quality product, many users make more than the number of spot welds specified, assuming that there might be some subquality welds produced. Automatic monitoring meters are used, which will warn the operator when any of the parameters exceed specific values. Another quality assurance method is an adaptive welding control system that uses special sensing devices. These systems are based on the motion of the electrode during the welding cycle to monitor the growth of the weld nugget. The adaptive controls will notify the operator to modify the welding parameters to correct potential weld problems.

Quality welds require continuous maintenance of the electrode tip contour. The tip or nose must be redressed often to maintain the proper shape. This must be done more often when welding highly conductive or coated materials. Special power tools (Figure 8-14) are available for redressing electrodes in the resistance welding machine. Another requirement is to check the tip

pressure using a portable tool such as that shown in Figure 8-15.

Spot welds are normally direct spot welds where the two electrodes are opposite each other, with the work to be welded between them. This may cause marking at the point where the electrode is in contact with the work. To avoid this, or where the back side of the joint is not accessible, the indirect spot is used (Figure 8-16). In this case, both electrodes are applied from one side and a large flat, or contoured, electrode is on the back side. This technique is used in the automobile industry to minimize metal finishing of exposed spot welds.

Projection Welding

Projection welding (PW) is "a resistance welding process that produces a weld by the heat obtained from the resistance to the flow of the welding current. The resulting welds are localized at predetermined points by projections, embossments, or intersections."

FIGURE 8–15   Electrode force meter.

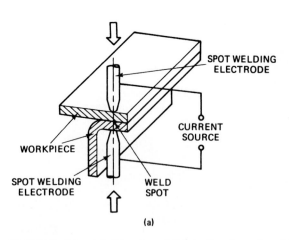

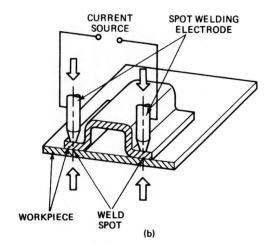

FIGURE 8–16 (a) Direct and (b) indirect spot welds.

Figure 8–17 shows the principles of projection welding. Localization of heating is obtained by a projection on one of the parts being welded. There are several types of projections: (1) the button or dome type, usually round; (2) elongated projections; (3) ring projections; (4) shoulder projections; (5) cross-wire welding; and (6) radius projection. The major advantage of projection welding is that electrode life is increased because larger contact surfaces are used. A very common use of projection welding is the use of special nuts that have projections on the portion of the part to be welded to the assembly. These are manufactured with the projections and assist in obtaining quality joints to the parts being welded. Projection dimensions must be properly designed since the height and area have optimum dimensions for welding to specific thicknesses of sheet metal. These data are in the "Resistance Welding Manual."[3] A press-type resistance welding machine is normally used. Flat nose or special electrodes are used.

One of the most common variations of projection welding is wire-to-wire welding, where they cross at approximately 90°. This is used for making gratings, wire racks, shelves, and similar items. A machine for making cross-wire welds is shown in Figure 8–18.

FIGURE 8–17 Projection welding.

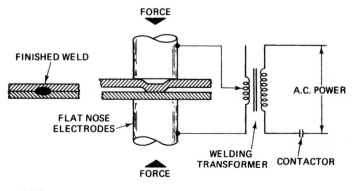

Resistance Seam Welding

Resistance seam welding (RSEW) is "a resistance welding process that produces a weld at the faying surfaces of overlapped parts progressively along a length of a joint. The weld may be made with overlapping weld nuggets, a continuous weld nugget, or by forging the joint as it is heated to the welding temperature by resistance to the flow of the welding current." The resulting weld is a series of overlapping spot welds made progressively along a joint by rotating the electrode. The resistance seam welding process is shown in Figure 8–19. A resistance seam welding machine is shown in Figure 8–20.

When the spots are not overlapped enough to produce gastight welds, it is a variation known as *roll resistance spot welding*. This process differs from spot welding, since the electrodes are wheels. Both the upper and lower electrode wheels are powered. Pressure is applied in the same manner as a press-type welder. The wheels can be either in line with the throat of the machine or transverse. If they are in line, it is normally called a longitudinal seam welding machine. Welding current is transferred through the bearings of the roller electrode wheels. Water cooling is not provided internally, and therefore the weld area is flooded with cooling water to keep the electrode wheels cool. In seam welding a complex control system is required. The welding speed, the spots per inch, and the timing schedule are dependent on each other. Welding schedules provide the pressure, the current, the speed, and the size of the electrode wheels. This process is quite common for making flange welds, for making watertight joints for tanks, and so on.

Another variation is *mash seam welding*, where the lap is fairly narrow and the electrode wheel is at least twice as wide as that used for standard seam welding. The pressure is increased to approximately 300 times normal pressure. The final weld mash seam thickness is only 25% greater than the original single sheet.

FIGURE 8–18 Cross-wire welding.

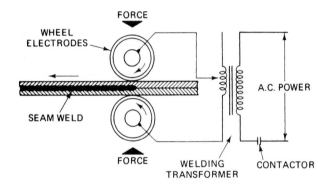

FIGURE 8–19 Resistance seam welding.

Another variation for welding coated steel utilizes a round copper wire that is fed between the electrode roll and the work. It is formed into an oval by the pressure in the machine. A wire is required for both wheel electrodes. It is the copper wire that is in contact with the work rather than the electrode. The continuously fed copper wire carries the melted coating away from the weld area and provides uniform resistance for consistent welds. The flattened copper wire is salvaged. It eliminates coated metal pickup on the roller electrodes and is claimed to give more consistent welding results.

Flash Welding

Flash welding (FW) is "a resistance welding process that produces a weld at the faying surfaces of a butt joint by a flashing action and by the application of pressure after heating is substantially completed (Figure 8–21). The flashing action, caused by the very high current densities at small contact points between the workpieces, forcibly

FIGURE 8–20 Resistance seam welding machine.

expels the material from the joint as the workpieces are slowly moved together. The weld is completed by a rapid upsetting of the workpieces." The flashing and upsetting are accompanied by expulsion of metal from the joint. This is very dramatic and is shown in Figure 8-22. After a

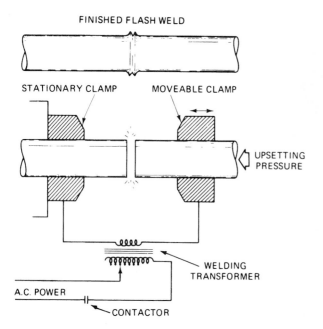

FINISHED FLASH WELD

STATIONARY CLAMP   MOVEABLE CLAMP

UPSETTING
PRESSURE

WELDING
TRANSFORMER

A.C. POWER

CONTACTOR

FIGURE 8–21   Flash welding process.

FIGURE 8–22   Flash weld being made.

predetermined time the two pieces are forced together and coalescence occurs at the interface; current flow is possible because of the light contact between the two parts being flash welded. The heat is generated by the flashing and is localized in the area between the two parts. The surfaces are brought to the melting point and expelled through the abutting area. As soon as this material is flashed away, another small arc is formed, which continues until the entire abutting surfaces are at the welding temperature. Pressure is then applied and the arcs are extinguished and upsetting occurs.

Flash welding can be used on most metals. No special preparation is required, except that heavy scale, rust, and grease must be removed. The joints must be cut square to provide an even flash across the entire surface. The material to be welded is clamped in the jaws of the flash welding machine with a high clamping pressure. The upset pressure for steel exceeds 10,000 psi (700 kg/cm^2). For high-strength materials these pressures may be doubled. For tubing or hollow members the pressures are reduced. As the weld area is more compact, upset pressures increase. If insufficient upset pressure is used, a porous, low-strength weld will result. Excess upset pressure will result in expelling too much weld metal and upsetting cold metal. The weld may not be uniform across the entire cross section, and fatigue and impact strength will be reduced. The speed of upset—that is, the time between the end of the flashing period and the end of the upset period—should be extremely short, to minimize oxidation of the molten surfaces. In the flash welding operation a certain amount of material is flashed or burned away. The distance between the jaws after welding compared to the distance before welding is known as the burn-off. It can be from 1/8 in. (3.2 mm) for thin material up to several inches for heavy material. Welding currents are high and are related to the following: 50 kVA per square inch of cross section at 8 seconds. It is desirable to use the lowest flashing voltage at a desired flashing speed. The lowest voltage is normally 2 to 5 V per square inch of cross section of the weld.

The upsetting force is usually mechanical cam action. The design of the cams is related to the size of the parts being welded. Flash welding is completely automatic and is an excellent process for mass-produced parts. It requires a machine of large capacity designed specifically for the parts to be welded. Flash welds produce a fin around the periphery of the weld, which is normally removed.

Upset Welding

Upset welding (UW) is "a resistance welding process that produces coalescence over the entire area of faying surfaces or progressively along a butt joint by the heat obtained from the resistance to the flow of welding current through the area where those surfaces are in contact. Pressure is used to complete the weld." Pressure is applied before heating is started and is maintained throughout the heating period (Figure 8–23). The equipment used for upset welding is very similar to that used for flash welding. It can be used only if the parts to be welded are equal in cross-sectional area. The abutting surfaces must be prepared very carefully to provide for proper heating. The difference from flash welding is that the parts are clamped in the welding machine and force is applied, bringing them tightly together. High-amperage current is then passed through the joint, which heats the abutting surfaces. When they have been heated to a suit-

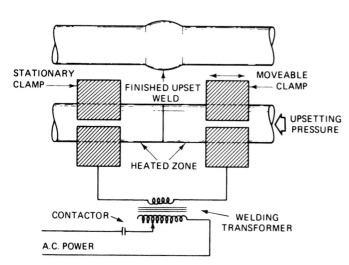

FIGURE 8–23 Upset welding process.

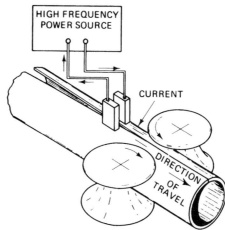

FIGURE 8–24 High-frequency seam welding.

able forging temperature, an upsetting force is applied and the current is stopped. The high temperature of the work at the abutting surfaces, plus the high pressure, causes coalescence to take place. After cooling, the force is released and the weld is completed. There is no arc or flash in upset welding. The area at the joint is usually enlarged over its original dimension. This process is used for welding small wires, tubing, piping, rings, strips, and so on, where the cross-sectional areas of both pieces are identical. If intimate contact is not obtained because of improper joint preparation, the weld will be defective.

High-Frequency Resistance Welding

High-frequency resistance welding is "a group of resistance welding process variations that use high-frequency welding current to concentrate the welding heat at the desired location." There are two variations.

High-frequency seam welding (RSEW-HF) is a resistance seam welding process variation in which high-frequency welding current is supplied through electrodes into the workpieces (Figure 8–24). *High-frequency upset welding* (UW-HF) is an upset welding process variation in which high-frequency welding current is supplied through electrodes into the workpieces. The systems are very similar, except that in one case the induction work coil generates the heat in the workpieces, and in the other case sliding electrodes are in contact with the workpieces. The frequency ranges from 10 to 500 kHz. The upsetting force is applied by rollers. These variations are ideally suited for making pipe, tubing, structural shapes, and other formed items made from continuous strip. In this process the high-frequency welding current is introduced into the metal at the surfaces to be welded but prior to their contact with each other. Current is introduced by means of sliding contacts or by induction coil. The high-frequency welding current flows along one edge of the seam to the welding point between

the pressure rolls and back along the opposite edge to the other sliding contact. The current is of such high frequency that it flows along the metal surface to a depth of several thousandths of an inch. Each edge of the joint is the conductor of the current and the heating is concentrated on the surface of these edges. At the area between the closing rolls the material is at the plastic temperature, and with the pressure applied, coalescence occurs. The surfaces must be reasonably true with respect to each other, and clean. No other special preparation is required. The process can be used to join most common metals and certain dissimilar metals. The process is entirely automatic and utilizes special control equipment. It is possible to make welds at extremely high speeds, approaching 500 ft/min (150 m/min) for thin-wall tubing.

Resistance Welding Safety

Only resistance welding equipment meeting the Resistance Welding Manufacturing Association (RWMA) standards should be utilized. All equipment must be installed in conformance with the *National Electrical Code*® and the local requirements.

Operating controls such as start buttons and foot switches must be guarded to prevent accidental startup of equipment. All chains, gears, linkages, belts, and so on, in the machine must be guarded in accordance with American National Safety Standards.

Fixed single-point equipment or single-ram equipment should be guarded or require two start buttons so that the operator's hands cannot be in the point of welding during operation. On multiple-point equipment interlocks, latches, barriers, or guards should be used. For portable equipment two handles are required so that the operator's fingers cannot be in the contact area.

All electrical controls must be enclosed in approved cabinets that should be grounded to earth. Stop buttons should be available at the operator's station to absolutely stop the welding sequence when they are pushed. If capacitors are involved in the welding machine, they should be properly enclosed and a positive device should be installed to discharge all capacitors whenever the enclosure is open.

Operators should wear face shields, spectacles, or goggles, depending on the type of work. Such devices are necessary to protect the face and eyes from flying sparks. Operators designated to operate resistance welding equipment must be properly instructed and judged competent to operate the equipment.

8-2 ELECTRON BEAM WELDING

Electron beam welding (EBW) is "a welding process that produces coalescence with a concentrated beam, composed primarily of high-velocity electrons, impinging on the joint. The process is used without shielding gas and without the application of pressure." It is a fusion welding process with the melting together of base metal, and possibly of filler metal, to produce a weld. Heat is generated in the workpiece as it is bombarded by a high-velocity electron beam. The kinetic energy, energy of motion, of the electrons is transferred to heat upon impact. It is a highly concentrated, high-powered source of heat and acts similar to the arc of gas tungsten arc welding in making welds.

The electron beam welding process was developed in France. J.A. Stohr, with the French Atomic Energy Commission, made the first public disclosure of this welding process at a symposium on fuel elements held in Paris on November 23, 1957.[4]

Principles of Operation

The original work was done in a high vacuum using an electron gun similar to an x-ray tube. In an x-ray tube the beam of electrons is focused on a target to give off x-rays; the target becomes very hot and requires water cooling. In electron beam welding, the target is the workpiece, which absorbs the heat to bring it to the molten stage to allow welding.

A modern electron beam welding machine consists of at least the following:

* Electron beam gun
* Power supply and control
* Gun and work motion equipment
* Welding chamber, with vacuum pumps
* Alignment and viewing system
* Miscellaneous auxiliary equipment

The electron beam gun is a device for producing and accelerating electrons. It consists of an emitter, which is a filament or cathode, a grid cup, an anode, and focus-

ing and deflection coils. It is housed in a hard vacuum. When the focus and deflection coils are included, this is called an electron beam gun column (Figure 8–25). The entire electron beam gun column is exhausted to a vacuum of 0.1 to 0.01 μm or 0.0001 torr (10^{-4}). One torr equals one millimeter of mercury. Atmospheric pressure is 760 torr or 760 mm hg, which is also 14.7 psi.

The emitter is either a tungsten filament or a tungsten rod heated by a filament. Electrons are freed from the tungsten when it is heated to a high temperature, causing thermionic emission. Electrons freed from the emitter are attracted to the anode, which is the positive pole. The beam is collected and partially focused and attracted to the anode, which has a hole. Beyond the anode hole, the beam is focused by means of magnetic forces generated by the focusing coil. Following this, the beam may be deflected by magnetic fields generated by deflection coils. The beam then leaves the electron beam gun through an exit port and impinges on the workpiece.

The next major component is the power supply and control. This unit takes power from the utility line and provides the beam current, normally less than 1 A, and the acceleration voltage, which is thousands of volts. The beam power is the product of the beam current and the acceleration voltage measured in kilovolts, and ranges from a few kilovolts up through 50 kV. The control system has total control of the electron beam welder system. It also controls relative motion between the gun and workpiece. It powers the vacuum pumps and other devices. Controls for electron beam welding machines must be very precise and are often computer driven. In many installations the electron gun is fixed and can be adjusted for specific targets. The work-handling equipment used to move the workpiece can be quite complex, ranging from single-axis motion to five or more axes of motion in three

---

FIGURE 8–25 Electron beam welding process.

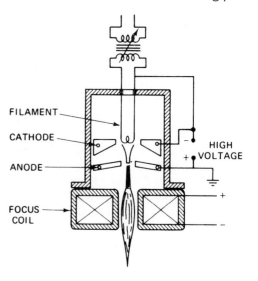

planes and with rotary motion. Equipment of extreme precision must be used. The travel mechanism must be designed for vacuum installations since lubricants and certain insulating varnishes in electric motors may volatilize in a vacuum. In some cases, motors and gearboxes are located outside the vacuum chamber, with shafts operating through pressure-sealed bearings.

The next major component is the welding chamber, which must be absolutely airtight. This container, which is evacuated to reduce the pressure to a high vacuum, must be extremely strong so that it will not crush under atmospheric pressure. It requires openings to allow the work to be enclosed and removed. The openings, doors, and so on, must be sealed to a vacuum tightness. The work chamber must be sufficiently large to enclose the parts to be welded, but should not be overly large because of the time and expense of evacuating it. Early chambers utilized a hard vacuum—the same as the vacuum in the electron beam gun column. As electron beam guns became more powerful, a second method of electron beam welding was developed. This allowed welding in a soft vacuum with a pressure of 0.1 torr (10^{-1} torr), known as soft-vacuum electron beam welding. This made larger work chambers possible, with quicker pump-down time.

The third method of electron beam welding, done in the open air, is known as nonvacuum electron beam welding. The electron beam gun is housed in the hard-vacuum chamber and there are several intermediate reduced-pressure chambers between the gun and the work. Each intermediate chamber has a reduced pressure with very small holes from one chamber to another so that the electron beam passes through—but they're too small for a volume of air to pass. This mode of operation eliminates the vacuum chamber for the work; however, certain sacrifices are made. Vacuum pumps are required to eliminate air in the electron beam gun column and in intermediate chambers between the gun and the work. Two vacuum pumps are required to produce a hard vacuum. A mechanical pump is used to eliminate the large volume of air and will pull a vacuum in the 10^{-1} torr range. To obtain the hard vacuum, a diffusion pump is required. The diffusion pumping does not remove large volumes of air and takes considerable time to reach a hard vacuum. The pumps are operated automatically by the control system.

The last component is an optical viewing system to line up the electron beam with the weld joint. This must be very accurate since the welding beam is small. The optical system is connected to the work motion device for precise alignment. Figure 8–26 shows a typical low-powered electron beam welding machine. Figure 8–27 shows a higher-powered machine with a larger work chamber.

FIGURE 8–26 Electron beam low-power welding equipment.

FIGURE 8–27 Electron beam high-power welding equipment.

Electron Beam Welding Equipment

There is a wide selection of electron beam welding machines, based on:

❋ The operating pressure of the work chamber
❋ The acceleration voltage
❋ The beam power level
❋ The complexity of control system

Electrons in the beam collide with molecules in air and lose velocity and direction. This causes scattering and dissipation of the beam strength. The hard-vacuum mode provides the most efficient welding operation. In the hard-vacuum mode the stand-off distance (the distance from the exit of the electron beam to the workpiece) can be as much as 30 in. (660 mm), and material up to 6 in. (150 mm) thick can be welded. Travel speeds will be the highest. On the negative side, the pump-down time is fairly long, based on chamber size, and the entire machine is expensive.

In the soft vacuum, sometimes called medium or partial vacuum, the pump-down time is greatly reduced and can be attained by using mechanical pumps without the diffusion pump. Even so, the electron beam gun column must be at a hard vacuum. With the soft vacuum, interlock doors can be used so that material can be introduced and taken out while the welding operation continues. Chambers can be larger. However, there is a sacrifice in operating conditions; the stand-off distance is reduced by approximately one-half, thickness is reduced to 2 in., and speed is also reduced with the same power.

In the nonvacuum mode, sometimes called out-of-vacuum or welding in the air, the workpiece is at atmosphere pressure. The stand-off distance is reduced to 1½ in. (37 mm), the maximum thickness is 2 in. (50 mm) and speed is reduced, again using the same power. A nonvacuum system requires a container to shield people from potential radiation. The size of work is unlimited and work motion is easier since the components are in the air. Even with the nonvacuum system, a hard vacuum is required for the electron beam gun, and intermediate reduced-pressure chambers are also required. Figure 8–28 shows electron beam welding of a part for the space shuttle engine.

There is a new method used experimentally that allows the welding to be done in a soft vacuum, which is contained in a small chamber attached to the work by means of seals. This allows the welding gun to move but requires that a seal be maintained between the welding apparatus and the workpiece. For some applications this is possible; for others it is not. This system is not popular.

The acceleration voltage is another way of specifying electron beam welding machines. There are low-voltage machines having an output of 15 to 60 kV, and high-voltage machines with an output of 100 to 200 kV. It is difficult to compare equipment on the basis of the accelerating voltage only since the basic design of the low- and high-voltage systems are radically different. From a safety standpoint, an accelerating voltage of less than 20 kV produces soft x-rays, while an accelerating voltage of over 20 kV produces hard x-rays. Shielding is more demanding against radiation as the acceleration voltage increases. The lower-voltage machines operate at a higher current; typically, 30- to 60-kV

FIGURE 8–28   Electron beam welding in a chamber.

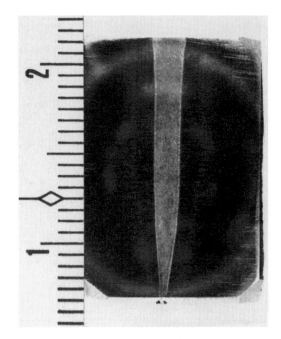

FIGURE 8–29   Electron beam weld in 1⅜-in.-thick aluminum.

machines operate at a 500-mA beam current. The high-voltage machines of 150 kV operate at 40 mA. The higher-voltage machines produce a greater depth-to-width ratio of the weld nugget. This could be the difference between a 12:1 depth-to-width ratio versus a 25:1 depth-to-width ratio. The higher-voltage machines can utilize a longer stand-off distance than can low-voltage machines; however, low-voltage machines are simpler in construction and less maintenance is required.

Electron beam machines are rated by their output power in kilowatts. They are available from approximately 1 kW to as high as 40 kW.

The ability of a machine to do work is based on its beam power, which is the product of the beam current and the accelerating voltage in kilovolts. Beam power relates to the power density of the electron beam. Power densities in the range 100,000 to 10,000,000 W/in.^{2} can be obtained. Temperatures are in the neighborhood of 25,000°F, which causes practically instantaneous vaporization of the surface of the workpiece. The depth of penetration is generally considered a function of the accelerating voltage. The accelerating voltage relates to the speed at which the electrons travel. The beam current, which relates to the number of electrons in the beam, influences the weld configuration.

The major advantage of electron beam welding is its tremendous penetration, which occurs when the highly accelerated electron hits the base metal. It will penetrate slightly below the surface and at that point release the bulk of its kinetic energy, which turns to heat energy. This brings about a tremendous temperature increase at the point of impact. The succession of electrons striking the same place causes melting and then evaporation of the base metal. This creates metal vapors, but the electron beam travels through the vapor much easier than solid metal. This causes the beam to penetrate deeper. The depth-to-width ratio can exceed 20:1. As the power density is increased, penetration is increased. An electron beam weld in 1⅜ in. aluminum is shown in Figure 8–29.

The heat input of electron beam welding is controlled by four variables: (1) the number of electrons per second hitting the workpiece or beam current; (2) the electron speed at the moment of impact, the accelerating potential; (3) the diameter of the beam at or within the workpiece, the beam spot size; and (4) the speed of travel, the welding speed. The first two variables, beam current and accelerating potential, are used in establishing welding parameters. The third factor, the beam spot size, is related to the focus of the beam, and the fourth factor is also part of the procedure. Normally, the electron beam current ranges from 250 to 1,000 mA; the beam currents can be as low as 25 mA. The accelerating voltage is within the two ranges mentioned previously. Travel speed can be extremely high and relates to the thickness of the base metal. The other parameter that must be controlled is the gun-to-work distance. It is difficult to establish welding schedules for electron beam welding because of the number of variables involved. However,

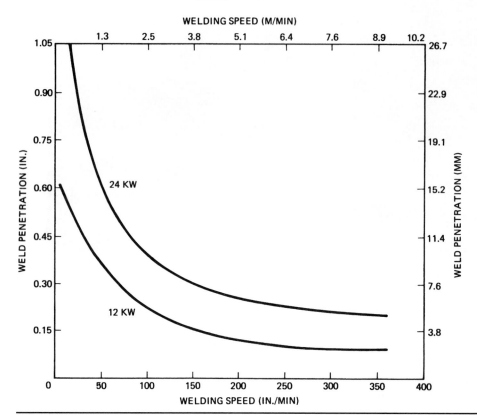

FIGURE 8–30 Travel speed versus penetration.

Figure 8–30 shows the relationship between travel speed and depth of penetration.

The beam spot size can be varied by the location of the focal point with respect to the surface of the work. Penetration can be increased by placing the focal point below the surface of the workpiece. As it is increased in depth below the surface, deeper penetration will result. When the beam is focused at the surface, there will be more reinforcement on the surface. When the beam is focused above the surface, there will be excessive reinforcement and the width of the weld will be greater.

Penetration is also dependent on the beam current. As beam current is increased, penetration is increased. The other variable, travel speed, also affects penetration. As travel speed is increased, penetration is reduced.

The power in an electron beam weld would be in the same relative amount as for a gas metal arc weld. The gas metal arc weld would require higher power to produce the same depth of penetration. The energy in joules per inch for the electron beam weld may be only one-tenth as great as the gas metal arc weld. The electron beam weld will be equivalent to the SMAW weld, with less power because of the tremendous penetration obtainable by electron beam welding. The power density is in the range of 100 to 10,000 kW/in^2.

Since the electron beam has tremendous penetrating characteristics, with the lower heat input, the heat-affected zone is much smaller than that of any arc welding process. In addition, because of the almost parallel sides of the weld nugget, distortion is greatly minimized. The cooling rate is much higher, and for many metals this is advantageous; however, for high-carbon steel this is a disadvantage and cracking may occur.

Some weld joint details for electron beam welding are shown in Figure 8–31. Welds are extremely narrow, and therefore penetration for welding must be extremely accurate.[6] The width of a weld in $\frac{1}{2}$-in. (12-mm)-thick stainless steel, for example, would only be 0.04 in. (0.10 mm), and for this reason a small misalignment would allow the electron beam to miss the joint completely. Special optical systems are used that enable the operator to align the work with the electron beam. The electron beam is not visible in the vacuum. The depth-to-width ratio allows for special lap-type joints. Where joint fitup is not precise, ordinary lap joints are used and the weld is an arc seam weld. Normally, filler metal is not used in electron beam welding; however, when welding mild steel, highly deoxidized filler metal is sometimes used to deoxidize the molten metal and produce dense welds.

Almost all metals can be welded with the electron beam welding process. The metals that are most often welded are the superalloys, the refractory metals, the reactive metals, and the stainless steels. Many combinations of dissimilar metals can also be welded.

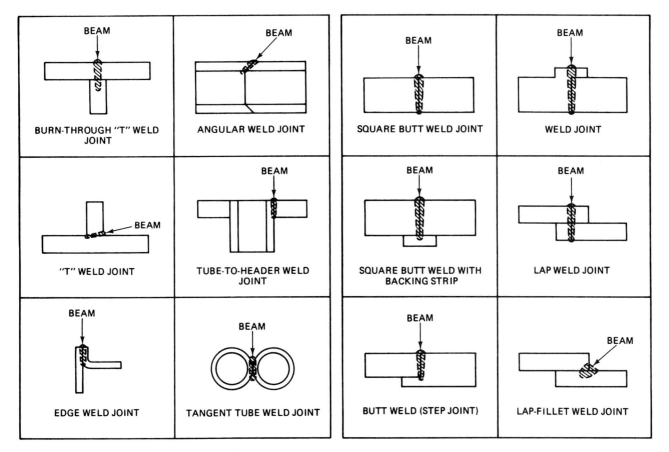

FIGURE 8–31 Weld joint types for EBW.

One of the disadvantages of the electron beam process is its high capital cost. The price of the equipment is very high, and it is expensive to operate due to the need for vacuum pumps. In addition, fitup must be precise and locating the parts with respect to the beam must be perfect.

Electron beam welding is not a cure-all; there are still the possibilities of defects of welds. A major problem is welding plain carbon steel in a vacuum. The melting of the metal releases gases originally in the metal and results in a porous weld. If deoxidizers cannot be used, the process is not suitable. It is expected that the electron beam process will become more popular for welding specialized metals where critical quality standards must be met.

8-3 LASER BEAM WELDING

Laser beam welding (LBW) is a welding process that uses the heat generated when a focused laser beam impinges on the joint. The process is used with or without a shielding gas and without the application of pressure. The laser is a device that produces a concentrated coherent light beam by stimulating electronic or molecular transitions to lower energy levels. The word *laser* is an acronym for "light amplification by stimulated emission of radiation." The laser beam is a highly concentrated source of energy that has many applications. It can be used for welding, for cutting metals and nonmetals, for surface heat treating of metals, and for cladding by fusing powders to base materials. It can also be used for brazing and soldering, and for drilling, machining, and marking. It is also used in other fields, including medicine, communications, marking, compact disc players, barcode reading, and surveying.

The laser was conceived by Charles H. Townes in 1951.[7] In 1960, T. H. Maiman of Hughes Aircraft Research Laboratories in California demonstrated a device, working in the visible region of the spectrum, utilizing a synthetic ruby crystal excited by a gas discharge flash tube and emitting short pulses of red coherent light.[8] In 1961, Ali Javan of Bell Labs produced a laser beam from a mixture of helium and neon gases excited directly by an electrical discharge. The CO_2 laser developed in 1964 by Patel of Bell Labs has become the industrial workhorse.[9] The focused laser beam has a very high energy concentration, on the same order as an electron beam in a hard vacuum. It is a source of electromagnetic energy, or light, that can be projected with low divergence and can be concentrated to a very small spot.

Light from an incandescent electric light bulb is "incoherent," which means out of phase, and as a result has a high divergence or is radiated in all directions from the source. It is not monochromatic, which means that it contains a wide spectrum of wavelengths (colors), from short to long. The radiation from a laser is monochromatic, which provides a single wavelength, which in turn allows for minimum beam divergence. The beam is also coherent, in that the light is all in phase. The laser beam has a high energy content; thus when it impinges on a surface, it creates heat. This heat can be used exactly as heat produced by an electron beam or a welding arc.

Laser Types

There are two basic types of lasers used in metalworking. The original types are the solid-state lasers, which use a solid medium. The second types are the gas lasers, which normally use a mixture of helium, nitrogen, and CO_2 gas in a tube. In either case, when the medium is sufficiently excited, it emits photons, which become the laser beam.

There are three types of solid-state lasers in commercial use: (1) the ruby laser, which uses a synthetic ruby with chromium in aluminum oxide; (2) the Nd:glass laser, which uses neodymium in glass; and (3) the Nd:YAG laser, which is a crystal doped with neodymium and made of yttrium, aluminum, and garnet. In the solid-state lasers, the Nd ions emit photons when their electrons are excited and then allowed to draw back to their original energy state.

The wavelength of the laser beam produced by solid-state lasers is much shorter than that for CO_2 lasers. A laser beam of this short wavelength is an eye hazard. In general, safe use of material working lasers requires eye protection, specifically blocking the wavelength of light that the laser produces. Unprotected eyes are at risk not only from direct laser energy but also from reflected energy. Therefore, eye protection is required when working around solid-state lasers. Beam-reflecting goggles coated with material that blocks or reflects radiation prevents the beam from passing through to the eyes.

The beam operating mode for the ruby and Nd:glass lasers is usually pulsed. The Nd:YAG laser can be of continuous-wave or pulsed mode. The average output pulse range for the ruby laser ranges from 10 to 20 W. Nd:YAG lasers are available to 3,000 W, with the power levels going up every year. The beam diameter is normally given only as a minimum, which is from $\frac{1}{16}$ in. (1.6 mm) for the ruby laser, from $\frac{1}{8}$ in. (3 mm) for the Nd:glass, and from 0.020 in. (0.05 mm) for the Nd:YAG laser, depending on whether it is continuous wave or pulsed. The upper diameter limit is the usable diameter of a laser beam and is dependent on the power limitations.

The solid-state lasers use a single crystal made into a round rod approximately $\frac{3}{4}$ in. (19 mm) in diameter and approximately 8 in. (200 mm) long, as shown in Figure 8–32.

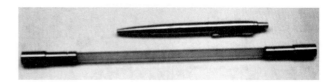

FIGURE 8–32 Nd:YAG crystal rod.

The end surfaces of the rod are ground flat and parallel and are polished to extreme smoothness. Both flat ends are covered with silver to reflect light; however, a small area in one end is left uncovered to allow the laser beam to exit from the rod. The solid-state rod is closely surrounded by a high-intensity light source, which is a flash tube with a xenon or krypton element. Figure 8–33 is a simplified diagram of a solid-state beam source. When the tube is flashed, it emits an intense pulse of light that lasts for approximately 2 msec. The high-intensity beam of coherent red light is emitted from the opening in the silver reflector on one end of the ruby rod. A burst of laser beam light, which lasts about 2 msec, occurs each time the flash tube is flashed. It is not possible to flash the ruby too often because of heat generated in the ruby crystal and in the flash tube. Thus it cannot operate continuously because of the heat buildup. The flash pulse durations are very short, and there is a relatively long period between pulses. The other two solid-state lasers operate in a similar manner.

The CO_2 laser is widely used for metalworking. The carbon dioxide laser uses gas that is a mixture of CO_2, helium, and nitrogen. Excitation of the gas laser is by means of high-voltage, low-current electric power. Lasers may use dc or ac, with ac being low or high frequency. The electrical discharge excites the CO_2 molecules, which on returning to their original energy state, emit photons. Mirrors are placed on both ends of the tube, one entirely reflective and the other with a small partially transmissive area to allow the beam to exit. This forms a cavity in which photons build up. The freed photons travel between the mirrors and excite the CO_2 molecules, starting a chain reaction of photon emissions. A stream of photons, the laser beam, exits through the unsilvered section of the one mirror.

The wavelength of the CO_2 laser beam is 10.6 μm. This wavelength is longer and does not pose quite the eye hazard of the shorter-wavelength solid-state lasers. Conventional safety eyewear can give satisfactory protection. The CO_2 gas laser can be operated in the continuous-wave or pulsed mode. A gas laser beam source is shown in Figure 8–34.

There are several types of CO_2 lasers. The early or lower-powered type used a "sealed tube" with power output of from 3 to 100 W. The fast axial-flow (FAF) type, which is a more complex method of producing laser beams, will produce from 50 W to 6 kW average output power. Another type, the high-power transfer-flow laser,

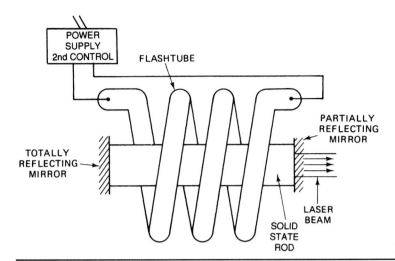

FIGURE 8–33 Solid-state laser.

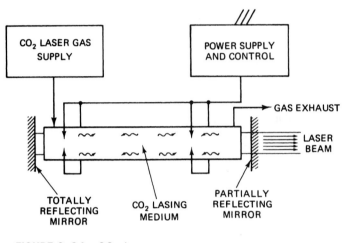

FIGURE 8–34 CO_2 laser.

is available up to 25 kW. The laser producing system is known as an oscillator. These different types of CO_2 lasers involve different designs of the beam-producing apparatus. The size of the apparatus and the efficiency of the system are different.

A summary of laser types and outputs is shown in Figure 8-35. There are two operating modes for lasers: the continuous-wave mode and the pulsing mode. These are similar to the operating modes of other welding processes.

Laser Beam Welding

The laser beam is very intense and unidirectional but can be focused and reflected in the same way as an ordinary light beam. The focus size is controlled by the choice of lenses and mirrors and the distance to the workpiece. The spot size can be made as small as 0.010 in. (0.2 mm) to up to $\frac{1}{2}$ in. (13 mm). The smaller focus spot size is used for cutting and welding, and the larger spot size is used for heat

treating. The laser beam can be used in open air and can be transmitted long distances with only minimal loss of power.

A block diagram of a **laser beam welding** system is shown in Figure 8-36. This shows the major components: the laser beam source (sometimes called the oscillator), the power supply, the cooling system, the gas supply for the laser beam source, the beam delivery system, the beam output coupling to the workpiece, the motion system for moving either the beam or the workpiece or both, the control system for the beam source and motion system and auxiliary systems, and the real-time monitor or feedback systems. Workpiece motion, parts handling, and workpiece motion feedback or monitoring are similar to those of other automatic welding systems, except that the accuracy of movement must be very precise.[10] The beam source power supply, cooling system, gas supply, and control system are particular to laser systems and are based on the types of laser, either solid state or gas, and the mode of operation, continuous or pulsed.

The beam delivery system must match the type of laser used. Fiber optics are used to transmit the short-wavelength laser beam from the solid-state lasers. The beam delivery system for the longer-wavelength laser, specifically the CO_2 laser beam, must use a lens and a mirror delivery system. The delivery system must match the type of laser and the particular application.

When using a laser beam for welding, the beam impinges on the surface of the base metal with such a concentration of energy that the surface is melted and volatilized. When the metal is raised to its melting temperature, the surface conditions have a minor effect on reflecting the beam.

The distance from the optical cavity to the workpiece has little effect on the laser. This is because it can be focused to the proper spot size at the work, with the same amount of energy available whether it is close or far away.

Use Factors	Solid-State Laser			Gas (CO_2) Laser
	Ruby	Nd:YAG (CW)	Nd:YAG (Pulsed)	
Wavelength (μm)	0.694	1.06	1.06	10.6
Continuous wave (CV)	No	Yes	No	Yes
Pulsed mode	Yes	No	Yes	Yes
Average power (W)	10–20	0.04–600	0.04–600	50–25,000
Beam diameter (mm)	1.5–25	1–6	5–10	1–10
Beam diameter (in.)	0.058–0.975	0.040–0.235	0.195–0.390	0.040–0.390
User for welding	Yes	Yes	Yes	Yes
Use for cutting	No	Yes	Yes	Yes

FIGURE 8–35 Summary of laser types for metalworking.

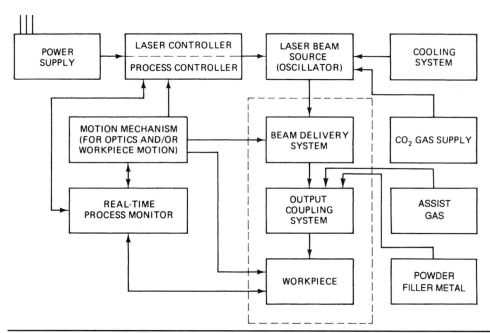

FIGURE 8–36 Block diagram of CO_2 laser system.

With laser welding, the molten metal takes on a radial configuration similar to plasma arc welding, known as "melt-in" or conduction mode welding. When the power density rises above a certain threshold level, keyholing occurs, the same as with plasma arc or electron beam welding. Keyholing provides for extremely deep penetration, which gives the weld a high depth-to-width ratio. Keyholing also minimizes the problem of beam reflection from the shiny molten metal surface since the keyhole behaves like a black body and absorbs the majority of the energy. For most applications, inert gas is used (the metal vapor in the weld area will ionize); an inert gas jet will minimize plasma formation. Plasma absorbs energy from the laser beam and can actually block the beam and reduce melting. This is overcome by using an inert-gas jet directed along the metal surface, which

eliminates the plasma buildup. It also shields the weld from the atmosphere.

The welding characteristics of the laser are similar to those of the electron beam. The laser can weld the same joint types. The concentration of energy by both beams is similar, with the laser having a power density on the order of 10^6 W $1 cm^2$. The power density of the electron beam is slightly greater.

The location of the focal point of the beam with respect to the surface of the workpiece is important. Maximum penetration occurs when the beam is focused slightly below the surface. Penetration is less when the beam is focused on the surface or deep within the surface. As power is increased, the depth of penetration increases.

Laser beam welding produces a tremendous temperature differential between the molten metal and the base metal immediately adjacent to the weld. Heating and cooling rates are much higher in laser beam welding than in arc welding, and the heat-affected zones are much smaller. Rapid cooling rates can create problems such as cracking in high-carbon steels.

The laser beam has been used to weld carbon steels, high-strength low-alloy steels, aluminum, stainless steel, and titanium. Laser welds are similar in quality to welds made by the electron beam process. Filler metal is used to weld metals that tend to show porosity when welded with either EB or LB welding.

Materials $\frac{1}{2}$ in. (12 mm) thick are being welded at a speed of 30 in. (760 mm) per minute. Figure 8-37 shows the top, underside bead, and the cross section of a laser weld made on stainless steel. This illustrates the characteristic cross section and the depth-to-width ratio of a laser weld. The welding speed for stainless steel of different thicknesses[11] welded with different-size power CO_2 laser welding machines is shown in Figure 8-39.

Laser beam welding for certain applications is very competitive. This is proven since LB welding is used for a special high-volume job in the automotive industry. Figure 8-38 shows a laser autogenous weld being applied by a robot on an auto muffler. The efficiency of laser beam welding equipment is 5% to 15%, but as new equipment becomes available this will increase. The cost of the equipment will decrease in the future.

8-4 HIGH ENERGY BEAM CUTTING

High energy beam cutting (HEBC). "A group of thermal cutting processes that severs or removes material by localized melting, burning, or vaporizing, of the workpieces using beams having high energy density." This is essentially electron beam cutting and laser beam cutting.

Electron Beam Cutting

Electron beam cutting (EBC) is "a thermal cutting process that severs metals by melting them with the heat from a concentrated beam composed primarily of high-velocity electrons impinging on the workpiece." The difference between electron beam welding and cutting is the heat input–heat output relationship. The electron beam generates heat in the base metal, which vaporizes the metal and allows it to penetrate deeper until the depth of the penetration, based on the power input, is achieved. In welding the electron beam actually produces a hole known as a keyhole. The metal flows around the keyhole and fills in behind. In the case of cutting, the heat input is increased so that the keyhole does not close.

All the metals that can be welded can also be cut. The quality of the cut surface is equal to the quality of a good oxyacetylene machine cut. The ability to shape a cut

---

FIGURE 8–37 Welds made in stainless steel.

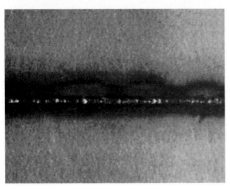
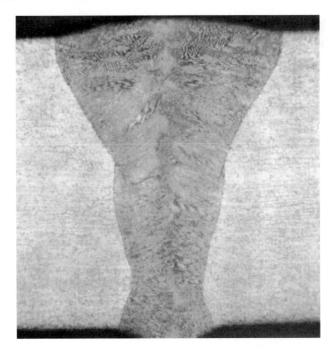

FIGURE 8–38 LB autogenous weld by a robot on auto muffler.

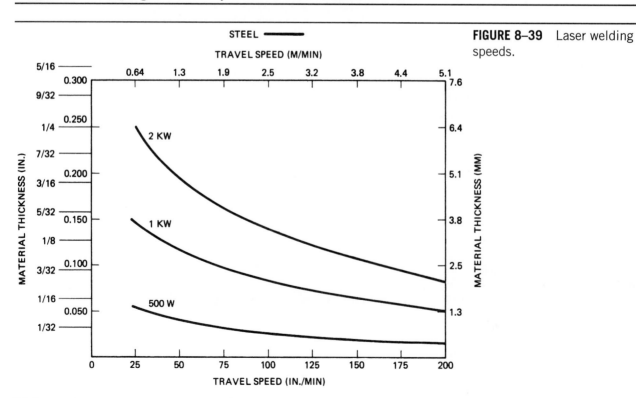

is limited only by the ability to move the work or the electron gun. The problems of electron beam cutting are greater than those of welding. The work must be in a vacuum. A large amount of volatilized metal will tend to plate out on the inside of the vacuum chamber. In view of these difficulties, the laser beam is replacing the electron beam for cutting.

Laser Beam Cutting

Laser beam cutting (LBC) is a thermal cutting process that severs metal by locally melting or vaporizing with the heat from a laser beam. The process is used with or without assist gas to aid removal of molten and vaporized material. There are several variations. In one case an inert gas jet assists the removal of molten and vaporized material. The other, laser beam oxygen cutting, is a variation that uses the heat from the chemical reaction between oxygen and the base metal at elevated temperatures. The necessary temperature is maintained with a laser beam. Other assist gases are compressed air and nitrogen. Figure 8-40 shows the Laser Beam Cutting Process.

The concentrated energy in the laser beam is only slightly less than the energy in the electron beam. The ability of both beams to cut materials is essentially the same. Laser beam cutting has many advantages over electron beam cutting. The laser beam can cut metal up to 1 in. (25 mm) in air. It can be used with automatic shape-cutting equipment at high travel speeds. The width of the laser beam cut is narrower and the angle of the cut is almost a perfect right angle. The quality of the cut surface is equal or superior to that of the best oxygen fuel gas cut surface. Figure 8-41 shows the materials that can be cut with a 400-W CO_2 laser and the speed of cutting. The CO_2

Material	Thickness		Cutting Speed	
	mm	in.	m/min	in./min
ABS plastic	4	0.157	4.5	177.1
Acrylic	6	0.236	1.7	66.9
Cardboard	0.1	0.004	96.0	3779.5
Ceramic tile	6.3	0.248	0.3	11.8
Formica	1.6	0.063	7.8	307.1
Galvanized steel	1	0.039	4.5	177.1
High-carbon steel	3	0.118	1.5	59.0
Mild steel	1	0.039	4.5	177.1
Plywood	18	0.708	0.5	19.7
Stainless steel	2.8	0.110	1.2	47.2
Titanium	3	0.188	4.1	161.4
Wool suit material	—	—	48.0	1889.7

FIGURE 8–41 Cutting speed of different materials using CO_2 laser. (From Ref. 12.)

Material	Laser type	
	1.06-µm Nd:YAG	10.6-µm CO_2
Mild steel	Excellent	Excellent
Stainless steel	Excellent	Excellent
Aluminum	Good	Good
Copper	Good	Difficult
Gold	Good	Not possible
Titanium	Good	Good
Ceramics	Fair	Good
Acrylics	Poor	Excellent
Polyethylene	Poor	Excellent
Polycarbonate	Poor	Good
Plywood	Poor	Excellent

FIGURE 8–42 Materials that can be cut with different types of lasers.

and the Nd:YAG lasers have different cutting abilities, depending on the absorption of their wavelength. Figure 8-42 describes the quality of cuts from the two types of lasers for different materials. Cutting power is usually between 500 and 2,000 W.

Precision laser cutting machines of various types are commercially available. The more common type uses a gantry frame that spans the workpiece. Others are similar to turret punches; in some cases the table moves, in others the head moves. They provide two axes of motion and are driven by a CNC controller. Software is available for nesting parts on material for minimum scrap loss. Figure 8-43 is a typical precision laser cutting machine.

Lasers are also used with robots. Figure 8-44 shows a Nd:YAG laser cutting thin sheet metal for a pickup truck

FIGURE 8–40 Process Diagram—Laser Cutting Beam.

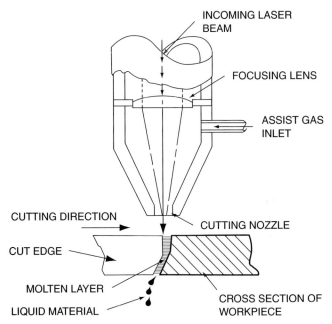

INCOMING LASER BEAM

FOCUSING LENS

ASSIST GAS INLET

CUTTING DIRECTION

CUT EDGE

CUTTING NOZZLE

MOLTEN LAYER

LIQUID MATERIAL

CROSS SECTION OF WORKPIECE

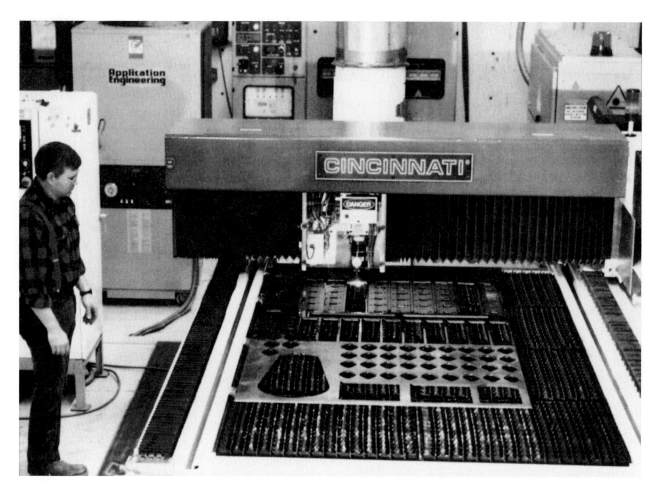

FIGURE 8–43 Typical laser shape-cutting machine.

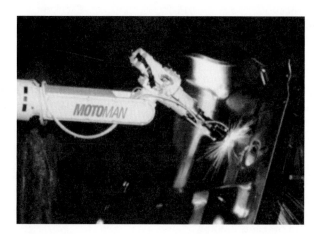

FIGURE 8–44 Robot cutting with Nd:YAG laser.

floor panel. In this case a proximity sensor is employed to maintain constant work-to-beam dimensions.

Typical laser-cut parts are shown in Figure 8–45. The dimensional accuracy is better than for oxyfuel gas cutting. The edges of a laser cut are square and sufficiently smooth that additional finishing is not necessary (Figure 8–46).

The power required for laser cutting is relatively low. In general, the continuous-wave CO_2 laser with up to 1 kW of power is sufficient to cut thin-gauge metals. The data given in Figure 8–47 show the travel speed of a laser beam for cutting different metals and thicknesses. This shows two different-size laser machines and uses a jet of oxygen to improve cutting speed.[13] Sharp corners, smooth surfaces, narrow cut width, minimum thermal damage, nonadherent dross, and 90° surfaces are all achieved with laser cutting. The laser beam can be used for drilling holes by using the cutting technique but without travel.

Laser applications in metalworking will continue to increase. As more powerful lasers are produced with higher efficiencies, there will be more welding applications for lasers. Lasers will be used for localized surface heat treating because of the accurate control possible. Lasers are being used for surfacing with the addition of powder as the filler metal instead of wire. In these applications, the depth of penetration can be very accurately controlled. The laser is also being used for fusing thermal spray surfaces to improve the density of the sprayed material and to create fusion with the substrate. It is expected that the use of lasers will greatly increase in the next few years.

FIGURE 8–45 Typical parts cut by a laser.

FIGURE 8–46 Surface of cut edges.

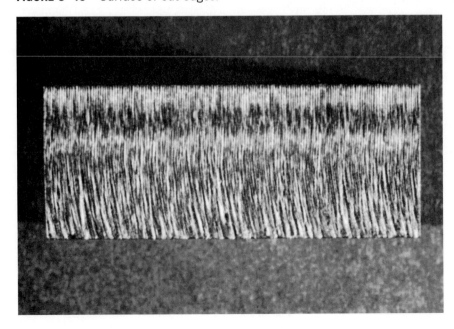

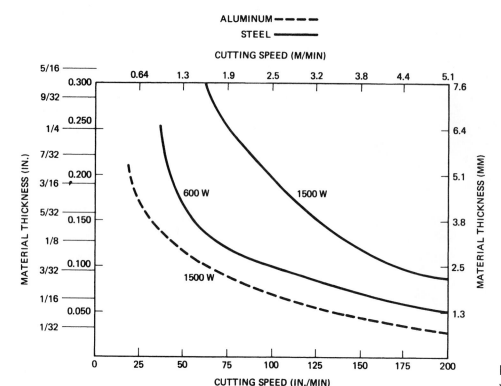

FIGURE 8-47 Cutting speed versus thickness.

8–1. What creates heat in a resistance weld?

8–2. Explain the difference between a resistance spot weld and a projection weld.

8–3. Resistance welding electrodes are usually made of what materials? Why?

8–4. Is filler metal required for making a resistance weld?

8–5. How does seam welding differ from spot welding? Is a resistance seam weld always watertight?

8–6. High-frequency resistance welding is regularly used for making what products?

8–7. What makes flash butt welding different from other resistance welding processes?

8–8. What is similar between an electron beam welding machine and a television set? Explain.

8–9. Why is electron beam welding recommended for hard-to-weld metals?

8–10. Explain the difference between electron beam welding in a hard vacuum, a soft vacuum, and in the atmosphere.

8–11. Why is precision joint preparation required for electron beam welding square butt joints?

8–12. What are the advantages and disadvantages of electron beam welding in the air?

8–13. What type of electron beam welding equipment generates x-rays?

8–14. What is the problem with electron beam cutting?

8–15. What does *laser* stand for?

8–16. What safety precautions should be taken when working around lasers?

8–17. Can all laser beams be transmitted by fiber optics?

8–18. What is the advantage of the CO_2 laser?

8–19. Can lasers cut nonmetals? What materials can be cut besides metals?

8–20. What is the advantage of laser welding over electron beam welding?

REFERENCES

1. R. D. Enquist, "How Easy Can You Join Metals by Resistance Spot Welding?" *Iron Age* (August 10, 1961).

2. "Resistance Welding Equipment Standards," Bulletin 16, Resistance Welders Manufacturers Association, Philadelphia, Pa.

3. "Resistance Welding Manual," Resistance Welders Manufacturers Association, Philadelphia, Pa.

4. J. A. Stohr and J. Eriola, "Vacuum Welding of Metals," *Welding and Metal Fabrication* (October 1958).

5. J.W. Meier, "High Power Density Electron Beam Welding of Several Materials," 2nd International Vacuum Congress, Washington, D.C., Oct. 16–19, 1961.

6. "Electron Beam Welding," Point Paper, U.S. Navy, Naval Air Station, North Island, San Diego, Calif.

7. A. L. Schawlow, *IEEE Transactions on Electron Devices,* Vol. ED23, 1976.

8. J. L. Bromberg, "The Construction of the Laser," *Laser Topics* (October 1985).

9. G. K. Klauminzer, "Twenty Years of Commercial Lasers—A Capsule History," *Laser Focus/Electro-Optics* (December 1984).

10. B. F. Kuvin, "Laser and Electron Beams for Deep, Fast Welding," *Welding Design and Fabrication* (August 1985).

11. D. A. Belforte, "Laser in Production Operations," Society of Manufacturing Engineers, Clearwater Beach, Fla.

12. D. A. Belforte, "Cutting with CO_2 Laser," *Metal Progress* (September 1974).

13. "CO_2 Laser Cutting," Technical Note, Spectra-Physics, San Jose, Calif.

WELDING-RELATED PROCESSES

9-1 OXYGEN CUTTING

Oxygen cutting (OC) is a group of thermal cutting processes that severs or removes metal by means of the chemical reaction between oxygen and the base metal at elevated temperature. The necessary temperature is maintained by the heat from an arc, an oxyfuel gas flame, or other source. In the case of oxidation-resistant metals the reaction is facilitated by the use of a chemical flux or metal power. Five basic processes are involved: (1) oxyfuel gas cutting, (2) metal powder cutting, (3) chemical flux cutting, (4) oxygen lance cutting, and (5) oxygen arc cutting. Each of these processes is different and will be described.

Oxyfuel Gas Cutting

Oxyfuel gas cutting (OFC) is "a group of oxygen-cutting processes that uses heat from an oxyfuel gas flame." The necessary temperature is maintained by means of gas flames obtained from the combustion of a fuel gas and oxygen. When an oxyfuel gas cutting operation is described, the fuel gas must be specified. There are a number of fuel gases used. The most popular is acetylene. Natural gas is widely used, as is propane, methyl-acetylene-propadiene stabilized, and various trade name fuel gases. Hydrogen is rarely used. Gasoline can even be used but is not popular. Each fuel gas has its particular characteristics and may require slightly different apparatus because of these characteristics. The characteristics relate to the flame temperatures, heat content, oxygen fuel gas ratios, and so on. The general concept of oxyfuel gas cutting is similar no matter what fuel gas is used. It is the oxygen jet that makes the cut in steel, and cutting speed depends on how efficiently the oxygen reacts with the steel. Oxygen for cutting must be 99% pure. If purity is less, cutting speed and efficiency will be reduced. For simplicity, we confine our discussion to the use of acetylene.

The generation of heat by combustion of acetylene and oxygen is used to bring the base metal steel up to its kindling temperature, where it will ignite and burn in an atmosphere of pure oxygen. The chemical formulas for three of the oxidation reactions are as follows:

$$2Fe + O_2 = FeO + heat\ (270\ kJ)$$

$$3Fe + 2O_2 = Fe_3O_4 + heat\ (1100\ kJ)$$

$$2Fe + 1.5O_2 = Fe_2O_3 + heat$$

At elevated temperatures all of the iron oxides are produced in the cutting zone.

The oxyacetylene cutting torch is used to heat the steel by increasing the temperature to its kindling point and then by introducing a stream of pure oxygen to create the burning or rapid oxidation of the steel. The stream of oxygen also assists in removing the material from the cut (Figure 9–1).

Steel and a number of other metals are flame cut with the oxyfuel gas cutting process. The following conditions must apply:

1. The melting point of the material must be above its kindling temperature in oxygen.
2. The oxides of the metal should melt at a lower temperature than the metal itself and below the temperature that is developed by cutting.
3. The heat produced by the combustion of the metal with oxygen must be sufficient to maintain the oxygen cutting operation.
4. The thermal conductivity must be low enough so that the material can be brought to its kindling temperature.
5. The oxides formed in cutting should be fluid when molten so as not to interrupt the cutting operation.

Iron and low-carbon steel fit all the listed requirements and are readily oxygen flame cut. Cast iron is not readily flame cut, because the kindling temperature is above the melting point. It also has a refractory silicate oxide, which produces a slag covering. Chrome-nickel stainless steels cannot be flame cut with the normal technique because of the refractory chromium oxide formed on the surface. Nonferrous metals such as copper and aluminum have refractory oxide coverings, which prohibit normal oxygen flame cutting; in addition, they have a high thermal conductivity.

When flame cutting, the preheating flame should be neutral or oxidizing. A reducing or carbonizing flame should not be used. Figure 9–2 shows the flame cutting operation being manually performed. The schedule for flame cutting clean mild steel is shown in Figure 9–3.

Torches are available for either welding or cutting. By placing the cutting torch attachment on the torch body, it is used for manual flame cutting. Figure 9–4 shows two styles of manual oxyacetylene flame cutting torches. Various sizes of tips can be used for manual flame cutting. The numbering system for tips is not standardized and most manufacturers use their own tip number system. Each system is, however, based on the size of the oxygen cutting orifice of the tip. These are related to drill sizes. Different tip sizes are required for cutting different thicknesses of carbon steel. The manual cutting torch and oxygen and acetylene cylinders and regulators are shown in Figure 9–5. Semiautomatic cutting is shown in Figure 9–6. Mechanized cutting is shown in Figure 9–7.

Many special cutting tips for specific applications are available. These are used for flame gouging, to remove weld defects, and for edge preparation. Other tips are designed

---

FIGURE 9–2 Manual oxyacetylene cutting operation.

FIGURE 9–1 Process diagram for oxygen cutting.

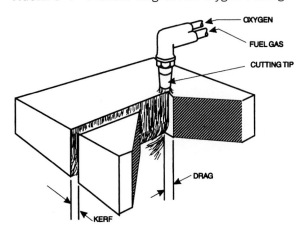

OXYGEN

FUEL GAS

CUTTING TIP

DRAG

KERF

Material Thickness			Cutting Orifice Diameter (Center Hole)		Approx. Gas Pressure (psi)		Travel Speed (in./min)	
in.	mm	Drill Size	in.	mm	Acetylene	Oxygen	Manual	Mechanized
$\frac{1}{8}$	3.2	60	0.040	1.0	3	10	20–22	22
$\frac{1}{4}$	6.4	60	0.040	1.0	3	15	16–18	20
$\frac{3}{8}$	9.5	55	0.052	1.3	3	20	14–16	19
$\frac{1}{2}$	12.7	55	0.052	1.3	3	25	12–14	17
$\frac{3}{4}$	19.0	55	0.052	1.3	4	30	10–12	15
1	25.4	53	0.060	1.5	4	35	8–11	14
$1\frac{1}{2}$	38.1	53	0.060	1.5	4	40	6–$7\frac{1}{2}$	12
2	50.8	49	0.073	1.9	4	45	$5\frac{1}{2}$–7	10
3	76.2	49	0.073	1.9	5	50	5–$6\frac{1}{2}$	8
4	101.6	49	0.073	1.9	5	55	4–5	7
5	127.0	45	0.082	2.1	5	60	$3\frac{1}{2}$–$4\frac{1}{2}$	6
6	152.4	45	0.082	2.1	6	70	3–4	5
8	203.2	45	0.082	2.1	6	75	3	4

FIGURE 9–3 Schedule for oxyacetylene flame cutting of carbon steels.

for rivet head removal, for removing risers from castings, and for shaping surfaces. A variety of tips are available, and they can be used with either manual or mechanized equipment. Surfaces of flame cut edges are shown by Figure 9–8.

Metal Powder Cutting

Metal powder cutting (POC) is "an oxygen cutting process that uses heat from oxyfuel gas flame, with iron or other metal powder to aid cutting." This process is used for cutting cast iron, chrome-nickel stainless steels, and some high-alloy steels. The process uses finely divided material, usually iron powder, added to the cutting-oxygen stream. The powder is heated as it passes through the oxyacetylene preheat flames and almost immediately oxidizes or ignites in the stream of the cutting oxygen.

The oxidation or burning of the iron powder provides a much higher temperature in the oxygen stream.

FIGURE 9–4 Two styles of manual cutting torches.

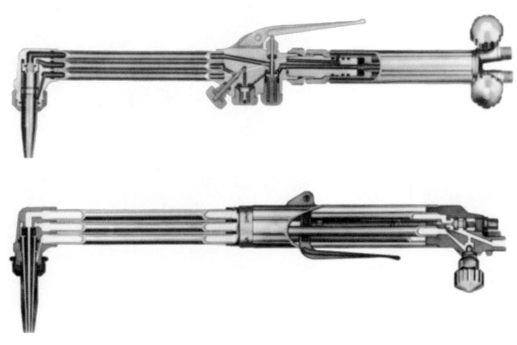

FIGURE 9–5 Complete manual oxygen cutting equipment.

FIGURE 9–6 Semiautomatic oxygen cutting equipment.

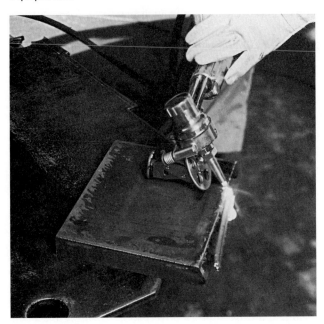

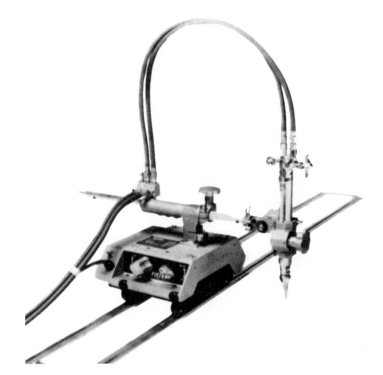

FIGURE 9–7 Mechanized oxygen cutting machine.

This plus the chemical reaction in the flame allows the cutting-oxygen stream to oxidize the metal being cut continuously in the same manner as when cutting carbon steels. The use of iron powder in the oxygen stream makes it possible to start cuts without preheating the base material. Powder cutting has found its broadest use in cutting cast iron and stainless steel. The same basic process can be used for scarfing to condition billets in steel mills. It creates large amounts of smoke.

Cutting speeds and cutting oxygen pressures are similar to those used when cutting carbon steels. For heavier material over 1 in. thick (25 mm), a nozzle one size larger should be used. Powder tends to leave a scale on the cut surface, which can easily be removed as the surface cools. This is a special application and is used only where required.

Flux Cutting

Flux cutting (FOC) is "an oxygen cutting process that uses heat from an oxyfuel gas flame, with a flux in the flame to aid cutting." Powdered chemicals are used in the same way as iron powder is used in the metal powder cutting process. This process is sometimes called flux injection cutting. It is of minor industrial significance.

Oxygen Lance Cutting

Oxygen lance cutting (LOC) is "an oxygen cutting process that uses oxygen supplied through a consumable lance. The preheat to start the cutting is obtained by other

CORRECT CUT

Cutting lines are almost vertical and not very pronounced. Edges are square. Little slag evident.

COMMON CUTTING MISTAKES

PREHEAT FLAMES TOO SMALL

Bottom half uneven and wavy.

OXYGEN PRESSURE TOO LOW

Top edge and cutting lines uneven.

PREHEAT FLAMES TOO LARGE

Top edge badly melted. Middle section is smooth. Slag evident at bottom.

OXYGEN PRESSURE TOO HIGH AND/OR NOZZLE SIZE TOO SMALL

Deep gouges into sides of cut due to lack of control.

CUTTING SPEED TOO SLOW

Upper edge melted. Cutting lines rather coarse.

TORCH TRAVEL UNSTEADY

Cutting lines erratic and uneven.

CUTTING SPEED TOO FAST

Cutting lines curve in opposite direction of travel. Cut edge irregular.

CUTTING WAS LOST

Cut stopped at gouged areas. Usually due to excess travel speed or insufficient preheating.

FIGURE 9–8   Surface of flame-cut edges.

means." This is sometimes called oxygen lancing. The oxygen lance is a length of pipe or tubing used to carry oxygen to the point of cutting. It uses a small ($\frac{1}{8}$ or $\frac{1}{4}$ in. nominal) black iron pipe connected to a suitable handle, which contains a shutoff valve. This handle is connected to the oxygen supply hose. The main difference between the oxygen lance and an ordinary flame cutting torch is that there is no preheat flame to maintain the material at the kindling temperature. The lance is consumed as it makes a cut. The principal use of the oxygen lance is the cutting of hot metal in steel mills. The steel is sufficiently heated so that the oxygen will cause rapid oxidation and cutting to occur. The end of the oxygen lance becomes hot and supplies iron to the reaction to maintain the high temperature.

9-2 ARC AND PLASMA CUTTING

The arc and plasma cutting processes are a group of thermal cutting processes that severs or removes metal by melting with the heat of an arc between an electrode and the workpiece. This group includes oxygen arc cutting, air carbon arc cutting, metal arc cutting, gas tungsten arc cutting, and plasma arc cutting. Most of these processes are applied manually, but some are used semiautomatically and others may be automated. Some of these processes may be used for underwater cutting.

Oxygen Arc Cutting

Oxygen arc cutting (AOC) is an oxygen cutting process that uses an arc between the workpiece and a consumable tubular electrode that directs oxygen to the workpiece. These are also known as oxyarc cutting processes. There are several variations. They all utilize a special electrode holder, compatible with the size of electrode used, that incorporates oxygen input and valve for controlling the oxygen. However, the different variations depend on the design of electrode utilized. The major difference is whether or not a power source is used to start and to maintain the arc. When a power source is used, it is a conventional ac or dc constant current welding machine. Different variations have different cutting abilities. In general, the high temperature heat source is an arc between the electrode and the metal to be cut. As soon as the arc is established, the valve in the electrode holder is depressed and oxygen is fed through the tubular electrode to the arc. The oxygen causes ferrous metals to oxidize (burn) and the stream of oxygen helps remove material from the cut. Steel from the electrode plus flux, when used, combine to create so much heat that thermal conductivity cannot remove the heat quickly enough to extinguish the oxidation reaction. This process will cut aluminum, copper, brasses, titanium, nickel alloys, cast iron, stainless, as well as carbon and low-alloy steels. The quality of the cut is not as good as the quality of the oxygen fuel gas cut on mild steel, but is sufficient for many applications. Material from $\frac{1}{4}$ to 3 inches (6.4 to 75 mm) can be cut. Electric current ranges from 150 to 250 amps and oxygen pressure from 3 to 60 psi (21 to 414 kPa) may be used. Electrodes are normally $\frac{3}{16}$ in. (4.8 mm) or $\frac{1}{4}$ in. (6.4 mm) in diameter and usually 18 in. (450 mm) long. The design of the electrode is the key to the cutting application. In all cases, when the flow of oxygen is stopped, the cut will stop. The process is used for salvage and repair work.

The main variations between the oxygen arc cutting processes are whether a power source is used or is not used continuously. Those electrodes that are used without the continuous power source are called exothermic electrodes. These electrodes usually consist of a tubular member normally made of steel and contain internal wires, which may be aluminum or magnesium or their alloys, or steel. The tube may be coated or not. In some cases the tubular electrode may be of a nonferrous metal. In operation, the oxygen flows through the tubular electrode and is ignited by an arc, a spark, or by a flame. The reaction of the pure oxygen with the sheath and the wires inside creates an exothermic reaction. This produces a very high temperature in the order of 10,000°F. A tremendous amount of heat produced is sufficient to melt all metals, composites, and nonmetals, such as concrete and bricks. This composite rod or tube can be used to cut concrete or masonry. It will also cut slag, rocks, and other nonmetals. It can also be used underwater. Rods of this type are known by various trade names such as Oxy-Lance, Hot-Rod, Slice, Prime Cut, and Burning Bar. Extra special safety precautions should be taken when using these type of rods due to high heat and smoke produced.

Air Carbon Arc Cutting

Air carbon arc cutting (CAC-A) is "a carbon arc cutting process variation that removes molten metal with a jet of air." It is also used for gouging.

A high-velocity air jet parallel to the carbon electrode strikes the molten metal puddle just behind the arc and blows the molten metal out of the immediate area. Figure 9–9 shows the arc between the carbon electrode

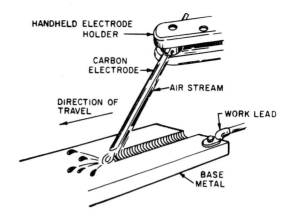

FIGURE 9–9 Process diagram for air carbon arc cutting or gouging.

and the work and the airstream parallel to the electrode from the special electrode holder.

The air carbon arc cutting process is used to cut metal, to gouge out defective metal, to remove inferior welds, for root gouging, and to prepare grooves for welding. Air carbon arc cutting is used when slightly ragged edges are not objectionable. The area of the cut is small, and, since the metal is melted and removed quickly, the surrounding area does not reach high temperatures. This reduces the tendency towards distortion and cracking.

Air carbon arc cutting and gouging is normally manually operated. The apparatus can be mounted on a travel carriage and considered machine cutting or gouging. Special applications have been made where cylindrical work has been placed on a lathe and rotated under the air carbon arc torch.

The air carbon arc cutting and gouging process can be used in all positions. The overhead position requires a high degree of skill. The process can be used for cutting or gouging most of the common metals.

The process is not recommended for weld preparation for stainless steel, titanium, zirconium, and other similar metals without subsequent cleaning. Cleaning, by grinding, must remove all of the surface carbonized material adjacent to the cut. The process can be used to cut most materials for scrap.

The circuit diagram for air carbon arc cutting or gouging is shown in Figure 9–10. Normally, conventional welding machines with constant current are used. Constant voltage can be used. When using a CV power source, precautions must be taken to operate it within its rated output. Alternating current power sources having conventional drooping characteristics can also be used for special applications. Ac carbon electrodes must be used.

Equipment required is shown by the block diagram. Special heavy-duty high-current machines have been made specifically for the air carbon arc process. This is because of extremely high currents used for the large carbon electrodes.

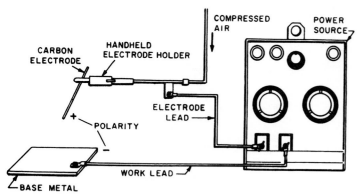

FIGURE 9–10 Circuit diagram, CAC-A.

The electrode holder shown in Figure 9–11 is designed for the air carbon arc process. The holder includes a small circular grip head, which contains the air jets for directing the compressed air along the electrode. It also has a groove for gripping the electrode. This head can be rotated to allow different angles of the electrode. A heavy electrical lead and an air supply hose are connected to the holder through a terminal block. A valve is included for turning the compressed air on and off. Holders are available in several sizes, depending on the duty cycle of the work performed, the welding current, and size of carbon electrode used. For extra-heavy-duty work, water-cooled holders are used.

The air pressure ranges from 80 to 100 psi (550 to 690 kPa). The volume of compressed air required ranges from as low as 5 ft^3/min (2.5 liters/min) up to 50 ft^3/min (24 liters/min) for the largest carbon electrodes. A 1-horse-power compressor will supply sufficient air for smaller electrodes and light duty. It will require up to a 10-horse-power compressor when using the largest electrodes. See Figure 9–12 for requirements.

The carbon graphite electrodes are made of a mixture of carbon and a graphite plus a binder that is baked to produce a homogeneous structure. Electrodes come in several types. The plain, uncoated electrode is less expensive, carries less current, and starts easier. The copper-coated electrode provides better electrical conductivity

between it and the holder. The copper-coated electrode is better for maintaining the original diameter during operation: It lasts longer and carries higher current. Copper-coated electrodes are of two types: the dc type and the ac type. The composition ratio of the carbon and graphite is slightly different for these two types. The dc type is more common. The ac type contains special elements to stabilize the arc. The ac electrode is used for direct current electrode negative when cutting cast irons. For normal use, the electrode is operated with the electrode positive. Electrodes range in diameter from $\frac{5}{32}$ in. (4.0 mm) to 1 in. (25.4 mm). Electrodes are normally 12 in. (300 mm) long; however, 6-in. (150-mm) electrodes are available. Copper-coated electrodes with tapered socket joints are available for automatic operation and allow continuous operation.

To make a cut or a gouging operation, the cutter starts the airflow and then strikes the arc. The electrode is pointed in the direction of travel with a push angle approximately 45° with the axis of the groove. The speed of travel, the electrode angle, and the electrode size and current determine the groove depth. Electrode diameter determines the groove width.

The safety precautions used for carbon arc welding and shielded metal arc welding apply to air carbon cutting and gouging. Two other precautions must be observed. First, the air blast will cause the molten metal to travel a very long distance. Metal deflection plates should be placed in front of the gouging operations. All combustible materials should be moved away from the work area. At high-current levels the mass of molten metal removed is quite large and will become a fire hazard if not properly contained.

Second is the high noise level. At high currents with high air pressure, a very loud noise occurs. Ear protection (earmuffs or earplugs) must be worn by the arc cutter.

The process is widely used for back gouging, for preparing joints, and for removing defective weld metal. It is also used in foundries for washing pads, removing risers, and removing defective areas of castings. Another major use is scrap preparation of metals to reduce them to proper size for handling. It is also used for maintenance and salvage.

FIGURE 9–11 Electrode holder for CAC-A.

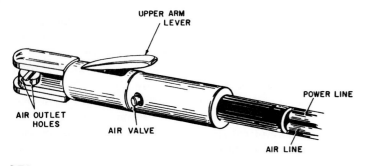

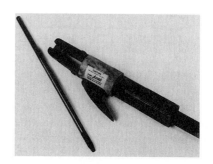

| Type of Torch | Air Hose Inside Diameter (in.) | Air Pressure | | Air Consumption (ft³/min) | Compressor Rating (hp) | |
		psi	kPa		Intermittent Use	Continuous Use
Light duty	¾	49	338	8	0.5	1.5
General duty	⅜	80	552	25	5	7.5
Heavy duty	½	80	552	33	7.5	10
Automatic	½	60	552	46	—	15

FIGURE 9–12 Recommended minimum compressed air requirements for cutting.

Carbon Arc Cutting

Carbon arc cutting (CAC) is "an arc cutting process that uses an arc between a carbon electrode and the weld pool." The process is similar to air carbon arc cutting except that the air blast is not employed. The process depends strictly on the heat input of the carbon arc to cause the metal to melt. The molten metal falls away by gravity to produce the cut. The process is relatively slow, a very ragged cut results, and it is used only when other cutting equipment is not available.

Shielded Metal Arc Cutting (SMAC)

Shielded metal arc cutting is an arc cutting process that uses a covered electrode.

The equipment required is identical to that used for shielded metal arc welding. When the heat input into the base metal exceeds the heat losses, the molten metal pool becomes large and unmanageable. If the base metal is not too thick, the molten metal will fall away and create a hole or cut. The cut produced by the shielded metal arc cutting process is rough and is not normally used for preparing parts for welding. The metal arc cutting process using covered electrodes is used only where a small cutting job is required and other means are not available, or in an emergency.

For metal arc cutting, an electrode with deep penetrating qualities such as an E6010 or an E6011 should be used. A relatively small electrode should be used with a dc electrode negative. The current should be set much higher than normally used for welding. This will create a maximum amount of heat in the weld pool, which will soon fall away making the cut. This technique can also be used for cutting cast iron. On thick material a sawing action is required to make the cut and to allow the molten metal to fall away. If the electrode coating is made wet by dipping in water, the electrode will melt more slowly so that more cut can be obtained per electrode.

The metal arc cutting technique can also be used for gouging when utilizing special electrodes. These special electrodes have an insulating coating that directs the arc. The technique is sometimes used for back gouging welds prior to making the backing weld.

Gas Tungsten Arc Cutting

Gas tungsten arc cutting (GTAC) is "an arc cutting process that uses a single tungsten electrode with gas shielding." The apparatus for gas tungsten arc cutting is identical to that used for gas tungsten arc welding. The gas tungsten arc is used to provide high heat input so that the molten pool will become large and unmanageable and will fall away. Gas tungsten arc cutting is used for thinner metals and can be used for cutting almost any metal. The smoothness of the cut is dependent on the skill of the arc cutter.

This process has largely been supplanted by plasma arc cutting and is of little industrial significance.

Plasma Arc Cutting

Plasma arc cutting (PAC) is "an arc cutting process that uses a constricted arc and removes the molten metal with a high-velocity jet of ionized gas issuing from the constricting orifice." It is shown by Figure 9–13.

FIGURE 9–13 Process Diagram—Plasma Arc Cutting.

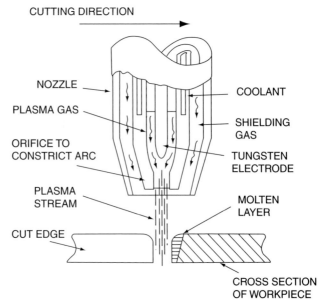

CUTTING DIRECTION

NOZZLE

PLASMA GAS

ORIFICE TO CONSTRICT ARC

PLASMA STREAM

CUT EDGE

COOLANT

SHIELDING GAS

TUNGSTEN ELECTRODE

MOLTEN LAYER

CROSS SECTION OF WORKPIECE

There are two major variations: (1) the low-current plasma cutting system, which normally uses air for the plasma and is usually manually applied; and (2) the high-current plasma cutting system, which normally uses nitrogen for the plasma and is usually applied automatically. A variation of the high-current plasma system utilizes water to improve the quality of the cut. Low- and high-current plasma cutting is shown in Figure 9–14.

The operation of plasma cutting is very similar to the keyhole mode of plasma welding. For cutting, the keyhole is not allowed to close. Heat input of the plasma arc is very high and the heat losses cannot carry the heat away quickly enough. The high-velocity jet of the plasma blows away the molten metal and produces the cut.

Plasma arc cutting has less of a detrimental metallurgical effect on the base metals than does oxygen cutting. Welding operations can often be made directly over plasma-cut edges in aluminum and stainless steel. The very thin oxide film on aluminum and on stainless steel edges is not detrimental and cannot be detected in the weld joint. In addition, it does not adversely affect the mechanical properties of the joint and cannot be seen in metallurgical examinations.

The plasma arc cutting process can be used in all positions. It can be used for piercing holes and for gouging, as shown in Figure 9–15. The piercing capacity is usually half the cutting thickness capacity. In addition, plasma arc cutting can be used to cut metals underwater.

The plasma arc welding torch described previously is modified for cutting. It uses the pilot arc or nontransferred mode as well as the transferred mode during cutting. It can be used to cut all metals. Due to its high tem-

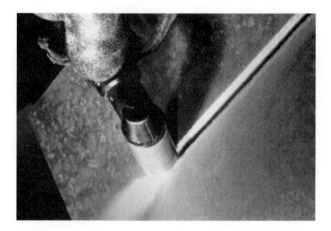

FIGURE 9–15 Plasma gouging steel.

perature, which approaches 30,000° F (16,650° C), high-melting-temperature oxides, which coat many metals, do not interfere with the operation. The torch will cut metals that cannot be cut with the oxygen cutting process since flame temperatures only approach 5600° F (3100° C). The metals cut with plasma arc cutting include aluminum, brass, bronze, copper, galvanized steel, coated or painted steel, mild steel, and stainless steel.

Plasma arc cutting can be used for interrupted cutting, contour cutting, and stack cutting. It is more efficient than oxyacetylene cutting for stack cutting.

The surface quality of the cut edge is equal to or better than that of oxyfuel gas cutting. The kerf is normally narrower than for oxyacetylene cutting and the angle of the cut is approximately 90°.

FIGURE 9–14 Plasma arc cutting: (a) manual, (b) mechanized.

(a)

(b)

Low-Powered Plasma Arc Cutting In the low-powered plasma arc cutting variation, the maximum current is normally 125 A. The torch is relatively small, usually air cooled, and manually operated. Air is usually used for the plasma. Low-powered units can be used with mechanized cutting systems and can utilize special devices to aid in manual cutting (Figure 9–16). This shows a small circle-cutting attachment which improves the quality of the cut surface. Straight-line attachments are also used. In all cases the cutting speed is greater than the cutting speed of oxyacetylene cutting. The cutting speed versus material thickness for low-powered plasma arc cutting is shown in Figure 9–17 using air.

Special plasma arc cutting equipment packages are available. These range in size from 30 to 120 A. Some of the power sources do not have current adjustments, but they all include power contactors since the maximum open-circuit voltage can be as high as 375 V. This keeps the open-circuit voltage from being present at the machine terminals or in the torch. The package may also include the torch and cable, plasma gas regulator, and solenoid valve. The torches are air cooled, and the machines may be adjusted for 20, 30, 40, 60 A, or higher current output.

The low-powered plasma systems are used in place of oxyacetylene cutting torches for manual cutting in maintenance and repair. One of the popular applications has been auto body repair and other sheet metal work.

High-Powered Plasma Arc Cutting High-powered plasma arc cutting uses 100 to 500 A of current. In high-powered cutting, automatic shape cutting machines are used. The equipment is similar to that used for oxygen flame cut-

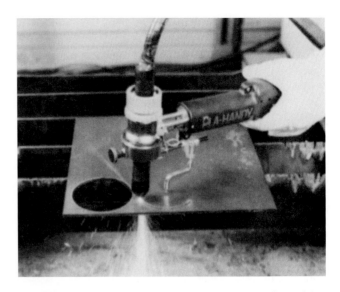

FIGURE 9–16 Plasma semiautomatic arc cutting with circle attachment.

ting. A higher travel speed is required. Motion is controlled in the same manner as in oxygen cutting systems.

The heavy-duty plasma torches are water cooled. They will fit in the same torch holders on automatic oxygen flame cutting machines. A water spray is sometimes used to surround the plasma to reduce smoke and noise. Worktables containing water, which is in contact with the underside of the metal being cut, also reduce noise and smoke. In some cases the water surface is above the top surface of the metal. The cutting speed versus material

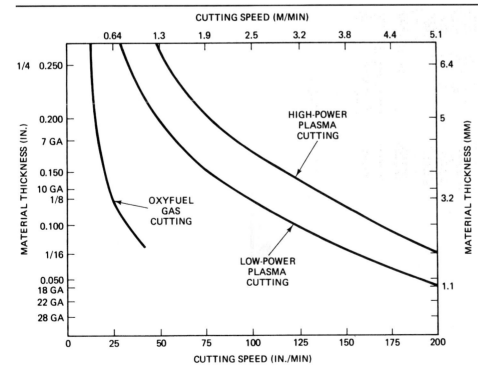

FIGURE 9–17 Plasma cutting speed on steel using air.

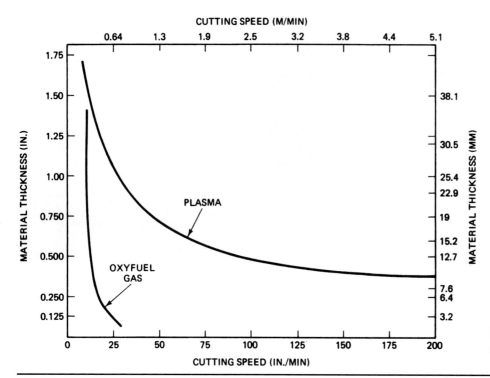

FIGURE 9–18 Plasma cutting steel plate material, air for plasma.

thickness for high-powered plasma arc cutting is shown in Figure 9–18. More information concerning this process variation is in the AWS "Recommended Practices for Plasma Arc Cutting."[1]

Control of the motion device, gas flow, and the power source is in a central control system, which may be computer driven. Nitrogen is usually used for high-current plasma systems; however, in some cases, 80% argon plus 20% hydrogen is used. The plasma gas should be matched to the work being cut.

Water Injection Variation The water injection variation is an effort to reduce fumes and smoke produced by the high-powered plasma arc cutting process. Use of water in the plasma improves the quality of the cut.

Safety Considerations The noise level generated by the high-powered equipment is uncomfortable. The cutter should wear ear protection. For low-power cutting, special ear protection is not required.

For high-powered cutting, local exhaust is required since large amounts of fumes or particulate matter are generated. High-powered cutting should be done over a water reservoir so that the molten metal removed from the cut will fall in the water and help reduce the amount of fumes released into the air.

The normal protective clothing to protect the cutter from arc and molten metal should be worn. The helmet should be supplied with a No. 9 filter glass lens for heavy-duty cutting, although lighter shades can be worn for low-powered cutting. Other safety considerations as outlined in Chapter 4 should be followed.

9-3 WATER JET CUTTING

Abrasive water jet cutting (AWJ) is not a thermal cutting process. It does not use the oxidation of metal similar to oxygen fuel gas cutting, nor does it use high-temperature melting. Instead, it cuts material by rapid erosion. It utilizes a very high-pressure jet of water, which attains a high speed of mach 2.5. When this high-speed jet of water, which sometimes includes finely ground abrasive materials, hits the material to be cut, it erodes the material rapidly and produces a cut very similar to a flame cut surface. The resistance to erosion is measured in hardness, toughness, and high tensile-strength. This high-speed jet of water operates at a pressure of up to 60,000 (4,140 bar) psi flowing through an orifice of from 0.003 in. (0.076 mm) up to 0.020 in. (0.508 mm). This jet of water is directed to the surface of the material to be cut in the same manner as an oxygen fuel gas cutting torch. Abrasive water jet cutting competes with the thermal processes and is being applied by automatic shape-cutting machines. The stream of water acts much in the same manner as the stream of oxygen in thermal cutting and it will produce contour-shaped cut parts.

Water jet cutting has many advantages over machining or contact cutting systems. No lateral force is involved and cutting tool wear is not a problem. It has some unique advantages over thermal cutting processes. It is unlike oxyfuel gas cutting, which depends on the oxidation (or burning) of steel. It is also unlike arc cutting, which depends on melting and blowing away the molten metal. It is also unlike high-energy beam cutting, which depends on vaporizing the material to be removed. Heat is not involved and therefore metallurgical changes do not occur on the cut surface. In addition, with no heat, thermal distortion is not a problem. It will cut all materials, not only metal but plastics, wood, cardboard, glass, cloth, rubber, and stone.

Abrasive water jet cutting heads are being used to replace torches on large automatic gantry-type shape-cutting machines. Since heat is not utilized, there is no heat-effected zone and it does not change the microstructure of the metal. The kerf, or width of cut, is very small and therefore parts can be nested closer, maximizing material utilization. High-precision cuts, plus or minus 0.03 in. (0.76 mm), are possible. The process will also pierce holes, eliminating the need to drill starter holes. The cut surface is very smooth, sometimes much better than oxygen fuel gas cutting and plasma cutting, and rivals that of the laser cut. No dross is produced, which eliminates secondary operations. In addition, there is no dust, smoke, or fumes produced. After the jet of water cuts the material, it takes about 24 to 30 in. (600 to 750 mm) of water to diffuse the jet stream. It does involve a fairly high level of noise, around 85 db. On thin materials very complex parts can be cut, which rival those made by a laser beam. It is obviously the recommended process to be used when cutting thick, nonferrous materials. The accuracy of the cut depends on the travel speed and the accuracy of the shape-cutting machine.

The heart of the water jet cutting system is the cutting head. This assembly consists of an orifice and focusing tube that produces a sharp and coherent water jet stream for cutting. The cutting head may include a diamond orifice, which assures proper orifice diameter after hours of cutting. The well-designed cutting head includes a mixing chamber, the orifice, and focusing tube. They are normally mounted with a Z-direction slide for proper alignment. A typical water jet cutting head is shown in Figure 9–19.

The second most important part of the system is the water jet pump or, as it is sometimes called, intensifier. These range in size from 25 hp to 100 hp and produce water jet pressures of up to 60,000 psi (4,140 bar). The water flow rate ranges from .25 gal/min (.95 L/min) to 2 gal/min (7.57 L/min). The pump, or intensifier, is designed to produce the ultrahigh pressure needed for water jet cutting.

A third component in the system is an abrasive hopper which holds the finely ground abrasive (usually garnet). The abrasive is used for hard metals, thick metals, and when required.

The final part of the total system is the high-pressure tubing used to connect the pump to the water jet.

Cutting conditions for cutting metals up to 4 in. thick are shown in Figure 9–20. In all cases the orifice size is 0.014 in. (0.36 mm). The nozzle size is 0.043 in. (1.1 mm), and the abrasive is added to the water jet. The pressure is 50,000 psi. Travel speed is in inches per minute. To improve the quality of the cut, the travel speed is re-

FIGURE 9–19 Water jet cutting head.

duced. If the quality of the cut surface is not important, the travel speed can be increased at least 50%.

The quality of a cut in 1-in.-thick steel is shown in Figure 9–21.

9-4 AUTOMATIC SHAPE CUTTING

Early efforts to mechanize shape cutting utilized pantograph machines using patterns that were the shape of the desired flame cut parts to guide the torch. These had a magnetic follower, which traversed the periphery of the pattern to make the torches follow the same path. Pantograph machines were replaced with cantilever or gantry bridge-type machines that allowed the cutting of large plates. The simple patterns were replaced by large templates made of aluminum strips attached to sheet metal that outlined the shape of the desired part. Mechanical following devices traversed these templates to guide the torch. As the electronics industry advanced, the track-type templates were replaced by electronic eye guidance systems that followed contrasting color patterns and guided the torches. Two torches were used to produce identical parts with one trip around the pattern, as shown in Figure 9–22. In this illustration the parts remain connected by a technique known as chain cutting. Chain cutting is accomplished by leaving small links of metal between adjacent pieces. This eliminates the need to start and stop the torch until the total cut

Material Type	Thickness		Travel Speed	
	in.	mm	in./min	mm/min
Mild steel	$\frac{1}{16}$	1.6	60–70	
	$\frac{1}{4}$	6.4	10–12	
	$\frac{1}{2}$	12.7	4–5	
	1	25.4	$1\frac{1}{2}$–2	
	4	101.6	$\frac{1}{4}$–$\frac{1}{2}$	
Stainless steel	$\frac{1}{16}$	1.6	50–60	
	$\frac{1}{4}$	6.4	8–10	
	$\frac{1}{2}$	12.7	3–5	
	1	25.4	1–2	
	4	101.6	$\frac{1}{4}$–$\frac{1}{2}$	
Aluminum	$\frac{1}{16}$	1.6	150–175	
	$\frac{1}{4}$	6.4	25–35	
	$\frac{1}{2}$	12.7	10–15	
	1	25.4	4–6	
	4	101.6	1	
Copper	$\frac{1}{16}$	1.6	75–85	
	$\frac{1}{4}$	6.4	12–16	
	$\frac{1}{2}$	12.7	5–7	
	1	25.4	$1\frac{1}{2}$–$2\frac{1}{2}$	
	4	101.6	$\frac{1}{2}$	
Titanium	$\frac{1}{16}$	1.6	75–85	
	$\frac{1}{4}$	6.4	14–16	
	$\frac{1}{2}$	12.7	5–7	
	1	25.4	2–$2\frac{1}{2}$	
	4	101.6	$\frac{1}{2}$	
Inconel	$\frac{1}{16}$	1.6	40–50	
	$\frac{1}{4}$	6.4	8–10	
	$\frac{1}{2}$	12.7	3–4	
	1	25.4	1–2	
	4	101.6	$\frac{1}{4}$	

Orifice 0.014 in., 0.36 mm; nozzle 0.043 in., 1.1 mm; pressure 50,000 psi

FIGURE 9–20 Water jet cutting conditions (with abrasive) from ESAB cutting systems.

FIGURE 9–21 The quality of cut by a water jet cutting head.

The method of guiding the torches has advanced to computer-controlled systems. These guidance systems also allow the nesting of parts with a minimum of scrap plate provided by software programs shown in Figure 9–25. These control systems have become very capable. They are user friendly with touch screen displays. They accept direct downloading and floppy disk input. They may contain memory libraries. They can start and adjust torches and control bevel cuts. Some units also control torch height. Some shape cutting control systems control the total cutting process.

These industrial cutting systems can be very large, ranging in width from 4 ft to 20 ft and even wider. The length is practically unlimited but is based on the length of the track. Long tracks are used to allow the gantry bridge carrying the torches to continue cutting, while the cut parts can be removed and cutting can begin on a new plate. In some cases, two gantry bridges are placed on the same tracks as shown by Figure 9–26. The accuracy of these large industrial gantry machines depends on the accuracy of the tracks, the motion motors, and the control system. They are approaching the accuracy of machine tools. The accuracy of the cut parts also depends on the torches and the cutting process.

The gantry bridge crane was originally designed to carry oxyfuel gas cutting torches. Most machines will accept plasma cutting torches, laser cutting torches, and water jet cutting nozzles. The same bridge can accommodate more than one type of torch. These machines are very precise, heavy duty, expensive, and are becoming very popular in industry.

For smaller requirements a small profile cutting machine is available with a 2-ft-wide capability and a 6-ft

is completed. The link is cut by a hand torch as the parts are removed from the cutting table.

To increase productivity, more than two torches may be used. The number is limited by the size of the parts and of the equipment. This is shown in Figure 9–23 using five torches. One of the newer advances in automatic flame cutting is the generation of bevel cuts. This breakthrough has made the use of computer-controlled oxygen cutting equipment even more productive. Figure 9–24 shows a cutting torch making bevel cuts.

The automatic shape cutting machines, sometimes called profiling machines, have become very big and complex, but very accurate. They were originally designed for the oxyfuel gas cutting processes. They have become automatic tools for accurately moving one or more torches in two directions over a wide area.

FIGURE 9–22 Electric eye template system (two torches).

FIGURE 9–23 Automatic shape-cutting machine using five torches.

FIGURE 9–24 Automatic bevel cut head.

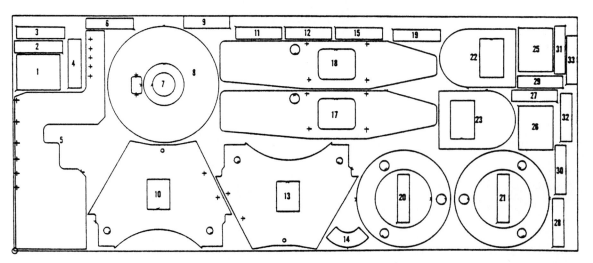

FIGURE 9–25 Computer-generated nesting program for automatic cutting.

or 12-ft length capability. A picture of this machine is shown in Figure 9–27. This machine travels on a track and has a cross rail that holds the torch. Its motion is programmed using a handheld terminal, or it can be programmed with a PC computer. Programs are available for cutting complex shapes and for manual or automatic nesting of the parts. Machines that use cutout templates and electric eye patterns are also available.

Dimensional tolerances for flame cutting depend largely on the thickness of the material and the size of the part. In medium thicknesses, the dimensions of small parts can be held within $\pm \frac{1}{16}$ in. (1.6 mm). Sprocket wheels for driving caterpillar-type treads are oxygen flame cut and used without further finishing. On larger parts the tolerance is not as close, and when chain burning is employed, warpage can create distortion and dimensional problems.

The American Welding Society has provided criteria for describing oxygen-cut surfaces, AWS C 4.1. This publication includes a plastic replica showing a surface reference guide for oxygen cutting at four different levels. It includes terms for describing oxygen-cut surfaces, including flatness, angularity, roughness, top edge rounding, notch, and slags. Figure 9–28 shows the correct and incorrect adjustments for the machine and the torch.

The introduction of the water table has improved large automatic shape-cutting operations. Use of the water table has several advantages. It greatly reduces the

FIGURE 9–26 Two Gantry Bridges on the same track.

FIGURE 9–27 Small profile cutting machine.

particulate matter released into the atmosphere by the oxygen cutting operation. It reduces distortion because the water in contact with the underside of the metal being cut eliminates the heat buildup in the metal. The water level can be raised or lowered, and it is raised during the cutting operation so that it is in contact with the metal being cut. Water tables include mechanisms for collecting the slag for easy disposal.

Stack cutting is the oxygen cutting of stacked metal sheets or plates arranged so that all the plates are severed by a single cut. In this way the total thickness of the stack is considered the same as the equivalent thickness of a solid piece of metal. When stack cutting, particularly thicker material, the cut is often lost because the adjacent plates may not be in intimate contact with each other. The preheat may not be sufficient on the lower plate to bring it to the kindling temperature and therefore the oxygen stream will no longer cut through the remaining portion of the stack. One way to overcome this problem is to use the metal powder cutting process. By means of the metal powder and its reaction in the oxygen, the cut is completed across separations between adjacent plates.

Automatic oxygen cutting machines are available for cutting pipe to fit other pipe at different angles and of different diameters. These are quite complex and have built-in contour templates to accommodate different cuts and bevels on the pipe.

9-5 THERMAL SPRAYING

Thermal spraying (THSP) is a group of processes in which finely divided metallic or nonmetallic surfacing materials are deposited in a molten or semimolten condition on a substrate or base metal to form a thermal spray deposit. The surfacing material may be in the form of powder, rod, cord, or wire.

There are three separate processes within this group: arc spraying, plasma spraying, and flame spraying. These three processes differ considerably since each uses a different source of heat and different apparatus. Thermal spraying was invented in 1913.

There are several variations of each process. A variation of flame spray is the detonation method, which uses combustible gases but attains a much higher temperature and particle velocity. A variation of the plasma spray process is the plasma transferred arc method, which provides higher temperatures and is more of a welding process. A summary of thermal spraying surfacing methods is given in Figure 9–29.

The selection of the spraying process depends on the properties desired of the coating. Thermal spraying is utilized to provide surface coatings of different characteristics, such as coatings to reduce abrasive wear, cavitation, or erosion. The coating may be either hard or soft. It may be used to provide thermal barriers for high-temperature protection. Thermal sprayed coatings improve atmosphere and water corrosion resistance. One of the major uses is to provide coating resistant to salt-water atmospheres. Another use is to restore dimensions to worn parts. The hardness and composition of the deposit are important and dictate whether the part will be machined or ground. Based on this decision, it is then necessary to determine the type of material that will be sprayed. If the spray material is available in wire form, the electric arc spray or the flame spray processes can be used. However, if it can be obtained only in a powder form, the flame spray or plasma spraying process can be used. The selection of materials for spraying is beyond the scope of this section. See the AWS "Thermal Spraying Practice, Theory and Application."[2]

Flame Spraying

Flame spraying (FLSP) is "a thermal spraying process in which an oxyfuel fuel gas flame is the source of heat for melting the surfacing material. Compressed gas may or may not be used for atomizing and propelling the surfacing material to the substrate." There are two major variations: One uses metal in wire form, and the other uses materials in powder form. The method of flame spraying that uses powder is sometimes known as powder flame spraying. The method of flame spraying using wire is known as metallizing or wire flame spraying.

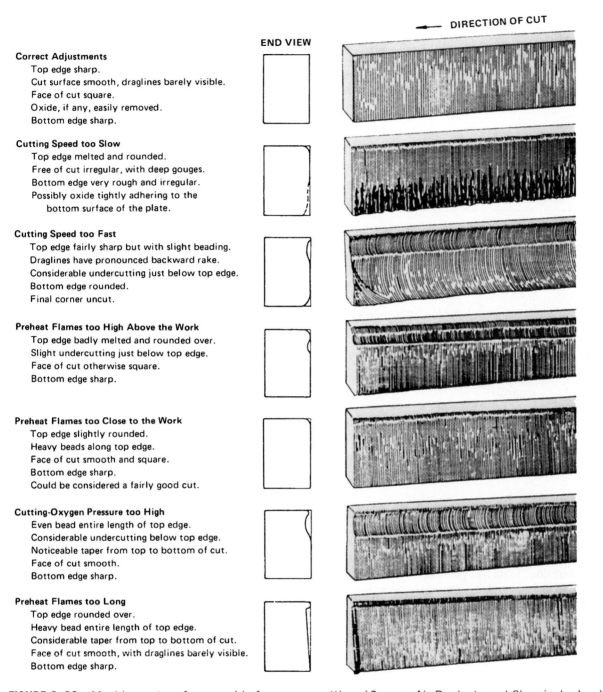

END VIEW

Correct Adjustments
    Top edge sharp.
    Cut surface smooth, draglines barely visible.
    Face of cut square.
    Oxide, if any, easily removed.
    Bottom edge sharp.

Cutting Speed too Slow
    Top edge melted and rounded.
    Free of cut irregular, with deep gouges.
    Bottom edge very rough and irregular.
    Possibly oxide tightly adhering to the
        bottom surface of the plate.

Cutting Speed too Fast
    Top edge fairly sharp but with slight beading.
    Draglines have pronounced backward rake.
    Considerable undercutting just below top edge.
    Bottom edge rounded.
    Final corner uncut.

Preheat Flames too High Above the Work
    Top edge badly melted and rounded over.
    Slight undercutting just below top edge.
    Face of cut otherwise square.
    Bottom edge sharp.

Preheat Flames too Close to the Work
    Top edge slightly rounded.
    Heavy beads along top edge.
    Face of cut smooth and square.
    Bottom edge sharp.
    Could be considered a fairly good cut.

Cutting-Oxygen Pressure too High
    Even bead entire length of top edge.
    Considerable undercutting below top edge.
    Noticeable taper from top to bottom of cut.
    Face of cut smooth.
    Bottom edge sharp.

Preheat Flames too Long
    Top edge rounded over.
    Heavy bead entire length of top edge.
    Considerable taper from top to bottom of cut.
    Face of cut smooth, with draglines barely visible.
    Bottom edge sharp.

FIGURE 9–28 Machine-cut surfaces: guide for oxygen cutting. (*Source:* Air Products and Chemicals, Inc.)

In both versions, the material is fed through a gun and nozzle and melted in the oxygen fuel gas flame. Atomizing, if required, is done by an air jet, which also propels the atomized particles to the workpiece. When wire is used for surfacing material, it is fed into the nozzle by a wire feeder and is melted in the gas flame. When powdered materials are used, they may be fed by gravity from a hopper, which is a part of the torch. In another system the powders are picked up by the oxygen fuel gas mix-

ture, carried through the gun where they are melted, and propelled to the surface of the workpiece.

Figure 9–30 shows the flame spray process using wire. This version can spray metals that are available in wire form. The variation that uses powder can feed normal metal alloys, oxidation-resistant metals and alloys, and ceramics. Ceramics can be provided in rod or cord form. It provides sprayed surfaces that can have many different characteristics.

	Flame: Oxyfuel Gas Combustion	Detonation: Oxyfuel Gas Pulsed Explosion	Arc: Electric Arc, Two Wires	Plasma: Nontransferred Arc	PTA: Plasma Transferred Arc
Heat source temperature (°F)	4700–5600	6000+	8000	15,000	15,000
Particle velocity (ft/sec)	800	2,500	800	1,800	—
Coating materials					
Wire–metal	Yes	No	Yes	Yes	Yes
Powder–metal	Yes	Yes	No	Yes	Yes
Powder–ceramic	Yes	Yes	No	No	No
Rod or cord					
ceramic and plastic	Yes	No	No	No	No

FIGURE 9–29 Thermal spray process for types of coating materials.

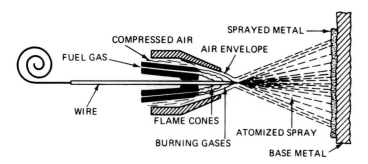

FIGURE 9–30 Flame spray process.

High-Velocity Oxyfuel (HVOF) Spraying This is a variation of flame spraying and is an internal combustion method that produces a high-speed jet. It utilizes the energy of rapid explosions of oxygen and fuel gas mixtures rather than a steadily burning flame. The powder is introduced into the combustion chamber. When the gas mixture is ignited, a controlled detonation wave or flame front accelerates and heats the powder particles as it moves down the length of the combustion chamber. Exit particle ve-

locities are extremely high, and the temperature is higher than the normal flame temperature. After each injection of powder has been discharged, a pulse of inert gas purges the barrel and chamber. Multiple detonations during each second build up the coating. Temperatures above 5000° F (2760° C) and velocities of 2500 ft/sec are attained. The density of the deposited coating is extremely high, and the bond with the workpiece is extremely good. Smooth deposits are achievable because of the high density of the deposit. This process is shown in Figure 9–31. There are several variations; they go by different trade names, such as Jet Coat, Diamond Jet, D-Gun, Top Gun.

Arc Spraying

Arc spraying (ASP) is "a thermal spraying process using an arc between two consumable electrodes of surfacing materials as a heat source and a compressed gas to atomize and propel the surfacing material to the substrate" (Figure 9–32). The two consumable electrode wires are fed, by a wire feeder, to bring them together at an angle of approximately 30° and to maintain an arc between

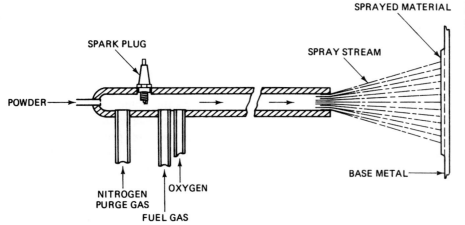

FIGURE 9–31 Detonation gun spray method.

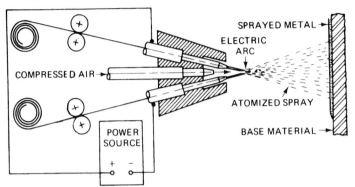

FIGURE 9–32 Arc spraying process.

them. A compressed air jet is located behind and directly in line with the intersecting wires. The wires melt in the arc, and the jet atomizes the melted metal and propels the fine molten particles to the workpiece. The power source for producing the arc is a dc constant voltage welding machine. The wire feeder is similar to that used for gas metal arc welding except that it feeds two wires. The gun can be handheld or mounted in a holder with a movement mechanism. The part or the gun is moved with respect to the other to provide a coating surface on the part.

The welding current ranges from 300 to 500 A direct current, with the voltage ranging from 25 to 35 V. This system will deposit from 15 to 100 lb/hr of metal. The amount of metal deposited depends on the current level and the type of metal being sprayed. Wires for spraying are sized according to the Brown and Sharp wire gauge system. Normally either 14 gauge (0.064 in. or 1.6 mm) or 11 gauge (0.091 in. or 2.3 mm) is used. Larger-diameter wires can be used. Flux cored wires can also be used.

The high temperature of the arc melts the electrode wire faster and deposits particles having a higher heat content and greater fluidity than in the flame spraying process. The deposition rates are from three to five times greater and the bond strength is greater. There is coalescence in addition to the mechanical bond. The deposit is more dense and coating strength is greater than when using flame spraying.

Dry compressed air is normally used for atomizing and propelling the molten metal. A pressure of 80 psi (552 kPa) and flow from 30 to 80 ft³/min (14 to 38 L/min) is

used. Almost any metal that can be drawn into a small wire can be sprayed. Figure 9-33 shows a list of metals that are arc sprayed.

Plasma Spraying

Plasma spraying (PSP) is "a thermal spraying process in which a nontransferred arc of the gun is used to create an arc plasma for melting and propelling the surfacing material to the substrate" (Figure 9-34). Plasma spraying is

FIGURE 9–33 Arc spraying: metals and spray rates.

Metal Sprayed	Spray Rate (lb/hr/100 A)
Aluminum	5–7
Babbitt	—
Brass	10–12
Bronze	10–12
Copper	12–15
Molybdenum	—
Monel	11–13
Nickel	9–11
Stainless steel	11–13
Carbon steel	10–14
Tin	—
Zinc	20–25

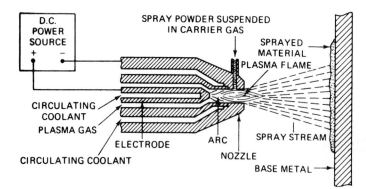

FIGURE 9–34 Plasma spray process.

sometimes called plasma flame spraying or plasma metallizing. It uses a nontransferred plasma arc, which is entirely within the plasma spray gun. The temperature is much higher than either arc spraying or flame spraying. Higher-temperature materials can be used for the coating. The material to be sprayed must be in a powder form since it is carried into the plasma spray gun suspended in a gas. The high-temperature plasma immediately melts the powdered material and propels it to the surface of the workpiece. Since inert gas and extra high temperatures are used, the mechanical and metallurgical properties of the coatings are generally superior to either flame spraying or arc spraying. This includes reduced porosity and improved bond tensile strengths. Coating density can reach 95%. The hardest metals known, and some with extremely high melting temperatures, can be sprayed with the plasma spraying process.

Preparation for Spraying

The most important aspect of thermal spraying is correct preparation of the workpiece. It must be clean. Machining is normally used to prepare round parts, such as shafting. When thermal spraying is used to correct a dimension, the part is usually machined undersize to allow a sufficient thickness of coating. For large flat areas, grit blasting is used. In any case, a roughened surface is preferred, but sharp corners should be avoided.

Spraying Operation

Spraying should be done immediately after the part is cleaned. If the part is not sprayed immediately, it should be protected from the atmosphere by wrapping with paper. If parts are extremely large, it may be necessary to preheat the part 200 to 400° F (95 to 205° C). Care must be exercised so that heat does not build up in the workpiece. This increases the possibility of cracking the sprayed surface. The part to be coated should be preheated to the approximate temperature that it would normally attain during the spraying operation. The distance between the spraying gun and the part is dependent on the process and material being sprayed. Recommendations of the equipment manufacturer should be followed and modified by experience. Speed and feed of spraying should be uniform. The first pass should be applied as quickly as possible. Additional coats may be applied slowly. It is important to maintain uniformity of temperature throughout the part. When there are areas of the part being sprayed where coating is not wanted, the area can be protected by masking it with tape. Relative motion is required. The spray gun may be handheld or mechanically held if the work is moving. This is common practice when building up shafting in a lathe. For larger areas the gun can be handheld, or mechanical motion devices may be employed. Robots are often used for thermal spray operations. Figure 9–35 shows the plasma spray system being applied by a robot. Note that the workpiece also rotates.

Quality of Coatings

Coatings must be inspected to determine that they are free of cracks, pinholes, blisters, voids, and so on. Coatings over sharp corners, such as keyways, require extra attention. The skill of the operator is a major factor in obtaining good-quality coatings. Written procedures are recommended for each type of application.[3],[4]

Posttreatment: Fusing or Recasting

Fusing or recasting is a posttreatment for a thermal spray coating. All thermal spray deposits have less than 100% density; they contain interparticle voids. The fusing or recasting operation melts and densifies the coating, eliminating most of the porosity. It increases the bond strength and improves the surface finish. In this operation the spray coating deposit is heated to a point between its solidus and liquidus where the surface attains a highly reflective glossy appearance called the *slick-up point*. This achieves an optimum combination of particle melting and closing the interparticle gaps. Shrinkage takes place in the deposited coating. Wetting and bonding occur between powder particles and base metal, resulting in a dense deposit with low-porosity levels.

Successful application of the fusion or recasting process requires careful heating. If the deposit is underheated, melting and fusing will be incomplete and result in poor bonding and undesired porosity. Heating above the proper temperature will produce a liquid phase and induce shrinkage, voids, distortions, melting, and running of the deposit and base metal dilution. The application of heating should be as rapid as possible. Correct temperature relates to the composition of the deposited metal. Temperature control is required for a high-quality fused coating.

Fusion or recasting can be performed with any heat source. One of the most popular is the oxyacetylene

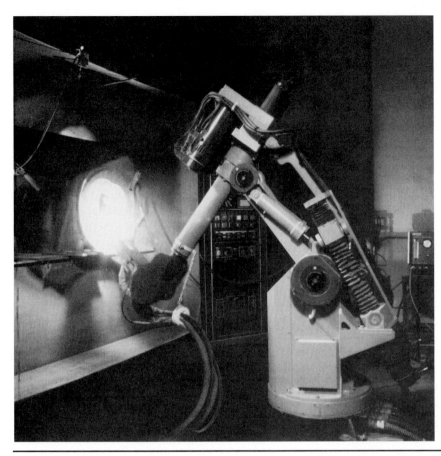

FIGURE 9–35   Robot spraying operation.

torch. It can also be accomplished by furnace heating, induction heating, or laser heating. Laser heating of deposited coatings provides a highly controlled heating and melting operation. Since the laser has such a wide range of energy densities, precise temperatures can be produced at specific depths in the coating. With certain high-temperature materials the fusion of a thin layer of the base metal at the bond line may be possible without melting the surface of the coating.

Safety

Potential hazards of thermal spraying include electric shock, arc radiation, fire, dust, fumes, gas, and noise. These subjects were all discussed in detail in Chapter 4. Thermal spray operators are required to wear protective clothing, including eye and ear protection as shown in Figure 9–36. OSHA sets permissible noise exposures. Ear protection should be provided for workers exposed to noise above 90 dBa. The noisier processes, plasma and detonation spray, require earphones. For these processes isolation booths are often employed.

Special attention should be paid to the content of sprayed materials. The spray powders must be handled carefully since particles should not be inhaled. When the spray material is handled, the best protection is to use breathing masks with filters. The spray operations must

FIGURE 9–36   Thermal spray operator wearing suitable protection.

be well ventilated, with air exhausting at 200 to 300 ft/min past the spray station, and passing through a wet collector to capture the spray dust.

AWS is developing a program to certify thermal spray operators and inspectors.

9-6 ADHESIVE BONDING

Adhesive bonding (ABD) is a material-joining process in which an adhesive is placed between the faying surfaces. The adhesive solidifies to produce an adhesive bond. The adhesive bond is the attractive forces between an adhesive and the base materials, or substrate. Two principal interactions that contribute to the adhesion are the van der Waals bond and permanent dipole bonds. The *van der Waals bond* is defined as a secondary bond arising from the fluctuating dipole nature of an atom with all occupied electron shells filled. The *dipole bond* is a pair of equal and opposite forces that hold two atoms together and results from a decrease in energy as two ends are brought closer to one another.

Adhesive bonding of metals, plastics, and composites to themselves and to each other is becoming more important. There are no industrial standards for adhesives. They are usually specified by proprietary trade names, which relate to manufacturers' specifications. There are a number of types of adhesives. We are interested primarily in structural adhesives that are capable of withstanding significant loads. However, there are others, such as holding adhesives, that cannot withstand a great deal of force: sealing adhesives used to prevent leakage; lock adhesives used to prevent the loosening of threaded parts; retaining adhesives used to prevent the twisting or sliding of nonthreaded parts; hot-melt adhesives that are applied in the heated state; pressure-sensitive adhesives used on self-sealing envelopes; instant adhesives, which cure within seconds; ultraviolet adhesives, which cure when exposed to ultraviolet light; and heat-cured adhesives, which require heat to cure. Adhesives can be classified according to their composition or other characteristic. This includes solvent cements, hot melts, silicones, urethanes, epoxies, anaerobics, cyanoacrylates, and acrylics. Each type has its advantages, weaknesses, and specific applications. Selection of the proper adhesive for specific applications is a technical subject beyond the scope of this book.

There are advantages to adhesive bonding. The stresses on adhesive-bonded parts are uniform over the entire bonded area, which may allow the use of thinner materials. Materials of all sizes and types can be bonded. If they have different coefficients of expansion, the adhesive bond will compensate since it retains some flexibility. Adhesive bonding has a high resistance to fatigue since it provides a damping action and retains some flexibility. There is no distortion with adhesive bonding since high temperatures are avoided and holes are not used. The properties of metals and other materials being joined are not affected. Adhesives have different characteristics—some may provide electrical insulation, while others will give electrical conduction depending on the design. Adhesive bonds provide seals to prevent leaks. Adhesive bonds eliminate bulges, gaps, projections, or indentations when compared to mechanical fasteners or resistance welds. Adhesive bonding provides cost savings when material thicknesses are reduced or when operations such as drilling or punching are eliminated, or when metal finishing operations can be reduced or eliminated.

There are problems with applications and with service life of adhesive bonding. Different adhesives require different handling and curing treatments. Some adhesives are toxic, some are flammable, some have a short shelf life, and so on. Curing of adhesives varies. Some require a long cure time with precise temperature and pressures. Others require ultraviolet radiation. High-temperature service will reduce bond strengths, and some adhesives are sensitive to environmental conditions such as humidity, temperature, and atmosphere. Some adhesives have more than one component, which complicates application.

The properties of adhesive-bonded joints are described differently from metalworking terms. The joints are judged by properties such as peel stress, shear stress distribution, tension and compression stress distribution, and cleavage and peel stress distribution. Testing methods have been established by the American Society for Testing and Materials (ASTM).[5] This includes tests for tensile strength, shear strength, cleavage strength, and peel resistance.

The joint types used for adhesive bonding are the lap/overlap; the joggle lap; the butt joint, which can utilize tongue-and-groove design; the scarf joint; and the strap joint (single or double). The mortise and tenon are used for corner joints as shown in Figure 9–37.

FIGURE 9–37   Joint types used for adhesive bonding.

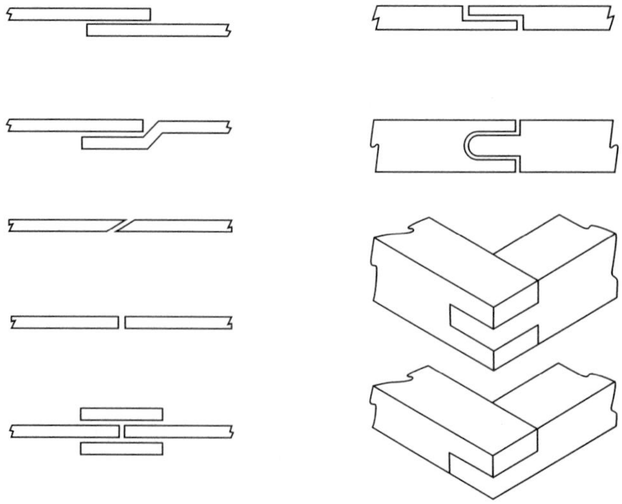

There are certain factors to consider when using adhesive bonding: (1) adhesive selection, (2) surface preparation, (3) application of the adhesive, and (4) curing. The successful performance of an adhesive bond depends on proper control of these four factors. They are equally important; however, the performance of an adhesive depends on surface preparation, and except for the adhesive for oily surfaces, the surface must be clean to produce an efficient joint. Adhesives are normally applied manually; however, robots are being used in the automotive and appliance industry. The application process can range from simple manual to sophisticated robotics and depends on the adhesive. For example, solvent cements may be sprayed on the surface; hot melts require heating applicators. Viscosity, whether thick or thin, dictates the application technique, as does the time and method of curing. There are single, double, or triple component adhesives and catalyst/accelerators.

The cost of different types of adhesives varies considerably and must be considered relative to alternative joining methods. In general, the adhesive that assures the most efficient joining, the least amount of preparation and curing, and requires minimum finishing yet provides the best strength is the proper selection.

Types of Adhesives for Bonding

The most commonly used structural adhesives are the acrylics, epoxies, urethanes, and cyanoacrylates. Each has its own characteristics.

Rubber-Base Adhesives   Rubber cements or solvent cements are adhesives that contain organic solvents rather than water. They are based on nitrocellulose or polyvinyl acetate, normally elastomeric products, dispersed in solvent. They are free flowing, thin-set materials that dry to hard, tack-free films. Some retain a soft, tacky, flexible film. They are used in pressure-sensitive labeling operations, in contact bonding for the woodworking industry, and in laminating applications, such as veneering. Mastic-type adhesives are a solvent cement used in the building and construction industry. They are used to bond wood and drywall to concrete and other vertical surfaces. Rubber and solvent cements can be sprayed or hand applied using a roller. They are usually flammable. The chemistry of the adhesive dictates its characteristics and recommended use.

Resin Adhesives   Synthetic resins are composed of synthetic organic materials and are relatively expensive. They are used when a high-quality bond is required, and they are relatively heat and moisture resistant. They can be applied by automatic or semiautomatic equipment, are used for sealing cardboard cartons and for wood, and for vinyl film laminations.

One of the major groups is the hot melts, which are combinations of waxes and resins that form a bond by applying heat and then cooling. Solid hot-melt synthetic resin adhesives contain neither water nor solvents and set up quickly. The hot-melt type must be heated to between 250 and 400° F (121 and 204° C) before they are applied. Hot-melt adhesives resist moisture and can be used on non-porous surfaces. They are used for bonding metal to various surfaces. They are used for packaging and in the furniture and woodworking industry.

Epoxy Adhesives   The epoxy adhesives can be used to bond metal to metal, metal to plastics, and plastics to plastics. They are a family of materials characterized by reactive epoxy chemical groups on the ends of resin molecules. They consist of two components, a liquid resin and the hardener to convert the liquid resins to solid. They may contain other modifiers to produce specific properties for special applications. Some epoxies will bond to concrete. One of the newer advances is the oily metal epoxy that bonds directly to oily metals "as received" with normal protective films on them. The oily coating need not be removed. Other epoxies can be a one-component type, but these require a heat-cure operation. Epoxies are good surface wetters. They achieve intimate molecular contact with the surface to be bonded and will achieve high adhesion on almost any surface. Epoxies are the most expensive of the adhesives; however, they offer more advantages.

Solvent Joining of Plastics

There are two basic methods of joining plastics by chemical action, the solvent joining method and the adhesive joining method. Solvent joining is applicable to the thermoplastics group of materials, which are readily dissolved in a solvent. Solvent joining cannot be applied to *inert* materials such as polyolefines. In the solvent cementing technique, the surfaces to be joined are coated with a solvent and then held together under pressure until the solvent has evaporated to form a seal. Butt, lap, and tongue-and-groove joints are employed. The joint strength is dependent on the material and the joint design. Improved bonds are achieved by using specially developed solvent cements, which contain a small quantity of the plastics of the type used in the components to be joined. The solvent cements have an advantage over straight solvent welding since they will fill voids in poorly fitted joints. Adhesive joining of plastics is similar to adhesive joining of metals.

9-7 JOINING PLASTICS

Plastics are being more widely used in the manufacture of more products. Plastics are replacing metals when their properties are appropriate, especially where corrosion resistance or weight restrictions are important. They are used for smaller parts or assemblies that are produced in high volume. Pipe and complex formed parts are often joined.

Plastic is defined as "a material that contains as an essential ingredient one or more polymeric substances of high molecular weight. It is solid in its finished state and at some stages in its manufacturing or processing can be shaped by flow into finished articles."[5] Plastics are organic, manmade materials. Plastic materials have a complex nomenclature based on their chemistry. Many have common names, trade names, or abbreviations, which may lead to confusion. Exact identification is important when working with plastics. Most plastic materials will burn, and they have a coefficient of expansion about four times that of steel.

All plastics fall within two categories, based on their chemical composition and on their elevated-temperature characteristics. They are classed as either thermoplastic or thermosetting.

Thermoplastic materials have long, chainlike polymer molecules held together by relatively weak van der Waals forces, hydrogen bonds, or the interaction of polar groups. When the plastic is heated, these forces are weakened so that the material becomes soft and flexible. At higher temperatures it becomes a viscous melt and can be molded or extruded into the required final shape. Thermoplastic materials can be repeatedly softened by heat and hardened by cooling. They can readily be welded by the application of heat, making a monolithic structure; fusing occurs across the bond line. Typical thermoplastics are polyethylene, polyvinyl chloride, polystyrene, polypropylene nylon, polycarbonate, and acetal.

Thermoset plastics are formed by a chemical reaction. Normally, the reaction occurs above room temperature and under pressure during the molding operation. During molding, polymer molecules capable of further reaction are chemically cross-linked into a close network structure. When cooled the resulting product is rigid. If extra heat is applied, the material will degrade. Thermoset plastics are not weldable by any method that involves heating. They can be joined by adhesive bonding. Typical thermoset plastics are phenolformaldehyde, melamine-formaldehyde, urea-formaldehyde, and epoxies.

The successful use of plastics often requires that parts be joined together securely. Mechanical fasteners can be used; however, for permanent joining a better method is desired. Permanent joining methods fall within two categories: welding and adhesive bonding. Welding produces a monolithic structure, but adhesive bonding does not.

Weldability of Plastics

Weldability of a thermal plastic depends on the welding method being used, the thickness of the materials being joined, and the joint design. There are two basic types of thermal plastics: amorphous and crystalline polymers.

The solid-state structure, that is, the manner in which the polymer molecules are arranged, determines its physical properties, melting, and welding characteristics.

Amorphous polymers have no orderly structure. The molecules are randomly arranged. Amorphous plastics do not have a definitely defined melting point. When heated they gradually soften when they pass from a rigid or solid state to a transition into a leathery and then a rubbery state. This is followed by a rubbery flow and then a liquid flow to a true molten state. Solidification is gradual, in the reverse process. The energy requirements remain relatively constant as the temperature changes.

Crystalline polymers have a very orderly molecular structure due to the chemical energy or interaction within each molecule. They can be considered as being like either flat or coiled springs. The higher the degree of crystallinity, the more complete the springlike structure. Crystalline polymers have a sharp melting point. The plastic remains rigid until it reaches its melting point, and then immediately becomes fluid. As the temperature of the crystalline materials approaches the melting point, a higher level of heat energy is required to continue to increase the temperature. Solidification occurs just as rapidly as melting, due to the sudden release of energy as chemical interaction or crystallization of the molecules takes place. In general, amorphous thermal plastics are easier to weld than the crystalline forms.

There are five steps to the thermal welding of thermoplastics:

1. Surface preparation
2. Heating
3. Application of pressure
4. Diffusion or welding
5. Cooling

Surface preparation is important since most molded plastics have a contaminated surface layer known as a mold release. This must be removed, and for certain processes the abutting surfaces must be absolutely flat for intimate contact. Heating is accomplished by different methods and is the basis for identifying the plastic welding methods.

The application of pressure is done in different ways. It can be done manually, in presses, or in automatic fixtures. It is often combined with tooling, which may include part of the heating apparatus as well as the pressure method.

Diffusion occurs once the liquid-to-liquid interface has been established. Diffusion occurs almost instantaneously with crystalline or semicrystalline materials. For amorphous materials heated only slightly above the melting point, diffusion takes longer.

The final step in making the weld is to cool the assembly and resolidify the joint. The load or pressure must be maintained until the resin has sufficient strength and stiffness to support the total weldment.

Welding Methods

There are a number of different welding methods.[6] They all require the use of heat; thus all of them can be called "thermal joining" methods. The heat for welding is produced in different ways. The heating method identifies the different welding/sealing/bonding methods:

- ✷ Electromagnetic (or induction)
- ✷ Friction (spin welding)
- ✷ Heated surface (heated tool/hot plate)
- ✷ High frequency (dielectric heating)
- ✷ Hot gas
- ✷ Implant
- ✷ Radiant
- ✷ Ultrasonic (sonic)
- ✷ Vibration

Electromagnetic Welding The electromagnetic bonding or magnetic heat sealing method uses induction heating for creating the weld. Induction heating utilizes high-frequency alternating current, which creates heat in magnetic particles in its field. In electromagnetic plastic welding, micron-sized magnetic particles are dispersed within a thermal plastic matrix. When this material is placed between the faying surfaces to be welded and exposed to the electromagnetic field, heat develops at the interface, causing melting and subsequent fusion of thermal plastic materials. It produces a polymer-to-polymer linkage between all compatible thermal plastics. It can be applied in hot melts or solvent binder systems, or as implants in the joint. Equipment required is a high-frequency power source from 2 to 20 kW output with a frequency of 3 to 30 mHz; 2.5 to 3.5 mHz is most often used. Work coils, usually water-cooled copper coils, produce the magnetic field in the workpiece. They can be incorporated into the fixtures. They can be used on thick or thin sections, irregular shapes, and with the right vehicle can be used to fill voids in the joints. They can be automated and are used to join plastics that are normally difficult to weld. They are relatively fast and are used for production applications.

Friction Welding Friction welding, sometimes called spin welding, uses heat that is developed at the interface of the parts being welded due to friction. The weld is made by holding one piece stationary and rotating the other piece against it under pressure. The rotating member is stopped as soon as melting occurs and the weld is consolidated under pressure while it cools. The typical joint strength on like plastics is 90% of the strength of the material. Strength is largely dependent on the joint area. Welding time is from 1 to 5 seconds. The method can be

used to produce welds in similar and dissimilar plastic materials. Equipment is similar to that used for metals except that it is lighter in design and construction. Friction welding is rapid and an efficient joining technique that can be applied to most thermal plastics. Linear motion can also be used for friction welding.

Heated Surface Welding Heated surface, sometimes called heated tool, hot plate, or heated bar welding, uses heat that is generated in the hot tool. Electrical resistance heating coils are normally used to heat the tool or bar. There are two variations. One is used for sealing plastic films or thin materials, and the other is for joining heavier pieces.

When joining thin materials, the parts are held together under pressure and heat is applied for a short period to produce the weld. There are several variations, one of which is known as impulse heat sealing, which uses the pressure bars but has a resistance heating element covered by fiberglass. A pair of bars produce pressure on the pieces being joined; short impulses of electric current provide the heating to complete the weld. Another variation uses circular bars, or wheels, which are heated and rotate and traverse the joint, providing heat and pressure. Circular bands are also used, but in all cases the bonding is accomplished by the direct application of heat.

For heavy materials, this technique requires that the two faces to be joined be flat. These two faces of the parts are held against a heated metal surface. When the plastic surface begins to melt, the heated metal plate is removed and the pieces are quickly brought together and held under pressure. In this application, a flash or reinforcement occurs at the joint. Both of these methods can be automated and programmed.

The hot plate method is widely used for plastic pipe welding for field installations. It is used for making butt welds and for branch joints. Portable semiautomatic equipment is used in the field (Figure 9–38).

High-Frequency Welding High-frequency sealing or bonding uses heat produced by dielectric heating. Dielectric heating occurs in insulating materials that possess electric dipole moments and exhibit polarization in a high-frequency electric field. Polarization and molecular agitation of the material creates heat. Nonresponsive materials can be bonded by using a film or coating between the two parts to be bonded, which generates the heat in the magnetic field. High-frequency generators with outputs up to 50 kW are used, with a frequency of 27.12 MHz. This frequency is used to avoid radio interference with communications. The power is programmed and this method is used primarily in automated systems, which may include a press or pressure-applying device. Special dies are often used, which may also emboss the joining material. In some applications, roll-type tooling is used to make a continuous weld.

FIGURE 9–38 Field welding plastic pipe.

Hot Gas Welding The hot gas welding method utilizes the heat transmitted by hot gases to melt the surface to be joined and also to melt a plastic welding rod to fill joint grooves. The apparatus for producing the hot gas is a torch, which normally uses resistance heating coils to heat the gas, which is directed to the welding area. It is used in the same manner as an oxyacetylene torch for welding metals.[7] Compressed air is most often used as the heated gas; however, nitrogen, or in some cases argon, is used to help eliminate problems associated with oxidation. Hot gas welding is normally a manual operation utilizing two hands. Manipulative skill is required. The hot gas welding method is used to produce large fabrications made of sheet materials. An example is ducting work, pipe work, and ventilator hoods for exhaust systems handling corrosive gases. The apparatus used to produce hot gas welding is relatively inexpensive, consisting of a hot gas torch, gas regulator, an air compressor when required, and the plastic filler rod.

The parts to be joined are beveled along the edges to provide a groove for filler material from the welding rod. The welding rod is of the same composition as the parts being welded; it is usually round and smaller than the groove. Multiple passes may be used. The plastic welding rod is not fluid but semifluid, and fairly easy to control. The application of hot gas welding is shown in Figure 9–39.

Implant Welding Implant welding utilizes heat generated by a metal implant or insert in or adjacent to the weld joint. There are two basic variations. One uses a molded-in resistance wire, which produces heat when it is connected to a source of electric power. The other utilizes a metal insert that is heated by induction via electromagnetic radiation. Both methods produce high-strength, high-quality welds in a variety of thermal plastic materials.

The best known application of hot wire welding is the joining of pipe to fittings. The electric resistance wire is molded into the fitting. Heat is produced by the resistance of the wire to electricity provided by a special power source usually operating at 24 V dc. Pressure is accomplished by the thermal expansion of the part that is heated. The disadvantage is the extra cost of the molded-in resistance wire, which remains in the joint. This process is commonly used for field welding of thermal plastic pipe.

FIGURE 9–39 Hot gas welding plastic assembly.

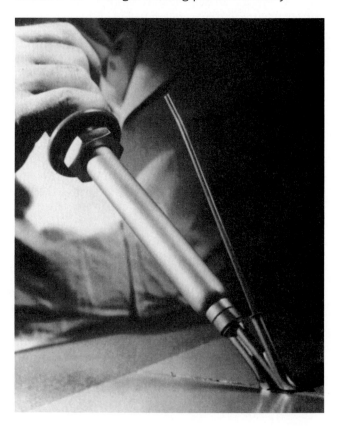

The other variation, metal insert welding, utilizes induction heating of the metal insert. This could be considered electromagnetic welding, but it does require that the implant use electromagnetic power, which generates heat in the implant which is placed at the interface of the parts to be joined. The implant can be a piece of iron or steel, or iron or stainless steel filings in a matrix. Screening is sometimes used. The high-frequency power source used to create the heat is similar to the equipment used for electromagnetic welding. During the heating period, pressure is applied to the joint. This method can be readily automated with the power source programmed. A successful application is the plastic membrane seal in bottle caps. In this case the metal component is a part of the product, and the welding occurs where the plastic membrane is in contact with the metal.

Radiant Welding   Radiant welding uses heat from an infrared heating source or from a laser beam. In either case, heating is due to electromagnetic waves being absorbed by the material surface. The weld surfaces are exposed to infrared lamps or to a laser, and are heated to melt the thermal plastic. When the thermal plastic is molten, the parts are pressed together until the material has cooled. This method is usually automated, and accurate dies are used to press the parts together and maintain dimensional accuracy. This process is rather critical from a time point of view and has not been extremely successful.

Ultrasonic Welding   Ultrasonic welding, sometimes called sonic welding, uses heat that is generated by vibration between the parts, causing the two surfaces to move relative to each other. This causes a temperature rise as mechanical energy is converted to heat. The heat generated is sufficiently high to melt the surfaces in contact. Pressure is required to hold the parts together during the welding operation. The workpieces transmit the ultrasonic vibration to the interface. The contact area with the sonic probe of the welding machines does not rise in temperature, and there is no marking. Frequencies of 20 to 30 kHz are used. Equipment consists of a power source and frequency converter and a piezoelectric crystal, which converts the electrical energy to mechanical energy in the form of vibrations. The frequency of 20 kHz is above the audible range for most people. Weld cycle time is very short, usually on the order of 1 to 2 seconds. Most thermal plastics can be joined by ultrasonics and the rigid plastics are the most easy to weld. Materials with a low modulus of elasticity attenuate vibrations and are more difficult to weld. Dissimilar plastics can be welded if their weld points are similar and there is chemical compatibility. Ultrasonic welding of plastic parts is shown in Figure 9–40.

Vibration Welding   Vibration welding uses heat that is generated by relative linear motion between the parts being welded, which are held together under pressure. The magnitude of movement is from $\frac{1}{16}$ in. (1.6 mm) to $\frac{1}{4}$ in. (3

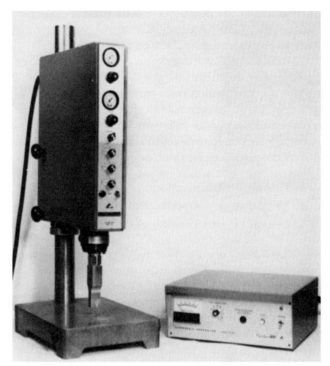

FIGURE 9–40   Ultrasonic welding of plastic parts.

mm). The frequency of movement is on the order of 100 to 300 Hz. This method is very similar to friction welding and is sometimes called linear friction welding. As soon as the generated heat causes melting of the surfaces at the interface, the vibration is stopped and the parts are aligned in final position and pressure is applied. Welding time ranges from 2 to 3 seconds, with a 1-second holding time for the joint to solidify. It can be used to weld most thermal plastic materials. It will weld numerous complex shapes together simultaneously. The equipment is designed especially for this welding and can be automated.

The AWS is currently developing a program to certify welders to join plastics and composites.

9-8   JOINING COMPOSITES AND CERAMICS

Composites are being used more and more by the automotive and aircraft industries. A composite material is composed of a combination of two or more constituents differing in forms that are essentially insoluble in each other. Generally, they combine high-strength reinforcing material, which is normally in a fiber form, and a holding material, or matrix. They offer certain advantages over metals and are replacing metals for some applications. The basic advantage of composites over metals is their high strength-to-weight ratio, their corrosion resistance, and the fact that they can be made either nonconductive or conductive to electrical current. They offer design flexibility, reduced finishing costs, and ease of parts assembly, and can be fabricated with less expensive tooling.[8]

Composites have disadvantages. They are more expensive than the metals they replace. They are difficult to join together, and certain types have specific disadvantages: Some are flammable, others do not have high-temperature properties, and so on. Even so, it is becoming necessary to join composites to make larger assemblies.

There are basically three categories of composites:

1. Polymer-matrix composites (plastics)
2. Metal-matrix composites (MMC) (metals)
3. Ceramic-matrix composites (ceramics)

The most widely used types are the polymer-matrix composites, which have been in use for many years and are finding wider applications. The least widely used are the ceramic-matrix composites, which are just now in research but will find industrial uses in the future. The metal-matrix composites (MMC) are currently being used; however, their price has prohibited widespread acceptance.

Composites are identified by the materials they contain, normally two. An example would be boron-aluminum, where boron fibers are in an aluminum matrix. The specification for a composite would include the names of the two components and a percentage of reinforcing materials by volume in the total composite. Normally, the reinforcing materials range from 10% to 60%. Often, the specification would include the analysis of the matrix and its treatment, such as heat treatment, and the type of reinforcement and its form. The reinforcing materials used in composites are either continuous or discontinuous. The continuous type would be wires or fibers that would be drawn, extruded, or spun. The discontinuous type would be chopped wires or fibers, whiskers, or particles. The reinforcing materials affect the engineering performance of the composite. It is important to know the orientation, length, shape, and composition of these materials. The matrix has two functions: (1) to hold the reinforcing materials or filaments in place; (2) as it deforms, it distributes the stress through the reinforcing materials. The matrix must be more ductile than the reinforcing material. It must transmit the forces to the reinforcing materials; thus there must be an intimate bond between the reinforcing material and the matrix. Joining composites depends entirely on the type of composite.

The polymer-matrix composites are joined by mechanical fasteners, adhesives, and, if the polymer matrix is a thermoplastic material, by welding.[9] For welding they are treated the same as a plastic material, based on the composition of the matrix. High-strength carbon fiber polymer composites can be joined by using resistance heating of the carbon filaments and pressure. Another system for producing lap welds is the use of induction heat-ing with a wire screen placed between the faying surfaces. An induction heating system introduces magnetic flux through the plastic to the wire screen, which becomes heated and creates melting of the surface and, by means of pressure, produces a weld. Fairly low power at a high frequency is used. Embedded resistance wires can also be used, as well as ultrasonic and friction welding. See Section 9-7 for details.

The metal-matrix composites (MMC) have the appearance of metals and are considered weldable with some of the welding processes. Welds have been made successfully using the gas tungsten arc welding process on titanium composites with metal reinforcing wires. Aluminum-boron composites have also been welded with gas tungsten arc welding. There can be severe damage to the boron filaments if the heat input is not accurately controlled. Research is ongoing. The plasma arc and electron beam processes are not acceptable since they usually cause melting of the boron filaments and result in metallurgical reactions that decrease the strength of the joints. Resistance welding, particularly spot welding, has been used for lap welding aluminum-boron composites. Resistance brazing has been used successfully, as well as weld bonding. Brazing and diffusion welding have also been applied successfully to the joining of metal-matrix composites. Ceramic-matrix composites are treated in the same way as ceramic materials.

Ceramics

Ceramic materials have been around for many years. They are commonly thought of as clay products made into dishes. Recently, engineered ceramic products have been considered for many applications, due to their high-temperature properties, low density, thermal insulation, wear characteristics, and corrosion resistance. Engineered ceramic products are made of high-purity raw materials so that their properties are consistent. Widespread use of ceramics require that they be joined together to metals and to nonmetals.

There are many different types of engineered ceramics. They fall into three general categories and are based on nitrides, carbides, and oxides. The two most widely used nitrides are silicone nitride and aluminum nitride. The most common oxides are alumina and zirconia. The most popular carbides are silicone carbide and boron carbide.

Ceramics are used by the aerospace industry, primarily in jet engines, the automotive industry, and the electronics industry. Much more use is planned as ceramic products are improved, have more consistent properties, and as additional data are acquired. Ceramics can be joined by means of adhesives and by means of cement-mortar-type inner layers. Unfortunately, adhesives do not have high-temperature properties and thus are limited to medium-temperature applications. Cement and mortar have higher-temperature capabilities but do not provide sufficient tensile strength for many applications. Fusion welding has not been applied successfully to joining ceramics. Successful joining has been accomplished with the use of metallic inner layers. The joining process usually involves brazing, soldering, or diffusion bonding. The inner layers are usually ductile metal foils placed between the parts to be joined. High-quality joints have been produced with metallic interlayers using the diffusion bonding procedure. The joining of ceramics to metals has been more successful. Mechanical joints are widely used. This is the familiar method of manufacturing spark plugs, which is done by crimping the metallic portion around the ceramic portion of the spark plug. Adhesive bonding has also been used and has many applications; however, it is severely limited due to the lack of high-temperature properties.

The joining of ceramics to metals seems best when a metallic interliner is used. Interliner metal must be selected so that it will "wet" both the metal and the ceramic and should have a melting temperature close to that of the metal. The diffusion bonding technique seems to be the most successful, and with a voltage applied across the joint, the bonding seems even more successful. Silicone nitride has been joined to various metals using a copper-titanium intermetallic or filler alloy. One of the major problems with joining ceramics to metals is their difference of thermal expansion, which limits the possibilities. The intermetallic layer seems to be an assist in this regard, such as a titanium layer of filler metal in austinetic stainless steel. Much research, and ultimately better solutions, will become available.

9-9 PREHEAT AND POSTHEAT TREATMENT

Preheating is the application of heat to the base metal immediately before welding, brazing, soldering, or cutting. Postheating is the application of heat to an assembly after welding. This might be better described as a postweld heat treatment, which includes any heat treatment following welding.

All the arc welding processes and many of the other welding processes use a high-temperature heat source. A steep temperature differential occurs between the localized arc heat source and the cool base metal. This temperature difference causes differential thermal expansion and contraction and high stresses. By reducing the temperature differential, these can be minimized. This will reduce the danger of weld cracking, reduce the maximum hardness, minimize shrinkage stresses, lessen distortion, and help gases—particularly hydrogen—escape from the metal. Preheating will reduce the temperature differential, which reduces the other problems. Preheating is also done on highly conductive metals in order to maintain sufficient heat at the weld area. The preheat temperature depends on factors, such as the composition and mass of the base metal, the ambient temperature, and the welding procedure.

The interpass temperature should also be considered. It is involved in multiple-pass welds and is the temperature, both minimum and maximum, of the deposited weld metal and adjacent base metal before the next pass is started. Usually, the minimum interpass temperature will be the same as the preheat temperature. The weldment temperature should never be allowed to become lower than the preheat or the interpass temperature. If welding is interrupted, the interpass temperature must be attained before welding is started again. Preheat and interpass temperatures must be completely through the thickness of the metal. The interpass temperature may also be specified as a maximum temperature. When welds are made on a small weldment, its temperature will usually increase due to the heat input from welding. Under certain conditions it is not desirable to allow the heat to build up and exceed a specific temperature; therefore, a maximum interpass temperature will be specified. When heat buildup becomes excessive, the weldment must be allowed to cool, but not below the minimum interpass temperature. The temperature of the welding area must be maintained within the minimum and maximum interpass temperature.

There are different ways for preheating, including use of gas torches, gas burners, heat-treating furnaces, electrical resistance heaters, low-frequency induction heating, and temporary furnaces. The choice of the preheating method depends on many factors, such as the preheat temperature, the length of preheating time, the size and shape of the parts, and whether it is one-of-a-kind or a continuous-production type operation. Any method used for postheating can also be used for preheating.

On critical work the preheat temperature must be precisely controlled. In these cases, controllable heating systems are used, and thermocouples are attached directly to the part being heated. The thermocouple will measure the exact temperature of the part and will provide a signal to a controller, which regulates the fuel or electrical power required for heating. The temperature of the part being heated can be held to close limits. Most code work requires precise heat temperature control.

A common method of preheating is by torches or burners utilizing flames. Figure 9–41 shows a natural gas burner used to preheat the lip of a powershovel dipper. This sub-assembly is a critical part made of quenched and tempered high-alloy steel. The preheat and interpass temperature must be maintained throughout the entire welding cycle. Open flames such as this cannot be as precisely controlled as other methods and are therefore used only for preheating and maintaining interpass heat.

FIGURE 9–41   Preheating with open gas flame.

Postweld Heating

There are a number of postweld heat treatments for weldments but stress relieving is the most widely used. Some of these other heat treatments are annealing, normalizing, drawing, and fusing. *Annealing,* or *full annealing* as it is sometimes called, is a heat treatment that increases the temperature of steel above the critical or recrystallization temperature, followed by slow cooling. It is normally heated to a point about 100° F (38° C) above the critical temperature line of the steel. Cooling is usually done in a furnace to provide a substantially stress-free condition. Figure 9–42 shows this in a diagram.

Normalizing is similar to annealing. The heating rate and holding periods are identical, but in normalizing the cooling rate is faster and is usually done by allowing the part to cool in still air rather than in the furnace. Due to the higher cooling rate, normalizing usually provides a structure with greater strength and less ductility than annealing.

Tempering is a heat treatment done at a much lower temperature than annealing, normalizing, or stress relieving. It is an operation that often follows a quenching operation. It tends to reduce the hardness, and strength of a steel, but it improves ductility and toughness.

FIGURE 9–42   Heat-treatment cycles.

Fusing is a specialized process of heating a thermal spray deposit to cause it to coalesce, solidify, and bond metallurgically to the base material. This can be done by almost any heating method.

Stress relieving is of major importance to weldments. It is similar to normalizing except that it is done at a temperature below the critical temperature, usually in the range of 1050 to 1200° F (566 to 649° C).

Both annealing and normalizing relieve residual stresses better than stress relieving. They are carried out above the critical temperature. They involve changes in grain structure and tend to produce heavy scale. They may also produce serious dimensional changes and require that complex large structures are braced to avoid sagging.

Stress relieving is required by some codes. Refer to the specific portion of the code that is applicable to decide on the stress-relieving schedule. Many products not built under code are stress relieved for the following reasons:

1. To reduce the residual stresses inherent in any weldment, casting, or forging
2. To improve the resistance to corrosion and caustic embrittlement
3. To improve the dimensional stability of the weldment during machine operations
4. To improve the service life of the weldment

Stress relieving should be performed if a weldment is subjected to impact loading or to low-temperature service, or if it is exposed to repetitive or fatigue loading. In all the heat treatments, heating rates and times may be specified. The maximum temperature is related to the composition of the steel; the holding time, at the maximum temperature, is related to the material thickness; and the cooling rate is related to the particular treatment and to the code. The rate of heating is usually in the 300 to 350° F (149 to 177° C) per hour rate. The holding temperature is usually 1 hour for each inch of maximum thickness in order to provide for uniform heating throughout. The cooling rate is also in the range 300 to 350° F (149 to 177° C) per hour down to a specific temperature. In some cases the cooling rate can be increased when the part has cooled to 500 to 600° F (260 to 316° C). These rates of heating, holding, and cooling are usually a part of the specification and must be followed explicitly.

Temperature indicators and controlling equipment must be used for postweld heat treatment. Figure 9–43 shows a typical car bottom furnace loaded with weldments that have been stress relieved. These furnaces are natural gas or fuel oil fired. The complete heat treating cycle may require up to 24 hours. Thermocouples are attached to the weldment and are used to signal the controller. Automatic controllers can be set for a specific rate of heating, a specific holding time, and a specific rate of cooling.

FIGURE 9–43 Car bottom furnace for stress relieving.

Accurate control is also possible with induction and resistance heating. Figure 9–44 shows the use of resistance heating coils for both pre- and postheating. Figure 9–45 shows the use of low-frequency induction heating coils (400 Hz) for pre- and postheating welds made on large pipe. Both resistance and induction heating are very popular for preheating and stress relieving weld joints in power plant piping welds. Temperature record charts are required by some codes for each weld or weldment.

A field method for preheating and stress-relieving pipe weld joints has recently been introduced. This method utilizes exothermic material, which is fitted around the welded pipe joint. An insulating nonflammable material is placed between the metal and the exothermic material. This is followed by another layer of insulating material, which is secured to hold the exothermic material in place. The exothermic material is then ignited by a torch or fuse. It burns, creates heat, and heat treats the pipe joint. Temperatures can be checked by means of temperature-measuring crayons, and the results of the heat treatment are checked by making hardness measurements after the heat treatment. This is acceptable by certain codes. The advantage of this system is that additional electric power and skilled personnel are not required.

It is important to check the temperature of items being preheated. An effort should be made to obtain and maintain uniform heating throughout the entire part. The surface temperature should be the same on the front and back. This can be checked by using a portable pyrometer, which shows the temperature based on touching the part being heated. This is shown in Figure 9–46. Another way is

FIGURE 9–44 Resistance heating for pre- and postheat.

FIGURE 9–45 Induction heating for pre- and postheat.

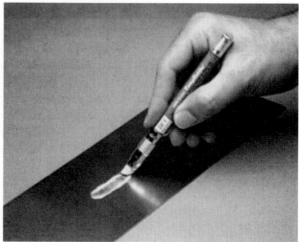

FIGURE 9–47 Melting of temperature-indicating crayon.

FIGURE 9–46 Portable pyrometer.

the use of temperature-indicating crayons.[10] They melt at specific temperatures, as shown in Figure 9–47. By selecting the crayon with the specific melting temperature, an accurate measurement can be made. Crayons are available in ratings starting at 100° F up to 2500° F. An example is the use of crayons for checking the temperature of a cast ringear being preheated, shown in Figure 9–48. They make positive contact so that the temperature being measured is the temperature of the workpiece. Crayons can be certified as to their melting temperatures. Certain temperature-indicating crayons contain sulfur, inorganic sulfate, or halogens and should not be used for certain materials. Contact the crayon manufacturer for details.

FIGURE 9–48 Use of temperature-indicating crayons.

9-10 MECHANICAL STRESS RELIEF

As described in the preceding section, thermal stress relief is used to relieve yield point stresses inherent in weld-

ments. In some cases thermal stress relief is impractical or extremely expensive due to rising energy costs. If this is the case, mechanical stress relief can be used. There are three basic types of mechanical stress relief:

1. Overstressing, creating plastic deformation
2. Surface treatment by hammering or shotpeening
3. Mechanical vibrations: vibratory stress relief

All three methods are used successfully and each has specific application advantages and disadvantages.

The overstressing method can be accomplished in several different ways. The objective is to stress the weldment beyond the yield point stress of the metal and cause plastic deformation. Overstressing loads can be applied slowly or rapidly. When the load is removed, the maximum stresses remaining will be below yield point and under certain conditions may become residual compressive stresses. One way of accomplishing this is by proof testing the weldment with 150% to 200% of the maximum design load. This will reduce the yield point stresses when the load is eliminated. Another method of overstressing is to thermally expand the metal adjacent to the weld. This is done by means of gas heating torches that are moved parallel to the weld joint. The expansion overstresses the weld and the heat-affected zone, which reduces yield point stresses. This method has been used successfully to reduce residual stresses in welded ships.

Surface treatment by hammering or shotpeening is widely used in many metalworking applications other than welding. This plastically deforms the surface, which reduces the surface yield point stresses. This technique can provide surface compression stresses and is widely used to increase fatigue life. The peening and hammering are difficult to control but widely used for local stress relief.

Stress relief by exposure to mechanical vibration is becoming more widely used.[11] It is performed by attaching a mechanical vibrator, usually an eccentric rotating weight, to the weldment. An example is shown in Figure 9-49. The vibrating device is firmly attached to the weldment and is shown attached to a large pipe weldment assembly. The speed of the drive motor that controls the frequency of vibration is adjusted until it matches the resonant frequency of the total weldment or of a specific area of the weldment. This is controlled by a speed regulator in the control panel shown to the left in the picture. According to some experts, resonance is not required for stress relief. However, vibrations of sufficient power at or near the resonant frequency of the weldment reduce yield point stresses. This occurs due to the alternating stresses that cause slip in individual grains. The critical cycle strain amplitude must be exceeded, and the weldment must be allowed to deform freely during treatment. Frequencies up to 100 Hz are used. One of the advantages of vibratory stress relief is the reduction of distortion during machining. This is due to the removal of residual stress in the surface layers of the weldment and its effect on the weldment when machined away.

The vibratory stress relief technique can be used while welding on complex weldments. This technique results in minimum distortion of the weldment during welding and during machining. Vibratory or mechanical stress relief has not been accepted by code-making bodies, and there seems to be considerable technical skill involved in accomplishing the desired results.

FIGURE 9-49 Vibratory stress relief apparatus.

QUESTIONS

9–1. What five conditions must apply for successful oxy-fuel gas cutting?

9–2. What is the best way to compare various makes of cutting tips?

9–3. What methods are used to guide an automatic flame cutting machine?

9–4. How is the nesting program for automatic cutting machines generated?

9–5. What is stack cutting? What problem is sometimes encountered?

9–6. Explain the oxygen arc cutting process. What metals can be cut?

9–7. Explain the difference between air carbon arc cutting and carbon arc cutting.

9–8. What is the advantage of plasma arc cutting?

9–9. Which process is fastest for cutting $\frac{1}{4}$-in. (6.4-mm)-thick material?

9–10. What is the advantage of using water with plasma arc cutting?

9–11. Is semiautomatic plasma arc cutting possible?

9–12. What are the three thermal spraying methods?

9–13. Are powders used for the electric arc spraying processes? If so, why?

9–14. Explain how parts are prepared for spraying.

9–15. Explain the difference between flame spraying and the detonation spray method.

9–16. What are the two major categories of plastics?

9–17. What are the major types of composites?

9–18. What is interpass temperature?

9–19. Explain the difference between annealing, stress relieving, and normalizing.

9–20. How can you measure the temperature of a heated weldment?

REFERENCES

1. "Recommended Practices for Plasma Arc Cutting," AWS C5.2, American Welding Society, Miami, Fla.

2. "Thermal Spraying Practice, Theory and Application," American Welding Society, Miami, Fla., 1985.

3. "Recommended Practices for Fused Thermal Spray Deposits," AWS C2.15, American Welding Society, Miami, Fla.

4. "Guide for Thermal Spray Operators and Equipment Qualifications," AWS C2.16, American Welding Society, Miami, Fla.

5. M. N. Watson and R. M. Rivet, "Welding of Plastics," Materials Engineering '85 Conference, London, Nov. 1985; The Welding Institute, Cambridge, England.

6. J. Agranoff, ed., *Modern Plastics Encyclopedia,* (New York: McGraw-Hill, published annually).

7. R. L. Miller and D. R. Winkleman, "Welding Plastics with Rod: A Downstream Operation for Custom Molders," *Plastics Engineering* (April 1980): 38.

8. M. M. Schwartz, *Composite Materials Handbook* (New York: McGraw-Hill, 1983).

9. A. Benatar and T. G. Gutowski, "A Review of Methods for Fusion Bonding Thermoplastics Composites," *SAMPE Journal* (Jan.–Feb. 1987).

10. "Temperature Indicators—Useful Tools in High-Tech Welding," *The Fabricator* (Nov 1989).

11. A. G. Hebel, Jr., "Subresonant Vibration Relieves Residual Stress," *Metal Progress* (Nov 1985).

10

POWER SOURCES FOR ARC WELDING

OUTLINE

10-1 ARC WELDING ELECTRICITY

The electrical arc welding circuit is the same as any electrical circuit. In the simplest electrical circuits, there are three factors:

1. *Current:* flow of electricity
2. *Pressure:* force required to cause the current to flow
3. *Resistance:* force used to regulate the flow of current

Current is a rate of flow. Current is measured by the amount of electricity that flows through a wire in one second. One *ampere* (A) is the amount of current per second that flows in a circuit. The letter I is used to designate current in amperes.

Pressure is the force that causes a current to flow. The measure of electrical pressure is the *volt*. The voltage between two points in an electrical circuit is called the *difference* in potential. This force or potential is called *electromotive force* (EMF). The difference of potential or voltage causes current to flow in an electrical circuit. The letter E is used to designate voltage or EMF.

Resistance is the restriction to current flow in an electrical circuit. Every component in the circuit, including the conductor, has some resistance to current flow. Current flows more easily through some conductors than others; that is, the resistance of some conductors is less than others. Resistance depends on the material, the cross-sectional area, and the temperature of the conductor. It is designated by the letter R. The unit of electrical resistance is the *ohm*. Copper is widely used for conductors since it has the lowest electrical resistivity of common metals. Insulators have a very high resistance and will not conduct current.

The simple electrical circuit shown in Figure 10-1 includes two meters for electrical measurement, a voltmeter and an ammeter. It also shows a symbol for a battery. The longer line of the symbol represents the positive terminal. The electron current flows from the negative (−) to the positive (+). The arrow shows the direction of current flow. Conventional current flows in the opposite direction.

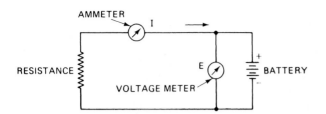

FIGURE 10–1 Simple electrical circuit.

The ammeter is a low-resistance meter, shown by the round circle and arrow adjacent to the letter *I*. The pressure or voltage across the battery can be measured by a voltmeter. The voltmeter is a high-resistance meter, shown by the round circle and arrow adjacent to the letter *E*.

The resistance in the circuit is shown by a zigzag symbol. The resistance of a resistor can be measured by an ohmmeter. An ohmmeter is *never* used to measure resistance in a circuit when current is flowing.

The relationship of these three factors is expressed by *Ohm's law* as follows:

$$current = \frac{pressure}{resistance}$$

or

$$amperes = \frac{volts}{ohms} \quad or \quad I = \frac{E}{R}$$

where I = current in amperes (flow)
E = pressure in volts (EMF)
R = resistance in ohms

Ohm's law can also be expressed as

$$E = IR \quad or \quad R = \frac{E}{I}$$

By simple arithmetic, if two values are known or measured, the third value can be determined.

A few changes to the circuit can be made to represent an arc welding circuit. Replace the battery with a welding machine, since they are both a source of EMF (or voltage), and replace the resistor with a welding arc, which is also a resistance to current flow (Figure 10–2).

FIGURE 10–2 Welding electrical circuit.

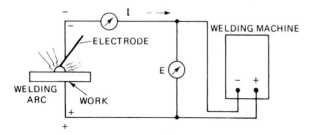

The electron current will flow from the negative terminal through the resistance of the arc to the positive terminal.

In the early days of arc welding, using bare metal electrodes, it was normal to connect the negative side of the generator to the electrode and the positive side to the workpiece. This was known as **straight polarity.** When deeper penetration was required on the base metal, the polarity would be reversed. This connected the electrode to the positive pole of the generator and the workpiece to the negative pole. The welder could quickly change the polarity of the welding current by means of a polarity switch. In marking welding machines and polarity switches, these old terms were used and indicated the polarity as *straight* when the electrode is negative, and *reverse* when the electrode is positive. In this book, to avoid confusion, whenever polarity is discussed the term **electrode negative** (DCEN) is used instead of *straight polarity* (DCSP) and **electrode positive** (DCEP) is used instead of *reverse polarity* (DCRP).

The ammeter used in a welding circuit is a millivoltmeter calibrated in amperes connected across a shunt in the welding circuit. The shunt is a calibrated, very low-resistance conductor. The voltmeter shown in the figure will measure the welding machine voltage output and the voltage across the arc, which are essentially the same. Before the arc is struck, the voltmeter will read the voltage with no current flowing in the circuit. This is known as the *open circuit voltage* and is usually higher than the arc voltage or voltage across the machine when current is flowing.

Another unit in an electrical circuit, and important to welding, is the unit of *power.* The rate of producing, or of using, electrical energy is called power and is measured in watts. Power in a circuit is the product of the current in amperes times the pressure in volts, or

$$power = current \times pressure$$

or

$$watts = amperes \times volts$$

or

$$P = I \times E$$

where P = power in watts
I = current in amperes
E = pressure in volts

When welding using a $\frac{1}{8}$-in. electrode at 100 A and an arc voltage of 25, the power would be 2500 watts (W); 2500 W can be expressed as 2.5 kilowatts (kW). Power is measured by a wattmeter, which is a combination of an ammeter and a voltmeter.

In addition to power, it is necessary to know the amount of work involved. Electrical *work* or energy is the product of power times time and is expressed as watt-seconds or joules or kilowatt-hours.

$$\text{Work} = \text{power} \times \text{time} \qquad \text{or} \qquad W = Pt$$

where W = work in watt-seconds or joules or kilowatt-hours

P = power in watts or kilowatts

t = time in seconds or hours

Cost-of-welding calculations involve these work units since the watt-hour or kilowatt-hour are commercial units of work and are the basis of charges by the electric utility companies. So far, we have dealt exclusively with direct current electricity, electricity that flows continually through the circuit in the same direction. Alternating current electricity is also important since it is the power furnished by utility companies.

Alternating current is an electrical current which flows back and forth at regular intervals in a circuit. When the current rises from zero to a maximum, returns to zero and increases to a maximum in the opposite direction, and finally returns to zero again, it is said to have completed one cycle. For convenience, a cycle is divided into 360 degrees. Figure 10–3 is a graphical representation of a cycle and is called a *sine wave*. It is generated by one revolution of a single loop coil armature in a two-pole alternating current generator. The maximum value in one direction is reached at the 90° point and in the other direction at the 270° point. The number of times this cycle is repeated in one second is called the frequency and is measured in hertz. When a current rises to a maximum in each direction 60 times a second, it completes 60 cycles per second or has a frequency of 60 hertz (Hz). The frequency of electrical power in North America and other parts of the world is 60 hertz. Fifty hertz is used in Europe, Africa, part of Asia, and South America.

The principle of electrical generation states that "when a conductor moves in a magnetic field so as to cut lines of force, an electromotive force is generated." The lines of force run between the north and south magnetic poles of the generator. The single turn coil rotates within these lines of force or magnetic field, and as the conductor cuts the lines of force at right angles, the maximum voltage is generated (i.e., at 90° and at 270°). When no lines of force are being cut as at positions 0°, 180°, and 360°, there is no EMF generated.

Alternating current (ac) for arc welding has the same frequency as the line current. The voltage and current in the ac welding arc follow the sine wave and go through zero twice each cycle. The frequency is so fast that the arc appears continuous and steady to the naked eye. The sine wave is the simplest form of alternating current. It is always assumed that alternating current has a sine wave shape unless otherwise stated.

Alternating current and voltage are measured with ac meters. An ac voltmeter measures the value of both the positive and negative parts of the sine wave. It reads the effective voltage, called the root-mean-square (rms) voltage. The effective direct-current value of an alternating current or voltage is 0.707 times the maximum value.

An alternating current has no unit of its own, but is measured in terms of direct current, the ampere. The ampere is defined as a steady rate of flow, but an alternating current is not a steady current. An alternating current is said to be equivalent to a direct current when it produces the same average heating effect under exactly similar conditions. This is used since the heating effect of a negative current is the same as that of a positive current. Therefore, an ac ammeter will measure a value called the effective value of an alternating current, which is shown in amperes. All ac meters, unless otherwise marked, read effective values of current and voltage.

Ohm's law also applies to ac circuits. This is because Ohm's law deals only with voltage, current, and resistance. In ac welding circuits there are other factors, and one of the most important is inductance. To understand inductance we must refer to magnetism.

A magnet has a north pole and a south pole, which have identical strength. Between these poles there are lines of force. This effect can be shown by sprinkling iron filings on a sheet of paper and placing it over a magnet. The distinct pattern shows these lines of force running from one pole to the other. Similar lines of force exist around electric conductors that carry direct current. This can be proven by placing a small compass near a current-carrying wire. The needle will deflect when the current is turned off and on. Magnetic lines of force create physical forces between magnets or magnetic fields around current-carrying wires. This is the principle of operation of an electric motor. The magnetic properties of a ferromagnetic material such as iron when wrapped with a coil of wire are such that the combination will produce a much stronger magnetic field than the magnetic field produced by the coil alone. The coil of wire around an iron core is a magnetic circuit. Magnetic circuits will have a specific inductance. Inductance expresses the results of

FIGURE 10–3 Sine-wave generation.

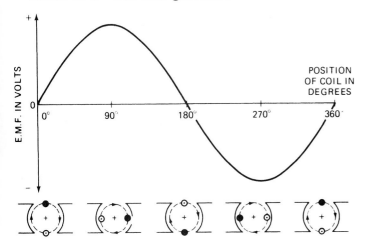

a certain arrangement of conductors, iron, and magnetic fields. Inductance involves change since it functions only when magnetic lines of force are cutting across electrical conductors. Inductance is important only in ac circuits or in dc circuits when they are connected or disconnected. When the current is turned off, the magnetic field collapses and the lines of force cut across the wires and induce current in the wires in the same direction as it had been flowing. If the coil is connected to alternating current, the lines of force build up to the maximum and then collapse and then build up in the opposite direction to a maximum and collapse each cycle. If another coil is placed on the same iron core and close to the first coil, the magnetic lines of force will cut across the second coil and induce the EMF in it. The closer the coils or the stronger the magnetic lines of force, the greater will be the induced EMF. This is the principle of the transformer and is shown in Figure 10–4. By changing the magnetic coupling of the two coils, we can control the output of the second coil (the secondary) and thus the output of the welding transformer. This coupling can be changed by moving the coils closer together or by increasing the strength of the magnetic field between them. The strength of the magnetic field can be changed by putting more or less iron in the area between the coils or by adjusting the availability of the magnetic field in other ways.

The output of a transformer welding machine is alternating current of the same frequency as the input power. A rectifier is a device that conducts current easier in one direction than the other. It has a high resistance to current flowing in one direction and a low resistance to current flowing in the opposite direction. A diode vacuum tube is an efficient rectifier but will not carry sufficient current for welding. Another type, the dry disk rectifier, employs layers of semiconductors such as selenium between plates. The newest and most popular rectifier is the silicon diode. These are made of thin wafers of silicon that have had small amounts of impurities added to make them semiconductors. The wafers are specially treated and then assembled in holders for mounting in welding machines. The diodes are connected to the output of a welding transformer to produce a rectifier welding machine with dc output.

FIGURE 10–4 Transformer principle.

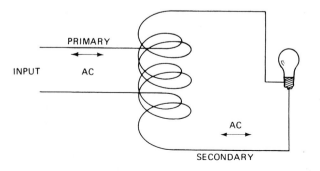

10-2 WELDING ARC REQUIREMENTS

An arc is a sustained luminous discharge of electricity across a gap in a circuit due to the incandescence of the conducting vapors. A welding arc is a controlled, sustained, and stable electrical discharge between the electrode and the workpiece. It is formed and sustained by the establishment of a gaseous conducted medium called an *arc plasma*. This is done by means of a welding power source.

There are two basic types of welding arcs: the *nonconsumable electrode arc,* which includes gas tungsten arc welding, carbon arc welding, and plasma arc welding, and the *consumable electrode arc,* which includes shielded metal arc welding, gas metal arc welding, flux cored arc welding, submerged arc welding, and electrogas arc welding. These two arcs are entirely different. The nonconsumable electrode does not melt in the arc, and filler metal is not carried across the arc gap. It was described in Section 5-1. The consumable electrode melts in the arc and the molten metal is carried across the arc gap. It was described in Section 6-1.

To provide a stable arc and to maintain a controllable molten weld metal pool, it is necessary to match the output characteristics of the power source to the welding process. Figure 10–5 shows the basic welding machine static volt–ampere characteristic curves and variations. It shows actual true constant current and true constant voltage, typical output curves, and slope curves. The volt-ampere curve is plotted for a machine by placing a voltmeter across the output terminals and an ammeter in the resistance load circuit attached to the output terminals. The voltage is plotted on the vertical left-hand scale and the current in amperes is plotted along the horizontal scale. The open-circuit voltage is obtained without any load attached to the output terminals. Resistance load is placed across the output terminal, and the output voltage along with the current is plotted on the chart. The load is changed gradually and points are plotted periodically until the maximum load rating of the power source or a short circuit is obtained. A short circuit occurs at zero output voltage. These curves are plotted for different types of machines using a resistance load bank. Of course, in addition, welding machines are designed to produce the specific output characteristic curve that is desired. This curve shows the normal arc voltage range for consumable electrodes. Special processes such as pulse welding may require unique volt–ampere characteristics.

Historically, the first welding power source was a battery with a resistor in series with the arc. To start the arc, the battery, or open-circuit voltage, would be approximately 90 V. The adjustable ballast resistors would allow the welder to vary the welding current. The static volt-ampere curve could be approximately a straight line

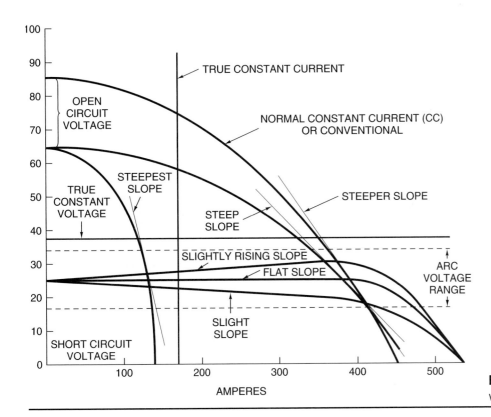

FIGURE 10-5 Basic static volt-ampere characteristic curves.

on an angle, depending on the resistance. The voltage drop across the ballast resistor and across the arc would equal the total voltage of the battery. Today this is an impractical solution due to the power dissipation and low efficiency, but it did provide a stable arc, with the welder having the ability of lengthening and shortening the arc to control the welding current. This first welding arc was between a carbon electrode and a metal workpiece. Electrical engineers studied the arc and the characteristics and designed the necessary parameters for a practical generator welding machine.

The first machines designed for welding were used for manual bare metal arc welding. This employed a welding arc between a bare or lightly coated metal electrode and the work. It was used without shielding and required a very good power source to provide a stable arc. Bare metal arc welding was soon replaced by the shielded metal arc welding process manually using covered electrodes. In the early days the welding power source was designed for use with bare electrodes and had drooping arc characteristics. Now known as a conventional welding power source, it is still used for shielded metal arc (stick) welding.

Constant Current System

The conventional or drooping constant current power source was designed to give the manual welder maximum flexibility. Figure 10-6 shows a typical characteristic output curve for a conventional drooping welding power source. It also shows curves for welding arcs. The top arc curve is for a long arc, the middle for a normal arc, and the lower curve is for a short arc. The intersection of a curve for an arc and a characteristic curve of a welding machine is known as an *operating point*. The operating point is changing continually during welding since the welder cannot maintain an absolutely perfect arc length. While welding, and without changing the controls on the machine, the welder can lengthen or shorten the arc and change the arc voltage from 35 to 25 V. The curve shows that without changing the machine setting, the short arc (lower voltage) produces a higher-current arc. Conversely, a long arc (higher voltage) is a lower-current arc. This allows the welder to control penetration of the arc. For example, during the welding operation, conditions in the joint may change. The root face might temporarily become thicker or the root opening smaller, or they could become thinner and the root opening wider. Higher current would be required in the first case, and lower current in the second case. By lengthening or shortening the arc, the welder can control penetration to obtain a good-quality weld. The longer arc will reduce the current, which in turn reduces penetration and allows for a smaller weld pool. The shorter arc (lower voltage) increases the welding current and increases penetration. Thus, by lengthening or shortening the arc, the welder can control penetration during welding without touching the machine controls. This is particularly important for vertical welding. When the welder lengthens the arc briefly, the current is reduced, the arc spreads out, and molten metal

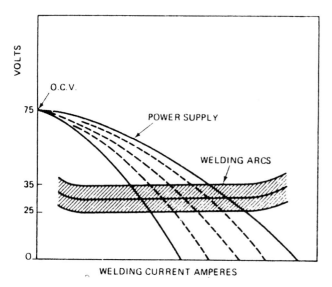

FIGURE 10–6 Volt-ampere curve, with single-control machine.

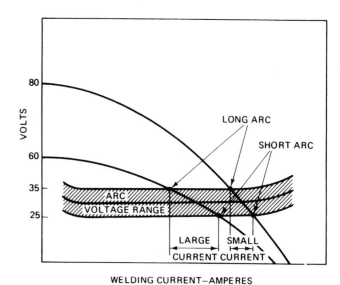

FIGURE 10–7 Volt-ampere slope versus welding operation, with dual control machine.

freezes more quickly. The amount or volume of molten metal is reduced, which allows the welder to control the molten metal in vertical welding. The slope of the characteristic curve through the arc voltage range determines the change in welding current for a specific change in arc voltage. This led in the early days to the development of the dual-controlled generator, to give the operator more control.

The single-control conventional machine is most common. It has only one adjustment, a range switch or plug-in connectors, which changes the current output from minimum to maximum. By adjusting the current control, a number of output curves are obtained. The dashed lines show intermediate adjustments of the machine. On a range switch or plug-in machine, the number of curves will correspond to the number of taps or plug-in combinations available. Most transformers and transformer-rectifier machines are single-control welding machines.

Later, dual-control machines, normally generator welding machines, were developed to provide more flexibility for different welding applications. These machines have both current and voltage control, normally two different adjustments: one for coarse current control and the other for fine current control, which also provides adjustment of the open-circuit voltage. The welder can adjust the machine for more or less change of current for a given change of arc voltage. Figure 10–7 shows two characteristic curves obtained on a dual-control machine by adjusting the fine control knob to 80-V or 60-V open-circuit voltage. When using the 80 V curve, a steeper slope is produced than when the adjustment is for 60 V. Through the arc voltage range a flatter or steeper slope is obtained. The flatter sloped curve provides a digging arc with an equal change in arc voltage. The steeper slope has less

change for the same change in arc length and provides a softer arc. Dual-range machines allow the welder to control the current over a fairly wide range. Dual-range machines are no longer popular.

Later, the ac transformer welding machine was developed for shielded metal arc welding. The static volt–ampere characteristic curve is as shown in Figure 10–6. Some transformer welding power sources have fine and coarse adjustment knobs but are not dual-control machines. Alternating current welding differs from direct current welding since voltage and current pass through zero at each current reversal according to the line frequency, 100 or 120 times per second. Reactance designed into the machine causes a phase shift between the voltage and current, so they do not both go through the zero at the same instant. When the current goes through zero, the arc is extinguished, but because of phase differences there is voltage present that helps to reestablish the arc quickly. The ionization in the arc stream affects the voltage required to reestablish the arc and overall stability of the arc. Arc stabilizers (ionizers) are included in the coatings of electrodes designed for ac welding to provide a stable arc.

Constant current welding machines are used for some automatic welding processes. The wire feeder and control must duplicate the manual motions of the welder to start and maintain an arc. This requires a complex system with feedback from the arc voltage to compensate for changes in arc length. These are known as *voltage-sensitive control wire feeders.* Constant current power supplies are rarely used for very small-diameter electrode wire welding applications.

As gas tungsten arc welding became more popular, there was a demand for a power source that had little or

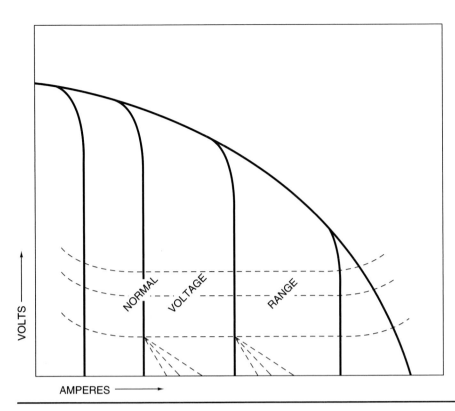

FIGURE 10–8 Characteristic curve for constant current with arc force.

no current change with arc length change. The welder could not maintain a perfectly consistent arc length, and this caused the welding current to change when it was not desired. This led to the development of a power source with a true constant current volt-ampere curve within the arc voltage range, shown in Figure 10-8. The welding current remains the same whether the arc is short or long. This led to the development of an *arc force* to avoid shorting, which kicks in at a value of or slightly below the normal low arc voltage. It provides higher current at a low voltage, which improves arc stability by providing quick recovery during partial arc outages. Arc force is adjustable, as shown by dashed lines in Figure 10-8. This is a great advantage of gas tungsten arc welding since the working arc length of the tungsten arc is limited. It provides a more driving arc for shielded metal arc welding.

Constant Voltage System

The constant voltage (CV) welding machine and the CV system of automatic arc length control was introduced at the same time as semiautomatic gas metal arc welding. This combination has made gas metal arc welding extremely popular. Prior to the introduction of the CV welding machine, the conventional drooping characteristic type of power source was employed with voltage-sensing electrode wire feed systems. The reaction time of the voltage sensing system was not fast enough to avoid burnback and stubbing when using small-diameter electrode

wire. The voltage sensing system is still widely used for automatic submerged arc welding.

The CV electrical system is not new to the electric power industry. It is the basis of operation of the entire commercial electric power distribution system. The electric power delivered to your house and available at every receptacle has a constant voltage, normally 115 V. This is shown in Figure 10-9. This voltage is maintained at each outlet, whether a small night-light with a very low wattage rating or an electric heater with a high wattage rating is connected. The current that flows through each circuit will be different based on the resistance of the particular appliance, in accordance with Ohm's law. For example, the small lightbulb will draw less than 0.01 A of current, whereas the electric heater may draw over 15 A. The voltage remains constant, but the current flowing through each appliance depends on its resistance or electrical load. The same principle is utilized by the CV welding system.

This relationship is the basis of the simplified control for electrode wire feeding using a CV power source. Instead of regulating the electrode wire feed speed rate to maintain the constant arc length, the electrode wire is fed into the arc at a fixed speed and the power source is designed to provide the necessary current to melt off the electrode wire at the same rate. This concept prompted development of the CV welding power source.

The volt–ampere characteristics of the CV power source shown in Figure 10-10 were designed to produce

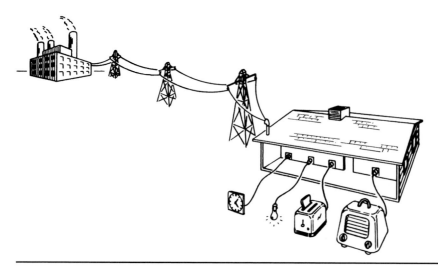

FIGURE 10-9 Constant voltage system.

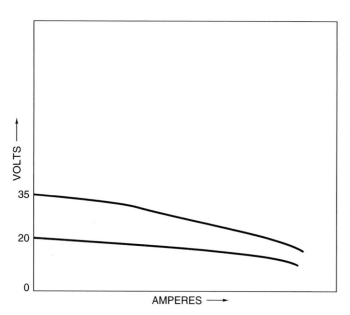

FIGURE 10-10 Static volt-ampere characteristic curve of CV machine.

substantially the same voltage at no load and at rated or full load. Most CV machines do not produce a true flat curve. Experience has shown that a slight droop is preferable. If the load in the circuit changes, the power source adjusts its current output automatically to satisfy this requirement and maintains essentially the same voltage across the output terminals. This system assured a self-regulating arc based on a fixed rate of wire feed speed and a CV power source. The arc voltage is controlled by adjusting the output voltage of the power source. The welding current is controlled by the wire speed of the constant (but adjustable) speed of the wire feeder.

Another way to understand the CV system is to consider the electrical resistance load in the welding circuit. These resistances or voltage drops occur in the welding arc, welding cables and connectors, welding gun, and electrode length beyond the current pickup tip to the arc. These voltage drops add up to the output voltage of the welding machine and represent the electrical resistance load on the welding power source. When the resistance of any component in the external circuit changes voltage, the balance will be achieved by changing the welding current in the system. The greatest voltage drop occurs across the welding arc. The other voltage drops are relatively small and constant. The voltage drop across the welding arc is dependent on the arc length. A small change in arc volts results in a relatively large change in welding current. Figure 10-11 shows that if the arc length is shortened by 2 V, the welding current increases by approximately 100 A. This change in arc length greatly increases the melt-off rate and quickly brings the arc length back to normal.

The CV power source is changing its current output continually to maintain the voltage drop in the welding circuit. Changes in wire feed speed that occur when the welder moves the gun toward or away from the work are compensated by momentarily changing the current and the melt-off rate until equilibrium is reestablished. The same corrective action occurs if the wire feeder has a temporary reduction in speed. The CV power source and fixed wire feed speed system is self-regulating. It is an excellent wire feed system, especially for semiautomatic welding, since movement of the cable assembly often changes the drag or feed rate of the electrode wire. The CV welding power source provides the proper current so that the melt-off rate is equal to the wire feed rate. The arc length is controlled by setting the voltage on the power source. The welding current is controlled by adjusting the wire feed speed.

The characteristics of the welding power source must be designed to provide a stable arc when welding with different electrode sizes and metals and in different atmospheres. Some CV welding machines have a means of adjusting the slope of the volt–ampere curve. Slope control provides control of the magnitude of the short

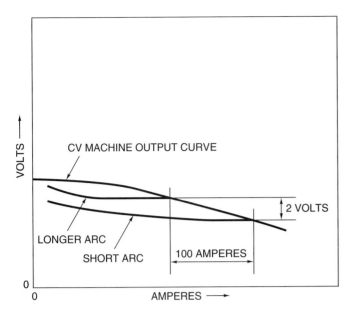

FIGURE 10–11 Static volt-ampere curve with arc range.

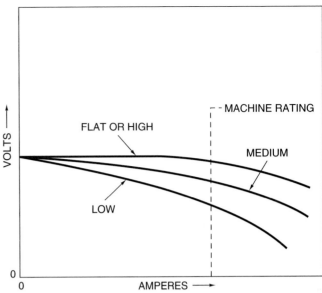

FIGURE 10–12 Various slopes of characteristic curves.

circuiting current: the steeper the slope, the smaller the short circuiting current. Experience indicates that a curve having a slope of $1\frac{1}{2}$ to 2 V per 100 A is best for gas metal arc welding with nonferrous electrodes in inert gas, for submerged arc welding, and for flux cored arc welding with relatively larger-diameter electrode wires. A curve having a medium slope of 2 to 3 V per 100 A is preferred for welding with CO_2 gas shielded metal arc welding and for smaller flux cored electrode wires. A steeper slope of 3 to 4 V per 100 A is recommended for short circuiting arc transfer. These three slopes are shown in Figure 10–12. The flatter the slope of the curve, the more the current changes for an equal change in arc voltage.

The dynamic characteristics of the power source must be carefully engineered. If the voltage changes abruptly with a short circuit, the current will tend to increase quickly to a very high value. This is an advantage in starting the arc but will create unwanted spatter if not controlled. It is controlled by adding reactance or inductance in the circuit. This changes the time factor or response time and provides a stable arc. In most machines a different amount of reactance, which can be resistance or inductance, is included in the circuit for the different slopes. In the new inverter power sources, the characteristics are controlled electronically.

Two factors determine the arc stability: the static volt–ampere characteristic curve, and the dynamic characteristics of the machine. Changes in voltage and current provide the dynamic characteristics, which relate primarily to the two extreme conditions of short-circuit and open-circuit voltage. At short circuit the voltage drops to approximately zero. When the short circuit clears, the voltage will rise quickly, followed by stabilization of the voltage to a value below the open-circuit voltage, called the *re-*

covery voltage. The current rises rapidly during the short circuit period. When the short is removed, the current quickly stabilizes to the normal value. Arc stability is related to the ability of the power source to reignite the arc smoothly following the short circuit. The quick-rising voltage must exceed the arc voltage if arc reignition is to occur. Excessive overshoot of welding current cannot be tolerated. An excessively fast increase in current during the period of initial shorting results in *overshoot* and causes excessive weld metal spatter. Additionally, an excessively fast decrease in the current results in instability, which might extinguish the newly formed arc. Current recovery should occur within one or two electrical cycles. The dynamic characteristic of the machine called the *response time* is the time required to return to steady or average conditions, shown in Figure 10–13. The conventional

FIGURE 10–13 Dynamic characteristic versus response time.

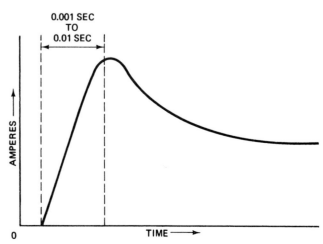

welding machine must contain the proper balance of impedance or inductance to stabilize the output of the machine quickly. The entire concept of machine control to match the physics of the dynamic welding arc is changing with the design of the new inverter power sources.

10-3 TYPES OF WELDING MACHINES

A special kind of electrical power is required to make an arc weld. The special power is provided by a *welding machine,* also known as a *welding power source.* Power directly from the utility line *cannot* be used for arc welding. A welding machine must deliver controllable current at a voltage according to the requirements of the welding process. Normally, the power required is from 10 to 35 V and from 5 to 500 A. The various welding processes and procedures have specific arc characteristics that demand specific outputs of the welding machine.

Electric power for arc welding is obtained in two ways: (1) generated at the point of use, or (2) converted from power available from utility lines. There are many ways of describing electric power used for welding. It can be direct current (dc) or alternating current (ac). Requirements of the welding process may be supplied by a constant current (CC) with a drooping characteristic, or it may require constant voltage (CV), which is a flat characteristic output. Figure 10–14 shows the principal types

of power sources and a classification system. Power sources can be described as rotating machines, static machines, engine-driven machines, transformer–rectifier machines, and so on. All of the machines are for a single operator; that is, they are designed to deliver current to only one welding arc.

The first way of classifying a welding machine is by the basic type, rotating or static. Rotating machines generate power for arc welding at the point of use and are usually run by an internal combustion engine. Static power sources have no moving parts and convert power available from utility line to power required for arc welding. There are three basic types: the transformer, the rectifier, and the inverter.

The second way of classifying a welding power source is by the type of welding current provided, whether ac or dc, pulsing, or a combination output. In connection with the type of current produced, the static volt–ampere characteristic output curve must be analyzed. There are two basic types of output characteristic curve. One is the conventional or CC welding machine that has a drooping volt–ampere characteristic curve. The other is the CV, sometimes called constant potential (CP) power source, which has a relatively flat volt–ampere characteristic curve. The actual output characteristic curves are slightly different. The terms *constant current* and *constant voltage* are widely used, even though the actual machine output curve is not true CC or CV but is

FIGURE 10–14 Simplified system of classifying welding machines.

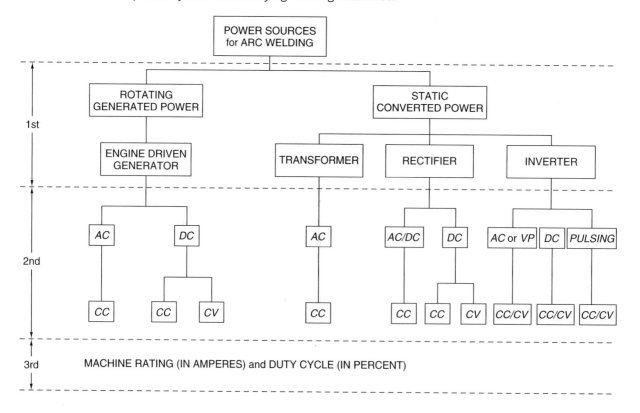

fairly close. Constant current or conventional machines are sometimes referred to as variable voltage (VV) machines. Pulsed-current welding has been adopted for other processes. It provides more precise control in mechanized welding systems.

A third way to categorize a power source is by the output rating. This is the machine rating or the rated output current in amperes, which is usually given as a maximum, and the duty cycle, which is based on the time limit of operating the machine at a specific current output. The rating is the load current produced by the welding machine without creating excessive temperature rise within the machine. All arc welding power sources are designed and rated in accordance with IEC 974-1, "Arc Welding Equipment, Part 1: Welding Power Sources."[1] This is an international standard that is accepted or approved by national standard societies in most of the industrialized countries. In the United States it is the National Electrical Manufacturers Association (NEMA), which formerly provided an electric arc welding apparatus standard for the United States. This IEC standard is a performance standard that provides definitions, environmental test types and conditions, designs for protection against electrical shock, thermal requirements including heating tests, overload protection, and design for safe operation. It also covers the design of the machine and auxiliary apparatus as well. Most countries have adopted this standard.

One of the ways to determine the capacity of a machine in addition to its rated output is by its *duty cycle,* the ratio of arc time to total time in a 10-minute period. The machine must operate properly without overheating during this period. For example, a 60%-duty-cycle machine must not overheat when providing rated output for 6 out of 10 minutes.

A number of terms used to describe power sources require explanation. The term *slope* relates to the slope of the static volt–ampere characteristic curve of the machine. It is defined as the output voltage change to the change in output current expressed in volts per 100 A. The term applies to both flat characteristic machines and drooping characteristic machines. Slope is important within the arc voltage range since it relates to the stability of the arc.

The term *power factor* also relates to welding power sources and is an electrical term of great interest to the electrical utility company and to plant engineers. In direct current, power is expressed in watts or kilowatts (kW). In a dc circuit the product of voltage and amperage in the circuit is all usable and it all registers on the electric power meter. With alternating current, however, kilowatts are used to indicate usable power, and *kilovolt-amperes* (kVA) are used to indicate the total product of amperes times volts delivered by the utility company. The power factor (PF) is the ratio of usable power (kW) to total power (kVA). When ac voltage and current are in

phase, the power factor is said to be unity or 1; thus, with unity power factor, kilowatts (kW) equals kilovolt-amperes (kVA).

The power factor of an industrial user company is rarely unity, however, since most electrical equipment in a factory generally consists of motors that tend to cause the volts and amperes to become out of phase. A single-phase ac transformer welding machine has an inductive electrical load. This causes the current curve of the alternating cycle to lag the voltage curve by a number of degrees. This creates a *lagging power factor.* The electrical utility company monitors industrial customers and establishes a power factor for the company. This is entered into a formula so that the factory will pay a penalty when the power factor is less than unity. The power factor of an industrial plant can be improved or moved toward unity by correction devices such as power factor–correcting capacitors. In the case of a single-phase transformer welding machine, a power factor–correcting capacitor can be built into the machine to provide a correction factor, bringing the power factor close to unity, and is normally recommended.

Common usage has generally matched the power source to the welding procedure and welding process. Figure 10-15 shows the more popular arrangements. Direct current electrode positive (DCEP) is used for gas metal arc welding. When dc electrode negative (DCEN) is used, the arc is erratic and produces an inferior weld. DCEN can be used for submerged arc welding, flux cored arc welding, and shielded metal arc welding.

The CV principle of welding with alternating current is normally not used. It can be used for submerged arc welding and for electroslag welding, but is not popular. A CV power system should not be used for shielded metal arc welding. It may overload and damage the power source by drawing too much current too long. It can be used for carbon arc cutting and gouging with small electrodes.

In the final analysis, the welding power source must provide a stable arc, instant arc starting, and a high degree of controllability. No matter what welding process is used, the power source must provide usability for the welder and must provide welder satisfaction. A power source that will enable the welder or the automatic welding machine to produce high-quality welds using the process and procedures specified would be the best welding power source for the purpose.

10-4 ROTATING WELDING MACHINES

The dc generator, which is a rotating machine, was the first type of power source designed for arc welding. At one time the motor generator was the most popular welding machine used. The first machine with the motor

Arc Welding Process	Welding Machine Output Characteristics			
	Direct (Dc):		Alternating (Ac):	Pulsed:
	CC Drooping	CV Flat	CC Drooping	CV or CC Flat or Drooping
Nonconsumable electrode process				
Gas tungsten arc welding (GTAW)	Yes	No	Yes	Yes
Plasma arc welding (PAW)	Yes	No	No	Yes
Carbon arc welding (CAW)	Yes	No	Two carbons	Not used
Stud welding (SW)	Yes	Possible	No	Not used
Consumable electrode processes				
Shielded metal arc welding (SMAW)	Yes	No	Yes	Not used
GMAW inert gas, nonferrous MIG	Possible	Yes	No	Yes
GMAW spray arc transfer MIG	Possible	Yes	No	Yes
GMAW globular transfer MIG	Possible	Yes	No	No
GMAW short circuiting transfer MIG	No	Yes	No	Yes
GMAW pulsed arc transfer MIG, pulsed	Special	Special	Possible	Yes
Flux cored arc welding (FCAW)	Yes	Yes	Experimental	Yes
Submerged arc welding (SAW)	Yes	Yes	Yes	Not used
Electrogas welding (EGW)	Possible	Yes	No	Not used
Electrogas welding (EW)	Possible	Yes	Yes	Not used

FIGURE 10–15 Arc welding process versus power source output.

and generator on a single shaft is shown in Figure 10–16. It was invented by E. A. Hobart in 1930 (Patent No. 1825064). Despite its flexibility and ruggedness, the motor generator welding machine is no longer produced in North America. It has been replaced by static machines, which are more efficient electrically, require less maintenance, and are less expensive.

The generator is still very popular but now as a portable power source driven by an internal combustion engine. The engine can be fueled by gasoline, diesel oil, liquid petroleum gas, or natural gas. The engine can be air- or water-cooled. For all-around portability and versatility, the lightweight air-cooled gas engine–driven generator is most popular. For heavy-duty field welding, an air- or water-cooled diesel engine–driven power source is preferred. Diesel-powered machines are required for welding in areas where gasoline-powered equipment is not permitted, such as on off-shore drill platforms. Many different types of engine-driven welding machines are available. Most provide constant-current conventional welding power with dc or ac output for SMAW (stick) welding. Other engine-driven machines provide CV output direct current for semiautomatic welding, while other machines provide both CC and CV power. Engine-driven welding machines can be *insonarized*—provided with soundproof enclosures for use in urban areas.

The principle of an electric welding generator power source is based on the principle that "when a conductor moves in a magnetic field so as to cut lines of force, a voltage is generated." A welding generator is very complex since there are multiple poles and hundreds of loops of wire that cut the magnetic fields.

The earlier design of generators used a stationary frame or stator for the magnetic field and a rotating ar-

FIGURE 10–16 First single-shaft motor generator welding machine.

mature that included many loops of wire connected to a commutator. The commutator rectified the power and carried it through brushes to the output terminals. This meant that the welding current had to be extracted from the rotating armature through the copper segmented commutator and large carbon brushes. The commutator requires periodic maintenance and the brushes need periodic replacement.

A newer design of rotating power source is known as a *revolving field* or *alternator-type generator.* In this machine the welding power is generated in the stationary windings of the generator's stator. A small amount of the stator power is fed to a solid-state diode rectifier bridge. The rectifier produces direct current used for the fields of the generator, which revolves with the main shaft. The revolving field produces the magnetic lines of force, which generate alternating current in the stator windings of the generator. This ac output may be used directly for ac welding or rectified by solid-state diodes to become dc output for welding. The advantage of the revolving field or alternator–generator machine is that only a small amount of current is carried by the brushes and slip rings. This results in a lower-cost, more reliable generator power source that requires less maintenance. Figure 10–17 shows a block diagram of a revolving field welding generator.

A special type of alternator–generator design is one that is totally *brushless.* This means that no electric current passes from the rotating part to the stationary part of the machine. This is accomplished in a variety of ways.

One method is to utilize a second generator as an exciter. The exciter consists of a stationary field which causes ac voltage to be generated in the rotating armature. The ac voltage is rectified by solid-state rectifier diodes mounted on the rotating shaft. The direct current from the rectifiers is then used to supply current to the revolving fields of the main generator. These, in turn, produce ac voltage in the stator windings. Some of the ac voltage from the main generator stator is rectified to dc to power the stationary fields of the exciter. Only magnetic flux crosses the interface between the stationary and revolving portions of the generator. A block diagram of a brushless type of machine is shown in Figure 10–18.

The speed of rotation of a generator welding power source with four poles is 1800 rpm. A four-pole revolving field or rotor is shown in Figure 10–19. This produces 60-Hz power from the exciter generator, which can be used to run power tools, electrode wire feeders, and so on. To produce 50-Hz power, the generator would be operated at 1500 rpm. Two-pole generators operate at 3600 rpm for 60 Hz or at 3000 rpm for 50 Hz. A two-pole revolving field or rotor has two windings instead of four. Engine-driven equipment will run at a slightly higher speed without load, but when under load should approximate 1800 rpm. Some engine-powered generators have idling devices that reduce the rotating speed to conserve fuel when not welding.

On dc commutator generators it is possible to reverse the polarity output. The terminals on a generator are normally marked positive and negative, or electrode and

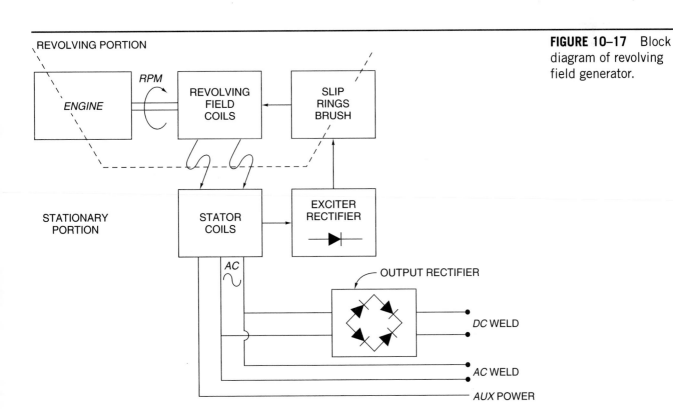

FIGURE 10–17 Block diagram of revolving field generator.

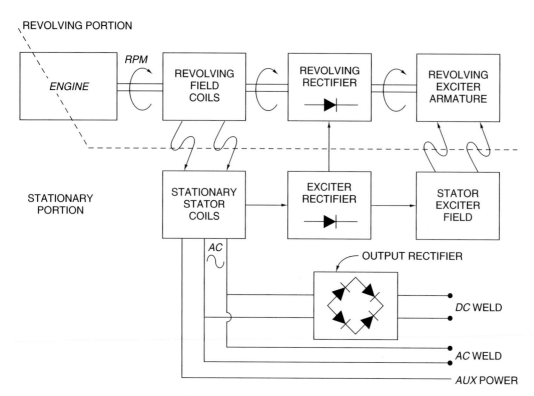

FIGURE 10–18 Block diagram of brushless generator design.

FIGURE 10–19 Four-pole revolving field or rotor.

work. By utilizing a reversing switch between the output of the exciter, the field coils change their magnetic polarity. When the polarity of the field coils is changed, the polarity of the output of the commutator generator is also changed.

The output of the generator, which is alternating current, can be processed and controlled in the same manner as the transformer welding machine. If direct current is desired, the generator output can be processed and controlled in the same manner as the rectifier welding machine. It is possible for the ac generator to power

an inverter and provide the various output characteristics and procedure programs.

One of the more popular heavy-duty generators with a 60% to 100% duty cycle provides alternating current with conventional characteristics, direct current with conventional characteristics, and direct current with CV characteristics. Cable terminals are provided for the different outputs. These machines can be used for SMAW (stick) welding and for semiautomatic GMAW and FCAW welding. An air-cooled diesel engine–powered machine of this type is shown in Figure 10–20. The heavy-duty machines usually have electric motor starting systems. Most machines of this type include welding voltage and ammeters, engine hour meters, oil pressure gauges, and fuel gauges. Some machines provide single-phase 115- and 230-V ac receptacles with circuit breakers for auxiliary power. Most machines include automatic idle systems for fuel economy. Other optional equipment and accessories are normally available.

A lightweight light-duty air-cooled gasoline engine–powered generator provides the ultimate in portability. A machine of this type is shown in Figure 10–21. These machines have either electric key or recoil starting systems. Trailer mountings as well as other auxiliary items, such as wire feeders and spool guns, are normally available. A fuel consumption curve is provided for engine-powered welding machines. This provides fuel consumption at different current levels at a specific operator factor. A curve of this type is shown in Figure 10–22.

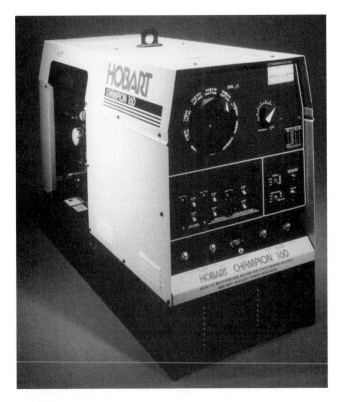

FIGURE 10–20 Air-cooled diesel-powered CC–CV welding generator.

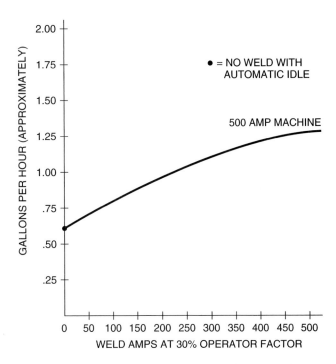

FIGURE 10–22 Fuel consumption based on welding current and operator factor.

FIGURE 10–21 Air-cooled gasoline-driven welding machine (front and back views).

10-5 TRANSFORMER WELDING MACHINES

Transformer welding power sources were developed by C. J. Holslag of New York (U.S. Patent 1,305,362). In 1919 Holslag discovered the phase-shifting aspects of the ampere and voltage curves, such that both will not go through zero at precisely the same instant. This helps stabilize the arc and make ac welding possible. However, not until the covered electrode became widely used did the transformer power source become popular. A very early transformer welding machine, built by Nels Miller in 1930, is shown in Figure 10–23. They are considered static electrical machines since motion is not used except for possible ventilating fans and control systems.

When a conductor cuts a line of force of a magnetic field, a voltage is generated. In alternating current the lines of force build up and collapse each half-cycle. There is no mechanical motion; it is the buildup and collapse of the lines of force that cuts through the conductors. This principle of the transformer is shown in Figure 10–24, where mutual induction is used between two coils in the same magnetic field. The ac power in the primary coil causes magnetic lines of force to build up to the maximum in one direction during one half-cycle and then collapse. Magnetic lines build up in the opposite direction during the other half-cycle of current flow and then collapse. The secondary coil is in the same magnetic field and the magnetic lines of force cut across the windings in the secondary coil and induce a voltage. A welding transformer is a step-down transformer. The voltage relationship between the primary and secondary is determined by the number of turns or loops in each coil. The primary coil has a large number of turns to accommodate the high voltage of the power lines. The secondary coil has fewer turns and a lower open-circuit voltage. This is known as the *turns ratio*. The power or watts value in both primary and secondary coils is approximately the same except for electrical losses. The product of the primary voltage times the primary current equals the product of the secondary voltage times the secondary current while welding, as shown by the formula $E_p \times I_p = E_s \times I_s$. The frequency of the alternating current is the same in both the primary and secondary. The waveshape of the secondary output is essentially the same as that of the primary output, normally sinusoidal. The magnetic core of a welding machine is made of thin iron laminations. Lines of force are concentrated in the iron core. The transformer isolates the incoming primary voltage, usually relatively high, from the secondary circuit of the welding machine. This is a safety feature. This principle is utilized in all transformer and transformer rectifier welding power sources.

Another important part of the transformer welding machine is the reactor. The reactor helps stabilize the output current of the welding transformer. It is often used to provide welding current adjustments. The reactor construction is similar to that of a transformer except that a reactor must have an air gap in the magnetic circuit. The reactor, which is connected in series with the welding current, supplies the lagging phase angle every half-cycle to smooth out the welding current output.

The output of the transformer welding machine can be adjusted in a number of ways. The open-circuit voltage of the machine depends on the turns ratio and the incoming power-line voltage. Changing the magnetic coupling of the two coils changes the current output of the welding transformer. There are various ways of changing the magnetic coupling to change the output of the machine.

The control system box, shown in Figure 10–24, indicates a means of changing the magnetic coupling of the two coils. This changes the output current of the welding transformer. There are numerous control systems. The more common systems are:

1. *Tapped resistance.* The amount of resistance in the secondary is changed by moving tap switches.
2. *Tapped secondary.* This changes the current's ratio between the primary and the secondary at different tap switches.
3. *Tapped reactor.* This is similar to the tap secondary but changes the reactor coil turns by moving tap switches.
4. *Moving coil.* This system moves one transformer coil with respect to the other, which changes the coupling.
5. *Moving reactor core.* This moves a core in the reactor coil.
6. *Moving shunt coil.* This moves a shunt to change the coupling.

FIGURE 10–23 Very early transformer welding machine.

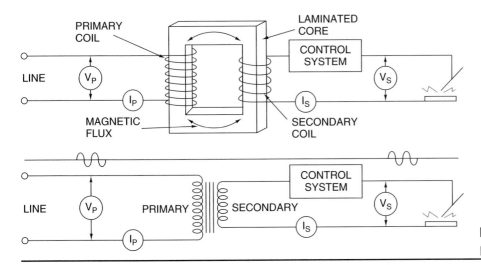

FIGURE 10–24 Transformer principle for alternating current.

The tap switch design does not allow continuous change from minimum to maximum current, which is a disadvantage. The moving core design allows continuous current adjustment but involves moving parts, which become loose and vibrate and create noise.

A certain amount of impedance must be built into the power source. This secondary impedance built into the machine is necessary; otherwise, it would have constant voltage (CV) characteristics. A CV or flat characteristic AC power source is not used for gas metal arc welding or flux cored arc welding.

The ac power source for shielded metal arc welding relies largely on ionizing elements in the arc atmosphere to provide a stable arc. Alternating current varies continuously from positive to negative. The arc goes out at each half-cycle as the voltage passes through the zero point, as shown by the current waveform in Figure 10–25. Covered electrodes designed for ac welding have ionizing elements added to the flux coating to help arc reignition each half-cycle and to stabilize the arc. It is very difficult

to use bare electrodes or bare solid electrode wire without some sort of arc stabilizer or stabilizing elements in the arc atmosphere.

To eliminate the moving parts and their possible service problems, a more modern method for controlling the output of the transformer evolved. This method utilized solid-state power electronic components. Figure 10–26 shows a simplified version of the saturable reactor method for controlling machine output electrically. This system uses a diode bridge rectifier along with a saturable reactor. Direct current from the rectifier flows through the reactor in the secondary circuit, which tends to saturate its magnetic field. When there is no dc current flowing in the reactor coil, it has its maximum impedance and thus minimum output of the transformer machine. As the dc current is increased by means of the rheostat, the magnetic field contains more direct current or more continuous magnetic lines of force in the magnetic circuit. Impedance of the reactors is decreased and the current output of the welding transformer is increased. This system is used widely in modern industrial transformer welding machines. This new way of controlling current output ushered in the use of power electronics, which revolutionized the design of welding power sources.

Static controls have a drawback. They cause a distortion of the waveshape of the secondary current, as shown in Figure 10–27. This makes the slope of the ac curve going through zero less steep and makes arc reignition more difficult. This distorted sine wave is common in most electronically controlled ac power sources.

One of the major problems in the early days of GTAW was arc rectification in the welding pool when welding on aluminum. Rectification occurred because the tungsten was a better emitter than the aluminum weld pool and caused the arc to miss half-waves of the ac cycle. This resulted in a very unstable arc. The dc or rectified component of the welding current created problems in the welding machine. This problem was overcome by

FIGURE 10–25 Current waveform for ac welding.

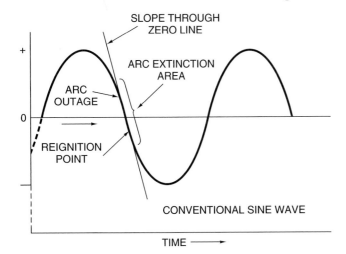

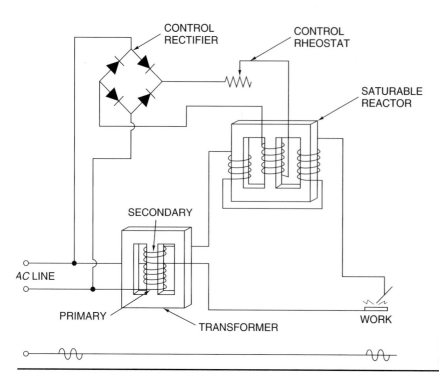

FIGURE 10–26 Saturable reactor control of transformer power source.

CONTROL RECTIFIER

CONTROL RHEOSTAT

SATURABLE REACTOR

SECONDARY

AC LINE

PRIMARY

TRANSFORMER

WORK

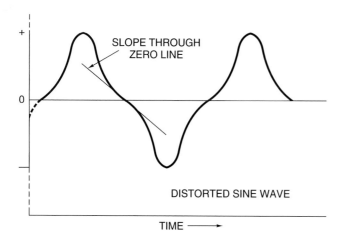

SLOPE THROUGH ZERO LINE

DISTORTED SINE WAVE

TIME →

FIGURE 10–27 Distorted sine wave.

superimposing a high-frequency stabilizing current on the welding current. This ionized the arc gap and partially corrected the rectification problem. It also aided arc starting. The high-frequency current ionized the arc gap so that when the welding current goes through zero, the arc would be reestablished instantaneously. This high-frequency current was provided by a high-frequency oscillator, which consisted of a high-voltage transformer, a spark gap, a high-voltage capacitor and resistance, and a coupling coil. A simplified circuit diagram for this device is shown in Figure 10–28. The high-frequency current is a broad-based signal with a fundamental frequency of about 2 MHz. The spectrum is very rich in harmonics, with frequencies to 20-MHz components. The frequency is determined by the charging time of the capacitor and

by the 50- or 60-Hz line frequency. High-frequency oscillators are included in the case of welding machines designed for GTAW. High-frequency oscillators, also available as stand-alone units external to the welding machine, were called the "missing link."

High-frequency stabilizing current must be carefully controlled. High-frequency current superimposed on the welding current can be radiated by the welding leads or can enter the welding machine's power input lines. Equipment must be installed correctly, and shielding is required to avoid radiation. Radiation of the high-frequency current can cause interference with aviation communications, broadcast, and TV stations. The length of the welding leads and their arrangement has an effect on radiation and on reducing the high-frequency current at the arc. The use of high-frequency spark gap oscillators is approved by the Federal Aviation Authority, provided that they are installed in accordance with manufacturers' instructions. In many cases, remote high-frequency units are installed near the arc to provide consistent arc starting.

Transformer welding machines are normally air cooled. Heavy-duty industrial transformer power sources have thermostat-controlled fans to cool the machine. Overheating may occur if the fan is not operating or if air passages are blocked. Oil-cooled transformer power sources that utilize circulating oil in a sealed circuit are available for special installations where dusty and corrosive environments are encountered. They are used mostly in Europe. The small, limited-input machines use convection cooling and rely on air that normally circulates through the louvers in the cabinet of the machine.

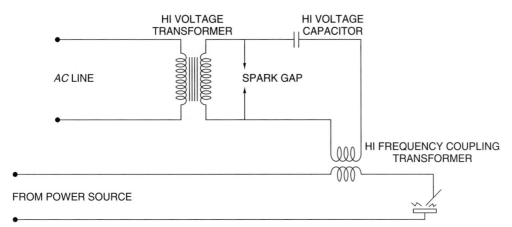

FIGURE 10–28 Simplified diagram of spark gap arc stabilizer.

The transformer welding machine is a very efficient welding machine from the electric power point of view. It has high efficiency, the lowest no-load losses, and, with proper power-factor correction, a good power factor. Figure 10–29 shows the performance curves of a typical 300-A machine.

The efficiency of a transformer welding machine operating on a 40-V load ranges from 80% to 90%. The no-load losses range from 150 to 375 W, depending on the machine. The power factor at a 40-V load is approximately 27% without correction, but can be improved by the use of capacitors. The correction action of capacitors should be based on the normal welding range. Too much correction may be as bad as none. The figure also shows the difference in power factor for a transformer welding machine with and without capacitor correction.

FIGURE 10–29 Performance curves of a transformer welding machine.

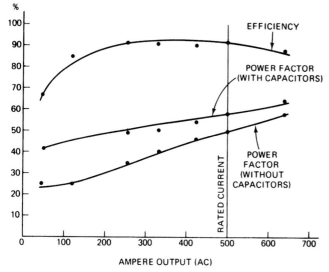

Single-operator CC welding transformer power sources are available, rated at different duty cycles. They are available with current outputs ranging from 50 to 1000 A. One of the most popular ac transformer welding machines is the *utility* or hobby type, commonly called *limited-input welding machines.* These were originally designed to have a limited input current at their rated output so that they could be used in rural areas on available utility power lines. They are rated with a 20% duty cycle and are used for light-duty work. They are also listed in Underwriters' Laboratories standards.[2]

Industrial welding power sources are rated at 60% or 100% duty cycle. The 60%-duty-cycle machines are used for manual shielded metal arc welding in industrial applications. The 100%-duty-cycle machines are used for automatic submerged arc welding.

Transformer power sources have several disadvantages. They normally operate on single-phase input power, which tends to unbalance utility power lines unless a sufficient number of ac transformer welding machines are used and balanced to the line. In addition, a limited number of types of covered electrodes are available for alternating current. Also, two transformer welding machines connected to the same weldment can become a potential safety hazard. The voltage difference between the electrode holders could be extremely high if the two machines were connected improperly to different phases of the incoming primary line. Alternating current is used for fewer welding applications.

Transformer power sources offer some advantages. They are the least expensive and very efficient. The maintenance expense is the lowest of any type of welding machine. The use of alternating current reduces the problem of arc blow on complex weldments.

Despite the many advantages of the transformer power source, they are becoming less popular. The limited input, hobby, or farm-type welding machine is now practically obsolete. It has been replaced by the dc light-duty rectifier or inverter power sources with built-in wire feeders.

10-6 RECTIFIER WELDING MACHINES

Direct current is used for most arc welding processes and for most welding applications. Alternating current is changed to direct current by means of a rectifier. A rectifier is a device that conducts current more easily in one direction than in the other. For many years the diode vacuum tube has been used in radio power-supply circuits to produce the required direct current. Large, heavy-duty vacuum-tube rectifiers were developed for radio transmitters and used in industrial battery chargers. Six mercury vapor vacuum-tube rectifiers were used in a three-phase circuit to power the first rectifier welding machine. This pioneer machine, developed in the mid-1930s by the Allis Chalmers Company, was called a Weld-O-Tron. It had a 75-A rating and is shown in Figure 10–30. Vacuum-tube rectifiers were not sufficiently rugged for industrial welding applications, so solid-state rectifiers were developed. The welding industry adopted the dry disk rectifier, which employs a layer of a semiconductor such as copper oxide or selenium between adjacent plates. Selenium rectifiers developed in the 1940s were used in many early rectifier welding power sources. The high-current solid-state silicon diode rectifier was developed in the mid-1950s and is used for most welding machines today.

A silicon diode rectifier is a semiconductor made of thin wafers of silicon that has small amounts of impurities added to make them semiconductors. The wafers are specially treated and encapsulated under pressure in a durable porcelain housing for welding machine requirements. Figure 10–31 shows a typical silicon diode. It enables current to flow in one direction while blocking it in the opposite direction. It has two terminals, an anode and a cathode. The internal parts of a silicon-controlled rectifier are shown in Figure 10–32. This rectifier is similar to a silicon diode but has a third terminal, a gate lead.

FIGURE 10–31 Silicon power diode rectifier.

FIGURE 10–32 Internal construction of high-current silicon-controlled rectifier.

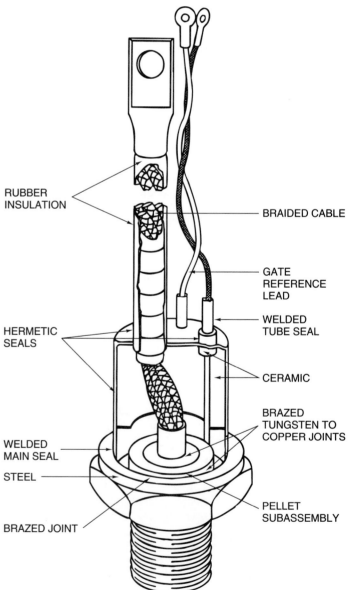

FIGURE 10–30 First rectifier welding machine.

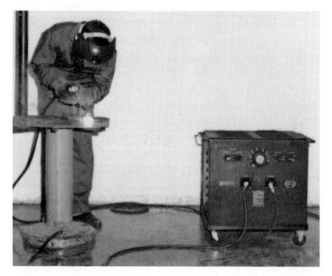

Rectifier electrical circuits to change alternating current to direct current were borrowed from the radio electronics technology. A simplified diagram of some basic rectifier circuits is shown in Figure 10-33. The smoothness of the dc output depends on the specific circuit. The simplest rectifier, known as a half-wave rectifier, will allow only the positive half of each sine wave through and blocks the other half of the cycle. Single-phase ac input and half-wave rectified dc output are shown by the top diagram. This circuit produces an output rising from zero to peak voltage and back to zero, then pausing and rising again. This waveshape is very rough. A circuit design that utilizes four rectifiers, known as a full-wave rectifier, provides a smoother output. This is shown on the second line. The output is considerably smoother but requires use of a filter. A filter consists of an inductor in series and capacitors across the output to smooth the roughness or ripple of the dc output. To provide smoother dc output, three-phase ac input is used. This is shown by the third line and provides smoother dc output. To provide even smoother output, twice as many power rectifiers are used. This is accomplished by using a full-wave rectifier, shown by the bottom line in the diagram. Most industrial rectifier welding power sources use three-phase input and full-wave rectification to provide

smooth dc output. These machines provide a balanced load on the three-phase utility power line, which is an advantage over the single-phase machines. Some rectifier welding machines do operate on single phase. These machines have a large filter circuit but do not provide an arc as smooth as that of a three-phase machine.

The availability of high-powered solid-state components contributed to the development of new welding power sources. The electronics industry has introduced solid-state components that include rectifiers, transistors, thyristors, high-speed switches, oscillators, inverters, and other devices. These components are small and require very little power to operate but control high currents. The symbols for these devices are shown in Figure 10-34. The welding industry uses them to provide solid-state contactors, solid-state timers, line voltage compensation control circuits, and high-current rectifiers. These devices are extremely rugged, fast, less heat sensitive, and more powerful.

An early industrial rectifier machine utilizing power solid-state rectifiers borrowed the circuit of the saturable reactor transformer welding machine. This transformer power source was discussed in Section 10-5. The power rectifiers were added to the basic transformer. Some machines remained with single-phase input and provides ac

FIGURE 10–33 Basic rectifier circuits.

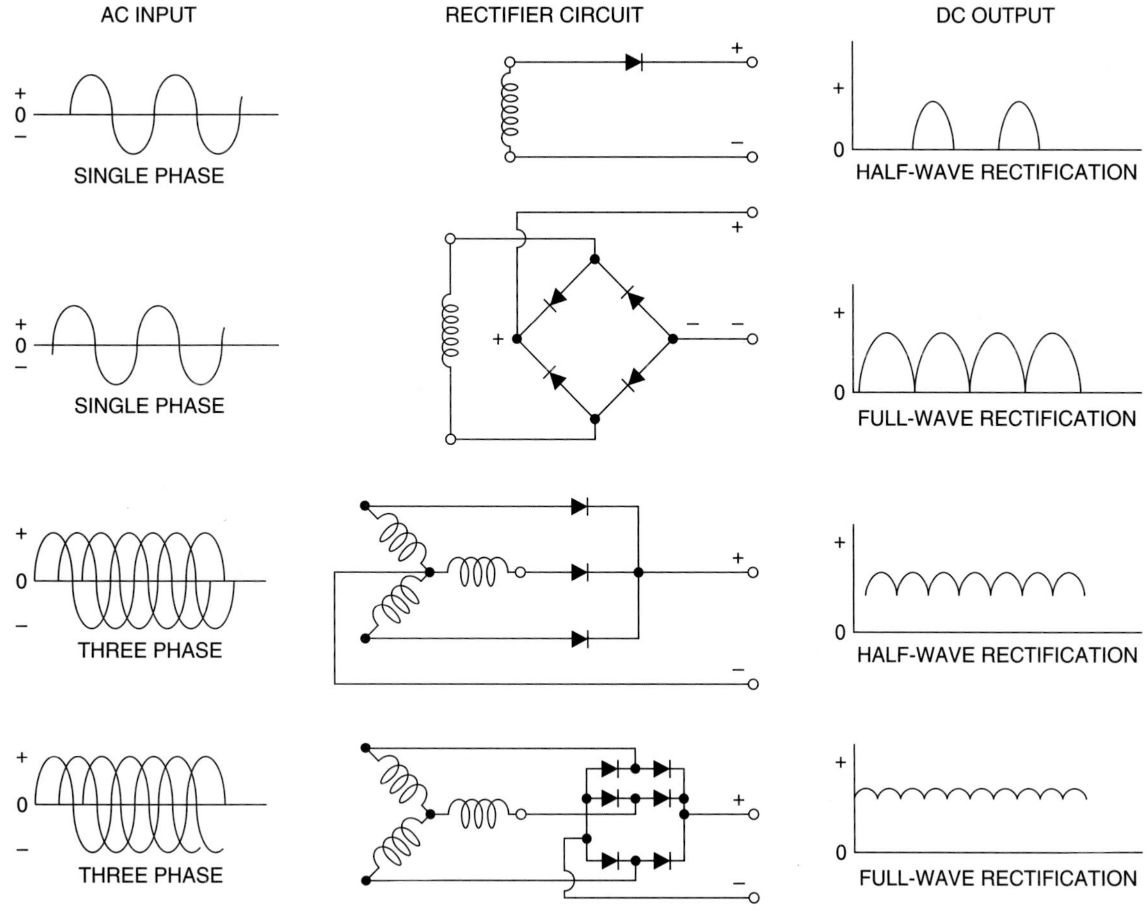

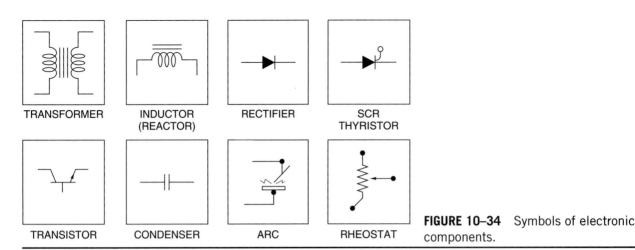

| TRANSFORMER | INDUCTOR (REACTOR) | RECTIFIER | SCR THYRISTOR |
| TRANSISTOR | CONDENSER | ARC | RHEOSTAT |

FIGURE 10–34 Symbols of electronic components.

or dc output with a conventional CC volt–ampere characteristic curve. The dc rectifier power source soon utilized three-phase input and produced only dc output. This machine was an advantage since it provided a balanced load on the utility lines and had a full-wave rectifier that provided a smoother arc. As gas metal arc welding became more popular, three-phase input machines were developed to provide CV output. Soon afterward, combination CC-CV machines appeared.

Another solid-state device used in welding machines was the silicon-controlled rectifier (SCR) shown in Figure 10-32. These devices were also called *thyristors,* which is a contraction of the word *thyraton,* a type of vacuum tube whose characteristics are similar to an SCR, and *transistor.* This semiconductor is a three-layer, three-terminal device that is very similar to the diode except for the addition of a very thin element called the gate, which is placed between the cathode and the anode. The output of the SCR is controlled by applying a small signal to the gate, which in turn controls a large amount of current. The SCR is a controlled diode, in that it has the capacity of blocking current flow in both directions. Its forward blocking action can be switched on or off. This switching function is accomplished by a signal fed to the third terminal, called the *gate.* Conduction starts when a controlled signal is applied to the gate. When the gate signal is applied, the SCR begins to conduct current. The SCR will continue to conduct the current after the gate signal stops, as long as the current continues to flow. If the current is turned off, another gate signal is required to start the current flowing again. The SCR turns off the current automatically when the alternating current goes through zero in a sinusoidal ac waveform. When the SCR falls out of conduction, it returns to a blocking state until the gate signal is again applied.

The phase-angle system has become very popular for controlling the output of the rectifier welding power source. This single-control machine uses SCRs to change the trigger point or angle of the ac cycle. The SCR can be made to become conductive at any specific point on the ac waveform. The SCR conducts current only a portion of the cycle based on the phase angle from 0 to 180° of the current. By turning on the SCRs full time, the maximum current will flow. By decreasing the ignition angle, a lesser current will flow. A simplified circuit diagram for the phase-angle-controlled machine is shown in Figure 10-36. The control circuit based on the current control knob of the machine will trigger the gate of the SCR at the appropriate angle. This provides one-knob control from minimum to maximum current output. This basic circuit can provide conventional output for SMAW, constant current for GTAW, or constant voltage for GMAW and FCAW. It is well controlled and has become the basis for most rectifier welding power sources.

FIGURE 10–35 Printed circuit board.

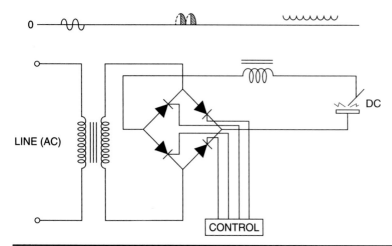

FIGURE 10–36 Simplified circuit for a phase-angle controlled rectifier power source.

Control circuits utilize another electronic device, known as an integrated circuit (IC). ICs are complete electronic circuits of miniature components mounted on printed circuit (PC) boards. Many different types of electrical circuits are available as integrated circuits. Since they are very small, they utilize low voltages and low currents. Each PC board plugs into a socket for ease of replacement. They are also used to control wire feeders and motion devices. A printed circuit board is shown in Figure 10-35.

The electrical efficiency and power factor performance curves for a typical rectifier welding power source are shown in Figure 10-37. A typical rectifier welding machine for CV use is shown in Figure 10-38. Many rectifier welding machines have additional features.

As rectifier welding machines became more popular, they acquired more features. Since they contained a power transformer, ac terminals were provided and

the transformer/reactor welding machine became available. This gave the welder one machine capable of providing power for SMAW (stick) welding, GTAW (TIG) welding, and GMAW (MIG) and FCAW. This led to many additional features because of the availability of high-powered electronic components. By means of integrated circuit and printed circuit boards, electrical engineers can control static characteristics of the power source, including waveform, and can change the welding characteristics to provide a soft arc or a harsh digging arc by a simple dial change. They provide electronic power contractors instead of mechanical contractors, which eliminates troublesome moving parts. They provide power line voltage compensation and can handle

FIGURE 10–38 Constant voltage solid-state dc rectifier welding machine.

FIGURE 10–37 Performance curves of a rectifier welding machine.

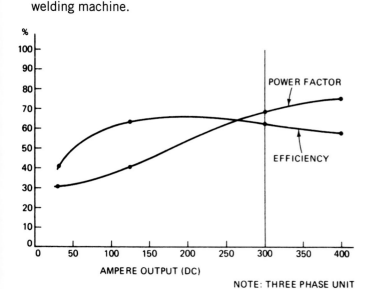

NOTE: THREE PHASE UNIT

variations of power line voltage of up to 10%, with only 2% variation of the output. They can provide a balance control to change the amount of ac positive cycles versus ac negative cycles for ac gas tungsten arc welding. They provide pulsed current of different frequencies and different waveshapes. They provide time-base programming by changing welding current output, sloping up or down, as required. They are used for remote control, which uses, for example, miniaturized potentiometers on welding guns. They greatly simplify all control aspects of semiautomatic and automatic welding. They also allow the control of welding equipment by computers for precise programming.

These features demanded by the welding industry are briefly described below.

Experience gained welding aluminum indicated that different power source output waveshapes were superior to the ac sinusoidal output of the normal transformer power source. The square wave output overcame the arc, extinguishing restricting problems by increasing the speed of polarity change through the zero point. Power electronics is used to vary the positive and negative output. The area above the zero point on the curve, the dc positive area, and the area below the curve, the negative area, can be equalized or balanced. A power source developed specifically for gas tungsten arc welding provides an approximate square-wave output form, but allows a balance or imbalance between the straight polarity and reverse polarity half-cycles of each cycle. This power source is not a transformer, but, instead, is similar to two rectifiers synchronized at 60 Hz. In welding aluminum, the electrode negative (straight polarity half-cycle) gives maximum penetration, whereas the electrode positive (reverse polarity half-cycle) provides cleaning action. It is advantageous to provide the most straight polarity current, and this is possible as shown in Figure 10–39.

Further research with plasma welding of aluminum revealed that a waveshape with very little electrode negative would improve the weld. This machine, known as a variable polarity power source, produced the waveform shown in Figure 10–40. It in effect was two SCR-type power sources connected by means of a high-speed electronic switch. This machine makes keyhole plasma welds in aluminum alloy having water clear x-rays.

Welders found that by changing the level of current rapidly they could improve the quality of the weld when welding thick-to-thin material. They found it helped when root opening varied or when the root face of the joint was different on each side. Using a foot rheostat, the welder could change the current level while welding. This change from a high current (HC) to a lower current (LC) on a repetitive basis became known as *pulsed welding* and is shown in Figure 10–41. In pulsed welding there are two current levels: high current, known as peak, and low current, called background current. By programming

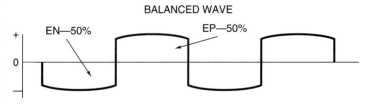

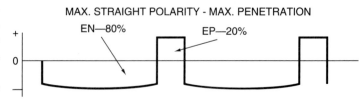

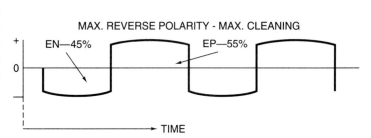

FIGURE 10–39 Semisquare-wave output, balanced and unbalanced.

a control circuit, the output of the machine can be made to switch from the high to the low current. The level of both high and low current is adjustable. In addition, the length of time is adjustable for high and low current pulses. This gives the welder better control over the arc and molten weld pool. Normally, the high current pulse is twice or 1½ times the normal steady current that would be employed for a similar application. The time period for the high current pulse may vary from 0.20 to 1.0 sec. The frequency of pulsing ranges from less than one to five pulses per second. The low current pulse has enough energy to sustain the arc properly. The time is adjusted to allow the molten pool to partially freeze. Figure 10–42 shows the type of weld produced, and Figure 10–43 shows a pulsed current weld on thin material. The values can be adjusted by the welder to provide optimum welding conditions.

Welding with GTAW-P allows very close control of the molten weld metal. GTAW-P can be used in both manual and automatic modes. It can provide increased or decreased penetration and can be coordinated with oscillation. It can be used to weld thick or thin materials and to provide high-quality welds where steady current systems cannot. It is claimed that it improves weld metallurgical characteristics, reduces warpage and distortion, and provides a weld with a lower heat input. It is commonly used on many automatic applications, including automatic tube head welding.

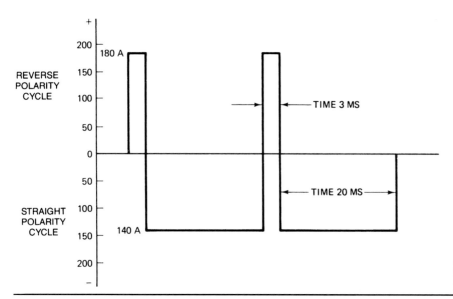

FIGURE 10–40 Waveform of variable polarity power source.

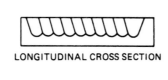

FIGURE 10–41 Pulsed current current–time relationship.

In addition to pulsing, programmers were developed to provide changes in current based on time. Programs allow the current to advance to a particular level, be tapered down as the part became heated, and then be tapered to a lower value before stopping the weld. These types of programmers became increasingly important for orbital tube welding machines. Power sources were developed with programmers built in. Figure 10–44 shows a typical programmable power source for GTAW. Various

FIGURE 10–44 Programmable ac–dc power source for GTAW.

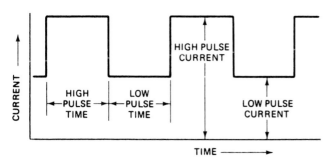

FIGURE 10–42 Pulsed current weld sectional view.

FIGURE 10–43 Pulsed current weld on thin material.

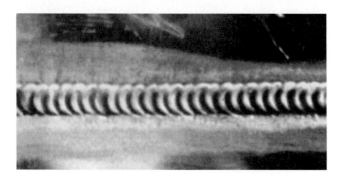

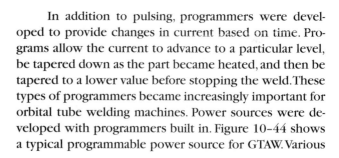

levels of complexity were designed into different programmers, which could be interchanged in the power source. These were based on adjustable rheostats and adjustable timers. Some of these became increasingly complex and included motor speed controllers to control automatic welding applications. Various types of controllers are shown in Figure 10–45. Machines of this nature have largely been replaced by inverter power sources utilizing complex electronic control systems, which in some cases involve microprocessors and computers.

Pulsed spray welding, a variation of gas metal arc welding, is known as GMAW-P. Pulsed spray metal transfer welding offers many advantages over spray or short circuit transfer welding. It provides a controllable molten weld pool and introduced many new procedure variables, which must be matched to the specific application. Spray transfer (see Chapter 6) is achieved when welding above a specific critical current value using argon-rich shielding gas. It has many advantages, such as minimum spatter, nice weld surface, and a larger electrode. Its major disadvantage is the large size of the molten weld pool, which makes it impossible to weld out of position and on thin material. Pulsed spray welding reduces the average current and reduces the size of the molten weld pool. Needham's original concept of one drop of molten metal crossing the arc at each current pulse was continued. The early version based on pulsing at line or double line frequency was abandoned since the power sources developed for this variation were very difficult to adjust. The ability of the new inverter power sources has greatly enhanced the pulsing mode by providing for a change of pulsing variables, current amplitude, frequency, and waveshape.

FIGURE 10–45 Popular program panels for GTAW.

DISTINCT CONTROL FUNCTIONS	APPLICATIONS	
Prepurge; pulsation and arc spot timer	Manual welding, TIG spot welding with or without pulsation. Used when slope control is not required.	
Upslope, downslope and pulsation weld time	Controlled slope rate and pulsation for manual or automatic welding. For pipe welding, thinwall tubing, and difficult to weld application and materials.	
Precision upslope, downslope, pulsation and weld taper with direct reading dials.	For machine TIG welding where pulsation and precision programming of upslope, downslope, and weld taper are required. Designed to perform the most difficult TIG welding jobs.	

10-7 INVERTER WELDING MACHINES

A new type of welding power source that became widely used in the early 1980s provides a small, lightweight welding machine with precise controllability. It is known as an inverter power source since its major electronic components act as an inverter. This circuitry was borrowed from TV technology. The inverter power source was originally requested by the U.S. Maritime Administration, representing the shipbuilding industry, which wanted a small, 40-lb, 300-A welding power source that could be taken through small hatches and manholes to any part of a ship. The resulting small portable inverter power source, developed by Jim Thommes[3] and built by Cyclomatic Industries, Inc.[4] in 1975, is shown in Figure 10–46. A recent Cyclomatic Industries (now PowCon) inverter machine is shown in Figure 10–47.

The inverter power source operates on an entirely different principle than those described previously. The inverter provides very high-frequency current, which can be controlled. This high-frequency current is transformed, rectified, smoothed, and programmed by small electronic components. A simplified circuit diagram of an inverter power source is shown in Figure 10–48. This is known as a primary inverter since the inverter is ahead of the primary winding of the main transformer. Power available from the utility line at normal line frequency goes first to a rectifier that changes the alternating sinusoidal current to direct current. The incoming power to an inverter power source need not be as smooth as required for a conventional power source. It can be single-phase or three-phase and can have different voltage variations without creating problems. A high-speed electronic switch called an inverter converts the dc power to very high-frequency ac power. The alternating current frequency is very high, in the range of 5 to 1000 kHz. This high-frequency power then goes through a

---

FIGURE 10–46 Early inverter built in 1975 power sources.

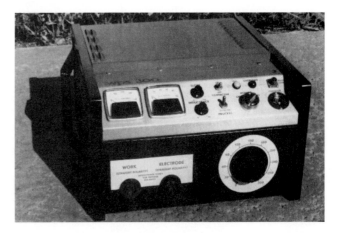

FIGURE 10–47 Current inverter power source.

transformer to lower the voltage and isolate the output from the utility line. This main transformer is very small due to the high frequency. The core is made of ferrite, which is a ceramic material. A conventional laminated steel core would overheat at the high frequencies. The transformer output, a low-voltage but high-frequency current, goes to a solid-state rectifier and changes back to direct current. This is filtered to produce smooth, low-ripple direct current for welding.

In recent years the inverter power source has become extremely popular, primarily due to its controllability. It has become economically feasible due to the availability of high-current, high-speed solid-state electronic components. Inverters have a higher electrical efficiency than rectifiers and a favorable power factor, as shown in Figure 10–49.

The reaction time of an inverter power source is much faster than that of a conventional power source. This is because a conventional power source operates on 50 or 60 Hz, and the reaction time may take one or two cycles, due to the time lag in iron-cored transformers and inductors. In the inverter the cycle time is so much faster and there are no iron-core transformers or inductors. As the frequency of the inverter is increased, the reaction time is even faster.

The inverter, the heart of this new type of power source, changes dc power into ac power with high-speed switching devices with switching speeds of 1 to 2 microseconds. The inverter can be made to control the machine output characteristics. Different types of inverters are used. The exact frequency depends on the type of circuit and the value of the components. These are very complex solid-state devices designed for specific uses. The more common types of inverters are the forward, full

FIGURE 10–48 Simplified circuit diagram of primary inverter power source.

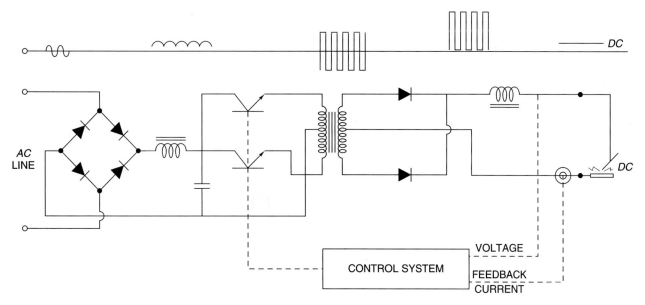

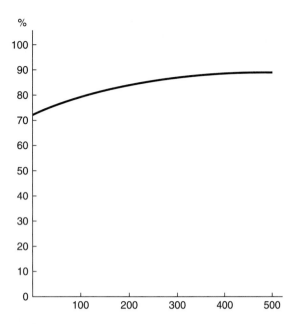

FIGURE 10–49 Electrical efficiency of inverter power source.

FIGURE 10–50 Comparison of main transformer conventional and inverter power sources.

or half bridge, and the series resonant and isolated gate bipolar transistor. Typically, inverters use transistors as power switch components and the series resonant inverters use thyristors. The thyristor changes the dc power to ac power; it does this by the on-off action of the high-power solid-state switches. This action alternately connects and disconnects the primary of the transformer and has the same basic effect as that of applying a regular sinusoidal waveform. The different types of inverters do this differently and are controlled differently to produce and control the output of the inverter. This very complex procedure is beyond the scope of this book.

Inverter power sources are about 25% the weight and size of a conventional rectifier of the same capacity. A comparison of the main transformer of a conventional and an inverter power source is shown in Figure 10-50. The conventional power source transformer, made of laminated silicon steel, is rated at 450 A, 100% duty cycle, and weighs 283 lb. The small one for the inverter power source using high frequency with the ferritic core is rated at 500 A, 100% duty cycle, and weighs 29 lb. The outward appearance of the inverter power source is very similar to that of a conventional rectifier power source except for the size. A typical inverter power source is shown in Figure 10-51.

The intermediate operating frequency of the inverter welding machine is very important. At the lower frequencies of approximately 5 kHz, audible noise is bothersome. This is also true at about 15 kHz. However, above 20 kHz the noise is inaudible except to people with very keen hearing. The trend is toward higher frequencies, up to 50 kHz.

FIGURE 10–51 Typical inverter power source.

Other than the small size and light weight, the major advantage of inverter power sources is their ability to control output, both static and dynamic. An inverter uses electronics to simulate changes in inductance and reacts exceptionally fast to dynamic changes. The static characteristic curve can have a drooping volt–ampere characteristic output for SMAW, a very steep output or CC output for precise GTAW or PAW, or it can have a flat characteristic output for GMAW or FCAW. The same basic machine can be programmed to provide any output characteristics. In this regard the inverter can be considered as a true multiprocess welding power source.

The inverter power source has opened the door to many different types of pulsing programs. It can provide dynamic output for pulsed welding of almost any type possible. Control systems for an inverter power source can provide various pulsing waveshapes. This has ushered in many different types of welding programs to provide specific arcs for very specific needs. Sensors and specialized circuits monitor the output and control it to match the needs of the welding process and procedure.

A recent development is known as *synergic pulsed-spray metal transfer* (the term *synergic* is a Greek word meaning "working together"). In synergic welding the shape of the pulse waveform can be changed. The duration of pulse and the amplitude of peak current and pulse frequency are matched to the electrode type and size and to the shielding atmosphere. The optimum welding procedure variables are built in to the software program that controls the power source microprocessor and in some cases the wire feeder processor. This results in a variable-frequency spray metal transfer mode that produces a very controllable weld pool and a smooth weld with virtually no spatter. There are two types of synergic systems: (1) manual control, and (2) one-knob control. With manual control the welder has independent control of the waveform, including pulse current, pulse time, and background current, as well as wire feed speed and pulse rate (frequency). As the wire feed rate (current output) is changed, the welder must change the frequency of the waveform to control and maintain a stable arc.

With one-knob control the waveform values and the wire feed rate versus frequency relationship is programmed into the system by means of the microprocessor. When the welder adjusts a single knob (average current or wire feed rate), the control automatically changes the other values to maintain the desired arc characteristics. For one-knob systems the pulse power source and the wire feeder are "electrically mated" and must be operated as a system. In the synergic mode the metal transfer is precisely regulated. It has a broader range of operating parameters. This mode of metal transfer is recommended for high-quality precision welding for semiautomatic, mechanized, or robotic welding applications.

There are several concepts of synergic welding using inverter power sources with matched wire feed systems. In one case, the frequency of pulsing remains the same but the maximum or peak pulsed current and the background current varies. In the other case, the frequency of the pulsing varies but the background pulsing current ratio remains the same. The wire feeder must match the output of the power source. Some power sources have a subprogram that provides the pulse waveform for each pulse. This waveform will change with different applications based on the filler metal type, size, and shielding gas atmosphere. Software in the microprocessor determines the waveform geometry of each pulse. In one case, the welder selects the desired program according to work to be performed and wire size and analysis, and by means of a single knob can adjust the machine from minimum to maximum output. Peak current, peak

time, background current, and frequency are adjusted automatically by the program selected. The knob adjusts average current from minimum to maximum. The microprocessor of the power source and its matched wire feeder automatically adjust wire feeder rate, pulse frequency, and other variables. There are two versions. One uses a "dumb" or standard conventional constant-feed wire feeder; the other uses a "smart" wire feeder interconnected to the software program of the power source microprocessor. These types of machines have become very popular. Synergic systems provide for a controllable weld pool, which results in a smooth weld with controlled penetration and virtually no spatter. Caution must be exercised to maintain a constant stickout distance while welding.

The newer control circuits for inverters utilize instantaneous feedback from the arc voltage and welding current. They feature closed-loop feedback monitoring of minute changes in the welding arc or current and adjust the output immediately. The feedback data are related to each pulse, where the microprocessor reads the welding conditions and alters the program to provide the output desired. These changes are made immediately and respond to fast-changing conditions in the arc. Microprocessors can be programmed with different software to provide different arcs for welding different metals of different thicknesses or to provide different characteristics to the arc. The inverter power source is changing the entire concept of arc physics and control of power sources.

As inverter power sources became more widely used, more complex monitoring and control circuits were introduced. The pulsing parameters are preset at the factory for different base metals and electrode sizes and provide optimum conditions. Figure 10–52 and Figure 10–53 show the control panel of a multiprocess machine. A single switch changes the output characteristics of the machine and allows the use of different welding

FIGURE 10–52 Control panel of a multiprocess machine.

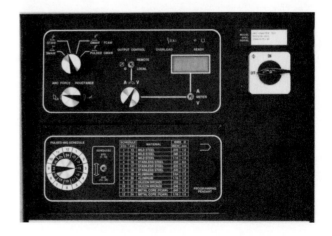

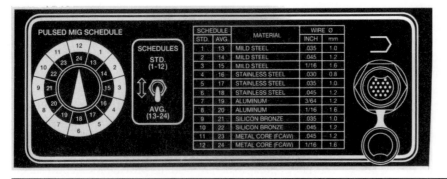

SCHEDULE		MATERIAL	WIRE Ø	
STD.	AVG		INCH	mm
1	13	MILD STEEL	.035	1.0
2	14	MILD STEEL	.045	1.2
3	15	MILD STEEL	1/16	1.6
4	16	STAINLESS STEEL	.030	0.8
5	17	STAINLESS STEEL	.035	1.0
6	18	STAINLESS STEEL	.045	1.2
7	19	ALUMINUM	3/64	1.2
8	20	ALUMINUM	1/16	1.6
9	21	SILICON BRONZE	.035	1.0
10	22	SILICON BRONZE	.045	1.2
11	23	METAL CORE (FCAW)	.045	1.2
12	24	METAL CORE (FCAW)	1/16	1.6

FIGURE 10–53   Details of the schedules.

processes. Different pulsing schedules can also be dialed in. Pulsing schedule changes can be made in the field using a programming pendant, shown in Figure 10-54. The pendant has the capability to provide a customized program by changing the starting current, peak time, peak pulse current, and background voltage and current.

The automobile industry had a need for welding extremely thin materials with a very flat bead shape and no

FIGURE 10–54   Programming pendant for developing specific programs.

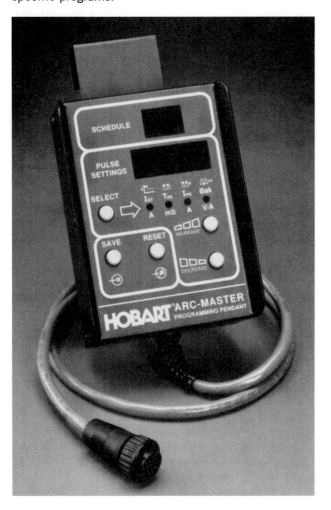

spatter. This required a nice-appearing weld with a minimum of finishing. This requirement, especially for aluminum welding, led to the development of a new pulse GMAW power source known as a twin-inverter system. The normal inverter power source has one inverter on the primary side of the main transformer. The twin inverter has a second inverter for polarity switching on the secondary side. This machine provides variable polarity output and controls penetration by changing the electrode polarity. It changes polarity output by switching from positive to negative and back again and controlling the ratio of reverse and straight polarity current. A simplified circuit diagram of a twin-inverter machine is shown in Figure 10-55. The microprocessor controls the power source precisely so that welds can be made at low ampere values with pulse rates required to provide the type of weld desired.

The need for more precise welding procedures continues. One specific desire is for welding extremely thin materials, including aluminum in automobile bodies. This brought about a power source that utilizes two inverters, known as a *twin inverter welding machine.* It provides variable-polarity welding at extremely high frequencies. The output waveform of this machine is shown in Figure 10-56. The length of the electrode negative time period and the amount of amplitude can be increased or decreased to change the penetration. This is necessary when welding thin aluminum. This system is known as ac pulse welding or dip pulse welding. The microprocessor in this case uses artificial intelligence to ensure proper pulse–wave control.

Another innovative welding system made possible by inverter technology is known as *surface tension transfer welding.* It is very similar to short-circuit transfer except that less spatter is produced. This is based on monitoring the individual pulses while the electrode is in contact with the work. The current is reduced for a very short time. The current is then increased until the shorted electrode wire separates from the deposited metal. The arc is monitored by the microprocessor, which controls the instantaneous current to reduce the spatter. This lower current level eliminates the spatter normally associated with fuse-effect arc restarting. It is claimed that this method will allow CO_2 shielded without spatter.

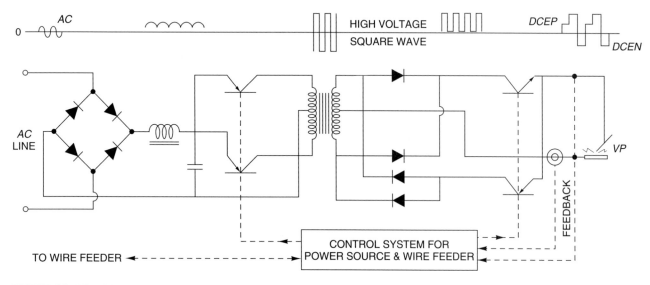

FIGURE 10–55 Simplified circuit diagram of twin-inverter power source.

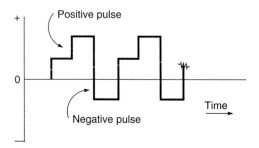

FIGURE 10–56 Variable-polarity pulsed waveform.

Inverter power sources allow more precise control and different waveforms, pulse mechanisms, and frequencies. It ensures perfect starts, reduced fumes, controlled penetration, improved appearance, and reduced spatter with both semiautomatic and automatic applications. It, along with the digitally controlled microprocessor, will be the power source of the future.

10-8 SELECTING AND SPECIFYING A POWER SOURCE

The type of welding to be done dictates the type of welding machine to be used. This requires an analysis of the weldments to be produced: factors such as thickness of members, size of the weldment, materials involved, method of application, and others. The following information should be considered to make an intelligent power source selection.

1. *Process selection.* The welding process to be used is based on the weldments to be produced.

2. *Welding current.* The welding work determines the welding current type: that is, dc, ac, steady state, pulsed, and type of pulsed current.

3. *Machine rating.* This determines the size of the power source. Heavier materials require larger electrodes, which indicates a higher welding current. Machines are rated in amperes at a given voltage.

4. *Type of power available.* If power is not available from a utility company, engine-driven generators are required. If utility power is available, the type, the number of phases of ac power, and the voltage are specified.

5. *Auxiliary devices.* This includes water and shielding gas control systems, current control systems, wire feeders, and any other factor required to allow manual, semiautomatic, or automatic welding.

6. *Duty cycle.* The duty cycle is a measure of the amount of work that the power source will do. Low-duty-cycle equipment is designed for light-duty work. High-duty-cycle work equipment is designed for semiautomatic or automatic welding. Duty cycle is the ratio of arc time to total time and is explained below.

Rating of the machine is determined by tests and is related to the static volt-ampere characteristic curves. Machines are rated according to the duty cycle at a specific load voltage. Load voltage standard changes from 28 to 44 V, depending on the size of the machine. Tests are run at the duty cycle specified to determine that specific temperatures within the machine are not exceeded. The method for testing and rating welding machines is in accordance with IEC 974-1 Standard.[1]

In general, 20%-duty-cycle machines are designed for light-duty work, 60%-duty-cycle work is designed for manual shielded metal arc work and for some semiautomatic welding, and 100%-duty-cycle machines are designed for automatic welding. Welding machines can be used at higher levels than their duty-cycle rating and/or welding current rating under specific conditions. It may be necessary to use a machine to weld automatically or for 100% of the 10-minute cycle, even though it has a 60% duty cycle. This is possible if the current is reduced below the rating. In other cases it may be necessary to use the machine at a higher current rating but for a short period of time.

Both of these situations can be resolved by use of the following formula:

desired duty cycle (%)

$$= \frac{(\text{rated current})^2}{(\text{desired current})^2} \times \text{rated duty cycle (\%)}$$

For example, a machine rated at 300 A and 60% duty cycle needs to produce 350 A. What is the maximum duty cycle that can be used?

$$\text{desired duty cycle (\%)} = \frac{(300)^2}{(350)^2} \times 0.60$$

$$= \frac{90,000}{122,500} \times 0.60 = 44\%$$

Thus, to use this machine at 350 A, the duty cycle would have to be reduced to 44%. This means welding 4.4 minutes out of every 10 minutes.

In the other situation, the same machine, a 300-A 60%-duty-cycle machine, must be used on an automatic welding application. It must run at 100% duty cycle, or for a full 10 minutes. What output current could safely be obtained from this machine?

$$1.00 = \frac{(300)^2}{(\text{desired current})^2} \times 0.60$$

$$(\text{desired current})^2 = \frac{(300)^2}{1.00} \times 0.60 = 90,000 \times 0.6$$

$$\text{desired current} = 54,000 = 232\,\text{A}$$

Thus, for an automatic operation running 10 minutes continuously, the machine output must not exceed 232 A without overloading the power source. These same determinations can be used without using the foregoing formula. Figure 10–57 is the duty cycle versus rated current plot of this formula. The sloping lines show typical machine ratings, and by drawing a sloping line in parallel to those shown, different duty cycles or different load current requirements can be determined. Manufacturers provide duty cycle versus rated current curves for their machines.

Specifying the Equipment

To specify a welding power source properly, the following data should be provided:

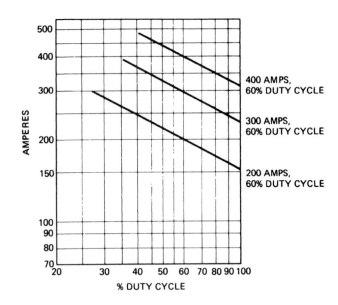

FIGURE 10–57 Duty cycle versus rated current curve.

1. *Manufacturer's machine designation.* This is determined by consulting the manufacturer's catalog or data sheets.

2. *Manufacturer's catalog number.* This is shown in the manufacturer's literature and is usually given as a model number.

3. *Rated load voltage.* Welding machines for production requirements are rated in accordance with the standard. This system shows that minimum load volts are related to the ampere output of the machine. For example, the 200-A machine has a minimum load of 28 V and this increases as the machine ratings are increased.

4. *Rated load amperes.* This is the rated current that the power source will deliver at the rated volts.

5. *Duty cycle.* Most production welding machines are rated at 60% or 100% duty cycle, in accordance with the standard. The manufacturer's data will provide this information.

6. *Voltage of incoming power.* The welding machine must match the power available at the fuse box. Most industrial welding machines can be reconnected for specific voltages. The voltage that is available must be specified.

7. *Frequency of incoming power.* This is the frequency of the power provided by the utility company. In North America this is normally 60 Hz. In some locations in the world it is 25, 50, or 60 Hz.

8. *Number of phases of incoming power.* For industrial equipment, three-phase power is normally provided. Single-phase power is used for limited input or low-duty-cycle welding machines. Some machines are capable of operating on either single- or three-phase power.

For engine-driven welding machines it is wise to specify the maximum rated speed in rpm at no load. The information outlined above permits accurate specification of the welding power source desired. Probably the most important factor is to specify the precise voltage of power that will be available at the fuse box.

10-9 INSTALLING AND MAINTAINING A POWER SOURCE

The installation, maintenance, and adjustment of welding machines and welding equipment are usually done by different people. Normally, the original installation is done by construction people, riggers, electricians, and so on. Maintenance is an ongoing operation and is usually done by plant maintenance electricians. They may also assist with troubleshooting when required. Minor adjustments of a routine nature are usually done by welders. This includes changing tips and nozzles, tungsten electrodes, wire drive rolls; blowing out cables; and other procedures.

Installation

Electric welding power sources must be installed with care and according to various codes and regulations and manufacturers' recommendations. The power source should be installed to avoid exposure to high ambient temperatures, high humidity, dust, or corrosive liquids or fumes. The case of the power source should be grounded to an earth ground. It should be installed in accordance with local or state codes. When state or local codes do not exist, it is recommended that the *National Electrical Code*® be followed. It should also be installed in accordance with the manufacturer's instructions and must comply with all rules and regulations of the owner. Power sources should be connected to the utility power using correct leads from the disconnect box to the welding machine. Manufacturers' instructions provide disconnect box sizes, fuse sizes, and cable size.

If the power source is utilized for GTAW, it may include a high-frequency arc stabilization unit. These are sources of EMI (electromagnetic interference). Machines with high-frequency stabilization must be grounded to earth in accordance with manufacturers' instructions. Newer-type all-solid-state power sources are susceptible to EMI. Machines of this type must also be grounded to earth. If automatic welding equipment involving motion is involved, emergency shutdown systems should be provided. In addition, load suppression relays should be installed. For details concerning this, see the manufacturer's instruction manual.

The power source must be matched to the primary utility power. Most industrial welding machines include voltage changeover links to accommodate different line voltages. The installation of welding machines must be inspected with a checklist utilizing the items mentioned above. These must be in accordance with all standards, codes, and instructions. Motor-generator welding machines must be checked for direction of rotation. This is easily done since arrows on the machine indicate the correct direction of rotation. Transformer welding machines must be installed to balance a three-phase power line. They must be phased with respect to adjacent units, especially if more than one transformer power source will be used on the same weldment.

Maintenance

Preventive maintenance is the routine scheduled maintenance performed on equipment while in service so that it does not deteriorate rapidly. Preventive maintenance for the different kinds of welding power sources is different. Specific types require specialized attention.

All welding machines should be kept dry and clean. Dirt and dust from the factory or the construction site are carried by ventilating air through the arc welding machine. This dirt collects in the internal parts of the machine and tends to build up on windings and prevent them from cooling efficiently. The dirt may build up to the point where it blocks the passage of cooling air. To counteract this, welding machines should be inspected every six months and cleaned. In extra-dirty environments, or when the machines are exposed to high humidity or corrosive fumes, the inspection period should be shortened. The electric power should be removed from the welding machine at the disconnect switch, which should be tagged, and the dirt should be removed from windings by blowing them out with dry compressed air at a pressure of 25 to 30 psi. High pressure should not be used since this will tend to drive the dust and dirt into crevasses, which will reduce the cooling efficiency of the machine. Vacuum cleaning can be used if metallic dust is present. In very dirty environments filters are recommended to keep dust from entering the machine. Look for internal corrosion and internal mechanical damage, and make sure that ventilating fans are operating properly.

Specific types of machines require specific attention. In the case of engine-driven generators, the engine requires much attention, such as changing oil after each week of continuous operation. Inspect and replace oil filters and fuel filters, check coolant and batteries each time fuel is added, check the radiator daily, and check idling devices weekly. The engine manufacturer's recommendation must be followed. The generators should be checked monthly. This includes windings, contact points, brushes, brushholders and commutator, control switches, and bearings. Brushes and commutators need special attention since they tend to wear. Bearings should be checked monthly to make sure that they are properly greased. If the generator is exposed to corrosive or salt atmospheres, special attention should be given to exposed metal parts. They should be cleaned and repainted and insulation replaced as necessary.

Most industrial inverter power sources have built-in protection features. Many machines have a green light which indicates that the machine is "ready to weld." This indicator light will turn off and a red indicator light will turn on if there are internal problems in the machine. A red indicator light may indicate one of the following problems: Input voltage is too high or too low, operation is overcapacity, or the machine is overheated. The machine can be restarted after corrections have been made.

Nonrotating welding machines require special attention. In most cases they include a ventilating fan which should be checked periodically to make sure it is operating correctly. All electrical contacts and terminals should be checked for tightness, but the most important thing is proper handling of printed circuit boards. Printed circuit boards and other devices of an electronic nature may be affected by static electricity. Special handling precautions are required. Most manufacturers have an exchange program since many users are unable to repair printed circuit boards. It is advisable to keep a detailed inspection record for each machine by serial number, inspection and periodic maintenance date, and extra activities or maintenance performed.

Troubleshooting

Troubleshooting is required on a welding machine when it does not operate satisfactorily. Troubleshooting is a matter of solving the problem of the machine and should be done only by trained and qualified personnel. Many of the newer machines are equipped with a fault-lamplight to indicate problems. The instruction manual will outline the solution. This type of work should be done without removing the machine's case. It is wise to do this type of work in connection with the welder, who understands the welding operation, and with maintenance personnel, who understand the equipment problems. Most static power sources are sufficiently complex that it is necessary to utilize the maintenance manual, which normally has troubleshooting instructions. They list various problems that may be encountered with a list of potential solutions. These get quite complex as the equipment gets more complicated. It is important to distinguish between a welding problem and an equipment problem. This is particularly apparent when the welder is utilizing unfamiliar equipment or when first using a different welding process. Welding processes have their own specific problem areas, and often the welder may tend to blame the equipment when it might be a problem with the process. Many difficulties are encountered when the equipment is not adjusted properly, cables are not tight, work is not properly connected to the work connection, and so on. Skillful diagnosis is required to determine the true cause of some welding problems. With complex equipment the owner's manual and a checklist should be employed. There should also be information related to the welding process, so that the distinction can be made between a welding problem and an equipment problem.

Repairing

If the diagnosis indicates that the machine must be repaired, it should be removed from service and taken to the maintenance repair shop. If in-house facilities are not available or if the diagnosis indicates problems with control circuits or printed circuit boards, the equipment should be sent to an authorized repair station or to the manufacturer. Local repair shops that have been approved by the equipment manufacturer can normally handle most of these problems. It is essential that genuine replacement parts and replacement PC boards be used for all repair work and that the repair mechanics have sufficient knowledge and skill to accomplish this work. After the machine has been repaired, it should be tested and checked to make sure it will fulfill its original function.

 **QUESTIONS**

10–1. What are the major two types of welding power sources? What are the differences?

10–2. What are the three types of static welding machines?

10–3. What are the static characteristics and dynamic characteristics of a power source?

10–4. What is a brushless generator?

10–5. What determines the open circuit voltage of a transformer welding machine?

10–6. How does the saturable reactor system work?

10–7. What are the principal methods of adjusting the output of a transformer welding machine?

10–8. What is the function of a rectifier?

10–9. How does an SCR work?

10–10. How does the phase-angle circuit adjust the output of a welding machine?

10–11. What is the advantage of an inverter type of power source?

10–12. Why is an inverter power source lighter in weight than a similar-sized rectifier power source?

10–13. Why is high frequency used in a power source?

10–14. What is a programmable power source?

10–15. What system can be used to control an inverter power source?

10–16. Explain a twin inverter machine.

10–17. What data are necessary to specify a welding power source?

10–18. Who should do electrical troubleshooting on an electrically hot power source?

10–19. What is preventive maintenance? How does it apply to welding machines?

REFERENCES

1. "Arc Welding Equipment, Part 1: Welding Power Sources," IEC 974-1. International Electrotechnical Commission, Geneva, Switzerland.

2. "Standard for Transformer-Type Arc-Welding Machines," ANSI/UL 551-1993, Underwriters' Laboratories, Chicago.

3. B. Irving, "What Welding Accomplished 'Way Back When,'" *Welding Journal* (Jan. 1994).

4. C. Shira, "Converter Power Supplies—More Options for Arc Welding," *Welding Design and Fabrication* (June 1985).

5. S. Barhorst and H. Cary, "Synergic Machines Simplify Pulsed Current Welding," *Welding Design and Fabrication* (November 1985).

6. H. R. Castner, "Gas Metal Arc Welding Fume Generation Using Pulsed Current," *Welding Journal,* Research Supplement (February 1995).

7. E. K. Stava, "The Surface-Tension Transfer Power Source: A New, Low-Spatter Arc Welding Machine," *Welding Journal* (January 1993).

8. T. L. Nancy, "Fourth-Generation Inverters Add Artificial Intelligence to the Control of GMA Welding," *Welding Journal* (January 1993).

9. W. J. Lester and F. L. Kaeser, "Todd Evaluates Welding Systems for Performance and Cost," *Marine Engineering/Log* (September 1961).

11 OTHER WELDING EQUIPMENT

OUTLINE

11-1 ARC WELDING GUNS AND TORCHES

All arc welding processes use a device to transmit welding current from a welding cable to the electrode. They also provide a means for shielding the weld area from the atmosphere, and for the consumable electrode processes they provide the metal for making the weld. These devices come in different types and sizes and are given different names. For shielded metal arc welding the device is called an *electrode holder.* It is designed to hold sticks of electrodes and is described in Section 6-3.

The carbon arc electrode holder is considerably different but holds the carbon electrodes securely and, in some cases, two electrodes. It is described in Section 5-4.

For gas tungsten arc welding the device is called a *welding torch* or *gun.* It holds the tungsten electrode and transfers the welding current to it. It also directs shielding gas to the arc area. It is described in Section 5-2.

The torch or gun for plasma arc welding is similar but contains an orifice that develops the plasma. This con-stricts the arc and creates high-temperature plasma. Plasma arc torches are described in Section 5-3. Both plasma arc and gas tungsten arc torches are available for manual and/or automatic use.

The guns for air arc cutting and gouging are de-signed specifically for this process. They contain a valve to actuate the flow of air. These are described in Section 5-4.

Stud welding uses a special gun that involves me-chanical action to carry out the stud welding operation. There are basically two types: the drawn arc type and the discharge type. Stud guns are described in Section 5-5.

Guns or torches for gas metal arc welding and flux cored arc welding are basically the same. They are used to direct the welding electrode into the arc, to transmit the welding current to the electrode, and to supply the shield-ing medium for the arc area. In general, they are called *guns* when used for semiautomatic applications since they are held manually. They are called *torches* when used for mechanized welding. There are two major variations: those that employ external shielding gases and those that do not. They are described in Sections 6-4 and 6-5.

The submerged arc welding process uses both man-ually held guns and automatic torches. There are two sys-tems for semiautomatic submerged arc welding. In one case, flux is carried in a hopper attached to the gun (Fig-ure 11-1), and in the other method compressed air feeds the flux through a cable assembly to the arc (Figure 11-2). In most cases, manually held guns are attached to cable assemblies, which are available in different lengths.

FIGURE 11–1 Hopper-feed semiautomatic gun for SAW.

FIGURE 11–2 Pressure-feed semiautomatic gun for SAW.

In others, the electrode cable is connected directly to the electrode holder.

A major function of the torch or gun using electrode wire systems is to deliver the welding current to the moving electrode wire. This is done by means of contact tips or contact jaws. The amount of current transmitted is a way of sizing welding guns and torches. This is the welding current rating, normally the maximum current that can be used with a particular gun or torch. Higher currents are generally used with larger-diameter electrode wires. Sliding contacts for transmitting large amounts of current generate heat. Heating of the gun occurs from this as well as from its closeness to the welding arc. This is another way of classifying guns and torches—the method of cooling—whether it is by circulating water or ambient air.

Another function of the gun or torch is to deliver shielding gas to the arc area. Shielding gas is not used with variations of flux cored arc welding; hence there are gas shielded and non–gas shielded types of welding guns and torches. They are called gas shielded or gasless guns.

Another way of classifying guns is by their shape, which relates to welding wire type and welding position, but also to the preference of the welder. The two most common are the curved head or gooseneck configuration, and the straight-line or pistol-grip variation. Automatic torches can also be either straight-line or bent with different angles for specific applications.

The handheld semiautomatic guns normally employ a trigger that activates the control circuit. Electrode touch starting is sometimes employed but is less popular. With a trigger switch, the voltage of the circuit must be supplied from an isolated voltage source and must not exceed 35 V ac or 50 V dc. There are no specifications in the United States for welding electrode holders, guns, or torches. In view of this, manufacturers' data should be consulted to specify the electrode holder, gun, or torch required. In each process section information is given about different sizes and types of guns and torches, different styles, and other details.

NEMA Standard EW3, "Semiautomatic Wire Feed Systems for Arc Welding,"[1] covers guns that use solid or flux cored electrode wire with or without gas shielding. It also covers GTAW torches. One provision is to list the maximum temperature of external surfaces of welding guns. The maximum temperature of a metal handle is 122° F (50° C). For a nonmetallic handle, the maximum temperature is 140° F (60° C). The maximum temperature of the nozzle is 158° F (50° C) if metallic. If nonmetallic, it is 203° F (95° C). The maximum temperatures are based on continuous use.

Welding Guns

The welding guns can be categorized as curved head or gooseneck and straight-line or pistol-grip guns. They are further subdivided into air-cooled or water-cooled guns. For flux cored arc welding they can be subdivided into those using external gas shielding and those using self-shielding electrodes. A variety of curved head welding guns is shown in Figure 11-3. A cross-sectional view of the curved head or gooseneck gun is shown in Figure 11-4. A pistol-grip water-cooled straight-line gun is shown in Figure 11-5. The gasless gun, for flux cored arc

FIGURE 11–3 Variety of curved-head welding guns.

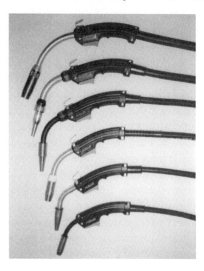

(a)

(b)

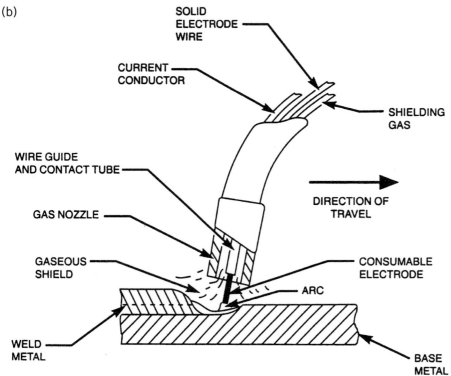

SOLID
ELECTRODE
WIRE

CURRENT
CONDUCTOR

SHIELDING
GAS

WIRE GUIDE
AND CONTACT TUBE

DIRECTION OF
TRAVEL

GAS NOZZLE

GASEOUS
SHIELD

CONSUMABLE
ELECTRODE

ARC

WELD
METAL

BASE
METAL

FIGURE 11–4 (a) Gooseneck gun for GMAW and FCAW. (b) Gas metal arc welding process.

welding, is shown in Figure 11–6. This also shows the details of the contact tube or tip for electrical stickout. This is often required since the self-shielding flux cored electrode wires normally operate more efficiently with extended stickout.

For low-current gas metal arc welding, the gooseneck-type air-cooled guns are usually employed. When CO_2 shielding is used, with larger-diameter electrodes at higher currents, an air-cooled gun can be used since the CO_2 is a cooling medium. When inert gas or argon–oxygen gas mixtures are used at higher currents, the guns must be water cooled to avoid overheating.

Handheld semiautomatic guns include the cable assembly, which attaches to the wire feeder. Cable assemblies are available in different lengths. The cable assembly includes the conduit tube for the electrode wire, a tube for supplying shielding gas, and two tubes for cooling water when used, plus the control cables to the trigger switch. It also includes the electrical conductor for the welding current. The size of this conductor is related to the rating of the welding gun. The welding guns allow the use of different nozzles and the replacement of the contact tube or tip.

The gun nozzle is usually identified by the inside diameter at the shielding gas discharge end. They are dif-

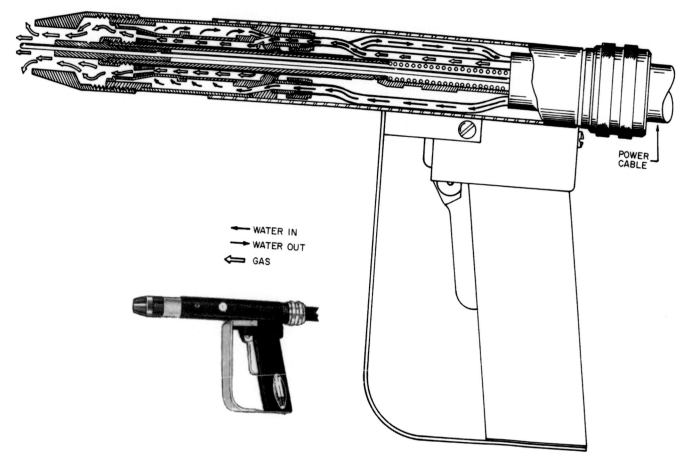

FIGURE 11–5   Pistol-grip or straight-line gun, water cooled.

→ WATER IN
→ WATER OUT
⇐ GAS

FIGURE 11–6   Gasless gun for FCAW, showing nozzle detail.

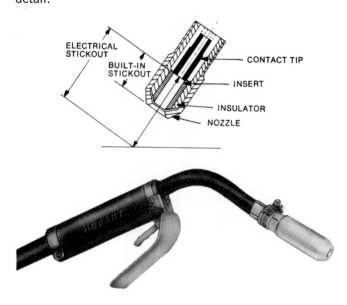

ELECTRICAL STICKOUT
BUILT-IN STICKOUT
CONTACT TIP
INSERT
INSULATOR
NOZZLE

ferent for each manufacturer and may fit only the guns of the same make.

The welding guns have replaceable contact tips or tubes for transferring current and guiding the electrode wire to the arc. These come with different inside diameters to accommodate different-diameter electrode wires. Manufacturers recommend specific tips for specific wire types and sizes. Efficient transfer of current from the cable to the electrode wire is necessary to avoid overheating. Contact tips are made of copper or copper alloys. Pure copper is very soft and the inside diameter will wear rapidly. When the inside hole becomes oversized, the welding current transfer efficiency diminishes and more heat will be generated. Hence, contact tips must be changed on a regular basis. Long-wearing contact tips are available and are made of special copper alloys. In some cases, special inserts are incorporated. The copper alloys are much harder than pure copper and will provide longer life, but are more expensive.

For welding aluminum, extra-long contact tubes are recommended because of the oxide coating on the aluminum electrode wire. The extra-long contact tubes provide more area to transfer the welding current to the electrode wire. To improve the current transfer, some contact

OTHER WELDING EQUIPMENT   **343**

tubes incorporate a slight bend to make sure that there is positive sliding contact between the electrode wire and the contact tip. Longer contact tubes usually mean straight-line pistol-grip guns.

The gooseneck type of air-cooled gun is most popular for welding steels, particularly using small-diameter electrode wire. The pistol-grip or straight-line gun is more often used with aluminum since the curve in the gooseneck gun tends to create resistance to the electrode wire, which may cause jamming of the cable assembly.

Welding guns and cable assemblies must be serviced regularly in order to provide efficient operation. The filler wire conduit must be replaced on a regular basis since it tends to fill with loose copper coating, metal shavings, and so on.

Adapters are available for attaching the gun cable assemblies of one manufacturer to the wire feeder of another. Quick-connection adapters are also available and are widely used.

There is no standard method of rating welding guns or torches or for measuring the angle of goosenecks, the weight or balance point, or the size of the guns. In view of this, it is necessary to use the manufacturers' data to specify guns and torches. Most manufacturers provide a duty cycle rating for use with CO_2 shielding gas and for inert shielding gases.

Automatic Welding Torches

Torches for mechanized welding are usually straight-line torches. Figure 11–7 shows a variety of torches for specific applications. The two torches on the left are used for smaller-diameter electrode wires. The torch on the extreme left provides concentric shielding gas delivery and the next one is for side delivery of CO_2 shielding gas since CO_2 gas is used at a higher flow rate. Side-delivery systems normally pick up less spatter and are widely used for automatic systems. Side-delivery torches should be used only when CO_2 is used for shielding. The other four torches are for larger-diameter electrode wires. The third and fourth torches are for concentric gas delivery. The larger one is water cooled and can be used for inert shielding gas. The fifth torch utilizes larger electrode wire and has the side delivery of CO_2 shielding gas. This same torch is used, without the side-delivery nozzle, for submerged arc welding and for gasless flux cored arc welding. There are exceptions to the straight-line design, shown by the extreme right-hand torch. This is curved for a specific automatic application. This type of torch is also used for robotic arc welding.

The automatic torch must be selected to fit the welding process. The torch must be selected with respect to the size of the electrode wire, the current range, and

FIGURE 11–7 Torches for automatic welding.

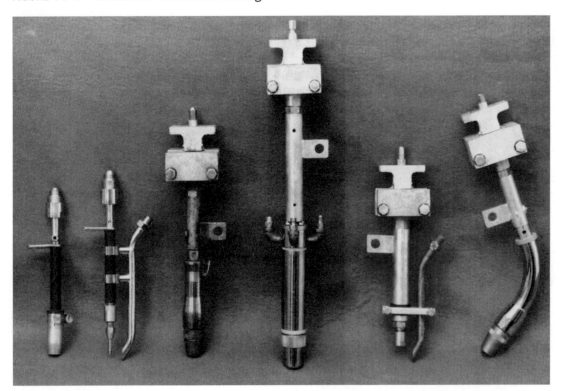

duty cycle of the operation. It is generally best to utilize a torch rated at a higher current level than will be employed. In addition, weight is less important since it is held by the machine rather than by the person. Torch current pickup tubes or tips can be selected to accommodate different wire sizes. For large electrodes, heavy-duty torches with spring-loaded current contact jaws are used. Spring-loaded contacts usually are made from special copper alloys. The jaws are loaded against the wire to provide efficient transmission of the current to the wire for cooler operation.

11-2 ELECTRODE FEED SYSTEMS

All of the continuous electrode wire arc welding processes require an electrode feeder of one type or another. It is used to feed the consumable electrode wire into the arc. There are many types of wire feeders. The most widely used is used with the consumable wire processes, where the electrode is part of the arc welding circuit. The other type of feeder, known as a "cold wire" feeder, is used with the arc welding processes, where the electrode is not part of the welding circuit. The basic requirement of the wire feeder is to feed the electrode continuously into the arc and to maintain a stable arc at the desired welding current and voltage. The basic requirement of the cold wire feeder is to feed the filler wire into the arc area at the correct rate to maintain proper melting and deposition. The components of a wire feeding system are the welding gun or torch, the wire drive mechanism, the control circuit, and the wire handling and dispensing system. See the NEMA standard "Semiautomatic Wire Feed Systems for Arc Welding"[1] for more details.

Wire Feeder Types

Welders need semiautomatic equipment that has the same flexibility and portability as shielded metal arc welding. Many developments have been made to provide this flexibility. It is now possible to weld almost anywhere and to make almost any kind of joint with semiautomatic equipment. The wire feeder is the heart of the system and has developed in many forms.

The most common type of wire feeder used for semiautomatic welding is shown in Figure 11–8. Feeders of this type carry the supply of electrode wire, the wire drive mechanism, the control circuit, and the adjustment for wire feed speed. They are sufficiently powerful to push electrode wires through a long cable assembly using gooseneck welding guns. These feeders usually carry 25-lb spools of electrode wire and have optional accessories such as wire covers, water valves, and mounting wheels.

FIGURE 11–8 Conventional wire feeder.

The control panel of a wire feeder includes a power on/off switch, wire feed speed control rheostat, and arc voltage control rheostat, as well as an inch/purge switch. It includes a LED meter, which shows either wire feed speed or voltage. Many feeders have an optional timer so that it can be used for arc spot welding. These types of wire feeders allow a variety of feed rolls to be used for different types of electrode wire.

Other wire feeders are designed to be used with conventional CV power sources and pulsing power sources, and in some cases with synergic welding. The wire feeder must be matched to the power source. In these cases the wire feeder may have its own microprocessor control. It is important that the wire feeder and power source be matched to obtain the ultimate in synergic welding control.

To provide better portability, smaller, enclosed or suitcase-type wire feeders are used. These weigh slightly over 20 lb and will pass through 14-in.-diameter manholes. They are totally enclosed and normally use a small spool of electrode wire that can carry 20 lb of steel or 5 lb of aluminum electrode wire. A typical example is shown in Figure 11-9.

Spool Guns

The spool gun is the ultimate in portability and is attained by combining the wire feeder and the gun. The wire feed motor is located in the gun, with the drive rolls just behind the gun nozzle. Electrode wire is pushed only several inches. The spool gun also contains the wire supply and uses a very small spool of electrode wire. Control is obtained in some cases by a separate control box, but the newer guns have the control system mounted in the gun. They were designed for pushing small, soft electrode wires and/or very small-diameter electrode wires. A typical spool gun is shown in Figure 11-10. These guns are

FIGURE 11-9   Portable wire feeder.

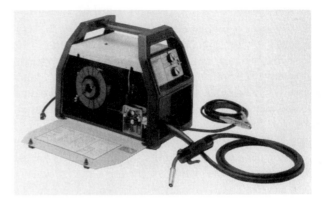

FIGURE 11-11   Combination power source and wire feeder.

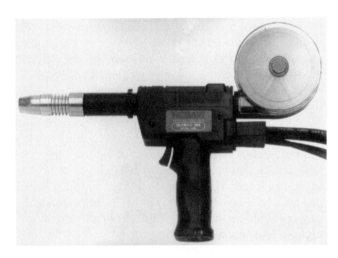

FIGURE 11-10   Spool gun.

lightweight and well-balanced for ease of manipulation. They are, however, more awkward than a gooseneck gun. They come in several sizes. A lightweight spool gun for aluminum will weigh approximately 1 lb. Spool guns are less rugged than pistol-grip or gooseneck guns. The wire feed motor is smaller and less robust; but, more importantly, the electrode wire, when purchased on small spools, is more expensive.

In some cases the wire feeder is built into the power source cabinet. This provides a single unit that is relatively portable. This kind of equipment can be taken into remote areas. It is also used for light production work and by hobbyists, and normally operates on 115 V ac. A typical example is shown in Figure 11-11.

When different welding parameters are used on the same job, several different types and sizes of electrode wires will be required. An example is the root-pass welding of a pipe joint with the filler passes made by larger-diameter flux cored wire. For this kind of work a dual

wire feeder is used, with two different coils of electrode wire and two gun-cable assemblies. Units of this type are very common in pipe fabrication shops. A typical example of a dual wire feeder is shown in Figure 11-12.

Equipment with flexibility that uses large packages of filler metal is the push-pull system. A drive motor included in the welding gun pulls the electrode wire, and another drive head placed at the wire supply pushes the electrode wire. These motors are designed so that the pull unit maintains a very slight tension on the wire as it passes through the flexible conduit and prevents kinking of the wire. These units, shown in Figure 11-13, are used for aluminum welding.

One of the newest types of wire feeder is the electronic control type, which allows presetting the wire feed speed and arc voltage. Some wire feeders have two or more schedules that can be preprogrammed into the controller memory circuit. This allows more control of the welding operation, but also allows the welder to select different schedules for different types of work. A typical example is shown in Figure 11-14.

Special wire feeders are sometimes required for pulsed arc welding and/or synergic welding. These re-

FIGURE 11-12   Dual-electrode wire feeder.

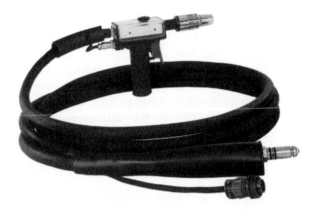

FIGURE 11-13   Twenty-five-foot extended wire feed system.

FIGURE 11-14   Wire feeder with electronic memory.

quire coordination controls between the wire feeder and the power source, but are easily adjusted by the welder.

There are features that should be incorporated in wire feeders for semiautomatic welding. These include at least the following. The wire feed motor and electrode

supply should be insulated from the cabinet so that the wire feeder can be placed on the work. The gun cable assembly must be easily attached and detached from the wire feeder. Certain controls should be at the wire feeder: the "inch" button to thread a new coil of wire, the inch/reverse switch to retract the wire, and the purge button. The control circuit should include dynamic breaking of the wire to prevent coasting, it should include preflow and postflow of shielding gas, and it should have a burnback control for crater filling.

Cold Wire Feeder

The cold wire feeder is an entirely different type of wire feeder, used to feed filler metal into the arc. It is used for gas tungsten, plasma arc welding, and for the high-energy beam welding processes. It is occasionally used with submerged arc welding. Normally, the filler metal does not carry current. An exception is a variation known as "hot-wire" welding, where the filler wire carries current to improve deposition rates. Current and voltage are sufficient to heat the filler wire but not to create a welding arc. The feed rate of the cold wire feeder must be very accurate, and the wire feed must have a continuously adjustable speed over a wide range. Cold wire feeders feed wire at a much lower rate than do consumable electrode wire feeders. A typical example of a cold wire feeder for a gas tungsten application is shown in Figure 11-15. Cold wire feeders are now available as cold wire spool guns.

Control Systems

The control system, or control circuits, is of different types, depending on the features required of the wire feeder and the type of power source involved. There are two basic types of systems, dictated by the type of power source involved. The most popular type is the constant feed speed system, which utilizes the fixed burnoff rate versus welding current relationship of the electrode wire. This system must be used with a constant voltage or "flat" characteristic power source. The electrode wire feed rate is set by the speed control of the wire feed motor. The CV power source automatically furnishes the correct amount of current to burn it off at the same rate that

FIGURE 11-15   Cold wire feeder.

it is fed into the arc. Thus the wire feed rate controls the welding current. The voltage at the arc is controlled by changing the output voltage of the power source. It is a self-regulating system and is popular for small-diameter electrode wires. It was originally developed for gas metal arc welding, to eliminate stubbing and burnback. The wire feed rates of constant-speed wire feeders are adjustable over a wide range of speeds. The range must include the welding conditions involved in the desired welding procedure. A constant-speed wire feeder may be used on a CC or "drooping" characteristic power source, but it is difficult to adjust the various controls properly; furthermore, sudden change in welding conditions may cause the system to go out of control.

The voltage-sensing wire feeder utilizes the voltage across the arc and regulates the wire feed speed to maintain a preset arc voltage. This type of unit is used with CC or drooping type of power source. The control system replaces the feeding action of the welder. It slows down or speeds up the feed rate of the electrode wire to maintain the set arc voltage. This system uses the arc voltage to control the wire feed motor. As the arc lengthens, the arc voltage increases, which increases the speed of the wire feed motor. This causes the electrode to feed faster and thus shorten the arc. Another system uses a feedback circuit, which takes the arc voltage and compares it to a standard, and the difference is used to vary the speed of the wire feed motor. The voltage-sensing system is self-regulating. Welding current is adjusted at the welding power source. A voltage-sensing wire feeder may be used for either dc or ac welding. It is most popular for feeding large-diameter electrode wires and was originally developed for submerged arc welding. The wire feed system may also include a retract circuit for automatically initiating the arc. Touch start and quick break of the arc are used for starting and stopping the arc in many semiautomatic systems; however, a trigger circuit is more popular. A comparison of

the electrode wire feeder types and power source types for different applications is shown in Figure 11–16.

The newer control circuits include memories for multiwelding schedules and also provide precision controls for pulsed MIG and GMAW welding, with particular emphasis on synergic systems. These are more complex control circuits, which are incorporated into the wire feeder controller.

Wire Drive Mechanisms

The wire drive mechanism, also known as the feedhead, consists of the drive motor, gearbox, and drive rolls assembly that actually feeds the wire. The most popular type of feeding system uses pinch rolls, which transform the feed motor rotary motion to the electrode wire to push it in a linear motion. The pinch rolls grip the wire on opposite sides, and by means of pressure provide positive linear motion (Figure 11–17). Two roll drives are most commonly used. Light-duty feeders use two drive rolls and only one roll is powered. Heavy-duty feeders use two drive rolls and both are powered. For special applications, four drive rolls are used and all four are powered. The advantage of the four drive rolls is that less pressure is required on the electrode wire. This is particularly important when feeding flux cored electrode wire since the sheath of the wire may collapse if the pressure is too great.

The design of the driving surface of the rolls is extremely important. Different-type rolls are used for different types and sizes of electrode wires. The drive rolls are steel and approximately 2 in. in diameter. The different driving surfaces are:

* Flat-smooth
* Flat-knurled
* Smooth V-groove

FIGURE 11–16 Power source type versus wire feeder type.

Power Source Type	Electrode Wire Feeder Type	
	Voltage Sensing	Constant Speed
CV, direct current	Difficult to adjust; seldom used; self-regulating within limits	Best for gas metal arc welding; best for flux cored arc welding; best for submerged arc when using small-diameter electrode wire; self-regulating
CC, direct current	Best for submerged arc when using large-diameter electrode wire; used for GMAW on aluminum; self-regulating	Difficult to control; not used for small wire GMAW; not self-regulating
CC, alternating current	Used for submerged arc (medium and large electrode diameters); used for flux cored arc welding; self-regulating	Difficult to control; not used for GMAW; not self-regulating

FIGURE 11–17   Pinch roll drive, two or four rolls.

* ✳ Knurled V-groove
* ✳ U-groove
* ✳ Cog wheel

The groove in the drive rolls is important and relates to wire drive efficiency. The U-shaped grooves are not recommended because of problems with the electrode diameter, which can vary by ±0.001 in. If the electrode wire is too large, it will not fit into the groove or may require too much force. If the wire is too small, it will slip in the groove and accurate feeding will not occur. The V-groove has advantages over the flat drive rolls in that there are four points of contact rather than two. This provides better control and better transfer of power to the electrode wire. In some cases the drive rolls are made as two individual pieces, which can be reversed to provide a new surface to replace worn surfaces. Figure 11–18 provides selection information for electrodes of hard wire, soft wire, and tubular wire of different sizes. Hard wires are steel, stainless steel, and nickel alloys. Soft wires are aluminum, magnesium, and copper. Tubular wires are flux cored electrode wires. The method of applying pressure to the wires should be positive but adjustable. Rolls should be adjusted so that they do not slip on the wire and do not deform the wire. Knurled rolls tend to indent the wire, which makes it more abrasive when going through conduits and current pickup tips. If too much pressure is used, it will deform the wire and possibly stall the drive motor. If too little pressure is used, slippage will occur.

The wire feed motors for heavy-duty wire feeders have up to $\frac{1}{4}$ hp. Smaller motors of $\frac{1}{8}$ or $\frac{1}{10}$ hp are often used. Smaller motors are used for the hand guns. Most feed motors are the dc shunt type; however, permanent-magnet motors, stepper motors, pancake motors, and print motors are all used. The resistance to the motor is the drag on the wire as it passes through the conduit, drive rolls, and current pickup tube. If there are kinks in the electrode wire, additional resistance is encountered. A problem with wire feed motors in automatic systems is the need to pull wire from large spools. This causes high-inertia starting loads, and the life of the wire feed motor may be shortened.

Each welding procedure has a specific electrode wire feed rate, given in inches per minute, millimeters per minute, and sometimes, meters per hour. Wire feeders have a range of wire feed speeds that can be adjusted. A typical minimum feed rate is 50 in. (127 cm) per minute and the maximum is 1000 in. (2540 cm) per minute. This speed range would be ample for most welding applications. This range is very broad and normally requires gearbox changes. Manufacturers' data sheets provide the maximum and minimum wire feed rates available for different-model wire feeders with different gearbox ratios, maximum and minimum size of electrode wire, and the different types that can be used with a particular wire feeder. The data sheets also provide the speed regulation and length of conduit that can be accommodated by a particular wire feeder. The wire feeder should have a speed range that includes the range of wire feed speeds for the welding conditions that are to be used. The speed regulation of the feeder indicates how much the wire feed motor will slow down when extra resistance is placed on the feeding of the electrode wire.

As an aid in selecting a wire feeder, see the charts showing the feed speeds of different electrode types used with different welding processes. Figures 11–19 to 11–21 show the wire feed speed versus welding current. Figure 11–19 is for the gas metal arc and flux cored arc welding process using solid and tubular small-diameter steel wires. Figure 11–20 is also for GMAW but for nonferrous electrode wires. Figure 11–21 is for steel electrode wires, used with submerged arc welding.

Wire drive mechanisms, or feedheads, are also used in automatic welding systems. Normally, the feedheads are the same as those in a heavy-duty semiautomatic system. For automatic welding a more complicated control system is used. Mounting hardware is available for attaching the feedhead to motion devices or in fixtures for specific applications.

Planetary Type Wire Feeders

Planetary electrode wire feeders, while not as popular as pinch roll feeders, are finding increasing use. They are sometimes known as linear feeders or concentric wire

Electrode Wire Diameter		Electrode Wire Types					
in.	mm	Hard Wire	Hard Wire	Hard and Tubular Wire	Soft Wire	Hard and Tubular Wire	Tubular Wire
0.024		X	X	—	X	—	—
0.030	0.75	X	X	—	X	—	—
0.035	0.9	X	X	—	X	—	—
0.045	1.1	X	X	—	—	—	—
3/64 (0.047)	1.2	—	—	—	X	—	—
0.052	1.3	X	X	—	X	—	X
1/16 (0.063)	1.6	—	—	X	X	X	X
5/64 (0.078)	2.0	—	—	X	X	X	X
3/32 (0.094)	2.4	—	—	X	X	X	X
7/64 (0.109)	2.8	—	—	X	X	X	X
1/8 (0.125)	3.2	—	—	X	X	X	X
5/32 (0.156)	4.0	—	—	—	—	X	X
3/16 (0.188)	4.8	—	—	—	—	X	X
7/32 (0.219)	5.6	—	—	—	—	X	X
1/4 (0.250)	6.4	—	—	—	—	X	X
Feed Rolls Selection		Flat - Smooth / Smooth V	Flat - Knurled / Smooth V	Smooth V / Smooth V	Smooth V / Smooth V	Knurled V / Smooth V	Cog / Cog

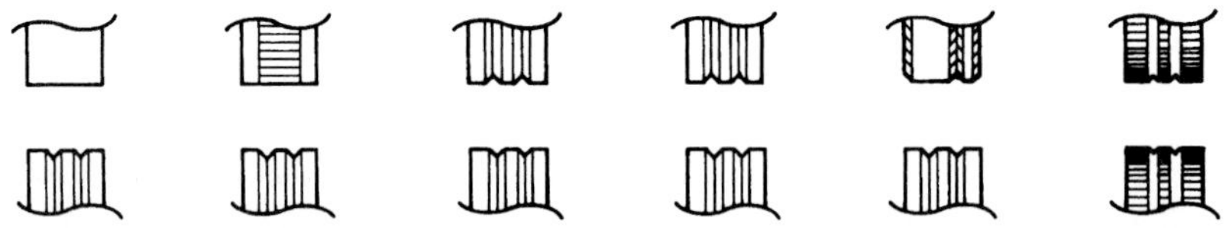

FIGURE 11–18 Drive roll selection chart.

feeders. This concept uses the planetary motion of two or three drive rollers. These rolls are mounted on an assembly attached to the drive motor that revolves around electrode wire. The electrode wire runs through the center of the feed motor shaft. Planetary feed rolls rotate 360° around the electrode wire. The rolls are skewed at an angle, so that with each revolution of the drive motor the wire is propelled by the amount of pitch or skew of the drive rolls. Two or three rollers are used. Figure 11–22 shows the principle of planetary linear wire feed operation. It is like revolving a nut to propel the bolt. The rollers have a special smooth, concave surface where they are in contact with the electrode wire. Roller pressure against the wire is from spring force, which can be adjusted. The system is self-regulating. The electrode wire rollers do not need to be changed for each electrode wire size. Planetary drive systems eliminate wire damage and deformation and virtually eliminate bird nesting. They also act as a wire straightener. They are synchronized with the wire feed speed and can be used in series to provide long feed systems. Planetary wire feeders are used for gas metal arc welding with solid, cored, or soft nonferrous wires, but can also be used as cold wire feeders, although the wire feed speed is greatly reduced. They have an advantage for feeding small soft wire.

The speed or wire feed rate is based on the rotational speed of the drive motor. The rotational speed is changed by the controller. Small versions of the planetary feeder can be placed in the handle of the welding gun. The larger models are placed at the wire supply spool and may be used as repeaters in a long cable assembly. They allow push-pull systems with truly synchronized wire speed.

Powder Feeders

Some welding processes use filler metal in the form of powder. Among them are thermal spraying, PTA, and laser

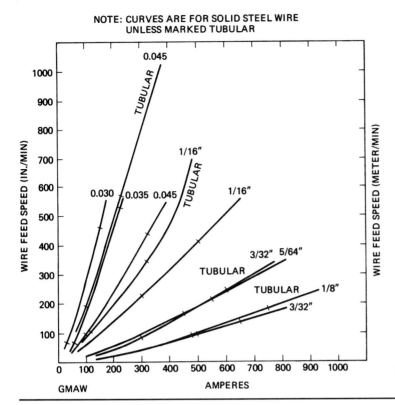

NOTE: CURVES ARE FOR SOLID STEEL WIRE
UNLESS MARKED TUBULAR

FIGURE 11–19 Steel electrode wire (GMAW and FCAW).

FIGURE 11–20 Nonferrous electrode wire (GMAW).

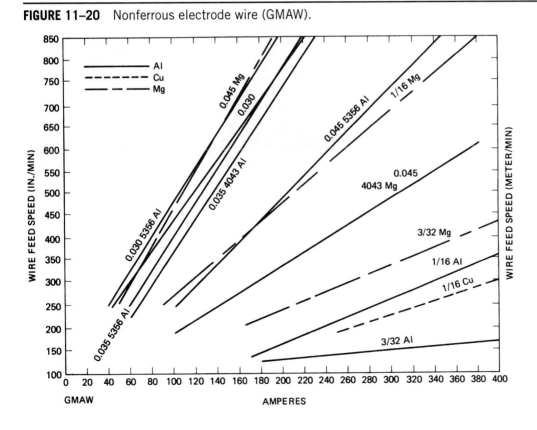

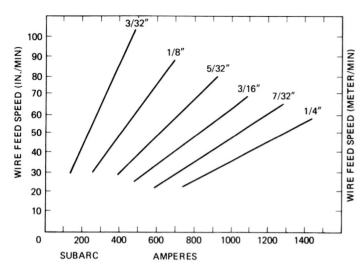

FIGURE 11-21 Steel electrode wire with submerged arc.

or 50 lb/hr. They must be precise and have very accurate feeding ability, must be adjustable to change the rate of feed, and must be programmable. Powder feeders must accommodate different particle sizes of powder and different shapes of particles and they must feed powder of different and varying densities, all at a very uniform rate. There are several types of powder feeders: the auger type, where a feed auger is rotated by a variable speed motor; the metering wheel type, where the motor rotates a wheel with holes or grooves; and the fluidized-bed type, which utilizes carrier gases. The carrier gas may operate at low or high pressure. Each type has its own advantages and must be matched to the application. In general, thermal spray uses lighter and smaller particle sizes, whereas PTA and the laser usually uses larger particles.

Powder feeders must be easily cleaned and should have hoppers or storage compartments easily changeable so that different powders can be fed for different applications. Powder feed systems are becoming more widely used. The type of powder feeder must be specified for the application in question. Specifications must be related to the rate of feed, the type of powder being fed, the size of the particles, the density of the powder, and the carrier gas, if used. A typical powder feeder is shown in Figure 11-23.

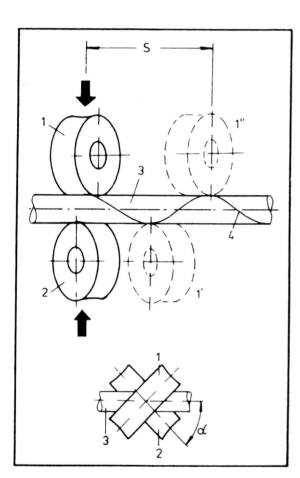

FIGURE 11-22 The planetary feeding principle: 1 and 2, rollers; 3, wire; 4, thread lines.

FIGURE 11-23 Powder feeder.

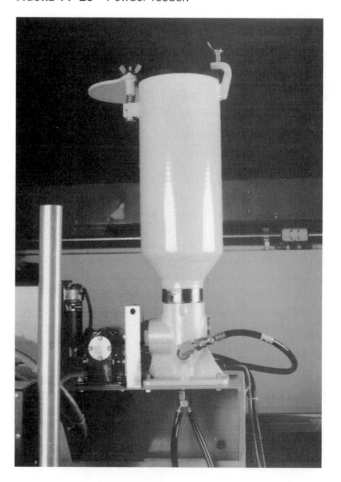

welding. The powder is fed into the arc or high-energy beam by a device called a powder feeder. Powder feeders are designed to feed metal and nonmetallic powders at a rate varying from a few grams per minute to as high as 25

Wire Handling and Dispensing Systems

The wire feeding or electrode wire dispensing equipment should accommodate the type of package of the electrode wires purchased. Welding electrode wire, solid or tubular, comes in different packages to suit the needs of the production operation. In general, they come packaged on small and medium-sized spools, small coils, reels, drums, or payoff packs and very large coils. Small spools are used on manually held welding guns. Medium-sized spools normally require a spool adapter, which is designed to fit the inside diameter of the spool and engage a hole in the spool. This will provide a braking function so that wire does not unwind from the spool. Small coils require an adapter or spider to center and retain the coil and allow uniform unwinding. These items are available from the wire feeder supplier.

Large reels require special dispensing equipment. These reels come with 250 to 1000 lb of electrode wire. The reels are made of wood with holes for an axle. The larger reels can be unwound with the axis horizontal, in which case an axle must be inserted in the reel and carried on a dereeling device. There are two general types of dereelers with a horizontal axis. One dereeler allows the wire feeder to pull the wire and rotate the reel. It may incorporate a brake to stop the reel when the wire feeder stops, to avoid unwinding and overlapping of loops, which may cause tangles. Wire feed motors are designed to utilize relatively small spools or coils that have relatively low inertia loads when starting to revolve. The larger reels with a large amount of electrode wire are quite heavy, and the inertia loads for starting a reel in rotation are high and cause premature failure of wire feed motors. A motorized dereeler (Figure 11-24) should be used. When the wire feed motor exerts a pull on the sup-

ply reel, the dereeler motor starts to rotate the reel. It has a variable-speed motor, which matches the speed of the wire feed motor. This reduces the load on the wire feed motor and improves its life. An extra load on the wire feeder can also occur if the electrode supply is remote from the wire feeder and the electrode wire is fed through conduits. This extra resistance load on the wire feed motor can be overcome by utilizing a motorized dispensing system. Motorized dispensing systems will also assist in arc starting, particularly when the reel is full.

Electrode wire comes in large coils of up to 1000 lb on a pallet. To feed wire from the coil, a dispenser with a rotating arm (Figure 11-25) is used. These rotary dispensers are used for $\frac{1}{16}$-in.-diameter and smaller solid or cored electrode wire. The arm rotates around the axis of the coil and there is no inertia load since the coil remains stationary. An adjustable drag brake prevents premature release of the wire, eliminating tangling. The wire may be carried in a conduit to the wire feeder. The problem with this type of dispenser is that it introduces one rotation of twist into the electrode wire for each revolution of the arm. This may cause wandering of the arc as the electrode wire twists when fed from the tip of the welding torch. This can be overcome with a rotary motorized wire straightener.

Another method of purchasing welding electrode wire is in a drum or payoff pack. The drums, made of heavy cardboard, will contain 250, 500, or over 700 lb of electrode wire. A special dereeling system for drums is shown in Figure 11-26. It sits on top of a drum and utilizes a rotating pickup arm and a pulley system to feed the wire to the wire feeder. There is no inertia since the supply of wire does not rotate. The electrode wire twists one revolution per loop as it is unwound. A rotary wire straightener will overcome the arc wandering problem.

FIGURE 11-24 Motorized dereeler.

FIGURE 11-25 Rotary dispenser for large coils.

FIGURE 11–26 Dispenser for wire in drum.

Another method of dispensing wire from a drum or payoff pack is by means of a rotating table, which revolves the wire as it is unwound. This eliminates the twist to the wire. A special control system and rotating device are required.

It is important to match the dispensing method to the automatic welding system. The expense of the system must be justified by the cost savings of purchasing the electrode wire in larger quantities and packages and the maintenance cost of the wire feed heads.

Wire Feeder Maintenance

Periodic inspection and preventive maintenance will reduce downtime and assure maximum service from the wire feed system. The control circuitry should be cleaned by blowing out with dry air at 25 to 30 psi every three months. Relay contacts and other sliding connections should be checked. The electrical connections, particularly plugs, that are connected daily should be inspected periodically. The grease in the gear case should be changed at least every 500 hours of operation. The housing should be flushed and new grease of the same type should be installed. The wire feed motors should be inspected and the brushes replaced according to wear. The commutating surface of the armature should also be checked, and if the surface appears rough or worn, it should be polished.

Gun cable assemblies, when used, should be blown out once a day so there is no accumulation of dirt in the electrode wire conduit. The gun, especially the electrode tip and nozzle, should be checked daily and replaced as required. Drive rolls should be checked weekly and aligned or replaced as required. Most manufacturers provide troubleshooting checklists for investigating equipment stoppages. Maintenance work should be done by qualified people.

11-3 WELDING CABLES AND CLAMPS

Welding cables are the electrical conductors, called the electrode lead and the work lead. These leads carry the welding current from the power source to the arc and back to the power source. The cables, along with the electrode holder and work connection, complete an electrical circuit.

The welding circuit can be a source of power waste and an economic loss, as well as a source of weld quality problems due to erratic operation. Power losses might result from the following:

1. Loose connections at the power source
2. Loose connections at the work connector or electrode holder
3. Poor-quality repair splices in the cable
4. Cable with broken strands and cable not properly repaired with a splice
5. Use of cable too small for the amperage or duty cycle being used
6. Enlarging the hole in a cable lug to fit a larger stud size
7. Use of excessively long cables, which cause abnormal voltage drop

As the price of electrical energy increases, it becomes extremely important to *inspect* and *maintain* welding conductors and the total circuit at peak operating efficiency. A hot point anywhere in the welding circuit is a source of high resistance, a point at which current is being wasted.

It has been found that if 10% of the strands of a cable are broken, the operating temperature of the cable can rise approximately 10° F (5.5° C). Cables will still carry welding current with up to 30% of the strands broken, but the operating temperature can rise 30° F (16.6° C). An easy way to check for damage and points where power is wasted is by feeling the cable for hot spots.

The work connection, erroneously called a ground clamp, is an important part of the welding circuit. There are several different types of connections available, ranging from spring clamps to actual welded-on connections. Figure 11-27 shows different types. Many styles come in different sizes and are rated according to current-carrying capacity. These connectors should be checked routinely to see that there is not an excessive voltage drop between the cable and the work. Figure 11-28 shows a rotary-type work connection.

Connectors for splicing additional lengths of cables are commercially available. These allow for quickly increasing or decreasing the lengths of leads. The connectors must be properly attached to the leads and must be well maintained to avoid excessive voltage drops at the connection. Connectors must be fully insulated.

Welding Cable Size Designation

In North America, electrical conductors and cables are specified in size by the American Wire Gauge (AWG). The AWG numbers range from a small size, such as number 54 gauge (ultrafine magnet wire), to cables so large that they are designated by MCM (thousands of circular mils). Wire such as number 10 or number 14 gauge is used for wiring a house. Welding cable sizes range from number 6 gauge through number 4/0 (pronounced "four ought"). The cable sizes for welding are shown in Figure 11-29. The two largest sizes are 250 MCM and 300 MCM. The term mil refers to 0.001 in. One circular mil equals the area of a circle whose diameter is 0.001 in.

Outside the United States, welding cable is specified in metric size. The relationship between metric and the American Wire Gauge is shown in Figure 11-29. Note that these are not soft conversions or exact comparisons since the nominal circular mils areas are not equal. The chart shows the metric size, which is its cross-sectional area in square millimeters. The overall diameter includes the jacket. The dimensions and characteristics given may vary between suppliers because of tolerance and differences in standards.

FIGURE 11-28   Magnetic and rotary work connections.

Welding cables are made of many strands of fine drawn, annealed copper. The copper in this form provides maximum flexibility of the welding cable. A separator, such as paper or Mylar foil, is placed between the copper strands and the insulation or jacket. This separator is an aid to jacket removal at terminations. Jacket compounds are designed to be flexible and to protect the copper conductor from the shop environment. These jackets are made of variations of synthetic rubber, which do not melt when they come into momentary contact with sparks or hot metals.

Welding cables are normally made of copper; however, aluminum welding cables are available. Aluminum cables have the advantages of being lighter in weight. Disadvantages are that to compensate for the lower conductivity of aluminum, welders normally must use two AWG sizes larger than would be used for copper. Terminations are more difficult and critical, and the flex life of aluminum cables is less than that of copper. Aluminum cables should be used for low-duty-cycle welding applications or where the cable is not normally flexed during the welding operation.

The arrangement of the strands within the cable has an influence on flexibility. Rope terminology is used to define these arrangements. A *rope lay* has all the strands, groups of strands, and group layers cabled in the same direction. There are seven groups of fine strands, six of which are cabled around a center group. This will produce a conductor having the correct combination of service life and flexibility at a reasonable cost. For extreme limpness a *hawser lay* is used. In this configuration each layer or group of strands is cabled in the direction opposite to the covering layer. Hawser lay cable provides greater flexibility and can be used for a short portion of the lead at the electrode holder. It is more expensive than the rope lay cable.

What Cable Size to Use

To determine what size of welding cable to use, refer to Figure 11-30. Three items must be considered:

1. Welding current
2. Duty cycle or operator factor
3. Total length of the welding circuit.

FIGURE 11-27   Variety of work connection clamps.

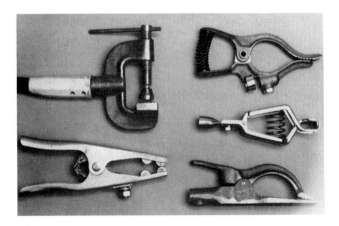

AWG Size	Nominal Overall Diameter (in.)	Approx. Weight per 1000 ft (lb)	Resistance dc per 1000 ft at 68° F (Ω)	Area (circ. mils)	Metric Nominal Cross-Sectional Area (mm²)	Overall Diameter (mm)	Approx. Weight (kg per 1000 m)	Resistance (per 1000 m)
8	0.340	121	0.688	16,510				
				19,740	10	10.5	130	1.75
6	0.390	137	0.435	26,240				
				31,580	16	11.5	235	1.09
4	0.440	194	0.272	41,740				
				49,350	25	13.0	330	0.70
2	0.550	306	0.173	66,360				
				69,100	35	14.5	440	0.50
1	0.600	376	0.137	83,690				
				98,700	50	17.0	610	0.35
1/0	0.660	464	0.109	105,600				
2/0	0.715	563	0.087	133,100				
				138,200	70	19.5	840	0.25
3/0	0.785	708	0.068	167,800				
				187,500	95	22.0	1120	0.18
4/0	0.875	884	0.054	211,600				
				237,000	120	24.0	1410	0.146
250 MCM	0.980	1070	0.045	250,000				
				296,000	150	26.5	1690	0.117
300 MCM	1.060	1260	0.038	300,000				
				365,000	185	29.0	2100	0.094

FIGURE 11–29 American and metric cable size comparison.

Weld Type	Welding Current (A)	Length of Welding Cable Circuit[a] Cable Size (AWG) for					
		50 ft	100 ft	150 ft	200 ft	300 ft	400 ft
Manual or	75	6	6	4	3	2	1
semiautomatic	100	4	4	3	2	1	1/0
welding	150	3	3	2	1	2/0	3/0
(up to 60%	200	2	2	1	1/0	3/0	4/0
duty cycle)	250	2	2	1/0	2/0	4/0	
	300	1	1	2/0	3/0	—	—
	350	1/0	1/0	3/0	4/0	—	—
	400	1/0	2/0	3/0	—	—	—
	450	2/0	3/0	4/0	—	—	—
	500	3/0	3/0	4/0	—	—	—
Semi or	400	4/0	4/0	—	—	—	—
automatic	800	2–4/0	2–4/0	—	—	—	—
welding	1200	3–4/0	3–4/0	—	—	—	—
(60% to	1600	4–4/0	4–4/0	—	—	—	—
100% duty cycle)							

[a]The length of the cable circuit is the length of the electrode lead plus the length of the work lead.

FIGURE 11–30 Copper welding cable size guide.

This means the total distance from the power source to the work and return. The 100-ft column means that the work is 50 ft from the power source. As the distance increases, the cable size should increase. This is to compensate for line loss within the cable due to increased length.

The chart also shows the duty cycle or operator factor that will be involved. This table assumes two categories; (1) up to 60% duty cycle, and (2) from 60% to 100%. Semiautomatic welding is in the top portion of the lower-duty-cycle range, whereas automatic welding is in the higher-duty-cycle range. The voltage drop in the welding circuit should not exceed 4 V.

There are three methods to determine the amount of power lost in the welding leads. In the first method, use an accurate voltmeter and measure the voltage at the welding machine terminals and the voltage between the electrode holder and the work connection while welding. Also, measure the welding current. The difference between the voltage at the power source terminals and at the electrode holder and work connection is the voltage lost in the leads. When multiplied by the welding current, this gives the amount of power lost in the leads. This is in accordance with the following formula:

$$\text{power loss} = V_1 \text{ (at terminals)} - V_2 \text{ (at holder)} \times I$$

or

$$PL = V_1 - V_2 \times I$$

An example would be 35 V measured at the terminals and 32 V measured between the electrode holder and the work connector or 3 V $\times$ welding current of 250 A = 750 W lost.

A second way to determine power loss is to find the resistance of the welding cables and multiply this by the welding current squared. The resistance of the different sizes of cables is shown in Figure 11-29. Modify data by total cable length. The formula is

$$PL = I^2 R$$

A third way is by the use of Figure 11-31. This provides the voltage drop for each cable size, based on a 100-ft-long circuit when welding at the current shown. In this case the power loss equals the welding current times the

voltage drop, or $PI = I \times VD$. These data would be factored according to the length of the cable circuit.

Termination Technique

The connection of the cable to the terminal lugs, the electrode holder, and to the work connector are potential sources of high resistance. Therefore, they should be made as efficient as possible. If any of these connections become hot, the joint should be reworked. Soldering the cable to the lugs is one way to achieve a highly efficient joint. The thermite welding process can also be used for joining cable to special lugs. Mechanical fastenings are also employed; however, these may become loose and must be retightened to maintain a low-resistance joint.

Power Cable

The power cable is the conductor used to carry the electrical power from the disconnect or fuse box of the building to the welding power source. Three-conductor cable is usually used for this application; however, four-conductor cable is sometimes used when the welding machine is on a portable mounting. The fourth wire is used to ground the case of the machine to earth.

The basis for determining the size of power conductor cables is the input power required by the welding machine. A factor to consider is whether the machine operates on single-phase or three-phase power. Figure 11–32 shows the three-conductor power cable size guide for welding machines. It provides size requirements for motor-driven three-phase welding machines and single-phase transformer-rectifier power sources. The power cables are rated at a higher voltage than welding cables since input power to machines can be 480 V or higher. The nameplate of the welding machine will provide the amperage drawn at the rated load and input voltage of the machine. This information is also shown on the data sheets of the machine, available from the manufacturer.

The normal color coding for three-conductor power cables is black, white, and green; for four-conductor cables, it is black, white, green, and red. The size of cables, their diameter, and weight are presented in Figure 11–33. These cables are flexible tinned-copper conductors with paper separators jacketed with insulation suitable for this voltage requirement.

Safety Considerations in the Use of Welding Cable

Welding cables are designed to be used only in conjunction with the relatively low voltages typical of welding equipment. Welding cable should not be used at power-line voltage or for other power applications. The Occupational Safety and Health Act contains specific requirements

FIGURE 11–31 Voltage drop for different cable size per 100 ft of welding cable.

Welding Current (A)	Voltage Drop per 100 ft of Lead for Cable Size (AWG):					
	2	1	1/0	2/0	3/0	4/0
50	1.0	0.7	0.5	0.4	0.3	0.3
75	1.3	1.0	0.8	0.7	0.5	0.4
100	1.8	1.4	1.2	0.9	0.7	0.6
125	2.3	1.7	1.4	1.1	1.0	0.7
150	2.8	2.1	1.7	1.4	1.1	0.9
175	3.3	2.6	2.0	1.7	1.3	1.0
200	3.7	3.0	2.4	2.0	1.5	1.2
250	4.7	3.6	3.0	2.4	1.8	1.5
300	—	4.4	3.4	2.8	2.2	1.7
350	—	—	4.0	3.2	2.5	2.0
400	—	—	4.6	3.7	2.9	2.3
450	—	—	—	4.2	3.2	2.6
500	—	—	—	4.7	3.6	2.8
550	—	—	—	—	3.9	3.1
600	—	—	—	—	4.3	3.4
650	—	—	—	—	—	3.7
700	—	—	—	—	—	4.0

Input of Welding Machine at Rated Output			
Motor-Driven, Three Phase	Rectifier or Transformer, Single Phase	Rectifier or Transformer, Three Phase	Three-Conductor Power Cable Wire Size (AWG)
Up–24	Up–30	Up–24	10
24–32	30–40	24–32	8
32–44	40–55	32–44	6
44–64	55–70	44–64	4
64–76	70–95	64–76	2
76–88	95–110	76–88	1
88–100	110–125	88–100	1/0
100–130	125–165	100–130	2/0
130–155	165–195	130–155	4/0

FIGURE 11–32 Copper power cable size guide.

Size (AWG)	Number of Conductors	Stranding (Number of Wires × Gauge)	Insulation Thickness (in.)	Sheath Thickness	Approx. Outside Diameter (in.)	IPCEA[a] Ampere Rating	Approx. Net Weight (lb per 1000 ft)
10	3	105 × 30	3/64	Type S	0.700	25	305
8	3	132 × 29	4/64	6/64	0.835	35	440
8	4	132 × 29	4/64	6/64	0.915	35	538
6	3	132 × 27	4/64	6/64	0.900	45	567
6	4	132 × 27	4/64	6/64	1.010	45	705
4	3	259 × 28	4/64	Type W	1.17	65	1050
4	4	259 × 28	4/64	Type W	1.27	55	1295
2	3	413 × 28	4/64	Type W	1.34	90	1275

[a]Insulated Power Cable Engineers Association.

FIGURE 11–33 Power cable size information.

that apply to the use of arc welding equipment. Some OSHA safety requirements:

1. Coiled welding cable must always be spread out before using to avoid overheating during use.
2. Cables must not be spliced within 10 ft of the holder.
3. Welding electrode cable must never be coiled or looped around the body of a welder.
4. Cables with damaged insulation must be repaired or replaced.
5. Welding cables must only be joined together by means of recommended connections.

11-4 AUXILIARY WELDING EQUIPMENT

There are a number of auxiliary devices employed in mechanized welding systems that greatly improve the operation of the system.

Wire Straighteners

Wire straighteners are often required for automatic systems. A wire straightener is used to remove the inherent cast and helix of the spooled or coiled electrode wire and make it straight. Two three-roll wire straighteners arranged in two planes will remove the majority of the cast and helix from the electrode wire. Wire straighteners of this type (Figure 11-34) are usually placed downstream from the wire feeder so that the electrode wire extending from the end of the torch contact tip will come out straight. This is to prevent arc wander of the wire after leaving the contact tip.

Rotary wire straighteners (Figure 11-35) are sometimes used. They must match the electrode size and type. They require a motor to provide rotational motion. In a rotary wire straightener the electrode wire runs through a bent tube that is rotating continuously.

Nozzle Cleaners

A torch cleaner, normally automatic, is often used in robot arc welding systems. The nozzle of the torch is close to

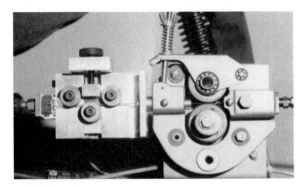

FIGURE 11–34 Three-roll wire straightener.

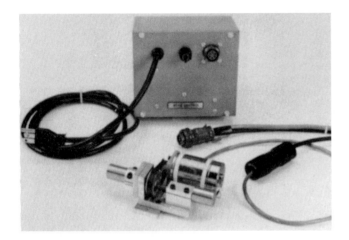

FIGURE 11–35 Rotary wire straightener.

the arc and will gradually pick up spatter. Spatter adheres to the nozzle and in time reduces the effectiveness of the nozzle to direct the shielding gas. The robot controller can be programmed to move the torch to the cleaner periodically and remove the accumulated spatter. There are also *blow down systems,* which attempt to remove spatter by an air blast. Some cleaners also will spray or dip the nozzle into antispatter material to reduce the frequency of cleaning required. These mechanical cleaners can be made automatic so that they operate only when the program calls for it.

Water Coolers/Circulators

Cooling water is commonly used for many heavy-duty welding operations. Plasma torches require cooling water; heavy-duty gas tungsten welding and high-current gas metal arc require water-cooled torches. Electroslag and electrogas retaining shoes are often water cooled. In addition, backing bars in seamers and heat sinks in fixtures often require water cooling. Water coolers/circulators are of two basic types. One system utilizes a pump that circulates water through the torch or item to be cooled to a reservoir. The volume of water in the reservoir is large

enough so that the torch is kept relatively cool. The circulator type of system is recommended for light-duty work only, since the water in the system will gradually rise in temperature until it reaches the boiling point. For certain types of work, particularly low-current plasma and gas tungsten arc welding, stainless steel tanks and tubing are required. In some cases deionized water must be used.

For heavy-duty work such as high-current welding or cooling retaining shoes, large-volume high-capacity heat exchangers are required. When a large amount of heat is generated over a long period, the heat must be extracted from the system and the water must be cooled by means of a heat exchanger or radiator. This is necessary to maintain a uniform cool operating temperature of the cooling water. Water cooler circulators are rated by the heat extraction rate, in Btu per hour. For light-duty welding in the medium-current range, a 25,000-Btu/hr unit is recommended. For heavy-duty work, a 50,000-Btu/hr unit is recommended. The circulator should be rated so that the temperature of the water does not exceed 150° F. The water-cooling circulator should have an adjustable flow rate control, a flow or pressure switch, an interlock circuit, and a fan circulating air through the radiator. It should also allow for adjustable pressure. All flow switches, pressure switches, and interlocks should be connected to the welding control circuit. For many applications the tank and piping system should be noncorrosive. The minimum flow rate should be at least $\frac{1}{2}$ gallon per minute and adjustable up to 4 gallons per minute. There should be sufficient water capacity in the system so that if a leak occurs, it will not immediately cause a burnout. Circulators are sometimes incorporated in the control cabinet or in the welding power source. It is important to specify the size heat exchanger required for a particular application. It is better to overspecify and have excess capacity than to underspecify. Tap water to be discharged is too expensive to use. Figure 11–36 shows a variety of water coolers/circulators of different capacities.

Smoke Exhaust Systems

Smoke exhaust devices are used in many semiautomatic and mechanized welding systems. This system is based on collecting the fumes as close as possible to the point of generation. The fumes collected in the immediate area of the arc are passed through a filter and then exhausted to the outside. In some cases, the cleaned air is returned to the welding shop. This is questionable practice since the filter system removes only the particulate matter of the fume and has no effect on gases. For gas metal arc or flux cored arc welding, a special nozzle is used on the welding gun that collects the fume from the arc area. The entire exhaust system consists of the smoke exhaust gun, cable assembly, vacuum blower, filter, and waste can. These systems greatly reduce the pollution in the air of a welding

FIGURE 11-36 Variety of water coolers/circulators.

shop. Different types of pickup devices are used for shielded metal arc welding. See Chapter 4.

Miscellaneous Arc Motion Devices

Welding oscillators or arc weavers are devices that provide transverse motion to an arc. These units provide a wider welding bead for surfacing applications and are sometimes used for wide joints. Oscillators are available as components that can be added to mechanized welding equipment. There are at least two types. One has a linear motion and the other a pivoting or swinging motion. Motion is controlled by mechanical devices such as lever arms or cams or by electronic devices containing timing circuits.

The mechanical type provides sinusoidal oscillation using an adjustable crank arm. Other types utilize cams, which can be shaped to provide a specific motion that can include a dwell time at the end of each oscillation stroke. The mechanical cam or crank arm types are difficult to adjust during operation. The newer types with electronic control can be changed easily. The dwell time at either end of the stroke can be lengthened or shortened, and the width of oscillation can be changed during operation. It is also possible to change the centerline of oscillation. The electronic control system can be used for pivoting or linear motion. If it is necessary to change any of the oscillation parameters during welding, the electronic controlled type is the best selection. If wide oscillation is required, the linear action is preferred to the pivoting action.

Portable Booms

There are a large number of devices that provide portability to the welding equipment. These are particularly popular for semiautomatic welding. These units usually include a boom or arm that supports the wire feeder so that the gun and its cable are usable over a wide area. This provides flexibility for semiautomatic welding, previously attained with shielded metal arc welding. Many of these units also carry the power supply and the electrode wire supply. Figure 11-37 shows two examples of a portable mounting for semiautomatic welding. The various types provide different features, and selection is largely a matter of personal preference.

11-5 WELD MONITORING

To collect information about a welding operation or procedure requires the use of many different instruments. They relate to the welding process and the degree of precision required. Originally, a procedure for shielded metal arc welding was monitored by a voltmeter, an amp meter, and a stopwatch to measure travel speed. Manual welding procedures were developed and monitored with panel meters of the welding machine; these procedures rarely used recording meters, and it is generally agreed that normal manual welding is based more on operator skill and ability than on following strict procedures.

As quality requirements became more stringent and as welding becomes more technically oriented, it

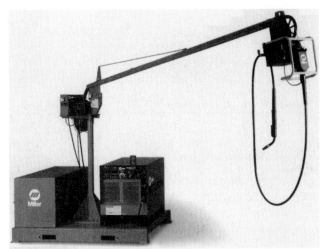

FIGURE 11–37   Portable booms for semiautomatic welding.

becomes necessary to accurately measure all the parameters involved. This requires the use of suitable, accurate measuring instruments. The panel meters on a welding machine are not acceptable for accurate measurements of arc voltage and welding current. Their accuracy is less than desired, the scale is too coarse, and they are too hard to read. Laboratory meters must be used. To qualify a welding procedure, it is necessary to measure and record all of the variables. To duplicate a welding procedure, it is necessary that these variables be the same as the original qualified procedure.

This equipment has brought about the need for more accurate measuring instruments and the need to measure variables heretofore ignored. Early methods to measure and record these parameters utilized recording meters with moving paper charts and ink pens. This type of equipment was satisfactory for many welding operations. However, high open-circuit voltage and use of high frequency in GTAW tended to complicate circuitry, which led to problems with automatic recording instruments. High-speed recording instruments are utilized to measure welding parameters and are often used for monitoring welding of critical welded products. Ac welding and pulsed-current welding adds to the complication. Alternating current welding normally assumes a sinusoidal wave form; ac meters are based on this wave form. If the wave form is not true sinusoidal, errors may be made. In

the case of pulsed-current welding, the wave shape of the pulse and the frequency of pulsing are extremely important to determine the total energy input. For this reason, an oscilloscope may be necessary to determine the wave shapes, the pulse shapes, and their frequency. In the case of pulsing, the parameters—including peak current, background current, peak current time, background current time, frequency, and shape—are important. In many automatic applications the ramping or change rates of current with respect to time are required. Other parameters must be measured and included in a specific procedure. Some procedures require base metal preheat and interpass temperatures. Some procedures require recording torch angle, work angle, tip-to-work distance, as well as the static and dynamic characteristics of the power source. Shielding gas flow rate, total elapsed welding time, are sometimes required.

In many cases the accuracy required must be verified by checking the meters to a national standard. This means also that meters and shunts must be checked for accuracy periodically. Current shunts and meter transformer leads must also be calibrated and cannot be spliced. Meters for welding should be damped for accurate reading. Care must be taken to attach meter leads to the proper location in the welding circuit. High-speed recording instruments such as the one shown in Figure 11–38 are often used. These instruments are much faster than the ink-type paper chart recorders.

FIGURE 11–38 High-speed recording instrument.

Remote Monitoring and Quality Control

Many new factors are dictating the need for remote-weld monitoring. These are:

1. The widespread use of automated welding systems
2. The increasing use of unmanned welding stations
3. Regulations that require verification that the weld was made properly

Automatic weld process monitoring systems that provide real-time data acquisition and recording are now available. These systems simultaneously collect and record all welding process data. In addition, they may provide for instant notification if a parameter exceeds specific limits or if the process goes out of control. These systems may send this data to remote locations, may summon help, or may shut down the operation. However, they may verify that the weld in question is satisfactory.

The monitor simultaneously measures all of the parameters in real time. They automatically read out this information visually, based on a microprocessor. These monitors may also interface with a computer and a printer to graphically print out these parameters. They may also be linked to an Internet system, which remotely provides the information to an office or may in fact connect to network webs that send the information remotely to the main office. Some systems summon help in case a parameter is out of control. It may, in fact, date stamp and approve the particular part, stating that it was made with the correct welding parameters. Graphic presentation of the welding parameters are made and recorded, and printouts are produced.

Monitors of this type can become very complex. They require numerous meters and sensors. For example, they sense the welding current, the wire feed speed, the arc voltage, the shielding gas flow rate, and the travel speed of the fixture or travel mechanism. They may sense the fact that the parts are in the fixture and that the fixture clamps are closed. They may utilize video monitors

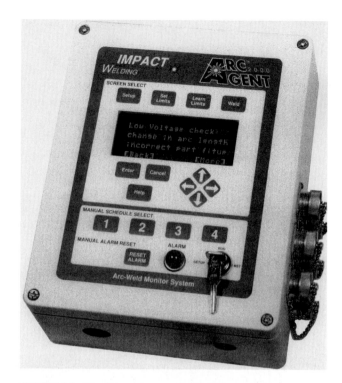

FIGURE 11–39 Remote monitoring and quality control.

and determine if the torch is misaligned. They may sense the presence of oil on the surface and the weld preparation. They may sense the temperature of the base metal and record the elapsed time for making a weld. In other words, statistical process control (SPC) data is gathered and analyzed by the quality control system. This will, in turn, activate an alarm such as a bell to obtain assistance, or a buzzer to quickly adjust parameters or to actually shut down the process if the parameters are beyond the control limit. A typical arc data monitor is shown in Figure 11–40. Printouts that provide high-resolution color graphics are shown in Figure 11–41.

FIGURE 11–40 Arc Data Monitor

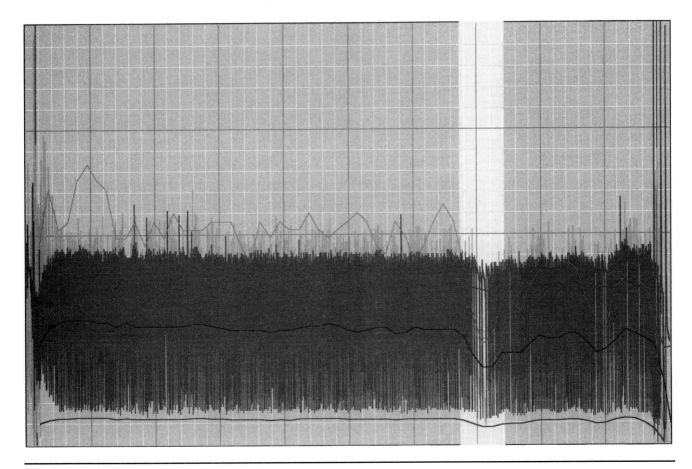

FIGURE 11–41   Printout of Arc Data Monitor

Systems of this sort are tailored to the particular requirements and may be as simple as a semiautomatic process monitor that provides signals on the visual instrumentation in the welder's helmet. They may monitor automated production welding operations that will provide quality control requirements for each part and provide information by broadcasting to wherever it is needed. It can also provide information for remote diagnosis and correction.

QUESTIONS

11-1. What are the main functions of a welding gun or torch?

11-2. What are the differences between a gooseneck and a pistol-grip gun?

11-3. What is the advantage of a pistol-grip gun?

11-4. Why is a water-cooled gun rarely used for CO_2 welding?

11-5. What is the importance of the gun contact tube or tip?

11-6. What type of contact type should be used for welding aluminum?

11-7. Explain the two types of wire feeder controls. Which type is used with the CV machine?

11-8. What is the disadvantage of spool guns?

11-9. What factors must be considered when selecting the speed range of a wire feeder?

11-10. What is the advantage of four drive rolls when using flux cored wire?

11-11. Why are V-groove drive rolls preferred over flat drive rolls?

11-12. Describe the characteristics of hard, soft, and tubular electrode wire.

11-13. What factors determine the size of the welding cable used?

11-14. What is the disadvantage of aluminum for welding cables?

11-15. What is indicated by a hot spot in a welding cable?

11-16. When should a heat exchanger be used in a water-cooling system?

11-17. What is the advantage of purchasing electrode wire in 1000-lb spools?

11-18. What is the disadvantage of dereeling 1000-lb spools?

11-19. What is the problem of using welding machine panel meters?

11-20. What types of meters will provide a permanent record of date?

REFERENCE

1. "Semiautomatic Wire Feed Systems for Arc Welding," NEMA Standards Publication EW3, National Electrical Manufacturers Association, Washington, D.C.

12

MECHANIZED, AUTOMATED, AND ROBOTIC ARC WELDING

12-1 AUTOMATION OF WELDING

The need to reduce the cost of welding is never ending. The use of larger-diameter electrodes and higher welding currents reduce costs somewhat, but not enough. The quest for improved productivity and lower production costs continues. This has brought about the transition from manual to semiautomatic to machine applications of welding continued until robotic welding with adaptive control became available. This has reduced the involvement of the individual welder and provides an improvement in operator factor, which has a major effect on the cost of welding.

Automation of welding became possible and practical when the continuous electrode wire arc welding processes became popular. The advantages of automatic welding are well known and include the following:

1. Increased productivity through higher operator factor
2. Increased productivity through higher deposition rates
3. Increased productivity through higher welding speeds
4. Good uniform quality that is predictable and consistent
5. Strict cost control through predictable weld time
6. Minimized operator skill and reduced training requirements
7. Operator removed from the welding arc area for safety and environmental reasons
8. Better weld appearance and consistency of product

The shift from manual to automatic welding and the resulting cost reduction has been known for many years. However, the automation of welding has lagged behind other metalworking operations. This is because arc welding is a much more complex process. Another reason is the lack of incentive to develop automatic welding since the welded product can still be produced by manual or semiautomatic welding.

The major deficiency of automatic welding is its inability to compensate for variations in welding joints in any but the simplest weldment designs. There are two potential solutions: (1) make the piece parts perfect in every respect; or (2) develop welding equipment that will compensate for these variations and still produce high-quality welds.

The first solution seems contrary to normal production operations. In the past, variations have been allowed to collect in the manufacturing processes, and the welder would overcome the accumulated tolerances and still produce a good-quality weldment. The welder would compensate for variations, utilizing the skill and attention of the human. This is a closed-loop welding system that overcomes the problems of variations in material, variations in piece part preparation, and so on. This is an expensive option.

Automatic welding is an open-loop system unable to make needed compensating changes. The solution is a closed-loop system to produce a good-quality weld in spite of variations. This requires a new method of application called adaptive control welding. It is a step beyond automatic welding since it involves complete control of the operation, including accommodations for poorly fit-

ted joints, for joint preparation errors, for warpage problems, and so on. The difference between adaptive control welding, which is a closed-loop system, and automatic welding, which is an open-loop system, is the use of feedback sensing devices and adaptive controls.

Use of the computer to control process motion and the retention of this in memory; the development of power electronics, making welding equipment computer controllable; the development of robot and precision motion devices; and the development of sensors that detect changes led initially to robotic arc welding, but also to the overall automation of welding.

The chart shown in Figure 12–1 describes the functions involved in making a weld. It also shows that manual, semiautomatic, and mechanized welding methods are closed-loop systems because of the human involvement. Automatic welding is not under constant supervision of an individual and so is an open-loop system. The functions involved in making an arc weld are expanded and show whether they are controlled by the individual or by the machine. These functions affect the level of fatigue of the individual. When more of these functions are taken over by the machine, fatigue levels are reduced and productivity is increased.

FIGURE 12–1 Person–machine relationship for arc welding with automation.

Method of Application / Arc Welding Elements/Function	MA Manual (closed loop)	SA Semiautomatic (closed loop)	ME Mechanized (closed loop)	AU Automatic (open loop)	RO Robotic (open or closed loop)	AD Adaptive Control (closed loop)
Starts, maintains, and controls the arc	Person	Machine	Machine	Machine	Machine	Machine
Feeds and directs the electrode into the arc	Person	Machine	Machine	Machine	Machine	Machine
Manipulates the arc to control the molten metal weld pool	Person	Person	Machine	Machine	Machine (robot) with or without sensor	Machine with sensor
Moves the arc along joint (travel)	Person	Person	Machine	Machine via prearranged path	Machine (robot) with or without sensor	Machine with sensor
Guides the arc along joint	Person	Person	Person	Machine via prearranged path	Machine (robot) with or without sensor	Machine with sensor
Corrects the arc to overcome deviations	Person	Person	Person	Does not correct hence potential weld imperfection	Machine (robot) only with sensor	Machine with sensor

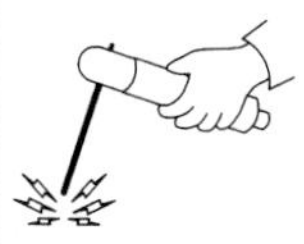

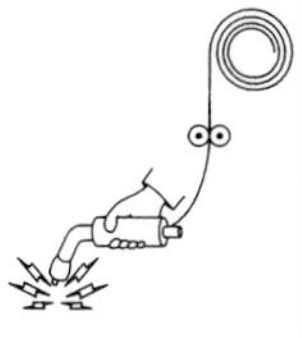

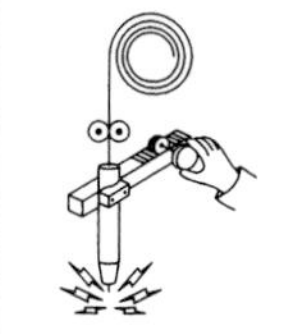

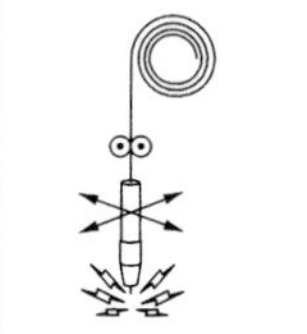

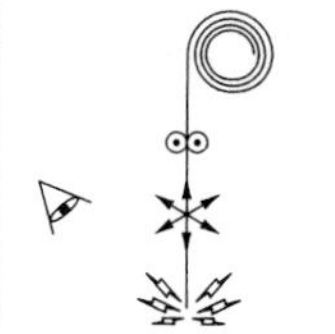

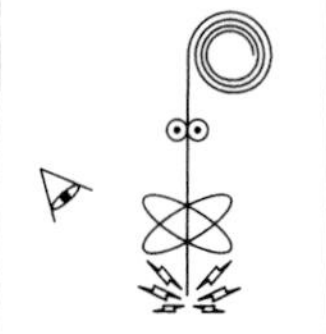

The functions are:

1. *Starts,* maintains, and controls the arc
2. *Feeds* and directs the electrode into the arc (to control the placement of the weld deposit and fill the joint)
3. *Manipulates* the arc to control the molten metal weld pool
4. *Moves* the arc along the joint (travels) (to provide motion at proper speed to make the weld joint)
5. *Guides* the arc along the joint (to track the weld joint)
6. *Corrects* the arc to overcome deviations (to compensate for improper fitup)

Closed loop means that real-time observations are made during welding and immediate corrections are made to compensate for deviations.

In "automatic" welding the welding apparatus is programmed to provide the exact taught motion patterns and the exact preset welding parameters. In many cases the weldment is simple and the parts are sufficiently accurate so that changes are not required in the welding conditions or the taught motion pattern. Good-quality welds will result since the inherent tolerance of the welding process will accommodate minor variations. If the joint location or geometry is beyond established variations, a defective weld may result.

An automatic or automated welding system consists of at least the following:

1. *Welding arc:* requires a welding power source and its control, an electrode wire feeder and its control, the welding gun assembly, and necessary interfacing hardware.
2. *Master controller:* controls all functions of the system. It can be the robot controller or a separate controller. It is the overall controller.
3. *Arc motion device:* can be the robot manipulator, a dedicated welding machine, or a standardized welding machine. It may involve several axes.
4. *Work motion device:* can be a standardized device such as a tilt-table positioner, a rotating turntable, or a dedicated fixture. It may involve several axes.
5. *Work holding fixture:* must be customized or dedicated to accommodate the specific weldment to be produced. It may be mounted on the work motion device.
6. *Welding program:* requires the development of the welding procedure and the software to operate the master controller to produce the weldment.
7. *Consumables:* includes the electrode wire or filler metal, the shielding media (normally gas), and possibly a tungsten electrode.

The "automatic-adaptive control" arc welding system is shown in Figure 12–2. Changing the top block from "master control with welding program plus human operation" to "master control with welding program and plus sensors and adaptive control," and changing the bottom right from human monitoring and supervising to multisensors, changes this from an automatic to an adaptive control welding system.

There are differences in the degree of automation of a welding system. This depends on the number of sensors employed to monitor conditions. Sensors are needed to find the joint, provide root penetration, provide bead placement, follow the joint, assure joint fill, and so on. Adaptive control requires sensing devices and computerized circuits that alter the motion and value of a particular variable in order to compensate and satisfy the new requirements. Many sensing devices will be required to provide total adaptive controlled or automated welding. Sensing devices and adaptive controls are expensive, and only a few "real-time" sensors are used.

The subject of automated welding is changing due to new developments. The methods of application of welding discussed in Chapter 2 are repeated here.

* *MA, Manual welding:* welding with the torch, gun, or electrode holder held and manipulated by hand.
* *SA, Semiautomatic welding:* manual welding with equipment that automatically controls one or more of the welding conditions.
* *ME, Mechanized welding:* welding with equipment that requires manual adjustment of the equipment controls in response to visual observation of the welding, with the torch, gun, or electrode holder held by a mechanical device.
* *AU, Automatic welding:* welding with equipment that requires only occasional or no observation of the welding and no manual adjustment of the equipment controls.
* *RO, Robotic welding:* welding that is performed and controlled by robotic equipment.
* *AD, Adaptive control welding:* welding with a process control system that determines changes in welding conditions automatically and directs the equipment to take appropriate action.

In robotic welding the weld sequence, and the path, once established, will always be followed identically. Robots will make weldments successfully with well-prepared piece parts over and over. This is an open-loop system. On the other hand, robotic welding equipment can utilize sensors with feedback capabilities so that the path can be followed even if it is not the original memorized path. Welding conditions can change if the sensors detect a change such as in the joint configuration. This is a closed-loop system and can be as complicated as necessary to control all parameters to make a perfect weld in

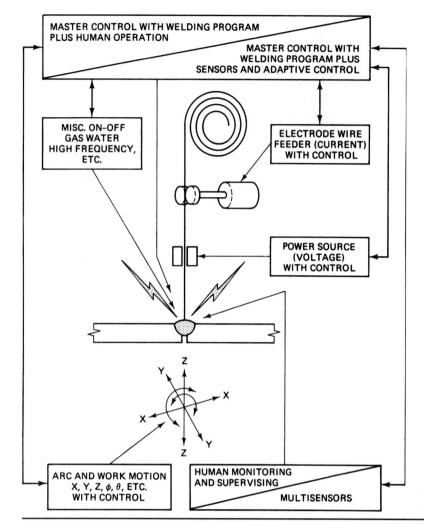

MASTER CONTROL WITH WELDING PROGRAM
PLUS HUMAN OPERATION

MASTER CONTROL WITH
WELDING PROGRAM PLUS
SENSORS AND ADAPTIVE CONTROL

MISC. ON–OFF
GAS WATER
HIGH FREQUENCY,
ETC.

ELECTRODE WIRE
FEEDER (CURRENT)
WITH CONTROL

POWER SOURCE
(VOLTAGE)
WITH CONTROL

ARC AND WORK MOTION
X, Y, Z, ϕ, θ, ETC.
WITH CONTROL

HUMAN MONITORING
AND SUPERVISING

MULTISENSORS

FIGURE 12–2 Automatic-adaptive control arc welding systems.

every situation. This system approaches the ability of a human welder to compensate for changes during the welding operation. The adaptive controller with appropriate sensors automatically determines changes in process conditions and directs the equipment to take appropriate action to ensure a high-quality weld. Automated welding is becoming widely used.

12-2 ARC MOTION DEVICES

Arc motion devices are required for mechanized welding. The machine moves the arc, torch, and welding head along the joint. The person or operator performs a supervisory role and may make adjustments to guide the arc, manipulate the torch, and change parameters to overcome deviations. Since the person is partially removed from the arc area, higher currents and higher travel speeds can be used. The fatigue factor is reduced and the operator factor is increased. Productivity is increased, with a resulting reduction of welding costs.

Arc motion devices fit into five categories:

1. Manipulator (boom and mast assembly)
2. Side beam carriages
3. Gantry or straddle carriages
4. Tractors for flat-position welding
5. Carriages for all-position welding

The arc motion devices carry the welding head and torch and provide travel or motion relative to the weld. They are used for the continuous wire processes, gas metal arc, flux cored arc and submerged arc welding, and also for gas tungsten arc and plasma arc welding. The motion device must be matched to the welding process. Gas tungsten and plasma arc welding require more accurate travel and speed regulation. This must be specified since tighter tolerances are used in manufacturing and the equipment will be more expensive.

Manipulator

A welding manipulator consists of a vertical mast and a horizontal boom that carries the welding head. They are sometimes referred to as boom and mast or column and boom positioners. Figure 12–3 shows a welding manipulator being used for submerged arc welding the circumferential joint of a large tank. Manipulators are specified by two dimensions: the maximum height under the arc from the floor and the maximum reach of the arc from the mast. The manipulator can be designated as a 6 × 6, which means that the height (z) weldable is 6 ft high and it can weld at a distance (y) 6 ft from the face of the mast. Other manipulators range from 4 × 4 to 12 × 12 and larger. A more detailed way of specifying manipulators is shown in Figure 12–4. Many companies supply manipulators. Manufacturers provide specification for their equipment, including the maximum weight that can be carried on the end of the boom and the maximum deflection.

There are many variations of manipulators. The assembly may be mounted on a carriage that travels on rails secured to the shop floor. The welding power source is usually mounted on the carriage. The length of travel can be unlimited; thus, the same welding manipulator can be used for different weldments by moving from one workstation to another. This is a light-duty model, to support the weight, electrode wire supply, flux supply, and welding head. Manipulators usually

FIGURE 12–3 Welding manipulator.

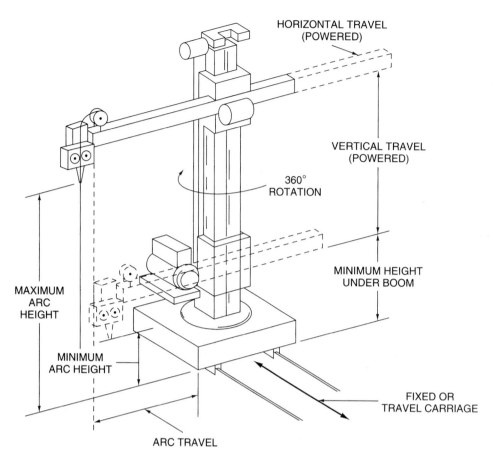

Labels in figure:
HORIZONTAL TRAVEL (POWERED)

VERTICAL TRAVEL (POWERED)

360° ROTATION

MINIMUM HEIGHT UNDER BOOM

MAXIMUM ARC HEIGHT

MINIMUM ARC HEIGHT

FIXED OR TRAVEL CARRIAGE

ARC TRAVEL

FIGURE 12–4 Welding manipulator work envelope.

have power for moving the boom up and down on the column. The boom may extend and move through the vertical adjusting assembly as shown, or the welding carriage head may move by power in and out along the boom to provide transverse motion. In some units the mast may rotate, but not with power. In selecting and specifying a welding manipulator, it is important to determine the weight to be carried on the end of the boom and how much deflection can be allowed. The welding torch should move smoothly at travel speed rates compatible with the welding process. The manipulator carriage must also move smoothly at the same speeds. Carriages should have high-speed return. Figure 12–5 shows a precision manipulator for gas tungsten arc welding. A precision manipulator, when specified, would have a "tracking" tolerance of 0.0015 in. per foot of reach runout. Standard manipulators would have a tolerance of $\frac{1}{32}$ in. per foot of reach runout. The quality of the hardware and adjusting devices largely determines the precision of the total machine.

Manipulators are one of the most versatile pieces of welding equipment available. They can be used for straight-line, longitudinal, and transverse welds, and for circular welds when a rotating device is used.

Side Beam Carriage

The side beam carriage is less versatile and less expensive than the boom and mast manipulator. The side beam carriage performs straight-line welds with longitudinal travel of the welding head. A side beam carriage using the flux cored arc welding process is shown in Figure 12–6. In this case the carriage is mounted on an I-beam modified with bars to provide for powered travel. Side beam carriages are available with high-precision motion, depending on the accuracy used in the manufacture of the beam and the speed regulation of the travel drive system. Figure 12–7 shows a precision side beam carriage for gas tungsten arc welding. The carriage will carry the welding head, wire supply, and so on, and the controls for the operator. The welding head on the carriage can be adjusted for different heights and for in-and-out variations. The welding arc is supervised by the welding operator, who makes adjustments to follow joints that are not in perfect alignment. The travel speed of the side beam carriage is adjustable to accommodate different welding procedures and processes. Rapid return speed should be available. A side beam carriage can be teamed up with a work-holding device or a rotating device.

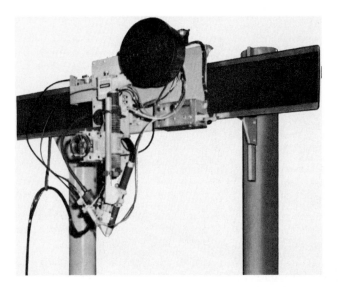

FIGURE 12–7 Precision side beam carriage.

Gantry Welding Machine

Gantry arc welding machines are motion devices that provide one or two axes of motion. The gantry consists of a horizontal beam supported at each end by a powered carriage (Figure 12-8). The gantry structure straddles the work to be welded, and the carriages run on two parallel rails secured to the floor. This provides the X or longitudinal motion and can be quite long. The length of the gantry bridge determines the width of the parts that can be welded. The torch or torches are mounted on carriages

FIGURE 12–5 Precision manipulator.

FIGURE 12–6 Side beam carriage.

FIGURE 12–8 Gantry welding machine with two heads.

that move along the gantry beam. This provides the *Y* or transverse motion. The travel speed of the carriages must be smooth and match the welding speed of the welding process. Rapid travel should be available for returning. It should go in either direction at welding speed and at high speed. The one or more welding heads on the gantry bridge will have power travel or will have adjusting devices to locate the head over the weld seam. Usually, a maximum of two torches is provided for transverse motion. The *X* and *Y* motions are not normally operated simultaneously. The *Z* or vertical motion should be available for adjustments.

Welding Tractor

A welding tractor is an inexpensive way of providing arc motion. Tractors are commonly used for mechanized flame cutting. Some tractors ride on the material being welded, while others ride on special tracks. The tractor should have sufficient stability to carry the welding head, the electrode wire supply, flux if used, and the welding controls. The welding tractor shown in Figure 12–9 rides on a track and has an adjustment so that the head will follow the weld joint. This type of equipment is extremely popular in shipyards and in plate-fabricating shops. The travel speed of the tractor must be closely regulated and smooth, and related to the welding process. It must have sufficient power to drag cables. A more specialized tractor carries two heads, straddles a stiffener, and makes double-fillet welds to a plate.

Another relatively inexpensive unit is known as an *automatic horizontal fillet welding carriage.* This lightweight carriage makes horizontal fillet welds and is used in shipyards and in plate fabricating shops. It is held in the

FIGURE 12–9 Welding tractor for SAW.

corner of the weldment by magnetics, has serrated rubber tires, and has sufficient power to pull the welding cable assembly. It travels in either direction and can be used for multipass fillets, which requires the adjustment of the torch position. It has an automatic stop function, which enables an operator to utilize several of these carriages. It is shown in Figure 12–10.

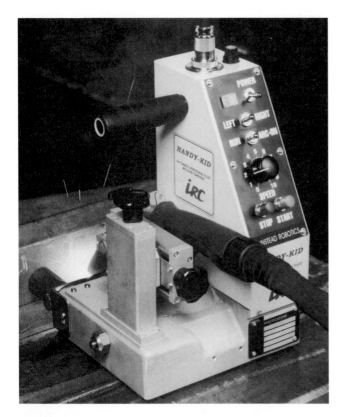

FIGURE 12–10 Automatic horizontal fillet welding carriage.

All-Position Welding Carriage

There are many requirements for mechanized vertical or horizontal position welding. A tractor that utilizes a special track is used. Figure 12–11 shows an all-position welding carriage that carries the welding gun. The gun is connected to the wire feeder by means of the standard cable assembly. In this case, an oscillator is employed to provide lateral arc motion. This type of welding carriage can be used in the flat, vertical, horizontal, or overhead positions. Adjustments can be made to align the torch to the joint and for maintaining this alignment. The track can be attached to the work with magnets or vacuum cups.

A special type of carriage is used for welding the horizontal girth welds on large storage tanks (Figure 12–12). The carriage may straddle the top course of plates and make welds simultaneously on the inside and the outside of the tank shell. Other carriages, designed for vertical welding, carry the welding head, controls, and the welding operator. They are used for the vertical joints of storage tanks and ships. A special carriage known as a skate welder is designed to follow irregular joint contours inside complex structures. Skate welder travel units are extremely compact and carry a miniaturized wire feeder or only a torch. Skate welders are used for welding inside aircraft assemblies.

(a)

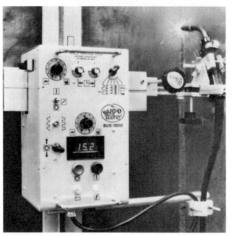

(b)

(c)

FIGURE 12–11 All-position welding tractor: (a) horizontal; (b) vertical; (c) overhead.

FIGURE 12–12 Tank construction welding carriage.

12-3 WORK MOTION DEVICES

A welding work motion device, commonly called a welding positioner, is a device that holds and moves a weldment to the desired location and angle for welding. The axis of the welds of a complex weldment are in many different angles. By means of a positioner, the weldment can be moved to put each weld in its most advantageous welding position. Flat-position welding is faster because higher welding current can be used and the weld can usually be made with fewer passes. It also has a higher quality level and will have a better appearance. A welder welding manually in the flat position will be more comfortable, will have less fatigue, and will have a higher percentage of arc time.

There are several negative aspects to weld positioning. Positioning equipment is relatively expensive, and to be cost-effective the savings of higher-current flat welding must pay back quickly. The weldment must be firmly attached to the positioner for safety reasons. The time required for loading and unloading the positioner must be considered in cost calculations justifying positioners.

The primary considerations for selecting a welding positioner are the size, shape, and weight of the weldment and the size, type, and quantity of welds. In addition, consideration must be given to the lot size of production and the number of arcs working simultaneously.

The type of production is important. For example, if like weldments are to be produced simultaneously, there must be a sufficient number of positioners so that all the weldments can be welded in the shortest cycle time. If there is only one positioner and each weldment requires a day on the positioner, it would take a week to produce five weldments. If there were five positioners, all five weldments could be welded simultaneously in one day. The cycle time versus the lot size of production must be considered carefully. In addition, there is the problem of accessibility for welding. The side attached to the positioner table is not accessible for welding, and this requires reattaching the weldment on the positioner. Most positioners hold the weldment above the floor so that ladders or scaffolding are required to bring the welder close to the arc. This extra equipment must be considered in the overall capital cost.

Four types of welding positioners are used for manual and mechanized welding, and one type is used for manual or semiautomatic welding only.

1. Universal or tilt-table positioner
2. Turning rolls
3. Head and tail stock positioners
4. Universal balanced positioners (used only for manual or semiautomatic welding)

There are specialized positioners designed to be used with robots. These are discussed later.

Universal or Tilt-Table Positioner

The universal tilt-table positioners are the most popular. An example is shown in Figure 12–13. Powered positioners are available for handling weldments weighing from 100 lb (+5.3 kg), known as bench positioners, to 150 tons (135 mg) or more for heavy-duty positioners. The principle of operation is a table that can be tilted from horizontal to the vertical position and beyond the vertical position by power. The table rotates about its center by power. The size of the table, which may be round or square, depends on the capacity of the positioners. Tables range from 12 in. in diameter up through 10 ft (3.05 m). Positioners must be securely anchored to the floor to avoid tipping over due to the strains imposed by heavy loads. The rotational and tilting speed of the table pertain to the size of the positioner. A much larger tabletop positioner is shown in Figure 12–14.

When selecting a positioner, two factors must be considered in specifying its size. A positioner is rated based on its ability to provide specific tilting and rotational torques. For tilting this is expressed in inch-pounds, which is the concentrated weight or center of gravity of the weldment times the distance from the face of the table when it is in the vertical position. This is

UNIVERSAL GEAR-DRIVEN POSITIONER

FIGURE 12–13 Table-type welding positioner.

FIGURE 12–14 Large weldment on bigger positioner.

shown in Figure 12–15 as weight W in pounds times distance D in inches. S is the inherent overhang designed into the machine. It is impossible for the centerline of gravity to be at the table face.

For rotation, the rated capacity is the product of the weight in pounds times the distance in inches from the center of gravity of the load to the centerline of table rotation. This is the distance in inches of eccentricity, shown by E in the figure. Each capacity is determined independently by the design of the yoke and frame members of the positioner. Positioner manufacturers provide load capacity charts that show safe operating limits based on the

weight of the load, the distance from the face of the table, and the eccentricity of the load.

It is extremely important to know the location of the center of gravity of weldments that are placed on positioners. This can be calculated accurately from the engineering drawing. However, its location can be approximated, in the shop, by balancing the weldment on an inverted angle iron in all three planes. Even with this information it is difficult to align the center of gravity of the weldment precisely with the center of rotation of the table.

The weldment must be very firmly attached to the positioner table. Most positioner tables have slots for bolts for this purpose. In some cases fixtures may be attached to the positioner table. Whenever fixtures are employed, the weight of the fixture must be included in establishing the total weight of the load.

Rotational speeds are important since positioners are often used for rotating parts under a fixed welding head. The circumferential speed must be within the parameters of the welding procedure. Manufacturers provide specifications showing tilt and rotational speeds.

Positioners must be well manufactured, with appropriate bearings and motors, so that travel speed is smooth and steady. A variation of the universal tilt-table positioner utilizes mechanical or hydraulic elevating mechanisms for raising or lowering the table. Another variation is the horizontal turntable, which is similar to the conventional turntable but without the tilt feature.

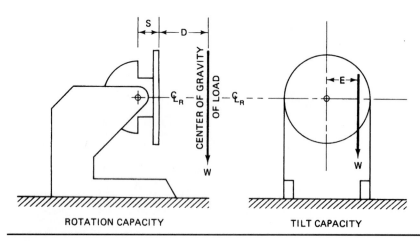

ROTATION CAPACITY   TILT CAPACITY

FIGURE 12–15 Center of gravity of load and location.

Turning Rolls

Turning rolls are ideally suited for cylindrical parts such as tanks. Turning rolls come in many sizes and ratings, based on the size and weight of the work to be rotated—ranging from 1 ton up through 250 tons. Usually, a powered set of rolls and idler sets are used together. Figure 12–16 shows the normal arrangement for turning rolls with a very large, heavy cylindrical load. For extremely heavy loads or long cylindrical members, additional sets of idler rolls can be used. For flexibility of operations, the sets of rolls can be placed on tracks so that they can be adjusted for long or short cylindrical weldments. The center-to-center spacing of the turning rolls shown by the figure relates to the diameter of the cylindrical part being rotated. Angle *A* should be 45°. Double sets of rollers are used for heavy loads. They distribute the load over more area of the rolls and the weldment. The surface of the turning rolls can be metal or composition. The composition rollers do not have as high a load capacity and should not be used when high-temperature preheating is used. The size of the drive motor is related to the size and weight capacity of the rolls. The off-center weight of the weldment must be considered. It also has an effect on the rotational speed of the weldment. The rotational speed of the rolls must be selected so that the speed of the weld will be compatible with the travel speed of the welding procedure. The speed of rotation must be smooth and steady. Turning rolls must be accurately aligned or else the cylindrical weldment will tend to move sideways during rotation. This is not acceptable when making circumferential welds.

Rolls are usually used with cylindrical parts; however, by the use of round fixtures or rings, noncylindrical parts can be rotated with conventional rolls. The rings are made in halves, which are clamped around the part to be rotated. With ring fixtures rectangular, square, or unusual-shaped weldments can be rotated for ease of welding (Figure 12–17). Note the adjustment screws, used so that different-sized rectangular weldments can be accommodated. This technique has been used for rotating small ships. Small turning rolls can be mounted on head and tail stock positioners for special applications.

Head and Tail Stock Positioners

Head and tail stock positioners (Figure 12–18) perform much the same function as turning rolls. They are similar to tilt-table positioners. A tilt-table positioner can be used for the head stock or powered member. The tilting device is deactivated so that only the power rotation is employed. Head stock positioners normally do not have the tilt feature. The capacity of head and tail stock positioners is similar to that of tilt-table positioners and ranges from 5 tons up to 50 tons. Since there are two units, the overhang weight problem is not important. The eccentricity weight must be considered. Head and tail stock positioners are used when long, irregular-shaped weldments are produced. Head and tail stock positioners are commonly used in railroad car building shops to rotate cars during the welding operation.

Universal Balanced Positioners

These positioners are relatively small compared to most powered tilt-table positioners. They are balanced and do not contain power; however, power rotation of the table is available. The universal balanced positioner is shown in Figure 12–19. The principle of this type of positioner is to determine accurately the center of gravity of the weldment and adjust the angles of the arms of the positioner so that the center of gravity is in line with the main axis of rotation of the arm and in line with the axis of rotation of the table. When the weldment is properly located and balanced on the positioner, it can be moved very easily to put it in any position required for welding. This type of unit is extremely popular for weldments that are too heavy to lift manually but do not require powered tilt-table positioners. When sufficient volume is to be produced, fixtures can be attached to the positioner so that when the parts are loaded they are accurately located at

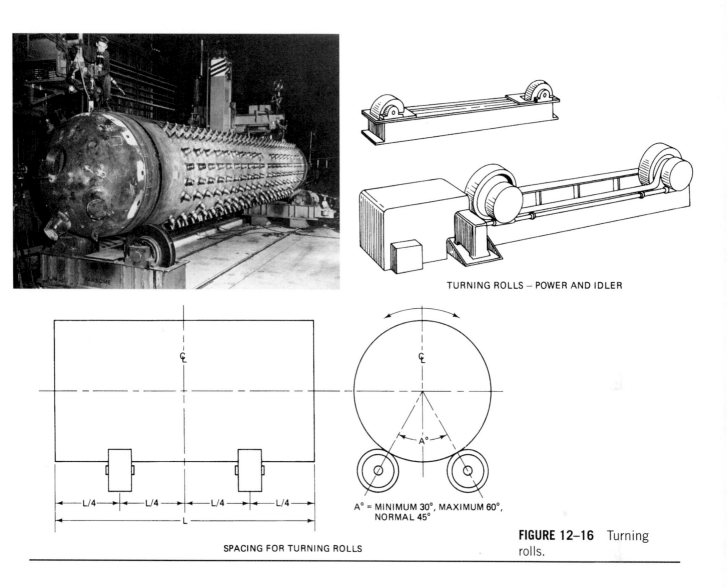

TURNING ROLLS — POWER AND IDLER

SPACING FOR TURNING ROLLS

A° = MINIMUM 30°, MAXIMUM 60°, NORMAL 45°

FIGURE 12–16 Turning rolls.

FIGURE 12–17 Ring fixtures holding rectangular parts on turning rolls.

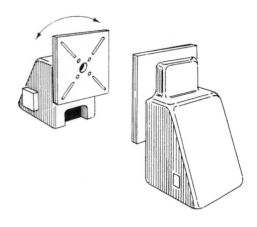

HEAD AND TAIL STOCK POSITIONER

FIGURE 12–18 Head stock/tail stock positioner.

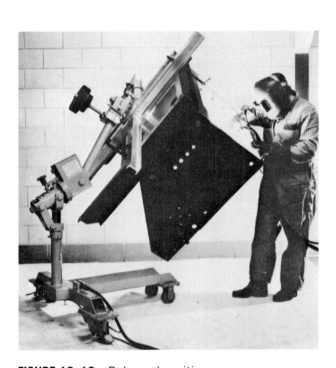

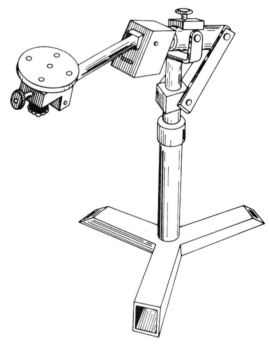

FIGURE 12–19 Balanced positioner.

the balance points. These positioners come in sizes ranging from 100 to 500 lb and have proven to be very efficient for smaller weldments. They are normally used with manual or semiautomatic welding.

There are many other types of positioners; some are made for special applications. One type, known as the cradle or drop center positioner, has two axes of rotation, but the overhang problem is less. They can be made in dif-

ferent sizes and weight capacities. A combination of arc and work motion is shown in Figure 12-20. With a computer program it can duplicate the work of a robot.

Robotic Arc Welding Positioners

Specialized positioners are used to improve the versatility and to extend the range of robotic arc welding sys-

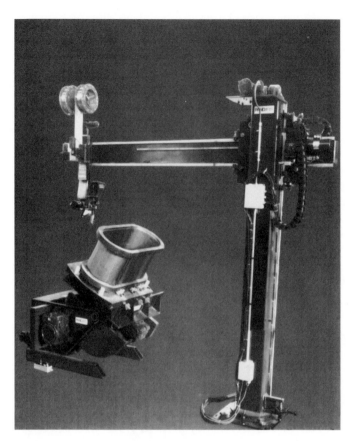

FIGURE 12–20 Combination of arc and work motion.

ing loaded or unloaded. It can have three axes of rotation at each side, and each side does not need to be the same. An example is shown in Figure 12–21b. These units do not normally have coordinated motion.

Single-axis rotational-type positioners that provide coordinated motion are shown in Figure 12–21a, c, and d. Simultaneous coordinated motion is obtained by the combination controller. The axis of motion can be vertical or horizontal. The "turn stock" positioner is used for two fixtures or long, slender weldments. Both sides have rotation, which can be with eight locked positions or with coordinated motion. This is a dual-station positioner with three axes of rotation. An example is shown in Figure 12–21e.

Another positioner, known as the drop center type, is called a turnover unit (Figure 12–21f). This unit has two axes of motion. Positioner motion is locked at specific points for welding.

A different type of positioner, also known as a turnover unit, is shown in Figure 12–21g. This unit has two axes of rotation and is designed for heavier loads. Positioner motion is locked at specific points for welding.

An item that expands the range of a robot is the shuttle carriage (Figure 12–22). This has a travel range of 6 or 12 ft. The work positioner can be mounted on the carriage or the robot can be mounted on the carriage, enabling it to weld at two workstations alternately. It could be considered an arc motion device. Other special robot positioners are available for special applications.

tems. The usable portion of a robot work envelope can be limited because the welding torch mounting method does not allow the torch to reach the joint properly. The lead and lag angles of the torch, which are controlled by the robot's wrist movements, reduce the distances that the torch tip can be extended. Thus the robot working range is limited since it is normally anchored at a fixed location. Special positioners eliminate some of these limitations by making the workpiece more accessible to the robot welding torch. They also provide additional axes of motion to the system. A variety of positioners is used with robots, including single- and dual-station types and the universal tilt-table unit manufactured to tight tolerances. Backlash is normally + 0.001 in. per inch (0.0025 mm per 25 mm). The positioners used with robots must be more accurate than required for manual or semiautomatic welding. In addition, the robot positioner controls must be compatible and controllable by the robot controller in order to have simultaneous coordinated motion of several axes while welding.

The following is a brief description of some of the more popular robot positioners, which are shown in Figure 12–21. The double-ended or twin-worktable indexing positioner, also called a *turnaround,* is very popular. This is a dual-station unit for small or medium-sized parts. While welding takes place at one station, the other is be-

12-4 STANDARDIZED AUTOMATIC ARC WELDING MACHINES

Standardized automatic arc welding machines are used for producing certain welded products or for making certain types of welds. They are considered standardized since they are adjustable to accommodate similar products of different sizes and different material thicknesses. They are tools for making specific items. They can make longitudinal seam welds on tanks, the weld joining of heads to the shell of a tank, the weld attaching small bosses or spuds to sheets, the welds to fabricate structural beams, pipe welds, and so on. They can be maintenance welding machines for building up tractor rollers, track pads, dipper teeth, and so on.

These machines are usually a combination of an arc motion device and a work motion device designed to work together to weld a family of products. Single or multiple arcs are involved and motion is relatively simple, usually only one axis. Work-holding devices are often included and may be customized for specific weldments. Many of these standardized machines have complex controls that are programmable to reduce the time involved in setup. The operator only loads and unloads these machines for maximum productivity.

(a)

(b)

(c)

(d)

(e)

(f)

(g)

FIGURE 12–21 Robot positioners.

Standardized automatic arc welding machines can be categorized as follows:

✳ Seamers, external and internal
✳ Circumferential welding machines
✳ Rotating head welding machines

✳ Beam fabricators
✳ Nozzle welding machines
✳ Internal bore buildup
✳ Rotary buildup machines
✳ Longitudinal welding machines

FIGURE 12–22 Shuttle carriage.

Machines of these types are available in different sizes with different configurations from numerous manufacturers. Users sometimes build their own machines using commercially available components. Many companies sell matched components for this purpose. Following is a brief discussion of some of the more popular standardized automatic arc welding machines.

The most popular is the external seamer (Figure 12-23). This machine utilizes a beam and carriage positioned over a hold-down fixture with a backup. It can be adjusted for different diameters and lengths of tank shells of different material thickness. The seamer may be adjustable for non-flat-position welding. Welding on an incline downhill improves weld travel speed. Internal seamers are constructed differently and are used for making the inside weld. An example is shown in Figure 12-24.

A companion piece is a tank head welding machine, sometimes called a weld lathe or circumferential welder. This machine will rotate the tank assembly under two welding heads and will make two cylindrical welds simultaneously (Figure 12-25). They can be used for different sizes, lengths, diameters, and material thickness. They are commonly used for making LNG tanks, hot water tanks, expansion tanks, and so on. The weld lathe can be used to weld other types of parts. Sometimes simple fixtures are required. See Figure 12-26 for examples.

Another standardized welding machine is the rotating head welding machine. It is also called a boss welder, or spud welding machine. This machine rotates a welding head or sometimes two heads around a relatively small diameter part such as the spud of a tank. The head rotates

FIGURE 12–23 External seamer.

FIGURE 12–24 Internal seamer.

the arc around the periphery of the part and makes a fillet weld joining it to a plate. These machines are used in the tank industry and can be used for welding small-diameter parts to flat or curved plates. When curved plates are involved, cams are used to follow the irregular seam. They are quickly adjustable for different sizes of spuds and they may clamp the parts as well as make the welds. A typical spud welder is shown in Figure 12–27.

Another name for the spud welder is the *weld-around machine.* Figure 12–28 shows a variety of parts welded with the weld all-around machine. One variation

for small parts uses a rotating fixture on the table. It is shown in Figure 12–29. Figure 12–30 shows a more complex version with two torches on the weld-around machine.

A similar circular welding machine, known as a *nozzle welder,* is used for attaching nozzles to tanks or boilers. This equipment can make a full-penetration groove weld or fillet weld on heavy material. The machine mounts on the nozzle and provides rotary motion for making the weld. It can use the submerged arc, flux cored, or gas metal arc welding process. Complex control and motion are required for thick material and for following complex joints on small-diameter tanks. The machine must be closely supervised because the welds must meet code requirements. It usually makes multipass welds. A typical nozzle welder is shown in Figure 12–31. It is much more complex than the spud welder since it produces a full-penetration groove weld, basically in the horizontal position, and makes multiple passes.

A very special machine has been developed for repairing incorrectly machined holes, rebuilding worn holes in machinery parts, and other uses. Known as a *bore welder,* it uses the gas metal arc welding process and makes welds on the inside diameter of holes by providing an all-position welding travel device. It is restricted to a minimum diameter based on the size of the machine. It can be used to weld the inside diameter of holes with the axis horizontal or vertical. It can accommodate large diameters and is shown in Figure 12–32.

An interesting type of specialized welding equipment is used for routine maintenance operations. This type of maintenance includes building up worn surfaces of tractor parts. This special machine is designed for

FIGURE 12–25 Circumferential welder-weld lathe.

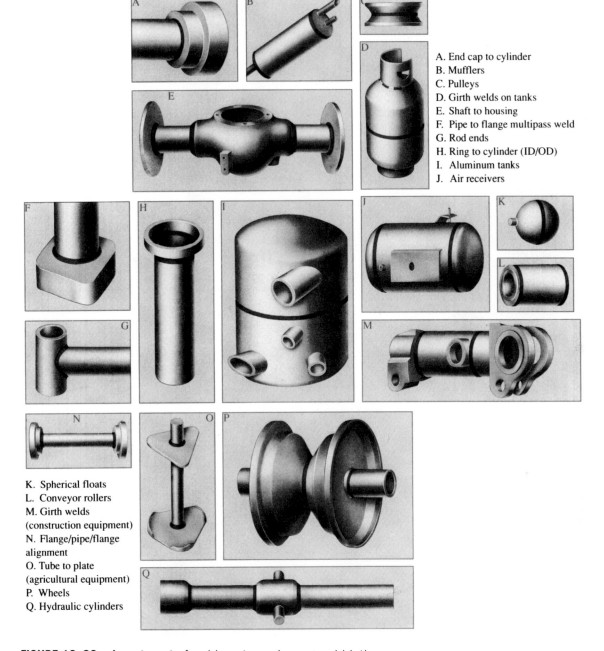

A. End cap to cylinder
B. Mufflers
C. Pulleys
D. Girth welds on tanks
E. Shaft to housing
F. Pipe to flange multipass weld
G. Rod ends
H. Ring to cylinder (ID/OD)
I. Aluminum tanks
J. Air receivers

K. Spherical floats
L. Conveyor rollers
M. Girth welds
(construction equipment)
N. Flange/pipe/flange
alignment
O. Tube to plate
(agricultural equipment)
P. Wheels
Q. Hydraulic cylinders

FIGURE 12–26   Assortment of weldments made on a weld lathe.

building up crawler tractor track shoe idlers. Worn rollers are rotated under the welding head to build them up to their original size or to give them a hard surface. A machine for building up rollers is shown in Figure 12-33.

Another wear item of a tractor is the track shoe. Figure 12-34 shows a weld buildup machine for track shoes. This machine includes several welding heads on a linear carriage with switches that start and stop the arc as the carriage progresses the length of the track shoe assembly. Special flux cored electrode wires are used to provide wear-resistant surfaces.

Other specialized standardized welding machines and equipment is used for routine maintenance and repair work. This includes welding the inside diameter of paper mill digesters, welding the seals of jet engines, welding jet engine blades, welding augers for material moving, and so on.

Orbital welding machines for welding pipe and tubing, considered to be standardized automatic welding machines, are described in Chapter 25. Standardized automatic welding machines greatly increase the productivity of welding. If machines of this type can be kept busy, they often are a more economical solution than a robot.

FIGURE 12–27   Rotary spud welder.

FIGURE 12–29   Variation of weld-around machine.

FIGURE 12–28   Variety of parts welded with a weld-around machine.

FIGURE 12–30 Two-torch variation of weld-around machine.

FIGURE 12–31 Nozzle welder.

12-5 DEDICATED AUTOMATIC ARC WELDING EQUIPMENT

A dedicated automatic arc welding machine is customized equipment designed to weld one specific part or a family of very similar parts on a high-volume production basis. Dedicated or customized machines are used whenever identical parts are manufactured in sufficient quantities or on a continuous basis. The automotive industry and the appliance industry are major users of dedicated automatic welding equipment. The very first automatic arc welding machine (Figure 12–35) was custom designed to automatically weld differential housings for automobiles. This machine utilized continuous bare electrode wire and produced good-quality parts in the early 1920s.

A dedicated or customized arc welding machine is designed to weld one specific part and that part only. This type of equipment is sometimes called *hard automation.* Some dedicated machines may allow for a family of parts that are very similar but vary in size or have only minor differences. They are quickly adjustable to allow for these differences. The customized dedicated welding machine incorporates arc motion, work motion, and work-handling equipment with appropriate controls and welding equipment.

The major disadvantage is the need to redesign or modify the dedicated machine when the design of the product is changed. Another disadvantage is the need to keep the customized equipment operating on a full-time continuous basis because it is expensive.

Despite this, automatic welding with customized machines is used in the high-volume production industries, because it is the most economical welding production method. It reduces labor requirements, produces consistent high-quality parts, maintains production schedules, and standardizes the cost of welded parts.

Customized or dedicated welding machines perform the entire welding operation automatically. The machine may be manually loaded with individual parts. Usually, it is manually unloaded but may be mechanically unloaded. The automated welding machine may be part of a total production line; it is integrated into the production operation.

Customized automatic welding machines are as varied as the weldments they produce, which range from tiny heart pacemakers to giant components for earthmoving equipment, and from small appliances to large bulldozer blades. Dedicated arc welding machines can be classified according to the number of welding arcs employed, the number of axes of motion, and the number of workstations. For example, a dedicated machine with a single weld process may be: single arc, single axis of motion, one workstation; single arc, multiple axis of motion,

FIGURE 12–32 Bore welder building up inside-diameter surface with axis horizontal or vertical.

one workstation; multiple arc, single axis of motion, one workstation; or multiple axis of motion and one workstation. It is also possible for a dedicated machine to utilize different welding processes at more than one workstation.

Machines that employ two or more arcs simultaneously would normally have higher welding productivity than a robotic that usually has only one arc. Multiple workstations using turnaround positioners with two identical holding fixtures will greatly increase productivity.

To obtain a better understanding of the variety of dedicated automatic welding stations, it is best to consider some typical applications. One early application of automatic and semiautomatic welding was the manufacture of domestic furnace heat exchangers (Figure 12–36). These are made in two halves pressed from thin-gauge plain carbon steel or aluminized steel. Baffle plates and

spacers are first tack welded to both the right- and left-hand press-formed pieces of the heat exchanger. Two welders using semiautomatic gas metal arc welding equipment do this work adjacent to the automatic welding machine. The right- and left-hand pieces are then placed manually in a fixture that clamps them together.

Only the very edges of the two halves to be welded are exposed. However, each corner is different. It was decided to simplify the automatic machine to make only the straight welds on each side. After loading, the unit is fully automatic. The weld produced is an edge weld on all four sides of the rectangular sections. Because of the different corner designs, each side of the workpiece has a different weld length. In operation, the machine automatically selects the weld length for the first side, makes the weld automatically, indexes the workpiece 90°, selects the next

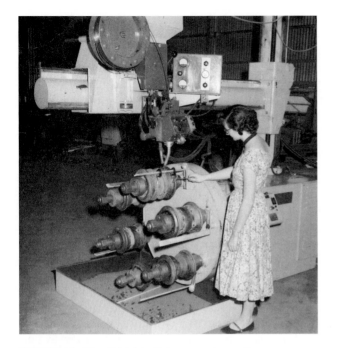

FIGURE 12–33   Weld buildup, tractor rolls.

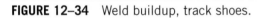

FIGURE 12–34   Weld buildup, track shoes.

FIGURE 12–35   First automatic welding machine.

FIGURE 12–36 Domestic furnace heat exchanger being automatically welded.

Each torch motion is controlled by a separate cam, with the cams mounted on a common shaft. This five-axis rotary motion provides a 360° circumferential weld even when the design of the part prevents full rotation. The 360° circumferential weld was made with only 270° of spindle rotation. Safety light curtains protected the operator from the motion of the machine. Loading and unloading is done manually, and the machine cycle time is 14 seconds.

The dedicated welding machine shown in Figure 12–38 is used to weld two separate models of an automotive catalytic converter assembly (shown in the foreground). It employs two arcs and nine axes of cam-controlled rotary motion. A two-piece rotating cradle is used to locate and clamp the converter body. Two different end pieces are located at the ends of the body section. Each torch is cam-controlled and welds across the body seam flanges and around the circumference of the pipes attached to each end. A single rotation cycle is used to weld both the large oval end and a smaller round end at the same time. It can be changed quickly to weld a different catalytic converter assembly. Two separate power sources and different weld schedules are provided by two separate welding controllers. The production rate for one converter model is 100 parts/hr and for the other is 180 parts/hr.

Automatic welding with dedicated machines is not restricted to the automotive and appliance industries. It is being used for producing large structural weldments. In this example, a dedicated machine is simultaneously making six welds to join three preformed stiffeners to the deck of the approach to a large bridge. These assemblies were fabricated of steel plates 1 in. (25 mm) thick by 10 × 50 ft (3 × 15 m), with three preformed stiffeners $\frac{1}{2}$ in. (12.5 mm) thick welded to them. These panels were then assembled and welded into a 40 × 50 ft (12 × 15 m) assembly and moved to the erection site. Six arcs are used simultaneously to weld the stiffener sections to the 1-in. steel deck plates. A gantry-type welding machine carries all the welding heads, controls, and electrode wire supply (Figure 12–39). The figure shows the welding with the six heads being used simultaneously. Flux cored arc welding utilizing CO_2 gas shielding is employed. High-quality full-penetration welds were required, which passed the necessary structural qualification tests.

Dedicated automatic welding machines are not all restricted to continuous electrode wire processes. The final example is a machine for producing electric motor parts. It is used for welding the stamped laminations together (Figure 12–40). These assemblies are made on equipment that includes two welding stations and two welding processes. The first machine, which has four positions, clamps and holds the lamination assembly in the proper position and moves it vertically in front of two gas tungsten arc welding torches. The workstation rotates

weld length, and repeats the sequence. This progresses automatically until all four sides are edge welded. This amounts to a total of 70 in. of weld made at a speed of 42 in./min, which greatly exceeds the 15 in./min attained with semiautomatic welding. Semiautomatic welding is still used to finish the corner joints adjacent to the openings. The semiautomatic welding operator removes the partially welded heat exchanger, inserts two new pieces, and then finishes welding the corners of each heat exchanger semiautomatically. This is done while the automatic welding machine is making the straight welds. This automatic machine increases production by 60% to 65% over semiautomatic welding.

Figure 12–37 shows a dedicated machine that uses a single torch but employs five axes of motion. It is used to weld a crossover pipe to a tubular automotive exhaust manifold assembly. The part in question is shown in front of the machine in the photograph. The cam-controlled rotary torch motion is the major feature of this machine.

FIGURE 12–37 Exhaust manifold assembly.

and makes two more welds. The final operation, done on a second machine, welds a cap ring to the stator. A gas metal arc plug weld is made through prepunched holes. The mandrel, which holds the stator, turns after each plug weld is made to bring the next hole in line with the torch. The total operation, which combines two welding processes and two stations, will produce hundreds of units per hour.

Many companies produce dedicated or hard automation welding equipment. These machines are relatively expensive since they are designed for a particular application. However, the payback period is short, provided that there is sufficient production volume. The disadvantage is that the machine must be rebuilt whenever the design of the product is changed.

12-6 FLEXIBLE AUTOMATION OF WELDING

Today there are greater demands on the manufacturing industry than ever before. Customers want shorter delivery time and require a greater variety of products. At the same time, the product's lifetime is shorter, and manufacturing batch or lot sizes are smaller. A response to these demands has been the development of **flexible manufacturing systems** (FMSs). Flexible manufacturing systems are being used to replace small batch and continuous manufacturing operations without losing the economies of volume production. The flexible manufacturing system was

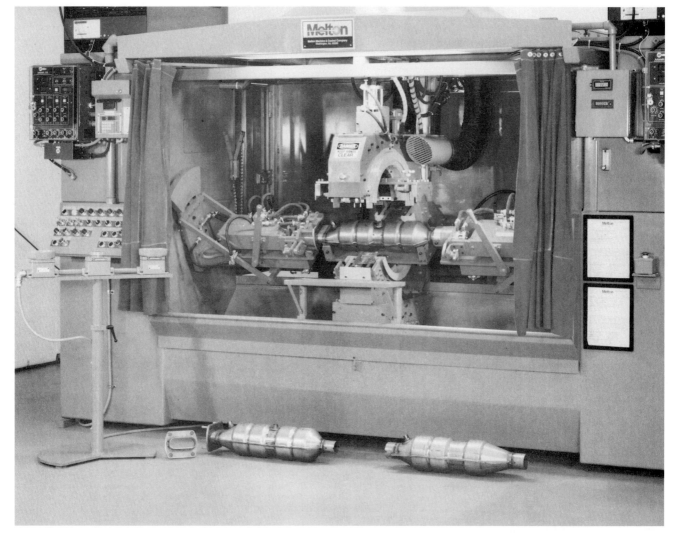

FIGURE 12–38   Welding a catalytic converter.

developed during the 1960s by the machine tool industry with government and customer assistance.

A study by the U.S. Congress, Office of Technology Assessment ("Computerized Manufacturing Automation: Employment Education and the Workplace" OTA CIT-235) indicated that discrete manufacturing could be divided into three categories, based on the volume and variety of products, shown by Figure 12–41.

1. Single-piece parts or an extremely low volume of similar items
2. Batch production of medium lot sizes
3. Continuous production or a high volume of similar parts

Job shop production is low-volume production with a lot size as small as one piece; that is, custom pro-

duction of 1 to 10 parts if it is a large complex part, or a volume of 1 to 300 units if it is a small simple part. *Batch production* is a moderate to medium lot size of 10 to 300 large, complex parts, or 300 to 15,000 small, simple parts. *Mass production* is high-volume production, usually over 300 large, complex parts, or over 15,000 small, simple parts. This usually means continuous operation of dedicated production equipment. These may be arbitrary figures, but are based on this study.

Job shop production involving single units or small lot sizes is very labor intensive. The parts produced are very expensive. There is insufficient volume to justify special machines or dedicated equipment.

Batch production involves medium-volume lot sizes. This type of production can justify simple fixturing and standardized machines to make weldments with less labor. This category is still relatively labor intensive but produces parts at a lower cost.

FIGURE 12–39   Welding stiffeners to bridge deck.

Mass production involves high-volume or continuous production. This type of production justifies customized welding equipment. The amount of labor per part is minimum, labor efficiency is maximum, and the end product is the least expensive.

The flexible automation of welding that makes use of flexible manufacturing systems can make batch manufacturing as efficient and productive as mass production. Carried to its ultimate, it could even make job shop production much more efficient and productive and greatly reduce the cost of "one-only" production.

Robotic arc welding is the obvious answer for flexible automation of welding. The use of a robot and a welding fixture that can be mounted on a work motion device is the key to reducing welding costs. A computer program is developed for each part. The program is placed in memory and used every time the particular part is manufactured. The setup time is minimal and the robot is kept busy welding many small lot sizes of parts all day long. It allows the capability of changing from one part to another quickly, and needs only a positive locating point to

align the robot's welding torch with the parts being welded. Extremely small or medium-sized lots can be processed economically in this manner. A different locating fixture is used for each part and for welding the lot size required.

Robots are expensive and must be kept busy on a fulltime basis to be economically acceptable. There is another way of accomplishing the economy of mass production of small parts produced in small lot sizes. Typical parts are shown in Figure 12–42. This can be done with a flexible welding system that is computer controlled. A flexible automatic welding station for welding small simple parts is shown in Figure 12–43. This workstation costs less than half that of a robot cell. The welding sequence for each workpiece is programmed and stored in the computer memory. A simple holding and locating fixture is provided for each weldment. When the part is to be produced, the operator places the fixture on the table and calls up the program from memory. This takes very little time. For production welding the operator loads the fixture and presses the start button, the machine makes

FIGURE 12–40 Welding motor laminations.

Type of Production	Job Shop	Batch	Mass

Lot size (volume)
  Large complex parts
  Small simple parts
	Low volume
	1–10
	1–300
		Medium volume
		10–300
		300–15,000
			High volume
			Over 300
			Over 15,000

Weld setup	Manual setup	Fixture, manual loading	Fixture, automatic loading
Welded production	Manual or semiauto weld	Standardized welding machine	Dedicated welding machine
Estimated percentage of U.S. production	10–20%	60–80%	20–30%

FIGURE 12–41 Characteristics of metal working production by lot size.

the weld, and the operator unloads the finished weldment. This machine can be programmed for linear arc motion using one or two torches. It can be arranged for rotating work or arc motion with the axis of rotation vertical or for head and tail stock rotary motion with the axis horizontal. Figure 12-44 shows the head-tail stock system with a bicycle fork being welded with rotation about the horizontal axis. Figure 12-45 shows modular components for a welding station and actual components.

The equipment is quickly changed over from one type of motion to another. It is easily adapted to product mix changes and to small batch sizes. This is flexible automation of welding. It is used for more and more short-run applications on simple parts. It will eliminate complex dedicated fixtures and is finding increasing acceptance in volume production plants. This equipment is much less expensive than a robot, yet will make the welds at the same production rate.

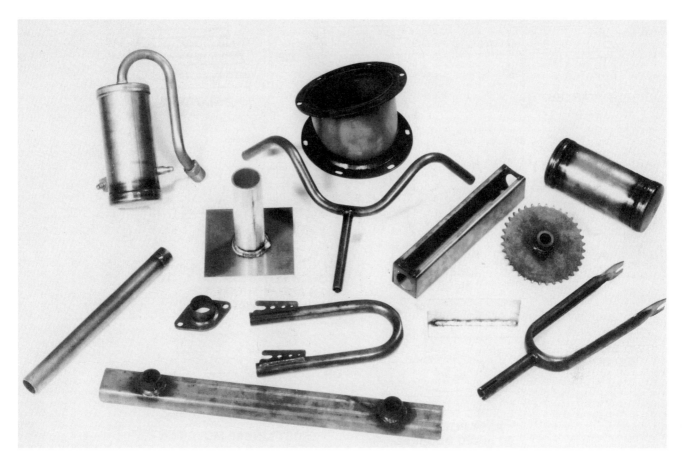

FIGURE 12–42 Typical parts manufactured with flexible welding equipment.

FIGURE 12–43 Flexible welding workstation with microprocessor controller.

FIGURE 12–44 Head-tail stock system welding bicycle fork.

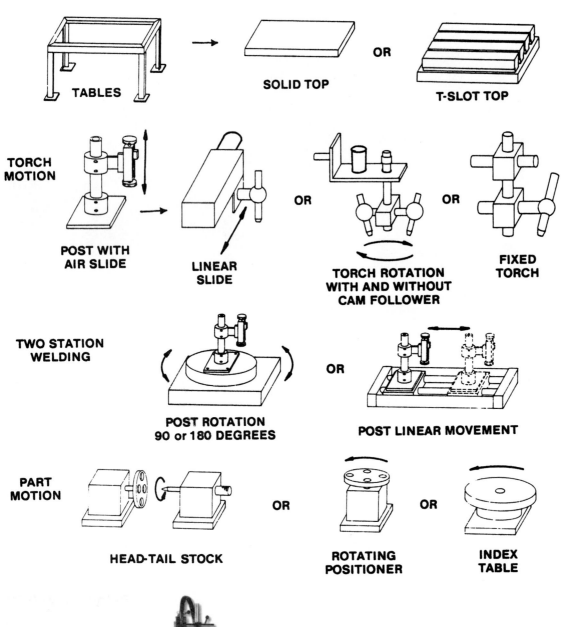

TABLES → SOLID TOP OR T-SLOT TOP

TORCH MOTION → POST WITH AIR SLIDE → LINEAR SLIDE OR TORCH ROTATION WITH AND WITHOUT CAM FOLLOWER OR FIXED TORCH

TWO STATION WELDING — POST ROTATION 90 or 180 DEGREES OR POST LINEAR MOVEMENT

PART MOTION — HEAD-TAIL STOCK OR ROTATING POSITIONER OR INDEX TABLE

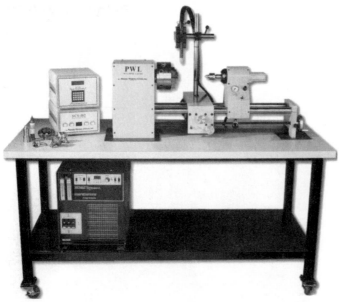

FIGURE 12–45 Modular components for welding system.

12-7 ARC WELDING ROBOTS

Arc welding robots have become very popular in the last few years; however, robots have been around for many years. Joseph Engleburger, the father of modern robots, developed a machine in the mid-1950s and gave it the name *robot*. This was based on the Czech word *robota,* which connotes forced labor that was depicted as a kind of automation in Karel Capek's 1920 play entitled *R.U.R.* Today the Robot Industries Association defines a robot as a "reprogrammable, multifunctional manipulator designed to move materials, parts, tools, or specialized devices, to variable programmed motions for the performance of a variety of tasks." The Japanese define the robot as a three-axis programmable tool. This difference in definition might explain why there are more robots in Japan than the United States.

Robots were introduced to North American industry in the early 1960s; however, it was not until the mid-1970s that robotic arc welding was used in production. Robotic welding grew rapidly in the early 1980s because of the emphasis by the automobile industry. In the mid-1980s other manufacturing companies started using welding robots, and today their use is widespread and growing.

The Robot Industries Association started keeping track of robots in early 1990. In 1994, approximately 8,000 robots were shipped; in 1995, approximately 9,000; in 1996, approximately 11,000; and in 1997, approximately 12,000. In 1998 there was a slight downturn to 11,000, but in 1999 it increased to over 15,000 units shipped. Today there is a current population of approximately 100,000 robots in North America. The pie chart shown in Figure 12–47 shows that most robots are used for material handling, including machine loading and unloading applications. However, spot welding robots, representing approximately 30%, are close behind. Arc welding robots represent 18% to 20%. Other applications, including painting, coating, and inspection, represent about 10%, whereas assembly and dispensing represent approximately 8% of the robot applications.

The automotive industry first used robots for spot welding (Figure 12–46). The robot replaced a person using spot welding guns. This has completely changed the automobile body production line. Today, almost every automobile body produced is spot welded with a robot.

All arc welding robot systems consist of a number of major components (Figure 12–48). The part referred to as the robot is known as the *manipulator* or *mechanical unit,* which performs the manipulative functions. The brain of the robot is the controller, and there are many auxiliary devices to make the robot more productive.

The robot manipulator is a series of mechanical linkages and joints capable of moving in various directions in order to provide motion. The mechanisms are

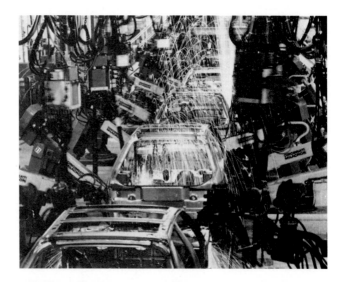

FIGURE 12–46 Robots handling spot welding guns on autobody line.

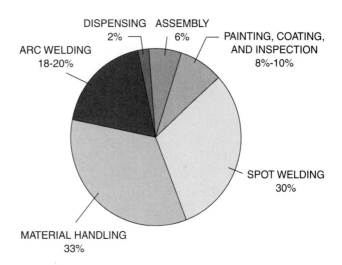

FIGURE 12–47 Robot applications.

driven by linear actuators, which may be hydraulic or pneumatic, or by rotary motors, which may be hydraulic or electric. They are coupled together by mechanical links and may be direct driven or driven indirectly through gears, chains, or screws. There are different designs of manipulators, ranging from three axes to multiple axes. The mechanical manipulator can be categorized by its general design. The more common types of manipulators are the (1) cartesian coordinate, (2) the cylindrical coordinate, (3) the spherical coordinate, (4) the anthropomorphic coordinate, (5) the gantry, and (6) the SCARA type. Each type has specific advantages and features, but all can be used to move a welding torch to make welds.

The complexity of the robot is usually described by the number of axes or *freedom of motion* that it is capable

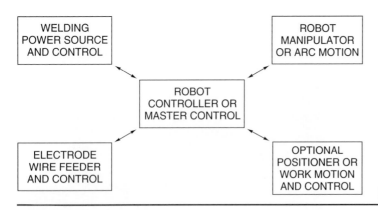

FIGURE 12–48 Robot arc welding system.

of providing. To provide more motion, most robots have a two- or three-axis-of-wrist motion in addition to their basic motions. In selecting robots it is important to understand the work envelope in which the robot can make welds. Each type of robot manipulator has a different work envelope configuration. The shape and size of the work envelope relates to the motions and size of linkages of the robot. Arc welding robots were originally designed to match the working area of a human being.

Robot Manipulator Configuration

Four of the six types of robots are shown in Figure 12–49. The first is the cartesian coordinate robot, based on the three-plane drawing system used for blueprints. It is often called the rectangular coordinate system since it moves within a box-shaped volume based on the x, y, and z directions. The direction X stands for longitudinal motion in a horizontal plane. Y stands for transverse or "in or out" motion in a horizontal plane, and Z stands for up-and-down motion in a vertical plane. It has sliding motion in all three directions. It has three motion axes: longitudinal, transverse, and vertical. Its work envelope is a rectangular box.

The second is the cylindrical coordinate robot. This robot type is similar since it uses sliding motion for two directions, the vertical and one extension, but has one rotational or swing motion. The work envelope is cylindri-

FIGURE 12–49 Four types of robots.

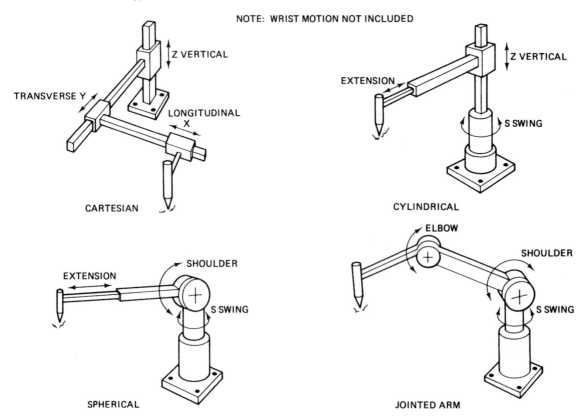

NOTE: WRIST MOTION NOT INCLUDED

cal in the plan view and rectangular in the elevation. The arm holding the welding torch moves up and down the mast and swings about the mast with less than a full circle. The torch extends and retracts.

The third is the spherical coordinate robot, also known as a polar coordinate robot. This robot type has one sliding motion and two rotational motions. One is around the vertical post, and the other is around a shoulder joint. The mechanism holding the arm swings about a vertical axis and rocks up and down about a horizontal axis. The arm slides to extend and retract. The work envelope is spherical, with a similar plan view as the cylindrical coordinate motion robot, but with an elevation view showing the rotational motions based on the shoulder rotation.

The fourth robot is the anthropomorphic robot, or revolute or jointed arm robot. The motions are all rotational with no sliding motion. The work envelope is irregularly shaped in the vertical plane and about two-thirds of a circle in the horizontal plane. This type of robot swings about its base to sweep the arm in a circle. It bends the upper arm forward and backward at the shoulder and raises and lowers the lower arm at the elbow.

The fifth robot, which can be considered as having a cartesian coordinate motion, is the gantry robot (Figure 12-50). The gantry is only part of the total motion since a jointed arm robot or a two- or three-axis wrist is attached to the gantry carriage to provide maximum movement within the work envelope. Its work envelope is a large rectangular box.

The sixth robot is the SCARA. SCARA is an acronym for *selection compliance assembly robot arm,* also known as a horizontal articulate robot. Some SCARA robots have all rotating axes and some have one sliding axis in combination with a rotating axis. SCARA robots have four axes of motion, but do not have much vertical travel. They are used for welding primarily in a single plane. Their work envelope is a flat rectangular box. The SCARA robot is not popular for welding.

There can be combinations of these types of motion systems for special applications. The work envelopes of different makes of robots of the same type are similar. The variations are due to different lengths of arms and links. The jointed arm or anthropomorphic robot is the most popular. The basic movements are shown in Figure 12-51. The work envelope of a typical jointed arm robot is shown in Figure 12-52.

The method for attaching a welding torch or gun is by means of an adapter attached to the wrist. Two different methods of attachment are shown in Figure 12-53. The adapter may have a breakaway feature, which avoids damage if the torch crashes into the work or fixtures. The wrist, which is attached to the work end of the robot upper arm, allows two or three additional axes of motion. They are very similar to the human wrist. These motions are known as pitch, roll, and yaw, which are boating

FIGURE 12–50 Gantry robot having eight axes of motion.

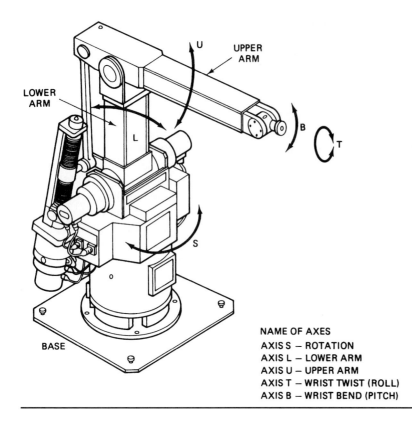

NAME OF AXES
AXIS S – ROTATION
AXIS L – LOWER ARM
AXIS U – UPPER ARM
AXIS T – WRIST TWIST (ROLL)
AXIS B – WRIST BEND (PITCH)

FIGURE 12–51 Basic movements of a jointed-arm robot manipulator.

terms, or bend, twist, and tilt. Figure 12–54 shows the wrist motions with a welding torch attached. Two- or three-axis wrists are used with arc welding robots.

The body motions and wrist motions allow the welding torch to be manipulated in space in almost the same fashion as a human being would manipulate it. This allows the torch angle and travel angle to change in order to make good-quality welds in all positions. They are also required in order to reach difficult-to-reach areas. Even so, a robot cannot provide the same manipulative motions as a human being, but it can come extremely close.

Additional axes are added when the robot is mounted on a moving carriage. This will add an additional axis of motion. A work-holding device can add an additional axis of motion. This is usually rotation and/or tilt, which will add two more axes of motion. A jointed arm robot with a three-axis wrist working with a two-axis manipulator would have eight axes of motion.

In selecting a robot it is necessary to determine its work envelope and reach and the number of axes of torch motion. This will allow you to determine if the robot will weld the weldment in question. This is difficult to determine without making tests; however, computer design programs are available that will help decide whether the robot can accommodate the weldment. An actual test is the positive method.

In selecting a robot it is important to determine the travel velocity while welding and while not welding, known as *air cut time*. The welding speed must be compatible with the welding process and procedures to be

used. The air cut time movement when not welding should be a minimum; the travel velocity when not welding should be high.

An important factor is the repeatability of the robot. This is the closeness of agreement of repeated position movements under the same conditions to the same location. This means to move the welding torch to the same point every time it goes through its program. Most electric robots provide a maximum variation of ±0.015 in. in robot movement for repeated returns to a programmed point. This is affected by operating speed and is acceptable for gas metal arc welding. For gas tungsten arc or plasma arc welding, a tighter tolerance is required and a repeatability of ±0.008 in. is desired. This information is provided by manufacturers' specifications; however, a test provides positive data.

Accuracy of robot movement is also very important. This is the degree to which the actual position corresponds to the desired command position. This is measured by comparing the command position to the actual position.

Resolution is also very important. This is a measure of the smallest possible increment of change in variable output of the robot. It is determined by the ability of the position feedback encoders or resolvers to determine the location of a particular joint and the position of the end-point, called the *tool center point.*

Another factor is the weight-carrying capacity of the robot. This is the weight that it will accommodate in its normal operating envelope at normal travel velocities

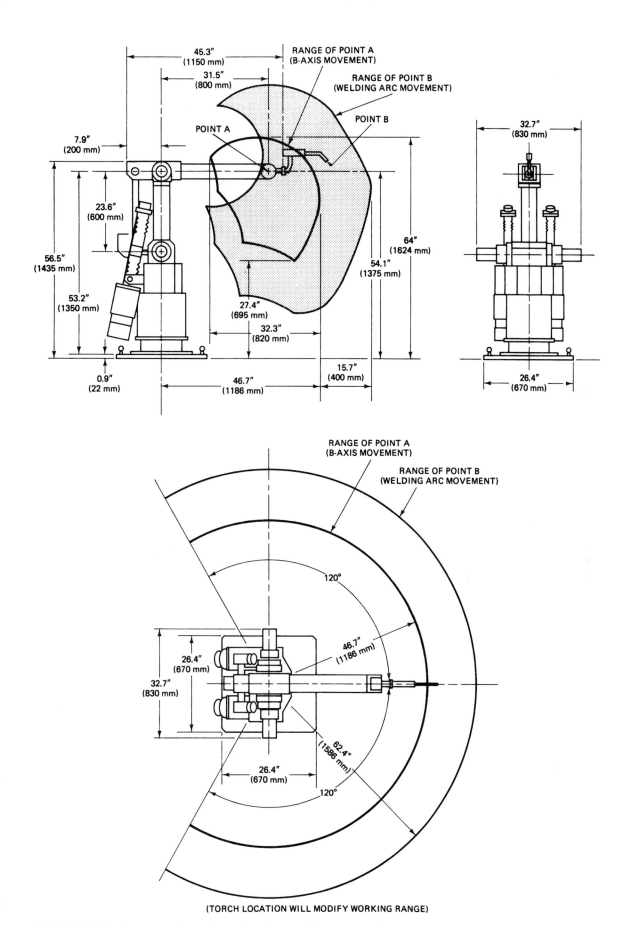

FIGURE 12–52 Work envelope of a jointed-arm robot manipulator.

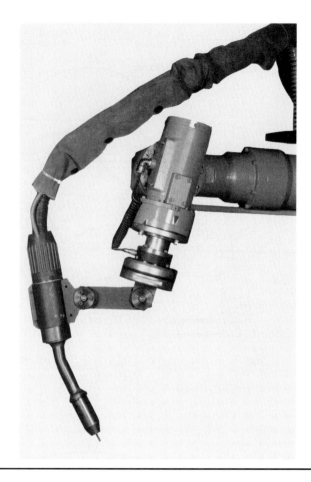

FIGURE 12–53 Methods of welding gun attachment.

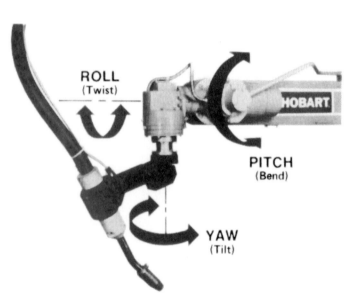

FIGURE 12–54 Wrist motions.

on the end of the wrist. Weight-carrying capacity should accommodate the welding torch, the torch breakaway devices, water and gas hoses, current-carrying cable, and in some cases the electrode wire feeder or feedhead and the electrode wire.

The type of motion drive system is extremely important, as well as the type of position feedback sensors. The motion should be smooth at all times in all positions. Electric drive robots are most widely used for arc welding. Hydraulic drives can be used for painting or spot welding since accuracy or repeatability is less critical. The electrical robots are more repeatable since hydraulic systems tend to drift during warm-up and during operation. In addition, hydraulic robots may have oil leaks. Finally, consideration should be given to mounting position, base height adjustment, manipulator weight, environmental limits, and approvals.

A typical robot cell installation is shown in Figure 12-55. This shows the robot manipulator, workpiece positioner, and necessary equipment in a safety enclosure.

Robot Welding Applications

Robots are welding many, many different products. They can weld just about anything that a human being can weld. An interesting application that shows the robot's capability of welding a complex structure can be seen in Figure 12–56A. This is the welded frame of a large motorcycle; a finished motorcycle is shown in Figure 12–56B. The following are examples to show the diversity of types of welds made by robots.

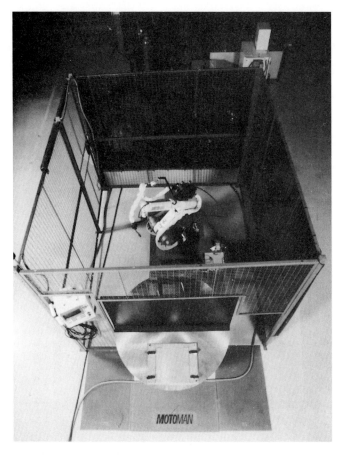

FIGURE 12–55 Typical robot cell with 180° index positioner.

bot. Five parts are required. The major part is thin-wall, small-diameter tubing. The other parts are sheet metal stampings. Tack welding is not used. The operator loads the fixture, and the frame is completely welded before removing from the fixture. One operator tends two robots loading and unloading workpieces on one positioner, while the robot welds on the other. Six robots are producing over 60,000 frames per month. Improved quality, reduced production costs, virtually eliminated scrap, and minimized inspection time resulted from changing from semiautomatic to robotic welding of this application, shown in Figure 12–57.

Case Study: Sheet Metal Assembly This manufacturer of health care equipment wanted to remove welders from routine work and from the welding environments and to reduce costs and improve quality. The weldment is a large sheet metal part (Figure 12–58). The work cell consisted of one robot and three workstations, all within the robot work envelope. One operator mans all three workstations; however, an operator start-and-stop control panel is located in each station. Each fixture is firmly fixed. The operation operates three shifts a day and has reduced production costs by using less electrode wire, less CO_2 shielding gas, and increased welding speed. The quality of welds is very good and consistent, so that fillet weld sizes can be reduced. The productivity has increased over 200%.

Case Study: Tubular Welded Product The need was to handle more different varieties of frames and to increase production. They did the following: Fixtures were modified to handle more than one size of frame. Robots and positioners were installed. The fixtures were mounted on positioners, and two positioners were used with each ro-

Case Study: Pipe Welded Assembly A manufacturer was seeking to reduce costs with small lot sizes and short production runs. Their work involved pipe fittings that required high-quality welds. Short runs are from 25 to 200 parts, and there are 35 different welded assemblies. The company installed a two-station, five-axis work positioner and elected to have different fixtures on each end of the turnaround positioner. An assembly is shown in Figure 12–59. This is a flange-to-T welded assembly. Switching to robotic welding has increased productivity, since the op-

FIGURE 12–56 Robot-welded motorcycle frame.

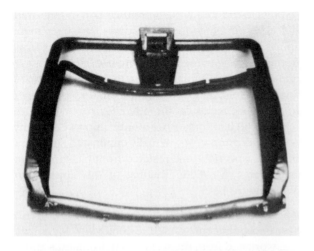

FIGURE 12–58   Sheet metal assembly.

FIGURE 12–57   Tubular product.

erator sets up workpieces of various size and types on one end of the turntable, while the robot is welding at the other end. The robot utilizes through-the-arc seam tracking, which provides good quality for every weld, even though fitup may not be perfect. The pipe assemblies are welded to code requirements.

Case Study: Gas Tungsten Arc Welds   Gas tungsten arc is being used for more and more applications. This aerospace supplier produces accessories for jet engines. The material is thin and medium-thick stainless steel and nickel alloys. This company installed a dual-station positioner and an inverter power source. The material thick-

FIGURE 12–59   Pipe welded assembly.

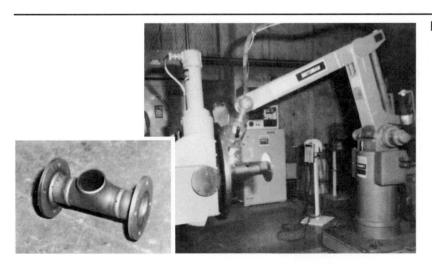

ness ranged from 0.032 to 0.215 in., which required a wide range of welding currents. The robot was equipped with an automatic arc length control (AVC) operating through the robot software. Precision and repeatability has been excellent and the resulting weldments are more consistent than those produced previously. Cold wire feed is used for heavier materials but not for the thin materials. This application is shown in Figure 12-60.

Case Study: Automotive Front Cross Member Automotive companies have scheduled model changeovers. Changeover expense can be minimized by using robots and dedicated holding fixtures rather than dedicated welding machines. The product is an automobile front cross member made from two heavy sheet metal stampings and miscellaneous smaller stampings. This assembly, shown in Figure 12-61, requires 56 in. of intricate curved welds. It

FIGURE 12–60   Gas tungsten arc welding.

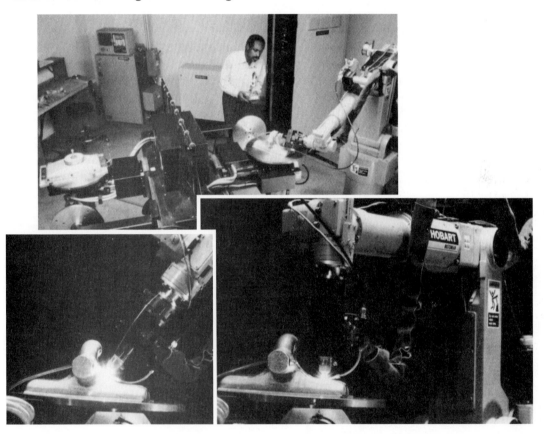

FIGURE 12–61   Auto front cross member.

is first spot welded together and then arc welded. The operation is completely automated, including transporting the workpiece from station to station. The conveyor transports the workpiece under the positioner, which clamps it and rotates it 180° for welding position. Following the weld cycle, the finished workpiece is automatically released onto the conveyor and transported to the next workstation.

Case Study: Aluminum Gas Metal Arc Welding   A supplier to the defense industry is producing aluminum louvered grill assemblies. The company selected a five-axis robot with a five-axis double-ended dual-station work positioner. The parts were self-jigging but required holding fixtures to keep parts in proper alignment. The holding fixtures were attached to the turntables on each end of the positioner. The system utilizes a push-pull wire feeding system with a water-cooled torch. Approximately 150 in. of weld is required to produce each louver. With the robot, weld quality has greatly improved, warpage has been greatly reduced, and productivity has increased. Welding these louvers is shown in Figure 12–62.

Buying a Welding Robot

Careful economic analysis is required to justify the purchase of automated equipment. This may be aggravated by the fact that it is difficult to hire qualified, skilled welders. Once the decision has been made to adopt automatic or robotic welding, the problem becomes how to select the most suitable arc welding system. There are only a handful of companies worldwide that produce arc welding robots and automatic arc welding equipment. However, there are many robot sales representatives and integraters that can assist you in selecting the best solution for your automated welding problem. In general, it might be advantageous to select a pre-engineered arc welding robot cell. These are engineered to use continuous wire welding systems, gas tungsten arc welding systems, or cutting or resistance welding. Pre-engineered robotic cells are available from basic units with minimum size capabilities to complex cells covering larger weldments. The advantage of the robot cell is that it can be installed and immediately will go to work to produce your parts. They are engineered with compatible equipment

FIGURE 12–62   Aluminum gas metal arc welding.

and built-in safety devices that meet national standards. Robot companies will also engineer special cells for your requirements—or you can purchase the necessary parts to construct your own cell for automated welding stations. When going this route, it is necessary that all of the components be matched for the ultimate requirement. For the do-it-yourself company it is recommended that you refer to American National Standard ANSI/AWS D16.2, entitled "Guide for Components of Robots and Automated Welding Installations." This document applies to the recommended design, integration, installation, and use of industrial welding robotic and automatic systems. It includes the various components such as the manipulator, power source, torch accessories, wire feed system, and shielding gas system. It does not contain information standardizing the control system. Control computer systems are designed and provided by the manipulator producer for their particular machine. Each control system and control software system are based on a particular manufacturer's robot and may not be compatible with other controllers. This problem is being considered by the robot industry, and, in fact, universal robot controllers are now available for reworked robots and to provide uniformity throughout the entire factory. They may allow a universal programming system that will standardize the programming of various makes of robots.

It is strongly suggested that the first robot installation be in the form of an integrated cell. It is important to make sure that the cell you order will perform the robotic welding on the products you manufacture.

Robot Safety

Robots were originally designed to duplicate the job functions of a human being. They were designed to relieve human beings of the drudgery of unpleasant, fatiguing, or repetitive tasks and to remove them from a potentially hazardous environment. In this regard, robots can replace people in the performance of dangerous jobs and are considered beneficial for preventing industrial accidents. On the other hand, robots have caused fatal accidents.

The best document relating to robot safety is the "American National Standard for Industrial Robots and Robot Systems—Safety Requirements ANSI/RIA R15.06." This standard provides guidelines for industrial robots. It covers manufacturing, remanufacturing, insulation, safeguarding, maintenance, testing, and startup requirements. A copy should be obtained from the Robot Industries Association, and it should be followed.

Robots work beyond their base area and have large work envelopes that may overlap with adjacent machinery. The travel speed is fast, and robots are multidirectional, operating with as many as six or more axes. Additionally, they start up suddenly and change direction abruptly during motion.

Due to the variable nature of robot applications, specific safety hazards for each installation must be studied on an individual basis. For a robot to be truly effective, it must maintain a high degree of flexibility. This implies that the working envelope must be unrestricted to allow for programs and path changes. Robots work best when they stand where a person once stood, next to other people. Unfortunately, a robot performing the same function as a human will occupy more space than the human. The primary safety rule is that the person and the robot should not occupy the same working space at the same time. The nature of arc welding requires a person to be close to the arc while programming or analyzing a program. To remove the person from the arc area limits the flexibility and accuracy of the robot. Hence one of the major problems associated with robotic arc welding is the presence of a human programmer in close proximity to the welding torch held by the robot. For small weldments, this is not a major problem. For very large weldments, the parts are heavy and a completed weldment will require cranes for loading and unloading. These introduce the normal problems of safety with respect to materials handling. However, they also introduce the problem of bringing heavy pieces of material into the robot's work envelope.

One of the best solutions for robot safety is to purchase the robot as a complete welding cell. A complete cell includes barriers, all necessary safety devices, and a method of loading and unloading the workstation. It is best used for the production of smaller weldments. In general, a turntable, turnover, or shuttle device is used for loading parts outside the robot's work envelope. The parts are then presented to the robot inside the barrier where the welding is performed. After the welding is completed, the parts are then transferred to outside the barrier, where they are unloaded and sent to their next destination. Figure 12-63 shows a pictorial view of a robot cell with a rotary 180° indexing table for the workstation. This cell is for small weldments that can be loaded and unloaded outside the barrier. Different index fixtures can be used for larger weldments. A robotic welding cell layout with two pneumatic indexing shuttle positioners and two robots is shown in Figure 12-64. Work-holding devices can also be loaded and unloaded automatically, in which case they must work with a material motion device that presents additional safety hazards. This becomes a special engineered project.

The robot industry has adopted special safety graphics. Although these have not yet been accepted as a national standard, they are used by many robot suppliers. They should be posted at appropriate locations. One warning sign that is agreed upon by all is shown in Figure 12-65. DANGER—DO NOT ENTER—MACHINE MAY START AT ANY TIME. This should be posted at the cell access door.

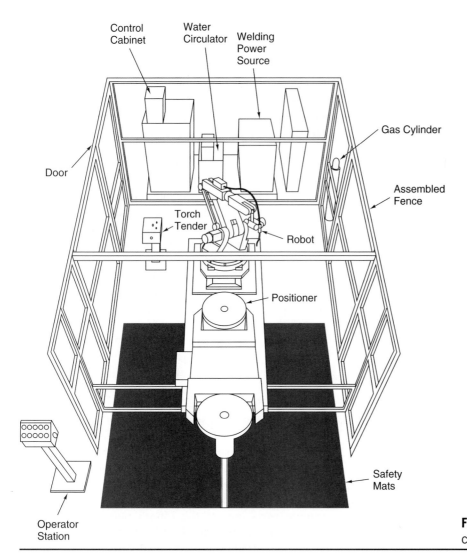

Control
Cabinet

Water
Circulator

Welding
Power
Source

Gas Cylinder

Door

Assembled
Fence

Torch
Tender

Robot

Positioner

Operator
Station

Safety
Mats

FIGURE 12-63 Layout of a robot
cell with 180° index positioner.

12-8 CONTROLS FOR AUTOMATIC ARC WELDING

When making a weld, it is always the intent to produce a perfect weld. In any of the methods of application, except manual welding, a control system is required to run the welding program. A welding program is always employed consciously or unconsciously whenever a weld is made. The program or welding procedure is the basis for making the weld. In manual welding these are established and controlled by the welder. In semiautomatic welding a control mechanism in the wire feeder actuates electrode wire feed, and starts the welding current and shielding gas flow when the welder presses the gun trigger.

Mechanized and automatic welding have more complicated programs and control additional functions, including travel or motion, torch position, and fixture motion. All motion functions are sequential. Adaptive welding, which varies weld parameters in accordance with ac-

tual conditions, has a complicated computer control system that includes sensing devices and adaptive feedback.

Automatic Welding Controllers

Programmers are designed to execute a welding program. As the welding program becomes more complex, the controller must include more electrical circuits. A typical program for gas metal arc welding is shown in Figure 12-66 (Figure 12-66). It can also be used for flux cored arc welding or submerged arc welding. The top three lines represent welding current (or wire feed speed) and arc voltage. The next two lines represent auxiliary activities shielding gas and cooling water flow. The bottom line represents travel or relative motion.

At the cycle start point, the operation begins and the specific activities occur. First is preflow of shielding gas and flow of cooling water. After a preset time period, the main contractor closes. The arc starts and the electrode wire feed begins. Single-axis travel, rotary or linear,

FIGURE 12–64 Robot cell with two shuttle index positioners.

FIGURE 12–65 Danger warning sign.

begins at this point or after a preset delay. Travel occurs until the weld is completed, but many end at different points, depending on the welding program. The travel or motion control circuit includes a motor speed control circuit. When the weld is completed, there is time for crater fill and time for burnback prior to terminating the weld. The welding circuit contractor will open, the arc stops, but shielding gas continues to flow during a preset postflow period. At the end of this period, the shielding gas and water cooling valves close and the welding cycle is completed. The cycle can be made to repeat for arc spot welds or for skip welds, or it can only repeat when new pieces are placed in the machine and the cycle reinitiated.

To fully understand a welding program, it is necessary to understand the terms used:

※ *Preflow time:* the time between start of shielding gas flow and arc starting (prepurge).

※ *Start time:* the time interval prior to weld time during which arc voltage and current reach a preset value greater or less than welding values.

※ *Start current:* the current value during the start-time interval.

※ *Start voltage:* the arc voltage during the start-time interval.

※ *Hot start current:* a very brief current pulse at arc initiation to stabilize the arc quickly.

※ *Initial current:* the current after starting but prior to upslope.

※ *Weld time:* the time interval from the end of start time or end of upslope to beginning of crater fill time or beginning of downslope.

※ *Travel start delay time:* the time interval from arc initiation to the start of work or torch travel.

※ *Crater fill time:* the time interval following weld time but prior to burnback time, during which arc voltage or current reach a preset value greater or less than welding values. Weld travel may or may not stop at this point.

※ *Crater fill current:* the arc current value during crater fill time.

※ *Burnback time:* the time interval at the end of crater fill time to arc outage, during which electrode feed is stopped. Arc voltage and arc length increase and current decreases to zero to prevent the electrode from freezing in the weld deposit.

※ *Downslope time:* the time during which the current is changed continuously from final taper current or welding current to final current.

※ *Upslope time:* the time during which the current changes continuously from initial current value to the welding value.

※ *Postflow time:* time interval from current shutoff to shielding gas and/or cooling water shutoff (postpurge).

※ *Weld cycle time:* the total time required to complete the series of events involved in making a weld from beginning of preflow to end of postflow.

The controller for running the program in Figure 12-66 is shown in Figure 12-67. Controllers of this type include meters for arc voltage, for welding current, and sometimes for electrode wire feed speed. It also includes pilot lights for other activities, such as an "arc on" signal to indicate that the arc has been established. This type of controller usually has input voltage compensation and will compensate for welding cable voltage drops, and so

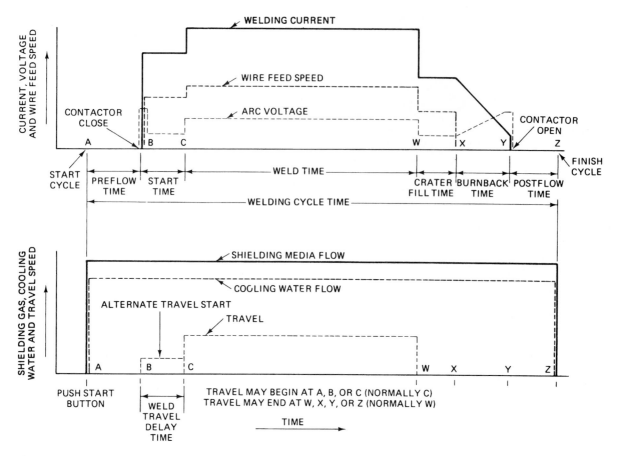

FIGURE 12–66 GMAW welding program.

FIGURE 12–67 Welding controller for single-axis motion.

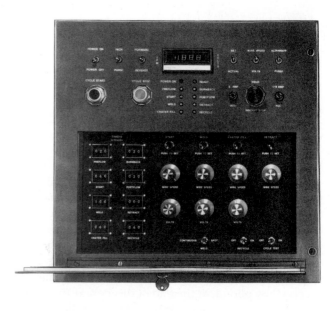

on. The controller includes motor speed control circuits, which accurately regulate the wire feed speed motor and the travel speed motor. Time-delay circuits are included whenever a delay period is required. Other activities, such as welding head positioning, fixture clamping, and so on, can be included. A relay controller of this type can only provide one function at a time in a prearranged sequence. Two on/off functions can be simultaneous, and sequenced functions can be in rapid order. This type of controller does not have the capability to ramp or gradually change the welding current during operation. This function is included, however, in controllers for gas tungsten welding and plasma arc welding. Adaptive feedback signals cannot be accommodated with this type of controller.

Controllers of this type can be preprogrammed to provide specific delays for shielding gas preflow and postflow time, travel start time, crater fill time, and burnback time. Welding current and arc voltage at different levels can be preprogrammed, as well as the total weld cycle time. This same type of programmer can be used with limit or proximity switches, to use motion as a control base rather than time.

This controller can control more than one axis of travel motion, and it has extra contacts so that it can control other motions, such as fixture clamp, torch advance, and so on. However, it cannot control coordinated or simultaneous motion of two or more axes. Motion must be sequential so that one activity immediately follows the previous one. Coordinated motion requires microprocessor-type controllers.

Controllers are simple or complex, depending on the number of activities that must be controlled. The one shown is for a semiautomatic wire feeder. A relay logic system requires some operator skill because the control provides the sequencing of operation but still requires the operator to establish parameters, delays, and decision-making capabilities to ensure a good-quality weld.

Controls and timers can be standard or precision, depending on the needs of the welding procedure program. Tachometer feedback of wire speed and travel speed motors can be included to provide for more precision and repeatability. The more precise controller ensures consistent high weld quality and repeatability.

The weld control systems described are relatively simple but are well suited for many, many applications utilizing arc motion and work motion devices. They can be used for standardized and dedicated automatic arc welding machines with not more than two axes of simultaneous (not coordinated) motion. Controllers such as those mentioned here and more complex control units are available from different manufacturers.

Robot Controllers

For robotic arc welding systems, a much more complex controller is required. Controllers include a high-speed

microprocessor since coordinated, simultaneous, continuous motion of up to eight axes and all welding parameters may be required. As the number of axes increases, the amount of computer capacity must increase.

The machine tool industry introduced numerical controls (NC) years ago. Automatic shape cutting machines use the same type of controller for directing the path of cutting torches. These are known as point-to-point (PTP) control systems. Points are locations in two dimensions in one plane. For arc welding robots the arc is moved from one point to the next in space. A typical robot arc welding controller is shown in Figure 12–68. The location of the arc is known as the *tool center point* (TCP). It is the path of the TCP that is programmed and stored in memory. For spot welding, pick and place, and machine loading, point-to-point playback is used. For arc welding, playback of the arc motion is a continuous path in space. The robot controller must be coordinated so that each axis movement begins and ends at the same time. It is the function of the programmer to accept the input of many point locations, relate welding parameters to the path taught, and to store this information in memory, then play it back to execute a welding program. It is beyond the scope of this book to explain its inner workings; however, we will explain how it is used to make welds. The major points of interest are the teach mode, the memory, and playback or execution.

FIGURE 12–68 Robot controller.

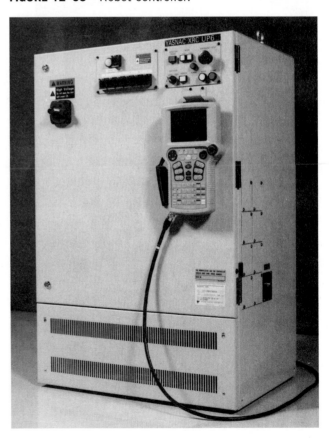

Teaching the Robot

There are at least four methods of teaching or programming a robot controller: manual methods, walk through, lead through, and off-line programming. The manual method is not used for arc welding robots. It is used mainly for pick-and-place robots.

The walk-through method requires the operator to move the torch manually through the desired sequence of movements. Each move is recorded into memory for play-back during welding. The welding parameters are controlled at appropriate positions during the weld cycle. This method was used in a few early welding robots.

The lead-through method is a popular way of programming a robot. The robot welding operator accomplishes this by using a teach pendant (Figure 12–69). By means of the keyboard on the teach pendant, the torch is power driven through the required sequence of motions. In addition, the operator inputs electrode wire feed speed, arc voltage, arc on, counters, output signals, job jump functions, and much more. All of these functions are related to a particular point along the taught path. In this way, if the robot speed is changed, it is not necessary to change the time for certain actions to happen. This means that actions are sequence and position related rather than time related. The travel speed of the torch is independently programmed between specific points by the keyboard.

The path of the arc or tool center point is taught by moving the TCP to a particular point using the teach pendant keyboard. The machine axes locate the torch, and the wrist axes control the angle of the torch. There is a control for each drive motor (i.e., one for each axis). When the desired position is reached, it is necessary to record the position by pushing the record button. This same operation is repeated for the next location point, until the complete path is taught. The robot controller must be coordinated to control all the axes simultaneously. The normal arc welding robot has five or six axes (including two or three in the wrist). The controller should have additional capacity to control the axes of positioning equipment. Robot positioners increase overall efficiency and the range of the robot and improve weld accessibility. The robot controller should be able to control the positioner and provides total coordinated motion.

In the playback mode the robot will follow the path between each point according to its interpolation function. Normally, linear interpolation is used, which means that the arc or TCP will move in a straight line between taught points. Circular interpolation means that the arc or TCP will move in a circle. Three points will designate and locate a circle. It is useful for developing a curved path and reduces the number of points required. The playback mode must be a continuous path.

The controller should allow revision of one taught point without reteaching the entire path. It should allow deletions or additions of taught points. Also, it should allow changes of travel speed or of welding parameters. The operator should be able to check the taught path and welding parameters without welding. The speed of the arc may be set in absolute values or by transverse run time (TRT) or time between points. The above is done in an edit mode so that the taught path can be modified or shortened, speed changed, or welding parameters changed.

Older robots require an interface panel between the robot controller and the welding power source and electrode wire heater. The panel allows the programmer to control the wire feeder and power source in exact volts and amperes. Some interface panels also provide subroutines such as weld termination. The controller usually has many steps from minimum to maximum to control current and voltage, and these must be converted to absolute current and voltage values to conform with the program. The newer robot controllers have more capabilities, and with newer power sources avoid the need for an interface panel.

The robot controller must program welding parameters in order to have a truly automatic welding system. They must be stored and retrieved the next time the job is run, without the necessity of adjusting the welding equipment.

FIGURE 12–69 Teach pendant.

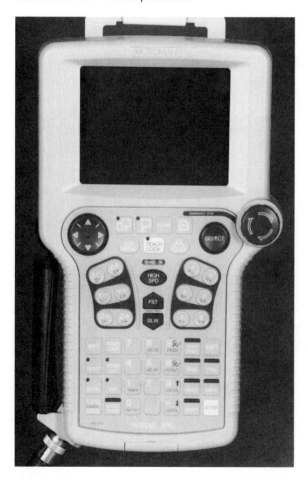

The robot controller must have a diagnostic system built in to allow a quick check when problems occur. Most robot controllers offer other features, which may be built in or optional. Linear and circular interpolation, mentioned previously, is important. Other features available as options could be:

- ✳ Automatic acceleration and deceleration
- ✳ Three-dimensional shift
- ✳ Simultaneous control of extra axes
- ✳ Scale-up and scale-down
- ✳ Mirror image
- ✳ Software weave

All of these are useful for an arc welding robot. The software weave is very useful since it allows the robot to manipulate the weld pool like a human welder. It allows a larger weld cross section, better bead contour, and enables the weld to bridge gaps. Different patterns can be programmed, from simple sideway oscillation to triangular patterns. This is taught in three steps. Frequency of the weaving oscillation, amplitude, and dwell at each end are taught. Once the weaving pattern is taught, welding will continue through changes of path in all planes without reteaching the weaving pattern. Other options include through-the-arc seam tracking and other tracking functions. A thorough study of the robot is necessary to determine and learn what these features include.

Off-line programming involves the preparation of the program on a computer. An appropriate language must be used. The program is entered into the robot memory very quickly. This increases the utilization of the robot, since lead-through teaching ties up the robot during programming. Off-line programming is becoming more widely used, but requires experienced personnel.

Robot Memory

The amount of memory of the controller is usually indicated by the number of steps and instructions that can be programmed with the number of axes involved. This is often described as having a memory capacity of 2,200 steps and 1,200 instructions. Memory should have at least 32K bytes with battery backup. There should be a programming terminal with keyboard and screen displays in addition to the teach pendant.

The controller usually has one or more microprocessors. Faster execution, response time to better input/output control, and overall flexibility is possible when two or more processors are used. Controller software that provides all the control features is stored in RAM (random-access memory) and ROM (read-only memory). Memory can be expanded with external cassette tapes, diskettes, or disk drives. External stored information must be read into the RAM prior to execution.

The computer must have communication ports so that it can talk to the overall controller. The robot memory should be selected based on the work to be done. Controllers allow feedback signals from various sensors.

Weld Execution

Welds can be made only when the power is on all components, electrode wire is installed, and the controller is in the playback or operate mode. The material must be in the fixture and ready. Pushing the start button will initiate the operation. The robot will move the torch to the start point. The welding equipment will begin its cycle of operation (i.e., gas preflow, start the arc, etc.). The robot controller will determine that the arc has started and then start motion. Points along the taught path will initiate other activities programmed. At the end of the taught path, the welding equipment will terminate the weld program and the robot controller will determine that the electrode wire has separated from the work. After this the robot will return to its home position, ready for another cycle. At this point the weld should be checked for quality. The program should be checked and edited to improve the weld if necessary and to minimize the air cut path and increase air cut speed. When the weld quality is acceptable and cycle time is at a minimum, it is time to freeze the program and start production.

12-9 SENSORS AND ADAPTIVE CONTROL

The ultimate automated welding system will simulate the human welder and provide a closed-loop system that compensates for all variations to produce a high-quality weld. This is true adaptive control welding. The components of an automatic welding system shown in Figure 12–2 are changed to adaptive control and are shown in Figure 12–70. Adaptive control welding can be applied to robotic welding systems or to complex automatic welding systems, which are open-loop systems until adaptive control is added.

True adaptive control for automatic or robotic welding closes the loop because sensing devices can replace the human operator for almost every function required, as shown in Figure 12–1 in the three right-hand columns. In practice, however, sensors should be provided only for the functions that require surveillance and control. The number of sensors indicates the level of completeness of a closed-loop system.

Two components must be added to a system to provide adaptive control:

1. A controller with a high-powered microprocessor that will accept signals from sensors and make the

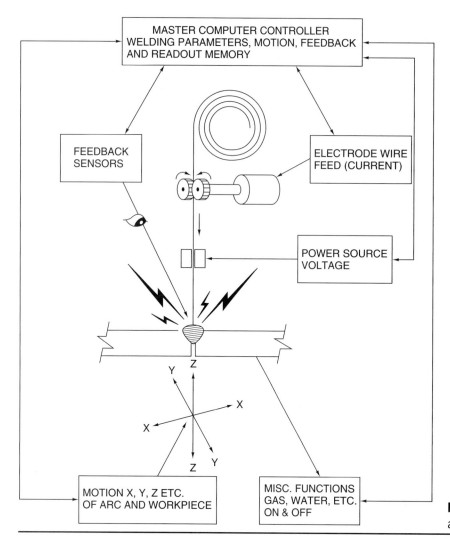

FIGURE 12–70 Adaptive control for an arc welding system.

necessary changes in the welding parameters in real time

2. Sensing devices that provide real-time information to cause parameter changes

The controller for a welding robot or for a multiaxis automated welding machine was described in Section 12-8. An advanced controller coordinates additional movements and has the ability to accept feedback sensor signals.

An adaptive controller for a robotic arc welding system must control all functions and accommodate feedback signals. The controller's central processing unit (CPU) controls and monitors arc motion and workpiece motion, torch location and angular position, welding parameters, parts location, shielding gas supply, and so on. The advanced controller eliminates the need for an interface panel and provides necessary analog-to-digital conversion for inputs and outputs. The processor must have very high speed with the ability to process vast amounts of data. The unit should contain at least 4 MB of RAM and a 20-MB hard disk for program storage and weld history.

A floppy drive or a tape drive should be included for making copies of welding procedures or taking historic real-time data. The robot controller should communicate with other factory processors and fit the CAD equipment used. It should provide a warning alarm system that is adjustable for variations in parameters if they exceed previously set tolerances.

Configuration editors should be included for different procedures and applications. The CPU should have a color display monitor with a touch-sensitive screen so that the operator can correct operations easily and accurately. The controller should have a logical menu-driven program with color icons and sound, written in plain English to lead the operator through the weld procedure. Only selections that make sense at a given time should be presented to the operator. If an error occurs, the controller should suggest solutions. The controller memory should contain hundreds of welding schedules in its library that can be recalled and applied as required. The controller should allow procedures to be checked by the system for logical consistency before the actual weld is made. Password routines should be used to allow a procedure to be viewed but changed only by personnel with the authority to do so. A lightweight portable pendant with a single connector to the main processor console should be provided. The pendant should contain an emergency stop button, individual trim knobs for each weld parameter, jog buttons for each motion axis, wire feed, and subroutines. It should have a display screen that displays messages and actual weld parameters. The programmer controller should be able to gather data on all activities, provide printouts of procedures and parameter values, and so on. Printouts should provide the date, time, operator identification, weld procedure, and part identification. Programmers should be sufficiently flexible that the operator working at the main menu can touch the screen and review the welding procedure or select a new one from the library. It should be possible to make a dry run to determine the procedure without the arc on. The operator's pendant, which displays parameter values, can be used to modify and control the weld. The master robot controller should be able to communicate with other computers in the factory. Two typical modern controllers are shown in Figure 12–71.

The other components necessary for adaptive control are sensors. A sensor is a device that determines or measures a function in real time during the welding operation. Sensors are used to determine actual conditions so that the welding procedure can be modified if necessary. They provide signals that are used to modify the motions of the arc as well as for changing welding parameters. Feedback of sensor variations causes the adaptive controller to change parameters and travel path to produce a quality weld despite problems that may be encountered. Sensors close the loop and make truly automatic welding possible.

A variety of sensing devices are commercially available. Special software or a special computer may be required to match a sensor to the robot controller. New and improved sensors are continually being developed, with their use becoming more widespread.

Contact Sensors

The two major categories of sensors for seam tracking are the contact (tactile) and the noncontact. Tactile sensors have been used for joint tracking for many years. They range from simple mechanical systems to complex electrical–mechanical contacting sensors. The simplest seam tracker is a spring-loaded roller with a floating welding torch. The roller fits against a reference surface and causes the head to maintain a specific dimensional relationship with the joint. The head will follow the motions generated by the roller.

The electromechanical system is more versatile. In this system, a wheel or a stylus probe will contact the surface, which can be the plate surface, the edge of a groove, the edge of a T-joint, or similar surface, and provide a signal that operates a motorized cross slide to adjust the torch for making the weld. A second axis can be provided to maintain accurate torch-to-work dimensions. The probe and torch are mounted on the carriage (Figure 12–72). This system is used for long, straight seams.

FIGURE 12–71 Advanced robot controllers.

FIGURE 12–72 Electromechanical contact-type seam tracker.

Probes wear and must be replaced. They are connected to switches that provide the correction signal. The probe must be sufficiently distanced from the arc to prevent spatter buildup. Tack welds and the start and end of welds pose a problem. This type of equipment is not suited for robotic arc welding.

The distance from the arc to the sensing location can pose a problem for a mechanical probe or wheel. If the distance is too great, deviations can occur; if it is too short, the arc will interfere with the probe and cause rapid wear and deterioration. These systems are not able to accommodate abrupt changes of direction at welding speeds.

A different type of touch system is employed in conjunction with through-the-arc tracking systems for a robot. This system uses the electrode wire, which protrudes beyond the current pickup tip, as the contact. The robot is programmed to move the electrode wire and touch the work surface at different points to determine the location of the start of the joint. It can also be programmed to measure the weld geometry and establish the size of the weld groove. It utilizes a complex motion system. This computer-driven system may employ an expert system with a memory. It is capable of sensing the joint path in three dimensions and storing it in memory. The calculation of the weld joint detail in connection with the expert data bank will establish a new welding procedure and modify the welding parameters.

Noncontact Sensor Systems

Noncontact sensor systems have become very popular. There are three basic types: (1) sensor systems that rely on physical characteristic of materials or energy output relationships, (2) through-the-arc systems that use electrical signals generated in the arc, and (3) optical–visual systems that attempt to duplicate the human eye.

Acoustics can be used to control the length of a gas tungsten arc and the standoff distance for laser heads. The sound energy is linearly proportional to arc voltage. An acoustical waveguide close to the arc leads to a microphone. The signal is amplified, filtered, and rectified and is used to control the torch movement and thus to control the standoff distance and the arc length. It is used for pulsed current gas tungsten arc welding and for laser cutting.

Capacitance is the property utilized by some proximity switches. The capacitance limit switch has been used in automatic equipment for years. It can be adjusted for different distances. It is also used to detect the presence or absence of material, such as if a clamp is closed or not.

Eddy currents are currents set up in the base metal by an adjacent ac field that is generated by a coil located close to the base metal. Another coil acts as the pickup and detects the eddy current. Electronic circuitry produces a voltage dependent on the distance from the base metal. The output changes when a joint interrupts the metal surface. The sensor is oscillated across the joint to produce control signals, which are processed to give the position of the joint centerline. Different types are required for ferrous metal and for nonferrous metals. Thickness is not a major factor. This system is for noncontact seam-tracking systems.

Inductance or induced current in the base metal can be detected and measured and used for seam tracking. In this case the sensor contains two coils, which scan the seam and provide signals that give information on the location of the joint. This is similar to the eddy current system. The sensor must be at a given distance above the base metal and placed ahead of the arc because of its sensitivity to heat, spatter, and so on.

Infrared radiation can be picked up by sensors that are used for penetration control. The infrared sensor is focused on the underside of the weld pool to detect the color of the metal under the weld. This system's accuracy is subject to surface conditions and exact target location. It is not considered extremely reliable as a penetration control system and has limited applications.

Through-the-arc seam tracking is a noncontact system with many advantages. It does not need accessory items attached to the torch. It is a real-time system that can be used for most types of welds. Monitoring occurs while the weld is being made. There are several types of through-the-arc systems and they are used both when metal crosses the arc and when metal does not cross the arc.

The earliest through-the-arc sensing system was the arc length control system for gas tungsten arc welding. Such a system is called an *arc voltage control* (AVC) *system;* however, *arc length control* is a more appropriate name. The starting mechanism of some AVC systems operates such that when the cold tungsten electrode touches the work, it initiates the arc and immediately withdraws to the preset voltage. Arc length control systems are very reliable and are widely used.

The major use of through-the-arc systems is for seam tracking. The welding torch is oscillated and the arc voltage and/or welding current are monitored before and after each oscillation. Mechanical oscillation is normally used, but magnetic oscillation can be used for gas tungsten arc welding but not for gas metal arc welding. Through-the-arc systems can be used for fillet or groove

welds. Figure 12-73 illustrates the principle of operation. Control circuits measure the voltage and/or current and reference the left- and right-hand values to equalize them. The control circuit moves the torch to the center point between the two equal points. This adjusts the path automatically. This system also has a corner recognition mode that allows tracking around a 90° change of direction and is capable of sensing the joint path in three dimensions. It can be used with all modes of metal transfer. Welding speeds of up to 40 in./min (1025 mm/min) can be attained. Oscillation can vary from $\frac{1}{8}$ in. (3.2 mm) to 1 in. (25 mm), and the frequency is from 1 to $4\frac{1}{2}$ Hz. The controls can be integrated into the controller. The final pass of a groove weld is attained by using the previous passes to establish the torch path in memory. This system can be coupled with the electrode contact system mentioned previously, where the electrode wire is used to find and measure the weld joint. If the root opening or gap in the groove joint is excessive, the machine can be programmed to select a different procedure from the memory bank and make alternate layers for each layer rather than a single pass.

Optical–visual sensor systems are based on an analysis of the manual welding operation, which states that the welder derives the bulk of the information required to make a high-quality weld through visual input. Optical–visual systems provide real-time signals for fully automated arc welding. Optical–visual systems find the seam, follow it, and identify and define the joint detail so that welding parameters can be adjusted to produce a high-quality weld. Optical–visual systems are extremely fast and do not become fatigued. However, they are extremely complex. A system flowchart is shown in Figure 12-74.

FIGURE 12–73 Through-the-arc guidance system.

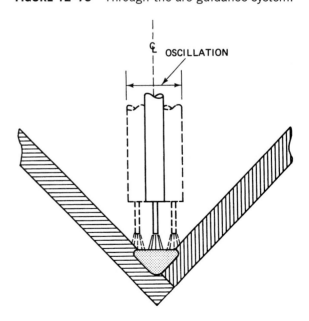

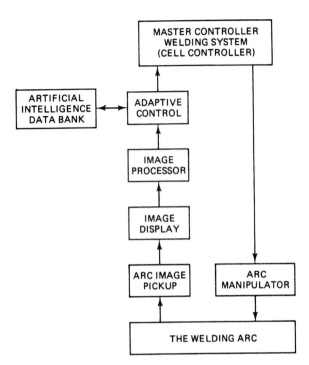

FIGURE 12-74 Block diagram of visual guidance system.

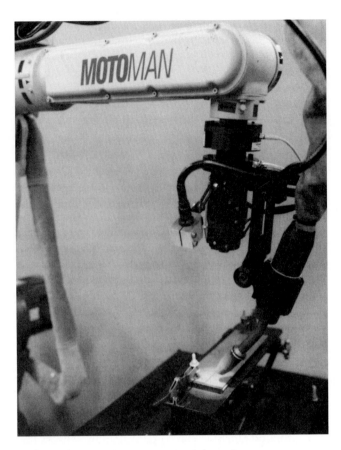

FIGURE 12-75 TV camera to pick up image.

Optical-visual systems have overcome the problem of viewing different colors and surface—bright, rusty, smooth, rough—that tend to confuse the sensor. They can pick up a very small joint in thin material, even when the joint separation is minimal. Many optical-visual systems are operating successfully, but there is no single system that can be applied universally to robotic welding applications. Different systems are designed for particular applications.

The image to be viewed can be the weld joint ahead of the arc, the arc itself, the weld pool under and behind the arc, or the light generated by the arc. The image selected depends on the viewing area and how it is lighted. The image can be picked up by means of a TV camera as shown in Figure 12-75 or by photodiodes arranged in a matrix array. The pickup method affects the image display and processing system. Two images are usually required. In some cases, images are triangulated to determine exact location. Fiber optics is used to transmit the image to the camera. The angle of viewing depends on the image processing method. Images can also be picked up by a system operating through the torch.

The picked-up image must be enhanced for better visibility. One method uses structured light, usually a pattern of bright and no light that can be directed from an oblique angle, as illustrated in Figure 12-76. The light source is usually a laser, which is more useful because it is monochromatic and can be highly focused. The incident arc light can be filtered out, which simplifies processing. If structured light from a point source is used, it is sometimes augmented by a beam from another direc-

tion, to facilitate triangulation for precise positioning. The image from the pickup device must be processed to provide a display. Digitizing the image is normal.

The most common image display device is the cathode ray tube, as shown by Figure 12-77. Image analysis requires the use of high-speed microprocessors. It also requires an extremely complex program to analyze all the data received and put them into a useful form so that the image can be used to make real-time changes based on variations in the weld.

Adaptive control systems require an interface between the sensor and the robot controller. It normally uses a database and an expert system to provide weld parameters when conditions change. The complete system will provide the necessary input and close the loop to produce the perfect weld.

Each optical-vision system has advantages and disadvantages. Each system is useful for certain applications. The following is a brief list of optical-visual systems:

* Reflecting light with the photodiode pickup
* Viewing the welding arc with a TV camera
* Viewing the molten weld pool either through the torch or adjacent to the torch
* Viewing the joint ahead of the arc
* Laser range-finding techniques (rastering)

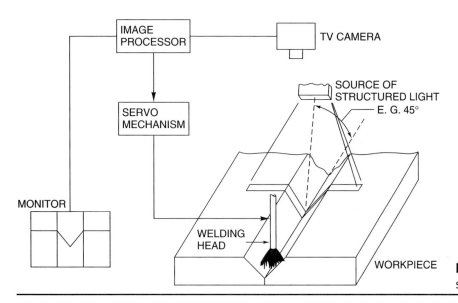

FIGURE 12–76 Structured light for seam tracking.

FIGURE 12–77 Image displayed on cathode ray tube.

Optical–visual sensing systems are continually being improved and are being more widely used for automated arc welding applications.

The economics of sensors must be considered. Only sensors needed to detect a repeatable problem should be used. For example, in a simple weldment, if the piece parts are always made accurately and holding devices are accurate, there is no need for a seam follower. If the location of the equipment, tooling, and piece parts are always accurate, there is no need for a seam-finding sensor. The more sensors involved, the more expensive and potentially troublesome the system becomes. Sensors must be small, robust, and durable. They must be able to withstand the hostile environment near the arc. They must be easy to connect to the controller. They must, im-

mediately and routinely, send a correction signal to the controller for immediate weld parameter correction.

The more commonly used systems are the electrode touch system for finding the joint combined with through-the-arc system for seam tracking. An optical–visual system is used for seam tracking when welding with GMAW. A system for maintaining the torch-to-work (standoff) distance with feedback from a capacitance sensor is used for laser cutting (see Figure 12–78). The robot control system must accommodate the different types of sensors. The system should be selected that fits the work.

12-10 TOOLING AND FIXTURES

A weldment is an assembly of piece parts. Automated assembly should be anticipated in the future. Parts should be designed so they can be inserted with a straight-line up-and-down motion. Robots with appropriate grippers will be used to assemble weldment parts in the properly designed holding fixture. If specific welds must be made prior to adding another piece part, consideration should be given to building the weldment in subassemblies and using a final welding operation to combine the subassemblies.

Fixturing for weldments should be coengineered by the product designer and the fixture designer. The weldment design should allow assembly of the weldment on a base with additional parts added to the top side. Self-jigging should be incorporated if at all possible. The addition of parts after a welding operation should be held to an absolute minimum.

Both designers must take into consideration and anticipate shrinkage and warpage inherent to weldments. When a weld cools, it shrinks and this causes warpage. One advantage of robotic welding is that distortion will

FIGURE 12-78 Torch-to-work distance for laser cutting.

normally be more uniform because the robot makes the weld in the same sequence every time. The designer should anticipate the pattern in which the welds will be made and attempt to balance welding to minimize warpage. The work-holding fixture for arc welding must accurately locate and hold the component parts of the weldment in their proper location for welding. It must locate the joints accurately and maintain the correct fitup. It speeds up the operation and improves the dimensional accuracy of the weldment. There are many types of welding fixtures, and there are many reasons for employing them.

Originally, welding fixtures were used for manual shielded metal arc welding (stick welding) to eliminate the time-consuming hand layout and tack welding of the parts. With manual welding, if a sufficient number of

weldments were made, the cost of the fixture would be recovered quickly due to the elimination of the setup and tack welding operation. The fixture also increased the accuracy of the weldment by eliminating the errors that could occur during the setup operation. With the advent of mechanized, automated, or robotic arc welding, the advantages of fixtures became even more pronounced. Efficient fixtures allow unattended welding. Once the parts are in the fixture and it is properly located, the automated or robotic welding process can start and operate without observation, monitoring, or supervision.

It is important to keep the fixture in operation as much of the time as possible to quickly recover its cost. In addition to the savings provided by the elimination of the layout operation, productivity is greatly increased because fixtures could be loaded and unloaded while the welding machine is making welds. Arc-on time is much higher, running as high as 90%. It also improves the safety of robotic welding since double-ended indexing positioners are normally used. Two fixtures are placed on the indexing positioner, which can rotate to position one fixture inside the welding cell, and the operator remain on the outside of the cell unloading and loading the other fixture.

There are basically two types of fixtures used for robotics arc welding: (1) those used for tack welding parts together, and (2) those that hold the weldment during the complete welding operation. The second type, sometimes called *strongbacks,* are heavier and more robust than tacking fixtures. They are used to hold the parts, maintain accurate alignment, and resist warpage of the weldment. The work-holding fixture is customized for each weldment. It is unique and must be reworked if the design of the weldment is changed.

For automated or robotic arc welding, the work-holding fixture is placed on an indexing positioner. This provides operator safety and increased productivity. Each end of the positioner may have two axes of motion, such as horizontal-vertical and/or tilt motion, which can be integrated by the controller. This allows the fixture and the work to move so that the welds can be made in the flat position and also maintains accessibility for the welding torch. The work-holding device on each end of an indexing positioner need not be for the same weldment. The robot can be programmed to weld different products on the two ends of the positioner. There is an exception to this. When rotary tables are used for loading, moving the work to the welding station, and moving the weldment to the unload station, the fixture must be for the same weldment.

The time required for unloading a finished weldment, loading the next weldment's piece parts, properly locating them, and clamping them must be less than the arc welding time for either weldment. This allows the operator time for inspecting, moving material, and so on. The total time from beginning to load parts to unloading the weldment is the factor that determines the robot system's production rate. Indexing fixtures change the index position when instructed by the operator rather than by the program of the robot.

Good fitup is required to obtain high-quality welds. The size of the root opening relates to the speed of welding. If the root opening is excessive, the root pass will burn through, resulting in the need to rework the joint. The piece part should be remade. Often a third type of fixture is used to attach the tack welded weldment to the positioner table to hold it in the proper location for the robotic program. It can also be used to facilitate quick attachment and easy removal. Clamping might not be necessary if the worktable of the positioner remains horizontal. Attention must be given to the positioner's weight capacity, which must also include the fixture.

In contrast to the previous information, keep in mind that extremely simple work-holding devices can be made quickly and will pay back after being used for a few batch runs of production. These fixtures can be built around a finished weldment. Assuming that the weldment is dimensionally accurate, the parts produced in the fixture will be accurate. Fixtures used for manual welding can be upgraded for automatic welding. They must be properly identified, stored, and called up again for the next production run of the same part. This keeps the automatic welding system running at full capacity, pays back quickly, and produces good-quality weldments.

It is essential that the weldment and the fixture provide accessibility for the welding gun to make the necessary welds. It may be necessary to redesign the fixture or weldment to allow weld location. This is why coengineering is essential.

Fixtures are normally purchased from a fixture builder or system company. The fixture builder or designer should be selected based on experience of building similar types of fixtures. Responsibility must be established and accepted. There must be complete understanding of the entire project by the fixture user and the fixture producer. This is done by having a meeting attended by the weldment designers, the welding production department, and the fixture designer or producer. It is necessary to agree on the productivity expected from the fixture, which would include welding time cycle, load time, unloading time, the annual quantity required, and the production lot size; that is, reach an agreement with all con-

cerned in order to obtain the desired fixture at a reasonable price. Written specifications are often used.

It is necessary to provide information concerning the weldment and its weight and size. If possible, show the exact weldment or a similar weldment to the fixture designer. It is necessary to agree on the dimensional tolerances that will be permitted. Show the dimensions and indicate which are critical and which are not. This allows the designer to determine how every piece part must be located and held, how much distortion can be allowed, how much material is allowed for finish machining, and so on. At the same time, it is worthwhile to review previous fixtures produced by the designer and producer.

The welding process to be used must be specified, as well as the size and type of each weld, the position of welding each joint, what type of work motion device will be used, the work envelope of the equipment or robot that is contemplated, the type of welding gun or torch, and whether multiple-pass welds will be required. Groove welds versus fillet welds should also be discussed, and in general, the weld details and the weld quality expected.

It is also desirable to indicate the target budget allowed for the fixture. Welding fixtures are expensive and can represent up to 50% of the total cost of the automatic welding cell. The fixture designer and producer should be able to provide an estimate of the cost of the fixture. Specifications should be understood and agreed to by all parties. Also, it is worthwhile to enter into a design-and-build contract between the parties. This would identify the fixture and weldment and finalize the specifications. The preliminary design should be reviewed and approved by the buyer. The fixture should then be manufactured and proven. This is done by making weldments with the mechanized equipment, and the resultant weldment must meet the specifications.

In view of the above, particularly on complex welding fixtures, complete trust must be established and responsibility accepted. Both parties must be satisfied. As the weldments become more complex, the fixture becomes more complex, and the cost goes up accordingly. Weld fixtures or work-holding devices should be designed and built by people with experience. Properly used, fixtures will pay for themselves quickly.

 QUESTIONS

12–1. What are the advantages of automatic arc welding?

12–2. Explain the man-machine relationship in arc welding.

12–3. What is an arc motion device? Name different types.

12–4. What is a work motion device? Name different types.

12–5. Discuss standardized arc welding machines. Name different types.

12–6. What is a dedicated automatic arc welding machine?

12–7. What was the product welded on the first dedicated arc welding machine?

12–8. What is lot size? What is the difference between job shop, batch, and mass production?

12–9. What machine provides for flexible manufacturing of weldments?

12–10. Define a robot.

12–11. What are the popular types of robots?

12–12. What is a popular application for robotic spot welding?

12–13. What is the robot work envelope?

12–14. How many body axes does a jointed robot have?

12–15. How many wrist axes can a robot have?

12–16. Discuss coordinated motion in a robot and in a robot and positioner.

12–17. What types of products can be welded on a robot? Give examples.

12–18. What is the disadvantage of a touch seam follower?

12–19. Explain how a through-the-arc sensor works.

12–20. What is coengineering? Why is this desirable when designing a new product to be robot welded?

ELECTRODES AND FILLER METALS

13-1 TYPES OF WELDING CONSUMABLES

There are many types of materials used when making welds. These welding materials are generally categorized under the term **filler metals,** defined as "the metal to be added in making a welded, brazed, or soldered joint." The filler metals are used or consumed and become a part of the finished weld. The definition has been expanded and now includes electrodes considered nonconsumable such as tungsten and carbon, and fluxes for brazing, submerged arc welding, electroslag welding, and so on. The term *filler metal* does not include electrodes used for resistance welding, nor does it include the studs involved in stud welding.

The American Welding Society has issued 31 specifications covering filler materials (Figure 13-1). This figure also shows the welding process for which each specification is intended. These specifications are periodically updated and a two-digit suffix indicating the year issued is added to the specification number. Additional specifications are added from time to time.

Most of the industrial countries issue filler metal specifications. A correlation of national filler metal specifications is shown in Figure 13-2.

In the United States, the American Welding Society provides filler metal specifications. They are approved by ANSI (the American National Standards Institute) and become an American national standard. The American Society of Mechanical Engineers (ASME) in their "Pressure Vessel and Boiler Code" issues filler metal specifications that are identical with AWS specifications. ASME adds the prefix letters SF to the specification number.

In Canada, the Canadian Standards Association (CSA) issues filler metal specifications that are in general agreement with the AWS specifications. However, Canada has switched to the metric system, and the familiar AWS specifications using E60XX, E70XX, and so on, which are related to the pounds per square inch strength levels, no longer apply. The Canadians changed the ksi two digits to a three-digit number standing for the tensile strength in MPa (megapascal). The three-digit classification number is slightly lower than the minimum megapascal strength requirement. The CSA also shows the diameter of the core

AWS Specification	Specification Title	OAW	SMAW	GTAW[a]	GMAW	SAW	Other
A5.1	Carbon-steel covered arc welding electrodes		X				
A5.2	Iron and steel gas welding rods	X					
A5.3	Aluminum and aluminum alloy arc welding electrodes		X				
A5.4	Corrosion-resisting chromium and chromium-nickel-steel covered welding electrodes		X				
A5.5	Low-alloy-steel covered arc welding electrodes		X				
A5.6	Copper and copper alloy covered electrodes		X				
A5.7	Copper and copper alloy welding rods	X		X			PAW
A5.8	Brazing filler metal						BR
A5.9	Corrosion-resisting chromium and chromium-nickel bare and composite metal cored and standard arc welding electrodes and rods			X	X	X	PAW
A5.10	Aluminum and aluminum alloy welding rods and bare electrodes	X		X	X		PAW
A5.11	Nickel and nickel-alloy covered welding electrodes		X				
A5.12	Tungsten arc welding electrodes			X			PAW
A5.13	Surfacing welding rods and electrodes	X		X			CAW
A5.14	Nickel and nickel alloy bare welding rods and electrodes	X		X	X	X	PAW
A5.15	Welding rods and covered electrodes for welding cast iron	X	X				CAW
A5.16	Titanium and titanium alloy bare welding rods and electrodes			X	X		PAW
A5.17	Bare carbon steel electrodes and fluxes for submerged arc welding					X	
A5.18	Carbon steel filler metals for gas shielded arc welding			X	X		PAW
A5.19	Magnesium alloy welding rods and bare electrodes	X		X	X		PAW
A5.20	Carbon steel electrodes for flux cored arc welding						FCAW
A5.21	Composite surfacing welding rods and electrodes	X	X	X			
A5.22	Flux cored corrosion-resisting chromium and chromium-nickel steel electrodes						FCAW
A5.23	Bare low-alloy steel electrodes and fluxes for submerged arc welding					X	
A5.24	Zirconium and zirconium alloy bare welding rods and electrodes			X	X		PAW
A5.25	Consumables used for electroslag welding of carbon and high-strength low-alloy steels						ES
A5.26	Consumables used for electrogas welding of carbon and high-strength low-alloy steels				X (EG)		FCAW (EG)
A5.27	Copper and copper alloy gas welding rods	X					
A5.28	Low-alloy steel filler metals for gas shielded arc welding			X	X		PAW
A5.29	Low-alloy-steel flux cored welding electrodes						FCAW
A5.30	Consumable inserts			X			
A5.31	Fluxes for brazing and braze welding						BR

[a]If GTAW is shown, the specifications will also apply to PAW even though not stated.

FIGURE 13–1   AWS filler metal specifications and welding processes.

Filler Metal Type	United States, AWS	Canada, CSA	International, ISO	United Kingdom, BS	Germany, DIN	Japan, JIS	U.S. Military, US-MIL
Covered electrodes Mild steel	A5.1	W48.1	670 R635 547	639 +1719	1913	Z3210 Z3211	E-15599 E-22200
Low-alloy steel	A5.5	W48.3	1045	2493 1719	1913	Z3212	E-22200/1
Stainless steel	A5.4	W48.2		2926	8556	Z3221	E-13080 E-22200/2A
Surfacing	A5.13				8555	Z3251	E-19141
For cast iron	A5.15		1163		8573	Z3252	
Aluminum	A5.3			1616	1732		E-15597
Copper	A5.6				1733		E-13191 E-21659
Nickel	A5.11				1736		E-21562
Bare, solid Stainless steel	A5.9		1159	2901	8556	Z3321	E-19933 R-5031
Steel	A5.17			4165		Z3311	E-18193
	A5.18	W48.4		1453	8559		E-23765
Flux cored steel	A5.20						
	A5.23	W48.4					E-24403

FIGURE 13–2 National specifications (International Filler Metal Specifications ANS IFS cross reference list).

wire and the length in millimeters. The Canadians have issued additional categories or classifications of covered electrodes. These are described in detail in the CSA filler metal specifications shown in Figure 13-2.

In Europe, the European Committee for Standardization (CEN) is working toward standards harmonization. They are providing European Community (EC) standards that will replace the individual standards of each of the 18 member states of the European Community and the European Free Trade Association (EFTA). Once these are completed, the British, German, French, and other European countries will discontinue publishing specific standards for welding consumables.

The International Standards Organization (ISO) also issues filler metal specifications. Many of the less industrialized nations utilize specifications of the industrialized countries or ISO standards. The ISO standards are available from the welding or standardization association of each country. AWS provides an "International Filler Metal Specification Cross Reference List" AWS-IFS.

Filler metals can be classified into four basic categories. These are:

1. Covered electrodes
2. Solid (bare) electrode wire or rod
3. Fabricated (tubular or cord) electrode wire
4. Fluxes for welding

The AWS specifications are written to provide specific chemical composition of the material and the mechanical properties of the deposited weld metal. The AWS specifications utilize similar methods and testing techniques so that there is consistency between all of the filler metal specifications. The mechanical properties of deposited weld metal are determined based on a standardized welding procedure, in a specified welding joint detail, to produce weld specimens for testing. Specifications may also require other properties, such as toughness, quality standards, and, in some cases, standards of porosity. Most specifications include usability factors showing the welding position that the electrode or filler metal is designed for, the welding current that should be used, and in the case of covered electrodes, the type of coating. Size and packaging information is also provided.

The American Welding Society does not test or approve filler metals. The society provides the specifications, which are voluntary conformance standards. It is the manufacturer of the filler material who guarantees that the product conforms to a specific AWS specification and classification. AWS provides charts showing comparison of brand names. AWS also provides "Filler Metal Procurement Guidelines" A5.01.

AWS also has available a CD "The Filler Metal Data Manager," which includes suppliers of filler metals, comparison charts, specifications, et al.

Weld Metal Certifications

For special applications, filler metals are tested and certified to be in conformance with a specific specification. In some cases, filler metals to be used for ships, nuclear reactors, vessels, highway bridges, and certain military products require certification. For military construction, approvals are granted when the specific electrodes are in conformance with the applicable military specification. Tests of this type are usually witnessed by a government inspector. Approved products are then placed on a qualified products list (QPL).

Approval of filler metals for ship construction is similar. A classification society, such as the American Bureau of Shipping, requires that one of their representatives, called a surveyor, witness the welding of test plates using electrodes selected at random. The surveyor also witnesses the testing of the weld specimens. Classification societies may have special mechanical property requirements; these usually include low-temperature impact data. Approvals are granted for different filler materials based on strength and impact requirements and for welding different classes of steels for ships. The classification society publishes lists of approved electrodes and filler metals manufactured by different companies. Retention of the filler metals on the approved list is subject to annual tests. Filler metal approvals include covered electrodes, submerged arc electrode wire with flux combinations, and flux cored arc welding electrodes with gas combinations.

Certification is handled differently from approvals. Certification is required for nuclear work and for certain types of military work. In these cases, a specific batch of electrodes or heat of wire made at one time is tested in accordance with either an AWS or MIL specification. The test results must then be certified by the manufacturer and provided to the user. A certification is used only for the specific batch or lot of filler materials made at one time and covered by the test result data. In many cases, the user must previously have approved the manufacturer prior to using filler materials produced by that manufacturer. This will require an audit by the user of the manufacturer to make sure that uniform quality manufacturing procedures are maintained and that quality control procedures provide for strict control and traceability of materials used in the manufacture of the filler material. After the producer has been approved, the certification test may still be required. Traceability of all items is necessary so that the user can provide traceability of all materials used to manufacture their products.

Filler metal specifications are of immense value to both producer and user. They allow the user to select the proper filler material to be used to manufacture all types of products. Specifications assure the user that the deposited weld metal, when normally applied, will provide the strength levels indicated by the specification and classification. Specifications are of value to the producer, since they provide standardization of testing methods and procedures and also since they provide categories and classifications to meet the needs of most users.

13-2 COVERED ELECTRODES

The covered electrode is a very popular type of filler metal used in arc welding. The identification of electrode types, the selection of electrodes for specific applications, and the usability of covered electrodes was discussed in Section 6-3.

The composition of the covering on the electrode determines the usability of the electrode, the composition of the deposited weld metal, and the specification of the electrode. The composition of coatings on covered arc welding electrodes has been surrounded in mystery, and little information has been published. The formulation of electrode coatings is very complex, and while it is not an exact science it is based on well-established principles of metallurgy, chemistry, and physics, tempered with experience.

The original purpose of the coating was to shield the arc from the oxygen and nitrogen in the atmosphere. It was subsequently found that ionizing agents could be added to the coating that helped stabilize the arc and made electrodes suitable for alternating current welding. It was found that silicates and metal oxides helped form slag, which would improve the weld bead shape because of the reaction at the surface of the weld metal. The deposited weld metal was further refined and its quality improved by the addition of deoxidizers in the coating. In addition, alloying elements were added to improve the strength and provide specific weld metal deposit composition. Finally, iron powder has been added to the coating to improve the deposition rate.

An electrode coating is designed to provide as many as possible of the following desirable characteristics. Some of these characteristics may be incompatible, and therefore compromises and balances must be designed into the coating. These desirable characteristics are:

1. Specific composition of the deposited weld metal
2. Specific mechanical properties of the deposited weld metal
3. Elimination of weld metal porosity
4. Elimination of weld metal cracking
5. Desirable weld deposit contour
6. Desirable weld metal surface finish (i.e., smooth, with even edges)
7. Elimination of undercut adjacent to the weld
8. Minimum spatter adjacent to the weld
9. Ease of manipulation to control slag in all positions

10. Stable welding arc
11. Penetration control (i.e., deep or shallow)
12. Initial immediate arc striking and restriking capabilities
13. High rate of metal deposition
14. Elimination of noxious odors and fumes
15. Reduced tendency of the coating to pick up moisture when in storage
16. Reduced electrode overheating during use
17. Strong, tough, durable coating
18. Easy slag removal
19. Ability to ship well and store indefinitely

These requirements must be achieved at the minimum possible cost. In addition, the formulation must be manufacturable with conventional extrusion equipment at high production rates. No single electrode type will meet all of the foregoing requirements, and there is no one single "universal electrode." Instead, there is a variety of electrode types, each having certain desirable characteristics.

The coatings of electrodes for welding mild and low-alloy steels may have from six to twelve ingredients, such as:

⁂ *Cellulose:* to provide a gaseous shield with a reducing agent. The gas shield surrounding the arc is produced by the disintegration of cellulose.

⁂ *Metal carbonates:* to adjust the basicity of the slag and to provide a reducing atmosphere.

⁂ *Titanium dioxide:* to help form a highly fluid but quick-freezing slag. It will also provide ionization for the arc.

⁂ *Ferromanganese and ferrosilicon:* to help deoxidize the molten weld metal and to supplement the manganese content and silicon content of the deposited weld metal.

⁂ *Clays and gums:* to provide elasticity for extruding the plastic coating material and to help provide strength to the coating.

⁂ *Calcium fluoride:* to provide shielding gas to protect the arc, adjust the basicity of the slag, and provide fluidity and solubility of the metal oxides.

⁂ *Mineral silicates:* to provide slag and give strength to the electrode covering.

⁂ *Alloying metals:* include nickel, molybdenum, chromium, and so on, to provide alloy content to the deposited weld metal.

⁂ *Iron or manganese oxide:* to adjust the fluidity and properties of the slag. In small amounts, iron oxide helps stabilize the arc.

⁂ *Iron powder:* to increase the productivity by providing additional metal to be deposited in the weld.

By using different amounts of these constituents, it is possible to provide an infinite variety of electrode coatings. The binder used for most electrode coatings is sodium silicate, which will chemically combine and harden to provide a tough, strong coating. The design of the coating provides the proper balance to give the electrode specific usability characteristics and to provide specific weld deposit chemistry and properties. In general, the different makes of electrodes that meet a particular classification have similar compositions.

Manufacturing

Covered arc welding electrodes are manufactured at machine-gun-like speeds. There are three basic parts of a covered electrode: the core wire, the chemicals and minerals that comprise the coating, and the liquid binder that hardens and holds it all together. The steps required to manufacture a covered electrode are shown in Figure 13–3, which is a flowchart showing each of the major steps involved.

The core wire for mild steel and low-alloy steel electrodes is normally a low-carbon steel having a carbon content of about 0.10% carbon, low manganese and silicon content, and the minimum amount of phosphorus and sulfur. Ingots of this composition are produced at the steel mill; they are hot rolled and reduced in size to billets. These are then taken to a bar mill and rolled into small-diameter rods that range in size from 1/4 in. (6.4 mm) to 3/8 in. (9.5 mm) in diameter. This product, which is known as *hotrolled wire rod,* is then taken to the wire drawing mill and drawn into the appropriate diameters for covered electrodes. After the wire has been drawn to the proper diameter, it is straightened and cut to the proper length. The lengths vary according to the size and range from 12 to 14 in. in the United States and from 200 to 500 mm in length elsewhere.

The coating is made of different chemicals and minerals obtained throughout the world. They are inspected and ground to the proper mesh size. The specific amount of each chemical is weighed and mixed together in the dry condition. After sufficient dry mixing, the proper amounts of liquid, binder, and water are added, and mixing is continued in the wet stage. Mixing is completed when it reaches the proper consistency. This material is then placed in a press, where it is formed into large briquets of moist flux coating material.

The briquets of coating material and the cut and straightened lengths of core wire are brought together at the extrusion press. The cut core wires are fed into the extrusion die by an automatic feeder. Simultaneously, the press feeds flux into a chamber and extrudes the coating onto the core wire as it passes through the chamber. The extrusion die holder must be adjusted with extreme accuracy so that the flux flows uniformly to make it concentric with the core wire. The coated electrode emerges

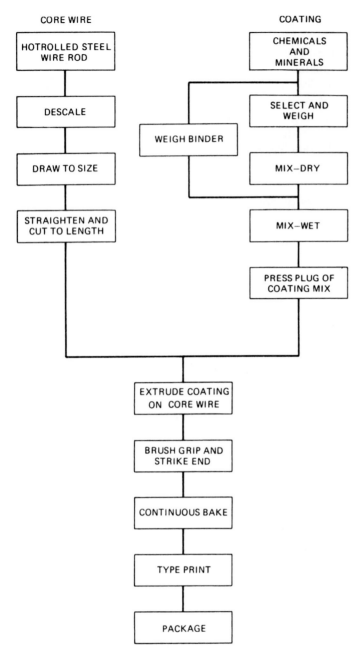

CORE WIRE

```
HOTROLLED STEEL
   WIRE ROD
        │
        ▼
     DESCALE
        │
        ▼
   DRAW TO SIZE
        │
        ▼
 STRAIGHTEN AND
 CUT TO LENGTH
```

COATING

```
  CHEMICALS
     AND
   MINERALS
        │
        ▼
  SELECT AND
    WEIGH
        │
        ▼
    MIX–DRY
        │
        ▼
    MIX–WET
        │
        ▼
  PRESS PLUG OF
  COATING MIX
```

WEIGH BINDER

```
EXTRUDE COATING
 ON CORE WIRE
        │
        ▼
 BRUSH GRIP AND
   STRIKE END
        │
        ▼
 CONTINUOUS BAKE
        │
        ▼
   TYPE PRINT
        │
        ▼
    PACKAGE
```

FIGURE 13–3 Flowchart for manufacturing covered electrodes.

FIGURE 13–4 Electrodes being manufactured.

from the die of the extrusion press at a rate of approximately 10 per second. They drop onto a conveyor where power brushes remove a portion of the coating at the grip end and clean the coating from the strike end. The electrodes move on the conveyor into a drying oven for a sufficient length of time for the coating to solidify and toughen. At the exit end of the oven, each electrode is individually printed with the AWS classification number and then inspected and placed in boxes and packed for shipment. The entire operation from start to finish is continuous. Figure 13–4 shows the electrodes during the manufacturing operation.

Care, Storage, and Reconditioning of Covered Electrodes

Covered electrodes can be easily damaged. Each electrode should be treated with care prior to its use. Rough handling in shipment or in storage can cause a portion of the coating to crack loose from the core wire and make the electrode unsuitable. Bending most electrode types will cause the coating to break loose from the core wire. The electrode should not be used where the core wire is exposed.

Electrodes may become unusable if they are exposed to moisture for an extended length of time. The coatings on some types of electrodes absorb moisture when exposed to humid atmospheres. Cellulose, rutile, and acid electrodes are fairly insensitive to moisture and can tolerate quite high moisture content without the risk of porosity in the weld. The coatings of low-hydrogen electrodes, particularly those of the EXX16 and EXX18 type, pick up moisture quickly when exposed to a high-humidity atmosphere. Since these electrodes are dried at high temperature in a low-moisture atmosphere, they are more sensitive to pickup. Stainless steel electrodes are in this same category.

If electrodes, even in unopened cardboard containers, are left outdoors, they will pick up moisture due to the change of temperature and humidity from day to night. The moisture is absorbed by the packing and in time is gradually absorbed by the coatings of the electrodes inside the package. Efforts to prohibit this are made by wrapping the electrodes in plastic liners or by using vaporproof or metal containers. These provide better protection for the electrodes.

Once the container is opened, the electrode should be stored in heated ovens, shown in Figure 13–5. These larger type ovens are kept in a storage room or electrode issuing area. Small or portable ovens, shown in Figure 13–6, are carried by the welder to the work site. These are powered by 115 V ac or by the output current of a conventional welding machine. These are useful for field welding. Nothing else should be stored in electrode ovens, especially those containing low-hydrogen electrodes. Food must not be placed in electrode ovens since the moisture given off during cooking would damage low-hydrogen coatings. Low-hydrogen electrodes must not be stored in ovens that hold electrodes of other classifications. The table shown in Figure 13–7 shows temperatures for storing electrodes in ovens and for reconditioning electrodes. Special ovens are available for storing submerged arc flux in the shop to keep it dry. Some codes have strict requirements for using, storing, and redrying low-hydrogen electrodes.

Damp electrodes are difficult to distinguish by the welder. It is easier to recognize the problem based on storage conditions. It is also easy to recognize the problem by reviewing x-rays of weld metal deposited by damp electrodes. The weld metal will be porous if the coatings are damp. It is sometimes possible to shake three or four low-hydrogen electrodes together and listen to the sound as they rattle against each other. If the electrode coatings are dry or contain only small amounts of moisture, a clear,

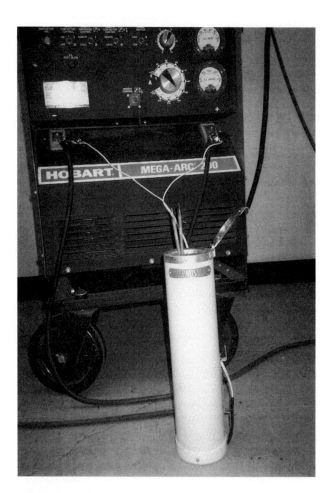

FIGURE 13–6 Portable electrode storage oven.

shrill metallic sound will be heard. Damp electrodes have a hollow sound, which is quite different. Experience in testing electrodes in this way will help to distinguish these two different sounds. When welding with an electrode with a damp coating, a fierce crackling or explosive sound may be heard. If the electrode is extremely damp, condensed vapor may be seen while welding. If the electrode is not completely consumed, the coating on the remaining part of the electrode will show longitudinal cracks.

Electrodes that are only slightly damp can be heated by shorting them against the work for a few seconds just before beginning to weld. For reconditioning electrodes, special ovens are available. These are set at specific temperatures for specific types of electrode coatings (Figure 13–7). The baking cycle for reconditioning electrodes should not exceed 4 hours. The heating rate in the oven is not critical. Electrodes can be taken from room temperature and placed in an oven without affecting the properties of the deposited weld metal. The maximum temperature for any low-hydrogen electrode is 800°F (427°C). Some ingredients in the coating tend to oxidize if the temperature is raised above this figure. The holding time at the maximum temperature should be at least 30 minutes. This ensures that the electrodes are up to the

FIGURE 13–5 Large electrode storage oven.

Electrode Classification	Recommended Storage			
	Unopened Boxes	Open Boxes	Holding Oven	Reconditioning
E-XX10	Dry at room temp.	Dry at room temp.	Not recommended	Not done
E-XX11	Dry at room temp.	Dry at room temp.	Not recommended	Not done
E-XX12	Dry at room temp.	Dry at room temp.	Not recommended	Not done
E-XX13	Dry at room temp.	Dry at room temp.	Not recommended	Not done
E-XX14	Dry at room temp.	150–200° F	150–200° F	250–300° F,
E-XX20	Dry at room temp.	150–200° F	150–200° F	1 hour
E-XX24	Dry at room temp.	150–200° F	150–200° F	
E-XX27	Dry at room temp.	150–200° F	150–200° F	
E-60 or 7015	Dry at room temp.	250–450° F	150–200° F	500–600° F,
E-60 or 7016	Dry at room temp.	250–450° F	150–200° F	1 hour
E-7018	Dry at room temp.	250–450° F	150–200° F	
E-7028	Dry at room temp.	250–450° F	150–200° F	
E-80 and 9015	Dry at room temp.	250–450° F	200–250° F	600 700° F,
E-80 and 9016	Dry at room temp.	250–450° F	200–250° F	1 hour
E-80 and 9018	Dry at room temp.	250–450° F	200–250° F	
E-90-12015	Dry at room temp.	250–450° F	200–250° F	650–750° F
E-90-12016	Dry at room temp.	250–450° F	200–250° F	1 hour
E-90-12018	Dry at room temp.	250–450° F	200–250° F	
E-XXX-15,16, or 17	Dry at room temp.	250–450° F	150–200° F	450° F,
Stainless	Dry at room temp.	250–450° F	150–200° F	1 hour

FIGURE 13–7 Storage and reconditioning electrodes.

oven temperature. The cooling rate is not critical; however, reconditioned electrodes should not be taken from the oven and allowed to cool until the oven has come down to approximately 300° F (149° C). Electrodes should not be reconditioned by heating more than three times. Going through the extra heating cycle tends to weaken the silicate binder and the coating will eventually become weak and fragile and will chip off easily.

Electrodes should be stored in a special storeroom with controlled atmosphere. The relative humidity should be maintained at 40% or less. This can be accomplished by sealing the room and installing a dehumidifier.

When low-hydrogen electrodes are issued from the controlled atmosphere storeroom, they should be used within 2 hours. When this cannot be done, individual ovens should be provided for each welder. They can then be left in the heated oven until the electrode is used. All low-hydrogen electrodes not used during a work shift should be returned to the holding oven. For critical work, special controls are instituted to maintain dry electrodes.

Electrodes can be damaged by aging. Very old electrodes of most types will have a furry surface on the coating, usually white. This is from the crystallization of the sodium silicate. This surface is normally harmless for mild steel low-hydrogen electrodes. They should not be used for extremely critical work. If iron powder–type electrodes are old, rust may form on the iron powder due to moisture absorbed in the coating. If the core wire is rusty,

it is evident that too much moisture may have been absorbed in the coating.

Deposition Rates

The different types of electrodes have different deposition rates, as a result of the composition of the coating. The electrodes containing iron powder in the coating have the highest deposition rates. The percentage of iron powder is confusing when comparing electrodes produced in Europe with those produced in the United States. In the United States the percentage of iron powder in a coating is in the range of 10% to 50%. This is based on the amount of iron powder in the coating versus the coating weight. This is shown in the formula

$$\% \text{ of iron powder} = \frac{\text{weight of iron powder}}{\text{total weight of coating}} \times 100$$

The percentages mentioned above are related to the requirements of the AWS specifications. The European method of specifying iron powder is based on the weight of deposited weld metal versus the weight of the bare core wire consumed, or

$$\% \text{ iron powder} = \frac{\text{weight of deposited metal}}{\text{weight of bare core wire}} \times 100$$

Thus if the weight of the deposit were double the weight of the core wire, it would indicate a 200% deposition efficiency even though the amount of the iron powder in coating represented only half of the total deposit. The 30% iron powder formula used in the United States would produce a 100% deposition efficiency using the European formula. The 50% iron powder electrode figured on U.S. standards would produce an efficiency of approximately 150% using the European formula.

Quality and Defects

Quality control in manufacturing of covered electrodes starts at the point of receiving chemicals and minerals, the binders, and the hotrolled wire rod. The chemicals must meet rigid specifications and are checked when they are received. The wire rod, which is checked on a continuous basis, must also meet stringent specifications. Grind sizes of chemicals, cleanliness of mixing containers, and so on, are routinely inspected. The adjustment of the extrusion die to maintain concentricity of the electrodes is checked often. Electrodes are checked after baking for coating concentricity. The surface and structure of the coating is also inspected, and each lot is checked by welding to determine that it meets the specifications.

A common complaint of quality of electrodes is *fingernailing*, which is the name given to the burning off of an electrode faster on one side than on the other. The welder assumes that fingernailing means a nonconcentric electrode; however, other factors can create the fingernailing. Fingernailing is most common when using direct current and is more evident with the smaller electrodes, $\frac{1}{8}$ in. (3.2 mm) and $\frac{5}{32}$ in. (4.0 mm), when used at low currents. This condition can be aggravated if the coating is not concentric with the core wire. This can be checked by removing the coating from the one side of the core wire and measuring the core wire and covering to the other side, and then removing coating on the opposite side of the electrode and measuring the electrode core wire and covering on the other side. Measuring should be done with a micrometer. Normally, electrodes are concentric within 0.002 to 0.003 of an inch (0.05 to 0.07 mm). Often fingernailing results from arc blow, welding current too low, incorrect electrode angle, unbalanced joint preparation, and in some cases, uneven moisture pickup in the coating, which might be greater on one side than the other. A quick check for fingernailing during welding is to stop welding when fingernailing is encountered and rotate the electrode in the holder 180°. Continue to weld and see if fingernailing continues on the same side of the electrode. If it does, the coating is probably off center. If it does not but instead fingernails off the other side, arc blow or one of the other factors already mentioned is the reason. Arc blow is a more frequent cause of fingernailing than off-center electrodes. When welding with lower than normal current, fingernailing will appear because there is insufficient arc force to overcome minor arc blow. The other factor is electrode angle,

which again can be checked by revolving the electrode 180° in the holder. Moisture can be checked as mentioned previously.

The welder or the welding supervisor may wish to compare different brands of electrodes. Many electrode manufacturers provide such charts; however, the most complete chart is published by the American Welding Society and is known as the "Filler Metals Comparison Chart" (AWS FMC).

13-3 SOLID ELECTRODE WIRES

Solid metal wires were first used for oxyfuel gas welding to add filler metal to the joint. These wires or rods were provided in straightened lengths approximately 36 in. (1m) long. Later, solid wire was provided in coils for "bare wire" automatic arc welding and later for submerged arc and electroslag welding. The newest process to use solid bare wire is gas metal arc welding, which uses relatively small-diameter electrode wires.

The manufacture of wire for welding electrodes or rod is essentially the same except that the straighten and cut operation is added for a welding rod. A simplified flowchart of the manufacturing operations for solid mild steel electrode wire is shown in Figure 13–8. The most complex portion is the drawing operation, shown partially in Figure 13–9. The drawing of steel wires and nonferrous wires is

---

FIGURE 13–8 Flowchart for manufacturing solid electrode

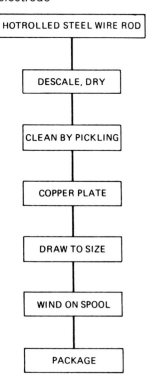

FIGURE 13-9 Wire drawing operation.

essentially the same; however, different amounts of reduction per drawing die, different drawing lubricants, different heat treatments, and so on, are involved.

The solid steel electrode wires may not be "bare." Most suppliers provide a very thin copper coating on the wire. The copper coating is for several purposes. It improves the current pickup between contact tip and the electrode. It aids drawing and helps prevent rusting of the wire when it is exposed to the atmosphere.

Solid steel electrode wires are available without the copper coating. Copper is undesirable in the welding shop atmosphere and therefore some suppliers provide gas metal arc electrode wires with an organic coating rather than copper. Non-copper-coated wire is often used for electroslag welding. Solid electrode wires are made of various stainless steel analyses, aluminum alloys, copper alloys, and other metals. Specifications for these electrodes are shown in Figure 13-1.

When the wire is cut and straightened, it is called a welding rod, which is a form of filler metal used for welding or brazing that does not conduct the electrical current. If the wire is used in the electrical circuit, it is called a welding electrode and is defined as a component of the welding circuit through which current is conducted. Rods are indicated by the prefix *R* for gas welding; rods for brazing have the prefix *RB*.

The American Welding Society's specifications are used for specifying bare welding rod and electrode wires. There are also military specifications such as the MIL-E or MIL-R types and federal specifications, and Aerospace Materials Specification (AMS) specifications. The particular specification involved should be used for specifying filler metals.

The most important aspect is the composition. The chemistry is the composition of the electrode or rod itself. The specifications provide the limits of composition and mechanical property requirements.

The bare electrode wires are identified in several different ways. When the wire is coiled on a spool, a label is placed on the spool identifying the size and type of the electrode. When it is in coils or in drums, the label is placed on the drum or on a liner on the inside diameter of the coil.

For straight lengths of welding rod, two systems are used. For large-diameter nonferrous rods, the classification number is stamped in the metal. When the diameter is too small, tags are stuck to each individual rod that shows the classification number, as shown in Figure 13-10. Color coding has been used but is being supplanted by tags. In all cases the container holding the rods carries an identification label.

In the case of coiled electrode wire, a maximum and minimum of cast and helix are specified. Normally, the cast is greater than the diameter of the package coil. This helps the electrode wire feed through cables and guns. The arc will tend to wander as the wire comes out of the welding gun tip if helix is too great.

Occasionally, on copper-plated wires, the copper may flake off in the feed roll mechanism and create problems. It may plug liners and contact tips. Therefore, a light

FIGURE 13–10   Identification of welding rods—tags.

copper coating is desirable. The electrode wire surface should be free of dirt and drawing compounds. This can be checked by using a white cleaning tissue and pulling a length of wire through it. Too much dirt will clog the liners, will reduce current pickup in the tip, and may create erratic welding operation.

Temper or strength of the wire can be checked in a testing machine. Wire of a higher strength will feed better through guns and cables. The minimum tensile strength recommended by the specification is 140,000 psi (98 kg/mm^2).

Feedability is a measure of the ease with which the wire can be fed through a gun and cable assembly. It depends on all the factors just mentioned.

13-4 CORED ELECTRODE WIRES

The outstanding performance of the flux cored arc welding process is made possible by the design of the cored electrode. This inside–outside electrode consists of a metal sheath surrounding a core of fluxing and alloying compounds. The compounds contained in the electrode perform essentially the same functions as the coating on a covered electrode (i.e., deoxidizers, slag formers, arc stabilizers, alloying elements, and may provide shielding gas).

There are three reasons why cored wires are developed to supplement solid electrode wires of the same or similar analysis.

1. There is an economic advantage. Solid wires are drawn from steel billets of the specified analyses. These billets are not readily available and are expensive. Also, a single billet might provide more solid electrode wire than needed.
2. Tubular wire production method provides versatility of composition and is not limited to the analysis of available steel billets.
3. Tubular electrode wires are easier for the welder to use than solid wires of the same deposit analysis, especially for welding pipe in the fixed position.

Figure 13–11 shows cross-sectional views of flux cored electrodes. The tubular type is the most popular. The sheath or steel portion of the flux cored wire comprises 75% to 90% of the weight, and the core material represents 10% to 25% of the weight of the electrode. For a covered electrode, the steel represents 75% of the weight and the flux 25%. This is shown in more detail in Figure 13–12.

More flux is used on covered electrodes than in a flux cored wire to do the same job. This is because the covered electrode coating contains binders to keep the coating intact and also contains agents to allow the coating to be extruded.

The manufacture of the flux cored electrode is an extremely technical and precise operation requiring specially designed machinery. Figure 13–13 shows the simplified flowchart of the manufacturing operation.

Figure 13–14 shows a simplified version of the apparatus for producing tubular cored electrodes. A thin, narrow, flat, low-carbon steel strip passes through forming rolls, which make it into a U-shaped section. This U-shaped steel is passed through a special filling device where a measured amount of the formulated granular core material is added. The flux-filled U-shaped strip then passes through closing rolls, which form it into a tube and tightly compress the core materials. This tube is then pulled through drawing dies which reduce its diameter and further compress the core materials. Drawing tightly seals the sheath and secures the core materials inside the tube, thus avoiding discontinuities of the flux. The electrode may or may not be baked during, or between, drawing operations. This depends on the type of electrode and the type of materials enclosed in the sheath.

Metal Cored Electrodes

Metal cored electrodes are composite tubular filler metal electrodes consisting of a metal sheath and a core of various powdered materials producing no more than slag islands on the face of the weld bead. These types of electrodes contain a very minor amount of fluxing ingredients

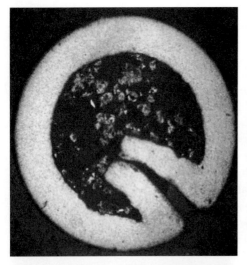

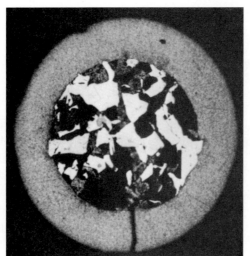

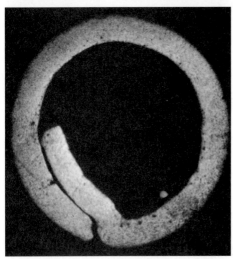

FIGURE 13–11 Cross-sectional views of various types of flux cored electrodes.

		Flux Cored Electrode Wire (E70T-1)	Covered Electrode (E7016)
By area:	Flux	25	55
	Steel	75	45
By weight:	Flux	15	24
	Steel	85	76

FIGURE 13–12 Summary of flux-to-steel ratios (percentages).

or no fluxing ingredients or gas formers. This is why external gas shielding is required. Metal core electrodes produce very high deposition efficiencies of 95% or greater. They provide high deposition rates and excellent operator appeal because of low spatter levels, low smoke level, and minimum slag coverage. They normally have spray transfer and have good mechanical properties.

The drawing operations produce the various sizes of electrodes. The normal diameters are $\frac{1}{8}$ in. (3.2 mm), $\frac{7}{64}$ in. (2.8 mm), $\frac{3}{32}$ in. (2.4 mm), $\frac{5}{64}$ in. (2.0 mm), $\frac{1}{16}$ in. (1.6 mm), 0.045 in. (1.1 mm), and now 0.035 in. (0.9 mm). The finished cored electrode is packaged as a continuous coil, on spools, or in round drums.

13-5 PACKAGING OF ELECTRODE WIRES

Filler materials are packaged in a variety of forms to meet the user's welding equipment, storage, and handling requirements. The American Welding Society has established standards for some spool and coil sizes, but there is no national standard for packaging of electrode wires. The industry has established various forms of packages, which are described. There are exceptions and additions to this compilation, but in general, filler wires can be obtained in the following basic packages: spools, small coils, reels, drums, or payoff packs and large coils. All packages

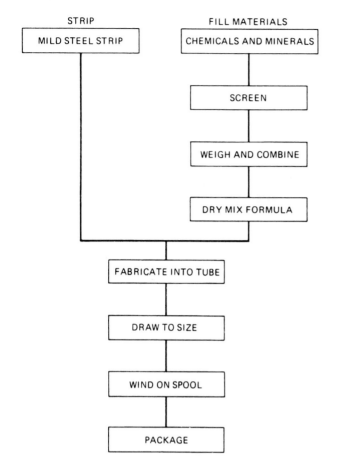

FIGURE 13-13 Flowchart for manufacturing tubular electrode wire.

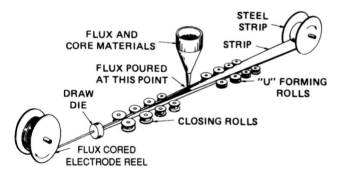

FIGURE 13-14 Simplified manufacturing operation to make tubular wire.

will carry the vendor's name, product size and specification, and the AWS safety warning label.

Spools

Spools made of plastic, formed steel wire, or composition wood are available in a variety of sizes and carry from 1 to 60 lb of electrode wire, depending on spool size and

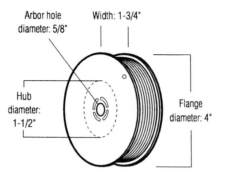

FIGURE 13-15 2-lb. plastic spool.

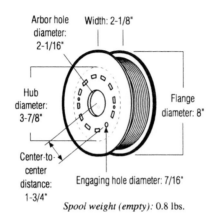

Spool weight (empty): 0.8 lbs.

FIGURE 13-16 10-lb fiber spool.

type of wire. In general, spools are wrapped in a thick plastic bag to provide maximum protection from moisture. Nonferrous wires are normally wrapped with protective paper. Standard 4-in. spool dimensions are shown in Figure 13-15. The 2-lb spools are usually level wound, individually boxed, and used for nonferrous wires. The small spools are used on spool guns or for orbital heads doing gas tungsten arc welding.

The 10-lb spools are wound transversely on 8-in. spools (Figure 13-16) and normally individually wrapped. They are usually for carbon steel or stainless steel electrode wires. Also, they are often used with portable wire feeders.

The 33- and 45-lb steel reels are transversely wound on the 12-in. reels and normally wrapped in plastic bags. This is the most popular package.

The 60-lb spools are transversely wound on 14-in. spools (Figure 13-17) and individually wrapped. These are normally used for carbon steel, stainless steel, and cored electrode wires. This is used on standard wire feeders. Spools, which are nonreturnable, should have one continuous length of electrode wire made from a single lot of material.

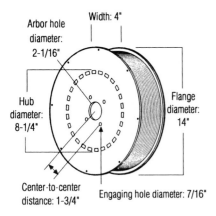

Spool weight (empty): 3 lbs.

FIGURE 13–17 60-lb fiber spool.

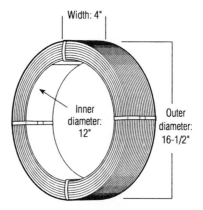

Coil weight (empty): 4 oz.

FIGURE 13–18 60-lb coil.

Small Coils

Coils are supplied with a cardboard inner liner to avoid the expense of spools. They require an adaptor to hold them on the dispensing equipment. The smaller coils come in 60-lb sizes. They are transversely wound, the ID of the core is $16\frac{1}{2}$ in., and they are 4 in. wide. Each coil is packed in a corrugated carton. The coil dimensions are shown in Figure 13–18. The weight of electrode wire in coils should not vary by more than 10%.

Reels

There are two types of reels, the steel reels and the flat reels. The steel reels are designed for small-, standard-, or portable-size wire feeders. They are made from wire and are recyclable with other scrap. They come in two sizes: the 33-lb. and the 45-lb. reel. They fit the standard spool hubs and are shown in Figure 13–19.

The large reels, sometimes called flat reels, are shown in Figure 13–20. These are designed for high-production automatic installations and contain 950 lb. of electrode wire. They may require special motorized dereelers. These are used to relieve the load on wire feeder motors. They are normally palletized and protected by shrink film wrapping. Reels are made of wood and are nonreturnable.

Drums/Payoff Packs

Another method of providing large quantities of electrode wire is by the use of drums or payoff packs. The drums shown in Figure 13–21 are made of heavy cardboard construction and will contain approximately 600 lb of electrode wire. These are normally used for solid carbon steel wire or flux cored electrode wires. Wire is placed in the drums to ensure a snarl-free payoff of the electrode with the drum usually rotating. Drums are palletized and covered with the polystyrene shrink film to provide protection during shipment and storage. Drums

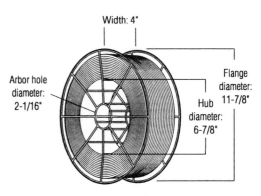

Steel reel weight (empty): 1.1 lbs.

FIGURE 13–19 Steel reel 33 & 45 lb.

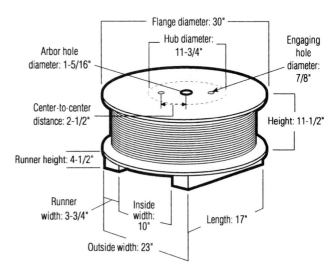

Reel weight (empty): 32.5 lbs.

FIGURE 13–20 950-lb reel.

are nonreturnable. Special dispensing equipment is available that eliminates problems related to out-of-spec cast or helix and eliminates wire flip. Speed nonflip packages are now available that supply straight wire.

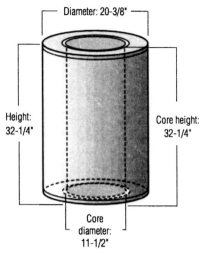

Diameter: 20-3/8"

Height: 32-1/4"

Core height: 32-1/4"

Core diameter: 11-1/2"

Drum weight (empty): 17.5 lbs.

FIGURE 13–21 Payoff pack.

Selection of Package

Figure 13-22 is a summary of the different types and packaging of electrode wires. This is not the total packaging method since some suppliers will provide special sizes of spools, coils, reels, and so on, for specialized equipment. Consult your supplier for the exact size of packages supplied.

In general, electrode wire is less expensive when ordered in larger packages. Additionally, the supply of electrode wire does not need to be changed as often. This is important when multiple welding heads are a part of the same system. This avoids frequent downtime to renew the electrode supply. However, larger packages require special dispensing equipment. These are different for the different packages. See the Section on electrode wire dispensing system for details. The wire supplier will normally be able to provide the types of dispensing equipment required for the package supplied.

Cast and Helix of Wire

AWS specifications require that cast and helix of wire on spools or coils must be suitable for feeding in an uninterrupted manner using automatic and semiautomatic welding equipment. The cast of a spooled electrode wire or filler metal wire wound on a spool is measured by removing several loops or rings of wire from the spool or drum. When cut from the spool and laid on a flat surface, it should form an unrestrained circle of not less than minimum diameter shown under "cast" in Figure 13-23.

The helix of coiled wire is measured with the loop or ring mentioned above. The loop or ring is placed on a flat surface without restraint. The maximum distance of any portion of the loop above the flat surface must not be greater than the dimension shown for the helix in Figure 13-23. The filler metal received from most manufacturers will meet these requirements.

FIGURE 13–22 Packaging of electrode size and type.

Wire Diameter		2-lb Spool	10-lb Spool	33-lb Spool	45-lb Spool	60-lb Spool or Coil	600-lb Pack	950-lb Reel
in.	mm							
.024	0.6	solid	solid	solid	—	—	—	—
.0.30	0.8	solid	solid	solid	—	—	—	—
.035	0.9	solid	solid	solid and cored	solid and cored	solid and cored	cored	cored
.045	1.2	—	solid	solid and cored	solid and cored	solid and cored	cored	cored
.052	1.3	—	—	cored	cored	cored	cored	cored
1/16	1.6	—	—	cored	cored	cored	cored	cored

FIGURE 13–23 Cast and helix requirements for electrode wires.

AWS Classification	Type of Package	Standard Size [in. (mm)]	Cast in.	Cast mm	Maximum Helix in.	Maximum Helix mm
All	4-in (100-mm) spools	0.045 (1.2) and less	4–9	100–230	$\frac{1}{2}$	13
All	All except 4-in. (100-mm) spools	0.030 (0.8) and less	12 min.	305 min.	1	25
		0.035 (0.9) and larger	15 min.	380 min.	1	25

13-6 WELDING FLUXES

There are many types of fluxes used in welding, brazing, and soldering. These include fluxes for oxyfuel gas welding, for oxygen cutting of certain hard-to-cut metals, for electroslag welding, and for submerged arc welding. The American Welding Society provides a specification for weld metal deposited by different combinations of steel electrodes and proprietary fluxes for submerged arc welding. The AWS provides a brazing flux specification that has recommended useful temperatures and the forms of the flux. There are also government specifications covering brazing fluxes. The manufacturers of brazing flux provide instructions for the application and use of their fluxes.

Fluxes for gas welding are not covered by the AWS. Manufacturers' recommendations should be followed.

Submerged Arc Flux

The function of the submerged arc flux is to produce a slag that will protect the molten metal from the atmosphere by providing a mechanical barrier. When it is molten, this slag should provide ionization to permit a stable arc. It should be fluid and of relatively low density so that it will float and cover the top of the weld metal. The melting temperature should be related to that of the molten weld metal, and it should have a different coefficient of expansion so that it can easily be removed after cooling. The slag should provide deoxidizers to help cleanse and purify the weld metal. It should also help reduce phosphorus and sulfur that might be present in the base metal. It should not introduce hydrogen into the weld. Finally, the flux should be granular and convenient to handle, should not provide noxious fumes, and should provide for a smooth weld surface.

Submerged arc fluxes consist of mixtures of chemicals and minerals in various combinations to provide properties just mentioned. Every grain of submerged arc flux should be similar in composition to the others and uniform in size. In the use of submerged arc flux, the granular material is placed over the welding joint and the heat of the arc causes it to melt and produce a molten slag. All of the flux placed over the weld does not melt, and the unmelted flux can be removed and reused. Upon cooling, the flux that melts transforms to a glasslike slag that must be removed from the weld deposit. The melted slag should not be used for welding since the deoxidizers and other cleansing and alloying elements in the flux are expended during melting.

There are three types of submerged arc welding fluxes based on the method of manufacturing: (1) fused flux, (2) agglomerated flux, and (3) bonded flux.

The ingredients for the flux must be ground, sized, and mixed prior to heating. In the case of *fused fluxes,* the mixture is melted in an electric or gas-fired furnace in a temperature range of 2912° F (1600° C). After melting, the molten flux is poured into water or onto a chilled plate to produce a glassy material. This material is then dried, crushed, and sized, by means of screens, and packaged; it is then ready for use.

The second method of manufacturing fluxes is the *agglomerated* method. Materials are dry mixed in the same way, except that a binder such as sodium or potassium silicate is added, after which the material is wet mixed. The mixture is then fed into a rotary kiln operating at approximately 1832° F (1000° C). Inside the kiln, by means of a tumbling process, the mixture forms into small balls and the ingredients tend to grow together and become larger. When they are properly heated, these balls become very tough. After cooling, the balls are ground in the same manner as already mentioned, sized, packaged, and ready for use.

The third method for making fluxes is the *bonded* method, and this is very similar to the agglomerated method, except that the mixture is bonded at a lower temperature. After the pellets are bonded, hardened, and cooled, they are ground, sized, packaged, and ready for use. Figure 13-24 is a simplified flowchart that applies to either agglomerated or bonded fluxes.

Each manufacturing method produces fluxes that are suitable for submerged arc welding and each has certain advantages and disadvantages. In the case of the fused flux, the high temperatures involved require considerably more energy for production. In addition, many of the elements used for deoxidation are partially expended at the high heat temperature and thus must be enriched to provide sufficient deoxidizing power for welding. The advantage is that all the grains of the flux are uniform in composition and in general the resulting flux is nonhygroscopic—that is, it will not pick up moisture.

In the case of the bonded and agglomerated fluxes, the temperatures are lower and therefore less energy is consumed. Additionally, the deoxidizers are not dissipated and are therefore active during the welding operation.

The composition of the welding fluxes can be varied to provide ranges in the melting and solidification temperatures, the viscosity, the current-carrying capacity, the arc stability, welding speed capacity, shape and appearance of the weld, and the ease of the slag removal. The slag metal reaction during the welding operation is extremely complex and beyond the scope of this book.

Submerged arc flux can be described as neutral, acidic, or basic while in the molten stage. If absolute neutrality cannot be obtained, it is best then for the flux to be basic, so that it will reduce impurities in the weld metal. The most ideal flux will be metallurgically inactive, which means that the composition of the deposited weld metal will be the same as the composition of the welding electrode. This is not possible over the entire range of welding electrode compositions available and thus a loss

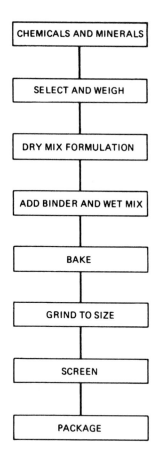

FIGURE 13–24 Flowchart for manufacturing submerged arc flux.

---

or buildup of certain elements may occur. The flux can be used to increase the amount of alloy added to the deposit. Alloying elements can be added via the electrode or the flux. In general, it is more economical to add alloying elements through the flux. This is particularly important when doing surfacing or when welding on alloy base metals.

Sizing is established by means of controlled mesh screens set to both the upper and lower limits of the size allowed for each particle. For example, a 12-mesh screen allows all particle sizes smaller than a certain size to fall through the screen. At the other end would be a 200-mesh screen, which would keep all the particles except the very fine particles from falling through. The resulting flux would have a maximum size allowed through a 12-mesh and a minimum size greater than that allowed to fall through the 200-mesh screen. If there are too many fines in the flux, it will tend to reduce the freezing period of the flux and it cannot be used for high-speed or circumferential welding. Larger sizes provide for higher-speed welding but allow arc flash through the layer of flux. Manufacturers provide fluxes between two limits since these sizes are recommended for most welding applications.

Submerged arc fluxes are classified according to the mechanical properties of the weld metal made with the specific electrode wires. This information is provided in the section on submerged arc welding. It is based on the AWS specification shown for submerged arc welding. The specific AWS specifications should be consulted since they are revised periodically.

Electroslag Fluxes

Electroslag fluxes are similar to submerged arc fluxes except that they are normally the fused type. Electroslag flux performs differently during the welding operation than submerged arc fluxes. The electrical conductivity of the flux makes the electroslag welding process operate. The flux becomes molten in a pool, and the electrode wire melts in the heated bath. It is the resistance of the bath to the welding current flowing between the electrode and the work that maintains the bath at the higher temperature. The flux is designed to provide a balance between conductivity and bath temperature for proper electroslag welding. In addition, the flux provides elements to purify and deoxidize the weld metal and also prohibits the oxygen and nitrogen of the air from coming in contact with the molten weld metal. The electroslag flux must have a lower density than steel so that it floats above the molten metal.

The criteria for selecting electroslag fluxes are based on the combination of flux and electrode wire. Tests must be made with standardized electrode wires and proprietary fluxes to qualify procedures.

13-7 OTHER WELDING MATERIALS

There are other filler metals and special items normally consumed in making welds. These include the nonconsumable electrodes—tungsten, carbon, and other materials, including backing tapes, backing devices, and flux additives. Another type of material consumed in making a weld is the consumable rings used for root pass welding and the guide tubes in the consumable guide electroslag welding method. Other filler materials are solders, brazing alloys, and powders.

Nonconsumable Electrodes

There are two types of *nonconsumable arc welding electrodes*. The carbon electrode is a nonfiller metal electrode used in arc welding or cutting, consisting of a carbon graphite rod that may or may not be coated with copper or other coatings. The second is the tungsten electrode, defined as a nonfiller metal electrode used in arc welding or cutting made principally of tungsten. The American Welding Society does not provide specification for carbon electrodes, but there is a military specification, MIL-E-17777C, entitled, "Electrodes Cutting and Welding Carbon-Graphite Uncoated and Copper Coated." This specification provides a classification system based on three grades: plain, uncoated, and copper coated (also copper coated with lock joint ends). It provides diameter information, length information, and requirements for size tolerances, quality assurance, sampling, various tests, and so on. Most manufacturers of carbon electrodes provide information indicating the type and size to be used for specific requirements. Applications including carbon arc welding, twin carbon arc welding, carbon cutting, and air carbon arc cutting and gouging.

The AWS specification for tungsten electrodes is entitled "Tungsten Arc Welding Electrodes." These electrodes are used for gas tungsten arc welding, plasma arc welding, and atomic hydrogen arc welding. This specification provides different classes of electrodes in various diameters and lengths and with two types of finish. The classifications relate to the composition of the tungsten, whether it is pure tungsten or tungsten with small amounts of rare earth elements added to improve electron emission. The information concerning tungsten electrodes was covered in Section 5-2.

Backing Materials

Backing materials are being used more frequently for welding. Special tapes exist, some of which include small amounts of flux, which can be used for backing the roots of joints. There are also different composite backing materials, for one-side welding. There are no specifications covering these materials, but more information about them is provided in Section 26-3. Consumable rings, which are considered a backing material, are specified by AWS A5.30, "Specifications for Consumable Inserts." More information is given in Section 25-2.

Submerged Arc Flux Additives

Metal powder is sometimes added to submerged arc flux. Additives are provided to increase productivity or enrich the alloy composition of the deposited weld metal. In both cases, the additives are of a proprietary nature and are described by their manufacturer, indicating the benefit derived by using the particular additive. Since there are no specifications covering these types of materials, the manufacturer's information is used.

Electroslag Guide Tubes

There are two types of guide tubes in common use for the consumable guide method of electroslag welding. Guide tubes may be bare or covered. Covered tubes are coated with a material that has a composition similar to that of the electroslag flux. Both types of tubes are consumed and the metal part of the tube becomes a portion of the deposited weld metal. The covered guide tube utilizes the flux coating to augment the flux used in the electroslag welding process. There is no AWS specification for guide tubes; however, they are normally seamless steel tubes of a low-carbon composition. Guide tubes are specified by the inside and outside diameter. The flux covering is proprietary and is compatible with the same manufacturer's electroslag flux. There are no specifications for electroslag fluxes.

Ceramic Ferrules

Ceramic ferrules are used in the stud welding process. These are small, short hollow cylinders that fit over the end of the stud and protect the molten metal from the atmosphere during welding. The ferrules also help mold the molten weld to an acceptable weld contour. Ceramic ferrules are available for all sizes of round studs and for many square or rectangular types. They are available from the stud manufacturer and are made to fit the stud sizes available. A ferrule is used only once and is easily broken away from the weld since it is very brittle. Manufacturers of studs provide the ceramic ferrules. No specifications exist for these items.

Solders

There are many different solder compositions and they are considered filler materials. Specifications for solder are issued by the American Society for Testing and Materials. The information about the different solders was summarized in Section 7-3.

Brazing Filler Metals and Fluxes

The brazing alloys are covered by a specification issued by the American Welding Society shown previously. The information about the different brazing alloys was summarized in Section 7-2. Brazing fluxes were also described in Section 7-2.

Powders

Filler metals used for some thermal spraying processes, for the PTA process, and sometimes for laser welding may be in the form of fine powder. The powder is fed into the thermal spray stream, the plasma arc stream, or the laser beam and is melted and carried to the work surface, where it is deposited. Powders are usually metal but may be ceramic or even plastic. These are used to provide a specific surface (i.e., to give the part necessary wear or corrosion resistance). Metal powders are commonly used for these applications. In many cases the metal for the desired surface cannot be drawn into a wire since it is non-ductile.

The powder is produced by atomization. This means that the metal is melted in a furnace with an air atmosphere or under gas protection or in vacuum. The molten metal is fed into a blast of water, air, or inert gas. This blast atomizes the liquid metal, which immediately solidifies and forms very small particles. The type of liquid or gas blast has an effect on the purity, shape, and nonmetallic content of the powder particles. The shape of each individual particle also relates to the atomizing technique. Inert gas atomization produces the highest-quality powder.

The individual particles are then screened by passing them through sieves of different sizes, where they are classified according to particle size. Classifications are based on the ASTM Standard E11, which relates mesh numbers to particle sizes. Different powder sizes are used for different processes and applications. For thermal spraying, including flame spraying and plasma spraying, the powder sizes are very small. Laser beam deposits usu-ally use larger particle sizes, due to high heat of the laser beam.

Powders are produced in many different analyses, such as stainless steel powders, nickel-base powders, cobalt-base powders, tool steel powders, nonferrous metal–base powders, and others. The application dictates the composition of the powder, and the process dictates the particle size to be utilized. There are no specifications for powder analysis; however, they follow normal specifications for stainless steel, tool steel, superalloys, and so on. ASTM provides specifications for some thermal spray powders.

Strip Electrodes

Strip electrodes are used for overlaying, usually stainless steels. They come in different thicknesses and widths. There are no specifications covering the size. The analysis is covered by the steel specifications.

 **QUESTIONS**

13–1. What are the four basic filler metals?

13–2. What is the difference between AWS and ASME specifications for welding filler metals?

13–3. Who tests and certifies welding filler metals?

13–4. List the major functions of a coating on a covered electrode.

13–5. What is a low-hydrogen coating?

13–6. How are electrodes reconditioned after they become damp?

13–7. Explain fingernailing and the reason for it.

13–8. How are bare solid steel electrodes specified? Give an example.

13–9. What is the purpose of a thin copper coating on a bare electrode wire?

13–10. In GMAW can argon–oxygen shielding gas be substituted for CO_2 shielding gas for welding carbon steel?

13–11. How are cut lengths of stainless steel rods identified?

13–12. What different packages are available for solid, bare electrode wires?

13–13. How are cast and helix measured? How do they affect welding?

13–14. What is the function of the core of a flux cored electrode?

13–15. Explain the specification system for flux cored steel electrodes.

13–16. Why is electrode wire less expensive when purchased in larger packages?

13–17. What welding processes normally use flux?

13–18. How is submerged arc flux made? Name two types.

13–19. What organization issues specifications for solder?

13–20. Explain the AWS classification system for tungsten electrodes.

14

GASES USED IN WELDING

14-1 SHIELDING GASES

The purpose of the shielding gas is to protect the arc area from the atmosphere. The shielding gas displaces the air and does not allow the atmospheric gases—nitrogen, oxygen, small amounts of argon, CO_2, and water vapor—to come into contact with the molten metal, the electrode, or the arc.

All the arc welding processes have some mechanism for shielding the arc area from the atmosphere. In shielded metal arc welding the disintegration of the coating creates gas that protects the molten metal from the atmosphere. In flux cored arc welding the disintegration of the core material, which may be supplemented by shielding gas, provides shielding from the atmosphere. In carbon arc welding the slow disintegration of the carbon electrode creates CO_2 gas, which shields the molten metal, and in submerged arc welding the granular flux performs this function. For gas metal arc welding and gas tungsten arc welding, the shielding gases must be supplied and directed around the arc area to provide protec-

tion from the atmosphere. The secondary purpose of the shielding gas is to establish the metal transfer mode and the deposited weld characteristics.

The shielding efficiency relates to how well the shielding gases displace the atmosphere from the arc area. This depends on the design of the nozzle, the distance from the nozzle to the work, the internal diameter or size of the nozzle, the gas flow rate, side winds, and the purity of the shielding gases.

Shielding gases are either inert or active. Inert gases will not combine chemically with other elements. There are only six inert gases: argon, helium, neon, crypton, xenon, and radon. All of these except argon and helium are too rare and expensive to be used for gas shielded welding. Inert gases must be used with the gas tungsten arc welding process and are normally used for welding nonferrous metals with gas metal arc welding. See ANSI/AWS A5.32 specifications for welding shielding gases.[1]

Active gases are either oxidizing or reducing. Active gases will combine with molten metal. Oxidizing gases are any gas that contains oxygen. Reducing gases are any gas that attracts oxygen. Following is a brief description of the different gases used for arc shielding. Pure gases and mixtures of two or three gases are employed. A better understanding of each gas will provide a basis for understanding the reasons for different gas mixtures. Properties of gases used are shown in Figure 14-1.

Several properties of gases have an effect on welding. Most of the gases are nontoxic but are an asphyxiant, meaning that a concentration of this gas will create suf-

Property	Air	Argon	Carbon Dioxide	Helium	Nitrogen	Oxygen	Hydrogen
International symbol and cylinder marking	Air	Ar	CO_2	He	N_2	O_2	H_2
Type of gas	Mixture oxidizing	Inert	Active oxidizing	Inert	Not true inert gas	Active oxidizing	Active reducing
Structure		Monatonic	Diatonic	Monatonic	Diatonic	Diatonic	Diatonic
Molecular weight	28.98	39.94	44.01	4.003	28.016	32.00	2.016
Boiling point (at 1 atm)							
°F	−317.8	−302.6	−109[a]	−452.1	−320.5	−297.3	−422.9
°C	−194	−184	−178	−269	−196	−182	−252
Specific volume (ft^2-lb) at 70° F, 1 atm	13.4	9.67	8.76	96.71	13.8	12.1	192.0
Density (lb/ft^3) at 70° F and 1 atm	0.0749	0.1034	0.1125	0.0103	0.0725	0.0828	0.0052
Specific gravity (air = 1)	1.000	1.380	1.530	0.137	0.967	1.105	0.069
Thermal conductivity (Btu/hr)	0.0140	0.0093	0.0085	0.0823	0.0146	0.0142	0.096
Ionization potential (electron volts)	—	15.7	14.4	24.5	15.51	12.5	15.6
Maximum allowable concentration	100%	Nontoxic asphyxiant	Nontoxic 5000 ppm	Nontoxic aphyxiant	Nontoxic, 82%	25%	Nontoxic asphyxiant

[1]Sublimes directly from a solid to a gas at −109° F (−178° C) and a pressure of 1 atm.
Note: The shipping containers of all these gases (except hydrogen) would be marked "Nonflammable Compressed Gas." The shipping containers of hydrogen would be marked "Flammable Compressed Gas."

FIGURE 14–1 Properties of inert and active gases.

focation due to the absence of oxygen. Too much oxygen or too much nitrogen in the breathing atmosphere will cause damage to humans.

The specific gravity relates to the weight of the gas with respect to air. The specific gravity of air is 1. Lightweight gases such as helium will float away and will not be an efficient shield. Heavier gases will displace air in enclosed areas.

Thermal conductivity relates to the heat in the arc column and whether it will create a small or a larger arc column; also, how fast the heat will travel in the gas.

Ionization potential establishes the ease of arc initiation and arc stability. The lower the ionization potential, the easier it is to start the arc. The higher the ionization potential, the hotter the arc.

Gases are diatomic or monatomic. A diatomic gas demonstrates disassociation of the molecules in the arc. This process absorbs heat energy, followed by recombination away from the arc, which releases latent heat. Monatomic gases do not disassociate in the arc.

The most important characteristic of a shielding gas is its purity. In all cases the purity must exceed 99%. This is governed by specifications that are shown for each gas.

Gas Shielding Efficiency

With gas shielded arc welding processes, specifically gas tungsten arc, gas metal arc, and flux cored arc welding with external gas shielding, the quality of the weld de-

pends on the efficiency of the atmospheric protection provided by the shielding gas. Figure 14-2 shows the surface of a gas tungsten arc weld on aluminum with proper gas shielding and with an inefficient shield. Figure 14-3 shows a gas metal arc weld on carbon steel with an efficient shield and with an inefficient shield. The following factors adversely affect the efficiency of the gas shielding:

1. Insufficient shielding gas due to breezes, low flow rates, high standoff distance

FIGURE 14–2 GTAW weld surfaces.

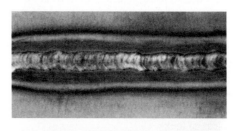

FIGURE 14–3 GMAW weld surfaces.

2. Insufficient shielding due to defects of the shielding gas delivery system
3. Impure shielding gas

The most common problem of inefficient shield is due to breezes in the weld area—caused by open windows and doors and the use of ventilating fans for the welder's comfort. The solution is to provide small windbreaks or shields in the arc area. In field welding the welder can use his body to shield the arc area from the breeze. Temporary enclosures are often used for welding high-rise buildings and for pipelines. Shielding gas flow can be increased; however, this can be expensive. The welder will immediately notice the deterioration of weld appearance that indicates poor shielding efficiency.

Gas delivery system problems affect the shielding efficiency. In gas metal arc welding the most prevalent problem is spatter buildup on the gas nozzle, which impairs gas flow. When using CO_2, freezing the regulator will stop gas flow. Other problems are broken hoses or loose hose connections within the wire feeder, the welding gun, or at the gas supply cylinder. Inoperative solenoid valves, gun switches, or control relays may also be a problem.

In gas tungsten arc welding, spatter buildup is not a problem; instead, leaking hoses, cracked tubing, and loose connections can be the problem. A specific problem is encountered with cable assemblies when water tubes and gas tubes are together; water leaks can get in the gas supply hose and create trouble. Certain kinds of tubing deteriorate in time and should be replaced. Maintenance of the gas supply system should be performed routinely or if there is a suspicion of leaks.

The third factor relates to the purity of shielding gas. This is rarely the problem since gas suppliers maintain constant checks on gas purity. If there is a suspicion that a cylinder has impure gas, it should be switched and a new cylinder used. If there is still a suspicion of the gas, the gas supplier can make measurements to determine the gas purity. In general, the purity problem relates to

moisture in the gas. The specifications for gases include the minimum dew point temperature, which relates to moisture in the gas. It can be measured with portable instruments and can be related to the standard. Dew point is covered more thoroughly in the following section on carbon dioxide gas. If this is a problem, and it can be in tropical countries, an in-line filter can be used. These have replaceable elements that must be replaced as they become saturated.

Gases for Shielding

Argon Argon has no color, odor, or taste, and is relatively plentiful compared to the other inert gases. A million cubic feet of air contain 93,000 ft^3 of argon. It is separated from air by liquefying the air under pressure and low temperatures and is then allowed to evaporate by raising the temperature. The argon boils off from the liquid at a temperature of $-302.6°$ F ($-184°$ C). For welding, the purity of argon is approximately 99.99%. Argon is relatively heavy, approximately 23% heavier than air. It is used as a shielding medium for gas tungsten arc welding and for gas metal arc welding of nonferrous metals. Argon has a relatively low ionization potential. The arc voltage of the gas tungsten arc in argon is lower than in helium. The welding arc tends to be stable in argon, and for this reason it is used in many shielding gas mixtures. Argon is nontoxic but can cause asphyxiation in confined spaces by replacing the air. Argon is specified by Military Specification MIL-A-18455B.

Helium Helium is the second lightest gas. It is one-seventh as heavy as air. It is inert, has no color, odor, or taste, and is nontoxic. In liquid form it is the only known substance to remain fluid at temperatures near absolute zero. Helium is obtained from natural gas, and in the Texas fields it represents 2% of the volume. It is found in natural gas in Canada and in Russia. Helium has the highest ionization potential of any shielding gas, and for this reason a gas tungsten arc in helium has an extremely high arc voltage. Because of this, arcs in an atmosphere of helium produce a greater amount of heat. Helium's light weight causes it to float away from the arc zone, producing an inefficient shield unless higher flow rates are employed. For overhead welding, this can be helpful. It is often mixed with other gases for gas metal arc welding. Helium is expensive for welding and is sometimes in scarce supply. Helium is specified by Federal Specification 88-H-11688.

Carbon Dioxide Carbon dioxide is a compound of about 27% carbon and 72% oxygen. It is made of two oxygen atoms joined with a single atom of carbon. At normal atmospheric temperature and pressure it is colorless, nontoxic, and does not burn. It has a faintly pungent odor and a slightly acid taste. It is about $1\frac{1}{2}$ times heavier than air, and in confined spaces it will displace the air. At elevated temperatures it will disassociate into oxygen and carbon monoxide. In the welding arc, disassociation takes place to the extent that 20% to 30% of the gas in the arc area is carbon monoxide and oxygen. Thus CO_2 has oxidizing characteristics in the welding arc. As the carbon monoxide leaves the arc area, it quickly recombines with oxygen to form CO_2. Extensive measurements have been made, and it has been found that the carbon monoxide level at a distance of 7 in. (175 mm) from the welding arc is 0.01% or 100 ppm, which is regarded as a safe limit for carbon monoxide gas. At a distance of 12 in. from the arc, the carbon monoxide concentration is 0.005%. A concentration of 5000 ppm of carbon dioxide is considered a safe level. Ventilation should be provided to keep the CO_2 level below this concentration.[2]

Carbon dioxide can exist simultaneously as a solid, a liquid, and a gas at its triple point. At atmospheric pressure, solid CO_2 (dry ice) transforms directly to a gas without passing through the liquid phase; that is, it sublimes. At temperatures and pressures above the triple point and below 87° F in a closed cylinder, carbon dioxide liquid and gas exist in an equilibrium. This is normally the way it occurs in high-pressure cylinders.

Carbon dioxide is manufactured from flue gases, given off by burning natural gas, fuel oil, or coke. It is also a by-product of the calcination operation of lime kilns, from the manufacturing of ammonia, and from the fermentation of alcohol. The CO_2 gas is cleaned, purified, and dried. The purity of carbon dioxide gas can vary considerably, depending on the process of manufacture. The federal specification covers two classifications of CO_2. Grade B, nonmedical, type 1, with very little moisture content for special uses, covers welding-grade CO_2. The purity specified for welding-grade CO_2 gas is a minimum dew point temperature of $-40°$ F ($-40°$ C). Figure 14–4 shows the dew point of CO_2 versus the percent of mois-

ture in the gas. The standard provides a minimum dew point of $-40°$ F ($-40°$ C); however, many manufacturers produce welding-grade CO_2 gas with a dew point temperature as low as $-70°$ F ($-57°$ C). This gas has a moisture content of 0.0091% by weight and/or 9 parts per million (ppm). Dew point can be measured using portable instruments. Too much moisture in the gas will cause weld porosity. CO_2 is covered by Federal Specification BBC-101A. Figure 14–5 shows a dew point testing instrument.

Welding with the Different Gases

The composition of the gas shielding envelope can be a single or pure gas; a mixture of two gases, known as duplex mixtures; or a mixture of three gases, known as trimix gases. Mixtures combine inert and active gases.

For GTAW, inert gases normally are used for shielding. Mixtures employing a small amount of an active gas are sometimes used. Mixtures of two inert gases are often used, and sometimes a reducing gas is included.

With GTAW, the pure inert gases do not provide good arc characteristics when welding steel. However, pure CO_2 does provide good arc characteristics. For GMAW, argon with small amounts of oxygen improves the penetration

FIGURE 14–5 Dew point testing instrument.

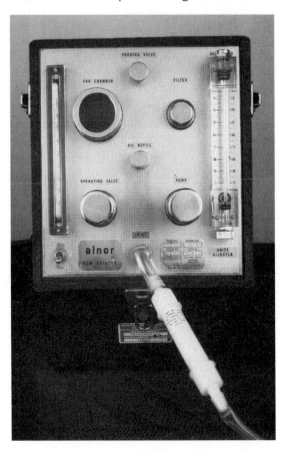

FIGURE 14–4 Dew point versus percentage of moisture in CO_2.

Dew Point		Moisture (% by weight)	Moisture in CO_2 (ppm)
°F	°C		
−90	−68	0.00021	2
−80	−62	0.00043	4
−70	−57	0.00091	9
−60	−51	0.00188	19
−50	−46	0.00365	36
−40	−40	0.0066	66
−30	−34	0.0120	120
−20	−29	0.0218	218
−10	−23	0.0354	354
0	−17.8	0.0590	590
10	−12.2	0.0980	980

pattern, bead contour, and eliminates undercut due to the wetting action. Argon with carbon dioxide is a popular mixture for welding steels. The triple-mix gases, usually argon with CO_2 and oxygen, or argon with CO_2 and helium, have specific advantages to be mentioned later.

It is important to select the correct gas mixture when using gas tungsten arc or GTAW and for welding a particular base metal. Following is a review of the gases and gas mixtures and their use for arc shielding.[2]

Argon Plus Oxygen For GTAW, very small additions of oxygen, less than 1%, help to stiffen the arc. Oxygen is used for dc electrode negative (DCEN) of aluminum. It is also used for thin steels, including stainless steels.

With GMAW, arc transfer characteristics are strongly influenced by the shielding gas composition. In mixtures, the amount of current needed to reach the transition point diminishes as the percent of CO_2 decreases. Poor bead contour and penetration pattern obtained with pure argon are improved with the addition of oxygen. Oxygen is normally added in amounts of 1% to 2%, or 3% to 5%. This provides for spray transfer. The amount of oxygen is limited to 5%. The weld bead profile is shown in Figure 14–6. The more oxidizing the shielding gas, the more important it is to select a welding electrode that contains sufficient deoxidizers to overcome the loss of silicon, manganese, and aluminum. More oxygen would lead to the formation of porosity in the deposit. Oxygen improves the penetration pattern by broadening the deep penetration finger at the center of the weld. It also improves the bead contour and eliminates the undercut at the edges of the weld, due to better wetting action.

Argon Plus Helium GMAW uses argon–helium mixtures for welding nonferrous metals. The addition of helium in percentages of 50% to 75% raises the arc voltage and increases the heat in the arc. It is useful for welding heavy thicknesses of aluminum, magnesium, and copper, and for overhead-position welding. With the higher percentages of helium, the speed and quality of ac welding of aluminum is improved. The 25% argon—75% helium mixture is used for the gas tungsten hot wire variation. The argon plus helium mixture is also used for GMAW of nonferrous metals.

Argon Plus Hydrogen Argon with the addition of small amounts of hydrogen increases the arc voltage and increases the heat in the arc. Mixtures of argon containing up to 5% hydrogen are used for welding nickel and nickel alloys and for welding heavier sections of austenitic stain-less steels. Mixtures of argon with up to 25% hydrogen are used for welding thick metals that have high heat conductivity, such as copper. It has an advantage in high-speed automatic welding. Hydrogen additions cannot be used for welding mild or low-alloy steels due to the problem of hydrogen pickup. Hydrogen should not be used with aluminum and magnesium.

Argon Plus Nitrogen In some countries pure nitrogen is used for GMAW of copper. The quality of the resulting welds is not as good as desired. Adding 50% to 75% argon to nitrogen produces a higher-quality weld. Nitrogen is not used as a shielding gas in North America.

Argon Plus Carbon Dioxide Argon plus carbon dioxide is not used for GMAW. For GMAW one of the most popular mixtures is 75% argon and 25% CO_2. However, outside North America the more popular mixture is 80% argon and 20% CO_2. It is widely used on thin steel, where deep penetration is not necessary and where bead appearance is important. It provides improved appearance over 100% CO_2. Spatter is reduced. It is also helpful for out-of-position welding, on thin sheet metal, and when fitup is poor.

Carbon Dioxide One hundred percent carbon dioxide shielding produces broad, deep-penetration welds. Bead contour is good and there is no tendency toward undercutting. Compared to inert gases, CO_2 is relatively inexpensive. The chief drawback of CO_2 shielding is the tendency for the arc to be somewhat violent. This can lead to spatter and makes welding of thin materials difficult. This is the reason for the argon—CO_2 mixtures. Carbon dioxide should not be used for GTAW. CO_2 is commonly used for flux cored arc welding.

Ternary Mixtures of Gases Commonly called tri-mix gas, shielding gases containing three gases are becoming more popular. Normally, the mixtures utilize argon with oxygen and CO_2, and sometimes argon, CO_2, and helium. In the liquefaction of argon, the raw argon contains about 2% oxygen before final purification. The impure argon is then mixed with CO_2, which provides a tri-mix of 70% argon, 2% oxygen, and the remainder CO_2. This mixture is popular for welding steels. Another tri-mix adds a small amount of helium to the argon—O_2 mixture. This tends to increase the arc voltage and provide higher deposition rates.

Various other mixtures of gases are becoming available that have specific features or advantages. Gas suppliers provide specific compositions and applications. This

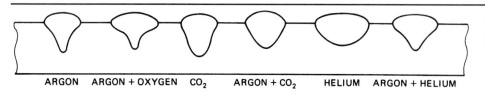

FIGURE 14–6 Shielding gas related to weld profile for DCEP.

ARGON   ARGON + OXYGEN CO_2 ARGON + CO_2 HELIUM   ARGON + HELIUM includes high-performance shielding gas mixtures that provide higher deposition rates or higher travel speeds. Most of these gases are three-component mixtures that contain helium. Use of these gases increases the arc voltage, and in many cases the user is expected to use a longer wire stick-out, increasing the I^2R heating of the welding electrode beyond the tip. The higher voltage and the extended electrode wire increase the energy in the arc and increase deposition rates. Proprietary shielding gases of this type are TIME gas, Stargon gas, and others. The gas suppliers claim improved weld deposit properties. It is wise to investigate these gases thoroughly under laboratory and production conditions. Laboratory tests should obtain all data and compare improved deposition or travel speed versus the extra cost of the gas. Figure 14–7 summarizes the more popular shielding gas mixtures.

14-2 FUEL GASES FOR WELDING AND CUTTING

Oxygen and sometimes air is used with various hydrocarbon fuel gases for producing heat by means of chemical combustion. These fuel gases, usually with oxygen, are used for soldering, brazing, welding, oxygen cutting, flame spraying, flame hardening, and flame straightening. The major fuel gases are acetylene, natural gas, liquid petroleum gases (propane and propylene), and synthetic gases such as methylacetylene propadiene. The only fuel gases used for welding are compounds of carbon and hydrogen that will react with oxygen to produce a flame having a temperature above the melting point of most metals. Nonhydrocarbon fuel gases should not be used for welding since their products of combustion are toxic.

The fuel gas–oxygen reaction is in two steps. The primary reaction produces carbon monoxide and hydrogen plus heat:

$$fuel\ gas + O_2 \rightarrow CO + H_2 + heat$$

The secondary reaction, which utilizes oxygen from the air, will oxidize the carbon monoxide and hydrogen to carbon dioxide and water vapor plus additional heat:

$$air + CO + H_2 \rightarrow CO_2 + H_2O + heat$$

This complete combustion reaction produces a large amount of heat known as the gross heat of combustion (heat of primary reaction plus heat of secondary reaction). This is given in Btu per pound of fuel gas or Btu per cubic foot of fuel gas or calories.

FIGURE 14–7 Summary of shielding gases and mixtures and their use (North America).

Shielding Gas	Gas Reaction	GMAW and FCAW	GTAW and PAW
Pure gases			
Argon, Ar	Inert	Nonferrous	All metals
Helium, He	Inert	Nonferrous	Al, Mg, and copper and alloys
Carbon dioxide, CO_2	Oxidizing	Mild and low-alloy steels some stainless steels	Not used
Two-component mixtures			
Argon mixtures			
Argon + 20–50% helium	Inert	Al, Mg, and Cu and alloys	Al, Mg, and Cu and alloys
Argon + 1–2% CO_2	Oxidizing	Stainless and low-alloy steels	Not used
Argon + 3–5% CO_2	Oxidizing	Mild, low-alloy, and stainless steels	Not used
Argon + 20–30% CO_2	Slightly oxidized	Mild and low-alloy steels some stainless steels	Not used
Argon + 2–4% He	Reducing	Not used	Nickel and alloy and austenitic stainless steel
Helium mixtures	—	Al and alloys, Cu and alloys	Al and alloys, Cu and alloys
Helium + 25% argon	Inert	Al and alloys, Cu and alloys	Al and alloys, Cu and alloys
CO_2 mixtures			
CO_2 + up to 20% O_2	Oxidizing	Mild and low-alloy steels (used in Japan)	Not used
CO_2 + 3–10% O_2	Oxidizing	Mild and low-alloy steels (used in Europe)	Not used
Three-component mixtures			
Helium mixtures			
Helium + 7.5% Ar + 2.5% CO_2	Inert	Stainless steel and low-alloy steels	Not used
Argon mixtures			
CO_2 + 3–10% O_2 + 15% CO_2	Oxidizing	Mild steels (used in Europe)	Not used

The properties of fuel gases are given in Figure 14-8. The figure shows the flame temperatures of each in oxygen and in air. Flame temperature in oxygen is always much higher than in air. The flame temperature and heat of combustion are indications of the amount of work that can be done by the fuel gases. However, when comparing the cost of using different gases it is important to consider the ratio of fuel gas to oxygen required for combustion. This is necessary so that the cost of both the fuel gas and the cost of oxygen are combined to obtain the total gas cost. These data are theoretical since in actual use a portion of the oxygen required for total combustion comes from the air surrounding the flame. For example, the combination of acetylene and oxygen for the primary reaction is a 1:1 ratio. The additional oxygen required for the secondary reaction requires 1.5 units of extra oxygen; therefore, the total ratio of oxygen to acetylene is 2.5 rather than 1. This is determined by working out the chemistry of both the primary and secondary reactions. The primary reaction is produced in the inner cone and the secondary reaction in the outer envelope of the flame.

Another important consideration is specific gravity. Hydrogen is the lightest of all gases. Propane, propylene, and methylacetylene are all heavier than air. Acetylene is slightly lighter than air. Methane and natural gas are slightly over half the weight of air. Some of the fuel gases would tend to float away into the atmosphere while others would collect in low spots, in enclosed areas of weldments, or in pits and bottoms of tanks. This is a very important safety consideration since fuel gas leakage can occur.

Another safety factor is the flammability limits in air. Acetylene will burn in air with a minimum of 2.5% to a maximum of 81%. This is the widest range of any fuel gas; however, hydrogen is almost as wide. The other gases are much lower. This means that acetylene is the most dangerous, since it will ignite in any percentage with air between these two limits. Figure 14-8 also shows the threshold limit values (TLV) of the different gases.

Acetylene

Acetylene (C_2H_2) is a compound of carbon and hydrogen. It is a colorless flammable gas slightly lighter than air. Acetylene of 100% purity is odorless, but the commercial grade has a distinctive garlic flavor. Acetylene burns in air with an intensely hot, yellow, luminous, and smoky flame. For safety reasons acetylene is never compressed above 15 psi (103 kPa). Acetylene cylinders are made safe by providing a porous mass of material inside the cylinder that is saturated with acetone. Acetylene dissolves in acetone and in this mode can be compressed to 250 psi (0.1750 kg/mm^2) without danger.

Acetylene with oxygen produces the highest flame temperature of any of the fuel gases. It also has the most concentrated flame, but it produces less gross heat of combustion than the liquid petroleum gases and the synthetic gases. Acetylene is manufactured by the reaction of water and calcium carbide. This is sometimes done at plant sites in acetylene generators. Acetylene is nontoxic; however, it is an anesthetic and if present in sufficiently high concentration it is an asphyxiant in that it replaces oxygen and will produce suffocation.

Hydrogen

Hydrogen (H_2) is the lightest gas and is present in the atmosphere in concentrations of about 0.01% at lower altitudes. Hydrogen may also be present in the arc area from water vapor resulting from the products of combustion and also from high temperature reaction with hydrocarbons that might be present. Hydrogen is soluble in molten steel but the solubility at room temperature is very low. As molten weld metal cools and solidifies, the hydrogen is rejected from the solution and becomes entrapped in the solidifying weld metal. It will collect at grain boundaries or at discontinuities of any type where it will create high pressures, and cause high stresses within the weld. These pressures and stresses lead to minute cracks in the weld metal that can develop into larger cracks. The small concentrations of hydrogen that appear on the fractured surface are known as fish eyes because of their characteristic appearance. Hydrogen also causes underbead cracking in the heat-affected zone. Hydrogen will, however, gradually escape from the solid steel over a period of time. This migration of hydrogen from the weld metal is accelerated if the temperature of the metal is increased.

Hydrogen can be used as a fuel gas and originally was an important commercial fuel gas. Its flammable limits in air range from 4% to 75%. When hydrogen is burned in either oxygen or air the flame temperature is lower than that of acetylene. It requires less oxygen for complete combustion but does not produce sufficient gross heat of combustion for industrial welding.

Methane

Methane (CH_4) is a colorless, odorless, tasteless, flammable gas. It is generally considered nontoxic, and concentrations of up to 9% can be inhaled without apparent ill effects. Methane is the major component of natural gas. It is separated from natural gas and can be obtained from petroleum. It is normally shipped and stored in high-pressure gas cylinders. It can be shipped in liquid form in special insulated tanks at temperatures below its boiling point. It acts in the flame similar to natural gas.

Natural Gas

Natural gas (essentially CH_4) has much of the same characteristics as methane. The composition of natural gas varies in different geographical locations, and the gross

Property	Acetylene	Hydrogen	Methane	Methyl Acetylene Propadiene	Propane	Propylene	Natural Gas
International symbol and cylinder marking	C_2H_2	H_2	CH_4	$CH_3C{:}CH$ (MPS)	C_3H_8 (LP gas)	C_3H_6 (PRY)	MET
Molecular weight	26.036	2.016	16.042	40.07	44.094	42.078	Similar to methane
Specific gravity of gas (air = 1)	0.91	0.069	0.55	1.48	1.56	1.48	0.56
Specific volume of gas at 60° F and 1 atm (ft³)	14.5	192.0	23.6	8.85	8.6	9.5	23.6
Specific gravity of liquid	—	—	—	0.576	0.507	0.527	—
Lb/gal of liquid at 60° F	—	—	—	4.80	4.25	4.38	—
Density of gas (lb/ft³)	0.0680	0.0052	0.0416	0.113	0.115	0.105	0.0424
Boiling point (at 1 atm) °F	−119.2	−422.9	−258.6	−9.6	−43.8	−53.9	−161
°C	−84	−252	−161	−23.1	−42.1	−47.7	−107
Flame temperature (neutral) In oxygen °F	5600	4800	5000	5300	4600	5250	4600
°C	3100	2650	2775	2925	2550	2900	2550
In air °F	4700	4000	3525	3200	3840	3150	3525
°C	2600	2200	1950	1760	2100	1730	1950
Ratio of oxygen to fuel gas required for combustion	1 to 1	0.5 to 1	1.75 to 1	2.5 to 1	3.5 to 1	4.5 to 1	2 to 1
Ratio of air to fuel gas required for combustion	11.9	2.38	9.52	21.83	24.30	21.83	10.04
Gross heat of combustion Btu/lb	21,600	52,800	23,000	21,000	21,500	22,000	24,000
Btu/ft³	1500	344	1000	2500	2500	2400	1000
Flammable limits in air by volume (%)	2.5–81	4–75	5.3–15	2.4–11.7	2.2–9.5	2.0–10.3	5.3–14
Max. allowable concentration in TLVs	Nontoxic asphyxiant	Nontoxic asphyxiant	Nontoxic up to 9	Nontoxic 1000 PPM	Nontoxic asphyxiant	Nontoxic	Nontoxic up to 25

Note: The shipping containers of these gases would all be marked "Flammable Compressed Gas."

FIGURE 14–8 Properties of fuel gases (From Ref. 3).

heat of combustion of natural gas varies from one locality to another; 1000 Btu per cubic foot is normally accepted as a minimum. Natural gas is used in oxygen flame cutting. Its flame temperature is relatively low and the gross heat of combustion is also relatively low. It is less expensive than other fuel gases and has become quite popular. It is not used for gas welding or flame hardening because of its lower flame temperature. It is normally supplied via pipeline to industrial sites and is sold by the cubic foot. It is usually compressed at the factory.

Liquefied Petroleum Gases

The liquefied petroleum (LP) gases are propane and propylene (propene) and butanes. They are by-products of oil refineries and are flammable, colorless, noncorrosive, and nontoxic. They have an anesthetic effect and when they displace oxygen in the air they act as asphyxiants. This is an important safety factor since they weigh approximately $1\frac{1}{2}$ to almost 2 times the weight of air.

Pure propane is odorless; however, it is given an artificial odorization. Propylene has an unpleasant odor characteristic of refineries. The flame temperature of propane is lower than that of acetylene but its gross heat of combustion is higher, more than $1\frac{1}{2}$ times that of acetylene. Propane is available in pure form and as mixtures. The mixtures contain additives such as ethylene, propylene, or ethyl ether, which increase the flame temperature and the heat of combustion. Additives also increase the price of the gas. Propane base gases are known as Acetogen, Chemi-gas, Flamex, Hy-Temp, Chem-O-Lene, and so on.

Propylene has a higher flame temperature than propane but not as high as acetylene. It also has a gross heat of combustion approximately $1\frac{1}{2}$ times that of acetylene. Propylene is also available as pure gas and with additives and is given such trade names as Apachi gas, HPG, B.T.U., and Liquifuel.

The liquefied petroleum gases require considerably more oxygen for combustion than acetylene. They are shipped and stored in the liquefied form in cylinders and tanks. They normally do not have pressures exceeding the 375 psi (2585 kPa). The liquefied petroleum vaporizes in the cylinder and is discharged as a gas. It is usually sold by weight. To determine the cubic feet of gas, multiply by the specific volume of the gas.

Synthetic Gases: Methylacetylene–Propadiene Stabilized

The most popular synthetic hydrocarbon fuel gas is methylacetylene plus propadiene (allene), sometimes called methylacetylene–propadiene stabilized (MPS) gas. It is a by-product of the chemical industry. It goes by several trade names, including MAPP gas and Fuel-gas. This

gas is colorless, flammable, and slightly toxic. The tentative maximum concentration of 1000 parts per million has been suggested for its TLV. Methylacetylene-propadiene stabilized has a flame temperature in oxygen higher than propane but less than acetylene. Its gross heat of combustion is over $1\frac{1}{2}$ times that of acetylene. It is stored and shipped as a liquefied gas in its own vapor pressure of about 60 psi (414 kPa) at 70° F (21.1° C). These gases are usually sold by weight and are available in cylinders and in bulk. When using these different fuel gases, different torches and tips are required.

Selecting Fuel Gases

The selection of a fuel gas should be based on the gas that will do the best job at the least cost. Necessary properties would include the flame temperature, the gross heat of combustion, and the oxygen-to-fuel gas ratio for combustion. This information is shown in Figure 14–8. Some of the fuel gases can be used only for heating, for oxygen cutting, or for soldering and brazing. They cannot all be used for gas welding or for flame hardening. The uses of a particular gas depend on its flame temperature, heat of combustion, heat distribution in the flame, and coupling distance. All fuel gases can be used for flame spraying; however, for spraying high-melting-temperature metals the higher-flame-temperature fuel gases must be used. All fuel gases can be used for heating, but the type of heating might dictate the fuel gas. For example, acetylene is a more concentrated heat source than the other gases. Figure 14–9 shows the flame temperatures for the common fuel gases versus the oxygen-to-fuel gas ratio. This is the cubic feet of pure oxygen per cubic foot of fuel gas. The dot on each curve is the temperature of the neutral flame. A neutral flame is used when welding steel. The temperature of all fuel gas flames increases when more oxygen is used.

For underwater oxygen flame cutting, acetylene can be used down to depths of 30 ft, but the methylacetylene-propadiene (MPS) gas can be used to depths of 100 ft. Special torches are required. Selection information is summarized in Figure 14–10.

The ratios of oxygen to fuel gases have an important bearing on the cost of the total operation since oxygen is expensive. The amount of oxygen required is difficult to determine since it depends on the type of torch employed. A single flame port torch is used for brazing or welding, and a multiflame port torch is used for flame cutting and heating. Multiport torches have a higher oxygen-to-fuel gas ratio since the inner flames are not able to obtain oxygen from the air. Another factor is the heat transfer or coupling. This is best done by practical tests.

The gross heat of combustion is an indication of how much work can be done by a given volume of fuel gas—hence, the amounts of oxygen and fuel gas that are required to do work. The measure of comparison here is

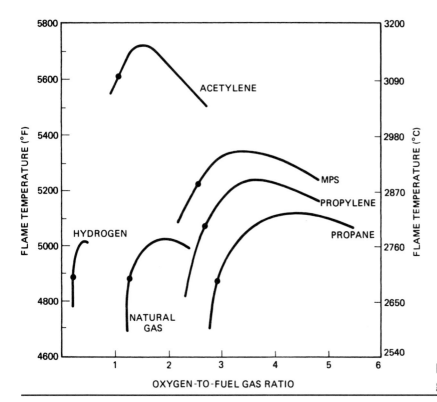

FIGURE 14–9 Flame temperatures of fuel gases.

Used for:	Acetylene	LPG Propane	Natural Gas or Methane	MPS	Propylene
Heating	Not preferred	Yes	Yes	Yes	Yes
Torch soldering	Yes (in air)	Yes (in air)	Yes (in air)	Yes (in air)	Yes (in air)
Torch brazing	Yes	Yes	Yes	Yes	Yes
Oxygen cutting	Yes	Yes	Yes	Yes	Yes
Flame spraying	Yes	Yes	Yes	Yes	Yes
Gas welding of steel	Yes	No	No	Marginal	No
Flame hardening	Yes	No	No	Yes	No

FIGURE 14–10 Uses of fuel gases with oxygen.

to establish welding or cutting procedures that will include gas usage and work travel speeds. This information is available from torch and gas manufacturers. Then calculate the time required for a specific cut or weld and compare the results. It is wise to confirm the calculated times by making tests under controlled conditions. In making cutting speed tests, make sure that the most efficient cutting tip is used for the conditions being tested. Differences in tips can be more of a determining factor than differences in gases. Remember, it is the oxygen jet that does the cutting.

It is becoming increasingly important to determine the availability of different fuel gases. As energy sources become more expensive, the cost-usefulness relationship can change.

14-3 ATMOSPHERE GASES

The atmosphere that surrounds the earth provides the air we breathe and supports life. At sea levels its pressure is approximately 14.7 psi (1 bar). Its composition is approximately 78% nitrogen, 21% oxygen, and 1% argon, with small amounts of carbon dioxide, hydrogen, and other gases. Nitrogen, oxygen, and argon are obtained by the liquefaction and distillation of air.

Oxygen

Oxygen is a colorless, odorless, tasteless gas that supports life and makes combustion possible. Oxygen combines with many elements to form oxides. Oxygen is very active and combines with most metals at room temperatures. Oxygen combines with iron to form compounds that can remain in the weld metal as inclusions. As the molten weld metal cools, free oxygen in the arc area will combine with carbon of the steel and form carbon monoxide. This may be trapped in the weld metal as it solidifies. The gases collect into pockets that cause pores or hollow spaces. This problem is often overcome by providing deoxidizers in the filler metal, such as manganese and silicon. These elements will combine with the oxygen to produce an oxide of manganese or silicon, which will float to the surface of the molten steel.

The purity of high-pressure oxygen supplied in a high-pressure cylinder is 99.6+ by volume. The oxygen used for flame cutting should have this purity. When the purity of oxygen is reduced, the oxidation of the metal being cut is retarded and more oxygen is consumed, cutting speed is reduced, and the cut quality is reduced. It is reported that a 1% decrease in oxygen purity decreases cutting speed by 10% to 15%. This reduction in purity also increases the consumption of oxygen by 25% to 35%. To compensate for reduced purity, the pressure is usually increased, which contributes further to poor flame cut surfaces. Oxygen with a purity below 97% should not be used.

Liquid oxygen is extremely cold, $-297°$ F $(-183°$ C) at atmosphere pressure. Accidental contact of liquid oxygen will cause severe frostbite to the eyes or skin. Protective clothing and safety goggles or face shield must be worn when handling liquid oxygen.

Combustibles must be kept away from oxygen. Many materials that do not normally burn in air, and other materials that are combustible in air, may burn violently in an atmosphere high in oxygen. All organic materials and flammable substances, such as oil, grease, kerosene, wood, paint, tar, and coal dust, must be kept away from oxygen. An accumulation of oxygen can be hazardous, and therefore proper ventilation is required. Oxygen should *never* be used in place of compressed air.

The Federal Specification BB-0-925A covers oxygen for industrial use. Purity must be 99.5% oxygen or greater.

Nitrogen

Nitrogen is the largest single element in the atmosphere. It is colorless, odorless, flavorless, nontoxic, and is almost an inert gas. Nitrogen does not burn or support combustion. In the arc, or at high temperatures, nitrogen will combine with other gases. It is soluble in molten iron, but at room temperature the solubility is very low. During the cooling and solidification process, the nitrogen collects in pockets or precipitates out as iron nitrites. In very small amounts, nitrites can increase the strength and hardness of steel. In larger amounts, nitrogen can lead to porosity in the weld deposit. The reduction of ductility due to the presence of iron nitrites may lead to cracking of the weld

metal. The typical purity of compressed nitrogen is 99.8% by volume. The dew point is approximately $-70°$ F $(-57°$ C).

Liquid nitrogen is very cold, $-320°$ F $(-196°$ C) at atmosphere pressure. Accidental contact of liquid nitrogen will cause severe burns to the eyes or skin. Protective clothing and safety goggles or face shields must be worn when handling liquid nitrogen. Nitrogen tends to vaporize very easily and an accumulation of nitrogen can be hazardous since it does not support life. The Federal Specification BB-N-411C covers nitrogen of three purities; the minimum is 99.50% nitrogen.

Nitrogen is not a true inert gas and should not be used as a shielding gas for welding steel. Nitrogen is used in some parts of the world for welding copper. It provides an extra-high-temperature arc that is useful in overcoming the high thermal conductivity of copper when using the gas tungsten arc process. Tungsten electrode erosion is very high when using nitrogen. Mixtures of argon and nitrogen produce higher-quality welds than nitrogen alone.

Nitrogen is often used for purging stainless steel pipe and tubing systems. It is much less expensive than argon, and it keeps the oxygen away from the root side of the weld. Nitrogen is also used for maintaining positive pressures in piping systems during testing and cleaning operations.

14-4 GAS CONTAINERS AND APPARATUSES

Shielding gases and fuel gases must be transported, stored, and available at the point of use. The most convenient way is by portable cylinders which are easily taken to the job site. For installations where a high volume of gas is required, the bulk storage system is used or the gas is manufactured at the site. This requires equipment to pipe the gas to the welding or cutting stations. The design of piping systems is complex and should be done only by experts who are familiar with safety regulations and codes. In bulk form, the gas is supplied as a liquid. The capacity of a bulk system is normally between 3000 and 1 million cubic feet (84,950 to 28,310,000 l).

Carbon dioxide can also be obtained in bulk containers. The bulk system is only used when supplying a large number of welding stations and where usage will justify the bulk system.

There are four basic types of cylinders used for transporting welding gases (Figure 14-11). In addition, these types of cylinders come in different sizes according to the gas producer. These high-pressure cylinders are commonly used for transporting and storing argon, oxygen, hydrogen, nitrogen, and helium. This same type of cylinder is used for mixtures of these gases and mixtures of argon with CO_2. The cylinder of this type is shown in

Cylinder Identification	Cylinder Type[a]	Cylinder Contents	Capacity[a] (ft³)	Full Cylinder Pressure at 70° F		Approximate Weight[a] (lb)	
				psi	kPa	Full	Empty
Nonflammable compressed gas	DOT type 3A or 3AA	Argon	244	2200	15,168	158	133
			330	2640	1820	177	143
	High pressure	Argon+ oxygen	330	2640	1820	177	143
	High pressure	Argon+ carbon dioxide	379	2640	1820	177	143
He nonflammable compressed gas	High pressure	Helium	213	2200	15,168	135	133
Hydrogen flammable compressed gas	High pressure	Hydrogen	191	2015	13,892	134	133
O_2 nonflammable compressed gas	High pressure	Oxygen	330	2640	1820	172	146
			244	2200	15,168	153	133
Co_2 non-flammable compressed gas	Medium pressure	Carbon dioxide liquid+ gas	435	1000	6894	183	133
LPG or LP gas or PRY or MAPP	DOT type B 240	Liquid under vapor pressure	Varies by gas and supplier	94	648	Varies by gas and supplier	Varies by gas and supplier
C_2H_2 flammable compressed gas	DOT type 8 AL	Acetylene disolvent in acetone	390	250	1724	207	180

[a]The cylinder capacity and weights vary by supplier.

FIGURE 14–11 High-pressure gas cylinder types and sizes.

Figure 14-12. These cylinders are made under very strict manufacturing procedures and are covered by various laws. In the United States, the Department of Transportation provides the regulations. They are made of manganese steel (3A) or chrome molybdenum steel (3AA) and each must be inspected, numbered, and reinspected at regular intervals, usually every five years.[4] Typical cylinder markings are shown in Figure 14-13.

There are no uniform standards for cylinder sizes even though different gas companies offer standards within their own organization. There is no standard color code in the United States for the industrial gases. Some gas producers have standardized cylinder color codes within their own organization, however. There is a standardized identification system.[5] Either the total name of the gas or the international symbol of the gas is required on each cylinder. Each cylinder must carry a label showing the hazardous classification of the gas.[6] This information is given in the two charts showing the properties of the gases.

The valve connections of the different gas cylinders have been standardized so that regulators for the same gas can be readily attached to cylinders supplied by different gas producing companies. These standards apply to North America only.

Carbon dioxide (CO_2) welding-grade gas is available in high-pressure steel cylinders. Cylinders containing CO_2 are always labeled "CO_2" and may be labeled "welding grade." They are usually aluminum colored but no standard color code exists.

The standard welding-grade carbon dioxide cylinder (Figure 14-14) contains approximately 50 lb (22.7 kg) or 435 ft³ (12,317 l) of carbon dioxide under a pressure of 1000 psi (0.7 kg/mm²). In the CO_2 cylinder, at 70° F (21.1° C) the carbon dioxide is in both a liquid and a vapor form. The liquid carbon dioxide takes up approximately two-thirds of the space in the cylinder. Above the liquid the CO_2 exists as a gas. As the gas is drawn from the cylinder, the liquid carbon dioxide vaporizes to replace it. The normal discharge rate of the CO_2 cylinder is from about 4 to 30 ft³/hr (2 to 14 l/min). However, a maximum discharge rate of 25 ft³/hr (12 l/min) is normally recommended when welding using a single cylinder. In cold weather the discharge rate is reduced. As the CO_2 vapor pressure drops from the cylinder pressure to discharge pressure through the CO_2 regulator, it absorbs heat. If

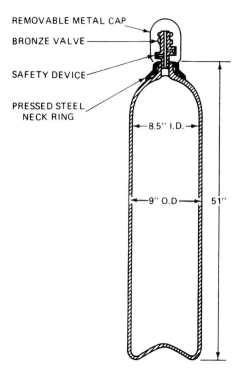

REMOVABLE METAL CAP
BRONZE VALVE
SAFETY DEVICE
PRESSED STEEL NECK RING

8.5" I.D.
9" O.D
51"

FIGURE 14–13 Typical number markings on a high-pressure cylinder.

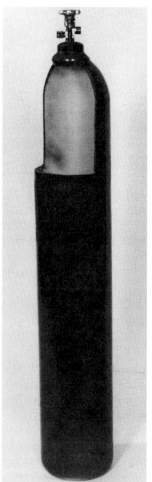

FIGURE 14–12 High-pressure gas cylinder.

FIGURE 14–14 Cylinder for carbon dioxide.

GAS

LIQUID

flow rates are too high, this absorption of heat can lead to freezing of the CO_2 regulator. When this happens, the gas shield is interrupted and weld porosity will result. When flow rates higher than 25 ft³/hr (12 l/min) are required, normal practice is to manifold two CO_2 cylinders in parallel or to place a heater between the CO_2 cylinder and the pressure regulator. As the liquid carbon dioxide is used, a drop in pressure will be indicated by the pressure gauge. When the pressure drops to 200 psi (1379 kPa), the cylinder should be replaced with a new one. A positive pressure should always be left in the cylinder to prevent moisture and other contaminants from entering. The valve should be closed.

Pressure is not an accurate measurement of cylinder contents. A partially used CO_2 cylinder should be weighed to determine how much CO_2 it still contains. To do this, weigh the cylinder then subtract the *tare weight* (weight of the cylinder when empty). Tare weight is usually stenciled on the cylinder neck. This gives the weight of the contents. At 70° F, there are 84.7 ft² of CO_2 per pound.

The liquefied petroleum gases are transported and stored in a different type of cylinder, one that is made to contain gas at a lower pressure. They are usually larger since the pressure is not so high. They are made to DOT specification B240 and are similar to the high-pressure cylinder except that they are usually larger in diameter.

When a gas is confined to a specific volume, the pressure exerted on the walls of the cylinder will vary in direct proportion with the temperature. Estimating the volume of gas remaining in a cylinder on the basis of gauge pressure is possible only within very broad limits. This is especially true of the liquefied gases.

Acetylene is transported and stored in a very special type of cylinder (Figure 14–15). This type of cylinder made to DOT specification 8AL is used only for acetylene. As mentioned previously, it is filled with a porous material soaked with acetone and the acetylene is dissolved in the acetone.

Where users have a shielding gas or oxygen demand of 10,000 to 20,000 ft³ per month, cryogenic liquid cylinders can be used (Figure 14–16). Cryogenic liquid cylinders are used for argon, carbon dioxide, nitrogen, and oxygen. The advantage of the cryogenic cylinder is that one cryogenic cylinder is equivalent to 15 to 24 high-pressure cylinders and they operate at a lower pressure. The cryogenic cylinders come in several sizes from different vendors but are all larger than high-pressure cylinders. They are approximately 20 or 24 in. in diameter and about 60 to 66 in. high. When empty they weigh between 250 and 280 lb; when filled they can weigh up to 700 lb. The cryogenic liquid cylinders are basically insulated vacuum bottles. The gas is stored as a liquid but is vaporized when it is withdrawn. The withdrawal rate can be as high as 250 ft³/hr. The service pressure is approximately 200

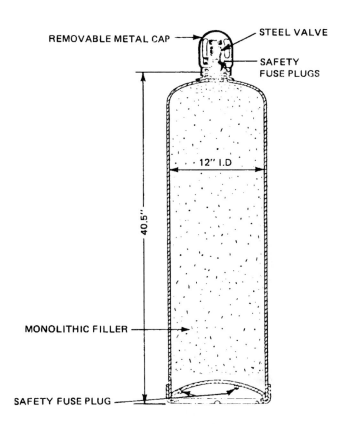

FIGURE 14–15 Cylinder for acetylene.

FIGURE 14–16   Cryogenic cylinder.

FIGURE 14–18   Manifolding cylinders for pipe distribution system.

psi. Cryogenic liquid cylinder data are shown in Figure 14-17. The hose connectors on cryogenic tanks are the same for the same gas as for high-pressure cylinders.

For fuel gas use exceeding the output of a single cylinder, the cylinders are manifolded together (Figure 14-18). Usually, flexible tubing is used to connect each cylinder to a manifold, which feeds the pipeline. In this way higher rates of gas can be supplied for multiple-torch flame cutting operations. For extremely large users of fuel gas, oxygen, or shielding gases, large liquid tanks or bulk storage tanks are used. These are filled from delivery trucks carrying liquefied gases. Typical insulated bulk storage tanks for oxygen and argon are shown in Figure 14-19 for liquid oxygen and in Figure 14-20 for liquid argon. The fuel gas tank shown in Figure 14-21 is not insulated. Heaters are sometimes used to convert the liquid to

a gas prior to piping it throughout the factory. Different-size tanks are available to satisfy the needs of a particular plant.

Apparatus

Various pieces of apparatus are required to utilize gas from high-pressure cylinders. These include regulators and flowmeters or combination units. The gas regulator was described previously. Its function is to reduce pressure and provide constant gas flow. Regulators must only be used for the gas for which they are designed.

The *flowmeter,* sometimes called a *rotometer,* contains two components: the adjustable needle valve, which allows for accurate control of gas flow, and a slightly tapered transparent tube that contains a float or indicator (Figure 14-22). The gas enters the flowmeter through the needle valve and then passes upward through the tapered tube. As it passes upward, the tapered tube is larger and the float is suspended in the stream of gas. The higher the flow rate, the higher the float will rise in the cali-

FIGURE 14–17   Cryogenic liquid cylinder data for 45-gallon type.

| Gas | Liquid Capacity[a] | | Product Weight[a] | | Gas Capacity (ft^3) |
	gal	l	lb	kg	
Argon	47.6	180	480	218	4.797 at 235 psi
Carbon dioxide	47.6	180	418	190	3.545 at 350 psi
Nitrogen	47.6	180	294	133	4.952 at 235 psi
Oxygen	47.6	180	415	188	5.010 at 235 psi

[a]Cylinder capacity and weight vary by supplier.

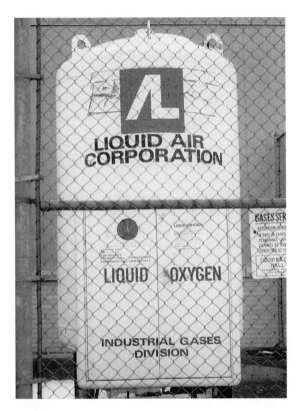

FIGURE 14–19 Oxygen liquid bulk storage tank.

FIGURE 14–20 Liquid argon bulk storage tank.

FIGURE 14–21 Liquid fuel gas bulk storage tank.

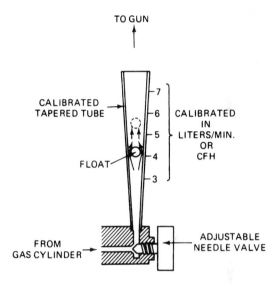

FIGURE 14–22 Diagram of gas flowmeter.

brated tube. The tapered tube is calibrated in either cubic feet per hour or liters per minute. It is important that the flowmeter is calibrated for gas being used. Different float weights are used for gases of different specific gravities. For extremely accurate work the discharge head or the resistance of the gas system beyond the flowmeter must be standardized and related to the calibration of the flowmeter. A typical flowmeter attached to a cylinder is shown in Figure 14-23. The flowmeter should be installed with the tube absolutely vertical for accurate measurement.

Orifice-type flowmeters can also be used; however, they are not adjustable and are installed in the line following the regulator to establish a specific flow rate of a specific gas. These are used when adjustments are not required.

Check valves are a safety device used to protect welding installations from the dangers of flashback and reverse flow. They prohibit a flashback in the torch from reaching the supply cylinders. These are spring-loaded

FIGURE 14–23 Gas flowmeter.

valves with rubber or neoprene actuators. With normal flow the check valve is open; however, if pressure from the downside exceeds flow pressure, the check valve will close. They are relatively inexpensive and should be used at every installation.

Gas Flow Rates

The flow rates required with the different shielding gases depend on the density or specific gravity of the gases. Welding procedures will provide the flow rates to be used.

When siphon tube CO_2 cylinders are used, external heaters are required. These cylinders are equipped with a plastic siphoning tube extending to the bottom of the cylinder that discharges liquid CO_2 rather than gaseous CO_2. The liquid must be vaporized by a heater. Siphon tube CO_2 cylinders are not popular in the United States, although they can be obtained by special order. They are used in Europe and in other parts of the world.

The final element in the system is the hose leading from the cylinder to the torch or to the control panel. Different types of hose are available, including rubber and various plastic materials. Each type has different characteristics with respect to pressure and the ability to maintain purity of the gas in the cylinder to the arc. It is best to check manufacturers' literature for this information.

Finally, there are such items as proportioners or mixers, which can be used to mix different gases together for a particular application. These are involved and should be used based on the manufacturers' recommendations.

QUESTIONS

14–1. What is the most common inert gas used for shielding?

14–2. Are mixed inert gases used for shielding?

14–3. What is the purpose of argon–oxygen for shielding gas?

14–4. What is the advantage of argon–CO_2 shielding gas on steel?

14–5. Why should active gases be used on aluminum?

14–6. Why is the density of a shielding gas important?

14–7. What happens to CO_2 in an arc? Away from the arc?

14–8. How is CO_2 gas specified? How is moisture checked?

14–9. What types of gases are used for GTAW welding?

14–10. Name the different fuel gases in use.

14–11. What fuel gas produces the highest flame temperature?

14–12. Does excess oxygen in the flame increase or decrease flame temperature?

14–13. What is the danger of fuel gases that are heavier than air?

14–14. What is the most abundant gas in the atmosphere?

14–15. Why is a higher-than-normal amount of oxygen dangerous?

14–16. What happens in oxygen flame cutting if the oxygen purity is low?

14–17. Describe the different types of tanks used for transporting welding gases.

14–18. Why are cryogenic cylinders of gas used?

14–19. Explain the principle of operation of a flowmeter?

14–20. What is the purpose of a check valve?

REFERENCES

1. Specifications for Welding Shielding Gases ANSI/AWS A5.32/A.32M American National Standards Institute/American Welding Society.

2. "Recommended Practices for Shielding Gases for Welding and Plasma Arc Cutting," C 5.10, American Welding Society, Miami, Fla.

3. "Threshold Limit Values for Chemical Substances and Physical Agents in the Workroom Environment," American Conference of Governmental Industrial Hygienists, Lansing, Mich.

4. "Code of Federal Regulations Support M—Compressed Gas and Compressed Air Equipment," Section 1910.166, "Inspection of Compressed Gas Cylinders," *Federal Register,* Vol. 36, No. 105, U.S. Government Printing Office, Washington, D.C.

5. "Registration Program for Cylinder Owner Symbols," CGAC-16-1991, Compressed Gas Association, Arlington, Va.

6. "American National Standard Method of Marking Portable Compressed Gas Containers to Identify the Material Contained," CGA C-4, Compressed Gas Association, Arlington, Va.

15

METALS AND THEIR WELDABILITY

OUTLINE

15-1 PROPERTIES OF METALS

When you look at a piece of metal, it may appear brownish, it may look bright or dull, it might appear gray; it has color. One of the physical properties of a metal is its color. When you lift a piece of metal, it may seem to be heavy or light; it has mass. When you bend a thin piece of metal, it may break or it may bend easily because it possesses ductility. If you attempt to melt the metal with a flame, it may become liquid quickly or it may not melt. Metals have different melting temperatures.

These are some of the many different properties of metals. They are used to help describe and specify them. Metals have physical properties such as density, melting point, color, conductivity, and others. They also have mechanical properties, which include strength, hardness, and ductility and all of them can be tested. Many mechanical and physical properties of metals determine

how they are used and how they can be welded. These properties determine how they will perform in service.

Physical Properties

The physical properties of metals are given in both the metric system and the conventional system. In time, we will all use the SI terms, which are based on international standards. Figure 15-1 shows the common metals and their physical properties. Each is briefly described.

Color   Color relates to the quality of light reflected from the metal.

Mass   Mass or density relates to mass with respect to volume. One of the more common ways of describing this property is by means of specific gravity, which is the ratio of the mass of a given volume of a metal to the mass of the same volume of water, at a specified temperature, usually 39° F (4° C). The mass of a volume of water is taken as unity, and the metals are related to it. In conventional terms this is taken as pounds per cubic foot of the metal or pounds per cubic inch. In the metric system this is taken as grams per cubic millimeter or centimeter.

Melting Point   The melting point is extremely important with regard to welding. A metal's fusibility is related to its melting point, the temperature when the metal changes from a solid into a molten state. Mercury is the only common metal that is in its molten state at normal room temperature. Metals having low melting temperatures can be

FIGURE 15–1 Physical properties of metals.

Base Metal or Alloy	Specific Gravity	Density lb/ft³	Density g/cm³	Melting Point (Liquidus) °F	Melting Point (Liquidus) °C	Boiling Point °F	Boiling Point °C	Relative Thermal Conductivity (Copper = 1)	Coefficient of Linear Expansion × 10⁻⁶ per Degree °F	Coefficient of Linear Expansion × 10⁻⁶ per Degree °C	Specific Heat (cal/g per °C)	Electrical Conductivity (%) (Copper = 100%)	Resistivity ($\mu\Omega$/cm)
Aluminum and alloys	2.70	166	2.7	1218	659	3270	2480	0.52	13.8	24.8	0.22	59.0	2.8
Brass, navy	8.60	532	8.6	1650	900	NA	NA	0.28	11.8	21.2	0.09	28.0	6.6
Bronze, aluminum (90 Cu–9 Al)	7.69	480	7.7	1905	1040	NA	NA	0.15	16.6	29.9	0.014	12.8	13.5
Bronze, phosphor (90 Cu–10 Sn)	8.78	551	8.8	1830	1000	NA	NA	0.12	10.2	18.4	0.09	11.0	16.0
Bronze, silicon (96 Cu–3 Si)	8.72	542	8.7	1880	1025	NA	NA	0.10	10.0	18.0	0.09	7.0	NA
Copper (deoxidized)	8.89	556	8.9	1981	1081	4700	2600	1.00	9.8	17.6	0.095	100.0	1.7
Copper nickel (70 Cu–30 Ni)	8.81	557	8.8	2140	1172	NA	NA	0.07	9.0	16.2	0.09	4.6	37.0
Everdur (96 Cu–3 Si–1 Mn)	8.37	523	8.4	1866	1019	NA	NA	0.09	10.0	18.0	0.095	NA	NA
Gold	19.3	1205	19.3	1945	1061	5380	2950	0.76	7.8	14.0	0.032	71.0	2.2
Inconel (72 Ni–16 Cr–8 Fe)	8.25	530	8.3	260	1425	NA	NA	0.04	6.4	11.5	0.109	1.5	98.1
Iron, cast	7.50	450	7.5	2300	1260	NA	NA	0.12	6.0	10.8	0.119	2.9	NA
Iron, wrought	7.80	485	7.8	2750	1510	5500	3000	0.16	6.7	12.1	0.115	15.0	NA
Lead	11.34	708	11.3	621	328	3100	1740	0.08	16.4	29.5	0.03	8.0	20.6
Magnesium	1.74	108	1.7	1202	650	2010	1100	0.40	14.3	25.7	0.246	37.0	5.0
Monel (67 Ni–30 Cu)	8.47	551	8.8	2400	1318	NA	NA	0.07	7.8	14.0	0.127	3.6	48.2
Nickel	8.8	556	8.8	2650	1452	5250	3000	0.16	7.4	13.3	0.105	23.0	7.9
Nickel silver	8.44	546	8.4	2030	1110	NA	NA	0.09	9.0	16.2	0.09	8.3	1.6
Silver	10.45	656	10.5	1764	962	4010	2210	1.07	10.6	19.1	0.056	106.0	—
Steel, high carbon	7.85	490	7.8	2500	1374	NA	NA	0.17	6.7	12.1	0.118	9.5	18.0
Steel, low alloy	7.85	490	7.8	2600	1430	NA	NA	0.12	6.7	12.1	0.118	14.5	12.0
Steel, low carbon	7.84	490	7.8	2700	1483	NA	NA	0.17	6.7	12.1	0.118	14.5	12.0
Steel, manganese (14 Mn)	7.81	490	7.8	2450	1342	NA	NA	0.04	6.7	12.1	0.210	NA	72.0
Steel, medium carbon	7.84	490	7.8	2600	1430	NA	NA	0.17	6.7	12.1	0.118	15.0	15.0
Steel, stainless (austenitic)	7.9	495	7.9	2550	1395	NA	NA	0.12	9.6	17.3	0.117	3.0	75.0
Steel, stainless (ferritic)	7.7	485	7.7	2750	1507	NA	NA	0.17	9.5	17.1	0.334	3.0	60.0
Steel, stainless (martensitic)	7.7	485	7.7	2600	1430	NA	NA	0.17	9.5	17.1	0.118	3.0	57.0
Tantalum	16.6	1035	16.6	5162	2996	7410	5430	0.13	3.6	6.5	0.052	13.9	12.5
Tin	7.29	455	7.3	449	232	4100	2270	0.15	12.8	23.0	0.125	13.5	11.0
Titanium	4.5	281	4.5	3031	1668	5900	3200	0.04	4.0	7.2	0.113	1.1	42.0
Tungsten	18.8	1190	19.3	6170	3420	10,600	5600	0.42	2.5	4.5	0.034	31.0	5.6
Zinc	7.13	442	7.1	788	419	1660	907	0.27	22.1	39.8	0.093	30.0	5.9

Note: NA, not available.

welded with lower-temperature heat sources. The soldering and brazing processes utilize low-temperature metals to join metals having higher melting temperature.

Boiling Point The boiling point is also an important factor. The boiling point is the temperature at which the metal changes from the liquid state to vapor state. In welding, some metals, when exposed to the heat of an arc, will boil and turn to vapor.

Thermal Conductivity The thermal conductivity of a metal is its ability to transmit heat throughout. It is of vital importance in welding since one metal may conduct heat from the welding area much more rapidly than another. It indicates the need for preheating and the size of heat source required. The thermal conductivity of metals is usually related to copper. Copper has the highest thermal conductivity of the common metals, exceeded only by silver. Aluminum has approximately half the thermal conductivity of copper, and steels have only about one-tenth the conductivity of copper. Some data use silver as the standard and rate the thermal conductivity with respect to silver. Thermal conductivity is measured in calories per square centimeter per second per degree Celsius. However, since we are using a relative figure these are not used.

Specific Heat Specific heat is a measure of the quantity of heat required to increase the temperature of a metal by a specific amount. Specific heat is important in welding since it is an indication of the amount of heat required to bring the metal to its melting point. A metal having a low melting point but having a relatively high specific heat may require as much heat to bring it to its point of fusion as a metal of a high melting point and low specific heat. Specific heat is the number of calories required to raise the temperature of 1 gram of metal 1 degree Celsius. It can be stated as a relative specific heat related to a standard. It is usually given at a standard temperature. Figure 15–1 provides the specific heat of the different metals based on the calories per gram per degrees C at 20° C.

Expansion The coefficient of linear thermal expansion is a measure of the linear increase per unit length based on the change in temperature of the metal. Expansion is the increase in the dimension of a metal caused by heat. The expansion of a metal in a longitudinal direction is known as the linear expansion. The coefficient of linear expansion is expressed as the linear expansion per unit length for 1 degree of temperature rise.

The expansion of the metal in volume is called the volumetric expansion. Linear expansion is most commonly used, and the data are available in both the conventional and metric values. The coefficient of linear expansion varies over a wide range for different metals. Aluminum has the greatest, expanding almost twice as much as steel for the same temperature change. This is important for welding with respect to warpage, warpage

control and fixturing, and for welding dissimilar metals together.

Electrical Conductivity Electrical conductivity is the capacity of metal to conduct an electric current. A measure of electrical conductivity is provided by the conductance of a metal to the passage of electrical current. The reciprocal of conductivity is resistivity. Electrical resistivity is measured in micro-ohms per cubic centimeter at a standardized temperature, normally 20° C. Electrical conductivity, however, is usually considered as a percentage and is related to copper or silver. Temperature bears an important part in this property. As the temperature of a metal is increased, conductivity decreases. This property is particularly important to resistance welding and to electrical circuits.

Mechanical Properties

The mechanical properties of metals determine the range of usefulness of the metal and establish the service that can be expected. Mechanical properties are also used to help specify and identify metals. They are vital to welding since the weld must provide mechanical properties in the same order as the base metals being joined. The adequacy of a weld depends on whether or not it provides properties equal to or exceeding those of the metals being joined.

The most common properties of the common metals—strength, hardness, ductility, and impact resistance—are shown in Figure 15–2.

Strength The strength of a metal is its ability to withstand the action of external forces without breaking. *Tensile strength,* also called *ultimate strength,* is the maximum strength developed in a metal in a tension test. The tension test is a method for determining the behavior of a metal under stretch loading. This test provides the elastic limit, elongation, yield point, yield strength, tensile strength, and the reduction in area. Tensile tests are normally taken at standardized room temperatures but may also be made at other temperatures. Figure 15–3 shows a tensile testing machine in operation. Many tensile testing machines are equipped to plot a curve that shows the load or stress and the strain or movement that occurs during the test operation. A typical curve for mild steel is shown in Figure 15–4. In the testing operation the load is increased gradually and the specimen will stretch or elongate in proportion to the tensile load. The specimen will elongate in direct proportion to the load during the elastic portion of the curve to point *A*. At this point, the specimen will continue to elongate but without an increase in the load. This is known as the yield point of the steel and is the end of the elastic portion. At any point up to point *A* if the load is eliminated, the specimen will come back to its original dimension. Yielding occurs from point *A* to point *B* and this is the area of plastic deformation. If the

Base Metal or Alloy	Yield Strength			Tensile Strength			Elongation % in 2 in. (50 mm)	Hardness (BHN)
	lb/in^2	MPa	kg/mm^2	lb/in^2	MPa	kg/mm^2		
Aluminum and alloys	5000	34.5	3.5	13,000	89.6	9.1	35	23
Brass, navy	30,000	206.8	21.0	62,000	427.4	43.6	47	89
Bronze, alum. (90 Cu–9 Al)	30,000	206.8	21.0	76,000	523.9	53.4	10	125
Bronze, phosphor (90 Cu–10 Sn)	28,000	193.0	19.7	66,000	455.0	46.4	35	148
Bronze, silicon (96 Cu–3 Si)	15,000	103.4	10.5	40,000	275.8	28.1	52	119
Copper (deoxidized)	10,000	68.9	7.0	33,000	227.5	23.2	40	30
Copper nickel (70 Cu–30 Ni)	20,000	137.9	14.0	55,000	379.2	38.6	45	95
Everdur (96 Cu–3 Si–1 Mn)	20,000	137.9	14.0	55,000	379.2	38.6	60	75
Gold	—	—	—	17,000	117.2	11.9	45	25
Inconel (72 Ni–16 Cr–8Fe)	35,000	241.3	24.6	85,000	586.0	59.7	45	150
Iron, cast	—	—	—	25,000	172.4	17.5	0.5	180
Iron, wrought	27,000	186.1	19.0	40,000	275.8	28.1	25	100
Lead	19,000	131.0	13.4	2500	17.2	1.7	45	6
Magnesium	13,000	89.6	9.1	25,000	172.4	17.5	4	40
Monel (67 Ni–30 cu)	35,000	241.3	24.6	75,000	517.1	52.7	45	125
Nickel	8500	58.6	6.0	46,000	317.1	32.3	40	85
Nickel silver	20,000	137.9	14.0	58,000	399.8	40.7	35	90
Silver	8000	55.2	5.6	23,000	158.6	16.2	35	90
Steel, high carbon	90,000	620.5	63.2	140,000	965.2	98.4	20	201
Steel, low alloy	50,000	344.7	35.1	75,000	517.1	52.7	28	170
Steel, low carbon	36,000	248.2	25.3	60,000	413.6	42.2	35	110
Steel, manganese (14 Mn)	75,000	517.1	52.7	118,000	813.5	82.9	22	200
Steel, medium carbon	52,000	358.5	36.5	87,000	599.8	61.2	24	170
Steel, stainless (austenitic)	40,000	275.8	28.1	90,000	620.5	63.2	23	160
Steel, stainless (ferritic)	45,000	310.2	31.6	75,000	517.1	52.7	30	155
Steel, stainless (martensitic)	80,000	551.5	56.2	100,000	68.9	70.3	26	250
Tantalum	—	—	—	50,000	344.7	35.1	40	300
Tin	1710	11.8	1.2	3130	21.6	2.2	50	5.3
Titanium	40,000	275.8	28.1	60,000	413.6	42.2	28	—
Tungsten	—	—	—	500,000	3447.0	351.5	15	230
Zinc	18,000	124.1	12.6	25,000	172.35	17.5	20	38

Note: Values depend on heat treatment or mechanical condition or mass of the metal.

FIGURE 15–2 Mechanical properties of metals.

load were eliminated at point *B,* the specimen would not go back to its original dimension but instead would take a permanent *set.* Beyond point *B* the load will have to be increased to stretch the specimen further. The load will increase to point *C,* which is the ultimate strength of the material. At point *C* the specimen will break and the load is no longer carried. The ultimate tensile strength of the material is obtained by dividing the ultimate load by the cross-sectional area of the original specimen. This provides the ultimate tensile strength in pounds per square inch or kilograms per square millimeter.

The *yield stress* or *yield point* is obtained by dividing the load at yield or at point *A* by the original area. This provides a figure in pounds per square inch or kilograms per square millimeter. Extremely ductile metals do not have a yield point. They stretch or yield at low loads. For

these metals the yield point is determined by the change in elongation. Two-tenths of 1% elongation is arbitrarily set as the yield point. The yield point is the limit upon which designs are calculated. Designs of weldments are expected to perform within the elastic limit, and the yield point is the measure of that limit.

Ductility The ductility of a metal is the property that allows it to be stretched or otherwise changed in shape without breaking and to retain the changed shape after the load has been removed. The ductility of a metal can be determined from the tensile test. This is done by determining the percent of elongation. Gauge marks are made 2 in. apart across the point where fracture will occur. The increase in gauge length related to the original length times 100 is the percentage of elongation. This is done by

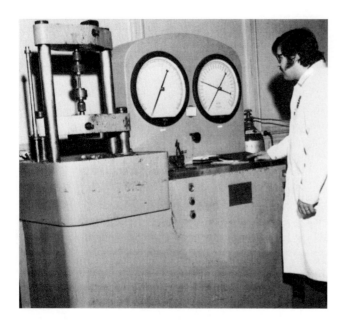

FIGURE 15–3   Tensile test machine.

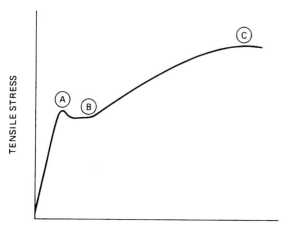

STRAIN OR UNIT DEFORMATION

FIGURE 15–4   Stress–strain curve.

making center punch marks 2 in. apart at the reduced section of the test coupon, testing the coupon, tightly holding the two pieces together and remeasuring the distance between the center punch marks. The original 2 in. is subtracted from the measured length, and the difference is divided by 2 and multiplied by 100 to obtain percentage of elongation.

Ductility of welds or of metals can also be measured by the bend test. In this case, gauge lines are drawn before testing, measured, and measured again after bending. The difference divided by the original length times 100 is the elongation in percentage.

The ductility is of extreme interest to welding since a higher ductility indicates a weld that would be less

FIGURE 15–5   Tensile specimens.

likely to crack in service. Figure 15–5 shows typical tensile specimens: unbroken, ductile, and brittle fracture.

Reduction of Area   Reduction of area is another measure of ductility and is obtained from the tensile test by measuring the original cross-sectional area of the specimen and relating it to the cross-sectional area after failure. For a round specimen the diameter is measured and the cross-sectional area is calculated. After the test bar is broken, the diameter is measured at the smallest point. The cross-sectional area is again calculated. The difference in area is divided by the original area and multiplied by 100 to give the percentage reduction of area. This figure is of less importance than the elongation but is usually reported when the mechanical properties of a metal are given.

The tensile test specimen also provides another property of metal known as its modulus of elasticity, also called *Young's modulus*. This is the ratio of the stress to the elastic strain. It relates to the slope of the curve to the yield point. For iron or steel the modulus of elasticity is approximately 30,000,000 psi, for aluminum it is approximately 10,300,000 psi, for copper 15,000,000 psi, and for magnesium 6,500,000 psi. The modulus of elasticity is important to designers and is incorporated in many design formulas.

Hardness   The hardness of a metal is defined as the resistance of a metal to local penetration by a harder substance. The hardness of metals is measured by forcing a hardened steel ball or diamond into the surface of the specimen, under a definite weight, in a hardness testing machine. Figure 15–6 shows the Vickers hardness testing machine. The Brinell is one of the more popular types of machines for measuring hardness. It provides a Brinell hardness number (BHN), which is in kilograms per square millimeter based on the load applied to the hardened ball in kilograms and divided by the area of the impression left by the ball in square millimeters. These metric factors

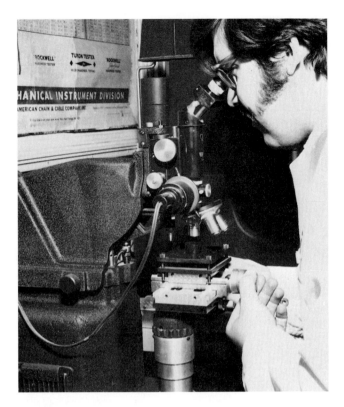

FIGURE 15–6 Vickers hardness testing machine.

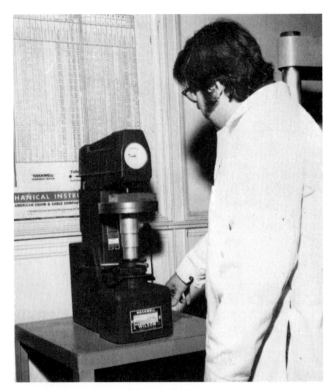

FIGURE 15–7 Rockwell hardness testing machine.

are disregarded and hardness is measured strictly as a Brinell hardness number. The BHN is obtained by an optical magnifier that reads the diameter of the impression and produces a hardness number. Various scales are used in the Brinell hardness testing system based on the load and the diameter of the ball, usually 10 mm using a 3000-kg load.

There are several other hardness measuring systems. A popular machine is the Rockwell hardness tester (Figure 15–7), which utilizes a diamond that is forced into the surface of the specimen. Different loads are used to provide different scales. Smaller loads are used for softer materials. The optical unit is not required since the hardness is read from a dial mounted on the machine that relates to the penetration. There is also the superficial Rockwell hardness tester for measuring surface hardnesses of metals. Another method is by means of the Vickers hardness machine, which reads directly as a diamond is pressed into the surface of the metal. Another way is the Shore scleroscope, which utilizes a small dropped weight that will bounce from the surface of the metal, providing a hardness measure. This device is used for field work and is not considered as accurate as those that make impressions. Figure 15–8 shows the hardness conversion numbers for these different hardness measuring systems. It is also possible to relate the approximate strength of a metal to its hardness as shown in the figure. Portable electronic units are now available for field use.

Impact Resistance Resistance of a metal to impacts is evaluated in terms of impact strength. A metal may possess satisfactory ductility under static loads but may fail under dynamic loads or impact. Impact strength is most often determined by the Charpy test. It is sometimes measured by the Izode test. Both types of tests use the same type of pendulum-testing machine. The Charpy test specimen is a beam supported at both ends and contains a notch in the center. The specimen is placed on supports and struck with a pendulum on the side opposite the notch. The accuracy and location of the notch is of extreme importance. There are several types of Charpy specimens; the V-notch type is the most popular. The specimen is standardized in metric dimensions. Figure 15–9 shows the impact testing machine in action.

The impact strength of a metal is determined by measuring the energy absorbed in the fracture. This is equal to the weight of the pendulum times the height at which the pendulum is released and the height to which the pendulum swings after it has struck the specimen. In conventional terms the impact strength is the foot pounds of energy absorbed. In metric practice, impact resistance is measured two ways: (1) the kilogram-meter based on energy absorbed and (2) the kilogram-meter per square centimeter of the area of the fractured surface or the cross-sectional area under the notch. Both terms are used, but care must be taken to determine which is appropriate. The SI system measures energy absorbed in

Brinell		Vickers or Firth Hardness No.	Rockwell		Scleroscope No.	Approximate Tensile Strength (1000 psi)
Diameter (mm): 3000-kg Load, 10-mm Ball	Hardness No.		C 150-kg Load 120° Diamond Cone	B 100-kg Load 1/16-in.-Diameter Ball		
2.05	898					440
2.10	857					420
2.15	817					401
2.20	780	1150	70		106	384
2.25	745	1050	68		100	368
2.30	712	960	66		95	352
2.35	682	885	64		91	337
2.40	653	820	62		87	324
2.45	627	765	60		84	311
2.50	601	717	58		81	298
2.55	578	675	57		78	287
2.60	555	633	55	120	75	276
2.65	534	598	53	119	72	266
2.70	514	567	52	119	70	256
2.75	495	540	50	117	67	247
2.80	477	515	49	117	65	238
2.85	461	494	47	116	63	229
2.90	444	472	46	115	61	220
2.95	429	454	45	115	59	212
3.00	415	437	44	114	57	204
3.05	401	420	42	113	55	196
3.10	388	404	41	112	54	189
3.15	375	389	40	112	52	182
3.20	363	375	38	110	51	176
3.25	352	363	37	110	49	170
3.30	341	350	36	109	48	165
3.35	331	339	35	109	46	160
3.40	321	327	34	108	45	155
3.45	311	316	33	108	44	150
3.50	302	305	32	107	43	146
3.55	293	296	31	106	42	142
3.60	285	287	30	105	40	138
3.65	277	279	29	104	39	134
3.70	269	270	28	104	38	131
3.75	262	263	26	103	37	128
3.80	255	256	25	102	37	125
3.85	248	248	24	102	36	122
3.90	241	241	23	100	35	119
3.95	235	235	22	99	34	116
4.00	229	229	21	98	33	113
4.05	223	223	20	97	32	110
4.10	217	217	18	96	31	107
4.15	212	212	17	96	31	104
4.20	207	207	16	95	30	101
4.25	202	202	15	94	30	99
4.30	197	197	13	93	29	97
4.35	192	192	12	92	28	95
4.40	187	187	10	91	28	93
4.45	183	183	9	90	27	91
4.50	179	179	8	89	27	89
4.55	174	174	7	88	26	87
4.60	170	170	6	87	26	85
4.65	166	166	4	86	25	83
4.70	163	163	3	85	25	82
4.75	159	159	2	84	24	80
4.80	156	156	1	83	24	78
4.85	153	153		82	23	76

FIGURE 15–8 Hardness conversion table.

| Brinell | | | Rockwell | | | |
Diameter (mm): 3000-kg Load, 10-mm Ball	Hardness No.	Vickers or Firth Hardness No.	C 150-kg Load 120° Diamond Cone	B 100-kg Load 1/16-in.-Diameter Ball	Scleroscope No.	Approximate Tensile Strength (1000 psi)
4.90	149	149		81	23	75
4.95	146	146		80	22	74
5.00	143	143		79	22	72
5.05	140	140		78	21	71
5.10	137	137		77	21	70
5.15	134	134		76	21	68
5.20	131	131		74	20	66
5.25	128	128		73	20	65
5.30	126	126		72		64
5.35	124	124		71		63
5.40	121	121		70		62
5.45	118	118		69		61
5.50	116	116		68		60
5.55	114	114		67		59
5.60	112	112		66		58
5.65	109	109		65		56
5.70	107	107		64		56
5.75	105	105		62		54
5.80	103	103		61		53
5.85	101	101		60		52
5.90	99	99		59		51
5.95	97	97		57		50
6.00	95	95		56		49

Note: Hardness conversion tables are approximate.

FIGURE 15–8 (*cont.*)

FIGURE 15–9 Making an impact test.

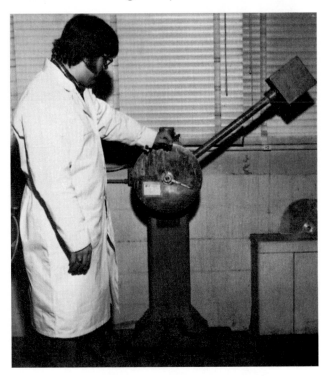

joules. Figure 15–10 shows V-notch Charpy impact bars before and after testing.

Impact tests are often made at different temperatures, since steels normally become more brittle or will absorb less energy at lower temperatures. Normally, seven specimens are broken at each test temperature and the high and low values are discarded. The reported value is the average of the remaining two specimens. Test temperatures are −60° F (−51° C), −50° F (−46° C), −40° F (−40° C), −20° F (−29° C), −14° F (−25° C), −4° F (−20° C), 0° F (−18° C), 14° F (−10° C), 32° F (0° C), 50° F (10° C), 68° F (20° C). All temperatures may not be used; however, usually five test temperatures are used so that a transition curve can be drawn. Figure 15–11 shows a temperature transition curve. At the point of *fall-off* the transition changes from ductile to brittle. This is known as the *transition temperature*. The change from ductile to brittle fracture can also be seen by the surface of the broken bars in the figure. Another measure of ductility is also utilized and this is the degree of lateral expansion of the bar at the fracture surface. The greater the degree of change, the more ductile the fracture. The fracture surface type is also reported for critical requirements. All these different tests and test specimens are standardized by the American Society for Testing and Materials.

FIGURE 15–10   V-notch Charpy impact bars.

FIGURE 15–11   Transition chart of impact tests.

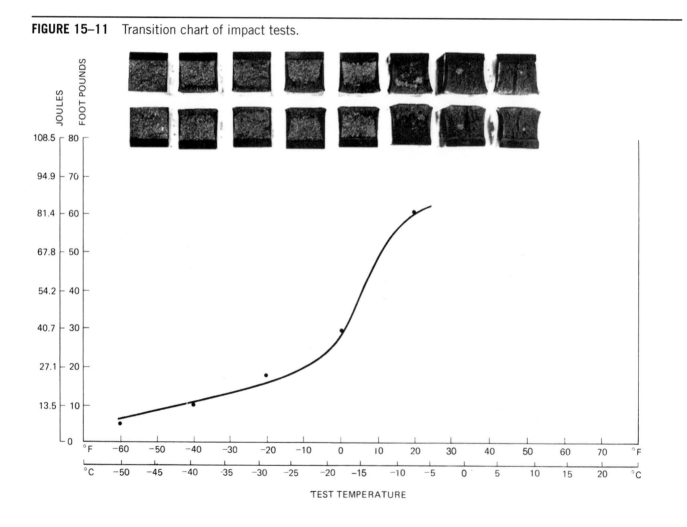

15-2 METAL SPECIFICATIONS AND STEEL CLASSIFICATIONS

There are many different ways that metals are identified and specified. These range from national specifications, professional society specifications, and trade association specifications to trade names of specific metals. In North America the most popular method of specifying a metal is by its ASTM number. ASTM stands for the American Society for Testing and Materials. ASTM, founded in 1898, is a scientific and technical organization that, among other things, produces standards on characteristics and performance of materials. It is the world's largest source of voluntary consensus standards. ASTM publishes over 60 volumes divided among 16 sections. There are three sections of interest to welding personnel:

* Section 1—Iron and Steel Products
* Section 2—Nonferrous Metal Products
* Section 3—Metals Test Methods and Analytical Procedures

Consult the ASTM standards book for a listing of specific volumes related to different metal specifications. Individual standards are available as separate copies.

Other sections of ASTM standards cover materials such as concrete, insulating materials, petroleum products, paint, textiles, plastics, and rubber.

The ASTM standards represent a consensus viewpoint of parties concerned with ASTM provisions: producers, users, and general interest groups. These are voluntary standards written by committees of the membership. The ASTM standards for metals provide the mechanical properties of the metal and in many cases chemical composition. Specifications for steels usually provide compositions that refer to either the analysis of the steel in the ladle or in its final form. The specifications also provide information concerning the form and size of the products, the size tolerance of products, testing procedures, inspection information, and so on.

The ASTM metal specifications are identified by prefix letters *A,* indicating ferrous materials, and *B,* indicating non-ferrous metals. This is followed by a one-, two-, or three-digit number indicating the exact specification number, which is then followed by a two-digit number indicating the year that the specification was formally adopted. A suffix letter *T,* when used, indicates that it is a tentative specification.

There are too many ASTM specifications to be listed here and it is therefore recommended that the sections or volumes be examined for a better understanding of the usefulness of the ASTM specifications.

Each part or volume is updated periodically and is available from the society. Many libraries have the ASTM volumes. Individual standards are available from the society.

The term *steel* encompasses many types of metals made principally of iron. Steel is an alloy of iron and carbon, but steels most often contain other metals such as manganese, chromium, and nickel, and nonmetals such as carbon, silicon, phosphorus, sulfur, and others. It is necessary to consider the different steels—how they are classified and identified.

There are so many different types and kinds of steels that it is sometimes confusing just to be able to identify the steel that is being used. For example, there are structural steels, cast steels, stainless steels, tool steels, hot rolled steel, reinforcing steel, low-alloy high-strength steel, and so on. Steels are sometimes given names based on their principal alloy, such as carbon steel, chrome-manganese steel, chrome-moly steel, and so on. Sometimes steels are identified by numbers, such as C-1020 steel, A36 steel, SAE 1045 steel, type 304 steel, and so on. In other cases, steels may be identified by letters, such as AR steels, T-1 steels, RQC steels, NAX steels, and so on. Steels are also called by a trade name given by their manufacturer. Examples of this are Mayari steel, Corten steel, Jalloy steel, and Naxtra steel. All these names tend to add to the confusion, but they are clues for finding the true identification of a steel.

The method of the manufacture of the steel also enters into the identification system. This would include cast steel, hot- or cold-rolled steels, forged steels, semikilled steels, and continuous cast steel.

The two best ways of identifying a steel are by its specification number and grade or trade name and number. In the case of the specification, it should be consulted to determine composition and properties, and it is necessary to determine the sponsoring group of the specification. When a trade name is used, the manufacturer's literature should be consulted for composition and properties. Of course, it is necessary to determine the manufacturer. When this is known, the identity of the steel can be determined.

A popular system for classifying steels is the American Iron and Steel Institute Numerical Designation of Standard Carbon and Alloy Steels.[1] This is known as the AISI designation system and sometimes known as the SAE system since it was originated by the Society of Automotive Engineers. The groupings of steels within this numerical system are shown in Figure 15-12. Numbers are used to designate different chemical compositions. A four-digit number series designates carbon and alloy steels according to the types and classes. This system has been expanded, and in some cases five digits are used to designate certain alloy steels. The last two digits are intended to indicate the approximate middle of the carbon range; for example, 0.21 indicates a range of 0.18% to 0.23% carbon. In a few cases, the system deviates from this rule and some carbon ranges relate to the ranges of manganese, sulfur, phosphorus, chromium, and other elements. Two letters are often used as a prefix to the numerals. The letter C

SAE Steel Specifications

The following numerical system for identifying carbon and alloy steels of various specifications has been adopted by the Society of Automotive Engineers.

COMPARISON
AISI-SAE Steel Specifications

The ever-growing variety of chemical compositions and quality requirements of steel specifications have resulted in several thousand different combinations of chemical elements being specified to meet individual demands of purchasers of steel products.

The SAE developed an excellent system of nomenclature for identification of various chemical compositions that have symbolized certain standards as to machining, heat treating, and carburizing performance. The American Iron and Steel Institute has now gone further in this regard with a new standardization setup with similar nomenclature but with restricted carbon ranges and combinations of other elements that have been accepted as standard by all manufacturers of bar steel in the steel industry, because it has become apparent that steel producers must concentrate their efforts on a smaller number of standardized grades. The Society of Automotive Engineers has, as a result, revised most of its specifications to coincide with those set up by the American Iron and Steel Institute.

PREFIX LETTERS

No prefix for basic open-hearth alloy steel.
(B) Indicates acid Bessemer carbon steel.
(C) Indicates basic open-hearth carbon steel.
(E) Indicates electric furnace steel.

NUMBER DESIGNATIONS—

(10XX series) Basic open-hearth and acid Bessemer carbon steel grades, nonsulfurized and nonphosphorized.
(11XX series) Basic open-hearth and acid Bessemer carbon steel grades, sulfurized but not phosphorized.
(1300 series) Manganese 1.60 to 1.90%.
(23XX series) Nickel 3.50%.
(25XX series) Nickel 5.0%.
(31XX series) Nickel 1.25%—Chromium .60%.
(33XX series) Nickel 3.50%—Chromium 1.60%.
(40XX series) Molybdenum.
(41XX series) Chromium—molybdenum.
(43XX series) Nickel—chromium—molybdenum.
(46XX series) Nickel 1.65%—molybdenum 0.25%.
(48XX series) Nickel 3.25%—molybdenum 0.25%.
(51XX series) Chromium.
(52XX series) Chromium and high carbon.
(61XX series) Chromium—vanadium.
(86XX series) Chrome—nickel—molybdenum.
(87XX series) Chrome—nickel—molybdenum.
(92XX series) Silicon 2.0%—chromium.
(93XX series) Nickel 3.0%—chromium—molybdenum.
(94XX series) Nickel—chromium—molybdenum.
(97XX series) Nickel—chromium—molybdenum.
(98XX series) Nickel—chromium—molybdenum.

Carbon Steels					
SAE Number	C	Mn	P Max.	S Max.	AISI Number
—	0.06 max.	0.35 max.	0.040	0.050	C1005
1006	0.08 max.	0.25–0.40	0.040	0.050	C1006
1008	0.10 max.	0.25–0.50	0.040	0.050	C1008
1010	0.08–0.13	0.30–0.60	0.040	0.050	C1010
—	0.10–0.15	0.30–0.60	0.040	0.050	C1012
—	0.11–0.16	0.50–0.80	0.040	0.050	C1013
1015	0.13–0.18	0.30–0.60	0.040	0.050	C1015
1016	0.13–0.18	0.60–0.90	0.040	0.050	C1016
1017	0.15–0.20	0.30–0.60	0.040	0.050	C1017
1018	0.15–0.20	0.60–0.90	0.040	0.050	C1018
1019	0.15–0.20	0.70–1.00	0.040	0.050	C1019
1020	0.18–0.23	0.30–0.60	0.040	0.050	C1020
—	0.18–0.23	0.60–0.90	0.040	0.050	C1021
1022	0.18–0.23	0.70–1.00	0.040	0.050	C1022
—	0.20–0.25	0.30–0.60	0.040	0.050	C1023

FIGURE 15–12 AISI-SAE numerical designation of carbon and alloy steels.

Carbon Steels					
SAE Number	C	Mn	P Max.	S Max.	AISI Number
1024	0.19–0.25	1.35–1.65	0.040	0.050	C1024
1025	0.22–0.28	0.30–0.60	0.040	0.050	C1025
—	0.22–0.28	0.60–0.90	0.040	0.050	C1026
1027	0.22–0.29	1.20–1.50	0.040	0.050	C1027
—	0.25–0.31	0.60–0.90	0.040	0.050	C1029
1030	0.28–0.34	0.60–0.90	0.040	0.050	C1030
1033	0.30–0.36	0.79–1.00	0.040	0.050	C1033
1034	0.32–0.38	0.50–0.80	0.040	0.050	C1034
1035	0.32–0.38	0.60–0.90	0.040	0.050	C1035
1036	0.30–0.37	1.20–1.50	0.040	0.050	C1036
1038	0.35–0.42	0.60–0.90	0.040	0.050	C1038
—	0.37–0.44	0.70–1.00	0.040	0.050	C1039
1040	0.37–0.44	0.60–0.90	0.040	0.050	C1040
1041	0.36–0.44	1.35–1.65	0.040	0.050	C1041
1042	0.40–0.47	0.60–0.90	0.040	0.050	C1042
1043	0.40–0.47	0.70–1.00	0.040	0.050	C1043
1045	0.43–0.50	0.60–0.90	0.040	0.050	C1045
1046	0.43–0.50	0.70–1.00	0.040	0.050	C1046
1050	0.48–0.55	0.60–0.90	0.040	0.050	C1050
—	0.45–0.56	0.85–1.15	0.040	0.050	C1051
1052	0.47–0.55	1.20–1.50	0.040	0.050	C1052
—	0.50–0.60	0.50–0.80	0.040	0.050	C1054
1055	0.50–0.60	0.60–0.90	0.040	0.050	C1055
—	0.50–0.61	0.85–1.15	0.040	0.050	C1057
—	0.55–0.65	0.50–0.80	0.040	0.050	C1059
1060	0.55–0.65	0.60–0.90	0.040	0.050	C1060
—	0.54–0.65	0.75–1.05	0.040	0.050	C1061
1062	0.54–0.65	0.85–1.15	0.040	0.050	C1062
1064	0.60–0.70	0.50–0.80	0.040	0.050	C1064
1065	0.60–0.70	0.60–0.90	0.040	0.050	C1065
1066	0.60–0.71	0.85–1.15	0.040	0.050	C1066
—	0.65–0.75	0.40–0.70	0.040	0.050	C1069
1070	0.65–0.75	0.60–0.90	0.040	0.050	C1070
—	0.65–0.76	0.75–1.05	0.040	0.050	C1071
1074	0.70–0.80	0.50–0.80	0.040	0.050	C1074

Alloy Steel									
AISI Number	C	Mn	P Max.	S Max.	Si	Ni	Cr	Other	SAE Number
1320	0.18–0.23	1.60–1.90	0.040	0.040	0.20–0.35	—	—	—	1320
1321	0.17–0.22	1.80–2.10	0.050	0.050	0.20–0.35	—	—	—	—
1330	0.28–0.33	1.60–1.90	0.040	0.040	0.20–0.35	—	—	—	1330
1335	0.33–0.38	1.60–1.90	0.040	0.040	0.20–0.35	—	—	—	1335
1340	0.38–0.43	1.60–1.90	0.040	0.040	0.20–0.35	—	—	—	1340
2317	0.15–0.20	0.40–0.60	0.040	0.040	0.20–0.35	3.25–3.75	—	—	2317
2330	0.28–0.33	0.60–0.80	0.040	0.040	0.20–0.35	3.25–3.75	—	—	2330
2335	0.33–0.38	0.60–0.80	0.040	0.040	0.20–0.35	3.25–3.75	—	—	—
2340	0.33–0.43	0.70–0.90	0.040	0.040	0.20–0.35	3.25–3.75	—	—	2340
2345	0.43–0.48	0.70–0.90	0.040	0.040	0.20–0.35	3.25–3.75	—	—	2345
E2512	0.09–0.14	0.45–0.60	0.025	0.025	0.20–0.35	4.75–5.25	—	—	2512
2515	0.12–0.17	0.40–0.60	0.040	0.040	0.20–0.35	4.75–5.25	—	—	2515
E2517	0.15–0.20	0.45–0.60	0.025	0.025	0.20–0.35	4.75–5.25	—	—	2517
3115	0.13–0.18	0.40–0.60	0.040	0.040	0.20–0.35	1.10–1.40	0.55–0.75	—	3115
3120	0.17–0.22	0.60–0.80	0.040	0.040	0.20–0.35	1.10–1.40	0.55–0.75	—	3120
3130	0.28–0.33	0.60–0.80	0.040	0.040	0.20–0.35	1.10–1.40	0.55–0.75	—	3130
3135	0.33–0.38	0.60–0.80	0.040	0.040	0.20–0.35	1.10–1.40	0.55–0.75	—	3135
3140	0.38–0.43	0.70–0.90	0.040	0.040	0.20–0.35	1.10–1.40	0.55–0.75	—	3140
3141	0.38–0.43	0.70–0.90	0.040	0.040	0.20–0.35	1.10–1.40	0.70–0.90	—	3141

FIGURE 15–12 (*cont.*)

Alloy Steel									
AISI Number	C	Mn	P Max.	S Max.	Si	Ni	Cr	Other	SAE Number
3145	0.43–0.48	0.70–0.90	0.040	0.040	0.20–0.35	1.10–1.40	0.70–0.90	—	3145
3150	0.48–0.53	0.70–0.90	0.040	0.040	0.20–0.35	1.10–1.40	0.70–0.90	—	3150
E3310	0.08–0.13	0.45–0.60	0.025	0.025	0.20–0.35	3.25–3.75	1.40–1.75	—	3310
E3316	0.14–0.19	0.45–0.60	0.025	0.025	0.20–0.35	3.25–3.75	1.40–1.75	—	3316
								Mo	
4017	0.15–0.20	0.70–0.90	0.040	0.040	0.20–0.35	—	—	0.20–0.30	4017
4023	0.20–0.25	0.70–0.90	0.040	0.040	0.20–0.35	—	—	0.20–0.30	4023
4024	0.20–0.25	0.70–0.90	0.040	0.035–0.050	0.20–0.35	—	—	0.20–0.30	4024
4027	0.25–0.30	0.70–0.90	0.040	0.20–0.35	0.20–0.35	—	—	0.20–0.30	4027
4028	0.25–0.30	0.70–0.90	0.040	0.040	0.035–0.050	—	—	0.20–0.30	4028
4032	0.30–0.35	0.70–0.90	0.040	0.040	0.20–0.35	—	—	0.20–0.30	4032
4037	0.35–0.40	0.70–0.90	0.040	0.040	0.20–0.35	—	—	0.20–0.30	4037
4042	0.40–0.45	0.70–0.90	0.040	0.040	0.20–0.35	—	—	0.20–0.30	4042
4047	0.45–0.50	0.70–0.90	0.040	0.040	0.20–0.35	—	—	0.20–0.30	4047
4053	0.50–0.56	0.75–1.00	0.040	0.040	0.20–0.35	—	—	0.20–0.30	4053
4063	0.60–0.67	0.75–1.00	0.040	0.040	0.20–0.35	—	—	0.20–0.30	4063
4068	0.65–0.70	0.75–1.00	0.040	0.040	0.20–0.35	—	—	0.20–0.30	4068
—	0.17–0.22	0.70–0.90	0.040	0.040	0.20–0.35	—	0.40–0.60	0.20–0.30	4119
—	0.23–0.28	0.70–0.90	0.040	0.040	0.20–0.35	—	0.40–0.60	0.20–0.30	4125
4130	0.28–0.33	0.40–0.60	0.040	0.040	0.20–0.35	—	0.80–1.10	0.15–0.25	4130
E4132	0.30–0.35	0.40–0.60	0.025	0.025	0.20–0.35	—	0.80–1.10	0.18–0.25	—
E4135	0.30–0.38	0.47–0.90	0.025	0.025	0.20–0.35	—	0.80–1.10	0.18–0.25	—
4137	0.35–0.40	0.70–0.90	0.040	0.040	0.20–0.35	—	0.80–1.10	0.15–0.25	4137
E4137	0.35–0.40	0.70–0.90	0.025	0.025	0.20–0.35	—	0.80–1.10	0.18–0.25	
4140	0.38–0.43	0.75–1.00	0.040	0.040	0.20–0.35	—	0.80–1.10	0.18–0.25	4140
4142	0.40–0.45	0.75–1.00	0.040	0.040	0.20–0.35	—	0.80–1.10	0.15–0.25	—
4145	0.43–0.48	0.75–1.00	0.040	0.040	0.20–0.35	—	0.80–1.10	0.15–0.25	4145
4147	0.45–0.50	0.75–1.00	0.040	0.040	0.20–0.35	—	0.80–1.10	0.15–0.25	—
4150	0.48–0.53	0.75–1.00	0.040	0.040	0.20–0.35	—	0.80–1.10	0.15–0.25	4150
4317	0.15–0.20	0.45–0.65	0.040	0.040	0.20–0.35	1.65–2.00	0.40–0.60	0.20–0.30	4317
4320	0.17–0.22	0.45–0.65	0.040	0.040	0.20–0.35	1.65–2.00	0.40–0.60	0.20–0.30	4320
4337	0.35–0.40	0.60–0.80	0.040	0.040	0.20–0.35	1.65–2.00	0.70–0.90	0.20–0.30	—
4340	0.38–0.43	0.60–0.80	0.040	0.040	0.20–0.35	1.65–2.00	0.70–0.90	0.20–0.30	4340
4608	0.06–0.11	0.25–0.45	0.040	0.040	0.25 Max	1.40–1.75	—	0.15–0.25	4608
4615	0.13–0.18	0.45–0.65	0.040	0.040	0.20–0.35	1.65–2.00	—	0.20–0.30	4615
—	0.15–0.20	0.45–0.65	0.040	0.040	0.20–0.35	1.65–2.00	—	0.20–0.30	4617
E4617	0.15–0.20	0.45–0.65	0.025	0.025	0.20–0.35	1.65–2.00	—	0.20–0.27	—
4620	0.17–0.22	0.45–0.65	0.040	0.040	0.20–0.35	1.65–2.00	—	0.20–0.30	4620
X4620	0.18–0.23	0.50–0.70	0.040	0.040	0.20–0.35	1.65–2.00	—	0.20–0.30	X4620
E4620	0.17–0.22	0.45–0.65	0.025	0.025	0.20–0.35	1.65–2.00	—	0.20–0.27	—
4621	0.18–0.23	0.70–0.90	0.040	0.040	0.20–0.35	1.65–2.00	—	0.20–0.30	4621
4640	0.38–0.43	0.60–0.80	0.040	0.040	0.20–0.35	1.65–2.00	—	0.20–0.30	4640
E4640	0.38–0.43	0.60–0.80	0.025	0.025	0.20–0.35	1.65–2.00	—	0.20–0.27	—
4812	0.10–0.15	0.40–0.60	0.040	0.040	0.20–0.35	3.25–3.75	—	0.20–0.30	4812
4815	0.13–0.18	0.40–0.60	0.040	0.040	0.20–0.35	3.25–3.75	—	0.20–0.30	4815
4817	0.15–0.20	0.40–0.60	0.040	0.040	0.20–0.35	3.25–3.75	—	0.20–0.30	4817
4820	0.18–0.23	0.50–0.70	0.040	0.040	0.20–0.35	3.25–3.75	—	0.20–0.30	4820
5045	0.43–0.48	0.70–0.90	0.040	0.040	0.20–0.35	—	0.55–0.75	—	5045
5046	0.43–0.50	0.75–0.100	0.040	0.040	0.20–0.35	—	0.20–0.35	—	4046
—	0.13–0.18	0.70–0.90	0.040	0.040	0.20–0.35	—	0.70–0.90	—	5115
5120	0.17–0.22	0.70–0.90	0.040	0.040	0.20–0.35	—	0.70–0.90	—	5120
5130	0.28–0.33	0.70–0.90	0.040	0.040	0.20–0.35	—	0.80–1.10	—	5130
5132	0.30–0.35	0.60–0.80	0.040	0.040	0.20–0.35	—	0.80–1.05	—	5132
5135	0.33–0.38	0.60–0.80	0.040	0.040	0.20–0.35	—	0.80–1.05	—	5135
5140	0.38–0.43	0.70–0.90	0.040	0.040	0.20–0.35	—	0.70–0.90	—	5140
5145	0.43–0.48	0.70–0.90	0.040	0.040	0.20–0.35	—	0.70–0.90	—	5145
5147	0.45–0.52	0.75–1.00	0.040	0.040	0.20–0.35	—	0.90–1.20	—	5147
5150	0.48–0.53	0.70–0.90	0.040	0.040	0.20–0.35	—	0.70–0.90	—	5150
5152	0.48–0.55	0.70–0.90	0.040	0.040	0.20–0.35	—	0.90–1.20	—	5152
E50100	0.95–1.10	0.25–0.45	0.025	0.025	0.20–0.35	—	0.40–0.60	—	50100

FIGURE 15–12 (*cont.*)

Alloy Steel									
AISI Number	C	Mn	P Max.	S Max.	Si	Ni	Cr	Other	SAE Number
E51100	0.95–1.10	0.25–0.45	0.025	0.025	0.20–0.35	—	0.90–1.15	—	51100
E52100	0.95–1.10	0.25–0.45	0.025	0.025	0.20–0.35	—	0.30–1.60	—	52100
								V	
6120	0.17–0.22	0.70–0.90	0.040	0.040	0.20–0.35	—	0.70–0.90	0.10 Min	—
6145	0.43–0.48	0.70–0.90	0.040	0.040	0.20–0.35	—	0.80–1.10	0.15 Min	—
6150	0.48–0.53	0.70–0.90	0.040	0.040	0.20–0.35	—	0.80–1.10	0.15 Min	6150
6152	0.48–0.55	0.70–0.90	0.040	0.040	0.20–0.35	—	0.80–1.10	0.10 Min	—
								Mo	
8615	0.13–0.18	0.70–0.90	0.040	0.040	0.20–0.35	0.40–0.70	0.50–0.60	0.15–0.25	8615
8617	0.15–0.20	0.70–0.90	0.040	0.040	0.20–0.35	0.40–0.70	0.50–0.60	0.15–0.25	8617
8620	0.18–0.23	0.70–0.90	0.040	0.040	0.20–0.35	0.40–0.70	0l.50–0.60	0.15–0.25	8620
8622	0.20–0.25	0.70–0.90	0.040	0.040	0.20–0.35	0.40–0.70	0.50–0.60	0.15–0.25	8622
8625	0.23–0.28	0.70–0.90	0.040	0.040	0.20–0.35	0.40–0.70	0.50–0.60	0.15–0.25	8625
8627	0.25–0.30	0.70–0.90	0.040	0.040	0.20–0.35	0.40–0.70	0.50–0.60	0.15–0.25	8627
8630	0.28–0.33	0.70–0.90	0.040	0.040	0.20–0.35	0.40–0.70	0.50–0.60	0.15–0.25	8630
8632	0.30–0.35	0.70–0.90	0.040	0.040	0.20–0.35	0.40–0.70	0.50–0.60	0.15–0.25	8632
8635	0.33–0.38	0.75–1.00	0.040	0.040	0.20–0.35	0.40–0.70	0.50–0.60	0.15–0.25	8635
8637	0.35–0.40	0.75–1.00	0.040	0.040	0.20–0.35	0.40–0.70	0.50–0.60	0.15–0.25	8637
8640	0.38–0.43	0.75–1.00	0.040	0.040	0.20–0.35	0.40–0.70	0.50–0.60	0.15–0.25	8640
8641	0.38–0.43	0.75–1.00	0.040	0.040–0.060	0.20–0.35	0.40–0.70	0.50–0.60	0.15–0.25	8641
8642	0.40–0.45	0.75–1.00	0.040	0.040	0.20–0.35	0.40–0.70	0.50–0.60	0.15–0.25	8642
8645	0.43–0.48	0.75–1.00	0.040	0.040	0.20–0.35	0.40–0.70	0.50–0.60	0.15–0.25	8645
8647	0.45–0.50	0.75–1.00	0.040	0.040	0.20–0.35	0.40–0.70	0.50–0.60	0.15–0.25	8647
8650	0.48–0.53	0.75–1.00	0.040	0.040	0.20–0.35	0.40–0.70	0.50–0.60	0.15–0.25	8650
8653	0.50–0.56	0.75–1.00	0.040	0.040	0.20–0.35	0.40–0.70	0.50–0.60	0.15–0.25	8653
8655	0.50–0.60	0.75–1.00	0.040	0.040	0.20–0.35	0.40–0.70	0.50–0.60	0.15–0.25	8655
8660	0.50–0.65	0.75–1.00	0.040	0.040	0.20–0.35	0.40–0.70	0.50–0.60	0.15–0.25	8660
8720	0.18–0.23	0.70–0.90	0.040	0.040	0.20–0.35	0.40–0.70	0.50–0.60	0.20–0.30	8720
8735	0.33–0.38	0.75–1.00	0.040	0.040	0.20–0.35	0.40–0.70	0.50–0.60	0.20–0.30	8735
8740	0.38–0.43	0.75–1.00	0.040	0.040	0.20–0.35	0.40–0.70	0.50–0.60	0.20–0.30	8740
8742	0.40–0.45	0.75–1.00	0.040	0.040	0.20–0.35	0.40–0.70	0.50–0.60	0.20–0.30	—
8745	0.43–0.48	0.75–1.00	0.040	0.040	0.20–0.35	0.40–0.70	0.50–0.60	0.20–0.30	8745
8747	0.45–0.50	0.75–1.00	0.040	0.040	0.20–0.35	0.40–0.70	0.50–0.60	0.20–0.30	—
8750	0.48–0.53	0.75–1.00	0.040	0.040	0.20–0.35	0.40–0.70	0.50–0.60	0.20–0.30	8750
—	0.50–0.60	0.50–0.60	0.040	0.040	1.20–1.60	—	0.50–0.80	—	9254
9255	0.50–0.60	0.70–0.95	0.040	0.040	1.80–2.20	—	—	—	9255
9260	0.55–0.65	0.70–1.00	0.040	0.040	1.80–2.20	—	—	—	9260
9261	0.55–0.65	0.75–1.00	0.040	0.040	1.80–2.20	—	0.10–0.25	—	9261
9262	0.55–0.65	0.75–1.00	0.040	0.040	1.80–2.20	—	0.25–0.40	—	9262
E9310	0.08–0.13	0.45–0.65	0.025	0.025	0.20–0.35	3.00–3.50	1.00–1.40	0.08–0.15	9310
E9315	0.13–0.18	0.45–0.65	0.025	0.025	0.20–0.35	3.00–3.50	1.00–1.40	0.08–0.15	9315
E9317	0.15–0.20	0.45–0.65	0.025	0.025	0.20–0.35	3.00–3.50	1.00–1.40	0.08–0.15	9317
9437	0.35–0.40	0.90–1.20	0.040	0.040	0.20–0.35	0.30–0.60	0.30–0.50	0.08–0.15	9437
9440	0.38–0.43	0.90–1.20	0.040	0.040	0.20–0.35	0.30–0.60	0.30–0.50	0.08–0.15	9440
9442	0.40–0.45	1.00–1.30	0.040	0.040	0.20–0.35	0.30–0.60	0.30–0.50	0.08–0.15	9442
9445	0.43–0.48	1.00–1.30	0.040	0.040	0.20–0.35	0.30–0.60	0.30–0.50	0.08–0.15	9445
9747	0.45–0.50	0.50–0.80	0.040	0.040	0.20–0.35	0.40–0.70	0.10–0.25	0.15–0.25	9747
9763	0.60–0.67	0.50–0.80	0.040	0.040	0.20–0.35	0.40–0.70	0.10–0.25	0.15–0.25	9763
9840	0.38–0.43	0.70–0.90	0.040	0.040	0.20–0.35	0.85–1.15	0.70–0.90	0.20–0.30	9840
9845	0.43–0.48	0.70–0.90	0.040	0.040	0.20–0.35	0.85–1.15	0.70–0.90	0.20–0.30	9845
9850	0.48–0.53	0.70–0.90	0.040	0.040	0.20–0.35	0.85–1.15	0.70–0.90	0.20–0.30	9850

FIGURE 15–12 (*cont.*)

indicates basic open hearth carbon steel, and E indicates electric furnace carbon and alloy steels. The letter H is sometimes used as a suffix to denote steels manufactured to meet hardenability limits. The first two digits indicate the major alloying metals in the steel such as manganese, nickel-chromium, chrome-molybdenum, and so on.

Other organizations that specify steels include the American Petroleum Institute (API), the American Association of Railroads (AAR), the American Bureau of Shipping (ABS), the Steel Founders Society of America (SFSA), the Society of Automotive Engineers (SAE), which include the AMS specifications, and many government agencies and private companies. In view of this, it is necessary to learn the different specifying groups and obtain copies of specifications or literature concerning the steels that are to be welded.

The AISI and SAE also collaborated on a system of identifying the corrosion-resistant or stainless steels. In this system, stainless and heat-resistant steels are classified into four general groups. They are identified by a three-digit number. The first number indicates the group and the last two numbers indicate the type. Modifications of types are indicated by suffix numbers. Information concerning the exact specifications for different steels will be given in Section 16-3.

Different branches of the U.S. federal government also write specifications for metals. The Department of Defense issues the MIL specifications, and the Department of Commerce issues the QQ specifications. Other groups may also issue metal specifications.

Professional societies and trade associations provide specifications for metals. For example, the American Petroleum Institute issues specifications covering the mechanical properties and composition of steel for pipe. The American Bureau of Shipping provides specifications for steels used in shipbuilding. The Association for American Railroads provides specifications for wrought and cast steels used by the railroad industry.

Steel castings are identified and specified by ASTM specifications and also by SAE, AAR, and ABS classes. These specifications provide for mechanical properties with various heat treatments and chemical compositions.

Specifications for materials used in the aircraft industry are designed as AMS, which stands for Aerospace Materials Specifications, a division of the Society of Automotive Engineers (SAE).

Most industrialized countries have national standards that provide specifications for metals. The British Standards (BS) issue specifications for many different types of metals, as do the German Standardization Institute (DIN) and the Japanese Standards Association (JIS). The standardization department of Russia issues GOST standards. Most of these standards are available in their original languages from the country's standardization society. The International Organization of Standards (ISO) also issues specifications for metals.

For nonferrous metals, several trade associations are involved. For example, the Aluminum Association provides a system of designating the compositions of aluminum alloys. This system uses four-digit numbers for wrought aluminum alloys. The AA also provides temper designations as suffix letters and numbers to indicate the temper condition. All of the aluminum producers in North America utilize the Aluminum Association alloy numbers for identifying their different products. More information is given in Section 17-1.

A standard designation for copper and copper alloys has been established by the Copper Development Association, Inc. Their system is used in North America and has been adopted by the U.S. government, ASTM, SAE, and nearly all producers of copper and copper alloy products. It is not a specification but rather an orderly method using a three-digit number of defining and identifying coppers and copper alloys. The system groups compositions into families, including the coppers, the high-copper alloys, the brasses, the bronzes, the copper-nickels, and the copper-zinc alloys. These numbers replace previous trade names such as copper-nickel, aluminum-bronze, phosphur-bronze, Naval-brass, and tough pitch copper.

The alloys of magnesium are identified by a special designation established by ASTM. Titanium, lead, tin, and other metals are specified by ASTM specifications. Specifications applicable to each metal will be provided in the chapter concerning each metal.

There are numerous alloys that are identified by trade names or numbers by their producer. Compositions of these alloys can be obtained from the producer. The Society of Automotive Engineers (SAE) also provides specifications for nonferrous metals.

Several books are available[1-4] that list different alloys and metals by trade name. For those who need to determine the analysis of different proprietary or trade name alloys, these books are of immense value.

15-3 IDENTIFICATION OF METALS

Often, particularly in repair work, it is necessary to weld on a material whose specification or trade name or composition is unknown. To produce a successful weld, it is necessary to know the composition of the metal being welded. There are a number of ways to determine the composition so that a welding procedure can be developed. The ability to make a rapid identification of the metal will reduce the time required to make a successful weld. If time permits, it is recommended that a piece of the metal be taken to a laboratory for analysis. Since time is rarely available, the following basis for identifying the metal should be followed. By using this technique and with experience, a fairly accurate identification can be made.

There are eight simple tests that can be performed to help identify metals. They will at least provide sufficient guidance to make a successful weld even though the exact composition may not be learned. Six of the different tests are summarized in Figure 15–13. This should be supplemented by Figures 15–1 and 15–2, which present physical and mechanical properties of metals.

Appearance Test

The first test is the appearance of the part. This includes features such as color and the appearance of the machined as well as unmachined surfaces. The shape can be descriptive; for example, shape includes such things as cast engine blocks, automobile bumpers, reinforcing rod,

FIGURE 15–13 Summary of identification tests of metals.

Base Metal or Alloy	Color	Magnet	Chisel	Fracture	Flame or Torch	Spark
Aluminum and alloys	Bluish-white	Nonmagnetic	Easily cut	White	Melts wo/col	
Brass, navy	Yellow or reddish	Nonmagnetic	Easily cut	Not used	Not used	
Bronze, aluminum (90 Cu–9 Al)	Reddish yellow	Nonmagnetic	Easily cut	Not used	Not used	
Bronze, phosphur (90 Cu–10 Sn)	Reddish yellow	Nonmagnetic	Easily cut	Not used	Not used	
Bronze, silicon (96 Cu–3 Si)	Reddish yellow	Nonmagnetic	Easily cut	Not used	Not used	
Copper (deoxidized)	Red; 1 cent piece	Nonmagnetic	Easily cut	Red	Not used	
Copper nickel (70 Cu–30 Ni)	White; 5 cent piece	Nonmagnetic	Easily cut	Not used	Not used	
Everdu (96 Cu–3 Si–1 Mn)	Gold	Nonmagnetic	Easily cut	Not used	Not used	
Gold	Yellow	Nonmagnetic	Easily cut	Not used	Not used	
Inconel (72 Ni–16 Cr–8 Fe)	White	Nonmagnetic	Easily cut	Not used	Not used	
Iron, cast	Dull gray	Magnetic	Not easily chipped	Brittle	Melts slowly	See text
Iron, wrought	Light gray	Magnetic	Easily cut	Bright gray fibers	Melts fast	See text
Lead	Dark gray	Nonmagnetic	Very soft	White; crystal	Melts quickly	
Magnesium	Silvery white	Nonmagnetic	Soft	Not used	Burns in air	
Monel (67 Ni–30 Cu)	Light gray	Slightly magnetic	Tough	Light gray	Not used	
Nickel	White	Magnetic	Easily cut	Almost white	Not used	See text
Nickel silver	White	Nonmagnetic	Easily chipped	Not used	Not used	
Silver	White; pre-1965 10¢ pc	Nonmagnetic	Not used	Not used	Not used	
Steel, high carbon	Dark gray	Magnetic	Hard to chip	Very light gray	Shows color	See text
Steel, low alloy	Blue-gray	Magnetic	Depends on composition	Medium gray	Shows color	See text
Steel, low carbon	Dark gray	Magnetic	Continuous chip	Bright gray	Shows color	See text
Steel, manganese (14 Mn)	Dull	Nonmagnetic	Work hardens	Coarse grained	Shows color	See text
Steel, medium carbon	Dark gray	Magnetic	Easily cut	Very light gray	Shows color	See text
Steel, stainless (austenitic)	Bright silvery	See text	Continuous chip	Depends on type	Melts fast	See text
Steel, stainless (ferritic)	Gray	Slightly magnetic	Continuous chip	Depends on type	Melts fast	See text
Steel, stainless (martensitic)	Bright silvery	Slightly magnetic		Depends on type		See text
Tantalum	Gray	Nonmagnetic	Hard to chip		High temperature	
Tin	Silvery white	Nonmagnetic	Usually as plating	Usually as plating	Melts quickly	
Titanium	Steel gray	Nonmagnetic	Hard	Not used	Not used	See text
Tungsten	Steel gray	Nonmagnetic	Hardest metal	Brittle	Highest temperature	
Zinc	Dark gray	Nonmagnetic	Usually as plating	Not used	Melts quickly	

I-beams or angle irons, pipes, and pipe fittings. Form should be considered and may show how the part was made, such as a casting with its obvious surface appearance and parting mold lines, or hot rolled wrought material, extruded or cold rolled with a smooth surface. Form and shape give definite clues. For example, pipe can be cast, in which case it would be cast iron, or wrought, which would normally be steel. Another example is a hot-rolled structural shape in a steel-framed building, which would be mild or low-alloy steel.

Color provides a very strong clue in metal identification. It can distinguish many metals such as copper, brass, aluminum, magnesium, and the precious metals. On oxidized metals, the oxidation can be scraped off to determine the color of the unoxidized metal. This helps to identify lead, magnesium, and even copper. The oxidation on steel, or rust, is usually a clue that can be used to separate plain carbon steels from the corrosion-resisting steels.

The use of the metal part is also a clue to identify it. Many machinery parts for agricultural equipment and light- and medium-duty industrial equipment are made of cast iron. For heavy-duty work such as brake presses, the castings would probably be steel. A railroad rail obviously can be identified by shape and this gives an immediate clue to its composition.

Hardness Test

A second test that should be used is the hardness test. Portable instruments are available that can be taken to the work. This gives a hardness indication, which helps determine the type of metal.

A less precise hardness test is the file test. A summary of the reaction to filing and the approximate Brinell hardness and the possible type of steel is given in Figure 15–14. A sharp mill file is used. It is assumed that the part is steel, and the file test will help identify the type of steel. Experience will help identify steel types with the file test.

Magnetic Test

The magnetic test can be quickly performed using a small pocket magnet. With experience it is possible to judge a strongly magnetic material from a slightly magnetic material. The nonmagnetic materials are easily recognized. The strongly magnetic materials include the carbon and low-alloy steels, iron alloys, pure nickel, and martensitic stainless steels. A slightly magnetic reaction is obtained from Monel and high-nickel alloys and the stainless steel of the 18% chrome–8% nickel type when cold worked, such as in a seamless tube.

The nonmagnetic materials are the copper-base alloys, aluminum-base alloys, zinc-base alloys, annealed 18% chrome–8% nickel stainless, the magnesiums, and the precious metals. With experience it is possible to use the magnetic test to help identify the various metals.

File Reaction	Brinell Hardness	Type of Steel
File bites easily into metal	100	Mild steel
File bites into metal with pressure	200	Medium-carbon steel
File does not bite into metal except with extreme pressure	300	High-alloy steel, high-carbon steel
Metal can only be filed with difficulty	400	Unhardened tool steel
File will mark metal but metal is nearly as hard as the file and filing is impractical	500	Hardened tool steel
Metal is harder than file	600+	

FIGURE 15–14 Approximate hardness of steel by the file test.

Chisel Test

The chip test or chisel test should also be used. The only tools required are a hammer and a cold chisel. The test is to use the cold chisel and hammer on the edge or corner of the material being examined. The ease of producing a chip is an indication of the hardness of the metal, and if the chip is continuous it is indicative of a ductile metal, whereas if chips break apart it indicates a brittle material. On such materials as aluminum, mild steel, and malleable iron the chips are continuous. They are easily chipped and the chips do not tend to break apart. The chips for gray cast iron are so brittle that they become small broken fragments. On high-carbon steel, the chips are hard to obtain because of the hardness of the material, but they can be continuous.

Fracture Test

This test is simple to use if a small piece of the metal being evaluated is available. The ease of breaking the part is an indication of its ductility or lack of ductility. If the piece bends easily without breaking, it is one of the more ductile metals. If it breaks easily with little or no bending, it is one of the brittle materials. The surface appearance of the fracture is also an indication. It will have the color of the base metal without oxidation. This will be true of copper, lead, and magnesium. In other cases, the coarseness or roughness of the broken surface is an indication of its structure. A careful study of known metals and how they appear at the fracture will help build experience to identify unknown specimens.

Flame or Torch Test

A high-temperature flame such as the oxyacetylene torch flame is used. The flame test should be used with discretion since it is possible that it will damage the part being investigated. If at all possible it should be used on a small piece of the metal being checked. The factors learned from this test are the rate of melting, the appearance of the molten metal and slag, and the action of the molten metal under the flame. All these factors provide clues that can aid in making the evaluation. When a sharp corner of a white metal part is heated, the rate of melting can be an indication. If the material is aluminum, it will not melt until sufficient heat has been used because of the high conductivity of aluminum. If the part is zinc, the sharp corner will melt quickly since zinc is not a good conductor. In the case of copper, if the sharp corner melts, it is normally deoxidized copper. If it does not melt until much heat has been applied, it is electrolytic copper. Also, with copper alloys, if lead is in the composition, it will boil, indicating a lead-bearing alloy. To distinguish aluminum from magnesium, apply the torch to filings. Magnesium will burn with a sparkling white flame. Steel will show characteristic colors before melting.

Spark Test

This is a very popular and reliable test for identification of different steels. The test requires the use of a high-speed grinding wheel, either fixed or portable. The grinding wheel should have a speed of at least 5000 surface feet per minute. The surface feet per minute equals the circumference in inches multiplied by the revolutions per minute divided by 12. Spark testing should be done in subdued light since the color of the spark is important. Spark testing is not used on nonferrous metals since they do not exhibit spark streams of any significance. This is one way to separate ferrous and nonferrous metals. For example, it can be used to separate stainless steel from high-nickel or copper-nickel materials, such as Monel, since sparks would be produced only by the stainless steel. It is advisable to have specimens or samples of known steels that can be sparked immediately before or after sparking the unknown material to help determine the unknown composition.

Spark testing has been thoroughly studied, and data are available with photographs showing the sparks that are produced. The spark resulting from the test should be directed downward and studied. Figure 15–15 shows the sparks resulting from a spark test. The color, shape, length, and activity of the sparks relate to characteristics of the material being tested. The spark stream has specific items that can be identified. The straight lines are called carrier lines. They are usually solid and continuous. At the end of the carrier line they may divide into three short lines called forks. If they divide into more lines at the end, they are called *sprigs*. Sprigs also occur at different places along the carrier line. These are sometimes called *bursts*, either star or fan bursts. In some cases, the carrier line will enlarge slightly for a very short

FIGURE 15–15 Spark test.

length, continue, and perhaps enlarge again for a short length. When these heavier portions occur at the end of the carrier line, they are called *spear points* or *buds*. High sulfur creates these thicker spots in carrier lines and the spearheads. Cast irons have extremely short streams, whereas low-carbon steels and most alloy steels have relatively long streams. Steels usually have white to yellow color sparks, while cast irons are reddish to straw yellow. By learning to identify the different portions of the spark, and by making tests on known samples, it is possible to acquire experience sufficient to make relatively accurate determinations of the metal being investigated. The following is a summary of some metals and the type of spark that is produced.

- *Cast iron:* dull red to straw yellow color, short spark stream, many small sprigs, short and repeating.
- *Wrought iron:* long straw-colored carrier lines, usually whiter away from the grinding wheel. Carrier lines usually end in spearhead arrows or small forks.
- *Low-carbon or mild steel:* long yellow carrier line; occasional forks and lines may end in an arrowhead.
- *Low-alloy steel:* each alloying element has an effect on the spark appearance and very careful observation is required. Type 4130 steel has carrier lines that often end in forks and sharp outer points with few sprigs.
- *High-carbon steel:* abundant yellow carrier lines with bright and abundant star bursts.
- *Manganese steel:* bright white carrier lines with fan-shaped bursts.
- *Stainless steels:* the chrome-nickel steels give off short carrier lines, sometimes making a dotted line without buds or sprigs.
- *Nickel:* extremely short spark stream. Carrier lines are orange. There are no forks or sprigs, and the sparks may follow the grinding wheel.

✳ *Titanium:* a very white spark stream. Carrier lines are uniform in size and terminate in forks and in arrowlike shapes at an angle from the carrier line.

Spark Atlas of Steels by Tschorn[5] is by far the most comprehensive authority on this subject. It relates to various national standards and has numerous pictures illustrating the spark stream produced by each type of material.

Chemical Test

There are numerous chemical tests that can be made in the shop for identifying some materials. Monel can be distinguished from Inconel by one drop of nitric acid applied to the surface. It will turn blue-green on Monel but will show no reaction on Inconel. A few drops of a 45% phosphoric acid will bubble on low-chromium stainless steels. These tests can become complicated and for this reason are not covered here. Metal identification kits, designed for portable use, are helpful (Figure 15–16). They electrically remove a minute amount of the test metal onto a filter paper. Reagents in the kit placed on the sample give distinct colors identifying metallic elements. For more information refer to ASTM booklet, STP 550.[6]

The use of these methods, coupled with samples of known metals and with experience, will enable you to make identifications sufficiently accurate for most welding requirements.

15-4 HEAT AND WELDING

Heat is employed in most welding processes. The heat source for welding may be generated in at least the following different ways:

1. An electric arc maintained between an electrode and the work
2. Resistance heating obtained by passing high current through the parts to be joined
3. A high-temperature flame obtained by burning a fuel gas with oxygen using a torch

FIGURE 15–16 Typical test kit.

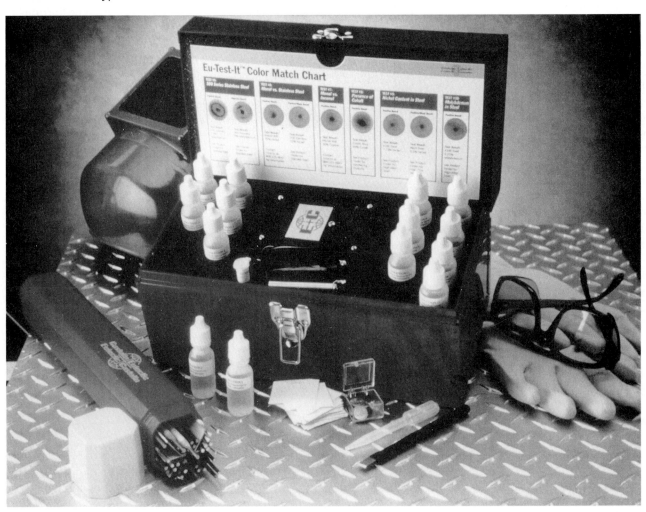

4. Mechanical sources from sliding friction, explosive impact, and ultrasonic vibrations

5. Exothermic chemical reaction producing super-heated liquid metal

6. Radiation from a focused high-energy beam of electrons

7. Radiation from a focused high-energy electromagnetic beam of coherent light

Heat is used to melt the surface of the metal to be welded so that coalescence, the growing together, can occur. Heat is also used to melt the filter metal that is added to the welding joint.

The most common source of heat for welding is the electric arc. The arc is a continuously moving heat source. Even though it moves, steady-state conditions are established and the temperature distribution relative to the heat source is relatively stable. The electric arc has a temperature of from 5000 to 20,000° C.

There are numerous detrimental effects of welding heat. Some of the disadvantages are:

1. High residual stresses from a localized heating cause differential shrinkage stresses, which may lead to warpage and distortion

2. A reduction of ductility or a degree of hardening in the heat-affected zone may lead to cracking

3. The deterioration of the toughness properties of the joint, primarily in the heat-affected zone

4. Loss of strength in the heat-affected zone of certain work-hardened quenched and tempered materials

The heat input–time–temperature relationship or thermal cycle of a weld cannot be precisely determined because there are so many variables involved. However, fairly accurate estimates can be made to predict or explain the effects of heat from a specific welding process on a given metal under practical conditions. The total heat input must be balanced to produce the desired weld. A complicating factor occurs when a *cold* filler rod is used to make the weld. Sufficient heat must be provided to melt the filler rod at the proper rate and add it to the molten pool. It is estimated that the temperature of the molten steel in the pool is 3500° F (1930° C). Extra heat is required, over and above the amount needed to melt the filler rod and the surfaces of the base metal, to compensate for the heat conducted away from the weld. It is necessary to control closely the amount of heat input while making a weld joint.

The heat developed by a moving arc can be calculated by the following:

$$HE = \frac{E \times I}{S} \times 60$$

where HE = energy input in joules per linear measure of weld (inches or millimeters)

E = arc voltage in volts
I = welding current in amperes
S = travel speed in lineal measure per minute (in./min or mm/min)

This energy input formula is used to calculate the heat developed in an arc and can be used in comparing welding procedures or for limiting heat input when welding quenched and tempered steels. Each welding process has a different thermal cycle. Figure 15–17 shows the time–temperature relationship of base metal taken immediately adjacent to welds made by two welding processes.[7] The rate of heat rise, the maximum temperature, the time at high temperature, and the rate of cooling of the metal are quite different for SMAW and ES welding. Processes with the highest concentration of heat cause the temperature to rise much more rapidly and to fall much more rapidly. The curve shown for the shielded metal arc weld rises almost instantaneously and the cooling rate of the base metal is a very steep slope, indicating quick cooling. The curve for electroslag welding rises more slowly, holds at a high temperature for a fairly long time, and then decreases slowly. The temperature changes that occur during an arc welding operation are much quicker and more abrupt than for most metallurgical processes. The metallurgical reactions from welding heat do not follow the normal heat-treating relationship. The temperature changes with electroslag welding are more similar to those encountered in foundry metallurgy.

The increase of heat in a metal increases the atomic mobility in the metal. When sufficient heat is absorbed by a solid, it will change to a liquid. In welding, it is necessary to produce a liquid at the surface of the parts being welded. The heat source is removed to allow cooling so that solidification or coalescence occurs and a weld is made. Heat moves rapidly in metal from one area to another area whenever there is a difference in temperatures.

FIGURE 15–17 Time–temperature relationship for various processes.

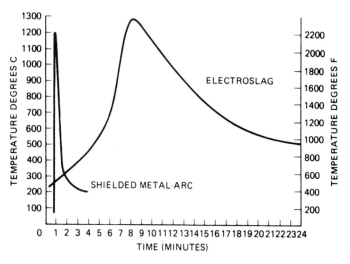

Heat will always move from the hot area to the cooler area. The welding heat source creates heat at a particular spot and is normally moving, at least in the arc processes. The heat is also moving so that a continually changing relationship occurs while welding. After an arc has stabilized, it will approach thermal equilibrium but never quite reach it. The rate at which the heat flows to the cold area depends upon the conductivity of the base metal. Heat also moves by means of convection, by radiation, and by absorption; however, for practical purposes most of the welding heat flows by means of conduction. Therefore the conductivity of the metal has a large influence on heat input-output time cycle relationship.

The temperature distribution around a point heat source can be shown by means of charts called *isotherms*.[8] These are lines that connect points of identical temperature. A typical temperature distribution curve made during the deposition of a shielded metal arc welding bead on thin material and on thick material is shown in Figure 15-18. This is the distribution of temperature around the arc with the arc at the highest temperature isotherm. The rise of the temperature, or the steepness of the curve, in front of the arc is much more

rapid than the fall of temperature, or the slope of the curve, behind the arc. This is due to the instantaneous heat transfer from the arc and the longer time for the heat to be removed. The influence of the thickness of the base metal is shown by the illustration. Using identical welding conditions, a much wider flow of heat is created in the thinner plate than in the thicker plate. This is due to the mass of the thicker plate and the fact that the heat flows in three directions in it rather than in two directions for the thin material.

Not all the heat generated by the heat source is used in making the weld. This is shown by relating the heat input, which is calculated by the formula shown, to the mass, or volume, of metal melted. The amount of heat required to melt a mass of metal is equal to the weight of the metal melted times the melting point (or degrees of temperature rise required) times the specific heat of the metal. It will be seen that from 20% to 75% of the energy available in the heat source is utilized in melting the metal. The percentage is different for different processes, procedures, base metals, base metal geometrics, and so on. For the shielded metal arc welding process, 70% to 85% of the heat is utilized in making the weld nugget. For

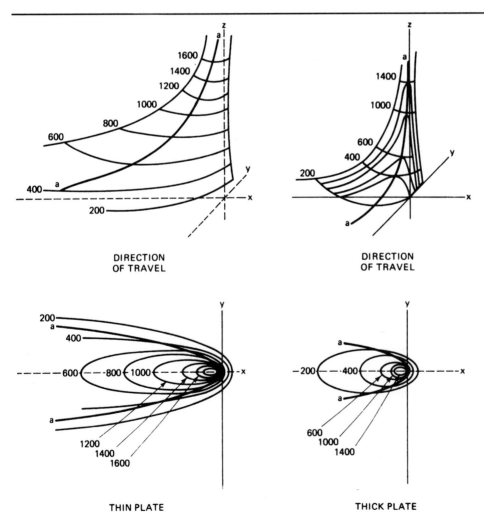

FIGURE 15–18 Temperature distribution around a moving arc.

the carbon arc welding process, 50% to 70% of the heat is utilized, while in submerged arc welding 80% to 90% of the heat generated by the arc is utilized in melting the weld metal. A major portion of this lost heat is used to raise the temperature of the base metal adjacent to the weld to near its melting point. Other losses come from weld spatter, heating the electrode and flux, and radiation and convection to the surrounding air.

In analyzing the effects of heat on a weld, a weld joint, or a weldment, it is necessary to determine:

1. The rate of heating
2. The maximum temperature attained
3. The length of time at temperature
4. The rate of cooling

These factors are difficult to determine; however, a good analysis of the potential damaging effects on the weld can be approximated. This allows precautions or procedure changes to minimize the harmful effects.

The *rate of heating* depends on a number of factors which include: the size and intensity of the heat source, the efficiency of the transfer of heat to the base metal, the utilization of heat in making the weld, the mass of the base metal, the joint geometry, and thermal conductivity. How much heat is in the heat source and how fast does it flow from the weld area? For example, a low-current gas tungsten arc can be struck on a large thick copper plate and it would be impossible to make a weld since the temperature of the copper plate would not rise sufficiently to cause melting. Not enough heat is produced by the tungsten arc to melt the copper since the conductivity of the copper and its mass cause the heat to flow away so rapidly. The thermal conductivity, the ability to transmit heat throughout its mass, is of major importance when considering the rate of heating. Figure 15-1 provides the relative thermal conductivity of the common metals. Numbers cannot be put into formulas to provide exact answers; however, the relationship of thermal conductivity of one metal versus another provides a clue. For example, steel has only about a tenth of the thermal conductivity of copper. With reference to the isotherm diagram, the conductivity of the metal has an effect on the steepness of the curve or change of temperature gradient. Extremely steep temperature gradients occur in welding as a result of the high temperatures of the heat source and the temperature of the base metal. Steepness of the rate of the heating curve is much different for the different processes.

The *maximum temperature* that will be attained in the base metal is also important. The base metal at the weld must be raised to its melting temperature and above. How *much* above is important and depends on the welding process. Efficient welding does not require the base metal to be raised much above its melting tempera-

ture. The welding processes that utilize extremely high temperature heat sources such as electron beam or laser beam can raise the base metal temperature so high that it will volatilize the metal. This is why some processes can be used for both welding and cutting, depending on the heat input. When welding thin sheet metal with too much heat input, the material becomes too hot. It melts rapidly and falls away, and holes are produced rather than welds. The maximum temperature reached by the base metal is related to the rate of heat input and to the rate of heat loss. As long as the heat input exceeds the rate of heat loss, the base metal will continue to get hotter. This relationship must continue until surface melting of the base metal occurs.

Another factor is the *specific heat* of the base metal. This is a measure of the quantity of heat required to increase the temperature of the metal. It relates to the amount of heat required to bring the metal to its melting point. A metal having a low melting temperature but with a relatively high specific heat may require as much heat to cause surface melting as a metal with a high-temperature melting point and a low specific heat. This can be seen by comparing aluminum to steel. The specific heat of the common metals is shown by Figure 15-1. The temperature required at the weld area should be only slightly greater than the melting temperature of the metal being welded. This is obtained in the base metal by balancing the heat input with the heat losses.

The length of time at the maximum temperature depends upon maintaining a heat balance between heat input and heat losses. There is rarely a true heat balance in any welding situation. During the arcing period, the heat input usually exceeds the heat losses and the base metal becomes hotter. Many times the welder must allow work to cool when the welding pool becomes too large and unmanageable. The current is reduced or the arc is broken, and the heat input is reduced or ceases. The heat losses continue, and the pool and base metal begin to cool. Normally the temperature of the work near the arc rises to a maximum. As soon as the arc moves on, the temperature adjacent to the weld begins to fall. The longer the base metal, adjacent to the weld, remains at a high temperature, the greater the possibility for grain growth in the weld metal and in the heat-affected area. The amount of metal melted, and the heat input and heat loss, affect this relationship.

The *rate of cooling* of the weld and adjacent metal is the rate of temperature change from welding temperatures to room temperature. The rate of cooling can be closely controlled and is governed by such conditions as heat transfer, heat losses, and thermal conductivity of the base metal. However, several factors must be considered since they can be used to regulate the cooling rate. The most important one is the initial temperature of the base metal before welding. A higher preheat or the more heat in the weldment, the slower it will cool.

The second important factor is the heat input that may be given to the weldment after the weld is made. It is usually desirable to reduce the cooling rate if metallurgical problems such as cracking or hard zones occur. Hard zones in or adjacent to the weld usually have lower toughness and ductility and tend to crack when thermal stresses are introduced. By reducing the cooling rate these can be eliminated and quality welds produced. Factors that increase the heat input or the mass of heat, or reduce the heat losses, will reduce the cooling rate, which will reduce the possibility of these defects.

15-5 WELDING METALLURGY

The science of joining metals by welding relates closely to the field of metallurgy. Metallurgy involves the science of producing metals from ores, of making and compounding alloys, and the reaction of metals to many different activities and situations. Heat treatment, steel making and processing, forging, and foundry all make use of the science of metallurgy. Welding metallurgy can be considered a special branch, since reaction times are in the order of minutes, seconds, and fractions of seconds, whereas in the other branches reactions are in hours and minutes.

Welding metallurgy deals with the interaction of different metals and the interaction of metals with gases and chemicals of all types. The welding metallurgist is also involved with changes in physical characteristics that happen in short periods. The solubility of gases in metals and between metals and the effect of impurities are all of major importance to the welding metallurgist.

In a general treatment of welding such as this, many metallurgical factors and practices are found throughout the entire book. This chapter presents a very brief coverage of welding metallurgy.

The structure of metals is complex. When metal is in a liquid state, usually hot, it has no distinct structure or orderly arrangement of atoms. The atoms move freely among themselves within the confines of the liquid. Their mobility allows the liquid metal to yield to the slightest pressure and to conform to the shape of the container. This high degree of mobility of the atoms is due to the heat energy involved during the melting process.

As molten metal cools, the heat energy of the atoms in the liquid state decreases and the atoms move with less mobility. As the temperature is further reduced and the metal cools, the atoms are no longer able to move and are attracted together into definite patterns. These patterns consist of three-dimensional lattices known as *space lattices,* which are made of imaginary lines connecting atoms in symmetrical arrangements. These imaginary lines are approximately the same distance from one another and limit the movement. Metals, in a solid state, possess this uniform arrangement, which is called *crystals.*

All metals and alloys are crystalline solids made of atoms arranged in a specific uniform manner.

There are over a dozen types of space lattices possible; however, the majority of common metals fall into only three: (1) the face-centered cubic lattice, (2) the body-centered cubic lattice, and (3) the hexagonal close-packed lattice. The metals and the form of the crystal lattice structure are shown in Figure 15–19. Note that iron has both the face-centered cubic structure and the body-centered cubic structure but at different temperatures. The change from one type of lattice structure to another takes place in the solid state with no change in specific gravity, but a small change in volume. This is known as an *allotropic change.*

The crystal lattices just mentioned are for pure metals that are composed of only one type of atom. Most metals in common use are alloys. In other words, they contain more than one metal. When more than one metal is present, the atoms making up the crystals will change. The atoms of the metal making up the minor portion of the alloy will at random replace some of the atoms of the metal making up the majority of the alloy. If the crystals are of essentially the same size, the minor metal will be considered to be dissolved in the major metal of the alloy. This condition is called a substitutional solid solution. A small amount of nickel added to copper will produce a substitutional solid solution.

If the atoms of the minor metal in the alloy are much smaller than those in the major lattice, they do not replace the atoms of the major metal in the lattice but rather locate in points between or in intervening spaces known as *interstices* in the lattice. This type of structure is called an *interstitial solid solution.* Very small amounts of carbon sometimes occur interstitially in iron.

If the minor metal atoms in the alloy cannot completely dissolve either interstitially or substitutionally, they will form the type of chemical compound the composition of which corresponds roughly to the chemical formula. This results in the formation of mixed kinds of atomic groupings consisting of different crystalline structures. These are referred to as *intermetallic compounds* and have a complicated crystal structure.

Each grouping with its own crystalline structure is referred to as a phase in the alloy, and the alloy is called a multiphase alloy. The individual phases may be seen and distinguished when examined under a microscope at extremely high magnification.

These different alloys, solid solutions, intermetallic compounds, and phases occur as the molten metal solidifies. Freezing or solidification of a liquid metal does not happen simultaneously throughout the entire melt. Freezing begins at the point of lowest temperature just below the liquids. At this point a small crystal forms, called a *nucleus.* Different nuclei may be formed almost simultaneously, and each is a point where solidification starts and the solidified metal grows from these points. The growth of solidification advances in all directions that are normal

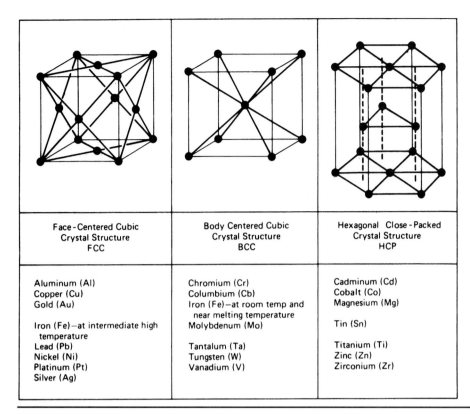

Face-Centered Cubic Crystal Structure FCC	Body Centered Cubic Crystal Structure BCC	Hexagonal Close-Packed Crystal Structure HCP
Aluminum (Al) Copper (Cu) Gold (Au) Iron (Fe)—at intermediate high temperature Lead (Pb) Nickel (Ni) Platinum (Pt) Silver (Ag)	Chromium (Cr) Columbium (Cb) Iron (Fe)—at room temp and near melting temperature Molybdenum (Mo) Tantalum (Ta) Tungsten (W) Vanadium (V)	Cadminum (Cd) Cobalt (Co) Magnesium (Mg) Tin (Sn) Titanium (Ti) Zinc (Zn) Zirconium (Zr)

FIGURE 15–19 Crystalline structure of common metals.

to the main axis of the nuclei crystal. Thus from a cubic crystal, growth progresses in six directions simultaneously. Growth is simply the adding on of additional crystals as temperature decreases. The growth continues and takes on a treelike pattern with branches and subbranches at right angles to one another. As solidification continues, the branches become thicker and larger and fill the spaces between additional branches, which are called *dendrites*. This continues until the entire mass has become solid. The dendritic growth of a crystal is shown in Figure 15–20.

The crystals that grow from one nucleus can grow only to the point where they come in contact with another crystal growing from a different nucleus. Since the nuclei occur randomly, growth of dendrites from nuclei crystals are at odd angles with one another and this does not permit the various crystals to merge into a single crystal. The completely solidified metal is made up of individual dendritic crystals that are oriented in different planes but held together by atomic attractive forces at the interface of adjacent dendrites. When this resultant structure is cut in a flat plane, the individual dendritic crystals, which grew until they met adjacent dendritic crystals, form an irregularly shaped area, which is known as a *grain*. The fitting together of the different grains is normally in an irregular outline shape and the interface between grains is known as grain boundaries (Figure 15–21). Grains are also very, very small but much larger than the individual crystals.

The size of the crystals and the grains depends on the rate of growth of the crystal. The rate of crystal growth depends on the rate of cooling of the molten solidifying metal. When the rate of cooling is high, the solidification process occurs more rapidly and the crystal size and grain size tend to be smaller. When the rate of cooling is slower, crystal and grain size tend to be larger. With extremely slow cooling or possibly with reheating, grains that have crystal axes almost parallel with one another will tend to grow together and it is possible that two grains will grow into one.

Grains can also grow relatively long and narrow because of the orientation of the nuclei. Grain growths that have proceeded in primarily one direction are those at the edge of a groove weld. The crystal is formed and

FIGURE 15–20 Dendritic growth from a nuclear crystal.

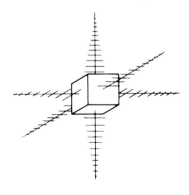

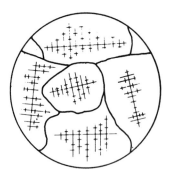

FIGURE 15–21 Grains are formed by dendrites growing together.

grows into elongated grains that produce a columnar structure. In restricted areas in which nuclei form close together, long grains are not possible and therefore more equiaxed grains result.

The overall arrangement of grains, grain boundaries, and phases present in an alloy is called its *microstructure*. The microstructure is largely responsible for the properties of the metal. It is affected by the composition or alloy content and by other factors such as hot or cold working, straining, heat treating, and so on. The microstructure of weld metal and adjacent metal is greatly influenced by the welding process and welding procedure, which influence the properties of the weld.

Anything done to the metal that will disturb or distort the lattice structure causes the metal to harden. Cold working of a metal distorts the structure and thereby hardens it. The presence of foreign atoms in the structure by alloy additions distorts the structure and tends to harden it. When atoms are dissolved in a solid state structure and are then precipitated out, the structure is distorted and thus hardened.

The grain boundaries contain lower melting point materials since the grain boundaries are the last portion to freeze or solidify. The strength of metals is sometimes determined by the grain boundaries. Grain boundaries increase the strength of some materials at room temperature by inhibiting the deformation of individual grains when the material is stressed. At elevated temperatures the atoms in the boundaries can move more easily and slide past one another, thus reducing the material strength. Fine-grained materials have better properties at room temperature. Metal structures can be characterized as having large grains (coarse grained) or small grains (fine grained) or a mixture of large and small (mixed grain size). The arrangement of atoms is irregular in the grain boundaries, and there are vacancies or missing atoms. The atom spacing may be larger than normal, and individual atoms can move more easily in the grain boundaries; because of this, the diffusion of elements, which is the movement of individual atoms through the solid structure, occurs more rapidly at grain boundaries. Odd-size

atoms segregate at the boundaries and this leads to the formation of undesirable phases that reduce the properties of a material by lowering the ductility while making it susceptible to cracking.

Phase Transformation

Some metals change their crystallographic arrangement with changes in temperature. Iron has a crystalline body-centered cubic lattice structure from room temperature up to 1670° F (910° C), and from this point to 2535° F (1388° C) it is face-centered cubic. Above this point, to the melting point of 2800° F (1538° C) it is again body-centered cubic. This change in crystalline structure is known as a *phase transformation* or an *allotropic transformation*. Other metals undergoing allotropic transformation at different temperatures are titanium, zirconium, and cobalt.

Another type of transformation occurs when the metal melts or solidifies. When the metal melts, the orderly crystalline arrangement of atoms disappears and there is then random movement of atoms. When the metal solidifies, the crystalline arrangement re-establishes itself. The change in crystalline structure or the change from liquid to solid is known as phase change. Pure metals melt or solidify at a single temperature, while alloys solidify or melt over a range of temperatures with a few exceptions.

The phase changes can be related to alloy composition and temperature when they are in equilibrium, and shown on a diagram. Such diagrams are called phase diagrams, alloy equilibrium diagrams, or constitution diagrams. Metallurgists have developed constitution diagrams for almost every combination of metal alloys.[9] By means of these diagrams, it is possible to determine the phases that are present and the percentage of each, based on the alloy composition at any specified temperature. In addition, it is possible to determine what phase changes tend to take place with increasing or decreasing temperature. Most constitutional phase diagrams describe alloy systems containing two elements. Diagrams of more than two elements are complex and difficult to interpret. Phase diagrams are based on equilibrium conditions. This means that the metal is stable at the particular point on the diagram based on relatively slow heating or cooling. In welding, this is not true since temperature changes are extremely rapid and equilibrium conditions rarely occur. Even so, the constitution diagram is the best tool available to determine phases.

The Iron-Carbon Diagram

An understanding of the iron–carbon equilibrium diagram shown in Figure 15–22 will provide an insight of the behavior of steels in connection with welding thermal cycles and heat treatment. This diagram represents the alloy of iron with carbon, ranging from 0% to 5% carbon.

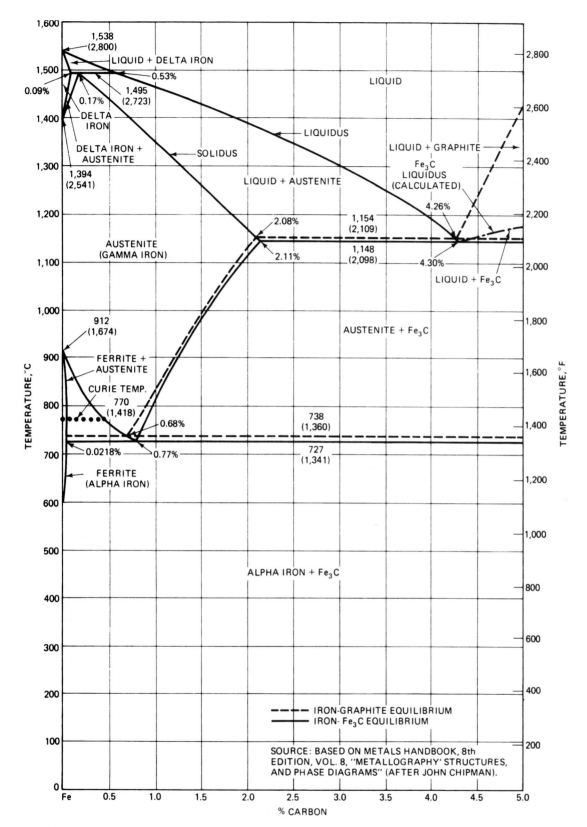

FIGURE 15–22 Iron–carbon equilibrium diagram. (After *Metal Progress Data Book,* © AMERICAN SOCIETY OF METALS, 1977.)

Pure iron is a relatively weak but ductile metal. When carbon is added in small amounts, the iron acquires a wide range of properties and uses and becomes the most popular metal, steel. It was previously mentioned that iron has either of two crystalline structures, depending on temperature. This can be shown by the iron–carbon diagram by considering pure iron or 0% carbon. Above 2800° F (1540° C) the iron is in liquid state and there is no crystalline structure. Below this temperature, it solidifies and has a body-centered cubic lattice which is known as *delta iron*. As the temperature is further reduced below 2540° F (1400° C), a transformation occurs and the crystalline structure changes to a face-centered cubic arrangement known as *gamma iron*. Below 1670° F (910° C) the iron transforms back to the body-centered structure, which is now known as *alpha iron*, and retains this structure down to room temperature. These transformation temperatures establish points on the iron–carbon diagram.

The other points and lines in the diagram show the percentage of carbon involved in solid solution. Iron and carbon form a compound known as iron carbide (Fe_3C) or cementite. When iron carbide or cementite is heated above 2100° F (1115° C), it decomposes into liquid iron saturated with graphite. Graphite is a crystalline form of carbon. Most metals have the ability to dissolve other elements in the solid state, and solid solutions are formed. Under suitable temperature and time conditions, the dissolved elements will diffuse and homogeneity will be obtained. A maximum solubility of carbon in alpha iron occurs at 1340° F (727° C) and decreases with lower temperature. This establishes a point on the diagram. A solid solution of carbon in alpha iron or delta iron (body-centered cubic) is known as ferrite. A solid solution of carbon in gamma iron (face-centered cubic) is known as austenite. As much as 2.1% carbon can be held in solid solution in gamma iron at a specific temperature and this establishes a point. In fact, the iron–carbon diagram can be divided at this point. Those alloys of iron and carbon less than 2.1% are considered steels, while those containing more than 2.1% are referred to as cast irons. Thus the line on the chart indicates whether or not the carbon is held in solid solution or whether it precipitates out.

To better understand the iron–carbon diagram, consider a steel with a composition of 0.25% carbon. This would be indicated by drawing a vertical line between the 0% and 0.5% carbon line. Considering this line, it will be seen that above approximately 2768° F (1520° C) the steel would be molten. As the temperature decreases, delta iron would start to form in the liquid. At just below 2732° F (1500° C) it would transform to austenite and molten metal. However, at about 2696° F (1480° C) all the liquid metal would be solidified and it would be austenite. At approximately 1500° F (815° C) the austenite commences to break down and form a new phase at the grain boundaries. This new phase is almost pure iron or ferrite. Ferrite formation would continue until a temperature of 1340° F

(727° C) was reached. At this point, the remaining austenite disappears completely, transforming to a structure known as pearlite plus ferrite. Pearlite is a mixture of ferrite and cementite and this structure would be retained down to room temperature. The microstructure of pearlite is shown in Figure 15-23. Only the room-temperature structures can be seen and analyzed through a microscope.

By means of alloy additions it is possible to have the other structures at room temperature. The microstructure of pearlite is a lamellar structure, which is relatively strong and ductile. The transformation during the cooling cycle will be reversed during the heating cycle. In welding the rise and fall of temperature or the rate of change of temperature is so fast that equilibrium does not occur. Therefore, some of the structures and temperatures mentioned here will be different. For example, if the cooling rate is faster, the austenite-to-ferrite transformation will be appreciably lower in temperature and this will also be true of pearlite. The pearlite will be more finely laminated since the transformation temperature is much lower. With extremely fast cooling rates the austenite might not have sufficient time to transform completely to ferrite and pearlite and will provide a different microstructure. In this case, some of the untransformed austenite will be retained and the carbon is held in a supersaturated state. This new structure is known as martensite and is shown in Figure 15-24. Martensite has a needlelike appearance, and if the cooling rate is sufficiently fast, the austenite

FIGURE 15–23 Pearlite microstructure.

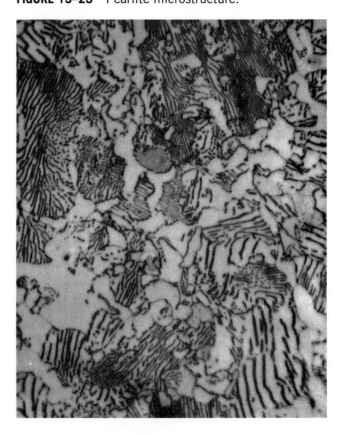

FIGURE 15–24 Martensite microstructure.

might transform completely to martensite. Martensite is harder than pearlite or the pearlite–ferrite microstructure and it has lower ductility. Its hardness depends on the carbon content. Thus, it can be seen that the cooling rate influences the microstructure and causes higher hardness. This is because the crystal lattice is changed or distorted and this in turn hardens the material.

By adding different alloys to the steel, the tendency of austenite to transform into martensite upon cooling increases. This is the basis of hardening steels. By proper use of different alloys, the amount of martensite produced can be changed. The rate of cooling changes depending on the method of quenching. The more severe quench will create more martensite, and the slower quench will create less martensite and thus a lower hardness. The amount of alloys and their power to create this microstructure transformation are known as *hardenability*. This is an advantage for heat treatment but can be detrimental to welding since high hardness is not desired in welds of softer materials.

The lines on the iron–carbon diagram show the different phases of iron–carbon alloy and indicate the microstructure of each phase. With respect to the iron–carbon alloy, only pearlite and ferrite and pearlite and cementite (Fe_3C) are found at room temperature. However, in other iron-alloy systems other microstructures exist at room temperature. Ferrite is an example. Ferrite is a solid solution of carbon in delta or alpha iron. It has a body-centered cubic structure and occurs when less than 0.08% carbon is dissolved in the iron. It is the softest constituent in steel, and as the amount of ferrite increases the steel is softer. In alloy steels, certain alloy elements may be dissolved in ferrite as a solid state solution so that it occurs at room temperature. The microstructure of ferrite is shown in Figure 15–25.

FIGURE 15–25 Ferrite microstructure.

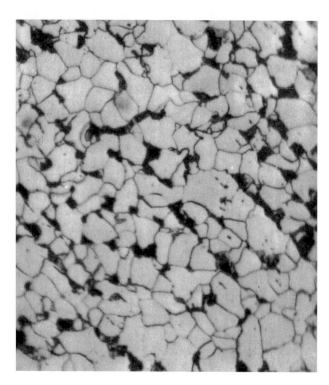

The microstructure of cementite (Figure 15–26) is hard and wear resistant, and the composition will be varied when other carbide-forming alloys are present. It appears in several different ways, sometimes as a network surrounding the grains in the region of grain boundaries and also within the grains. It may also appear as round, roughly globular-shaped particles in steel that has been specially heat treated.

Austenite is another important constituent. It is the face-centered cubic lattice form and occurs in low-carbon steels in temperatures above 1333° F (722° C). It is not stable at room temperatures in carbon steel; however, in highly alloyed steels and stainless steels it is stable at room temperature. It has good tensile strength and is ductile, but it has a strong tendency to work-harden and is shown in Figure 15–27. There are other microstructures in steels; however, the foregoing are the most important.

Hardenability

The heat treatment of steels to increase hardness and the metallurgy of welding have much in common. Heat treating to increase hardness is accomplished by heating followed by rapid cooling. The rapid cooling of metal in and adjacent to a weld is in this same order. An understanding of hardening by heat treating will make the metallurgical changes during welding clearer. Most steels possess the property of hardenability, which is defined as the property that determines the depth and distribution of hard-

---

FIGURE 15–26 Cementite microstructure.

ness induced by quenching. This property can be measured by the end quench test. In this test, a round bar is heated to a temperature in the austenite range and is then placed vertically in a fixture, and a jet of water is directed upwards to the bottom end. The bar rapidly cools to room temperature. Hardness measurements are then made along the bar from the quenched end to the unquenched end and plotted against distance. This produces a curve with the high hardness at the quenched end and dropping off to normal hardness at the unquenched end. This hardenability curve shows the maximum hardness, the depth of hardness, and so on, under standardized conditions. It is very useful in establishing heat-treating procedures. This information also provides data for welding since it indicates the effect of different alloying elements on the hardness of the quenched steel. The microstructure of the quenched steel can also be studied and related to the microstructure of welds.

Grain size and microstructure relate directly to hardness and strength. It was mentioned previously that as crystals deform they become harder, and it is the growth of crystals that dictates grain size. In heat treating, the steel is heated above the critical temperature, a phase change occurs, and new grains will nucleate and grow within the old grains. Since new grains of austenite were formed within each of the former grains, the steel now has more finer grains. Fine grain size promotes both increased strength and hardness.

Alloying elements are added to steel to increase its hardenability. Carbon is the most important and effective, and small amounts will greatly increase hardness up to about 0.65%. Manganese is the next most important. Chromium and molybdenum also increase hardenability. These alloys contribute to other properties as well.

The time required for transformation of steel to begin and end at any constant temperature is a useful measure of the heat-treating characteristics of the steel. This can be shown by diagrams known as isothermal transformation diagrams, which plot temperature against time in seconds. Diagrams are available for steels of different composition and they show the phases that occur at different temperatures and times. The lines are constructed from experimental data and show the process of transformation at each temperature. Isothermal diagrams help to explain the relationship between cooling rates and the microstructure of a specific steel composition. During cooling the steel remains a certain period of time in each phase, and the time period is inversely proportional to the cooling rate. With very slow cooling rates the phase changes take place near equilibrium values. As the cooling rate is increased, time is reduced so that there is not enough time for completion of the pearlite reaction and some austenite remains below the equilibrium transformation temperature, which creates hardening. The cooling rate must be so rapid that much of the austenite transforms to martensite rather than pearlite. These curves are

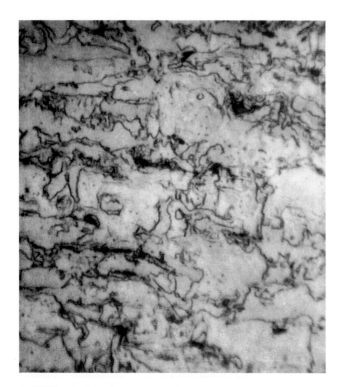
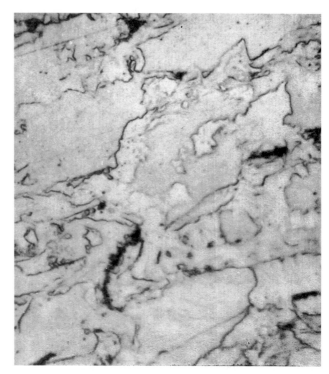

FIGURE 15–27 Austenite microstructure.

of limited use in welding but do show the types of transformation occurring at subcritical temperatures and their effects.

Welds

When a weld is made, all the factors just mentioned occur: the changes of temperature, the changes of dimensions, the growth of crystals and grains, the phase transformation, and so on. The type of welding process dictates, in general, how these will occur. In the arc welding processes the heat cycle is of primary importance. The previous section explained the heat input time-temperature relationship or thermal cycle. The rate of cooling or quench is of primary importance and this is controlled by the process, procedure, metal, and mass. The electroslag process has the slowest cooling rate, while the gas metal arc has a much faster cooling rate. The rate of change decreases as the distance from the center of the weld increases (Figure 15-28). It is obvious that many different cooling rates occur and that different microstructures will result. This is shown in Figure 15-29, which is a multipass groove weld. The different structures that occur in the weld are shown, as well as the different phases shown in the base metal adjacent to the weld.

With any arc process, where metal is transferred across the arc, the metal reaches a superheated temperature much above the melting temperature shown on the iron-carbon diagram. When it is deposited in the weld, it is molten or in the liquid phase. Immediately the weld

metal starts to freeze or solidify. The heat contained in the molten metal is transmitted to the base metal, and its temperature at the weld is raised to the molten stage. Away from the weld the metal is raised to a lower temperature. This creates a multitude of time-temperature curves based on location. As the weld metal freezes, the crystals form into grains that rapidly cool until there is no more liquid metal. The cooling rate is much faster than occurs in a casting or ingot and therefore, equilibrium, as represented in the iron-carbon diagram, really does not occur.

In addition to the complications created by the rapid cooling, there is also the complication of composition variations. As weld metal is deposited on base metal, some of the base metal melts and mixes with the weld metal, producing a dilution of weld metal. Unless the composition of the deposited filler metal and the composition of the base metal are identical, there will be a variation of composition of the metal at the interface. In multipass welds, the first pass will have a high dilution factor, the second pass less, and the third pass perhaps little or none. When welding on a base metal with a different composition from the deposited metal, this variation can be considerable. Variation in composition and the variation in the cooling rates will create variations in microstructure as shown in Figure 15-29. This is the reason that the microstructure of the weld is important and should be studied. The microstructure taken at different locations in the weld is shown in Figure 15-30. Each microstructure has its particular characteristics, as previously discussed. One of the important characteristics is

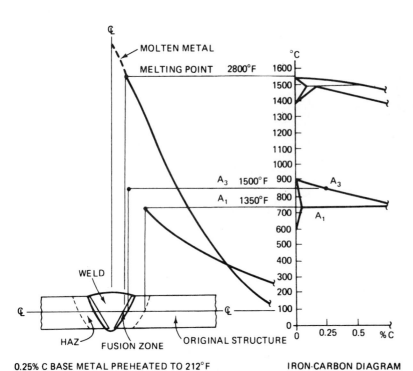

FIGURE 15–28 Temperature distribution at a weld.

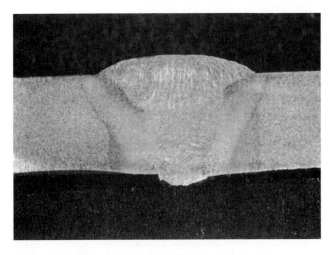

FIGURE 15–29 Cross section of a multipass arc weld.

the hardness of the microstructure throughout the weld area. The hardness should not vary over specific limitations. Figure 15–31 shows the macrostructure of a weld metal-base metal interface and the hardness at different points across this interface. Note the higher hardness of the weld metal compared to the base metal and that in the heat-affected zone the hardness is between these two values. This macrograph is of a low-carbon base metal joined with slightly alloyed weld metal.

The area between the interface of the deposited weld metal, and extending into the base metal far enough that any phase change occurs, is known as the *heat-affected zone* (HAZ). This is shown in several of the pic-

tures. The heat-affected zone, while part of the base metal, is considered to be a portion of the weld joint since it influences the service life of the weld. The heat-affected zone and the admixture zone are the most critical in many welds. For example, when welding a hardenable steel the heat-affected zone can increase in hardness to an undesirable level. On the other hand, when welding a hardened steel the heat-affected zone can become a softened zone since the heat of the weld has annealed the hardened metal. In the case of electroslag welding, the heat-affected zone will contain extremely large grains because of the long time at high temperature and the possibility for grain growth. The hardness of the weld metal, the base metal, and the heat-affected zone of an electroslag weld will be relatively uniform.

When the base metal and weld metal are of completely different analysis, the interface zone contains alloys that can be detrimental. Figure 15–32 shows the interface between a low-alloy high-strength steel and a stainless steel weld. Note the heat-affected zone of the base metal and the minute amount of mixing of the lighter colored stainless steel deposit with the base metal. In this case, the alloy mixture produced is of such a small amount that it does not have an appreciable effect on the overall properties of the weld joint.

Another problem of the weld is the segregation during the thermal cycle. Segregation relates to the solubility of elements in metals, particularly alloys. The composition of the first crystals that form as an alloy freezes is different from the composition of the liquid that freezes last. The purer metals have the higher melting point or freez-

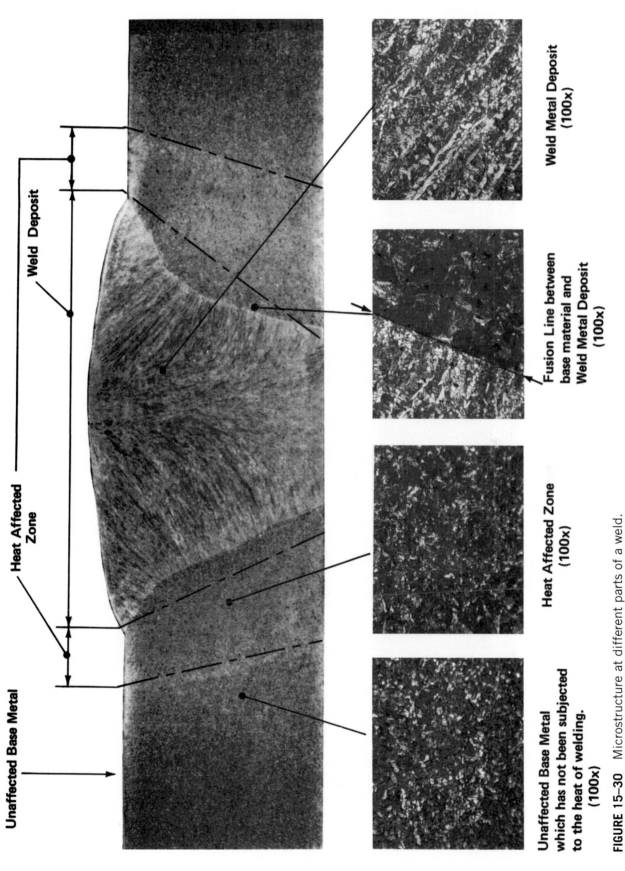

Unaffected Base Metal

Weld Deposit

Heat Affected Zone

Weld Metal Deposit
(100x)

Fusion Line between
base material and
Weld Metal Deposit
(100x)

Heat Affected Zone
(100x)

Unaffected Base Metal
which has not been subjected
to the heat of welding.
(100x)

FIGURE 15–30 Microstructure at different parts of a weld.

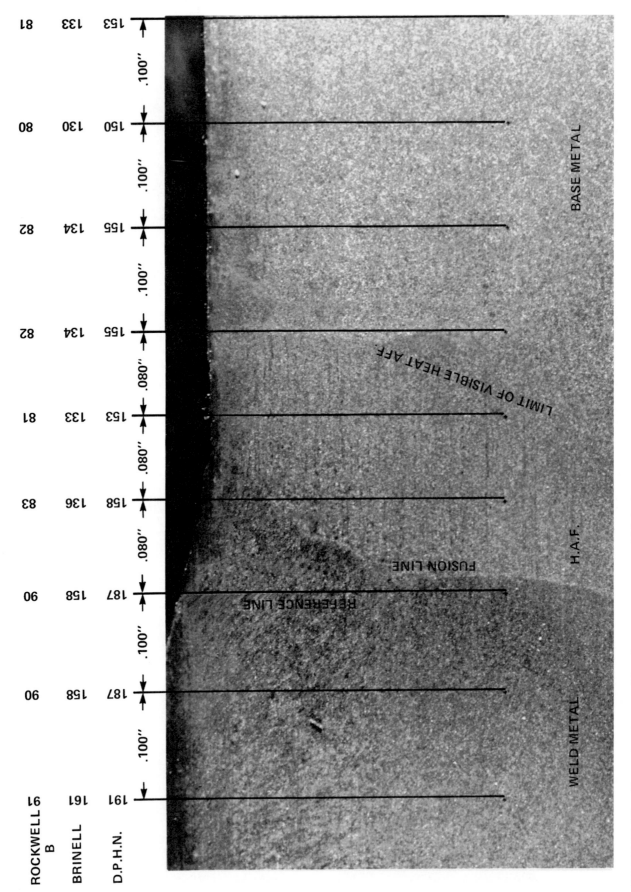

	ROCKWELL B	BRINELL	D.P.H.N.			
	81	133	153		.100"	
	80	130	150		.100"	
	82	134	155		.100"	
	82	134	155		.080"	
	81	133	153		.080"	
	83	136	158		.080"	
	90	158	187		.100"	
	90	158	187		.100"	
	91	161	191			

BASE METAL

LIMIT OF VISIBLE HEAT AFF.

H.A.F.

FUSION LINE

REFERENCE LINE

WELD METAL

FIGURE 15–31 Macrostructure and hardness across arc weld interface.

FIGURE 15–32 Interface of dissimilar weld joint.

ing point and therefore freeze first. Metals or elements with lower melting points freeze last. In addition, in weld metal, because of the rapidity of freezing time, very little diffusion occurs and there is a lack of homogeneity in the total weld. Carbon, phosphorus, sulfur, and sometimes manganese are frequently in the segregated state in steel. This can be determined by high-magnification study of the microstructure. Segregation, particularly of carbon, can be partially dispersed by means of heat treatment.

Molten metal has a relatively high capacity of dissolving gases in contact with it. As the metal cools it has less capacity for dissolved gases, and when going from liquid to solid state the solubility of gas in metal is much lower. The gas is rejected as the crystals solidify, but it may be trapped because of almost instantaneous solidification. Entrapment of the gas causes gas pockets and porosity in the weld. Carbon monoxide, which is present in many arc and fuel gas atmospheres, is sometimes trapped. Hydrogen that may be present in the arc atmosphere is also trapped. Hydrogen, however, will gradually disperse and escape from the weld metal over a period of time. High temperatures increase the speed for hydrogen migration and removal. The inert gases are not soluble in molten metal and for this reason are used in many gas-shielded applications.

The solubility of metals within metals is also of high interest, particularly for dissimilar metal welding. The solubility is determined by the equilibrium diagram of the alloys. The greater the degree of solubility, the better the success of welding dissimilar metal combinations.

It is impossible to provide a more thorough coverage of the metallurgy of steel welds in the space available. For more information, see References 10 to 12. The metallurgy of nonferrous metals resembles, but is quite different from, the metallurgy of iron and steel.

15-6 WELDABILITY OF METALS

The American Welding Society defines weldability as "the capacity of a material to be welded under the imposed fabrication conditions into a specific, suitably designed structure and to perform satisfactorily in the intended service." This definition includes many qualifying statements; for example, Is the design suitable? and Is the material suitable? And what about the welding process and the welding procedures? A more practical definition might be "the ease with which a satisfactory weld can be made that will produce a joint equal to the metal being welded."

It has been stated that all metals are weldable, but some are much more difficult to weld than others. In view of this, it is vitally important that the welding process and welding procedures be considered when determining the weldability of a particular metal.

In any of these definitions it is important to know all about the metals to be welded, the design of the welds and the weldment, and the service requirements including loadings and the environment to which it will be exposed. Perhaps the best definition is that a weldable material can be welded so that the joint is equal in all respects to the base metal—in other words, a 100% weld joint.

The base metal or metal to be welded must be considered from all points of view. This includes its physical properties, mechanical properties, and chemical composition and structure.

The physical properties are not always identical in materials of the same composition. This relates to the size of the test specimen, method of testing, and the type of microstructure. The mechanical properties can be different for different materials even though they may fall within the same specifications or class. For example, hardness is related to structure, which is affected by thermal history or heat treatment. The direction of testing has a large effect on strength levels, toughness, and ductility. In addition, composition and microstructure may vary. In heavy material, the composition may have more carbon or alloy to provide the strength called for by the specification, and the structure will change from the outside to the center based on different rates of cooling when the material was produced.

It becomes apparent that materials may not be as uniform or consistent as we have thought. Segregation and changes in metallurgical structure affect the properties. Despite this, we must still produce the weld or weldment that provides the required service performance. To better determine weldability, it becomes necessary to make several assumptions:

1. The material to be welded is suitable for the intended use. In other words, it will provide the proper and necessary properties to withstand its service requirements.
2. The design of the weldment is suitable for its intended use. In considering the design for the weldment, we should include the design of the welds.

Based on these assumptions, it is necessary then to study the weld joint. The desirable weld joint has uniform strength, ductility, notch toughness, fatigue strength, and corrosion resistance throughout the weld and adjacent material.

Most welds involve the use of heat and the addition of different metallurgical structure from the unaffected base metal. Welds may also include defects such as voids, cracks, and entrapped materials. Two types of problems may occur:

1. Problems of the weld metal deposit or heat-affected zone that occur in connection with or immediately following the welding operation, such as hot cracking, heat-affected zone cracking, hydrogen-induced cracking, and so on.
2. Problems in the weld or adjacent to the weld that occur any time during service of the weldment.

These can be any kind of defects that will reduce the efficiency of the weld joint under service conditions.

It is our objective to produce a weld that will avoid these problems. Hot cracking may result from any of the following four factors: restraint, weld shape, excessive heat input, or material composition. It can result from any one factor but is much more likely if two or more factors are present.

Restraint is always present in any weld because as the weld solidifies it acquires strength but continues to cool and shrink. It is the degree of restraint that becomes critical. Restraint relates to the weld design, the weldment design, and the thickness of the materials being joined.

Weld shape is also a function of weld design, weldment design, and welding procedure. Weld procedure relates to the placement of welds or beads in the weld, the shape of the beads, and the shape of the finished surface of the weld.

Another factor is the metal composition. Segregation is important, however, since impurities such as sulfur and phosphorus tend to form low-melting-point films between solidifying grains of the metal. These impurities relate to weld joint detail and the welding process, since they affect the amount of dilution. Lamellar tearing is also associated with base metal impurities and its through-directional strength. When the degree of restraint increases, as it does in thicker metals, this problem becomes more serious.

Hydrogen cracking is considered cold cracking since it occurs soon after the weld is completed, usually within 4 to 8 hours. It usually occurs in the heat-affected zone. It may occur while the weld is cooling down to room temperature. The four main factors that affect heat-affected zone cracking are:

* The thickness of the base metal and the type of weld
* The composition of the base metal
* The welding process and filler metal type
* Energy input and preheat temperatures

The effects of these four factors are interrelated. The thickness and composition of the base metal is established by the design. The weld joint configuration, the welding process, the type of filler metal, and the welding procedure can all contribute to the severity of the factors that cause HAZ cracking. The input energy can be modified by the welding process, welding procedure, and the welding preheat temperature. These can also be changed to reduce the cooling rate.

All of the factors above determine the type of microstructure that will occur in the heat-affected zone. The two factors most important to weldability are hardenability and the susceptibility of the hardened structure to cracking. Both are increased by using a higher carbon or higher alloy content in the base metal. Certain alloying elements increase hardenability without a significant increase in the susceptibility to cracking. In this regard the carbon equivalent of the base metal becomes important. The carbon equivalent formula[13] is shown in Figure 15–33.

Plain carbon steels, which have a carbon equivalent of not over 0.40%, are considered readily weldable. This carbon equivalent can be increased up to 0.45% provided that the carbon does not exceed 0.22%, the phosphorus does not exceed 0.06%, and the steel is not over $\frac{3}{4}$ in. (19.1 mm) thick.

Usually when the carbon equivalent exceeds 0.40%, special controls are required. The low-hydrogen type of filler metal should be employed. Higher heat input should be employed and preheat may be required. When the carbon equivalent exceeds 0.60%, low-hydrogen processes are required; preheating is required if the thickness exceeds $\frac{3}{4}$ in. (19.1 mm).

Hardenability is related to the cooling rate of metals. The faster cooling rate tends to produce higher hardnesses. The cooling rate depends on the mass of the metal, the welding process, the welding procedure, and preheat temperatures. The welding process and procedure influence the energy input used to make the weld. The greater the energy input, the slower the cooling rate. Heat input is a function of welding current, arc voltage, and travel speed. To increase the heat input, increase the welding current or reduce the travel speed. Welding current relates to process and to electrode size. Heat input is calculated by using the formula given previously.

Thus, by increasing amperes or voltage, heat input increases but by increasing travel speed heat input decreases. The voltage has a minor effect since it varies only slightly compared to the other factors. In general, higher heat input reduces cooling rate. This must be used with caution since on quenched and tempered steels, too high a heat input will tend to soften the heat-affected zone and its strength level will be reduced.

In steels of relatively low hardenability it is possible to produce an unhardened heat-affected zone by increasing the heat input. In higher hardenability steels, the tendency toward cracking, and the maximum hardness, will be reduced by a slower cooling rate. There are limits to the amount of heat input that can be used. In this case, preheating is used to reduce cooling rates.

There is an interplay of several variables in regard to hydrogen cracking. These are the base metal composition and the heat input, the preheat temperature, the rate of cooling, and restraint. On a nonhardenable thin material the control of hydrogen is not important. As carbon and alloy increase, to provide greater hardenability, and as thickness increases, the effect of hydrogen becomes of vital importance.

Heat-affected zone cracking depends on many of the same factors just mentioned. There are, however, general precautions that should be taken with certain types of steel to avoid HAZ cracking. Constructional steels can be grouped into five general classifications depending on whether or not they are hardenable and the nature of the hardened structure. These must be tempered with the effects of thickness, which increases restraint and adds to the cracking problem. The five classes are:

1. Nonhardenable low-carbon mild steel.
2. Low hardenability with low susceptibility of cracking when hardened, or low-alloy steels with a carbon equivalent of not over 0.20% maximum.
3. Low hardenability with high susceptibility of cracking when hardened, normally carbon manganese steels with less than 0.25% carbon and not over 1% manganese.
4. High-hardenable steels with low susceptibility of cracking when hardened. This includes most low-carbon low-alloy high-strength steels, usually with carbon less than 0.15%, manganese up to 1.5%, nickel up to 1.5%, chromium 0.25%, molybdenum up to 0.25%, and vanadium up to 0.20%.
5. High-hardenable steels with high susceptibility of cracking when hardened. This would include alloy steels with carbon not exceeding 0.25% but with alloys.

There are several precautions that should be taken with the five classifications:

1. No extraordinary precautions required when welding thin-to-medium-thickness materials.
2. Use low-hydrogen processes and filler materials and utilize preheat for thick sections or increase heat input.
3. Low-hydrogen processes are recommended but not essential. High heat input should be used and preheat is not required except in thicker materials and should be in the range 480 to 660° F (250 to 350° C).
4. Low-hydrogen processes are required, preheat and interpass are suggested, high-heat-input processes are recommended, and preheat is increased as thickness increases.

FIGURE 15–33 Carbon equivalent formula.

$$C.E. = C\% + \frac{Mn\%}{6} + \frac{Ni\%}{20} + \frac{Cr\% + Mo\%}{10} + \frac{Cu\%}{40}$$

5. Low-hydrogen processes are required, preheat and interpass temperatures in the range 300 to 480° F (150 to 250° C) are necessary, and a postweld heat treatment is required.

Recommended preheat and filler metals are discussed in Chapter 16. Weldability is extremely complex and all the welding factors interrelate. It is of vital importance for the success of welds to give consideration to all these factors.

QUESTIONS

15–1. What are some of the physical properties of a metal?

15–2. What are some of the mechanical properties of a metal?

15–3. What is ductility, and why is it important for welding?

15–4. Explain the different methods of measuring hardness. Is there a comparison?

15–5. What types of groups write metal specifications? What is the purpose of a specification?

15–6. How can you determine the metal composition if you know the specification number?

15–7. What are seven tests that can be made to identify a metal piece?

15–8. What can be determined by a metal's appearance?

15–9. Why is a magnet of value to help identify a piece of metal?

15–10. What is the spark test?

15–11. Why is it important to identify a metal before welding it?

15–12. What different ways is heat generated for welding?

15–13. What is the weld cooling rate and what importance is it to welding?

15–14. What is the major difference between welding metallurgy and casting metallurgy?

15–15. What are the three most common crystal structures of metals? Explain.

15–16. What is an equilibrium diagram?

15–17. What is the microstructure of a metal? Name some microstructures.

15–18. What is the HAZ? What causes it?

15–19. What is the carbon equivalent formula? How is it used?

15–20. What is weldability?

REFERENCES

1. H. David and C.T. Graff, "Alloys Index," American Society for Metals, Metals Park, Ohio, and The Metals Society, London, vol. 3 (1976).

2. E. M. Simons, *A Dictionary of Alloys* (London: Frederick Muller, 1969).

3. R. C. Gibbons, "Woldman's Engineering Alloys," 6th ed. American Society for Metals, Metals Park, Ohio, 1979.

4. W. F. Simmons and R. B. Gunia, "Compilation of Trade Names, Specification and Producers of Stainless Alloys and Super Alloys," ASTM Data Series DS45, American Society for Testing and Materials, Philadelphia, Pa., 1969.

5. G. Tschorn, *Spark Atlas of Steels* (translated from German) (New York: A Pergamon Press Book, Macmillan Company, 1963).

6. "Nondestructive Rapid Identification of Metals and Alloys by Spot Test," Special Technical Publication 550, American Society for Testing and Materials, Philadelphia, Pa.

7. I. Hrivnak, *The Theory of Mild Steel and Micro Alloy Steel's Weldability,* Bratislava, Czechoslovakia, 1969.

8. D. Rosenthal, "The Theory of Moving Sources of Heat and Its Application to Metal Treatment," *Transactions of the ASME* (November 1946), 849.

9. *Metals Handbook,* vol. 1, *Properties and Selection of Metals,* American Society for Metals, Metal Park, Ohio.

10. G. E. Linnert, "Welding Metallurgy," The American Welding Society, Miami, Fla., 1967.

11. A. Phillips, ed., "Introductory Welding Metallurgy," The American Welding Society, Miami, Fla., 1968.

12. *Metals Handbook,* vol. 6, *Welding and Brazing,* American Society for Metals, Metals Park, Ohio.

13. R. D. Stout, *Weldability of Steels,* 4th ed. (New York: Welding Research Council, 1987), 191.

16

WELDING STEELS

OUTLINE

16-1 WELDING CARBON AND LOW-ALLOY STEELS

Low-carbon steels are considered to be steels with a maximum of 0.15% carbon. Mild steels are those that contain 0.15% to 0.29% carbon. Low-alloy steels are those having a maximum of 0.29% carbon and with their total metal alloy content not exceeding 2%. Different groups have slightly different composition limits.

The E60XX and E70XX classes of electrodes provide sufficient strength to produce 100% weld joints in mild steels. The yield strength of electrodes, in these classes, will overmatch the yield strength of the mild and low-alloy steels. The E60XX class should be used for steels having yield strengths below 50,000 psi (344.7 MPa) and the E70XX class should be used for welding steels having a yield strength below 60,000 psi (414 MPa). Low-hydrogen electrodes should be used and preheat is suggested when welding heavier materials, or restrained joints. The electrode that provides the desired operational features should be selected. When welding the low-alloy steels, the operational characteristics of electrodes are ignored and only low-hydrogen electrodes or procedures are used.

Filler metal should be selected on the basis of strength and composition of the weld deposit. For low-alloy steels this means that the electrode class would be E70XX or higher. The strength level of bare solid and flux cored electrode wires is also contained in the classification number. Select the strength level to match or overmatch the base metal. The composition is described by the suffix letter in the filler metal classification. For the different electrodes the suffix is shown in Figure 16-1 along with the chemistry of the deposited metal. For low-alloy, bare solid, or flux cored electrodes, see the appropriate specification. The suffix is used to match the composition of the base metal. Electrodes and wire are not available to match the composition of every base metal, but an effort should be made to come as close as possible. For GMAW use 98% argon–2% oxygen, or 75% argon–25% CO_2, for shielding. For FCAW use 75% argon–25% CO_2, or 100% CO_2, for shielding. This allows the selection of the electrode to match not only the mechanical properties of the base metal, but also to approximately match the composition of the base metal.

The only E80XX or higher-strength electrodes that do not have low-hydrogen coverings are the EXX10 type electrodes, which are designed specifically for welding pipe. The deep penetrating characteristics of the cellulosic

Suffix	C	Mn	Si	Ni	Cr	Mo	Additional
A1	0.12	0.60–1.00	0.40–0.80	—	—	0.40–0.65	—
B1	0.05–0.12	0.90	0.60–0.80	—	0.40–0.65	0.40–0.65	—
B2	0.05–0.12	0.90	0.60–0.80	—	1.00–1.50	0.40–0.65	—
B2L	0.05	0.90	1.00	—	1.00–1.50	0.40–0.65	—
B3	0.05–0.12	0.90	1.00	—	2.00–2.50	0.90–1.20	—
B3L	0.05	0.90	1.00	—	2.00–2.50	0.90–1.20	—
B4L	0.05	0.90	1.00	—	1.75–2.25	0.40–0.65	—
B5	0.07–0.15	0.40–0.70	0.30–0.60	—	0.40–0.60	1.00–1.25	V–0.05
B6	0.05–0.10	1.0	0.90	0.40	4.0–6.0	0.45–0.65	—
B6L	0.05	1.0	0.90	0.40	4.0–6.0	0.45–0.65	—
B7	0.05–0.10	1.0	0.90	0.40	6.0–8.0	0.45–0.65	—
B7L	0.05	1.0	0.90	0.40	6.0–8.0	0.45–0.65	—
B8	0.05–0.10	1.0	0.90	0.40	8.0–10.5	0.85–1.20	—
B8L	0.05	1.0	0.90	0.40	8.0–10.5	0.85–1.20	—
B9	0.08–0.13	1.25	0.30	1.0	8.0–10.5	0.85–1.20	(1)
C1	0.12	1.25	0.60–0.80	2.00–2.75	—	—	—
C1L	0.05	1.25	0.50	2.00–2.75	—	—	—
C2	0.12	1.25	0.60–0.80	3.00–3.75	—	—	—
C2L	0.05	1.25	0.50	3.00–3.75	—	—	—
C3	0.12	0.40–1.25	0.80	0.80–1.10	0.15	0.35	V–0.05
C3L	0.08	0.40–1.40	0.50	0.80–1.10	0.15	0.35	V–0.05
C4	0.10	1.25	0.60–0.80	1.10–2.00	—	—	—
C5L	0.05	0.40–1.00	0.50	6.00–7.25	—	—	—
NM1	0.10	0.80–1.25	0.60	0.80–1.10	0.10	0.40–0.65	(2)
D1	0.12	1.00–1.75	0.60–0.80	0.90	—	0.25–0.45	—
D2	0.15	1.65–2.00	0.60–0.80	0.90	—	0.25–0.45	—
D3	0.12	1.00–1.80	0.60–0.80	0.90	—	0.40–0.65	—
G	—	1.00 MIN	0.80 MIN	0.50 MIN	0.30 MIN	0.20 MIN	(3)
M	0.10	0.60–2.25	0.60–0.80	1.25–3.80	0.15–1.50	0.20–0.55	V–0.05
P1	0.20	1.20	0.60	1.00	0.30	0.50	V–0.10
W1	0.12	0.40–0.70	0.40–0.70	0.20–0.40	0.15–0.30	—	V and Cu
W2	0.12	0.50–1.30	0.35–0.80	0.40–0.80	0.45–0.70	—	Cu

Note (1) Various amounts of V, Cu, AL, Nb(Cb) and N
Note (2) V-0.02, CU-0.10 & AL 0.05
Note (3) V-0.10 MIN Cu-0.20 MIN; see ANS A5.5 for details
See ANS A5.5 for exact analysis; this is a summary

FIGURE 16–1 Suffixes for low-alloy steel electrodes.

electrodes make them suitable for cross-country pipe welding. The theory and practice is that alloy steel pipe is relatively thin and it is welded with cellulosic electrodes at relatively high currents. In addition, each welding pass is very thin and the weld metal is aged for a considerable length of time prior to putting the pipeline into service. This allows for hydrogen, which might be absorbed, to escape from the metal and not adversely affect the service life of the pipeline.

The remainder of this section is related to specific types of steels and provides guidance in the selection of filler metal for joining them. For those steels that may not be specifically mentioned, it is possible to relate them to the chemistry of the deposited weld metal in order to establish the proper electrode class.

Low-Carbon (Mild) Steels

Low-carbon steels include those in the AISI series C-1008 to C-1025. Carbon ranges from 0.10% to 0.25%, manganese ranges from 0.25% to 1.5%, phosphorus is 0.40% maximum, and sulfur is 0.50% maximum. Steels in this range are most widely used for industrial fabrication and construction. These steels can be easily welded with any of the arc, gas, and resistance welding processes.

Medium-Carbon Steels

The medium-carbon steels include those in the AISI series C-1030 to C-1050. The composition is similar to low-carbon steels, except that the carbon ranges from 0.25%

to 0.50% and the manganese from 0.60% to 1.65%. With higher carbon and manganese the low-hydrogen electrodes are recommended, particularly in thicker sections. Preheating may be required and should range from 300 to 500° F (149 to 260° C). Postheating is often specified to relieve stress and help reduce hardness that may have been caused by rapid cooling. Medium-carbon steels are readily weldable provided the precautions are observed. They can be welded with all of the processes mentioned previously.

High-Carbon Steels

High-carbon steels include those in the AISI series from C-1050 to C-1095. The composition is similar to medium-carbon steels, except that carbon ranges from 0.50% to 1.03% and manganese ranges from 0.30% to 1.00%. Special precautions must be taken when welding steels in these classes. The low-hydrogen process or electrodes must be employed, and preheating of from 400 to 600° F (204 to 316° C) is necessary, especially when heavier sections are welded. A postheat treatment, either stress relieving or annealing, is usually specified. High-carbon steels can be welded with the same processes mentioned previously.

Low-Alloy High-Strength Steels

The low-alloy high-strength steels represent the bulk of the remaining steels in the AISI designation system. These steels are welded with the E80XX through E120XXX class of covered welding electrodes. It is for these types of steels that the suffix to the electrode classification number is used. These steels include the low-manganese steels, the low- to medium-nickel steels, the low-nickel–chromium steels, the molybdenum steels, the chromium–molybdenum steels, the nickel–chromium–molybdenum steels, and the other groups. Not included as an alloy steel but part of the AISI series are the sulfur steels. These are the series designated by 11XX, sometimes known as free-machining steels. Sulfur is 0.08% to 0.33%. These steels are difficult to weld, because of the high sulfur content, which has a tendency to produce porosity in the weld and cracking. Low-hydrogen electrodes of the minimum-strength level should be used. Welding is tedious on these steels and their use should be avoided when welding is required.

Low-Nickel Steels

These include those in AISI series 2315, 2515, and 2517. Carbon ranges from 0.12% to 0.30%, manganese from 0.40% to 0.60%, silicon from 0.20% to 0.45%, and nickel from 3.25% to 5.25%. If the carbon does not exceed 0.15%, preheat is not necessary except for extremely heavy sections. If the carbon exceeds 0.15%, preheat of up to 500° F (260° C), depending on thickness, is required. On thin material, $\frac{1}{4}$ in. (6.4 mm) or less, preheating is unnecessary. Stress relieving after welding is advisable. The electrode suffix C-1 or C-2 would be used, depending on the level of nickel in the base material. The strength level would be matched to the base metal. In all cases, the low-hydrogen coating is used.

Low-Nickel Chrome Steels

Steels in this group include the AISI 3120, 3135, 3140, 3310, and 3316. In these steels, carbon ranges from 0.14% to 0.34%, manganese from 0.40% to 0.90%, silicon from 0.20% to 0.35%, nickel from 1.10% to 3.75%, and chromium from 0.55% to 0.75%. Thin sections of these steels in the lower carbon ranges can be welded without preheat. A preheat of 200 to 300° F (93 to 149° C) is necessary for carbon in the 0.20% range; for higher carbon a preheat of up to 600° F (316° C) should be used. The weldment must be stress relieved or annealed after welding. The E80XX or E90XX electrodes should be used with the C-1 or C-2 suffix. There is no electrode type that will exactly provide a nickel-chrome deposit the same as the base metal.

Low-Manganese Steels

Included in this group are the AISI type 1320, 1330, 1335, 1340, and 1345 designations. In these steels, the carbon ranges from 0.18% to 0.48%, manganese from 1.60% to 1.90%, and silicon from 0.20% to 0.35%. Preheat is not required at the low range of carbon and manganese. Preheat of 250 to 300° F (121 to 149° C) is desirable as the carbon approaches 0.25%, and mandatory at the higher range of manganese. Thicker sections should be preheated to double the above figure. A stress relief postheat treatment is recommended. The E80XX or E90XX electrodes with the A-1, D-1, or D-2 suffix should be used.

Low-Alloy Chromium Steels

Included in this group are the AISI type 5015 to 5160 and the electric furnace steels 50100, 51100, and 52100. In these steels, carbon ranges from 0.12% to 1.10%, manganese from 0.30%, to 1.00%, chromium from 0.20% to 1.60%, and silicon from 0.20% to 0.30%. When carbon is at the low end of the range, these steels can be welded without special precautions. As the carbon increases and as the chromium increases, high hardenability results, and a preheat of as high as 750° F (399° C) will be required, particularly for heavy sections. The B suffix should be used. Match the suffix to the chromium content.

Welding procedure information is then determined by means of the carbon equivalent formula given in Chapter 15. The carbon equivalent should be calculated for the exact composition. When only a range of composition is known, use the maximum values to be on the safe side. When the carbon equivalent is 0.40% or lower, the material is considered readily weldable. Above 0.40% special controls are required. In all cases, low-hydrogen processes should be used and preheat may also be required. When the carbon equivalent exceeds 0.60%, preheating is required if the thickness exceeds 3/4 in. (19.1 mm). When the carbon equivalent exceeds 0.90%, preheat is absolutely required to a relatively high temperature on all except the thinnest material. This provides the guidance for establishing welding procedures using covered electrodes. For all but the simple work, the procedure should be qualified according to one of the standard tests to determine whether it produces the quality of weld expected.

When using the submerged arc welding process, it is also necessary to match the composition of the electrode with the composition of the base metal. A neutral flux that neither detracts nor adds elements to the weld metal should be used. In general, preheat can be reduced for submerged arc welding because of the higher heat input and slower cooling rates involved. To make sure that the submerged arc deposit is low hydrogen, the flux must be dry and the electrode and base metal must be clean.

When using the gas metal arc welding process, the electrode should be selected to match the base metal and the shielding gas should be selected to avoid excessive oxidation of the weld metal. Preheating with the gas metal arc welding process should be in the same order as with shielded metal arc welding since the heat input is similar.

When using the flux cored arc welding process, the deposited weld metal produced by the flux cored electrode should match the base metal being welded. Preheat requirements would be similar to gas metal welding.

When low-alloy high-strength steels are welded to lower-strength grades, the electrode should be selected to match the strength of the lower-strength steel. The welding procedure, that is, preheat, heat input, and so on, should be suitable for higher-strength steel.

Weathering Steels

Weathering steels are low-alloy steels that can be exposed to the weather without being painted. The steel protects itself by means of a dense oxide coating (patina) that forms naturally on the steel when it is exposed to the weather. This tight oxide coating reduces continuing corrosion. The corrosion resistance of weathering steels is four to six times that of normal structural carbon steels, and two to three times that of many of the low-alloy structural steels. The weathering steels are covered by the ASTM specifications A242 and A588. These steels have a minimum yield strength of 50,000 psi (345 MPa) with an ultimate tensile strength of 70,000 psi (482 MPa). Two

of the better known weathering steels are Corten and Mayari R.

The weathering steels can be welded by all the arc welding processes and by gas welding and resistance welding. To maintain the weather resistance characteristic of the steel, a special welding procedure should be employed. Use the E7018 class to within one layer of the top of the joint. The top layer should be made with an E7018-C1 electrode since the 2% nickel in the weld deposit will cause the weld metal to weather the same as the weathering steel. The C-1 suffix weld deposit should be used for the top layer of any multipass weld.

The same concept can be used for gas metal arc welding, flux cored arc welding, or submerged arc welding. The normal electrode used for a 50,000-psi steel would be employed, but this last layer would include alloy to provide weathering resistance.

16-2 WELDING ALLOY STEELS

Steel is considered to be an alloy steel when the maximum of the range given for the content of alloying elements exceeds one or more of the following limits: manganese 1.65%, silicon 0.60%, copper 0.60%, or in which a definite range or a definite minimum quantity of any of the following elements is specified or required within the limits of the recognized field of constructional alloy steels: aluminum, boron, chromium up to 3.99%, cobalt, columbium, molybdenum, nickel, titanium, tungsten, vanadium, zirconium, or any other alloying element added to obtain a desired alloying effect.[1]

Some of the steels mentioned in the preceding section fall within this range and are considered low-alloying steels. Many of the steels shown by the AISI steel classification system are also in the alloy range. If the chromium content exceeds 11%, it will be considered a stainless steel.

To weld alloy steels successfully, four factors must be considered.

1. Always use a low-hydrogen welding procedure, process, and filler metal.
2. Select a filler metal that matches the strength level of the alloy steel.
3. Select a filler metal that comes close to matching the composition of the alloy steel.
4. Match the welding heat input requirement to the alloy steel and its thickness, and use the proper welding procedure.

The AWS filler metal specifications for covered electrodes,[2] bare solid electrode wire,[3] and flux cored electrode wires[4] provide data to select filler metals that meet specific strength levels and provide the desired alloy composition.

The first one, two, or three digits of the electrode classification give the minimum tensile strength of the deposited weld metal. This is handled slightly differently in the three specifications. It provides the information necessary to select the strength level of the weld metal, which should meet or slightly overmatch the base metal.

All three specifications provide suffix letters that designate the chemical composition of the deposited weld metal. Unfortunately, the three specifications are not in exact agreement. In general, the suffix letters indicate the following chemistry (see Figure 16-1):

Suffix Letter	Chemistry
A	Carbon–molybdenum steel
B	Chromium–molybdenum steel
C or NI	Nickel–steel
D	Manganese–molybdenum steel
G, I, M, and W	Other low-alloy steels

The suffix letters are sometimes followed by a number that indicates a specific chemical composition range. The suffix numbers are a starting point for selecting filler metals. Refer to the specification for exact chemical composition. This information allows you to select the composition of the filler metal. The heat input requirement is discussed with the alloy steel information.

Quenched and Tempered Constructional Steels

The quenched and tempered constructional steels were developed in the early 1950s. These steels offer a number of advantages, which have made them extremely popular where a high strength-to-weight ratio is important. The unique properties of these steels are obtained from their chemical composition plus a quenched and tempered heat treatment. These steels have extremely high strength, in the order of 100,000 psi (689 MPa) yield strength, combined with good weldability. In addition, they possess good ductility, good notch toughness, good fatigue strength, and corrosion resistance. They can be welded successfully with relatively conventional welding procedures. Minimum preheat is used and for most applications the weldments are used in the as-welded condition.

ASTM Specifications A514 to A517 have been written to cover the quenched and tempered constructional steels produced by different steel manufacturers. The grades and compositions are shown in Figure 16-2. A different grade is given to each steel trade name. Several other ASTM specifications cover these types of steels, but the welding outlined here will apply to all when the composition is similar.

Steels of the same basic composition are also made into castings and forgings. These steels are water quenched by special techniques from a temperature of 1500 to 1600° F (816 to 871° C) and tempered at a temperature of from 1000 to 1110° F (538 to 593° C). This

FIGURE 16–2 Popular quenched and tempered steels, grades and composition per ASTM A514/517.

ASTM Grade	Proprietary Steels	C (max.)	Mn	Si	Cr	Ni	Mo	Cu (min.)	Other
A	NAXTRA 100	0.15–0.21	0.80–1.10	0.40–0.80	0.50–0.80	—	0.18–0.28	—	0.035-P, 0.040-S
B	T-1 Type A	0.15–0.21	0.70–1.00	0.20–0.30	0.40–0.65	—	0.15–0.25	—	0.035-P, 0.040-S, 0.01–0.03-Ti, 0.03–0.08-Va
C	Jalloy S-100	0.10–0.20	1.01–1.50	0.15–0.30	—	—	0.20–0.30	—	0.035-P, 0.040-S,
D	Armco SSS100A	0.13–0.20	0.40–0.70	0.20–0.35	0.85–1.20	—	0.20–0.25	0.20–0.40	0.035-P, 0.040-S, 0.04–0.10-Ti
E	Armco SSS100	0.12–0.20	0.40–0.70	0.20–0.35	1.40–2.00	—	0.40–0.60	0.20–0.40	0.035-P, 0.040-S, 0.04–0, 10-Ti
F	USS T-1	0.10–0.20	0.60–1.00	0.15–0.35	0.40–0.65	0.70–1.00	0.40–0.60	0.15–0.50	0.035-P, 0.040-S, 0.03-Va
G	PX-100	0.15–0.21	0.80–1.10	0.50–0.90	0.50–0.90	—	0.40–0.60	—	0.035-P, 0.040-S
H	T-1 Type B	0.12–0.21	0.95–1.30	0.20–0.30	0.40–0.65	0.30–0.70	0.20–0.30	—	0.035-P, 0.040-S, 0.03–0.08-Va
J	RQ 100A	0.12–0.21	0.45–0.70	0.20–0.35	—	—	0.50–0.65	—	0.035-P, 0.040-S
K	CHT-100	0.10–0.20	1.10–1.50	0.15–0.30	—	—	0.45–0.65	—	0.035-P, 0.040-S
L	Armco SSS100B	0.13–0.20	0.40–0.70	0.20–0.35	1.15–1.65	—	0.25–0.40	0.20–0.40	0.035-P, 0.040-S, 0.04–0.10-Ti
M	RQ-100B	0.12–0.21	0.45–0.70	0.20–0.35	—	1.20–1.50	0.45–0.60	—	0.035-P, 0.040-S
N	NAXTRA 100A								
P	RQC-100	0.12–0.20	0.45–0.70	0.20–0.35	0.85–1.20	1.20–1.50	0.45–0.60	—	0.035-P, 0.040-S

Note: Boron of 0.0005 to 0.005 added to each grade.

produces a microstructure of tempered, low-temperature transformation products that have an excellent combination of strength and toughness.

Shielded metal arc, gas metal arc, flux cored arc, and submerged arc welding processes can all be used. The gas tungsten arc welding process can be used but is restricted to the thinner sections. The electroslag welding process is not recommended because the long time at high temperature destroys the heat treatment. It is essential to keep the process a low-hydrogen process. This means dry electrode coating, dry flux, dry gas, clean joint preparation, and so on.

When using the shielded metal arc welding process, electrodes of the E11018M or E12018M classification should be used.

With the gas metal arc welding process the electrode should be an AWS Class E 120C-G, which is a proprietary electrode wire designed for this class of steel. The shielding gas should be 98% argon–2% oxygen or 75% argon–25% CO_2. Pure carbon dioxide can also be used. The electrode wire composition should be approximately the same as the composition of the base metal.

When welding with the flux cored process, use one of the AWS E110 or E120 type with T-1 through 5 usability class and with K number appropriate to the brand steel being used. The T number will indicate what shielding gas should be used. With the submerged arc welding process, a neutral flux should be used. The electrode wire should be approximately the composition of the base metal.

The last factor is to maintain the proper heat input. The heat input depends on the material thickness, the preheat employed, and the interpass temperature. For 1-in.-thick plate a minimum preheat of 50° F (10° C) is normally used. When the thickness is increased, a preheat of 200° F (93° C) is recommended. Higher preheats may be required for restrained joints. The allowable heat input is based on the joules per inch of weld joint given by the standard formula. The maximum heat input for different thickness of the different grades of the ASTM A514/A517 steels is given in Figure 16–3. These may vary from grade to grade or by different manufacturers. Consult the steel manufacturer's technical data to determine the recommended maximum heat input limit. When these heat inputs are exceeded, there will be a loss of strength of the weld joint. The toughness of the steel in the heat-affected zone is usually excellent, and the hardness of the heat-affected zone is normally lower than the base metal or the deposited weld zone. Producers of T-1 steel have developed a welding heat input calculator that relates the travel speed, welding current, and arc volts to the heat input in kilo-joules per inch. This calculator provides maximum heat inputs for different preheat and interpass temperatures based on different thicknesses of steel. As the plate thickness increases and with lower preheat temperatures, the maximum heat units are unlimited.

Plate Thickness (in.)	Preheat and Interpass Temperature (° F)			
	70	200	300	400
$\frac{3}{16}$	17,500	14,000	11,500	9,000
$\frac{1}{4}$	23,700	19,200	15,800	12,300
$\frac{1}{2}$	47,400	35,500	31,900	25,900
$\frac{3}{4}$	88,600	69,900	55,700	41,900
1	Any	110,000	86,000	65,600
$1\frac{1}{4}$	Any	154,000	120,000	94,000

Note: This applies to grade F. Check suppliers of other grades for values.

FIGURE 16–3 Suggested maximum heat input limits (joules per inch).

The stringer bead technique is preferred. The use of a full weave reduces the travel speed and increases the heat input above the maximum limits. If weaving is used, it should be restricted to two electrode diameters. The base metal should not be allowed to become overheated. When back gouging is required, it should be done with the air carbon arc process or by grinding. The oxyacetylene flame should not be used for back gouging.

Defects in welds made on quenched and tempered constructional steels are more serious than the same defects in mild or low-carbon steels. The weld surface should be smooth, with contours that are well blended into the pieces being joined. Each weld should be made so that there is good penetration into the previous weld and no undercut. Complete penetration is essential to utilize the full strength of the quenched and tempered steel.

High-Strength Low-Alloy Steels

The quenched and tempered steels just described give excellent service but have proven to be hydrogen sensitive, and for this reason preheating has been employed for welding. The Q and C steels have relatively large amounts of alloying elements and require heat treatment in manufacturing. It was felt that if preheat could be reduced or eliminated in thicker sections, considerable savings would result. In addition, heat input restrictions could be relaxed or eliminated.

The steel industry has developed a new family of weldable steels that overcome these problems. They are known as HSLA steels, microalloyed steel, and clean steel, and are identified by the ASTM A710 or ASTM 736 specification.[5] They have excellent notch toughness and good weldability due to the very low carbon content, which is 0.08% carbon maximum. Grain refinement is attained by microalloying with 0.01% to 0.05% columbium, and small additions of copper (1% to 1.3%), nickel (0.7% to 1%), chromium (0.60% to 0.90%), and molybdenum (0.15% to 0.25%) are used to achieve high strength. Small amounts of nitrogen are used. Each manufacturer has a different composition. The chromium and molybdenum optimize the precipitation of the copper. The nickel prevents hot shortness brought on by the copper and improves toughness. Aluminum is used for deoxidizing and grain refining. This microalloyed steel is extremely weldable without expensive preheat and heat input restrictions. This is due primarily to the low carbon content, which ranges from 0.04% to 0.08% carbon. It is also less sensitive to hydrogen. These steels are becoming very popular. They are available from sheet metal thicknesses to heavy plate.

There are three different classes of this steel based on three different methods of heat treatment. Plates rolled followed by an aging treatment constitute one class; plates rolled, normalized, and aged are covered by the second class; and the third class covers plates that are rolled, quenched, and aged. There are numerous producers, each with different composition and properties. The covered electrode used for welding this steel is the E10018-MI or the E12018-M2.

For GMAW use the electrode wire matching the strength level and alloy content of the HSLA steel you are using.[6] The shielding gas can be 98% argon–2%O_2, 75% argon–25% CO_2, or CO_2. For FCAW the same applies.

Special welding precautions are not required. The weld will have strength equal to the base metal and will have good toughness properties.

High-Nickel Steels

The 9% nickel steels are quenched and tempered but are considered separately because they are intended for different types of service. The 9% nickel steels were developed to provide high strength and extreme toughness at very low operating temperatures. The reason for developing the material was to provide a steel that could be used to build tanks and vessels for containing liquefied natural gas. The temperature of liquefied natural gas is −262° F (−160° C). The 9% nickel steel will provide good notch toughness at temperatures down to −320° F (−196° C). There is also a low-nickel steel in the 5% range, plus 0.25% molybdenum, which will provide good properties at temperatures as low as −275° F (−170° C). These steels are welded in the heat-treated condition and do not require a postweld heat treatment to obtain welds that provide properties essentially equal to the base metal.

The 9% nickel steel is supplied in the heat-treated condition. It is specified by ASTM A353 and A553. Two types of heat treatments may be used. One is known as the *double normalized and tempered condition* and the other is accomplished by normalizing at 1650° F (900° C), then normalizing at 1450° F (790° C) followed by a tempering at 1050° F (570° C). In the second way the steel furnished is water quenched and tempered with water quenching at 1470° F (800° C) and then tempering at 1050° F (570° C). The toughness of this steel is obtained by the small amount

of austenite, which is reformed during the tempering treatment. This phase is stable at subzero temperatures and contributes to the toughness of the steel.

The 9% nickel steel can be flame cut using normal oxygen fuel gas equipment. The cutting speed is slower than on mild steel. Flame-cut surfaces should be ground to remove any hardened metal and the oxide surface. Welding is done in the fully heat-treated condition and the heat-affected zone has a somewhat different microstructure than the base metal. Welding can be accomplished by SMAW submerged arc welding, GMAW, and flux cored arc welding.

When the SMAW process is used, the high nickel–chrome–iron electrodes are used. These are the AWS ENi-CrFe-2 type and EniCrFe-3 type. The higher nickel–chrome electrode will produce slightly higher strength welds that will match the base metal. A preheat or postheat is not required on material 2 in. (50 mm) thick or less. Before welding the base metal should be brought up to normal room temperature of 70° F (21° C). When making V- or bevel groove welds, the minimum included angle should be 70°. The high-nickel electrodes operate differently from mild or stainless steel electrodes. They have low penetration and do not flow or wash into the sidewall of the weld joint. The electrode should be pointed to place the deposited metal where it is desired.

When using submerged arc welding, the same analysis of electrode wire is used with a neutral-welding flux. For thinner materials, a room-temperature preheat of 70° F (21° C) is used. When welding material 2 in. (50 mm) and thicker, a preheat of 250 to 300° F (121 to 149° C) is recommended. The same temperature is used for the interpass temperature.

When using GMAW, the high-nickel–chrome electrode, AWS type ERNiCrFe-6, is used and a shielding gas of 90% helium and 10% argon is recommended. Short-circuiting transfer is used and the properties of the weld are essentially the same as the base metal. The pulsed mode of GMAW is widely used for welding 9% nickel steel. When using flux cored arc welding, special proprietary electrode wires are used. For difficult problems, consult the Nickel Development Institute in Toronto, Canada.

Chromium-Molybdenum Steels

The chromium-molybdenum steels (called chrome–moly) were developed for elevated-temperature service. They have been used extensively in power piping, where they operate at high pressures and temperatures between 700 and 1100° F (371 to 599° C). Popular Cr–Mo steels are shown in Figure 16-4.

The major reason for using chrome–moly steels is that they maintain their strength at high temperatures. They do not creep, which means that they do not stretch or deform under long periods of use at high pressures and temperatures. Also, they do not become brittle after

Popular Name	Composition (%)					Recommended Electrode Suffix
	C	Mn	Si	Cr	Mo	
$\frac{1}{2}$ Cr–$\frac{1}{2}$ Mo	0.10–0.20	0.30–0.60	0.10–0.30	0.50–0.81	0.44–0.65	B1
1 Cr–$\frac{1}{2}$ Mo	0.15 max.	0.30–0.60	0.50 max.	0.80–1.25	0.44–0.65	B2L
1$\frac{1}{4}$ Cr–$\frac{1}{2}$ Mo	0.15 max.	0.30–0.60	0.50–1.00	1.0–1.50	0.44–0.65	B2L
2 Cr–$\frac{1}{2}$ Mo	0.15 max.	0.30–0.60	0.50 max.	1.65–2.35	0.44–0.65	B4L
2$\frac{1}{4}$ Cr–1 Mo	0.15 max.	0.30–0.60	0.50 max.	1.90–2.60	0.87–1.13	B3

FIGURE 16–4 Composition of popular chrome–molybdenum steels.

extended periods of high-temperature service. Carbon steels, on the other hand, do tend to stretch at high-temperature service and will become brittle in time.

The chrome–moly steels are used in the normalized and tempered condition and in the quenched and tempered condition. The type of heat treatment dictates the strength level of the steel. The strength levels extend from 85,000 psi (586 MPa) to 135,000 psi (930 MPa). There are a number of compositions that have become popular. These are the 1% Cr–$\frac{1}{2}$% Mo, 1$\frac{1}{4}$% Cr–$\frac{1}{2}$% Mo, the 2% Cr–$\frac{1}{2}$% Mo, the 2$\frac{1}{4}$% Cr–1% Mo, and the 5% Cr–$\frac{1}{2}$% Mo.

SMAW, GTAW, and GMAW are widely used for joining the chrome–moly steels. Submerged arc welding and flux cored arc welding are also used. It is necessary to match the weld metal deposit analysis closely with the composition of the base metal.

For SMAW, the electrode class suffix ranging from B1 with $\frac{1}{2}$% chrome–$\frac{1}{2}$% moly up through the B4 for the 2$\frac{1}{2}$% chrome–$\frac{1}{2}$% moly identifies the composition. The higher levels of chromium are not specified by means of a suffix system. Proprietary electrodes are available for the higher chrome–moly steels.

The AWS specifications for chrome–moly bare solid and flux cored electrodes go up to the 2$\frac{1}{2}$% chrome–1% moly analysis. Above this proprietary electrodes are available that match many compositions.

- For GTAW use argon for shielding.
- For GMAW use CO_2 or argon–CO_2 mixture for shielding.
- For FCAW use CO_2 or argon–CO_2 mixture for shielding.

Much of the welding on these steels is done on pipe. For pipe welding, the gas tungsten arc welding process is often used for making the root pass. The SMAW process, gas metal arc, or flux cored arc welding can be used for the remainder of the weld joint. The submerged arc welding process would be used for roll welding of pipe subassemblies.

The chrome–moly steels are hardenable steels; therefore, it is necessary to provide a welding procedure that includes preheating and postheating. Preheat temperatures range from a minimum of 100° F (37.8° C) to as high as 700° F (371° C). The preheat temperature is de-pendent on the carbon content and the thickness of the material being welded. If the carbon content is below 0.20% and the thickness is less than $\frac{3}{8}$ in. (9.5 mm), the minimum 100° F preheat can be used. However, if carbon is above this figure and the wall thickness is greater, the temperature should be increased to 200° F and up to 400° F. For the higher chrome–molys and thicker sections, the preheat will extend up to 700° F (371° C); however, if thickness is less than $\frac{3}{4}$ in. (19 mm), the preheat can be reduced to half this value. Details of welding procedures for welding piping is given by the AWS "Recommended Practices for Welding of Chromium–Molybdenum Steel Piping and Tubing."[7] Specific preheat values are given for different types and wall thicknesses of chrome–moly pipe.

SMAW electrodes for chrome–moly steels are always of the low-hydrogen type. Low-hydrogen electrodes are difficult to use with open root joints; therefore, the GTAW process is used for making the root pass. Backup rings are not used for welding high-pressure, high-temperature steam pipe.

A postheat treatment is required when the carbon content exceeds 0.20% or the wall thickness is over $\frac{1}{2}$ in. (12 mm). The heat-treatment temperature is from 1150 to 1300° F (621 to 704° C). The lower temperatures are used with the thinner material and the higher temperatures for the heavier wall thickness. Specific recommendations are provided in the above-mentioned AWS booklet.

Where different grades of chrome–moly steels are welded together, the preheat and postheat temperatures should be based on the higher-alloy material, but the welding electrode can be based on the lower-alloy material.

Steel Castings

The welding of steel castings is important since they are often incorporated into weldments. Steel castings may have foundry defects that are repaired by welding. Steel castings are made in many different analyses, and it is necessary to know the composition of the casting in order to select the proper filler metal. Castings are easily identifiable since they carry an imprint of the foundry where they are made. By checking with the foundry it is possible to determine the exact analysis of the casting.

In general, steel castings have higher amounts of carbon than rolled steel. Many steel castings are heat treated to obtain desired properties. When welding heat-treated castings, one of the problem areas is that the weld metal deposit usually has a lower carbon content than the casting, and the heat treatment may not produce the same mechanical properties in the weld metal as in the casting. It is best to overmatch the analysis of the weld deposit over the composition of the casting. This will tend to produce a hardness level in the weld metal similar to that in the casting.

When using gas metal arc, flux cored arc, or submerged arc welding, this problem is reduced because of the higher penetration of these processes. There is more dilution of the weld metal from the base metal, and a higher carbon deposit will result. This provides a weld metal deposit more similar to the casting and will provide comparable heat-treated properties.

The flux cored arc welding process is extremely popular for weld repairing castings. Flux cored electrode wires are now available with weld metal deposits matching many steel casting compositions.[4]

Welding procedures for castings should be developed based on the casting analysis. All other factors concerning weldability must be considered in developing the procedure, including preheat, heat input, and postheat requirements.

16-3 WELDING STAINLESS STEELS

Stainless steels or, more precisely, corrosion-resisting steels, are a family of iron-base alloys having excellent resistance to corrosion. These steels do not rust and strongly resist attack by a great many liquids, gases, and chemicals. Many of the stainless steels have good low-temperature toughness and ductility. Most stainless steels exhibit good strength properties and resistance to scaling at high temperatures. All stainless steels contain iron as the main element and chromium in amounts ranging from about 11% to 30%. Chromium provides the basic corrosion resistance to stainless steels. A thin film of chromium oxide forms on the surface of the metal when it is exposed to oxygen in the air. This film acts as a barrier to further oxidation, rust, and corrosion. Steels that contain only chromium or chromium with small amounts of other alloys are known as straight chrome types. There are about 15 types of straight chrome stainless steels. The straight chrome steels are the 400 series of stainless steels, which are highly magnetic.

Nickel is added to certain of the stainless steels, which are known as *chrome-nickel* stainless steels. The addition of nickel reduces the thermal conductivity and decreases the electrical conductivity. The chrome-nickel steels are in the 300 series of stainless steels. They have austenitic microstructure and are nonmagnetic.

The chrome-nickel stainless steels contain small amounts of carbon. Carbon is undesirable particularly in the 18% chrome-8% nickel group. Carbon will combine with chromium to form chromium carbides, which do not have corrosion resistance. Chromium carbides are formed when the steel is held in the temperature range of 800 to 1600° F (427 to 871° C) for prolonged periods, which can happen during welding with slow cooling. The chemical reaction of carbon with chromium to form chromium carbide is called *carbide precipitation.* Carbon can be controlled, however, by the use of stabilizing elements. Carbide precipitation can be reduced or prevented in two ways. The first way is to keep the carbon level at 0.03% or less, which eliminates the formation of chromium carbides. Stainless steels with low carbon in this range are commonly referred to as ELC (extra low carbon) types. The other way of preventing carbide precipitation is to use a stabilizing element. The most popular stabilizers are titanium and columbium (niobium). These elements will combine with carbon to form titanium or columbium carbides, which have corrosion resistance. Both types of stainless steels have equivalent corrosion resistance. These types of stainless steels are identified as L type or stabilized type.

Manganese is added to some of the chrome-nickel alloys. Usually, these steels contain slightly less nickel since the chrome-nickel-manganese alloys were developed originally to conserve nickel. In these alloys, a small portion of the nickel is replaced by the manganese, generally in a 2:1 relationship. The 200 series of stainless steels is the chrome-nickel-manganese series. These steels have an austenitic microstructure and are nonmagnetic. The 201 and 202 types are used as alternates for 301 and 302.

Molybdenum is also included in some stainless steel alloys. Molybdenum is added to improve the creep resistance of the steel at elevated temperatures. It will also increase resistance to pitting and corrosion in many applications. The different alloy groupings are shown in Figure 16-5.

Stainless steels are sometimes identified by numbers that refer to the principal alloying elements, such as 18/8, 25/20, and so on. This identification system has been supplanted by the American Iron and Steel Institute system, which utilizes a three-digit number (Figure 16-6). The first digit indicates the group and the last two digits indicate specific alloys. The AISI numbers refer to the alloys as chrome-nickel stainless steels and chromium stainless steels. They are, however, also identified according to their microstructure, which can be austenitic, martensitic, or ferritic. The austenitic chrome-nickel-manganese (200 series) and austenitic chrome-nickel (300 series) steels are shown in the upper portion of the table. The martensitic types are shown in the center part of the table and represent a portion of the 400 series; the ferritic types are shown in the lower portion of the table and are the remaining alloys in the 400 series. The duplex stainless steels are covered later.

Series Designation	Metallurgical Group	Principal Elements	Hardenable by Heat Treatment	Magnetic
2xx	Austenitic	Chromium–nickel–manganese	Nonhardenable	Nonmagnetic
3xx	Austenitic	Chromium–nickel steels	Nonhardenable[a]	Nonmagnetic
4xx	Martensitic	Chromium steels	Hardenable	Magnetic
4xx	Ferritic	Chromium steels	Nonhardenable	Magnetic
5xx[b]	Martensitic	Chromium–molybdenum steels	Martensitic	Magnetic

[a]Will work harden.
[b]Not stainless.

FIGURE 16–5 Groups of stainless steels.

The three most popular processes for welding stainless steels are shielded metal arc, gas tungsten arc, and gas metal arc welding; however, almost all the welding processes can be used.

Stainless steels are slightly more difficult to weld than mild carbon steels. The physical properties of stainless steel are different from mild steel and this makes it weld differently. These differences are:

1. Lower melting temperature
2. Lower coefficient of thermal conductivity
3. Higher coefficient of thermal expansion
4. Higher electrical resistance

The properties are not the same for all stainless steels, but they are the same for those having the same microstructure. In view of this, stainless steels of the same metallurgical class have similar welding characteristics and are grouped according to the metallurgical structure with respect to welding.

Austenitic Types

The austenitic stainless steels have about 45% higher manganese, are not hardenable by heat treatment, and are nonmagnetic in the annealed condition. They may become slightly magnetic when cold worked or welded. This helps identify this class of stainless steels. All the austenitic stainless steels are weldable with most of the welding processes, with the exception of type 303, which contains high sulfur, and type 303Se, which contains selenium to improve machinability.

The austenitic stainless steels have about 45% higher thermal coefficient of expansion, higher electrical resistance, and lower thermal conductivity than mild-carbon steels. High travel speed welding is recommended, which will reduce heat input, reduce carbide precipitation, and minimize distortion. The melting point of austenitic stainless steel is slightly lower than mild-carbon steel. Because of lower melting temperature and lower thermal conductivity, welding current is usually lower. The higher thermal expansion dictates that special precautions should be taken with regard to warpage and distor-

tion. Tack welds should be twice as often as normal. Any of the distortion reducing techniques such as back-step welding, skip welding, and wandering sequence should be used. On thin materials it is very difficult to completely avoid buckling and distortion.

Ferritic Stainless Steels

The ferritic stainless steels are not hardenable by heat treatment and are magnetic. All the ferritic types are considered weldable with the majority of the welding processes except for the free-machining grade of 430F, which contains high sulfur. The coefficient of thermal expansion is lower than the austenitic types and is about the same as mild steel. Welding processes that tend to increase carbon pickup are not recommended. This would include the oxyfuel gas process, carbon arc process, and gas metal arc welding with CO_2 shielding gas. The ferritic steels in the 400 series have a tendency for grain growth at elevated temperatures. Grain growth occurs at about 1600° F (871° C) and increases rapidly at higher temperatures. The lower chromium types show tendencies toward hardening with a resulting martensitic type structure at grain boundaries of the weld area. This lowers the ductility, toughness, and corrosion resistance at the weld. For heavier sections a preheat of 400° F (204° C) is beneficial. To restore full corrosion resistance and improve ductility after welding, annealing at 1400 to 1500° F (760 to 816° C), followed by a water or air quench, is recommended. Large grain size will still prevail, however, and toughness may be impaired. Toughness can be improved only by cold working such as peening the weld. If heat treating after welding is not possible and service demands impact resistance, an austenitic stainless steel filler metal should be used. Otherwise, the filler metal is selected to match the base metal.

Martensitic Stainless Steels

The martensitic stainless steels are hardenable by heat treatment and are magnetic. The low-carbon types can be welded without special precautions. The types with over 0.15% carbon tend to be air hardenable and, therefore,

AISI No.	Composition (%)					
	Carbon	Manganese	Silicon	Chromium	Nickel	Other Elements
Chromium–nickel–magnesium–austenitic, nonhardenable						
201	0.15 max.	5.5/7.5	1.0	16.0/18.0	3.5/5.5	$N_2$0.25 max.
202	0.15 max.	7.5/10.	1.0	17.0/19.0	4.0/6.0	$N_2$0.25 max.
Chromium–nickel–austenitic, nonhardenable						
301	0.15 max.	2.0	1.0	16.0/18.0	6.0/8.0	—
302	0.15 max.	2.0	1.0	17.0/19.0	8.0/10.0	—
302B	0.15 max.	2.0	2.0/3.0	17.0/19.0	8.0/10.0	—
303	0.15 max.	2.0	1.0	17.0/19.0	8.0/10.0	S 0.15 min.
303Se	0.15 max.	2.0	1.0	17.0.19.0	8.0/10.0	Se 0.15 min.
304	0.08 max.	2.0	1.0	18.0/20.0	8.0/12.0	—
304L	0.03 max.	2.0	1.0	18.0/20.0	8.0/12.0	—
305	0.12 max.	2.0	1.0	17.0/19.0	10.0/13.0	—
308	0.08 max.	2.0	1.0	19.0/21.0	10.0/12.0	—
309	0.20 max.	2.0	1.0	22.0/24.0	12.0/15.0	—
309S	0.08 max.	2.0	1.0	22.0/24.0	12.0/15.0	—
310	0.25 max.	2.0	1.50	24.0/26.0	19.0/22.0	—
310S	0.08 max.	2.0	1.50	24.0/26,0	19.0/22.0	—
314	0.25 max.	2.0	1.5/3.0	23.0/26.0	19.0/22.0	—
316	0.08 max.	2.0	1.0	16.0/18.0	10.0/14.0	Mo 2.0/3.0
316L	0.03 max.	2.0	1.0	16.0/18.0	10.0/14.0	Mo 2.0/3.0
317	0.08 max.	2.0	1.0	18.0/20.0	11.0/15.0	Mo 3.0/4.0
321	0.08 max.	2.0	1.0	17.0/19.0	9.0/12.0	Ti 5 × C min.
347	0.08 max.	2.0	1.0	17.0/19.0	9.0/13.0	Cb + Ta10 × C min.
348	0.08 max.	2.0	1.0	17.0/19.0	9.0/13.0	Ta 0.10 max.
Chromium–martensitic, hardenable						
403	0.15 max.	1.0	0.5	11.5/13.0	—	—
410	0.15 max.	1.0	1.0	11.5/13.5	—	—
414	0.15 max.	1.0	1.0	11.5/13.5	1.25/2.5	—
416	0.15 max.	1.25	1.0	12.0/14.0	—	S 0.15 min.
416Se	0.15 max.	1.25	1.0	12.0/14.0	—	Se 0.15 min.
420	Over 0.15	1.0	1.0	12.0/14.0	—	—
431	0.20 max.	1.0	1.0	15.0/17.0	1.25/2.5	—
440A	0.60/0.85	1.0	1.0	16.0/18.0	—	Mo 0.75 max.
440B	0.75/0.95	1.0	1.0	16.0/18.0	—	Mo 0.75 max.
440C	0.95/1.2	1.0	1.0	16.0/18.0	—	Mo 0.75 max.
Chromium–ferritic, nonhardenable						
405	0.08 max.	1.0	1.0	11.5/14.5	—	A1 1.1/0.3
430	0.12 max.	1.0	1.0	14.0/18.0	—	—
430F	0.12 max.	1.25	1.0	14.0/18.0	—	S 0.15 min.
430Se	0.12 max.	1.25	1.0	14.0/18.0	—	Se 0.15 min.
446	0.20 max.	1.50	1.0	23.0/27.0	—	N 0.25 max.
Martensitic*						
501	Over 0.10	1.0	1.0	4.0/6.0	—	Mo 0.40/0.65
502	0.10 max.	1.0	1.0	4.0/6.0	—	Mo 0.40/0.65

*Chromium–molybdenum steel, not stainless.

FIGURE 16–6   AISI stainless steel classification system. (Courtesy of the American Iron and Steel Institute.)

preheat and postheat of weldments are required. A preheat temperature range of 450 to 550° F (232 to 288° C) is recommended. Postheating should immediately follow welding and be in the range of 1200 to 1400° F (649 to 760° C), followed by slow cooling.

If preheat and postheat are not possible, an austenitic stainless steel filler metal should be used. Type 416Se is the free-machining composition and should not be welded. Welding processes that tend to increase carbon pickup are not recommended. Increased carbon content increases crack sensitivity in the weld area.

Duplex Stainless Steels

A new class of stainless steels has been developed that combines the best properties of austenitic and ferritic stainless steels. They combine the ductility and corrosion resistance of the austenitic types and strength and resistance to corrosion cracking of the ferritic types. The name *duplex* indicates that their characteristic microstructure is typically 50% ferrite and 50% austenite. This is done by the adjustment of the chromium and nickel contents and by the addition of nitrogen or copper to stabilize the austenite. The duplex stainless steels are widely utilized by the petrochemical, pulp and paper, and oil and gas industries. One of the major uses is for pipe and tubing systems. This is to provide better corrosion resistance to sulfuric acid corrosion and better resistance to pitting in seawater, and to provide good resistance to stress corrosion cracking in these types of environments.

There are three basic classes based on the percent chromium: the 18%, 22%, and 25% chromium-containing alloys (see ASTM A789 for chemical requirements). Some are nitrogen containing, which improves mechanical properties and provides better resistance to general corrosion and pitting. Some contain small amounts of copper, which promotes better corrosion resistance in polluted seawater and improved resistance to sulfuric acid corrosion, as well as high mechanical properties. The duplex alloys also have a much lower carbon level than regular stainless steel compositions. The duplex stainless steels are susceptible to embrittlement if used for prolonged periods at elevated temperatures.

The physical and mechanical properties of the duplex stainless steels affect welding. The yield strength is typically about double that of type 316L, and the tensile properties are considerably higher than standard austenitic grades. The thermal conductivity of duplex stainless is approximately half that of carbon steels, but about 25% more than most austenitic stainless steels. The coefficient of thermal expansion of the duplex stainless steel is approximately the same as carbon steel, and about 40% less than that of the austenitic stainless steels. Duplex stainless steels are magnetic. Because of these factors, duplex stainless steels are easier to weld than austenitic stainless steels.

The normal arc welding processes, shielded metal arc, gas tungsten arc, gas metal arc, plasma arc, and submerged arc welding can all be used. In addition, electron beam and laser welding are used, as well as resistance welding. The joint details can be the same as those employed for austenitic stainless steels. The welding parameters would be essentially the same. Preheat should not be used and the interpass temperature should not exceed 300° F (150° C). Heat input should be on the low side. Surface cleanliness is a must when welding duplex stainless steels. It is necessary to eliminate any source of hydrogen in the welding operation. For the gas-shielded processes, particularly on pipe, argon purge gas should be used.

The filler metals to be used for welding duplex stainless steels should approximately match or overmatch the base metal composition. Filler metal and electrodes are available to match each of the three grades. Normally, there is no need for a postweld heat treatment. It is essential that thorough cleaning, chemical or mechanical, be used after welding.

The selection of the filler metal alloy for welding the stainless steels is based on the composition of the stainless steel. The various stainless steel filler metal alloys (Figure 16-7) are normally available as covered elec-

FIGURE 16–7   Stainless steel filler metal alloys. (From Ref. 8.)

AWS Class	Typical Composition (%)						
	C	Cr	Ni	Mo	Mn	Si	Others
E308	0.08	19.5	10.5	—	2.5	0.90	—
E308L	0.04	19.5	10.5	—	2.5	0.90	—
E309	0.15	23.5	13.5	—	2.5	0.90	—
E39Cb	0.12	23.5	13.5	—	2.5	0.90	Cb + Ti − 0.85
E309Mo	0.12	23.5	13.5	2.5	2.5	0.90	—
E310	0.20	26.5	21.5	—	2.5	0.75	—
E310Cb	0.12	26.5	21.5	—	2.5	0.75	Cb + Ti − 0.85
E310Mo	0.12	26.5	21.5	2.5	2.5	0.75	—
E312	0.15	30.0	9.0	—	2.5	0.90	—
E316	0.08	18.5	12.5	2.5	2.5	0.90	—
E316L	0.04	18.5	12.5	2.5	2.5	0.90	—
E317	0.08	19.5	13.0	3.5	2.5	0.90	—
E318	0.08	18.5	12.5	2.5	2.5	0.90	—
E320	0.07	20.0	34.0	2.5	2.5	0.60	—
E330	0.25	15.5	35.0	—	2.5	0.90	—
E347	0.08	19.5	10.0	—	2.5	0.90	—
E410	0.12	12.5	0.60	—	1.0	0.90	—
E430	0.10	16.5	0.60	—	1.0	0.90	—

Note: Remainder is iron.

trodes and as bare solid wires. Flux cored electrode wires are available for welding most stainless steels.[9]

Figure 16-8 gives the recommended filler metal alloy for welding the various stainless steel base metals. The table also shows some alternate alloys. Alternates are provided since there are so many different stainless steel types and there are not electrodes of each type. It is possible to weld several different stainless base metals with the same filler metal alloy.

For SMAW, there are three types of electrode coatings. These are the lime type indicated by the suffix 15, the titania type designated by the suffix 16, and the new type indicated by the suffix 17. The 17 classification is a titania type with silica replacing some titania to provide a true spray transfer. The Lime type (−15) is used on DCEP only. The Titania types are used with AC or DCEP. The type 15 produces a convex weld and can be used in all positions. The type 16 produces welds that are smooth and uniform with a profile that is flat to slightly convex and can be used in all positions. The type 17 provides spray transfers in the flat and horizontal positions. It provides a smooth weld with a concave profile. Diameters up to and including $\frac{5}{32}$ in. can be used in all positions. The procedure schedule for using the SMAW process is given

FIGURE 16–8 Recommended filler metals for stainless steels (use E or R prefix).

AISI No.	Recommended Filler Metal		Popular Name	Remarks
	First Choice	Second Choice		
Cr–Ni–Mn				
201	308	308L		Substitute for 301
202	308	308L		Substitute for 302
301	308	308L		
302	308	308L		
302B	308	309		High silicon
303	—	—		Free machining, welding not recommended, 312
303Se	—	—		Free machining, welding not recommended, 312
Cr–Ni–austenitic				
304	308	308L	18/8	
304L	308L	347	18/8 Elc	Extra low carbon
305	308	—		
308	308	—	19/9	
309	309	—	25/12	
309S	309	—		Low carbon
310	310	—	25/20	
310S	310	—		Low carbon
314	310	—		
316	316	309Cb	18/12 Mo	
316L	316L	309Cb	18/12 Elc	Extra low carbon
317	317	309Cb	19/14 Mo	
321	347	308L		
347	347	308L	19/9 Cb	Difficult to weld in heavy sections
348	347	—	19/9 CbLTa	
Cr–martensitic				
403	410	—		
410	410	430	12 Cr	
414	410	—		
416	410	—		Use 410-15
416Se	—	—		Free machining, welding not recommended
420	410	—	12 CrHC	High carbon
431	430	—		
440A	—	—		High carbon, welding not recommended
440B	—	—		High carbon, welding not recommended
440C	—	—		High carbon, welding not recommended
Cr–ferritic				
405	410	405Cb		
430	430	309	16 Cr	
430F	—	—		Free machining, welding not recommended
430FSe	—	—		Free machining, welding not recommended
446	309	310		
501	502	—	5 Cr–$\frac{1}{2}$ Mo	Chrome–moly steel
502	502	—	5 Cr–$\frac{1}{2}$ Mo	Chrome–moly steel

in Figure 16–9. The width of weaving should be limited to $2\frac{1}{2}$ times the diameter of the electrode core wire.

Covered electrodes for SMAW must be stored at normal room temperatures in dry areas. These electrodes are of the low-hydrogen type and are susceptible to moisture pickup. Once the electrode box has been opened, the electrodes should be kept in a dry box until used. If the electrodes are exposed to moisture, they should be reconditioned in accordance with procedures that were presented in Chapter 13.

The GTAW process is widely used for thinner sections of stainless steel. The 2% thoriated tungsten is recommended, and the electrode should be ground to a taper. Argon is normally used for gas shielding; however, argon–helium mixtures are sometimes used for automatic applications. Figure 16–10 shows the welding procedure schedule for the GTAW process for stainless steel.

The GMAW process is widely used for thicker materials. The spray transfer mode is used for flat-position welding and this requires the use of argon for shielding with 2% or 5% oxygen or special mixtures. The oxygen helps produce better wetting action on the edges of the weld and stabilize the arc. Figure 16–11 shows the welding procedure schedule for GMAW. The short-circuiting transfer can be used on thinner materials. In this case, CO_2 shielding or the 25% CO_2 plus 75% argon mixture is used. The argon–oxygen mixture can also be used with small-diameter electrode wires. With extra-low-carbon

FIGURE 16–9 Welding procedure schedule for SMAW of stainless steel.

Material Thickness			Electrode Diameter (in.)	Welding Current DCEP		
Gauge	Fraction	in.		Flat	Vertical	Overhead
26	—	0.018	$\frac{5}{64}$	20–35	20–25	20–30
22	—	0.030	$\frac{5}{64}$	30–45	30–40	30–40
18	—	0.048	$\frac{3}{32}$	50–70	40–55	50–60
14	—	0.075	$\frac{3}{32}$	60–90	50–65	60–95
11	—	0.120	$\frac{1}{8}$	90–120	75–90	90–110
—	$\frac{3}{16}$	0.188	$\frac{5}{32}$	120–150	90–110	120–140
—	$\frac{1}{4}$	0.250	$\frac{3}{16}$	150–200	100–125	—

FIGURE 16–10 Welding procedure schedule for GTAW of stainless steel.

Material Thickness (or Fillet Size)			Type of Weld	Tungsten Electrode Diameter		Filler Rod Diameter		Nozzle Size, Inside Diameter (in.)	Shielding Gas Flow (ft³/hr)	Welding Current (amps DCEN)	Number of Passes	Travel Speed per Pass (in./min)
Gauge	in.	mm		in.	mm	in.	mm					
24			Square groove	0.040	1.0	$\frac{1}{16}$	1.6	$\frac{1}{4}$	10	20–50	1	26
18			Square groove	$\frac{1}{16}$	1.6	$\frac{1}{16}$	1.6	$\frac{1}{4}$	10	50–80	1	22
$\frac{1}{16}$ in.	0.062	1.6	Square groove	$\frac{1}{16}$	1.6	$\frac{1}{16}$	1.6	$\frac{1}{4}$	12	65–105	1	12
$\frac{1}{16}$ in.	0.062	1.6	Fillet	$\frac{1}{16}$	1.6	$\frac{1}{16}$	1.6	$\frac{1}{4}$	12	75–125	1	10
$\frac{3}{32}$ in.	0.093	2.4	Square groove	$\frac{1}{16}$	1.6	$\frac{3}{32}$	2.4	$\frac{1}{4}$	12	85–125	1	12
$\frac{3}{32}$ in.	0.093	2.4	Fillet	$\frac{1}{16}$	1.6	$\frac{3}{32}$	2.4	$\frac{1}{4}$	12	95–135	1	10
$\frac{1}{8}$ in.	0.125	3.2	Square groove	$\frac{1}{16}$	1.6	$\frac{3}{32}$	2.4	$\frac{5}{16}$	15	100–135	1	12
$\frac{1}{8}$ in.	0.125	3.2	Fillet	$\frac{1}{16}$	1.6	$\frac{3}{32}$	2.4	$\frac{5}{16}$	15	115–145	1	10
$\frac{3}{16}$ in.	0.188	4.8	Square groove	$\frac{3}{32}$	2.4	$\frac{1}{8}$	3.2	$\frac{5}{16}$	15	150–225	1	10
$\frac{3}{16}$ in.	0.188	4.8	Fillet	$\frac{1}{8}$	3.2	$\frac{1}{8}$	3.2	$\frac{3}{8}$	18	175–250	1	8
$\frac{1}{4}$ in.	0.25	6.4	V-groove	$\frac{1}{8}$	3.2	$\frac{3}{16}$	4.8	$\frac{3}{8}$	18	225–300	2	10
$\frac{1}{4}$ in.	0.25	6.4	Fillet	$\frac{1}{8}$	3.2	$\frac{3}{16}$	4.8	$\frac{3}{8}$	18	225–300	2	10
$\frac{3}{8}$ in.	0.375	9.5	V-groove	$\frac{3}{16}$	4.8	$\frac{3}{16}$	4.8	$\frac{1}{2}$	25	220–350	2–3	10
$\frac{3}{8}$ in.	0.379	9.5	Fillet	$\frac{3}{16}$	4.8	$\frac{3}{16}$	4.8	$\frac{1}{2}$	25	250–350	3	10
$\frac{1}{2}$ in.	0.50	12.7	V-groove	$\frac{3}{16}$	4.8	$\frac{1}{4}$	6.4	$\frac{1}{2}$	25	250–350	3	10
$\frac{1}{2}$ in.	0.50	12.7	Fillet	$\frac{3}{16}$	4.8	$\frac{1}{4}$	6.4	$\frac{1}{2}$	25	250–350	3	10

Notes: 1. Increase amperage when backup is used.
2. Data are for flat position. Reduce amperage 10%–20% when welding is horizontal, vertical, or overhead position.
3. For tungsten electrodes: first choice, 2% thoriated EWTh2; second choice, 1% thoriated EWTh1.
4. Argon is used for shielding. The 75% helium–25% argon mixture is used for heavier thickness.

| Material Thickness (or Fillet Size) | | | Type of Weld | Electrode Diameter | | Welding Power | | Wire Feed Speed (in./min) | Shielding Gas Flow (ft³/hr) | Number of Passes | Travel Speed per Pass (in./min) |
Gauge	in.	mm		in.	mm	Current (A dc)	Arc Volts EP				
16	0.063	1.6	Square groove and fillet	0.035	0.9	60–100	15–18	90–190	12–15	1	15–30
13	0.093	2.4	Square groove and fillet	0.035	0.9	125–150	18–21	230–280	12–15	1	20–30
				0.045	1.1	125–150	18–21	130–160			20–30
11	0.125	3.2	Square groove and fillet	0.035	0.9	130–160	19–24	250–280	12–15	1	20–25
				0.045	1.1	150–225	19–24	160–260			20–30
$\frac{5}{32}$ in.	0.156	3.9	V-groove and fillet	0.045	1.1	190–250	22–26	200–290	15–20	1	25–30
$\frac{1}{4}$ in.	0.250	6.4	V-groove and fillet	0.045	1.1	225–300	24–30	260–370	25–30	2	25–30

Notes: 1. Data are for flat position. Reduce current 10%–20% for other positions.
2. Gas selection: argon–oxygen (1% to 2% oxygen + argon) for flat position and horizontal fillets.
   argon–CO_2 (75% argon–25% CO_2) for all-position, some carbon pickup.
   helium–argon–CO_2 (90%–7.5%–2.5%) for all-position welding.
   carbon dioxide (CO_2) where carbon pickup can be tolerated.

FIGURE 16–11 Welding procedure schedule for GMAW of stainless steel.

electrode wires and CO_2 shielding the amount of carbon pickup will increase slightly. This should be related to the service life of the weldment. If corrosion resistance is a major factor, the CO_2 gas or the CO_2–argon mixture should not be used.

Stainless steel can be welded with the submerged arc welding process. In this case, the electrode wire would be the same as shown in the selection guide table. The submerged arc flux must be selected for stainless steel welding.

For all welding operations, the weld area should be cleaned and free from all foreign material, oil, paint, dirt, and so on. The welding arc should be as short as possible when using any of the arc processes.

Problem Areas

The most serious problem when welding stainless steels is avoiding carbide precipitation. As mentioned previously, this may occur when the material is held at a high temperature for a long period. Electrodes containing ex-

tra low carbon, indicated by the suffix L, should be used. Most electrodes are stabilized with columbium or titanium. This also helps to minimize the carbide precipitation problem. These are definitely required when the weldment will be subjected to high temperature service.

Another factor that affects the quality of austenitic weld joints is the control of ferrite content in the microstructure. Austenitic weld deposits may develop microcracks during welding if ferrite is not controlled. The composition of the filler metal should be selected based on the deposit containing a small percentage of ferrite. The ferrite content should not be too high or the weldment will have lower than desired impact strength. For low-temperature service the weld metal should have the ferrite in the range 4% to 6%. The ferrite content of the weld deposit depends on the composition of the base metal as well as the composition of the deposited filler metal. A special constitution diagram for stainless steel weld metal has been designed by Schaeffler and modified by DeLong.[10] The diagram in Figure 16–12 relates the nickel and chromium equivalents to lines that show the

FIGURE 16–12 Constitution diagram for stainless steel weld metal. (From *Metal Progress Data Book*, ©American Society for Metals, 1977.)

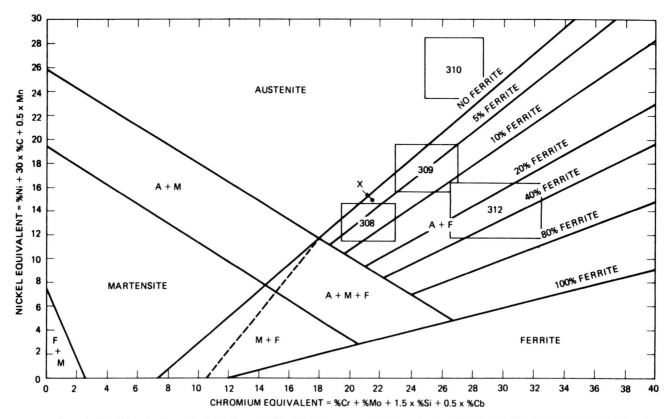

Example: Point X on the diagram indicates the equivalent composition of a type 318 (316 CB) weld deposit containing 0.07 C, 1.55 Mn, 0.57 Si, 18.02 Cr, 11.87 Ni, 2.16 Mo, 0.80 Cb. Each of these percentages was multiplied by the "potency factor" indicated for the element in question along the axes of the diagram, in order to determine the chromium equivalent and the nickel equivalent. When these were plotted, as point X, the constitution of the weld was indicated as austenite plus from 0 to 5% ferrite; magnetic analysis of the actual sample revealed an average ferrite content of 2%.

For austenite-plus-ferrite structures, the diagram predicts the percentage ferrite within 4% for the following stainless steels: 308, 309, 309 Cb, 310, 312, 316, 317, 318 (316 Cb), and 347.

Dashed line is the martensite/M + F boundary modification by Eberhard Leinhos, "Mechanische Eigenschaften und Gefugeausbildung von mit Chrom und Nickel legiertem Schweissgut," VEB Deutscher Verlag fur Grundstoffindustrie, Leipzig, 1966.

percentage of ferrite. This diagram is useful for estimating the microstructure of the weld deposit and the filler metal composition required to produce the prescribed amount of ferrite in the deposit. The diagram shows how the microstructure of the weld deposit is affected by the alloying elements in the stainless steel, based on those that act like nickel and those that act like chromium. The nickel equivalent group includes nickel and the effect of carbon and manganese. The chromium equivalent group includes chromium and the effect of molybdenum, silicon, and columbium. To estimate the microstructure of a deposit, the nickel and the chromium equivalents are calculated using the following formulas:

$$\text{nickel equiv.} = \%Ni + 30\% \text{ C} + 0.5\% \text{ Mn}$$
$$\text{chromium equiv.} = \%Cr + \%Mo + 1.5\% \text{ Si} + 0.5\% \text{ Cb}$$

The values obtained are marked on the coordinates of the diagram and a point is located. The microstructure at that point is the one predicted for a deposit of that composition. It is possible to plot the composition of the filler wire and the composition of the base metals and connect them with a line, and a resulting weld would be along this line. By the use of the Schaeffler-DeLong diagram, it is possible to select a filler metal that will avoid ferrite or martensite in the stainless steel weld deposit. The diagram can also be used to predict weld deposit composition when welding dissimilar stainless steels.[11] Instruments are available to measure the delta ferrite content of austenitic stainless steel weld metal.

Numerous computer software programs based on the constitution diagrams for stainless steel weld metal are available. These can be used to predict the weld deposit position with different welding procedures when welding different stainless steels. One program assists in the selection of electrodes for welding stainless steel. It uses the Schaeffler-DeLong diagrams to calculate ferrite percentage and number (FN). It is used for complex dissimilar welds between different stainless steels. It calculates FN by applying complex mathematical formulas. Different programs are referenced in the Appendix.

The stainless steels can be welded by resistance welding and by many of the other specialty welding processes. Stainless steels can also be soldered and brazed.

Removal of Weld Discoloration of Stainless Steel

It is sometimes difficult to remove heat discoloration from weld seams, especially on inside corners. It can be removed by grinding, but this may be impractical or expensive. Commercial stainless steel chemical cleaners are available and are widely used to clean stainless steel to prevent corrosion and contamination, and are used on food-processing equipment, chemical-processing equipment, metal furniture, tanks, transportation equipment, and so on. Most systems use the paste, which is painted on the surface and allowed to soak for 10 to 15 minutes, after which it is removed by using a stainless steel wire brush. It is usually available from local supply houses. Electro-chemical devices are also available for removing discoloration. It uses a cleaning solution and a low voltage wand to wipe over the affected area.

16-4 WELDING ULTRAHIGH-STRENGTH STEELS

The term *high-strength steel* is often applied to all steels other than mild low-carbon steels. The steels under discussion in this section are those that have a yield strength of at least 80,000 psi (552 MPa). These are sometimes called the ultrahigh-strength steels or super alloys.

The groups of steels that fall into this category are:

1. Medium-carbon low-alloy hardenable steels
2. Medium-alloy hardenable or tool and die steels
3. High-alloy hardenable steels
4. High-nickel maraging steels
5. Martensitic stainless steels
6. Semiaustenitic precipitation-hardenable stainless steels

Each of these groups will be briefly described and welding information will be presented. Nominal compositions of steels in each of these groups is shown in Figure 16-13.

Medium-Carbon Low-Alloy Hardenable Steels

The best known steels in this class are the AISI 4130 and AISI 4140 steels. Also in this class are the higher-strength AISI 4340 steel and the AMS 6434 steel. These steels obtain their high strength by heat treatment to a full martensitic microstructure, which is tempered to improve ductility and toughness. Tempering temperatures greatly affect the strength levels of these steels. The carbon is in the medium range and as low as possible but sufficient to give the required strength. Impurities are kept to an absolute minimum because of high-quality melting and refining methods. These steels are available as sheets, bars, tubing, and light plate. The steels in this group can be mechanically cut or flame cut. However, when they are flame cut they must be preheated to 600° F (316° C). Flame-cut parts should be annealed before additional operations in order to reduce the hardness of the flame-cut edges.

Welding is usually done on these steels when they are in the annealed or normalized condition. They are then heat treated to obtain the desired strength. The gas tungsten arc, the gas metal arc, the shielded metal arc, and the gas welding process are all used for welding these

	Composition (wt %)						
Designation	C	Mn	Si	Cr	Ni	Mo	Other
Medium-carbon low-alloy hardenable steels							
AISI 4130	0.28/0.33	0.40/0.60	0.20/0.35	0.80/1.10	—	0.15/0.25	—
AISI 4140	0.38/0.43	0.75/1.0	0.20/0.35	0.80/1.10	—	0.15/0.25	—
AISI 4340	0.38/0.43	0.60.0.80	0.20/0.35	0.70/0.90	1.65/2.00	0.20/0.30	—
AMS 6434	0.31/0.38	0.60/0.80	0.20/0.35	0.65/0.90	1.65/0.90	0.30/0.40	0.17/0.23V
Medium-alloy hardenable or tool and die steels							
5Cr–Mo–V aircraft steel	0.37/0.43	0.20/0.40	0.80/1.20	4.75/5.25	—	1.20/1.40	0.4/0.6V
H-11 tool steel	0.30/0.40	0.20/0.40	0.80/120	4.75/5.50	—	1.25/1.75	0.30/0.50V
H-13 tool steel	0.30/0.40	0.20/0.40	0.80/1.20	4.75/5.50	—	1.25/1.75	0.80/1.20V
High-alloy hardenable steels							
HP 9-4-20	0.20	0.30	0.10 max.	0.75	9.0	0.75	0.01S, 0.01P, 0.10V, 4.50Co
HP 9-4-30 (Cr, Mo)	0.30	0.20	0.10 max.	1.00	7.5	1.00	0.01S, 0.01P, 0.10V, 4.50Co
High nickel maraging steels							
18Ni	0.03 max.	0.10 max.	0.10 max.	—	18.0	3.25	0.01S, 0.01P, 8.5Co, 0.20Ti, 0.10Al
18Ni	0.03 max.	0.10 max.	0.10 max.	—	18.0	4.90	0.01S, 0.01P, 8.0Co, 0.40Ti, 0.10Al
18Ni	0.03 max.	0.10 max.	0.10 max.	—	17.5	4.90	0.01S, 0.01P, 9.0Co, 0.65Ti, 0.10Al
18Ni	0.01 max.	0.10 max.	0.10 max.	—	17.5	3.75	0.01S, 0.01P, 12.5Co, 1.80Ti, 0.15Al
Martensitic stainless steels							
AISI 420	0.15 max.	1.0 max.	1.0 max.	13	—	—	—
AISI 431	0.20 max.	1.0 max.	1.0 max.	16	2.0	—	—
12Mo–V	0.25	0.5	0.5	12	0.5	1.0	0.3V
17-4PH	0.07 max.	1.0 max.	1.0 max.	16.5	4.0	—	4.0Cu, 0.3Cb
PH13–8Mo	0.05 max.	0.1 max.	0.1 max.	12.5	8.0	2.5	1.1Al
Pyromet X-15	0.03 max.	0.1	0.1 max.	15	—	2.9	20.0Co
Custom 455	0.05 max.	0.5 max.	0.5 max.	12	8.5	0.5	2.0Cu, 0.3Cb, 1.1Ti
AFC 77	0.15	—	—	14.5	—	5.0	4.0Cu, 13.5Co, 0.5V, 0.5N
Semiaustenitic precipitation-hardenable stainless steels							
17-7PH	0.09 max.	1.0 max.	1.0 max.	17.0	7.0	—	1.0Al
PH15–7Mo	0.09 max.	1.0 max.	1.0 max.	15.0	7.0	2.5	1.1Al
PH14–8Mo	0.05 max.	0.1 max.	0.1 max.	15.0	8.5	2.5	1.0Al
AM 350	0.12 max.	0.90	0.5 max.	16.5	4.5	3.0	0.10N
AM 355	0.15 max.	0.95	0.5	15.5	4.5	3.0	0.09N

FIGURE 16–13 Nominal composition of ultrahigh-strength steels.

steels. The composition of the filler metal is designed to produce a weld deposit that responds to a heat treatment in approximately the same manner as the base metal. In order to avoid brittleness and the possibility of cracks during welding, relatively high preheat and interpass temperatures are used. Preheating is on the order of 600° F (316° C). Complex weldments are heat treated immediately after welding.

Aircraft engine mounts, aircraft tubular frames, and racing car frames are made from AISI 4130 tubular sections. These types of structures are normally not heat treated after welding.

Medium-Alloy Hardenable Steels

These steels are used largely in the aircraft industry for ultra-high-strength structural applications. They have carbon in the low to medium range and possess good fracture toughness at high strength levels. In addition, they are air hardened, which reduces the distortion that is encountered with more drastic quenching methods. Some of the steels in this group are known as hot work die steels and another grade has become known as 5Cr-Mo-V aircraft quality steel. There are proprietary names for other steels in this class. The steels are available as forging billets, bars, sheet, strip, and plate.

There is another type of steel in this general class, which is a medium-alloy quenched and tempered steel known as high-yield or HY 130/150. This type of steel is used for submarines, aerospace applications, and pressure vessels, and is normally available as plate. This steel has good notch toughness properties at 0° F and below. These types of steels have much lower carbon than the grades mentioned previously.

When flame cutting or welding the aircraft quality steels, preheating is absolutely necessary since the steels are air hardening. A preheat of 600° F (316° C) is used before flame cutting and then annealed immediately after the flame cutting operation. This will avoid a brittle layer at the flame-cut edge, which is susceptible to cracking. This type of steel should only be welded in the annealed condition. The steel should be preheated to 600° F (316° C) and this temperature must be maintained throughout the welding operation. After welding, the work must be cooled slowly. This can be done by postheating, or by furnace cooling. The weldment is then stress relieved at 1300° F (704° C) and air cooled to obtain a fully tempered microstructure suitable for additional operations. It is usually annealed, after all welding is done, prior to final heat treatment. The filler metal should be of the same composition as the base metal. The gas tungsten arc and gas metal arc processes are most widely used. However, shielded metal arc welding, plasma arc, and electron beam welding processes can be used.

The medium-alloy quenched and tempered high-yield strength steels are usually welded with the shielded metal arc, gas metal arc, or the submerged arc welding process. The filler metal must provide deposited metal of a strength level equal to the base material. In all cases, a low-hydrogen or no-hydrogen process is required. For shielded metal arc welding the low-hydrogen electrodes of the E-13018 type are recommended. Electrodes must be properly stored. In the case of the other processes, precautions should be taken to make sure that the gas is dry and that the submerged arc flux is dry. By employing the proper heat input–heat output procedure, yield strength and toughness are maintained. Preheating should be at least 100° F (38° C) for thinner materials and double that for heavier materials. The heat input should be such that the adjacent base metal does not become overheated. The heat input is sufficient to maintain the proper microstructure in the heat-affected zone. There may be some softening in the intermixing zone. The properties of welded joints that are properly made will be in the same order as the base metal. Subsequent heat treating is usually not required or desired.

High-Alloy Hardenable Steels

The steels in this group develop high strength by standard hardening and tempering heat treatments. The steels possess extremely high strength in the range of 180,000 psi yield and have a high degree of toughness. This is obtained with a minimum carbon content usually in the range of 0.20%; however, these steels contain relatively high amounts of nickel and cobalt, and they are sometimes called the 9% Ni–4% Co steels. These steels also contain small amounts of other alloys. They are normally welded in the quenched and tempered condition by the gas tungsten arc welding process. No postheat treatment is required. The filler metal must match the analysis of the base metal.

High-Nickel Maraging Steels

This type of steel has a relatively high nickel content but is a low-carbon steel. It obtains its high strength from a special heat treatment called *maraging*. These steels possess an extraordinary combination of ultra high-strength and fracture toughness and at the same time are formable, weldable, and easy to heat treat. There are three basic types: the 18% nickel, the 20% nickel, and the 25% nickel types. These steels are available in sheet, forging billets, bars, strip, and plate. Some are available as tubing.

The extra-special properties of these steels are obtained by heating the steel to 900° F (482° C) and allowing it to cool to room temperature. During this heat treatment all of the austenite transforms to martensite, which is of the very tough massive type. The time at the 900° F temperature is extremely important and usually is in the range of three hours. The steels derive their strength while aging at this temperature in the martensitic condition and for this reason are known as maraging steels.

These steels are supplied in the soft or annealed condition. They can be cold worked in this condition and can be flame cut or plasma arc cut. Plasma arc cutting is preferred. These steels are usually welded by the gas tungsten arc or the GMAW process. The shielded metal arc and submerged arc process can also be used with special electrode–flux combinations. The filler metal should have the same composition as the base metal. In addition, the filler metal must be of high purity with low carbon. Preheat or postheat is not required; however, the welding must be followed by the maraging heat treatment, which produces weld joints of an extremely high strength.

Martensitic Stainless Steels

These steels are of the straight chromium type, essentially the AISI 420 classification. These steels contain 12% to 14% chromium and up to 0.35% carbon. This composition combines corrosion resistance with high strength. Numerous variations of this basic composition are available, all of which are in the martensitic classification. This type of steel has been used for compressor and turbine blades of jet engines and for other applications in which moderate corrosion resistance and high strength are required. The strength level of these steels is obtained by a quenching and tempering heat treatment. They can be obtained as sheet, strip, tubing, and plate. The compositions are also used for castings. These steels can be heat treated to strengths as high as 250,000 psi (175 kg/mm^2) yield strength.

These stainless steels can be flame cut by the powder cutting system normally used for flame cutting stainless steels. They can also be cut with the oxyarc process. Flame cutting should be done with the steel in the annealed condition. Most grades should be preheated to 600° F (316° C) because they are air hardenable. They should be annealed after cutting to restore softness and ductility. These materials can also be cold worked in the annealed condition.

The martensitic stainless steels can be welded in the annealed or fully hardened condition, usually without preheat. The GTAW and GMAW processes are normally used. The filler metal must be of the same analysis as the base metal. Following welding the weldment should be annealed and then heat treated to the desired strength level.

Semiaustenitic Precipitation-Hardenable Stainless Steels

The steels in this group are chrome–nickel steels that are ductile in the annealed condition but can be hardened to high strength by proper heat treatment. In the annealed condition the steels are austenitic and can be readily cold worked. By special heat treatment the austenite is transformed to martensite and later a precipitant is formed in the martensite. The outstanding extra-high strength is obtained by a combination of these two hardening processes. The term *semiaustenitic type* was given these steels to distinguish them from normal stainless steels. They are also called precipitation-hardening (PH) steels. The heat treatment for these steels is based on heating the annealed material to a temperature of 1700 to 1750° F (927 to 954° C) followed by a tempering or aging treatment in the range 850 to 1100° F (454 to 593° C). These steels are available as billets, sheet, tubing, and plate.

These steels are normally not flame cut. Welding is performed using the gas tungsten arc or the gas metal arc welding process. The shielded metal arc welding process is not used. The filler metal should have the same composition as the base metal. No preheat is required if the parts are welded in the annealed condition. Following welding it should be annealed and then heat treated to develop optimum strength levels.

It is possible to weld the PH steels in the heat-treated condition using an arc welding process. However, there is a loss of joint strength due to heating of the heat-affected zone above the aging temperature. In view of this, it is not possible to produce a 100% efficient joint. Extra reinforcing must be utilized to develop full-strength joints. These steels are also brazed using a wide range of filler metal alloys.

When welding on any of these high-strength steels, weld quality must be of the highest degree. Root fusion must be complete, and there should be no undercut or any type of stress risers. The weld metal should be free of porosity and any weld cracking is absolutely unacceptable. All precautions must be taken in order to produce the highest weld quality. Arc strikes should be the basis for rejection.

QUESTIONS

16–1. What is the maximum carbon content for low-carbon steels?

16–2. Why are the free-machining steels difficult to weld?

16–3. Can the E60XX and E70XX electrodes be used to weld all carbon and mild steels?

16–4. What is the significance of the covered electrode class suffix letter?

16–5. Why can a cellulosic electrode be used to weld pipelines made of high-strength steels?

16–6. How is the electrode class suffix letter used to select electrodes for alloy steels?

16–7. How are the first two (or three) digits useful in selecting electrodes for alloy steels?

16–8. What are weathering steels? Name two popular brands.

16–9. Why is carbon–moly steel selected for high-temperature service? How is it welded?

16–10. What is the advantage of high-nickel steel for low-temperature service? How is it welded?

16–11. What are the advantages of quenched and tempered steels for construction equipment?

16–12. Why is heat input important when welding quenched and tempered steels?

16–13. What are the three types of stainless steel? Are any magnetic? Which?

16–14. Why is the welding of austenitic stainless steel different from mild steel?

16–15. Are the stainless steel covered electrodes of the low-hydrogen type? Why?

16–16. What is a duplex stainless steel?

16–17. What use is made of the Schaeffler diagram?

16–18. Can any of the ultrahigh-strength steels be welded after heat treatment?

16–19. What is maraging steel? What processes are used for welding?

16–20. What is a PH steel? Can it be welded with SMAW?

REFERENCES

1. "Steel Products Manual," American Iron and Steel Institute, New York.

2. "Specifications for Low Alloy Steel Covered Arc Welding Electrodes," AWS A5.5, American Welding Society, Miami, Fla.

3. "Specification for Low Alloy Steel Filler Metals for Gas Shielded Arc Welding," AWS A5.28, American Welding Society, Miami, Fla.

4. "Specifications for Low Alloy Steel Electrodes for Flux Cored Arc Welding," AWS A5.29, American Welding Society, Miami, Fla.

5. Robert Irving, "Micro Alloying the Route to Stronger, Tougher Steels," *Iron Age* (February 16, 1983).

6. "Selecting Welding Wire for HSLA Steels," *Fabricator* (May–June 1984).

7. "Recommended Practices for Welding of Chromium–Molybdenum Steel Piping and Tubing," AWS D10.8, American Welding Society, Miami, Fla.

8. "Specifications for Corrosion-Resisting Chromium and Chromium Nickel Steel Covered Welding Electrodes," AWS A5.4, American Welding Society, Miami, Fla.

9. "Specifications for Flux-Cored Corrosion-Resisting Chromium and Chromium-Nickel Steel Electrodes," AWS A5.22, American Welding Society, Miami, Fla.

10. A. L. Schaeffler, "Constitution Diagram for Stainless Steel Weld Metal," *Metal Progress Data Book,* American Society for Metals, Metals Park, Ohio, Mid-June 1977.

11. C. Long and W. DeLong, "The Ferrite Content of Austenitic Stainless Steel Weld Metal," *Welding Journal,* Research Supplement (July 1973).

17

WELDING NONFERROUS METALS

OUTLINE

17-1 ALUMINUM AND ALUMINUM ALLOYS

The unique combination of light weight and relatively high strength makes aluminum the second most popular metal that is welded. Aluminum is not difficult to join, but aluminum welding is different from welding steels.

Many alloys of aluminum have been developed, and it is important to know which alloy is to be welded. A system of four-digit numbers has been developed by the Aluminum Association, Inc., and adopted by the American Society for Testing and Materials (ASTM) to designate the wrought aluminum alloy types.[1] UNS designations are also shown; this adds A9 ahead of the AA number. This system of alloy groups (Figure 17–1) is as follows:

❊ *1XXX series.* These are aluminums of 99% or higher purity. They are used primarily in the electrical and chemical industries.

❊ *2XXX series.* Copper is the principal alloy in this group. This group provides high strength when properly heat treated. These alloys do not produce as good corrosion resistance and are often clad with pure aluminum or special-alloy aluminum. These alloys are used in the aircraft industry.

❊ *3XXX series.* Manganese is the major alloying element in this group. These alloys are non-heat-treatable. Manganese is limited to about 1.5%. These alloys have moderate strength and are easily worked.

❊ *4XXX series.* Silicon is the major alloying element in this group. It can be added in sufficient quantities to reduce the melting point and is used for brazing alloys and welding electrodes. Most of the alloys in this group are non-heat-treatable.

❊ *5XXX series.* Magnesium is the major alloying element of this group. These alloys are of medium strength. They possess good welding characteristics, good resistance to corrosion, but the amount of cold work should be limited.

❊ *6XXX series.* Alloys in this group contain silicon and magnesium, which make them heat treatable. These alloys possess medium strength and good corrosion resistance.

❊ *7XXX series.* Zinc is the major alloying element in this group. Magnesium is also included in most of these alloys. Together they result in a heat-treatable alloy of high strength. This series is used for aircraft frames.

FIGURE 17–1 Designation of aluminum alloy groups.

Major Alloying Element	Designation
99.0% minimum aluminum and over	1xxx
Copper	2xxx
Manganese	3xxx
Silicon	4xxx
Magnesium	5xxx
Magnesium and silicon	6xxx
Zinc	7xxx
Other elements	8xxx

✳ *8XXX series.* Other elements such as iron, nickel, or lithium.

The composition of the wrought aluminum alloys is shown in Figure 17–2.

Aluminum alloy casting alloys are also designated by the Aluminum Association. Figure 17–3 shows the nominal chemical composition of casting alloys. With respect to welding, it is the composition that is important rather than how the part was made. Castings as well as wrought forms are heat treated, which must be considered. Otherwise, the welding procedures can be essentially the same.

Temper Designation System

The Aluminum Association and ASTM provide a temper designation system used for wrought and cast aluminum alloys. It is based on the sequence of treatments to produce various tempers. In specifying an alloy, the temper designation follows the alloy designation separated by a dash. Basic temper designations consist of letters. Subdivisions of the basic tempers, when required, are indicated by one or more digits following the letter.

The basic temper designations and subdivisions are as follows:

F As fabricated.

O Annealed, recrystallized (wrought products only); applies to the softest tempers of the wrought products.

H Strain hardened (wrought products only). This applies to products whose strength is increased by strain hardening with or without supplementary treatment. The H is always followed by two or more digits. The first digit indicates the specific combination of basic operations as follows:

H-1 Strain hardened only.

H-2 Strain hardened and then partially annealed.

H-3 Strain hardened and then stabilized.

The digit following the designation H-1, H-2, and H-3 indicates the final degree of strain hardening. Tempers between 0 (annealed) and 8 (full hard) are designated by numbers 1 through 7. Numeral 2 indicates quarter hard, the numeral 4 indicates half hard, the numeral 6 indicates three-quarters hard, etc. The numeral 9 indicates extra hard temper.

The third digit, when used, indicates a variation of the two-digit H temper number.

W Solution heat treated. This is an unstable temper applied only to alloys that are age hardened at room temperatures after solution heat treatment.

T Thermally treated to produce stable tempers other than F, O, or H. The T is always followed by one or more digits as follows:

T-1 Cooled from an elevated temperature shaping process and naturally aged to a substantially stable condition.

T-2 Annealed (cast products only).

T-3 Solution heat treated and then cold worked.

T-4 Solution heat treated and naturally aged to a substantially stable condition.

T-5 Cooled from an elevated temperature shaping process and then artificially aged.

T-6 Solution heat treated and then artificially aged.

T-7 Solution heat treated and then stabilized.

T-8 Solution heat treated and then heat treated, cold worked, and then artificially aged.

T-9 Solution heat treated, artificially aged, and then cold worked.

T-10 Cooled from an elevated temperature shaping process, artificially aged, and then cold worked.

An additional digit may be used that indicates the variation and treatment that significantly alters the characteristics of the product. For example, TX indicates stress relieving by some process such as stretching, compressing, or thermal treatment.

The temper designations are important from a welding point of view since welding, which is normally a thermal process, can change the characteristics of the metal in the heat-affected zone. Care must be taken when welding on the H, W, or T designations. Get metallurgical advice to determine treatment required to obtain original properties.

The different temper designations are used for different products such as sheet, plate, pipe, shapes, rod, bar, etc. In addition, the different alloys are available in certain types of mill products. In other words, all products are not available in all compositions or in all of the different tempers.

The heat-treatable alloys that contain copper or zinc are less resistant to corrosion than the non-heat-treatable alloys. To increase the corrosion resistance of these alloys in sheet and plate, they are sometimes clad

FIGURE 17-2 Nominal chemical composition of aluminum-wrought alloys. (Courtesy of the Aluminum Association.)

UNS Number	AA Designation	Composition (%)									Others		Al
		Si	Fe	Cu	Mn	Mg	Cr	Ni	Zn	Ti	Each	Total	(Min.)
—	1050	0.25	0.40	0.05	0.05	0.05	—	—	0.05	0.03	0.03	—	99.50
A91060	1060	0.25	0.35	0.05	0.05	0.03	—	—	0.05	0.03	0.03	—	99.60
A91100	1100	1.0	Si + Fe	0.05–0.20	0.05	—	—	—	0.10	—	0.05	0.15	99.00
A91145	1145	0.55	Si + Fe	0.05	0.05	—	—	—	0.05	0.03	0.03	—	99.45
—	1175	0.15	Si + Fe	0.10	0.10	0.02	—	—	0.04	0.02	0.02	—	99.75
—	1200	1.0	Si + Fe	0.05	0.05	—	—	—	0.10	0.05	0.05	0.15	99.00
A91230	1230	0.7	Si + Fe	0.10	0.05	0.05	—	—	0.10	0.03	0.03	—	99.30
A91235	1235	0.65	Si + Fe	0.05	0.05	0.05	—	—	0.05	0.03	0.03	—	99.35
—	1345	0.30	0.40	0.10	0.05	0.05	—	—	0.05	0.03	0.03	—	99.45
—	1350	0.10	0.40	0.05	0.01	—	0.01	—	0.05	0.03	0.03	0.10	99.50
A92011	2011	0.40	0.7	5.0–6.0	—	—	—	—	0.30	—	0.05	0.15	Remainder
A92014	2014	0.50–1.2	0.7	3.9–5.0	0.40–1.2	0.20–0.8	0.10	—	0.25	0.15	0.05	0.15	Remainder
A92017	2017	0.20–0.8	0.7	3.5–4.5	0.40–1.0	0.40–0.8	0.10	—	0.25	0.15	0.05	0.15	Remainder
A92018	2018	0.9	1.0	3.5–4.5	0.20	0.45–0.9	0.10	1.7–2.3	0.25	—	0.05	0.15	Remainder
A92024	2024	0.50	0.50	3.8–4.9	0.30–0.9	1.2–1.8	0.10	—	0.25	0.15	0.05	0.15	Remainder
A92025	2025	0.50–1.2	1.0	3.9–5.0	0.40–1.2	0.05	0.10	—	0.25	0.15	0.05	0.15	Remainder
—	2036	0.50	0.50	2.2–3.0	0.10–0.40	0.30–0.6	0.10	—	0.25	0.15	0.05	0.15	Remainder
A92117	2117	0.8	0.7	2.2–3.0	0.20	0.20–0.50	0.10	—	0.25	—	0.05	0.15	Remainder
A92124	2124	0.20	0.30	3.8–4.9	0.30–0.9	1.2–1.8	0.10	—	0.25	—	0.05	0.15	Remainder
A92218	2218	0.9	1.0	3.5–4.5	0.20	1.2–1.8	0.10	1.7–2.3	0.25	0.15	0.05	0.15	Remainder
A92219	2219	0.20	0.30	5.8–6.8	0.20–0.40	0.02	—	—	0.10	0.02–0.10	0.05	0.15	Remainder
—	2319	0.20	0.30	5.8–6.8	0.20–0.40	0.02	—	—	0.10	0.10–0.20	0.05	0.15	Remainder
A92618	2618	0.10–0.25	0.9–1.3	1.9–2.7	—	1.3–1.8	—	0.9–1.2	0.10	0.04–0.10	0.05	0.15	Remainder
A93003	3003	0.6	0.7	0.05–0.20	1.0–1.5	—	—	—	0.10	—	0.05	0.15	Remainder
A93004	3004	0.30	0.7	0.25	1.0–1.5	0.8–1.3	—	—	0.25	—	0.05	0.15	Remainder
A93005	3005	0.6	0.7	0.30	1.0–1.5	0.20–0.6	0.10	—	0.25	0.10	0.05	0.15	Remainder
A93105	3105	0.6	0.7	0.30	0.30–0.8	0.20–0.8	0.20	—	0.40	0.10	0.05	0.15	Remainder
A94032	4032	11.0–13.5	1.0	0.50–1.3	—	0.8–1.3	0.10	0.50–1.3	0.25	—	0.05	0.15	Remainder
—	4043	4.5–6.0	0.8	0.30	0.05	0.05	—	—	0.10	0.20	0.05	0.15	Remainder
—	4045	9.0–11.0	0.8	0.30	0.05	0.05	—	—	0.10	0.20	0.05	0.15	Remainder
—	4047	11.0–13.0	0.8	0.30	0.15	0.10	—	—	0.20	—	0.05	0.15	Remainder
—	4145	9.3–10.7	0.8	3.3–4.7	0.15	0.15	0.15	—	0.20	—	0.05	0.15	Remainder
—	4343	6.8–8.2	0.8	0.25	0.10	—	—	—	0.20	—	0.05	0.15	Remainder
—	4643	3.6–4.6	0.8	0.10	0.05	0.10–0.30	—	—	0.10	0.15	0.05	0.15	Remainder
A95005	5005	0.30	0.7	0.20	0.20	0.50–1.1	0.10	—	0.25	—	0.05	0.15	Remainder
A95050	5050	0.40	0.7	0.20	0.10	1.1–1.8	0.10	—	0.25	—	0.05	0.15	Remainder
A95052	5052	0.25	0.40	0.10	0.10	2.2–2.8	0.15–0.35	—	0.10	—	0.05	0.15	Remainder
A95056	5056	0.30	0.40	0.10	0.05–0.20	4.5–5.6	0.05–0.20	—	0.10	—	0.05	0.15	Remainder
A95083	5083	0.40	0.40	0.10	0.40–1.0	4.0–4.9	0.05–0.25	—	0.25	0.15	0.05	0.15	Remainder
A95086	5086	0.40	0.50	0.10	0.20–0.7	3.5–4.5	0.05–0.25	—	0.25	0.15	0.05	0.15	Remainder
A95154	5154	0.45	Si + Fe	0.10	0.10	3.1–3.9	0.15–0.35	—	0.20	0.20	0.05	0.15	Remainder
—	5183	0.40	0.40	0.10	0.50–1.0	4.3–5.2	0.05–0.25	—	0.25	0.15	0.05	0.15	Remainder
A95252	5252	0.08	0.10	0.10	0.10	2.2–2.8	—	—	0.05	—	0.05	0.15	Remainder
A95254	5254	0.45	Si + Fe	0.05	0.01	3.1–3.9	0.15–0.35	—	0.20	0.05	0.05	0.15	Remainder
—	5356	0.50	Si + Fe	0.10	0.05–0.20	4.5–5.5	0.05–0.20	—	0.10	0.06–0.20	0.05	0.15	Remainder

FIGURE 17–2 (cont.)

UNS Number	AA Designation	Composition (%)									Others		(Min.)
		Si	Fe	Cu	Mn	Mg	Cr	Ni	Zn	Ti	Each	Total	
A95454	5454	0.45	0.4	0.10	0.50–1.0	2.4–3.0	0.05–0.20	—	0.25	0.20	0.05	0.15	Remainder
A95456	5456	0.40	Si + Fe	0.10	0.50–1.0	4.7–5.5	0.05–0.20	—	0.25	0.20	0.05	0.15	Remainder
A95457	5457	0.08	0.10	0.20	0.15–0.45	0.8–1.2	—	—	0.05	—	0.03	0.10	Remainder
—	5554	0.40	Si + Fe	0.10	0.50–1.0	2.4–3.0	0.05–0.20	—	0.25	0.05–0.20	0.05	0.15	Remainder
—	5556	0.40	Si + Fe	0.10	0.50–1.0	4.7–5.5	0.05–0.20	—	0.25	0.05–0.20	0.05	0.15	Remainder
A95652	5652	0.40	Si + Fe	0.04	0.01	2.2–2.8	0.15–0.35	—	0.10	—	0.05	0.15	Remainder
—	5654	0.45	Si + Fe	0.05	0.01	3.1–3.9	0.15–0.35	—	0.20	0.05–0.15	0.05	0.15	Remainder
A95657	5657	0.08	0.10	0.10	0.03	0.6–1.0	—	—	0.05	—	0.02	0.05	Remainder
A96003	6003	0.35–1.0	0.6	0.10	0.08	0.8–1.5	0.35	—	0.20	0.10	0.05	0.15	Remainder
A96005	6005	0.6–0.9	0.35	0.10	0.10	0.40–0.6	0.10	—	0.10	0.10	0.05	0.15	Remainder
A96053	6053	—	0.35	0.10	—	1.1–1.4	0.15–0.35	—	0.10	—	0.05	0.15	Remainder
A96061	6061	0.40–0.8	0.7	0.15–0.40	0.15	0.8–1.2	0.04–0.35	—	0.25	0.15	0.05	0.15	Remainder
A96063	6063	0.20–0.6	0.35	0.10	0.10	0.45–0.9	0.10	—	0.10	0.10	0.05	0.15	Remainder
A96066	6066	0.9–1.8	0.50	0.7–1.2	0.6–1.1	0.8–1.4	0.40	—	0.25	0.20	0.05	0.15	Remainder
A96070	6070	1.0–1.7	0.50	0.15–0.40	0.40–1.0	0.50–1.2	0.10	—	0.25	0.15	0.05	0.10	Remainder
A96101	6101	0.30–0.7	0.50	0.10	0.03	0.35–0.8	0.03	—	0.10	—	0.03	0.10	Remainder
A96151	6151	0.6–1.2	1.0	0.35	0.20	0.45–0.8	0.15–0.35	—	0.25	0.15	0.05	0.15	Remainder
A96162	6162	0.40–0.8	0.50	0.20	0.10	0.7–1.1	0.10	—	0.25	0.10	0.05	0.15	Remainder
A96201	6201	0.50–0.9	0.50	0.10	0.03	0.6–0.9	0.03	—	0.10	—	0.03	0.10	Remainder
A96253	6253	—	0.50	0.10	—	1.0–1.5	0.15–0.35	—	1.6–2.4	—	0.05	0.15	Remainder
A96262	6262	0.40–0.8	0.7	0.15–0.40	0.15	0.8–1.2	0.04–0.14	—	0.25	0.15	0.05	0.15	Remainder
A96351	6351	0.7–1.3	0.50	0.10	0.40–0.8	0.40–0.8	—	—	0.20	0.20	0.05	0.15	Remainder
A96463	6463	0.20–0.6	0.15	0.20	0.05	0.45–0.9	—	—	—	—	0.05	0.15	Remainder
A96951	6951	0.20–0.50	0.8	0.15–0.40	0.10	0.40–0.8	—	—	0.20	—	0.05	0.15	Remainder
—	7001	0.35	0.40	1.6–2.6	0.20	2.6–3.4	0.18–0.35	—	6.8–8.0	0.20	0.05	0.15	Remainder
A97005	7005	0.35	0.40	0.10	0.20–0.7	1.0–1.8	0.06–0.20	—	4.0–5.0	0.01–0.06	0.05	0.15	Remainder
A97008	7008	0.10	0.10	0.05	0.05	0.7–1.4	0.12–0.25	—	4.5–5.5	0.05	0.05	0.15	Remainder
A97011	7011	0.15	0.20	0.05	0.10–0.30	1.0–1.6	0.05–0.20	—	4.0–5.5	0.05	0.05	0.15	Remainder
A97072	7072	0.7	Si + Fe	0.10	0.10	0.10	—	—	0.8–1.3	—	0.05	0.15	Remainder
A97075	7075	0.40	0.50	1.2–2.0	0.30	2.1–2.9	0.18–0.35	—	5.1–6.1	0.20	0.05	0.15	Remainder
—	7079	0.30	0.40	0.40–0.8	0.10–0.30	2.9–3.7	0.10–0.25	—	3.8–4.8	0.10	0.05	0.15	Remainder
A97178	7178	0.40	0.50	1.6–2.4	0.30	2.4–3.1	0.18–0.35	—	6.3–7.3	0.20	0.05	0.15	Remainder

Notes: 1. Composition is percent maximum unless shown as a range or a minimum.
2. There are sometimes other minor elements present.

FIGURE 17–3 Nominal chemical composition of aluminum-casting alloys. (Courtesy of the Aluminum Association.)

AA Number	Former Designation	Product[a]	Composition (%)										Others	
			Si	Fe	Cu	Mn	Mg	Cr	Ni	Zn	Sn	Ti	Each	Total
208.0	108	S	2.5–3.5	1.2	3.5–4.5	0.50	0.10	—	0.35	1.0	—	0.25	—	0.50
213.0	C113	P	1.0–3.0	1.2	6.0–8.0	0.6	0.10	—	0.35	2.5	—	0.25	—	0.50
222.0	122	S&P	2.0	1.5	9.2–10.7	0.50	0.15–0.35	—	0.50	0.8	—	0.25	—	0.35
242.0	142	S&P	0.7	1.0	3.5–4.5	0.35	1.2–1.8	0.25	1.7–2.3	0.35	—	0.25	0.05	0.15
295.0	195	S	0.7–1.5	1.0	4.0–5.0	0.35	0.03	—	—	0.35	—	0.25	0.05	0.15
B295.0	B195	P	2.0–3.0	1.2	4.0–5.0	0.35	0.05	—	0.35	0.50	—	0.25	—	0.35
308.0	A108	P	5.0–6.0	1.0	4.0–5.0	0.50	0.10	—	—	1.0	—	0.25	—	0.50
319.0	319, Allcast	S&P	5.5–6.5	1.0	3.0–4.0	0.50	0.10	—	0.35	1.0	—	0.25	—	0.50
328.0	Red X-8	S	7.5–8.5	1.0	1.0–2.0	0.20–0.6	0.20–0.6	0.35	0.25	1.5	—	0.25	—	0.50
A332.0	A132	P	11.0–13.0	1.2	0.50–1.5	0.35	0.7–1.3	—	2.0–3.0	0.35	—	0.25	0.05	—
F332.0	F132	P	8.5–10.5	1.2	2.0–4.0	0.50	0.50–1.5	—	0.50	1.0	—	0.25	—	0.50
333.0	333	P	8.0–10.8	1.0	3.0–4.0	0.50	0.05–0.50	—	0.50	1.0	—	0.25	—	0.50
355.0	355	S&P	4.5–5.5	0.6	1.0–1.5	0.50	0.40–0.6	0.25	—	0.35	—	0.25	0.05	0.15
C355.0	C355	S&P	4.5–5.5	0.20	1.0–1.5	0.10	0.40–0.6	—	—	0.10	—	0.20	0.05	0.15
356.0	356	S&P	6.5–7.5	0.6	0.25	0.35	0.20–0.40	—	—	0.35	—	0.25	0.05	0.15
A356.0	A356	S&P	6.5–7.5	0.20	0.20	0.10	0.20–0.40	—	—	0.10	—	0.20	0.05	0.15
357.0	357	S&P	6.5–7.5	0.15	0.05	0.03	0.45–0.6	—	—	0.05	—	0.20	0.05	0.15
360.0	360	D	9.0–10.0	2.0	0.6	0.35	0.40–0.6	—	0.50	0.50	0.15	—	—	0.25
A360.0	A360	D	9.0–10.0	1.3	0.6	0.35	0.40–0.6	—	0.50	0.50	0.15	—	—	0.25
380.0	380	D	7.5–9.5	2.0	3.0–4.0	0.50	0.10	—	0.50	3.0	0.35	—	—	0.50
A380.0	A380	D	7.5–9.5	1.3	3.0–4.0	0.50	0.10	—	0.50	3.0	0.35	—	—	0.50
384.0	384	D	10.5–12.0	1.3	3.0–4.5	0.50	0.10	—	0.50	1.0	0.35	—	—	0.50
413.0	13	D	11.0–13.0	2.0	1.0	0.35	0.10	—	0.50	0.50	0.15	—	—	0.25
A413.0	A13	D	11.0–13.0	1.3	1.0	0.35	0.10	—	0.50	0.50	0.15	—	—	0.25
B443.0	43 (0.15 max. cu)	S&P	4.5–6.0	0.8	0.15	0.35	0.05	—	—	0.35	—	0.25	0.05	0.15
C443.0	43	D	4.5–6.0	2.0	0.6	0.35	0.10	—	0.50	0.50	0.15	—	—	0.25
514.0	214	S	0.35	0.50	0.15	0.35	3.5–4.5	—	—	0.15	—	0.25	0.05	0.15
A514.0	A214	P	0.30	0.40	0.10	0.30	3.5–4.5	—	0.50	1.4–2.2	—	0.20	0.05	0.15
B514.0	B214	S&P	1.4–2.2	0.6	0.35	0.8	1.8–2.4	0.25	—	0.35	—	0.25	0.05	0.15
518.0	218	D	0.35	1.8	0.25	0.35	7.5–8.5	—	0.15	0.15	0.15	—	—	0.25
520.0	220	S	0.25	0.30	0.25	0.15	9.5–10.6	—	—	0.15	—	0.25	0.05	0.15
535.0	Almost 35	S	0.15	0.15	0.05	0.10–0.25	6.2–7.5	—	—	—	—	0.10–0.25	0.05	0.15
705.0	603, Ternalloy 5	S&P	0.20	0.8	0.20	0.40–0.6	1.4–1.8	0.20–0.40	—	2.7–3.3	—	0.25	0.05	0.15
707.0	607, Ternalloy 7	S&P	0.20	0.8	0.20	0.40–0.6	1.8–2.4	0.20–0.40	—	4.0–4.5	—	0.25	0.05	0.15
A712.0	A612	S	0.15	0.50	0.35–0.65	0.05	0.6–0.8	0.40–0.6	—	6.0–7.0	—	0.25	0.05	0.15
D712.0	D612, 40E	S	0.30	0.50	0.25	0.10	0.50–0.65	0.40–0.6	—	5.0–6.5	—	0.15–0.25	0.05	0.20
713.0	613, Tenzaloy	S&P	0.25	1.1	0.40–1.0	0.6	0.20–0.50	0.35	0.15	7.0–8.0	—	0.25	0.10	0.25
771.0	Precedent 71A	S	0.15	0.15	0.10	0.10	0.8–1.0	0.06–0.20	—	6.5–7.5	—	0.10–0.20	0.05	0.15
850.0	750	S&P	0.7	0.7	0.7–1.3	0.10	0.10	—	0.7–1.3	—	5.5–7.0	0.20	—	0.30
A850.0	A750	S&P	2.0–3.0	0.7	0.7–1.3	0.10	0.10	—	0.30–0.7	—	5.5–7.0	0.20	—	0.30
B850.0	B750	S&P	0.40	0.7	1.7–2.3	0.10	0.6–0.9	—	0.9–1.5	—	5.5–7.0	0.20	—	0.30

Notes:
1. Composition is percent maximum unless shown as a range. Aluminum is the remainder.
2. There may be minor elements present.
[a]Product: S, sand cast; P, permanent mold cast; D, die cast.

with high-purity aluminum, usually $2\frac{1}{2}$% to 4% of the total thickness on each side. These are known as *alclad* products.

Welding Aluminum Alloys

Aluminum possesses a number of properties that make welding different than welding steels. These are:

1. Aluminum oxide surface coating
2. High thermal conductivity
3. High thermal expansion coefficient
4. Low melting temperature
5. The absence of color change as temperature approaches the melting point

Aluminum is an active metal and it reacts with oxygen in the air to produce a thin hard film of aluminum oxide on the surface. The melting point of aluminum oxide is approximately 3600° F (1926° C), which is almost three times the melting point of pure aluminum, 1220° F (660° C). This aluminum oxide film, particularly as it becomes thicker, will absorb moisture from the air. Moisture is a source of hydrogen, which is the cause of porosity in aluminum welds. Hydrogen may also come from oil, paint, and dirt in the weld area. It also comes from the oxide and foreign materials on the electrode or filler wire, as well as from the base metal. Hydrogen will enter the weld pool and is soluble in molten aluminum. As the aluminum solidifies it will retain much less hydrogen, and the hydrogen is rejected during solidification. With a rapid cooling rate, free hydrogen is trapped in the weld and will cause porosity.

The aluminum oxide film must be removed prior to welding. If it is not all removed, small particles of unmelted oxide will be entrapped in the weld and will cause a reduction in ductility, lack of fusion, and may cause weld cracking. Anodized coatings must be removed before welding.

The aluminum oxide can be removed by mechanical, chemical, or electrical means. Mechanical removal involves scraping with a sharp tool, sandpaper, wire brush (stainless steel), filing, or any other mechanical method. Chemical removal can be done in two ways. One is by use of cleaning solutions, either the etching types or the nonetching types. The nonetching types should be used only when starting with relatively clean parts. They are used in conjunction with other solvent cleaners. For better cleaning the etching type solutions are recommended but must be used with care. When dipping is employed, hot and cold rinsing is recommended. The etching type solutions are alkaline solutions. The time in the solution must be controlled so that too much etching does not occur.

Chemical cleaning includes the use of welding fluxes. Fluxes are used for gas welding, brazing, and soldering. The coating on covered aluminum electrodes also contains fluxes for cleaning the base metal. Whenever etch cleaning or flux cleaning is used, the flux and alkaline etching materials must be completely removed from the weld area to avoid future corrosion.

The electrical oxide removal system uses cathodic bombardment. Cathodic bombardment occurs during the half cycle of alternating current gas tungsten arc welding when the electrode is positive (reverse polarity). This is an electrical phenomenon that actually blasts away the oxide coating to produce a clean surface. This is one of the reasons why ac gas tungsten arc welding is so popular for welding aluminum.

The oxide film will immediately start to reform. The time of buildup is not extremely fast, but welds should be made after aluminum is cleaned within at least 8 hours for good quality welding.

Aluminum conducts heat from three to five times as fast as steel, depending on the specific alloy. This means that more heat must be put into the aluminum even though the melting temperature of aluminum is less than half that of steel. Because of the high thermal conductivity, preheat is often used for welding thicker sections. If the temperature is too high or the period of time is too long, it can be detrimental to weld joint strength in both heat-treated and work-hardened alloys. The preheat for aluminum should not exceed 400° F (204° C), and the parts should not be held at that temperature longer than necessary. Because of the high heat conductivity, procedures should utilize higher-speed welding processes using high heat input.

The high heat conductivity of aluminum can also be helpful, since if heat is conducted away from the weld extremely fast the weld will solidify very quickly. This, with surface tension, helps hold the weld metal in position and makes all-position welding practical.

The thermal expansion of aluminum is twice that of steel. In addition, aluminum welds decrease about 6% in volume when solidifying from the molten state. This change in dimension or attempt to change in dimension may cause distortion and cracking.

Aluminum does not exhibit color as it approaches its melting temperature. Aluminum will show color above the melting point, at which time it will glow a dull red. When soldering or brazing aluminum with a torch, flux is used and the flux will melt as the temperature of the base metal approaches the temperature required. The flux first dries out and then melts as the base metal reaches the correct working temperature. When torch welding with oxyacetylene or oxyhydrogen, the surface of the flux will melt first and assume a characteristic wet and shiny appearance. (This aids in knowing when welding temperatures are reached.) When welding with gas tungsten arc or gas metal arc, color is not too important because the weld is quickly completed before the adjoining area would melt.

When the factors above are taken into consideration, it will allow making welded joints in aluminum with little or no more trouble than when welding steels.

With either gas metal arc or gas tungsten arc welding, the selection of filler metal is the same. The base metal composition or alloy must be known. Figure 17–4 provides the nominal composition of the different aluminum filler metals. Refer to AWS specifications A5.3 and A5.10 for details. These provide for bare, solid, straightened electrode wires, coiled wires, and covered electrodes. It may not be necessary to make the comparison or selection of the filler metal to weld the different aluminum alloys since this has been standardized. Figure 17–5 is a guide to the choice of filler metals for aluminum welding established by the American Welding Society and is recommended.

Gas Tungsten Arc Welding

The GTAW process is used for welding the thinner sections of aluminum and aluminum alloys. Alternating current is recommended for general-purpose work since it provides the half-cycle of cleaning action. Figure 17–6 provides welding procedure schedules for using the process on different thicknesses. Ac welding, usually with high frequency, is widely used with manual and automatic applications. Procedures should be followed closely and special attention should be given to the type of tungsten electrode, size of welding nozzle, gas type, and gas flow rates. When manual welding, the arc length should be kept short and equal to the diameter of the electrode. The tungsten electrode should not protrude too far beyond the end of the nozzle. The tungsten should be kept clean, and if it does accidentally touch the molten metal it must be redressed.

Welding power sources designed for the gas tungsten arc welding process should be used since they provide for programming, pre- and postflow of shielding gas, pulsing, and special wave shapes.

For automatic or machine welding, direct current electrode negative (straight polarity) can be used. Cleaning must be extremely efficient since there is no cathodic bombardment to assist. When dc electrode negative is used, extremely deep penetration and high speeds can be obtained. Cleanliness is an absolute necessity. Figure 17–7 provides welding procedure schedules for dc electrode negative welding.

The gases are either argon or helium or a mixture of the two. Argon is the most popular and is used at a lower flow rate. Helium will increase penetration but a higher flow rate is required.

When filler wire is used, either manually or automatically, it must be clean. If the oxide is not removed from the filler wire, it may include moisture that will produce porosity in the weld deposit.

Gas Metal Arc Welding

The gas metal welding process is applicable to heavier thicknesses of aluminum. It is much faster than gas tungsten arc welding.

Several factors should be mentioned with respect to GMAW welding aluminum. The electrode wire must be clean. If porosity occurs, it is possible that it came from moisture absorbed in the oxide coating of the electrode wire.

Pure argon is normally used for gas metal arc welding of aluminum. On occasion, leaks in the gas system, in the gun or cable assembly, will allow air to be drawn into the argon, which will cause porosity. Gas purge control and post-gas flow should be used. The angle of the gun or torch is critical. A 30° leading travel angle is recommended. The electrode wire tip should be oversize for aluminum. Figure 17–8 provides welding procedure schedules for gas metal arc welding of aluminum.

FIGURE 17–4 Composition of aluminum filler metals.

	Nominal Composition (%)								
Type	Si	Fe	Cu	Mn	Mg	Cr	Zn	Ti	Al
1100	1.0 Si + Fe		0.05–0.20	0.05	—	—	0.10	—	99.0 min.
4043	4.5–6.0	0.80	0.30	0.05	0.05	—	0.10	0.20	Remainder
5154	0.45 Si + Fe		0.10	0.10	3.1–3.9	0.15–0.35	0.20	0.20	Remainder
5254	0.45 Si + Fe		0.05	0.01	3.1–3.9	0.15–0.35	0.20	0.05	Remainder
5652	0.40 Si + Fe		0.04	0.01	2.2–2.8	0.15–0.35	0.10	—	Remainder
5554	0.40 Si + Fe		0.10	0.50–1.0	2.4–3.0	0.05–0.20	0.25	0.05–0.20	Remainder
5356	0.50 Si + Fe		0.10	0.05–0.20	4.5–5.5	0.05–0.20	0.10	0.06–0.20	Remainder
5183	0.40	0.40	0.10	0.50–1.0	4.3–5.2	0.05–0.25	0.25	0.15	Remainder
5556	0.40 Si + Fe		0.10	0.50–1.0	4.7–5.5	0.04–0.35	0.25	0.05–0.20	Remainder
6061	0.40–0.80	0.70	0.15–0.40	0.15	0.80–1.2	0.04–0.35	0.25	0.15	Remainder

Note: Per AWS specification A.510, "Aluminum and Aluminum Alloy Welding Rods and Bare Electrodes."

FIGURE 17–5 Guide to the choice of filler metal for welding aluminum.

Base Metal	201.0 206.0 224.0	319.0, 333.0 354.0, 355.0 C355.0	356.0, A356.0 357.0, A357.0 413.0, 443.0 A444.0	511.0, 512.0, 513.0, 514.0, 535.0	7004, 7005, 7039, 710.0, 712.0	6009 6010 6070	6005, 6061, 6063, 6101, 6151, 6201, 6351, 6951	5456	5454
1060, 1070, 1080, 1350	ER4145	ER4145	ER4043[a,b]	ER5356[c,d]	ER5356[c,d]	ER4043[a,b]	ER4043[b]	ER5356[d]	ER4043[b,d]
1100, 3003, Alc 3003	ER4145	ER4145	ER4043[a,b]	ER5356[c,d]	ER5356[c,d]	ER4043[a,b]	ER4043[b]	ER5356[d]	ER4043[b,d]
2014, 2036	ER4145[e]	ER4145[e]	ER4145	—	—	ER4145	ER4145	—	—
2219	ER2319[a]	ER4145[e]	ER4145[b,c]	ER4043	ER4043	ER4043[a,b]	ER4043[a,b]	—	ER4043[b]
3004, Alc 3004	—	ER4043[b]	ER4043[b]	ER5356[f]	ER5356[f]	ER4043[b]	ER4043[b,f]	ER5356[d]	ER5356[f]
5005, 5050	—	ER4043[b]	ER4043[b]	ER5356[f]	ER5356[f]	ER4043[b]	ER5356[c,f]	ER5356[f]	ER5356[f]
5052, 5652[i]	—	ER4043[b]	ER4043[f]	ER5356[f]	ER5356[f]	ER4043[b]	ER5356[c,f]	ER5356[f]	ER5356[f]
5083	—	—	ER5356[c,d]	ER5356[d]	ER5183[d]	—	ER5356[d]	ER5183[d]	ER5356[d]
5086	—	—	ER5356[c,d]	ER5356[d]	ER5356[d]	—	ER5356[d]	ER5356[d]	ER5356[d]
5154, 5254[i]	—	—	ER4043[f]	ER5356[f]	ER5356[f]	—	ER5356[f]	ER5356[f]	ER5356[f]
5454	—	ER4043[b]	ER4043[f]	ER5356[f]	ER5356[f]	ER4043[b]	ER5356[c,f]	ER5356[f]	ER5554[c,f]
5456	—	—	ER5356[c,d]	ER5356[d]	ER5556[d]	—	ER5356[d]	ER5556[d]	
6005, 6061, 6063, 6101, 6151, 6201, 6351, 6951	ER4145	ER4145[b,c]	ER4043[b,f,g]	ER5356[f]	ER5356[c,f]	ER4043[a,b,g]	ER4043[b,f,g]		
6009, 6010, 6070	ER4145	ER4145[b,c]	ER4043[a,b,g]	ER4043	ER4043	ER4043[a,b,g]			
7004, 7005, 7039, 710.0, 712.0	—	ER4043[b]	ER4043[b,f]	ER5356[f]	ER5356[d]				
511.0, 512.0, 513.0, 514.0, 535.0	—	—	ER4043[f]	ER5356[f]					
356.0, A356.0, 357.0, A357.0, 413.0, 443.0, A444.0	ER4145	ER4145[b,c]	ER4043[b,h]						
319.0, 333.0, 354.0, 355.0, C355.0	ER4145[e]	ER4145[b,c,h]							
201.0, 206.0, 224.0	ER2319[a,h]								

Base Metal	5154 5254[i]	5086	5083	5052 5652[i]	5005 5050	3004 Alc 3004	2219	2014 2036	1100 3003 Alc 3003	1060 1070 1080 1350
1060, 1070, 1080, 1350	ER5356[c,d]	ER5356[d]	ER5356[d]	ER4043[b,d]	ER1100[b,c]	ER4043[b,d]	ER4145[b,c]	ER4145	ER1100[b,c]	ER1188[b,c,h,j]
1100, 3003, Alc 3003	ER5356[c,d]	ER5356[d]	ER5356[d]	ER4043[b,d]	ER1100[b,c]	ER4043[b,d]	ER4145[b,c]	ER4145	ER1100[b,c]	
2014, 2036	—	—	—	—	ER4145	ER4145	ER4145[e]	ER4145[e]		
2219	ER4043	—	—	ER4043[b]	ER4043[a,b]	ER4043[a,b]	ER2319[a]			
3004, Alc 3004	ER5356[f]	ER5356[d]	ER5356[d]	ER5356[c,f]	ER5356[c,f]	ER5356[c,f]				
5005, 5050	ER5356[f]	ER5356[d]	ER5356[d]	ER5356[c,d]	ER5356[c,f]					
5052, 5652[i]	ER5356[f]	ER5356[d]	ER5356[d]	ER5654[c,f,i]						
5083	ER5356[d]	ER5356[d]	ER5183[d]							
5086	ER5356[d]	ER5356[d]								
5154, 5254[i]	ER5654[f,i]									

Notes:

1. Service conditions such as immersion in fresh or salt water, exposure to specific chemicals, or a sustained high temperature [over 150° F (66° C)] may limit the choice of filler metals. Filler metals ER5183, ER5356, ER5556, and ER5654 are not recommended for sustained elevated temperature service.

2. Recommendations in this table apply to GSAW processes. For oxyfuel gas welding, only ER1188, ER1100, ER4043, ER4047, and ER4145 filler metals are ordinarily used.

3. Where no filler metal is listed, the base metal combination is not recommended for welding.

[a] ER4145 may be used for some applications.

[b] ER4047 may be used for some applications.

[c] ER4043 may be used for some applications.

[d] ER5183, ER5356, or ER5556 may be used.

[e] ER2319 may be used for some applications. It can supply high strength when the weldment is postweld solution heat treated and aged.

[f] ER5183, ER5356, ER5554, ER5556, and ER5654 may be used. In some cases, they provide (1) improved color match after anodizing treatment, (2) higher weld ductility, and (3) higher weld strength. ER5554 is suitable for sustained elevated temperature service.

[g] ER4643 will provide high strength in $\frac{1}{2}$ in. (12 mm) and thicker groove welds in 6XXX base alloys when postweld solution heat treated and aged.

[h] Filler metal with the same analysis as the base metal is sometimes used. The following wrought filler metals possess the same chemical composition limits as cast filler alloys: ER4009 and R4009 as R-C355.0; ER4010 and R4010 as R-A356.0; and R4011 as R-A357.0.

[i] Base metal alloys 5254 and 5652 are used for hydrogen peroxide service. ER5654 filler metal is used for welding both alloys for service temperatures below 150° F (66° C).

[j] ER1100 may be used for some applications.

FIGURE 17-6 Welding procedure schedules for ac-GTAW of aluminum.

Material Thickness (or Fillet Size)			Type of Weld	Tungsten Electrode Diameter		Filler Rod Diameter		Nozzle Size, Inside Diameter	Shielding Gas Flow (ft^3/hr)	Welding Current (A ac)	Number of Passes	Travel Speed per Pass (in./min)
Gauge	in.	mm		in.	mm	in.	mm	in.				
$\frac{3}{64}$ in.	0.046	1.2	Square groove and fillet	$\frac{1}{16}$	1.6	$\frac{1}{16}$	1.6	$\frac{1}{4}-\frac{3}{8}$	20	40–60	1	14–18
$\frac{1}{16}$ in.	0.063	1.6	Square groove and fillet	$\frac{3}{32}$	2.4	$\frac{3}{32}$	2.4	$\frac{5}{16}-\frac{3}{8}$	20	70–90	1	8–12
$\frac{3}{32}$ in.	0.094	2.4	Square groove and fillet	$\frac{3}{32}$	2.4	$\frac{3}{32}$	2.4	$\frac{5}{16}-\frac{3}{8}$	20	95–115	1	10–12
$\frac{1}{8}$ in.	0.125	3.2	Square groove and fillet	$\frac{1}{8}$	3.2	$\frac{1}{8}$	3.2	$\frac{3}{8}$	20	120–140	1	9–12
$\frac{3}{16}$ in.	0.187	4.7	Fillet	$\frac{5}{32}$	3.9	$\frac{5}{32}$	3.9	$\frac{7}{16}-\frac{1}{2}$	25	160–200	1	9–12
$\frac{3}{16}$ in.	0.187	4.7	V-groove	$\frac{5}{32}$	3.9	$\frac{5}{32}$	3.9	$\frac{7}{16}-\frac{1}{2}$	25	160–180	2	10–12
$\frac{1}{4}$ in.	0.250	6.4	Fillet	$\frac{3}{16}$	4.8	$\frac{3}{16}$	4.8	$\frac{7}{16}-\frac{1}{2}$	30	230–250	1	8–11
$\frac{1}{4}$ in.	0.250	6.4	V-groove	$\frac{3}{16}$	4.8	$\frac{3}{16}$	4.8	$\frac{7}{16}-\frac{1}{2}$	30	200–220	2	8–11
$\frac{3}{8}$ in.	0.375	9.5	V-groove	$\frac{3}{16}$	4.8	$\frac{3}{16}$	4.8	$\frac{1}{2}$	35	250–310	2–3	9–11
$\frac{1}{2}$ in.	0.500	12.7	V- or U-groove	$\frac{1}{4}$	6.4	$\frac{1}{4}$	6.4	$\frac{5}{8}$	35	400–470	3–4	6

Notes: 1. Increase amperage when backup is used.
2. Data are for all welding positions. Use low side of range for out of position.
3. For tungsten electrodes: first choice, pure tungsten EWP; second choice, zirconated EWZr.
4. Normally, argon is used for shielding; however, mixtures of 10% or more helium with argon are sometimes used for increased penetration in aluminum $\frac{1}{4}$ in. thick and over. The gas flow should be increased when helium is added. A mixture of 75% He + 25% argon is popular. When 100% helium is used, gas flow rates are about twice those used for argon.

FIGURE 17-7 Welding procedure schedules for dc-GTAW of aluminum.

Material Thickness (or Fillet Size)			Type of Weld	Tungsten Electrode Diameter		Filler Rod Diameter		Nozzle Size, Inside Diameter	Shielding Gas Flow (ft³/hr)	Welding Current (A ac)	Number of Passes	Travel Speed per Pass (in./min)
Gauge	in.	mm		in.	mm	in.	mm	in.				
20	0.032	0.8	Square groove and fillet	$\frac{3}{32}$	2.4	None		$\frac{3}{8}$	30	65–70	1	52
18	0.046	1.2	Square groove and fillet	$\frac{3}{64}$	1.2	$\frac{3}{64}$	1.2	$\frac{3}{8}$	30	35–90	1	45
16	0.063	1.6	Square groove and fillet	$\frac{3}{64}$	1.2	$\frac{3}{64}$	1.2	$\frac{3}{8}$	30	45–120	1	36
13	0.094	2.4	Square groove and fillet	$\frac{1}{16}$	1.6	$\frac{1}{16}$	1.6	$\frac{3}{8}$	30	90–185	1	32
11	$\frac{1}{8}$	3.2	Square groove and fillet	$\frac{1}{8}$	3.2	$\frac{1}{8}$	3.2	$\frac{3}{8}$	30	120–220	1	20
11	$\frac{1}{8}$	3.2	Square groove and fillet	$\frac{1}{8}$	3.2	None		$\frac{3}{8}$	30	180–200	1	24
—	$\frac{1}{4}$	6.4	Square groove and fillet	$\frac{1}{8}$	3.2	$\frac{1}{8}$	3.2	$\frac{1}{2}$	40	230–340	1	22
—	$\frac{1}{4}$	6.4	Square groove and fillet	$\frac{1}{8}$	3.2	None		$\frac{1}{2}$	40	220–240	1	22
—	$\frac{1}{2}$	12.7	V-groove	$\frac{3}{16}$	4.8	$\frac{1}{8}$	3.2	$\frac{1}{2}$	40	300–450	1	20
—	$\frac{1}{2}$	12.7	Square groove	$\frac{5}{32}$	3.9	None		$\frac{1}{2}$	40	260–300	2	20
—	$\frac{3}{4}$	19.1	V-groove	$\frac{3}{16}$	4.8	$\frac{1}{8}$	3.2	$\frac{1}{2}$	40	300–450	2	6
—	$\frac{3}{4}$	19.1	Square groove	$\frac{3}{16}$	4.8	None		$\frac{1}{2}$	40	450–470	2	6
—	1	25.4	V-groove	$\frac{3}{16}$	4.8	$\frac{1}{8}$	3.2	$\frac{5}{8}$	40	300–450	2	5

Notes: 1. Normally for automatic travel.
2. Use helium or 75% helium–25% argon.

FIGURE 17–8 Welding procedure schedules for GMAW of aluminum.

| Material Thickness (or Fillet Size) | | | Type of Weld | Electrode Diameter | | Welding Power | | Wire Feed Speed (in./min) | Shielding Gas Flow (ft³/hr) | Number of Passes | Travel Speed per Pass (in./min) |
| | | | | | | Current (dc) | Arc Volt EP | | | | |
Gauge	in.	mm		in.	mm						
—	0.050	—	Square groove and fillet	0.030	0.8	50	12–14	268–308	30	1	17–25
—	0.062	1.6	Square groove and fillet	0.030	0.8	55–60	12–14	295–320	30	1	17–25
—	0.062	1.6	Square groove and fillet	$\frac{3}{64}$	1.2	110–125	19–21	175–185	30	1	20–27
—	0.093	2.4	Square groove and fillet	0.030	0.8	90–100	14–18	330–370	30	1	24–36
—	0.125	3.2	Fillet	0.030	0.8	110–125	19–22	410–460	30	1	20–24
11	0.125	3.2	Square groove	$\frac{3}{64}$	1.2	110–125	20–24	175–190	40	1	20–24
$\frac{3}{16}$ in.	0.187	4.7	Square groove and fillet	$\frac{3}{64}$	1.2	160–195	20–24	215–225	40	1	20–25
$\frac{1}{4}$ in.	0.250	6.4	Fillet	$\frac{3}{64}$	1.2	160–195	20–24	215–225	40	1	20–25
$\frac{1}{4}$ in.	0.250	6.4	V-groove	$\frac{1}{16}$	1.6	175–225	22–26	150–195	40	3	20–25
$\frac{3}{8}$ in.	0.375	9.5	V-groove and fillet	$\frac{1}{16}$	1.6	200–300	22–26	170–275	40	2–5	25–30
$\frac{1}{2}$ in.	0.500	12.7	V-groove and fillet	$\frac{1}{16}$	1.6	220–230	22–27	195–205	40	3–8	12–18
$\frac{1}{2}$ in.	0.500	12.7	Double v-groove	$\frac{3}{32}$	2.4	320–340	22–29	140–150	45	2–5	15–17
$\frac{3}{4}$ in.	0.750	19.0	Double v-groove	$\frac{1}{16}$	1.6	255–275	22–27	230–250	50	4–10	8–18
$\frac{3}{4}$ in.	0.750	19.0	Double v-groove	$\frac{3}{32}$	2.4	355–375	22–29	155–160	50	4–10	14–16
1 in.	1.000	25.4	Double v-groove	$\frac{1}{16}$	1.6	255–290	22–27	230–265	50	4–14	6–18
1 in.	1.000	25.4	Double v-groove	$\frac{3}{32}$	2.4	405–425	22–27	175–180	50	4–8	8–12

Notes: 1. For groove and fillet welds, material thickness also indicated fillet weld size. Use V-groove for $\frac{3}{16}$ in. and thicker.
2. Use argon for thin and medium material; use 50% argon and 50% helium for thick material. Increase gas flow rate 10% for overhead position.
3. Increase amperage 10%–20% when backup is used.
4. Decrease amperage 10%–20% when welding out of position.

The wire feeding equipment for aluminum welding must be in good adjustment for efficient wire feeding. Nylon liners should be used in cable assemblies. Proper drive rolls should be selected for the aluminum wire and for the size of the electrode wire. It is difficult to push extremely small diameter aluminum wires through long gun cable assemblies. The spool gun is used for the small-diameter electrode wires. Water-cooled guns are required except for low-current welding.

Both the constant-current power source with matching voltage-sensing wire feeder and the constant-voltage power source with constant-speed wire feeder are used for welding aluminum. The CV system is preferred when welding on thin material and using small-diameter electrode wire. It provides better arc starting and regulation. The CC system is preferred when welding thick material using larger electrode wires. The CC power source with a moderate droop of 15 to 20 V per 100 A and with a constant-speed wire feeder provides the most stable power input and provides the highest weld quality.

The recent availability of lithium aluminum alloys has excited the aerospace industry. These alloys are approximately 10% stronger than existing alloys. This allows the use of thinner sections for weight reduction or higher-strength parts of the same weight. The problem: lithium is very active and the lithium aluminum alloys are difficult to weld. The variable-polarity power source described in the power source section with the keyhold plasma arc process is being used to weld this material in thicknesses up to $\frac{1}{2}$ in. The quality of the welds exceeds the quality of multipass gas tungsten arc welds. The VPPA welding process is able to make welds in one pass; however, to provide reinforcing a second pass is made.

Other Welding Processes

The shielded metal arc welding process can be used for welding aluminum. The covering on the electrodes is hydroscopic and must be protected from the atmosphere. Welds made with covered electrodes must be cleaned since the residue that remains on the weld will cause corrosion. Arc stability is rather poor, and there are a limited number of electrode types. This process is not popular.

Gas welding has been done using both oxyacetylene and oxyhydrogen flames. In either case, neutral flame is required. Flux is used as well as a filler rod. The process is not too popular because of low heat input and the need to remove flux.

Electroslag welding is used for joining pure aluminum. So far it has not been successful for welding the aluminum alloys. Submerged arc welding has been used in some countries where inert gas is not available. It is not used in North America.

All of the resistance welding processes are used for welding aluminum. In the case of spot and seam welding, extreme cleanliness of surface is required. Different types

of power are used but the process is extremely efficient and is widely used in the aircraft industry.

Most of the solid-state welding processes, including friction welding, ultrasonic welding, and cold welding, are used for aluminums. In addition, the stud welding process is used for aluminum. Aluminum can also be joined by soldering and brazing. Brazing can be accomplished by most brazing methods. A high-silicon alloy filler material is used.

The electron beam welding process is also used for aluminum welding, as are the plasma process and laser welding. These processes have not been used to a very great degree.

All the major aluminum companies provide welding manuals and data for the welding of aluminum.[2–4] If more detail is required, contact the Aluminum Association.

17-2 COPPER AND COPPER-BASE ALLOYS

Copper and copper-base alloys have specific properties that make them widely used. Their high electrical conductivity makes them widely used in the electrical industries and corrosion resistance of certain alloys makes them very useful in the process industries. Copper alloys are also widely used for friction or bearing applications.

There are over 300 different alloys commercially available. All of these different alloys have been used for many years. The Copper Development Association, Inc., has established an alloy designation system that is widely accepted in North America. It is not a specification system but rather a method of identifying and grouping different coppers and copper alloys. This system has been updated so that it now fits the unified numbering system (UNS). It provides one unified numbering system that includes all of the commercially available metals and alloys. The UNS designation consists of the prefix letter C followed by five digits without spaces (the final digits may or may not be zeroes). The compositions of each UNS number or copper alloy number and its common name are published by the association.[5]

Figure 17–9 shows the grouping of these copper alloys by common names that normally include the constituent alloys. There may be alloys within a grouping that have a composition sufficiently different to create welding problems. These are the exception and the data presented will provide starting point guidelines. There are two categories, wrought materials and cast materials. The welding information is the same whether the material is cast or rolled.

Copper shares some of the characteristics of aluminum, but it is weldable. Attention should be given to its properties that make the welding of copper and copper alloys different from the welding of carbon steels. Copper

FIGURE 17–9 Copper and copper alloy designation system.

Copper Number	Wrought Alloys—Groups
C11X00	Oxygen-free, high-conductivity copper (99.95 + %)
C11X00	Tough pitch copper (99.88+%)
C12X00	
C13X00	
C19X00	High-copper alloys (96+% copper)
C2XX00	Copper-zinc alloys (brasses)
C3XX00	Copper-zinc-lead alloys (leaded brasses)
C4XX00	Copper-zinc-tin alloys (tin brasses)
C50X00	Copper-tin alloys (phosphor bronzes)
C51X00	
C52X00	
C53X00	Copper-tin-lead alloys (leaded phosphor bronzes)
C54X00	
C61X00	Copper-aluminum alloys (aluminum bronzes)
C62X00	
C63X00	
C64X00	Copper-silicon alloys (silicon bronzes)
C65X00	
C66X00	Copper-zinc alloys (misc. brasses and bronzes)
C67X00	
C68X00	
C69X00	
C70X00	Copper-nickel alloys
C71X00	
C72X00	
C73X00	Copper-nickel-zinc alloys (nickel-silvers)
C74X00	
C75X00	
C76X00	
C77X00	
C78X00	
C79X00	

	Cast Alloys—Groups
C80X00	Copper alloys (99 + % copper)
C81X00	High-copper alloys (beryllium copper)
C82X00	
C83X00	Copper-tin-zinc + copper-tin-zinc-lead alloys (red brasses and leaded red brasses)
C84X00	Semired brasses and leaded semired brasses
C85X00	Yellow brasses and leaded yellow brasses
C86X00	Manganese and leaded manganese bronze alloys
C87X00	Copper-zinc-silicon alloys (silicon bronzes and brasses)
C90X00	Copper-tin alloys (tin bronzes)
C91X00	
C92X00	Copper-tin-lead alloy (leaded tin bronze)
C93X00	Copper-tin-lead alloy (high-leaded tin bronze)

alloys possess properties that require special attention when welding:

1. High thermal conductivity
2. High thermal expansion coefficient
3. Relatively low melting point
4. *Hot short* (i.e., brittle at elevated temperatures)
5. Very fluid molten metal
6. High electrical conductivity
7. Much of its strength due to cold working

Copper has the highest thermal conductivity of all commercial metals, and the comments made concerning thermal conductivity of aluminum apply to copper, to an even greater degree.

Copper has a relatively high coefficient of thermal expansion, approximately 50% higher than carbon steel, but lower than aluminum. One of the problems associated with copper alloys is the fact that some of them, such as aluminum bronze, have a coefficient of expansion over 50% greater than that of copper. This creates problems when making generalized statements about the different copper-based alloys.

The melting point of the different copper alloys varies over a relatively wide range, but is at least 1000° F (538° C) lower than carbon steel. Some of the copper alloys are *hot short,* meaning they become brittle at high temperatures. This is because some of the alloying elements form oxides and other compounds at the grain boundaries, embrittling the material.

Copper does not exhibit heat colors like steel, and when it melts it is relatively fluid. Copper's high electrical conductivity is a problem with the resistance welding processes.

All copper alloys derive their strength from cold working. The heat of welding will anneal the copper in the heat-affected area adjacent to the weld and reduce the strength provided by cold working. This must be considered when welding high-strength joints.

There is one other problem associated with the copper alloys that contain zinc. Zinc has a relatively low boiling temperature, and the heat of an arc will tend to vaporize the zinc. The arc processes must be used with care for the alloys containing zinc.

The grouping of the copper alloys in Figure 17–9 is for convenience; however, there may be certain alloys within the grouping that are different from the others. In view of these, it is difficult to make generalized statements that apply to all the alloys in a particular grouping. For best results it is wise to know the exact composition of the alloy being welded. If it fits within a particular grouping, the recommended filler metal can be checked by referring to Figure 17–10, which gives the nominal composition of the copper alloy filler metals. The data shown here are for the filler metal, whether

FIGURE 17–10 Composition of copper alloy filler metals. (From AWS specifications A5.6, A5.7, and A5.8.)

AWS Class[a]	Nominal Composition (%)										
	Cu	Al	Fe	Mn	Ni	Si	Sn	Pb	Ti	Zn	Other
Cu	98	0.01	—	0.5	—	0.50	1.0	0.22	—	—	0.50
CuAl–A1	Remainder	6.0–9.0	—	—	—	0.10	—	0.02	—	0.20	0.50
CuAl–A2	Remainder	9.0–11.0	1.5	—	—	0.10	—	0.02	—	0.02	0.50
CuAl–B	Remainder	11.0–12.0	3.0–4.25	—	—	0.10	—	0.02	—	0.02	0.50
CuNi	Remainder	—	0.40–0.75	1.00	29	0.50	—	0.02	0.15–1.00	—	0.50
CuSi	Remainder	0.01	0.5	1.5	—	2.8–4.0	1.5	0.02	—	—	0.50
CuSi–A	94	0.01	0.5	1.5	—	2.8–4.0	1.5	0.02	—	1.5	0.50
CuSn–A	Remainder	0.01	—	—	—	—	4.0–6.0	0.02	—	—	0.50
CuSn–C	Remainder	0.01	—	—	—	—	7.0–9.0	0.02	—	—	0.50
CuZn–A	57–61	0.01	—	—	—	—	0.25–1.0	0.05	—	Remainder	0.50
CuZn–B	56–60	0.01	0.25–1.2	0.01–0.50	0.2–0.8	0.04–0.15	0.8–1.1	0.05	—	Remainder	0.50
CuZn–C	56–60	0.01	0.25–1.2	0.01–0.50	—	0.04–0.15	0.8–1.1	0.05	—	Remainder	0.50
CuZn–D	46–50	0.01	—	—	9.0–11.0	0.04–0.25	—	0.05	—	Remainder	0.50

[a]AWS class filler metals may have the prefix letter E, electrode; R, rod; RB, rod or brazing; B, brazing.

it is an electrode, a rod, or wire, or for brazing. Refer to AWS specifications A5.6, A5.7, and A5.8 for details. The composition of the filler material should be chosen to match the base metal as closely as possible.

The GMAW and GTAW processes are the most popular for welding copper and copper alloys. The GTAW process normally uses dc electrode negative (straight polarity), but in some cases alternating current with high frequency is recommended. GTAW is best for welding the thinner gauges. It is also recommended for repairing copper alloy castings.

The GMAW process is used for welding thicker materials. It is faster, has a higher deposition rate, and usually results in less distortion. It can produce high-quality welds in all positions. It uses dc electrode positive, and the CV power source is recommended.

Copper

There are three basic groups in the C100 series of copper designation. The C10X is the oxygen-free type, which has a copper analysis of 99.95% or higher. This high-conductivity copper contains no oxygen and is not subjected to grain boundary migration. Adequate gas coverage should be employed to avoid oxygen of the air coming into contact with the molten metal. Welds should be made as quickly as possible since too much heat or slow welding can contribute to oxidation. The deoxidized coppers are preferred because of their freedom from embrittlement by hydrogen. Hydrogen embrittlement occurs when copper oxide is exposed to a reducing gas at high temperature. The hydrogen reduces the copper oxide to copper and water vapor. The entrapped high-temperature water vapor or steam can create sufficient pressure to

cause cracking. In common with all copper welding, preheat should be used and can run from 250 to 1000° F (121 to 538° C), depending on the mass involved.

The second subgroup is the tough pitch coppers, which have a copper composition of 99.88% or higher, and some high-copper alloys that have 96% or more copper. The ECu or RCu class filler metal is recommended.

The tough pitch electrolytic copper is difficult to weld because of the presence of copper oxide within the material. During welding the copper oxide will migrate to the grain boundaries, which reduces ductility and tensile strength. The gas-shielded processes are recommended since the welding area is more localized and the copper oxide is less able to migrate in appreciable quantities.

The third copper subgroup is the high-copper alloys that may contain deoxidizers such as phosphorus. The ECuSi filler wires are used with this material. The preheat temperatures needed to make the weld quickly apply to all three grades.

Copper–Zinc Alloys (Brasses)

These are the C2XX family of copper alloys. Within this group there are many different types of brasses. These alloys contain zinc, and zinc vaporization can be reduced by decreasing preheat and by using lower welding currents. The various filler metals, copper–silicon, copper–tin, and aluminum–bronze, can all be used.

For lighter sections argon shielding is used. For thicker sections preheat should be employed at approximately 400° F (200° C). Helium and helium gas mixtures are recommended. The weld joint should be opened up sufficiently to allow root penetration.

Copper-Zinc-Lead Alloys (Leaded Brasses)

The leaded alloys in the C3XX group are not suitable for welding since the lead will create excessive porosity and promote cracking in the weld area.

Copper-Zinc-Tin Alloys (Tin Brasses)

This is the C4XX subgroup. These are the yellow brasses and are generally welded with the CuAl-A2 aluminum bronze filler metal. The same comments made concerning the copper–zinc alloys apply here.

Copper-Tin Alloys (Phosphor Bronzes)

This is the C5XX series, which also includes the leaded phosphor bronzes. Except for the leaded bronzes, both the GMAW and GTAW processes can be used with the normal recommendations concerning thickness. These alloys have a tendency to be hot short. High current density and a high travel speed should be used. Helium is recommended for shielding with GTAW. The level of tin in the filler wire should be selected to match the tin in the base metal. Preheat should be used in the range of 300 to 400°F (150 to 200°C). When groove angles are used, wide angles should be used. To reduce stresses and distortion, hot peening of the weld deposit is recommended. The CuSn filler metal should be used with the higher amount of tin to match the higher amount of tin in the base metals. The leaded phosphor bronzes should not be welded.

Copper-Aluminum Alloys (Aluminum Bronzes)

These are a subgroup in the C6XX class representing the lower numbers. Both the GMAW and GTAW processes are used, with GMAW used for the heavier thicknesses. Filler metal should be the CuAl-A2 type. Argon–helium mixtures are recommended. Preheat is required only for the heavier thicknesses. Full-penetration welds are recommended.

Copper-Silicon Alloys (Silicon Bronzes)

This is also a subgroup of the C6XX series. Both GMAW and GTAW can be used for this family of copper alloys. This alloy is free of volatile alloying elements and has a lower conductivity. Preheat is recommended for heavier thicknesses. The leaded grade in this class is not suitable for welding.

Copper-Nickel Alloys

These alloys are in the low C7XX class of alloys. Both GMAW and GTAW processes can be used for these alloys. The filler metal should be the CuNi 70/30 type. Argon is normally used, but for heavy thicknesses argon and helium mixtures can be employed. Preheating is normally not used, and the interpass temperatures should not be allowed to rise above 150° F (65° C).

Copper-Nickel-Zinc Alloys (Nickel-Silver)

These alloys are in the high C7XX class of alloys. These alloys are not normally arc welded and should be joined by brazing. This is because of the relatively high amount of zinc included in these compositions.

The analysis of cast alloys similar to wrought alloys would be welded the same way. The filler metal should be selected to most closely approximate the analysis of the base metal.

Figure 17-11 provides recommended welding conditions for both GMAW and GTAW of copper alloys. This is a summary of material just covered and is a starting point for establishing a welding procedure. Welding procedure schedules are provided for welding the different copper alloys with both GTAW and GMAW (Figures 17-12 and 17-13).

Other Welding Processes

Many welding processes can be used to join copper and copper alloys. Soldering is widely used for joining most of the copper alloys; however, the high aluminum content and aluminum–manganese bronzes are not readily soldered. Both corrosive and resin fluxes are used for soldering copper. There is one precaution. Solders containing more than 1.0% antimony or more than 0.02% arsenic should not be used to solder the copper–zinc alloys. They will produce brittle joints or have poor bonding. The soldering process does not utilize sufficiently high heat to cause annealing of the copper-base alloys.

Brazing is widely used. The copper-phosphorus filler material (BCuP) and some of the silver alloy (Bag) types are used. The copper-phosphorus is much less expensive but is not used for copper alloys that contain more than 10% nickel. In addition, the copper-phosphorus alloy does not provide as high an electrical conductivity as the silver alloys.

Several of the other arc welding processes can be used. Plasma arc welding is becoming more popular for welding copper alloys. The same comments made concerning GTAW apply. The submerged arc welding process has been used for copper alloy welding and overlaying. Specialized fluxes for copper alloys must be used.

The cold welding process is widely used on coppers; so are high-frequency welding and the electron and laser beam welding processes. Additional information for the welding of copper and copper alloys can be obtained from the various bulletins published by the Copper Development Association, Inc.

FIGURE 17–11 Recommended welding conditions for GMAW and GTAW of copper alloys.

Material Type	GTAW			GMAW			Notes
	Filler Metal	Shielding Gas	Welding Current	Electrode Type	Electrode Class	Shielding Gas	
Copper[a] (E1xx)	RCu	Helium, argon, or mixture	DCEN AC-HF	EWTh-2	ECu	Argon + helium	Preheat, higher temperature for thicker materials
Brasses[a] (C–2xx) Copper–zinc)	RCuZn-B RCuZn-C RCuZn-D	Argon– helium mixture	DCEN AC-HF	EWTh-1	—	—	Preheat, open up joint, do not weld leaded types
Tin brasses[a] (C4xx)	RCuZn-A RCuZn-C	Argon– helium mixture	DCEN AC-HF	EWTh-1	ECuA1–A2	Argon + helium	Preheat, open up joint, higher temperature for thicker materials
Phosphor bronze (C5xx) (copper–tin)	RCuSn-A RCuSn-C RCuSn-D	Argon– helium mixture	DCEN	EWTh-2	ECuSn-A ECuSn-C	Argon	Weld quickly, hot short, do not weld leaded types
Aluminum bronze (C61x, 62x and 63x) (copper– aluminum)	RCuA1–A2 RCuA1-B	Argon– helium mixture	AC-HF DCEN	EWTh-1	ECuA1–A1 ECuA1–A2 ECuA1–B	Argon	Relatively easy to weld
Silicon bronze (C64x and 65x) (copper– aluminum)	RCuSi-A	Argon	DCEN	EWTh-1	ECuSi	Argon	Relatively easy to weld; do not weld leaded types
Copper– nickel (C7xx)	RCuNi	Argon	DCEN	EWTh-1	ECuNi	Argon	Relatively easy to weld

[a]Preheat.

17-3 MAGNESIUM-BASE ALLOYS

Magnesium is the lightest structural metal. It is approximately two-thirds as heavy as aluminum and one-fourth as heavy as steel. Magnesium alloys containing small amounts of aluminum, manganese, zinc, zirconium, and so on, have strengths equaling that of mild steels. They can be rolled into plate, shapes, and strip. Magnesium can be cast, forged, fabricated, and machined. As a structural metal it is used in aircraft. It is used by the materials-moving industry for parts of machinery and for handpower tools due to its strength-to-weight ratio. Magnesium can be welded by many of the arc and resistance welding processes, as well as by the oxyfuel gas welding process, and it can be brazed.

The more popular magnesium alloys are shown in Figure 17–14. This chart shows the ASTM designations per ASTM B275 and the UNS designation. Magnesium, like aluminum, is produced with different tempers. These are based on heat treatment and work hardening. They are listed following the alloy classification and use the prefix letter T followed by a number ranging from 1 to 10, the higher numbers indicating the higher hardness. The letter F is also used indicating as fabricated. The letter H is used to indicate the heat treat condition. The strength of a weld joint is lowered in base metal, in the work-hardened condition, as a result of recrystallization and grain growth in the heat-affected zone. This effect is minimized with gas metal arc welding because of higher welding speed. This is not a factor in the base metals that are welded in the soft condition.

Welding Magnesium Alloys

Magnesium possesses properties that make welding it different than welding steel. Many of these are the same as for aluminum. These are:

1. Magnesium oxide surface coating
2. High thermal conductivity

FIGURE 17-12 Welding procedure schedules for GTAW of copper alloys.

Material Thickness (or Fillet Size)				Tungsten Electrode Diameter		Filler Rod Diameter		Nozzle Size, Inside Diameter	Shielding Gas Flow	Welding Current (amps	Number of	Travel Speed per Pass
Gauge	in.	mm	Type of Weld	in.	mm	in.	mm	in.	(ft³/hr)	DCEN)	Passes	(in./min)
16	0.063	1.6	Square groove	$\frac{1}{16}$	1.6	$\frac{1}{16}$	1.6	$\frac{3}{8}$	20	100–150	1	10–12
16	0.063	1.6	Fillet	$\frac{1}{16}$	1.6	$\frac{1}{16}$	1.6	$\frac{3}{8}$	20	85–125	1	10–12
11	0.125	3.2	Square groove	$\frac{3}{32}$	2.4	$\frac{3}{32}$	2.4	$\frac{3}{8}$	20	170–235	1	8–11
11	0.125	3.2	Fillet	$\frac{3}{32}$	2.4	$\frac{3}{32}$	2.4	$\frac{3}{8}$	20	115–165	1	10–12
$\frac{3}{16}$ in.	0.187	4.7	Square groove	$\frac{1}{8}$	3.2	$\frac{1}{8}$	3.2	$\frac{5}{8}$	35	185–255	1	8–12
$\frac{3}{16}$ in.	0.187	4.7	Fillet	$\frac{3}{32}$	2.4	$\frac{3}{32}$	2.4	$\frac{3}{8}$	25	170–230	1	8–12
$\frac{1}{4}$ in.	0.250	6.4	Fillet	$\frac{1}{8}$	3.2	$\frac{1}{8}$	3.2	$\frac{1}{2}$	40	220–275	1	7–10
$\frac{1}{4}$ in.	0.250	6.4	Single V-groove	$\frac{1}{8}$	3.2	$\frac{1}{8}$	3.2	$\frac{1}{2}$	40	220–275	2	7–10
$\frac{1}{4}$ in.	0.250	6.4	Edge	$\frac{1}{8}$	3.2	$\frac{1}{8}$	3.2	$\frac{1}{2}$	25	160–225	1	7–10
$\frac{1}{4}$ in.	0.250	6.4	Double V-groove	$\frac{1}{8}$	3.2	$\frac{1}{8}$	3.2	$\frac{1}{2}$	20	180–220	3	8–12
$\frac{3}{8}$ in.	0.375	9.5	Fillet	$\frac{3}{16}$	4.8	$\frac{3}{16}$	4.8	$\frac{1}{2}$	45	275–325	3	8–12
$\frac{3}{8}$ in.	0.375	9.5	Single V-groove	$\frac{1}{8}$	3.2	$\frac{1}{8}$	3.2	$\frac{1}{2}$	25	225–290	3	8–12
$\frac{3}{8}$ in.	0.375	9.5	Double V-groove	$\frac{5}{32}$	3.9	$\frac{1}{8}$	3.2	$\frac{1}{2}$	20	200–250	3	8–12
$\frac{1}{2}$ in.	0.500	12.7	Fillet	$\frac{1}{4}$	6.4	$\frac{1}{4}$	6.4	$\frac{5}{8}$	45	370–500	4	8–12
$\frac{1}{2}$ in.	0.500	12.7	Single V-groove	$\frac{1}{8}$	3.2	$\frac{1}{8}$	3.2	$\frac{1}{2}$	30	280–330	7	7–10
$\frac{1}{2}$ in.	0.500	12.7	Double V-groove	$\frac{5}{32}$	3.9	$\frac{1}{8}$	3.2	$\frac{1}{2}$	30	180–250	4	7–10

Notes:
1. Increase amperage 100% when backup is used.
2. Data are for flat position. Reduce amperage 10%–20% when welding in horizontal, vertical, or overhead position.
3. For tungsten electrodes: first choice, 1% thoriated EWTh1; second choice, 2% thoriated EWTh2.
4. For copper, use helium for shielding; however, a mixture of 75% He + 25% argon is very popular on copper and some copper alloys. Argon is usually for bronzes.
5. Preheat $\frac{3}{16}$ in. copper 200° F, $\frac{1}{4}$ in., 300° F, $\frac{3}{8}$ in., 500° F; preheat $\frac{1}{4}$ in. and up 900° F.
6. Deoxidized copper and copper alloys use DCEN; aluminum bronze uses ACHF and argon for shielding.

3. Relatively high thermal expansion coefficient
4. Relatively low melting temperature
5. Absence of color change as temperature approaches the melting point

Magnesium is a very active metal, and the rate of oxidation increases as the temperature is increased. The melting point of magnesium is very close to that of aluminum, but the melting point of the oxide is very high. In view of this, the oxide coating must be removed.

Magnesium has high thermal heat conductivity and a high coefficient of thermal expansion. The thermal conductivity is not as high as aluminum but the coefficient of thermal expansion is very nearly the same. The absence of color change is not too important with the arc welding processes.

The welds produced between similar alloys will develop the full strength of the base metals; however, the strength of the heat-affected zone will be reduced slightly. In all magnesium alloys the solidification range increases and the melting point and the thermal expansion decrease as the alloy content increases. Aluminum added as an alloy up to 10% improves weldability since it tends to refine the weld grain structure. Zinc of more than 1% increases hot shortness, which can result in weld cracking. The high-zinc alloys are not recommended for arc welding because of their cracking tendencies. Magnesium, containing small amounts of thorium, possesses excellent welding qualities and freedom from cracking. Weldments of these alloys do not require stress relieving.

Certain magnesium alloys are subject to stress corrosion. Weldments subjected to corrosive attack over a period of time may crack adjacent to welds if the residual stresses are not removed. For weldments intended for this type of service, stress relieving is required.

GTAW and GMAW are recommended for joining magnesium. Gas tungsten arc is recommended for thinner materials and gas metal arc is recommended for thicker materials; however, there is considerable overlap.

The filler metal alloys used for joining magnesium are shown in Figure 17-15. They are based on AWS specification A5.19. The composition of filler metals should match the composition of the base materials; however, there are many cases in which this cannot be done. In all cases the recommendations shown in Figure 17-16 should be used.

FIGURE 17–13 Welding procedure schedules for GMAW of copper alloys.

Material Thickness (or Fillet Size)			Type of Weld	Electrode Diameter		Welding Power		Wire Feed Speed (in./min)	Shielding Gas Flow (ft³/hr)	Number of Passes	Travel Speed per Pass (in./min)
Gauge	in.	mm		in.	mm	Current (A dc)	Arc Volt EP				
16	0.063	1.6	Deoxidized copper	$\frac{3}{64}$	1.2	150–170	22–24	210–220	35	1	20–23
14	0.078	1.9	Square groove and fillet	$\frac{3}{64}$	1.2	180–200	22–25	240–270	40	1	20–25
12	0.109	2.8	Square groove and fillet	$\frac{3}{64}$	1.2	200–230	23–27	270–290	40	1	20–25
11	0.125	3.2	Square groove and fillet	$\frac{3}{64}$	1.2	210–240	23–27	280–300	40	1	20–25
$\frac{1}{4}$ in.	0.250	6.4	Square groove and fillet	$\frac{1}{16}$	1.6	380–410	23–29	260–270	40	1	12–15
$\frac{1}{4}$ in.	0.250	6.4	V-groove and fillet	$\frac{1}{16}$	1.6	300–330	23–27	190–210	40	1–3	14–17
$\frac{3}{8}$ in.	0.375	9.2	V-groove and fillet	$\frac{1}{16}$	1.6	340–360	24–28	220–240	40	1	12–15
$\frac{1}{2}$ in.	0.500	12.7	Double V-groove	$\frac{3}{32}$	2.4	400–440	24–30	270–290	50	2	8–10
$\frac{3}{4}$ in.	0.750	19.0	Double V-groove	$\frac{3}{32}$	2.4	420–460	24–30	290–315	50	3	7–9
1 in.	1.000	25.4	Double V-groove	$\frac{3}{32}$	2.4	420–460	24–30	270–300	50	4	7–9
$\frac{1}{8}$ in.	0.125	1.2	Silicon bronze	$\frac{3}{64}$	1.2	130–160	25–28	220–230	35	1	25–32
$\frac{1}{4}$ in.	0.250	6.4	Fillet and V-groove	$\frac{1}{16}$	1.6	270–290	27–30	170–190	40	1–3	26–33
$\frac{1}{4}$ in.	0.250	6.4	Fillet and V-groove	$\frac{1}{16}$	1.6	450–465	25–28	220–250	50	1	30–34
$\frac{1}{2}$ in.	0.500	12.7	Fillet and V-groove	$\frac{1}{16}$	1.6	335–350	27–30	180–200	50	3–5	15–20
$\frac{1}{8}$ in.	0.125	3.2	Aluminum bronze	$\frac{3}{64}$	1.2	190–225	22–25	280–300	40	1	18–24
$\frac{1}{4}$ in.	0.250	6.4	V-groove and fillet	$\frac{1}{16}$	1.6	275–300	23–29	170–190	50	2	16–22
$\frac{3}{8}$ in.	0.375	9.2	V-groove and fillet	$\frac{1}{16}$	1.6	300–340	23–29	190–210	50	3–6	16–22
$\frac{1}{2}$ in.	0.500	12.7	Double V-groove	$\frac{1}{16}$	1.6	320–350	23–29	200–220	50	6–8	11–15
$\frac{5}{8}$ in.	0.625	15.9	Double V-groove	$\frac{1}{16}$	1.6	320–345	23–29	220–240	50	6–8	9–13
$\frac{3}{4}$ in.	0.750	19.0	Double V-groove	$\frac{1}{16}$	1.6	340–370	23–29	220–240	50	6–8	9–12

Notes: 1. If preheating is required, a range of 500 to 900° F may be used for aluminum, bronze, and deoxidized copper and 400 to 600° F for silicon bronze.
2. Argon is normally used. If porosity is encountered, it can be eliminated by adding an equal amount of helium to the argon flow.
3. Speeds and currents for fully automatic welding are approximately 15% higher.

FIGURE 17-14 Composition of magnesium alloys. From ASTM 8275, magnesium alloys (abridged).

ASTM Alloy	Nominal Composition (%)						
	Aluminum	Manganese	Zinc	Zirconium	Rare Earths	Thorium	Magnesium
Sand and permanent mold castings							
AX92A	9.0	0.15	2.0	—	—	—	Remainder
AZ63A	6.9	0.25	3.0	—	—	—	Remainder
AZ81A	7.6	0.13 min.	0.7	—	—	—	Remainder
AZ91C	8.7	0.20	0.7	—	—	—	Remainder
EK30A	—	—	—	0.35	3.0	—	Remainder
EK41A	—	—	—	0.6	4.0	—	Remainder
EZ33A	—	—	2.7	0.7	3.0	—	Remainder
HK31A	—	—	—	0.7	—	3.0	Remainder
HZ32A	—	—	2.1	0.7	—	3.0	Remainder
Die castings							
AZ91A	9.0	0.20	0.6	—	—	—	Remainder
AZ91B							
Extrusions							
AZ31B	3.0	0.45	1.0	—	—	—	Remainder
AZ31C							
AZ61A	6.5	0.30	1.0	—	—	—	Remainder
M1A	—	1.50	—	—	—	—	Remainder
AZ80A	8.5	0.25	0.5	—	—	—	Remainder
ZK60A	—	—	5.7	0.55	—	—	Remainder
Sheet and plate							
AZ31B	3.0	0.45	1.0	—	—	—	Remainder
HK31A	—	—	—	0.7	—	3.0	Remainder

FIGURE 17-15 Composition of magnesium filler metals per AWS A5.19.

AWS Classification[a]	Nominal Composition (%)										
	Mg	Al	Be	Mn	Zn	Zi	Rare Earths	Cu	Fe	Ni	Si
AZ61A	Remainder	5.8–7.2	0.0002–0.0008	0.15	0.40–1.5	—	—	0.05	0.005	0.005	0.05
AZ101A	Remainder	9.5	0.0002–0.0008	0.13	0.75	—	—	0.05	0.005	0.005	0.05
AZ92A	Remainder	8.3–9.7	0.0002–0.0008	0.15	1.75	—	—	0.05	0.005	0.005	0.05
EZ33A	Remainder	—	—	—	2.0–3.1	0.45–1.0	2.5–4.0	—	—	—	—

[a]Use suffix letter E, electrode, or R, rod.

Gas Tungsten Arc Welding

Welding procedure schedules for gas tungsten arc welding of magnesium are given in Figure 17-17. All the precautions mentioned for welding aluminum should be observed. A short arc should be used and the torch should have a slight leading travel angle. The cold wire filler metal should be brought in as near to horizontal as possible (on flat work). The filler wire is added to the leading edge of the weld puddle. High-frequency current should be used for starting the direct current arc. With alternating current, high frequency should be used continuously. Runoff tabs are recommended for welding any except the thinner materials. Uniform travel speed and weld beads are recommended. The shielding gas is normally argon.

However, a mixture of 75% helium and 25% argon is used for thicker materials. For heavy thicknesses 100% helium can be used; more helium is required than argon to do the same job.

Direct current with electrode positive can be used for machine or automatic welding. In this case, materials must be perfectly clean prior to welding. Additional details are given by the welding procedure schedule.

Gas Metal Arc Welding of Magnesium

Gas metal arc welding is used for the medium to thicker sections. Special high-speed gear ratios are required in the wire feeders since the magnesium electrode wire has an extremely high melt-off rate. The normal wire feeder and power supply used for aluminum welding will be suitable for welding magnesium. The different types of arc transfer can be obtained when welding magnesium. This is primarily a matter of current level or current density and voltage setting. The short-circuiting transfer and the spray transfer are recommended. Argon is usually used for gas metal arc welding of magnesium; however, argon–helium mixtures can be used. In general, the spray transfer should be used on material $\frac{3}{16}$ in. and thicker and the short-circuiting arc used for thinner metals. Figure 17-18 provides welding procedure schedules.

Welding Problems

Magnesium is usually delivered with an oil preservative. The oil must be removed with a solvent and the material should be cleaned either by mechanical or chemical methods. Scraping and brushing is often used, and a stainless steel brush should be used.

If cracking persists in magnesium welds, check the welding technique. Craters must always be filled and runoff tabs should be used. Preheating in the range of 200 to 400° F (93 to 204° C) is recommended for complex weldments. Stress relieving is recommended when the weldment is exposed to corrosion.

Other Welding Processes

Resistance welding can be used for welding magnesium, including spot welding, seam welding, and flash welding. Magnesium can also be joined by brazing. In all cases, brazing flux is required and the flux residue must be completely removed from the finished part. Soldering is not popular since the strength of the joint is relatively low.

Magnesium can be stud welded, gas welded, and plasma welded. Finely divided pieces of magnesium such as shavings, filings, and so on, should not be in the welding area since they will burn. Magnesium castings or wrought materials do not create a safety hazard since the possibility of fire caused by welding on these sections is very remote. The producers of magnesium provide additional data for welding magnesium.[6] Consult the International Magnesium Association for additional information.

17-4 NICKEL-BASE ALLOYS

Nickel and high-nickel alloys are commonly used when corrosion resistance is required. They are used in the chemical industry and the food industry. Nickel and nickel alloys are also widely used as filler metals for joining dissimilar materials and cast iron.

Nickel and nickel alloys are identified by trademark names and suffix numbers. They are also specified by ASTM and AMS numbers and by others. UNS designations use the letter N followed by five digits. The trademark names are popular and are:

* *Monel:* a nickel–copper alloy
* *Inconel:* a high nickel–chromium alloy with iron
* *Incoloy:* a nickel–iron–chromium alloy
* *Hastealloy:* a nickel–molybdenum–iron alloy

There are other nickel alloys. The trademark names of the International Nickel Company or the Huntington Alloy Products Division will be used. The more common nickel alloys are shown in Figure 17-19, which also shows UNS numbers.

When welding, the nickel alloys can be treated much in the same manner as austenitic stainless steels with a few exceptions. These exceptions are:

1. The nickel alloys will acquire a surface oxide coating that melts at a temperature approximately 1000° F (538° C) above the melting point of the base metal.
2. The nickel alloys are susceptible to embrittlement at welding temperatures by lead, sulfur, phosphorus, and some low-temperature metals and alloys.
3. Weld penetration is less than expected with other metals.

A weld metal flow is quite sluggish. When adjustments are made for these three factors, the welding procedures used for the nickel alloys can be the same as those used for stainless steel. This is because the melting point, the coefficient of thermal expansion, and the thermal conductivity are similar to austenitic stainless steel.

It is necessary that each of these precautions be considered. The surface oxide should be completely removed from the joint area by grinding, abrasive blasting, machining, or by chemical means. When chemical etches are used, they must be completely removed by rinsing prior to welding. The oxide, which melts at temperatures above the melting point of the base metal, may enter the weld as a foreign material, or impurity, and will greatly reduce the strength and ductility of the weld.

The problem of embrittlement at welding temperatures also means that the weld surface must be absolutely clean. Paints, marking crayons, grease, oil, machining lubricants, and cutting oils may all contain the ingredients that will cause embrittlement. They must be completely removed from the weld area to avoid embrittlement.

With respect to the minimum penetration, it is necessary to increase the opening of groove angles and to provide adequate root openings when full-penetration welds are used. The bevel or groove angles should be increased to approximately 40° over those used for carbon steel.

FIGURE 17-16 Guide to the choice of filler metal for welding magnesium. (From Ref. 6.)

Base Alloy	AM100A	AZ10A	AZ31B&C	AZ61A	AZ63A	AZ80A	AZ81A	AZ91C	AX92A	EK41A	EZ33A
AM100A	1. AZ92A 2. AZ101										
AZ10A	AZ92A	1. AZ61A 2. AZ32A									
AZ31B&C	AZ92A	1. AZ61A 2. AX92A	1. AZ61A 2. AZ92A								
AZ61A	AZ92A	1. AZ61A 2. AZ92A	1. AZ61A 2. AZ92A	1. AZ61A 2. AZ92A							
AZ63A	X	X	X	X	AZ92A						
AZ80A	AZ92A	1. AZ61A 2. AZ92A	1. AZ61A 2. AZ92A	1. AZ61A 2. AZ92A	X	1. AZ61A 2. AZ92A					
AZ81A	AZ92A	AZ92A	AZ92A	AZ92A	X	AZ92A	1. AZ92A 2. AZ101				
AZ91C	AZ92A	AZ92A	AZ92A	AZ92A	X	AZ92A	AZ92A	1. AZ92A 2. AZ101			
AZ92A	AZ92A	AZ92A	AZ92A	AZ92A	X	AZ92A	AZ92A	AZ92A	AZ101		
EK41A	AZ92A	AZ92A	AZ92A	AZ92A	X	AZ92A	AZ92A	AZ92A	AZ92A	EZ33A	
EZ33A	AZ92A	AZ92A	AZ92A	AZ02A	X	AZ92A	AZ92A	AZ92A	AZ92A	EZ33A	EZ33A
HK31A	AZ92A	AZ92A	AZ92A	AZ92A	X	AZ92A	AZ92A	AZ92A	AZ92A	EZ33A	EZ33A
HM21A	AZ92A	AZ92A	AZ92A	AZ92A	X	AZ92A	AZ92A	AZ92A	AZ92A	EZ33A	EZ33A
HM31A	AZ92A	AZ92A	AZ92A	AZ92A	X	AZ92A	AZ92A	AZ92A	AZ92A	EZ33A	EZ33A
HZ32A	AZ92A	AZ92A	AZ92A	AZ92A	X	AZ92A	AZ92A	AZ92A	AZ92A	EZ33A	EZ33A
K1A	AZ92A	AZ92A	AZ92A	AZ92A	X	AZ92A	AZ92A	AZ92A	AZ92A	EZ33A	EZ33A
M1A MG1	AZ92A	1. AZ61A 2. AZ92A	1. AZ61A 2. AZ92A	1. AZ61A 2. AZ92A	X	1. AZ61A 2. AZ92A	AZ92A	AZ92A	AZ92A	AZ92A	AZ92A
ZE41A	0	0	0	0	X	0	0	0	0	EZ33A	EZ33A
ZX21A	AZ92A	1. AZ61A 2. AZ92A	1. AZ61A 2. AZ92A	1. AZ61A 2. AZ92A	X	1. AZ61A 2. AZ92A	AZ92A	AZ92A	AX92A	AZ92A	AX92A
ZH62A ZK51A ZK60A ZK61A	X	X	X	X	X	X	X	X	X	X	X

Note: 0, no data available for welding this combination; x, welding not recommended.

Almost all the welding processes can be used for welding the nickel alloys. In addition, they can be joined by brazing and soldering. The filler metals to be used for joining nickel alloys are shown in Figure 17-20. These are based on AWS filler metal specifications A5.11, A5.14, and A5.15 and include covered electrodes as well as bare solid wire for GMAW or for filler wire with other processes. The recommended filler metals for joining different alloys are shown in Figure 17-21.

Welding Nickel Alloys

The most popular processes for welding nickel alloys are SMAW, GTAW, and GMAW. When shielded metal arc weld- ing is used, the procedures are essentially the same as those used for stainless steel welding.

The welding procedure schedule for using gas tung- sten arc welding is shown in Figure 17-22. The welding procedure schedule for gas metal arc welding is shown in Figure 17-23. This procedure information on these tables will provide starting points for developing the welding procedures. The submerged arc welding process is used with proprietary fluxes manufactured by the nickel pro- ducer.

No postweld heat treatment is required to maintain or restore corrosion resistance of the nickel alloys. Heat treatment is required for precipitating hardening alloys and stress relief may be required to meet certain specifi-

FIGURE 17–16 *(cont.)*

HK31A	HM21A	HM31A	HZ32A	K1A	LA141A	M1A MG1	QE22A	ZE10A	ZE41A	ZX21A	ZH62A ZH51A ZK60A ZK61A
EZ33A											
EZ33A	EZ33A										
EZ33A	EZ33A	EZ33A									
EZ33A	EZ33A	EZ33A	EZ33A								
EZ33A	EZ33A	EZ33A	EZ33A	EZ33A							
AZ92A	AZ92A	AZ92A	AZ92A	AZ92A	0	1. AZ61A 2. AZ92A					
EZ33A	EZ33A	EZ33A	EZ33A	EZ33A	0	0	EZ33A	0	EZ33A		
AZ92A	AZ92A	AZ92A	AZ92A	AZ92A	0	1. AZ61A 2. AZ92A	AZ92A	1. AZ61A 2. AZ92A	AZ92A	1. AZ61A 2. AZ92A	
X	X	X	X	X	X	X	X	X	X	X	EZ33A

cations to avoid stress corrosion cracking in applications involving hydrofluoric acid vapors or caustic solutions. Additional welding information is available from the Nickel Development Institute and from producers.[7]

17-5 REACTIVE AND REFRACTORY METALS

The reactive and refractory metals were originally used in the aerospace industry and are now being welded for more and more requirements. These metals share many common welding problems and are, therefore, grouped together in this section. Reactive metals have a strong affinity for oxygen and nitrogen at elevated temperatures.

At lower temperatures they are highly resistant to corrosion. Refractory metals have extremely high melting points. They may also exhibit some of the same characteristics of reactive metals.

The reactive metals are:

- Zirconium
- Titanium
- Beryllium

The refractory metals are:

- Tungsten
- Molybdenum
- Tantalum
- Columbian (niobium)

FIGURE 17–17 Welding procedure schedule for GTAW of magnesium.

Material Thickness (or Fillet Size)			Type of Weld	Tungsten Electrode Diameter		Filler Rod Diameter		Nozzle Size, Inside Diameter	Shielding Gas Flow (ft³/hr)	Welding Current (amps DCEN)	Number of Passes	Travel Speed per Pass (in./min)
Gauge	in.	mm		in.	mm	in.	mm	in.				
20	0.038	0.9	Square groove	$\frac{1}{16}$	1.6	$\frac{3}{32}$	2.4	$\frac{1}{4}$	15	25–40	1	20
20	0.038	0.9	Fillet	$\frac{1}{16}$	1.6	$\frac{3}{32}$	2.4	$\frac{1}{4}$	15	30–45	1	20
16	0.063	1.6	Square groove	$\frac{1}{16}$	1.6	$\frac{3}{32}$	2.4	$\frac{1}{4}$	15	45–60	1	20
16	0.063	1.6	Fillet	$\frac{1}{16}$	1.6	$\frac{3}{32}$	2.4	$\frac{1}{4}$	15	45–60	1	20
14	0.078	1.9	Square groove	$\frac{1}{16}$	1.6	$\frac{3}{32}$	2.4	$\frac{1}{4}$	15	60–75	1	17
14	0.078	1.9	Fillet	$\frac{1}{16}$	1.6	$\frac{3}{32}$	2.4	$\frac{1}{4}$	15	60–75	1	17
12	0.109	2.8	Square groove	$\frac{3}{32}$	2.4	$\frac{1}{8}$	3.2	$\frac{5}{16}$	15	80–100	1	17
12	0.109	2.8	Fillet	$\frac{3}{32}$	2.4	$\frac{1}{8}$	3.2	$\frac{5}{16}$	15	80–100	1	17
11	0.125	3.2	Square groove	$\frac{3}{32}$	2.4	$\frac{1}{8}$	3.2	$\frac{5}{16}$	25	95–115	1	17
11	0.125	3.2	Fillet	$\frac{3}{32}$	2.4	$\frac{1}{8}$	3.2	$\frac{5}{16}$	25	95–115	1	17
$\frac{3}{16}$ in.	0.187	4.7	V-groove	$\frac{1}{8}$	3.2	$\frac{1}{8}$	3.2	$\frac{3}{8}$	25	95–115	2	26
$\frac{1}{4}$ in.	0.250	6.4	V-groove	$\frac{1}{8}$	3.2	$\frac{3}{16}$	4.8	$\frac{1}{2}$	25	110–130	2	24
$\frac{3}{8}$ in.	0.375	9.5	V-groove	$\frac{1}{8}$	3.2	$\frac{3}{16}$	4.8	$\frac{1}{2}$	30	135–165	2	20

Notes: 1. Increase amperage when backup is used.
2. Data are for flat position. Reduce amperage 10%–20% when welding in horizontal, vertical, or overhead position.
3. Tungsten electrode: first choice, zirconated EWZr; second choice, pure tungsten EWP.
4. Select filler metal in accordance to selection chart.
5. Shielding gas is normally argon. A mixture of 75% helium + 25% argon is used for heavier thickness. For heavy thickness 100% helium is used. Gas flow rates for helium are approximately twice those used for argon.

FIGURE 17–18 Welding procedure schedule for GMAW of magnesium.

Material Thickness (or Fillet Size)			Type of Weld	Electrode Diameter		Current (amps DCEP)	Arc Volts	Wire Feed Speed (in./min)	Shielding Gas Flow (ft³/hr)	Number of Passes	Travel Speed per Pass (in./min)
Gauge	in.	mm		in.	mm						
0.025 in.	—	—	Square groove and fillet	0.040	1.0	26–27	13–16	180	40–60	1	24–36
0.040 in.	—	—	Square groove and fillet	0.040	1.0	35–50	13–16	250–340	40–60	1	24–36
0.063 in.	$\frac{1}{16}$	1.6	Square groove and fillet	0.063	1.6	60–75	13–16	140–170	40–60	1	24–36
0.090 in.	$\frac{3}{32}$	2.4	Square groove and fillet	0.063	1.6	95–125	13–16	210–280	40–60	1	24–36
0.125 in.	$\frac{1}{8}$	3.2	Square groove and fillet	0.094	2.4	110–135	13–16	100–130	40–60	1	24–36
0.160 in.	$\frac{5}{32}$	3.9	Square groove and fillet	0.094	2.4	135–140	13–16	130–140	40–60	1	24–36
0.190 in.	$\frac{3}{16}$	4.8	V-groove and fillet	0.094	2.4	175–205	13–16	160–190	40–60	2	24–36
0.250 in.	$\frac{1}{4}$	6.4	V-groove and fillet	0.063	1.6	240–290	24–30	550–660	50–80	2	24–36
0.375 in.	$\frac{3}{8}$	9.5	V-groove and fillet	0.094	2.4	320–350	24–30	350–385	50–80	2	24–36
0.500 in.	$\frac{1}{2}$	12.7	V-groove and fillet	0.094	2.4	350–420	24–30	385–415	50–80	2	24–36
1.00 in.	1	25.4	V-groove and fillet	0.094	2.4	350–420	24–30	385–415	50–80	4	24–36

Notes: 1. Values are for flat position.
2. For groove and fillet welds, material thickness also indicates fillet weld size. Use V-groove for $\frac{1}{4}$ in. and thicker.
3. Shielding gas is argon or for heavier thicknesses use helium–argon mixtures.
4. Above 200 A and 20 V, metal transfer is spray type; below 200 A and 20 V, metal transfer is short-circuiting type.

A summary of the physical properties of these metals is given in Figure 17-24. The refractory metals all have relatively high density and thermal conductivity. The reactive metals have lower melting points, lower densities, and, except for zirconium, have higher coefficients of thermal expression.

The reactive metals are becoming increasingly important because of their use in nuclear and space technology. They are considered in the difficult-to-weld category. These metals have a high affinity for oxygen and other gases at elevated temperatures. They cannot be welded with any process that utilizes fluxes or where

FIGURE 17–19 Composition of nickels and nickel alloys.

Alloy Designation	UNS Number	Nominal Composition (%)											
		Ni	C	Mn	Fe	S	Si	Cu	Cr	Al	Ti	Cb[a]	Others
Nickel 200	N02200	99.5[b]	0.08	0.18	0.2	0.005	0.18	0.13	—	—	—	—	—
Nickel 201	N02201	99.5[b]	0.01	0.18	0.2	0.005	0.18	0.13	—	—	—	—	—
Nickel 205	N02205	99.5[b]	0.08	0.18	0.10	0.004	0.08	0.08	—	—	0.03	—	Mg 0.05
Nickel 211	—	95.0[b]	0.10	4.75	0.38	0.008	0.08	0.13	—	—	—	—	—
Nickel 220	N02220	99.5[b]	0.04	0.10	0.05	0.004	0.03	0.05	—	—	0.03	—	Mg 0.05
Nickel 230	N02230	99.5[b]	0.05	0.08	0.05	0.004	0.02	0.05	—	—	0.003	—	Mg 0.06
Nickel 270	N02270	99.98	0.01	<0.001	0.003	<0.001	<0.001	<0.001	<0.001	—	<0.001	—	Mg <0.001, Co <0.001
Duranickel alloy 301	—	96.5[b]	0.15	0.25	0.30	0.005	0.5	0.13	—	4.38	0.63	—	—
Permanickel alloy 300	—	98.5[b]	0.20	0.25	0.30	0.005	0.18	0.13	—	—	0.40	—	Mg 0.35
Monel alloy 400	N04400	66.5[b]	0.15	1.0	1.25	0.012	0.25	31.5	—	—	—	—	—
Monel alloy 401	—	42.5[b]	0.05	1.6	0.38	0.008	0.13	Bal.	—	—	—	—	—
Monel alloy 404	N04404	54.5[b]	0.08	0.05	0.25	0.012	0.05	44.0	—	0.03	—	—	—
Monel alloy R–405	N04405	66.5[b]	0.15	1.0	1.25	0.043	0.25	31.5	—	—	—	—	—
Monel alloy K–500	N05500	66.5[b]	0.13	0.75	1.00	0.005	0.25	29.5	—	2.73	0.60	—	—
Monel alloy 502	N05502	66.5[b]	0.05	0.75	1.00	0.005	0.25	28.0	—	3.00	0.25	—	—
Inconel alloy 600	N06600	76.0	0.08	0.5	8.0	0.008	0.25	0.25	15.5	—	—	—	—
Inconel alloy 601	N06601	60.5	0.05	0.5	14.1	0.007	0.25	0.50	23.0	1.35	—	—	—
Inconel alloy 617	—	54.0	0.07	—	—	—	—	—	22.0	1.0	—	—	Co 12.5, Mo 9.0
Inconel alloy 625	N06625	61.0[b]	0.05	0.25	2.5	0.008	0.25	—	21.5	0.2	0.2	3.65	Mo 9.0
Inconel alloy 671	—	Bal.	0.05	—	—	—	0.25	—	48.0	—	0.35	—	—
Inconel alloy 702	N07702	79.5[b]	0.05	0.50	1.0	0.005	0.35	0.25	15.5	3.25	0.63	—	—
Inconel alloy 706	N09706	41.5	0.03	0.18	40.0	0.008	0.18	0.15	16.0	0.20	1.75	2.9	—
Inconel alloy 718	N07718	52.5	0.04	0.18	18.5	0.008	0.18	0.15	19.0	0.50	0.90	5.13	Mo 3.05
Inconel alloy 721	—	71.0[b]	0.04	2.25	6.5	0.005	0.08	0.10	16.0	—	3.05	—	—
Inconel alloy 722	N07722	75.0[b]	0.04	0.50	7.0	0.005	0.35	0.25	15.5	0.70	2.38	—	—
Inconel alloy X–750	N07750	73.0[b]	0.04	0.50	7.0	0.005	0.25	0.25	15.5	0.70	2.50	0.95	—
Inconel alloy 751	—	72.5[b]	0.05	0.5	7.0	0.005	0.25	0.25	15.5	1.20	2.30	0.95	—
Incoloy alloy 800	N08800	32.5	0.05	0.75	46.0	0.008	0.50	0.38	21.0	0.38	0.38	—	—
Incoloy alloy 801	N08801	32.0	0.05	0.75	44.5	0.008	0.50	0.25	20.5	—	1.13	—	—
Incoloy alloy 802	—	32.5	0.35	0.75	46.0	0.008	0.38	—	21.0	0.58	0.75	—	—
Incoloy alloy 804	—	41.0	0.05	0.75	25.4	0.008	0.38	0.25	29.5	0.30	0.60	—	—
Incoloy alloy 825	N08825	42.0	0.03	0.50	30.0	0.015	0.25	2.25	21.5	0.10	0.90	—	Mo 3.0
Ni–span–C alloy 902	N09902	42.25	0.03	0.40	48.5	0.02	0.50	0.05	5.33	0.55	2.58	—	—

[a]Cobalt included.
[b]Not for specification purposes.

AWS Class[a]	UNS Number	Trade Name	Nominal Composition (%)											
			Ni	C	Mn	Fe	S	Si	Cu	Cr	Al	Ti	Cb	Mo
ERNi-3	N02061	Nickel 61	96.0	0.06	0.30	0.10	0.005	0.40	0.02	—	—	3.0	—	—
ERNi-1	—	Nickel 141	96.0	0.03	0.30	0.05	0.005	0.60	0.03	—	0.25	2.5	—	—
ERNiCu-7	—	Monel 60	65.0	0.03	3.5	0.20	0.005	1.00	27.0	—	—	2.2	—	—
ERNiCu-8	—	Monel 64	65.0	0.15	0.60	1.00	0.005	0.15	30.0	—	2.8	0.5	—	—
ERCuNi	—	Monel 67	31.0	0.02	0.75	0.50	0.005	0.10	67.5	—	—	0.30	—	—
ENiCuAl-1	—	Monel 134	64.0	0.20	2.50	1.00	0.005	0.30	30.0	—	1.8	0.75	—	—
ECuNi	—	Monel 187	32.0	0.02	2.00	0.60	0.01	0.15	65.0	—	—	—	—	—
ENiCu-2	—	Monel 190	65.0	0.01	3.10	0.30	0.007	0.75	30.5	—	0.15	0.55	—	—
ERNiCrFe-5	N06062	Inconel 62	74.0	0.02	0.10	7.50	0.005	0.10	0.03	16.0	—	—	2.25	—
ERNiCrFe-7	N07069	Inconel 69	73.0	0.04	0.55	6.50	0.007	0.30	0.05	15.2	0.70	2.5	0.85	—
ERNiCr-3	N06082	Inconel 82	72.0	0.02	3.00	1.00	0.007	0.20	0.04	20.0	—	0.55	2.55	—
ERNiCrFe-6	N07092	Inconel 92	71.0	0.03	2.30	6.60	0.007	0.10	0.04	16.4	—	3.2	—	—
		Inconel 601	60.5	0.05	0.50	14.1	0.007	0.25	0.50	23.0	1.35	0.2	—	—
		Inconel 625	61.0	0.05	0.25	2.5	0.008	0.25	—	21.5	0.2	0.2	3.65	9.0
		Inconel 718	52.5	0.04	0.20	18.5	0.007	0.30	0.07	18.6	0.40	0.90	5.0	3.1
		Inconel 112	61.0	0.05	0.3	4.0	0.010	0.40	—	21.5	—	—	3.6	9.0
ENiCrFe-1	N06132	Inconel 132	73.0	0.04	0.75	8.5	0.006	0.20	0.04	15.0	—	—	2.1	—
ENiCrFe-3	—	Inconel 182	67.0	0.05	7.75	7.5	0.008	0.50	0.10	14.0	—	0.40	1.75	—
ENiCrFe-2	—	Inco-weld A	70.0	0.03	2.0	9.0	0.008	0.30	0.06	15.0	—	—	2.0	1.5
		Inco-weld B	70.0	0.13	2.0	9.0	0.008	0.30	0.06	15.0	—	—	2.5	2.0
	N08065	Incoloy 65	42.0	0.03	0.70	30.0	0.007	0.30	1.70	21.0	—	1.0	—	3.0
		Incoloy 135	36.0	0.05	2.00	26.0	0.008	0.40	1.80	29.0	—	—	—	3.75
ENi–C1	—	Ni–rod	95.0	1.00	0.20	3.0	0.005	0.70	0.10	—	—	—	—	—
ENiFe–C1	—	Ni–rod 55	53.0	1.50	0.30	45.0	0.005	0.50	0.10	—	—	—	—	—

[a]E, electrode; R, rod.
From AWS specifications A5.6, A5.7, A5.11, A5.14, and A5.15.

FIGURE 17–20 Composition of nickel alloy filler metals.

FIGURE 17–21 Guide for selecting filler metal for welding nickel alloys.

Alloy Designation	Electrode		Rod	
	Inco Name	AWS—Spec	Inco Name	AWS—Spec
Nickel 200	141	ENi–1	61	ERNi–3
Nickel 201	141	ENi–1	61	ERNi–3
Monel alloy 400	190	ENiCu–2	60	ERNiCu–7
Monel alloy 404	190	ENiCu–2	60	ERNiCu–7
Monel alloy K–500	190[a]	ENiCu–2	60[a]	ERNiCu–7
Monel alloy 502	190[a]	ENiCu–2	60[a]	ERNiCu–7
Inconel alloy 600	132 (182)	ENiCrFe–1	62 (82)	ERNiCrFe–5
Inconel alloy 601	132 (182)	ENiCrFe–1	601 (82)	ERNiCrFe–5
Inconel alloy 625	112	—	625	—
Inconel alloy 706	NA	—	718	—
Inconel alloy 718	NA	—	718	—
Inconel alloy 722	NA	—	69	ERNiCrFe–7
Inconel alloy X–750	NA	—	69	ERNiCrFe–7
Incoloy alloy 800	182 (A)	ENiCrFe–3	82 (625)	ERNiCr–3
Incoloy alloy 801	182 (A)	ENiCrFe–3	82 (625)	ERNiCr–3
Incoloy alloy 825	135 (112)	—	65 (625)	—

Notes: 1. Electrodes are covered electrode for SMAW. Rods are solid bare for GTAW, GMAW, and SAW.
2. NA, not available; use GTAW only.
[a]Will not age harden.

FIGURE 17–22 Welding procedure schedules for GTAW of nickel alloys.

Material Thickness (or Fillet Size)				Tungsten Electrode Diameter		Filler Rod Diameter		Nozzle Size, Inside Diameter	Shielding Gas Flow	Welding Current (amps	Number of	Travel Speed per Pass
Gauge	in.	mm	Type of Weld	in.	mm	in.	mm	in.	(ft³/hr)	DCEN)	Passes	(in./min)
24	0.024	0.6	Square groove and fillet	$\frac{1}{16}$	1.6	None		$\frac{3}{8}$	15	8–10	1	8
16	0.063	1.6	Square groove and fillet	$\frac{3}{32}$	2.4	$\frac{1}{16}$	1.6	$\frac{1}{2}$	18	25–45	1	8
$\frac{1}{8}$ in.	0.125	3.2	Square groove and fillet	$\frac{1}{8}$	3.2	$\frac{3}{32}$	2.4	$\frac{1}{2}$	25	125–175	1	11
$\frac{1}{4}$ in.	0.25	6.4	V-groove and fillet	$\frac{1}{8}$	3.2	$\frac{1}{8}$	3.2	$\frac{1}{2}$	30	125–175	2	8

Notes: 1. Tungsten used; first choice, 2% thoriated EWTh2; second choice, 1% thoriated EWTh1.
2. Adequate gas shielding is a must not only for the arc but also for heated metal. Backing gas is recommended at all times. A trailing gas shield is also recommended. Argon is preferred, but for higher heat input on thicker material use argon–helium mixture.
3. Data are for flat position. Reduce amperage 10%–20% when welding is horizontal, vertical, or overhead position.

FIGURE 17–23 Welding procedure schedules for GMAW of nickel alloys.

Material Thickness (or Fillet Size)				Electrode Diameter		Welding Power		Wire Feed Speed	Shielding Gas Flow	Number of	Travel Speed per Pass
Gauge	in.	mm	Type of Weld	in.	mm	Current (dc)	Arc Volt EP	(in./min)	(ft³/hr)	Passes	(in./min)
$\frac{1}{16}$ in.	0.062	1.6	Square groove and fillet	$\frac{3}{64}$	1.2	200–250	23–27	200–250	50	1	55–65
$\frac{1}{8}$ in.	0.125	3.2	Square groove and fillet	$\frac{1}{16}$	1.6	290–340	25–35	150–175	60	1	30–35
$\frac{1}{4}$ in.	0.250	6.4	Double V-groove and fillet	$\frac{1}{16}$	1.6	300–350	28–38	170–200	80	3	20–25

Notes: 1. Use 50% helium and 50% argon for thin metal and 100% helium for thick; higher voltage is for helium.
2. Increase amperage 10%–20% when backup is used.
3. Data are for flat position. Reduce current 10%–20% for other positions.

FIGURE 17–24 Physical properties of the refractory and reactive metals.

Element	Crystal Structure	Melting Point		Density		Thermal Conductivity (cal/cm^2/sec/° C)	Thermal Expansion (μ in./in./° F)
		F°	C°	lb/ft^3	gr/cc		
Tungsten	BCC	6170	3410	1190	19.3	0.397	2.55
Tantalum	BCC	5425	2996	1035	16.6	0.130	3.6
Molybdenum	BCC	4730	2610	650	9.0	0.34	2.7
Columbium	BCC	4474	2567	524	8.4	0.125	4.06
Zirconium	HCP	3366	1796	402	6.4	—	3.2
Beryllium	HCP	2332	1377	114	1.8	0.35	6.4
Titanium	HCP	3035	1668	281	4.5	—	4.67

heated metal is exposed to the atmosphere. Minor amounts of impurities cause these metals to become brittle.

Most of these metals have the characteristic known as the *ductile–brittle* transition. This refers to a temperature at which the metal breaks in a brittle manner rather than in a ductile fashion. The recrystallization of the metal during welding can raise the transition temperature. Contamination during the high-temperature period and impurities can raise the transition temperature so that the material is brittle at room temperatures. If contamination occurs so that transition temperature is raised sufficiently, it will make the weldment worthless. Gas contamination can occur at temperatures below the melting point of the metal. These temperatures range from 700° F (371° C) up to 1000° F (538° C).

At room temperature the reactive metals have an impervious oxide coating that resists further reaction with air. The oxide coatings melt at temperatures considerably higher than the melting point of the base metal. The oxidized coating may enter molten weld metal and create discontinuities that reduce the strength and ductility of the weld. Of the three reactive metals, titanium is the most popular and is routinely welded with special precautions.

All the refractory metals incur internal contamination or surface erosion when exposed to the air at elevated temperatures. Molybdenum has an extremely high rate of oxidation at high temperatures above 1500° F (816° C). Tungsten is much the same. Tantalum and columbium form pentoxides that are not volatile below 25° F (−3.9° C), but these provide little protection because they are nonadherent. Molybdenum and tungsten both become embrittled when a minute amount of oxygen or nitrogen is absorbed. Columbium and tantalum can withstand larger amounts of oxygen and nitrogen.

Titanium can withstand much more oxygen or nitrogen before becoming embrittled; however, small amounts of hydrogen will cause embrittlement. Zirconium can withstand about as much oxygen but much less

nitrogen or hydrogen. Beryllium is similar to zirconium in this regard.

Welding Refractory Metals

These metals must be perfectly clean prior to welding, and they must be welded in such a manner that air does not come into contact with the heated material. Cleaning is usually done with chemicals. A water rinse is necessary to remove all traces of chemicals from the surface. After the parts are cleaned, they must be protected from reoxidation. This is best done by storing in an inert gas chamber or in a vacuum chamber.

Molybdenum is welded by the gas tungsten arc welding process and the electron beam process. The gas metal arc process can be used, but sufficient thickness of molybdenum is rarely available to justify this process. It has been welded by other arc processes but results are not too satisfactory. Welding with the gas shielding processes is accomplished in an inert gas chamber or dry box. This is a chamber that can be evacuated and purged with inert gas until all active gases are removed. A typical dry box welding chamber is shown in Figure 17–25. Welding is done in the pure inert atmosphere. The filler metal compositions should be the same as the base metal. The base metal in the heat-affected zone becomes embrittled by grain growth and recrystallization as a result of the welding temperatures. Recrystallization raises the transition temperature so that molybdenum welds tend to be brittle. Molybdenum is highly notch sensitive; craters and notch effects such as undercutting must be avoided. Molybdenum can also be welded with the resistance welding processes and by diffusion welding.

Tungsten is welded in the same manner as molybdenum and has the same problems, only more intensely so. It has greater susceptibility to cracking because the ductile-to-brittle transition temperatures are higher. The preparation of tungsten for welding is more difficult. The GTAW process is used with direct current electrode negative. Welding should be done slowly to avoid cracking. Preheat-

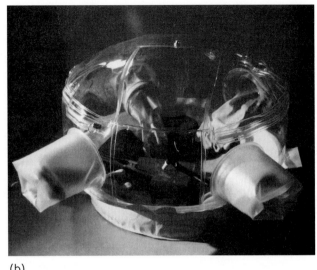

FIGURE 17-25   (a) Welding in a dry box. (b) Inflatable purge chamber.

ing may assist in reducing cracking but must be done in the inert gas atmosphere.

Commercially pure tantalum is soft and ductile and does not seem to have a ductile-brittle transition. There are several alloys of tantalum commercially available. Even though the material is easier to weld, it should be well cleaned and for best results should be welded in the inert gas chamber. The gas tungsten arc welding process is recommended. Some tantalum products are produced by powder metallurgy technology, and this may result in porosity in the weld. The arc cast product does not have porosity. Filler wire is normally not used when welding tantalum, and for best results direct current electrode negative is used. High frequency should be used for initiating the arc. Helium is recommended for welding tantalum to provide for maximum penetration since joints are designed to avoid using filler metal.

There are several different alloys of columbium (niobium) available. Some are ductile and others brittle since the transition temperature is near room temperature. The gas tungsten arc welding process is used for pure columbium and for the lower strength commercial alloys. In certain alloys the welding can be done outside an inert gas chamber, but special precautions should be taken to provide extremely good inert gas shielding coverage. In certain of the alloys preheating is recommended to provide for a crack-free weld. Electron beam welding is used, and columbium can be resistance welded.

Reactive Metals

Beryllium has been welded with the gas tungsten arc welding process and with the gas metal arc welding process. It is also joined by brazing. Beryllium should not be welded without expert technical assistance. Beryllium is a toxic metal, and extra special precautions should be provided for proper ventilation and handling.

Zirconium and zirconium–tin alloys are ductile metals and can be prepared by conventional processes. Cleaning is extremely important; chemical cleaning is preferred over mechanical cleaning. Both GTAW and GMAW are used for joining zirconium. The inert gas chamber should be employed to maintain an efficient gas shield. Argon or argon–helium mixtures are used. The zircalloys are alloys of zirconium, which contain small amounts of tin, iron, and chromium. These alloys can be welded in the open in much the same manner as titanium. Electron beam and resistance welding have been used for joining zirconium.

The secret to the successful welding of titanium is *cleanliness*. Small amounts of contamination can render a titanium weld completely brittle. Contamination from grease, oils, paint, fingerprints, or dirt can have the same effect. If the material is cleaned thoroughly before welding and well protected during welding, there is little difficulty in the welding of titanium.

GTAW and GMAW can be used for welding titanium. Special procedures must be employed that include the use of large gas nozzles and trailing shields to shield the face of the weld from air. Backing bars that provide inert gas to shield the back of the welds from air are used. Not only the molten weld metal, but the material heated above 1000° F by the weld must be adequately shielded in order to prevent embrittlement.

When using GTAW, a thoriated tungsten electrode should be used. The electrode size should be the smallest diameter that will carry the welding current. The electrode should be ground to a point. The electrode may extend $1\frac{1}{2}$ times its diameter beyond the end of the nozzle. Welding is done with direct current electrode negative (straight polarity).

Selection of the filler metal will depend on the titanium alloys being joined. When welding pure titanium, pure titanium should be used. When welding a titanium alloy, the next lower strength alloy should be employed as a filler wire. Due to the dilution that will take place during welding, the weld deposit will pick up the required strength. The same considerations are true when GMAW is used.

Argon is normally used with the gas-shielded process. For thicker metal use helium or a mixture of argon and helium. The purity of welding grade gases is normally satisfactory; however, tests can be made before welding. A simple test is to make a bead on a piece of clean titanium scrap and notice its color. The bead should be shiny. Any discoloration of the surface indicates a contamination.

Extra gas shielding provides protection for the heated solid metal next to the weld metal. This shielding is provided by special trailing gas nozzles (Figure 17–26) or by chill bars laid immediately next to the weld. Backup gas shielding should be provided to protect the underside of the weld joint. Protection of the back side of the joint can also be provided by placing chill bars in intimate contact with the backing strips. If the contact is close enough, backup shielding gas is not required. For critical applications use an inert gas welding chamber. These can either be flexible, rigid, or vacuum purge chambers.

To guarantee that embrittlement of the weld will not occur, proper cleaning steps must be taken. Solvents containing chlorine should not be used. Recommended solvents would be trialcohol or acetone. Titanium can be ground with disks of aluminum oxide or silicon carbide. Wet grinding is preferred; however, if wet grinding cannot be used, the grinding should be done slowly to avoid overheating the surface of the titanium.

Figure 17–27 gives procedure schedules for welding titanium. Joint types that are satisfactory for stainless steel should be used. Consult the International Titanium Association regarding difficult welding problems (address in Appendix).

17-6 OTHER NONFERROUS METALS

This section covers the welding of the low-melting-point metals, lead and zinc, and the precious metals, silver, gold, and platinum. Lead has one of the lowest melting temperatures. It melts at 621° F (328° C) and boils at 3092° F (1700° C). It is very soft and very ductile. When freshly cut its surface has a bright silvery luster that almost immediately oxidizes to a dull gray. It is available in sheet form and is used as a liner for tanks because of its corrosion resistance, particularly to sulfuric acid. It is also available in pipe.

Lead is normally joined by the oxyacetylene or oxyfuel gas torch, which for this reason has erroneously been called lead burning. Lead can also be welded with the gas

---

FIGURE 17–26 (a) Trailing shield on welding torch. (b) Visible nozzle with improved gas shielding.

(a)

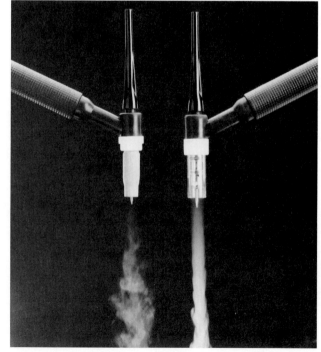

(b)

FIGURE 17–27 Welding procedure schedules for GTAW of titanium.

Material Thickness (or Fillet Size)			Type of Weld	Tungsten Electrode Diameter		Filler Rod Diameter		Nozzle Size, Inside Diameter (in.)	Shielding Gas Flow (ft³/hr)	Welding Current (amps DCEN)	Number of Passes	Travel Speed per Pass (in./min)
Gauge	in.	mm		in.	mm	in.	mm					
24	0.024	0.6	Square groove and fillet	$\frac{1}{16}$	1.6	None		$\frac{3}{8}$	18	20–35	1	6
16	0.063	1.6	Square groove and fillet	$\frac{1}{16}$	1.6	None		$\frac{5}{8}$	18	85–140	1	6
$\frac{3}{32}$ in.	0.093	1.6	Square groove and fillet	$\frac{3}{32}$	2.4	$\frac{1}{16}$	1.6	$\frac{5}{8}$	25	170–215	1	8
$\frac{1}{8}$ in.	0.125	3.2	Square groove and fillet	$\frac{3}{32}$	2.4	$\frac{1}{16}$	1.6	$\frac{5}{8}$	25	190–235	1	8
$\frac{3}{16}$ in.	0.188	4.8	Square groove and fillet	$\frac{3}{32}$	2.4	$\frac{1}{8}$	3.2	$\frac{5}{8}$	25	220–280	2	8
$\frac{1}{4}$ in.	0.25	6.4	V-groove and fillet	$\frac{1}{8}$	3.2	$\frac{1}{8}$	3.2	$\frac{5}{8}$	30	275–320	2	8
$\frac{3}{8}$ in.	0.375	9.5	V-groove and fillet	$\frac{1}{8}$	3.2	$\frac{1}{8}$	3.2	$\frac{3}{4}$	35	300–350	2	6
$\frac{1}{2}$ in.	0.50	12.7	V-groove and fillet	$\frac{1}{8}$	3.2	$\frac{5}{32}$	3.9	$\frac{3}{4}$	40	325–425	3	6

Notes:
1. Tungsten used; first choice, 2% thoriated EWTh2; second choice, 1% thoriated EWTh1.
2. Use filler metal one or two grades lower in strength than the base metal.
3. Adequate gas shielding is a must not only for the arc but also heated metal. Backing gas is recommended at all times. A trailing gas shield is also recommended. Argon is preferred for higher heat input; on thicker material use argon–helium mixture.
4. Without backup or chill bar, decrease current 20%.

tungsten arc welding process. Since lead is usually in thin sheets, the square butt and lap joints are the most commonly used. The surface of the lead at the welding area should be cleaned. Filler metal can be obtained by shearing strips of the base metal.

Sufficient heat is obtained to melt the surface of the base metal and the filler rod when used. When using gas tungsten arc a long arc is recommended to reduce the actual heat in the joint. There are a few difficulties that can be encountered when welding lead. Its popularity and use is declining, and for this reason the joining of lead is of minor importance. Proper ventilation is required.

Zinc is very similar to lead except that it is not as heavy. Zinc is widely used as a coating on steel, which is called galvanized steel. Zinc also has a tough oxide coating that must be removed prior to welding. The major use of zinc where welding is involved is as high-zinc die castings. Die castings of this type are commonly used for automotive grills and for decorative trim. For decorative trim and grills, the zinc may be chromium plated. It is important to distinguish zinc die castings from aluminum and magnesium alloy castings. Zinc has a weight in the same order as steel. In addition, zinc has a light gray appearance that is rather easy to identify.

The oxyacetylene or oxyfuel gas torch method is used to weld zinc alloys, particularly the die castings. The technique is essentially the same as for lead. The filler metal for matching zinc die castings is difficult to obtain and for this reason it may be necessary to manufacture it. One way is to melt down the bodies of carburetors and pour into a groove formed when an angle iron is positioned with the point down. These alloys contain approximately 10% aluminum, 1% to $2\frac{1}{2}$% copper, a trace of magnesium, and the remainder zinc.

For welding die castings the joint must be prepared with an extra wide root angle approaching 90°. The part should be positioned and braced in position. If the part is chrome plated, the plating should be removed adjacent to the welding area. The surfaces to be welded should be cleaned by wire brushing, sanding, filing, and so on. In addition, the filler metal should be cleaned by sanding. A small torch should be used and only sufficient heat to melt the surface and the filler rod. The filler rod is handled in the same manner as with lead welding.

The GTAW process can also be used for welding zinc die-casting alloys. When using the gas tungsten torch, a small tungsten diameter and low current should be employed. Sufficient heat should be maintained to melt the surface of the work being welded. The filler rod is moved in and out of the molten pool, normally on the leading edge. If the part is unusually shaped, it may be necessary to back it up to maintain the shape of the part. Proper ventilation is required.

Welding the Precious Metals

Silver is welded in much the same manner as copper since it has a high conductivity and a low affinity for oxygen and nitrogen. Filler metal normally used for brazing can be used. Silver is used for jewelry and tableware, but also for industrial applications where tanks are lined with sheet silver for the chemical industry.

The oxyacetylene or oxyfuel gas torch has been widely used for welding silver. Silver is also clad to other metals for chemical vessels. The gas tungsten arc welding process can also be used for welding silver and in this case direct current electrode negative is employed. The torch or tungsten must be sufficiently small to match the welding job. Procedure data for welding pure copper can be used to establish starting points for a silver welding procedure. Silver is also brazed and soldered.

Gold is one of the most expensive metals and, therefore, parts to be welded are unusually thin or of a small intricate shape. Gold can be soldered or brazed, and it may be cold or pressure welded. It is normally welded with the oxyfuel gas process using the small torch.

Platinum is used in the chemical industry and in the glass industry for making filaments for fiberglass. Welding is often required and is normally accomplished by the oxyacetylene or oxyfuel gas and the gas tungsten arc welding process. The other precious metals of the platinum group can be welded in the same manner. All the precious metals can be resistance welded, in spite of their high conductivity. The plasma arc welding process can also be used.

 # QUESTIONS

17–1. Explain the aluminum designation system of alloys and tempers.

17–2. What properties of aluminum make it different from welding steel? Explain.

17–3. What are the two most popular processes for welding aluminum? Where is each used?

17–4. Of all of the factors involved in welding aluminum, which one is the most important?

17–5. What are the properties of copper and its alloys that make welding difficult?

17–6. What is the effect of zinc in copper alloys on welding?

17–7. What problem does magnesium oxide present when welding magnesium?

17–8. Welding nickel and high-nickel alloys is similar to welding what other metals?

17–9. Why should the bevel or groove angles be increased for nickel alloys?

17–10. What welding process is most widely used for welding nickel alloys?

17–11. What are the reactive metals? Why are they difficult to weld?

17–12. What precautions should be taken when welding beryllium or its alloys?

17–13. What welding processes are used to join titanium *in the open*?

17–14. What is the most important factor to consider when welding titanium?

17–15. What precautions must be taken when welding lead?

17–16. What filler metal is used for lead? Where is it obtained?

17–17. Can zinc die castings be welded? How?

17–18. What precautions must be taken when welding zinc?

17–19. Silver is welded in much the same way as what other metal?

17–20. What processes are commonly used to join gold?

REFERENCES

1. "Aluminum Standards and Data," the Aluminum Association, New York.
2. "Welding Kaiser Aluminum," Kaiser Aluminum and Chemical Sales, Oakland, Calif.
3. "Welding Alcoa Aluminum," Aluminum Company of America, Pittsburgh, Pa.
4. "Welding Aluminum," Reynolds Metals Company, Richmond, Va.
5. "Standards Handbook Wrought Copper and Copper Alloy Mill Products, Part 2: Alloy Data," Copper Development Association, New York.
6. "Joining Magnesium," The Dow Metal Products Company, Midland, Mich.
7. "Joining Huntington Alloys," The International Nickel Company, Huntington, W. Va.

18

WELDING SPECIAL AND DISSIMILAR METALS

OUTLINE

18-1 CAST IRON AND OTHER IRONS

The term *cast iron* is a rather broad description of many types of irons that are castings but may have different properties and serve different purposes. In general, a cast iron is an alloy of iron, carbon, and silicon in which more carbon is present than can be retained in solid solution in austenite at the Eutectic temperature. The amount of carbon is usually more than 1.7% and less than 4.5%. There are many types of cast iron. The most widely used type of cast iron is known as gray iron. Its tonnage production exceeds that of any other cast metal.

Gray iron has a variety of compositions but it is usually such that the matrix structure is primarily pearlite with many graphite flakes dispersed throughout. It is these graphite flakes that provide the characteristic "gray" appearance of the fracture. The graphite flakes pro-

mote machinability, which is one of the primary advantages of gray cast iron. Another advantage is the ability to cast material into complex shapes with relatively thin walls. Gray cast iron is the least expensive metal for making many parts. Gray cast iron also has good resistance to wear and has a damping effect for vibration.

Gray cast iron is used in the automotive industry for engine blocks and heads, automatic transmission housings, differential housings, water pump housings, brake drums, and engine pistons. There are exceptions but the exceptions are usually aluminum, which is readily identifiable from cast iron.

There are also alloy cast irons that contain small amounts of chromium, nickel, molybdenum, copper, or other elements added to provide specific properties. These usually provide higher-strength cast irons. One of the major uses for the higher-strength irons is casting automotive crankshafts. These are sometimes called semi-steel or proprietary names.

Another alloy iron is the austenitic cast iron, which is modified by additions of nickel and other elements to reduce the transformation temperature so that the structure is austenitic at room or normal temperatures. Austenitic cast irons have a high degree of corrosion resistance.

Another type of cast iron is known as white cast iron, in which almost all the carbon is in the combined form. This provides a cast iron with higher hardness, which is used for abrasion resistance.

One class of cast iron is called malleable iron. This is made by giving white cast iron a special annealing heat treatment to change the structure of the carbon in the iron. By so doing, the structure is changed to pearlitic or ferritic, which increases its ductility.

There are two other classes of cast iron that are more ductile than gray cast iron. These are known as nodular iron and ductile cast iron. These are made by the addition of magnesium or aluminum, which will either tie up the carbon in a combined state or will give the free carbon a spherical or nodular shape rather than the normal flake shape in gray cast iron. This structure provides a greater degree of ductility or malleability of the casting.

Cast irons are used in many industries and they are popular because of their ease of casting, the ease of machining, and the relative low cost. They are widely used in agricultural equipment, for bases, brackets, covers, and so on, on machine tools, and for pipe fittings and for cast iron pipe. Cast iron is not used in structural work except for compression members.

Cast irons, particularly gray cast irons, are covered by ASTM specification A48, which establishes seven classes based on the tensile strength of the material. These range from 20,000 psi (14.5 kg/mm^2) tensile and 150 Brinell hardness number to 40,000 psi (28.5 kg/mm^2) and 250 Brinell hardness number. The cast irons above 40,000 psi (28.5 kg/mm^2) tensile strength are considered high-strength irons and are more expensive and more difficult to machine. Cast iron is twice as strong in compression than in tension. Other ASTM specifications are used to describe other classes of cast iron, but these are primarily with respect to the end use of the material.

Welding the Cast Irons

Gray cast iron has a very low ductility. Possibly a maximum of 2% ductility will be obtained in the extreme low carbon range. The low ductility is due to the presence of the graphite flakes, which act as discontinuities. In most welding processes the heating and cooling cycle creates expansion and contraction, which sets up tensile stresses during the contraction period. For this reason, gray cast iron is difficult to weld without special precautions. The ductile cast irons such as malleable iron, ductile iron, and nodular iron, can be welded successfully. For best results, these types of cast irons should be welded in the annealed condition.

Welding is used to salvage new iron castings, to repair castings that have broken in service, and to join castings to each other or to steel parts in manufacturing operations. Figure 18–1 shows the welding processes that can be used for welding cast, malleable, and nodular irons. The selection of the welding process and the welding filler metals depends on the type of weld properties desired and the service life that is expected. The filler metal will have an effect on the color match of the weld compared to the base material. The color match can be a determining factor in the salvage or repair of castings, where a difference of color would not be acceptable.

No matter which welding process is selected, certain preparatory steps should be made. It is important to determine the exact type of cast iron to be welded. If exact information is not known, it is best to assume that it is gray cast iron with little or no ductility. This would err on the side of safety so that a successful repair weld would be obtained.

In general it is not recommended to weld-repair gray iron castings that are subjected to repeated heating and cooling in normal service. At least it should not be used when heating and cooling vary over a range of temperatures exceeding 400° F (204° C). The reason is that unless cast iron is used as the filler material, the weld metal and base metal may have different coefficients of expansion and contraction. This will contribute to internal stresses that cannot be withstood by gray cast iron. Repair of these types of castings can be made, but the reliability and service life of such repairs cannot be predicted with accuracy.

Preparation for Welding

In preparing the casting for welding, it is necessary to remove all surface contaminants to completely clean the casting in the area of the weld. This means removing paint, grease, oil, and other foreign material from the weld zone. It is desirable to heat the weld area for a short time to remove entrapped gas from the weld zone of the base metal. Additionally, the skin or high-silicon surface should be removed adjacent to the weld area on both the face and root side.

Where grooves are involved, a V-groove from a 60 to 90° included angle should be used. Complete penetration welds should always be used since a crack or defect not removed completely may quickly reappear under service conditions.

Preheating is desirable for welding. It can be reduced when using extremely ductile filler metal. Preheating will reduce the thermal gradient between the weld and the remainder of the cast iron. Preheat temperatures should be related to the welding process, the filler metal type, the mass, and the complexity of the casting.

Preheating can be done by any of the normal methods. Torch heating is normally used for relatively small castings weighing 30 lb (13.6 kg) or less. Larger parts may be furnace preheated and in some cases temporary furnaces are built around the part rather than taking the part to a furnace. Preheating should be general since it helps to improve the ductility of the material and will spread shrinkage stresses over a large area to avoid critical stresses at any one point. It tends to help soften the area adjacent to the weld; it assists in degassing the casting and this in turn reduces the possibility of porosity of the deposited weld metal; and it also increases welding speed.

FIGURE 18–1 Welding processes and filler metals for cast iron.

Welding Process and Filler Metal Type	Filler Metal Spec[a]	Filler Metal Type[a]	Color Match	Machinable Deposit
MAW (stick)				
Cast iron	E-CI	Cast iron	Good	Yes
Copper–tin[b]	ECuSn A and C	Copper–5% or 8% tin	No	Yes
Copper–aluminum[b]	ECuA1–A2	Copper–10% aluminum	No	Yes
Mild steel	E-St	Mild steel	Fair	No
Nickel	ENi-CI	High-nickel alloy	No	Yes
Nickel–iron	ENiFe-CI	50% nickel plus iron	No	Yes
Nickel–copper	ENiCu-A and B	55% or 65% Ni + 40% or 30%	No	Yes
Oxyfuel gas				
Cast iron	RCI & A and B	Cast iron, with minor alloys	Good	Yes
Copper–Zinc[b]	RCuZn B and C	58% copper–zinc	No	Yes
Brazing[c]				
Copper–zinc	RBCuZn A & D	Copper–zinc and copper– zinc–nickel	No	Yes
GMAW (MIG)				
Mild steel	E60S-3	Mild steel	Fair	No
Copper base[c]	ECuZn-C	Silicon bronze	No	Yes
Nickel–copper	ENiCu-B	High-nickel alloy	No	Yes
FCAW				
Mild steel	E70T-7	Mild steel	Fair	No
Nickel type	No spec	50% nickel plus iron	No	Yes

[a]See "Specification for Welding Electrodes and Rods for Cast Iron." AWS A5.15
[b]Would be considered a brass weld.
[c]Heat source, any for brazing; also carbon arc, twin carbon arc, gas tungsten arc, or plasma arc.

Slow cooling or postheating improves the machinability of the heat-affected zone in the cast iron adjacent to the weld. The postcooling should be as slow as possible. Often this is done by covering the casting with insulating materials to keep the air or breezes from it.

Arc Welding

The shielded metal arc welding process can be utilized. There are four types of filler metals that may be used: cast iron-covered electrodes, covered copper-base alloy electrodes, covered nickel-base alloy electrodes, and mild steel-covered electrodes. There are reasons for using each of the different types: the machinability of the deposit, the color match of the deposit, the strength of the deposit, and the ductility of the final weld.

When arc welding with cast iron electrodes (ECI), preheat to between 250 and 800° F (121 and 425° C), depending on the size and complexity of the casting and the need to machine the deposit and adjacent areas. The higher the degree of heating, the easier it will be to machine the weld deposit. In general, it is best to use small electrodes and a relatively low current setting. A medium arc length should be used, and if at all possible welding should be done in the flat position. Wandering or skip welding pro-

cedures should be used, and peening will help reduce stresses and will minimize distortion. Slow cooling after welding is recommended. These electrodes provide an excellent color match on gray iron. The strength of the weld will equal the strength of the base metal.

There are two types of copper-base electrodes, the copper–tin alloy (ECuSn-A and C) and the copper–aluminum (ECuA1–A2) types. The copper–zinc alloys cannot be used for arc welding electrodes because of the low boiling temperature of zinc. Zinc will volatilize in the arc and will cause weld metal porosity. The copper tin electrodes will produce a bronze weld having good ductility. The ECuSn-A has less amount of tin. It is more of a general-purpose electrode. The ECuSn-C provides a stronger deposit with higher hardness. The copper–aluminum alloy electrode (ECuA1–A2) provides much stronger welds and is used on the higher-strength alloy cast irons.

When the copper-base electrodes are used, a preheat of 250 to 400° F (121 to 204° C) is recommended and small electrodes and low current should be used. The welding technique should be to direct the arc against the deposited metal or puddle to avoid penetration and mixing the base metal with the weld metal. Slow cooling is recommended after welding. The copper-base electrodes do not provide a color match.

There are three types of nickel electrodes used for welding cast iron. The ENiFe-CI contains approximately 50% nickel with iron, the ENiCI contains about 85% nickel, and the ENiCu type contains nickel and copper. The ENiFeCI electrode is less expensive and provides results approximately equal to the high-nickel electrode. These electrodes can be used without preheat; however, heating to 100° F (38° C) is recommended. These electrodes can be used in all positions; however, the flat position is recommended. The welding slag should be removed between passes. The nickel and nickel-iron deposits are extremely ductile and will not become brittle with the carbon pickup. The hardness of the heat-affected zone can be minimized by reducing penetration into the cast iron base metal. The technique mentioned above, that is, playing the arc on the puddle rather than on the base metal, will help minimize dilution. Slow cooling and, if necessary, postheating will improve machinability of the heat-affected zone. The nickel-base electrodes do not provide a close color match.

The copper–nickel type comes in two grades: the ENiCu-A with 55% nickel and 40% copper and the ENiCu-B with 65% nickel and 30% copper. Either of these electrodes can be used in the same manner as the nickel or nickel-iron electrode with about the same technique and results. The deposits of these electrodes do not provide a color match.

Mild steel electrodes (E St) or other low H_2 mild steel are not recommended for welding cast iron if the deposit is to be machined. The mild steel deposit will pick up sufficient carbon to make a high-carbon deposit which is difficult to machine. Additionally, the mild steel deposit will have a reduced level of ductility as a result of increased carbon content. This type of electrode should be used only for small repairs. Small electrodes at low current are recommended to minimize dilution and to avoid the concentration of shrinkage stresses. Short welds using a wandering sequence should be used and the weld should be peened as quickly as possible after welding. The mild steel electrode deposit provides a fair color match.

Oxyfuel Gas Welding

The oxyfuel gas process is often used for welding cast iron. The flame should be neutral to slightly reducing. Flux should be used. Two types of filler metals are available: the cast iron rods (RCI and A and B) and the copper zinc rods (RCuZn-B and C).

Welds made with the proper cast iron electrode will be as strong as the base metal. The RCI classification is used for ordinary gray cast iron. The RCI-A has small amounts of alloy and is used for the high strength alloy cast irons and the RCI-B is used for welding malleable and nodular cast iron. Good color match is provided by all these welding rods. The optimum welding procedure should be used with regard to joint preparation, preheat, and postheat.

The copper–zinc rods produce bronze welds. There are two classifications: RCuZn-B, which is a manganese bronze, and RCuZn-C, which is a low-fuming bronze. The bronze deposited has relatively high ductility but will not provide a color match.

Brazing and Braze Welding

Brazing is used for joining cast iron to cast iron and steels. In these cases, the joint design must be selected for brazing so that capillary action causes the filler metal to flow between closely fitting parts. The torch method is normally used; however, any of the other heating methods can be used. In addition, the carbon arc, the twin carbon arc, the gas tungsten arc, and the plasma arc can all be used as sources of heat. Two brazing filler metal alloys are normally used, both copper–zinc alloys; see Figure 18–1 for specification for the brazing alloys.

Braze welding can also be used to join cast iron. In braze welding the filler metal is not drawn into the joint by capillary attraction. This is sometimes called bronze welding. The filler material having liquidus above 850° F (454° C) should be used. Braze welding will not provide a color match.

Braze welding can also be accomplished by the shielded metal arc and the gas metal arc welding processes. High-temperature preheating is not usually required for braze welding unless the part is extremely heavy or complex in geometry. The bronze weld metal deposit has extremely high ductility, which compensates for the lack of ductility of the cast iron. The reason for not requiring high-temperature preheat is the desire to avoid intermix of base metal with the filler metal. The heat of the arc is sufficient to bring the surface of the cast iron up to a temperature at which the copper-base filler metal alloy will make a bond to the cast iron. Since there is little or no intermixing of the materials, the zone adjacent to the weld in the base metal is not appreciably hardened. The weld and adjacent area are machinable after the weld is completed. In general, a 200° F (93° C) preheat is sufficient for most applications. The cooling rate is not extremely critical and a stress relief heat treatment is not usually required. This type of welding is commonly used for repair welding of automotive parts, agricultural implement parts, and even automotive engine blocks and heads. It can be used only when the absence of color match is not objectionable.

Gas Metal Arc Welding

The gas metal arc welding process can be used for making welds between malleable iron and carbon steels. Several types of electrode wires can be used, including:

❋ Mild steel (E7OS-3) using 100% CO_2, 75% argon + 25% CO_2 for shielding

* Nickel–copper (ENiCu-B) using 100% argon for shielding
* Silicon bronze (ECuZn-C) using 50% argon + 50% helium for shielding
* Nickel–copper–iron (ER Ni-CI) using argon for shielding

In all cases, small-diameter electrode wire should be used at low current. With the mild steel electrode wire the argon–CO_2 shielding-gas mixture is used to minimize penetration. In the case of the nickel-base filler metal and the copper-base filler metal, the filler metal deposited is extremely ductile. The mild steel provides a fair color match. A higher preheat is usually required to reduce residual stresses and cracking tendencies.

Flux Cored Arc Welding

This process can be used for welding cast irons. The more successful application has been using a nickel-base flux cored wire, which produces a weld metal deposit very similar to the 50% nickel deposit provided by the ENiFeCI covered electrode. This electrode wire is normally operated with CO_2 shielding gas, but when lower mechanical properties are not objectionable it can be operated without external shielding gas. The minimum preheat temperatures can be used. The technique should minimize penetration into the cast iron base metal. Postheating is normally not required. A color match is not obtained.

Flux cored self-shielding electrode wires (E60T-7), operating with electrode negative (straight polarity), have also been used for certain cast iron to mild steel applications. In this case, a minimum penetration-type weld is obtained and by the proper technique penetration should be kept to a minimum. It is not recommended for deposits that must be machined or use ENCFe T-3-CI.

Figure 18-1 provides a summary of welding processes for joining cast iron. For manufacturing operations it is highly recommended that a welding procedure be developed utilizing the process selected and that service-type tests be made prior to using the process on a particular product. In general, those cast irons having maximum ductility are those that can be most successfully welded. Thus malleable iron, nodular iron, and ductile iron can be welded for many applications. The most successful welds would be those that provide an extremely ductile weld deposit.

Other welding processes can also be used for cast iron. Thermite welding has been used for repairing certain types of cast iron machine tool parts. The procedure is identical to that used for welding steel except that a special thermite mixture is required. Flash welding can also be used for welding cast iron. See "Guide for Welding Iron Castings" AWS D11.2.

18-2 TOOL STEELS

Steels used for making tools, punches, and dies are perhaps the hardest, the strongest, and toughest steels used in industry. It is obvious that tools used for working steels must be stronger and harder than the steels they cut or form. The metallurgical characteristics or compositions of tool steels are extremely complex and beyond the scope of this book. Tools and dies wear and are damaged but they can be repaired and returned to service. In addition, certain tools and dies can be fabricated by welding. Repairing damaged tools and dies and the fabrication by welding of dies will save money.

There are hundreds of different makes and types of tool steels available, and each has a specific composition and end use. The Society of Automotive Engineers, in co-operation with the American Iron and Steel Institute, has established a classification system that relates to the use of the material and its composition or type of heat treatment (Figure 18-2). In general, tool steels are basically medium-to high-carbon steels with specific elements in different amounts to provide special characteristics. The carbon in the tool steel is provided to help harden the steel for cutting and wear resistance. Other elements are added to provide greater toughness or strength.

The addition of alloys produces different effects. Chromium produces deeper hardness penetration in heat treatment and contributes wear resistance and toughness. Cobalt is used in high-speed steels and increases the red hardness so that they can be used at higher operating temperatures. Manganese in small amounts is used to aid in making steel sound, and further additions help steel to harden deeper and more quickly in heat treatment. It also helps to lower the quenching temperature necessary to harden steels. Larger amounts of manganese in the range 1.20% to 1.60% allow steels to be oil quenched rather than water quenched. Molybdenum increases the hardness penetration in heat treatment and reduces quenching temperatures. It also helps increase red hardness and wear resistance. Nickel adds toughness and wear resistance to steel and is used in conjunction with hardening elements. Tungsten added to the steel increases its wear resistance and provides red hardness characteristics: Approximately 1.5% increases wear resistance, and about 4% in combination with high carbon will greatly increase wear resistance. Tungsten in large quantities with chromium provides red hardness. Vanadium in small quantities increases the toughening effect and reduces grain size. Vanadium in amounts over 1% produces extreme wear resistance especially to high-speed steels. Smaller amounts of vanadium in conjunction with chromium and tungsten aid in increasing red hardness properties.

The tool steels are designed for special purposes that are dependent on composition. Certain tool steels

FIGURE 18–2 Abridged chart of tool steel types.

AISI-SAE Type	Classification of Tool Steels	Composition (%)					
		C	Cr	V	W	Mo	Other
W1	Water hardening	0.60	—	—	—	—	—
W2		0.60	—	0.25	—	—	—
S1		0.50	1.50	—	2.50	—	—
S5	Shock resisting	0.55	—	—	—	0.40	0.80 Mn 2.00 Si
S7		0.50	3.25	—	—	1.40	—
O1	Oil hardening	0.90	0.50	—	0.50	—	—
O6		1.45	—	—	—	0.25	1.00 Si
A2	Cold work	1.00	5.00	—	—	1.00	—
A4	Medium-alloy air hardening	1.00	1.00	—	—	1.00	2.00 Mn
D2	Cold work, high carbon, high chromium	1.50	12.00	—	—	1.00	—
M1	Cold work	0.80	4.00	1.00	1.50	8.00	—
M2	Molybdenum	0.85	4.00	2.00	6.00	5.00	—
M10		0.90	4.00	2.00	—	8.00	—
H11	Hot work	0.35	5.00	0.40	—	1.50	—
H12	Chromium	0.35	5.00	0.40	1.50	1.50	—
H13		0.35	5.00	1.00	—	1.50	—
P20	Die-casting mold	0.35	1.25	—	—	0.40	—

are made for producing die blocks; some are made for producing molds, others are made for hot working, and still others for high-speed cutting applications. The other way for classifying tool steels is according to the type of quench required to harden the steel. The most severe quench is the water quench (water-hardening steels). A less severe quench is the oil quench obtained by cooling in oil baths (oil-hardening steels). The least drastic quench is cooling in air (air-hardening steels).

Tool steels can also be classified according to the work that is to be done by the tool. This is based on class numbers. Class I steels are used to make tools that work by a shearing or cutting action, such as cutoff dies, shearing dies, blanking dies, trimming dies, and so on. Class II steels are used to make tools that produce the desired shape of the part by causing the material being worked, either hot or cold, to *flow* under pressure. This includes drawing dies, forming dies, reducing dies, forging dies, and so on. This class also includes plastic molds and die cast molding dies. The class III steels are used to make tools that act on the material being worked by partially or wholly reforming it without changing the actual dimensions. This includes bending dies, folding dies, twisting dies, and so on. Class IV steels are used to make dies that work under heavy pressure and that produce a flow of metal or other material compressing it into the desired form. This includes crimping dies, embossing dies, heading dies, extrusion dies, staking dies, and so on. It is im-

portant to understand and have sufficient information concerning the composition of the tool or die, the type of heat treatment that it has received, and the type of work that it performs.

As far as welding is concerned, there are four basic types of die steels that are weld repairable: water-hardening dies, oil-hardening dies, air-hardening dies, and hot work tools. High-speed tools can also be repaired. In tool and die welding it is not always necessary for the filler metal used to provide a deposited weld metal that exactly matches the analysis of the tool steel being welded. It is necessary, however, that the weld metal deposited match the heat treatment of the tool or die steel as closely as possible. Thus, selection of the proper electrode is based on matching the heat treatment of the tool or die steel.

There are no specifications covering the composition of tool and die welding electrodes. However, all manufacturers of these electrodes provide information concerning each, showing the type of tool or die steels for which it is designed. They also provide the properties of the weld metal that is deposited. Welding filler metal is not available to match the composition of each and every tool steel composition or to match the specific heat treatment of each tool or die steel. However, tool and die electrodes are available that match the different categories of tool and die steels. Assistance can be obtained from the catalogs of electrodes for tool and die welding or by consulting with representatives of the companies that manufacture them. If the identification of the electrodes is lost, it is possible to use the spark test in matching the electrode to the tool steel. A comparison is made of the sparks from the tool or die steel to be welded and compared with the spark pattern of the welding electrode. The matching spark patterns will be the guide or basis for selection of the electrode.

Successful tool and die welding depends on the selection or development of a welding procedure and welding sequence. It is beyond the scope of this chapter to outline specific welding procedures to be used for each classification of tool steel using each type of tool and die welding filler metal. Normally, the manufacturer of filler metal will provide specific procedure sheets pertaining to the different products they offer. These should be carefully followed.

In general, weld deposits of tool and die electrodes are sufficiently hard in the as-welded condition. If the welded tools or dies lend themselves to grinding, treatment other than tempering is not required. However, if machining is required, the weld deposits should be annealed and heat treated after machining. The hardness of the weld deposit will vary in accordance with the following:

⁕ Preheat temperature if used

⁕ Welding technique and sequence

⁕ Mixture or dilution of the weld metal with base metal

⁕ Rate of cooling, which depends on the mass of the tool being welded

⁕ Tempering temperature of the welded tool or die after welding

Uniform hardness of the as-welded deposit is obtained if the temperature of the tool or die is maintained constant during the welding operation. The temperature of the tool being welded should never exceed the maximum of the *draw range* temperature for the particular class of tool steel being welded. The manufacturer's recommendation should be followed with respect to those temperatures.

The welding procedure for repair welding of tools and dies should consist of at least the following factors:

⁕ Identification of the tool steel being welded

⁕ Selection of the filler metal to match the same class of material or heat treatment

⁕ Establishing the correct joint detail for the repair and preparing the joint

⁕ Preheating the workpiece

⁕ Making the weld deposit in accordance with manufacturer's recommendations

⁕ Postheating to temper the deposit or the repaired part

One of the major problems is proper preparation of the part for repair welding. When making large repairs to worn cutting edges or surfaces, the damaged area should be ground sufficiently under size to allow a uniform depth of finished deposit of at least 1/8 in. (3.2 mm). In some cases very small weld deposits are made using the gas tungsten arc welding process to build up a worn or damaged edge or corner. It is important to provide a uniformly thick weld deposit that will be refinished to the original dimensions. This ensures a more uniform hardness throughout the deposit. For inlay deposit or other overlay work a thickness of 3/16 in. (4.8 mm) is required.

When preheating the part to be repaired, observe the "draw temperature range" of the base metal. The preheat temperature should slightly exceed the minimum of the draw range and the interpass temperature should never exceed the maximum of the draw range of the particular tool steel. Exceeding the maximum draw range will reduce the hardness of the tool by softening it. Figure 18-3 provides recommended preheat temperatures to be employed on the popular types of tool steels.

Most of the tool and die SMAW welding electrodes are used with dc electrode positive or with alternating current. The recommended currents for each different size should be provided with the electrode manufacturer's technical data. For making tool and die welds, a slow travel is recommended in order to maintain an even deposit and to assure uniform weld penetration. The work should be positioned for flat-position welding except that it is recommended that welding be done with the work positioned for slightly uphill travel, on the order of 5 to 15°. This causes the deposit to build up evenly and

FIGURE 18-3 Preheat and interpass temperatures when welding tool steels.

AISI-SAE Type	Temperature (°F)		
	Preheat	Interpass	Draw or Tempering
W1	250/450	250/450	300/650
W2	250/450	250/450	300/650
S1	300/500	300/500	300/500
S5	300/500	300/500	500 min.
S7	300/500	300/500	425/400
O1	300/400	300/400	300/450
O6	300/400	300/400	300/450
A2	300/500	300/500	350/400
A4	300/500	300/500	350/400
D2	700/900	700/900	925/900
M1	950/1100	950/1100	1000/1050
M2	950/1100	950/1100	1000/1050
M10	950/1100	950/1100	950/1050
H11	900/1200	900/1200	1000/1150
H12	900/1200	900/1200	1000/1150
H13	900/1200	900/1200	1000/1150
P20	400/800	400/800	1000

helps keep the slag free of the weld pool. Uniform motion without weaving is recommended. When welding on tool cutting edges, position the work so that the deposit will flow or roll over the cutting edge.

Peening should be done immediately on all weld deposits. Peening should be controlled and used to provide sufficient mechanical work to help improve the properties of the deposit and help refine the metallurgical structure. It will also assist in relieving shrinkage stresses and possibly assist in correcting distortion. Peening can be done manually, or small air power hammers can be used. The welding technique should avoid craters. In all cases, craters should be filled by reversing the direction of travel and pausing slightly. This will ensure a more uniform deposit.

When welding deeply damaged cutting edges that require multiple passes, it is necessary to start at the bottom and gradually fill up damaged areas. The current for the first or second beads can be higher than used on the final bead. It is important to peen the weld metal while hot to help eliminate shrinkage, warpage, and possibly cracks. The random or wandering welding technique should be used when welding circular parts, such as on the inner edge of a die. Warpage or distortion can be reduced by preheating, which expands the part, and peening during the contraction period, which will reduce stresses. On parts such as a long shear blade where welding is done all on one side, it is recommended that the parts be reverse formed. This will help keep the part straight during welding. It is recommended to weld only short lengths of 2 to 3 in. and then to peen to reduce stresses and warpage.

After the repair welds are completed, the part should be allowed to cool to room temperature. It is then tempered by reheating to the recommended temperature, as specified by the type of tool steel being welded or by the welding filler metal manufacturer's technical data. The draw temperature would always be used. For small or light-duty work parts, the draw temperature should be on the minimum side of the draw range. On larger or heavy-duty parts, the draw temperature should be the maximum.

The fabrication of composite trim and blanking dies is becoming popular. The tool can be completely fabricated using a low-alloy steel base and then building up the cutting edge with tool steel welding filler metal having the desired characteristics. Following normal wear, the cutting edge can be rewelded with the tool steel welding electrode appropriate to the application. The base steel for the composite tool, or die, must possess the required mechanical properties for the specific application. Normally, low-alloy steels are suitable; however, if the tool operates at elevated temperatures, an alloyed steel must be utilized. Any heat treatment needed by the tool must be provided prior to welding. The welding filler metal should be selected to provide a deposit having the characteristics suited for the type of work the tool will

do. Resistance to heat, abrasion, shock, and so on, should all be considered. The size of the SMAW welding electrode depends on the amount of welding to be done and the type of preparation. Dilution of the base metal must be considered. A preheat in the range of 200 to 400° F is recommended. The larger the unit, the higher the preheat. The welding procedure should be similar to that used when making repairs to similar tools. After welding the composite tool should be tempered, and the tempering temperature should be that recommended by the filler metal manufacturer and should not exceed the one specified for the base material.

Experience with tool and die welding is very helpful and will avoid the possibility of failures. The procedure development, including identification of material, selection of filler metal, and welding techniques should follow the tool steel manufacturer's data and the welding filler metal manufacturer's information.

18-3 REINFORCING BARS

Concrete reinforcing bars, or as they are more technically known, *deformed steel reinforcing bars,* are used in reinforced concrete construction. This includes buildings, bridges, highways, locks, dams, docks, piers, and so on. The principal applications of reinforcing bars include reinforcement of columns, girders, beams, slabs, and pavements, as well as precast and prestressed concrete structures. Concrete is strong in compression and shear but is weak in tension. By using deformed steel reinforcing bars embedded in the concrete, tensile stresses can be accommodated; thus, reinforced concrete provides compression strength of concrete and tensile strength of steel. The concrete and steel must work together. This is accomplished by a bond between the bar and the concrete, which is achieved by means of deformations that are rolled into the bars. These deformations keep the bars from slipping through the concrete.

Concrete reinforcing bars come in different sizes. There are 11 standard sizes, known as *reinforcing bar* No. 3 through No. 11 and No. 14 and No. 18. The numbers assigned to bars are based on the number of 1/8 in. included in the nominal diameter. The nominal diameter of a deformed reinforcement bar is equivalent to the diameter of a plain steel bar having the same weight per foot as the deformed bar. Lengths up to 60 ft are available.

There are three ASTM specifications for reinforcing bars: A615, plain billet steel bars; A616, rail steel reinforcing bars; and A617, axle steel reinforcement bars. Information concerning these different specifications is shown in Figure 18–4.

All of the reinforcing bars produced in the United States are identified by markings rolled into the bar. These markings will show the code for the manufacturer

ASTM Specification	Specification Identification	Grades Produced	Grade Identification	Size Designation Bar Number	Strength		Composition (1)	
					Tensile min psi	Yield min psi	Carbon	Manganese
A-615 (New billet steel)	N	40	Blank	#3 thru #11 and #14 and #18	70,000	40,000	– –	– –
		60	60		90,000	60,000	– –	– –
		75	75		100,000	75,000	– –	– –
A-616 (Made from A-1)	(rail symbol)	50	Blank	#3 thru #11	80,000	50,000	0.55-0.82	0.60-1.00
		60	60		90,000	60,000		
A-617 (Made from A-21)	A	40	Blank	#3 thru #11	70,000	40,000	0.40-0.59	0.60-0.90
		60	60		90,000	60,000		

Note: See specific ASTM specification for additional information—
(1) composition based on A-1 and A-21

FIGURE 18–4 Summary of information for reinforcing bars.

of the steel bar. The different code letters have been identified by the Concrete Reinforcing Steel Institute. This is then followed by the letter identifying the specific steel mill where the bar was produced, based on standard designations. The next symbol indicates the bar size by the bar number. The next symbol indicates the type of steel as follows: N indicates new billet steel, A indicates axle steel, and the third symbol which is a cross section of a railroad rail indicates that the bar was rerolled from used railroad rails.

The next identification symbol is a number indicating the grade of steel. If there is no number, it normally means that it is the minimum grade within the specification. Grades are also identified by a single or double continuous longitudinal line through at least five spaces offset from the center of the bar. A single line indicates the middle-strength grade and a double line indicates the highest-strength grade. It is important to determine the type of steel and the grade since this will be valuable information in establishing the welding procedure.

The specifications do not include chemical requirements for the different classes; however, when bars are purchased from the mill, the mill will provide a chemical analysis report of the bars, if requested. The grade number is the indication of the strength of the bars and the numbers indicate the yield point in thousand pounds per square inch minimum. All bar sizes are not made in all grades and it is only the specification A615 that provides the large No. 14 and No. 18 size bars.

It is necessary to splice concrete reinforcing bars in all but the most simple concrete structures. In the past, splicing was done by overlapping the bars from 20 to 40 diameters and wiring them together and relying on the surrounding concrete to transmit the load from one bar to the other. This method is wasteful of the steel and is sometimes impractical. Welding is now used for splicing concrete reinforcing bars. Three welding processes are used for the majority of welding splices; however, several

of the other processes can be used. There is a mechanical splice similar to welding that utilizes medium-strength metal cast (*metallic grout*) around the ends of the bars enclosed within a steel sleeve having internal grooves. The welding processes most commonly used are shielded metal arc, gas metal arc, flux cored arc, and thermite welding. It was mentioned previously that there are no chemistry requirements for the three ASTM specifications. However, the reinforcing bars of specification A616 rail steel are produced from used railroad rails that were originally made to specification A1. Old railroad rails are salvaged, heated, and cut into three parts, the flange, the web, and the head. The heads are then rolled into the deformed reinforcing bars. ASTM specification A1 has chemical requirements for steel rails, and they contain relatively high amounts of carbon and manganese.

The reinforcing bars made to specification A617 are made from salvaged carbon steel axles used for railroad cars. These axles when originally produced were made to specification A21. In this ASTM specification the carbon and manganese is relatively high. These are both considered in the hard-to-weld category of steels. The bars produced to A615 have only a maximum for phosphorus content; however, based on the strength level of steels, the alloy content should not be too high. For quality welding it is best to assume that they, too, are in the hard-to-weld category. If at all possible, the analysis of the reinforcing bars should be determined from mill reports. If this is not possible, the bars could be analyzed for exact composition. In lieu of this, it is recommended that the bars be considered to have a carbon equivalent of 0.75, thus in the hard-to-weld category.

The American Welding Society has provided a specification entitled "Reinforcing Steel Structural Welding Code." D1.4. This code provides a table of carbon equivalents that relates to the bar size and then presents recommended preheat and interpass temperatures. The standard formula for the determination of carbon equivalent is used. There are six carbon equivalents, which can be calculated only if the analysis of the reinforcing bars is known.

The code also provides joint design information for making direct butt splices, for making indirect butt splices, and for making lap splices (Figure 18-5). A butt splice is a direct end-to-end splice of bars with their axes approximately in line and of approximately the same size. A split pipe is often used for backing. An indirect butt splice is one in which an intermediary piece such as a steel plate or rolled angle is used with each reinforcing bar welded directly to the same piece. The lap welded splice is made by overlapping the two bars alongside each other and welding together. Direct splices can be made between bars of different sizes by providing a transition-type configuration to aid stress flow. For butt splices when the bars are in the horizontal position, the single groove weld is most often used with a 45 to 60° in-cluded angle. Double-groove welds can be made in the larger bars. When the bars are to be welded with the axis vertical, a single or double bevel groove weld is used with the flat side or horizontal side on the lower bar. On occasion, the reinforcing bar may need to be welded to other steel members and a variety of weld joints can be used.

This code provides filler metal selection information based on the grade number of the steel. When using the shielded metal arc welding process, grade 40, the AWS E-7018, is recommended, for grade 50 the AWS E-8018 is recommended, for grade 60 and the low-alloy A706 the AWS E-9018 electrode is recommended, and for the grade 75 the AWS E-10018 electrode is recommended. If the XX18 is not available, the XX16 can be used. In the case of gas metal arc welding, the E-70S electrode would be used; for flux cored arc welding, the E70T type would be used when welding grade 40 bars. If these processes are used, the filler metal must meet the same mechanical

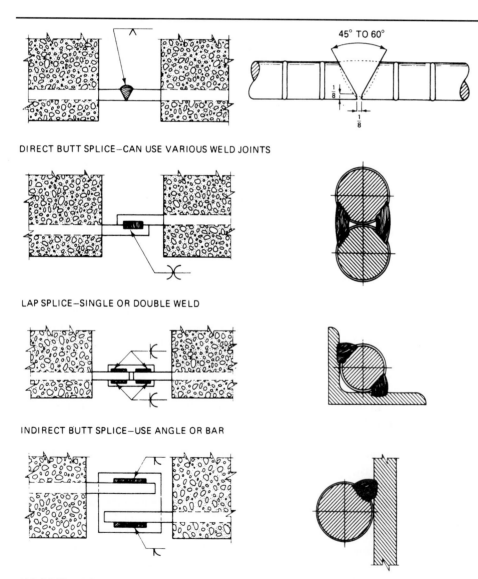

45° TO 60°

$\frac{1}{8}$

$\frac{1}{8}$

FIGURE 18-5 Types of reinforcing bar splices.

DIRECT BUTT SPLICE—CAN USE VARIOUS WELD JOINTS

LAP SPLICE—SINGLE OR DOUBLE WELD

INDIRECT BUTT SPLICE—USE ANGLE OR BAR

INDIRECT LAP SPLICE—AVOID EXCESSIVE ECCENTRICITY

properties as the equivalent shielded metal arc welding electrode mentioned.

The code also provides minimum preheat and inner-pass temperatures based on the carbon equivalent of the reinforcing bars. It also relates to the size of the bar. It is important to determine the composition of the bar so that the carbon equivalent can be determined. This establishes the heat requirement, which ranges from 50° F (10° C) up through 500° F (260° C) based on the size of the bar and the carbon equivalent. In the case of large bars and if the carbon equivalent is not known, the 500° F preheat would be recommended. Consult the code for further information.

The code further requires that joint welding procedures should be established based on the welding process, filler metal type and size, and welding technique, which involves position, joint detail, and so on. Welders must be qualified. A direct butt splice or indirect butt splice specimen is used. The test bars are tested in tension. Figure 18–6 shows a reinforcing bar being welded using the gas metal arc welding process.

The gas metal arc and flux cored arc welding processes will make the weld in approximately one-half the time required for shielded metal arc. Welding is highly recommended as the way to splice reinforcing bars. The welded splices will exceed the strength of lapped and wired splices. It will also exceed a strength level of the cast metal splices, which are sufficiently strong to withstand the strength level of the reinforced concrete composite structure.

FIGURE 18–6 Joining bars with GMAW.

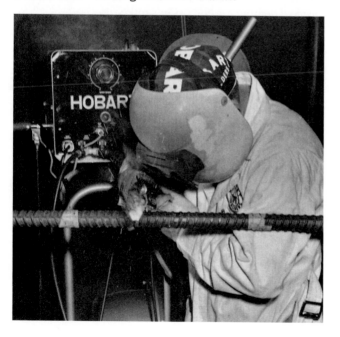

18-4 COATED STEELS

The coated steel that will be discussed in detail is galvanized or zinc-coated sheet steel. Galvanized steel is widely used and is becoming increasingly important. Manufacturers of many items such as truck bodies, buses, and automobiles are increasingly concerned with corrosion, particularly when chemicals are used on roads for ice control. Galvanized metal is also used in many appliances, such as washing machines and dryers, and in many industrial products, such as air conditioning housings and processing tanks. Other uses for galvanized products are for high-tension electrical transmission towers, highway sign standards, and protective items.

There are two methods of galvanizing steel. One is by coating sheet metal and the other is by hot dipping the individual item. The coated sheet metal is produced by the continuous hot dip process. The continuous hot-dip or zinc-coated sheet comes in different classes based on the thickness of the zinc coating. The coating varies from 1 to 1.75 oz of zinc per square foot of the surface, based on coating both sides of the sheet. One-side-coated steel is also available. Hot-dipped individual parts have coatings exceeding the thickness mentioned above. Welding of zinc-coated steel can be done with specific precautions. When galvanized steel is arc welded, the heat of the welding arc vaporizes the zinc coating in the weld area. The zinc volatilizes and leaves the base metal adjacent to the weld. The extent to which the coating is disturbed depends on the heat input of the arc and the heat loss from the base metal. The disturbed area is greater with the slower welding speed processes.

When galvanized sheet is resistance welded, the welding heat causes less disturbance of the zinc coating than the arc processes. The resistance to corrosion, or rather the protection by the zinc, is not disturbed since the zinc forced from the spot weld will solidify adjacent to the spot weld and protect the weld nugget. Resistance welding of galvanized steel is more of a problem because of the zinc pickup of the welding tips and tools.

Weld Quality

The zinc in the gaseous state may become entrapped in the molten weld metal as it solidifies. If this occurs, there will be porosity in the weld metal and if sufficient zinc is available it will cause large voids in the surface of the deposit. The presence of the zinc in stressed welds can cause cracking and it may also cause delayed cracking due to stress corrosion. To eliminate this, the weld joint must be designed to allow the zinc vapor to completely escape from the joint. Fixturing, backing straps, and so on, should be arranged to allow for the zinc to escape completely. Other ways to avoid zinc entrapment in weld metal is to use sufficient heat input when making the weld. It is also important to secure complete and full penetration of the joint. The best precaution is to remove the zinc from the weld area.

When welding on galvanized steel or any coated steel, particularly those with coatings that produce noxious fumes, positive ventilation must be provided. Positive ventilation involves the use of a suction hose at the weld area. When using the gas metal arc or the flux cored arc process, suction-type gun nozzles should be used. Welding on zinc or other coated steels should never be done in confined areas.

For corrosion resistance of the weld it is advisable to use a corrosion-resistant weld metal, such as a copper–zinc alloy or a stainless steel. In any case, when arc welding is used the area adjacent to the weld will lose the protective zinc coating, which must be replaced.

Arc Welding

The electrode selection should be based on the thickness of metal and the position that will be used when welding galvanized steel. The EXX12 or 13 will be used for welding thinner material; the EXX10 or 11 will be used for welding galvanized pipe and for welding hot-dipped galvanized parts of heavier thickness. The low-hydrogen electrodes can also be used on heavier thickness. The welding technique should utilize slow travel speed to permit degassing of the molten metal. The electrode should point forward to force the zinc vapor ahead of the arc. The quality of welds will be equal to those of bare metal, assuming the weldability of the steel is equal.

The GMAW process is widely used for joining galvanized steel. For the thinner gauges the fine-wire short-circuiting method is recommended. In this case, the technique would be similar to that used for bare metal. The shielding gas can be 100% CO_2 or the 75% argon and 25% CO_2 mixture. The selection is dependent on the material thickness and position of welding. For certain applications, the argon–oxygen mixture is used. The amount of spatter produced when welding galvanized steel is slightly greater than when welding bare steel. The gun tip and nozzle should be cleaned more often. However, a stainless steel or bronze type can be employed. This will produce a weld deposit that will be corrosion resistant.

The flux cored arc welding process can be used for galvanized steel. It is recommended for the heavy gauges and on hot-dipped galvanized parts. The highly deoxidized type of welding electrode should be used.

The GTAW process is not popular since it causes a larger area of zinc adjacent to the weld to be destroyed. In addition, the volatilized zinc will contaminate the tungsten electrode and require frequent redressing. To overcome this, extra high gas flow rates are used, which can be expensive. If a filler rod is used, it should be of either the highly deoxidized steel type or bronze. In this case the arc is played on the filler rod and zinc contamination

of the tungsten electrode is reduced.

The carbon arc welding process can be used for welding galvanized steel. Both the single carbon torch and twin carbon torch can be used. When the single carbon is used, the arc is played on the filler rod and high rates of speed can be attained. Normally in this situation the filler rod is the 60% copper–40% zinc alloy, type RBCuZn-A. By directing the arc on the filler rod, it melts and sufficient heat is produced in the base metal for fusion without destroying the zinc coating. This process is used in the sheet metal industry.

Torch Brazing

The oxyacetylene torch is used for brazing galvanized steel. The technique is similar to the carbon arc. The flame is directed toward the filler rod, which melts and then fills the weld joint. A generous quantity of brazing flux is used to help reduce the zinc loss adjacent to the weld.

Repairing the Zinc Coating

The area adjacent to the weld may lose zinc because of the high temperature of the arc. To produce a corrosion-resistant joint, the zinc must be replaced. There are several ways of replacing the zinc. One is by the use of zinc base sticks sometimes called zinc sticks or galvanized sticks sold under different proprietary names. These sticks are wiped on the heated bare metal. With practice a very good coating can be placed that will blend with the original zinc coating. This coating may be thicker than the original coating. Another way of replacing the depleted zinc coating is by flame spraying using a zinc spray filler material. This is a faster method and is used if there is sufficient zinc coating to be replaced. The coating should be two to two-and-one-half times as thick as the original coating for corrosion protection.

Other Coated Metals

One other coated metal that is often welded is known as tern plate. This is sheet steel hot dipped with a coating of a lead–tin alloy. The tern alloy is specified in thicknesses based on the weight of tern coating per square foot of sheet metal. This ranges from 0.35 to 1.45 oz per square foot of sheet metal based on both sides being coated. Tern plate is often used for making gasoline tanks for automobiles. It is welded most often by the resistance welding process. If it is arc welded or oxyacetylene welded, the tern plating is destroyed adjacent to the weld and it must be replaced. This can be done similar to soldering.

Aluminized steel is also widely used in the automobile industry, particularly for exhaust mufflers. In this case, a high-silicon–aluminum alloy is coated on both sides of the sheet steel by the hot dip method. There are two common weights of coating: the regular is 0.40 oz/ft^2

and the light-weight coating is 0.25 oz/ft² based on coating both sides of the sheet steel. If an arc or gas weld is made on aluminum-coated steel, the aluminum coating is destroyed. It is relatively difficult to replace the aluminum coating; therefore, painting is most often used.

18-5 OTHER METALS

This section includes special steels not covered previously. These metals are abrasion-resisting steel, free-machining steel, manganese steel, silicon steel, and wrought iron.

Abrasion-Resisting Steel

Abrasion-resisting (AR) steel is carbon steel usually with a high-carbon analysis, used as liners in material-moving systems and for construction equipment, where severe abrasion and sharp hard materials are encountered. Abrasion-resisting steels are often used to line dump truck bodies for quarry service, for lining conveyors, chutes, bins, and so on. The abrasion-resisting steel is not used for structural strength purposes, but only to provide lining materials for wear resistance. Steel companies make different proprietary alloys that all have similar properties and, in general, similar compositions. Most AR steels are high-carbon steel in the range 0.80% to 0.90% carbon; however, some are low carbon with multiple alloying elements. These steels are strong and have a hardness up to 40 Rockwell C or 375 BHN. Abrasion-resisting bars or plates are welded to the structures, and when worn out are removed by oxygen cutting or air carbon arc and new plates installed by welding.

Low-hydrogen welding processes are required. Local preheat of 400° F (204° C) is advisable to avoid underbead cracking of the base metal or cracking of the weld. In some cases this can be avoided by using a preheat weld bead on the carbon steel structure and filling in between the bead and the abrasion-resisting steel with a second bead in the groove provided. The first bead tends to locally preheat the abrasion-resisting steel to avoid cracking, and the second bead is made having an oversized throat (see Figure 18–7). Intermittent welds are made since continuous or full-length welds are usually not required. Efforts should be made to avoid deep weld penetration into the abrasion-resisting steel so as not to pick up too much carbon in the weld metal. If too much carbon is picked up, the weld bead may crack.

When using shielded metal arc welding, the EXX16 or EXX18 electrodes are used. When using gas metal arc welding, low penetrating-type shielding gases such as the 75% argon–25% CO_2 mixture should be used. When the flux cored arc welding process is used, self-shielding electrodes are preferred. During cold weather applications, it

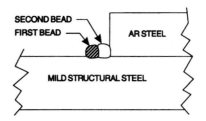

FIGURE 18–7 Preheat bead technique.

is recommended that the abrasion-resistant steel be brought up to 100° F (38° C) temperature prior to welding.

Free-Machining Steels

The term *free machining* is normally associated with steel and brass. Free machining is the property that makes machining easy because small cutting chips are formed. This characteristic is given to steel by sulfur and in some cases by lead. It is given to brass by lead. Sulfur and lead are not considered alloying elements. In general, they are considered impurities in the steel and are purposely added to give free-machining properties. They are difficult to weld.

Free-machining steels are usually specified for parts that require a considerable amount of machine tool work. The addition of the sulfur makes the steel easier to turn, drill, mill, and so on, even though the hardness is the same as a steel of the same composition without the sulfur.

The sulfur content of free-machining steels will range from 0.07% to 0.12% to as high as 0.24% to 0.33%. The amount of sulfur is specified in the AISI specifications for carbon steels. Sulfur is not added to any of the alloy steels. Leaded grades comparable to 12L14 and 11L18 are available. Unless the correct welding procedure is used, the weld deposits on free-machining steel will be porous, may crack, and will not provide properties expected.

The welding procedure for free-machining steels is the same as for carbon steels of the same analysis. These steels usually run from 0.010% carbon to as high as 1.0% carbon. They may also contain manganese ranging from 0.30% to as high as 1.65%. In the case of shielded metal arc welding, use a low-hydrogen electrode of the EXXX18 classification. In the case of gas metal arc or flux cored arc welding, the same type of filler metal is specified as is normally used. Submerged arc and gas tungsten welding is not used on free-machining steels.

The welding procedure should minimize dilution of base metal with the filler metal. Efforts should be made to reduce penetration so as to melt less sulfur or lead.

Free-machining steel can be successfully welded and limited quality welds made; however, the procedures are slower. For this reason, free-machining steel should not be specified for weldments.

Manganese Steel

Manganese steel is sometimes called austenitic manganese steel because of its metallurgical structure. It is also called Hadfield manganese steel after its inventor. It is an extremely tough, nonmagnetic alloy. It has an extremely high tensile strength, a high percentage of ductility, and excellent wear resistance when work hardened. It also has a high resistance to impact and is difficult to machine.

Hadfield manganese steel is widely used as castings but is also available as rolled plate. Manganese steel is popular for impact wear resistance. It is used for railroad frogs, for steel mill coupling housings, pinions, spindles, and for dipper lips of power shovels operating in quarries. It is also used for power shovel track pads, drive tumblers, and dipper racks and pinions.

The composition of austentic manganese is 12% to 14% manganese and 1% to 1.4% carbon. The composition of cast manganese steel would be 12% manganese and 1.2% carbon. Nickel is often added to the composition of the rolled manganese steel.

A special heat treatment is required to provide the superior properties of manganese steel. This involves heating to 1850° F (1008° C) followed by quenching in water. In view of this heat treatment and the material toughness, special attention must be given to welding and to any reheating of manganese steel.

Manganese steel can be welded to itself, and defects can be weld repaired in manganese castings. Manganese steel can also be welded to carbon and alloy steels, and weld surfacing deposits can be made on manganese steels.

Manganese steel can be prepared for welding by flame cutting; however, every effort should be made to keep the base metal as cool as possible. If the mass of the part to be cut is sufficiently large, it is doubtful if enough heat will build up in the part to cause embrittlement. However, if the part is small, it is recommended that it be frequently cooled in water, or, if possible, partially submerged in water during the flame cutting operation. For removal of cracks, the air carbon arc can be used. The base metal must be kept cool. Cracks should be completely removed to sound metal prior to rewelding. Grinding can be employed to smooth the surfaces.

There are two types of manganese steel electrodes available. Both are similar in analysis to the base metal but have added elements that maintain the toughness of the weld deposit without quenching. The EFeMn-A electrode is known as the nickel–manganese electrode and contains from 3% to 5% nickel in addition to 12% to 14% manganese. The carbon is lower, ranging from 0.50% to 0.90%. The weld deposits of this electrode on large manganese castings will result in a tough deposit due to the rapid cooling of the weld metal.

The other electrode used is a molybdenum-manganese steel type EFeMn-B. This electrode contains 0.6% to 1.4% molybdenum instead of the nickel. This electrode is less often used for repair welding of manganese steel or for joining manganese steel itself or to carbon steel. The nickel manganese steel is more often used as a buildup deposit to maintain the characteristics of manganese steel when surfacing is required.

Stainless steel electrodes can also be used for welding manganese steels and for welding them to carbon and low-alloy steels. The 18-8% chrome–nickel types are popular; however, in some cases when welding to alloy steels the 29-9% type is used. These electrodes are more expensive than the manganese steel electrode and are not popular.

When welding manganese steel with manganese type electrodes, the welds should be made with relatively low current and they should be peened with an air hammer as quickly as possible. This helps spread or deform the deposited weld metal and avoids triaxial shrinkage stresses, which can cause cracking. The base metal should be kept cool. Small parts must be cooled frequently or partially submerged in water. The manganese steel electrodes are available as covered electrodes and as bare electrodes. Bare electrodes are not popular. The electrodes are operated with the electrode positive on direct current. Preheating is never employed when welding manganese steels, and should the part become heated to over 500° F (260° C) it must be reheat-treated to retain its toughness.

Silicon Steel

Silicon steels, or, as they are sometimes called, electrical steels, are steels that contain from 0.5% to almost 5% silicon but with low carbon and low sulfur and phosphorus. Silicon steel is provided as sheet or strip. The silicon steels are designed to have lower hysteresis and eddy current losses than plain steel when used in magnetic circuits. Their magnetic properties make silicon steels useful in direct-current fields for most applications. Silicon steel stampings are used in the laminations of electric motor armatures, rotors, and generators. They are widely used in transformers for the electrical power industry and for transformers, chokes, and other components in the electronics industry.

Welding is important to silicon steels since many of the laminations are assembled in packs that are welded together. Figure 18-8 shows an example of welding a stack of laminations. Welds are made on the edge of each sheet to hold the stack together. Welding is done instead of punching holes and riveting the laminations to reduce manufacturing costs. Almost all the arc welding processes are used. The more popular processes are gas metal arc using CO_2 for gas shielding and gas tungsten arc and

FIGURE 18–8 Welded laminations stack of silicon steel.

plasma arc. When the consumable electrode processes are used, the stampings are usually indented to allow for deposition of filler metal. For gas tungsten arc and plasma arc the filler metals are not used and the edges are fused. The size of the weld bead should be kept minimum so that eddy currents are not conducted between laminations in the electrical stack.

One precaution that should be taken in welding silicon steel laminations is to make sure that the laminations are tightly pressed together and that the oil used for protection and in manufacturing is at a minimum. Oil can cause porosity in the welds, which might be detrimental to the lamination assembly.

Wrought Iron

Wrought iron is a ferrous material made of highly refined iron with slag minutely and uniformly distributed throughout. The slag is a form of stringers that are in a longitudinal arrangement in the finished product. Wrought iron has been used for structural applications and for pipe. It provides good corrosion resistance and has been used for piping systems such as hot water coils for radiant heating and brine coils, for cooling ice rinks, and so on. It is also used for certain architectural applications.

Many applications of so-called "wrought iron" are actually made of mild low-carbon steel. Very little wrought iron is manufactured today; however, welding is sometimes required for repair or modifications of existing systems. Wrought iron should be treated exactly the same as low-carbon mild steel and the same welding processes, procedures, and filler metals should be employed. For welding small wrought iron pipe, the oxyacetylene process should be used.

18-6 CLAD METALS

Most clad metals have a cladding metal such as stainless steel, nickel and nickel alloys, or copper and copper alloys welded to a backing material of either carbon or alloy steel. The two metals are welded together at a mill in a roll under heat and pressure. The clad composite plates are usually specified in a thickness of the cladding, which ranges from 5% to 20% of the total composite thickness. The advantage of composite material is to provide the benefits of an expensive material that can provide corrosion resistance and other benefits with the strength of the backing metal. Clad metals were developed in the early 1930s and one of the first used was nickel bonded to carbon steel. The composites are used in the construction of tank cars, heat exchangers, tanks, processing vessels, storage equipment, and so on.

Clad or composites can be made by several different welding manufacturing methods. The most widely used process is roll welding, which employs heat and roll pressure to weld the clad to the backing steel. Explosive welding is also used and weld surfacing or overlay is another method of producing a composite material.

Clad steels can have as the cladding material chromium steel in the range 12% to 15%, stainless steels primarily of the 18–8% and 25–12% analysis, nickel–base alloys such as Monel and Inconel, copper–nickel, and copper. The backing material is usually high-quality steel of the ASTM A285, A212, or similar grade. The tensile strength of clad material depends on the tensile strength of its components and their ratio to its thickness. The clad thickness is uniform throughout the cross section, and the weld between the two metals is continuous throughout.

A different procedure is used for oxygen cutting of clad steel. All of the clad metals mentioned can be oxygen flame cut with the exception of the copper-clad composite material. The normal limit of clad plate cutting is when the clad material does not exceed 30% of the total thickness. However, a higher percentage of cladding may be cut in thicknesses of 1/2 in. (12 mm) and over. The oxygen pressure is lower when cutting clad steel; however, larger cutting tips are used. The quality of the cut is very similar to the quality of the cut of carbon steel. When flame cutting clad material, the cladding material must be on the underside so that the flame will first cut the carbon steel. The addition of iron powder to the flame will assist the cutting operation. Schedules of flame cutting are provided by clad steel producers as well as flame cutting equipment producers. For oxygen flame cutting copper and copper-nickel clad steels, the copper clad surface must be removed and the backing steel cut in the same fashion as bare carbon steel. Copper and brass clad plate can be cut using iron powder cutting. Clad steels can be fabricated by bending and rolling, shearing, punching, and machining in the same manner as the equivalent carbon steels. Clad materials can be preheated and given stress relief heat treatment in the same manner as carbon steels. However, stress-relieving temperatures should be verified by consulting with the manufacturer of the clad material.

Welding Clad Steels

Clad materials can be successfully welded by using special joint details and following special welding procedures. Since the clad material is utilized to provide special properties, it is important that the weld joint retain these same properties. It is also important that the structural strength of the joint be obtained with the quality welds of the backing metal.

The normal procedure for making a butt joint in clad plate is to weld the backing or steel side first with a welding procedure suitable for the base material being welded. Then the clad side is welded with the suitable procedure for the material being joined. This sequence is preferable to avoid the possibility of producing hard brittle deposits, which might occur if carbon steel weld metal is deposited on the clad material. Different joint preparations can be used to avoid the possible pickup of carbon steel in the clad alloy weld. Any weld joint made on clad material should be a full-penetration joint. When designing the joint details, it is wise to make the root of the weld the clad side of the composite plate. This may not always be possible; however, it is more economical since most of the weld metal can be of the less expensive carbon steel rather than the expensive alloy clad metal. This is shown in Figure 18-9 as the preferred type of joint. If the material is of sufficient thickness that a double-groove weld is required, it is recommended that the smaller groove side be the clad side. For material 3/16 in. (4.8 mm) or thinner, a square groove joint detail should be used.

The selection of the welding process or processes used would be based on welding the material in the thickness and position required. Shielded metal arc welding is probably used more often; however, submerged arc welding is used for fabricating large thick vessels, and the gas metal arc welding process is used for medium thicknesses; the flux cored arc welding process is used for the steel side, and gas tungsten arc welding is sometimes used for the thinner materials, particularly the clad side. It is important to select a process that will avoid penetrating from one material into the other. The welding procedure should be designed so that the clad side is joined using the appropriate process and filler metal to be used with the clad metal and the backing side should be welded with the appropriate process and filler metal recommended for the backing metal. Exceptions will be covered later. For code work, the welding procedure must be qualified in accordance with the specification requirements.

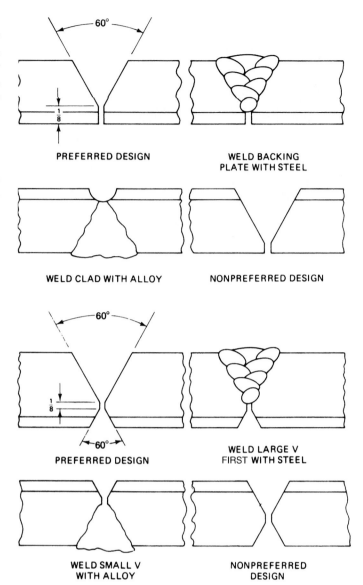

FIGURE 18-9 Weld joint design for clad plate.

The backing side or steel side would be welded first. The depth of the penetration of the root pass must be closely controlled. It is desirable to produce a root pass that will penetrate through the root of the backing metal weld joint into the root face area, yet not come in contact with the clad metal. If penetration is excessive and the root bead melts into the clad material because of poor fitup or any other reason, the deposit will be brittle. If this occurs, the weld will have to be removed and remade. However, if the penetration of the backing steel root bead is insufficient, the amount of back gouging will be excessive and larger amounts of the clad material weld metal will be required. The steel side of the joint should be welded at least halfway prior to making any of the weld on the clad side. If warpage is not a factor, the steel side weld can be completed before welding is started on the clad side.

The clad side of the joint is prepared by gouging to sound metal or into the root pass made from the backing steel side. This can be done by air carbon arc gouging plasma or by gouging chipping. The gouging should be sufficient to penetrate into the root pass so that full penetration of the joint will result. This will determine the depth of the gouging operation. It is also a measure of the depth of penetration of the root pass. Grinding is not recommended since it tends to wander from the root of the joint and may also cover up an unfused root by smearing the metal. If the depth of gouging is excessive, weld passes made with the steel electrode may be required to avoid using an excessive amount of clad metal electrode.

On thin materials the gas tungsten arc welding process may be used; on thicker materials shielded metal arc or gas metal arc may be used. The filler metal must be selected to be compatible with the clad metal analysis. There is always the likelihood of diluting the clad metal deposit by too much penetration into the steel backing metal. Special technique should be used to minimize penetration into the steel backing material. This is done by directing the arc on the molten pool instead of on the base metal. When welding copper or copper–nickel clad steels, a high-nickel electrode is recommended for the first pass (ECuNi or ENi-1). The remaining passes of the joint in the clad metal should be welded so that the copper or copper–nickel electrode matches the composition of the clad metal.

When the clad metal is stainless steel, the initial pass, which might fuse into the carbon steel backing, should be of a richer analysis of alloying elements than necessary to match the stainless cladding. This same principle is used when the clad material is Inconel or Monel. The remaining portion of the clad side weld should be made with the electrode compatible with or having the same analysis as the clad metal. The procedure should be designed so that the final weld layer will have the same composition as the clad metal.

On heavier thicknesses, where the weld of the backing steel is made from both sides, it is important to avoid allowing the steel weld metal to come in contact or to fuse with the clad metal. This will cause a contamination of the deposit, which may result in a brittle weld.

When welding thinner gauge clad plate and inside clad pipe, it may be more economical to make the complete weld using the alloy weld metal compatible with the clad metal instead of using two types of filler metal. The alloy filler metal must be compatible with the steel backing metal. The expense of the welding filler metal may be higher, but the total weld joint may be less expensive because of the more straightforward procedure. Joint preparation may also be less extensive using this procedure. For medium thickness, the joint preparation is a single V or bevel without a large root face. The root face is obtained by grinding the feather edge to provide a small root face. If possible, the face of the weld will be the steel or backing side of the joint. The backing side or steel side is welded first using the small-diameter electrode for the root pass to ensure complete penetration. The remainder of the weld is made on the steel side. The weld is completed by chipping the back side or clad side of the joint and making a final pass from that side (Figure 18–10).

If the composite is a pipe or if it must be welded from one side, the buttering technique should be used. In this case the filler metal must provide an analysis equal to the clad metal and be compatible with the backing steel. Weld passes are made on the edge of the composite to butter the clad and backing metal. The buttering pass must be smoothed to the design dimensions prior to fitup. The same electrode can be used to make the joint.

When the joint is welded from the clad side, the procedure is the same. The filler metal deposit must match the clad metal but be compatible with the backing metal (Figure 18–11). The same basic joint procedure can be used for flat, vertical, horizontal, or overhead weld joints.

When welding heavy, thick composite plate, the U-groove weld joint design is recommended instead of the V-groove, to minimize the amount of weld metal.

When the submerged arc welding process is used for the steel side of the clad plate, caution must be exercised to avoid penetrating into the clad metal. This same

---

FIGURE 18–10 Alloy weld from both sides.

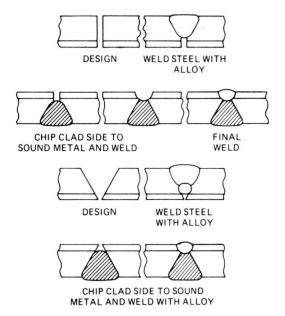

DESIGN   WELD STEEL WITH ALLOY

CHIP CLAD SIDE TO SOUND METAL AND WELD   FINAL WELD

DESIGN   WELD STEEL WITH ALLOY

CHIP CLAD SIDE TO SOUND METAL AND WELD WITH ALLOY

FIGURE 18–11 Alloy weld from clad side.

ALLOY WELD METAL TO MATCH CLAD

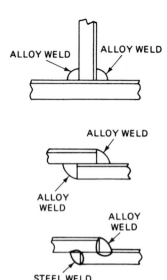

ALLOY WELD   ALLOY WELD

ALLOY WELD

ALLOY
WELD

ALLOY
WELD

STEEL WELD

FIGURE 18–12 T- and lap joints.

caution applies to automatic flux cored arc welding or gas metal arc welding. A larger root face is required, and fitup must be very accurate to control root bead penetration.

The submerged arc process can also be used on the clad side when welding stainless alloys. However, caution must be exercised to minimize dilution of a high-alloy material with the carbon steel backing metal. The proper filler metal and flux must be utilized. To minimize admixture of the final pass, it is recommended that the clad side be welded with at least two passes so that dilution would be minimized in the final pass. When making T-joints, corner joints, or lap joints, special precautions must be taken so as to have the weld metal in proper relationship to the clad and the backing material (Figure 18–12).

Special quality control precautions must be established when welding clad metals so that undercut, incomplete penetration, lack of fusion, and so on, are not allowed. In addition, special inspection techniques must be incorporated to detect cracks or other defects in the weld joints.

18-7 DISSIMILAR METALS

There are many applications where weldments are made from metals of different compositions. A successful weld between dissimilar metals is one that is as strong as the weaker of the two metals being joined. It must possess sufficient tensile strength and ductility so that the joint will not fail. Such joints can be accomplished in a variety of different metals and by many of the welding processes. The problem of making welds between dissimilar metals relates to the transition zone between the metals and the intermetallic compounds formed in this transition zone. For

the fusion welding processes it is important to investigate the phase diagram of the two metals involved. If there is mutual solubility of the two metals, the dissimilar joints can be made successfully. If there is little or no solubility between the two metals to be joined, the weld joint will not be successful. The intermetallic compounds that are formed between the dissimilar metals must be investigated to determine their crack sensitivity, ductility, susceptibility to corrosion, and so on. The microstructure of this intermetallic compound is extremely important. In some cases, it is necessary to use a third metal that is soluble with each metal in order to produce a successful joint.

Another factor is the coefficient of thermal expansion of both materials. If these are widely different, there will be internal stresses set up in the intermetallic zone during any temperature change of the weldment. If the intermetallic zone is extremely brittle, service failure may soon occur.

The difference in melting temperatures of the two metals that are to be joined must also be considered. This is of primary interest when a welding process utilizing heat is involved since one metal will be molten long before the other when subjected to the same heat source. When metals of different melting temperatures and thermal expansion rates are to be joined, the welding process with a high heat input that will make the weld quickly has an advantage.

The difference between the metals on the electrochemical scale is an indication of their susceptibility to corrosion at the intermetallic zone. If they are far apart on the scale, corrosion can be a serious problem.

In certain situations, the only way to make a successful joint is to use a transition material between the two dissimilar metals. An example of this is the attempt to weld copper to steel. The two metals are not mutually soluble, but nickel is soluble with both of them. Therefore, by using nickel as an intermediary metal the joint can be made. Two methods are used: (1) use a piece of nickel, or (2) deposit several layers of nickel alloy on the steel (i.e., butter or surface the steel with a nickel weld metal deposit). The nickel or nickel deposit can be welded to the copper alloy using a nickel filler metal. Such a joint will provide satisfactory properties and will be successful.

Another method of joining dissimilar metals is the use of a composite insert between the two metals at the weld joint. The composite insert consists of a transition joint between dissimilar metals made by a welding process that does not involve heating. Following is a brief description of some of the welding processes that can be used for making composite inserts.

Explosion welding is used to join many so-called incompatible metals. In explosion welding the joint properties will be equal to those of the weaker of the two base materials. Since minimum heat is introduced there is minor melting and no thermal compounds are formed. The characteristic sine wave pattern of the interface greatly increases the interface area. Composites containing a transition joint are commercially available between aluminum and steel, aluminum and stainless steel, aluminum and copper, and other materials.

Cold welding is used for making dissimilar metal transition joints. This process does not use heat and thus avoids the heat-affected zone and the intermetallic fusion alloy. Little or no mixing of the base metals takes place. It is commonly used to join aluminum to copper.

Ultrasonic welding is used for welding dissimilar metals since very little heat is developed at the weld joint. Ultrasonic welding can be used only for very thin materials or small parts.

Friction welding is used for joining dissimilar metals and for making composite transition inserts. Various dissimilar combinations have been welded, including steel to copper-base alloys, steel to aluminum, stainless to nickel-base alloys, and so on. In friction welding only a very small amount of the base metal is heated and that which is melted is thrown from the joint; therefore, the intermetallic material is kept to the very minimum. The heat-affected zone is also minimal.

The high-frequency resistance process is also widely used for dissimilar metal welding. Here the heat is concentrated on the very surface of the parts being joined and pressure applied is sufficient to make welds of many dissimilar materials. It can be used for joining copper to steel at very high speeds.

Diffusion welding is widely used for aerospace applications of dissimilar metals welding. Percussion welding

is also used but this process is restricted to wires or small parts. The laser beam welding process has also been used.

The electron beam and laser welding has had wide application for joining dissimilar metals. They use high-density energy and fast welding speed. This seems to overcome the difference of thermal conductivity when welding metals together having wide variation of thermal conductivity. In addition, the weld zone is extremely small and filler metal is not introduced. Since there is such a small amount of intermetallic compound formed, they offer an advantage for many dissimilar combinations.

The flash butt welding process will make high-quality welds between copper and aluminum. With proper controls all or most of the molten metal is forced out of the joint and the weld is complete as a solid-state process. Flash butt welds are made in rods, wires, bars, and tubes.

Arc Welding Dissimilar Metals

The three popular arc welding processes are most often utilized: shielded metal arc, gas tungsten arc, and gas metal arc welding. The popular combinations of dissimilar metals that are joined are shown in Figure 18–13. The table summarizes the requirement to join aluminum to different metals, copper and copper alloys to different metals, nickel alloys to different metals, stainless steel to carbon steels, and the welding together of various types of steels. All these combinations can be successfully welded using the correct procedures.

| Base Metal Combinations | Welding Process and Filler Metal | | |
	SMAW	GTAW	GMAW
Aluminum to mild and low-alloy steel	Use a transition insert of these metals or coat the surface of the steel with GTAW		
Aluminum to stainless steel	Use a transition insert of these metals or coat the surface of the SS with GTAW		
Aluminum to copper	Use a transition insert of these metals		
Copper to mild and low-alloy steel	ECu	RCu	ECu
Copper to stainless steel	ECuA1–A2	RCuA1–A2	ECuA1–A2
Brass to mild and low-alloy steel	ECuA1–A2	RCuA1–A2	ECuA1–A2
Aluminum bronze to low-alloy steel	ECuA1–A2	RCuA1–A2	ECuA1–A2
Inconel to mild and low-alloy steel	ENiCrFe–3	RNiCrFe–3	ENiCrFe–3
Inconel to austenitic stainless steel	ENiCrFe–3	RNiCrFe–3	ENiCrFe–3
Inconel to ferritic stainless steel	ENiCrFe–3	RNiCrFe–3	ENiCrFe–3
Monel to mild and low-alloy steel	ENiCu–2	RNiCu–2	ENiCu–2
Monel to austenitic stainless steel	ENiCu–2	RNiCu–2	ENiCu–2
Monel to ferritic stainless steel	ENiCu–2	RNiCu–2	ENiCu–2
Ferritic stainless steel to mild and low-alloy steel	ENiCrFe–3 E 309	RNiCrFe–3 E 309–XX	ENiCrFe–3 E 309
Austenitic stainless steel to mild and low-alloy steel	ENiCrFe–3	RNiCrFe–3	ENiCrFe–3
Alloy steel to mild and low-alloy steel	E7018	E70S–X	E70S–X
Q and T steel to mild and low-alloy steel	E7018	E70S–X	E70S–X

FIGURE 18–13 Popular dissimilar metal combinations.

Welding Aluminum to Different Metals

There is a wide difference between the melting temperature of aluminum, approximately 1200°F (649° C), and of steel, approximately 2800° F (1538° C). The aluminum will melt and flow away well before the steel has melted. The aluminum iron phase diagram shows that a number of complex brittle intermetallics are formed. It is found that iron–aluminum alloys containing more than 12% iron have little or no ductility. There is wide difference in the coefficient of linear expansion, in thermal conductivity, and in specific heats of aluminum and steel. This will introduce thermal stresses of considerable magnitude.

The most successful method is to use an aluminum–steel transition insert with each metal welded to its own base metal using any of the three arc welding processes.

The other way is to coat the steel surface with a metal compatible with aluminum. A coating of zinc on steel can be used and the aluminum welded to it by the gas tungsten arc welding process. A high-silicon-aluminum filler wire should be used. Direct the arc toward the aluminum; pulsing will assist the welder.

For welding aluminum to stainless steel, transition inserts are available. It is also possible to use the coating technique. A coating for the stainless steel is pure aluminum coating, which can be applied by dipping clean stainless steel into molten aluminum. Another way to obtain a compatible coating is by tinning the stainless steel with a high-silicon aluminum alloy. The aluminum surface can then be gas tungsten arc-welded to the aluminum. The arc should be directed toward the aluminum; pulsing will assist the welder. The welding of aluminum to copper is accomplished by using a copper–aluminum transition insert piece.

Welding Copper to Various Metals

Copper and copper-base alloys can be welded to mild and low-alloy steels and to stainless steels. For thinner sections, the gas tungsten arc welding process can be used with a high-copper-alloy filler rod. The pulsed mode makes it easier to obtain a good-quality weld. The arc should be directed to the copper section to minimize pickup of iron. In the heavier thicknesses, first overlay or butter the steel with the same filler metal and then weld the overlaid surface to the copper. It is important to avoid excessive penetration into the steel portion of the joint since iron pickup in copper alloys creates a brittle material. The copper must be preheated.

Another method is to overlay the copper with a nickel-base electrode. A second layer is recommended on thicker materials. When making the overlay welds on thick copper, the copper should be preheated to 1000° F (538° C). The overlay or buttered surface of the copper part should be smoothed to provide a uniform joint preparation. Effort should be made to minimize dilution of the copper with the nickel electrode. Copper can also be joined to stainless steel, and brass can be joined to mild and low-alloy steels.

Welding Nickel-Base Alloys to Steels

Nickel-base alloys such as Monel and Inconel can be successfully welded to low-alloy steel by using the Monel filler material and any of the arc welding processes. In the case of Inconel to mild or low-alloy steel, the Inconel-base electrode would be used. The same situation applies also to the welding of Inconel or Monel to stainless steels.

Welding Stainless Steels to Various Metals

Most stainless steels can be successfully welded to mild and low-alloy steels. Consideration must be given to the effects of dilution of the weld metal with the two base metals and the different coefficients of thermal expansion of stainless steel and mild or low-alloy steels. The weld metal deposited will tend to pick up alloys from both parts of the joint. The effect of this dilution can be controlled by buttering or overlaying the surface of one of the metals being joined and by selecting the correct electrode or filler material. The weld joint between a slightly ferritic stainless steel and a mild or low-alloy steel would be hard and brittle if made with a slightly ferritic stainless steel electrode. However, if a fully austenitic stainless steel electrode or filler rod is used the amount of ferrite in the weld metal would be reduced to a tolerable level. The electrode or filler wires normally used would be an E310 electrode or filler rod corresponding to the 310 composition. The stabilized stainless steel electrodes and filler wires should be used. The best selection would be an ENiCr-1, which is a 15% chromium–high-nickel composition. This analysis has an austenitic composition, and the weld deposit can tolerate considerable dilution before becoming crack sensitive.

Austenite stainless steel has a coefficient of thermal expansion about twice that of mild or low-alloy steel. During the cooling cycle of the weld, the stainless steel side will tend to contract more than the mild steel or low-alloy steel side. This difference in contraction will set up stresses in the weld joint. If the weld joint is subjected to repetitive thermal cycles, the resulting stress cycling could cause premature failure similar to fatigue fracture. The use of stainless steel for buttering will not solve this situation. The buttering technique is not recommended for those situations in which repetitive thermal cycles are involved. The best solution is to use a high-nickel electrode such as the ENiCr-1 or the ENiCrFe-2 electrodes or the ERNiCrFe-5 or ERNiCr-3 filler wires. The relatively low iron content of these alloys allows minimum iron in the weld, which reduces cracking. High-nickel deposits have a thermal expansion coefficient similar to that of the low-alloy steel. Thus the thermal stress will be set up in the weld metal and stainless steel rather than the mild or low-alloy steel. Both the weld metal and the stainless steel have good ductility and can absorb the stress cycles without premature failure. When using the stainless steel or the high-nickel overlay, the joint design must be altered to provide space for the buttering layer.

Welding Steel to Different Steels

Steels that have similar metallurgical structures are normally welded with electrodes matching the composition of the lower-strength material. This applies not only to various grades or strength levels of carbon and low-alloy steels but also to various grades of stainless steels. For example, a 316 stainless steel should be welded to a 304 stainless steel with a 308 composition filler metal. The 308 would slightly overmatch the 304 composition.

Another example would be the welding of a quenched and tempered steel to a low-alloy high-strength steel. The electrode normally used for joining the low-alloy high-strength steel to itself should be used for welding it to the quenched and tempered steel. The heat input requirements of the quenched and tempered steel should be followed and in all cases a low-hydrogen deposit is required. Another example is the welding of a low-chrome–moly steel to a plain carbon mild steel. In this case, the standard E7018 electrode or filler metal designed for the carbon or low-alloy steel would be used.

Conclusions

When welding dissimilar metals it is important to consider the problem areas. These relate to the solubility of the metals with one another and the formation of brittle alloys. Second, they relate to the difference in thermal expansion and contraction and the recommended use of ductile weld metals to help absorb these stresses. The buttering technique is most often used. Other techniques involve the plating or coating of one of the base materials with a material compatible to both metals and then making the weld. Bimetallic, composite transition inserts, which can be welded to each type of base metal, are used for certain combinations. Information about joining less common metals can be found in the Welding Research Bulletin, "The Fabrication of Dissimilar Metal Joints Containing Reactive and Refractory Metals" No. 210, Oct 1975.

 **QUESTIONS**

18–1. What property of cast iron makes it difficult to weld? Why?

18–2. Why is a full-penetration weld necessary for repairing cast iron?

18–3. Identify the four types of covered electrodes for cast iron welding. What is the advantage of each?

18–4. Why is preheat usually specified for welding cast iron?

18–5. Tool and die welding is very complex. What are the important factors?

18–6. How are deformed reinforcing bars identified? Why is this important?

18–7. Describe the different types of splices of reinforcing bars.

18–8. Low-hydrogen electrodes are required. What else is required for successfully welding rebars?

18–9. What safety precaution should be taken when welding galvanized steel?

18–10. Why is the galvanized coating damaged adjacent to the weld?

18–11. Why is GTAW not recommended for welding galvanized steel? How can the difficulty be reduced?

18–12. How can the galvanized coating be repaired alongside the weld?

18–13. What welding technique should be utilized when welding free-machining steel?

18–14. Why shouldn't preheat be used when welding Hadfield manganese steel?

18–15. When welding clad metals, which side should have the V-groove? Why?

18–16. Which side should be welded first, the clad side or the steel side?

18–17. When making a T-joint of clad metal, should alloy or steel electrode be used?

18–18. What makes welding different types of metal together more difficult? Why?

18–19. What is a transition piece? Where is it placed in the joint?

18–20. What welding processes are used to produce transition pieces?

19

DESIGN FOR WELDING

OUTLINE

19-1 ADVANTAGE OF WELDED CONSTRUCTION

A weldment is an assembly whose component parts are joined by welding. A weldment can range from a huge structure such as an all-welded ship, the world's tallest structure, the Sears Tower, or an all-welded long-span bridge, to a relatively small item such as a bicycle frame or a coffeepot. Weldments offer many advantages over other design concepts.

1. A weldment is lighter in weight than cast or mechanically fastened structures; it requires less material.

2. The weldment design can readily be modified to meet changing product requirements.

3. The production time for a weldment is less than that of other manufacturing methods.

4. The weldment will be more accurate than a casting with respect to dimensional tolerances.

5. Weldments are more easily machined than castings.

6. Weldments are tight and leakproof and will not shift.

7. The capital investment for producing weldments is lower than that for producing castings. Environmental controls are more easily adapted to the welding shop than to the foundry.

8. Weldments can be more pleasing to the eye than castings. They are cleaner in their lines and usually smoother and more easily prepared for final use.

To join two members by bolting or riveting requires holes in the parts to accommodate the bolts or rivets. These holes reduce the cross-sectional area of the members to be joined by up to 10%. The joint may also require the use of one or two gusset plates, thus increasing the weight of material required and the cost. This expense can be eliminated by the use of a weld. The greatest economy of a welded design will be obtained if the cross-sectional area of the entire structural member is reduced by the amount of the bolt holes. This can be done since the entire cross section of a member of a welded design is utilized to carry the load. The amount of material required is reduced (Figure 19-1) as well as its cost. This same design concept applies to joining plates used to build a ship or a container. In view of this material savings, ships and storage tanks are no longer riveted.

Pipes joined by welding offer major economies. The wall thickness of a pipe should be heavy enough to carry the required load. However, if the pipe is joined by screw threads, a heavier wall thickness is used to allow for cutting away a portion of the thickness for the threads (Figure 19-2). A thinner pipe wall thickness is used for the entire welded pipe system. This reduces the amount of metal required and the cost. The inside surface of the welded joint is smoother. Large-diameter pipes are no longer connected using screw threads and pipe fittings.

Converting castings to weldments allows the designer to reduce weight by reducing metal thickness. Welding is a design concept that allows freedom and flexibility not possible with cast construction. Heavy plates can be used where strength is required and thin ones where possible. The uniform thickness rule and minimum thickness required for foundry practice are not necessary for weldments. Additionally, high-strength materials can be used in specific areas.

The weldment usually offers advantages over a casting for the same function.[1] Figure 19-3 shows a welded and cast truck brake shoe. The weight of the rough casting is 23 lb (10.5 kg) and of the weldment is 13 lb (5.8 kg).

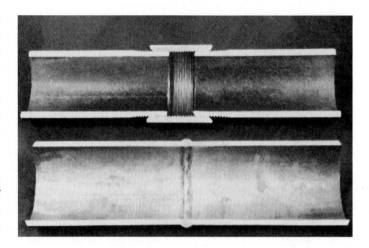

FIGURE 19–2 Pipe joints welded and threaded.

FIGURE 19–3 Welded and cast truck brake shoe.

FIGURE 19–1 Comparison of welded and riveted structural joints.

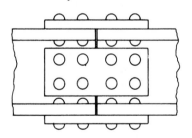
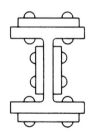

RIVETED SPLICE IN WIDE FLANGE MEMBER

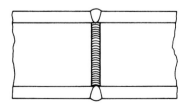

WELDED SPLICE IN WIDE FLANGE MEMBER

The extra material and machine work penalizes the casting design. Quantity production is required to justify the blanking dies of the web plates and for automating the welding operation.

Additional advantages of weldments include the fact that welding is the lowest-cost joining method, joins all commercial metals, and can be used anywhere. It provides design flexibility; that is, the correct metals can be used where required.

Limitations of weldments include the fact that some welding depends on the human factor and welding often requires internal inspection. These limitations can be overcome by good controls and supervision and nondestructive evaluation.

19-2 WELDMENT DESIGN FACTORS

Designers of weldments must have knowledge of the various welding processes and welds and welding joint designs. They need to know how materials are prepared for welding, as well as the actual making of welds. This is complicated because of the many types of designers that utilize welding. For example, the civil engineer designs bridges and similar structures; the structural engineer designs high-rise buildings and factories; the mechanical engineer designs pressure vessels, piping systems, and machinery components; the chemical engineer designs vessels, process components, and piping systems; the marine architect designs ships and boats; and the electrical engineer designs electrical generating and electrical machinery. Each of these design specialists must know the details of weldment design. In addition, many other products that are designed require knowledge of welding. The primary objective of all designers is to design the part to fulfill its design function, but also to design a weldment that is an efficient, cost-effective part that fulfills the design requirements.

The economical solution to a design requirement will be accomplished if the designer takes into consideration the following points:

1. The total service requirements
2. The types of loadings and methods of accurately calculating stresses
3. The allowable working stresses
4. The mechanical and physical properties of materials employed
5. The capabilities of the welding processes and the weld deposit properties
6. Joint types and weld design
7. Fabrication methods available
8. The cost of welding using the variety of processes available
9. Clear communication of design
10. Quality specifications and inspection techniques

The intelligently designed weldment will always be less expensive than the mechanically assembled parts or castings designed for the same function. If this is not true, it is an indication that the weldment design is poor or that other factors are involved. The best design involves ingenuity, but the resultant welded designs are worth this effort.

The designer is expected to design a product that will function properly under the service conditions encountered. The designer must be completely aware of the properties of the material involved and how they must be treated in fabricating and welding. The design, the materials, and the production procedures are interrelated and must all be considered when making a design. Codes or specifications must also be considered when they are involved. Weldment design is a particularly difficult process since weldments cover such a wide variety of parts with such a wide variety of service requirements. In addition, the economic factors must be considered. The first cost of the weldment may not always be the determining factor since maintenance and repair costs over the service life may be more important than the initial costs. Weight reduction is desirable, especially for moving structures, since this can provide for increased payloads and less energy requirement.

The total service requirements must be known. It is absolutely essential that the design meet all these requirements. This can include dynamic loading, fatigue loading, or impact loading, the service temperature (whether hot or cold), and the thermal temperature changes in the service. Another factor is corrosion resistance under normal and extended use and resistance to abrasion of all types.

A major problem is the type of loading and methods of accurately calculating stresses. It is beyond the scope of this book to develop how loads are imposed on structures and how they are calculated. Special design procedures provide this information. It is necessary to consider both static and dynamic loads, including impact and fatigue. The term *loads* is identified by referring to Figure 19–4, the five basic types of loads. A weldment is a monolithic structure, and therefore any load imposed anywhere on the weldment will be transmitted throughout the weldment.

The allowable working stress, sometimes called *allowable unit stress,* is the maximum stress level that is allowed anywhere within a weldment. Stress is calculated using standard engineering principles based on the type of product or structure being designed. The mechanical and physical properties of materials are also important. The designer has the responsibility to select materials to meet the design requirements. This includes materials with the correct properties to withstand loads but also to withstand corrosion, abrasion, and other factors.

The capabilities of welding processes and of the weld deposit must be known. In general, most weldments are designed to be manufactured by the use of an arc welding process. It is normal practice to assume that the weld joint of a full-penetration weld will equal or exceed the properties of the base material employed. Codes and specifications may apply and provide specific factors that must be used.

Joint types and weld designs must be established. It is essential that they be specified to carry the loads. Loadings greatly influence the type of weld joint that should be employed, specifically full-penetration welds versus partial-penetration welds. It is common practice to assume that the weld is stressed evenly over its entire area. This is not always true. Fillet welds are the most widely

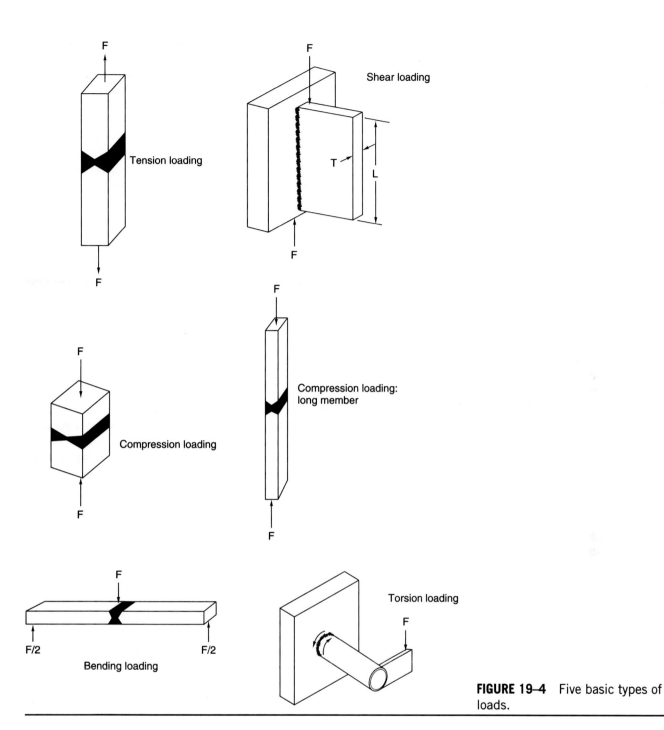

FIGURE 19–4 Five basic types of loads.

used welds, and they usually fail through the throat. Stresses in a fillet weld can be longitudinal or transverse shear, as shown in Figure 19-5. Unfused sections of the joint are stress risers.

The designer must understand the cost factors when welding with different processes. This relates to the method of application and thus the labor cost. The more mechanized the process, the less labor is involved. However, this depends on the quantity of production and the capital investment allowed for automation. Different processes also have different productivity factors. The

most productive process that is practical for the part should be employed.

Fabrication facilities in the factory have an influence on design. Most parts for many weldments are flame cut or blanked from hot-rolled steel plate. They are sometimes then bent or rolled to design requirements. The designer must know about the facilities that are available, whether the work must be done hot or cold, and so on.

Clear communication of designs is essential. It is absolutely necessary that the designer provide complete drawings and procedures for manufacturing a product.

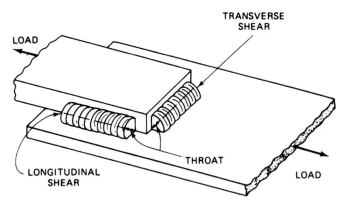

FIGURE 19–5 Stresses in a fillet weld.

The most common method for welding design information interchange is by the use of welding symbols. Weld symbols also indicate the edge preparation of parts entering the weldment. This involves bevels of different types. The type of equipment used to provide joint preparations is important and must be known by the designer and communicated to the shop. There should be sufficient information so that each part will be prepared properly before going into the weldment. Standardized data and supplementary information are often required to provide clear communication for parts preparation for welding and for continuing operations.

Quality specifications and inspection techniques must be designated. The designer must indicate any particular joints where high quality must be assured by means of nondestructive testing. Welding symbols can be combined with nondestructive examination symbols for this purpose. The designer must also have the knowledge of the materials being used and if they need further heat treatment following the welding operation. This can include stress relieving, normalizing, or heat treatments of various types required to provide the properties required.

One of the most important functions of the designer is to design for dynamic or low-temperature or impact loadings. The designer must also indicate the life of the product based on service requirements. As far as welding is concerned, the most important factor is the use of full-penetration welds versus partial-penetration welds. For static loading, partial-penetration welds are ample. However, for other types of loading it is important to design for full-penetration welds based on possible stress concentrations in the design.

There are at least four types of stress concentrations that must be avoided. There are structural notches in the overall total design of the weldment (see the ship failure report described in Section 24-1). There are weld joint stress concentrations designed into the joint. This includes partial penetration welds, two fillet welds with an unfused area in a T-joint, and so on. There are workmanship stress concentration notches such as incomplete penetration welds where it was designed for full pene-

tration, sharp corners of fillets, undercut, and so on. Finally, we note one of less significance: stress concentrations at metallurgical notches when materials of different properties are brought together.

Stress Concentrations

When considering the design of a weldment, special emphasis must be placed on stress concentrations—that is, the distribution of stress at notches or discontinuities. When the parts of a weldment are welded together, they act as a monolithic structure. This means that stresses are spread throughout the entire structure when transmitting a force from one point to another within the structure. Once a piece is welded into the weldment it becomes a part of the weldment and will carry a portion of the load. Tensile stresses are normally thought to be uniformly distributed throughout the entire cross section of a member. This is usually a safe assumption for simple statically loaded structures. However, complex structures exposed to dynamic loading, repetitive loading, and impact loading will not have a uniform stress pattern across the cross section of the structure. The effect of the nonuniformity is more important in fatigue and cold weather service.

Under some service requirements external loads may fluctuate or may be applied *repetitively* thousands of times. Most metals exhibit a lower ultimate strength under the application of repetitive loads, and it is normal practice to reduce the working unit stress when cyclic or repeated loading is applied. This involves the fatigue life of the metal and depends on the stress cycle imposed. The types of stress cycle refer to whether there is a complete reversal of stress from tension to compression or whether it is a loading from a minimum to a maximum amount of either tension or compression. Most codes specify an allowable fatigue stress.

Suddenly applied loads are called *impact loads,* and the suddenness of the application of the load is a matter of the degree of impact. The effect of impact is to immediately increase the internal stress in the weldment. These internal stresses may be localized and cause problems. The design calculations are made on the basis of equivalent static load conditions. It is normal practice to allow for the effect of impact-producing stresses by the use of factors based on static loading. The factor is an across-the-board cut in the allowable unit stress of the total structure or weldment.

A stress concentration is a point within the structure at which the stresses will be more concentrated than throughout the remaining cross-sectional area of the weldment. Stresses at a notch, for example, can be two to four times as great as the uniform stresses throughout the remaining portion of the structure. For weldments loaded dynamically, repetitively, or by impact, it can create a point of premature failure. A simple example of stress concentration based on the presence of a notch is shown in Figure 19-6. On the left, the bar stressed in tension will

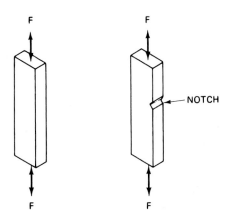

FIGURE 19–6   Bar with and without notch.

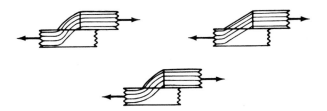

FIGURE 19–7   Single fillet lap weld in tension.

have uniform stress throughout its cross section. The bar on the right has a sharp notch in the edge, and the stresses will be concentrated at the root of the notch; even though the stresses were distributed uniformly at the ends of the bar, they cannot be transmitted through the notch. The bar with the notch will have service life considerably less than that of the bar not containing the notch.

A design notch is designed into a weldment as in an abrupt change of section. A practical example is a square hatch opening in the deck of a ship. The cross section at the point where the hatch opening occurs changes abruptly and there is a concentration of stresses. The cross section changes drastically at the point where the deckhouse is welded to the deck. It might be argued that the deckhouse is not expected to carry any of the load or stress. It will, since it is welded and becomes integral to the other parts of the hull. The intersection of members, reentrant angles, abrupt reinforcing structures, and square holes are all examples of notches in the design of weldments.

Notches also occur in welds.[2] Any weld joint design that provides partial penetration includes a notch. A design notch in a weld joint would be the incomplete fusion area at the center or root of a weld. Fillet welds used for lap joints are notch prone. See Figure 19-7 for examples of stress concentration when the joint is loaded in tension. T-joints made with two fillet welds include a notch. The worst notch is the single fillet welded T-joint. Figure 19-8 shows these T weld joint details and how different designs contain notches. This figure shows three joint designs, the stress paths, and a comparison of their relative static tensile strength, resistance to fatigue, and impact strength. The static strength is determined by the area of the weld, the other factors by tests. Butt joints are less notch prone because of their geometry. Full-penetration welds produce highly efficient butt joints. However, partial-penetration welds contain notches at the center of the weld or at the outer surface. Four examples are shown in Figure 19-9. It is easy to determine whether a notch is designed into the joint by drawing a cross section of the weld joint indicating the paths that the lines of stress must follow. Notches can also occur in weld joints when the section changes abruptly. Figure 19-10 shows butt joints between a thick and a thin member. There are several ways to provide a smooth stress flow through the weld. It is also possible to make two stress concentration points. The left joint would create two points of stress concentration, one at the toe of the fillet

FIGURE 19–8   T-joints: stress pattern.

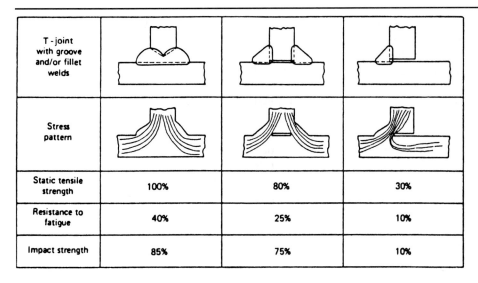

T - joint with groove and/or fillet welds			
Stress pattern			
Static tensile strength	100%	80%	30%
Resistance to fatigue	40%	25%	10%
Impact strength	85%	75%	10%

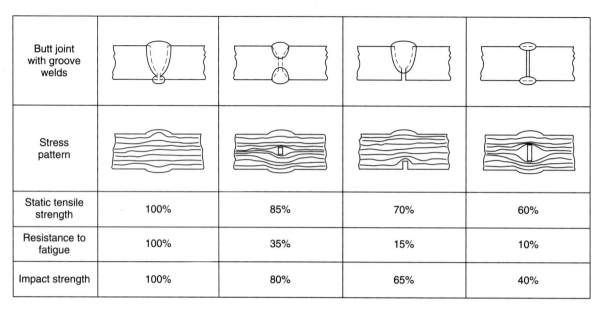

Butt joint with groove welds				
Stress pattern				
Static tensile strength	100%	85%	70%	60%
Resistance to fatigue	100%	35%	15%	10%
Impact strength	100%	80%	65%	40%

FIGURE 19–9 Butt joints: stress pattern.

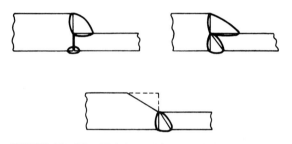

FIGURE 19–10   Thick-to-thin butt joint.

and one at the unfused root. The right joint would be a better solution and would provide a relatively smooth stress flow. The joint on the bottom would also provide a relatively smooth stress flow and would be less expensive since less weld metal is required.

Workmanship notches can be troublesome and difficult to control. These occur when the welds are not full-penetration welds, even though they are designed to be. The root opening may have been eliminated by an accumulation of tolerances, or back gouging and back welding may have been omitted. A fillet weld at the point of joining a thin section to a thick section may not be full size and will create a notch.

The fourth type of notch is in the minority and creates the least trouble. These are metallurgical notches that may be caused by joining metals of different yield strengths or by welding on hardenable steels and creating extremely hard spots in welds. Abrupt change in strength along a cross section can have the effect of a notch even though there is no abrupt change in geometry.

In the design and construction of weldments subjected to dynamic loads and low-temperature service, every effort must be taken to provide a smooth flow of

stress lines throughout the weldment. This becomes more important when high-strength materials are used.

Fatigue of Welded Structures

Fatigue failures in engineered structures usually occur at connections where a change of section or shape occurs. They typically initiate at discontinuities or stress risers associated with joints. They are the result of design or workmanship errors. Major design errors can be corrected by experienced designers or by fatigue element analysis. Welding design errors may not be as evident. For example, fatigue cracks may initiate at the toe of fillet welds or from a partial-penetration joint. These factors were discussed in the last section and should be corrected. Faulty workmanship is another source of discontinuities that may initiate failure. These include lack of penetration or lack of fusion, porosity or slag inclusions, or poor weld geometry, especially in fillets if the weld joint is oriented perpendicular to the cyclic stresses. Close visual inspection and nondestructive examination will help alleviate the problem. Another factor is residual stresses that are balanced in the weldment. Weldments almost always contain residual stresses.[3]

Fatigue failure is divided into initiation, propagation, and terminal fracture. Total fatigue life is governed by the applied loads, geometry of the part, and the joint design. It also includes workmanship faults, environment, and the fracture characteristics of the material. The magnitude of tensile residual stresses should be controlled, especially in thick, highly constrained weldments made of low fracture toughness material. To greatly reduce stresses, one solution is stress relief heat treatment, which is recommended if possible for critical components.

Rigid-Frame Structures

Rigid-frame construction, also called *continuous construction,* is a design system incorporating *plastic analysis.* This type of design and construction is specifically suited to welded construction. The members are welded directly together rather than through connection plates, gussets, or filler plates. The material in a rigid or continuous welded frame is utilized more efficiently because the bending moments are distributed better. This type of construction provides the maximum degree of end restraint. Rigid frame design allows large savings of steel since it eliminates the connecting plates, gussets, and so on. It requires a thorough analysis and should be done only by designers well qualified to perform this type of work.

In machine members the rigid connection concepts or plastic analysis provide greater dimensional accuracy. The weldment will maintain correct alignment and dimensional accuracy throughout its service life. This type of design and construction should be used for all welded machine parts unless they must be disassembled.

19-3 WELDING POSITIONS AND WELD ACCESSIBILITY

Most welds made in the field must be made in the position in which they are found. Other weldments are too large to be moved and the welds must be made in the position they are found. This involves welding on the ceiling, on the corner, or on the floor—in other words, in the position in which the part will be used. These welding positions must be accurately described. AWS has defined the four basic welding positions as follows:

1. *Flat welding position:* the welding position used to weld from the upper side of the joint at a point where the weld axis is approximately horizontal and the weld face lies in an approximately horizontal plane

2a. *Horizontal welding position, fillet weld:* the welding position in which the weld is on the upper side of an approximately horizontal surface and against an approximately vertical surface

2b. *Horizontal welding position, groove weld:* the welding position in which the weld face lies in an approximately vertical plane and the weld axis at the point of welding is approximately horizontal

3. *Overhead welding position:* the welding position in which welding is performed from the underside of the joint

4. *Vertical welding position:* the welding position in which the weld axis, at the point of welding, is approximately vertical and the weld face lies in an approximately vertical plane

There are several foreign definitions that are different from those of the AWS but are commonly used. The British and others use the term *downhand* to describe the flat position. They also use the term *horizontal-vertical* to describe welds between two plates, one approximately in the horizontal plane and one in the vertical plane, when the axis of the weld is approximately horizontal. The positions are identified further in Figure 19-11, which shows welding positions for fillet and groove welds. Here F represents fillets; 1, flat; 2, horizontal; 3, vertical; and 4, overhead. The same numbers apply to groove welds, which are designated with the letter G.

Pipe weld joints are a special case and are identified in Figure 19-12. They are normally groove welds and they are indicated by the letter G. Position 1G is roll welding with the axis of pipe horizontal, but with the welding done in the flat position with the pipe rotating under the arc. Position 2G is known as horizontal welding; the axis of the pipe is in the vertical position with the axis of the weld in the horizontal position. There is no 3G or 4G position on pipe welding. Position 5G is known as the multiple position. The axis of the pipe is horizontal, but the pipe is not turned or rolled during the welding operation. Position 6G for pipe has the axis of the pipe at 45° and the pipe is not turned while welding. For qualification work a 6G restricted position is often used. Restricted accessibility is provided by a restriction ring placed near the weld. It is called 6GR. The axis of the pipe may vary +15° for the 1G, 2G, and 5G test positions, but only +5° for the 6G position. Square and rectangular tubing is accommodated in the 2G and 5G test positions.

The official AWS diagrams for welding positions are precise. They utilize the angle of the axis of the weld, which is "a line through the length of the weld perpendicular to the cross section at its center of gravity." Figure 19-13 shows the fillet weld and the limits of the various positions. It is necessary to consider the inclination of the axis of the weld as well as the rotation of the face of the fillet weld.

Figure 19-14 shows the groove weld positions in the same manner. The inclination of the axis of the groove and fillet weld is the same as far as limits are concerned. For the flat position the rotation of the face of the weld is the same for fillet and groove welds. However, the rotation of the face of the horizontal, vertical, and overhead groove welds are different. The design of a joint is normally changed whenever the welding position or type of backing is changed. In general, narrower included angles are used for other than flat-position groove welds. Welds made in the horizontal position usually have a flat face on the bottom member and a beveled face on the upper member. When backing strips are used, the root opening is usually wider. Specific joint details for different thicknesses and positions are shown in the next section.

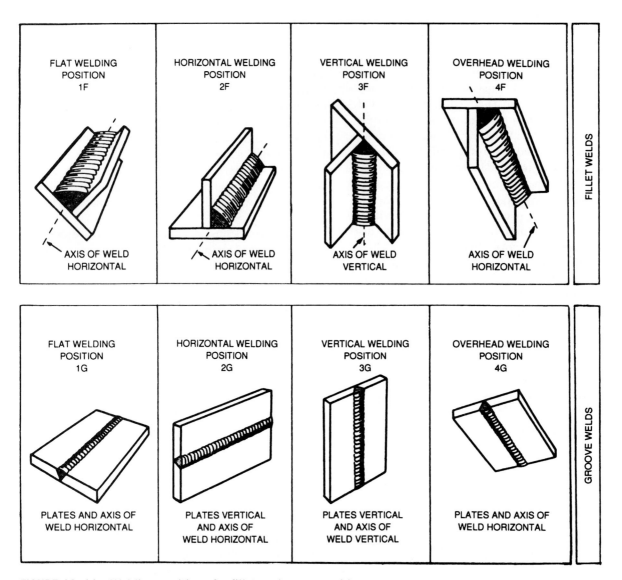

FLAT WELDING POSITION 1F · AXIS OF WELD HORIZONTAL

HORIZONTAL WELDING POSITION 2F · AXIS OF WELD HORIZONTAL

VERTICAL WELDING POSITION 3F · AXIS OF WELD VERTICAL

OVERHEAD WELDING POSITION 4F · AXIS OF WELD HORIZONTAL

FILLET WELDS

FLAT WELDING POSITION 1G · PLATES AND AXIS OF WELD HORIZONTAL

HORIZONTAL WELDING POSITION 2G · PLATES VERTICAL AND AXIS OF WELD HORIZONTAL

VERTICAL WELDING POSITION 3G · PLATES VERTICAL AND AXIS OF WELD VERTICAL

OVERHEAD WELDING POSITION 4G · PLATES AND AXIS OF WELD HORIZONTAL

GROOVE WELDS

FIGURE 19–11 Welding positions for fillet and groove welds.

The welding position must always be accurately described. It is an important variable in any welding procedure. It is especially important with respect to training and qualifying welders and must always be given consideration when selecting a welding process. The position must be considered in the design of a joint. A good example of this is the design of weld joints to splice columns in steel frame buildings. It is normal practice to have the bottom side of the joint flat with the bevel on the upper piece. When backing strips are used, the root opening is usually wider. Another important factor that designers must consider is the accessibility for making the weld. The weld joint must be accessible to the welder. In other words, it must be possible to make the weld. Figure 19–15 shows several examples of inaccessible welds. Welds cannot be made on the inside of small-diameter

pipe or inside box columns. Welds shown in the structural detail through a channel would be impossible to make as shown. There are other examples of inaccessible welds that can be made on paper but not in the shop.

The sequence of assembly of a weldment has an effect on accessibility. The welding area can be covered up as additional parts are assembled in the weldment. It is advantageous if the weldment can be designed so that all weld joints are accessible for welding after the weldment has been completely assembled. Missed welds of this type can lead to premature failures of the weldment.

It is also important to consider the back welding that may be required for a particular joint. In some cases back gouging and welding is required for a fully penetrating high-quality joint. Accessibility must be available to the back side of the joint. When the back side of the

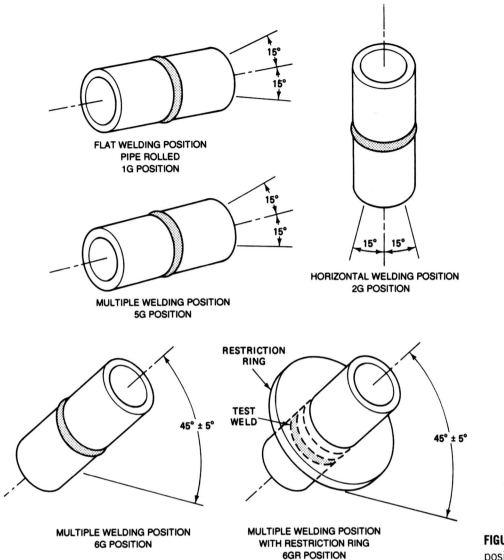

FLAT WELDING POSITION
PIPE ROLLED
1G POSITION

15°
15°

MULTIPLE WELDING POSITION
5G POSITION

15°
15°

15° 15°

HORIZONTAL WELDING POSITION
2G POSITION

45° ± 5°

MULTIPLE WELDING POSITION
6G POSITION

RESTRICTION
RING

TEST
WELD

45° ± 5°

MULTIPLE WELDING POSITION
WITH RESTRICTION RING
6GR POSITION

FIGURE 19–12 Welding positions for pipe welds.

weld is not accessible, the designer must consider *one-side welding*. This technique is used when the weld must be made completely from one side of the joint. An example of a one-side weld is a weld on small-diameter pipe. It is impossible to get to the back side of the joint; therefore, the joint must be made completely from the outside of the pipe. That is usually done with an open root. A one-side weld usually requires some type of backing. This is covered more completely in Section 26–3. Briefly, there are different types of backing and techniques for making complete penetration welds from one side. Consumable inserts are in common use for pipe welding. Backing straps that become part of the joints are common. Fluxes, flux-coated tape, and flux-filled bars are all used, and water-cooled copper bars are widely used in automatic applications.

Another issue that designers must consider is the practical problem of weld distortion. Distortion can occur in any weldment and can create a problem since moving parts may close up root openings and joints. The welders must be careful to avoid unfused roots in joints. This is particularly troublesome when the design requires full-penetration welds. To understand this problem completely, it is advantageous if designers have some shop experience. Bracing of the weldment and distortion correction have their place. However, it is sometimes possible to reduce distortion by means of double welds or welds made on opposite sides of the centerline of the joint or weldment, to minimize angular distortion. Weld sizes also affect distortion and it is advisable to use the smallest welds possible. This subject is covered more completely elsewhere in this book.

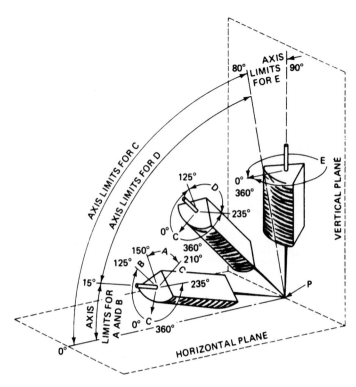

TABULATION OF POSITIONS OF FILLET WELDS			
POSITION	DIAGRAM REFERENCE	INCLINATION OF AXIS	ROTATION OF FACE
FLAT	A	0° TO 15°	150° TO 210°
HORIZONTAL	B	0° TO 15°	125° TO 150°
			210° TO 235°
OVERHEAD	C	0° TO 80°	0° TO 125°
			235° TO 360°
VERTICAL	D	15° TO 80°	125° TO 235°
	E	80° TO 90°	0° TO 360°

FIGURE 19–13   Welding positions for fillet welds.

FIGURE 19–14   Welding positions for groove welds.

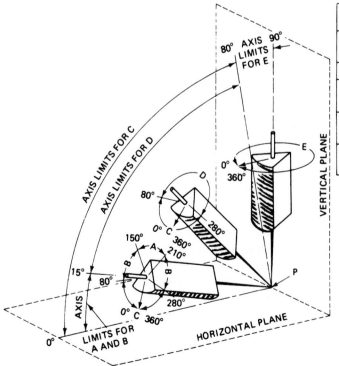

TABULATION OF POSITIONS OF GROOVE WELDS			
POSITION	DIAGRAM REFERENCE	INCLINATION OF AXIS	ROTATION OF FACE
FLAT	A	0° TO 15°	150° TO 210°
HORIZONTAL	B	0° TO 15°	80° TO 150°
			210° TO 280°
OVERHEAD	C	0° TO 80°	0° TO 80°
			280° TO 360°
VERTICAL	D	15° TO 80°	80° TO 280°
	E	80° TO 90°	0° TO 360°

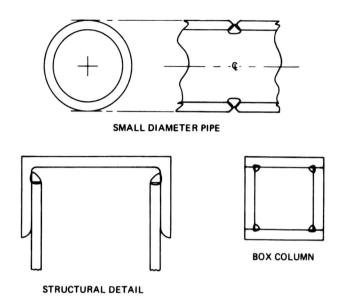

FIGURE 19–15 Inaccessible welds (impossible to make).

SMALL DIAMETER PIPE

STRUCTURAL DETAIL

BOX COLUMN

19-4 DESIGN OF WELD JOINTS AND WELDS

Welds are made at the junction of all of the pieces that make up the weldment. The junction of parts is called a joint, defined as "the junction of members or edges of members which are to be joined or have been joined." Parts being joined to produce a weldment may be hot-rolled plates, structural shapes, pipe, castings, or forgings. It is the placement of these members that creates the joints. Five basic joints are used for bringing two members together for welding. Definitions of the terms used to describe these joints are similar to those used by various crafts. The figure shows the relationship between the two members as well as the location of the joint between them. They are shown in Figure 19–16 and are defined as follows:

A. *Butt joint:* a joint between two members aligned approximately in the same plane

B. *Corner joint:* a joint between two members located approximately at right angles to each other in the form of an L

C. *T-joint:* a joint between two members located approximately at right angles to each other in the form of a T

D. *Lap joint:* a joint between two overlapping members located in parallel

E. *Edge joint:* a joint between the edges of two or more parallel or nearly parallel members.

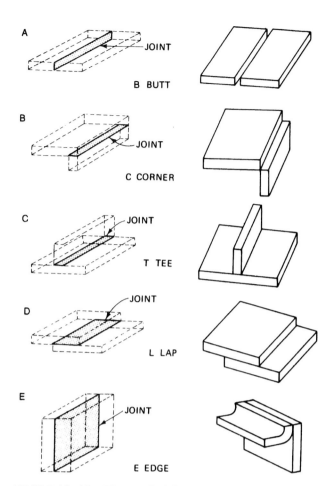

FIGURE 19–16 Five basic joint types.

When more than two members are brought together, the joint is a combination of one of the five basic joints. The most popular joint is the *cross* or *cruciform joint,* a joint between three members at right angles to each other in the form of a cross. It is actually a double-T joint.

Weld Types

There are eight separate and distinct welds, shown in Figure 19–17, some of which have variations. In addition, welds can be combined.

1. *Fillet weld.* This is the most commonly used weld. The fillet weld is so named because of its cross-sectional shape. The fillet is regarded as being *on the joint* and is defined as a "weld of approximately triangular cross section joining two surfaces approximately at right angles to each other." Details of the fillet weld are shown in Figure 19–18. Variations of the fillet are shown in Figure 19–19.

2. *Groove weld.* This is the second most popular weld. It is defined as a "weld made in the groove between

FILLET
Most popular of all welds
(may be single or double)

GROOVE
Second most popular—may
be single or double—has
many variations

Eleven
types

BACK OR BACKING WELD
Bead type back or backing
welds of single groove welds

PLUG OR SLOT WELD
Used with prepared holes

SPOT OR PROJECTION WELD
Used without prepared holes
Use arc or resistance

SEAM WELD
Continuous—use arc or
resistance

STUD WELD
Special application
welding process

Stud

SURFACING WELD
Surface built up by welding

FIGURE 19–17   Eight basic welds.

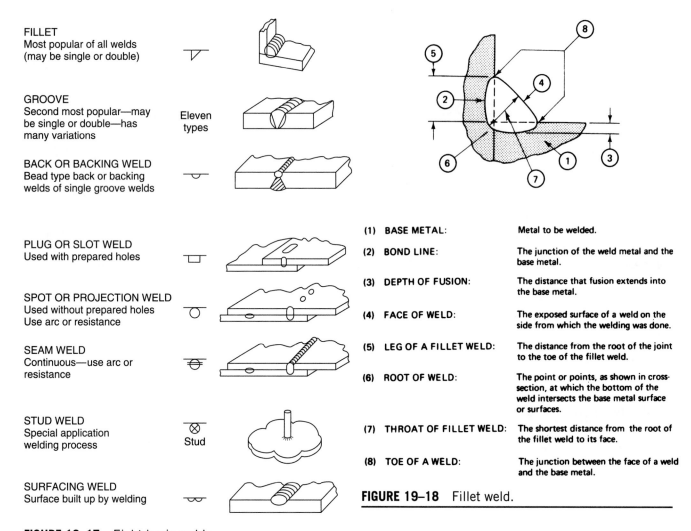

(1)	**BASE METAL:**	Metal to be welded.
(2)	**BOND LINE:**	The junction of the weld metal and the base metal.
(3)	**DEPTH OF FUSION:**	The distance that fusion extends into the base metal.
(4)	**FACE OF WELD:**	The exposed surface of a weld on the side from which the welding was done.
(5)	**LEG OF A FILLET WELD:**	The distance from the root of the joint to the toe of the fillet weld.
(6)	**ROOT OF WELD:**	The point or points, as shown in cross-section, at which the bottom of the weld intersects the base metal surface or surfaces.
(7)	**THROAT OF FILLET WELD:**	The shortest distance from the root of the fillet weld to its face.
(8)	**TOE OF A WELD:**	The junction between the face of a weld and the base metal.

FIGURE 19–18   Fillet weld.

two members to be joined." The groove weld is re-garded as being *in the joint*. There are 11 basic groove weld designs, and they can be used as single or double welds. The details of the groove weld are shown in Figure 19-20. Different types of groove welds are shown in Figure 19-21.

3. *Back or backing weld.* This is a special weld made on the back side or root side of a previous weld. The root of the original weld is gouged, chipped, or ground to sound metal before the backing weld is made. This will improve the quality of the weld joint by assuring complete penetration. It, in itself, cannot make a joint.

4. *Plug or slot welds.* These are made using prepared holes. They are considered together since the welding symbol to specify them is the same. The important difference is the type of prepared hole in one of the members being joined. If the hole is round, it is a plug weld; if it is elongated, it is a slot weld.

FIGURE 19–19   Fillet weld throat dimension.

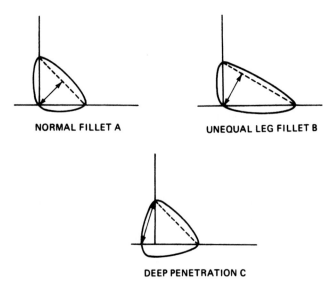

NORMAL FILLET A

UNEQUAL LEG FILLET B

DEEP PENETRATION C

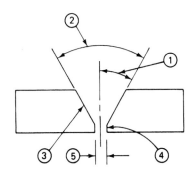

(1) BEVEL ANGLE: The angle formed between the prepared edge of a member and a plane perpendicular to the surface of the member.

(2) GROOVE ANGLE: The total included angle of the groove between parts to be joined by a groove weld.

(3) GROOVE FACE: The surface of a member included in the groove.

(4) ROOT FACE: That portion of the groove face adjacent to the root of the joint.

(5) ROOT OPENING: The separation between the members to be joined at the root of the joint.

FIGURE 19–20 Groove weld.

FIGURE 19–21 Types of groove welds.

I			SQUARE
Y			SINGLE-V
X			DOUBLE-V
			SINGLE-BEVEL
K			DOUBLE-BEVEL
			SINGLE-U
			DOUBLE-U
			SINGLE-J
			DOUBLE-J
			FLARE-V (MAY BE DOUBLE)
			FLARE-BEVEL (MAY BE DOUBLE)

5. *Spot or projection weld.* This is shown by the same weld symbol. These welds can be applied by different welding processes, which change the actual weld. For example, when the resistance welding process is used, the weld is at the interface of the members being joined. If the electron beam, laser, or arc welding process is used, the weld melts through one member into the second member.

6. *Seam weld.* This weld in cross section looks similar to a spot weld. The weld geometry is influenced by the welding process employed. With resistance welding, the weld is at the interface between members being joined. With the electron beam, laser, or arc welding process, the weld melts through the one member to join it to the second member. There are no prepared holes in either the spot or the seam weld.

7. *Stud weld.* This is a special type of a weld produced by a stud welding process, used for joining a metal stud or similar part to a workpiece.

8. *Surfacing weld.* This weld is composed of one or more stringer or weave beads deposited on base metal as an unbroken surface. It is not used to make a joint. It is used to build up surface dimensions, to provide metals of different properties, or to provide protection of the base metal from a hostile environment.

Weld Joints

To produce weldments it is necessary to combine the joint with the weld to produce weld joints for joining the separate members. Each weld cannot always be combined with each joint to make a weld joint. Figure 19–22 shows the welds applicable to the basic joints. Since fillet welds and groove welds can be combined, they have more possibilities and their use is more complex.

Weld Joint Design

The purpose of the weld joint is to transfer the loads and stresses between the members and throughout the weldment. The type of loading and service of the weldment have a great bearing on the joint design that should be selected. All weld joints are either full- or partial-penetration joints. A *full-penetration weld joint* has weld metal throughout the entire cross section of the weld joint. A *partial-penetration joint* is designed to have an unfused area; that is, the weld does not penetrate the joint completely. The rating of the joint is based on the percentage of weld metal to the total joint. If the weld metal penetrated a quarter of the way from both sides, it would leave half of the joint unfused. A 50% partial-penetration joint would have weld metal halfway through the joint. Weldments subjected to static loading need only sufficient

WELD	Symbol	THE FIVE BASIC JOINTS				
		A Butt	B Corner	C Tee	D Lap	E Edge
Fillet		Special	Yes	Yes	Yes	Special
Square-groove		Yes	Yes	Yes	—	Yes
V-groove		Yes	Yes	Yes	—	Yes
Bevel-groove		Yes	Yes	Yes	Yes	Yes
U-groove		Yes	Yes	—	—	Yes
J-groove		Yes	Yes	Yes	Yes	Yes
Flare V-groove		Yes	Yes	—	—	—
Flare-bevel-groove		Yes	Yes	Yes	Yes	—
Backing weld		Combin.	Combin.	Combin.	—	—
Plug or slot		—	—	Yes	Yes	—
Spot or projection		—	—	Special	Yes	—
Seam		—	Special	Special	Yes	—

FIGURE 19–22 Welds applicable to the basic joints.

weld metal to transfer the static loads. When joints are subjected to dynamic loading, reversing loads, impact loads, or cold-temperature service, the weld joint must be more efficient.

The strength of the weld joint depends on the size of the weld and the strength of the weld metal. When using mild and low-alloy steels, the strength of the weld metal is normally stronger than that of the materials being joined. The yield strength of normal structural steel is a minimum of 36,000 psi (248 MPa). The yield strength for a normal E60XX type of electrode deposit is about 50,000 psi (345 MPa); thus the weld metal is stronger than the base metal. When this joint is pulled in tension, it will always break outside the weld since the base metal will yield first, because weld metal strength over-matches the base metal yield strength. Weld reinforcement should be kept to a minimum since it is not required, it is wasteful, and it may act as a stress riser.

In welding high-alloy steel, heat-treated steels, or other high-strength metals, this situation does not apply. Many materials obtain their strength by heat treatment. The weld metal does not have this same heat treatment; therefore, it might have lower strength properties. The welding operation might nullify the heat treatment of the base metal, causing it to revert to a lower strength adjacent to the weld. When welding high-alloy or heat-treated materials, the filler metal must be properly matched to the base metal.

There are at least three factors that must be considered in designing a weld joint. They have an influence on the economics of the weldment design as well as on the strength of the weld joint and the ability of the welder to make it. The designer must first consider the strength re-

quirements and the penetration requirements dictated by loading and service. The joint should be made in the most economical way. The weld joint should be designed so that its cross-sectional area is the minimum possible. The cross-sectional area is a measure of the amount or weight of weld metal needed to make the joint. The welding process, the base metal, and its thickness dictate, and the welding position indicates, the joint design, which dictates the type of preparation tools to be used.

The edge preparation required to produce the particular weld joint design is an economic factor. Weld joints are normally prepared either by shearing, thermal cutting, or machining. Shearing is the most economical way to cut metals; however, there are limitations to thickness, and the sheared edge is a square-cut edge without bevels. Thermal cutting is the most popular method of preparation and is used for most material above the sheet metal thicknesses. It can be used for cutting square edges and for bevels. The root face or square edge, the bevel—both top and bottom—can all be accomplished with one passage of a torch assembly, especially for straight-line cuts. The machining method of preparation is more expensive. However, normally it is used for J and U edge preparation. It is popular for preparing weld joints on circular parts. The edge shapes for most joints are shown in Figure 19–23.

Fillet Welds

The fillet weld is the most popular of all welds because there is normally no preparation required. The fillet weld might be the least expensive, even though it may require more filler metal than a groove weld but the preparation

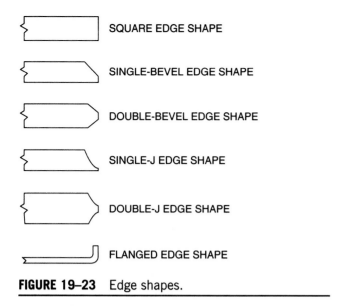

SQUARE EDGE SHAPE

SINGLE-BEVEL EDGE SHAPE

DOUBLE-BEVEL EDGE SHAPE

SINGLE-J EDGE SHAPE

DOUBLE-J EDGE SHAPE

FLANGED EDGE SHAPE

FIGURE 19–23 Edge shapes.

cost would be less. It can be used for lap, T, and corner joints without preparation. On corner joints the double fillet can actually produce a full-penetration weld joint. The use of the fillet for making the basic joints is shown in Figure 19–24. Fillet welds are also used in conjunction with groove welds, particularly for corner and T-joints. The fillet weld is expected to have equal-length legs, and thus the face of the fillet is on a 45° angle. A fillet can be designed to have a longer base than height, in which case it is specified by the two leg lengths (see Figure 19–19). On the 45° or normal type of fillet, the strength of the fillet is based on the shortest or throat dimension, which is 0.707 times the leg length. In North America, fillet welds are specified by the leg length; in many European countries, they are specified by the throat dimension. For fillets having unequal legs, the throat length must be calculated and is the shortest distance between the root of the

fillet and the theoretical face of the fillet. In calculating the strength of fillet welds, the reinforcement is ignored. The root penetration is ignored unless a deep penetrating process is used. If semi- or fully automatic application is used, the extra penetration allows the size of the fillet to be reduced, yet provides equal strength. Such reductions can be utilized only when strict welding procedures are enforced. Figure 19–19 shows details about the throat of a fillet.

The strength of the fillet weld is determined by its failure area, which is the throat dimension. Doubling the size (leg length) of a fillet will double its strength, since it doubles the throat dimension. However, doubling the fillet size will increase its cross-sectional area and weight four times. This is illustrated in Figure 19–25, which shows the relationship of strength that is the throat-versus-cross-sectional area or weight of a fillet weld. For example, a $\frac{3}{8}$-in. (20-mm) fillet is twice as strong as a $\frac{3}{16}$-in. (10-mm) fillet; however, the $\frac{3}{8}$-in. (20-mm) fillet requires four times as much weld metal.

With a T-joint, plates being joined may not always be at 90° with each other. These are known as *skewed T-joints*. The angle of approach can range from a minimum of 60° to a maximum of 135°. The required leg size for the fillet weld in a skewed joint is normally based on a layout of the joint and measurement of the effective throat of the weld. The size would be related to the throat dimension required.

In design work the fillet size is sometimes governed by the thickness of the metals joined. In some situations the minimum size of the fillet must be based on practical reasons rather than on the theoretical need of the design. Intermittent fillets are sometimes used when the size is minimal, based on code or for practical reasons, rather than because of strength requirements. Many intermittent welds are based on a pitch and length such that the weld metal is reduced in half. The minimum length of

FIGURE 19–24 Fillet welds to make the five basic joints.

FILLET WELDS	THE FIVE BASIC JOINT DESIGNS				
	BUTT (B) A	CORNER (C) B	TEE (T) C	LAP (L) D	EDGE (E) E
SINGLE	F ← OR ↑ F	↓ F OR F	F →	↑ F	Not Applicable
DOUBLE	F ← OR ↑ F	F	F →	↑ F	Not Applicable

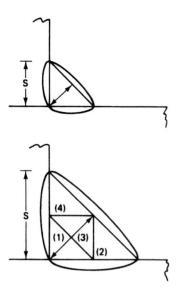

FIGURE 19–25 Fillet weld size versus strength.

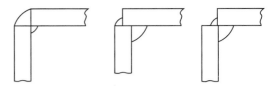

FIGURE 19–26 Fillet welds used for corner joints.

each intermittent weld is normally $1\frac{1}{2}$ in. (38 mm). Large intermittent fillets are not recommended because of the volume–throat dimension relationship mentioned previously. For example, a $\frac{3}{8}$-in. (20-mm) fillet 6 in. (150 mm) long on a 12-in. (300-mm) pitch (center to center of intermittent welds) could be reduced to a continuous $\frac{3}{16}$-in. (10-mm) fillet and the strength would be the same, but the amount of weld metal would be only one-half as much. The minimum length of a fillet should be at least four times its size.

The butt, corner, T, and lap joints can all be made with fillet welds, as shown in Figure 19–24. Single fillet welds are extremely vulnerable to cracking if the root of the weld is subjected to tension loading. Notice the F arrowhead, which stands for *force*. The simple remedy for such joints is to make double fillets, which prohibit the tensile load from being applied to the root of the fillet.

The corner joint, a very popular joint, can be welded with fillets. Figure 19–26 shows variations of the corner joint design and the use of the double fillet. This can be handy in allowing for accumulated tolerances, which make exact fitup of the joint difficult.

Groove Welds

There are seven basic groove welds: the square, V, bevel, U, J, flare V, and flare bevel, shown in Figure 19–27. They can all be used as single or double welds. They can be used to make butt, corner, or T-joints. The dimensions shown are commonly used for arc welding processes. Some dimensions are changed to meet the requirements for the various processes. Three of them—the square-groove, the flare-V, and the flare-bevel-groove weld—can be made without extra preparation of the joint detail. The square groove is simplest, since it requires only a square

edge. The flare-V and flare-bevel welds are normally used for thinner materials in which a bent section joins another section. It may involve a round member. The bevel- and the J-groove welds require preparation on only one of the members of the joint. The remaining two, the V-groove and the U-groove welds, require preparation of both members of the joint.

There are several names given to groove weld designs that are not standard with AWS. These names, *X-, K-* and *Y-groove,* used mostly in Europe, are descriptive. The letters describe the weld in cross section. The X-weld is a double-V weld without a root face, the K-weld is a double-bevel weld, and the Y-weld is a single V with a relatively large root face.

Joint preparation cost is a factor in deciding which groove type to use. Preparation requires shearing, bending, thermal cutting, or machining. Shearing is used if the metal is relatively thin, but square groove preparation is also used on thick materials for the electroslag process and other *narrow-groove processes.* When the thickness is greater, preparation requires thermal cutting or flame cutting. The flare-V joints can be made with thin material flanged by a brake press. These joints are also used when round sections are welded. An example is the welding of reinforcing bars together. The bevel- and V-groove welds are normally used for medium-to-thicker materials and thermal cutting is required. The choice between single- and double-groove welds is shown in Figure 19–27. The double-groove design uses less weld metal, but the single-groove design is less expensive to prepare. In the V-groove and bevel-groove a root face may or may not be involved. The normal practice is to have a small root face, which helps provide dimensional control of the parts during preparation operations. The J-groove design requires J-type preparation on one of the parts, whether a single or a double J is used. In the case of the U-groove, both members must have the special curved shape, which involves either machining or special gouging and cutting. The other groove designs are also easier to make on circular parts.

The root face is used primarily to assure dimensional control of the parts during the parts preparation operation. When large plates are flame cut to feather edges, either V or bevel joint preparations, it is difficult to hold dimensions if a root face is not involved. The root face should be kept to a minimum where full joint penetration is required. When a partial penetration joint is re-

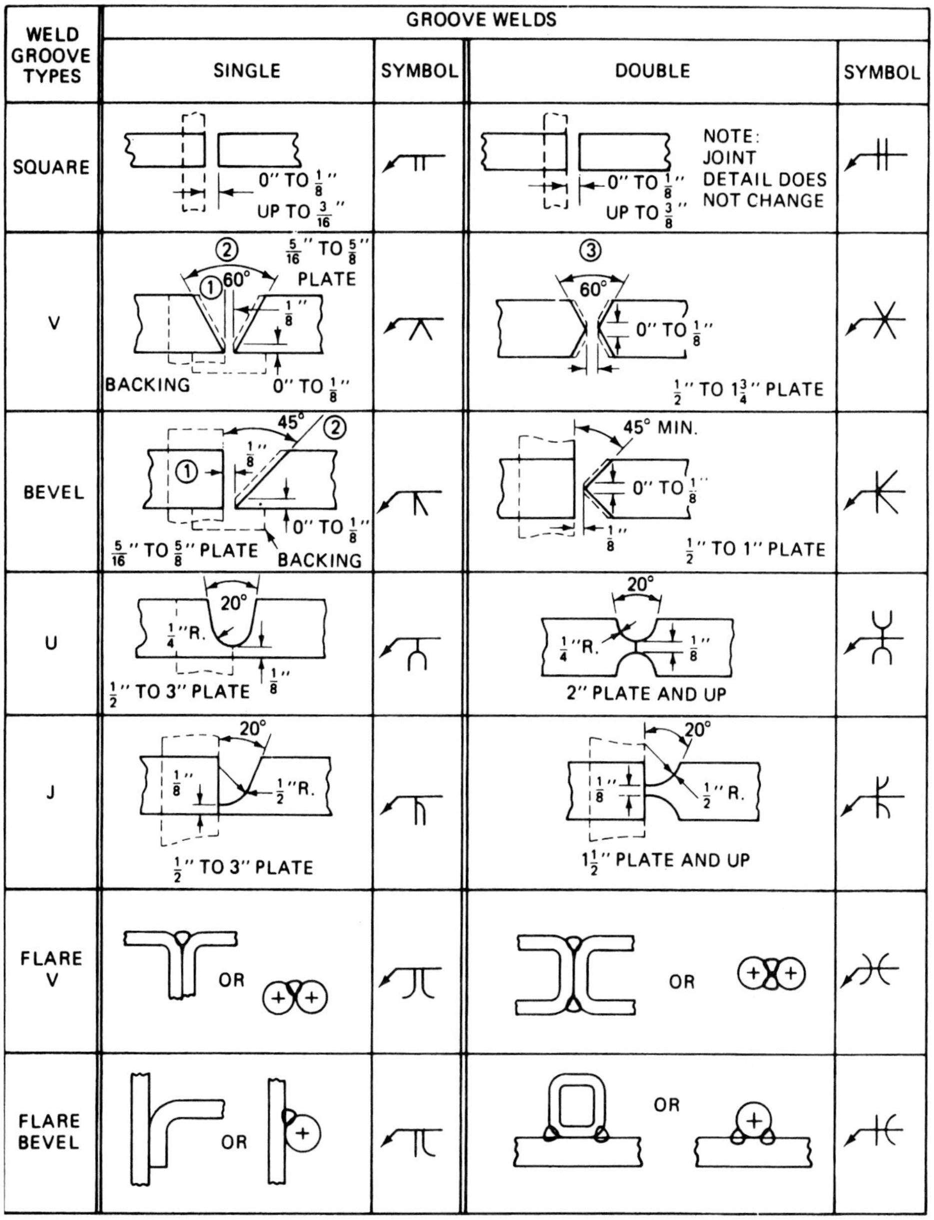

GROOVE WELDS

WELD GROOVE TYPES	SINGLE	SYMBOL	DOUBLE	SYMBOL
SQUARE	0" TO $\frac{1}{8}$" UP TO $\frac{3}{16}$"		NOTE: JOINT DETAIL DOES NOT CHANGE 0" TO $\frac{1}{8}$" UP TO $\frac{3}{8}$"	
V	② $\frac{5}{16}$" TO $\frac{5}{8}$" PLATE ① 60° $\frac{1}{8}$" BACKING 0" TO $\frac{1}{8}$"		③ 60° 0" TO $\frac{1}{8}$" $\frac{1}{2}$" TO $1\frac{3}{4}$" PLATE	
BEVEL	45° ② ① $\frac{1}{8}$" 0" TO $\frac{1}{8}$" $\frac{5}{16}$" TO $\frac{5}{8}$" PLATE BACKING		45° MIN. 0" TO $\frac{1}{8}$" $\frac{1}{8}$" $\frac{1}{2}$" TO 1" PLATE	
U	20° $\frac{1}{4}$" R. $\frac{1}{8}$" $\frac{1}{2}$" TO 3" PLATE		20° $\frac{1}{4}$" R. $\frac{1}{8}$" 2" PLATE AND UP	
J	20° $\frac{1}{8}$" $\frac{1}{2}$" R. $\frac{1}{2}$" TO 3" PLATE		20° $\frac{1}{8}$" $\frac{1}{2}$" R. $1\frac{1}{2}$" PLATE AND UP	
FLARE V	OR		OR	
FLARE BEVEL	OR		OR	

FIGURE 19–27 Seven basic types of groove welds.

quired, the root face can be larger but rarely over 50% of the thickness of the part being beveled. In weldments for which stiffness and weight are the primary criteria, and if a dynamic load is not involved, large root faces will save a considerable amount of weld metal and make the joint less expensive.

There are two other factors that must be considered with respect to the V- and bevel-groove welds. They must be considered together since they do affect the welder's ability to make or place a weld bead at the root of the joint. These are the included angle and the root opening. In full-penetration welds it is absolutely necessary that the welder have sufficient room and accessibility to place the weld at the root of the joint. If the root opening is too tight or if the included angle is too narrow, it will be impossible for the welding electrode to deposit the weld metal at the root of the joint, as shown in Figure 19-28. The smaller-diameter electrode used by GMAW or FCAW will better reach the root of the weld. It is obvious that one or the other must be widened to allow the root weld to be made. The illustration shows what is accomplished by increasing the included angle, but the best solution is accomplished by increasing the root opening. There are optimum included angles and root openings for SMAW and for gas metal and flux cored arc welding. The designs are based on producing a completely penetrated root pass. The sample joint designs referenced utilize these optimum or standardized dimensions. They vary for the different processes. Each welding process section contains design information relating to the process.

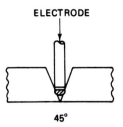

ELECTRODE

45°

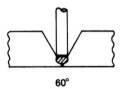

60°

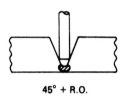

45° + R.O.

FIGURE 19–28 Groove weld root opening–included angle relationship.

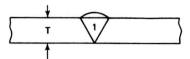

SINGLE - V GROOVE

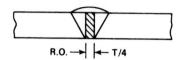

R.O. →| |← T/4

WITH AND WITHOUT
ROOT OPENING

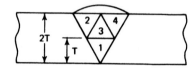

SINGLE - V IN 2T

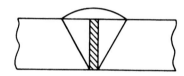

FOUR TIMES AS MUCH
WELD METAL

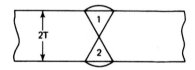

DOUBLE - V IN 2T

FIGURE 19–29 Summary of groove weld design dimensions.

FIGURE 19–30 Reentrant angle of weld.

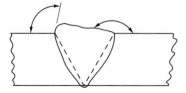

The J- and U-groove welds have been fairly well standardized in design. This means that the radius at the root and the included angle have been optimized, and these are shown by the designs discussed. Finally, for certain metals and for certain applications, special joint details are used. For example, for aluminum pipe a broad single-U preparation is used so that the root is similar to thinner members that can readily be fused together when making the root pass. In heavy-wall pipe, compound angles are often used when the joints are prepared by machining. Figure 19–29 shows the relationship of cross-sectional area for single and double welds and when the root opening is increased. This requires more weld metal but will assure a complete penetration weld. Finally, with groove welds on corner and T-joints, fillets are used for reinforcement and to avoid sharp changes of direction and stress concentrations.

The reinforcement on a groove weld is not a design factor since the designer prefers a small reinforcement. It should be considered a workmanship factor since the welder often adds to the reinforcement to the point where it becomes excessive. Different codes specify the maximum reinforcement allowable. Often, the maximum reinforcement is removed by grinding after inspection. This is time consuming, expensive, and should be avoided, but is not the problem. The reentrant angle where the reinforcement meets the original surface is the problem. This is shown in Figure 19–30. If this angle is sharp, as shown on the left, it is a stress riser and can contribute to fatigue

CLOSED   HALF OPEN   FULL OPEN

FIGURE 19–31 Corner joints.

cracks. It should be a very large angle, as shown on the right, or as smooth a surface as possible where the weld meets and blends into the original base metal.

The joint designs can sometimes be a problem in the weldment if tolerances are allowed to accumulate. In the shop situation, root openings are sometimes used to accommodate tolerances of parts. If the part is made larger than it should be, the root opening will disappear and the ability of the welder to make a full-penetration weld is reduced. It is necessary to trim the part to its designed dimension or to remake the part so that the weld joint geometry is according to the design. If the parts are too small, the root openings will be excessively large and extra weld metal is required to make the weld joint. This is expensive and sometimes it is more economical to remake a piece than to fill in a larger-than-designed weld groove.

There are some cases where tolerances can be accepted in a corner joint. A corner joint can be closed, half open, or fully open, as shown in Figure 19-31, or designs in between. The weld at the corner must accomplish the strength requirements and must vary with fill-up. This requires the shop to obtain engineering approval for the joint design used.

Overwelding Overwelding is an all too-common problem in many shops. The designer and the welder must have confidence in each other. Neither should increase the size of a weld to compensate for assumed deficiencies of the other. Oversize welds are expensive, especially fillets. Figure 19-25 shows the size versus strength of the fillet; the amount of weld metal is four times as much when the size is doubled. Just slightly oversized fillets increase the weld metal 30%-70%. This is expensive. Welders may increase the reinforcement of a groove weld to insure quality. This is not required and increases cost.

Just as important, overwelding increases distortion of the weldment and may require extensive rework.

Every effort should be made to produce welds at the size that they are designed.

An example of weld and weld joint selection is provided by the design for closing the end of a small circular tank. Figure 19-32 shows a variety of designs. The

FIGURE 19–32 End closure of small tank.

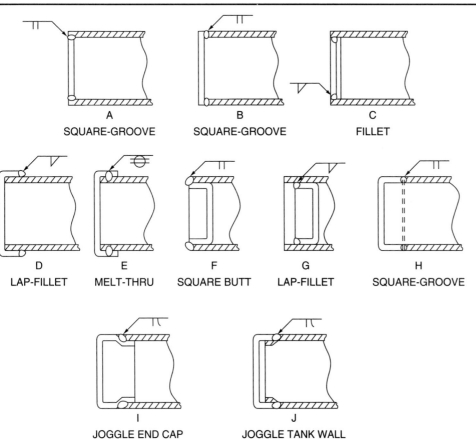

A
SQUARE-GROOVE

B
SQUARE-GROOVE

C
FILLET

D
LAP-FILLET

E
MELT-THRU

F
SQUARE BUTT

G
LAP-FILLET

H
SQUARE-GROOVE

I
JOGGLE END CAP

J
JOGGLE TANK WALL

simplest from the point of view of piece-part preparation is to use a flat plate, shown by details A, B, and C. What welding procedure to use then becomes the question. Both B and C use the square-groove weld, but this is difficult to accomplish because fitup must be perfect if penetration is expected to be uniform, or a backing ring must be used, which increases cost. Also, a fixture must be used to hold the end plate in place prior to welding. Additionally, in detail B the weld is made in the horizontal position, which is more difficult. In detail C, a fillet can be used, which requires less precise fitup. Also, both ends can be welded simultaneously, which increases productivity. The formed head is a better solution. Of course, the formed head is more expensive than the flat plate; however, a fixture is not required to hold it in place, assuming that parts are to size—friction should hold it in place. Details D and G require fillet welds, and both ends can be welded simultaneously. Detail E can be used with thinner material, and a laser beam could be used for welding. For thinner material GMAW would be used for a melt-through weld. Detail F has the same problem as detail A, and detail H has the same problem as detail B. Backing rings should be avoided. They add to the cost and may increase corrosion problems, depending on the contents of the tank. The joggled end design has many advantages but requires special tooling, which might affect the cost adversely. The welding procedure is the easiest. This type of analysis can be applied to the selection of all joint designs.

Other Weld Designs

Corner joints, T-joints, and lap joints can be made with plug and slot welds. Seam welds and spot welds can be made with resistance welding processes as well as with arc welding processes. In this chapter we are interested in the weld joints designed for the arc welding processes. Seam welds and spot welds are made by melting through the top member of the joint into other members when using arc processes. Sometimes called *burn-through welds*, these are restricted to thin materials and are largely dependent on the depth of penetration of the process involved. They are quite popular for the CO_2 welding process, which has deep penetrating qualities. The joint strength is based on the area of the weld at the interface between the two members. The way these and the other weld types are applied to the five basic joint types is shown in Figure 19–33.

Plug and slot welds have holes prepared in one member so that the welding can be done through the hole into the other member. Plug welds are round and slot welds are elongated. The holes are filled when making the plug or slot welds. In early design, plug welds were transitional from riveted structures to welded structures but are no longer popular. Standard dimensions have been established for plug or slot welds and are shown in Figure 19–34. The strength of these welds is obtained by calculating the area of the plug hole or slot hole where it interfaces the other member.

FIGURE 19–33 Other types of welds related to the five basic joints.

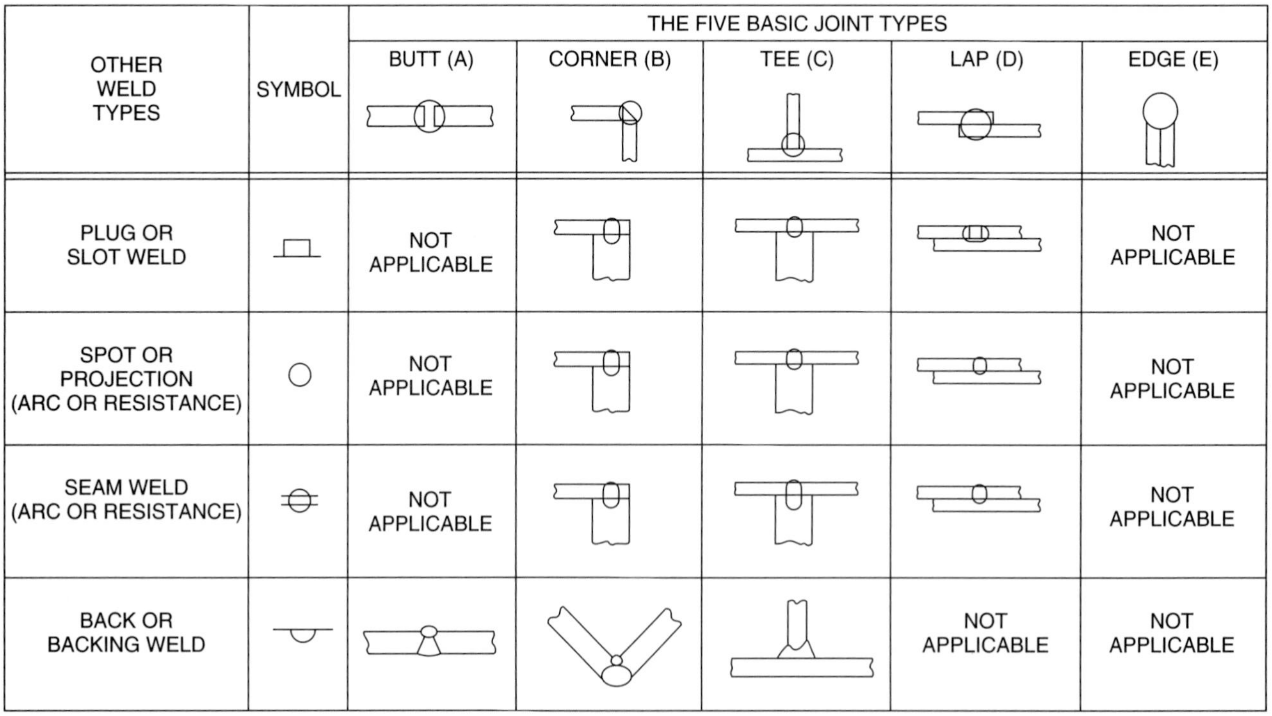

OTHER WELD TYPES	SYMBOL	THE FIVE BASIC JOINT TYPES				
		BUTT (A)	CORNER (B)	TEE (C)	LAP (D)	EDGE (E)
PLUG OR SLOT WELD		NOT APPLICABLE				NOT APPLICABLE
SPOT OR PROJECTION (ARC OR RESISTANCE)		NOT APPLICABLE				NOT APPLICABLE
SEAM WELD (ARC OR RESISTANCE)		NOT APPLICABLE				NOT APPLICABLE
BACK OR BACKING WELD					NOT APPLICABLE	NOT APPLICABLE

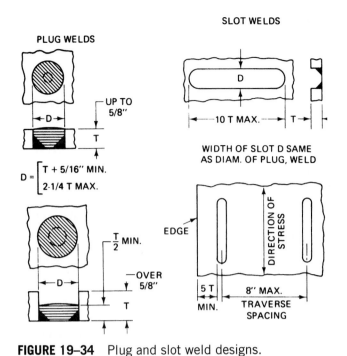

FIGURE 19–34 Plug and slot weld designs.

Surfacing welds are normally welds made on a surface to provide special properties or dimensions of that surface. They are commonly used to build up areas for remachining, to provide corrosion-resistant surfaces, or to provide abrasion-resistant or hard surfaces for wear. Surfacing welds can be made using most of the arc processes and with strip electrode to provide wider beads. They do not involve making joints.

Flange joints are used primarily for lighter-gauge metals, usually called *sheet metals.* Flanging of parts at welds is common practice to improve stiffness, reduce distortion, and provide an area for welding. The only preparation involved is shearing and bending. A flare groove weld is used with flanged preparation.

The last of the weld types is the back or backing weld, which is used for improving the properties of the root of single-groove welds. To perform a back or backing weld, the back side of the weld must be accessible. It is used to ensure that root fusion is complete and that potential stress risers are eliminated. Gouging, chipping, or grinding is performed before the backing weld is made.

Weld Joint Details

Weld joint details have been standardized by the AWS "Structural Welding Code—Steel."[4] The weld joint details are indexed by a code system, shown in Figure 19–35, which identifies the type of joint, the material thickness limitation, the welding process, the welding position, joint penetration requirement, and the type of weld. Similar weld designs are then expanded alphabetically to provide variations. These designs can be used as a standardized system for a specific company. This system is similar to that of "Welded Joint Design," MIL-STD-0022B (SHIPS), but sufficiently different so that the numbering system is not interchangeable.[5] A similar system is presented by the

FIGURE 19–35 Weld joint identification system.

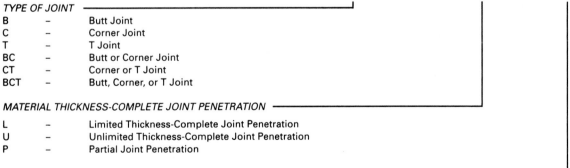

TYPE OF JOINT
B — Butt Joint
C — Corner Joint
T — T Joint
BC — Butt or Corner Joint
CT — Corner or T Joint
BCT — Butt, Corner, or T Joint

MATERIAL THICKNESS-COMPLETE JOINT PENETRATION
L — Limited Thickness-Complete Joint Penetration
U — Unlimited Thickness-Complete Joint Penetration
P — Partial Joint Penetration

TYPE OF WELD
1. Square Groove
2. Single-V Groove
3. Double-V Groove
4. Single Bevel Groove
5. Double Bevel Groove
6. Single-U Groove
7. Double-U Groove
8. Single-J Groove
9. Double-J Groove
10. Flare Groove

American Institute of Steel Construction.[6] This weld joint identification system presents a series of weld joint details. In this identification system, the joint type is first. The types of joints are listed alphabetically; thus the butt joints (B) come first, followed by corner joints (C), and so on. Combination joints are also included, since the weld detail can be the same for more than one type of joint.

An important factor is the material thickness and penetration requirements. There are three categories. With L, which indicates *limited thickness,* a maximum nominal thickness is shown for each joint and must be adhered to. U, which indicates *unlimited thickness,* would be used for materials thicker than the L category. In P, which indicates *partial penetration,* a sufficiently large root face is used to avoid complete penetration. Caution must be exercised when using partial-penetration joints in which dynamic loading and cold-temperature service are involved.

19-5 INFLUENCE OF SPECIFICATIONS ON DESIGN

All codes and specifications that apply to weldments include design information. In some cases detailed information is presented that must be followed to meet the code or specification. These include overall general design guidelines, minimum thicknesses of particular materials, minimum weld sizes, and so on. Different specifications have been mentioned throughout this book; however, the most complete listing of codes and specifications and the products they cover are given in Section 22-2. Please refer to this section for the type of products involved. This provides the particular code or specification name and the organization that issues it.

The designer is responsible for obtaining and studying the correct edition of the specifications covering the product being designed. The code rules and conditions must apply. Information in this chapter cannot be a substitute for a specification.

Twenty Welding Joint Design Guidelines

Many products are not covered by codes or specifications. When designers are not governed by a specific code or specification, the following welding joint design guidelines may be used for mild and low-alloy steel fabrication.

1. *Designed strength.* Each weld joint shall be designed to meet the strength requirements for the intended application. Consideration should be given to stress concentrations due to abrupt changes in section, especially when impact, fatigue loads, or low-temperature service is involved. The weld metal should equal the strength of the metal being welded.

2. *Standardized joints.* Use the standard welding joints shown by the AWS Structural Code. These have been designed to require the least amount of weld metal. The design of additional joints not covered should be designed on the same basis.

3. *Complete penetration joints.* Highest efficiencies for all types of loadings result from joints designed for full penetration. Joints requiring the highest efficiency must specify CP in the tail of the weld symbol.

4. *Joint preparation time.* Weld joints should require the least amount of edge preparation with respect to the welding time required to fill the joint. It is generally less expensive to bevel and weld thinner plates from one side; however, for thicker plate, it is less time-consuming to bevel and weld from both sides.

5. *J- and U-groove preparation.* J- and U-grooves shall be used only on parts that are readily prepared by machining. Machining is more expensive than flame cutting.

6. *Reduce overwelding.* Overwelding increases welding costs and causes extra distortion. Joints designed for 100% efficiency may be subjected to all types of loadings; however, when stiffness is the principal requirement, joints with efficiencies as low as 50% may be satisfactory.

7. *Fillet weld size.* The designed size might be smaller than shown, but for general appearance and for fabricating reasons, the minimum fillet weld size to be used on a given thickness of plate for T-joints is as follows (the design size might be larger in some cases):

Thickness of Thinner Plate (in.)	Minimum Fillet Weld Size (in.)
Up to $\frac{1}{4}$ incl.	$\frac{1}{8}$
$\frac{3}{8}$ to $\frac{3}{4}$ incl.	$\frac{1}{4}$
$\frac{7}{8}$ to $1\frac{1}{4}$ incl.	$\frac{1}{2}$
$1\frac{3}{8}$ to 2 incl.	$\frac{1}{2}$
$2\frac{1}{4}$ to 4 incl.	$\frac{3}{4}$

8. *Intermittent fillet welds.* Intermittent fillet welds should only be used for strength when the smallest fillet sizes, as given in 7 above, are too large for continuous welds. Exceptions are metallurgical or warpage reasons. Intermittent welds should be used whenever possible on sheet metal and structural parts when stiffness is their prime purpose.

9. *Length of intermittent fillet welds.* For material $\frac{1}{4}$ in. (12.5 mm) thick and greater, the minimum length of intermittent fillet welds should be 8 times their nominal size, but not less than 2 in. (50 mm): the maximum length should be 16 times their nominal size, but not more than 6 in. (150 mm).

10. *Pitch of intermittent fillet welds.* For material $\frac{1}{4}$ in. (12.5 mm) thick and greater, the maximum center to center dimension should be 32 times the thickness of the thinner plate, but in no case should the clear spacing between intermittent fillets be greater than 12 in. (300 mm).

11. *Reduce welds.* Eliminate a weld joint by making simple bends wherever possible.

12. *Butt joints.* For butt joints of unequal thickness, smooth the transition by removing metal rather than by adding weld metal.

13. *Double-T joints.* Avoid double-T or cruciform joints whenever possible. Such joints have maximum locked-up stresses and are suspect due to possible laminar tearing.

14. *Corner joints.* For corner joints when bevels are used, prepare the thinner member whenever possible.

15. *Plug and slot welds.* Plug or slot welds, or fillet welds in holes or slots, should not be used in highly stressed members unless absolutely necessary. They should be used where subjected principally to shearing stresses or where needed to prevent buckling of lapped parts.

16. *Groove weld preparation.* Whenever possible, require only one member of a joint to have bevel preparation.

17. *Welding position.* Weldments should be designed so that the position in which the welds are made should have the following order of preference:

    (a) Fillet welds, flat, horizontal fillet, horizontal, vertical, overhead

    (b) Groove welds, flat, vertical, horizontal, overhead

18. *Enclosed welding.* Whenever welding is required in enclosed areas or pockets, the enclosed area must contain sufficient openings for access and ventilation for the welder.

19. *Accessibility.* All joints should be located so that the welder will have sufficient room to weld, gouge, peen, and clean slag. There should be no obstructions that prevent the welder from seeing to the root of the joint.

20. *Weld symbols.* All weld joints should be specified by weld symbols. Symbols should be conspicuously placed on all drawings and should refer to views that show the most joint detail, which is normally the profile view.

19-6 DESIGN CONVERSION TO WELDMENTS

The redesign from other manufacturing methods to weldments is being done for many different reasons. These reasons can be classified into the following categories:

- Economics or reduced cost
- Quality improvement
- Appearance improvement
- Design or product improvement
- Easier to machine
- Reduced production cycle time
- Environmental and other reasons

An investigation into these reasons for redesign soon shows the basic facts that are involved. By taking each of the reasons above and investigating them, we are able to establish the following factors that should assist in making a decision to redesign the present part or structure to a weldment. The economic or *cost reduction* reasons consist of at least the following.

One of the primary cost reduction reasons is the ability to make the product lighter. This occurs since the metal thickness of castings is often thicker than the stresses require but is needed to assist metal flow in the mold. Parts with excessive thickness due to the metal casting flow requirements are wasteful and would be reduced in the weldment. Extremely generous radiuses are provided at intersecting planes to assist in metal flow during the casting process. Many surfaces are made extra thick to allow for machining, and the machining allowance may be oversize to allow for potential warpage, pattern shifts, and so on. The lap joint used for riveting is also excess metal that can be eliminated with the welding operation. This applies not only to structural work but to tanks and vessels of all types. Weight can be reduced. This reduces the initial cost of the weldment because less metal is involved. This also reduces other charges, for example, shipping charges of the finished product. Reducing the weight allows for more payload in many types of structures.

A cost problem that plagues users of cast parts is the detection of internal defects in the casting after it has been partially or completely machined. Sometimes these castings can be salvaged by welding; many times they must be scrapped with the consequent loss of the time spent machining them prior to finding the defect. Another important cost factor is that weldments frequently simplify and streamline production flow of manufactured parts. Many companies acquire their castings from outside sources. A long time is usually required to place the order and get the casting produced. This cycle time can be reduced by having an in-plant welding department. In comparing the cost of a foundry and cost of a welding shop and metal-preparation shop, studies show that the capital investment to produce the same quantity of product is much lower for the weldment. The direct labor required to produce the same part in the foundry is usually higher. Many structures made of castings are so large that the castings must be connected together. This requires machined surfaces and bolt holes for fitted bolts. This extra machining can run up the cost of the final product.

Many foundries do not want short-run jobs. Many foundries also do not have their own pattern shops and this adds more cycle time whenever changes are involved or patterns need to be repaired. Pattern damage and damage to core boxes are relatively common for large castings, and these items are expensive to maintain and repair. Many foundries are discontinuing business due to the air pollution problems associated with older iron foundries.

The *quality* of castings is another reason why companies are changing to weldments. There are many designs that are extremely difficult to cast from the foundry's point of view, and these usually contribute to high casting costs. This difficulty creates a high scrappage rate and consequently higher prices.

Mechanical fasteners often contribute to other problems such as leaks of threaded screw joints in piping systems. In large structures in which castings and structural members are bolted or riveted together, the joints become loose and start to flex. In time the rivets shear, the holes elongate, and the working stresses cause the parts to fail. The service problems of composite structures are the reason why they are being changed to weldments.

Design improvement is another reason for redesign. This can be not only for more reliability and reduced cost, but also for several other reasons. One of the most important is the design freedom that welding provides the designer. Changes in the ultimate product are relatively easy to make if the product is designed of weldments. Design changes are more often required in low-volume production equipment than in mass-produced items. On some types of machinery, the production volume is low and pattern costs are absorbed by only a limited number of pieces and thus become an excessive cost. Another reason for changing is to design the product to reduce space requirements. Weldments can be designed to be more compact than castings. This is particularly important in the transportation industry. Another design improvement is the ability for composite weldment construction. Composite construction is a weldment produced by joining steel castings, hot-rolled plate, rolled shapes, stamped items, forged, and formed pieces. Composite construction can also involve the welding together of parts of different types of steel. The type of metal may be placed at a specific spot to provide the necessary strength, corrosion resistance, abrasion resistance, and so on, whereas a casting or forging must be completely of the same material. Design improvement can also be made with regard to deflection resistance or stiffness and vibration control. In general, since the modulus of elasticity of steel is at least double that of cast iron, the weight of items of similar stiffness can be reduced.

Improved appearance is a reason for redesign. This is very similar to design improvement but can be considered as styling, which is becoming increasingly important for industrial machinery. Weldments have cleaner, crisper lines than castings. Figure 19–36 illustrates this point. It shows the top and bottom view of a cast food machinery base and the part redesigned as a weldment. The weldment utilized square mechanical tubing and thin plate. The cost was reduced and cycle time reduced. A casting gives an appearance of an old-fashioned design and detracts from its appeal to the buyer. Weldments eliminate the cracks and crevices in riveted and bolted construction, and eliminate the problem of rivets, rivet heads, and so on. The surface of hot-rolled steel is vastly superior to that of cast steel or cast iron and allows reduced finishing costs. Cold-rolled steel is as easy to plate with a minimum amount of finishing and polishing. It is sometimes difficult to get a cast surface free enough of porosity to meet sanitation standards. The welds when made with semiautomatic or automatic equipment have a smooth appearance and a surface that can be painted with little or no finishing.

Another reason for the change to a weldment has to do with manufacturing or machining requirements. Machining a large casting requires large machine tools that are extremely expensive. In many cases small castings can be incorporated into composite design weldments. The small castings can be premachined on small machine tools. By proper fixturing and using balanced procedures, the parts can be incorporated into weldments with minimum warpage, which will eliminate much of the machining requiring large machine tools. The use of small castings eliminates much of the foundry quality problems since they have simpler designs and are easier to produce. Scrappage rates and machining time are reduced.

An important factor has to do with maintenance or rebuilding of equipment. In repair work parts must often be replaced. If they are castings that are no longer available or require excessive procurement time, it is wise to redesign the part as a weldment. The weldment can be made immediately and machined so that the equipment can be returned to service more quickly. In preparing welded steel versus cast iron parts, the ease of weld repair of the steel is much greater than the repair by welding of cast iron parts. Another reason to use welded steel is the fact that cast iron parts may without warning suddenly fail in a brittle manner. This rarely occurs with welded steel structures.

A cost-savings reason for redesign is the reduced cycle time required by the weldment versus other methods of manufacture. It is extremely important in the maintenance of equipment and machinery. An effective way for a company to reduce costs is to reduce the cycle time from start to finish of the manufactured product. With an in-house welding shop and parts preparation department, this is obtainable. Reduced cycle time is also important for companies supplying spare parts for existing machinery. Most often when a spare part is ordered, the user is experiencing an emergency and needs assistance to return to production as quickly as possible. If the extra

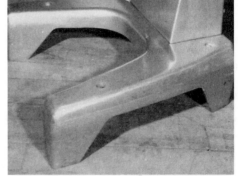

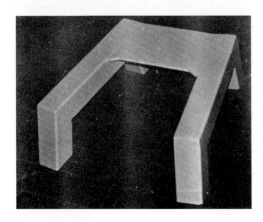

FIGURE 19–36   Cast and welded food machinery base, top, and bottom views.

time required by the foundry is included, the time to supply the spare part becomes excessive and the user may decide to fabricate the part.

These are valid reasons for changing to weldments from other types of construction. If you make an analysis of some of the castings or riveted or other mechanically fastened products in your plant, you might find that some of these reasons apply and that you can reap some of the economic advantages by making the change.

Redesign to a Weldment

There are three basic methods for converting from the existing design to a weldment. These are known as (1) direct copy redesign, (2) redesign from the existing part, and (3) new design based on loads and stresses. The optimum economic advantage is obtained when the part is designed from the beginning as a weldment based on the loads and stresses applied to it. The second best method, from an economic point of view, would be the redesign based on the existing part. The quickest method, but the least economical, is to copy the existing part as the new weldment. The latter two methods will be discussed briefly.

Direct Copy

Direct copy redesign is done quickly, usually when time is not available for a complete study. Storage tanks, vessels, and similar items can be very easily redesigned as a direct copy of an existing structure. The lap joints for the riveted design are eliminated and butt joints are substituted. The full strength of the metal should be utilized. A weld joint having full penetration should be used. The direct copy redesign will produce a stronger tank as well as a leakproof tank. It will be stronger unless the thickness of the metal is reduced. An analysis should be made before reducing metal thickness. If the thickness can be safely reduced, it should be done by compensating for the areas used for the rivet holes. This can be done by determining the vertical percentage of the area through the metal and rivet holes compared to the area through the metal alone. The rivet holes can reduce this area by as much as 25%. One-half of this percentage can be used to reduce the thickness of the plate. If the plates become

too thin, corrosion or pitting may be the controlling factor. Before reducing the plate thickness, a thorough analysis should be made by a qualified designer.

Piping work can also be redesigned by the direct copy method. This is simply a matter of eliminating screw-threaded joints and substituting welded joints or welding fittings. If the same size and wall thickness of pipe are used, the welded design will have a higher safety factor than the screw-threaded design. This is because of the material cut away when making the threads on the pipe. The pipe wall thickness or schedule should not be reduced to the thickness at the thinnest point, which is at the root of the threads. This could shorten the life of the piping system as a result of corrosion, erosion, and so on. If wall thickness is reduced, it should be done by a qualified designer after reviewing the entire design. A conservative way would be to reduce wall thickness by one schedule rather than attempting to provide a wall thickness equal to the thickness at the root of the thread. In any event, a leakproof system will result, which will be superior to the threaded system.

In structural work the direct copy conversion can be used for certain types of structures. Riveted roof trusses, for example, can be changed to a welded roof truss merely by eliminating the rivets in certain of the members and replacing them with plug welds. The number of holes required would be cut in half. However, the section sizes would remain the same and gusset plates, spacers, and so on, would still be required. Another way for the direct copy method would be to eliminate all holes and use fillet welds at the toes and heels of angles to join the members to gusset plates rather than using plug welds. The equivalent length of weld to provide the same strength as rivets is shown in Figure 19–37. The direct copy method can be used for many other similar type structures.

Castings can be redesigned as weldments utilizing the direct copy method. To make these conversions intelligently it is necessary to consider the metal from which the casting is made. In the case of redesigning a cast iron part, the section thicknesses can be reduced by one-half and still produce a weldment stronger than the replaced cast iron part. This is because the strength of the steel is at least twice as great as the strength of the cast iron. The

Rivet Diameter (in.)	Rivet Shear Value at 15,000 psi Shear Stress (lb)	Equivalent Length of Fillet Weld Based on 19,000 psi Deposit (in.)		
		$\frac{1}{4}$ Fillet	$\frac{3}{8}$ Fillet	$\frac{1}{2}$ Fillet
$\frac{1}{2}$	2,950	1	$\frac{5}{8}$	$\frac{1}{2}$
$\frac{5}{8}$	4,600	$1\frac{3}{8}$	1	$\frac{3}{4}$
$\frac{3}{4}$	6,630	2	$1\frac{3}{8}$	1
$\frac{7}{8}$	9,020	$2\frac{3}{4}$	$1\frac{7}{8}$	$1\frac{3}{8}$
1	11,780	$3\frac{1}{2}$	$2\frac{3}{8}$	$1\frac{3}{4}$

FIGURE 19–37 Length of fillet weld (AWS E70XX electrode) to replace rivets (ASTM A502-1 rivets). From Ref. 5.

stiffness of the steel part will be as stiff as the cast iron part even though the section thicknesses are reduced by one-half. Stiffness is related to deflection, and the deflection formulas involve the modulus of elasticity of the material. If all other factors are equal, the deflection of the steel part would be only one-half the deflection of the cast iron part. There is sometimes the question of vibration of a steel weldment versus a cast iron part. The damping capacity of steel and cast iron are about the same. In steel plate construction, the design should be such that the natural frequencies that might occur are outside of the operating range. This is a complex problem and should be placed in the hands of an experienced designer.

In the case of redesigning steel castings to steel weldments, the dimensions of the sections should be left the same. It is difficult to determine when sections of the steel casting are heavier than they need to be unless the design history for the casting is known. Since this is normally not known, it is wise to make the sections of the same thickness. The welds that are taking the full load should be designed for complete penetration. Various gussets and large radiuses can often be eliminated or reduced without affecting the service capabilities of the weldment.

Aluminum sand castings are also redesigned as weldments. The ratio of strength of steel versus strength of aluminum should be considered. However, there are many different strength levels of aluminum; therefore, it is impossible to provide ratios unless tests are performed. Aluminum castings can often be replaced by steel stampings of fairly thin parts. An example of this type of redesign is shown in Figure 19–38. This is almost a direct copy of the casting, although several dimensions are changed. The weldment made of the stampings is about the same weight as the aluminum casting, but is much less expensive.

FIGURE 19–38 Redesign aluminum casting to steel stampings.

Analysis of the Part

The redesign by analysis of the existing part to determine its strength is more economically advantageous than the direct copy technique. In this case efforts are made to analyze the part that is being redesigned without actually analyzing the loads and stresses that are imposed on it. In this type of redesign it is usually possible to reduce the size of members or thickness of sections appreciably to reduce the total weight of the weldment. An example of the redesign by part analysis would be the redesign of a structural member such as a roof truss. In a roof truss there are certain members subjected to tensile loading and other members subjected to compressive or possibly column loading. In the members loaded in tension, the part can be reduced in cross-sectional area by the amount or by the size of the rivet holes that were punched in the member. The replacement member would have the same strength since its entire cross-section would be utilized. In such a substitution it is necessary to utilize standard rolled sections and the section that provides the cross-sectional area equal to that of the original less the rivet hole area should be used. For safety's sake it might be wise to go up one size for the member. This technique must not be used for long members subjected to compressive loads. It cannot be used because such members are designed based on the slenderness ratio and the possibility of buckling rather than the unit stress produced in compression. For those members subjected to bending, particularly outer fibers where the bending load produces tensile stresses, the size of the members should not be reduced; the reason is that the stress is not consistent through the entire length of the member, and the stress level could be extremely low at the joint or where the rivet holes are located. The allowable stresses may be proper at some other point; therefore, this member should not be reduced in size. A designer with experience in this field should design the product.

For maximum cost savings, the complete redesign based on the fundamental factors involved should be done. This would mean following the general concepts outlined in the earlier part of this chapter.

19-7 COMPUTER-AIDED DESIGN (CAD)

Computer-aided design (CAD) and computer-aided manufacturing (CAM) are now the industry standard for product design and manufacture. Computers are being used to design and analyze weldments and welded structures. Computers are used for other activities within the manufacturing operation, such as programming, tools, automatic shape-cutting machines, automatic welding machines, and welding robots. The computer has replaced the drafting board for making drawings used in the shop to manufacture weldments. Drawings are still made in the usual fashion—that is, front, side, and top views with joint details—but on the computer screen. The resultant drawing is printed by the computer printer or, in the case of a large drawing, is made by electronic plotters. In many cases, the computer-generated information is fed into the controller of manufacturing process tools being used. These CAD and CAM systems greatly reduce the time required in the shop and greatly reduce—and in some cases eliminate—the amount of paper involved.

Computer-aided design does not alter the responsibility of the engineer. It is still necessary to establish the functions of the weldment, the load transfer, the environmental conditions, the code restrictions, and the constraints of the welding process. The computer hardware required ranges from PCs with advanced microprocessors (CPU) to the supercomputers. The type of hardware used depends on the weldment's size and complexity and the speed needed to get the job done in a timely manner.

Many software programs are available for PCs. They cover various aspects of welding design and analysis, and are available from various organizations that provide computer-aided design and drafting (CADD) software. These programs must be carefully researched to determine the best one for your product. They allow the designer to create accurate three-dimensional (3-D) drawings of the desired part, which render images that appear similar to photographs of the finished product.

The many advantages of computer-aided design are illustrated by the following example of a very simple weldment. It is manually drawn as the front, side, and top view, as shown in Figure 19-39. Usually this is drawn on paper, with separate detail drawings of each component so that the shop can produce the weldment. When using CAD, it is drawn electronically on the computer screen. By means of the computer, it can be reviewed from many different points of view. An isometric view is a typical printout, shown in Figure 19-40. More details are shown in Figure 19-41, which shows each joint with the weld symbol and the joints in cross section. The weldment is

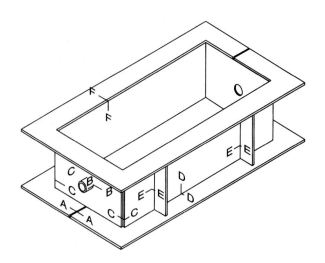

FIGURE 19–40   Isometric view original design.

FIGURE 19–41   Cross-section views.

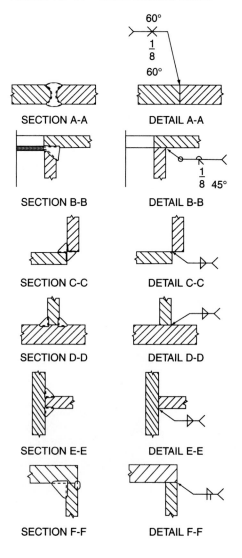

FIGURE 19–39   Side, top, and front views. Original design.

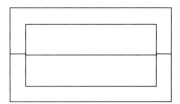

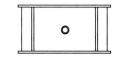

then studied by looking at different views that are provided by the computer program. An example is shown in Figure 19–42, which is an isometric view of the bottom of the tank. This and other views show the designer that specific changes could be made to improve the weldment. For example, the joint should be eliminated on the bottom plate, saving considerable welding. Two corner joints should be eliminated by a 90-degree bend, which would also reduce the amount of welding. Using bars for the top of the tank would eliminate much flame cutting. This would require four extra small welds but would be a net cost savings. Once the design is finalized and approved, the computer program will produce detailed drawings of each component part. The final design is shown in Figure 19–43, which should be compared with Figure 19–40 to show the improvement.

Another advantage of CAD is the ability to modify the design when one factor changes. This can be accomplished very quickly and easily by skilled people.

FIGURE 19–42 Isometric view of bottom. Original design.

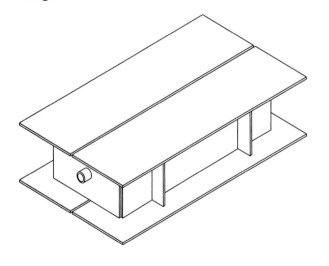

FIGURE 19–43 Final design.

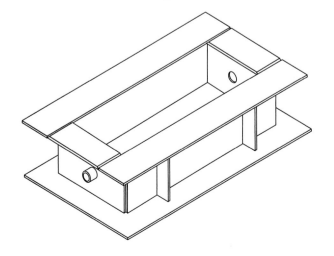

Various software packages are available to provide thorough analysis. Most of them are based on finite element analysis (FEA). These programs are available as packages that link with other finite element analysis programs or stand alone. In this way CAD allows for the assessment of a weldment's fitness for service.

These studies may need to be conducted in several phases, each of different complexities. Some analysis programs highlight stress concentrations and predict fatigue failures based on different loadings. Some predict residual stresses by analyzing mechanical, electrical, thermal, and metallurgical factors; others provide welding thermal analysis showing heat flow. Software packages might provide adaptive meshing that refines or simplifies the operation to improve accuracy.

These programs provide a multitude of information such as weld joint details, properties and stresses, and joint cross-sectional data. Other programs provide welding procedures based on process materials, position, and so on. The analysis may also include welding parameters, welding sequence, and joint geometry. Some provide cost estimates for making the weld. Thus, a complete study of the weldment is performed via computer analysis.

Computers are helping to optimize weldment design. They aid in synthesizing the knowledge of welding experts, textbook information, and problem-solving tactics. They provide the ability to easily build extensive databases for expert systems. They are paving the way for welding technological advancements through improved modeling and data analysis. Taking advantage of these developments requires computer-literate personnel.

19-8 WELDMENT REDESIGN TO REDUCE COST

Value engineering programs generate large cost savings from welded product analysis. Value engineering groups review all practices within a company to find a better way to provide the same value for less cost. The value engineering technique should be used for evaluating welded products and will usually suggest changes that will save money.

Value engineering of welded products may take several paths: Analyze the weldment design, the welding procedure, materials preparation, and the welding operation. Since the design is the basis for the product, it is good to relate all factors to the design except those that must be accomplished by other groups.

The first step is to involve welding shop personnel. Ask them if the weldment is easy to weld. Are there any difficult or inaccessible weld joints? Are welds covered up by parts added later? Do the parts fit together without trimming or create large gaps? The investigation team should also check other manufacturing personnel and everyone involved in the manufacture of the weldment. Critical items should be investigated and solutions found. The following items should be checked and corrected.

Designers and Engineers

1. Investigate the field service record. If there have been no field failures, recheck the load calculations; the weldment may be overdesigned.

2. Investigate field failure reports. The redesign should correct design weakness but not overloading.

3. Eliminate weld joints whenever possible. Use rolled sections, use small steel castings for complex area, and use formed or bent plates.

4. Reduce the cross-sectional area of all welds. Utilize small root openings, small groove angles, double- instead of single-groove welds, and so on. The weld metal is the most expensive metal on the weldment. The amount of weld metal should be kept to a minimum (see Figure 19–44).

5. Utilize fillet welds with caution. If the size is doubled, the strength is doubled but the cross-sectional area and the weight increase four times.

6. Intermittent fillet welds should be studied. It may be possible to reduce the fillet size and make the weld continuous and use less weld metal.

7. Where extra thickness is required, use flame-cut heavy plates or billets or small simple castings.

8. Pre-machine parts wherever possible for bearings, housings, and so on. This avoids large machine tools for completed weldments.

9. Provide easy accessibility for all welds. Otherwise, special attention and time are required (see Figure 19–15).

10. Select weldable materials: mild steel and low-alloy steels. Difficult-to-weld metals require preheat and complex expensive procedures.

11. Design the weldment with the least number of thicknesses of steel. This reduces stocking extra thicknesses.

12. Balance using heavier materials without reinforcements against the cost of extra pieces of gussets and welding.

13. Use weld symbols with size notations for all welds.

Process or Manufacturing Engineers

1. Make sure written welding procedure or job sheets are provided to the welding departments for all jobs.

2. Use the semiautomatic or automatic or robotic method of application of welding.

3. Use high-deposition-rate welding processes and filler metal.

4. Consider stress relieving. Is it required for service or for machining stability? Can the vibratory technique be used?

5. Provide fixtures when economically possible; use simple fixtures for setup operations.

6. Use subassemblies where possible.

7. Utilize positioning equipment when economies are provided.

Parts Preparation Personnel

1. Shear or blank parts whenever possible—use automatic back gauges.

2. Use automatic shape-cutting equipment. Use the optimum process, fuel gas, or cutting tips.

3. Utilize stops, gauges, and so on in metal-forming operations to increase accuracy of parts.

4. Check with the welding department for fitup problems and fix part drawings.

Welding Department Personnel

1. Provide accurate fitup of all weldments. Use locating fixtures for setup. Tack weld only in weld setup fixtures. Inspect fitup prior to welding. Figure 19–44 shows the effect of fitup on extra weld metal and time required.

2. Do not overweld. Make welds the size shown on weld symbols. Figure 19–45 shows the additional weld metal and time required when overwelding (see also Figure 19–24).

FIGURE 19–44 Extra weld metal required for poor fitup.

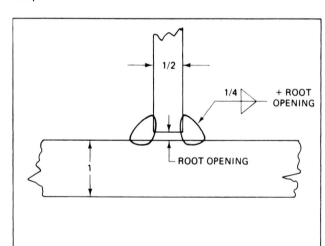

ROOT OPENING	FILLET SIZE REQUIRED	WELD METAL REQUIRED LB/FT.	% MORE WELD METAL REQUIRED
0	1/4"	0.212	–
1/16"	5/16'	0.334	157%
1/8"	3/8"	0.476	224%
3/16"	7/16"	0.652	308%

Fillet Size Overwelded Size		Theoretical wt/ft	Overwelding Requires This Much More Weldmetal
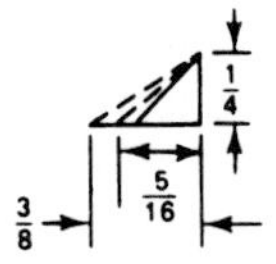	1/4 × 1/4 design 1/4 × 5/16 1/4 × 3/8	0.106 0.133 0.159	— 25% 50%
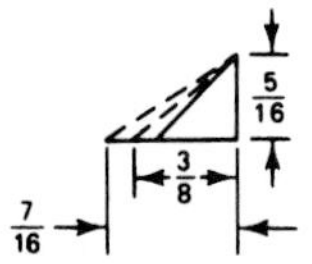	5/16 × 5/16 design 5/16 × 3/8 5/16 × 7/16	0.166 0.199 0.232	— 20% 40%
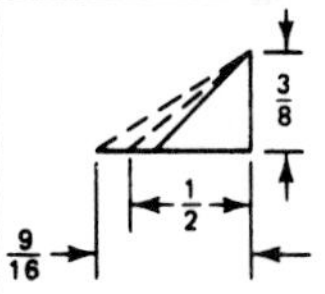	3/8 × 3/8 design 3/8 × 1/2 3/8 × 9/16	0.239 0.318 0.358	— 33% 50%
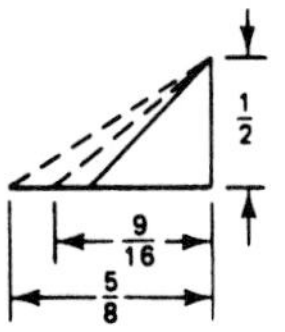	1/2 × 1/2 design 1/2 × 9/16 1/2 × 5/8	0.425 0.477 0.531	— 12% 25%

FIGURE 19–45 Extra weld metal required for overwelded fillets.

3. Eliminate excessive reinforcing on all welds. Reinforcing is not required to obtain weld strength.

4. Utilize subassemblies whenever possible. This will minimize distortion and reduce cycle time.

5. Follow welding procedures to minimize distortion. Distortion will create poor fitup, which may require additional welding.

6. Use positioning equipment and flat-position welding in all production.

7. Follow welding procedures. Use proper arc length and welding current.

8. Use all the filler metal purchased. Do not discard long stubs. Use large electrode sizes. Purchase filler metals in large lot sizes.

9. Provide power tools for slag removal and for weld finishing.

10. Provide for welder comfort and safety by using scaffolding, worker positioners, guard rails, and so on.

11. Maintain welding equipment efficiency through routine maintenance procedures.

12. Check for cable and connector efficiency in the welding circuit. Hot spots waste power.

Involvement of all personnel will turn up unexpected cost savings. Be flexible.

Finally, take an overall view of the weldment. The design must be functional and improve every time it is changed.

19-9 WELDING SYMBOLS

The American Welding Society developed and established a system of welding symbols in the 1930s. The purpose was to identify the location of welds and transmit this information on engineering drawings from the designer to the welding shop. Since that time, numerous revisions have been made. The original purpose is being fulfilled and welding symbols are being used increasingly by progressive fabricators and users of welding. In the recent past, through the efforts of the International Institute of Welding and the International Standards Organization, the welding symbols of different countries are being unified so that there will be a common system of weld symbols throughout the world. The latest edition of the American Welding Society's "Standard Symbols for Welding, Brazing, and Nondestructive Examination"[7] is in substantial agreement within the ISO standards and those of many countries throughout the world. Increased international trade makes the standardization of welding symbols an important objective. For universal international use, two more steps are required. This includes the conversion of the American measuring system to the metric system, and the second is to resolve the differences between drafting-room practices in North America and in Europe. In North America, the third-angle projection is normally used. Europe uses the first-angle projection. In the first-angle projection, the side view is the left-side view, and in the third-angle projection, the side view is the right-side view. This is not too important with respect to weld symbols but does cause confusion in interpreting drawings and symbols.

The purpose of welding symbols is to describe the desired weld accurately and completely. The welding symbol can also be used to transmit other information, such as specifications and procedures. This is done by means of special information in the tail of the arrow. Welding symbols can also be combined with nondestructive examination symbols.

The welding symbol consists of eight elements that may or may not all be used in each symbol.

1. Reference line
2. Arrow
3. Basic weld symbol
4. Dimension and other data
5. Supplementary symbol
6. Finish symbol
7. Tail
8. Specification, process, or reference

The first and second elements and either the third or seventh must be used to make an intelligible welding symbol. The others may or may not be used, in accordance with the necessity of passing along the information or the standard practice of the organization that is using them.

The foundation for constructing a welding symbol is the reference line. The reference line is always shown in the horizontal position, and it should be drawn near the weld joint that it is to identify. The other

parts of the welding symbol are constructed on the reference line. Each of the other elements of the welding symbol must be placed in proper location with respect to the reference line and in accordance with the standard location (Figure 19-46). The elements that describe the basic weld, the dimensions and other data, and the supplementary and finish symbols are always located with the same relationship to the reference line no matter which end of the reference line carries the arrow.

The next important element of the welding symbol is the arrow. This is a line from one end of the reference line to an arrowhead to the arrow side or arrow side member of the weld joint. When the symbol is used for joints that require the preparation of one member, only the arrowhead should point with a definite break to the member of the joint that is to be prepared.

The other end of the reference line carries the tail of the arrow. The area in the tail can be used to provide references to specifications, processes, or other specific information; when there are no such references, the tail is omitted.

The most important part is the weld symbol, which is used to indicate the desired type of weld. The basic symbols are shown in Figure 19-47.

If the basic weld symbol is placed under the reference line, that symbol is to define the weld on the arrow side of the joint or the arrow side member of the joint. If the weld symbol is placed above the reference line, it is to define the weld made on the other side or other side member of the joint. When the symbol is placed on both sides, it would indicate that the weld is made on both sides. Figure 19-48 shows the identification of the arrow

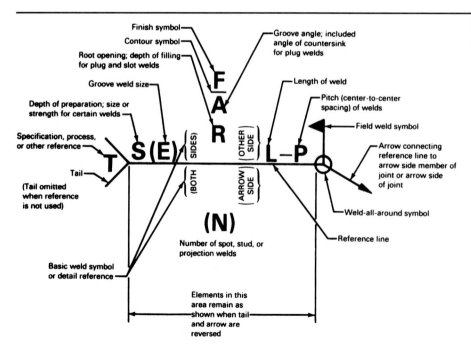

FIGURE 19–46 Standard location of elements of a welding symbol.

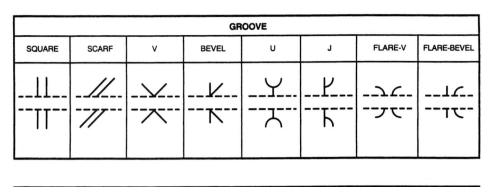

NOTE: THE REFERENCE LINE IS SHOWN DASHED FOR ILLUSTRATIVE PURPOSES.

FIGURE 19–47 Basic weld symbols.

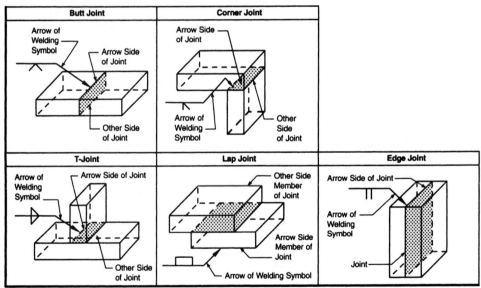

FIGURE 19–48 Identification of arrow side and other side.

side and the other side of the joint and the arrow side and other side member of the joint.

The various dimensions that help describe and define the weld have a specific location relationship to the weld symbol. The size of the weld is to be placed at the left of the weld symbol. In groove welds, if the size is not shown, it indicates complete joint penetration. The root opening or depth of filling for plug and slot welding is to be placed directly in the weld symbol. The groove angle, that is, the included angle, for groove welds and the included angle of countersink for plug welds are placed above or below the weld symbol. This dimension is often omitted if there is a company standard or all-inclusive note on the drawing. To the immediate right of the weld symbol will be the dimension indicating the length of the weld, and if required, a dash and the next number will indicate the pitch, which is the center-to-center spacing of intermittent welds. These are all given by Figure 19–49, showing the standard location of elements of a welding symbol.

More than one basic weld symbol can be used to specify a weld joint. For example, in a T-joint a fillet weld may be included in addition to a groove weld. In this case, the basic groove weld symbol would be made to touch the reference line and the basic fillet weld symbol would be added on top.

The next element of the welding symbol consists of the supplementary symbols. These are to be used in conjunction with welding symbols and have a specific location. The supplementary symbols are shown in Figure 19–50. The supplementary symbols are used for situations that require them. Supplementary symbols include the weld-all-around symbol, the field weld symbol, symbols indicating the contour of the finished weld, and others. Surface finish symbols are used for very exacting requirements.

FIGURE 19–49 Basic welding symbols and their location.

Basic Welding Symbols and Their Location Significance								
Location Significance	Fillet	Plug or Slot	Spot or Projection	Stud	Seam	Back or Backing	Surfacing	Edge
Arrow Side								
Other Side				Not Used			Not Used	
Both Sides		Not Used	Not Used	Not Used	Not Used	Not Used	Not Used	
No Arrow Side or Other Side Significance	Not Used	Not Used		Not Used		Not Used	Not Used	Not Used
Location Significance	Groove							Scarf for Brazed Joint
	Square	V	Bevel	U	J	Flare-V	Flare-Bevel	
Arrow Side								
Other Side								
Both Sides								
No Arrow Side or Other Side Significance		Not Used	Not Used	Not Used	Not Used	Not Used	Not Used	Not Used

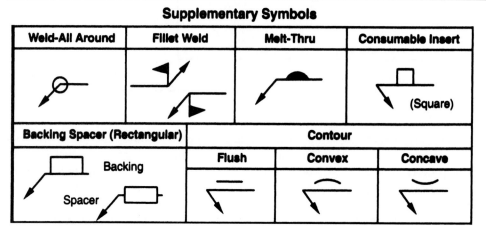

Supplementary Symbols

Weld-All Around	Fillet Weld	Melt-Thru	Consumable Insert
			(Square)
Backing Spacer (Rectangular)	Contour		
Backing / Spacer	Flush	Convex	Concave

FIGURE 19–50
Supplementary symbols.

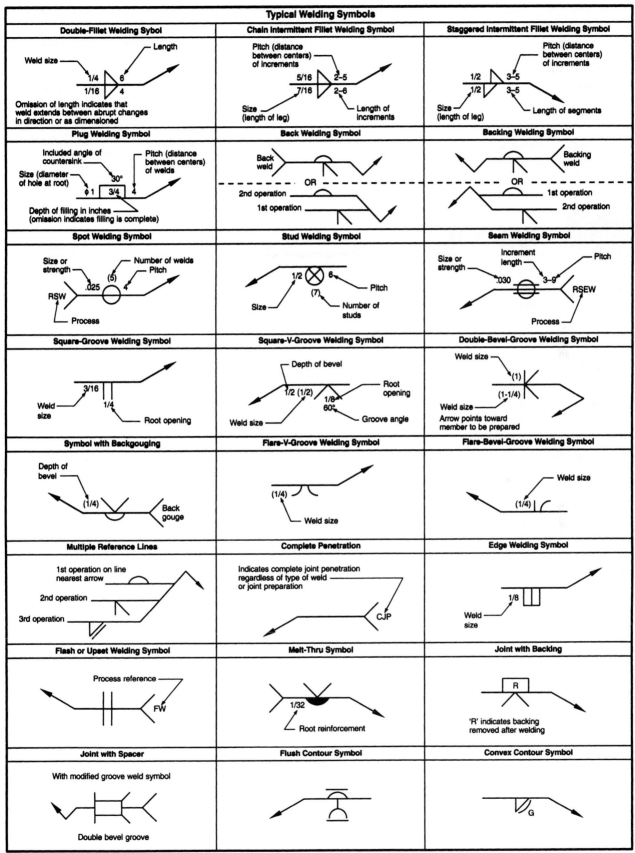

'It should be understood that these charts are intended only as shop aids. The only complete and official presentation of the standard welding symbols is in A2.4.

FIGURE 19–51 Typical welding symbols including use of supplementary symbols.

The following letters are used to indicate the method of finishing but not the degree of finish.

- ※ C Chipping
- ※ G Grinding
- ※ H Hammering
- ※ M Machining
- ※ R Rolling

A number of typical welding symbols are shown in Figure 19-51. Basic weld symbols, as well as the standard location of elements of a welding symbol, should be learned by all who use weld symbols: the designer, the draftsman, the detailer, the layout man, the setup man, the welder, and the welding inspector.

The AWS "Standard Symbols for Welding, Brazing, and Nondestructive Examination" should be available in every drafting room, engineering department, and weld shop. Training programs are available for learning welding symbols. AWS will soon have a recommended practice for detailing welding symbols for computer-aided drafting.

QUESTIONS

19–1. List all the advantages of welded construction.

19–2. The designer is responsible for weldment design. List the factors that the designer must consider.

19–3. Define and explain the factor of safety.

19–4. Explain tensile loading, compression loading, bending loading, torsion loading, and shear loading.

19–5. What is a stress concentration? What causes it? How can it be eliminated?

19–6. How does a stress concentration affect the life of a weld?

19–7. What are the five basic types of joints? Describe each.

19–8. What are the eight types of welds? Describe or sketch each.

19–9. Define the six parts of a fillet weld.

19–10. Define the five parts of a groove weld.

19–11. If the fillet weld size is doubled, how much more weld metal is required?

19–12. What are the seven types of groove welds? Are they used as single or double?

19–13. Is the surfacing weld used for making corner welds or T-welds?

19–14. Name the fields of welding in which weldment designs are governed by specifications.

19–15. What is weldment conversion? How is it done?

19–16. Is the welder responsible for not making an inaccessible weld? Why?

19–17. Which is the arrow side of a weld symbol reference line?

19–18. What does the triangle symbol indicate? How do you specify a double one?

19–19. What is the symbol for *weld all around?* What does it mean?

19–20. What symbol indicates welding at field erection? Describe it.

REFERENCES

1. "Welded Brake Shoes: End of the Road for Castings," *Welding Design and Fabrication* (October 1963).

2. "Steel Plates and Their Fabrication," Lukens Steel Company, Coatsville, Pa.

3. "Fatigue of Welded Structures," Welding Research Bulletin No. 422 June 1997.

4. "Structural Welding Code—Steel," AWS D1.1, American Welding Society, Miami, Fla.

5. "Welded Joint Design," Military Standard MIL-STD-0022B (SHIPS), U.S. Naval Ship Engineering Center, Hyattsville, Md.

6. "Manual of Steel Construction," American Institute of Steel Construction, New York.

7. "Standard Symbols for Welding, Brazing, and Nondestructive Examination," AWS A2.4, American Welding Society, Miami, Fla.

20

COST OF WELDING

OUTLINE

20-1 WELDMENT COST ELEMENTS

The cost of welding, like the cost of any industrial process, includes the cost of labor, materials, and overhead. Welding costs are used to make cost estimates for bidding on welding work, for setting rates for incentive programs, and for comparing welded construction and competing processes.

It is necessary to differentiate between the cost of weldments versus the cost of a specific weld. For example, in pipe welding, we are interested in the cost of making each weld. This is done in order to determine the cost advantage of one process versus another or between one joint design versus another.

The cost of the weldment is of major importance. This includes the cost of the weld, the cost of the material required, the preparation of the parts, and the post-weld treatment required. The cost of the weldment is required to determine the competitive advantage of a weldment versus a casting or some other type of construction.

The costing of welds and weldments must fit into the company's accounting practices. This can be done since all cost systems include labor, material, and overhead. This chapter presents methods that allow welds to be costed when using any of the arc welding processes. These methods can be used to compare the cost of competing welding processes. The basis for this system is the amount of weld metal deposited to produce the weld joint. It requires the input of data concerning labor costs related to the time required to accomplish the weld, plus the materials required, plus the overhead involved. Normally, additional overhead such as factory administrative costs and company overall sales and administrative costs are prorated according to direct labor involved in manufacturing the part.

The basis for labor costs is time: time per weld, time per increment of the length of a weld, or the time required to weld a part. Time is universal, and in many situations time represents the cost of the labor. When labor costs are prorated in another fashion, this can quickly be related in time per part or parts per unit of time. Welding materials are also related to the total cost. For those processes where weld metal is deposited in the joint, the amount of metal deposited becomes the basis of the calculation of material costs. The amount of metal required is then related to process to determine the amount of filler metal required to accomplish the weld joint. The rate of depositing weld metal relates to time, which is the basis for labor costs. This same basis can be used for establishing time standards and for constructing charts of data that can be used for estimating purposes, rate-setting purposes, and so on.

The welding procedure is the starting point for establishing welding costs. Welding procedure schedules presented throughout this book can all be used in establishing welding costs. To be of value, the welding procedure schedule must include the welding joint details, which in turn establish the amount of weld metal required to accomplish a weld joint. The procedure must also include the welding process and the type of filler metal involved since this relates to the cost of the filler metal. The welding procedure must specify the welding current since welding current, filler metal, and process relate to the rate of depositing weld metal and the utilization of filler metal purchased and deposited in the weld. The welding procedure must include the method of application since this influences the operator factor or duty cycle, which in turn determines the amount of labor used in depositing weld metal. Finally, the travel speed should be included in the procedure since it determines the amount of time required to produce weld joints. Travel speed is vitally important when weld metal is not deposited in the joint (i.e., autogenous welds).

The procedure provides the data needed to calculate the cost of a weld. The labor rate and the cost of filler metals must be included to get actual costs. In this regard, filler metals include fluxes, shielding gases, and any other material consumed in making the weld.

The cost of the weldment includes the cost of welding; however, closely related to the cost of welding is the cost of joint preparation. This is included in total weldment cost since joint preparation is required and it adds to the total weldment cost. Joint preparation varies according to material thickness and joint design. To minimize welding cost it is wise to consider alternate joint designs and alternate welding processes. Less expensive joint preparation, based on joint design, can be used with certain processes; for example, with the electroslag welding process a square cut or a mill edge can be used for the joint preparation. For deep groove welds, one or two cuts are required for the submerged arc welding process. The tradeoff involves the amount of weld metal required to produce the joint. In some cases more weld metal might be required, but it can be deposited at a higher rate. This might offset a different joint design that uses less weld metal but is applied with a slower process. These factors must all be studied in order that the optimum weld joint can be made for the least cost.

The cost of postweld treatment will be considered but without the same detail. Postweld treatment includes

final machining, grinding or polishing, heat treating, shot blasting, and possibly straightening. They are important since some welding processes require more or less postweld treatment, which influences the total cost of the weldment.

Welding, particularly manual welding, is a highly labor-intensive manufacturing process. As plants change from manual to semiautomatic, the labor input is reduced, and in changing to automatic it is further reduced. In view of this, calculations for weld costs must be extremely accurate and based on practical procedure data. Comparisons of different welding processes or methods of application can be the basis for investments in automatic equipment. Data in the book and this chapter provide the basis for making these cost determinations, which can greatly affect capital expenditures.

Field welding costs more than shop welding. Welding in the horizontal, vertical, or overhead position costs more than welding in the flat position. Local working conditions in the shop or on the job affect costs. Available equipment for fitting up or handling the work, experience and skill of the welders, power rates, special code requirements, weather and temperature conditions, industrial regulations, and many other variables affect costs; tables, graphs, charts, and so on, can be offered only as general guidelines applying to average conditions.

The tables and graphs that provide final costs when labor rates are inserted should be used with caution. These data must be based on many assumptions that might not apply in your situation. For example, the weld designs can vary in root opening, groove angle, and so on. The reinforcement may be different. The filler metal yield varies from electrode to electrode, and the operator factor can also vary widely. In addition, the data might be slanted to favor one process over another. The data presented in this chapter will enable you to construct your own tables and charts based on your standard weld designs, the processes you employ, the filler metals used, and the operator factors you expect. When these data are modified by your pay rates, overhead rates, and materials prices, they will apply to your operation. It is wise to include only the standard data in charts, since rates and prices continuously change. Computer programs are available for making cost calculations (see Section A-3 in the Appendix).

20-2 WELD METAL REQUIRED FOR JOINTS

The filler metal cost is based on the amount of weld metal deposited in the joint. The exception is the autogenous welds when weld metal is not deposited. The same method of calculating weld metal in joints can also be used for surfacing and overlay applications. The procedure involved applies to all the arc welding and other welding processes in which metal is deposited.

This system utilizes standardized joint designs. Most welding organizations' welding codes and standards provide standardized designs. Standardized designs are also presented in Chapter 19. The same information is presented here with area and weight calculated for these joint designs in various thicknesses of material. This information is based on the use of steel base metal and weld metal. However, the information is presented so that data for other metals can be easily calculated. The cross-sectional area is related to the standard joints and can be modified for different metals based on their density. In this chapter only the conventional measurements are provided since to include metric measurements would unduly complicate the tables.

Every weld has a cross-sectional area that can be determined by straightforward geometric calculations. If weld details are standardized, it is a simple matter to calculate the cross-sectional area. The formulas for the different welds are shown in Figure 20–1. In these figures the letter designations for the different parts of the weld are the same as Chapter 19. These are as follows:

- *A*   Groove angle
- *CSA*   Cross-sectional area
- *D*   Diameter plug weld, arc spot weld
- *L*   Length of slot weld
- *R*   Radius (used in J- and U-grooves)
- *RF*   Root face
- *RO*   Root opening
- *S*   Fillet size, bead size, or size of groove weld if not complete penetration
- *T*   Thickness
- *W*   Width of surfacing

In addition, the weld designs are the same as those in Chapter 19.

Information shown in Figure 20–1 can be applied in different ways. For example, the bead weld can be used for a backing bead or a sealing bead or for thinner metals for the flange, edge, and corner weld. The amount of metal required for the bead weld may sometimes be added to the full-penetration groove welds for root reinforcing.

When welds of different design configuration are used, the cross-sectional area is calculated using geo-

metric area formulas. The formulas for each weld provide theoretical cross-sectional area values with a flush surface. Weld reinforcement is not included in these values. For practical purposes, however, reinforcement must be added to a weld. The standardized weld designs are shown in Figure 20–2. In this figure, the different weld designs are related to material thicknesses. The theoretical cross-sectional area in square inches is shown for each weld in its normal range thickness. In addition, the theoretical weight of the weld deposit is shown related to the design and thickness. The weight is based on 12 in. or 1 ft of joint of steel weld metal in pounds per linear foot of weld. This is calculated using Equation (20-1).

The constant 0.283 is the weight of a cubic inch of steel in pounds per cubic inch. The data can be used for any metal by using its density or weight per cubic inch.

To make this datum more practical, a reinforcement factor is added. A value of reinforcement of 10% is added to single-groove welds and 20% reinforcement is added to the double-groove welds. Ten percent is also added as reinforcement for fillet welds. These are arbitrary figures, but they are sufficiently accurate for most calculations. For greater accuracy it may be desirable to make representative welds, cross section them, and measure the reinforcement. This would be accurate for a particular weld, but since reinforcement varies it may not be worth the effort. Figure 20–2 therefore includes two additional columns. One provides the cross-sectional area of the welds in square inches in the different material thicknesses without reinforcement. The final column provides the weight of the weld deposit with reinforcement in pounds per foot of weld.

Another value of the data presented in Figure 20–2 is their usefulness in visualizing how welding costs are related to weld designs. They illustrate the amount of weld metal required for different weld designs. For example, they will show the increase in cross-sectional area or weight of weld metal required when the size of a fillet weld is increased. Other comparisons can be made such as the difference in the cross-sectional area or weld metal required between a bevel to a V-groove weld or between a single- and a double-groove weld. They can also be used for braze welding and for oxy-fuel gas welding. These data can be the basis for a standard cost system when standard weld design as shown is used. If the weld designs are different from those used in the charts, the data must be recalculated to reflect these changes.

$$\text{weight of deposit (lb/ft)} = \text{cross-sectional area (in}^2) \times 0.283 \text{ (lb/in}^3) \times 12 \text{ (in/ft)} = 3.396 \text{ (lb/ft)} \qquad (20\text{-}1)$$

WELD	DESIGN	FORMULA FOR CROSS-SECTIONAL AREA
FILLET (EQUAL LEGS)	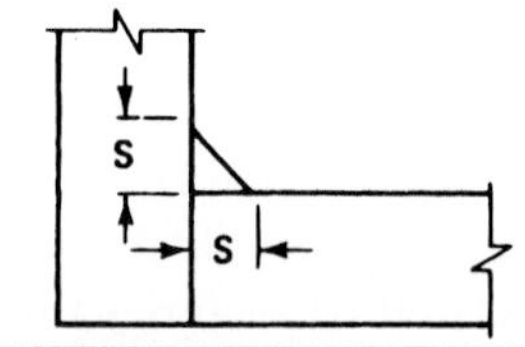	$CSA = 1/2(S)^2$
FILLET (UNEQUAL LEGS)	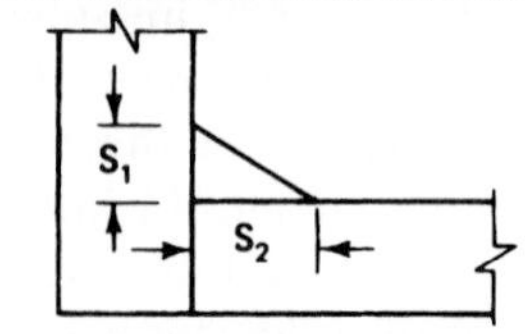	$CSA = 1/2 (S_1 \times S_2)$
SURFACE	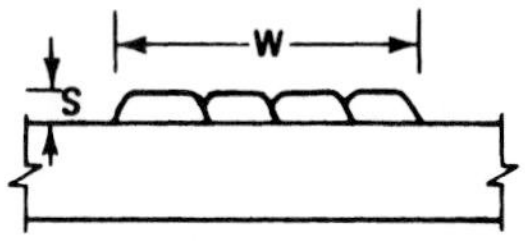	$CSA = S \times W$
PLUG	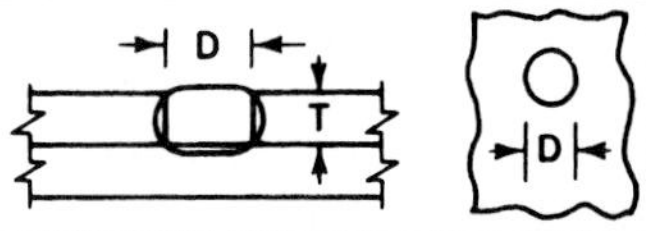	$VOL = \pi \left(\dfrac{D}{2}\right)^2 \times T$ FORMULA PROVIDES VOLUME OF WELD METAL PER WELD
SLOT	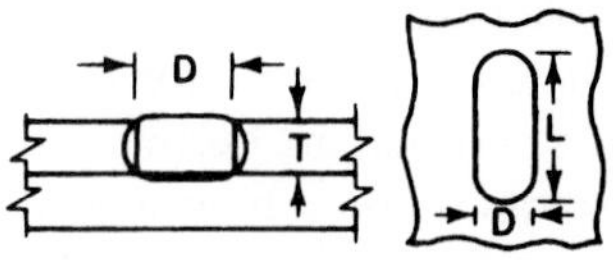	$VOL = \left[\pi \left(\dfrac{D}{2}\right)^2 + (L - D)D\right] T$ FORMULA PROVIDES VOLUME OF WELD METAL PER WELD
ARC SPOT	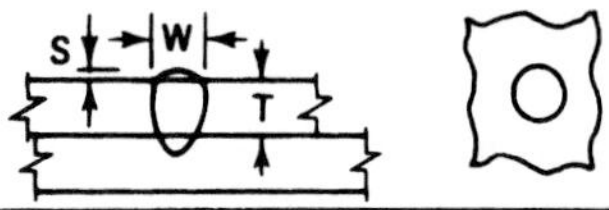	$VOL = 1/2S \left(\pi \dfrac{W}{2}\right)^2$ (VOLUME PER WELD) FORMULA PROVIDES VOLUME OF WELD METAL PER WELD
ARC SEAM	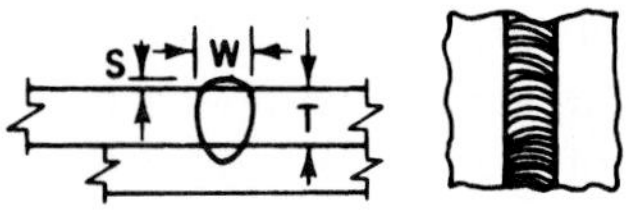	$CSA = 1/2WS$
BEAD	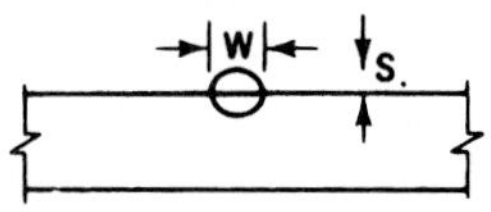	$CSA = 1/2WS$
SQUARE	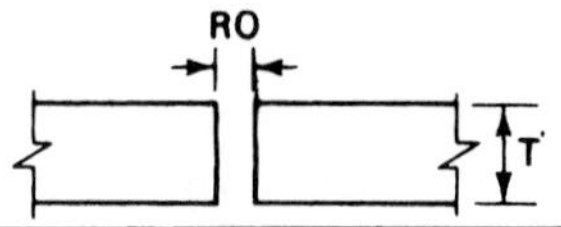	$CSA = RO \times T$
SINGLE-V	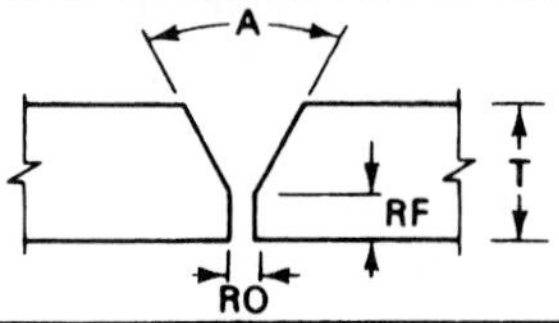	$CSA = (T - RF)^2 \tan \left(\dfrac{A}{2}\right) + RO \times T$
DOUBLE-V	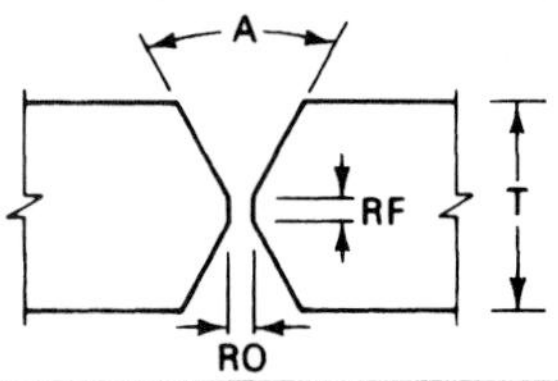	$CSA = 1/2(T - RF)^2 \tan \left(\dfrac{A}{2}\right) + RO \times T$

FIGURE 20–1   Cross-sectional area of welds.

608

WELD	DESIGN	FORMULA FOR CROSS-SECTIONAL AREA
SINGLE-BEVEL	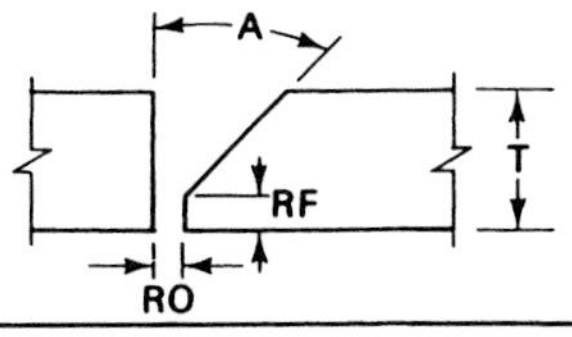	$CSA = 1/2(T - RF)^2 \tan A + RO \times T$
DOUBLE-BEVEL	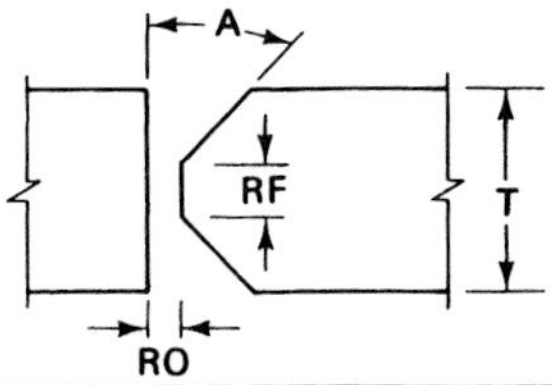	$CSA = 1/4(T - RF)^2 \tan A + RO \times T$
SINGLE-U	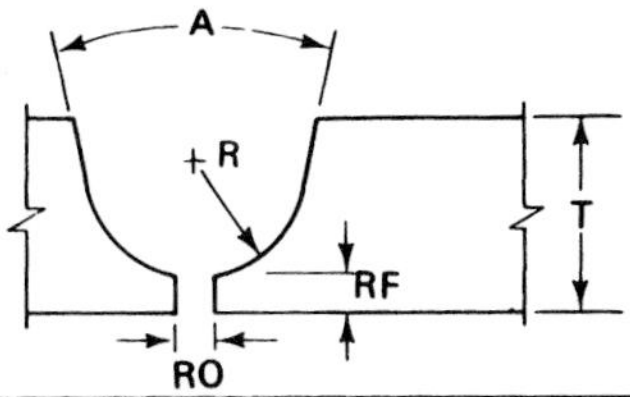	$CSA = (T - R - RF)^2 \tan \left(\dfrac{A}{2}\right) + 2R(T - R - RF)$ $+ 1/2\pi R^2 + RO \times T$
DOUBLE-U	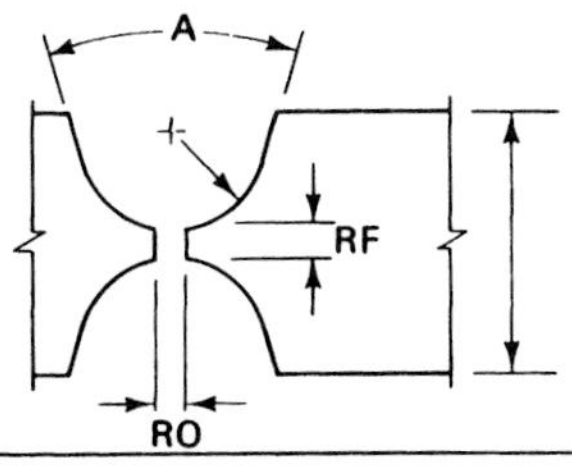	$CSA = 1/2(T - 2R - RF)^2 \tan \left(\dfrac{A}{2}\right) + 2R(T - 2R - RF)$ $+ \pi R^2 + RO \times T$
SINGLE-J	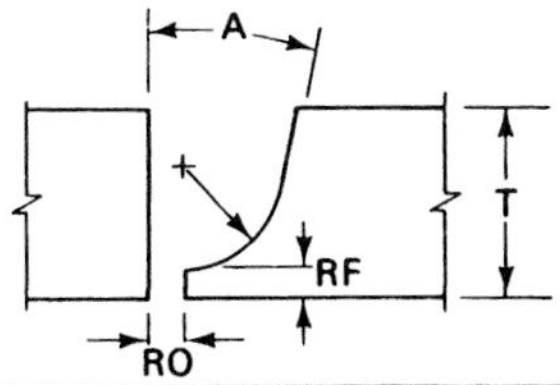	$CSA = 1/2(T - R - RF)^2 \tan A +$ $R(T - R - RF) + 1/4\pi R^2 + RO \times T$
DOUBLE-J	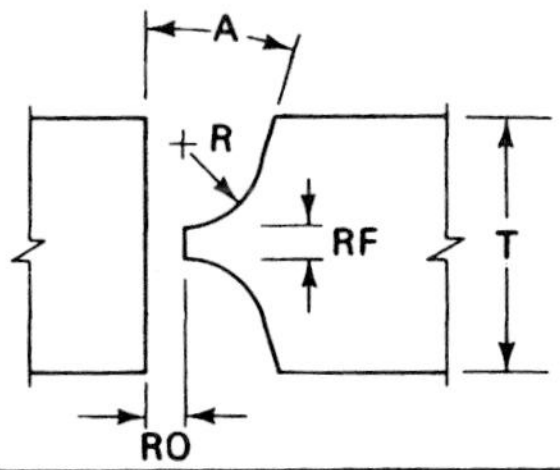	$CSA = 1/4(T - 2R - RF)^2 \tan A +$ $R(T - 2R - RF) + 1/2\pi R^2 + RO \times T$
FLARE-V	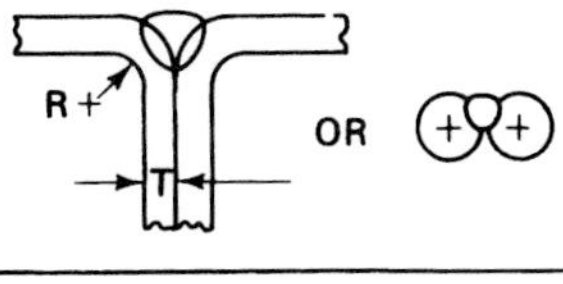	$CSA = \dfrac{(2 \times R + T)^2 - \pi(R + T)^2}{2}$
FLARE-BEVEL	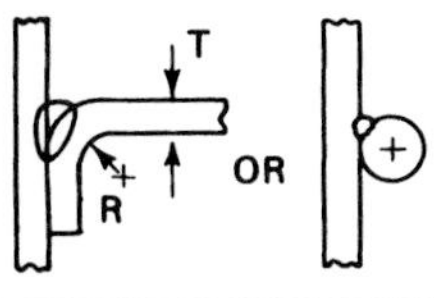	$CSA = \dfrac{(2 \times R + T)^2 - \pi(R + T)^2}{4}$

FIGURE 20–1 *(cont.)*

609

Weld	Design	T in inch	CSA Theor. in.²	Weld Deposit Theoretical lb/ft	CSA w/ref. in.²	Weld Deposit w/ Reinforcement lb/ft
Fillet (equal legs)	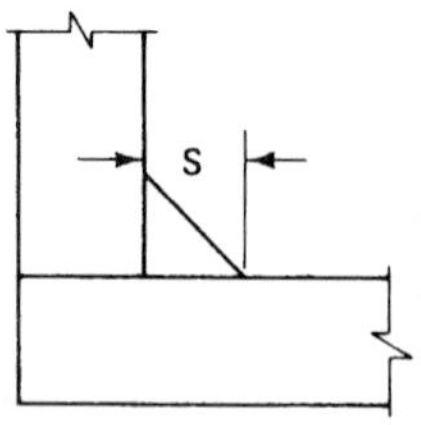	1/8	0.008	0.027	0.009	0.030
		3/16	0.018	0.061	0.020	0.067
		1/4	0.031	0.106	0.034	0.117
		5/16	0.049	0.167	0.054	0.184
		3/8	0.070	0.238	0.077	0.262
		7/16	0.096	0.326	0.106	0.360
		1/2	0.125	0.425	0.138	0.468
		9/16	0.158	0.537	0.174	0.591
		5/8	0.195	0.663	0.215	0.729
		3/4	0.281	0.956	0.309	1.052
		7/8	0.383	1.503	0.421	1.653
		1	0.500	1.700	0.550	1.876
Fillet (unequal legs)	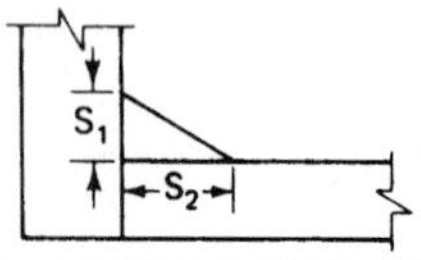	1/4 × 3/8	0.047	0.160	0.052	0.176
		3/8 × 1/2	0.094	0.319	0.103	0.351
		1/2 × 5/8	0.156	0.530	0.172	0.583
		5/8 × 3/4	0.234	0.795	0.258	0.875
		3/4 × 1	0.375	1.274	0.413	1.401
Square	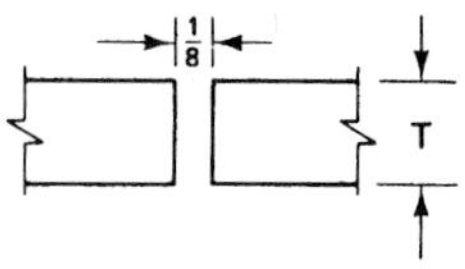	1/8	0.016	0.054	0.019	0.065
		5/32	0.019	0.065	0.023	0.078
		3/16	0.023	0.078	0.027	0.094
		7/32	0.027	0.092	0.032	0.110
		1/4	0.031	0.105	0.037	0.126
		9/32	0.035	0.119	0.042	0.143
		5/16	0.039	0.132	0.047	0.158
Single-V	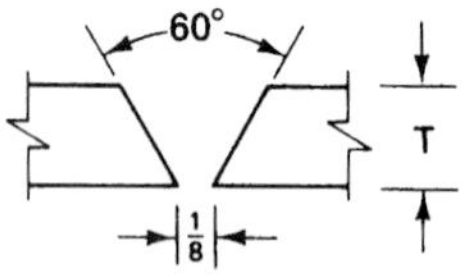	1/4	0.067	0.228	0.074	0.251
		3/8	0.128	0.384	0.141	0.422
		1/2	0.206	0.702	0.227	0.772
		5/8	0.305	1.040	0.336	1.144
		3/4	0.418	1.430	0.460	1.573
		1	0.702	2.395	0.772	2.635
Double-V	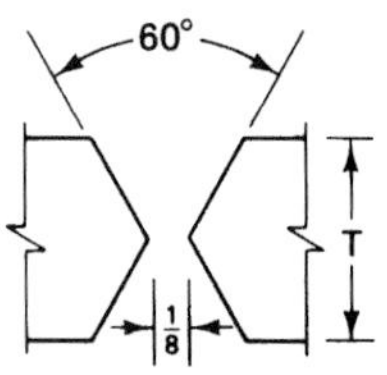	3/4	0.256	0.874	0.307	1.049
		1	0.414	1.420	0.497	1.704
		1-1/4	0.608	2.075	0.730	2.490
		1-1/2	0.838	2.860	1.006	3.432
		1-3/4	1.105	3.765	1.326	4.518
		2	1.405	4.780	1.686	5.736
		2-1/4	1.742	5.945	2.090	7.134
		2-1/2	2.210	7.530	2.652	9.036
		2-3/4	2.530	8.620	3.036	10.344
		3	2.978	10.150	3.574	12.180
		3-1/2	3.970	13.530	4.764	16.236
		4	5.620	19.130	6.744	22.956
Single-bevel	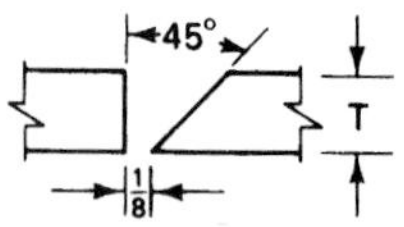	1/4	0.063	0.215	0.069	0.237
		3/8	0.117	0.364	0.129	0.400
		1/2	0.188	0.641	0.207	0.705
		5/8	0.301	1.025	0.331	1.128
		3/4	0.375	1.280	0.413	1.408
		1	0.625	2.135	0.687	2.349
Double-bevel	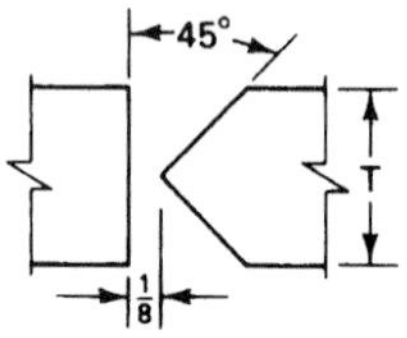	5/8	0.176	0.600	0.211	0.720
		3/4	0.234	0.798	0.281	0.958
		7/8	0.301	1.025	0.361	1.230
		1	0.375	1.279	0.450	1.535
		1-1/4	0.547	1.862	0.656	2.234
		1-1/2	0.750	2.560	0.900	3.072
		1-3/4	0.984	3.360	1.181	4.032
		2	1.250	4.260	1.500	5.112
		2-1/2	1.875	6.398	2.250	7.677
		3	2.625	8.950	3.150	10.740

FIGURE 20–2 Area and weight of weld metal deposit.

Weld	Design	T in inch	CSA Theor. in.²	Weld Deposit Theoretical lb/ft	CSA w/ref. in.²	Weld Deposit w/ Reinforcement lb/ft
Single-U	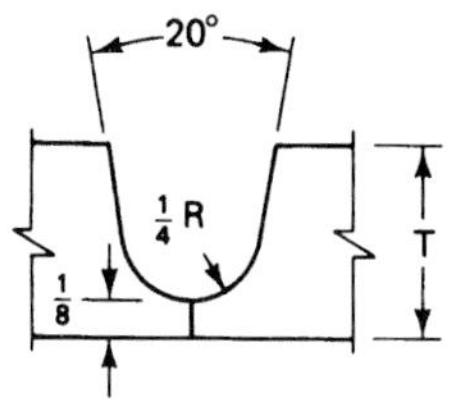 (20° groove, 1/4 R, 1/8)	1/2	0.163	0.555	0.179	0.611
		3/4	0.310	1.058	0.341	1.164
		7/8	0.392	1.338	0.431	1.472
		1	0.479	1.635	0.527	1.799
		1-1/4	0.671	2.288	0.738	2.517
		1-1/2	0.885	3.020	0.974	3.322
		1-3/4	1.120	3.820	1.232	4.202
		2	1.376	4.680	1.514	5.148
		2-1/2	1.961	6.680	2.157	7.348
		3	2.631	8.960	2.894	9.856
Double-U	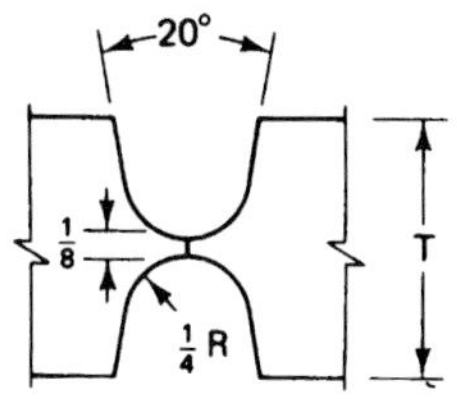 (20° groove, 1/8, 1/4 R)	1	0.396	1.350	0.475	1.620
		1-1/4	0.543	1.852	0.652	2.222
		1-1/2	0.701	2.390	0.841	2.868
		1-3/4	0.870	2.968	1.044	3.562
		2	1.151	3.922	1.381	4.706
		2-1/4	1.242	4.235	1.490	5.082
		2-1/2	1.444	4.925	1.732	5.910
		2-3/4	1.658	5.650	2.000	6.780
		3	1.879	6.410	2.255	7.692
		3-1/2	2.389	8.150	2.867	9.780
		4	2.951	10.070	3.541	12.084
		4-1/2	3.459	11.790	4.151	14.148
Single-J	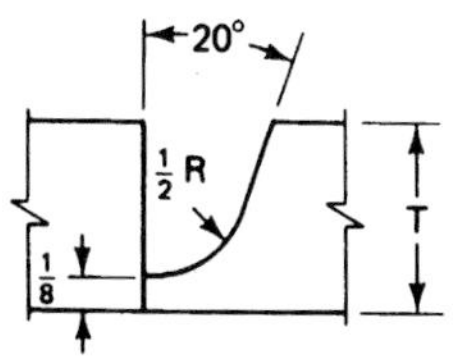 (20°, 1/2 R, 1/8)	1/2	0.180	0.614	0.198	0.675
		3/4	0.261	0.890	0.287	0.979
		1	0.409	1.395	0.450	1.535
		1-1/4	0.580	1.978	0.638	2.176
		1-1/2	0.774	2.640	0.851	2.904
		1-3/4	0.989	3.375	1.088	3.713
		2	1.229	4.190	1.352	4.609
		2-1/4	1.491	5.080	1.640	5.589
		2-1/2	1.774	6.050	1.957	6.655
Double-J	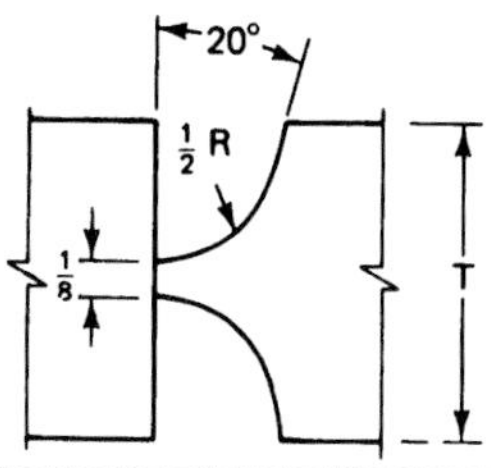 (20°, 1/2 R, 1/8)	1	0.360	1.228	0.432	1.474
		1-1/4	0.437	1.490	0.524	1.788
		1-1/2	0.589	2.010	0.707	2.412
		1-3/4	0.728	2.482	0.874	2.978
		2	0.875	2.983	1.050	3.580
		2-1/4	1.029	3.510	1.235	4.212
		2-1/2	1.191	4.065	1.429	4.878
		2-3/4	1.360	4.640	1.632	5.568
		3	1.535	5.235	1.842	6.282
		3-1/2	1.909	6.510	2.291	7.812
		4	2.313	7.880	2.776	9.456

FIGURE 20-2 (cont.)

20-3 FILLER METAL AND MATERIALS REQUIRED

Electrodes

Section 20-2 provides information necessary to calculate the "weight of weld metal deposited" in a weld joint or on a weldment. It also provides tables that give the weight of weld metal deposit for standard weld designs made in different thicknesses of material. The total weight of weld metal deposited in the joint or required to produce the weldment can easily be calculated using these tables.

The weight of filler metal purchased to make the weld or the weldment is greater than the weight of the weld metal deposit. This is true for most of the arc welding processes, but not for all. Stated another way, it means that more filler metal must be purchased than is deposited because of stub end losses, coating or slag losses, spatter losses, and so on. This can be shown by Equation (20-2). These losses are sometimes represented as a ratio and called deposition efficiency, filler metal yield, or recovery rate.

Deposition efficiency is the ratio of the weight of deposited weld metal in the weld divided by the net weight of filler metal consumed, exclusive of stubs: As

$$\text{weight of weld metal deposited (lb)} = \frac{\text{weight of weld metal deposited (lb)}}{1 - \text{total electrode loss}} \qquad (20\text{-}2)$$

shown by Equation (20-3), filler metal yield is the ratio of the weight of deposited weld metal divided by the gross weight of the filler metal used. Thus yield relates to the amount of filler metal purchased. The filler metal yield for the different types of electrodes and filler metals will range from 50% to 100%. Yield is the better term to use since stubs occur only when using covered electrodes. When the equation calls for a % figure, use the decimal equivalent.

The covered electrode has the lowest yield, that is, it has the highest losses. These losses are made up of the stub end loss, the coating or slag loss, and the splatter loss. Considering a 2-in. stub, the 14-in.-long electrode has a 14% stub loss, the 18-in. electrode has an 11% stub loss, and the 28-in. electrode has a 7% stub loss. Electrodes are not always melted to a 2-in. stub. The coating or slag loss of a covered electrode can range from 10% to 50%. The thinner coatings on an E6010 electrode are at the lower end of the scale approximating 10%, while the heavy coating on an E7028 electrode will approach 50%. This can apply even when iron powder is incorporated in the coating. For accurate results, measure this factor for the electrodes to be used. The spatter loss depends on the welding technique, but normally ranges from 5% to 15% loss.

The solid bare electrode or filler rod has the highest yield since the losses are minimized. Normally, in the continuous electrode wire processes the entire spool or coil of electrode wire is consumed in making the weld. Scrap ends of coils are discarded, but this is usually negligible compared to the total weight. The spatter loss relates to welding process and welding technique. In the submerged arc and electroslag welding processes, virtually all of the electrode is deposited in the weld. There is little spatter and, therefore, deposition efficiency or yield approaches 100%. In gas metal arc welding there is a loss from spatter and this amounts to approximately 5% of the electrode melted, which provides a deposition efficiency or yield of 95%. In the case of the cold wire processes, gas tungsten arc, plasma arc, and carbon arc, the *cold wire* filler metal is completely used and has a 100% yield.

The flux cored electrode has slightly higher losses because of the fluxing ingredients within the tubular wire that are consumed and lost as slag. The fluxing materials in the core represent from 10% to 20% of the weight of the electrode. Different types and sizes have different core-to-steel weight ratios. For accurate results, measure yield of the electrode to be used. There can be up to 5% loss as spatter; therefore, the deposition efficiency or yield for flux cored electrodes ranges from 75% to 85%. The deposition efficiency or yield of filler material and the process have an important bearing on the cost of the deposited weld metal.

The filler metal cost can be calculated several different ways. The most common is based on cost per foot of weld, as shown by Equation (20-4).

The electrode price is the delivered cost to your plant. Electrode price can be reduced by purchasing in large lot sizes. In this formula the yield can be taken from Figure 20-3. These are average figures and should be sufficient for most calculations; however, for more accuracy actual measurements should be made using the filler metals that are employed.

A different method can be used for calculating the amount of weld metal required when the continuous wire processes are used. It is particularly advantageous for single-pass welding. Three simple calculations are required, but the end result is the electrode cost per foot of weld. The first step is to determine the amount of elec-

FIGURE 20–3 Filler metal yield for various types of electrodes.

Electrode Type and Process	Yield (%)
Covered Electrode for:	
SMAW 14-in. manual	55–65
SMAW 18-in. manual	60–70
SMAW 28-in. automatic	65–75
Solid bare electrode for:	
Submerged arc	95–100
Electroslag	95–100
Gas metal arc welding	90–95
Cold wire use	100
Tubular flux cored electrode for:	
Flux cored arc welding	80–85
Cold wire use	100

Note: Does not include shielding except for covered electrodes.

$$\text{weight of filler metal required (lb)} = \frac{\text{weight of weld metal deposited (lb)}}{\text{filler metal yield (\%)}} \qquad (20\text{-}3)$$

$$\text{electrode cost (\$/ft)} = \frac{\text{electrode price (\$/lb)} \times \text{weld metal deposited (lb/ft)}}{\text{filler metal yield (\%)}} \qquad (20\text{-}4)$$

trode used, expressed as pounds per hour using Equation (20-5).

The pounds per hour of electrode used disregards the yield or deposition rate factor since we are measuring the actual filler material consumed. The factor 60 is the minutes in an hour, which converts minutes to hours. The weight per length of the electrode wire is a physical property of wire based on the size of the wire and the density of the metal of the wire (Figure 20-4). The wire feed speed in inches per minute is the same as the melt-off rate of an electrode wire. It is not a true deposition rate since the spatter losses and slag losses are not considered. The wire feed speed can be determined from charts that relate the welding current to the wire feed speed, according to the size of the electrode wire, the composition of the electrode wire, and the welding process. Charts showing these data are given in Section 11-2. To use this formula it is essential to know the welding current or to measure the wire feed speed. In some instances wire feed speed is called for in the welding procedure. For very accurate work it is best to make a measurement of wire feed speed. This can be done simply by setting the wire feeder, making a test weld to determine that the weld procedure is satisfactory, and then without welding, actually allow the electrode wire to feed through the gun and measure the amount of wire fed per minute. Since this can be an extremely large amount of wire, it can be simplified by feeding for 5 seconds and multiplying the amount of wire fed by 12 to relate it to inches per minute. Instruments are available for making this measurement.

The second part of this calculation is to determine or measure the weld travel speed and arrive at this rate in feet per hour. Normally, welding procedures provide weld travel speed in inches per minute. If these data are not in the welding procedure, tests should be made to determine travel speed while making the required weld. This is then converted to feet per hour by using Equation (20-6). The 60 represents minutes in an hour and the 12 represents inches in a foot, which is the conversion factor 5.

The third part of this calculation is to determine the weight of weld metal required per foot of weld, as shown by using Equation (20-7).

This information would then be multiplied by the electrode price ($/lb) to obtain electrode cost in $/ft.

This system can also be used for flux cored arc welding, but in this case the length of electrode wire per pound is a little more difficult to determine since different types of flux cored wire have different amounts of core material of different densities (see Figure 20-5). The length per pound of steel flux cored electrode wire can be used for normal calculations. For better accuracy, actual tests should be made to determine the number of inches of the wire that weighs a pound.

Flux

When flux is used, the cost of flux must be included in the cost of materials used. The cost of flux in submerged arc welding and in electroslag welding and even in oxyfuel gas welding can be related to the weight of weld metal deposited and may be calculated using Equation (20-8).

In the submerged arc welding process, normally one pound of submerged arc flux is used with each pound of electrode wire deposited. This is a flux-to-steel ratio of 1. This ratio may change for different welding procedures and for different types of flux.

The flux ratio of 1 can be used for costing; however, for more accuracy, tests should be run with the particular flux used. The flux ratio can rise as high as 1.5.

For electroslag welding the ratio of flux to electrode wire deposited is 5 to 10 lb of flux per 100 lb of electrode consumed. This is a flux-to-steel ratio of 0.05 to 0.10. The exact amount of flux used is based on the surface area of

$$\text{weight of filler metal required (lb/hr)} = \frac{\text{wire feed speed (in./min} \times 60 \text{ (min/hr)}}{\text{length of wire per weight (in./lb)}} \qquad (20\text{-}5)$$

$$\text{travel speed (ft/hr)} = \frac{\text{travel speed (in./min)} \times 60 \text{ (min/hr)}}{12 \text{ (in./ft)}} = \text{travel speed (in./min)} \times 5 \qquad (20\text{-}6)$$

$$\text{weight of filler metal required (lb/ft)} = \frac{\text{deposition rate (lb/hr)}}{\text{weld travel speed (in./min)} \times 5} \qquad (20\text{-}7)$$

$$\text{flux cost (\$/ft)} = \text{flux price (\$/lb)} \times \text{weld metal deposit (lb/ft)} \times \text{flux ratio} \qquad (20\text{-}8)$$

FIGURE 20–4 Length versus weight (inches per pound) of bare electrode wire of type and size shown.

Wire Diameter		Inches of Metal or Alloy Per Pound								
Decimal in.	Fraction in.	Aluminum	Alum. 10% Bronze	Silicon Bronze	Copper (Deox.)	Copper– Nickel	Magnesium	Nickel	Steel, Mild	Steel, Stainless
0.020		32,400	11,600	10,300	9800	9950	50,500	9900	11,100	10,950
0.025		22,300	7960	7100	6750	6820	34,700	6820	7680	7550
0.030		14,420	5150	4600	4360	4430	22,400	4400	4960	4880
0.035		10,600	3780	3380	3200	3260	16,500	3240	3650	3590
0.040		8120	2900	2580	2450	2490	12,600	2480	2790	2750
0.045	$\frac{3}{64}$	6410	2290	2040	1940	1970	9990	1960	2210	2170
0.062	$\frac{1}{16}$	3382	1120	1070	1020	1040	5270	1030	1160	1140
0.078	$\frac{5}{64}$	2120	756	675	640	650	3300	647	730	718
0.093	$\frac{3}{32}$	1510	538	510	455	462	2350	460	519	510
0.125	$\frac{1}{8}$	825	295	263	249	253	1280	252	284	279
0.156	$\frac{5}{32}$	530	189	169	160	163	825	162	182	179
0.187	$\frac{3}{16}$	377	134	120	114	116	587	115	130	127
0.250	$\frac{1}{4}$	206	74	66	62	64	320	63	71	70

FIGURE 20–5 Length per pound of steel flux cored electrode wires.

Electrode Diameter (in.)	Length by Weight (in./lb)
0.045	2400
$\frac{1}{16}$	1250
$\frac{5}{64}$	1000
$\frac{3}{32}$	650
$\frac{7}{64}$	470
0.120	380
$\frac{1}{8}$	345
$\frac{5}{32}$	225

retaining shoes exposed to the flux pool. The rule of thumb is $\frac{1}{4}$ lb of flux per vertical foot of joint height. The 10% ratio mentioned above is ample to cover the cost of electroslag flux.

In oxyfuel gas welding and torch brazing the amount of flux used can also be related to the amount of filler wire consumed. This ratio is usually on the order of 5 to 10%. Accurate checks can be made for a more precise flux ratio if desired.

Shielding Gas

When shielding gas is used, it must be included in material cost. The cost of the gas is related to the time required to make the weld. Shielding gas is normally used at a specified flow rate and measured in cubic feet per hour. The amount of shielding gas used would be the time required to make the weld times the rate of gas usage. The cost of shielding gas can be figured two ways. Normally, the gas cost is based on the cost per foot of the weld and it is calculated using Equation (20-9).

The gas flow rate is provided in the welding procedure or it can be measured with a flow meter. The price of gas is the delivered price at the welding station.

Shielding gas cost per minute of operation is used when calculating the cost of making an arc spot weld, a small joint, or a small part. This is based on the time required to make the weld, and it may be calculated using Equation (20-10).

$$\text{gas cost (\$/ft)} = \frac{\text{price of gas (\$/ft}^3) \times \text{flow rate (ft}^3/\text{hr})}{\text{weld travel speed (in./min)} \times 5} \qquad (20\text{-}9)$$

$$\text{gas cost (\$/weld)} = \frac{\text{price of gas (\$/ft}^3) \times \text{flow rate (ft}^3/\text{hr}) \times \text{weld time (min)}}{60 \text{ (min/hr)}} \qquad (20\text{-}10)$$

Miscellaneous Material Costs

To obtain a total cost of a particular weld, other items should be included. These include the guide tube used in the consumable guide electroslag welding process, the cost of ferrules and studs in arc stud welding, and so on. In stud welding the price of each stud must be considered, even though studs are not strictly filler metals. They are related to the number of welds made and can be calculated in this manner. Also, since a ceramic ferrule is required for each weld, it must also be included in the cost of making each stud weld.

20-4 TIME AND LABOR REQUIRED

The cost of the labor required to make a weld is the single greatest factor in the total cost of a weld. Section 20-3 provided the weight of metal deposit for different welds of different sizes. With this information, the cost of the filler metals required is determined. The amount of weld deposit or the amount of filler metal required is one basis for determining the amount of time required to make a weld or weldment. Time is normally the basis for pay for welders since many are paid by the hour. The data that follow will be used to determine the cost of welding when welders are paid an hourly rate.

Welders are sometimes paid on the basis of welds made. This may be on a per footage basis for different size welds or on the number of pieces welded per hour. This type of pay is usually involved with incentive systems. In order to establish costs, on this basis it may be necessary to determine time required for making welds of different types or, conversely, the speed of making certain welds. In some cases, time studies are used to determine the normal welding time for making a particular weld or weldment. In other cases, standard cost data are used, and quite often these are based on weight of weld metal deposited. In developing the cost of welds, these data can be used in many different ways according to the pay systems and accounting systems employed. The basis for accurate weld costing is a welding procedure. The welding procedure may not be available at the time of setting cost standards or estimating costs; therefore, welding procedure schedules like those presented throughout this book should be employed.

The basis for calculating the cost of labor on a dollars-per-foot basis is given in Equation (20-11). The operator factor shown here is the same as duty cycle, which is the percentage of arc time against the total paid time.

Each of the elements of this formula requires considerable analysis. The welder's hourly pay rate can be entered into the formula; however, in most cases companies prefer to factor the pay rate to cover fringe benefits such as cost of insurance, cost of holidays, cost of vacations, and so on. This is a factor that must be determined and must be in line with the company accounting policies. It should be the same as the method used for determining machining costs and other direct labor costs used in the plant.

This assumption applies primarily to single-pass welds, since weld travel speed is available in the welding procedure schedules. It should be used when metal is not deposited, specifically GTAW and PAW. Travel speeds relate to the welding job, the weld type, and the welding process employed. This is relatively easy to determine when single-pass welds are made, but it is more difficult with multipass welds and for this reason a different system is used for large multipass welds.

Duty cycle or operator factor also requires analysis since it varies considerably from job to job and from process to process. The shielded metal arc welding process has the lowest operator factor, while semiautomatic welding is much higher, often double. The type of work dictates the operator factor. Construction work, in which small welds are made in scattered locations, has a low operator factor. Heavy production work, in which large welds are made on heavy weldments, can have much higher operator factors. Time study is sometimes used to determine operator factor based on work similar to the job being costed. Automatic time recorders are sometimes used to accurately determine operator factor on repetitive jobs. The results of this calculation provide labor costs in dollars per foot of weld joint. This can be added to the cost per foot of materials, filler metals, and so on.

As an estimate for operator factor, when other data are not available, refer to Figure 20-6. This shows the operator factor related to the method of applying the weld. The arrangement of the work, the use of positioners and fixtures, whether the work is indoors or outdoors, all have an effect on the operator factor. Construction would tend to be the lower end of the range, with heavy production toward the high end of the range. The operator factor will vary from plant to plant, and it will also vary from part to part if the amount of weld is much different. The operator factor is higher for the continuous electrode wire processes since the welder does not have to stop every time a covered electrode is consumed. Slag chipping, electrode changes, and moving from one joint to another all reduce the operator factor.

$$\text{labor cost (\$/ft)} = \frac{\text{welder pay rate (\$/hr)}}{\text{travel speed (in./min)} \times \text{operator factor (\%)} \times 5} \qquad (20\text{-}11)$$

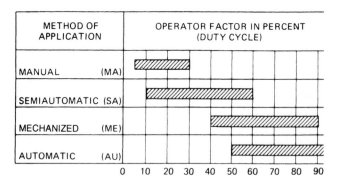

FIGURE 20–6 Welder operator factor related to method of application.

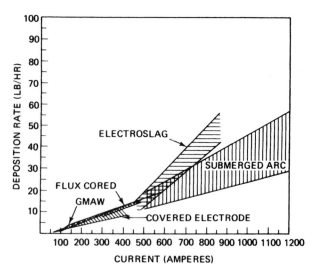

FIGURE 20–7 Deposition rates for various processes with steel electrodes.

When the welding procedure schedule is not available or when travel speed involves more than one pass, Equation (20-12) is used.

The weight of weld metal deposit in pounds per foot is the datum presented in Figure 20–2. As mentioned previously, this can be changed for different metals if the density of the metal is used in the formula. This datum is universal when the proper density factor is used.

The new factor introduced by this formula is deposition rate in pounds per hour. Deposition rate is the weight of the filler metal deposited in a unit of time. It is expressed as pounds per hour. Charts and graphs presented in the sections for particular processes provide deposition rate information. Figure 20–7 is a composite chart covering most of these processes using steel electrodes. The deposition rate has a tremendous effect on welding costs. The greater the deposition rate, usually the less time required to make a weld. This should be tempered, however, since certain high-deposition-rate processes cannot be applied to smaller jobs. For accurate results it is necessary to calculate deposition rates for specific weld procedures. This can be done by weighing the filler metal used, weighing the weld metal deposited, and measuring the arc time. This will provide the deposition rate and also the filler metal yield or utilization.

Deposition rate can also be calculated. This requires the use of Figures 11–19 to 11–21. These curves show the electrode wire feed speed per minute versus welding current. The melt-off rate and the wire feed speed are the same. The curves can be used for these calculations. The melt-off rates are given in inches per minute based on the

type and size of electrode, the welding process, and current. Figure 20–4 shows the length of bare electrode wire per pound of different metals. The relationship between melt-off rate and weight of filler metal melted can be determined by using Equation (20-13).

Use the wire feed speed graphs for melt-off rate and use Figure 20–4 for the length and weight of electrode wire.

The deposition rate, which is pounds of metal deposited per hour, is related to the melt-off rate by dividing melt-off rate by the filler metal yield or deposition efficiency. For solid wire systems the yield is high, as shown in Figure 20–3.

20-5 POWER AND OVERHEAD COSTS

Overhead cost consists of many, many things, both in the factory and in the office. It consists of the salaries of plant executives, supervisors, inspectors, maintenance personnel, janitors, and others whose time cannot be directly charged to the individual job or weldment. These costs are apportioned pro rata among all work going through the plant. Another important overhead cost is rent or depreciation of the plant and the general maintenance of

$$\text{labor cost (\$/ft)} = \frac{\text{welder pay rate (\$/hr)} \times \text{weight of weld metal deposit (lb/ft)}}{\text{deposition rate (lb/hr)} \times \text{operator factor (\%)}} \qquad (20\text{-}12)$$

$$\text{deposition rate (lb/hr)} = \frac{\text{melt-off rate (in./min)} \times 60 \times \text{filler metal yield (\%)}}{\text{length of electrode wire per weight (in./lb)}} \qquad (20\text{-}13)$$

the building, grounds, and so on. Depreciation of plant equipment includes welding machines, materials-handling equipment, overhead cranes, and all other equipment that is not charged directly to individual or specific weldments. In addition, all the taxes on the plant, the real estate, equipment, and payroll, and any other taxes applying to the operation of the plant are considered as overhead. Another item of overhead is the small tools such as chipping hammers, electrode holders, safety equipment, and so on, that are not charged to specific jobs. In addition, in most plants the cost of heat, light, maintenance, and repair of the building and equipment are also charged to overhead accounts.

Almost all plants or factories have similar systems for handling overhead expenses, but they vary in detail and for this reason are not described here. In all cases, however, the overhead charges must be distributed to the welding jobs in one manner or another. In some cases this is based on per ton of steel fabricated or on the weight of steel consumed. Usually, however, the overhead costs are prorated in accordance with the direct labor charges against the different welding jobs. When this system is used, accurate labor costs for weldments are essential.

Overhead rates are sometimes separate and not included in the welder pay rate. When this is the case, the same formulas as described before [i.e., Equations (20-10) and (20-11)] are used, but the overhead rate is substituted for the welder's pay rate. Both are in dollars per hour.

For single-pass welding, use Equation (20-14). Duty cycle and operator factor are the same. For multipass welding, Equation (20-15) should be used.

The cost of electrical power is sometimes considered part of the overhead expense. On the other hand, when it is necessary to compare competing manufacturing processes or competing welding processes, it is wise to include the cost of electric power in the calculations. In some plants, electric power is considered a direct cost and is charged against the particular job. This is more often the case for field welding than for large production weldment shops. In this case, Equation (20-16) is used.

The local power rate is based on the rate charged to the factory by the local utility company. If time penalties, power factor penalties, and so on, are involved, they should be included. The volts and amperes are the values used when making the weld. The weight of the weld metal is the weight of the weld metal deposited. The deposition rate is that used for a particular weld, as is the operator factor. The final factor is the efficiency of the power source, and this can be found from the machine performance curve. Performance curves were presented in Chapter 10.

Special fixtures for tack welding and holding may be considered an indirect labor cost and are classed as overhead. These items are considered as capital expenditures and are depreciated over a period of five years or so. These costs are also added to the welding department or plant overhead. When this information is calculated, it is added to overhead costs, the cost of labor, and the cost of filler metal to arrive at the total weld cost.

20-6 WELD COST FORMULAS AND EXAMPLES

Welding costs are obtained by adding the major cost elements:

✳ *Materials cost:* filler metals, flux, gas, etc.
✳ *Labor cost:* direct labor.
✳ *Overhead cost:* normally prorated to direct labor

One method of calculating cost is using the basis of cost per weld. This is used for arc spot welds (sometimes cost is based on per 100 arc spot welds), for plug welds, and for small welds used to make a small part.

The more common method of calculating welding cost is on the basis of cost per foot of weld. Each of the cost elements mentioned above can be determined on this basis. There are several ways of determining the cost

$$\text{overhead cost (\$/ft)} = \frac{\text{overhead rate (\$/hr)}}{\text{travel speed (in./min)} \times \text{operator factor (\%)} \times 5} \qquad (20\text{-}14)$$

$$\text{overhead cost (\$/ft)} = \frac{\text{overhead rate (\$/hr)} \times \text{weight of weld metal deposit (lb/ft)}}{\text{deposition rate (lb/hr)} \times \text{operator factor (\%)}} \qquad (20\text{-}15)$$

$$\text{power cost (\$/ft)} = \frac{\text{local power rate (\$/kWh)} \times \text{volts} \times \text{amperes} \times \text{weld metal deposit (lb/ft)}}{1000 \text{ (w/kw)} \times \text{deposition rate (lb/hr)} \times \text{operator factor (\%)} \times \text{power source efficiency (\%)}} \qquad (20\text{-}16)$$

of the principal elements. One method is best for single-pass welding with continuous wire processes, and the other is best for multipass welding.

Materials Cost

For all welding processes that utilize deposited metal, the weight of deposited metal is the basis for material costs. The weight of each type of weld in the weldment is calculated, and the results are added together to determine the total weight. Use Equation (20-3). The weight of weld metal deposit for each weld type is obtained from Figure 20-2. Filler metal yield is based on the type of filler metal and the data of Figure 20-3. These results can be factored by the electrode price to obtain the cost of filler metal required.

To determine the electrode cost per foot, use Equation (20-4). Obtain the weld metal deposit from Figure 20-2. The electrode price is the delivered cost to the plant, and the filler metal yield is from Figure 20-3.

Another method of determining the weight of filler metal required, when using a continuous electrode wire process, is to use Equation (20-5). The wire feed speed is taken from the welding procedure, or can be obtained from the process schedule charts based on welding current, or it can be measured. The length of wire per weight is taken from Figure 20-4.

A third method of determining filler metal required is to use Equation (20-7). The weld travel speed can be obtained from the welding procedure, or from process schedule charts, or by measurement. The deposition rate is arrived at from process deposition rate charts based on the current or by Equation (20-13).

The wire feed rate is obtained from the welding procedure, from charts based on welding current, or by measurement. The other two factors were mentioned earlier.

The cost of welding flux used for SAW is determined by using Equation (20-8). The flux price is the delivered cost to the plant. The weld deposit metal deposit is computed using Equation (20-1) or from Figure 20-2. The flux ratio is 1, 1.1, or 1.2. For accuracy, make a test and measure the weld deposit and flux consumed and establish the ratio.

The cost of shielding gas per foot for GMAW, FCAW, or GTAW is determined by using Equation (20-9). The price of gas is the cost of the shielding gas delivered to the plant. The flow rate is the usage and is from the welding procedure, from process schedule charts, or by measurement. The welding travel speed is obtained the same way.

The cost of shielding gas per weld is used when costing arc spot welds, plug welds, or small weldments and is determined by using Equation (20-10). The price of gas and the flow rate are obtained as mentioned. The weld time is established by the welding procedure, or from process schedule charts, or by measurement.

Labor and Overhead Cost

The labor and overhead cost for single-pass welding is determined by using Equation (20-11). The welder pay rate is normally factored to cover fringe benefits. The travel speed is taken from the welding procedure, process schedule charts, or is measured. The operator factor, or duty cycle, is obtained from Figure 20-5; however, for better accuracy it should be measured or related to similar jobs.

The labor and overhead cost for multipass welding is determined by using Equation (20-12). The data for each factor involved in this formula have been explained.

The overhead cost must be calculated separately if all of the overhead is not included in the welder pay rate. The equations to use are similar to Equations (20-14) and (20-15). The only difference is that the overhead rate is substituted for the welder pay rate, and it should be obtained from the accounting calculations, which is beyond the scope of this book. For simplicity, the overhead rate can be assumed to be equal to the welder pay rate. In some plants it is two times the pay or even higher.

There are a variety of computer software programs that will help provide cost information for welding. In general they are mostly based on the principles outlined in this chapter: They are based on calculated weld metal requirements to produce the weld joint. These different programs are referenced in the Appendix.

Sample Cost Calculations

The following will show how the different equations are used. The following rates will be used in these examples.

$$
\begin{aligned}
\text{Welder pay rate} &= \$15.00 \text{ per hour} \\
\text{Overhead rate} &= \$30.00 \text{ per hour} \\
\text{Power cost} &= \$0.06 \text{ per kWh} \\
\text{Argon gas cost} &= \$0.21 \text{ per cubic feet} \\
CO_2 \text{ gas cost} &= \$0.10 \text{ per cubic feet} \\
\text{Covered electrode price} &= \$0.48 \text{ per pound} \\
\text{Steel electrode wire price} &= \$0.55 \text{ per pound} \\
\text{Flux cored electrode price} &= \$0.65 \text{ per pound}
\end{aligned}
$$

All other data are taken from information referenced elsewhere in this book.

EXAMPLE 1 In this example we wish to establish the cost of a weldment that contains 120 ft of $\frac{1}{4}$-in. fillet, 300 ft of $\frac{3}{16}$-in. fillet, 40 ft of $\frac{1}{2}$-in. V-groove, and 60 ft of $\frac{3}{8}$-in V-groove weld. To determine the amount of weld metal deposited, consult Figure 20-2. Obtain the weight of weld metal of each type of weld, multiply it by the length of that type of weld, and add the total:

$$
\begin{aligned}
120 \text{ ft of } \tfrac{1}{4}\text{-in. fillet} &= 12.7 \text{ lb} \\
300 \text{ ft of } \tfrac{3}{16}\text{-in. fillet} &= 18.3 \text{ lb} \\
40 \text{ ft of } \tfrac{1}{2}\text{-in. V-groove} &= 28.1 \text{ lb} \\
60 \text{ ft of } \tfrac{3}{8}\text{-in. V-groove} &= 23.0 \text{ lb} \\
\text{Total of weld metal deposited} &= 82.1 \text{ lb}
\end{aligned}
$$

What is the cost of the filler metal required when covered electrodes are used? Use Equation (20-4) (modified) and a filler metal yield of 60% from Figure 20–3:

$$\text{electrode cost (\$)} = \frac{\text{(price\$/lb) 0.48} \times \text{(total wt lb) 82.1}}{\text{(yield \%) 0.60}} = \$65.68$$

What is the cost of filler metal required when solid-steel electrodes are used? Use Equation (20-4) (modified) and a filler metal yield of 92.5% from Figure 20–3:

$$\text{electrode cost (\$)} = \frac{\text{(price\$/lb) 0.48} \times \text{(total wt lb) 82.1}}{\text{(yield \%) 0.925}} = \$42.60$$

Note that the total cost is less when using a higher-speed electrode.

EXAMPLE 2   What is the cost of a foot of $\frac{1}{4}$-in. fillet weld made manually with SMAW using E6024 electrode $\frac{3}{16}$-in. size? The operator factor is 30% and is from Figure 20-6. The filler metal yield is 55% and is from Figure 20–3. The weight of weld metal deposited is 0.117 lb/ft and is from Figure 20–2. Use Equation (20-4):

$$\text{electrode cost (\$/ft)} = \frac{\text{(price\$/lb) 0.48} \times \text{(wt lb/ft) 0.117}}{\text{(yield \%) 0.55}} = 0.102 \ \$/\text{ft}$$

The labor cost is calculated using Equation (20-11). The travel speed of 15 in./min is from the process schedule chart:

$$\text{labor cost (\$/ft)} = \frac{\text{(pay \$/hr) 15.00}}{\text{(speed in./min) 15} \times \text{(factor \%) 0.30} \times 5} = \$0.666 \ \$/\text{ft}$$

The overhead cost is double the labor cost. The total cost is the total of: $0.102 + 0.666 + 1.332 = 2.10$ \$/ft of weld.

EXAMPLE 3   What is the cost of a foot of $\frac{1}{4}$-in. fillet weld made semiautomatically using GMAW with CO_2 gas shielding using an E70S-1 electrode of 0.035 in. size? The operator factor is 50% and is from Figure 20-6. The filler metal yield is 95% and is from Figure 20–3. Use Equation (20-4):

$$\text{electrode cost (\$/ft)} = \frac{\text{(cost \$/lb) 0.55} \times \text{(wt ft/lb) 0.117}}{\text{(yield \%) 0.95}} = 0.067 \ \$/\text{ft}$$

The gas cost is calculated using Equation (20-9). The shielding gas flow rate is 25 cubic feet per minute and the travel speed is 15 in./min obtained from the process schedule chart:

$$\text{gas cost (\$/ft)} = \frac{\text{(gas \$/ft}^3\text{) 0.10} \times \text{(flow ft}^3\text{/min) 25}}{\text{(speed in./min) 15} \times 5} = 0.033 \ \$/\text{ft}$$

The labor cost is determined using Equation (20-11) (all of the factors involved were mentioned above):

$$\text{labor cost (\$/ft)} = \frac{\text{(pay \$/hr) 15.00}}{\text{(speed in./min) 15} \times \text{(factor \%) 0.5} \times 5} = 0.400 \ \$/\text{ft}$$

The overhead cost is determined using Equation (20-14) (all of the factors involved were mentioned above):

$$\text{overhead cost (\$/ft)} = \frac{\text{(rate \$/hr) 30}}{\text{(speed in./min 15} \times \text{(factor \%) 0.5} \times 5}$$
$$= 0.80 \ \$/\text{ft}$$

The total cost is the total of: $0.067 + 0.033 + 0.400 + 0.80 = 1.90$ \$/ft of weld. Note that the gas metal arc is less expensive than shielded metal arc welding for the same weld size.

EXAMPLE 4   How many pounds of E70S-1 electrode wire 0.035 in. in diameter should be purchased to make a tank requiring 25,000 ft of $\frac{1}{8}$-in. square-groove butt weld? The weld deposit, with reinforcement, is 0.065 lb/ft from Figure 20–2, times 25,500 ft. which is 1657.5 lb. To determine the pounds of filler metal needed, use Equation (20-3) (the filler metal yield of 90% is from Figure 20–3):

$$\text{weight of filler metal required (lb)} = \frac{\text{(wt lb) 1657.5}}{\text{(yield \%) 0.90}} = 1842 \ \text{lb}$$

These computations are quickly made using an electronic calculator.

QUESTIONS

20-1.   Why is the cross-sectional area of a weld joint important to cost?

20-2.   What is the formula for the cross-sectional area of a regular fillet weld?

20-3.   If the cross-sectional area is known, how do you determine the weight of a foot of weld?

20-4.   What is filler metal yield? Why is the yield lower for covered electrodes?

20-5.   Which has the better yield, a 14-in. electrode or an 18-in. electrode? Why?

20-6.   Why is deposition rate so important for welding costs?

20-7.   What different materials are included in gas metal arc welding? In submerged arc welding? In shielded metal arc welding?

20-8.   What is the operator factor when calculating welding cost?

20-9.   Why is operator factor so important?

20-10.   What major cost elements must be totaled to obtain total welding cost?

20-11.   How does poor fitup affect welding costs?

20-12.   How does one's welding affect welding costs? How does excessive reinforcement affect costs?

20-13.   Why is a positioner used to reduce welding costs?

20-14. What is the most expensive metal of a weldment? Why?

20-15. What is the formula for the cross-sectional area of a corner joint utilizing a single bevel weld reinforced by a fillet weld?

20-16. What is the theoretical weld deposit for a $\frac{3}{8}$-in. fillet weld 72 in. long?

20-17. What is the weight of weld deposit for a single-V butt joint in a 30-in.-diameter pipe having a $\frac{3}{8}$-in. wall thickness?

20-18. How many pounds of E6010 electrodes should be purchased for making a 5-mile-long pipeline using the pipe size in Question 20-17? Assume that each length of pipe is 20 ft.

20-19. A weldment requires 40 ft of $\frac{1}{2}$-in. V, 60 ft of $\frac{3}{8}$-in. V, 120 ft of $\frac{1}{4}$-in. fillet, and 300 of $\frac{3}{16}$-in. fillet welds. What is the weight of weld metal deposited, including reinforcement?

20-20. In Question 20-19, how many pounds of electrode should be purchased to make 10 weldments using the shielded metal arc process? How many pounds of electrode wire should be purchased for the 10 weldments using the flux cored arc welding process?

21

QUALITY CONTROL AND EVALUATION OF WELDS

21-1 QUALITY CONTROL PROGRAM

A principal factor in the performance of an organization is the quality of its products. There is a worldwide trend toward more stringent customer expectations with respect to quality. There is a growing realization that continual improvements in quality are necessary to maintain success of an organization.[1] A need for more reliable products, increasingly complex technology, and the need to conserve resources makes weld quality increasingly important. This has prompted many organizations to provide special specifications; however, these will not in themselves guarantee that the quality requirements will be consistently met. This has led to the development of quality system standards and guidelines that complement relevant product requirements. The international community has established such a requirement, known as ISO 9000.[2] This is similar to a U.S. military specification "Quality Program Requirements."[3] This manufacturing system provides management support through policy and delegated authority. The system includes documentation to establish designs, manufacturing techniques, and quality control methods. From a welding point of view, it includes welding procedure qualifications, welder performance qualifications, and an overall total welding quality control program. It is a "welding manufacturing system." As part of a total manufacturing system it will provide good-quality welded products by establishing the necessary engineering capabilities.

These different quality control programs are all slightly different. It is important to study the particular specification involved. Review your program with the representative of the accrediting organization. Make sure that your approach is the approach they desire and that your documents will meet their approval. In general, all quality control programs are similar, but the accrediting agencies may have different requirements.

The remaining portion of this section concerns the quality control plan and offers a suggested quality control program. It can be adopted by companies desiring to improve weld quality and is very similar to the program established by the nuclear code.[4]

For certain classes of work, quality control requirements are well established. Strict requirements, which are found in the nuclear code, require a quality assurance program that is based on the technical and manufacturing aspects of the product. The program must ensure adequate quality from the design, acquisition, and manufacturing to final shipment. The program must define authority and responsibility for each portion of the work. The quality assurance plan must include the following:

1. *Organization.* The organization for quality must be clearly prescribed. It should define and show charts for responsibility and authority and the organizational freedom to identify and evaluate quality problems. Quality control personnel should not report to production personnel.

2. *Quality assurance program.* The producer must conduct a review of the requirements of the quality required of the product. The various factors, such as specialized controls, processes, testing equipment, and skills, for assuring product quality must be identified. This program must be documented by written policies, procedures, and instructions.

3. *Design control.* The design must provide for verifying the adequacy of the design, via performance testing and independent review. It should include qualification and testing of prototypes. Measures must be established to ensure that the design specifications and code requirements are correctly translated into drawings, procedures, and instructions.

4. *Procurement documents.* The program requires that specifications be written for each item purchased and that the specification ensure the quality required by the end product. These specifications also require quality assurance programs from vendors.

5. *Instructions, procedures, and drawings.* The quality program must ensure that all work affecting quality must be prescribed in clear and complete documented instructions of a type appropriate to the work. Compliance with instructions must be monitored.

6. *Document control.* The quality program must include a procedure for maintaining the completeness and correctness of drawings and instructions, showing dates, control, effective point, and so on. These drawings, procedures, and instructions must be maintained and continuity explained by change notices.

7. *Control of purchased material, equipment, and services.* The program must include a control system for purchasing from qualified vendors. This means that vendors must have acceptable quality programs for producing their items. A qualified product list is required, and only those vendors having adequate quality programs and providing quality products will be included. The program requires receiving inspection systems so that purchased items can be checked against the specifications. Raw materials and purchased parts will be inspected by means of instruments, laboratory procedures, and so on, to ensure that the products meet the specifications.

8. *Identification and control of materials.* The program must provide for identification of all parts, materials, components, and so on, from receipt throughout all processing to the final item. Records shall provide traceability of all materials and components. A checklist shall be established for reporting all characteristics and recording test reports that have been received, reviewed, and found acceptable.

9. *Control of special processes.* The quality program must ensure that all manufacturing operations, including welding, are accomplished under controlled conditions. These controlled conditions involve the use of documented work instructions, drawings, special equipment, and so on. It further requires that such instructions be provided, with space for reporting results of inspection by the manufacturer and the inspector, including the date and initials.

10. *Inspection.* The quality assurance program should ensure a system of inspection and testing for all products. Such testing should stimulate the product service, and records must be maintained on the adequacy of the product to meet these specifications.

11. *Test control.* The program must assure that all tests are performed according to written instructions. Instructions must provide requirements and acceptance limits. Test results must be documented and evaluated to assure that test requirements are met.

12. *Control of measuring and testing equipment.* The program should provide for methods of maintaining the accuracy of gauges, testing devices, meters, and other precision devices, showing that they are calibrated against certified measurement standards on a periodic basis.

13. *Handling, storage, and delivery.* The program should provide adequate instructions for handling, storage, preservation, packaging, shipping, and so on, so that the product is protected from its time of manufacture until its time of use.

14. *Inspection test and operating status.* The program must include methods of identifying parts to determine its status as far as inspection and approval are concerned.

15. *Nonconforming materials, parts, or components.* There should be a procedure established to maintain an effective and positive system for controlling nonconforming material. It may include and allow for rework; however, records must be maintained of such

work. Resolution of nonconformities should be in conformance with paragraph 7 of this program.

16. *Corrective actions.* The quality program must establish methods of dealing promptly with any conditions that are adverse to quality, including design, procurement, manufacturing, testing, and so on. The program should also include methods of overcoming defects, taking corrective action to produce a product that meets the required quality.

17. *Quality assurance records.* The program requires that records be maintained, including all data essential to the economical and effective operation of the quality program. Records must be complete and reliable and include measurements, inspections, observations, and so on, and these records must be available for review.

18. *Cost related to quality.* The program should allow for maintenance and use of cost data for identifying the cost of the program and for the prevention and correction of defects encountered.

19. *Production tooling and inspection equipment.* Various items of tooling, including fixtures, templates, patterns, and so on, may be used for inspection purposes, provided that their accuracy is checked at periodic intervals.

20. *Audits.* The program must include a system of planned and periodic audits to verify compliance with all aspects of the quality assurance program. The audit must be done by personnel not normally involved in the areas being audited. Audits must be documented and reviewed, and action must be taken to correct any deficiencies found.

The preceding list is an abbreviated outline of the requirements of a quality assurance program necessary for critical products. It can be modified and expanded and used by companies where a code of specification is not imposed. When applying for approval of any type, it is absolutely necessary to review the specification or standard that is being applied. In addition, consult with the accrediting agency that you are dealing with to make sure that your input is in agreement with their requirements.

21-2 DESTRUCTIVE TESTING

Welds and weld metal are subjected to more different types of tests than any other metal produced. Weld metal can be tested in the same manner as any other form of metal. Mechanical tests are used to qualify welding procedures, welders, and welding processes, and to determine if electrodes and filler metals meet the requirements of the specification. Welds in weldments are often tested for soundness, strength, and toughness by mechanical tests.

Mechanical tests are destructive tests since the weld joint is destroyed in making the test. They are also expensive since they involve preparation of material, making welds, cutting and often machining, and destructive testing the specimens.

Procedure Qualification

To qualify a welding procedure specification (WPS), specific welds are made, cut into standardized sizes and shapes, and tested to destruction. The tests are spelled out in detail by the code or specification. This involves making test plates according to the welding procedure specification and then mechanically testing the welds. The welding process, filler metals, and the welding schedule are selected to make the weld in the position required on the base metal that is to be used. Welding joint details and material thickness may be specified and may not be exactly as used in making the production weldment. Requirements vary from code to code. The test specimens are not always exactly the same, nor are they taken from the same positions of a test weld. It is essential that the proper edition of the code or specification in question be checked when making test welds, test specimens, and the tests of the weld.

The American Petroleum Institute standard 1104, "Standard for Welding Pipeline and Related Facilities," is extremely popular and used for pipeline welding. Figure 21–1 shows the location of test specimens based on the size of pipe. It also indicates the number of test specimens that should be taken. The detail of the nick break test specimen is shown in Figure 21–2. The bend test specimens are shown in Figures 21–3 and 21–4. Note that root and face bend specimens are used for thinner material and the side bend specimen is used when the wall thickness of the pipe is over $\frac{1}{2}$ in. (12.7 mm).

The API tests are often made in the field, and it is permissible to use flame-cut edges and grinding to prepare specimens. This eliminates machining and makes it possible to quickly prepare test specimens and to make tests with a portable bend test machine. The basis for acceptance of the test welds is spelled out in detail in the code.

The American Society of Mechanical Engineers Boiler and Pressure Vessel Code Section IX, "Welding and Brazing Qualifications," is widely used for pressure vessel work and is the reference specification for pressure piping work. This code makes use of the guided bend test and the fillet weld test. The ASME uses reduced section tensile test for certain requirements. The location of test specimens and the details of them are shown in the code. For the guided bend test, special dimensions are provided for different thicknesses of test specimens. The American Welding Society's "Structural Welding Code," D1.1, is widely used for qualifying procedures and welders. One of the tests, which is often used for preliminary evaluation, is the fillet weld break specimen used for qualifying

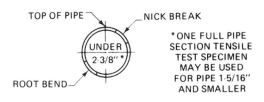

TOP OF PIPE · NICK BREAK

UNDER 2-3/8" *

ROOT BEND

*ONE FULL PIPE SECTION TENSILE TEST SPECIMEN MAY BE USED FOR PIPE 1-5/16" AND SMALLER

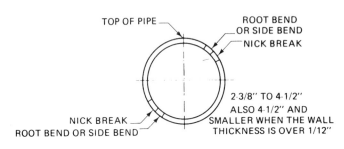

TOP OF PIPE

ROOT BEND OR SIDE BEND
NICK BREAK

2-3/8" TO 4-1/2" ALSO 4-1/2" AND SMALLER WHEN THE WALL THICKNESS IS OVER 1/12"

NICK BREAK
ROOT BEND OR SIDE BEND

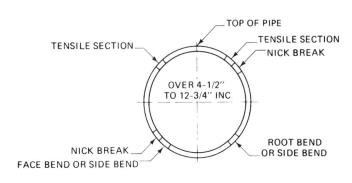

TOP OF PIPE

TENSILE SECTION
TENSILE SECTION
NICK BREAK

OVER 4-1/2" TO 12-3/4" INC

NICK BREAK
FACE BEND OR SIDE BEND

ROOT BEND OR SIDE BEND

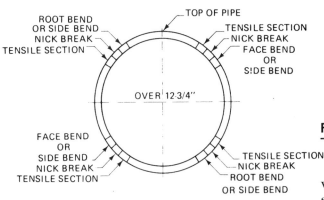

ROOT BEND OR SIDE BEND
NICK BREAK
TENSILE SECTION

TOP OF PIPE

TENSILE SECTION
NICK BREAK
FACE BEND OR SIDE BEND

OVER 12-3/4"

FACE BEND OR SIDE BEND
NICK BREAK
TENSILE SECTION

TENSILE SECTION
NICK BREAK
ROOT BEND OR SIDE BEND

NOTE: AT THE COMPANY'S OPTION, THE LOCATIONS MAY BE ROTATED 45 DEGREES COUNTERCLOCKWISE OR THEY MAY BE EQUALLY SPACED AROUND THE PIPE EXCEPT SPECIMENS SHALL NOT INCLUDE THE LONGITUDINAL WELD. ALSO, AT THE COMPANY'S OPTION, ADDITIONAL SPECIMENS MAY BE TAKEN.

FIGURE 21–1 API test specimen locations.

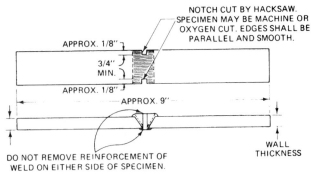

NOTCH CUT BY HACKSAW. SPECIMEN MAY BE MACHINE OR OXYGEN CUT. EDGES SHALL BE PARALLEL AND SMOOTH.

APPROX. 1/8"
3/4" MIN.
APPROX. 1/8"
APPROX. 9"
WALL THICKNESS

DO NOT REMOVE REINFORCEMENT OF WELD ON EITHER SIDE OF SPECIMEN.

FIGURE 21–2 API code nick break test specimen.

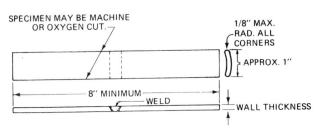

SPECIMEN MAY BE MACHINE OR OXYGEN CUT.

1/8" MAX. RAD. ALL CORNERS
APPROX. 1"

8" MINIMUM
WELD
WALL THICKNESS

WELD REINFORCEMENT SHALL BE REMOVED FROM BOTH FACES FLUSH WITH THE SURFACE OF THE SPECIMEN. SPECIMEN SHALL NOT BE FLATTENED PRIOR TO TESTING

FIGURE 21–3 API code root and face bend specimen.

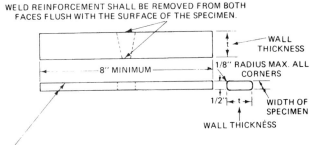

WELD REINFORCEMENT SHALL BE REMOVED FROM BOTH FACES FLUSH WITH THE SURFACE OF THE SPECIMEN.

WALL THICKNESS
1/8" RADIUS MAX. ALL CORNERS
8" MINIMUM
1/2" t
WIDTH OF SPECIMEN
WALL THICKNESS

SPECIMENS MAY BE MACHINE CUT TO 1/2 INCH WIDTH OR THEY MAY BE OXYGEN CUT TO APPROXIMATELY 3/4 INCH WIDE AND THEN 1/8 INCH REMOVED BY MACHINING OR GRINDING. CUT SURFACES SHALL BE SMOOTH AND PARALLEL.

FIGURE 21–4 API code root bend specimen.

weld tackers. This specimen, and the way it is fractured, is shown in Figure 21–5. The thickness of the plates shall be $\frac{1}{2}$ in. (12.6 mm) and the fillet weld should be $\frac{1}{4}$ in. (6.3 mm). Variations of this specimen use thinner plates and smaller fillet sizes. This specimen can be used for each welding position and can be used for all of the arc welding processes and any electrode type.

The structural code includes another fillet test that has been used for welder qualification. In this test, fillet welds are made between two plates and a backing bar. The plates are separated $\frac{15}{16}$ in. (24 mm), and fillet welds

TEST SPECIMEN DETAIL:

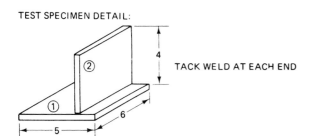

TACK WELD AT EACH END

4

② ①

6

5

BILL OF MATERIAL
Item 1 – 1 piece 3/8 x 5 x 6 Mild Steel – ASTM A-37
Item 2 – 1 piece 3/8 x 4 x 6 Mild Steel – ASTM A-37

WELDING PROCEDURE:

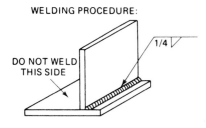

1/4 ⌵

DO NOT WELD
THIS SIDE

Position	Electrode size (in)	Fillet size (in)
Flat	3/16	1/4
Horizontal	3/16	1/4
Vertical	5/32	1/4 and 3/8
Overhead	5/32	1/4 and 3/8

TEST BAR PREPARATION: None

TEST SPECIFIED

Force

PLACE IN PRESS OR HIT
WITH SLEDGEHAMMER UNTIL
BROKEN OR FLATTENED.

STANDARD OF ACCEPTABILITY:
(a) Contour–The exposed face of the weld shall be reasonably smooth and regular. There shall be no overlapping or undercutting. The weld shall conform to the required cross section for the size of weld specified per gauge.
(b) Extent of Fusion–There shall be complete fusion between the weld and base metal and full penetration to the root of the weld.
(c) Soundness–The weld shall contain no gas pocket, oxide particle or slag inclusion exceeding 3/32 in. in greatest dimension. In addition, no square inch of weld metal area shall contain more than 6 gas pockets exceeding 1/16 inch in greatest dimension.

FIGURE 21–5 AWS fillet break test.

are made between each plate and the backing bar. The remaining area between the fillet welds is filled in like a groove weld. The difficulty with this test is that the backing bar must be removed and this requires machining. This specimen can be tested with the root in tension or the face in tension. This test can be very critical when it is tested with the root in tension, since this shows root penetration of the fillets. The details of this test are shown in Figure 21-6. The examination of this test and the requirements are listed in the code. For welder qualification, the code also provides for groove weld test specimens. The weld joint detail is a single V-groove weld with a 45° included angle and a $\frac{1}{4}$-in. (6.3-mm) root opening. A backup bar is used. The specimen can be welded in any position. However, when it is used in the horizontal position, it is usually a single bevel weld with the bottom horizontal and the 45° bevel on the top piece. These specimens are shown in Figure 21-7. These specimens are

then cut and given a guided bend test. For plate heavier than $\frac{3}{8}$ in. (9.5 mm), the side bend specimen is required. Since different strength-level materials have different beading characteristics, the design detail of the guided bend fixture is altered. The details of the guided bend fixture are shown in Figure 21-8. Testing a specimen in a guided bend fixture is shown in Figure 21-9.

Other tests are required for other reasons. For example, to check the compliance of deposited weld metal, a special joint design and *all-weld* metal test specimen is required. This joint design and test specimen are specified in the filler metal specifications. Figure 21-10 shows the joint detail for making *all-weld* metal test specimens. In some cases impact properties are also specified. When required, impact test specimens must be prepared. They are made with the same joint detail as shown in Figure 21-10. The detail of the *all-weld* metal test specimen Type 505 is shown in Figure 21-11 and the detail of the Charpy

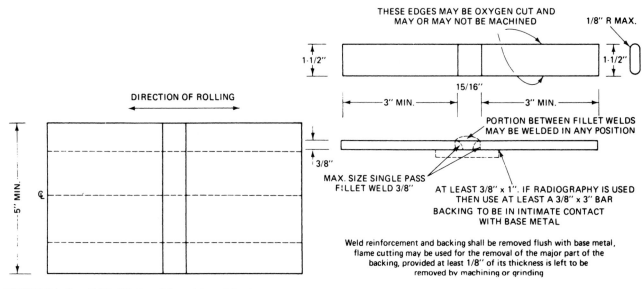

DIRECTION OF ROLLING

5" MIN.

3/8"

THESE EDGES MAY BE OXYGEN CUT AND
MAY OR MAY NOT BE MACHINED

1/8" R MAX.

1-1/2"

1-1/2"

15/16"

3" MIN.

3" MIN.

PORTION BETWEEN FILLET WELDS
MAY BE WELDED IN ANY POSITION

MAX. SIZE SINGLE PASS
FILLET WELD 3/8"

AT LEAST 3/8" x 1". IF RADIOGRAPHY IS USED
THEN USE AT LEAST A 3/8" x 3" BAR
BACKING TO BE IN INTIMATE CONTACT
WITH BASE METAL

Weld reinforcement and backing shall be removed flush with base metal,
flame cutting may be used for the removal of the major part of the
backing, provided at least 1/8" of its thickness is left to be
removed by machining or grinding

FIGURE 21–6 AWS fillet weld root bend test.

FIGURE 21–7 AWS groove weld.

TEST SPECIMEN DETAIL:

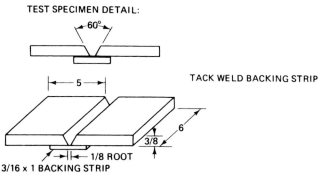

60°

TACK WELD BACKING STRIP

5

6

3/8

1/8 ROOT

3/16 x 1 BACKING STRIP

BILL OF MATERIAL: 2 pieces 5 x 6 x 3/8 and 1 piece 1 x 3/16 Mild Steel

WELDING PROCEDURE:

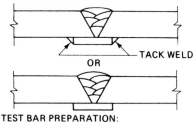

TACK WELD

OR

Position	Electrode size (in)	No. of Passes
Horizontal	1/8	6 to 7
Vertical	1/8	6 to 7
Overhead	1/8	6 to 7

TEST BAR PREPARATION:

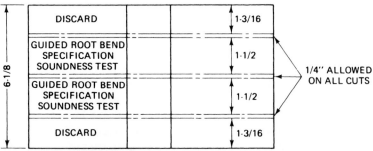

6-1/8

DISCARD	1-3/16
GUIDED ROOT BEND SPECIFICATION SOUNDNESS TEST	1-1/2
GUIDED ROOT BEND SPECIFICATION SOUNDNESS TEST	1-1/2
DISCARD	1-3/16

1/4" ALLOWED
ON ALL CUTS

STANDARD OF ACCEPTABILITY:

(a) Contour—The exposed face of the weld shall be reasonably smooth and regular. There shall be no overlapping or undercutting.

(b) Extent of Fusion—There shall be complete fusion between the weld and base metal and full penetration to the root of the weld.

(c) Soundness—The weld shall contain no gas pocket, oxide particle or slag inclusion exceeding 1/8" in greatest dimension. In addition, no square inch of weld metal area shall contain more than 6 gas pockets exceeding 1/16" in greatest dimension.

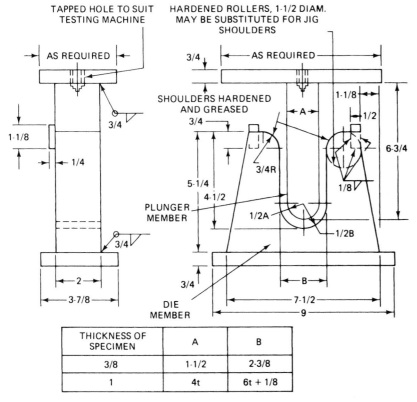

TAPPED HOLE TO SUIT
TESTING MACHINE

HARDENED ROLLERS, 1-1/2 DIAM.
MAY BE SUBSTITUTED FOR JIG
SHOULDERS

THICKNESS OF SPECIMEN	A	B
3/8	1-1/2	2-3/8
1	4t	6t + 1/8

ALL DIMENSIONS ARE IN INCHES.

NOTES:

1. The ram shall be fitted with an appropriate base and provision for attachment to the testing machine. The ram shall also be designed to minimize deflection and misalignment.

2. The specimen shall be forced into the die by applying the load on the plunger until the curvature of the specimen is such that a 1/8 in. (3.2 mm) diam. wire cannot be placed between the specimen and any point in the curvature of the plunger member of the jig.

FIGURE 21-8 AWS guided bend test jig.

FIGURE 21-9 Making a guided bend test.

FIGURE 21-10 AWS all-weld metal test.

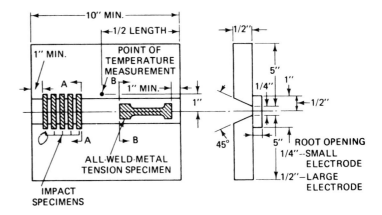

V notch impact test specimen is shown in Figure 21-12. These test specimens are universally used, and the detail dimensions of them are identical in most specifications.

There are several other weld test specimens that may be used, including the transverse fillet weld specimen

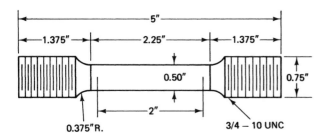

FIGURE 21–11 Tensile test specimen: 505.

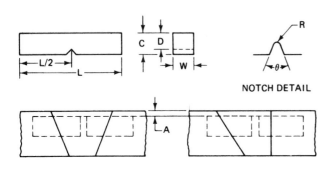

NOTCH DETAIL

Dimension	in.	mm
A—Distance of sample below specimen	0.06 ± 0.002	1.6 ± 0.05 – 0
L—Length of specimen	2.165 ± 0.002	55.0 ± 0, –2.5
L/2—Location of notch	1.082 ± 0.002	27.5 ± 1.0
C—Cross section (depth)	0.394 ± 0.001	10.000 ± 0.025
W—Cross section (width)	0.394 ± 0.001	10.000 ± 0.025
D—Bottom of notch to base	0.315 ± 0.001	8.000 ± 0.025
R—Radius of notch	0.010 ± 0.001	0.250 ± 0.025
θ—Angle of notch 45 deg ± 1 deg	Adjacent sides shall be 90 deg ± 10 minutes	

FIGURE 21–12 Impact test specimen: Charpy V notch.

(Figure 21-13) and the longitudinal fillet weld shear specimen (Figure 21-14). Many other weld specimens are used for the development and research work. For further information refer to the AWS "Standard Methods for Mechanical Testing of Welds."[5] This document shows the detail dimensions of most weld specimens.

21-3 VISUAL INSPECTION

Visual examination is the most widely used nondestructive testing technique. It is extremely effective and is the least expensive inspection method. The welding inspector can utilize visual inspection throughout the entire production cycle of a weldment. It is an effective quality control method that will ensure procedure conformity and will catch errors at early stages. The work of the welding inspector utilizing visual inspection methods can be subdivided into three divisions: (1) visual examination prior to welding, (2) visual examination during welding, and (3) visual examination of the finished weldment.

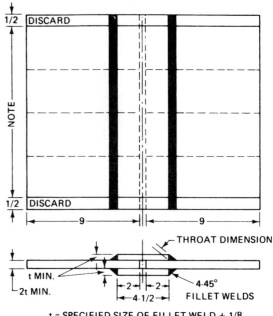

t = SPECIFIED SIZE OF FILLET WELD + 1/8

ALL DIMENSIONS ARE IN INCHES.

NOTE: SEE SPECIFICATION FOR ADDITIONAL DETAILS.

FIGURE 21–13 Transverse fillet weld test.

Visual Examination Prior to Welding

There are a great many items that must be reviewed and checked prior to welding. These include:

1. Review all applicable drawings, specifications, procedures, welder qualifications, and so on. This helps the inspector to become familiar with the job and all specifications that apply to it.
2. Review the material specifications of the parts comprising the weldment and determine that the materials are according to specifications.
3. Compare the edge preparation of each joint with the drawings. At the same time, check edge preparation for surface conditions.
4. Check the dimensions of each item since they will affect weldment fitup.
5. At the fitup operation check assembly dimensions and fitup, with special emphasis on root openings of the weld joints.
6. At the fitup operation check the backing, bars, rings, flux, and so on, to be sure that they are in accordance with requirements.
7. At the fitup operation, check the cleanliness of the welding joints and the conditions of tack welds.

At the fitup and tack weld operation, many weldments are completely fitted and ready for production

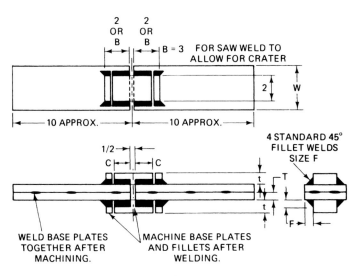

DIMENSION	IN.	MM
SIZE OF WELD, F	1/8	3.2
	1/4	6.4
	3/8	9.5
	1/2	12.7
THICKNESS, t, MIN.	3/8	9.5
	1/2	12.7
	3/4	19.1
	1	25.4
THICKNESS, T, MIN.	1/4	6.4
	3/8	9.5
	1/2	12.7
	5/8	15.9
WIDTH, W	3	76.2
	3	76.2
	3	76.2
	3-1/2*	88.9
FILLET LENGTH, C	1-1/2	38.1
	1-1/2	38.1
	1-1/2	38.1
	1	25.4

NOTE: SEE SPECIFICATION FOR ADDITIONAL DETAILS.

FIGURE 21–14 Longitudinal fillet weld test.

Visual Examination during Welding

When welding begins, there are several items that should be checked, including the welding procedures. Make sure that they are in order, applicable to the weldment, and available to the people doing the welding. Items that must be checked are:

1. Determine that the designated welding process and method of application are in accordance with procedures.

2. Determine that the designated electrode is proper for the base metals. Determine the storage facilities, the condition of electrodes, and for critical work, record the heat numbers of the electrodes used in specific joints or weldments.

3. Survey the welding equipment to make sure that it is in good operating condition. This should include clamping devices, fixtures, and locating devices, as well as the electrical machinery.

4. Determine that the correct welding current and the proper polarity are being used. Use portable meters.

5. Determine that preheat requirements are adhered to at the time of welding. Determine that base metal temperatures are heat-soaking temperatures instead of merely surface heat. The time of preheating can help establish whether through-heating is accomplished.

6. Identify all welders assigned to the weldment or job or joint in question. Their qualification level must be in accordance with the requirements of the job. Qualification papers should be reviewed to determine that they are in order.

7. Observe welders making welds. This has a rather startling effect on welders, especially when they know that their welds are being watched as they are being made. If a welder does not appear to have the necessary skill for the job in question, the inspector can, in consultation with the supervisor, request that the welder make requalification tests. This requirement may not be in the code but is common practice for high-quality work.

8. Determine that interpass temperatures are being maintained. If welding operations are discontinued for a period, the interpass temperatures must be obtained before welding is resumed.

9. Determine that interpass cleaning by chipping, grinding, gouging, and so on, is being done in accordance with the procedure or specification or in accordance with good practice.

10. Watch out for slugging (i.e., adding rods or metal to a weld groove that does not belong and weakens the joint).

welding. In other cases certain welds may later be hidden, and these welds must be completed before fitup is finished. It is recommended good practice for the fitter to mark in chalk the symbols showing sizes for all welds to be made by the production welders. On high-volume production work this may not be necessary, especially if samples of production parts are available for reference.

At the fitup station, the welding inspector should check the tack welds to determine that the correct electrode types are being used for the base metal and to see if any special precautions such as preheat are required. If preheating is specified, local preheating may be used.

With the welding inspector on the site during welding operations, it is possible for any unusual activities or repair to be noticed. This type of work is often required but should have special attention and supervision to determine that the quality requirements are maintained. In many situations repair work must be described and approved prior to doing the work.

1. The inspector must document all repair welding, why it was required, the extent of the work, and how it was done. This should be recorded in an inspection notebook or on applicable report forms, whichever is required.

2. The inspector should determine that any type of postheat treatment performed is done in accordance with the procedure or other requirements.

3. Finally, the inspector should check any warpage corrective activities, such as press work or thermal bending, that might be employed. It is also necessary that these types of activities be recorded in the inspection notebook.

Visual Examination after Welding on the Completed Weldment

The inspector is expected to determine that the weldment conforms to the drawings and specifications for which it is designed and constructed. This includes many items with respect to the weldment but more importantly to the welds. The welds must all be made to the size specified.

1. It is important to check the weld size of all welds. The size of fillet welds can easily be determined by means of weld gauges. Figure 21–15 shows the use of a standard fillet weld gauge used in North Amer-

ica. Figure 21–16 shows the size of fillet welds and the method of checking fillet welds to determine that they do meet the size specifications. There are many other types of gauges. Figure 21–17 shows the use of a U.S. Navy gauge for checking fillets. There are other gauges, primarily from Europe, that are used throughout the world, including North America. One of the most popular is the British gauge, shown in Figure 21–18. It has capabilities for checking the sizes of different types of welds.

2. All welds should be inspected to see that they do not have any of the defects listed below.
   Surface cracks (including toe cracks)
   Crater cracks (or unfilled crater)
   Surface porosity
   Incomplete root penetration
   Undercut

FIGURE 21–16   Fillet weld size and method of checking.

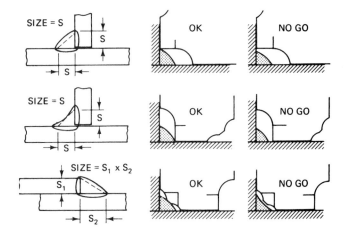

FIGURE 21–17   U.S. Navy weld gauge.

FIGURE 21–15   Use of standard fillet gauge.

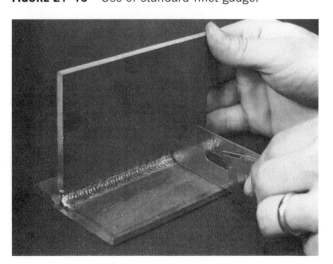

FIGURE 21–18 British welding gauge.

Underfill on face, groove, or fillet (concave)
Underfill of root (suck-back)
Excessive face reinforcement, groove, or fillet (convex)
Excessive root reinforcement (or drop-through)
Overlap
Misalignment (high-low)
Arc strikes
Excessive spatter
Each of these defects is described in detail in Section 21-5.

3. The weldment must also be checked by the inspector. The following are considered weldment defects.

Warpage Warpage of weldments can be a reason for rejection or repair work. If warpage is beyond the allowable or acceptable limits, corrective action should be initiated. This can include mechanical methods, such as the use of clamps, strong backs, and presses, or thermal methods, such as the use of torches. Judgment of the inspector is important. If it is felt that damage can be done to the weldment by means of these remedial actions, records of action taken must be made in the inspection notebook.

Base Metal Defects The inspector must also be on the lookout for these, which can appear as lamination in edges of steel plates. They can also be scabs or seams in the base metal. Caution is particularly important when steel plate is stressed in the through or *Z* direction.

Backing Welds These must be utilized whenever there is a question about the quality of root fusion of groove welds and corner type fillet weld joints.

Nondestructive Examination Recommendations

The welding inspector should determine if nondestructive testing symbols are on weldment drawings. If so, it is important that the appropriate joints be marked for nondestructive testing (NDT) and the inspector determine that such tests are made. These tests can involve the use of ultrasonic inspection, magnetic particle inspection, radiographic inspection, or dye-penetrant inspection. Dye-penetrant inspection is sometimes used for root-pass inspection, especially on high-quality pipe weldments.

It is the privilege of the welding inspector, utilizing visual inspection, to call for any of the nondestructive techniques if there is reason to be suspicious of a specific joint, welder, or weldment. Good practice allows such weldments to be taken and subjected to at least one of the NDT methods. It is expected that the inspector will have reason for making such requests. If such weldments, joints, and so on, do not show defects, it is then wise to determine the competence of the inspector. Visual inspection of the surface or welds or of any metal part will not reveal internal flaws or problems. Surface inspection cannot show the lack of fusion at the root of the weld if the root of the weld is inaccessible for visual inspection. Internal porosity cannot be seen from the surface, nor can internal cracks and other internal defects. It is therefore necessary to allow the inspector the privilege of requiring occasional internal examinations in order for the inspector to maintain credibility with welders.

The welding inspector must have certain qualifications. Under certain circumstances on some types of work, it is necessary that the welding inspector be qualified and certified. The American Welding Society tests and certifies welding inspectors (CWI).[6] It is necessary that the inspector have good eyesight corrected to 20-20 vision. The inspector must also be given the necessary tools and instruments for making inspections. This includes welding gauges, measuring instruments, temperature indicators, welding shields, flashlights, notebooks for recording data, and a marking crayon, normally yellow or red, which can be used to mark those welds or weldments that must be repaired or rejected. Finally, it is necessary that the welders cooperate with the inspector, that weld slag be removed for adequate inspection, and that critical welds are never covered over before they can be inspected. This type of cooperation will ensure that quality weldments are produced by the team of welders, supervisor, and inspector. For more information, see AWS "Guide for the Visual Inspection of Welds."[7] See also the AWS book, "Welding Inspection."[8]

21-4 NONDESTRUCTIVE TESTING

Nondestructive testing (NDT) is also known as nondestructive examination or evaluation (NDE) and nondestructive inspection (NDI). These techniques use the application of physical principles for the detection of flaws or discontinuities in materials without impairing their usefulness. There are a number of examination methods or techniques. The growth of nondestructive testing has been greatly accelerated by the need for higher quality and better reliability of manufactured products.

In the field of welding, four nondestructive tests are widely used: dye-penetrant testing and fluorescent-penetrant testing, magnetic particle testing, ultrasonic testing, and radiographic testing. Each of these techniques has specific advantages and limitations. A comparison of the different techniques compared with visual testing is shown to provide guidance in selection of the different tests. For more information see AWS "Guide for the Nondestructive Inspection of Welds."[9]

Penetrant Examination

Liquid-penetrant examination (PT) is a highly sensitive, nondestructive method for detecting minute discontinuities (flaws) such as cracks, pores, and porosity, which are *open to the surface* of the material being inspected. This method may be applied to many materials, such as ferrous and nonferrous metals, glass, and plastics. Although there are several types of penetrants and developers, they all employ common fundamental principles, as shown in Figure 21–19.

One of the most important aspects of liquid-penetrant examination is the preparation of the part before the penetrant is applied. The surface must be cleaned with a solvent to remove any dirt or film. Discontinuities must be free from dirt, rust, grease, or paint, to enable the penetrant to enter the surface opening. The solvent cleaner, used to remove the excess penetrant, is excellent for precleaning of the part surfaces.

A liquid penetrant is applied to the surface of the part to be inspected. The penetrant remains on the surface and seeps into any surface opening. The penetrant is drawn into the surface opening by capillary action. The parts may be in any position when tested. After sufficient penetration time has elapsed, the surface is cleaned and excess penetrant is removed. When the surface is dry, a powdered, absorbent material or a powder suspended in a liquid is applied. The result is a blotting action that draws the penetrant from any surface opening. The penetrant is usually a red color; therefore, the indication shows up brilliantly against the white background of the developer. The indication is larger than the actual defect. Thus even small defects may be located.

Equipment   Portable inspection kits are available for visible dye penetrant. Some of the kits use pressurized cans so that liquids may be sprayed on the parts to be inspected. The use of pressurized liquids is shown in Figure 21–20.

Stationary equipment usually consists of a tank in which the penetrant is applied either by dipping, pouring, or by brushing. Other tanks provide drain, wash, and developer stations. The size of the tanks and individual stations is governed by the sizes and quantities of the parts to be examined. Specialized, high-volume units are available that provide for parts examination at production rate speeds. This equipment utilizes moving conveyors. The operator places the parts on the conveyor and they proceed unattended throughout the entire processing operation. At the end, an inspector examines the parts and interprets the indications.

Applications   In the field of welding, liquid-penetrant examination is used to detect surface defects in aluminum, magnesium, and stainless steel weldments when the magnetic particle examination method cannot be used. It is very useful for locating leaks in all types of welds. Welds in pressure and storage vessels and in piping for the petroleum industry are examined for surface cracks and for porosity, using this method.

Fluorescent-Penetrant Examination

The fluorescent-penetrant examination (FPT) technique is almost identical with the dye-penetrant technique. There are two basic differences, however. The penetrant is fluorescent, and when it is exposed to ultraviolet or black light it shows as a glowing fluorescent type of readout. It provides a greater contrast than the visible dye penetrants. It is considered to have greater sensitivity.

FIGURE 21–19   Principle of penetrant examination.

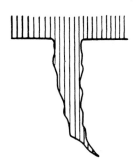

A. LIQUID PENETRANT
APPLIED

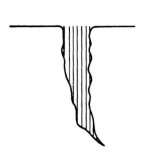

B. EXCESS PENETRANT
REMOVED

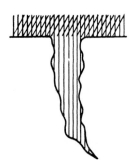

C. DEVELOPER DRAWS PENE-
TRANT FROM CRACK.

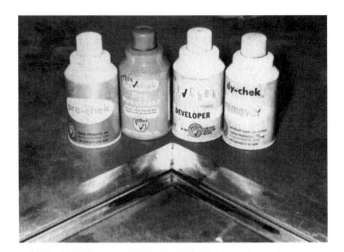

FIGURE 21–20 Using penetrant on finished weld.

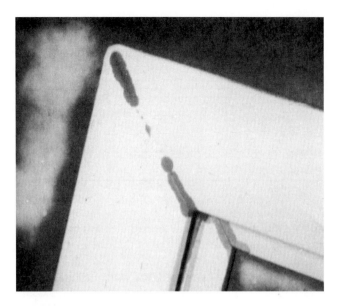

FIGURE 21–21 Penetrant indication.

For the use of the fluorescent-penetrant system, the other extra piece of equipment required is the ultraviolet or black light source. It is recommended that the inspection be done in a darkened area with the black light as the predominant source of light available.

Kits similar to the dye-penetrant equipment are also available for fluorescent-penetrant examination; however, it is probably more popularly used with liquid penetrants in tanks similar to that mentioned for dye penetrant but with the fluorescent-penetrant material and the black light for examination.

Examination is made using the ultraviolet or the black light. Sound areas appear deep violet, while the defects will glow a brilliant yellowish green. The width and brightness of the fluorescent indication depends on the size of the crack or defect.

Applications   Probably one of the most useful applications of fluorescent-penetrant examination is for leak detection in magnetic and nonmagnetic weldments. A fluorescent penetrant is applied to one side of the joint and a portable ultraviolet light (black light) is then used on the reverse side of the joint to examine the weld for leaks. Fluorescent-penetrant examination is also widely used to inspect the root pass of highly critical pipe welds.

Interpretation   When visible dye penetrants are used, defects are indicated by the presence of a red color against the white background of the developer. A crack appears as a continuous line indication. The width and brightness of the dye indication depend on the volume of the crack or defect. Figure 21–21 shows a typical indication.

A *cold shut* (lap), which is caused by imperfect fusion, is smooth in outline and continuous. Penetrant indications of gas holes appear round with definite color contrast.

The most effective aid for identifying and recognizing defects is a collection of parts containing defects. These can be compared to parts containing unknown in-

dications. Extreme care and judgment must be exercised in interpreting indications. Consult the specification involved for standards of acceptability and qualification of operators.

Magnetic Particle Examination

Magnetic particle examination (MT) is a nondestructive method of detecting cracks, porosity, seams, inclusions, lack of fusion, and other discontinuities in ferromagnetic materials. Surface discontinuities and shallow subsurface discontinuities can be detected by using this method. There is no restriction as to the size and shape of the parts to be inspected; *only ferromagnetic materials can be examined by this method.*

This examination method consists of establishing a magnetic field in the test object, applying magnetic particles to the surface of the test object, and examining the surface for accumulations of the particles that are the indications of defects.

Ferromagnetism is the property of some metals, mainly iron and steel, to attract other pieces of iron and steel. A magnet will attract magnetic particles to its ends or poles, as they are called. Magnetic lines of force or flux flow between the poles of a magnet. Magnets will attract magnetic materials only where the lines of force enter or leave the magnet at the poles.

If a magnet is bent and the two poles are joined so as to form a closed ring, no external poles exist and hence it will have no attraction for magnetic materials. This is the basic principle of magnetic particle inspection. As long as the part is free of cracks or other discontinuities, magnetic particles will not be attracted. When a crack is present, north and south magnetic poles are set up at the

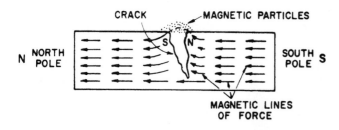

FIGURE 21–22 Principle of magnetic particle examination.

edge of the crack. The magnetic particles will be attracted to the poles that are the edges of the crack or discontinuity (Figure 21–22).

Electric currents are used to induce magnetic fields in ferromagnetic materials. An electric current passing through a straight conductor creates a circular magnetic field. For reliable examination, the magnetic lines of force should be at right angles to the defect to be detected. Hence, for a straight conductor with a circular field, any defect parallel to the conductor will be detected.

If the part is too large to run current through it, the part can be circularly magnetized by using probe contacts (Figure 21–23). Direct current is the most desirable type of current for subsurface discontinuities. It is most commonly used for wet magnetic particle inspection. For dry particles, pulsating dc is used for both surface and subsurface defects. This current causes the particles to pulse, which gives them mobility and aids in the formation of indications. Alternating current tends to magnetize the cracks of the metal only, and hence it is used only

for surface discontinuities such as fatigue or cracks caused by grinding.

Ferromagnetic parts that have been magnetized retain a certain amount of residual magnetism. Certain parts may require demagnetization if they are to function properly. The attraction of small chips or particles caused by the residual magnetism may cause excessive wear and failures with rotating parts such as bearings and bearing surfaces.

Equipment The most necessary piece of equipment for magnetic particle examination is the *specialized power source.* Small portable units are available that supply ac while operating from 115-V ac power lines. These units generally use dry powder, but portable magnetic particle units that employ a pressurized spray may also be used.

Stationary units are widely used for examination of small manufactured parts. These units usually contain a built-in tank with a pump that agitates the wet particle bath and pumps the fluid through a hose for application to the test parts. These stationary units are usually provided with an inspection hood; ultraviolet or black light can be used so that fluorescent particles can be used and viewed.

Applications The iron particles can be applied as dry powder or suspended in a liquid. Magnetic particle examination may be applied to all types of weldments. On multipass welds, it is sometimes used to examine each pass immediately after it has been deposited. An indication using the dry powder method is shown in Figure 21–24.

The majority of steel weldments in the aircraft industry are examined by the magnetic particle method. If

FIGURE 21–23 Using magnetic particle examination.

FIGURE 21–24 Magnetic powder indication.

ergy remain lighter. Therefore, areas of the material where the thickness has been changed by discontinuities, such as porosity or cracks, will appear as dark outlines on the film. Inclusions of low density, such as slag, will appear as dark areas on the film, while inclusions of high density, such as tungsten, will appear as light areas. All discontinuities are detected by viewing shape and variations in the density of the processed film.

The x-ray or gamma-ray source and penetrameter are placed above the piece to be radiographed and the film is placed on the opposite side of the part (Figure 21–25). Figure 21–26 shows an oil-cooled and shielded

FIGURE 21–25 Principle of radiographic examination.

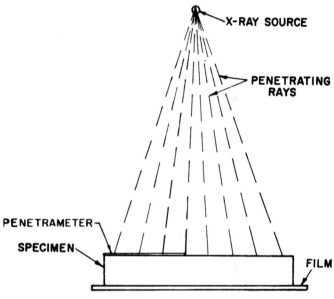

the weldments are thin enough, this method may provide sufficient sensitivity to detect any subsurface defects. Consult the specification involved for standards of acceptability and qualifications of equipment and operators. Parts may have to be demagnetized after testing.

Radiographic Examination

Radiography is a nondestructive examination method that uses invisible, x-ray, or gamma radiation to examine the interior of materials. Radiographic examination (RT) gives a permanent film record of defects that is relatively easy to interpret. Although this is a slow and expensive method of nondestructive examination, it is a positive method for detecting porosity, inclusions, cracks, and voids in the interior of castings, welds, and other structures.

X-rays, generated by electron bombardment of tungsten, and gamma rays emitted by radioactive elements are penetrating radiation whose intensity is modified by passage through a material. The amount of energy absorbed by a material depends on its thickness and density. Thus a thinner part will absorb less energy than a thick part, and a heavy dense metal, such as steel, will absorb more energy than a light metal such as aluminum. Energy not absorbed by the material will cause exposure of the radiographic film. These areas will be dark when the film is developed. Areas of the film exposed to less en-

FIGURE 21–26 Setting up to take radiograph.

head, which encloses the x-ray tube, being positioned to make a radiograph of a weld test plate.

Equipment X-rays are produced by electrons hitting a tungsten target inside an x-ray tube. In addition to the x-ray tube, the apparatus consists of a high-voltage generator with necessary controls. X-rays are produced when a high-speed stream of electrons collides with a piece of tungsten. The electrons are produced in the x-ray tube by a hot cathode. They are accelerated towards the anode by means of an electron gun in the vacuum of the x-ray tube. The anode is a piece of tungsten, and when the electrons hit it, x-rays are produced which are directed through a window to the part being inspected.

Gamma rays are produced by radioactive decay of certain radioisotopes. The radioisotopes normally used are cobalt-60, iridium-192, thulium-170, and cesium-137. These isotopes are contained in a lead or spent uranium vault or capsule to provide safe handling. They have a relatively short half-life and the strength of the radiation decreases with time.

Isotopes must be handled in such a way that radiographic sources can be positioned and yet produce minimum radiation hazards to operating personnel. Remote handling equipment is employed when the radioactive source is drawn from the shielded container to the material to be radiographed. In the United States a license from the U.S. Department of Energy is required for use of radioisotopes.

The radiation intensity or output from an x-ray machine or from radioisotope sources will vary. Common materials such as concrete and steel are used to house the x-ray machine and protect the operator from exposure. The thickness of the shielding enclosure walls should be sufficient to reduce exposure in all occupied areas to a minimum value. If the work is too large or too heavy to be brought into the shielded room, special precautions such as lead-lined booths and portable screens are used to protect personnel. In the field, radiography protection is usually obtained from distance alone, since radiation intensity decreases as distance increases.

Penetrameters are used to determine the sensitivity of the radiograph. They are made of the same material that is being inspected and are usually 2% of the thickness of the part being tested. Therefore, if the penetrameter can be seen clearly on the radiograph, any change in thickness of the part (2% or more) will be seen clearly.

Radiographic film consists of a transparent plastic sheet coated with a photographic emulsion. When x-rays strike the emulsion, an image is produced. The image is made visible and permanent by a film-processing operation. Most processing equipment consists of tanks that contain a developer, a fixer, and rinse solutions. Film-processing operations are just as critical as the film exposure. Unsatisfactory radiographs can sometimes be attributed to errors in the processing technique or from mishandling of materials.

Applications Radiography is one of the most popular nondestructive examination methods for locating subsurface defects. It is used for examination of weldments in all types of materials: steel, aluminum, magnesium, and so on. Radiography is used in the pipeline industry to ensure proper weld quality.

Interpretation Most indications will show up as dark regions against the light background of the sound weld. Radiographs should be examined with a film illuminator providing a strong light source. Figure 21–27 shows a radiograph being examined.

It is essential that qualified personnel conduct x-ray interpretations since false interpretation of radiographs causes a loss of time and money. Radiographs for reference are extremely helpful in securing correct interpretations. Consult the specification involved for standards of acceptability and qualification of equipment and operators.

Ultrasonic Examination

Ultrasonic examination (UT) is a nondestructive examination method that employs mechanical vibrations similar to sound waves but of a higher frequency. A beam of ultrasonic energy is directed into the specimen to be examined. This beam travels through a material with only a small loss, except when it is intercepted and reflected by a discontinuity or by a change in material.

Ultrasonic examination is capable of finding *surface and subsurface* discontinuities. The ultrasonic contact pulse reflection technique is used. This system uses a transducer, which changes electrical energy into mechanical energy. The transducer is excited by a high-frequency voltage that causes a crystal to vibrate mechanically. The crystal probe becomes the source of ultrasonic mechanical vibrations. These vibrations are transmitted into the test piece through a coupling fluid, usually a film of oil, called a *couplant*. When the pulse of

FIGURE 21–27 Examination of a radiograph.

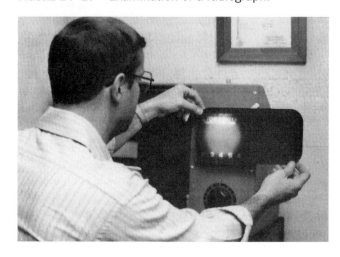

ultrasonic waves strikes a discontinuity in the test piece, it is reflected back to its point of origin. Thus the energy returns to the transducer. The transducer now serves as a receiver for the reflected energy. The initial signal or main bang, the returned echoes from the discontinuities, and the echo of the rear surface of the test material are all displayed by a trace on the screen of a cathode-ray oscilloscope. Videotapes may be used for permanent records.

The basic principles of ultrasonic examination are shown in Figure 21-28. The transducer is sending out a beam of ultrasonic energy. Some of the energy is reflected by the internal flaw, and the remainder is reflected by the back surface of the specimen. Figure 21-29 shows the equipment in use.

Figure 21-30 shows a typical display as presented on the oscilloscope screen. Signal strength is indicated by a vertical deflection of the pip on the screen, and transmitted time is indicated by horizontal deflection. By measuring the height of the pip, the size of the flaw can be determined. The depth of the flaw from the surface is

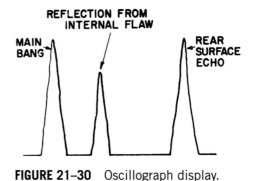

FIGURE 21–30   Oscillograph display.

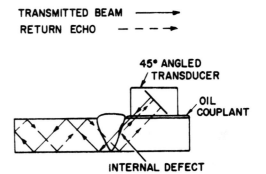

FIGURE 21–31   45° use of ultrasonic inspection.

FIGURE 21–28   Principle of ultrasonic examination.

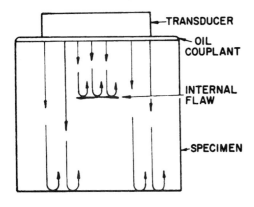

FIGURE 21–29   Making an ultrasonic examination of a weld.

found by use of horizontal base. The front reflection and the rear reflection are at the extreme ends of the screen. The echo from the flaw is between the two.

To determine the size and depth of flaws, calibration techniques and references standards must be used. A set of test blanks that have holes of known diameter and depth is used to calibrate the test instrument. The reference standard is described in detail by the AWS structural code.

Testing can be done with one transducer that serves both as a transmitter and receiver, or with two transducers, one transmitting only and the other receiving only. The single transducer is used for portable equipment.

For examining irregularly shaped parts, the immersion testing method is often used. With this method, the part and the transducer are submerged in water. The water transmits and couples the ultrasonic beam and the part. Actual contact is not required, thus irregular surfaces can be scanned.

The application of ultrasonic examination to weldments is shown in Figure 21-31. A 45° angled beam transducer is used to inspect the weld area. This search unit directs the beam toward the weld from a position on one side of the weld.

Butt joint welds in plate are usually examined with the angled search unit. The reflection obtained by the use of the 45° head is similar to Figure 21-30. The joint welds of heavy plate are examined with a straight beam transducer through the top of the joint, or with an angled beam transducer from one side of the bottom. Fillet welds are more difficult to examine.

Equipment Equipment required for this process consists of a transducer, pulse rate generator, amplifier, timer, and cathode ray oscilloscope. These devices are electronic, quite small in size for portability, yet rugged. Instant cameras are available to photograph oscilloscope displays for permanent records. Tapes can also be used.

Applications Ultrasonic examination can be used to test practically any metal or material. Its use is restricted only by very complex weldments. The process is increasing in popularity and widely used. Consult the specification involved for standards of acceptability and qualification of equipment and operators.

Leak Testing

Leak testing (LT) can be accomplished in many different ways. It can be used only when the weldment can be made to contain a gas or liquid. The most common leak test is the soap bubble test, which can be applied to external joints if internal gas pressure is present. Another is the use of internal liquids and the maintaining of high pressure over an extended period. The same test can utilize a vacuum. Halogen gases with sensitive detection meters are used to inspect production parts.

Do not use toxic or flammable gases or air for internal pressure testing. The internal pressure will store energy and if the part should fail it could cause an explosion that might cause bodily harm. Use internal liquids instead, or test inside a safety chamber.

Proof Testing

Proof testing (PRT) is controversial. In the past, proof testing was used on vessels and mechanical components. It consisted of loading the part 50% or 100% greater than the designed load. It was felt that if the part passed this test and if it was never loaded above its designed load that it would never fail in service. It is now thought that proof testing could cause internal damage that might reduce service life. Also, proof testing cannot provide assurance if the part is subjected to corrosion, fatigue, low-temperature impacts, and so on. Consider proof testing but only in light of the above. Its use is declining.

Guide to NDT Techniques

Figure 21–32 is a guide to welding quality control comparing the different nondestructive examination techniques. This table shows the equipment required, the defects that can be detected, the advantages and disadvantages of each technique, and other factors.

21-5 CORRECTIVE ACTIONS FOR WELD DEFECTS

The problem of weld defects has become complex, partially because of the great number of different words and definitions that are being used. For example, a welding flaw is a synonym for discontinuity; *discontinuity* is the preferred term. A discontinuity is "an interruption of the typical structure of a material, such as a lack of homogeneity in its mechanical, metallurgical, or physical characteristics." A discontinuity is not necessarily a defect. A *defect* is "a discontinuity or discontinuities that by nature or accumulated effect render a part or product unable to meet minimum applicable acceptance standards or specifications." This term designates rejectability. A *defective weld* then becomes "a weld containing one or more defects." For our purposes, we will consider defects as anything undesirable in a weld. It may or may not be cause for rejection or cause for repair. This is normally a matter left up to the specifications or codes involved. It is important that we learn to recognize the different types of weld defects and learn enough about them so that we can recognize them, repair them, and avoid them.

Upcoming figures will show photographs and drawings of the possible different welding defects that can occur. It is an effort to present, in an organized manner, the various defects that can occur, the description of the particular defect or problem, and an indication of how or what caused the problem and the action that should be taken to correct the specific problem. The presentation may not describe all possible weld defects. There are others, and some of the defects described may occur on different types of welds, but they would generally resemble those presented.

In this collection, an effort is made to indicate the responsibility for each of the defects. This breakdown is broad, but will indicate whether it is the fault of the welder—that is, a problem of welding technique—the fault of the designer—that is, a drawing or design error—or a fault of some manufacturing or shop function, such as materials preparation. In the figures these are designated as a *welder* responsibility, *designer* responsibility, or a *shop* responsibility.

An indication of how the particular defect was detected will also be given. This is based on the inspection techniques used to find the defect. The five most popular nondestructive examination techniques are as follows:

* VT   Visual examination
* MT   Magnetic particle examination
* PT   Dye-penetrant examination (including fluorescent)
* RT   Radiographic examination
* UT   Ultrasonic examination

Examination Technique	Equipment	Defects Detected	Advantages	Disadvantages	Other Considerations
Visual: VT	Pocket magnifier, welding viewer, flashlight, weld gauge, scale	Weld preparation, fitup, cleanliness, roughness, spatter, undercuts, overlaps, weld contour and size; welding procedures	Easy to use; fast; inexpensive; usable at all stages of production	For surface conditions only; dependent on subjective opinion inspector	Most universally used examination method
Dye penetrant or fluorescent: DPT, FPT	Fluorescent or visible penetrating liquids and developers; ultraviolet light for the fluorescent type	Defects open to the surface only; good for leak detection	Detects very small, tight, surface imperfections, easy to apply and to interpret; inexpensive; use on magnetic or nonmagnetic materials	Time consuming in the various steps of the process; normally no permanent record	Often used on root pass of highly critical gas welds if material improperly cleaned; some indications may be misleading
Magnetic particle: MT	Iron particles, wet or dry, or fluorescent; special power source; ultraviolet light for the fluorescent type	Surface and near-surface discontinuities, cracks, etc.; porosity, slag	Indicates discontinuities not visible to the naked eye; useful in checking edges prior to welding, also, repairs; no size restriction	Used on magnetic materials only; surface roughness may distort magnetic field; normally no permanent record	Examination should be from two perpendicular directions to catch discontinuities that may be parallel to one set of magnetic lines of force
Radiographic: RT	X-ray or gamma ray; source: film processing equipment, film viewing equipment, penetrameters	Most internal discontinuities and flaws; limited by direction of discontinuity	Provides permanent record; indicates both surface and internal flaws; applicable on all materials	Usually no suitable for fillet weld inspection; film exposure and processing critical; slow and expensive	Most popular technique for subsurface inspection; required by some codes and specifications
Ultrasonic: UT	Ultrasonic units and probes; reference and comparison patterns	Can locate all internal flaws located by other methods with the addition of exceptionally small flaws	Extremely sensitive; use restricted only by very complex weldments; can be used on all materials	Demands highly developed interpretation skill	Required by some codes and specifications

FIGURE 21–32 Guide to weld quality control techniques.

The corrective action for the particular defect is briefly mentioned. Corrective action means the way to correct the specific defect in order to make the weldment suitable for service. Efforts will also be made to provide an explanation to prevent a recurrence of the same type of defect. This may be covered in the general information concerning the different classifications of defects.

The defects are arranged according to the classification of weld defect system established by Commission V of the International Institute of Welding. Their document IIS/11W-340-69 and Commission V Document[10] classify the defects into six groups:

- ※ *Series 100, Cracks:* including longitudinal, transverse, radiation, crater
- ※ *Series 200, Cavities:* including gas pockets, internal porosity, surface porosity, shrinkage
- ※ *Series 300, Solid inclusions:* including slag, flux, metal oxides, foreign material
- ※ *Series 400, Incomplete fusion or penetration:* including incomplete fusion, incomplete penetration
- ※ *Series 500, Imperfect shape or unacceptable contour:* including undercut, excessive reinforcement, underfill, fillet shape, overlap
- ※ *Series 600, Miscellaneous defects not included above:* including arc strikes, excessive spatter, rough surface

This International Institute of Welding document catalogs all welding defects and provides index numbers of three digits for groups and four digits for specific types of defects within the groups. It also cross-indexes with the International Institute of Welding Radiographic Reference Radiographics.

The examples of defects presented here are shown to illustrate the problems, and no reference is made to what is acceptable or allowable. To determine the acceptability limits of these different defects you must refer to the specification or code that is involved. Certain defects of a minor nature may be acceptable under specific conditions in some codes, whereas they may be unacceptable in other codes. In general, defects that can propagate under stress are not acceptable in any code; for example, cracks are not allowed in any of the major codes but porosity of a specific magnitude and spacing may be acceptable.

There are other problem areas encountered in welding that are not specifically included. These would include such factors as distortion and warping, brittle heat-affected zones, brittle weld metal, arc blow, and so on. These topics are covered elsewhere in the book.

Series 100: Cracks

Cracks are the first category of weld defects. A crack is "a fracture-type discontinuity characterized by a sharp tip and high ratio of length and width to opening displacement." Cracks are perhaps the most serious of the defects that occur in the welds or weld joints in weldments. However, cracks are defects that can be found in other metal products such as forgings, castings, and even hotrolled steel products. Cracks are considered dangerous because they create a serious reduction in strength. They can propagate and cause sudden failure. They are most serious when impact loading and cold-temperature service are involved. Cracks must be repaired.

There are many different types of cracks. One way of categorizing them is as surface or subsurface cracks. *Surface cracks* can be seen on the surface of the weld using the visual testing technique. There are several types of surface cracks—transverse, longitudinal, and crater cracks (Figure 21–33). There are also toe cracks in adjacent parent metal that normally come to the surface. The subsurface or internal cracks are also of many types. Some may be in the weld, some in the heat-affected zone—sometimes called *underbead cracks* (Figure 21–34), at the interface between weld metal and base metal, and sometimes called *lamanar tearing,* which is completely

FIGURE 21–33 Types of surface cracks of welds.

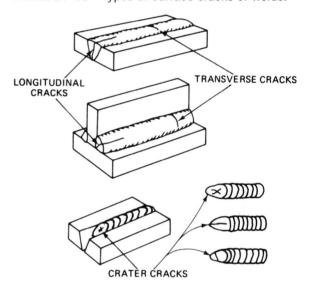

FIGURE 21–34 Weld toe cracks and underbead cracks.

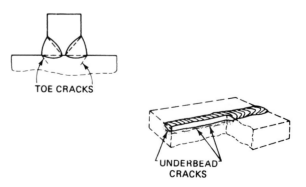

in the base metal. There can also be microsize cracks as well as macrosize cracks. Sometimes the smaller cracks are called *fissures,* or if the cracks are extremely small they are called *microfissures,* which require special techniques to find.

There is another way of classifying cracks and this is with respect to temperature. Hot cracks occur during or immediately after the weld is made or during the cooling cycle. Cold cracks occur sometime after the weld is finished, after it has cooled to room temperature. Cold cracks may be delayed hours or even days after the weld is finished. There are also fatigue-type cracks that may occur months or years after fabrication as a result of an initiating point and fatigue loading. There are also stress corrosion cracks that are caused by a corrosive atmosphere and a high stress condition.

In general, cracks in welds or cracks adjacent to welds indicate that the weld metal or the base metal has low ductility and that there is high restraint. Any factor that contributes to low ductility of the weld and adjacent metal and high restraint will contribute to cracking. Some of these factors are rapid cooling, high-alloy composition, insufficient heat input, poor joint preparation, and incorrect electrode type. The examples that are shown in Figure 21–35 will help further explain the different types of weld cracks, the probable cause, and corrective action.

Series 200: Cavities

Cavities are the second category of defects. The most common type of cavity is called porosity, defined as "cavity type discontinuities formed by gas entrapment during solidification." Specific defects can be called *gas pockets,* which are cavities caused by entrapped gas. These are sometimes called *blow holes.* Porosity can also be divided into two types: surface porosity, which can be seen by the naked eye and detected by visual inspection technique; and subsurface porosity, which can be found only by the internal detecting techniques. The gas pockets can occur as extremely large holes in the weld metal or extremely

FIGURE 21–35 Collection of weld defects: cracks.

Weld Defect Class — Cracks Appearance or Cross Section or Radiograph Class No. 100		How Defected	Responsibility	Probable Cause	Corrective Action
Longitudinal Crack A	VT	X	design	General	1. Use proper or matched electrode.
	MT	X	X	1. Incorrect electrode.	2. Reduce rigidity of weldment or change welding sequence. Use higher ductility welding filler metal.
	PT	X	welder	2. High restraint of joint.	
	RT	X	X		
	UT	X	shop	3. Rapid cooling of weld.	3. Use preheat and/or inner pass heat to reduce cooling rate.
Longitudinal Crack B	VT		design		4. Use proper joint for welding process.
	MT	X	X	4. Improper joint preparation.	5. Change center line of weld to avoid interface between parts.
	PT		welder		
	RT	X	X	5. Fillet weld longitudinal crack.	
	UT	X	shop		
Crater Crack C	VT	X	design	Crater Crack	1. Filler crater with proper technique.
	MT	X		1. Unfilled crater.	2. Utilize run-out-tab.
	PT	X	welder	2. Crater crack in submerged arc welding.	
	RT	X	X		
	UT	X	shop		
Transverse Crack D	VT	X	design	Transverse Crack	1. Use proper electrode.
	MT	X		1. Incorrect electrode.	2. Use larger electrode, higher welding current, or preheat.
	PT	X	welder	2. Rapid cooling.	
	RT	X	X	3. Welds too small for size of parts joined.	3. Use larger weld, possibly larger welding electrode.
	UT	X	shop X		

small holes scattered throughout the cross section of a weld. Some types of porosity are called *worm holes* when they are long and continuous. Others are called piping, usually long in length and parallel to the root of the weld. Some types may occur exclusively at the root and others almost at the surface. Porosity is not as serious a defect as cracks primarily because porosity cavities usually have rounded ends and will not propagate like cracks. Many codes and specifications provide comparison charts showing the amount of porosity that may be acceptable. Figure 21-36 shows an example of comparison charts used by the API 1104 code. The AWS structural welding code has taken a slightly different point of view and has a sliding scale (Figure 21-37). This takes into account size and spacing of the porosity and relates to the size of the weld. For more information refer to these codes.

There are other types of cavities and some are called shrinkage voids, which are defined as "a cavity discontinuity normally formed by shrinkage during solidification." Cavities, voids, and porosity are caused by gases that are present in the arc area, or may be present in base metal, that are trapped in the molten weld during the solidification process. Common causes for porosity are high sulfur in the base metal, hydrocarbons such as paint on the surface of the metal, water, oil, moisture from damp electrodes, wet submerged arc flux, or wet shielding gas. When the porosity exceeds that acceptable by the code, it must be removed and repair welds made. In general, surface porosity is an indication that subsurface porosity may have been in the weld before it became noticeable as surface porosity. In these cases extra inspection should be done to determine the extent of the subsurface porosity. The following examples of cavities will help further explain this type of weld defect. These are shown in Figure 21-38.

Series 300: Inclusions

Solid inclusions are the next type of defect. Solid inclusions are normally expected to be a subsurface type of defect and would include any foreign material entrapped in the deposited weld metal. The most common type of solid inclusion is a slag inclusion defined as "nonmetallic solid material entrapped in weld metal or between weld metal and base metal." Another and very similar type of inclusion is a flux inclusion, which is an entrapment of flux from an electrode, from submerged arc flux, or from another source of flux that for one reason or another did not float out of the weld metal as it solidified. Slag inclusions and flux inclusions can be continuous, intermittent, or very randomly spaced. In general, flux or slag inclusions are rounded and do not possess sharp corners like cracks and for this reason are not quite as serious as cracks. The applicable code or specification indicates how much entrapped slag or flux is acceptable.

On certain metals, particularly those that have high-temperature oxide coatings, there is the possibility of oxide inclusions in the weld metal. This is a troublesome problem when welding aluminum. Aluminum oxide will form rapidly in the atmosphere and can be entrapped in the weld metal very easily if cleaning and other precautions are not taken. Oxide inclusions are detected by the internal inspection techniques. There are also other metallic inclusions such as tungsten inclusions that can only be found by internal inspection techniques, particularly radiographic testing. Oxide inclusions or tungsten inclusions are not acceptable for high-quality work. When copper backing bars are used, local melting may occur and copper can be entrapped in the weld metal. This can be detected from the underside surface of a weld or by internal detection techniques. All such inclusions are de-

FIGURE 21-36 Porosity chart for pipe welds.

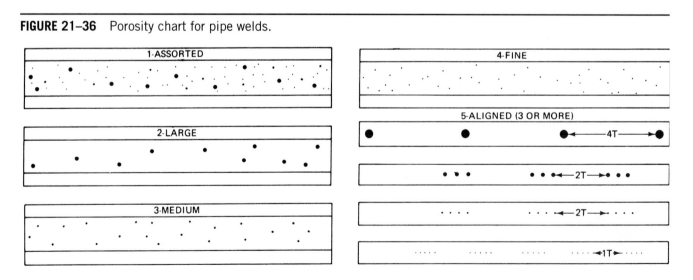

WALL THICKNESS 1/2" OR LESS
MAXIMUM DISTRIBUTION OF GAS POCKETS

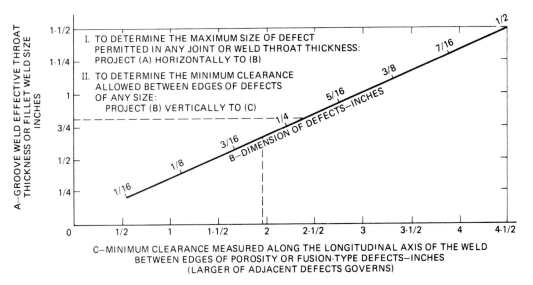

FIGURE 21–37 AWS structure code weld quality requirements.

FIGURE 21–38 Collection of weld defects: cavities.

Weld Defect Class – Appearance or Cross Section or Radiograph Class No. 200		How Defected		Respon-sibility	Probable Cause	Corrective Action
A		VT	X	design	**General** Surface Porosity 1. Welding over foreign material on surface such as rust, oil, moisture, paint, etc. 2. Damp electrodes. 3. Improper base metals such as free machining or high sulphur. 4. Welding current too low.	1. Clean weld bevels and area adjacent to weld and keep clean. 2. Use fresh dry electrodes or rebake electrodes that have been exposed to dampness. 3. Utilize correct base metal possibility to use low hydrogen-type electrodes. 4. Increase welding current.
		MT				
		PT	X	welder		
		RT		shop		
		UT	X			
B		VT	X	design	**Gas Shielded Welding Processes** 1. Incorrect shielding gas type. 2. Incomplete gas coverage due to breeze, defective gas system, clogged nozzle, etc. 3. Moisture in the shielding system. 4. Poor gas coverage. 5. Welding over tack weld made with shielded metal arc process.	1. Use specified shielding gas. 2. Provide windshields, check efficiency of gas system such as broken hoses, gas valves, empty tanks, clean nozzle. 3. Check to make sure gas is dry or welding grade. Check for water leaks in water cooled systems. 4. Use proper nozzle to work distance. Check gas flow rate, may be too high or too low. 5. Utilize gas metal arc for tack welding.
		MT				
		PT	X	welder		
		RT		shop		
		UT	X			
C		VT	X	design		
		MT	X			
		PT	X	welder		
			X			
		RT	X	shop		
		UT	X	X		
D		VT	X	design	**Submerged Arc Welding** 1. Damp submerged arc flux. 2. Contaminated surface of electrode wire, dirt and/or moisture. 3. Too many fines in flux.	1. Utilize fresh dry flux or dry damp flux. 2. Clean surface of electrode wire. 3. Use fresh flux and discard fines.
		MT				
		PT		welder		
			X			
		RT	X	shop		
		UT	X	X		

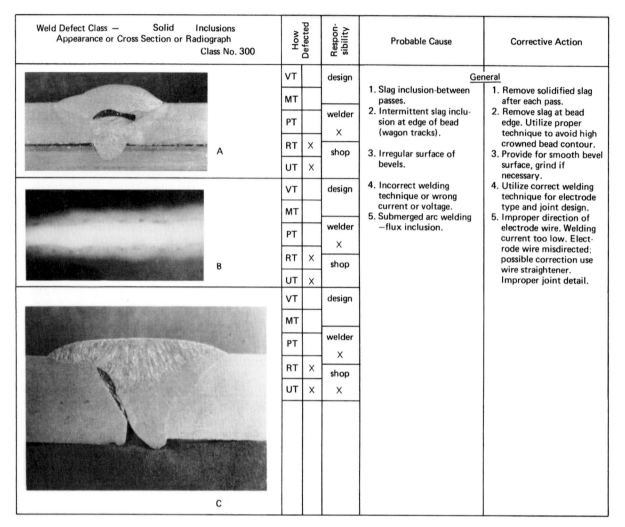

Weld Defect Class — Solid Inclusions Appearance or Cross Section or Radiograph Class No. 300	How Defected		Respon-sibility	Probable Cause	Corrective Action
A	VT		design	**General**	
	MT			1. Slag inclusion-between passes.	1. Remove solidified slag after each pass.
	PT		welder	2. Intermittent slag inclusion at edge of bead (wagon tracks).	2. Remove slag at bead edge. Utilize proper technique to avoid high crowned bead contour.
	RT	X	X	3. Irregular surface of bevels.	3. Provide for smooth bevel surface, grind if necessary.
	UT	X	shop	4. Incorrect welding technique or wrong current or voltage.	4. Utilize correct welding technique for electrode type and joint design.
B	VT		design	5. Submerged arc welding —flux inclusion.	5. Improper direction of electrode wire. Welding current too low. Electrode wire misdirected; possible correction use wire straightener. Improper joint detail.
	MT				
	PT		welder		
	RT	X	X		
	UT	X	shop		
C	VT		design		
	MT				
	PT		welder		
	RT	X	X		
	UT	X	shop X		

FIGURE 21–39 Collection of weld defects: solid inclusions.

fects that must be evaluated in accordance with the code or specification in question or with respect to good practice. The examples shown in Figure 21-39 will help further explain these types of defects.

Series 400: Incomplete Fusion

Incomplete fusion or *penetration* is the next defect category. This is sometimes called lack of fusion or lack of penetration; however, the preferred term is *incomplete fusion*, defined as "a weld discontinuity in which fusion did not occur between weld metal and fusion faces on adjoining weld beads." It is shown in Figure 21-40. This can be inadequate joint penetration, defined as "joint penetration which is less than specified." The word *penetration* is not preferred; the term should be *joint penetration*, defined as "the distance the weld metal extends from its face into a joint, exclusive of weld reinforcement," or root penetration. *Root penetration* is defined as "the distance the weld metal extends into the joint root"

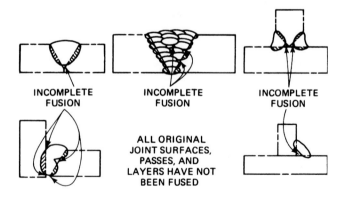

FIGURE 21–40 Incomplete fusion.

and is shown in Figure 21-41. These illustrations help show the difference between complete and incomplete fusion and complete joint versus partial joint penetration and root penetration. Incomplete fusion as a defect means that the weld deposited did not completely fill the

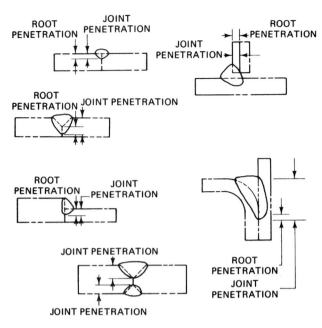

FIGURE 21–41 Root penetration and joint penetration.

joint preparation or there is space in between the beads or passes or a space at the root of the joint. *Penetration* is a slightly different term. The term *joint penetration* is the minimum depth of the joint; the groove or flange weld extends from its face into the root, exclusive of reinforcement. The term *penetration* means the depth that the groove weld extends into the root of a joint measured on the centerline of the cross section. These terms are often used interchangeably but do have a different meaning. The defect is the absence of complete fusion of a joint and this provides a stress riser, which is undesirable for welds loaded in fatigue or subject to impacts or low-temperature service. Figure 21–42 helps illustrate different types of this defect. The cause of such defects can be dirty surfaces such as heavy mill scale, heavy rust, or grease; failure to remove slag from previous beads; the fact that the root opening may not be sufficiently large; or unsatisfactory welding technique. The danger of the defect is the serious reduction in static strength and the production of a stress riser.

FIGURE 21–42 Collection of weld defects: incomplete fusion.

Weld Defect Class — Incomplete Fusion Appearance or Cross Section or Radiograph Class No. 400		How Defected	Responsibility	Probable Cause	Corrective Action
A	VT	X	design	**General** 1. Welding speed too fast. 2. Electrode too large for joint detail. 3. Welding current too low. 4. Improper joint design such as excessive root face or minimum root opening. 5. Improper joint fitup such as root opening too small.	1. Reduce welding speed. 2. Utilize correct size electrode. 3. Increase welding current for more penetration. 4. Utilize correct joint detail. 5. Make setup correct to agree with joint design detail.
	MT		X		
	PT	X	welder		
	RT	X	X		
	UT	X	shop		
B	VT	X	design	**Shielded Metal Arc Welding** 1. Irregular travel speed. 2. Irregular arc length	1. High speed will reduce complete fusion, lower speed will cause complete fusion. 2. Maintain proper arc length.
	MT	X	X		
	PT		welder		
	RT	X	X		
	UT	X	shop		
C	VT	X	design	**Gas Metal Arc Welding** 1. Incomplete fusion (cold shut).	1. Direct arc at leading edge of puddle. Current too low, voltage too low, adjust for proper procedure. Pause too short at dwell when weaving. Increase pause to allow melting of base metal.
	MT	X	X		
	PT		welder		
	RT	X	X		
	UT	X	shop		
D	VT	X	design	**Submerged Arc Welding—Semiautomatic** 1. Incomplete root fusion.	1. Failure to direct welding electrode to root of weld joint.
	MT				
	PT	X	welder		
	RT	X	X		
	UT	X	shop		

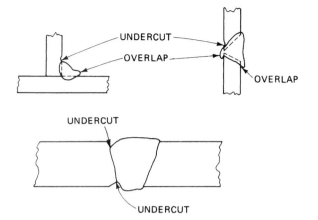

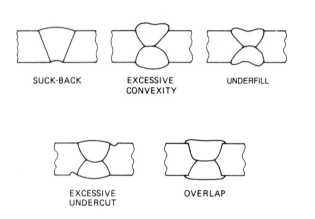

FIGURE 21-43 Undercut fillet and groove and overlap.

SUCK-BACK   EXCESSIVE CONVEXITY   UNDERFILL

EXCESSIVE UNDERCUT   OVERLAP

FIGURE 21-44 Groove welds and various defects.

Series 500: Imperfect Shape

Imperfect shape, or *unacceptable contour,* is the next category of defect. One of the most serious of these defects is undercut (Figure 21-43). Undercut occurs not only on fillet welds but on groove welds as well. Undercut also produces stress risers that create problems under impact, fatigue, or low-temperature service. It is normally caused by excessive currents, incorrect manipulation of electrode, incorrect electrode angle, or type of electrode. Undercut actually refers more to the base metal adjacent to the weld, whereas imperfect shape is a defect of the weld itself. This can include such things as excessive reinforcement on the face of a weld, which can occur on groove welds as well as fillet welds. There is also the problem of excessive reinforcement on the root of the weld, primarily open root groove welds. Excessive reinforcement is an economic waste. It can also be a stress riser and is objectionable from an appearance point of view. It is normally a factor involved with fitup, welder technique, welding current, type of electrode, and so on. A similar flaw is the concave type contour or lack of fill on the face of the weld or a suck-back on the root of a groove weld. The proper term in both cases is *underfill,* defined "as a depression on the weld face or root surface extending below the adjacent surface of the base metal" (Figure 21-44). Underfill does reduce the cross-sectional area of the weld below the designed amount and therefore is a point of weakness and potentially a stress riser where failure may initiate. The fillet weld is particularly vulnerable to the problem of imperfect shape. Figure 21-45 shows some different fillet weld contours, both acceptable and unacceptable. Those that reduce the throat of the fillet weld actually reduce the strength of the weld so that premature failure may occur. Examples shown in Figure 21-46 help show the different types of possible defects in this series.

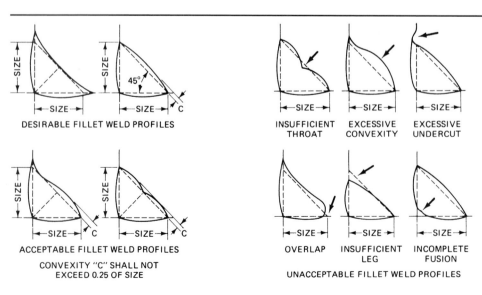

FIGURE 21-45 Contour of fillet welds.

DESIRABLE FILLET WELD PROFILES

ACCEPTABLE FILLET WELD PROFILES

CONVEXITY "C" SHALL NOT EXCEED 0.25 OF SIZE

INSUFFICIENT THROAT   EXCESSIVE CONVEXITY   EXCESSIVE UNDERCUT

OVERLAP   INSUFFICIENT LEG   INCOMPLETE FUSION

UNACCEPTABLE FILLET WELD PROFILES

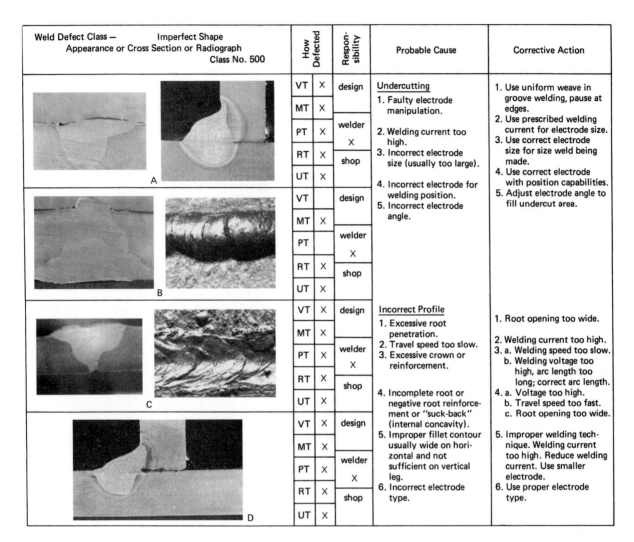

Weld Defect Class — Imperfect Shape Appearance or Cross Section or Radiograph Class No. 500	How Defected		Responsibility	Probable Cause	Corrective Action
A	VT	X	design	**Undercutting** 1. Faulty electrode manipulation. 2. Welding current too high. 3. Incorrect electrode size (usually too large). 4. Incorrect electrode for welding position. 5. Incorrect electrode angle.	1. Use uniform weave in groove welding, pause at edges. 2. Use prescribed welding current for electrode size. 3. Use correct electrode size for size weld being made. 4. Use correct electrode with position capabilities. 5. Adjust electrode angle to fill undercut area.
	MT	X			
	PT	X	welder X		
	RT	X	shop		
	UT	X			
B	VT		design		
	MT	X			
	PT		welder X		
	RT	X	shop		
	UT	X			
C	VT	X	design	**Incorrect Profile** 1. Excessive root penetration. 2. Travel speed too slow. 3. Excessive crown or reinforcement. 4. Incomplete root or negative root reinforcement or "suck-back" (internal concavity). 5. Improper fillet contour usually wide on horizontal and not sufficient on vertical leg. 6. Incorrect electrode type.	1. Root opening too wide. 2. Welding current too high. 3. a. Welding speed too slow. b. Welding voltage too high, arc length too long; correct arc length. 4. a. Voltage too high. b. Travel speed too fast. c. Root opening too wide. 5. Improper welding technique. Welding current too high. Reduce welding current. Use smaller electrode. 6. Use proper electrode type.
	MT	X			
	PT	X	welder X		
	RT	X	shop		
	UT	X			
D	VT	X	design		
	MT	X			
	PT	X	welder X		
	RT	X	shop		
	UT	X			

FIGURE 21–46 Collection of weld defects: incorrect shape.

Series 600: Miscellaneous Defects

Miscellaneous defects are the final category considered and actually cover all defects that may not be categorized in any of the other classifications given above. One type of defect in this classification is *arc strikes,* which are unacceptable in certain types of work. These are defects where the welder accidentally struck the electrode on the base metal adjacent to the weld. This creates problems particularly on hardenable steel and on critical types of applications and is not acceptable. For certain types of work, protective wrappings are made around the part, especially pipe adjacent to the weld, to avoid stray arc strikes. *Excessive spatter* adjacent to the weld is also a defect and is unacceptable. This may be caused by arc blow, by the selection of the incorrect electrode or welding current, or the technique of the welder. There are several other types of miscellaneous defects, and those defects that apply specifically to particular processes are covered in the examples shown in Figure 21-47.

The explanations and illustrations of different weld defects are based on opinions of various experts in the welding industry. Unfortunately, it is not universally possible to indicate which are acceptable and which are unacceptable. Some defects can be critical under certain service conditions or on specific metals. The same defect on less-demanding service on mild steel would be acceptable. There is no common agreement as to what is acceptable and unacceptable in different types of service.

A discontinuity is any flaw, crack, or unwanted imperfection in the material or weld metal. Some discontinuities are acceptable and some are not. Acceptable discontinuities do not exceed the limits of the standard of the welding code being used as an acceptance criterion. A discontinuity that exceeds the limits of the working standard or code is called a *defect* and is unacceptable. It is therefore necessary to refer to the code or standard that is being used.

Weld Defect Class — Miscellaneous Defects Appearance or Cross Section or Radiograph Class No. 600	How Defected		Responsibility	Probable Cause	Corrective Action
	VT	X	design	**Poor Appearance** 1. Welding current too high or too low. 2. Improper technique. 3. Faulty electrode. 4. Irregular travel speed.	1. Use prescribed procedure. 2. Provide additional welding training. 3. Use fresh electrode or determine correct type to be used. 4. Allow for additional practice and experience.
	MT				
	PT		welder X		
	RT	X			
	UT		shop		
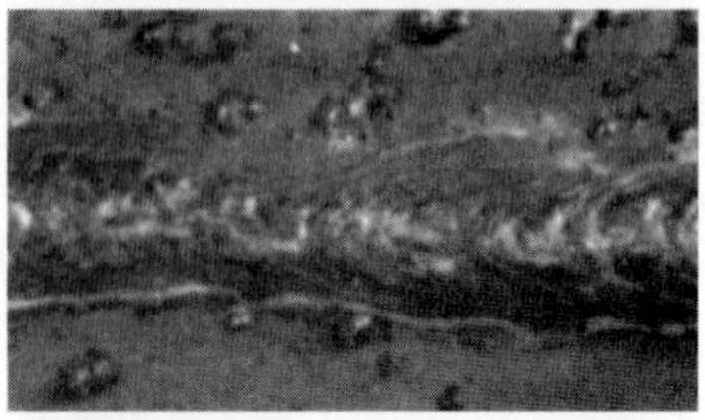	VT	X	design	**Excessive Weld Spatter** 1. Arc blow—see welding current. 2. Excessive welding cur for type and size electrode. 3. Excessive long arc—high voltage. 4. Improper electrode type.	1. Reduce arc blow—use alternating current. 2. Adjust for proper welding current for size electrode used. 3. Hold proper arc length and use correct arc voltage. 4. Utilize proper electrode type for location.
	MT				
	PT		welder X		
	RT				
	UT		shop		
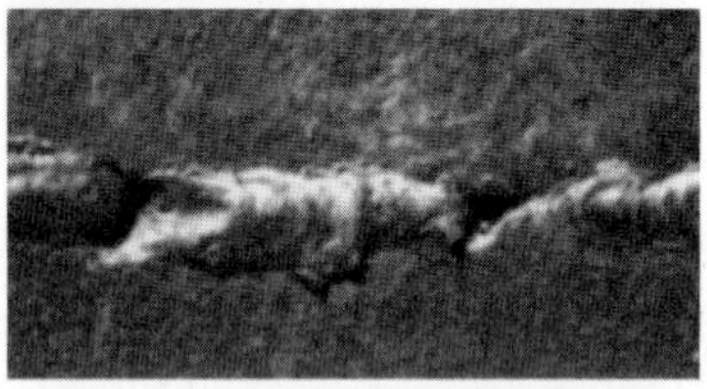	VT	X	design	**Poor Tie-In** 1. Incorrect electrode angle. 2. Improper technique for restriking arc.	1. Use correct electrode angle. Improve training. 2. Provide additional training and experience.
	MT				
	PT		welder X		
	RT	X			
	UT		shop		
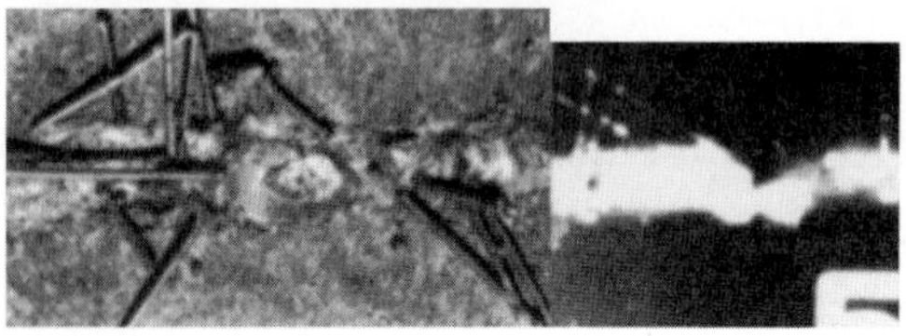	VT	X	design	**Whiskers** 1. Root opening too wide.	1. Use correct root opening. Use weaving motion and direct arc on weld puddle.
	MT				
	PT		welder X		
	RT	X			
	UT	X	shop		

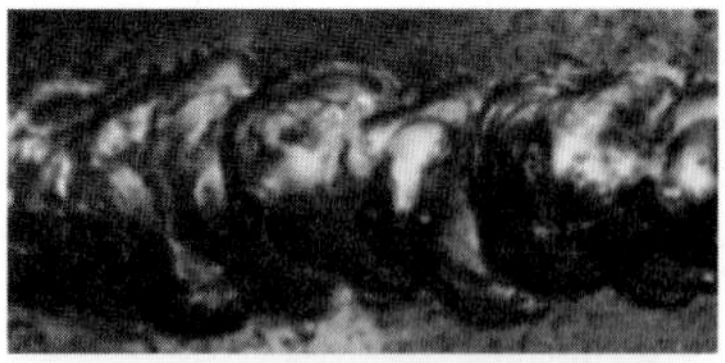

FIGURE 21–47 Collection of weld defects: miscellaneous.

21-6 WORKMANSHIP SPECIMENS AND STANDARDS

A *workmanship specimen* is an actual weld with each weld bead showing for a short length. Such a specimen provides the welder with an example of what is expected and also gives the welding supervisor and the inspector an example. Figure 21–48 shows a vertical up V-groove weld made of medium thickness plate. The workmanship specimen is a quality control tool and also an instruction device. The workmanship specimen concept originated with the U.S. Army Ordnance Department to ensure the quality of weldments produced during World War II. Workmanship specimens were made for each of the weld joints of an ordnance weldment. Cross-sectional samples were cut from the weld joint, polished, etched, and tack welded to the specimen to provide additional information. A weld schedule or procedure chart shows the welding joint design, welding conditions, current, voltage, and so on, for each layer of the particular joint made. The schedules were followed in making the specific workmanship specimens. This technique provides a good tool for quality control. The information should be posted and made available to welders, welding supervisors, inspectors, engineers, and others at the point where the welding is done. Workmanship specimens can be made for any weld joint welded in any position using any of the arc welding processes. The principle is to show the joint fitup detail and each bead or weld as it is made in producing the total weld joint.

Complete parts are sometimes used instead of welding drawings or blueprints in the manufacturing department. These welded parts sometimes are posted in the welding booth where that part is manufactured. For companies making relatively small parts on a production basis, that is an excellent technique of informing the welder and others of exactly the type of welds expected on the finished weldment. Many times these parts are painted with the welds highlighted in a different color of

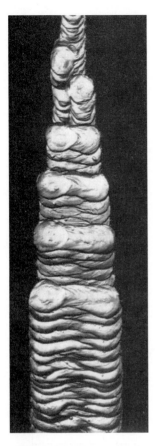

FIGURE 21–48 Workmanship specimen: vertical up SMAW.

paint to make them stand out. Welding schedule and weld size information can be posted adjacent to the weldment specimen.

The concept of workmanship specimens is used extensively by major structural steel contracting companies. Many of these companies operate erection crews in widely separated areas, yet expect to have welds made the same way by the different crews. These companies produce workmanship specimens and provide the welding schedules for producing these workmanship specimens. They go a step further and run qualification tests on each of the different joints that are normally employed.

The tail of the arrow of the welding symbol will show a specific joint detail specification. This detail specification will refer to a particular weld schedule and workmanship specimen that has been qualified in accordance with the code. By this means, the welding crew will always make the weld joint in the same manner and will utilize the procedure that is known to produce quality welds. This assists in consistency of welding but also in consistency of weld quality and further provides cost control since the weld joint is always made the same way. Figure 21–49 shows a typical welding specimen and the test bars produced from the specimen. The schedule for making this weld is shown in Figure 21–50.

Companies producing heavy weldments can also use workmanship specimens, but in these cases they may only show the different joints in a small section. These joints serve to show a proper way for making a particular

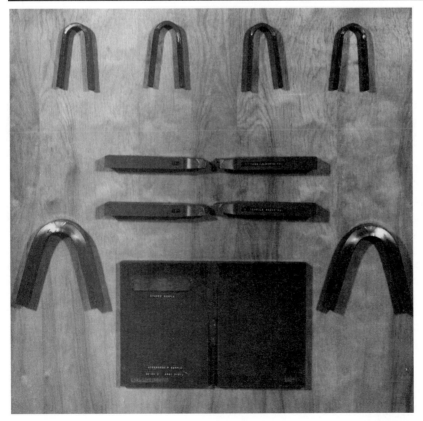

FIGURE 21–49 Workmanship specimen and test bars.

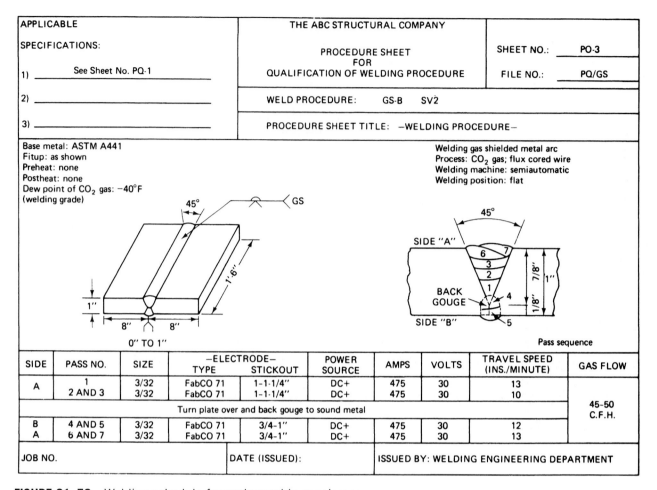

APPLICABLE SPECIFICATIONS:	THE ABC STRUCTURAL COMPANY							
1) See Sheet No. PQ-1	PROCEDURE SHEET FOR QUALIFICATION OF WELDING PROCEDURE		SHEET NO.: PQ-3 FILE NO.: PQ/GS					
2)	WELD PROCEDURE: GS-B SV2							
3)	PROCEDURE SHEET TITLE: —WELDING PROCEDURE—							

Base metal: ASTM A441
Fitup: as shown
Preheat: none
Postheat: none
Dew point of CO_2 gas: −40°F (welding grade)

Welding gas shielded metal arc
Process: CO_2 gas; flux cored wire
Welding machine: semiautomatic
Welding position: flat

SIDE	PASS NO.	SIZE	—ELECTRODE— TYPE	STICKOUT	POWER SOURCE	AMPS	VOLTS	TRAVEL SPEED (INS./MINUTE)	GAS FLOW
A	1	3/32	FabCO 71	1-1-1/4"	DC+	475	30	13	
	2 AND 3	3/32	FabCO 71	1-1-1/4"	DC+	475	30	10	45-50 C.F.H.
	Turn plate over and back gouge to sound metal								
B	4 AND 5	3/32	FabCO 71	3/4-1"	DC+	475	30	12	
A	6 AND 7	3/32	FabCO 71	3/4-1"	DC+	475	30	13	

JOB NO.	DATE (ISSUED):	ISSUED BY: WELDING ENGINEERING DEPARTMENT

FIGURE 21-50   Welding schedule for workmanship specimens.

weld and also provide the appearance expected for making these welds. Companies that manufacture construction equipment often use these types of welding workmanship specimens.

Other companies have used the same concept to produce examples of welds that are acceptable and unacceptable. Originally the construction equipment company produced good and questionable welds of different types and posted the actual welds in the manufacturing department. However, they found that there was sufficient variation in the different specimens to create confusion. To overcome this, they produced one set of acceptable welds and one set of unacceptable welds and then made plastic replicas of each weld from these molds. The plastic weld replicas were posted in the department and in this case each and every impression was identical. Figure 21-51 shows the plastic replicas of acceptable and unacceptable welds. These types of exhibits are also very useful in the training of welders. The Department of Defense has used a similar technique with the military specification, which covers the smoothness of flame-cut surfaces.[11] To actually portray the surface as expected and surfaces not acceptable, they have made

a plastic replica of the actual flame cut surfaces. This is reproduced and is a part of a military specification (Figure 21-52). For the purpose of welder training, workmanship specimens have been used and specimens have been developed that include typical problem areas for training welders. Some of the problem areas are welding into and out of corners, changing electrodes, and making fillet welds on outside corners. These welding problems can be put together into one workmanship specimen and displayed in the welding department.

The use of *workmanship standards* is another technique for maintaining weld quality and can be used as a quality control tool. Workmanship standards are a list of requirements that should be followed in producing quality weldments. A list of 20 requirements has been compiled and is shown in Figure 21-53. This list was made by reviewing the major welding codes and specifications and adopting those rules that repeat in the different codes. In many cases, welding codes and specifications are written in legal terms that may not be easily understood by welders. In compiling these 20 requirements an effort has been made to utilize terminology more understandable by welders. It is suggested that

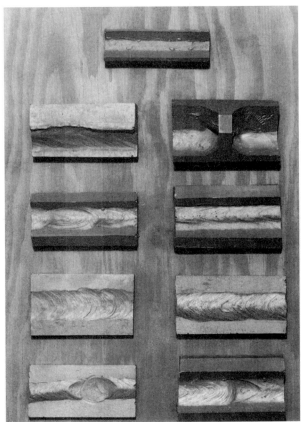

FIGURE 21–51 Plastic replicas of acceptable and unacceptable welds.

companies that do not utilize established codes adopt these 20 requirements as their own company standard. The basis for this standard is stated in the first sentence: Welders are responsible for their own work.

1. "Surfaces to be welded must be reasonably free from scale, paint, grease, water, etc." This is qualified by the word *reasonably* to allow welders and inspectors to exercise judgment. For example, mill scale on new steel just received is usually tight and it would be needless to remove it; if the steel has excessive mill scale or rust, it should be removed. The basis here is to provide a good surface for welding. The surface must be sufficiently clean so that there is nothing that might contain hydrocarbons, which break down in the heat of the arc providing hydrogen and can be absorbed in the weld and cause cracks. Welding over paint is a special case and is discussed elsewhere.

2. "If it is necessary to trim adjacent parts for proper fitup, the designed bevel and root opening must be maintained." Too often tolerances accumulate when setting up a weldment, and these are absorbed in weld joints. This can eliminate root openings so that incomplete penetration will result. On the other

hand, it can result in excessively large gaps that will require too much weld metal to make the joint and may also introduce undesired warpage. Welders should be cautioned to review the requirements of the weld joint and take corrective action when required.

3. Where the spacing between members to be joined by a T-type fillet welded joint is greater than $\frac{1}{16}$ in. (1.6 mm), the size of the fillets should be the size specified plus the amount of the opening. This is shown in Figure 21-54. A weld fillet size as specified can be made and the inspection after the weld is made would indicate that it is acceptable. In view of the opening between the parts, the effective area of the fillets would be much less than designed. This problem is relatively common, particularly when fixtures are used to locate premachined parts such as lugs, gussets, and so on. Premature failure will occur if these lugs are load-carrying parts.

4. The preheat or interpass temperatures specified must be adhered to. It may be that the preheat is only a surface heat instead of a through heat. If at all possible, the temperature of the part should be checked on the side opposite that to which heat is being applied. Specifications recommend a soak heating, not merely a surface heating.

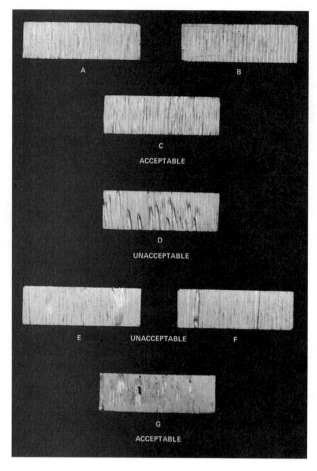

A

B

C
ACCEPTABLE

D
UNACCEPTABLE

E   UNACCEPTABLE   F

G
ACCEPTABLE

FIGURE 21–52 Plastic replicas of flame cut surfaces. (From Ref. 11.)

5. "The specified electrode or electrode matching the base metal must be used." On some weldments two, three, or more types of electrodes may be employed. It is extremely important that the proper electrode be used where specified. The quality of deposited weld metal from different shielded metal arc welding electrodes varies, and the properties may not be satisfactory for the service life.

6. The written welding procedure provided must be followed explicitly. However, in construction jobs and in large manufacturing plants the written procedures may not be available to the welders. The cost of producing written qualified procedures is expensive. It is prudent to provide the information to those who must follow it.

7. "Cracked or defective tack welds must be removed before the weld is made." Tack welds frequently crack. They crack because they are small with respect to the loads that can be imposed on them. Some welders feel that they have sufficient skill to burn through or melt out a cracked tack weld. This may be so, but the average welder normally does not have this skill and the crack is still in the final weld where it can cause problems.

8. "Cracked welds or welds having surface irregularities must be repaired before welding continues." This is sometimes ignored with the feeling that the crack or the irregular weld metal can be covered up with the next pass. This might be so, but it is poor practice since the crack is still in the weld joint and may propagate and create problems later. The best policy is to repair a weld at its earliest possible time of repair.

9. "Welds showing excessive surface porosity must be removed and rewelded." Surface porosity is usually an indication that there is subsurface porosity. Porous welds should be removed and a determination made if subsurface porosity occurs. The cause of the porosity should be corrected before welding continues.

10. All weld craters should be completely filled before depositing the next weld bead or pass. A crater is the depression at the end of the weld. If welding is to be immediately continued, after changing an electrode, the crater is filled quickly with the next electrode when the bead is continued. If the crater is at the end of a weld or at a change of direction or at a corner, it must be filled. This can be done by the welder by hesitating shortly before breaking the arc. Craters are prone to produce cracks that might propagate.

11. "Each bead or pass of a multipass weld must be cleaned before the next bead is made." This requirement is for shielded metal arc welding. It applies also to submerged arc welding and flux cored arc welding. It is important to remove the slag since it can be trapped in an undercut area and will reduce the strength of the joint. In the case of horizontal multipass fillets, judgment can be exercised and cleaning may not be required if, in the opinion of the inspector, the welder has sufficient skill to avoid entrapping the slag.

12. The specified size and length of welds as shown on blueprints and drawings is the minimum acceptable. Weld size tolerances should be -0 and $+\frac{1}{16}$ in. (1.6 mm). Many drawings have standardized tolerances printed on the drawing form, stating that all fractional dimensions must be $+$ or $-\frac{1}{16}$ in. or similar. If there are many $\frac{3}{16}$ in. fillets, this would mean that fillet welds could all be $\frac{1}{8}$ in., which would in effect greatly weaken them.

13. Weld reinforcement or crown should not exceed $\frac{1}{16}$ in. (1.6 mm) for manual welds or $\frac{1}{8}$ in. (3.2 mm) for automatic welds. The extra reinforcement is an economic waste since it is not required. Some welders may put extra reinforcement on a weld to camouflage internal flaws in a groove weld. Additionally, extra reinforcement causes stress concentrations. It is absolutely unnecessary for mild steel and low-alloy steels; the weld metal is stronger because it has a higher yield strength than the base metal.

Every welder is responsible for welding done. The following workmanship standards must be followed to produce high-quality welds. They may be employed as a company standard. (Specific codes or specifications take priority over these standards.)

1. Surfaces to be welded must be reasonably free from scale, paint, grease, water, etc.
2. If it is necessary to trim adjacent parts for proper fitup, the designed bevel and root opening must be maintained.
3. Where the spacing between members to be joined in a T-joint is greater that $\frac{1}{16}$ in. the size of the fillet weld shall be the size specified plus the amount of the opening.
4. Preheat or interpass temperature requirements must be adhered to.
5. The specified electrode or electrode matching the base metal must be used.
6. Welding procedures must be followed explicitly.
7. Cracked or defective tack welds must be removed before weld is made.
8. Cracked welds or welds having surface irregularities must be repaired before welding continues.
9. Welds showing excessive surface porosity must be removed and rewelded.
10. All weld craters shall be filled completely before depositing the next weld pass.
11. Each bead or pass of a multipass weld must be cleaned before the next bead is made.
12. The specified size and length of weld as shown on drawings are the minimum acceptable. Weld size tolerances should be $+\frac{1}{16}$ in.–0.
13. Weld reinforcement or crown shall not exceed $\frac{1}{16}$ in. for manual welds.
14. Undercut is not permissible on highly stressed or dynamically loaded members.
15. Root fusion must be complete on all joints designed with a root opening.
16. Welds showing subsurface slag or voids, by nondestructive inspection, must be gouged out to sound metal and rewelded.
17. Welds showing subsurface cracks by NDT must be gouged out to sound metal and rewelded.
18. All work should be positioned for flat-position welding whenever possible.
19. Specific welds may be taken at random and submitted to a 100% visual inspection or to any of the nondestructive testing methods.
20. Welders may be required to requalify if, in the opinion of the inspector and the supervisor, the work is of questionable quality.

FIGURE 21–53 Workmanship standards.

14. "Undercut is not permissible on highly stressed or dynamically loaded members." Undercut is a defect that is not allowed by most codes. Some codes allow a small amount of undercut, provided that it is within specified limits. Undercut areas tend to concentrate stresses and will create field problems. This can occur on both tensile and compressive loaded weld joints.

15. "Root fusion must be complete on all joints designed with a root opening." Root fusion is absolutely necessary for any weld with a root opening. The root face of both sides of the joint must be fused into the weld.

16. Welds showing subsurface slag or voids by nondestructive testing must be removed to solid metal and rewelded. Small amounts of slag or voids may be permitted. Some codes allow small amounts of slag or porosity provided they are small and not continuous. Usually, slag and porosity voids have rounded edges and will not propagate under designed loads.

17. Welds showing subsurface cracks by nondestructive testing must be removed to sound metal and rewelded. This is absolutely necessary since cracks have sharp corners or sharp ends and propagate under load. They are stress risers and will cause pre-mature failure or shortened service life. Internal cracks are not acceptable by any code.

18. "All work should be positioned for flat-position welding whenever possible." This is good economical practice. A welder who is working in a comfortable position will produce higher-quality welds.

19. Weldments or specific welds can be taken at random and submitted to 100% visual inspection or to any of the nondestructive testing methods. This rule and the next are provided to give the inspector the necessary authority to maintain quality.

20. Welders may be required to requalify if, in the opinion of the inspector and the welder's supervisor, the work produced is of questionable quality. A welder's work may deteriorate for one reason or another. This rule provides the mechanism of having the welder retested to determine the cause of the problem. This is a tool that the inspector can use whenever a questionable weld quality situation results from the work of a specific welder.

Companies that do not produce weldments by a national code should adopt this set of workmanship standards. This will provide a better understanding between the designer, the welder, the welding supervisor, and the inspector to maintain weld quality.

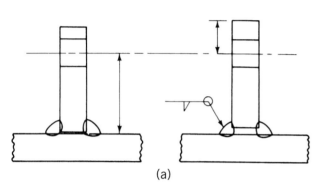

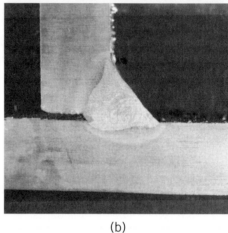

(a)

(b)

FIGURE 21–54 Equal-size fillets with different strength.

21-7 NONDESTRUCTIVE EXAMINATION SYMBOLS

Nondestructive examination symbols have been established by the American Welding Society.[12] They are used by the designer to convey information to the inspector concerning joints, welds, or weldments that need special attention. These symbols are very similar to welding symbols and can be used in conjunction with welding symbols. Figure 21-55 shows the elements of the examination symbol and the standard location with respect to each other, and only those elements required to provide the needed information are used. The examination symbol or designated letters are shown in Figure 21-56. One special symbol is used to show the direction of radiation that is used in conjunction with the radiographic examination symbol (Figure 21-57). Typical testing symbols are shown in Figure 21-58. For complete information on nondestructive examination symbols it is recommended that the reader consult the AWS standard.

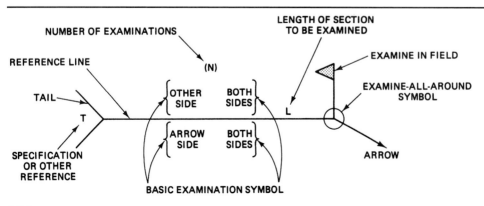

FIGURE 21–55 Standard location of elements.

Examination Method	Letter Designation
Acoustic emission	AET
Electromagnetic	ET
Leak	LT
Magnetic particle	MT
Neutron radiographic	NRT
Penetrant	PT
Proof	PRT
Radiographic	RT
Ultrasonic	UT
Visual	VT

FIGURE 21–56 Examination methods: letter designation.

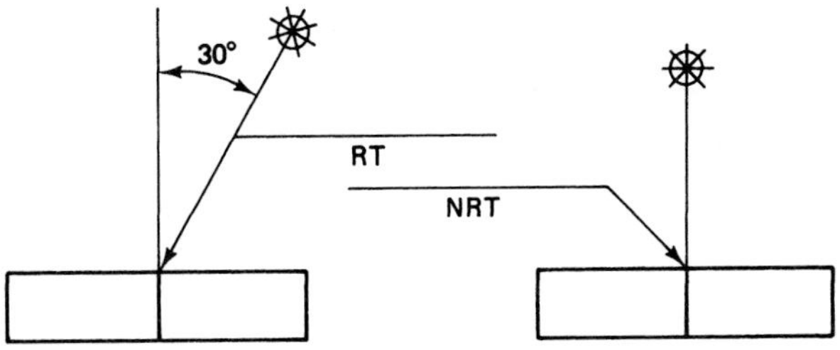

FIGURE 21–57 Radiation location symbol.

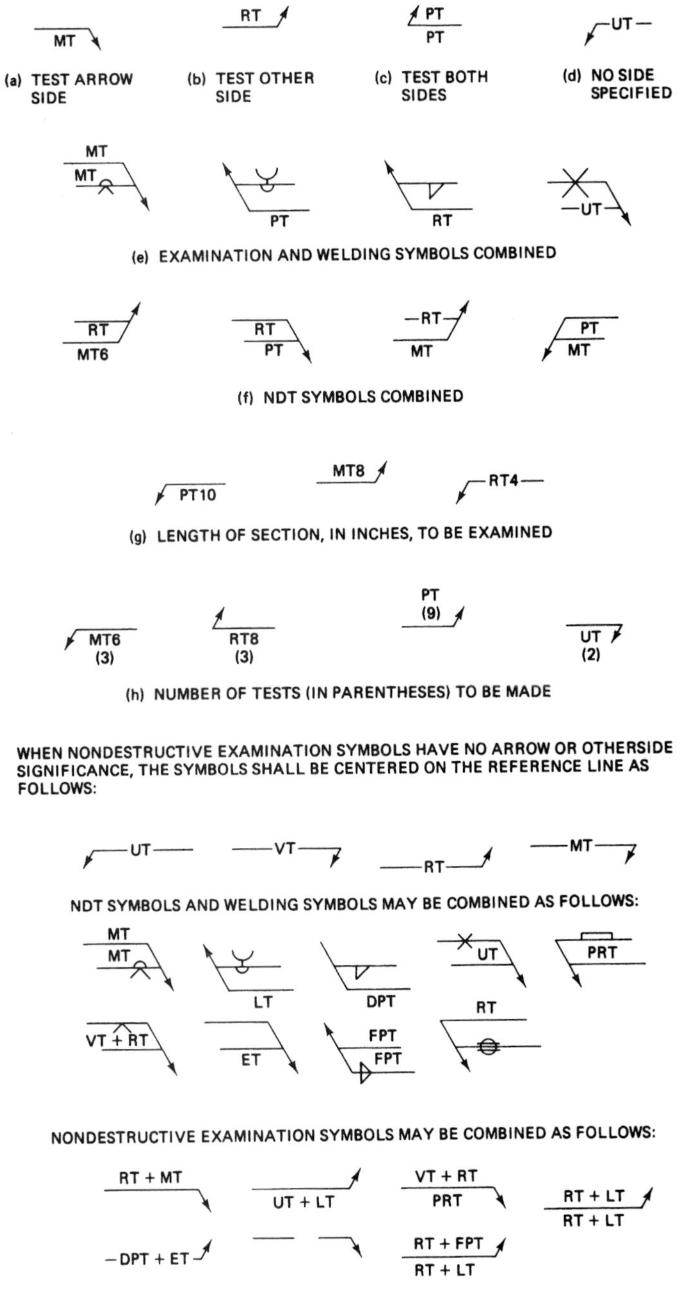

FIGURE 21–58 Typical examination symbols.

(a) TEST ARROW SIDE

(b) TEST OTHER SIDE

(c) TEST BOTH SIDES

(d) NO SIDE SPECIFIED

(e) EXAMINATION AND WELDING SYMBOLS COMBINED

(f) NDT SYMBOLS COMBINED

(g) LENGTH OF SECTION, IN INCHES, TO BE EXAMINED

(h) NUMBER OF TESTS (IN PARENTHESES) TO BE MADE

WHEN NONDESTRUCTIVE EXAMINATION SYMBOLS HAVE NO ARROW OR OTHERSIDE SIGNIFICANCE, THE SYMBOLS SHALL BE CENTERED ON THE REFERENCE LINE AS FOLLOWS:

NDT SYMBOLS AND WELDING SYMBOLS MAY BE COMBINED AS FOLLOWS:

NONDESTRUCTIVE EXAMINATION SYMBOLS MAY BE COMBINED AS FOLLOWS:

QUESTIONS

21–1. There are 20 factors in a quality assurance plan. Name as many as you can.

21–2. What is the difference between a root bend, a face bend, and a side bend test? When is the side bend test used instead of root or face bend tests?

21–3. When is a fillet break test used?

21–4. What is different about the guided bend test jig for testing $\frac{3}{8}$-in. specimens or 1-in. specimens?

21–5. What is the reason for the 0.505-in. diameter of a tensile test specimen?

21–6. What is the most widely used nondestructive evaluation technique? What are its three divisions?

21–7. Explain the use of a fillet weld gauge. Show typical fillet size problems.

21–8. Does AWS qualify and certify welding inspectors? How is this done?

21–9. Explain the principle of dye-penetrant testing. What are its limitations?

21–10. Can penetrant testing be used on plastic parts?

21–11. Explain the principle of magnetic particle testing. Why is it not used on aluminum?

21–12. Is ac or dc current used for MT examinations?

21–13. Explain the principle of radiographic testing. How are radiographs developed?

21–14. What is a penetrameter? Explain.

21–15. Can permanent records be made of UT results?

21–16. Explain the principle of ultrasonic testing. Can it be used for field inspection?

21–17. Why is leak testing with compressed air dangerous?

21–18. Name the six classes of defects.

21–19. Why are cracks more dangerous than porosity to weld service life?

21–20. Can testing symbols be combined with weld symbols?

REFERENCES

1. "Guidelines for Quality Assurance in Welding Technology," 11S/11W 902–86, International Institute of Welding, AWS, Miami, Fla.

2. "Quality Management and Quality Control Assurance Standards—Guidelines for Selection and Use," ISO 9000, International Organization for Standardization, ANSI, New York.

3. "Quality Program Requirements," Military Specification MIL-Q-9858. U.S. Department of the Navy, Washington, D.C.

4. "Quality Assurance Criteria for Nuclear Power Plants and Fuel Reprocessing Plants," Code of Federal Regulations, Section 10, Energy Part 50, Appendix B (10 CFR50-B).

5. "Standard Methods for Mechanical Testing of Welds," AWS B4.0, American Welding Society, Miami, Fla.

6. "Standard for Qualification and Certification of Welding Inspectors," AWS QC1, American Welding Society, Miami, Fla.

7. "Guide for the Visual Inspection of Welds," ANSI/AWS B1.11, American Welding Society, Miami, Fla.

8. "Welding Inspection," AWS W1-80, American Welding Society, Miami, Fla.

9. "Guide for the Nondestructive Inspection of Welds," AWS B1.10, American Welding Society, Miami, Fla.

10. "Classification of Defects in Metallic Fusion Welds with Explanation," IIS/11W-340-69 (ex doc V-360-67), *Metal Construction* and *British Welding Journal* (February 1970).

11. "Acceptance Standards for Surface Finish on Flame or Arc-Cut Material," Navships 0900-999-9000, U.S. Department of the Navy, Naval Engineering Center, Washington, D.C.

12. "Standard Symbols for Welding, Brazing, and Nondestructive Examination," AWS A2.4, American Welding Society, Miami, Fla.

22

WELDING SPECIFICATIONS, PROCEDURES, AND QUALIFICATIONS

OUTLINE

22-1 WELD RELIABILITY

Reliable welds are required on every weldment produced. Many different techniques are used to assure weld quality or reliability. Almost all welding codes, standards, and specifications include a plan for assuring weld quality. Most of these documents prescribe a method of assuring that the welding procedure will provide the quality required for the product. They almost always provide a method of testing the welders who apply the welds to determine if they have the skill required to provide the desired quality. Close adherence will assure that the product as designed will satisfy the service requirements.

The purpose of a welding procedure qualification is to show that the proposed weld joint or weldment will have the required properties for its intended application. The document that does this is the record of the actual welding variables used to produce the test weld and the results of the tests conducted. It is often called the *procedure qualification record.*

The purpose of the welder performance qualification is to determine the ability of the welder or welding operator to produce sound weld metal following a welding procedure specification. This qualifies the individual welder or welding operator for specific processes made on specific materials under identified conditions. It is often called the *performance qualification test record.*

Neither the welding procedure qualification nor the welder performance qualification establishes the capabilities of an organization to continuously make an acceptable welded product. A quality control program, which was described in detail in Chapter 21, is employed to assure adherence.

Quality welding on a product must be judged with respect to a specific quality standard that is based on its intended service. It must be a balance between the service requirements and the consequences of failure versus economic factors. For many products, weld quality requirements are controlled by codes, specifications, and standards. However, when codes or specifications do not apply, the producer must make products to a high quality standard in order to survive. The success of maintaining the balance between high quality and high cost is decided in the field and in the marketplace, where quality and price determine the producer's continuing success.

Weldments in space vehicles and nuclear plants are exposed to environments unheard of in the past. The weld perfection obtained by this class of work is obtained due to excellent procedures, extensive training, and stringent quality assurance methods. It is obtained because of extensive preparation and time-consuming procedure testing and qualifications that contribute to high cost. Perfect welds are not required on all weldments. The industry must guard against establishing super-quality requirements when they are not necessary.

Responsibility for producing high-quality products rests on many people. It is management's responsibility to create the proper cooperative spirit among designers, welders, supervisors, and inspection personnel to make sure that the quality requirement is reasonable and in agreement with the service life expected. The responsibility for producing high-quality welds rests with the welder. Each welder must accept this responsibility. The welding supervisor has the responsibility for welders and their performance. The welding inspector must verify that quality standards are met. The welding standards, specifications, and procedures are the basis for weld quality, and these factors, coupled with weldment design, are the responsibility of designers, welding engineers, material managers, and quality assurance personnel. It is a total responsibility with a very complex interrelationship.

Designers, material specifiers, and others must keep close contact with field service personnel with respect to field requirements and problems. They must be sensitive to needs for change and they must respond as required. Welding supervisors and production managers must be continually alert for evidence of substandard workmanship.

The need to differentiate between the adequate and the perfect weld has led to research concerning the acceptability of weld imperfections and how they affect service life. The term *fitness for service* is the result of studies relating to degrees of imperfection and reliability in service. Years of investigation of this information has allowed the establishment of fitness for service data. These data have been translated into codes and standards for different types of equipment based on different types of service. This knowledge gained from field experience and investigation is reflected in the revision of codes and standards.

A major problem encountered in weldment production is the suspicion of the designer that the shop will not manufacture the weldment as designed. This suspicion occurs when designers consider welding workmanship factors that are seemingly beyond the designer's control. They feel that the welder can produce welds that meet the design requirements under ideal conditions. They also feel that welders produce good-quality welds when the performance qualification tests are passed. However, they want assurance that every weld in the weldment will be of high quality. This depends on the quality assurance program, which saves money in the long run since they eliminate premature field failures, catastrophic disasters, and the cost of overwelding in order to overcome suspected shop malpractice.

22-2 WELDING CODES AND SPECIFICATIONS

Many codes and specifications cover welding applications. Having evolved over years of experience based on specific industries or types of products, codes and specifications change continually as the technology changes and as the demands for the products covered change. They are written by different organizations, including professional societies and trade associations. They are known as consensus standards, codes, and specifications. Federal government departments, state governments, provincial governments, and cities either adopt these consensus documents or write standards for application to items under their jurisdiction.

The standards adopted or issued by government bodies are enforceable by law and quite often become contractual requirements. Trade association and professional engineering society specifications are usually involved in purchase agreements as standards of quality acceptance. Similar products have very similar welding specifications. In some cases the qualification of a procedure to one specification may be acceptable by another specification for the same or similar products. Figure 22–1 shows some typical specifications.

Code-writing bodies are continually upgrading specifications and attempting to make them more uniform and to make interchange of qualifications easier. Many codes and specifications are approved by the American National Standards Institute (ANSI).

The following products utilize welding specifications:

1. Pressure vessels
2. Nuclear reactors

FIGURE 22–1 Popular codes and specifications.

3. Piping and pipelines
4. Buildings and bridges
5. Storage tanks and vessels
6. Ships, barges, offshore drill rigs
7. Railroad rolling stock
8. Aircraft and spacecraft
9. Automotive and trucks
10. Construction and agricultural equipment
11. Industrial machinery
12. Shipping containers
13. Ordnance material
14. Nonstructural sheet metal components and systems
15. Miscellaneous equipment

Each is explained briefly and the specific codes or specifications are identified.

Pressure Vessels and Boilers

The manufacturer of boilers and pressure vessels, and all other items defined as pressure vessels, come under the specification of the ASME "Boiler and Pressure Vessel Code."[1] This code consists of 11 sections:

* Section I Power boilers
* Section II Material specifications—ferrous
  Material specifications—nonferrous
  Material specifications—welding rods, electrodes, and filler metals
* Section III Nuclear power plant components
* Section IV Heating boilers
* Section V Nondestructive examination
* Section VI Recommended rules for care and operation of heating boilers
* Section VII Recommended rules for the care of power boilers
* Section VIII Pressure vessels, divisions I and II
* Section IX Welding qualifications
* Section X Fiberglass-reinforced plastic pressure vessels
* Section XI Rules for in-service inspection of clear reactor coolant system

Products manufactured under the requirements of these codes may also be manufactured under the varying rules and regulations of states and provinces that follow the boiler and pressure vessel code. In general, Section IX is used throughout North America and in other parts of the world as the method of qualifying procedures and welders for work on boilers, pressure vessels, and pressure piping.

Code Symbol Stamps Manufacturers or contractors who regularly build or install pressure vessels or pressure pip-

ing will usually have an ASME *symbol stamp*. This means that they have been approved by the American Society of Mechanical Engineers as an authorized manufacturer or installer of the type of equipment specified. Different stamps are used to make the installation or the product manufactured. Some of the symbol stamps are:

* N Nuclear vessel
* PP Pressure piping
* U Pressure vessel
* S Power boilers
* H Heating boilers

To obtain an ASME symbol stamp, a manufacturer or contractor must contact the American Society of Mechanical Engineers, Boiler and Pressure Code Committee, and apply for the code symbol stamp required. The actual mechanics are quite involved but include obtaining a contract with an authorized inspection agency. The American Society of Mechanical Engineers will advise of the exact requirements. The requirements include at least the need to prepare a written quality control manual describing a controlled manufacturing system for the scope of the proposed ASME certificates of authorization. The ASME will send a survey team to inspect the facilities and review the quality control manual and witness a demonstration of all items affecting quality within the scope of the certificate. If everything is satisfactory, ASME will issue a certificate of authorization and the applicable code symbol stamp.

The National Board Inspection Code provides rules for repairs and alterations to boilers and pressure vessels by welding after such vessels have been in service. This is in contrast to the ASME code rules, which apply to new construction (except Section XI). Repair organizations that make repairs or alterations will usually have an "R" or "NR" code symbol stamp issued by the National Board of Boilers and Pressure Vessel Inspectors.

Nuclear Reactors

The nuclear reactors, components, and material used in nuclear power plants are covered by the provisions of Section III of the ASME pressure vessel code and the Nuclear Regulatory Commission Specification.[2] Any part that is utilized in a nuclear plant must be manufactured under the jurisdiction of these codes. The exceptions are those components for navy ship use that are covered by a similar but different code issued by the Department of Defense Naval Ship Division. This is known as "Standard for Welding of Reactor Coolant and Associated Systems and Components for Naval Nuclear Power Plants."[3] The above-mentioned codes require the certification of materials and traceability of all materials, including welding filler metals to the point of origin. They also include strict control systems of inspection during the manufacture of the nuclear power plant components.

Pressure Piping

Code specifications and procedures for welding piping are covered in Chapter 25.

Buildings and Bridges

Steel buildings welded in most cities in North America are covered by city code and specifications. The larger cities may publish their own codes, while others follow the AWS "Structural Welding Code—Steel"[4] and AISC "Manual of Steel Construction,"[5] which includes the AISC Specification for Structural Steel Buildings and the AISC Code of Standard Practice. Some cities require qualification of welders and certification of electrodes for structural applications.

Welding on highway bridges is under the jurisdiction of the state or provincial department of transportation. Basis for these codes, either by reference or by direct copy, is the "Bridge Welding Code" of AASHTO/AWS D1.5[6] This code incorporates the requirements of the U.S. Department of Transportation, Bureau of Public Roads. Some states publish their own bridge codes, which reference the AASHTO/AWS "Bridge Welding Code." Many states supplement the AASHTO/AWS requirements. Some require welders to be examined yearly and certified by the state to work on bridges. Some states maintain rosters of certified welders. Some highway departments also require yearly certification of welding electrodes. In addition, AWS also provides a welding code for aluminum structures[7] and a code for welding reinforcing steel.[8]

Storage Tanks and Vessels

There are two major codes for welding storage tanks. One is for the welding of elevated storage tanks and is published by AWS and the American Water Works Association, "Standard for Welded Steel Elevated Tanks, Standpipes, and Reservoirs for Water Storage."[9] The other one is for oil or petroleum product storage tanks, published by the American Petroleum Institute, entitled "Standard for Welded Steel Tanks for Oil Storage."[10] Both of these codes refer to Section IX of the ASME boiler code for welding qualification.

Ships

Welding on ships is covered by different specifications and codes. In the United States, all federal government vessels are covered by codes issued by the U.S. Coast Guard[11] or the Navships Division of the Department of Defense.[12] These requirements are nearly identical as to welding procedure qualification and welder qualification. They are also very similar to the requirements of the Maritime Administration for commercial ships.[13] Qualification of welders is usually transferable among these

three organizations. The American Bureau of Shipping has similar requirements for welding on ships that they survey.[14] Lloyds and other classification societies also publish specifications that cover welding. Certification of filler metal is required. The American Welding Society publishes two guides related to ship welding: "Guide for Steel Hull Welding"[15] and "Guide for Aluminum Hull Welding."[16]

Railroad Rolling Stock

In the United States, specifications for rolling stock for North American railroads is under the jurisdiction of the federal Department of Transportation. However, as far as welding qualification and welding design requirements are concerned, the controlling specifications are issued by the Association of American Railroads. Two specifications are involved: "Specifications for Tank Cars"[17] and "Specifications for Design, Fabrication, and Construction of Freight Cars."[18] These specifications provide information concerning the design of welds and the qualifications of welders. They are in substantial agreement with requirements of the AWS "Railroad Welding Specification."[19]

Aircraft and Spacecraft

Weldments intended for use in aircraft and spacecraft are welded to the requirements of U.S. government specifications. There are other groups that write specifications for materials that might be utilized, including the Society of Automotive Engineers[20] and the Aerospace Industries Association of America.[21] Welding codes or requirements are covered by specifications of the National Aeronautics and Space Administration (NASA) and of the Department of Defense Military (MIL) Standards and Specifications. The one pertaining primarily to welding on aircraft is "Qualification of Aircraft, Missile and Aerospace Fusion Welders."[22] This standard covers many welding processes, metals, and levels of proficiency for testing welders and must be adhered to when welding on aircraft. Qualification under this standard is done under the supervision of government inspectors.

Automotive and Trucks

There is no national specification for welding on automobiles or trucks; however, AWS has issued two documents relating to the welding of automobiles and trucks: "Specification for Automotive Frame Weld Quality—Arc Welding"[23] and "Recommended Practices for Automotive Weld Quality—Resistance Spot Welding."[24]

Construction Equipment

Most manufacturers of construction, earth moving, and agricultural equipment have their own specifications that cover welding. These have been in use for years and are acceptable based on field experience. Recently, the American Welding Society has issued a specification that covers all structural welds used in this type of equipment. In addition, it covers weld joints, welding procedures, welder and procedure qualification tests, and the inspection of welds used in this type of equipment. This is known as "Specification for Welding Earth Moving and Construction Equipment."[25] This specification also accepts welders tested to the AWS structural code. Rollover protective structures (ROPS) on construction, mining, and agriculture equipment may be field welded using welders qualified to AWS D1.1 or MIL-STD 248 or their equivalent.[26]

Industrial Machinery

Such a wide variety of industrial machinery is produced by welding that it is impossible to cover each and every type. However, the American Welding Society has issued standards covering some types. They also issued a standard with a classification and application of welded joints for all types of machines. The AWS specs are "Specification for Welding Industrial and Mill Cranes and Other Material Handling Equipment,"[27] "Specification for Metal Cutting Machine Tool Weldments,"[28] "Specification for Welding Earth Moving and Construction Equipment,"[25] "Classification and Application of Welded Joints for Machinery and Equipment,"[29] "Specification for Welding of Presses and Press Components,"[30] and "Specification for Rotating Elements of Equipment."[31] In general, these are minimum requirements for welded fabrication of the types of equipment covered. They cover the specifications, joint types, documentation, procedure qualification, and welder qualification. Refer to the specific specification with respect to welder qualification.

Shipping Containers

Strict welding specifications are required for shipping containers used for transporting gas under high pressure and for tanks carrying liquid petroleum and similar products. These codes and specifications are used by the U.S. Department of Transportation. Additional detail is given in Section 14-4. See also the Code of Federal Regulations, Title 49.[32,33]

Ordnance Material

Ordnance material is produced for the government, normally the U.S. Army, and includes such items as combat tanks, gun carriers and mounts, personnel carriers, and retrieval vehicles. These are covered by MIL Specifications and Standards issued by the Department of Defense. Performance, procedure, and welder qualification are usually covered by MIL-STD-2219,[34] but each spe-

cific MIL specification should be referenced to determine the exact requirements.

Nonstructural Sheet Metal Components and Systems

The AWS "Sheet Metal Welding Code"[35] provides qualifications, workmanship, and inspection requirements for both arc welding and braze welding as they apply to the fabrication, manufacture, and erection of nonstructural sheet metal components and systems, such as ductwork.

Miscellaneous Equipment

Undoubtedly, there are other specifications and codes covering items not mentioned here. It is important to determine the code or specification involved. These can be obtained from the buyer or purchaser of the weldments in question.

22-3 WELDING PROCEDURES AND QUALIFYING THEM

Qualifying welding procedures is complicated because of slightly different terminology and definitions of the different codes and standards. It is extremely important to consult the latest or specific edition of the code or standard involved and follow it in detail. A welding procedure is "the detailed methods and practices involved in the production of a weldment." This broad definition covers two types. The first is the legal requirement of a code or specification. The second is the directions for making a specific weldment. Procedures of this type are written to show how a weldment must be built to maintain consistency.

The *welding procedure specification* (WPS) required by many codes describes the step-by-step directions for making a specific weld and proof that the weld is acceptable. All welding codes and specifications require qualified procedures. It is necessary to write a welding procedure and then to prove or qualify it by making welds that are then tested for acceptability. All codes require proof that welders and welding operators have the skill and ability to follow the welding procedures successfully. They must make specific welds to prove that they can provide the quality required. The codes and standard requirements are somewhat different, and this is covered in the next section.

Qualifying a procedure by one manufacturer under a specific code normally will not qualify the procedure for another manufacturer or contractor. There are a few exceptions. In the piping field, procedures are qualified in the name of the National Certified Pipe Welding Bureau or a local contractor's association. The procedures may be used by all of the association members, and welders qualified to the procedures can transfer from one member company to another without requalification. This saves the expense of continually requalifying. The welders are hired from a labor pool and may work for different contractors on each new job. With this arrangement they are covered by the association-qualified procedures and need not retest for each job. The contractor/employer is still responsible for the procedures and for the welders and for the quality of the work.

Boiler Code Welding

Probably the most widely used code for procedure and performance qualifications is Section IX of the ASME "Boiler and Pressure Vessel Code." It is entitled "Qualification Standard for Welding and Brazing Procedures—Welders, Brazers, and Welding and Brazing Operators." It will be explained in detail with examples. The other popular codes will be mentioned briefly since the boiler code explanation will provide the background information needed.

The boiler code makes the following statement concerning welder responsibility: "Each manufacturer or contractor is responsible for the welding done by his organization and shall conduct the tests required to qualify the welding procedures he uses in the construction of the weldments built under this code, and the performance of welders and welding operators who apply these procedures." It further states: "Each manufacturer, or contractor, shall maintain a record of the results obtained in welding procedures and welder and welding operator performance qualifications. These records shall be certified by the manufacturer or contractor and shall be accessible to the authorized inspector." Computer software programs are available for recording this information and should be used; see Appendix A-3.

The ASME code calls the welding procedure a welding procedure specification (WPS). This document provides in detail the required conditions for specific applications to assure repeatability by properly trained welders and welding operators. A WPS is a written welding procedure prepared to provide direction for making production welds to code requirements. The ASME provides a sample form, which may be used or modified provided that it covers all information. The completed WPS provides directions to the welder or welding operator to assure compliance with the code requirements. The completed WPS describes all of the essential, nonessential, and supplementary essential (when required) variables for the welding process. The WPS should reference the supporting *procedure qualification record* (PQR). A PQR is a record of all of the welding conditions that were used when welding the test coupons and the actual results of the tested specimens. The completed PQR should record

all essential and supplementary essential (when required) variables for the welding process used to weld the test coupon. Nonessential or other variables used during the welding of test coupons need not be recorded. The PQR should be certified accurate by the manufacturer or contractor. This certification is the manufacturer's or contractor's verification that the information is a true record of the variables that were used during the welding of the test coupon and that the test results are in compliance with Section IX of the code. The manufacturer or contractor cannot subcontract this certification function.

There are three types of variables for WPS. *Essential variables* are those in which change is considered to affect mechanical properties of the weld joint or weldment. *Supplementary essential variables* are required for metals for which notch toughness tests are required. *Nonessential variables* are those in which a change may be made in the WPS without requalification. The variables for each welding process are listed in detail in Section IX. For this reason it is necessary to refer to the code when writing, testing, or certifying the welding procedures.

Welding Procedure Specification

An example of a WPS is shown in Figures 22-2, 22-3, and 22-4. In this example the ABC Pressure Vessel Company is using gas metal arc welding, semiautomatically applied, to weld P-1 grade steel pipe in the horizontal fixed and vertical positions. Each entry will be explained.

Joints The joint design is a single V-groove with a 60 to 70° included angle. A sketch is drawn on the form under details. If more space is needed, use a third sheet, such as Figure 22-4 (sheet 3 of 3) in the example. The welding variables are placed in the table provided. Backing is not used. However, if backing is used, it must be described.

Base Metals To reduce the number of WPSs required, P numbers are assigned to base metals depending on characteristics such as composition, weldability, and mechanical properties. Groups within P numbers are assigned for ferrous metals for the purpose of procedure qualifications where notch toughness requirements are not specified. The same P numbers group the various base metals having comparable characteristics. The P numbers and groupings of most of the different steels are given in Section IX of the code. If a P number is not available for the material involved, its ASTM specification number may be used. If an ASTM specification number is not available, the chemical analysis and mechanical properties can be used. Under base metals the thickness range must be shown, and if it is of pipe, the pipe diameter range must be shown.

Filler Metals Electrodes and welding rods are grouped according to their usability characteristics, which deter-

QW-482 SUGGESTED FORMAT FOR WELDING PROCEDURE SPECIFICATION (WPS)
(See QW-201.1, Section IX, ASME Boiler and Pressure Vessel Code)

Company Name: ABC Pressure Vessel Company By: Frank Jones Weld. Engr.
Welding Procedure Specification No. 1 Date Aug. 11, 1999 Supporting PQR No.(s) 101
Revision No. ~ Date ~
Welding Process(es) Gas Metal Arc Welding ~ Short Circuiting Type(s) Semi-Automatic
(Automatic, Manual, Machine, or Semi-Auto.)

JOINTS (QW-402)

Joint Design: Single Vee
Backing (Yes) ____ (No) X
Backing Material (Type) ~
(Refer to both backing and retainers.)

☐ Metal ☐ Nonfusing Metal
☐ Nonmetallic ☐ Other

Sketches, Production Drawings, Weld Symbols or Written Description should show the general arrangement of the parts to be welded. Where applicable, the root spacing and the details of weld groove may be specified.

(At the option of the Mfgr., sketches may be attached to illustrate joint design, weld layers and bead sequence, e.g. for notch toughness procedures, for multiple process procedures, etc.)

Details

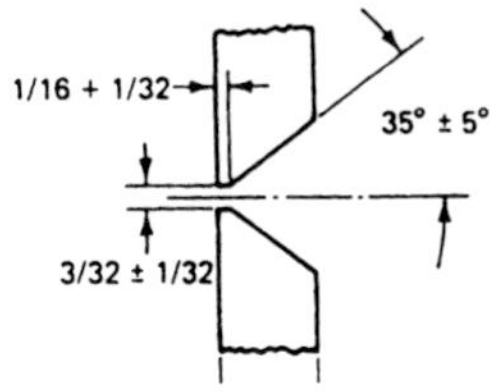

1/16 + 1/32
35° ± 5°
3/32 ± 1/32

***BASE METALS (QW-403)**

P-No. 1 Group No. 1 to P-No. 1 Group No. 1
OR
Specification type and grade ~
to Specification type and grade ~
OR
Chem. Analysis and Mech. Prop. ~
to Chem. Analysis and Mech. Prop. ~
Thickness Range:
Base Metal: Groove 3/16 to 1 1/8 Fillet ~
Pipe Dia. Range: Groove unlimited Fillet ~
Other

***FILLER METALS (QW-404)**

Spec. No. (SFA) 5.18		
AWS No. (Class) ER 70S-3		
F-No. 6		
A-No. 1		
Size of Filler Metals 0.35-in		
Deposited Weld Metal ~		
Thickness Range: ~		
Groove 1/8-inch		
Fillet ~		
Electrode-Flux (Class) None		
Flux Trade Name ~		
Consumable Insert None		
Other ~		

*Each base metal-filler metal combination should be recorded individually.

FIGURE 22–2 ASME welding procedure specifications (WPS), sheet 1.

POSITIONS (QW-405)

Position(s) of Groove _2G Pipe Axis Vertical_

Welding Progression: Up _~_ Down _~_

Position(s) of Fillet _~_

PREHEAT (QW-406)

Preheat Temp. Min. _100° F_

Interpass Temp. Max. _200° F_

Preheat Maintenance _100° F_

(Continuous or special heating where applicable should be recorded)

POSTWELD HEAT TREATMENT (QW-407)

Temperature Range _None_

Time Range _~_

GAS (QW-408)

Percent Composition

	Gas(es)	(Mixture)	Flow Rate
Shielding	CO_2	welding grade	20 CFH
Trailing	None	~	~
Backing	None	~	~

ELECTRICAL CHARACTERISTICS (QW-409)

Current AC or DC _DC_ Polarity _Electrode Positive_

Amps (Range) _150-170_ Volts (Range) _21-23_

(Amps and volts range should be recorded for each electrode size, position, and thickness, etc. This information may be listed in a tabular form similar to that shown below.)

Tungsten Electrode Size and Type _None_

(Pure Tungsten, 2% Thoriated, etc.)

Mode of Metal Transfer for GMAW _Short Circuiting Mode_

(Spray arc, short circuiting arc, etc.)

Electrode Wire feed speed range _230 to 300 ipm_

TECHNIQUE (QW-410)

String or Weave Bead _See Details Sketch →_

Orifice or Gas Cup Size _1/2 - inch I.D._

Initial and Interpass Cleaning (Brushing, Grinding, etc.) _Brush to clean metal_

1/16 MAX →

1/16 MAX

Method of Back Gouging _None_

Oscillation _As required_

Contact Tube to Work Distance _1/2 to 3/4 - inch_

Multiple or Single Pass (per side) _Multiple_

Multiple or Single Electrodes _Single_

Travel Speed (Range) _21 to 26 ipm_

Peening _None_

Other _All tack welds to be ground to feather edge. All starts and stops to be ground to sound metal. All surface cracks or holes to be removed before continuing._

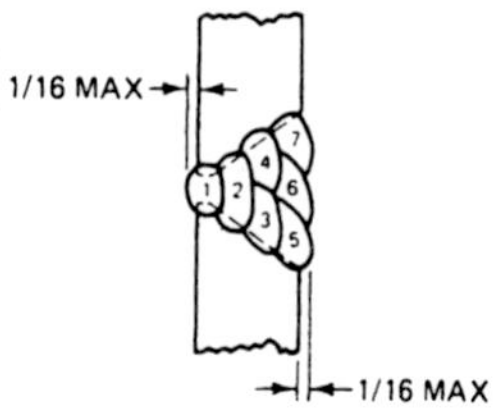

| Weld Layer(s) | Process | Filler Metal | | Current | | | | Travel | Other |
		Class	Dia.	D.C. Type Polar.	Amp. Range	Volt Range		Speed Range	(e.g., Remarks, Comments, Hot Wire Addition, Technique, Torch Angle, Etc.)
1	GMAW	ER 70 S-3	0.035-in	Elec +	150-170	21-23		22-25	Increase shielding gas flow 50% when welding outdoors.
2	"	"	"	"	"	"		"	
3	"	"	"	"	"	"		"	
4	"	"	"	"	"	"		"	
5	"	"	"	"	"	"		"	
6	"	"	"	"	"	"		"	
7	"	"	"	"	"	"		"	

FIGURE 22-3 ASME welding procedure specifications (WPS), sheet 2.

POSITIONS (QW-405)

Position(s) of Groove _5 G Pipe Axis Flat-Fixed_

Welding Progression: Up _____ Down _X_

Position(s) of Fillet _~_

PREHEAT (QW-406)

Preheat Temp. Min. _100° F_

Interpass Temp. Max. _200° F_

Preheat Maintenance _100° F_

(Continuous or special heating where applicable should be recorded)

POSTWELD HEAT TREATMENT (QW-407)

Temperature Range _None_

Time Range _~_

GAS (QW-408)

	Gas(es)	Percent Composition (Mixture)	Flow Rate
Shielding	CO_2	welding grade	20 CFH
Trailing	None	~	~
Backing	None	~	~

ELECTRICAL CHARACTERISTICS (QW-409)

Current AC or DC _DC_ Polarity _Electrode Positive_

Amps (Range) _150-170_ Volts (Range) _21-23_

(Amps and volts range should be recorded for each electrode size, position, and thickness, etc. This information may be listed in a tabular form similar to that shown below.)

Tungsten Electrode Size and Type _None_

(Pure Tungsten, 2% Thoriated, etc.)

Mode of Metal Transfer for GMAW _Short Circuiting arc_

(Spray arc, short circuiting arc, etc.)

Electrode Wire feed speed range _230 to 300 ipm_

TECHNIQUE (QW-410)

String or Weave Bead _See Details Sketch →_

Orifice or Gas Cup Size _1/2 - inch I.D._

Initial and Interpass Cleaning (Brushing, Grinding, etc.) _Brush to clean metal_

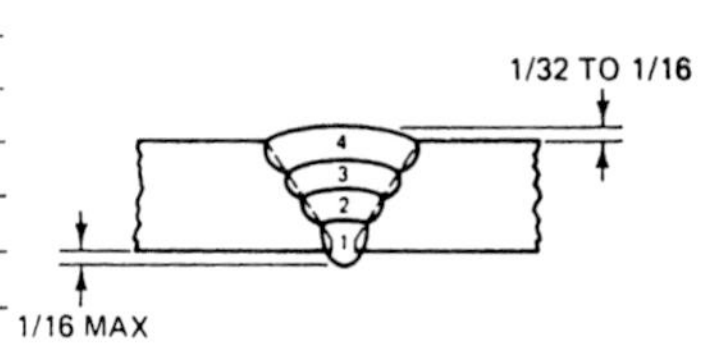

Method of Back Gouging _None_

Oscillation _As required_

Contact Tube to Work Distance _1/2 to 3/4 · inch_

Multiple or Single Pass (per side) _Multiple_

Multiple or Single Electrodes _Single_

Travel Speed (Range) _21 to 26 ipm_

Peening _None_

Other _All tack welds to be ground to feather edge. All starts and stops to be ground to sound metal. All surface cracks or holes to be removed before continuing_

1/32 TO 1/16

1/16 MAX

Weld Layer(s)	Process	Filler Metal		Current			Travel Speed Range	Other (e.g., Remarks, Comments, Hot Wire Addition, Technique, Torch Angle, Etc.)
		Class	Dia.	D.C. Type Polar.	Amp. Range	Volt Range		
1	GMAW	ER70S-3	0.035-in	Elec +	150-170	21-23	22-25	Increase shielding gas flow 50% when welding outdoors.
2	"	"	"	"	"	"	"	
3	"	"	"	"	"	"	"	
4	"	"	"	"	"	"	"	

FIGURE 22–4 ASME welding procedure specifications (WPS), sheet 3.

mines the ability of the welders to make satisfactory welds with a given filler metal. The groups are given F numbers, which relate to the composition and usability. This is filled in on the form. This block also requires the ASME specification number and the AWS classification number of the filler metal used. The ASME specification numbers are the same as the AWS specification numbers with the addition of the letters SF. These data are given in Section IX. The AWS classification number of the filler metal specification is also given on the label on the electrode boxes. For example, A-1 is a mild steel weld metal deposit. The size of the filler metal, which is its diameter, must be shown as well as deposited weld metal thickness range for groove or fillet welds. In the case of submerged arc welding, electrode flux class and the flux trade name must be shown. For gas tungsten arc welding, the consumable insert analysis should be shown. Other information relating to filler metals not mentioned above should be given when available.

Position The welding position of the groove or fillet weld must be described according to AWS terminology. If vertical welding is involved, it should be mentioned whether progression is upward (uphill) or downward (downhill).

Preheat A minimum temperature shall be given as well as the maximum interpass temperature. Preheat maintenance temperature should be given. Where applicable, special heating should be recorded.

Postweld Heat Treatment If a postweld heat treatment is used, it must be described. This includes the temperature range and the time at temperature. If there is no postweld heat treatment, write in "none."

Gas The shielding gas, if used, must be identified, and if it is a mixture, should be described. The shielding gas flow rate should be given. If backing gas or trailing shield gas is used, the gas composition and flow rate should be given.

Electrical Characteristics The welding current should be shown as alternating (ac) or direct current (dc). If direct current is used, the polarity of the electrode must be shown. The amperes, voltage, and travel speed range must be shown for each electrode class, size, and position. This is presented in tabular form, as shown on Figure 22-4 (sheet 3 of 3). In the case of GTAW, the tungsten electrode size and type should be described. For GMAW the mode of metal transfer must be described. The electrode wire feed speed range and travel speed range should be shown.

Technique Under technique, describe the weld as made with stringer or weave beads. Oscillation should be used to make weave beads. This should show in the sketch. Often, both techniques are used in the same weld. For the gas-shielded process the nozzle inside diameter should

be shown. The method of cleaning before welding and between passes must be shown. If back gouging is employed, it should be described. The contact tip-to-work distance should be described as a minimum-maximum dimension. It should be stated whether multiple- or single-pass technique is used. It is also necessary to indicate whether a single electrode or multiple electrodes are used. The travel speed range should be described. Peening, if used, must be described and any other pertinent information should be mentioned. For example, pulsing, if employed, would need to be described.

Procedure Qualification Record

To support the welding procedure specification (WPS), it is necessary to test and certify the weld results. This is done by making the welds described in the WPS, machining them, and testing the specimen in accordance with the code. The data are entered on the procedure qualification record (PQR), which is defined as a document providing the actual welding variables used to produce the acceptable test weld, and the results of tests conducted on the weld for the purpose of qualifying a welding procedure specification (WPS). It must reference a specific WPS. An example of a PQR is shown in Figures 22-5 and 22-6. This sample PQR is a record of actual conditions used to weld the coupons made in accordance with WPS No. 1, the example shown previously. Many of the data required by the PQR are the same as the information on the referenced WPS. In fact, the data on the front sheets are almost identical. The back, Figure 22-6 (sheet 2 of 2) of the PQR, is straightforward and is a record of the mechanical tests, the tension test, the guided bend test, the toughness test when required, and the fillet weld test, when used. A toughness test, either impact or drop weight, is not required by Section IX of the ASME code. These tests may be required by other sections of the code and must be made according to the provisions of the code. The example shows typical data that would be entered. If the test data meet the requirements of the code, the form is then signed by the manufacturer's representative, certifying that the statements in the record are correct and that the test welds were prepared, welded, and tested in accordance with requirements of Section IX. The test record of the PQR qualifies the WPS and fulfills the requirements for the code. All changes to a PQR require recertification by the manufacturer or contractor.

It is necessary to have specific WPSs and PQRs to cover all the weld processes, combination of welding processes, different P groupings of base materials, and so on, to comply with the variables involved. Every process and base metal used in producing the product must be covered by a WPS, which must be qualified by a PQR.

Keeping track of all of your WPSs can be difficult, but it is necessary to avoid duplicating them. It is recom-

PROCEDURE QUALIFICATION RECORD (PQR)
(See QW-201.2, Section IX, ASME Boiler and Pressure Vessel Code)
Record Actual Conditions Used to Weld Test Coupon.

Company Name __ABC Pressure Vessel Co.   Anytown U.S.A 12345__

Procedure Qualification Record No. __101__ Date __Aug 8 1999__

WPS No. __1__

Welding Process(es) __GMAW (Gas Metal Arc Welding)__

Types (Manual, Automatic, Semi-Auto.) __Semi-automatic__

JOINTS (QW-402)

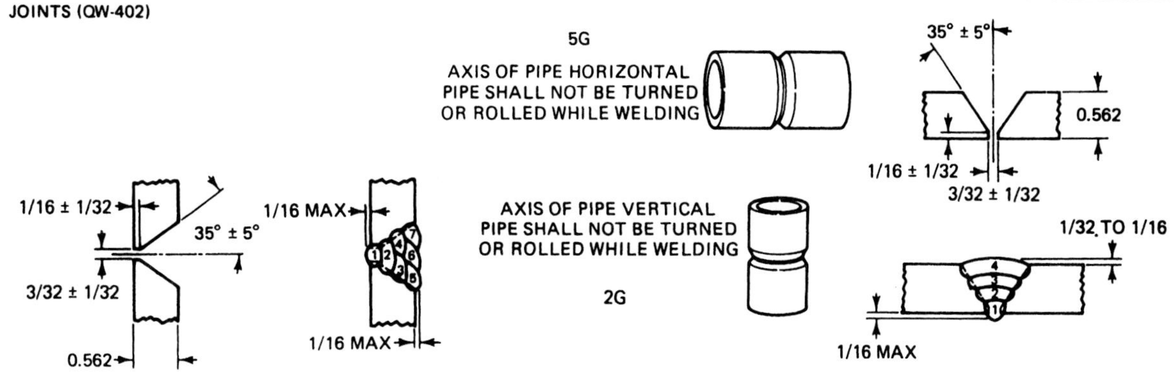

5G

AXIS OF PIPE HORIZONTAL
PIPE SHALL NOT BE TURNED
OR ROLLED WHILE WELDING

35° ± 5°

0.562

1/16 ± 1/32

3/32 ± 1/32

1/16 ± 1/32

35° ± 5°

1/16 MAX

AXIS OF PIPE VERTICAL
PIPE SHALL NOT BE TURNED
OR ROLLED WHILE WELDING

2G

1/32 TO 1/16

3/32 ± 1/32

0.562

1/16 MAX

1/16 MAX

Groove Design of Test Coupon

(For combination qualifications, the deposited weld metal thickness shall be recorded for each filler metal or process used.)

BASE METALS (QW-403)	POSTWELD HEAT TREATMENT (QW-407)
Material Spec. __ASTM A53 pipe__	Temperature __None__
Type or Grade __A__	Time __~__
P-No. __1__ to P-No. __1__	Other __~__
Thickness of Test Coupon __0.562 - inch__	
Diameter of Test Coupon __24 - inch O.D.__	
Other __~__	

GAS (QW-408)

Percent Composition

	Gas(es)	(Mixture)	Flow Rate
Shielding	CO_2	welding grade	20 CFH
Trailing	None	~	~
Backing	None	~	~

FILLER METALS (QW-404)

SFA Specification __5.18__

AWS Classification __ER 70S-3__

Filler Metal F-No. __6__

Weld Metal Analysis A-No. __1__

Size of Filler Metal __0.035-in__

Other __~__

Deposited Weld Metal

ELECTRICAL CHARACTERISTICS (QW-409)

Current __D.C.__

Polarity __Electrode Positive__

Amps. __150-180__ Volts __21-23__

Tungsten Electrode Size __None__

Other __Short circuiting mode__

POSITION (QW-405)

Position of Groove __2G and 5G__

Weld Progression (Uphill, Downhill) __Downhill__

Other __~__

TECHNIQUE (QW-410)

Travel Speed __21-26 ipm__

String or Weave Bead __as required - see sketch__

Oscillation __as required - see sketch__

Multipass or Single Pass (per side) __Multiple__

Single or Multiple Electrodes __Single__

Other __~__

PREHEAT (QW-406)

Preheat Temp. __100°F__

Interpass Temp. __200°F Max__

Other

FIGURE 22–5   ASME procedure qualification record (PQR), sheet 1.

Tensile Test (QW-150)

Specimen No.	Width	Thickness	Area	Ultimate Total Load lb	Ultimate Unit Stress psi	Type of Failure & Location
2 G 1	0.752	0.377	0.283	26,000	95,500	ductile - BM
2 G 2	0.754	0.377	0.282	25,400	89,500	"
2 G U 2	0.753	0.378	0.284	21,800	77,000	"
2 G U 4	0.754	0.378	0.284	25,000	88,000	"

Guided-Bend Tests (QW-160)

Type and Figure No.	Result
Side bend Q7.1	No defect
Side bend Q7.1	No defect
Side bend Q7.1	No defect
Side bend Q7.1	No defect

Toughness Tests (QW-170)

Specimen No.	Notch Location	Notch Type	Test Temp.	Impact Values	Lateral Exp. % Shear	Mils	Drop Weight Break	No Break
None								

Fillet-Weld Test (QW-180)

Result — Satisfactory: Yes X   No   Penetration into Parent Metal: Yes X   No

Macro—Results _Normal_

Other Tests

Type of Test ... None

Deposit Analysis   None

Other   None

....

Welder's Name   Peter J. Arc   Clock No. 3506   Stamp No. 506

Tests conducted by: Hobart Procedure Laboratory   Laboratory Test No. T-376

We certify that the statements in this record are correct and that the test welds were prepared, welded, and tested in accordance with the requirements of Section IX of the ASME Code.

Manufacturer   ABC Pressure Vessel Co.

Date   Aug. 11, 1999   By   John Dough

(Detail of record of tests are illustrative only and may be modified to conform to the type and number of tests required by the Code.)

FIGURE 22–6   ASME procedure qualification record (PQR), sheet 2.

mended that all WPSs be recorded and documented in a computer program. A successful computer program virtually eliminates cross-referencing. However, most programs are related to a particular specification and are designed to accommodate the more popular specifications. See Appendix A-3 for more information.

Structural Welding

Requirements for the AWS "Structural Welding Code-Steel" D1.1 are different from the Pressure Vessel Code. However, "Each manufacturer or contractor shall conduct the tests required by this code to qualify the welding procedures." In addition, "The engineer, at his direction, may accept evidence of previous qualification of welders, welding operators and tackers to be employed." Thus, as with the Pressure Vessel Code, the manufacturer or contractor is totally responsible for qualification of procedures and personnel.

The AWS structural code covers the welding requirements applicable to welded structures. It allows the use of prequalified welding procedures. Prequalified procedures are exempt from tests provided that they conform in all respects to code requirements, which are described completely. The use of prequalified joint welding procedures is not intended as a substitute for engineering judgment or the suitability of applications to a welded assembly or connection. The code requires that the manufacturer or contractor prepare a written procedure specification for the joint welding procedure to be used. This is a record of materials and welding variables showing that the joint welding procedure meets the requirements for prequalified status. It is therefore necessary to prepare welding procedure specifications that cover the work to be done under the requirements of the AWS Structural Welding Code.

The forms used by the code are shown by Figure 22-7. The front side gives the details of the welding procedure specification; the data asked for are almost the same as utilized for the ASME code. This can be prequalified or qualified by testing. If the procedure is qualified by testing, it is necessary to use Figure 22-8. This form is in reality the procedure qualification record (PQR) and shows the results of destructive testings, visual inspection, and radiographic tests. This shows that the test specimens were acceptable and therefore provides the support data to qualify the WPS. For information and limits of variables and acceptability of tests, refer to the code.

The AWS "Structural Welding Code—Sheet Steel"[36] has slightly different provisions. It does not accord prequalified status to any welding procedures for sheet steel. On the other hand, at the engineer's discretion, evidence of previous qualification of the welding procedure to be employed on production work may be accepted. It further states that each welding procedure shall be prepared as a specification for each type of weld and shall be qualified by the manufacturer or contractor. An independent laboratory or testing agency may do the testing. Further welding procedures shall be qualified for each change of essential variables as listed in the code. Once the procedure is qualified, it shall be considered qualified for that contractor's use indefinitely. The forms used by the sheet metal code are different from the standard structural code.

Other AWS Requirements

The AWS "Standard for Welding Procedure and Performance Qualification" (B2.1)[37] covers the requirements for qualifications of welding procedures for weldments other than structural for buildings and bridges. The requirements of the B2.1 standard are very similar to those of Section XI of the Boiler Code. The forms are similar, and the information required to fill in the forms for welding procedure specification (WPS) is essentially the same as the information required by the ASME welding procedure specification.

The welding procedure specification is qualified by making specific welds as described and testing them. This information is documented on the procedure qualification record (PQR) form, which is also very similar to the Boiler Code procedure qualification record. There are minor differences in that AWS uses the American Welding Society specification numbers for filler metals and the listing of base metals is slightly different. In addition, the variables are slightly different from ASME. For this reason, the B2.1 document must be referred to.

Cross-Country Pipelines

The API Standard 1104[38] for welding pipelines and related facilities requires procedure and welder qualification. This code, which is used worldwide, was designed so that qualification tests can be made in the field. The procedure specification includes the process, the base metal, the size of pipe, diameter and wall thickness, the joint detail, the filler metal type, size and number of passes, and the electrical characteristics utilized. There are also provisions for qualifying welding equipment and operators even if more than one operator is required. If any of the essential variables are changed, including a change in the welding process, a change in the pipe material or size, a change in the joint design, a change in the position, a change in filler metal, a change in filler metal size, and so on, the welding procedure must be reestablished and completely requalified. These requirements are described in detail. The code must be referred to in writing in a qualified welding procedure. An example of an API-qualified welding procedure specification is given in Figures 22-9 and 22-10. The coupon test report form example is given in Figure 22-11. This form is used for both procedure qualification and welder qualification. Always refer to the latest edition of the standard.

22-4 STANDARD WELDING PROCEDURE SPECIFICATIONS (SWPS)

A major expense to producers of code-welded products is the necessity of writing, preparing, testing, and qualifying welding procedures. This becomes excessive because of the need to requalify the same procedure over and over. Requalification is due to code requirements when customers or products are changed or for legal reasons.

WELDING PROCEDURE SPECIFICATION (WPS)   Yes (X)
PREQUALIFIED _____   QUALIFIED BY TESTING YES _____
or PROCEDURE QUALIFICATION RECORD (PQR)   Yes (X)

Identification # _300_
Revision __—__ Date _____ By _____
Authorized by _____ Date _____

Company Name _XYZ Structural Company_
Welding Process(es) _FCAW_
Supporting PQR No.(s) _2_

Type - Manual ( )   Semi-Automatic (X)
Machine ( )   Automatic ( )

JOINT DESIGN USED
Type _Vee and bevel groove_
Single (X) ·   Double Weld ( )
Backing   Yes (X)   No ( )
    Backing Material _steel_
Root Opening _3/16_   Root Face Dimension _0_
Groove Angle _30°-5-0_   Radius (J-U) _None_
Back Gouging: Yes ( )   No (X)   Method _⌣_

BASE METALS
Material Spec. _A-441_
Type or Grade _⌣_
Thickness   Groove _1-inch_   Fillet _—_
Diameter (Pipe) _⌣_

FILLER METALS
AWS Specification _A-5.20_
AWS Classification _E 70T-5_

SHIELDING
Flux _None_   Gas _CO2_
                Composition _Welding grade_
Electrode-Flux (Class)   Flow Rate _30-35 CFH_
            _⌣_   Gas Cup Size _3/4 inch_

PREHEAT
Preheat Temp., Min. _50°F_
Interpass Temp., Min _50°F_   Max _⌣_

POSITION
Position of Groove _Flat & Horiz._ Fillet _—_
Vertical Progression: Up (←) Down (↙)

ELECTRICAL CHARACTERISTICS
Transfer Mode (GMAW)   Short-Circuiting ( )
  Globular ( )   Spray (X)
Current: AC ( ) DCEP (X) DCEN ( ) Pulsed ( )
Other _____

TECHNIQUE
Stringer or Weave Bead _stringer_
Multi-pass or Single Pass (per side) _multi-pass_
Number of electrodes _one_
Electrode Spacing   Longitudinal _⌣_
                    Lateral _⌣_
                    Angle _⌣_

Contact Tube to Work Distance _5/8 to 3/4 inch_
Peening _none_
Interpass Cleaning: _yes_

POSTWELD HEAT TREATMENT
Temp. _None_
Time _____

WELDING PROCEDURE

Pass or Weld Layer(s)	Process	Filler Metals		Current			Travel Speed	Joint Details
		Class	Diam.	Type & Polarity	Amps or Wire Feed Speed	Volts		
All	FCAW	E 70T-5	3/32-in	D.C. +	460-450	28-30	10-14 IPM	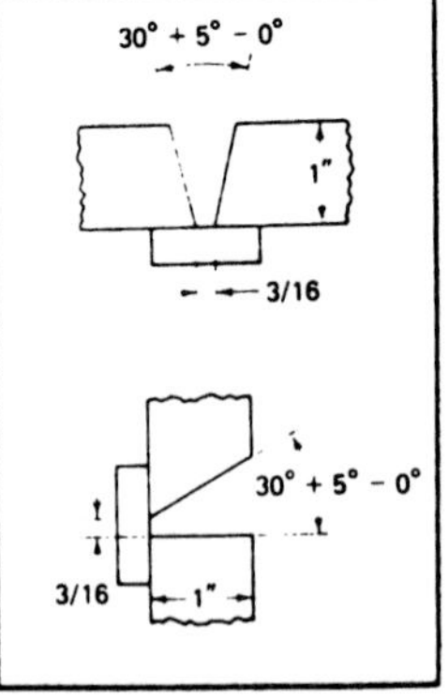

FIGURE 22-7 AWS structural welding procedure specification.

Standard welding procedures supported by adequate test data and that satisfy the technical requirements for the commonly used welding codes and specifications have been needed for many years.

Through the cooperation of the Welding Research Council and the American Welding Society, thousands of qualified welding procedures and procedure qualification records have been collected and analyzed. Based on these data, a number of standard welding procedure specifications (WPSs) have been written. This is an ongoing activity, and so far over 50 standard welding procedure specifications have been established, approved, and published by

TENSILE TEST

Specimen no.	Width	Thickness	Area	Ultimate tensile load. lb	Ultimate unit stress. psi	Character of failure and location
F { M-8751-C1	0.990	0.746	0.7385	56,700	76,773	Base metal
M-8751-C2	0.994	0.746	0.7415	57,275	77,240	Base metal
H { M-8751-E1	0.991	0.790	0.7829	68,000	86,857	Base metal
M-8751-E2	0.993	0.789	0.7835	69,850	89,153	Base metal

GUIDED BEND TEST

Specimen No.	Type of bend	Result	Remarks
M-8751	Side bend	Passed	Flat position
M-8751	Side bend	Passed	Flat position
M-8751	Side bend	Passed	Horizontal position
M-8751	Side bend	Passed	Horizontal position

VISUAL INSPECTION
Appearance _acceptable_
Undercut _None_
Piping porosity _None_
Convexity _Acceptable_
Test date _Nov. 12, 1991_
Witnessed by _John Jones_

Radiographic-ultrasonic examination
RT report no: _None_ Result _~_
UT report no: _None_ Result _~_

FILLET WELD TEST RESULTS
Minimum size multiple pass   Maximum size single pass
Macroetch _None_   Macroetch _None_
1. _~_   3. _~_   1. _~_   3. _~_
2. _~_   2. _~_

All-weld-metal tension test _None_

Tensile strength, psi _~_
Yield point/strength, psi _~_
Elongation in 2 in., % _~_
Laboratory test no. _~_

Other Tests _Charpy V-notch 10mm x 10mm_

Welding position	Specimen location	Test temp	Energy absorbed	Lateral expansion
Flat	WM	-20°F	95 ft-lbs	80 mils
Flat	WM	-40°F	82 ft-lbs	76 mils
Flat	WM	-60°F	65 ft-lbs	70 mils
Flat	WM	-80°F	50 ft-lbs	70 mils

Welder's name _John Doe_   Clock no. _123-45-4321_   Stamp no. _4321_

Tests conducted by _Hobart Technical Center_   Laboratory

Test number _TW B-707-C_

Per _John Jones_

We, the undersigned, certify that the statements in this record are correct and that the test welds were prepared, welded, and tested in accordance with the requirements of section 5, Part B of ANSI/AWS D1.1. (_1990_) Structural Welding Code-Steel.
year

Signed _XYZ Structural Company_
Manufacturer or Contractor

By _Tom Brown_

Title _Gen. Manager_

Date _Nov. 12, 1999_

FIGURE 22–8   AWS structural procedure qualification record.

the American Welding Society. They are also approved as national standards by the American National Standards Institute (ANSI). The current standard WPSs based on ANSI/AWS B2.1, "Standard for Welding Procedure and Performance Qualifications,"[37] are shown in Figure 22-12.

Many more standard welding procedure specifications are needed. In the meantime it is important to determine if a particular need can be satisfied by an existing standard welding procedure specification or if it will be necessary to generate a welding procedure specification.

Reference: API Standard 1104, 2.2

PROCEDURE SPECIFICATION NO. 3

For S.M.A.W. Welding of X-52 Pipe and Fittings
Process Shielded Metal Arc Welding
Material A.P.I. std Line Pipe X-52
Diameter and wall thickness 4 1/2 inch to 12 3/4 inch by 3/16 inch to 3/4 inch
Joint design Single vee groove 60° to 75° 1/16 R.O. + 1/16 R.F.
Filler metal and no. of beads See sketch and schedule
Electrical or flame characteristics D.C. electrode positive
Position Horizontal fixed 5G
Direction of welding Downhill
No. of welders One
Time lapse between passes Unlimited
Type and removal of lineup clamp None
Cleaning and/or grinding Mechanical to remove all slag
Preheat stress relief None
Shielding gas and flow rate None
Shielding flux None
Speed of travel
Sketches and tabulations attached

Tested Nov. 30, 1999 Welder P. L. Welder
Approved Yes Welding supervisor W.S. Jones
Adopted Yes Chief engineer C.E. Smith

FIGURE 22–9 Sample procedure specification form.

FIGURE 22–10 Sample procedure specification form (cont.)

Reference: API Standard 1104, 2.2

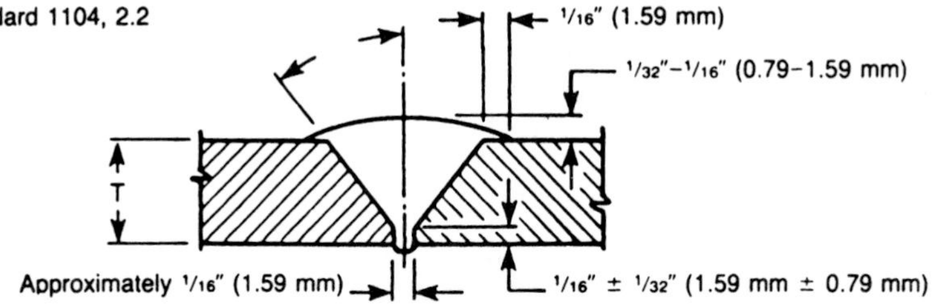

1/16" (1.59 mm)
1/32"–1/16" (0.79–1.59 mm)
T
Approximately 1/16" (1.59 mm)
1/16" ± 1/32" (1.59 mm ± 0.79 mm)

Standard V-Bevel Butt Joint

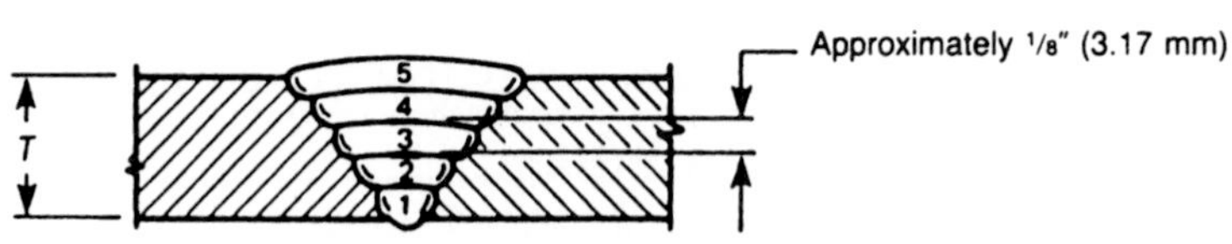

Approximately 1/8" (3.17 mm)
T
5
4
3
2
1

Sequence of Beads

Note: Dimensions are for reference only.

ELECTRODE SIZE AND NUMBER OF BEADS

Bead Number	Electrode Size and Type	Voltage	Amperage and Polarity	Speed
1	1/8" E 6010	27	110 EP	
2	1/8" E 7010	27	110 EP	
3	5/32" E 7010	26	130 EP	

COUPON TEST REPORT

Date _Nov. 30, 1999_   Test No. _T 242_

Location _Troy_

State _Ohio_   Weld Position: ___ Roll ☐   Fixed ☐

Welder _John C. Hickman_   Mark _3509_

Welding time _25 mins_   Time of day _10:00 AM_

Mean temperature _70° F_   Wind break used _None_

Weather conditions _indoors_

Voltage _25-28_   Amperage _110-130_

Welding machine type _Hobart D.C. generator_   Welding machine size _300 Amp._

Filler metal _Hobart # 10 E 6070_

Reinforcement size _1/32 to 1/16_

Pipe type and grade _5LX X-52_

Wall thickness _.203 inch_   Outside diameter _8 inch_

	1	2	3	4	5	6	7
Coupon stenciled	1	2					
Original specimen dimensions	0.937 x 0.203	1.015 x 0.203					
Original specimen area	0.190	0.206					
Maximum load	15,200	16,300					
Tensile strength per square inch of plate area	79,911	79,108					
Fracture location	base metal	base metal					

☒ Procedure   ☒ Qualifying test   ☒ Qualified
☒ Welder   ☐ Line test   ☐ Disqualified

Maximum tensile _80,000_   Minimum tensile _79,200_   Average tensile _79,500_

Remarks on tensile-strength tests
1. _Failed in base metal 2" from weld_
2. _Failed in base metal 1½" from weld_
3. _____
4. _____

Remarks on bend tests
1. _Root bend, no defects, passed_
2. _Root bend, one minor defect, passed_
3. _Face bend, no defects, passed_
4. _Face bend, minor defects, passed_

Remarks on nick-break tests
1. _No defects_
2. _No defects_
3. _____
4. _____

Test made at _Hobart Technical Center_   Date _Nov. 30, 1999_

Tested by _W.S. Jones_   Supervised by _C.E. Smith_

Note: Use back for additional remarks. This form can be used to report either a procedure qualification test or a welder qualification test.

FIGURE 22–11   Sample coupon test report form.

A welding procedure selection criteria chart is shown in Figure 22–13. This should be used to determine whether an AWS Standard Welding Procedure Specification (SWPS) is available for the specific application or whether it will be necessary to develop a customized welding procedure specification. Some of the standard WPSs have been approved by the National Board for use with the "National Board Inspection Code."[39] The ASME Boiler Code Committee has approved specific SWPSs.

The *standard welding procedure specifications* provide directions for making acceptable welds with specific processes on specific metals in the size range and positions covered. They provide ranges of welding variables that will be practical for the applications. It is the

Base Metal table (WPS data)

WPS Number	Welding Process[a]	Variation[b]	Method of Application[c]	M Class[d]	Base Metal		Backing[e]	Position[f]	Heat Treatment[g]	Miscellaneous Information
					Type	Thickness (in.)				
001	SMAW	—	MA	1	Carbon steel	$\frac{3}{16}$–$\frac{3}{4}$	Yes	All (UH)	AW	low hydrogen electrode
002	GTAW	—	MA	1	Carbon steel	$\frac{3}{16}$–$\frac{3}{8}$	Opt.	All	AW	Argon shielding
003	GMAW	SC	SA		Galvanized steel	Sheet metal	Opt.	All	AW	Argon + CO_2 shielding
004	GMAW	SC	SA	1	Carbon steel	Sheet metal	Opt.	All	AW	Argon + CO_2 shielding
005	GMAW	SC	SA	8	Austenitic SS	Sheet metal	Opt.	All	AW	90% He + 7% argon + 2% CO_2 shielding
006	GMAW	SC	SA	1	CS to austenitic SS	Sheet metal	Opt.	All	AW	90% He + 7% argon + 2% CO_2 shielding
007	GTAW	—	MA		Galvanized steel	Sheet metal	Opt.	All	AW	Argon shielding
008	GTAW	—	MA	8	Caron steel	Sheet metal	Opt.	All	AW	Argon shielding
009	GTAW	—	MA	1,8	Austenitic SS	Sheet metal	Opt.	All	AW	Argon shielding
010	GTAW	—	MA		CS to austenitic SS	Sheet metal	Opt.	All	AW	Argon shielding
011	SMAW	—	MA	1	Galvanized steel	10-18 gauge	Opt.	All	AW	E6010 or E6013
012	SMAW	—	MA	8	Carbon steel	10-18 gauge	Opt.	All	AW	E6010 or E6013
013	SMAW	—	MA	1,8	Austenitic SS	10-18 gauge	Opt.	All	AW	
014	SMAW	—	MA	22	CS to austenitic SS	10-18 gauge	Opt.	All (UH)	AW	
015	GTAW	—	MA	1	Aluminum		Opt.	All (UH)	AW or PW	Argon shielding
016	SMAW	—	MA	1	Carbon steel	$\frac{3}{16}$–$1\frac{1}{2}$	Opt.	All (UH)	AW or PW	E7018
017	SMAW	—	MA	1	Carbon steel	$\frac{1}{8}$–$1\frac{1}{2}$	Opt.	All (UH)	AW or PW	E6010
018	FCAW	SS	SA	1	Carbon steel	$\frac{1}{8}$–$1\frac{1}{2}$	Opt.	All (UH)	AW	E71 T-8
019	FCAW	GS	SA	1	Carbon steel	$\frac{1}{2}$	Opt.	All (UH)	AW	100% CO_2 shielding
020	FCAW	GS	SA	1	Carbon steel	$\frac{1}{8}$–$1\frac{1}{2}$	Opt.	All (UH)	AW or PW	75% Ar + 25% CO_2
021	GTAW, SMAW	—	MA	1	Carbon steel	$\frac{1}{8}$–$1\frac{1}{2}$	Opt.	All (UH)	Aw or PW	ER 70S-2, E7018
022	SMAW	—	MA	1	Carbon steel	$\frac{1}{8}$–$1\frac{1}{2}$	No	All (UH)	AW or PW	E6010, E7018
023	SMAW	—	MA	8	Austenitic SS	$\frac{1}{8}$–$1\frac{1}{2}$	Opt.	All (UH)	AW	Various F.M.
024	GTAW	—	MA	8	Austenitic SS	$\frac{1}{8}$–$1\frac{1}{2}$	Opt.	All (UH)	AW	Various F.M.
025	GTAW, SMAW	—	MA	8	Austenitic SS	$\frac{1}{8}$–$1\frac{1}{2}$	Opt.	All (UH)	AW	Various F.M.
026	SMAW	—	MA	1	Carbon steel	$\frac{1}{8}$–$1\frac{1}{2}$	No	All (DH)	AW	E6010 downhill, E7018
027	FCAW	SS	SA	1,2	Carbon steel	$\frac{1}{8}$–1	Opt.	All (opt)	AW	E71T-1
201	SMAW	—	MA	1	Carbon steel	$\frac{1}{2}$–3	No	All (R-UH)	AW	E6010 root uphill, E7018
202	SMAW	—	MA	1	Carbon steel	$\frac{1}{8}$–3	No	All (R-DH)	AW	E6010 root downhill, E7018
203	SMAW	—	MA	1	Carbon Steel	$\frac{3}{4}$	Opt.	All (UH)	AW	E6010 uphill
204	SMAW	—	MA	1	Carbon steel	$\frac{1}{8}$–3	No	All (R-DH)	AW	E6010 root downhill, balance uphill

No.	Process		Base metal class[d]	Base metal	Thickness	Consumable insert[e]	Position[f]	AW/PW[g]	Comments
205	SMAW	—	1	Carbon steel	$\frac{1}{8}-1\frac{1}{2}$	No	All (R-UH)	AW or PW	E6010 root uphill E7018
206	SMAW	—	1	Carbon steel	$\frac{1}{8}-1\frac{1}{2}$	No	All (R-DH)	AW or PW	E6010 root downhill, E7018
207	GTAW	—	1	Carbon steel	$\frac{1}{8}-1\frac{1}{2}$	No	All (UH)	AW or PW	Argon, ER 70S-2
208	SMAW	—	1	Carbon steel	$\frac{1}{8}-1\frac{1}{2}$	Yes	All (UH)	AW or PW	E7018
209	GTAW, SMAW	—	1	Carbon steel	$\frac{1}{8}-1\frac{1}{2}$	Opt.	All (UH)	AW or PW	GTW root, E7018
210	GTAW	—	1	Carbon steel	$\frac{1}{8}-1\frac{1}{2}$	CI	All (UH)	AW or PW	Argon, ER 70S-2
211	GTAW, SMAW	—	1	Carbon steel	$\frac{1}{8}-1\frac{1}{2}$	CI	All (UH)	AW or PW	GTAW root, E7018
212	GTAW	—	4	Chromium–molybdenum steel	$\frac{1}{8}-1\frac{1}{2}$	No	All (UH)	AW or PW	Preheat, Argon, ER 80S-B2
213	SMAW	—	4	Chromium–molybdenum steel	$\frac{1}{8}-1\frac{1}{2}$	Yes	All (UH)	AW or PW	Preheat, steel backing, E8018-B2
214	GTAW, SMAW	—	4	Chromium–molybdenum steel	$\frac{1}{8}-1\frac{1}{2}$	No	All (UH)	AW or PW	Preheat, GTAW root, ER80S-B2
215	GTAW	—	4	Chromium–molybdenum steel	$\frac{1}{8}-1\frac{1}{2}$	CI	All (UH)	AW or PW	Preheat, Argon IN515, ER80S-B2
216	GTAW, SMAW	—	4	Chromium–molybdenum steel	$\frac{1}{8}-1\frac{1}{2}$	CI	All (UH)	AW or PW	Preheat, GTAW root, IN515, ER80S-B2, E8018-B2
217	GTAW	—	4	Chromium–molybdenum steel	$\frac{1}{8}-\frac{3}{4}$	Opt.	All (UH)	AW	E880S-B2 Argon
218	SMAW	—	4	Chromium–molybdenum steel	$\frac{1}{8}-1\frac{1}{2}$	Yes	All (UH)	AW or PW	E8018-B2
219	GTAW, SMAW	—	4	Chromium–molybdenum steel	$\frac{1}{8}-1\frac{1}{2}$	Opt.	All (UH)	AW or PW	E880S-B2 E8018-B2
220	GTAW	—	4	Chromium–molybdenum steel	$\frac{1}{8}-\frac{3}{4}$	CI	All (UH)	AW	ER80S-B2 Argon
221	GTAW, SMAW	—	4	Chromium–molybdenum steel	$\frac{1}{8}-1\frac{1}{2}$	CI	All (UH)	AW or PW	ER80S-B2 E9018-B2
222	GTAW	—	5A	Chromium–molybdenum steel	$\frac{1}{8}-\frac{3}{4}$	Opt.	All (UH)	AW or PW	FR905-B3
223	SMAW	—	5A	Chromium–molybdenum steel	$\frac{1}{8}-1\frac{1}{2}$	Yes	All (UH)	AW or PW	FR905-B3
224	GTAW, SMAW	—	5A	Chromium–molybdenum steel	$\frac{1}{8}-1\frac{1}{2}$	Opt.	All (UH)	AW or PW	FR905-B3
225	GTAW	—	5A	Chromium–molybdenum steel	$\frac{1}{8}-\frac{3}{4}$	CI	All (UH)	AW	FR905-B3
226	GTAW, SMAW	—	5A	Chromium–molybdenum steel	$\frac{1}{8}-1\frac{1}{2}$	CI	All (UH)	AW	FR905-B3

[a]SMAW, shielded metal arc welding (stick); GTAW, gas tungsten arc welding (TIG); FCAW, flux cored arc welding; GMAW, gas metal arc welding (MIG).
[b]SC, short-circuiting, SS, self-shielded; GS, gas-shielded.
[c]MA, manual; SA, semiautomatic.
[d]Base metal class is M numbers of material.
[e]No, none; Yes, metal backing bar; CI, consumable insert; Opt., optional.
[f]UH, uphill; DH, downhill.
[g]AW, as welded; PW, postwelded.

FIGURE 21–12 Index of standard welding procedure specifications for sheet metal and structural applications.

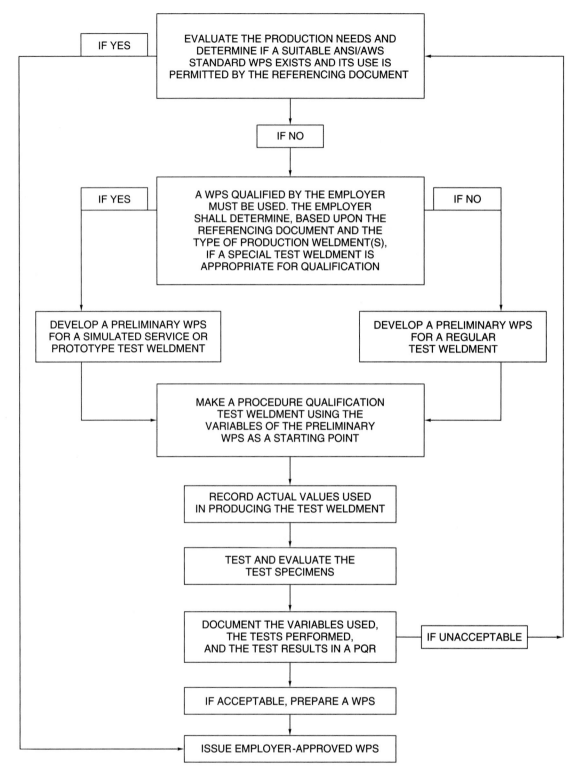

FIGURE 22–13 Welding procedure selection criteria.

policy of the AWS B2 Committee that the range of conditions and variables listed in the standard WPS be significantly restrictive to ensure a high probability of successful application by all users. Each standard welding procedure specification (WPS) used by a company must be accepted with a company official signing to the effect that "I accept full responsibility for the application of this standard WPS for use with the particular product or project under construction." At the option of the accepting company, the variables can be limited or restricted.

The standard welding procedure specifications are permitted to be used on work covered by the code or specification for which it has been approved. They do not require further testing or qualification work by the producer company. Thus it will no longer be necessary for the company to develop and qualify its own welding procedure specifications for standardized requirements. New welding processes, special techniques, materials, or filler metals must still be tested and qualified as in the past.

The standard WPSs do not replace or substitute for fabrication documents (codes, specifications, or contract requirements). They are an alternative to the individual company's own WPSs and PQRs. The standard WPS is not valid using conditions and variables outside the ranges listed. The user organization must have personnel with significant knowledge of welding. They must provide the engineering capability, training and qualified personnel, and proper equipment to implement the standard welding procedure specifications. The ability to make production welds having properties suitable for application depends on supplementing the standard WPS with appropriate performance qualification tests and sound engineering judgment. A standard WPS must be supplemented by information and instructions for use by the organization and for information to the welders.

The index for standard WPSs (Figure 22–12) shows the standard WPS number, the base metal it is designed to weld, the type of base metal by M number, the geometry (whether structural plate or shapes), sheet metal or pipe, the thickness range of the base metal, the welding process employed with variations, whether backing is used or not, the type of backing, the type of filler metal, the welding position, whether the vertical progression is uphill or downhill, and whether it is designed for "as welded" use or whether it will be "postweld heat treated." The format for each welding procedure specification is very similar to those required by AWS B2.1 or ASME Section IX.

Additional WPSs will be generated in the future. It is anticipated that the standard WPSs will save the welding industry tremendous sums of money and enable it to become more competitive throughout the world. The standard WPSs can be obtained from the American Welding Society headquarters in Miami, Florida.

22-5 QUALIFYING AND CERTIFYING WELDERS

All codes and specifications require proof that welders and welding operators have the necessary skill and ability to follow a qualified welding procedure successfully. This is a legal requirement since the specifications state that the manufacturer, contractor, owner, or user is responsible for the welding done. They are unable to personally observe the work of each and every welder; therefore, they rely on the fact that each welder and welding operator has passed qualification tests. Contractors, manufacturers, owners, and users must maintain complete records of qualified procedures and test results.

Computer software programs are available that allow you to create welder performance qualification records (WPQR). These programs allow you to keep close track of every qualified welder on the payroll. Each welder's record is entered into the program, which can be used to monitor that welder's test record for various processes and which may provide warnings of test expiration dates. Some programs provide for welder inspection results and rejection rates.

To be qualified, welders and welding operators must make specific welds that are then tested to prove they are of the quality required. These tests may be different for different codes. Qualification under one code will not necessarily qualify the welder to weld under a different code even though the tests are similar. There is one exception: this involves certain Navy codes, the Coast Guard code, and the American Bureau of Shipping Specifications for the welding of ships. These codes are very similar, and qualifications for one are usually accepted for the others.

It is absolutely essential that the applicable code be used in qualifying welders and welding operators. Qualification of welders is an extremely technical subject and carries with it contractual responsibilities. In addition, different codes and specifications have slightly different definitions. The following are used in this section.

- ✳ *Welder's performance qualifications* means "the demonstration of a welder's ability to produce welds meeting prescribed standards."
- ✳ *Welder certification* means "written verification that a welder has produced welds meeting a prescribed standard of welder performance."
- ✳ *Qualification* is described differently in different specifications, but in general qualification means the ability of an individual to perform to a required standard. It is the demonstration of a welder's ability to produce welds meeting prescribed standards or that welds made to a specific procedure can

meet prescribed standards. It involves taking and passing a practical welding test.

⁂ *Certification* is a written statement to the fact that the welder has produced welds meeting a prescribed standard of welding performance. It implies that a testing organization, a manufacturer, a contractor, or an owner or user has witnessed the preparation of test welds, has conducted the prescribed testing of the welds, and has recorded the successful results of a test in accordance with prescribed accepted standards.

⁂ *Welder registration* is "the act of registering a welder certification or a photostatic copy of the welder's certification." This is done by an appropriate or authoritative organization. Welders and others sometimes become confused when they encounter the need for a "qualified welder." This confusion is the result of the great number of different codes, specifications, and government regulations with seemingly different requirements that apply.

The following is a brief review of welder qualifications under ASME Section IX of the "Boiler and Pressure Vessel Code." ASME Section IX is used for qualifying welders to work on some other products; consult the latest edition of the code.

The welder who prepares the procedure qualification record (PQR) specimens that pass code requirements is personally qualified within his or her performance qualification variables. All other welders and welding operators are qualified by specific welding tests required by the WPS (welding procedure specification) that will cover the work. An example of the Record of Welder or Welding Operator Qualification Test (WPQ) is shown in Figure 22-14. This record should include the essential variables, the type of test and test results, and the ranges qualified for each welder and welding operator. Each welder and welding operator must be assigned an identifying number, letter, or symbol. It is used to identify the work of that person. The tests assigned are in accordance with the code, and the mechanical tests should meet the requirements applicable by the code. Radiographic examination may be substituted for mechanical tests except for GMAW using short circuiting metal transfer. The radiographic technique and acceptance criteria should be in accordance with the code. In general, welders who meet the code requirements for groove welds are also qualified for fillet welds, but not vice versa. A welder qualified to weld in accordance with one qualified WPS is also qualified to weld in accordance with other qualified WPSs using the same welding process, within the limits of essential variables according to the code.

If a welder has not welded for a period of six months or more, his or her qualifications shall be expired.

If there is reason to question the welder's ability to make welds that meet the specifications, his or her qualification shall be considered expired. There are various other conditions relative to welder qualifications listed in the code. Performance testing can be done by independent testing laboratories. The code requires welding of performance qualification tests to be witnessed by the certificate holder. The preparation and mechanical testing may be subcontracted, but the full responsibility for all requirements must be borne by the manufacturer, contractor, owner, or user. The code must be consulted for this information.

The following is a brief review of welder qualification under AWS "Standard for Welding Procedure and Performance Qualifications" (B2.1). This code is referenced by many AWS specifications and should be used except for qualifying welders, welding operators, or tackers to the Structural Code. "Standard for Welding Procedure and Performance Qualifications" is in many respects very similar to the ASME Section IX requirements. The terminology is essentially the same and information required is almost identical. The variables may vary slightly and the forms are slightly different. The recording of test results made by welders or welding operators is recorded on a form called the Performance Qualification Test Record, which is similar to the ASME form. Upon completion of the tests and if they meet the code requirements, this record is then signed by the qualifier, which qualifies the welder or welding operator. Qualification record applies continuously unless no welding is done for a period of six months or if there is reason to question the ability of the welder. Refer in detail to this specific document.

The following is a brief review of the qualification of welders, welding operators, and tackers under the requirements of AWS "Structural Welding Code—Steel." This code is different from the Boiler Code and AWS B2.1. The structural code provides for the use of welding procedure specifications as mentioned previously but also allows the use of prequalified joint welding procedures. Prequalified welding procedures must be used with caution, and also standard welding procedure specifications may be used. Consult the code for specific information. The form used for performance qualification is entitled Welder and Welding Operator Qualification Test Record, shown in Figure 22-15. The specific tests that must be taken are covered in the code. It is further stated that at the engineer's discretion proper documented evidence of previous qualifications of welders, welding operators, and tack welders may be accepted. Visual inspection, guided bend test, fillet test, and radiographic test results are employed.

Welder performance qualification for the "Structural Welding Code—Sheet Steel" AWS *D1.3* is different from the Structural Code. Qualification is established for any one of the steels permitted by the specification, and the welder shall be considered as qualified to weld on any

QW-484 SUGGESTED FORMAT FOR MANUFACTURER'S RECORD OF WELDER OR WELDING OPERATOR QUALIFICATION TESTS (WPQ)
(See QW-301, Section IX, ASME Boiler and Pressure Vessel Code)

Welder Name Peter J. Arc Check No. 3506 Stamp No. 506

Using WPS No. 1 Rev. ~ Date 8/11/99

the above welder is qualified for the following ranges.

Variable	Record Actual Values Used in Qualification	Qualification Range
Process	GMAW	GMAW
Process Type	semi-automatic	semi-automatic
Backing (metal, weld metal, flux, etc. (QW-402))	None	None
Material Spec. (QW-403)	P-1 to P-1	P-1 to P-1
Thickness		
Groove	0.562-in	All
Fillet	~	~
Diameter		
Groove	24-in	No maximum
Fillet	~	~
Filler Metal (QW-404)		
Spec. No.	5.18	5.18
Class	ER70S-3	ER70S-3
F-No.	6	6
Deposited Weld Metal Thickness		
Groove X Fillet		
Position (QW-405)	2G and 5G	All position
Weld Progression	Downhill	~
Gas Type (QW-408)	CO2	CO2
Backing Gas (QW-408) None		
Electrical Characteristics (QW-409)		
Current	D.C.	D.C.
Polarity	Electrode +	Electrode +

Guided Bend Test Results QW-462.2(a), QW-462.3(a), QW-462.3(b)

Type and Fig. No.	Result
Side bend Q 7.1	No defects
Side bend Q 7.1	No defects
Side bend Q 7.1	No defects
Side bend Q 7.1	One minor defect
Side bend Q 7.1	No defects
Side bend Q 7.1	No defects

Radiographic Test Results (QW-304 & QW-305)

For alternative qualification of groove welds by radiography

Radiographic Results: None

Fillet Weld Test Results [See QW-462.4(a), QW-462.4(b)]

Fracture Test (Describe the location, nature and size of any crack or tearing of the specimen)
None

Length and Per Cent of Defects ~ inches ~ %

Macro Test—Fusion None

Appearance—Fillet Size (leg) ~ in. X ~ in. Convexity ~ in. or Concavity ~ in.

Test Conducted by Hobart Welding Procedure Lab Laboratory—Test No. 1065

We certify that the statements in this record are correct and that the test welds were prepared, welded and tested in accordance with the requirements of Section IX of the ASME Code.

Organization ABC Pressure Vessel Co.

Date Aug. 18, 1999 By Dick Brown

(Detail of record of tests are illustrative only and may be modified to conform to the type and number of tests required by the Code.)
NOTE: Any essential variables in addition to those above shall be recorded.

FIGURE 22–14 ASME record of welder qualification tests (WPQ).

WELDER AND WELDING OPERATOR QUALIFICATION TEST RECORD

Welder or welding operator's name __John Doe__   Identification no. __123-45-4321__

Welding process __FCAW__   Manual _____   Semiautomatic __X__   Machine _____

Position __Flat and horizontal__

(Flat, horizontal, overhead or vertical — if vertical, state whether upward or downward)

In accordance with procedure specification no __200__

Material specification __ASTM A-441__

Diameter and wall thickness (if pipe) — otherwise, joint thickness __1-inch__

Thickess range this qualifies __3/16 inch to unlimited__

FILLER METAL

Specification no. __A 5.20__   Classification __E 70T-5__   F no. __10__

Describe filler metal (if not covered by AWS specification) __—__

Is backing strip used? __YES__

Filler metal diameter and trade name __3/32-inch__   Flux for submerged arc or gas for gas metal arc or flux cored arc welding __CO2 welding grade__

VISUAL INSPECTION (9.25.1)

Appearance __Good__   Undercut __None__   Piping porosity __None__

Guided Bend Test Results

Type	Result	Type	Result
Flat side bend	Passed	Horizontal side bend	Passed
Flat side bend	Passed	Horizontal side bend	Passed

Test conducted by __Hobart Technical center__   Laboratory test no. __IWB-707-C__

per __John Jones__   Test date __Nov. 17, 1991__

Fillet Test Results

Appearance __None__   Fillet size __~__

Fracture test root penetration __~__   Macroetch __~__

(Describe the location, nature, and size of any crack or tearing of the specimen.)

Test conducted by __Hobart Technical Center__   Laboratory test no. __IWB-707-C__

per __John Jones__   Test date __Nov. 17, 1991__

RADIOGRAPHIC TEST RESULTS

Film identification	Results	Remarks	Film identification	Results	Remarks
None					

Test witnessed by __John Jones__   Test no. __IWB-707-C__

per _____

We, the undersigned, certify that the statements in this record are correct and that the welds were prepared and tested in accordance with the requirements of section 5, Part C or D of ANSI/AWS D1.1, (__1990__) Structural Welding Code-Steel.

year

Manufacturer or contractor __XYZ Structural Co.__

Authorized by __Tom Brown__

Date __Nov. 17, 1999__

FIGURE 22–15 Welder and welding operator qualification test record.

other steels permitted by the specification with the exception of coated steels. If coated steels are involved, the qualification must be on coated steels. The welder should be qualified for each welding process used and in each position used. In the case of vertical, it relates to uphill or downhill travel. Welders are qualified for all electrodes within a group designation. Different combinations of electrode and shielding gas must be qualified. There is a differentiation made between different types of welds, such as fillet welds, flare groove welds, and so on. The manufacturer or contractor must keep track of the qualification records. The qualification is considered indefinite unless the welder does not weld for a period of six months or there is some reason to question the welder's ability. The form used is different from the structural code form. Refer to AWS *D1.3* for details.

The qualification of welders for cross-country pipeline work, according to API Standard 1104, is different from the others just mentioned. The cross-country pipeline qualification work is usually done in the field, and welds are tested in the field. The tests require tensile, bend tests, and nick break tests. The welders can be qualified for single qualification or multiple qualification, depending on taking one or two tests. The Coupon Test Report form is used for performance and for procedure qualification and is shown in Figure 22–11. The API standard should be consulted for details including variables, test results, and so on.

Welders or welding operators cannot be qualified or certified on their own. Normally, manufacturers, contractors, owners, or users certify that a welder is qualified based on successful completion of specific test welds. This means that recertification or qualification is normally required when welders change employers. The American Welding Society, in an effort to reduce the expense of requalifying welders for code work, has announced the "Standard for AWS Certified Welders" (AWS QC-3)[40] program, which promises to save the industry millions of dollars annually. Welders who complete a series of standardized skill tests successfully and pass a visual acuity examination become AWS Certified Welders. Certification remains in effect indefinitely as long as they weld every six months, provide employment verification to AWS headquarters, and pass the annual eye examination.

A test facility accredited by the American Welding Society[41] will provide the skill tests to welders based on their application being accepted by AWS headquarters. The tests will be in accordance with the Certified Welder Program. Accreditation of the independent test laboratories is handled by the American Bureau of Shipping, which performs on-site test assessments of the laboratory. AWS will provide a list of accredited test facilities upon request.

QUESTIONS

22–1. Why is welding covered in so many codes?

22–2. What is the purpose of a procedure qualification record (PQR)?

22–3. What is the purpose of a welder performance qualification (WPQ) test?

22–4. Why are perfect welds required for some classes of work and not others?

22–5. Who is responsible for a good-quality product?

22–6. Who is responsible for a good-quality weld?

22–7. What is a certified welder? What is welder registration?

22–8. Codes and specifications are related to industries. What industries use welding specifications?

22–9. What is a welding procedure specification (WPS)?

22–10. What is a welding procedure? Name two types.

22–11. Is the ASME Section IX, welding qualifications, enforceable by law?

22–12. Can welders be certified by a contractors' association? If so, how is it done?

22–13. What is an ASME symbol stamp? Who can use it?

22–14. What welded products are covered by the AWS structural code? What materials?

22–15. What are prequalified welding procedures? Explain.

22–16. What welded products are covered by API Standard 1104?

22–17. How are automatic welding equipment and operators qualified?

22–18. What method does API 1104 use for fracture toughness testing?

22–19. How will standard WPSs reduce the cost of weldments?

22–20. Who makes standard welding procedure specifications available to the industry?

REFERENCES

1. "Boiler and Pressure Vessel Code," American Society of Mechanical Engineers, New York.

2. "Quality Assurance Criteria for Nuclear Power Plants and Fuel Reprocessing Plants," Code of Federal Regulations, Section 10, Energy Part 50, Appendix B (10 CFR50-B).

3. "Standard for Welding of Reactor Coolant and Associated Systems and Components for Naval Nuclear Power Plants," Navships 250-1500-1, U.S. Department of the Navy, Naval Engineering Center, Washington, D.C.

4. "Structural Welding Code—Steel," ANSI/AWS D1.1, American Welding Society, Miami, Fla.

5. "Manual of Steel Construction—Allowable Stress Design," American Institute of Steel Construction, Chicago.

6. "Bridge Welding Code," AASHTO/AWS D1.5, American Welding Society, Miami, Fla.

7. "Structural Welding Code—Aluminum," ANSI/AWS D1.2, American Welding Society, Miami, Fla.

8. "Structural Welding Code—Reinforcing Steel," ANSI/AWS D1.4, American Welding Society, Miami, Fla.

9. "Standard for Welded Steel Elevated Tanks, Standpipes, and Reservoirs for Water Storage," AWS D1.1, American Welding Society, Miami, Fla.

10. "Standard for Welded Steel Tanks for Oil Storage," API Standard 650, American Petroleum Institute, Washington, D.C.

11. U.S. Coast Guard, Department of Transportation, "Marine Engineering Regulations," Sub Chapter F, Part 57, "Welding and Brazing," Code of Federal Regulations, Washington, D.C.

12. "Fabrication, Welding and Inspection of Ship Hulls," Navships 0900-000-1000, U.S. Department of the Navy, Naval Ship Systems Command, Washington, D.C.

13. "Standard Specification for Merchant Ship Construction," U.S. Maritime Administration, Washington, D.C.

14. "Rules for Building and Classing Steel Vessels," American Bureau of Shipping, New York.

15. "Guide for Steel Hull Welding," ANSI/AWS D3.5, American Welding Society, Miami, Fla.

16. "Guide for Aluminum Hull Welding," ANSI/AWS D3.7, American Welding Society, Miami, Fla.

17. "Specifications for Tank Cars," Association of American Railroads, Washington, D.C.

18. "Specifications for Design, Fabrication, and Construction of Freight Cars," Association of American Railroads, Chicago.

19. "Railroad Welding Specification," ANSI/AWS D15.1, American Welding Society, Miami, Fla.

20. "Aerospace Material Specifications," Society of Automotive Engineers, Warrendale, Pa.

21. "National Aerospace Standards," Aerospace Industries Association of America, Washington, D.C.

22. "Qualification of Aircraft, Missile and Aerospace Fusion Welders," Military Standard MIL-STD-1595A (Notice 1), U.S. Department of Defense, Washington, D.C.

23. "Specification for Automotive Frame Weld Quality—Arc Welding," ANSI/AWS D8.8, American Welding Society, Miami, Fla.

24. "Recommended Practices for Automotive Weld Quality—Resistance Spot Welding," ANSI/AWS D8.7, American Welding Society, Miami, Fla.

25. "Specification for Welding Earth Moving and Construction Equipment," AWS D14.3, American Welding Society, Miami, Fla.

26. "30 CFR Roll-Over Protective Structures," *Federal Register* 39, no. 207.

27. "Specification for Welding Industrial and Mill Cranes and Other Material Handling Equipment," ANSI/AWS D14.1, American Welding Society, Miami, Fla.

28. "Specification for Metal Cutting Machine Tool Weldments," ANSI/AWS D14.2, American Welding Society, Miami, Fla.

29. "Classification and Application of Welded Joints for Machinery and Equipment," AWS D14.4, American Welding Society, Miami, Fla.

30. "Specification for Welding of Presses and Press Components," AWS D14.5, American Welding Society, Miami, Fla.

31. "Specification for Rotating Elements of Equipment," ANSI/AWS D14.6, American Welding Society, Miami, Fla.

32. "General Design and Construction Requirements," Code of Federal Regulations, Title 49, Transportation Section 178.340, Part D, U.S. Government Printing Office, Washington, D.C.

33. "Specification of Cargo Tanks," ML.331, Code of Federal Regulations, Title 49, Transportation Section 178.337, U.S. Government Printing Office, Washington, D.C.

34. "Fusion Welding for Aerospace Welders," Military Standard MIL-STD-2219, U.S. Department of Defense, Washington, D.C.

35. "Sheet Metal Welding Code," AWS D9.1, American Welding Society, Miami, Fla.

36. "Structural Welding Code—Sheet Steel," AWS D1.3, American Welding Society, Miami, Fla.

37. "Standard for Welding Procedure and Performance Qualifications," ANSI/AWS B2.1, American Welding Society, Miami, Fla.

38. "Welding of Pipelines and Related Facilities," API Standard 1104, American Petroleum Institute, Washington, D.C.

39. "The National Board Inspection Code," National Board of Boiler and Pressure Vessel Inspectors, Columbus, Ohio.

40. "Standard for AWS Certified Welders," AWS QC-3, American Welding Society, Miami, Fla.

41. "Standard for Accreditation Test Facilities for AWS Certified Welder Program," AWS QC-4, American Welding Society, Miami, Fla.

23

WELDING PROBLEMS AND SOLUTIONS

OUTLINE

23-1 ARC BLOW

Arc blow is the deflection of a welding arc from its normal path because of magnetic forces.[1] Deflection of the arc can be extremely frustrating to a welder. It will adversely affect the appearance of the weld, will cause excessive spatter, and can impair the quality of the weld. It is well known to the welder using the shielded metal arc welding process. It is also a factor in semiautomatic and automatic arc welding applications. Arc blow occurs primarily when welding steel or ferromagnetic materials, but it can be encountered when welding nonmagnetic materials. The welding arc is usually deflected forward or backward of the direction of travel; however, it may be deflected from one side to the other. It can become so severe that it is impossible to make a satisfactory weld. Arc blow is one of the most troublesome problems encountered by the welder and one that is the least understood.[1]

The laws of electricity and magnetism can help explain the problem of arc blow. When an electric current passes through an electrical conductor, it produces a magnetic flux in circles around the conductor in planes perpendicular to the conductor and with their centers in the conductor. The *right-hand rule* is used to determine the direction of the magnetic flux. It states that when the thumb of the right hand points in the direction the current flows (conventional flow) in the conductor, the fingers point in the direction of the flux. The direction of the magnetic flux produces polarity in the magnetic field, the same as the north and south poles of a permanent magnet. The rules of magnetism, which state that like poles repel and opposite poles attract, apply in this situation. Welding current is much higher than the electrical current normally encountered. Similarly, the magnetic fields are much stronger.

The welding arc is an electrical conductor and the magnetic flux is set up surrounding it in accordance with the right-hand rule. The important magnetic field in the vicinity of the welding arc is the field produced by the welding current that passes through it from the electrode and to the work. This is a self-induced circular magnetic field that surrounds the arc and exerts a force on it from all sides according to the electrical-magnetic rule. As long as the magnetic field is symmetrical, there is no unbalanced magnetic force and there is no arc deflection. The arc is parallel or in line with the centerline of the electrode, and it takes the shortest path to the base plate. If the symmetry of this magnetic field is disturbed, the forces on the arc are no longer equal and the arc is deflected by the strongest force.

This electrical-magnetic relationship is used in certain welding applications for magnetically moving or oscillating the welding arc. The gas tungsten arc is easily deflected by a magnetic flux. It can be oscillated by transverse magnetic fields or it can be made to deflect in the direction of travel. To move the arc is to rely on the flux field surrounding the arc and to move this field by introducing an external polarity field. Oscillation is obtained by reversing the external transverse field. As the self-induced field around the arc is attracted and repelled, it tends to move the arc column. Magnetic oscillation of the gas tungsten welding arc is used to widen the weld deposit. Long arcs are more easily moved than short arcs. The amount of magnetic flux to create the movement must be of the same order as the flux field surrounding the arc column.

Whenever the symmetry of the field is disturbed by another magnetic force, it will tend to move the self-induced field surrounding the arc and deflect the arc.

Except under the most simple conditions, the self-induced magnetic field is not symmetrical throughout the entire electric circuit and changes direction at the arc. There is always an imbalance of the magnetic field around the arc because the arc is moving and the current flow pattern through the base material is not constant.

There is another factor that helps produce the nonsymmetrical or unbalanced relationship of the magnetic forces. The magnetic flux will pass through a magnetic material such as steel much easier than it will pass through air. The magnetic flux path will tend to stay within the steel and will be more concentrated and stronger than in air. Welding current passes through the electrode lead, the electrode holder to the welding electrode, then through the arc into the base metal (assume steel). At this point the current changes direction to pass to the work lead connection, then through the work lead back to the welding machine (Figure 23–1). At the point the arc is in contact with the work, the change of direction is relatively abrupt and the lines of force are perpendicular to the path of the welding current, which creates a magnetic imbalance at this point. The lines of force are concentrated on the inside of the angle of the current path through the electrode and the work and are spread out on the outside angle of this path. Consequently, the magnetic field is much stronger on the side of the arc toward the work lead connection than on the other side. This imbalance produces a force on the stronger side, which deflects the arc to the left. This is toward the weaker force and is opposite the direction of the current path. The direction of this force is the same whether the current is flowing in one direction or the other. If the welding current is reversed, the magnetic field is also reversed

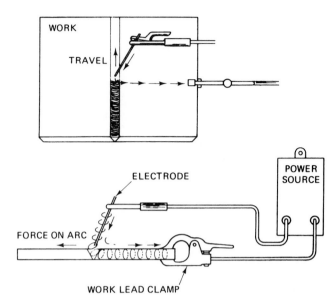

FIGURE 23–1 Unbalanced magnetic force due to current direction change.

but the direction of the magnetic force acting on the arc is in the same direction.

The second factor that keeps a magnetic field from being symmetrical is the fact that the arc is moving and depositing weld metal. As a weld is made joining two plates, the arc moves from one end of the joint to the other. The magnetic field in the plates will constantly change. If we assume the work lead is immediately under the arc and moving with the arc, the magnetic path in the work will not be concentric about the point of the arc. This is because the lines of force do not take the shortest path but the easiest path. Near the start end of the joint the lines of force are crowded together since they will tend to stay within the steel. Toward the finish end of the joint the lines of force will be separated since there is more area (Figure 23–2). In addition, where the weld has been made the lines of force must cross the air gap or

FIGURE 23–2 Unbalanced magnetic force due to unbalanced magnetic path.

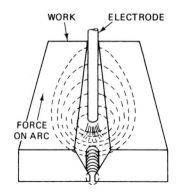

root opening. The magnetic field is more intense on the short end and the imbalance produces a force that deflects the arc to the right or toward the long end.

When welding with direct current, the total force tending to cause the arc to deflect is a combination of these two forces. Sometimes these forces add and sometimes they subtract from each other, and at times they may meet at right angles. The polarity or direction of flow of the current does not affect the direction of these forces nor the resultant force. By analyzing the path of the welding current through the electrode and into the base metal to the work lead, and analyzing the magnetic field within the base metal, it is possible to determine the resultant forces and predict the resulting arc deflection.

The use of alternating current greatly reduces the magnitude of the deflection. Alternating current does not completely eliminate arc blow. The reason for the reduction of arc blow is that the alternating current sets up other currents that tend to neutralize the magnetic field or greatly reduce its strength. Alternating current varies between maximum value of one polarity and the maximum value of the opposite polarity, and the magnetic field surrounding the alternating current conductor does the same. The alternating magnetic field is a moving field that induces current in any conductor through which it passes, according to the basic laws of electricity, the *induction principle*. This means that currents are induced in nearby conductors in a direction opposite to that of the inducing current. These induced currents are called eddy currents. They, in turn, produce a magnetic field of their own that tends to neutralize the magnetic field of arc current. These currents are alternating currents of the same frequency and are in the part of the work nearest the arc. They always flow from the opposite direction (Figure 23–3). When alternating current is used, eddy currents are induced in the workpiece, which produce magnetic fields that reduce the intensity of the field acting on the arc. Unfortunately, alternating current cannot be used for all welding applications, and changing from direct current to alternating current may not always be possible to reduce arc blow.

With an understanding of the factors that affect arc blow, it is possible to explain the practical factors and provide solutions for overcoming them. Arc blow is caused by magnetic forces. The induced magnetic forces are not symmetrical about the magnetic field surrounding the path of the welding current. One factor is the nonsymmetrical location of magnetic material with respect to the arc. This creates a magnetic force on the arc, which acts toward the easiest magnetic path and is independent of electrode polarity. The location of the easiest magnetic path changes constantly as welding progresses; therefore, the intensity and the direction of the force change.

The second factor is the change in direction of the welding current as it leaves the arc and enters the workpiece. Welding current will take the easiest path but not always the most direct path through the work to the work lead connection. The resultant magnetic force is opposite in direction to the current from the arc to the work lead connection. It is independent of welding current polarity.

The third factor explains why arc blow is much less with alternating current. This is because the induction principle creates current flow within the base metal, which creates magnetic fields that tend to neutralize the magnetic field affecting the arc.

The greatest magnetic force is caused by the difference in resistance of the magnetic path in the base metal around the arc. The location of the work lead connection is of secondary importance. It is best to have the work lead connection at the starting point of the weld. On occasion, the work lead can be changed to the opposite end of the joint. In some cases, leads can be connected to both ends.

The conditions that affect the magnetic force acting on the arc vary so widely that it is impossible to do more than make generalized statements regarding them. The following suggestions may help reduce arc blow.

The magnetic forces acting on the arc can be modified by changing the magnetic path across the joint. This can be accomplished by runoff tabs, starting plates, large tack welds, and backing strips, as well as the welding sequence. An external magnetic field produced by an electromagnet may be effective. This can be accomplished by wrapping several turns of welding lead around the workpiece. Arc blow is usually more pronounced at the start of the weld seam. In this case a magnetic shunt or runoff tab will reduce the blow. It is wise to use as short an arc as possible so that there is less of an arc for the magnetic forces to control.

The welding fixture can be a source of arc blow; therefore, an analysis with respect to fixturing is important. The hold-down clamps and backing bars must fit tightly to the work. Copper or nonferrous metals should be used. Magnetic structure of the fixture can affect the magnetic forces affecting the arc.

A major problem results from magnetic fields already in the base metal. This happens when the base metal has been handled by magnet lifting cranes. Residual magnetism

FIGURE 23–3 Reduction of magnetic force due to induced fields.

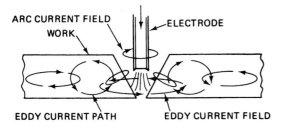

in thick plates handled by magnets can be of such magnitude that it is impossible to make a weld. The solution here is to demagnetize the part, wrap the part with welding leads, or if this fails, stress relieve or anneal the parts.

23-2 WELDING DISTORTION AND WARPAGE

The arc welding processes involve heat. High-temperature heat is largely responsible for welding distortion, warpage, and stresses. When metal is heated it expands, and it expands in all directions. When metal cools it contracts in all directions. To understand this, consider an extremely small piece of metal the shape of a cube (Figure 23-4). When it is exposed to a temperature increase, it will expand in all three directions, x, y, and z. There is a direct relationship between the amount of temperature change and the change in dimension, based on the coefficient of thermal expansion. This is a measure of the linear increase per unit length based on the change in temperature of the material. The coefficient of expansion is different for different metals. Aluminum has one of the greatest coefficient of expansion ratios, and changes in dimension almost twice as much as steel. The coefficient of expansion of the common metals is given in Figure 15-1. A metal expands or contracts by the same amount when heated or cooled by the same temperature, if it is not restrained. In the case of welding, the metals that are heated and cooled are restrained. They are restrained because they are a part of a larger piece of metal that is not heated to the same temperature. This is the problem. Within uniform heating and unrestrained parts, the heating and cooling are relatively distortion free. In actual practice, the heating is not uniform across the cross section of a part. There is always restraint, because the parts not heated or heated to a lesser amount tend to restrain the portion of the same piece that is heated to a higher temperature. This differential or nonuniform heating and the partial restraint is the cause for thermal distortion and warpage that occur in welding.

The coefficient of expansion is important when considering warpage. It is the factor responsible for the different degrees of warpage between different metals. Of the structural metals, aluminum has the highest coefficient of expansion; it is approximately twice as great as plain carbon steel. This is significant when relating the warpage that occurs when welding steel is compared to welding aluminum. The coefficient of expansion for steel is 6.7×10^{-6}; written out, this would be 0.0000067, which is the amount in inches that steel expands for every degree Fahrenheit the temperature rises. As an example, if a piece of steel is taken from 100° F (38° C) up to a dull red heat of 1100° F (593° C), there would be temperature change of 1000° F (538° C). Multiplying these would give a change in dimension per inch of 0.0067 in. (0.17 mm), or 6.7 thousandths of an inch. This is a small change in dimension, but it is significant. If aluminum is heated for the same temperature change, the results would be larger. Aluminum has a coefficient of expansion of 13.8×10^{-6} in. per degree change of temperature. Exposing it to 1000° F (538° C) change in temperature (which, incidentally would almost cause aluminum to melt) would cause a change of 0.0000138 × 1000, or a dimensional change of 0.0138 in. (0.35 mm), which is a change of almost 1/64 in. (0.4 mm) per inch. To make this a little more meaningful, take a 10-in. (250-mm)-long round rod. Each inch of the aluminum will expand the 0.0138 in. (0.35 mm), but the entire bar, which is 10 in. (250 mm) long, would expand 0.138 in. (3.5 mm) or slightly over 1/8 in. (3.2 mm), and this is significant with respect to warpage.

In practical application, this is not free expansion and contracting and uniform heating. Consider a round rod placed in a vise or in some device that is absolutely unmovable (Figure 23-5A). With this rod in between two

FIGURE 23-5 Round rod in vise.

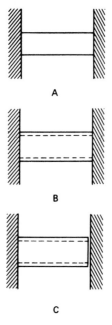

A

B

C

FIGURE 23-4 Cube of metal showing expansion.

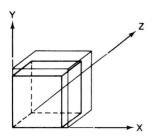

unmovable surfaces, uniformly heat the bar 1000° F (538° C) or the difference above room temperature to, say, 1070° F (577° C). In the case of the steel bar it would try to expand 0.067 in. (1.7 mm) per inch, or in the case of aluminum would try to expand 0.138 in. (3.5 mm) per inch. However, the restraining surfaces will not move. The bar is restrained from expanding in the length or *x* direction. However, each of the small cubes within this bar still will expand according to the laws of the temperature and expansion. All the expansion will be in the *y* and *z* directions because it will be unable to expand in the *x* direction. This means that the bar will become slightly larger but not longer (Figure 23-5B). This is the principle of upsetting. Molecules have rearranged themselves and have expanded in two directions, but not in the third direction. Now allow the bar to cool down to room temperature. The small cubes within the bar will tend to contract in the *x*, *y*, and *z* directions and will contract the same amount that they expanded. After cooling back to room temperature the bar will be slightly shorter than it was originally (Figure 23-5C). It will be slightly larger from the upsetting force that occurred during the heating cycle. This illustrates the effect of restraint and shows that the heated portion will not return to its original shape. This may be too elementary, at least when considering differential heating with respect to welding. To be more practical, assume the same round rod between two immovable surfaces, but in this case include a compression spring (Figure 23-6A). Now go through the same heating cycle. Heat the rod the same 1000° F (538° C), but in this situation it will expand in length (*x* direction) as well as in the *y* and *z* directions; however, there is restraint in the *x* direction. The rod actually becomes longer and will cause the spring to compress slightly. Because of the re-

straint, by the spring, the bar length increase will be less than if it were unrestrained. The spring exerts force against the rod, which restrains the rod from expanding as much as when free. There will be expansion in the *x*, *y*, and *z* directions, so that the bar will actually become somewhat larger as well as longer (Figure 23-6B). When the heat is removed and the part returns to its original room temperature, it will be slightly shorter than originally and somewhat larger in diameter (Figure 23-6C). The deformation will not be quite so great because the restraint was less. This is more similar to what happens to weld metal in a differentially heated cycle.

Another example: Again use a round rod, but in this case considerably longer, and place it between the immovable surfaces (Figure 23-7A). In this situation, heat only the center portion of the rod. In the heated area, the temperature rise will cause expansion in all three directions. Restraint will be exercised by jaws and the compressibility of unheated metal. In the heated area there will be expansion in the *y* and *z* directions so that the diameter of the bar in the heated area will become larger (Figure 23-7B). This is an example of plastic deformation or upsetting. When the bar is allowed to cool, contraction will occur uniformly in the *x*, *y*, and *z* directions and the length of the bar will be slightly reduced when it returns to room temperature (Figure 23-7C).

One final example to bring the problem of differential heating and upsetting into focus with regard to practical applications is illustrated by the use of a piece of flat rectangular bar stock fairly thin and fairly wide: $\frac{1}{4}$ in. (6.4 mm) thick, 2 in. (50 mm) wide, and 12 in. (300 mm) long. Pass a high-temperature heat source such as a gas tungsten arc along one edge (Figure 23-8A). This creates differential heating across the width of the bar. The top edge of the bar will be heated almost to the molten stage. Approximately $\frac{1}{2}$ in. (13 mm) below the edge in the bar it will remain at room temperature. At the bottom edge of

FIGURE 23-6   Round rod in vise with spring.

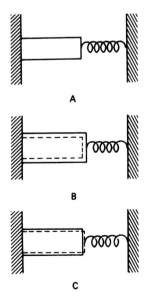

A

B

C

FIGURE 23-7   Long rod in vise.

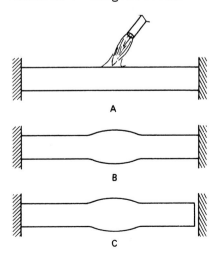

A

B

C

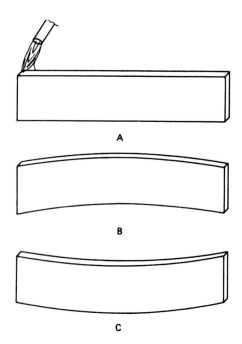

FIGURE 23–8 Long rectangular bar heated on one edge.

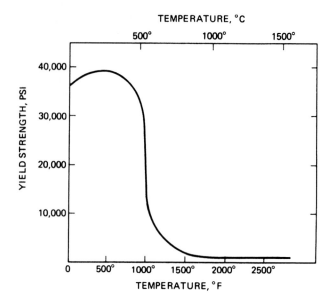

FIGURE 23–9 Temperature–yield strength relationship.

the bar it will also be at room temperature. Each small increment of metal in the upper edge of the bar is being heated to approximately 2000° F (1093° C) and it will expand in all three directions. Slightly below the top edge, the bar will be heated to a lesser degree or to a lower temperature. Even at this point there will be a degree of expansion in all three directions. Farther down with little or no change of temperature there will be little or no change in dimension. This will cause the top edge of the bar to expand (Figure 23–8), but it is intimately a part of the lower portion of the bar, which has no tendency to expand because it is not heated. The restraint is from the lower portion of the bar, which will restrain the upper portion from expanding to the amount to which it would expand if it were free. Plastic deformation will occur and the bar will become slightly thicker at the heated edge. When the bar cools, contraction will occur in all three directions and this will cause the upper edge of the bar to shorten. Shortening the upper edge of the bar without shortening the center or lower edge of the bar will cause warpage (Figure 23–8C). Shortening one edge of a bar and not shortening the other edge will create a curved bar.

Another factor must be considered. This is the fact that metals have lower strengths at high temperatures. As the temperature of a metal increases, its strength decreases. This can be shown by a curve plotting the yield strength against temperature. The case of steel results in a curve shown in Figure 23–9. As the temperature rises the strength decreases at approximately 1000° F (538° C) to 1500° F (816° C), depending on its composition. For low-carbon mild steel, when the temperature is above

1500° F (816° C) the strength is reduced drastically. This factor is involved because in arc welding a portion of the base metal goes above this temperature since surface melting is involved.

In the case of the rectangular bar in the previous example, the bar was made of steel. When the temperature at the top edge was practically at the molten stage, a quarter of the way down it would be at a relatively low temperature, and halfway down the width of the bar and at the bottom edge it would be at room temperature. If the bar is copper or aluminum, *thermal conductivity* becomes important. The property of conductivity is shown in Figure 15–1 for common metals. The thermal conductivity of copper is the highest, that of aluminum is approximately half, but that of steels is only about one-fifth as much. If the bar were made of copper, the temperature at the top would be practically at the molten stage. The heat would quickly move within the bar to the lower portions so that the temperature differential would be minimal. This would be true of high-thermal-conductivity metals. The higher the thermal conductivity the less effect differential heating will have. This physical property should be considered, since arc temperatures are similar but metal melting points are greatly different.

When making a weld, take into consideration all the factors mentioned above and determine how each one reacts alone but also how they react with one another. Consider a weld bead made longitudinally on a relatively thin rectangular plate (Figure 23–10). When making a weld bead on the plate, the deposited weld metal is momentarily at a temperature of about 3000° F (1649° C), slightly above its melting point. The base metal immediately under the arc is also brought to the molten stage. As the weld metal cools and fuses to the base metal, it takes

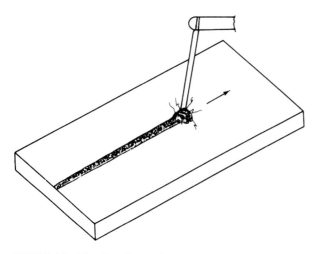

FIGURE 23-10   Bead on plate.

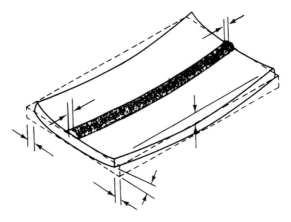

FIGURE 23-11   Warpage produced by bead on plate.

shape and forms a bead. The weld bead freezes. At the point of solidification the molten metal has little strength. As it cools it acquires strength (Figure 23-9). It is also in its expanded form because of its high temperature. The weld metal is now intimately fused to the base metal and they work together. As the metal continues to cool, it acquires higher strength and contracts in all three directions. These factors are further complicated because the arc depositing molten metal is moving. In addition, the cooling differential is also moving, but follows the travel of the arc. With the temperature further reducing and each small increment of heated metal tending to contract, contracting stresses occur and there will be movement in the metal adjacent to the weld. The unheated metal tends to resist the cooling dimension changes. Temperature differential has an effect on this. If it is a low-conductivity metal, the changes will occur over a relatively small distance. If the metal has high thermal conductivity, the heating differential will be less and the change in dimensions will be spread over a larger area. In this bead-on-plate example, the cooling shrinkage changes are above the centerline of the thickness of the plate. The tendency of the weld bead to shorten in length, in thickness, and width tends to warp the plate by shortening the weld area and shortening the top surface in both directions. An exaggeration of how a plate of this type will warp is shown in Figure 23-11.

Running a weld bead on the edge of a bar would be similar to the example of a heat source on the edge of the bar. The effect might be slightly more since additional molten or high-temperature metal is deposited on the edge of the bar. The deposited metal becomes integral with the bar and provides a greater mass of metal solidifying, gaining strength and cooling. This shortening of the heated edge without a similar change in the nonheated edge would create the warpage that would be the same or similar to that shown in Figure 23-8.

When making a weld joint, specifically a butt joint, between two narrow, thin plates, another factor becomes important. This has to do with heat input or the travel speed of welding. Refer to Figure 23-12, which shows the weld joint partially made. If the same joint is welded with covered electrodes, the unwelded end of the joint tends to close or the parts become closer together. If the same joint is welded with submerged arc, the unwelded end of the joint tends to open or the parts become farther apart. The explanation is complicated since it involves the geometry of the pieces being joined, the thermal coefficient of expansion, the thermal conductivity, the mass of molten metal, but most importantly, the travel speed of the arc. If the travel speed is relatively fast, the effect of the heat of the arc will cause expansion of the edges of the plates and they will bow outward and open up the joint. If the travel speed is relatively slow, the effect of the arc temperature and the cooling will cause contraction of the edges of the plates and they will bow inward and close up the joint. This is the same as running a bead on the edge of the plate. In either case it is a momentary situation that continues to change as the weld progresses.

FIGURE 23-12   Butt joint showing warpage.

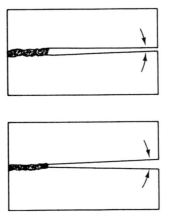

By experimenting with current and travel speed, the exact speed can be found for a specific joint so that the root will neither open up nor close. This is one of the advantages of fine wire gas metal arc welding of sheet metal. The heat input balance, using normal procedures, approaches this travel speed relationship and warpage is minimized. Conversely, when using the gas tungsten arc process on sheet metal, the travel speed is slower and greater warpage results.

All these factors must be considered, yet they are not the entire story for more complicated welds. For example, in a large fillet weld there are all the factors mentioned previously, plus more. Consider a single-pass fillet weld to make a corner joint with the fillet weld on the inside (Figure 23–13). A fillet weld, by definition, has a triangular cross section. In making the weld, the base metal immediately under the fillet is molten. The fillet is completely molten and soon after it is deposited it begins to freeze. Initially it has little strength, but the strength rapidly increases as the temperature decreases. From the geometry of the joint we have absolute restraint where the edge of the vertical plate is in contact with the surface of the horizontal plate. The metal deposited is now integral with the material of the two parts; its strength is increasing but is decreasing in volume. Each and every small increment of the heated metal is decreasing volume in its x, y, and z directions. As each increment shrinks and as the surfaces between the two plates are restrained and cannot shrink, warpage occurs. Normally, the fillet weld freezes from the root to the face and the strength would increase in this same relationship. One reason why fillet welds are a special case is the fact that at the root there are less small increments of metal contracting. But toward the face of the fillet there are more increments that are contracting. There is more shrinkage at the face of the fillet than there is at the root, not only because of the larger volume of metal at the face but also because of the restraint between the plates. If the plates had not been tightly placed together, there would be less total angular warpage at the joint. The fillet weld is particularly vulnerable to warpage because of these two factors. Figure 23–13 shows how this warpage occurs. Note that all the warpage occurs on one side of the centerline of the vertical plate and on the top side of the centerline of the horizontal plate.

One practical method of eliminating the angular distortion produced by a single-fillet weld is to utilize the T-type joint or the corner joint with the one plate inset so that double-fillet welds would be made. If the fillet welds could be made on both sides simultaneously, it would greatly reduce the angular warpage since the two would be working against each other and their shrinkage. Longitudinal shrinkage will still occur and result in highly stressed welds that might be stressed to the yield point of the weld metal.

If one fillet is made and then the second fillet is made, there will be a small amount of angular distortion because the initial fillet will freeze first and create distortion. However, if the second fillet is made larger than the first fillet, this can be overcome. In common practice it is normal to place one pass of the fillet on one side, then to place two passes of the second fillet on the other side and finish up on the first side with its second pass. This type of procedure tends to reduce angular warpage of fillet welded T-joints.

The cross-section geometry of other welds and the technique of making them contribute to warpage. Consider a single V-weld joint. A single-V-weld in thin material can be made with one pass. Figure 23–14 shows a single-pass V-groove butt joint and the warpage that would occur. As the weld is deposited in the groove, the adjacent base metal is raised to its melting temperature and becomes molten. The entire weld is molten but immediately begins to freeze. As it freezes it increases its strength, but it still has considerable temperature, on the order of 1000° F (538° C). At the root of the V-groove, each increment of weld metal shrinks in all three directions. A weld in thinner material may freeze throughout its cross section almost simultaneously. Since there is more volume of metal freezing and contracting at the top of the joint, there will be more total dimensional contraction at this point. Angular distortion will occur because of the greater amount of weld metal at the top of the joint.

A change in weld joint design has an effect on warpage. If the root opening is increased and the bevel angle is reduced, the difference in the amount of weld metal at the root and at the face of the weld would be more similar and therefore angular distortion would be less. Another approach to reducing angular distortion is to use the double-V preparation. One portion of the weld would be made on one side of the plate and the other portion on the other side. This is applicable only to

FIGURE 23–13 Inside fillet corner joint.

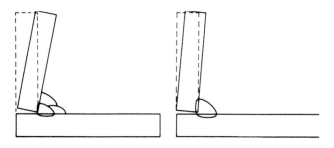

FIGURE 23–14 Single-pass V-groove butt joint.

thicker plates. On a thinner material the square groove weld has an advantage since the shrinkage dimension at the root of the weld would be the same as that at the face of the weld. The differential in the cooling rate would lead to some angular distortion. In these cases, the longitudinal shrinkage will still occur and create the shortening of the total weld and its effect on distortion.

The cross section of an EB weld is almost rectangular, since the width at the face and the width close to the root are almost the same. Angular distortion is very small since the sides of the weld are nearly parallel.

The problem of distortion in fillet welds and in V-groove welds increases as the sizes of the welds increase. The groove weld shown in Figure 23–15 will require many passes to complete the joint. With arc welding the first or root pass would be placed in the joint and in effect would create a homogeneous structure between the two parts. Very little angular distortion would result from the root pass weld. The next pass would cause some angular distortion. In the second pass the first pass acts as a restraining force, assuming that it is not completely melted by the second pass. The freezing of the second pass would create shrinkage of the deposited metal but the root pass would offer restraint and there would be shrinkage closer to the upper surface of the plates and less at the bottom surface. When the next pass is made, there would be more restraint at the root and more shrinkage at the surface of the weld. Successive passes are larger and wider, and there is a greater mass of weld metal shrinking. This condition continues until the joint is completed. Each new pass creates its heating and cooling and shrinking cycle, with previous passes acting as restraint. It is like a hinge with the root acting as the hinge pin, and each additional pass tending to bring the edges of the joint closer together. Multipass single-V-groove welds are particularly susceptible to angular distortion.

The larger number of passes used increases the angular distortion in single-groove welds. This disproves the theory that by making many small passes with low currents the distortion will be reduced. The mass of molten metal and the mass of metal restraining the shrinkage does affect the relationship. The additional applications of heat make for more angular distortion of multipass single fillets and multipass single-V-groove welds. A solution to this is the use of larger passes, larger electrodes, or processes that provide larger passes.

One solution is to use the double-weld, double-V, or double-bevel preparation. The double fillet should be used if at all possible; if not, use a combination of fillet and groove welds. The principle of using double welds is to equalize the shrinkage on both sides of the centerline of the weld joint. If one side of the centerline contracts more than the other, it will create angular distortion.

Angular distortion is greatly reduced by balancing the welding on either side of the centerline of the joints being made. The distortion that occurs is based on the shrinkage that follows thermal cooling versus restraint conditions previously described. Even though there may be no visible evidence of warpage, there are high stresses approaching yield points of the metals in and near the joint. If a square-groove butt weld were made between two edges of two flat plates and if the welds were made from both sides properly, there would be little evidence of warpage. However, if we saw through the throat of the weld its entire length, the warpage would become immediately evident. The result would be the same as two bars with welds on one edge, as described previously (Figure 23–16).

The principle of balanced welding must be considered in designing welds. In the case of the bar, the centerline of the bar is actually its neutral axis. The neutral axis is the center of gravity of the cross section of the part. Weldments also have a center of gravity or neutral axis that can be calculated based on the thickness and size of the component parts. An example of a symmetrical cross-sectional part is a fabricated wide-flange or H-type beam (Figure 23–17). To build such a beam will require the welding of two flange plates to a web plate. If the parts are all equal size and thickness, the center of

---

FIGURE 23–16 Square-groove butt weld cut apart.

FIGURE 23–17 Fabricated beam.

FIGURE 23–15 Multiple-pass single-V-groove butt joint.

gravity would be the exact center of this assembly. The welds to join the flanges to the web would be equally spaced about the center of gravity. If all four of these fillet welds could be made simultaneously, it would be possible to produce the beam without longitudinal distortion. The edges of the flanges will pull in slightly because of the angular distortion produced by each fillet weld. If the welding can be balanced around the neutral axis of the weldment, the distortion will be reduced. If each application of heat is done in a logical manner about the center of gravity of the weldment, it tends to keep the weldment to true shape. A box section, such as a box column, presents the same opportunity. The weldment is symmetrical around its neutral axis and the applications of the welds are also symmetrical. If all four welds could be made simultaneously, a straight part would result. In practice it is rarely possible to make all four welds simultaneously. Box structures have been welded in the vertical position to make the welds simultaneously. A more common practice, however, is to make two of the welds at one time and then make the remaining two welds. Sometimes it is possible to vary the sizes of the welds to produce balanced stresses to create a straight member. Unfortunately, most weldments are not symmetrical around their neutral axis, and more often than not most welding has to be done on one side or another of the neutral axis, which creates warpage.

The following factors should be taken into consideration in order to reduce welding warpage.

1. The location of the neutral axis and its relationship in both directions.
2. The locations of welds, size of welds, and distance from the neutral axis in both directions.
3. The time factor of welding and cooling rates when making the various welds
4. The opportunity for balancing welding around the neutral axis
5. Repetitive identical structures and varying the welding technique based on measurable warpage
6. The use of procedures and sequences to minimize weldment distortion

It is impossible to provide more than general guidelines in warpage control. When identical parts are made with identical procedures, the warpage of each assembly may be different. This can result from stresses within parts themselves prior to welding or insignificant changes in technique by different welders. However, based on these rules it is possible to minimize distortion on most weldments. When making welds on a large box structure such as a ship hull, the sequence of welding can be varied from side to side and from top to bottom to minimize distortion. At shipyards, precision measuring

devices are used to determine the amount of distortion. Corrective action is taken to maintain straightness of the structure. This same technique can be used for any large engineering structures. If the structure tends to warp to one side, welding can be increased on the other side to compensate; by measuring continuously and altering procedures, the structure can be made to come true. On large weldments it is important to establish a procedure to minimize warpage. The order of joining plates in a deck or on a tank will affect stresses and distortion. As a general rule, transverse welds should be made before longitudinal welds. Figure 23–18 shows the order in which the joints should be welded.

Platens and bases that are large and thin and usually made of "egg crate" design can also be regulated by making certain joints first. The joints joining the web sections together should be made prior to joining them to the top and bottom or flange sections. The joining of the web sections to the flanges should be regulated and constant measurements made to determine that the weldment remains true.

The design and type of weldment have much to do with warpage. Large weldments made of relatively thin materials have a tendency to warp. Control is possible with proper procedures. Large weldments made of extremely heavy or thick members have a greater degree of restraint and the amount of warpage may be less. Balancing the welding and monitoring the effect of changes in technique and procedure are the best ways of controlling warpage in large structures.

FIGURE 23–18 Order of making weld joints.

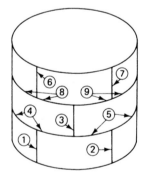

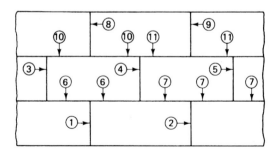

Warpage can be minimized in smaller structures by:

1. The use of restraining fixtures, strong backs, or many tack welds
2. The use of heat sinks or the fast cooling of welds
3. The predistortion or prebending of parts prior to welding
4. Balancing welds about the weldment neutral axis or using wandering sequences or back-step welding
5. Intermittent welding to reduce the volume of weld metal
6. Proper joint design selection and minimizing of size
7. As a last resort, use of preheat or peening

Each technique has advantages and can be used in certain applications. No one of them is a cure-all for the problem of weld warpage. Fixtures can be very effective, but fixtures must be extremely strong if they are to resist warpage. Even with fixtures, balanced welding is necessary. Strong backs or face plates can be used to physically restrain the parts. Massive tack welds and heavy bracing can also be used. If the weldment is stress relieved with bracing in place, warpage will be minimized. Figure 23–19 shows a dipper for a large power shovel with massive bracing. Each half was stress relieved separately, because of furnace size limitations, with the bracing in place. Distortion was minimized.

Prebending or predistortion of parts or providing special dimension for warpage, or prewarping with the hope the welds will bring the parts into proper alignment, can be helpful. This technique is most useful when repetitive products are being welded and a history of the amount of warpage can be determined. This technique is particularly useful in field erection of large structures. It is also used for welding premachined parts together to avoid finish machining of a total weldment.

Rapid cooling by means of heat extractors or heat sinks has been used successfully in the aircraft industry. By means of hydraulic or pneumatic clamps the parts of the weldment are put in intimate contact with large masses of highly conductive metal. These are known as heat sinks, which pull the heat away from the weld quicker than normal. This creates a more uniform heat distribution and reduces the heat differential and distortion. Some weld fixtures have water-cooled heat sinks to help reduce distortion.

Use of the backstep technique shown in Figure 23–20 has an advantage in that each small increment will have its own shrinkage pattern, which then becomes insignificant to the pattern of the total weldment. This technique may be rather time consuming and questionable for certain types of applications.

Intermittent welding can be helpful; however, intermittent welding is merely a method of reducing the amount of weld metal in certain areas to avoid warpage at that specific area. Using the smallest possible weld size helps reduce distortion.

FIGURE 23–19 Heavy braces in weldment.

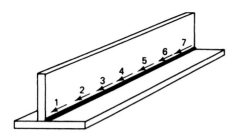

FIGURE 23–20 Backstep technique.

As a last resort use peening. Peening works the weld metal and expands it, which counteracts the shrinkage that occurred upon freezing. Peening is hard to regulate, noisy, and usually not the best solution. If the weldment is warped so that it cannot be used, it may be salvageable by mechanical or thermal methods. Mechanical methods involve the use of force, such as a straightening press, jacks, clamps, and so on. This can be expensive but justifiable. Thermal methods involve local heating to relieve stresses in some cases or to cause corrective warpage. Heat is usually applied by torches. Caution should be exercised not to overheat the metal, especially heat-treated materials. Local torch heating should not be used on highly thermal conductive metals such as aluminum or copper. The heat will be conducted away so quickly that local upsetting will not occur. This was covered in an earlier section.

Finally, one of the best approaches to weldment distortion is the intelligent design of the weldment itself. With intelligent review it is possible to design or redesign a weldment to better place the welds in a balanced geometry around the neutral axis of the weldment. Many times this can change the location of a weld. It might involve the use of a bend or it might involve the use of two welds; but if it can be accomplished, it will greatly enhance the distortion control of the weldment.

23-3 HEAT FORMING AND STRAIGHTENING

Heat, normally applied by an oxyfuel gas torch, can be used to bend or straighten metal parts. In the previous section we discussed how high-temperature heat from welding operations is largely responsible for welding distortion and warpage and how this occurs. Briefly, all metal expands when heated and contracts when cooled. The amount of expansion depends on the temperature increase, the coefficient of the expansion of the metal heated, and the size of the heated area. Unrestrained metal expands in all three directions, but metal is normally restrained due to unequal heating. Rapid heating causes plastic deformation or upsetting, which creates dimensional changes upon cooling. The strength of metal decreases significantly as the temperature of the metal is increased. Distortion occurs as a result of forces created by differential heating and restraint. These forces can be used to bend or to straighten metal pieces. This principle was put to practical use by Joseph Holt, who published the book *Contraction as a Friend in Need* in 1938. This principle was perfected by his son, Richard Holt, in the 1960s; practiced throughout the world by Ray Stitt; has now been given a scientific approach by Richard Avent[2]; and may soon become a recommended practice document by AWS.

Heat forming and straightening, the application of these principles, can be used to correct distortion from welding or from accidents. To understand the application of heat to bend or straighten members, start with a relatively simple example. Consider a flat bar 1/4 in. (6.4 mm) thick by 2 in. (50 mm) wide and 24 in. (650 mm) long. Heat with any oxyacetylene welding torch using a medium-size single-orifice tip. Propane can also be used as the fuel gas. Adjust the torch for a neutral flame. Before beginning the heat pattern use a piece of soapstone and mark a triangular area from point *A,* located one-eighth of the width of the bar toward edge *D,* and mark a pie-shaped V-shaped area to points *B* and *C* on the *E* edge of the bar (Figure 23–21). The angle formed by *B, C* to *A* should be approximately 30° toward edge *E.* Holding the flame steady with the flame pointed slightly toward edge *E* until point *A* becomes heated to a dull red color, start to move the torch as shown in the figure. Move the torch slowly in a zigzag fashion, bringing it up to the same dull-red temperature. Continue traveling, making sure that the part comes up to temperature before moving farther toward the far edge, *E.* Continue the zigzag line of travel until you have reached points *B* and *C* on edge *E.* This will produce a V-shaped section that has been heated progressively from near edge *D* to edge *E.* Cooling will also be progressive in the same direc-

FIGURE 23–21 Application of heat to V-shaped areas.

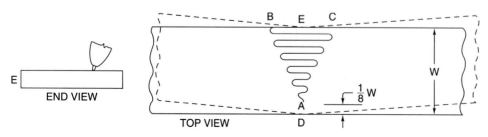

tion. After the bar has cooled, it will take the shape shown by the dashed lines.

The heat at point *A* causes the metal to expand in all three directions; however, it is restrained by the adjacent cold metal and it will therefore upset or become slightly thicker. The localized heating continues as the flame is moved from point *A* toward *B* and *C*. The heated area is increased as the flame moves away from point *A*. The metal at point *A* begins to cool and contract as the temperature falls. Normally, points *B* and *C* are reached before the temperature at point *A* cools appreciably. The widened heated area and the contracting metal behind it create the forces that cause the bending. As the cooling continues from point *A* toward edge *E*, the metal contracts. There is more metal to contract as the heated area becomes wider. This contributes to creating more motion or bending as cooling continues. Point *D* acts as the hinge pin; the material between *B* and *C* contracts the most. The amount of distortion depends on the angle between *B* and *C*. If more motion is required, additional heated Vs can be made adjacent to the first one.

The same application of V-shaped heating areas can be used to bend or straighten other types of structural shapes. Figure 23–22 shows the heat application that can be used for other shapes. The large end of the V will be the portion where the maximum contraction occurs. The amount of contraction can actually be calculated; however, since temperature control is not precise, the amount of metal heated is not known exactly, due to conduction of heat to the cooler areas. Formulas have been worked out and are available. The best method of understanding the amount required for shrinkage is by experience.

When straightening shapes other than flat bars, it is necessary to consider the relationship between webs and flanges. For example, in the figure the rolled angle can be bent with the vertical leg acting as the hinge point. All the heating is done on the horizontal leg. If the vertical leg of the angle must be shortened, however, it should be done with thorough heating of the vertical leg when the wide portion of the V comes to that leg. Progressive heating should not be done in the vertical leg. Progressive heating is used only when a V-shaped area is to be heated. By following the figure closely, an angle can be bent in either direction. The same applies to channels, Ts, and I or wide-flange beams. The basic principles can also be used for box sections and pipe. In all cases the starting point should be closest to the edge that is to be the hinge point.

The same technique can be used to correct warpage of structures that involve T-joints utilizing double fillet welds. In this case the application of heat is linear rather than V-shaped. Warpage is often encountered when double fillets are made on one side of a member forming a T-joint. Applying the torch flame to the center of the back side of the T will create shrinkage at this point, which will tend to bring the top of the T-joint back into a flat plane (Figure 23–23). The same technique of straight-line heat application can be used to shorten parts.

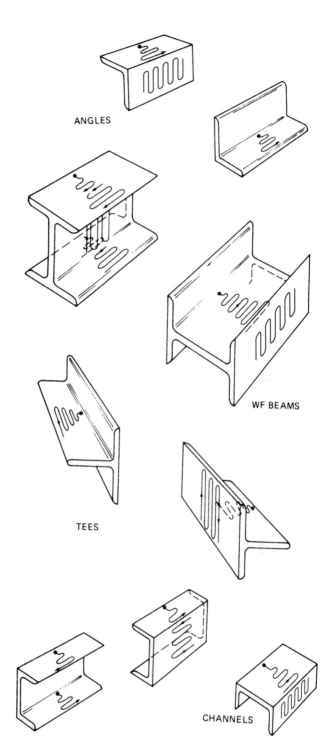

FIGURE 23–22 Application of heat to bend rolled shapes.

Diaphragms, bulkheads, or flat plates in welded assemblies sometimes buckle as a result of weld distortion. One method of reducing buckling is to create round heated areas approximately 3 in. (75 mm) in diameter across the surface of the buckled plate. Each individual heated spot will upset and as it cools will create shrinkage in all directions. By adding enough of these round

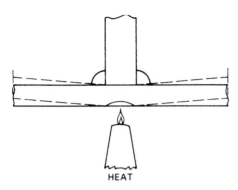

FIGURE 23–23 Application of heat to correct warpage.

heated spots, buckling can be completely eliminated from the flat plate.

The technique of heat flame straightening and bending can be used to salvage members damaged by accident, as shown in Figure 23–24. It has been applied suc-

cessfully to structural work of bridges, large buildings, and offshore drilling platforms. In these cases, special precautions must be exercised. Force is sometimes applied to assist the heat-forming operation. A careful analysis must be made before attempting such jobs, and it must be determined whether the part being straightened is stressed in tension or in compression. When the temperature of the member is increased by flame heating, its strength is greatly decreased. To create shrinkage action, it is beneficial to have the members loaded in compression. This will assist upsetting and will help create the favorable direction of shrinkage to straighten the member. If the member is stressed in tension, compression loading should be added, by means of temporary bracing, to accomplish the heat-straightening operation.

The flame bending system is also used for creating camber in beams. Samples of wide-flange beams formed to large radiuses for roof structures are given in Reference 3. Here wide-flange beams 80 ft (24 m) long and 24 in. (288 mm) wide were formed to curvatures of 135 ft (41.1

FIGURE 23–24 Salvage of I beam after accident.

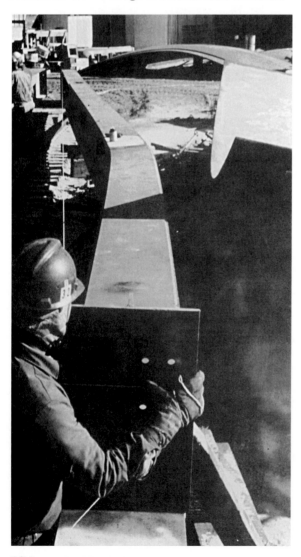

m).These were erected and became the arch of a gymnasium. With skill and experience this type of heat forming can be very precise. Heat forming is expensive compared to forming in large beam bending rolls.

Certain precautions should be taken when heat forming or bending. Although used on low-carbon steels, these procedures should be performed with caution on medium- or high-carbon or quenched and tempered steels because of possible local hardening or effects on earlier heat treatment. The temperature of the heated area must be controlled carefully. Precautions should be used in cooling the heated zones. Forced-air cooling or water cooling will not appreciably increase the amount of distortion. It will merely decrease the time for the shrinkage action to take place.

To avoid metallurgical damage to the steel, it is recommended that 1200° F (649° C) be the maximum temperature utilized. This temperature produces a dark red color on steel in a subdued light. It has been found that reheating the same spot does not affect the steel adversely, provided that it is not heated above 1200° F (649° C). Heat forming can be used on stainless steels. For stainless steels the maximum temperature should be 800° F (427° C). Rapid cooling is encouraged to minimize chromium carbide formation.

For thicker materials, larger torch sizes are required. For extremely heavy materials, multiorifice torches can be used. Propane can be used as well as acetylene for heat forming. It is important to maintain the maximum heat differential between the heated area and the adjacent cool area. This provides for more contraction during the cooling period.

In straightening materials in two dimensions, it must be remembered that the torch side of the part being heated will be heated to a higher temperature than the underside. For this reason, the torch side will normally have greater reaction than the underside. Thorough heating is required, and often it is possible to judge the heating by the color on the underside. With sufficient practice and experience, some rather amazing feats can be done utilizing the oxyfuel gas torch for heat forming and straightening.

23-4 WELD STRESSES AND CRACKING

The subjects of weld stresses, cracking, weld distortion, lamellar tearing, brittle fracture, fatigue cracking, weld design, and weld defects are so interrelated that it is impossible to treat them separately. All of these factors relate to weldment failure, and it is weldment failure that should be eliminated. For clarification and ease of understanding, these factors are discussed in an orderly fashion. In this section the problem of welding stress and its effect on weld cracking is explained.

Residual Stresses

It was pointed out in the previous sections that metals expand and contract the same amount when heated and cooled the same amount, if the metal is not restrained. It was also explained that the heating and cooling that occur in welding are not uniform and there is a temperature difference between the weld and areas adjacent to the weld. The amount of nonuniform heating and the partial restraint creates stresses in the weld area, including the weld metal. If further temperature change occurs, the stresses will be greater than the yield point of the metal. Yielding will occur so that the retained or residual stress will be at the yield point of the metal. This means that yield point stresses within the weldment may occur in all three directions simultaneously. These internal or remaining stresses are known as *residual stress*,[4] the "stress present in a joint, member, or material that is free of external forces or thermal gradients."

When stresses applied to a member exceed the yield strength, the member will yield in a plastic fashion so that the stresses will be reduced to yield point. This is normal in simple structures with stresses occurring in one direction on parts made of ductile materials. Shrinkage stresses due to normal heating and cooling do occur in all three dimensions, however. For example, in a thin flat plate there will be tension stresses at right angles, in other words, in the x and y directions. As the plate becomes thicker the stresses occur in the x, y, and z—or through—directions as well.

When simple stresses are imposed on a thin brittle material, the material will fail in tension in a brittle manner; the fracture will exhibit little or no ductility. In such cases there is no yield point for the material since the yield strength and the ultimate strength are practically the same. The failures that occur without plastic deformation are known as brittle failures. When two or more stresses occur in a ductile material and particularly when three stresses occur in the x, y, and z directions in a thick material, brittle fracture may occur, which is similar to the fracture of a brittle material.

Residual stresses are not peculiar to weldments. They occur in other types of metal structures, such as castings and forgings, and even hotrolled shapes. Several examples of spontaneous fracturing of rolled structural shapes under conditions of zero external load have been reported.[2] In one case an I beam fractured spontaneously under a condition of zero external load when the beam was lying flat on the ground and under normal temperature conditions. The failure occurred through the center of the web splitting the beam its entire length. High residual stresses also occur in castings and forgings as a result of differential cooling. The outer portion of the part cools first and the thicker and the inner portion considerably later. As the parts cool, they contract and pick up strength. The earlier portions that cool go into a compressive load and the latter portions go into a tensile stress mode. In complicated parts the stresses may cause warpage.

Residual stresses are not always detrimental. They may have a beneficial effect on the service life of parts. Normally, the outer fibers of a part are subjected to tensile loading and with residual compression loading there is a tendency to neutralize the stress in the outer fibers. One example of the use of residual stress is in the shrink fit assembly of parts. An example is the cooling of sleeve bearings to insert them into machined holes, then allowing them to expand to their normal dimension. Sleeve bearings are used for heavy slow machinery and are subjected to compressive residual loading. One of the most dramatic uses of shrink-fit assembly for heavy-duty service is the shrink fitting of steel tires on wheels for railroad locomotives. The tire is made of relatively high-carbon steel. These steel tires are heated and then placed on the locomotive wheel and allowed to cool around the wheel and make a very strong mechanical connection. Even with the tremendous loads encountered in use, the residual stresses continue to hold the tire on the wheel. There is no relaxing of the stresses from mechanical working. It seems certain that normal operating loads do not reduce the magnitude of internal residual stresses.

Many investigations have been made and techniques established to measure residual stresses. Residual stresses occur in all arc welds but only in the more simple joints have accurate measurements been made. The most common method of measuring has been to produce weld specimens and then to machine away specific amounts of metal adjacent to the weld and measure the movement that occurs. This is done to produce data showing the magnitude of the residual stresses. Another method is the use of grid marks or data points on the surface of weldments that can be measured in multiple directions. Cuts are made to reduce or release residual stresses from certain parts of the weld joint and the measurements are taken again. The amount of movement relates to the magnitude of the stresses. Another technique is to utilize extremely small strain gauges and gradually mechanically cut the weldment from adjoining portions to determine the change in internal stresses. In this way, experts have been able to establish patterns and determine amounts of stress within parts that were caused by the thermal effect of welds.

Based on these data, it is possible to establish a pattern of residual stresses that occurs in a simple weld.

The residual stresses in an edge weld are shown in Figure 23–25. When the weld metal starts to cool, the upset area attempts to contract but is restrained by cooler metal. This results in the heated zone (upset zone) becoming stressed, in tension. When the weld has cooled to room temperature, the weld metal and the adjacent base metal are under tensile stresses close to the yield strength. There is a portion that is compressive and beyond this another tensile stress area. The two edges are in tensile stress with the center in compressive residual stress, as shown by the figure.

The residual stresses in a butt weld joint made of relatively thin plate are more difficult to analyze. This is because stresses occur in the longitudinal direction and perpendicular to the axis of the weld. The residual stresses within the weld are tensile in the longitudinal direction of the weld and the magnitude is at the yield strength. The base metal adjacent to the weld is also at yield stress parallel to the weld and along most of the length of the weld. When moving away from the weld into the base metal, the residual stresses rapidly fall to zero, and to maintain equilibrium, change to compression (Figure 23–26). The residual stresses in the weld at right angles to the axis of the weld are tensile at the center of the plate and are compressive at the ends. For thicker materials, when the welds are made with many passes, the relationship is different because of the multiple passes of the heat source. Except for a single-pass simple joint, the compressive and tensile residual stresses can only be estimated. To determine at least generally the mode and type of stresses, it is important to remember that as each weld is made it will contract as it solidifies and gain strength as the metal cools. As it contracts it tends to pull, and this creates tensile stresses at and adjacent to the weld. Farther from the weld the metal must remain in equilibrium and therefore compressive stresses occur. In heavier weldments when restraint is involved, movement is not possible and therefore residual stresses are of a higher magnitude. For example, in a multipass single-groove weld the first weld or root pass originally created a tensile stress. The second, third, and fourth passes contract and cause a compressive load in the root pass. As more passes are made until the weld is finished, the top

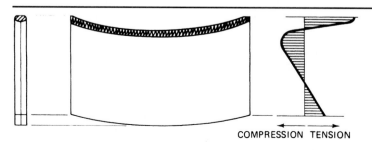

COMPRESSION   TENSION

FIGURE 23–25 Edge-welded joint: residual stress pattern.

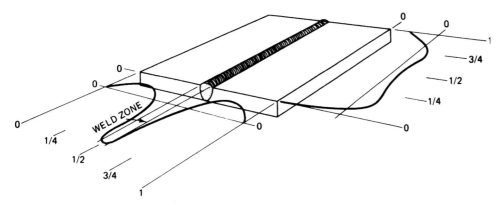

FIGURE 23–26 Butt-welded joint: residual stress pattern.

passes will be in tensile load, the center of the plate in compression, and the root pass will have tensile residual stress again.

Residual stresses can be decreased in several ways. If the weld is stressed beyond its yield strength, plastic deformation will occur and the stresses will be made more uniform but still at the yield point of the metal. This will not eliminate residual stresses but at least will create a more uniform stress pattern. Another way to reduce high or peak residual stresses is by means of loading or stretching the weld by heating adjacent areas, causing them to expand. The heat reduces the yield strength of the weld metal and the expansion will tend to reduce peak residual stresses within the weld. This technique will also make the stress pattern at the weld area more uniform. The more positive way of reducing high residual stresses is by means of the stress relief heat treatment. Here the weldment is uniformly heated to an elevated temperature at which the yield strength of the metal is greatly reduced. The weldment is then allowed to cool slowly and uniformly, so that the temperature differential between parts is minor and the cooling will be uniform and a uniform low-stress pattern will develop within the weldment. High-temperature preheating will also reduce residual stresses since the entire weldment is at a relatively high temperature and will cool more uniformly from that temperature and thus reduce peak residual stresses.

Weld Cracking

Residual stresses contribute to weld cracking. Weld cracking may occur during the manufacturing operation or shortly after the weldment is completed. Cracking may occur for many reasons and may occur years after the weldment is completed. Weld cracking that is a result of the residual stresses and weld cracking that occurs during the manufacturing operation or shortly thereafter will be described here.

Cracks are the most serious defect that occurs in welds. Cracks are not permitted in weldments, particu-

larly those subjected to low-temperature service, impact loading, reversing stresses, or when the failure of the weldment will endanger life. It is important to understand the mechanism of weld cracking so as to avoid cracks in all welds. Weld cracking that occurs during or shortly after the fabrication of the weldment can be classified as hot cracking or cold cracking. In addition, welds may crack in the weld metal or in the base metal adjacent to the weld metal, usually in the heat-affected zone. Welds crack for a variety of reasons:

* Insufficient weld metal cross section to sustain the loads involved
* Insufficient ductility of weld metal to yield under stresses involved
* Under-bead cracking due to hydrogen pickup in a hardenable base material
* Chemistry, i.e., sulfur or phosphorus
* Poor width-to-depth profile

Restraint and residual stresses are among the main reasons for weld cracking during fabrication. Weld restraint can come from several factors. The most important is the rigidity of the weldment. For example, if the weldment is made of thick material and it is of a highly restrained nature, there will be little chance for yielding or movement in the weld joint. If the weld metal does not have sufficient ductility, cracking will occur. Weld metal shrinks as it cools, and if the parts being welded cannot move with respect to one another, and if the weld metal has insufficient ductility, a crack will result. Additionally, movement of welds may impose high loads on other welds and cause them to crack during fabrication. The best solution is to use a more ductile weld metal or make the weld with sufficient cross-sectional area that will have enough strength to withstand cracking tendencies. A weld crack can occur in the root pass when the parts are unable to move.

Another possible factor is rapid cooling of the weld deposit. If the base metal is cold and the weld is relatively small, it will cool extremely rapidly. Shrinkage will occur quickly and cracking can occur. If the parts being joined are preheated, the cooling rate will be slower and cracking can be eliminated. One of the reasons for preheat is to reduce the cooling rate of the weld. In addition, if the base metal is at an elevated temperature, it will have lower yield strength and will not be as restrictive as far as its restraint on the weld.

Another reason for this type of cracking is the alloy or carbon content of the base material. When a weld is made with higher-carbon or high-alloy base material, a small amount of base material is melted and mixed with the electrode to produce the weld metal. The resulting weld metal has higher carbon and alloy content; it may have a higher strength, but it has lower ductility and may not have sufficient ductility for plastic deformation and cracking may occur. This can be eliminated by using a more ductile weld metal or by reducing the cooling rate of the weld and also by reducing the amount of base metal picked up and mixed in with the weld metal.

Another factor involved can be hydrogen pickup in the weld metal and in the heat-affected zone. When using cellulose-covered electrodes or when hydrogen is present because of damp gas, damp flux, hydrocarbon surface materials, and so on, the hydrogen in the arc atmosphere will be absorbed in the molten weld metal and the adjacent base metal. As the metal cools it will reject the hydrogen, and if there is sufficient restraint cracking will occur. This type of cracking can be reduced by increasing preheat, reducing restraint, and eliminating the hydrogen from the arc atmosphere.

As a general rule, to eliminate weld cracking during fabrication it is wise to follow these principles:

* Use ductile weld metal
* Avoid extremely high restraint
* Adjust welding procedures to reduce restraint
* Utilize low-alloy and low-carbon materials
* Reduce the cooling rate by use of preheat
* Utilize low-hydrogen welding processes and filler metals

When cracking is in the heat-affected zone or if cracking is delayed, the culprit is possibly hydrogen pickup in the weld metal and heat-affected zone. Another important factor is the presence of higher-carbon materials or high alloy in the base metal. It is important to utilize low-hydrogen filler metals, to reduce cooling rates, by means of preheat, and to use ductile filler materials.

One solution when welding high-alloy or high-carbon steels is to use the buttering technique. This involves surfacing the weld face of the joint with a weld metal that is much lower in carbon and alloy content than the base metal. The weld is then made between the deposited surfacing material and avoids the carbon and alloy pickup in the weld metal and allows a more ductile weld metal deposit. Care must be used so that the total joint strength is sufficient to meet design requirements. Underbead crack-

ing is greatly reduced by use of low-hydrogen-type processes and filler metals. Reducing the cooling rate will materially reduce the chance of cracks. When welds are too small for the service intended, they will probably crack. This is common in tack welds where a small weld is expected to carry extreme loads. Many specifications list minimum size of fillet welds that can be used to join different thicknesses of steel sections. If these minimum sizes are used, cracking will be eliminated.

23-5 IN-SERVICE CRACKING

The objective is always to design and build weldments that perform adequately in service. The risk of failure of a weldment is relatively small, but it can occur in structures, mechanical parts, tanks, and so on.[5] Welding has sometimes been blamed for the failure, but failures have occurred in riveted and bolted structures in castings, forgings, hotrolled plate and shapes, as well as other types of construction. It is important to make weldments and welded structures as safe against premature failure as we possibly can. There are at least four types of failures that we should be aware of so that proper steps can be taken to avoid them. These are:

* Brittle fracture
* Fatigue fracture
* Lamellar tearing
* Stress corrosion cracking

Each of these failure modes will be covered in detail.

Brittle Fracture

The fracture of metals is a very complex subject and is beyond the scope of this book; however, fracture can be classified into two general categories: *ductile* and *brittle*.

Ductile fracture occurs by deformation of the crystals and slip relative to each other. There is a definite stretching or yielding. There is a reduction of cross-sectional area at the fracture (Figure 23–27). Brittle fracture

FIGURE 23–27 Ductile fracture surface.

occurs by cleavage across individual crystals and the fracture exposes the granular structure; there is little or no stretching or yielding. There is no reduction of area at the fracture (Figure 23–28).

It is possible that a broken surface will display both ductile and brittle fracture over different areas of the surface. This means that the fracture that propagated across the section changed its mode of fracture. There are four factors that should be reviewed when analyzing a fractured surface: (1) growth marking, (2) fracture mode, (3) fracture surface texture and appearance, and (4) amount of yielding or plastic deformation at the fracture surface. Growth markings are one way to identify the type of failure. Fatigue failures are characterized by a fine texture surface with distinct markings produced by erratic growth of the crack as it progresses. The *chevron* or *herringbone* pattern occurs with brittle or impact failures. The apex of the chevron appearing on the fractured surface always points toward the origin of the fracture and is an indicator of the direction of crack propagation. The second factor is the fracture mode. Ductile fractures have a shear mode of crystalline failure. The surface texture is silky or fibrous in appearance. Ductile fractures often appear to have failed in shear as evidenced by all parts of the fracture surface assuming an angle of approximately 45° with respect to the axis of the load stress. Brittle or cleavage fractures have either a granular or a crystalline appearance. There is usually a point of origin of brittle fractures. The chevron pattern will help locate this point. The necking down of the surface of the fractured part is an indication of the amount of plastic deformation. There is little or no deformation for a brittle fracture and usually a considerably necked down area in the case of a ductile fracture.

One characteristic of brittle fracture is that the steel breaks quickly and without warning. The fractures propagate at very high speeds, and the steels fracture at stresses below the yield strength normal for the steel. Mild steels, which show a normal degree of ductility when tested in tension as a test bar, may fail in a brittle manner. In fact, mild steel may exhibit good toughness characteristics at room temperature. Brittle fracture is therefore more similar to the fracture of glass than fracture of normal ductile materials. A combination of conditions must be present simultaneously for brittle fracture to occur. This is reassuring since some of these factors can be eliminated and reduce the possibility of brittle fracture. The following conditions must be present: (1) low temperature, (2) a notch or defect, (3) a relatively high rate of loading, (4) triaxial stresses normally due to thickness or residual stresses, and (5) the microstructure of the metal.

Temperature is an important factor. However, temperature must be considered in conjunction with microstructure of the material and the presence of a notch. Impact testing of steels using a standard notched bar specimen at different temperatures shows a transition from a ductile-type failure to a brittle-type failure based on a lowered temperature. The change from ductile to brittle fracture is known as the transition temperature. Unfortunately, notched specimens are different from large weldments. However, notched specimen results do provide a correlation that is useful in selecting the better material.

A notch can result from faulty workmanship or from improper design that produces an extremely high stress concentration that prohibits yielding in the normal sense. A crack, for example, will not carry stress across it, and the load is transmitted to the end of the crack. It is concentrated at this point, and little or no yielding will occur. Metal adjacent to the end of the crack that does not carry load will not undergo a reduction of area since it is not stressed. It is in effect a restraint that helps set up triaxial stresses at the base of the notch or the end of the crack. Stress levels much higher than normal occur at this point and contribute to starting the fracture.

The rate of loading is the time versus strain rate. The high rate of strain, which is a result of impact or shock loading, does not allow sufficient time for the normal slip process to occur. The material under load behaves elastically, allowing a stress level beyond the normal yield point. When the rate of loading from impact or shock stresses occurs near a notch in heavy thick material, the material at the base of the notch is subjected very suddenly to very high stresses. The effect is rapid failure of the structure. This is what makes brittle fracture so dangerous.

Triaxial stresses are more likely to occur in thick material than in thin material. The z direction acts as a restraint at the base of the notch, and for thicker material the degree of restraint in the through direction is higher. This is why brittle fracture is more likely to occur in thick plates or complex sections than in thinner materials. In addition, thicker plates usually have less mechanical working during manufacture than thinner plates and may have lower ductility in the z axis. The microstructure and chemistry of the material in the center of thick plates have poorer properties than thin plates.

The microstructure of the material is important with respect to the fracture behavior and transition temperature range. Microstructure of a steel depends on the

FIGURE 23–28 Brittle fracture surface.

chemical composition and production processes used in manufacturing it. A steel in the "as-rolled" condition will have a higher transition temperature or lower toughness than the same steel in a normalized condition. Normalizing produces a grain refinement, which provides higher toughness. Unfortunately, fabrication operations on steel such as hot and cold forming, punching, and flame cutting affect the original microstructure. This raises the transition temperature of the steel.

Welding tends to accentuate some of the undesirable characteristics that we wish to avoid. The thermal treatment resulting from welding will tend to reduce the toughness of the steel or possibly to raise its transition temperature in the heat-affected zone. The monolithic structure of a weldment means that more energy is locked up and there is the possibility of residual stresses that may be at yield point levels. Additionally, the monolithic structure causes stresses and strains to be transmitted throughout the entire weldment. Defects in welds can be the nucleus for the notch or crack that will cause fracture initiation.

The problem of brittle fracture can be greatly reduced in weldments by selecting steels that have sufficient toughness at the service temperatures. The transition temperature should be below the service temperature to which the weldment will be subjected. Heat treatment or normalizing or any method of reducing locked-up stresses will reduce the triaxial yield strength stresses within the weldment. Design notches must be eliminated, and notches resulting from poor workmanship must not occur. This requires the elimination of internal cracks within the welds and of unfused root areas either by design or by accident. By closely following these conditions, the possibility of brittle fracture will be eliminated or greatly reduced.

Fatigue Failure

Structures sometimes fail at nominal stresses considerably below the tensile strength of the materials involved. The materials involved were ductile in the normal tensile tests but the failures generally exhibited little or no ductility. Most of these failures developed after the structure had been subjected to a large number of cycles of loading. This type of failure is called a fatigue failure.[6] Fatigue failure is the formation of and development of a crack by repeated or fluctuating loading. When sudden failure occurs, it is because the crack has propagated sufficiently to reduce the load-carrying capacity of the part. Fatigue cracks may exist in weldments, but they will not fail until the load-carrying area is sufficiently reduced. Repeated loading causes progressive enlargement of the fatigue crack. The rate at which the fatigue crack propagates depends upon the type and intensity of stress and a number of other factors involving the design, the rate of loading, type of material, and so on.

The fracture surface of a fatigue failure has a typical characteristic appearance. It is generally a smooth surface and frequently shows concentric rings or areas spreading from the point where the crack initiated. These rings show the propagation of the crack, which might be related to periods of high stress followed by periods of inactivity. The fracture surface also tends to become rougher as the rate of propagation of the crack increases. Figure 23–29 shows the characteristic fatigue failure surface.

Most structures are designed to a permissible static stress based on the yield point of the material in use and the safety factor selected. This is based on statically loaded structures, the stress of which remains relatively constant with respect to time. Many structures, however, are subjected to other than static loads in service. They are loaded by various live loads applied in different ways—for example, cyclic loading in the case of a rotating device or of a bridge carrying varying traffic, or dynamic loads from machinery, or loads based on temperature changes, vibrations, and so on. These changes range from simple cyclic fluctuations to completely random variations. In this type of loading the structure must be designed for dynamic loading and considered with respect to fatigue stresses.

The varying loads involved with fatigue stresses can be categorized. These can be alternating cycles from tension to compression. They can be pulsating loads with pulses from zero load to a maximum tensile load, or from a zero load to a compressive load; or loads can be high and rise higher, either tensile or compressive. It is important to consider the number of times the weldment is subjected to the cyclic loading. For practical purposes, loading is considered in millions of cycles. Fatigue is a cumulative process and its effect is in no way healed during periods of inactivity. Testing machines are used for loading metal specimens to millions of cycles, and the results are plotted on stress versus cycle curves. These show the relation between the stress range and the number of cycles for the particular stress used. Fatigue test specimens are machined and polished, and the results obtained may

FIGURE 23–29 Fatigue failure fracture surface.

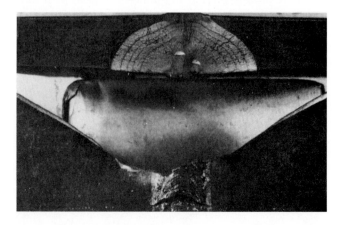

not correlate with actual service life of a weldment. It is important to determine those factors that adversely affect the fatigue life of a weldment.

The possibility of a fatigue failure depends on four factors:

1. Material used
2. Number of loading cycles
3. Stress level and nature of stress variations
4. Total design and design details

This last factor is controllable in the design and manufacture of the weldment. Weld joints can be designed for uniform stress distribution utilizing a full-penetration weld, but in other cases joints may not have full penetration because of an unfused root that prohibits uniform stress distribution. With a full-penetration weld, if the reinforcement is excessive, a portion of the stress will flow through the reinforced area and will not be uniformly distributed. Welds designed for full penetration might not have complete penetration because of workmanship factors such as cracks, slag inclusions, incomplete penetration, and so on, and therefore contain a stress concentration. One reason fatigue failures in welded structures occur is because the welded design can introduce severe stress concentrations. The weld defect, including excessive reinforcement, undercut, or negative reinforcement, will contribute to the stress concentration factor. In addition, a weld forms an integral part of the structure, and when parts are attached by welding they may produce sudden changes of section that contribute to stress concentrations.

Anything that can be done to smooth out the stress flow in the weldment will reduce stress concentrations and make the weldment less subject to fatigue failure. Total design and careful workmanship will greatly eliminate this type of a problem. See Reference 6 for more information on fatigue failure.

Lamellar Tearing

Lamellar tearing has come into prominence because of failures in structural steel work in buildings and in offshore drill rigs and platforms. Lamellar tearing is cracking that occurs beneath welds and is found in rolled steel plate weldments. The tearing always lies within the base metal, usually outside the heat-affected zone and generally parallel to the weld fusion boundary. This type of cracking has been found in corner joints where the shrinkage across the weld tends to open up the steel similar to laminations. In these cases, the lamination-type crack is removed and replaced with weld metal. It is only when welds subject the base metal to tensile loads in the z or through direction of the rolled steel that the problem is encountered. For many years the lower strength of rolled steel in the through direction was recognized, and

the structural code prohibited z-directional tensile loads on steel spacer plates. Figure 23–30 shows how lamellar tearing will come to the surface of the metal. Figure 23–31, showing a T-joint, is a more common type of lamellar tearing, which is much more difficult to find. In this case, the crack does not come to the surface and is under the weld. This type of crack can only be found with ultrasonic testing, or if failure occurs, the section can actually come out and separate from the main piece of metal.

Three conditions must occur to cause lamellar tearing:

1. Strains must develop in the through direction of the plate. These strains are caused by weld metal shrinkage and can be increased by residual stresses and by loading.
2. The weld orientation must be such that the stress acts through the joint across the plate thickness or in the z direction. The fusion line beneath the weld is roughly parallel to the lamellar separation.
3. The material will have poor ductility in the through or z direction.

Lamellar tearing can occur during flame cutting operations and in cold-shearing operations. It is primarily the low strength of the material in the z or through direction that contributes to the problem. It is a stress placed in the z direction that triggers the tearing. The thermal heating and the stresses resulting from weld shrinking create the fracture. Lamellar tearing is not the same as under-bead hydrogen cracking. It can occur soon

FIGURE 23–30   Corner joint.

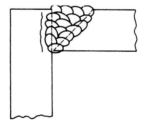

FIGURE 23–31   T-joint.

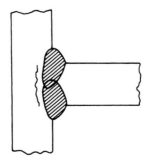

after the weld has been made but on occasion will occur much later. Also, the tears are under the heat-affected zone and it is more apt to happen in thicker materials and in higher-strength materials.

Only a very small percentage of steel plates are susceptible to lamellar tearing. There are only certain plates where the concentration of inclusions is coupled with the unfavorable shape and type that presents the risk of tearing. These conditions rarely occur with the other two factors mentioned previously. In general, the three situations must occur in combination: structural restraint, joint design, and the condition of the steel. The experience gained to date indicates that joint details can be changed to avoid the possibility of lamellar tearing. In the case of T-joints, double-fillet weld joints are less susceptible than full-penetration welds. In addition, balanced welds on both sides of the joint seem to present less risk than large single-sided welds.

Corner joints are common in box columns. Lamellar tearing at the corner joints is readily detected on the exposed edge of the plate. The easy way of overcoming the problem of corner joints is to place the bevel for the joint on the edge of the plate that would have tearing, rather than on the other plate (Figure 23–32).

Butt joints rarely have a problem with lamellar tearing since the shrinkage of the weld does not set up a tensile stress in the through thickness direction of the steel.

Experience indicates that higher heat input welds are less likely to create lamellar tearing. This because of the fewer number of applications of heat and the lesser number of shrinkage cycles involved in making a weld. It is also found that the deposited filler metal with lower yield strength and high ductility reduces the possibility of lamellar tearing. It does not appear that preheat is specifically advantageous. Also, stress-relief heat treatment does not appear to have any beneficial effect. The buttering technique of laying one or more layers of low-strength high-ductility weld metal deposit on the surface of the plate stressed in the z direction will reduce the possibility of lamellar tearing. This is perhaps an extreme solution and should only be used as a last resort. Steel companies are making improvements in steel processing to avoid lamellar tearing. By using the design factors just mentioned, the lamellar tearing problem is reduced. More information regarding lamellar tearing can be found in Reference 7.

FIGURE 23–32 Redesigned corner joint to avoid lamellar tearing.

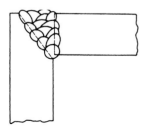

Stress Corrosion Cracking

Stress corrosion cracking and delayed cracking due to hydrogen embrittlement is troublesome when the weldment is subjected to an environment that accentuates this problem. Delayed cracking is caused by hydrogen absorbed in the weld metal at high temperatures. Molten steel will absorb large quantities of hydrogen. As the metal solidifies it rejects the hydrogen that is forced out of solution. The hydrogen coming out of the solution sets up high stresses, and if a sufficient amount is present it will cause cracking in the weld or the heat-affected zone. These cracks develop over a period of time. The concentration of hydrogen and the stresses resulting from it when coupled with residual stresses promote cracking. Cracking will be accelerated if the weldment is subjected to thermal stresses due to repeated heating and cooling.

Stress corrosion cracking in steels is sometimes called *caustic embrittlement*. This type of cracking takes place when hot concentrated caustic solutions are in contact with steel that is stressed in tension to a high level. The high level of tension stresses can be created by loading or by high residual stresses. Stress corrosion cracking will occur if the concentration of the caustic solution in contact with the steel is high and if the stress level in the weldment is sufficiently high. This situation can be reduced by reducing the stress level. Various inhibitors can be added to the solution to reduce the concentration. A practical solution is to reduce the tensile stress of the area in contact with the corrosive solution. On piping this can be done by making weld beads on the outer surface, which causes compressive stresses on the inside-diameter surface.

Another type of cracking is called *graphitization*. This is caused by long service life exposed to thermal cycling—that is, repeated heating and cooling. This may cause a breakdown of carbides in the steel into small areas of graphite and iron. This formation of graphite, in the edge of the heat-affected area, exposed to the thermal cycling causes cracking. It will more often occur in carbon steels deoxidized with aluminum. The addition of molybdenum to the steel tends to restrict graphitization and for this reason carbon molybdenum steels are normally used in high-temperature power plant service. These steels must be welded with filler metals of the same composition.

23-6 WELDING-PAINTING

There are two other welding problems that require explanation and solution. These are welding over painted surfaces and painting of welds.

The practice of welding over paint should be discouraged. In every code or specification it is specifically stated that welding should be done on clean metal. In some industries, however, welds are made on painted base metal.

In the shipbuilding industry and in others, steel, when it is received, is shot blasted, given a coat of prime paint, and then stored out of doors. Painting is done to preserve the steel during storage and also to identify it. In some shipyards a different color paint is used for different classes of steel. When this practice is used, every effort should be made to obtain a prime paint that is compatible with welding. There are three factors involved with the success of the weld when welding over painted surfaces:

* Compatibility of the paint with welding
* Dryness of the paint
* Paint film thickness

Paint compatibility varies according to the composition of the paint. Certain paints contain large amounts of aluminum or titanium dioxide and are usually compatible with welding. Other paints may contain zinc, lead, vinyls, and other hydrocarbons, and are not compatible with welding. The paint supplier should be consulted. Anything that contributes to deoxidizing the weld such as aluminum, silicon, or titanium will be compatible. Anything that is a harmful ingredient such as lead, zinc, and hydrocarbons will be detrimental. The fillet break test can be used to determine compatibility. The surfaces should be painted with the paint under consideration. The normal paint film thickness should be used, and the paint must be dry. The test should be run using the proposed welding procedure on the painted surface. It should be broken and the weld examined. If the weld breaks at the interface of the plate with the paint, it is obvious that the paint is not compatible.

The dryness of the paint should be considered. Many paints employ an oil base that is a hydrocarbon. These paints dry slowly since it takes considerable time for the hydrocarbons to evaporate. If welding is done before the paint is dry, hydrogen will be in the arc atmosphere and can contribute to underbead cracking. It will also cause porosity if there is sufficient oil present. Water-based paints should also be dry prior to welding.

The thickness of the paint film is another factor. Some paints may be compatible if the thickness of the film is in the neighborhood of 3 to 4 mils. If the paint film thicknesses are double that amount, which occurs at an overlap, there is the possibility of weld porosity. Paint films that are to be welded over should be of the minimum thickness possible.

Tests should be run with the maximum film thickness to be used, but dry, to determine which paint has the least harmful effect on the weld deposit.

Cutting painted surfaces should be done with caution. Demolition of old structural steel work that had been painted many, many times with flame- or arc-cutting techniques can create health problems. Cutting through many layers of lead paint will cause an abnormally high lead concentration in the immediate area and will require extra ventilation or personnel protection.

Painting over welds is also a problem. The success of the paint film depends on its adherence to the base metal and the weld. This is influenced by surface deposits left on the weld. Paint failure occurs when the weld and the immediate area are not properly cleaned prior to painting. Deterioration of the paint over the weld also seems to be dependent upon the amount of spatter present. Spatter adjacent to the weld leads to rusting of the base material. The paint does not completely adhere to spatter and some spatter falls off in time, leaving bare spots in the paint coating.

The success of the paint job can be ensured by preweld and postweld treatment. Preweld treatment found most effective is to use antispatter compounds, as well as cleaning of the weld area, before welding. The antispatter compound extends the paint life because of the reduction of spatter. The antispatter compound must be compatible with the paint to be used.

Postweld treatment for ensuring paint film success consists of mechanical and chemical cleaning. Mechanical cleaning can consist of hand chipping and wire brushing or power brushing, or sand or grit blasting. Sand or grit blasting is the most effective mechanical cleaning method. If the weldment is furnace stress relieved and then grit blasted, it is well prepared for painting. When sand or grit blasting cannot be used, power wire brushing is the next most effective method. In addition, a chemical bath washing is recommended. Slag coverings on weld deposits must be thoroughly removed from the surface of the weld and adjacent area. Different types of coatings create more or less problems in their removal and also with respect to paint adherence. Weld slag of many electrodes is alkaline in nature and must be neutralized to avoid chemical reactions with the paint. The weld should be scrubbed with water, which will usually remove the residual coating slag and smoke film. If a small amount of phosphoric acid up to a 5% solution is used, it will be more effective in neutralizing and removing the slag. However, if this is used it should be followed by a water rinse. If water only is used, it is advisable to add small amounts of phosphate or chromate inhibitors to avoid rusting.

It has been found that the method of applying paint is not an important factor. The type of paint employed must be suitable for coating metals and proper for the service intended.

Successful paint jobs over welds can be obtained by:

1. Removing weld spatter
2. Mechanically cleaning the weld and adjacent area
3. Washing the weld area with a neutralizing bath and rinse

The American Welding Society is preparing a standard practice document on this subject.

QUESTIONS

23-1. What causes arc blow?

23-2. Why is ac welding less likely to have arc blow?

23-3. What is the best solution for arc blow when parts are magnetized?

23-4. If a metal piece is not restrained, will it come back to its original dimension after heating?

23-5. Is heating uniform in metal during welding? Does this cause distortion?

23-6. Is a weldment restrained? Does this cause warpage?

23-7. What is plastic deformation? How is it affected by heat?

23-8. How can angular distortion be reduced?

23-9. Explain the reason for the special order of making weld joints on a tank.

23-10. What is the stress level of residual stresses?

23-11. How do residual stresses build up and change in a multipass groove weld?

23-12. How can residual stresses be reduced?

23-13. What is the danger of brittle fracture?

23-14. How is the type of failure determined?

23-15. What is the characteristic of a fatigue fracture? What four factors are involved?

23-16. Describe lamellar tearing.

23-17. What causes stress corrosion cracking?

23-18. How can stress corrosion cracking be overcome in piping?

23-19. What is the hazard of flame cutting old structures that are covered with many layers of paint?

23-20. What is required to obtain a good paint job over welds?

REFERENCES

1. C. H. Jennings and A. B . White, "Magnetic Arc Blow," *Welding Journal* (October 1941): page?

2. R. R. Avent, "Engineered Heat Straightening Comes of Age," *Modern Steel Construction* 35, no. 2 (February 1995): 32–39.

3. J. R. Stitt, "Distortion Control during Welding of Large Structures," SAE-ASME (April 1964): 844B.

4. F. Campus, "Effects of Residual Stresses on the Behavior of Structures," in *Residual Stresses in Metals and Metal Construction,* W. R. Osgood, ed. (New York: Reinhold, 1954).

5. M. E. Shank, "Control of Steel Construction to Avoid Brittle Failure," Welding Research Council, New York, 1957.

6. W. H. Munse, "Fatigue of Welded Steel Structures," Welding Research Council, New York, 1964.

7. J. C. M. Farrar and R. E. Dolby, "Lamellar Tearing in Welded Steel Fabrication," The Welding Institute, Cambridge, England, 1966.

24

FAILURE ANALYSIS, REPAIR WELDING, AND SURFACING

OUTLINE

24-1 WELD FAILURE ANALYSIS

Failures of large engineered welded structures are very rare. Catastrophic failures of major structures and results of investigations are usually reported. These reports are useful since they provide information that is helpful in avoiding future similar problems.

It is important to make an objective study of failure of parts of structures to determine the cause. This is done by investigating the service life, the conditions that led up to the failure, and the actual mode of the failure. The study should utilize every bit of information available, investigate all factors, and evaluate this information to arrive at the reason for the failure.

Failure investigation will uncover facts that will lead to changes in design, manufacturing, or operating practice that will eliminate similar failures in the future.

Each failure and subsequent investigation will lead to changes that will assure a more reliable product. The investigators must use extreme care and should present the facts in a logical order. The following four areas should be investigated to determine the cause of the failure and the interplay of factors involved.

1. *Initial observation.* Investigators should make a detailed study of the actual component that failed. This should be made at the failure site as quickly as possible. Photographs should be taken, in color, of all parts, structures, failure surfaces, fracture texture appearance, final location of component debris, and all other factors. Witnesses to the failure should all be interviewed and all information should be recorded.

2. *Background data.* Investigators should gather all information concerning specifications, drawings, component design, fabrication methods, welding procedures, weld schedules, repairs in and during manufacturing and in service, maintenance, and service use. Particular attention should be given to environmental details, including operating temperatures, service loads, overloads, cyclic loading, abuse, and so on.

3. *Laboratory studies.* Investigators should make laboratory tests to verify that the materials in the failed parts have the specified composition, mechanical properties, dimensions, and so on. Micrographic

studies should be made. Each failed part should be thoroughly investigated to determine what bits of information it can add to the total picture. Fracture surfaces can be extremely important. Original drawings should be obtained and marked showing failure locations. This should be coupled to design stress data used in designing the product. Other defects in the structure that are apparent, even though they might not have contributed to the failure, should be noted and investigated.

4. *Failure assumptions.* Investigators should list not only all positive facts and evidence that may have contributed to the failure, but also all negative responses that may be learned about the failure. It is sometimes as important to know what specific things did not happen or what evidence did not appear to help determine what happened. These data should be tabulated. The actual failure should be synthesized to include all available evidence. This might lead to the need for collecting additional data or asking more questions.

The true cause of failure will emerge from this study. Assumptions must be challenged by every bit of information available until it stands up as the one and only plausible cause for the failure. Failure cause can usually be classified as one of the following:

1. Failure due to faulty design or misapplication of material
2. Failure due to improper processing or improper workmanship
3. Failure due to deterioration during service

The following is a summary of these three situations.

Failure due to faulty design or misapplication of the material involves failure due to inadequate stress analysis or a mistake in design such as incorrect calculations on the basis of static loading instead of dynamic or fatigue loading. Ductile failure can be caused by a load too great for the strength of the material. Brittle fracture may occur from stress risers inherent in the design, or the wrong material may have been specified for the part, or the weld joint is improper.

Failures can be due to faulty processing or poor workmanship. The quality of the weld may be substandard. Failures can be attributed to poor fabrication practice such as the elimination of a root opening, which may cause incomplete penetration. There is also the possibility that incorrect filler metal was used.

One of the major problems is the problem of overload. Normal wear and abuse to the equipment may have resulted in reducing sections to the degree that they no longer can support the load. Corrosion due to environmental conditions and accentuated by stress concentrations will contribute to failure. There may be other situations such as poor maintenance, poor repair techniques, and accidental conditions beyond the user's control. Or, the product might be exposed to an environment for which it was not designed.

Failure Analysis Examples

Thorough investigations and reports are always made of major failures, especially if there is a loss of life. The Federal Aviation Agency always investigates aircraft accidents and publishes a report. The Office of Pipeline Safety determines the cause of pipeline failures, and the U.S. Department of Transportation investigates ship, railroad, and highway catastrophes and serious bridge failures. Engineering magazines summarize and publish these reports. Investigation reports are always interesting and informative. A study of these reports will reduce the likelihood of future failures of similar types. The following is a review of some studies that relate to welding.

The report of the inquiry into the accidents of the Comet airplane failures[1] was extremely informative. This study represented an outstanding example of investigation and experimentation necessary to track down the cause of failure. It provided knowledge regarding the fatigue problem of aircraft structures and changed the design concepts for all large aircraft with respect to fatigue loading.

Another important failure analysis report is the "Brittle Failure in Carbon Plate Steel Structures Other Than Ships."[2] This investigation provided insight into the failure mode of large weld structures. It emphasized the fact that weldments are monolithic structures, and that a welded structure is one piece of metal that may have designed into it internal and external notches, stress risers, and crack starters. It also emphasized that stresses are distributed throughout the weldment because of its monolithic structure, whether the designer had this in mind or not. It helped clarify the design concepts of large welded structures.

Figure 24–1 shows the result of a weld failure, a fire at a large refinery. It was apparently caused when corrosion attacked a part, allowing a leak. The part was of the wrong material, and the result was a catastrophic multi-million dollar explosion and fire that required many months to repair.

The "Report of the Royal Commission into the Failure of Kings Bridge"[3] in Melbourne, Australia, is a classic. The fractured beam is shown in Figure 24–2. The report indicated that the cause of failure was due to the combined effects of local stress risers due to poor design of cover plates, improper welding procedure for high-strength steel, and the failure to control low-hydrogen electrodes, which resulted in many cracks in the welds.

Another famous failure is shown in Figure 24–3, which was a large pressure vessel made of $5\frac{3}{4}$-in. (150-mm)-thick material with an inside diameter of $5\frac{1}{2}$ ft (1.67 m) and nearly 60 ft (18.3 m) long.[4] The failure occurred

FIGURE 24–1 Refinery fire.

FIGURE 24–2 Fractured beam.

FIGURE 24–3 Failed pressure vessel.

during the hydrostatic proof test at a temperature of 50° F (10° C) and a pressure of 5000 psi with a designed maximum test pressure of 5100 psi. The fracture was of a brittle type. There were two fracture initiation sites located in the heat-affected zone of the submerged arc weld be-tween the rolled plate and the end forging. The steel involved banding and segregation, and failure resulted from high residual stresses around the weld and the fact the base metal had low toughness.

The failure of the Kansas City Hyatt Regency Hotel pedestrian walkways resulted in a large loss of life.[5] The investigation revealed that the design had been changed during construction. This design change using two rods instead of one essentially doubled the stress on the hangers and this, in addition to the extra heavy human load, caused the catastrophic failure.

Welded ships have been plagued with problems due to at-sea failures. Figure 24–4 shows a 32,000-ton vessel in two pieces in the North Sea.[6] The break occurred nearly amidships and ran approximately in line with the transverse bulkhead at the end of the longitudinal stiffeners, which were interrupted at the bulkhead. This design created discontinuities and stress risers, which caused the failure in heavy seas.

Another example of a failure on a machinery part is shown in Figure 24–5. A diesel power shovel with a 3-cubic-yard dipper was digging taconite at an open pit mine in the Mesabi Range in northern Minnesota. On a mid-January day when the temperature was −20° F (−29° C), the boom failed without warning. The point of the boom broke off completely about midway between the shipper shaft bearing and the point of the boom (Figure 24–5). This was a brittle failure since there was no deformation or necking down of the material thickness. The working stresses on this portion of the boom are compressive. When the machine rotates, there is a bending moment on the boom point section. The severest load is caused by the pull of the dipper as it goes through the heavy ore. During this period the boom shakes and vibrates violently. The calculated stresses when combined were within the allowable limit.

The boom is a rectangular cross section about 16 in. (424 mm) by 20 in. (526 mm) at the location of the

FIGURE 24–4 Ship broken in two.

fracture. It was made of low-carbon mild steel whose notch bar impact properties were poor. The boom is made of two formed half-sections joined longitudinally by submerged arc welds. A $\frac{3}{8}$-in. (9.5-mm) by 1-in. (25.4-mm) backup bar was used. The point section was made as a subassembly and butt welded to the main section with shielded metal arc process using E6012 electrodes. A diaphragm was located very near this butt welded joint. The boom was not stress relieved.

The fracture was initiated at the fillet weld attaching the diaphragm to the box section and at a point of poor root fusion in the transverse butt weld. It was concluded that the stresses were concentrated at this area because of the abrupt end of the backing bars, the fillet weld joining the diaphragm to the box section, and the unfused root of the butt transverse weld. The base metal and the E6012 weld metal both have relatively poor low-temperature impact resistance. It was concluded that the low ambient temperature, the shock loading, stress concentration, and poor low-temperature toughness of the steel caused the failure.

Early in World War II, welded merchant vessels built in the United States experienced difficulties in the form of fractures that could not be explained. Many fractures occurred with explosive suddenness and exhibited a quality of brittleness that was not associated with the behavior of normally ductile materials used. Immediate steps were taken to investigate and solve the problem. A

board was appointed to make a complete investigation and report the facts. "The Design and Methods of Construction of Welded Steel Merchant Vessels "[7] is the comprehensive report produced. The investigation took over three years, starting in April 1943 and concluding in July 1946. The investigation involved a total of 4,694 ships, of which 970 sustained some type of structural casualty. Eight ships were lost at sea, four others broke in two but were not lost. Twenty-six lives were lost.

The study involved design studies of each type of vessel involved, loading and ballasting conditions, convoy routes with accompanying sea and weather conditions, and extensive laboratory research aimed at studying fabrication and materials used in construction. The results of this effort to eliminate the occurrence of hull fractures were successful. The number of fractures decreased sharply after remedial measures were taken based on the findings of the board of inquiry.

The following is a summary of the findings:

1. The highest incidence of fracture occurred under the combination of low temperatures and heavy seas.
2. The age of the vessel had no appreciable influence on the tendency to fracture.
3. The loading and ballasting system did not create abnormal bending moments.
4. There was no marked correlation between the incidence of fracture on the ships and the construction practices of shipyards. It was found, however, that ships constructed in yards utilizing subaverage construction practices showed a higher-than-average incidence of fractures.
5. The bulk of failures were reported on Liberty ships, with only relatively fewer serious fractures on the Victory ships. (Victory ships were designed with fewer structural notches.)
6. The steel supplied for ship construction complied with the applicable specification for ship steel.
7. Locked-in stresses in the decks of completed vessels were not appreciably reduced in service.
8. Welding sequence in general had no effect upon the magnitude of residual welding stresses.
9. Every fracture examined started at a geometrical discontinuity or notch resulting from unsuitable design or poor workmanship.
10. There is a large variation in the notch sensitivity of steels used in ship construction. Steel removed from fractured vessels showed high notch sensitivity.

The investigators researched failures of riveted ships previously reported. They found that when a crack starts in a riveted ship, it generally progresses only to the first break, that is, a riveted seam. There it awaits reload-

(a)

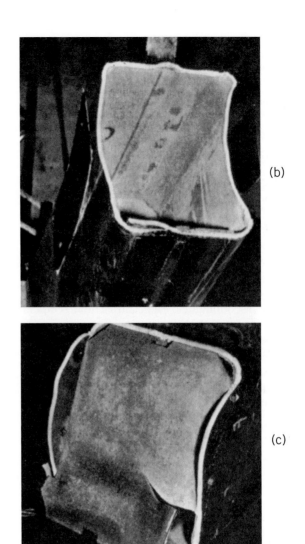

(b)

(c)

FIGURE 24–5   Power shovel boom failure.

ing to a stress that will give it a fresh start. In a welded structure, the crack will continue to propagate as long as sufficient energy is available.

A particularly bewildering phenomenon was the appearance and nature of the fracture. It had been generally believed that ship steel would deform elastically when loaded within the elastic limit, and that if it were loaded beyond that point, plastic flow would take place and a permanent deformation would result, as evidenced by a reduction in thickness. It was previously believed that if the load were increased sufficiently, material would fail only after considerable elongation. It was found on examining the ship fractures that the fractured surface appeared crystalline, rather than silky as it would in a ductile failure. The break was square and the line of separation normal to the surface of the plate. Very little ductility occurred as indicated by practically zero reduction in the thickness at the fracture. This type of fracture is termed a cleavage fracture, denoting a separation of the surface of the crystal lattice rather than sliding action along slip planes.

In the investigation, the designs were recalculated. These calculations showed that the hull girder strength was ample and that the margin of strength in the structure exceeded that required by the design standards. The monolithic character of the welded ship produced specific areas with high stress concentrations and severe restraint. This condition did not exist in riveted ships. The danger of high concentration at points of structural discontinuities in the welded ship is further aggravated by the welds present at such points. The welding produces a complex metallurgical condition, which is sometimes aggravated by discontinuities in the form of weld defects. A ship hull requires numerous openings, machinery foundations, deck houses, and so on. At each of these points the section changes abruptly, and when under a bending load, a stress concentration occurs. Stress concentrations of dangerous magnitude exist at structural discontinuities including hatch corners, shear strake cutouts, and where foundations and deck houses are welded to the deck. Investigation found that most of the serious fractures started at hatch corners and many started in the shear strake cutout for the accommodation ladder. This indicated that insufficient attention was paid to discontinuities or notches, whether they were large or small. At the time the investigation started the mechanism of metal fracture was not well understood. The incidence of serious failures of large welded steel structures, both during construction and during service, indicated the need for a better understanding of the fundamental factors affecting steel performance. Lack of reliable information had led designers to overdesign in the interest of safety, which in some cases enhanced the possibility of failure. Impact tests of steel samples taken from vessels that suffered fractures indicated that the steel was notch sensitive—that is, its ability to absorb energy in the notched condition at low temperatures was low. The investigators explored the behavior of ship steel in the welded and unwelded condition and under the influence of multiaxial stress in the presence of discontinuities, especially at low temperatures. These studies found that *notch sensitivity* was an important factor in the occurrence of *brittle* failures.

The welding subcommittee made a survey of shipyards and found varying degrees of quality workmanship. The analysis did not indicate a marked correlation between the incidence of fractures in welded ships in shipyard construction practice; however, ships produced in yards utilizing below-average practices showed a higher-than-average incidence of failure. It was concluded that high-quality workmanship is important in welded ships. They felt that welds should be identified as to who made the weld and that there should be an improvement of welder training. They also felt that welding sequences and procedures must be prepared and followed. They concluded that evidence was found to indicate that residual welding stresses were important in causing the fractures.

The board concluded that the fractures in welded ships were caused by notches and by steel that was notch sensitive at operating temperatures.

Figure 24-6 shows the Liberty ship and details the abnormal frequency of fractures. Figure 24-7 shows the tragic results of the S.S. *Schenectady* breaking in two at the outfitting dock prior to being placed in service.

The epidemic of fractures was greatly reduced through the combined effect of corrective measures taken on the structure of the ship during construction and after completion, improvements in new design, and improved construction practices in the shipyards. The

FIGURE 24–6 Liberty ship: location of fractures.

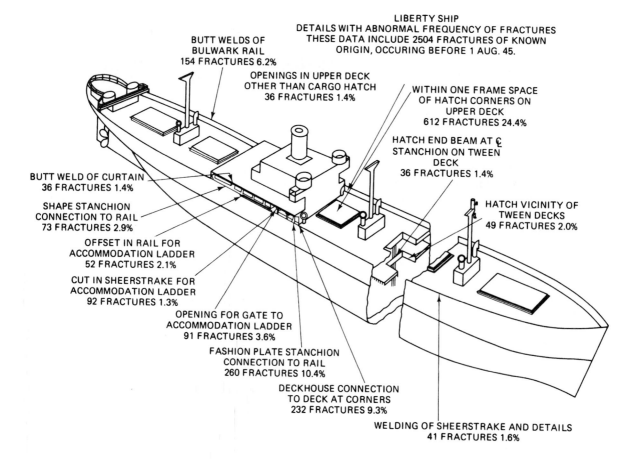

LIBERTY SHIP
DETAILS WITH ABNORMAL FREQUENCY OF FRACTURES
THESE DATA INCLUDE 2504 FRACTURES OF KNOWN
ORIGIN, OCCURING BEFORE 1 AUG. 45.

BUTT WELDS OF
BULWARK RAIL
154 FRACTURES 6.2%

OPENINGS IN UPPER DECK
OTHER THAN CARGO HATCH
36 FRACTURES 1.4%

WITHIN ONE FRAME SPACE
OF HATCH CORNERS ON
UPPER DECK
612 FRACTURES 24.4%

HATCH END BEAM AT ₵
STANCHION ON TWEEN
DECK
36 FRACTURES 1.4%

HATCH VICINITY OF
TWEEN DECKS
49 FRACTURES 2.0%

BUTT WELD OF CURTAIN
36 FRACTURES 1.4%

SHAPE STANCHION
CONNECTION TO RAIL
73 FRACTURES 2.9%

OFFSET IN RAIL FOR
ACCOMMODATION LADDER
52 FRACTURES 2.1%

CUT IN SHEERSTRAKE FOR
ACCOMMODATION LADDER
92 FRACTURES 1.3%

OPENING FOR GATE TO
ACCOMMODATION LADDER
91 FRACTURES 3.6%

FASHION PLATE STANCHION
CONNECTION TO RAIL
260 FRACTURES 10.4%

DECKHOUSE CONNECTION
TO DECK AT CORNERS
232 FRACTURES 9.3%

WELDING OF SHEERSTRAKE AND DETAILS
41 FRACTURES 1.6%

FIGURE 24–7 S.S. *Schenectady* after splitting in two at dock.

first remedial step taken was to eliminate stress concentration of cargo hatch openings. This was done by modifying the corners to provide rounded corners rather than square corners. *Crack arrestors* were installed.

This exhaustive examination of failures led designers to appreciate the fact that weldments are monolithic in character; that anything welded onto a structure will carry part of the load whether intended or not; and that abrupt changes in section, because of adding a deck house or removing a portion of the deck for a hatch opening, create stress concentration. Under normal loading, if the steel at the point of stress concentration is notch sensitive at the service temperature, failure can result.

It was reported by the board that the results of the investigation vindicated the all-welded ship. The statistics show that the percentage of vessels sustaining serious fracture is small. With proper design, high-quality workmanship, and steel with good notch sensitivity at operating temperatures, a satisfactory all-welded ship can be obtained. This was reinforced by the fact that the Victory ships, which were designed to reduce stress concentration, sustained fewer and less serious fractures. Novels have even been written about failure investigations. One of the most famous was about an airplane crash.[8]

A thorough analysis as just described is not required in most situations. This is due to experience gained in analyzing jobs, making repairs, and then checking on the service life of the repaired part. As experience is gained, shortcuts can be taken. The reason for an investigation is to establish the cause of the failure.

24-2 DEVELOPING A REWORK PROCEDURE

The success of a repair or surfacing job depends on the thought and preparation prior to doing the work. Metal surfaces deteriorate from corrosion, abrasion, erosion, and so on, until the part is no longer serviceable. Repairs are required, and welding is the quickest and most reliable method for returning the part to service. Weld repair is often the most economical solution, especially when the "out of service" time depends on obtaining a new part. Replacement parts are often not immediately available, and if they are, they are very expensive. The economies of weld repairing are very favorable. Some weld repair jobs may only take a few hours, and some complex structures may require weeks for preparation and welding.

Once the decision has been made to make a weld repair, it is necessary to review why the part failed or wore out. This relates to the type of repair job since it indicates if reinforcing is required. Reasons for the part to fail or wear out are determined by the failure analysis.

If the part failed because of an accident or an overload, it can be returned to service with the weld repair made to bring it back to its original strength. This applies also if the part has been abused or misapplied. It may be necessary to reinforce the part so that it will stand temporary overloads, misapplication, or abuse.

In the case of poor workmanship, the weld repair should rework the poor workmanship responsible for the failure. The part would be returned to its original condition. If failure is due to poor design, design changes are required and reinforcement may be added. In a case of incorrect material, it is assumed that the material was of a lower strength level, which contributed to the failure. In this case reinforcing would be required. If the repair is to alter the part, it is necessary that the modification is approved by an experienced designer. Drawings and additional parts will be required. It is important that the repaired or reworked part meets or exceeds the strength of the original part.

An important factor is the type of repair work required. It can be a standardized, repetitive job such as the resurfacing of dipper teeth of an excavator or the rebuilding of track shoes of a crawler tractor. These are parts that routinely wear and must be repaired by welding on a scheduled basis. Or it can be a weld repair because of a breakdown, which is a one-of-a-kind job and often an emergency. An example would be the broken power shovel boom mentioned previously. Emergency repair work must be analyzed quickly and a procedure established as soon as possible. Large machines, when down, create delays in an entire operation, such as a mine, and cost extremely large amounts of money while they are out of operation. This also applies to oil-drilling operations, offshore platforms, steel rolling mills, electric power generators, and other production equipment. These are the types of repair work where "return to service" is most important and there is no time to obtain a replacement part.

Investigate before Repairing

There are certain situations and certain types of equipment where repair welding should not be done or may be done only with prior approvals. It may be uneconomical to repair some parts. An example is the weld repair of a cast iron part that is repeatedly heated and cooled. Weld repairs on cast iron parts subjected to repeated heating and cooling may not provide adequate service life. The problem is that cast iron parts such as machinery friction brakes, furnace sections, and so on, failed originally from this kind of service. The metallurgical changes involved with the weld may not be able to withstand repetitive heating and cooling cycles. Such repairs should only be made on an emergency basis until replacement parts are available.

If a failure occurs when equipment is new and within the manufacturer's warranty, it is necessary to contact the manufacturer of the equipment. The manufacturer must be made aware of the problem and the repair that is planned. Failure to do this will cancel the machine's warranty.

Aircraft may be repaired by welding but only under stringent controls. The welder doing repair welding on aircraft should be qualified in accordance with MIL-T-5021D or latest "Tests; Aircraft and Missile Welding Operators Certification," on the type of metal being welded, using the process for which the welder is qualified and on the category of parts involved. Furthermore, the welder should be certified in accordance with requirements of the Department of Transportation Federal Aviation Administration. The FAA issues two documents, "Acceptable Methods, Techniques and Practices—Aircraft Inspection and Repair"[9] and the "Air Frame and Power Plant Mechanics Air Frame Handbook."[10] Both provide precautionary information techniques, practices, and methods that may be used for repair welding. Techniques, practices, and methods other than those prescribed may be used provided that they are acceptable to the FAA administrator. Extensive damage must not be weld repaired on items such as engine mounts, landing gear, or fuselage components unless the method of repair is specifically approved by an authorized representative of the FAA or the repair is accomplished in accordance with the FAA-approved instructions furnished by the aircraft manufacturer. The reason for these regulations is that many parts are made of high-strength material, and the strength is obtained by postweld heat treatment. Certain parts are not to be welded if the damage is beyond a specific amount. Consult with Federal Aviation Administration authorities or the manufacturer of the aircraft. For safety reasons welding must not be done on aircraft inside hangers, unless all fuel is completely removed and the aircraft is made inert.

Certain types of containers and transportation equipment must not be weld repaired or may be welded only with special permission and approval. These include railroad locomotive and railroad car wheels, high-alloy high-strength truck frames, and compressed gas cylinders.

Most power-generating machinery, including turbines, generators, and large engines, are covered by casualty insurance. Weld repair can be done only with the prior approval of the welding procedure by the insurance company. Approval may not be granted. In many cases weld repairs can be made, but it is necessary to develop a written procedure that must be approved in writing by the insurance company representative.

Alterations of bridges, large steel frame buildings, and ships may be done only with special authorization. The alteration work must be designed and approved. The welders must be qualified according to the code used and the work must be inspected. Written welding procedures are required.

Repairs to boilers and pressure vessels require special attention. The ASME codes are for new construction. Repair, maintenance, or alterations are a jurisdictional responsibility. The "National Board Inspection Code"[11] has been adopted by most jurisdictional authorities in North America (cities, states, and provinces) to provide rules for inspection, maintenance, repairs, and alterations to boilers and pressure vessels. To maintain reliability and insurability and ASME stamping on boilers and pressure vessels, the rules of National Board of Inspection Code must be met and an authorized inspector must be involved during repairs and alterations.

The National Board Inspection Code defines basic and routine repairs and alteration. A repair is the work necessary to restore a boiler or pressure vessel to its original or safe and satisfactory operation condition. Alterations are any changes that affect the operation of the boiler or pressure vessel from its original design.

A written procedure is required for doing either repair work or alteration. All work must be performed in a manner to maintain the original integrity of the ASME code vessel. All welding procedures and welders must be qualified in accordance with ASME Section IX.

There are companies that specialize in the repair and alterations of boilers and pressure vessels. These companies have authorization from the appropriate jurisdictional authority, or they possess a current ASME Code Symbol Stamp covering the scope of the repair work, or they possess a current National Board "R" Repair Code Symbol Stamp. In any case, when repair welding is performed, the authorized inspector must be involved. The repair firm will contact the jurisdictional authority, the insurer, and the owner of the boiler or pressure vessel to assure that the method and extent of repair or alteration is given proper prior approval. This is required to ensure the proper continued use of the boiler or pressure vessel that is repaired or altered.

Alterations to boilers and pressure vessels require special attention. A statement must be obtained from an ASME certificate holder with the appropriate stamp, certifying that the redesigned portion of the alteration is correct. This certification of the design is made on the R-1 Alteration Form and must be accepted and signed off by the ASME certificate holder's authorized inspector. This authorizes the repair company to proceed with the alteration. The alteration must include the involvement of the authorized inspector. In any case, the repair company must be an ASME certificate holder with the appropriate stamp covering the alterations or must have a National Board "R" stamp following the rules published in the National Board Inspection Code.

Rework Procedure

A written repair procedure is required for all but the simplest jobs. The composition of the material being welded must be known. If this is not possible, particularly in the field, look for clues as to the metal involved. Refer to Section 15-3 for help. As a final resort, obtain a laboratory analysis of the metal. Filings or a piece of the metal must be sent to a laboratory for analysis.

The normal method of selecting the welding process based on the type and thickness of the metal, the position of welding, and so on, should be followed. This aids selection of filler metal, which involves matching composition and properties to provide weld metal that will withstand the service involved.

In surfacing, the desired surface characteristics depend on the service to which it will be exposed. Surfaces can be rebuilt many times without reducing the strength of the part, and the service life will be greatly extended.

The repair procedure should be complete. It should include the process, filler metal, and the technique to be used in making welds. The format utilized by Section IX of the pressure vessel code can be utilized for repair procedures. The procedure, for complex jobs, should be qualified to determine that it will provide a repair weld that is equal in strength to the original part. This is done in the same manner as qualifying a welding procedure by a code. The repair procedure should be approved by the proper authority. This could be the inspector of a casualty insurance company, the inspector of the National Board, the representative of the manufacturer of the original equipment, or a governmental representative, such as the state piping or boiler inspectors. For work on ships, the ship's rating agency should be consulted. In every case consider the specification or code under which the product was built. In case of extensive repairs on critical items, make sure that the procedure is practical and will provide the necessary strength for the service intended. Assume that written procedures and approvals are required prior to making any repair welds. Only after the procedure has been approved by all necessary parties is it time to make the repair weld.

24-3 MAKING THE REPAIR WELD

Once all factors have been reviewed and analyzed, the decision should be made to repair by welding. The analysis indicates the cause of the failure. The material composition is known. A repair welding procedure has been prepared and has been approved. The weld repair may be as simple as the removal and replacing of a body panel in an automobile or as complex as the repair of a rolling mill frame (Figure 24-8). In any case, there are three separate phases to the job:

1. Preparation for welding
2. Repair welding
3. Postweld operation

The amount of detail that must be considered depends on the complexity of the job.

Preparation for Welding

A large number of factors should be considered and decisions made before starting to weld.

FIGURE 24–8 Complex weld repair: rolling mill frame.

1. *Safety.* The repair welding location must be surveyed and all safety considerations satisfied. This includes posting the area, required by certain regulations; removing all combustible materials from the area; draining fuel tanks of construction equipment, aircraft, boats, trucks, and so on; and removing or inerting fuel pipelines, tanks, blind compartments, and so on. If electrical cables are involved, they should be removed or made inactive. Other precautions include the elimination of toxic materials such as thick coats of lead paint, plastic coverings, and so on. If heights are involved, proper scaffolding with safety devices should be used. If welding is enclosed, preparations for proper ventilation and personnel removal should be made. If these hazards cannot all be removed, special safeguards should be established such as fire watch, wetting down, or protecting combustible wooden floors. Traditionally, repair welding creates more safety problems than production welding; extra special precautions must be taken.

2. *Cleaning.* The immediate work area must be clean. This includes removal of dirt, grease, oil, rust, paint, plastic coverings, and so on, from the surface of the parts being welded. The method of cleaning depends on the material to be removed and the location of the workpiece. For most construction and production equipment, steam cleaning is recommended. When this is not possible, solvent cleaning can be used. Blast cleaning with abrasives is also used. For small parts, pickling or solvent dip cleaning can be used. Power tool cleaning with brushes, grinding wheels, disk grinding, and so on, can be employed. The time spent cleaning a weld repair area will pay off in the long run.

3. *Disassembly.* Except for the most simple repair jobs, disassembly may be required. This applies to lubrication lines, instrument tubing, wiring, and so on. Sometimes it is necessary to disassemble major components. Experience with similar jobs is important, since it is expensive to disassemble and remove machinery when not required.

4. *Protection of adjacent machinery and machined surfaces.* When repair welding is done on machinery, parts that are not removed should be protected from weld spatter, flame cutting sparks, and foreign material generated by the repair process. Sheet metal baffles are used to protect adjacent machinery. For machined surfaces, cloth can be employed. Secure protective material with wire, clamps, tape, or temporary bracing.

5. *Bracing and clamping.* On complex repair jobs bracing or clamping may be required. This is because of the heavy weight of parts or the fact that

loads may be exerted on the part being weld repaired. If main structural members are to be cut, the load must be carried by temporary braces. The braces can be temporarily welded to the structure being repaired. The braces can be strong backs or pieces welded on both sides of the repair area to maintain alignment of the part while the repair weld is being made. If strong backs or bracing are used, they should be located so that they do not interfere with the repair welding.

6. *Layout repair work.* In most repair jobs it is necessary to remove metal so that a full-penetration weld can be made. A layout should be made to show the metal that is to be removed by cutting or gouging. The minimum amount of metal should be removed to obtain a full-penetration weld. The layout should be selected so that welding can be balanced, if possible, and that the bulk of the welding can be made from the more comfortable welding position. The root opening should be specified, and if the welding can be done on the back side it should be gouged for full-penetration welding. If the back side cannot be reached for welding, backing straps should be employed. The groove angle should be the minimum possible for use but should be sufficient so that the welder has room to manipulate the arc at the root.

7. *Preheating.* Preheating and flame cutting or gouging is part of the preparation for welding but can be considered part of the welding operation. When flame cutting or gouging is required, preheat temperature should be the same as for welding. It is wise to preheat prior to cutting or gouging to at least one-half the temperature that will be used for the repair welding operation. Preheating should be based on the mass of the metal involved. If the mass is great, heating should be slow so that thorough heating occurs. Surface heating is not acceptable. Preheating can be done by any of the normal methods; however, the slower processes would be advantageous. The equipment for preheating and sufficient fuel should be available prior to starting.

8. *Cutting and gouging.* The oxygen fuel gas cutting torch is most often used for this application. Special gouging tips should be selected based on the geometry of the joint preparation. It is possible, by closely watching the cut surface, to find and follow cracks during the flame-gouging operation. The edges of the cracks will show since they become slightly hotter. The air carbon arc gouging process is also widely used for weld repair preparation. Proper power sources and carbons should be selected for the volume of metal to be removed. For some metals the torch or carbon arc might not be appropriate and in these cases mechanical chipping and grinding may be employed. Chipping is preferable

to grinding, and power tools should be employed. The resulting groove should be smooth without reentrant gouges or notches.

9. *Grinding and cleaning.* The resulting surfaces may not be as smooth as desired and may include burned areas, oxide, and so on. Grind the surfaces to clean bright metal prior to starting to weld. For critical work or where there is a suspicion of additional cracks, it is wise to inspect the surface by magnetic particle examination to make sure that all defects have been removed.

The nine steps above should be followed for weld preparation. Some of these may be eliminated but they should all be considered to properly prepare the joint for welding.

Repair Welding

Successful repair welding involves following a logical sequence to make sure that all factors are considered and adequately provided for.

1. *Welding procedure.* The welding procedure must be available for the use of the welders. It must include the process to be used, the specific filler metals, the preheat required, and any other specific information concerning the welding joint technique. This procedure must be understood by all concerned. It should be written.

2. *Welding equipment.* All welding equipment should be available so that there will be no delays. Standby equipment might also be required. This should include sufficient electrode holders, grinders, wire feeders, cables, and so on. Sufficient power must be available at the site to run all of the equipment required. In addition, if the job runs around the clock, provisions for lighting and for personnel comforts such as wind breaks or covers should be provided.

3. *Materials.* Sufficient materials must be available for the entire job. This includes the filler metals, insert pieces, reinforcing pieces, and so on; fuel for preheat and interpass temperature; shielding gases; and fuel for engine-powered welding machines. If inspection equipment is required for intermediate checking, this equipment must also be available.

4. *Alignment markers.* Prior to making the weld alignment, markers are sometimes used. These can be center punch marks made across the joint in various locations. With precise measuring equipment such marks are useful in maintaining dimensional control and alignment during the welding operation. This is more important when repairing mechanical equipment than for structural applications.

5. *Welding sequences.* The welding sequence should be well described in the welding procedure and can include block welding, backstep sequence welding, wandering sequence welding, and peening. These techniques are useful to reduce distortion and to help maintain alignment and dimensional control. By making precision measurements from the checkpoints, the technique can be varied to maintain alignment.

6. *Personnel.* There should be sufficient welders assigned to the job so that it can be completed quickly. Welders should be rotated so that they will be able to produce good quality welds. Welders should not work excessive hours on precise jobs. Many jobs require three shifts of welders when the need to return to service is paramount. Supervision must be around the clock.

7. *Safety.* Safety cannot be compromised throughout the entire operation. For example, ventilation must be provided when fuel gases are used for preheating.

8. *Weld quality.* The quality of the weld should be continually checked. The final weld should be smooth, there should be no notches, and reinforcing, if used, should fair smoothly into the existing structure. If necessary, grinding should be done to maintain smooth flow contours.

Postweld Operation

After the weld has been completed, it should be allowed to slow cool. It should not be exposed to winds or drafts, nor should the machinery loads be placed on the repaired part until the temperature has returned to the normal ambient temperature.

1. *Inspection.* The finished weld should be inspected for smoothness and quality. This can include nondestructive examination. The repair weld should be of high quality since it is replacing original metal of high quality.

2. *Cleanup operation.* This includes the removal of strong-backs and the smooth grinding of the points where they were attached. It also involves the removal of other bracing and protective cover, and so on. In addition, all weld stubs, weld spatter, weld slag, and other residue should be removed from the repair area to make it cleaner than it was originally. Grinding dust is particularly troublesome, and every effort should be made to remove it entirely since it is abrasive and can get into working joints, bearings, and so on, and create future problems.

3. *Repainting.* After the weld and adjacent repair area have been cleaned, they should be repainted and other areas should be regreased in preparation for the reoperation of the machinery.

4. *Reassembly.* The pieces of machinery that were taken away are returned for reassembly. Particular attention should be paid to the fit of machinery. If necessary, remachining or redressing should be done to assure proper fit. All other items, such as lubrication lines, cables, conduits, and so on, should be reassembled; once this has been done, the machinery should be tested prior to operation.

There is a gratifying sense of accomplishment of a successful repair job. Most complex repair jobs are done under much pressure due to the short time out of service.

Weld repair is an extremely technical subject and must be properly handled. On the other hand, it is a time saver and an economic advantage. It requires greater-than-normal skills for successful jobs, but will pay off in the long run.

Remote Welding

Remote welding is automatic welding since the welding is done without the presence of a human welding operator. The operator may be a short distance from the welding operation, but could be far away.

Remote welding is often done for maintenance operations where each weld is different. Remote welding is performed where humans cannot be present because of a hostile atmosphere, such as a high level of radioactivity. A hostile atmosphere requires special protective equipment for the welder, which seriously hampers visibility and flexibility. Radiation greatly reduces the time that a person can work and allows time only for setting up the equipment.

Pipe welds are made remotely in radioactive atmospheres. The joints are properly prepared and aligned. The pipe welding head is attached to the joint. The power source and operator control pendant are remote from the welding operation and shielded from radioactivity. Remote welds are made in pipe and tubing, as they would be made with the equipment under normal conditions. To observe the weld being made, a TV camera is mounted on the welding head and monitored remotely. Adjustments can be made if required. Figure 24–9 shows equipment of this type in operation.

The more difficult jobs are the one-of-a-kind type, where standard equipment cannot be used. The ingenuity of the personnel and the use of gadgetry and special welding equipment is a must. More and more of this type of work is being done as repair welding in radioactive hot areas increases.

Remote cutting using thermal processes is also done. The plasma arc cutting process is used for remote preparation in a radioactive atmosphere. Another requirement is the necessity to cut under water for certain maintenance operations. This is used when the area is flooded.

FIGURE 24–9 TV viewing remote welding.

24-4 REBUILDING AND OVERLAY WELDING

Rebuilding and overlaying with weld metal or thermal spray metal are both considered surfacing operations. Surfacing is the deposition of metal on a base metal to obtain desired dimensions or properties. Overlay is considered to be a weld or spray metal deposit that has specific properties sometimes unlike the original surface. Rebuilding is used on worn shafts, on parts that were machined undersize, and so on. Overlay surfacing is used to return the part to original dimensions but with the deposited metal having specific properties to reduce wear, erosion, corrosion, and so on.

Rebuilding and overlay, or the all-embracing term *surfacing,* can be done by many of the welding processes and by the thermal spraying processes. In some situations thermal spray should be selected. The thermal spray processes do not introduce as much heat into the work as the welding processes. It is possible to thermal spray materials that cannot be deposited with welding, such as ceramic coatings.

The selection of the welding process and the welding procedure and technique is as important as the selection of the deposit alloy. The factors discussed previously should be considered; however, there are additional factors. Whether the job is to be done in the field or in the shop has a definite bearing on process selection. In addition, if it cannot be moved and must be welded in place, the use of some processes is prohibited. The properties and analysis of the base metal has an important bearing, as does the cost factor.

Shielded metal arc welding is commonly used for hardfacing. Figure 24-10 shows the shielded metal arc process being used to surface a dredge cutter head.

Submerged arc welding is used for plant operations. It is used for repeat applications when the same part is surfaced on a routine basis. Rollers, track shoes, and drums are commonly hardfaced with submerged arc welding. Figure 24-11 shows a power shovel ring gear being resurfaced. Over 400 lb of weld metal was deposited in this operation.

Flux cored arc welding is popular and is not restricted to the flat position. Figure 24-12 shows the process being used in the field to build up a dipper lip.

Gas metal arc welding is also used, but there is not as wide a selection of solid electrode wires available for hard surfacing applications. It is used for buildup application, either semiautomatic or fully automatic. Figure 24-13 shows the process being used on a small shaft.

FIGURE 24-10 Surfacing a dredge cutter head with SMAW.

FIGURE 24–11 Surfacing a ring gear with submerged arc.

FIGURE 24–12 Surfacing a dipper lip with flux cored arc welding.

FIGURE 24–13 Surfacing a shaft with gas metal arc.

Gas tungsten arc welding is used for many smaller applications. It is more expensive and is widely used for nonferrous metals.

Plasma arc welding is also used in much the same manner as gas tungsten arc welding.

The electroslag welding process is used for special applications such as for rebuilding crusher hammers. These can be rebuilt with special fixturing and done quite rapidly.

Oxyacetylene welding is used for certain applications such as bronze bearing overlay.

In general, the process is selected based on normal process selection factors and modified by the comments above. Once the process is selected, the next requirement is the selection of the deposited metal to provide the necessary properties.

Salvaging of Shafts

Rebuilding round shafting is an important application of surfacing. Worn or mismachined shafts can be salvaged by surfacing with many different processes. The same procedure can be used to provide an overlay with specific properties to improve its service life. Shafts exposed to corrosive atmospheres can be surfaced with stainless steel to improve service performance.

Arc welding processes are preferred to thermal spraying since they produce a weld. This is important since splines and keyways can be cut in weld deposits without harming the overlay.

Close adherence to the procedure is important when the diameter of the part is small. Figure 24–13 shows the GMAW process surfacing a relatively small diameter shaft. An old lathe or similar device can be used to provide rotation. Precautions should be taken to avoid welding current from passing through roller bearings. A rotary connection should be used. The welding gun can be mounted in the lathe toolholder. It should be offset approximately one-fourth of the diameter of the part being welded or at the 1:30 or 2-o'clock position. The offset is always toward the direction of rotation, and the electrode should point to the centerline of the rotating part. The longitudinal travel of the gun should be fast enough so that each weld bead blends smoothly into the preceding one. If the offset distance is insufficient, the molten metal will not solidify before it reaches the top or 12-o'clock position and may form a high crown bead. If the offset distance is too large, the molten weld metal may run down the shaft ahead of the arc. The angle of the gun can be adjusted to point slightly ahead up to 5°, to improve shielding gas coverage. Experience will assist in setting these exact distances based on different diameter parts.

The welding procedure must include the welding travel speed. On rotating parts this is surface inches per minute. The correct speed will ensure a smooth weld deposit that will require a minimum amount of machining. Figure 24–14 provides the surface speed based on the diameter of the part and the rotational speed. To determine the revolutions needed for the desired travel speed, draw a straight line between the diameter of the shaft and the desired travel speed. The desired travel speed is the 20 in./min. Where the line intersects revolutions per minute, read the rpm required. For example, with a 2-in.-diameter shaft, draw the line through 20 of the surface travel speed to the intersection with revolutions per minute, which would be 3.3. The 20-in./min surface speed is proper when using the 0.0035-in.-diameter electrode wire in the range 120 to 150 A. For bigger jobs, larger electrodes can be used and the procedure variables will be different.

24-5 SURFACING FOR WEAR RESISTANCE

Wear

Wear is the result of impact, erosion, metal-to-metal contact, abrasion, oxidation, and corrosion, or a combination of these. The effects of wear can be repaired by welding. Surfacing with special welding filler metals is used to replace worn metal. Hardfacing applies a coating to reduce

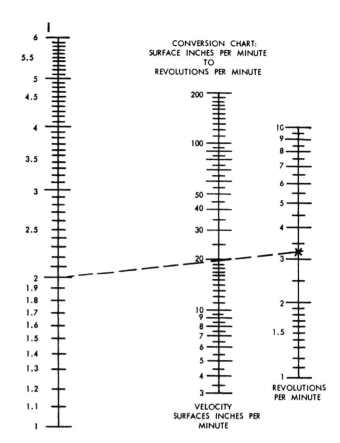

FIGURE 24–14 Conversion chart: in./min to rpm.

wear. It can be used to extend the usable life of wear parts. It will save money since the replacement of worn parts is costly, particularly when the downtime and repair labor is considered.

To select the proper hardfacing alloy, it is necessary to understand the wear that caused the metal wear. The types of wear can be categorized as follows:

Impact wear is the striking of one object against another. It is a battering, pounding type of wear that breaks, splits, or deforms metal surfaces. A good example is the impact encountered by a tamper.

Abrasion is the wearing away of surfaces by rubbing, grinding, or friction. It usually occurs when a hard material is used on a softer material. It is caused by the scouring action of sand, gravel, slag, earth, or gritty material on machinery.

Erosion is the wearing away of materials by the abrasive action of a liquid. This type of action gouges out metal surfaces. This is also caused by steam and slurries that carry abrasive materials. Pump parts are subject to this type of wear.

Compression is a deformation type of wear caused by heavy static loads or by slowly increasing pressure on metal surfaces. Compression wear causes metal to move and lose its dimensional accuracy.

Cavitation wear results from turbulent flow of liquids, which carry suspended abrasive particles.

Metal-to-metal wear is a seizing and galling type of wear that rips and tears out portions of metal surfaces. It is often caused by metal parts seizing together because of lack of lubrication. Frictional heat helps create this type of wear.

Corrosion wear is the gradual eating away of metal surfaces by the effects of the atmosphere, acids, gases, alkalis, and so on. This type of wear creates pits and perforations and may eventually dissolve metal parts.

Oxidation is a type of wear indicated by the flaking off or crumbling of metal surfaces. This takes place when unprotected metal is exposed to a combination of heat, air, and moisture. Rust is an example of oxidation.

Thermal shock is a problem indicated by cracking or splintering, which is caused by repeated rapid heating and cooling. Although not exactly a wear problem, it is a deterioration problem and is considered here.

Many of these types of wear occur in combination. It is wise to look for a combination of factors that create the wear problem in order to best determine the type of hardfacing material to apply. This is done by studying the worn part, the job it does, how it works with other parts of the equipment, and the environment in which it works.

Hardfacing Alloy Selection

There is no standardized method of classifying and specifying the different surfacing weld rods and electrodes. The American Welding Society has issued two specifications, A5.13, "Specifications for Surfacing Weld Rods and Electrodes," and A5.21, "Specifications for Composite Surfacing Weld Rods and Electrodes." There is some overlap between these two specifications and with A5.6 and A5.7, "Copper and Copper Alloy Welding Electrodes and Rods." Many of the hardfacing electrodes commercially available are not covered by these specifications. Filler metal suppliers provide data establishing classes of service and have categorized their products within these classes. Suppliers also provide complete information for using their products for various applications and for different industries, such as quarrying, steel mills, foundries, and so on. This information is extremely valuable and should be consulted.

A good system of classification has been established by the American Society for Metals Committee on Hardfacing. These data are provided in the *Metals Handbook*,[12] Volume 6, "Hardfacing by Arc Welding." This system has five major groups classed according to alloy content, with subdivisions based on the major alloying elements. These data have been abridged and simplified by L. F. Spencer, who added the AWS specifications where they apply[13] (Figure 24-15). Most of these alloys are available as solid bare filler rod in straightened lengths or

in coils or covered electrodes. Some of the materials are available as powder for special applications. Following is a brief description of the five major groups, what they contain as alloys, and where they are recommended.

Group 1 is the low-alloy steels that, with few exceptions, contain chromium as the principal alloying element. The subgroup 1A has from 2% to 6% alloy including carbon. These alloys are often used as buildup materials under higher-alloy hardfacing materials. The group 1B is similar except that they have a higher-alloy content ranging from 6% to 12%. Several alloys in the group have higher carbon content exceeding 2%, and include several alloy cast irons. The alloys of group 1 have the greatest impact resistance of all hardfacing alloys except the austenitic manganese steels (group 2D) and have better wear resistance than low or medium carbon steels. They are the least expensive of the alloy surfacing materials and are extremely popular. They are machinable. They have a high compressive strength and fair resistance to erosion and scratch abrasion.

Group 2 contains higher alloyed steels. Group 2A has chromium (Cr) as the chief alloying element, with a total alloy content of 12% to 25%. Many of these alloys also contain molybdenum. Those with over 1.75% carbon are medium-alloy cast irons. Group 2B has molybdenum (Mo) as the principal alloying element but many also contain appreciable amounts of chromium. The hardfacing alloys of groups 2A and 2B are more wear resistant, less shock resistant, and more expensive than those in group 1.

Groups 2A and 2B are quite strong and have relatively high compressive strengths. They are effective for rebuilding severely worn parts and are used for buildup prior to using higher-alloy facing materials. They provide high impact resistance and good abrasion resistance at normal temperatures.

Group 2C contains tungsten and modified high-speed tool steels. They are excellent choices at service temperatures up to 1100° F (593° C) and when good resistance coupled with toughness is required. They are not considered as good high abrasion-resistant types but are resistant to hot abrasion up to 1100° F and exhibit good metal-to-metal wear at elevated temperatures.

Group 2D are the austenitic manganese steels, which contain either nickel or molybdenum as stabilizers. The alloys in group 2D are highly stock resistant but have limited wear resistance unless subjected to work hardening. The total alloy content ranges from 12% to 25%. This group is excellent for metal-to-metal wear and impact when the deposit is work hardened in use. The as-welded deposit hardness is low, from 170 to 230 BHN, but will work harden to 450 to 550 BHN. The deposit may deform under battering but it will not crack. The deposit should not be heated to above 500° F (260° C), which would cause embrittlement.

Group 3 contains higher-alloyed compositions ranging from 25% to 50% total alloy. They are all high-

FIGURE 24–15 Typical composition of hardfacing materials.

Group	AWS Class	Composition (%)								
		C	Mn	Si	Cr	Ni	Mo	W	V	Co
1A	—	0.10	1.3	0.75	2.0		1.0			
		0.25	0.8	0.50	0.4	0.70	0.6			
		0.20	0.25	0.40	3.25		1.0			
		0.35	1.2	0.10	4.0		0.5			
		0.55	1.0		1.8			2.25		
1B	—	0.70	0.9	0.3	6.5		0.8			
		0.70	1.0	0.7	3.0		4.0			
		0.70	1.2	1.0	5.0		0.5			
		2.2	0.4	0.5	5.0				5.0	
		3.0	0.7	1.0	3.0					
		3.4			4.8				2.4	
2A	—	0.6	0.4	0.7	7.0		0.9	3.5	1.0	
		0.5	2.0	1.0	9.0		1.7			
		3.0	2.5	1.0	12.0		1.5			
		1.0	4.0		12.0					
		3.8			15.0	2.0	8.0			
		3.0			16.0	6.0	8.0			
2B		0.80			4.0		9.0		1.5	
		1.0			0.9		15.3			
		1.4			4.2		9.7			
		3.5			5.0		4.0			
	RFeMoC	3.6					10.0			
2C	EFe5A	0.85	0.5	0.7	4.0		5.0	6.0	2.0	
	EFe5B	0.70	0.5	0.7	4.0		8.0	2.0	1.0	
	EFe5C	0.40	0.5	0.7	4.0		8.0	2.0	1.0	
2D	EFeMnA	0.80	16.0	0.3	0.4	4.0				
	EFeMnB	0.80	14.0	0.8	0.5		1.0			
		1.2	12.0	0.6		4.8				
3A		2.7	1.0	1.0	26.0					
		3.0			18.0		16.0		1.5	6.0
		3.7	1.0	0.8	28.0					3.0
	EFeCrA1	4.0	6.0	1.7	29.0					
	EFeCrA2	4.0	1.0	1.3	29.0	3.5				
3B	—	2.5			25.0	12.0	8.0			
		4.0		4.5	16.0				0.5	
		4.0		1.0	17.0	6.0			0.5	
		3.4	4.5	0.8	30.0					
3C	—	2.3			16.0	6.0				20.0
		3.6	0.6	1.6	15.5		3.0			23.5
4A	ECoCrA	1.0	2.0	0.5	29.0	3.0	1.0	4.0		Remainder
	ECoCrB	1.3	2.0	1.0	29.0	3.0	1.0	8.0		Remainder
		2.5			32.0		17.0			Remainder
	ECoCrC	2.5	2.0	1.0	30.0	3.0	1.0	12.0		Remainder
		0.3			27.0	2.7	5.0			Remainder
4B	ENiCrA	0.35		3.5	12.0	Rem.				
	ENiCrB	0.40		4.0	15.0	Rem.				
		0.10			16.0	Rem.	17.0	4.5		
	ENiCrC	0.75		4.5	15.0	Rem.				1.0
4C	—	2.5			29.0	39.0		14.0		8.0
		2.5			25.0	15.0	8.0			25.0
		3.7			16.0	4.0	6.5			20.0
5	EWC	Tungsten carbide particles (38% to 60+% encased in matrix) most wear resistant of materials.								
	RWC									

chromium alloys and some contain nickel, molybdenum, or both. The carbon can range from slightly under 2% to over 4%. The alloys in this group exhibit better impact, erosion resistance, metal-to-metal wear, and shock resistance than the previous groups. The 3B grouping will withstand elevated temperatures of up to 1000° F (538° C). The 3C group is high in cobalt, which improves high-temperature properties. The group 3 alloys are more expensive than groups 1 and 2.

The compositions within group 4 are nonferrous alloys—either cobalt base or nickel base with total content of nonferrous metals from 50 to 99%.

The group 4A alloys are the high-cobalt-based alloys with a high percentage of chromium. These alloys are used exclusively for applications subjected to a combination of heat, corrosion, erosion, and oxidation. They are considered the most versatile of the hardfacing materials. The alloys with higher carbon are used for applications requiring high hardness and abrasion resistance but when impact is not as important. These alloys are excellent when service temperatures are above 1200° F (649° C). They resist oxidation temperatures of up to 1800° F (982° C).

The group 4B alloys are the nickel-base alloys, which contain relatively high percentages of chromium. This group of alloys is excellent for metal-to-metal resistance, exhibits good scratch abrasion resistance, and corrosion resistance. They will retain hardness to 1000° F (538° C). The alloys with higher carbon content provide higher hardnesses but are more difficult to machine and provide for less toughness. These alloys show good oxidation resistance up to 1750° F (954° C).

The group 4C alloys are the chrome-nickel cobalt alloys and all are recommended for elevated temperatures. The high-nickel alloy has excellent resistance to hot impact, abrasion, and corrosion, and moderate resistance to wear and deformation at elevated temperatures. The medium-nickel alloy has high-temperature wear resistance and impact resistance. It also provides resistance to erosion, corrosion, and oxidation. The low-nickel alloy is used for moderate high temperatures and provides good edge strength, corrosion resistance, and moderate strength.

The group 5 alloys provide a tungsten carbide weld deposit. This deposit consists of tungsten carbide particles distributed in a metal matrix. The matrix metals may be iron, carbon steel, nickel-base alloys, cobalt-based alloys, and copper-base alloys. The tungsten carbide particles are crushed to mesh sizes varying from 8 to 10 down to 100 and have excellent resistance to abrasion and corrosion, and moderate resistance to impact. The matrix material determines the resistance to corrosion and high-temperature resistance. The finish of the deposit depends on the tungsten carbide particle size. The finer the particles the smoother the finish. The deposits are not machinable and are very difficult to grind.

There is another class of surfacing materials used to provide corrosion or oxidation resistance surfaces that will be covered in Section 24-6. Figure 24–16 shows the hard-surfacing alloy classes just mentioned and provides properties and the welding processes that can be used. The method of finishing and the application for the different alloy classes is shown. This is generalized information and is presented as a starting point for making the final selection. A welding procedure should be established for the successful hard surfacing or overlaying operation. The procedure should relate to the particular part being surfaced. It should specify the welding process, the method of application, and the preweld operations. The welding procedure should give the preheat and interpass temperature and any special techniques that should be employed, such as the pattern of hard surfacing, whether beading or weaving, the interface between adjacent beads, and finally, any postwelding operations such as peening and the method of cooling. When a properly developed procedure is followed, the service life of the job will be predictable.

In many cases two separate materials may be required—the buildup alloy, which is used when the part is to be reclaimed or is excessively worn, and the hardfacing alloy. In general, over three layers of hardfacing alloys are not deposited. The hard-surfacing alloys are considerably more expensive than buildup alloys. The hard surfacing should be replaced when the hardfacing alloy is worn away. When deposit exceeds three layers, other problems may be encountered such as cracking.

A major consideration is the location of finished surface with respect to the worn surface. In many cases, the first layer of surfacing may have sufficient dilution of base metal so that it is unsuitable for the desired service. In this case, the worn surface should be further removed so that there is sufficient room for two layers of surfacing metal, which will provide a better service life. Where the part is to be remachined after surfacing, the machining surface should not be at the interface between weld surfacing metal and the base metal. Premachining may be required. This is important when the base metal is of hardenable material.

Preheating, interpass temperature, and cooling of the part being surfaced are important. The factors that apply to welding the base metal in normal fabrication should be followed when overlaying. Preheating is used to minimize distortion, to avoid thermal shock, and to prevent surface cracking. The preheat temperature depends on the carbon and the alloy content of the base metal and the mass of the part being surfaced. A soak-type preheat should be used. If it is extremely complex in shape, preheat should be increased. The preheat temperature should be maintained throughout the entire welding operation and should then be allowed to slow cool. The base metal composition must be known to determine preheat temperatures. Welding should be done in the flat position if at all possible.

FIGURE 24–16 Hardfacing alloys versus surface conditions.

Class	Cold Abrasion	Impact	Erosion	Metal-to Metal Wear	Corrosion	Hot Abrasion	Rockwell Hardness, Layers, and Process	Finishing	Applications
	Characteristic[a]								
1A	F	Ex	No	No	No	No	RC 30–40; two or three layers; FCAW, SAW, SMAW	Machinable, with carbide tools	Used as built up for hardfacing or by itself; fair-to-good strength, toughness, and moderate abrasion resistance
1B	F	G	F	F	No	No	RC 50–57; two layers; SAW, SMAW, FCAW	use carbide tools or grind	Hardfacing or heavy-duty built-up application involving heavy impact
2A	F–G	F–G	F	F	No	No	RC 50–55; two layers; SMAW, SAW	Use grinding practices	High-strength, low-crack-sensitivity deposits for severe abrasion and compression, moderate to heavy impact, good resistance to erosion, and mild corrosion.
2B	Ex	No	Ex	F	No	No	RC 65; gas	Use grinding practices	Excellent for cold abrasive wear, also for metal-to-metal wear and mild impact.
2C	F	F	G	F	F	Ex	RC 55–60; anneal to RC 30	Machines if softened	Hardness up to 1100° F (593° C); good wear resistance and toughness use for tools
2D	F–G	Ex	F–G	G–Ex	No	F–G	gas SMAW	Use grinding practices	Work hardness—built-up and hard-facing—austenitic manganese, steel
3A	Ex	F–G	G–Ex	F–G	No	No	RC 47–62; SMAW, gas	Machine if softened or grind	Holds hardness up to 800° F (427° C) or 1150° F (621° C) depending on alloy, good wear, and oxidation resistance; moderate impact and severe abrasion
3B	G–Ex	F–G	F–G	G–Ex	F–G	F	RC 35–65; SMAW, gas	Use carbide tool or grind	Excellent abrasion resistance and metal-to-metal wear at moderate temperatures
3C	Ex	Ex	F	Ex	G	F	RC 45–55; SMAW, gas	Machinable with carbide tools	Good edge strength—use for tools
4A	V	V	G	G	G	G	RC 35–50; SMAW, gas	Use carbide tools or grind	Very good for metal-to-metal wear; good for hot and cold abrasion—depends on specific alloy; impact varies
4B	V	V	G	G	G–E	G	RC 30–40; SMAW, gas	Machinable with carbide tools	Best for corrosion or erosion; also good for metal-to-metal wear; other properties depend on specific alloy
4C	Ex	G	F	F	G	Ex	SMAW, gas	Use carbide	Excellent wear resistance; good high-temperature properties
5	Ex	F	G	No	V	V	RC 90–95 SMAW, gas	Use grinding practices	Severe abrasion; limited to 1200° F (649° C); moderate resistance to impact

[a]Ex, excellent; G, good; F; fair; No, do not use; V, varies

The thickness of the surfacing deposit is extremely important. If the deposit is too thick, problems can be encountered. Hardfacing alloys should be restricted to two layers. The first will include dilution from the base metal, but the second layer should provide the properties expected. Some types of alloys can be used in three layers. Consult the manufacturer's data for the particular product involved.

The technique for buildup should be to within $\frac{1}{4}$ in. (6 mm) of the final surface. This will then allow two layers of surfacing material to bring the part to final dimension. A weaving technique is recommended instead of stringer bead welding. The pass thickness or layer thickness should not exceed $\frac{3}{16}$ in. (5 mm). The adjacent beads must fair into the previous bead to provide as smooth a surface as possible. There is controversy concerning the pattern of welds that should be made when applying the surfacing deposits. In general, the direction of welding should not be transverse to the load on the part. This can create stress concentrations and may affect service life. In certain types of metal, peening is recommended but this is based on the metal. The manufacturer's instructions should be followed.

Hardfacing by welding is an excellent method of reclaiming parts and will save time and money. It is becoming popular for original equipment manufacturers to hardface wear parts on new equipment to provide better service life.

Pickup and Dilution

In any surfacing operation, be it for corrosion resistance, wear resistance, or whatever, the analysis of the exposed deposit is of major concern. The composition of the deposit is designed to withstand the service environment and be compatible with the base metal. The selection of the overlay material must take into account the metallurgical interrelationship between the deposit and the base metal. The resulting deposit analysis depends on the dilution and pickup, which relate to the welding process and the welding procedure.

Pickup and *dilution* are terms used in analyzing surfacing of overlay deposits. Their meanings are different and should not be confused. *Pickup* is defined as an addition or increase in any alloying element in a weld deposit by virtue of melting and incorporating some of the base metal. The mixture of base metal with filler metal forms the deposit. The first bead or layer in contact with the base metal has the greatest amount of pickup or of alloying elements, depending on penetration. The second layer has much less pickup and is more the analysis of the filler metal. A greater depth of penetration has more effect on the alloy content of the deposit weld. On the other hand, the pickup from the base metal might reduce the alloy content of the deposit weld and reduce its ability to function as desired.

Dilution is "the change in composition of a welding filler metal caused by the admixture of the base metal, or previous weld metal, in the weld deposit. It is measured by the percentage of base metal or previous weld metal in the weld bead." Dilution is commonly considered a reduction in alloy content of the weld deposit by virtue of melting and incorporating melted base metal of lower alloy content. It will reduce the effectiveness of the hardfacing or corrosion resistance of the overlay. It is measured as a percent of base metal that enters into the weld deposit. If the composition of the filler metal is known and the composition of base metal or previous metal is known, then by knowing the percent dilution the weld deposit composition can be calculated. Dilution or alloy pickup can change the composition of the weld deposit considerably. The amount of mixing or dilution is measured as the percentage of the entire cross section of weld metal, as shown by a photomicrograph of the cross section that shows the outlines of the deposit and the base metal. The cross section is polished and etched to reveal the boundary of the deposit and the junction between the weld and base material. This is shown by Figure 24–17. The calculation of dilution from the measured cross sectional area of weld bead is shown by the formula.

$$\% \text{ dilution} = \frac{B}{A + B} \times 100$$

The penetration or dilution percentage increases with the penetration of the weld into the base metal. Minimum penetration is desired to reduce the amount of dilution; however, a penetration of 5% to 8% is required for a good quality weld. Uniform dilution is preferred. The welding process and procedure controls penetration. Multipass buildup is required if the deposit is expected to have zero pickup or dilution.

FIGURE 24–17 Dilution of weld deposit.

$$\% \text{ dilution} = \frac{B}{A + B} \times 100$$

24-6 SURFACING FOR CORROSION RESISTANCE

The corrosion of metals causes premature failures of many objects, from automobile bodies to chemical plant equipment. Corrosion can be prevented or at least substantially reduced. One of the best ways to reduce corrosion is to protect the metal with an overlay or surface of a material less susceptible to corrosion in a specific environment. Galvanized steel and clad metals with nonferrous facings have long been used to reduce corrosion.

The deterioration of metal surfaces is caused by a combination of factors, such as corrosion and oxidation, corrosion and erosion, or cavitation. Before repairing corroded or deteriorated surfaces, it is necessary to determine the reason. These factors should be considered in selecting a material for overlays for specific types of service.

Once the surfacing material or composition has been selected, it is important to select the method of repairing the corroded surface. There are two different methods. One is to use corrosion-resistant filler metal and deposit the desired analysis on the corroded surface. This technique is also used to rebuild wall thickness to original dimensions. The second method is known as *wallpapering*. This is done by applying small thin pieces of the corrosion-resistant material and welding them to the corroded surface. Each technique will be described briefly.

The selection of the overlay composition is based on the service environment to which the surface is exposed. There are many alloys that are commonly used for corrosion and oxidation resistance. They are summarized as follows:

* The copper-base alloys, including the copper silicon alloys and the copper tin alloys, are used for specific corrosion resistance requirements.
* The austenitic stainless steels, including types 308, 309, 310, 316, and 347, are used for corrosion-resistant surfaces. These alloys exhibit moderate resistance to high-stress abrasion and have excellent oxidation resistance with impact properties.
* The nickel-base alloys, including 100% nickel, Monel, and Inconel, are frequently used for specific applications.
* The high-cobalt chromium alloys are used where corrosion is a major problem. They are used in refineries where high pressures and high temperatures are encountered.

These and special alloys including titanium are applied as weld surfacing deposits or using the wallpapering technique.

The weld surfacing technique is used for new products as well as for resurfacing older items that have corroded. In either case, the weld surface must be clean. Normally, on corroded surfaces the base metal is prepared for welding by grit blasting the entire surface. This is followed by an acid wash and water rinse. Quite often it is the inside diameter of tanks, vessels, digesters, and even pipe that is overlaid. New items that are being manufactured are rotated so that welding processes are utilized, and it is economically important to utilize high deposition processes. Submerged arc welding is often used and often multiple electrode wires are employed. The flux may contain metal additives that will enhance the composition of the deposit. Strip overlay, shown in Figure 24–18, is a popular overlay method. GMAW is also used, as well as gas tungsten arc and plasma arc. In the case of GTAW and PAW, the hot wire technique is often employed. Electroslag welding is also used for surfacing.

Normally, single layers of weld metal are used; however, where dilution is a problem or where thicker layers are required, such as pump linings and high-wear areas, a second layer of surfacing is applied. The second layer can be made with an electrode of lower alloy content since the dilution factor is reduced.

Much of this type of work is performed in situ (in the field). Special equipment has been developed to repair the inside diameter of round pressure vessels such as paper mill digesters. The internal surface of digesters corrode at a high rate due to the chemical process and the

---

FIGURE 24–18 Submerged arc strip overlay.

high pressures and temperatures involved. Automatic machines with one or two gas metal arc welding heads deposit metal on the vertical inside surface of these digesters. The welding heads are mounted on a boom that rotates around the centerline of the tank and makes the metal deposit on the ID or the corroded surface. These operations start at the bottom and automatically move vertically at each revolution as they revolve around the tank. The procedures have been developed so that a smooth internal surface of the weld deposit results.

The second method of providing corrosion protection, known as wallpapering,[14] is done by applying small thin [usually $\frac{1}{16}$-in. (1.6-mm)] pieces of corrosion-resistant material and welding them to the corroded surface. The corrosion-resistant sheets can be of any material listed previously; however, high-nickel alloys and titanium are popular. The problem of dilution and pickup should be considered in selecting the filler material. The material must be firmly attached with intimate contact between the corrosion resistance sheets and the corroded surface. The most popular welding process is gas metal arc with the short-circuiting or pulse mode. A special technique is used, starting with intermittent welds. However, eventually a seal fillet weld is required around the entire outer edge of the corrosion-resistant thin sheet (see Figure 24–19; the reference provides details). For sheets larger than 1 ft square, arc spot welds or plug welds are recommended in the middle of the sheet. Adjacent sheets should slightly overlap previous sheets, and a fillet weld should be made between the new sheet and the sheet

previously installed. Caution should be exercised so that there is no possible leakage of the corroding material through the thin sheet overlay. Wallpapering is replacing nonmetallic linings in many situations.

24-7 OTHER SURFACING APPLICATIONS

There are other surfacing requirements that are best done by welding.

One popular use is the overlaying of metal parts with bronze to provide wearing surfaces for metal-to-metal contact. This includes guides and ways for reciprocating and sliding motion. In order to avoid making a part completely of brass or bronze, it is possible that it can be made of steel and then overlaid with bronze. The copper-base alloys provide a relatively soft deposit for metal-to-metal wear. The AWS Class ECuAl types, which are the aluminum bronzes, are well suited for overlay for bearing surfaces. The different classes such as ECuAl, A-2, B, C, D, and E can all be used. The hardness is greater with the higher suffix letters.

Aluminum bronze overlays can be used for plungers in pumps, rams in extrusion presses, and for rings on hydraulic rams. They will wear faster than the hardened steel with which they make contact. It is advantageous to concentrate the wear on the bronze that can be replaced rather than cause wear on the inside diameter of a hardened steel cylinder.

The aluminum bronzes can also be used for repair welding of worn bronze bearings used in heavy slow moving machinery, and for overlaying worn cast iron gears and sheaves. By overlaying with bronze and remachining, the part can be made better than new.

Overlay of bronze is sometimes used for decorative purposes, particularly for architectural metal application. By judicious use of the bronze and stainless steel, the color contrast can be made very attractive.

Another use for bronze surfacing is for projectiles where a brass overlay is welded around the projectile. This causes it to fit the rifling of the gun barrel tightly to avoid loss of pressure and also to give the desired spin to the projectile.

The use of weld surfacing can be extended to provide safety surfaces. When metal flooring becomes smooth from wear, it is advisable to run stringer beads on the smooth surface to produce a rougher surface. The weld surface will be safer and eliminate slipping. This is used on treads of metal steps, walkways, and other places where smooth metal surfaces can be hazardous. Undoubtedly, there are other applications for surfacing and overlays since it is an ideal method to utilize less expensive materials and provide specialized materials for specific surfaces.

FIGURE 24–19 Wallpapering technique.

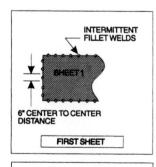

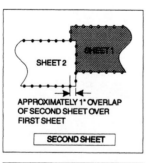

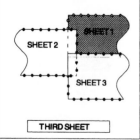

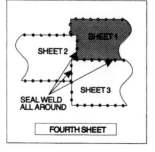

QUESTIONS

24–1. What are the four areas of interest when making a failure analysis?

24–2. What is a monolithic structure? Is a weldment a monolithic structure?

24–3. What are the four points of a failure analysis?

24–4. When should repair welding *not* be performed?

24–5. Approval of a repair procedure is required on what type of products?

24–6. Why is it wise to prepare a written procedure for repair welds?

24–7. List the factors involved in preparation for repair welding.

24–8. List the factors involved in repair welding.

24–9. List the factors involved in repair postweld treatment.

24–10. Explain rebuilding, overlaying, and surfacing. How do they differ?

24–11. What is the best process to use to rebuild small shafts?

24–12. Define and give an example of impact wear.

24–13. Define and give an example of abrasion wear. How is it different from erosion?

24–14. Define and give an example of corrosion wear. How is it different from oxidation?

24–15. What is the basis for selecting hardfacing alloys?

24–16. What is a buildup surfacing material? Where is it used?

24–17. What is a corrosion-resistant weld-overlay cladding? Where is it used?

24–18. What is the advantage of stainless steel cladding over carbon steel?

24–19. Why are wear parts surfaced with bronze or brass?

24–20. How is weld surfacing used to provide a safety surface?

REFERENCES

1. "Civil Aircraft Accident Report of the Court of Inquiry into the Accidents to the Comet G," Her Majesty's Stationery Office, London, 1955.

2. M. E. Shank, "Brittle Failure in Carbon Plate Steel Structures Other Than Ships," Bulletin 19, Welding Research Council, New York, January 1954.

3. A. C. Brooks, "Report of the Royal Commission into the Failure of Kings Bridge," Government Printer, Melbourne, Australia.

4. N. Smith and I. G. Hamilton, "Failures in Heavy Pressure Vessels during Manufacture and Hydraulic Testing." *Journal of West Scotland Iron and Steel Institute* 76 (1968–1969), Glasgow, U.K.

5. "Connection Cited in Hyatt Collapse," *Engineering News Record,* March 4, 1982, McGraw-Hill, New York.

6. S. J. Garwood and J. D. Harrison, "The Use of Yielding Fracture Mechanics in Post Failure Analysis," The Welding Institute, Cambridge, England.

7. "The Design and Methods of Construction of Welded Steel Merchant Vessels," First Report of a Board of Investigation, *Welding Journal* (July 1974).

8. N. Shute, *No Highway* (London: Pan Books, 1948).

9. "Acceptable Methods, Techniques and Practices—Aircraft Inspection and Repair," AC43.13-1A-1972, Federal Aviation Administration, U.S. Department of Transportation, Washington, D.C.

10. "Air Frame and Power Plant Mechanics Air Frame Handbook," AC65-15, Federal Aviation Administration, U.S. Department of Transportation, Washington, D.C.

11. "National Board Inspection Code," 12th Edition, National Board of Boiler and Pressure Vessel Inspectors, Columbus, Ohio.

12. *Metals Handbook,* 8th ed., vol. 6, *Welding and Brazing,* Meals Park, Ohio (*American Society for Metals*).

13. L. F. Spencer, "Hardfacing, Picking the Proper Alloy," *Welding Engineer* (November 1970).

14. B. Irving, "Wallpapering: Another Growth Market for Welding Fabricators," *Welding Journal* (July 1991).

25

WELDING PIPE AND TUBING

OUTLINE

25-1 TUBULAR PRODUCTS

Tubular products, known as pipe or tubing, are hollow items, normally circular, used for transmitting gases or liquids, or for structural, mechanical, or decorative functions. They can range from the smallest to over 60 in. in diameter and with wall thicknesses from very thin to thick. Tubular products are manufactured as seamless or welded. Welded tubular products are the most popular and are the only ones considered here. There are many different ways of classifying pipe and tubing, usually based on shape and intended use. General classifications are as follows:

1. *Standard pipe:* used for transmission of low-pressure air, steam, other gases, water, oil, and/or other fluids. Used primarily in buildings, sprinkler systems, irrigation systems, and in machinery.

2. *Line pipe:* used for the transportation of gas, oil, water, and so on, in cross-country pipelines and for utility distribution systems.

3. *Oil country goods:* tubular products used by the oil and gas industries with three subdivisions; casings for well walls, tubing used within the casings, and drill pipe used to carry rotary drilling tools.

4. *Pressure tubing:* used to transmit fluids or gases at elevated temperatures or pressures or both.

5. *Mechanical tubing:* used to manufacture industrial, construction, and agricultural equipment.

6. *Structural pipe and tube:* used for structural or load-bearing purposes, for architectural or structural purposes, and can be of different shapes.

7. *Thin-wall tubing:* used for instrument tubing, aircraft control tubing, air conditioning, and miscellaneous applications; can be of different sizes and of stainless steels and nonferrous metals.

Each classification can be made of different materials. Standard pipe is normally made of carbon steel. It may be uncoated, galvanized, or plastic coated, and is made in different wall thicknesses, known as standard, extra-strong, double-extra strong, and others. Wall thickness may be indicated by schedule number (Figure 25–1). Schedule 40 is standard-weight pipe.

Line pipe is usually made of carbon steel or of low-alloy high-strength steel. They are made of weldable steels since line pipe is joined by welding. Special pipelines

730

FIGURE 25–1 Standard pipe size and wall thickness.

Nominal Pipe Size (in.)	Outside Diameter	Nominal Wall Thickness for:													
		Sched. 5	Sched. 10	Sched. 20	Sched. 30	Standard	Sched. 40	Sched. 60	Extra Strong	Sched. 80	Sched. 100	Sched. 120	Sched. 140	Sched. 160	xx Strong
1/8	0.405	—	0.049	—	—	0.068	0.068	—	0.095	0.095	—	—	—	—	—
1/4	0.540	—	0.065	—	—	0.088	0.088	—	0.119	0.119	—	—	—	—	—
3/8	0.675	—	0.065	—	—	0.091	0.091	—	0.126	0.126	—	—	—	—	—
1/2	0.840	—	0.083	—	—	0.109	0.109	—	0.147	0.147	—	—	—	0.187	0.294
3/4	1.050	0.065	0.083	—	—	0.113	0.113	—	0.154	0.154	—	—	—	0.218	0.308
1	1.315	0.065	0.109	—	—	0.133	0.133	—	0.179	0.179	—	—	—	0.250	0.358
1 1/4	1.660	0.065	0.109	—	—	0.140	0.140	—	0.191	0.191	—	—	—	0.250	0.382
1 1/2	1.900	0.065	0.109	—	—	0.145	0.145	—	0.200	0.200	—	—	—	0.281	0.400
2	2.375	0.065	0.109	—	—	0.154	0.154	—	0.218	0.218	—	—	—	0.343	0.436
2 1/2	2.875	0.083	0.120	—	—	0.203	0.203	—	0.276	0.276	—	—	—	0.375	0.552
3	3.5	0.083	0.120	—	—	0.216	0.216	—	0.300	0.300	—	—	—	0.438	0.600
3 1/2	4.0	0.083	0.120	—	—	0.226	0.226	—	0.318	0.318	—	—	—	—	—
4	4.5	0.083	0.120	—	—	0.237	0.237	—	0.337	0.337	—	0.438	—	0.531	0.674
5	5.563	0.109	0.134	—	—	0.258	0.258	—	0.375	0.375	—	0.500	—	0.625	0.750
6	6.625	0.109	0.134	—	—	0.280	0.280	—	0.432	0.432	—	0.562	—	0.718	0.864
8	8.625	0.109	0.148	0.250	0.277	0.322	0.322	0.406	0.500	0.500	0.593	0.718	0.812	0.906	0.875
10	10.75	0.134	0.165	0.250	0.307	0.365	0.365	0.500	0.500	0.593	0.718	0.843	1.000	1.125	—
12	12.75	0.156	0.180	0.250	0.330	0.375	0.406	0.562	0.500	0.687	0.843	1.000	1.125	1.312	—
14	14.0	—	0.250	0.312	0.375	0.375	0.438	0.593	0.500	0.750	0.937	1.093	1.250	1.406	—
16	16.0	—	0.250	0.312	0.375	0.375	0.500	0.656	0.500	0.843	1.031	1.218	1.438	1.593	—
18	18.0	—	0.250	0.312	0.438	0.375	0.562	0.750	0.500	0.937	1.156	1.375	1.562	1.781	—
20	20.0	—	0.250	0.375	0.500	0.375	0.593	0.812	0.500	1.031	1.281	1.500	1.750	1.968	—
22	22.0	—	0.250	0.375	—	0.375	—	—	0.500	—	—	—	—	—	—
24	24.0	—	0.250	0.375	0.562	0.375	0.687	0.968	0.500	1.218	1.531	1.812	2.062	2.343	—
26	26.0	—	—	0.500	—	0.375	—	—	0.500	—	—	—	—	—	—
30	30.0	—	0.312	0.500	0.625	0.375	—	—	0.500	—	—	—	—	—	—
34	34.0	—	—	—	—	0.375	—	—	0.500	—	—	—	—	—	—
36	36.0	—	—	—	—	0.375	—	—	0.500	—	—	—	—	—	—
42	42.0	—	—	—	—	0.375	—	—	0.500	—	—	—	—	—	—

have been made of corrosion-resistant steels and stainless steels. Line pipe is made to API specifications.

Oil country goods are made of carbon steel and alloy steels, and some items are made of extremely high-alloy, high-strength materials.

Pressure tubing, which is made to exact dimensions of outside diameter and wall thickness, is made of carbon steels, alloy steels, creep-resisting steels, heat-resisting steels, and stainless steels of different types.

Structural steel pipe and tube is made of low-carbon weldable steels. The analysis of the steel used for making the pipe is normally specified by the producer or by specifications for the material.

Thin-wall tubing is made of low-alloy steels and stainless steels. Stainless steel tubing is available in almost any alloy of stainless available. In addition to steels, tubing is available in aluminum, copper, titanium, and nickel alloys.

The dimensions used for pipe and tubing depend on the product classification and the country of origin. See standard pipe sizes already mentioned and metric sizes shown in Figure 25–2.

In specifying pipe and tubing, it is necessary to provide exact dimensions and the material classification or composition in order to obtain the type requested.

Methods for Manufacturing

Welded tubing is preferred over seamless tubing since it has more uniform wall thickness and is less expensive. The following welding processes are used:

1. Continuous butt welding process
2. Resistance welding processes
3. Arc welding processes
4. High-energy beam (electron and laser) processes

Two types of weld joints are used. The most common is the straight, longitudinal joint from end to end of pipe, used for all sizes from smallest to largest. The spiral joint is used for medium- and larger-sized tubular products.

The continuous mill for making tubular products, when the weld joint is longitudinal, is similar for all of the welding processes.[1] A continuous mill (Figure 25–3) consists of the following:

1. Coil of strip or skelp
2. Splicing operation for the skelp
3. Strip flattening and trimming station (optional)
4. Multiple forming rolls, including closing rolls
5. Welding station, including the squeeze or pressure rolls
6. Sizing rolls, or die
7. Cut off operation

The number and size of rolls, number of stations, and so on, will vary depending on the manufacturer of

FIGURE 25–2 Metric pipe size and wall thickness.

Nominal Pipe Size		Schedule Number	Outside Diameter		Wall Thickness	
in.	mm		in.	mm	in.	mm
$\frac{1}{8}$	3	10	0.405	10.3	0.049	1.2
		40			0.068	1.7
		80			0.095	2.4
$\frac{1}{4}$	6	10	0.540	13.7	0.065	1.7
		40			0.088	2.2
		80			0.119	3.0
$\frac{3}{8}$	10	10	0.675	17.1	0.065	1.7
		40			0.091	2.3
		80			0.126	3.2
$\frac{1}{2}$	13	5	0.840	21.3	0.065	1.7
		10			0.083	2.1
		40			0.109	2.8
		80			0.147	3.7
$\frac{3}{4}$	19	5	1.050	26.7	0.065	1.7
		10			0.083	2.1
		40			0.113	2.9
		80			0.154	3.9
1	25	5	1.315	33.4	0.065	1.7
		10			0.109	2.8
		40			0.133	3.4
		80			0.179	4.5
$1\frac{1}{4}$	32	5	1.660	42.2	0.065	1.7
		10			0.109	2.8
		40			0.140	3.6
		80			0.191	4.9
$1\frac{1}{2}$	38	5	1.900	48.3	0.065	1.7
		10			0.109	2.8
		40			0.145	3.7
		80			0.200	5.1
2	51	5	2.375	60.3	0.065	1.7
		10			0.109	2.8
		40			0.154	3.9
		80			0.218	5.5
$2\frac{1}{2}$	64	5	2.875	73.0	0.083	2.1
		10			0.120	3.0
		40			0.203	5.2
3	76	5	3.500	88.9	0.083	2.1
		10			0.120	3.0
		40			0.216	5.5
$3\frac{1}{2}$	89	5	4.000	101.6	0.083	2.1
		10			0.120	3.0
		40			0.226	5.7
4	102	5	4.500	114.3	0.083	2.1
		10			0.120	3.0
		40			0.237	6.0
6	152	5	6.625	168.3	0.109	2.8
		10			0.134	3.4
8	203	5	8.625	219.1	0.109	2.8
		10			0.148	3.8
10	254	5	10.750	273.1	0.134	3.4
		10			0.165	4.2
12	305	5	12.750	323.9	0.156	4.0
		10			0.180	4.8
14	356	5	14.000	355.6	0.156	4.0
		10			0.188	4.8

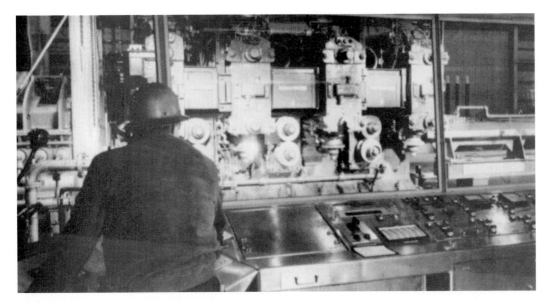

FIGURE 25–3 Continuous mill for making tubular products.

the mill and the size and type of tubular products being produced.

The welding station produces a high-quality weld with full penetration, minimum root and face reinforcement, and minimum bead widths. The weld must be smooth, uniform, and clean without cracks and without undercutting, and the reinforcement of the weld should not exceed 10% of the wall thickness.

The so-called butt welding process, commonly called the CW (continuous welding) process, is the oldest welding process for welding pipe. It is forge welding in which the flat stock, known as skelp, is formed into a tubular shape while very hot and pulled through a die. This causes the abutting edges to come together under very high pressure and high temperatures in a continuous-welding mill, to make a forge weld. This process is used to manufacture standard pipe of $\frac{1}{8}$ to 4 in. nominal diameter at a high rate of speed on a continuous welded pipe mill.

There are three different electric resistance welding processes employed for continuous mill welding. The choice of the welding process variation depends on the diameter of the tubular product, the wall thickness, and the production rate. In all three methods the power for welding is provided either by low-frequency current through revolving electrode wheels, or by radio-frequency current through sliding contacts or induction coils.

GTAW is popular for thin-wall stainless steel tubing and tubing made of nonferrous alloys. Plasma arc welding is finding increased use for stainless steel tube mills. Gas metal arc welding is often employed where thickness is greater and filler metal is required. For thick-wall pipe, flux cored arc welding or submerged arc welding can be used. These processes are also used for making spiral joint pipe.

The GTAW process produces high-quality welds in tube mills in square groove welds from 0.020 in. (0.5 mm) to 0.118 in. (3 mm) thick without the addition of filler wire. The wall thickness of the tubing and the metal composition greatly influence the welding parameters. Most procedure tables for mechanized welding relate to average conditions. Tubular product mills use welding conditions modified for maximum travel speed. The welding parameters must be analyzed and each adjusted to provide the maximum travel speed. The following suggestions are made to speed up the welding station of the mill.

The primary variables with the arc processes are travel speed, welding current, and arc voltage. This is the heat input into the weld. The secondary adjustable variables include the torch travel angle and the arc direction when a magnetic arc deflecting system is used. The distinct level variables include the electrode size, type and point geometry, the composition of the shielding gas (and trailing gas shield if used), the use of more than one torch and their spacing, and the use of oscillation, either mechanical or magnetic.

The fixed conditions include thickness and composition of the metal. Increasing productivity means increasing the speed of the tube, which can be done by increasing the energy employed in making the weld. A combination of improvements can increase the welding speed of the tube mill. Revise the welding procedure by examining each variable and adjusting them independently until the right combination has been obtained. Figure 25–4 gives parameters for single-arc gas tungsten arc welding for stainless steel of the wall thicknesses shown.

Adjust the primary variables. As the welding current increases, the travel speed must increase, to avoid burn-through. The travel speed must be continuously adjustable so that it can be changed as the current is increased. The top limit of current with the GTAW process seems to be 250 to 300 A.

FIGURE 25–4 Welding schedule for GTAW welding stainless steel tubing.

Wall Thickness		Travel Speed		Current (A DCEN)	Arc Voltage (V)	Shield Gas Type	Joint Detail Type
in.	mm	in./min	mm/min				
0.078	1.98	12	305	180	10	Argon	Square butt
0.078	1.98	20	508	200	12	Argon + 5% h$_2$	Square butt
0.109	2.77	16	406	220	10	Argon	Square butt
0.126	3.18	10	254	230	11	Argon	Square butt
0.154	3.91	12	305	245	12	95% argon + 5% H$_2$	60° V-groove
0.216	5.48	8	205	260	13	95 argon + 5% H$_2$	60° V-groove
0.226	5.74	7	178	240	13	95 argon + 5% H$_2$	60° V-groove
0.237	6.02	6	167	280	13	95 argon + 5% H$_2$	60° V-groove

Arc voltage relates to arc length; it can be varied between narrow limits. The minimum arc length should not be less than one diameter of the electrode. The maximum arc length should not be more than twice the diameter of the electrode. Many GTAW mills use automatic arc length control, which allows setting the torch to a specific arc voltage.

The practical maximum travel speed [approximately 1 m (39 in.) per minute] is limited by the quality of the weld. As travel speed increases beyond this rate, undercutting will occur. The weld bead may be high and crowned, and there will be a depression in the center of the bead that reduces the cross section along the weld centerline (Figure 25-5).

These defects occur because of a lagging arc (Figure 25-6). This makes the arc longer as travel speed increases. As the arc length increases it flares, is less concentrated, and has a higher voltage. Giving the torch a lead angle overcomes the lagging arc, reduces the arc length, and generally allows travel speed to be increased without undercutting. A push angle of up to 20° will move the undercutting occurrence to a higher speed and tends to flatten the weld bead. This can also be accomplished by a magnetic arc deflection system, which corrects for the arc lag and reduces arc length. This system is adjusted to cause the arc to lead, which preheats the weld area and allows higher travel speeds before undercutting occurs.

The optimum arc length is $1\frac{1}{2}$ times the electrode diameter, which provides optimum arc voltage. Torch position and angle adjustment is required for the best welding conditions.

FIGURE 25–6 Arc length factors.

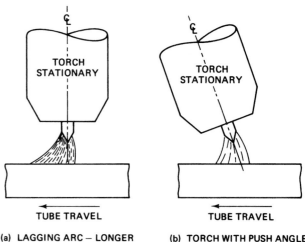

(a) LAGGING ARC – LONGER

(b) TORCH WITH PUSH ANGLE

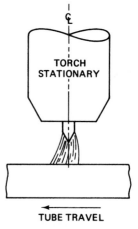

(c) WITH MAGNETICS PROVIDING LEADING ARC

Another way to increase the heat input in the arc is to use helium shielding gas. Helium provides more heat in the arc, hence travel speed can be increased. Increased production must be related to the higher cost and greater flow rate of helium gas. A mixture of 50:50 argon—

FIGURE 25–5 Undesirable weld cross section.

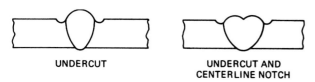

UNDERCUT

UNDERCUT AND CENTERLINE NOTCH

FIGURE 25-7 Welding schedule for plasma arc welding stainless steel tubing.

Wall Thickness		Travel Speed		Current (A DCEN)	Orifice Size		Gas Shield and Plasma Type
in.	mm	(in./min)	mm/min		in.	mm	
0.062	1.57	14	355	65	0.081	2.05	A + 5% H_2
0.078	1.98	14	355	70	0.081	2.05	A + 5% H_2
0.093	2.36	12	305	85	0.081	2.05	A
0.109	2.77	16	406	85	0.081	2.05	A + 5% H_2
0.126	3.18	10	254	100	0.081	2.05	A + 5% H_2
0.154	3.91	16	406	100	0.081	2.05	A + 5% H_2
0.187	4.75	7	178	100	0.081	2.05	A + 5% H_2

helium can be used to reduce gas cost. Another way to increase the heat of the arc is to use hydrogen in the argon shielding gas. Up to 10% hydrogen can be used for welding nickel and nickel alloys and some stainless steels. In general, hydrogen mixtures should not be used for welding carbon and low-alloy steels.

The use of helium or hydrogen mixtures will increase travel speed up to 50%. Special nozzles are required to shield the longer molten weld pool. Heavy-duty, water-cooled, automatic torches should be used. The torch rating should be at least 50% greater than the welding current.

For welding ferrous metals, dc electrode negative is used. The 2% thoria type (EWTh2) tungsten electrode should be used. The ground finish should be specified since this improves heat transfer and increases electrode life and time between regrinding the point. The size of the electrode should be the largest for the welding current to be used. The electrode should be precision ground to a point of 30° included angle, but the end of the point should be flattened.

The use of an additional GTAW torch ahead of the welding arc will allow increased speed since more energy is being put into the material to preheat it. The use of an additional torch following the welding torch will reduce the undercut problem. The use of three torches will increase tube travel speed by up to 100% while producing a good-quality weld. The leading (preheat) torch should operate at about 50% of the current of the welding torch. The trailing torch will operate at about 33% of the welding torch. The spacing between the torches should be the minimum.

The plasma arc welding keyhole process can be used in place of the GTAW to increase the production rate. Speed increase of from 33% to 100% is possible, with the greatest improvement on thicker metal. Increased production is due to the higher temperature plasma and the constricted stiffer arc, which improves heat transfer to the work. The welding schedule shown in Figure 25-7 shows the productivity improvement. Analysis and adjustment of all variables would be similar to that described for GTAW.

In submerged arc, gas metal, or flux cored arc welding applications, multiple torches can be used. Submerged arc using ac and three torches is often used. Electron beam and laser beam welding processes are both used for welding specialty-type tubular products.

25-2 PIPE AND TUBE WELDING

In the United States, approximately 10% of the steel produced is made into tubular products, essentially pipe. Except for the small size, the majority of pipe is installed by welding.

The piping industry is roughly divided into three major categories:

* Pressure or power piping
* Transmission and distribution piping
* Noncritical piping

The welding of pressure piping used in thermal and nuclear power stations, refineries, chemical plants, on ships, and so on, is done in accordance with the ASME code for pressure piping.[2] All of the ASME piping codes of B31 are shown in Figure 25-8. The pipe employed normally has a medium to thick wall thickness in medium to large sizes. Welding procedures, and qualifications, are largely in accordance with Section IX of the ASME pressure vessel code.

Transmission and distribution pipelines transmit gas and petroleum products from the producing fields to the consumers. Welding this type of pipe utilizes special techniques and procedures, and is governed by API Standard 1104.[3] This specification is in general agreement with B31.8, "Gas Transmission and Distribution Piping Systems." The pipe employed is usually of medium to high strength and has relatively thin walls in medium to large diameters. Distribution piping is normally carbon steel standard-size pipe of smaller size.

The noncritical piping field includes many different systems: domestic hot water supply, sprinkler, sanitary, gas

FIGURE 25–8 ASME code for pressure piping.

B31.1	Power piping
B31.2	Fuel gas piping
B31.3	Chemical plant and petroleum refinery piping
B31.4	Liquid transportation systems for hydrocarbons, liquid petroleum gas, anhydrous ammonia, and alcohols
B31.5	Refrigeration piping
B31.8	Gas transmission and distribution piping systems
B31.9	Building services piping
B31.11	Slurry transportation piping systems
B31G	Manual for determining the remaining strength of corroded pipelines

and air lines, and many other applications. Welding has not been universally adopted. Screw-thread connections, soldered copper tubing, and plastic pipe are used for many applications. The steel pipe is usually standard wall thicknesses in the small and medium sizes. Qualification tests may not be required, although qualified welders and qualified procedures are often used.

Welding pipe and tubing is normally done in accordance with established written procedures. The specific code involved must be consulted. Welding procedures are based on the pipe material, pipe diameter, and wall thickness. Welding position depends on the job and the code, but the procedure must indicate the welding process and progression of travel. The method of application depends on the process and equipment available. The filler metal is selected based on the composition of the pipe material and the mechanical properties requirement. A listing of pipe welding procedure schedules is given in Figure 25–9. This list is based on the pipe or tubing size, which is categorized as small [4 in. (100 mm) and smaller], medium [4 in. (100 mm) to 12 in. (300 mm)], and large [12 in. (300 mm) and larger]. The wall thickness is categorized as thin (less than standard), standard (Schedule 40), and heavier (greater than standard). This is followed by the welding position, the welding process, and the method of application. In some cases, combinations of welding processes and methods of application are used.

With these data, welding procedures can be designed to meet the job requirements based on the specification involved. Welding procedure specifications must be qualified to meet the code requirements.

FIGURE 25–9 Pipe welding procedure schedules.

Tube and Pipe Diameter and Wall Thickness	Welding Position	Welding Process	Method of Applying
Small tubing, thin wall	All position	GTAW	MA
	All position	PAW	MA
	All position	GTAW, no FM	Au
Small pipe, standard wall	All position	OFG	MA
	All position	GTAW with FM	AU
Small to medium pipe, standard wall	All position	GMAW	SA
	All position	SMAW	MA
Medium pipe, standard and heavy wall	All position	Comb. GTAW and SMAW	MA
Medium and large pipe, thin and standard wall	All downhill	SMAW	MA
	All up or down	GMAW	SA
	All up or down	FCAW	SA
Medium and large pipe, all wall	All uphill	SMAW	MA
	Flat roll (1G)	SAW	AU
	Flat roll (1G)	Comb. GMAW and SAW	SA or AU and AU
	Flat roll (1G)	Comb. GMAW and FCAW	SA or AU and AU
Small, medium, and large pipe, standard to thick wall	All position	FCAW	SA
	All position	Comb. GTAW and FCAW	MA and SA
	All position	Comb. GTAW and GMAW	MA and SA

Joint Design

The joint designs for pipe welding have been fairly well standardized and are shown in Figure 25–10. For thinner-wall pipe the joint design is the square-groove weld. As thickness increases, a single-V-pipe joint is used. The included angle of the V-groove has been standardized at 60 and 75°. The 75° included angle is more common in pressure piping, and the 60° included angle is common in cross-country transmission-line piping. The root face and root opening are approximately the same. As the wall thickness increases, the joint design will change so that less weld metal will be required. The included angle changes to a narrower angle partially up the joint. These joint designs are commonly used in power plant piping, where heavy wall thickness pipe is used. Other variations in joint design depend on the composition of the pipe. Some automatic procedures require special joint designs. For aluminum pipe special joint details have

been developed, and these are normally welded with combination process procedures. This allows a root weld to be made in much the same manner as the weld on thin-wall tubing.

Consumable insert rings are often used for making critical welds in pipe and tubing. There are five classes of rings called *consumable inserts* and specified by AWS A5.30, Consumable Inserts. They are shown in Figure 25–11. They are given class numbers. Class 1 is called A-shaped or inverted T. Class 2 is called J-shaped, classes 3 and 5 are rectangular and sometimes called K-shaped, and class 4, is called Y-shaped. They are placed in the root of the weld joint when the joint is tacked up. GTAW is used for the root pass, and the rings are fused into the root of the joint. A skilled welder can make a very smooth root that gives a smooth inside diameter of the pipe. The rings are available in different sizes and of different analysis so that they can match the pipe composition and size. The composition of the consumable insert is covered by the specification.

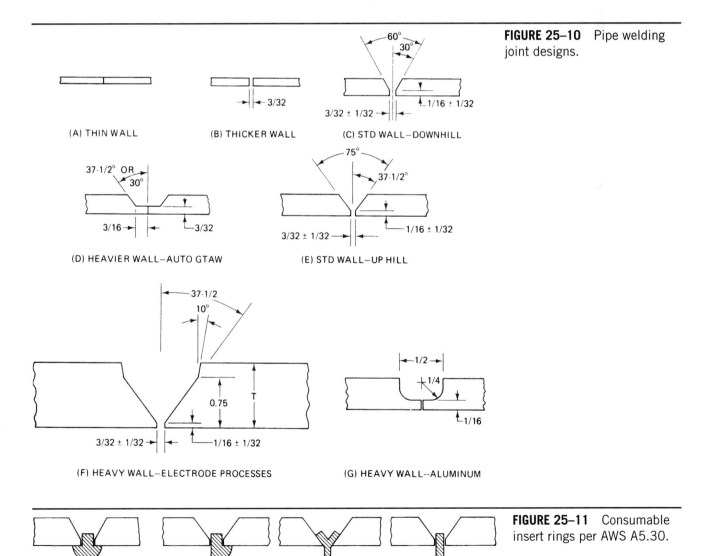

FIGURE 25–10 Pipe welding joint designs.

(A) THIN WALL

(B) THICKER WALL

(C) STD WALL—DOWNHILL

(D) HEAVIER WALL—AUTO GTAW

(E) STD WALL—UP HILL

(F) HEAVY WALL—ELECTRODE PROCESSES

(G) HEAVY WALL—ALUMINUM

FIGURE 25–11 Consumable insert rings per AWS A5.30.

CLASS 1   CLASS 2   CLASS 4   CLASSES 3 AND 5

FIGURE 25–12 Flame-cutting bevel on pipe.

Rectangular backing rings are rarely used when gas or fluids are transmitted in the piping system, since the rings reduce the inside diameter of the pipe. They may be used for structural applications where tubular members are used to transmit loads rather than materials. They are useful in helping align the pipe assembly.

For critical pipe welding, internal gas purging is often used. Special dams of soluble paper or balloons are used to contain the purge gas in the area of the pipe joint. The purge gas is argon, but nitrogen may be used for stainless steel tubing.

Internally clad pipe is being used increasingly. This is normally carbon steel pipe clad on the inside with stainless steel. The internal clad thickness usually is on the order of $\frac{1}{8}$ in. (3 mm) thick where the carbon steel pipe is of normal diameter and wall thickness. Internally clad pipe is used for carrying corrosive materials, usually oil and gas products with corrosive characteristics or accompanying gases. Special precautions are used to weld clad pipe, in that the internal clad must be welded with a filler metal of approximately the same composition as the clad material in the remainder of the pipe welded with filler metals comparable to those of the pipe.

Joint Alignment and Fitup

An important requirement in obtaining a high-quality pipe joint is to make sure that the fitup of the joint prior to welding is as good as possible. It must meet the joint design detail and be uniform around the circumference of the joint. This is sometimes difficult to obtain because pipe is not always exactly round and the diameter may vary within the limit of the pipe size. Tubing is manufactured closer to size and is easier to fit up. The nonroundness or ovality of pipe can present a welding problem, especially with large-diameter pipe.

Pipe joints, particularly for structural applications, are varied and complex. The most simple type is the butt joint, which requires square cutoffs and single bevels. Portable machines with mechanized torches that revolve around the pipe are widely used and are shown in Figure 25–12. Complex joints require detailed layout and accurate cutting. Up to five-axis computer-controlled machine tool machines are used in shops. The operator uses dedicated software that calculates the bevel angle of intersections. These, coupled with critical weld preparation angles, provide nearly perfect fitup. Flame-cutting torches, rotary cutters, abrasive wheels, and grinders can all be used.

Pipe received from the steel mill usually has a standard end preparation used for butt joints. Forged pipe fittings commonly used in complex assemblies have joint preparation prepared at the manufacturer's plant.

Fitting the pipe is the most difficult part of piping installations. For small assemblies, this is relatively easy and much of this work is done in pipe-fabricating shops. Setup and welding subassemblies with different kinds of

FIGURE 25–13 Shop fabrication of pipe subassembly.

fittings and pipe sections are welded there under ideal conditions. A typical shop fabrication welded with semi-automatic equipment is shown in Figure 25–13. These assemblies are then transported to the erection site for field welding. Assembly in the field is usually more difficult since dimensional variations are more difficult to control.

A variety of alignment devices are used for pipe installations. For small-diameter pipe, external-type clamps are normally employed. Figure 25–14 shows an assortment of these. Some of these clamps have sufficient force to re-form the pipe into a perfect circle to facilitate fitup. This is possible on the thin-wall pipe, but becomes increasingly difficult as the pipe wall thickness increases.

FIGURE 25–14 Variety of external clamps.

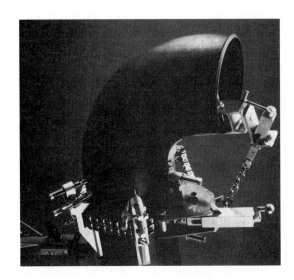

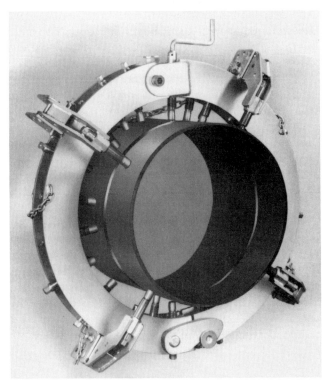

Cross-Country Pipelines

The cross-country transmission pipeline welding techniques have become extremely sophisticated. Normally, the "stove pipe" method of installing pipe is used. This means that each section or length of pipe is added on to the existing pipe installation. The crew for doing this moves along the right-of-way from the beginning to the end of the pipeline. Welding procedures and techniques vary based on the diameter of the pipe.

The SMA stick electrode process has been used and still is the predominant welding process for field girth welding. However, the use of semiautomatic and automatic GMAW is increasing steadily. Self-shielding FCAW is also used. All procedures use the downhill technique. Figure 25–15 shows a field welding spread for welding large-diameter pipe with automatic equipment in hilly country.

For the large-diameter pipe welds an internal lineup clamp is utilized (Figure 25–16). The clamp is inserted in the end of the last section of the pipeline and is operated remotely by air pressure. The air pressure clamps the internal lineup clamp to the section already welded to the pipeline and then as the new section is being placed in position, it clamps, locates, and spaces the new section. It helps round out the pipe due to the strength of the clamp. Some of these clamps include a copper backing ring. The clamps are left in the pipe joint until the first or stringer pass is made. Usually, the second pass is also made before the clamp is released and removed. In welding large-diameter pipes there is one welding crew that makes the root pass and second pass, commonly known as the stringer and hot pass. They move on with the lineup clamp crew and work on the next pipe joint, and other welding crews come in to finish the weld. These crews make the filler passes, which are those that fill the weld joint; the stripper passes, which are usually made in the vertical portion of the pipe joint; and the last pass, known as the cap pass. The stringer crew and other crews represent three or four pipe welding groups that are progressing along the pipeline during its construction. However, the production rate is determined by the front-end crew depositing the root and hot pass.

When semiautomatic welding GMAW is used for cross-country pipe welding, the welding equipment is placed on a flatbed truck or a tractor with a boom supporting the welding cables and guns over the pipe to be welded. This technique has almost doubled the production rates over manual SMAW and has become very popular in many parts of the world. Figure 25–17 shows the semiautomatic welding of a large cross-country pipe.

GMAW will meet the requirements of the API 1104 specification for medium and large pipe with relatively thin wall.

The construction of offshore pipelines differs considerably from that of cross-country pipelines. The girth welding of a 40-ft length of pipe (80 ft if double-jointed)

FIGURE 25–15 Pipeline spread—large pipe.

is done on lay barges such as the one shown in Figure 25–18. Lay barges are 400 to 500 ft long. Welding stations are located 40 ft apart along the centerline of the vessel.

The welding, inspection, and field coating of the weld joint area are carried out simultaneously at these workstations. Station 1 is the head station, where the root and hot passes are deposited. Subsequent weld passes are made at later stations. Each welder deposits the same weld pass in each joint and does not make the entire weld. The welding procedure and the welder have been qualified, and each welder uses the same welding variables for a particular operation. Weld inspection (usually radiographic) is done in the station following the cap pass. Any weld repairs are made at this or the next station. Finally, the weld area is coated at the last station. After each station has completed its particular task, the barge is advanced 40 ft by wenches taking up cables attached to anchors. The completed section is then lowered into the water. The rate of pipe laying (i.e., the productivity) is determined by the speed of the slowest station.

FIGURE 25–16 Pneumatic internal lineup clamps.

FIGURE 25–17 Semiautomatic welding of a large-diameter pipeline.

FIGURE 25–18 Air view of pipe-laying barge.

Quality Assurance

The quality of butt welds in piping systems must be closely monitored. Visual inspection is always used; however, there is increasing use of ultrasonic inspection. Traditionally, x-ray inspection has been employed. The quality level, level of acceptable defects, and so on, are established by the code involved. Accidents or any failure of a pipeline must be reported to the office of pipeline safety in Washington, D.C.

25-3 MANUAL AND SEMIAUTOMATIC PIPE WELDING

Neither screw-thread joints nor mechanical joints develop the full strength of pipe; hence one of the earliest applications of welding was to join pipe. The oxyacetylene welding process was used for many years to make pipe welds. Oxyacetylene welds develop the full strength of the pipe. Oxyacetylene welding is slow, so the time involved for making heavy-wall large-diameter pipe welds was excessive. However, welding of 2 in. and smaller standard wall pipe is done by oxyacetylene welding. It is used for radiant heating systems, cooling systems for ice rinks, and similar applications. Figure 25–19 shows the oxyacetylene welding of small-diameter pipe.

Electric arc welding has been used for pipe joining for many years. Initially, bare or lightly coated electrodes were used, and the welds developed the full strength of the pipe. Recently, pipelines welded with bare electrodes over 50 years ago have been uncovered, and inspection indicated that the welds were still good. The covered electrode made manual shielded metal arc welding of pipe a very popular process.

For medium-diameter, heavy-wall pressure piping, the uphill technique is used. This technique meets the requirements of the ASME piping codes, and literally thou-

FIGURE 25–19 Oxyacetylene welding of small pipe.

sands of procedures have been qualified using E6010 electrodes.

With low-alloy, high-strength steels for powerhouse construction, a new type of electrode is used. These are low-hydrogen types with low-alloy deposited metal matching the analysis of the pipe. An illustration showing this type of application is shown in Figure 25–20.

For critical piping the root pass is made with the gas tungsten arc welding process. This is done using an open root or using consumable insert rings and fusing them to the roof of the joint. A second pass is usually made with GTAW and the remaining weld by SMAW. This produces an excellent weld joint with an extremely smooth inner surface. Procedures have been developed for most low-alloy steels, including the chrome-molybdenum steels.

GTAW and SMAW applied manually are relatively low-production welding methods. When GMAW was developed, it was soon applied to pipe welding. It was used for roll welding (IG) with the handheld gun, and for fixed position (5G) welding at construction sites. Figure 25–21 shows semiautomatic GMAW in a factory installation. Both the small wire short-circuiting technique and the spray technique are employed. Electrode wire compositions must match the composition of the base metal. Flux cored electrode wires are available to match the composition of pressure piping.

This work is governed by codes and specifications. A summary of pipe welding schedules is given in Figure 25–22. Recommended practices are given in Figure 25–23. These are available from AWS, Miami, Florida. A

FIGURE 25–21 Semiautomatic welding of power plant piping.

high level of skill is required to make pipe welds manually or with semiautomatic equipment either uphill or downhill. Stringent qualification tests apply to pipe work.

25-4 MECHANIZED PIPE AND TUBE WELDING

Mechanized welding systems are used for welding pipe and tubing. Two basic types of procedures are used. One is *roll welding* pipe in the flat or downhand position. This is the IG position when the joint is rotated under the welding head. The second, known as *orbital welding*, is used when the pipe is in a fixed position and the machine rotates around the pipe to make the weld. A further subdivision is for thin- or heavy-wall pipe, or for small- or large-diameter pipe or tubing. A further subdivision relates to the weld process. Submerged arc welding (SAW) is used for roll welding only; other processes—GMAW, FCAW, and GTAW—are used for making pipe welds in any position.

Roll welding was an early application of machine or automatic welding. In the field this is known as "double jointing" (Figure 25–24). This means the welding together of two sections of straight pipe for cross-country pipelines. Double jointing is done at the pipe storage yard where standard lengths of pipe are welded and then transported to the construction site. Roll welding is more productive than fixed position welding and reduces the welding hours to construct a pipeline.

SAW has historically been used for roll welding. An internal lineup clamp, containing a backup bar, is used. Roll welding is also done in the fabrication shop

FIGURE 25–20 SMAW used for heavy-wall pipe.

FIGURE 25-22 Pipe weld schedules.

Process	Joint Design (See Figure 25-10)	Method of Application	Pass	in.	mm	Amperes DC	Voltage	Shielding Details	Travel	Other Information
GTAW	A	MA	1	$\frac{1}{16}$	1.6	35–45	10–12 EN	Argon at 12–15 ft³/hr	Downhill	Use purge gas inside for high quality
PAW	A or D	MA	1	None	None	60–70	9–10 EP	Argon at 12–15 ft³/hr	10 in./min	Plasma gas is 95% argon + 5% H_2 at 1 ft³/hr
GTAW	A or D	AU	1	None	None	40	10 EN	Argon at 20 ft³/hr	$4\frac{1}{4}$ in./min either	$\frac{1}{16}$-in. Tungsten use purge gas
OFG	B or E	MA	1 2	$\frac{1}{8}$	3.2	—	—	Product of Combustion	3 in./min	Forehand—oxygen and acetylene
GTAW	E	MA	1 2 +	$\frac{3}{32}$ $\frac{1}{8}$	2.4 3.2	90–100 100–140	EN EN	Argon at 15–20 ft³/hr Argon at 20–25 ft³/hr	Uphill Uphill	Use purge gas No purge gas
GMAW	C or E	SA	1 2 +	0.035 0.035	0.8 0.8	150–170 120–130	22 EP 20 EP	CO_2 at 12–15 ft³/hr CO_2 at 12–15 ft³/hr	Downhill Up or down	Travel 11 in./min Travel 4 in./min
SMAW	C	MA	1 2 +	$\frac{1}{8}$ $\frac{1}{8}$	3.2 3.2	90–100 110–140	EP EP	Coating Coating	Uphill Uphill	E6010 E6010
SMAW	E	MA	1 2 +	$\frac{1}{8}$ $\frac{1}{8}$	3.2 3.2	80–120 150–200	EP EP	Coating Coating	Downhill Downhill	E6010 E6010
GTAW SMAW	E or F	MA	1 + 2 3 +	$\frac{3}{32}$ $\frac{5}{32}$	2.4 4.0	90–100 120–200	EN 20–22 EP	Argon at 15–20 ft³/hr Coating	Uphill Uphill	Alt. use insert Low hydrogen
SMAW	C	MA	1 2 +	$\frac{5}{32}$ $\frac{3}{16}$	4.0 4.8	140–160 160–190	24–26 EP 24–28 EP	Coating Coating	Downhill Downhill	E6010 or E7016 E6010 or E7016
GMAW	C or E	SA	1 + 1 +	0.035 0.035	0.8 0.8	180–200 120–150	21–23 EP 19–20 EP	CO_2 at 20–30 ft³/hr CO_2 at 20–30 ft³/hr	Downhill Uphill	Less passes for downhill
FCAW	C or E	SA	1 2 +	0.045 $\frac{3}{32}$	0.9 2.4	135–145 325–350	20–21 EP 25–28 EP	CO_2 at 20–30 ft³/hr CO_2 at 25–30 ft³/hr	Uphill Uphill	Use purge for high quality
SMAW	E or F	MA	1 2 +	$\frac{1}{8}$ $\frac{5}{32}$	3.2 4.0	90–100 120–140	EP EP	Coating Coating	Uphill Uphill	EXX10 electrode Low-hydrogen elec.
SAW	E or F	AU	1 2 +	$\frac{5}{32}$ $\frac{5}{32}$	4.0 4.0	375–425 480–505	23–25 EP 26–27 EP	Sub arc flux Sub arc flux	20 in./min 26 in./min	Backup ring rqd. Flat-roll
GMAW SAW	C,E,F	SA or AU AU	1 2 +	0.035 $\frac{5}{32}$	0.8 4.0	150–170 480–505	20–22 EP 26–27 EP	CO_2 at 20–35 ft³/hr Sub arc flux	12 in./min 26 in./min	Double ending Double ending
GMAW FCAW	C,E,F	SA or AU AU	1 2 +	0.035 $\frac{3}{32}$	0.8 2.4	150–170 350–400	20–22 EP 25–28 EP	CO_2 at 20–25 ft³/hr CO_2 at 30–35 ft³/hr	12 in./min 10 in./min	Flat-roll Double ending
FCAW	C or E	SA	1 2 +	0.045 $\frac{3}{32}$	0.9 2.4	135–145 325–350	20–21 EP 25–28 EP	CO_2 at 20–30 ft³/hr CO_2 at 25–30 ft³/hr	Uphill Uphill	Use purge for high quality
GTAW FCAW	C or E	MA SA	1 + 2 3 +	$\frac{3}{32}$ $\frac{3}{32}$	2.4 2.4	90–100 325–350	EN 25–28 EP	Argon at 15–20 ft³/hr CO_2 at 25–30 ft³/hr	Uphill Uphill	Alt. use insert
GTAW GMAW	C or E	MA SA	1 + 2 3 +	$\frac{3}{32}$ 0.035	2.4 0.8	90–100 120–150	EN 19–20 EP	Argon at 15–20 ft³/hr CO_2 at 20–30 ft³/hr	Uphill Either	Alt. use insert

FIGURE 25–23 AWS recommended practices for pipe welding.

D10.4	Austenitic chromium–nickel, stainless steel piping and tubing
D10.6	Gas tungsten arc welding of titanium piping and tubing
D10.7	Gas shielded arc welding of aluminum and aluminum alloy pipe
D10.8	Chromium–molybdenum steel piping and tubing
D10.9	Qualification of welding procedures and welding for piping and tubing
D10.10	Local heating of welds in piping and tubing
D10.11	Root-pass welding without bacing
D10.12	Welding low-carbon steel pipe

FIGURE 25–25 Roll welding in pipe fabrication shop.

FIGURE 25–24 Double joining pipe by roll welding.

FIGURE 25–26 Orbital head welding tubing.

on subassemblies. Normally straight pipe sections are joined to ells, flanges, and so on (Figure 25-25). This provides higher efficiency since welds can be made more rapidly in the flat or roll position. In some cases the root pass, and even the second pass, is made by GMAW or FCAW. FCAW or GMAW can be used for subsequent passes, as well as submerged arc welding.

Orbital welding of thin-wall tubing and standard wall pipe is being done with the GTAW process. Mechanized orbital tube and pipe welding systems are used (Figure 25-26). They are available as complete systems consisting of the power source, programmer, welding head, and so on (Figure 25-27). Remote control pendants or controls on the head allow operation at the point of welding. This equipment can weld tubes with an outside diameter from $\frac{1}{4}$ in. (6.35 mm) to over 8 in. (200 mm), with wall thicknesses from 0.015 in. (0.35 mm) up to $\frac{1}{2}$ in. (6.35 mm). Exact capabilities depend on the welding head design as well as the pipe material. The head shown in Figure 25-28 is designed with a minimum radial clearance of 1 13/16 in. (46 mm), so that it can be used to weld pipe in clusters. These mechanized orbital

heads are compact and rugged and clamp on the pipe or tube. A family of heads is required to weld the smallest to the larger tubes. The welding torch rotates around the pipe and carries the tungsten electrode. In some designs, slip rings are used to avoid rotating or twisting cables and hose. Other heads, which do not include slip rings, allow the cable and hose to wrap around the pipe. Three revolutions are usually the maximum used. A clamshell head design (Figure 25-29) is used for smaller tubes.

The three joint types commonly used are shown in Figure 25-30. This includes the square-groove joint,

FIGURE 25–27 Complete system for welding small-diameter tubing.

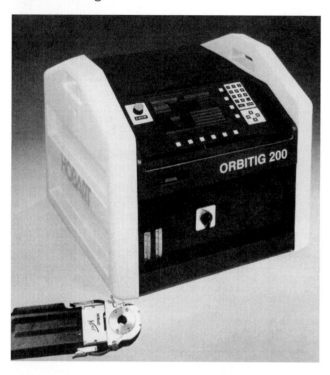

FIGURE 25–28 Tube-to-tube orbital head for GTAW.

FIGURE 25–29 Clamshell orbital head.

FIGURE 25–30 Three weld or joint types.

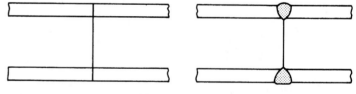

(a) SQUARE GROOVE – BUTT JOINT

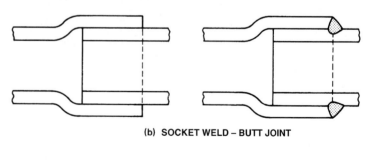

(b) SOCKET WELD – BUTT JOINT

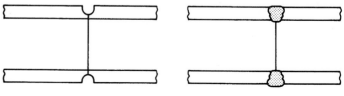

(c) U GROOVE WELD – BUTT JOINT

socket joint, and the U-groove joint. The square groove joint is used for thin-wall tubing and only a single pass is used. Socket joints provide easy fitup and a fillet weld is used. Groove joints, U and V, are used for thicker-wall tubing where full-penetration welds are required. Multiple passes with filler metal are used.

Specialized programmers having upslope and downslope of welding current plus control of rotation speed, preflow and postflow of gas, and high frequency for arc initiation are all included. Pulsing is used for most mechanized welding procedures.

For larger-diameter pipe with thicker wall, a different type of welding head is used. This type rotates around the pipe but is held to the pipe by means of a split-ring or chain-drive assembly (Figure 25–31). Some units have a low profile and can weld pipe with minimum clearance. This machine includes a wire feeder and various collets to weld different sizes of pipe. It can also be used for welding pipe when it is in the vertical position (Figure 25–32). Groove joints are normally employed.

Similar machines have been developed that utilize a combination of welding processes. The first pass will use GTAW and subsequent passes may use the GMAW. Torches are changed for making the total weld. This equipment is becoming increasingly popular for welding pressure piping. Complex controllers and inverter power sources with CC and CV characteristics are used with pulsing.

Mechanized welding machines for welding large-diameter pipe, cross-country pipelines, use the GMAW welding process. These are large machines that fit around the circumference of the pipe and will make gas metal arc welds in the field, on lay-barges, or in the shop. Different types are available that utilize different weld joint details. In some cases, automatic welds are made on the inside diameter of the pipe as well as the outside diameter. A special joint design is used. The inside welding head is combined with a lineup clamp and is shown in Figure 25–33. The equipment for the outside weld is also shown.

FIGURE 25–31   Mechanized pipe welding head for heavy-wall pipe.

FIGURE 25–32 Mechanized pipe welding with mechanized GMAW equipment.

25-5 AUTOMATED PIPE WELDING

Efforts to reduce the time of pipe installation have resulted in fully automated pipe welding systems that are computer driven. Figure 25–34 shows the system for making all-position GTAWs on small-diameter pipe. The cabinet includes the microprocessor controller, computer keyboard and display screen, and a 150-A inverter power source. This equipment includes a remote teaching pendant, shown with the automatic head in Figure 25–35. The welding head on the pipe is shown in Figure 25–36. This head weighs approximately 25 lb and will weld pipe sizes from $1\frac{3}{4}$ to $2\frac{1}{2}$ in. standard wall to the heaviest pipe available. This head has minimum radical clearance so that welds can be made when the pipe is separated by only $2\frac{1}{2}$ in. The head hinges in the middle in a clamshell fashion. Slip rings are used for welding current, control signals, and shielding gas so that cables and hose do not wrap around the pipe during operation. Functional motors for torch rotation, oscillation, tungsten-to-work distance, and cold wire feeder are all built into the housing of the head.

The heart of this automated welding system is the microprocessor controller. The input is by means of the keyboard, and the readout is on the monitor screen. The microprocessor controls all functions. The heart of the operation is the extremely complex software program, which is user friendly. The initial readout is the mode of operation and menu (Figure 25–37). The operator selects the teach mode, which is the next display. Specific instructions for welding are given. The operator will then key in the welding parameters as requested and they will appear on the next display, which is the operating mode. This input provides welding parameters for making the weld utilizing a specific procedure based on pipe size, wall thickness, joint details, pipe analysis, and so on. This procedure can be modified and the procedure can be recorded by means of a hard-copy printout at any step. Welding operators learned to program this equipment in a very short time.

In operation, the head is clamped on the pipe and lined up with the joint. The root pass does not require oscillation; subsequent passes utilize oscillation. Oscillation is programmed with the exact dimensions, which change for each layer. Dwell time is programmed for each end of the stroke, and for each layer the speed of oscillation is

FIGURE 25–33 GMAW automatic welding machine for cross-country pipe.

also programmed and welding current pulsing is synchronized with oscillation. When the second pass is completed, the programmer automatically changes to the third pass without stopping the weld. Welding parameters can be changed each 10° around the pipe. The welding is uninterrupted from root to final pass and provides 100% arc time. When the final pass is completed, the controller turns off the machine. Welds produced by this automated system meet the requirements of the most stringent codes. Radiographs are water clear.

The teach pendant is used to input information to the microprocessor to establish the total welding procedure. Arc sensing is used to control oscillation. This allows the head to mechanically oscillate during setup to determine the centerline of the weld joint prior to striking an arc. The arc will sense the joint at each end of the oscillation stroke. The controller will reverse the stroke and keep the weld head centered on the joint. This can be modified for a split weave technique and can be different for each layer.

FIGURE 25–34   Automated pipe welding system.

FIGURE 25–35   Automated head and remote pendant.

FIGURE 25–36   Welding head on pipe.

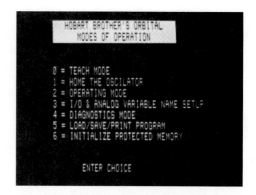

```
HOBART BROTHER'S ORBITAL
MODES OF OPERATION

0 = TEACH MODE
1 = HOME THE OSCILATOR
2 = OPERATING MODE
3 = I/O & ANALOG VARIABLE NAME SETUP
4 = DIAGNOSTICS MODE
5 = LOAD/SAVE/PRINT PROGRAM
6 = INITIALIZE PROTECTED MEMORY

          ENTER CHOICE
```

FIGURE 25–37   Program menu.

Practical application in a fabrication shop utilizes two heads with one controller power source panel. While one head is making a weld, the other head is being attached and aligned to another joint. When the first joint is completed, the controller switches to the second head and makes the weld. Meanwhile the first head is removed and attached to a new joint.

This machine can be utilized for remote welding in dangerous or radioactive atmospheres. The controllers can also collect data during welding. This includes all parameters. It can select data when the parameters are outside defined limits. It can also transfer data.

Automated welding has been applied to the plasma arc welding, keyhole mode process for roll welding. In this case the torch is stationary, but adjustable, while the pipe joint rotates under the arc. The welding head has automatic X, Y, and Z adjustments. The angle of the torch is preset. The microprocessor initiates the plasma arc and all other functions, including torch adjustment, pipe rotation, and gas coverage. Welding parameters are programmed to initiate the keyhole and to provide filler metal. This equipment is designed for high-alloy steel piping and is normally used with single-pass operation. Upon completion of the weld, the computer programs the closing of the keyhole, which involves simultaneous changing of four variables. This equipment is shown in Figure 25–38.

25-6 TUBE TO SHEET WELDING

Mechanized equipment is widely used for welding tubes to tube sheets or heads. This equipment is popular in the heat-exchanger industry. A heat exchanger consists of many tubes between two headers, where the tubes are attached to the headers with perfect, leakproof connections. A heat exchanger may have hundreds of tube-to-header joints. Previously, these were mechanically connected or manually welded with GTAW, which was a tedious job. The change from manual to machine welding has improved quality and reduced the cost per weld.[4]

Complete welding systems are available, including the mechanized orbital welding head, the welding power

FIGURE 25–38 Roll pipe welding equipment: plasma.

FIGURE 25–39 Tube-to-sheet mechanized welding machine.

FIGURE 25–40 Tube-to-tube sheet welding head.

source and the programmer, which completely mechanize the welding operation.

The welding head is a compact, lightweight device that rotates the gas tungsten arc welding torch around the periphery of the tube to sheet joint. The head includes a mandrel, which fits inside the tube to be welded and locates the torch. The head will rotate the torch 360° plus overlap in each direction. Slip rings are incorporated so that the hose and cables do not twist. Figure 25–39 shows this equipment in use welding a small heat exchanger. This photograph shows the tube sheet in the vertical position; however, the equipment can be used if the tube sheet is horizontal and the weld is flat or even if the weld is overhead.

Heads of this type can weld tubes with outside diameter from $\frac{3}{8}$ in. (16 mm) to 6 in. (150 mm) with a tube wall thickness of 0.015 in. (0.4 mm) and larger. With the proper attachments the head can also make welds in the internal surface or bore. These are used for retubing heat exchangers.

A close-up of the welding head is shown in Figure 25–40. The head includes the GTAW torch, rotation motor and filler wire, and drive motor. The tungsten electrode is the 2% thoriated type and $\frac{3}{32}$ in. diameter is normally used. Filler metal can be added for some types of joints. When filler metal is added, the arc length should be slightly greater than the electrode diameter. When filler metal is not added, the arc length is slightly less than the electrode diameter.

The tungsten electrode position is adjustable and critical for tube-to-header welds. Figure 25–41 shows the electrode position for the most common joint designs.

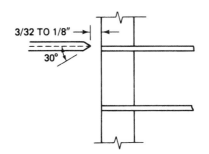

FLUSH TUBE WELD

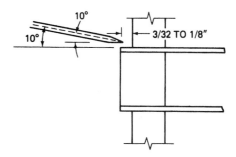

FIGURE 25–41 Tungsten electrode position for welds.

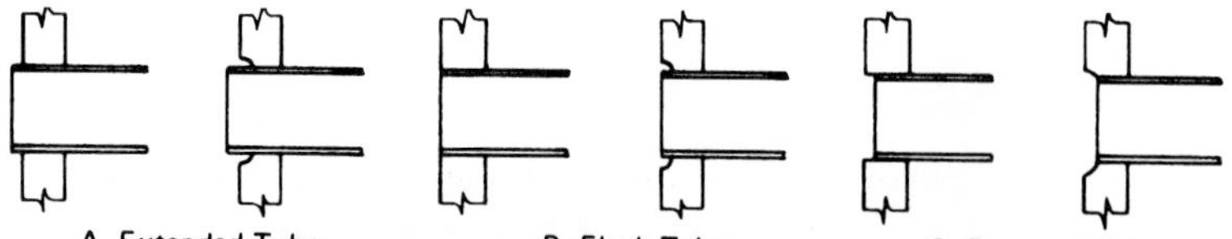

A. Extended Tube   B. Flush Tube   C. Recessed Tube

Tube O.D.		Wall Thickness		Joint Type	Weld Current (Amperes)	Filler Rod Type	Weld Time (sec)
in.	mm	in.	mm				
Stainless steel tube-to-stainless steel tube sheet							
0.75	19.1	0.062	1.6	A	140	E304	18
0.75	19.1	0.080	2.0	A	130	None	18
0.75	19.1	0.090	2.3	A	140	E304	31
Mild steel to mild steel							
0.75	19.1	0.062	1.6	B	140	E705-3	28
1.00	25	0.125	3.2	B	140	E705-3	184
1.00	25*	0.125	3.2	C	180	E705-3	49
2.50	62	0.187	4.7	B	175	E705-3	245
Stainless steel to mild steel							
0.62	15.7	0.062	1.6	A	120	E309	14
1.00	25	0.083	2.1	A	155	E309	40
CuNi-CuNi							
0.75	19.1	0.062	1.6	A	160	ERCuNi	28
CuNi—mild steel							
0.75	19.1	0.062	1.6	A	160	ERNiCrFe-6	28
Cu—mild steel							
0.62	15.7	0.025	0.6	A	140	ERCuSi-A	16

*2 passes

FIGURE 25-42 Tube-to-tube sheet welding schedule.

The programmer is the same as used with mechanized tube-to-tube welding heads. It starts the gas preflow, torch rotation, high frequency, and welding current, which changes during the weld cycle. Pulsing is normally used for making tube-to-tube sheet welds. The controller has various delays and ends with postflow of shielding gas. The equipment may also include a weld control pendant, which is used when remote welding is required.

The joint detail used for this type of welding is shown in Figure 25-42. The three most common joint designs are the extended tube, flush tube, and recessed tube. There are variations of each design. Some applications require only a seal bead between the tube and the tube sheet. Filler metal is normally added for the ex-

tended tube or the recessed tube joint design. Nuclear specifications require that the weld metal thickness be equal to the thickness of the tube. The fillet weld design will not produce the desired cross-section dimension since the throat dimension is less than the thickness of the tube. The design selected must have sufficient filler metal so that the weld is stronger through its shortest dimension than the thickness of the wall of the tube. Joint design is based on the specifications involved. The welding procedure shown is with the tube sheet vertical. With the tube sheet flat, higher currents can be used. Any metal welded by the gas tungsten arc process can be welded with mechanized GTAW tube-to-tube sheet welding heads.

QUESTIONS

25–1. What are the seven different classifications of pipe and tubing?

25–2. What schedule number pertains to standard wall pipe?

25–3. Briefly describe a continuous pipe mill.

25–4. What arc welding process is used to make spiral joint pipe?

25–5. What percentage of steel produced is made into tubular products?

25–6. What code applies to most high-pressure pipe welding? Cross-country pipe?

25–7. Can more than one welding process be used to make a pipe weld? Explain.

25–8. Why are different pipe weld joint designs used? Where is each used?

25–9. Explain the difference between internal and external line-up clamps. What determines the type to be used?

25–10. Explain the difference between uphill and downhill pipe welding.

25–11. What type of covered electrode is widely used on cross-country pipe welding?

25–12. What is stove pipe welding?

25–13. What is double jointing?

25–14. What is the difference between roll welding and fixed-position welding?

25–15. Can submerged arc welding be used for fixed-position welding? For roll welding?

25–16. Are low-hydrogen welding electrodes used for pressure piping?

25–17. What is the advantage of consumable inserts for pipe welding?

25–18. Can gas metal arc welding be used on pipelines?

25–19. What is the advantage of mechanized orbital welding of tubes?

25–20. What are the three joint types for tube-to-tube sheet welds?

REFERENCES

1. H. E. McGannon, ed., "The Making, Shaping and Treating of Steel," (Pittsburgh, Pa.: Association of Iron and Steel Engineers, 1995).

2. "Power Piping," ASME Code for Pressure Piping B31, American Society of Mechanical Engineers, New York.

3. "Standard for Welding Pipe Lines and Related Facilities," API Standard 1104, American Petroleum Institute, Washington, D.C.

4. W. Hebert, "Mechanized Tube Welding Speeds Heat Exchanger Fabrication," *Welding Journal* (May 1986).

26

SPECIAL WELDING APPLICATIONS

26-1 ARC SPOT WELDING

An arc spot weld is a spot weld made by an arc welding process. A spot weld is "a weld made between or upon overlapping members in which coalescence may start and occur on the faying surfaces or may proceed from the outer surface of one member. The weld cross section is approximately circular." Figure 26-1 shows the different arc spot welds. An arc spot weld differs from a resistance spot weld since in resistance welding coalescence is located at the faying surfaces of the parts being joined.[1] The arc spot weld does not require a hole in either member. It differs from the plug weld, which requires a hole to be prepared for the weld.

Arc spot welding is performed by melting through the top member. The thickness of this member is the limiting factor, which also depends on the welding process used.

The gas tungsten arc and the gas metal arc welding process are most commonly employed for making arc spot welds. Flux cored arc welding, shielded metal arc welding using covered electrodes, and the plasma arc process can also be used.

The principal operation of arc spot welding is to strike and hold an arc without travel at a point where the two parts are held tightly together. The heat of the arc melts the surface of the top member, and the depth of melting is dependent on the welding process, the electrode size and type, the welding current and the time, and, in the case of gas metal arc welding, the shielding gas employed. Timing is done automatically by means of a timer. With the exception of the timer and special electrode holder or gun, the process utilizes the same equipment that is normally used for that process.

Arc spot welding offers many metal joining advantages. One advantage is that it can be used without manipulative skills. The training required to make arc spot welds is minor. In addition, the welder making arc spot welds does not have to use a welding helmet since the arc is contained within a gun nozzle. Arc spot welding is extremely fast, can be used for a variety of joining requirements, and can be fully automated.

Arc spot welds can be made in the flat position and in the horizontal position, but are almost impossible to make in the overhead position. The reason is that the member closest to the welding gun must melt completely through and results in a fairly large mass of molten metal that falls away from the weld.

Material	Mild Steel	Mild Steel	Mild Steel	Stainless	Aluminum
Thickness	20 ga (0.0359 in.)	11 ga (0.1196 in.)	3/16 in.	22 ga (0.0299 in.)	11 ga (0.0907 in.)
Process	GMAW (CO_2)	GMAW (CO_2)	FCAW (CO_2)	GTAW (Argon)	GMAW (Argon)
Electrode size	0.035 in.-.9 mm	1/16 in.-1.6 mm	7/64 in.-2.8 mm	– – –	3/64 in.-1.2 mm
Electrode type	E70S-3	E70S-3	E70T-1	– – –	5356
Top side	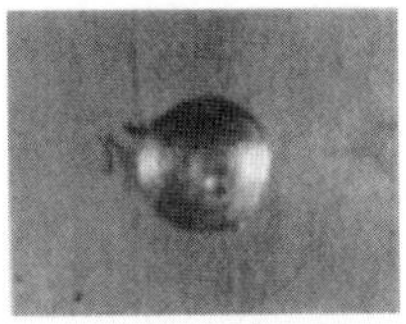	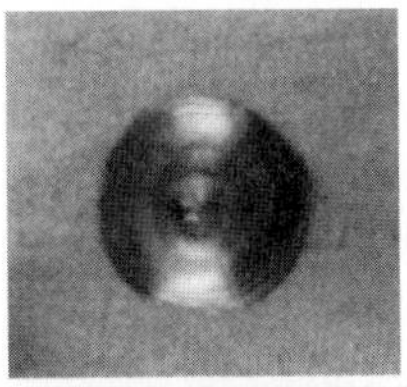	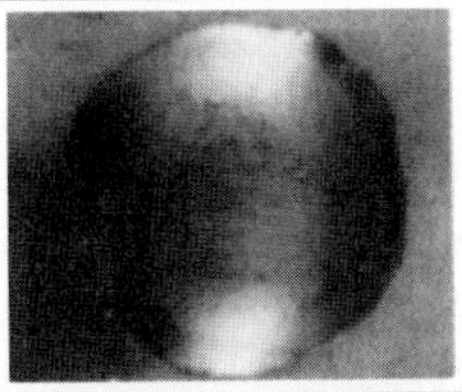	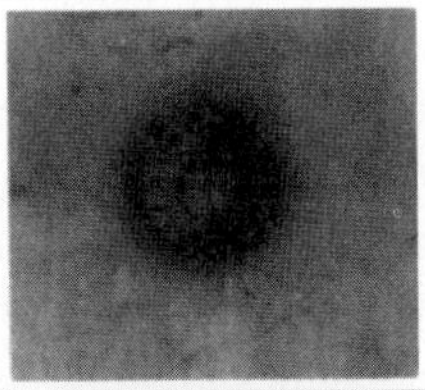	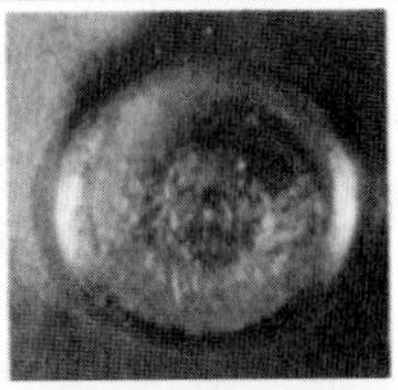
Bottom side	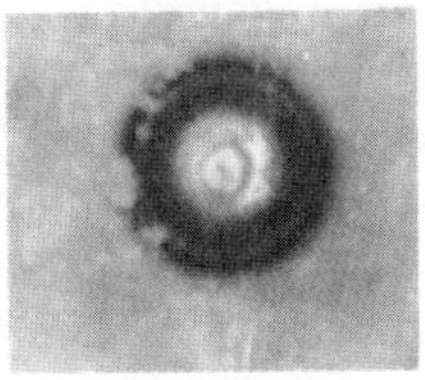	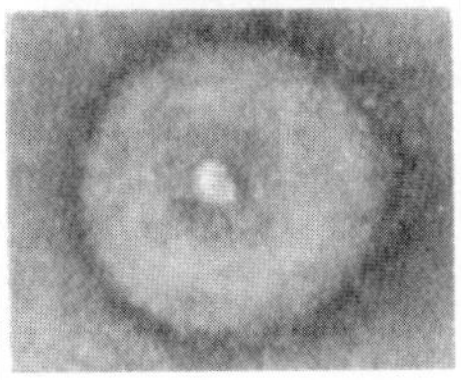	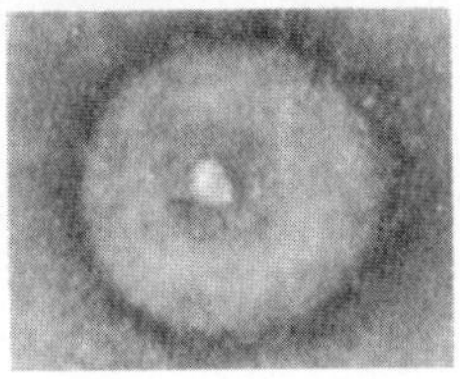	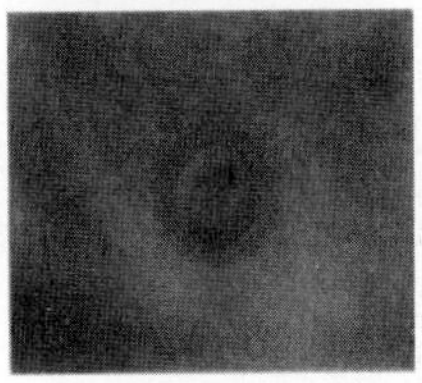	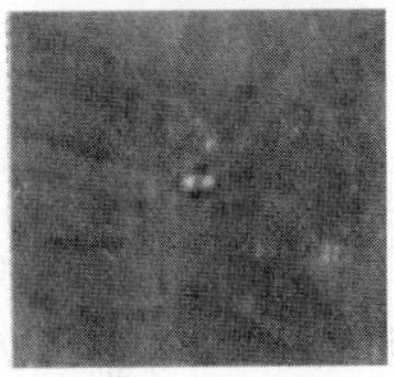
Cross section	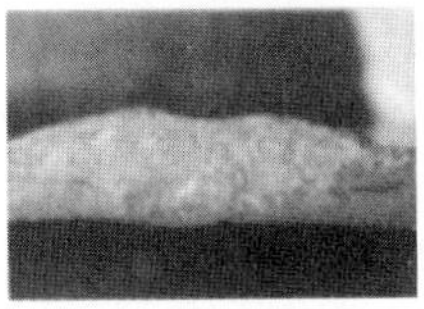	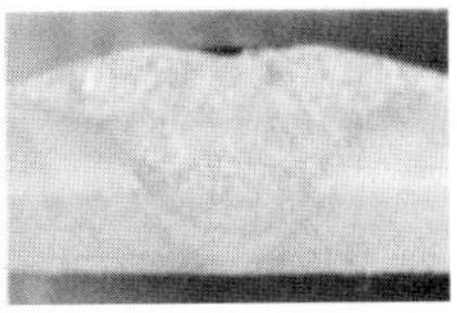	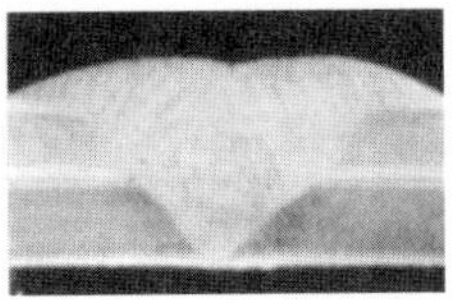		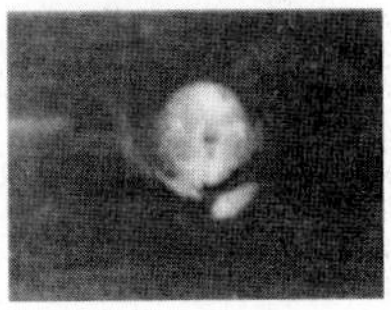

FIGURE 26–1 Spot welds: top, bottom, and cross section.

The metals normally welded with arc spot welding are the mild, low-alloy, and stainless steels. With special precautions other metals can be welded. Aluminum can be arc spot welded if a clean interface between the parts to be joined is maintained.

Thickness range of metals that can be welded by the different arc processes is shown in the schedules for GTAW and GMAW welding. This involves the thickness of the top member that must be melted through. The thickness of the other member is not important. Lap joints are the most common types of joints used for arc spot welding; however, T-joints can also be made.

Gas Tungsten Arc Spot Welding

Arc spot welding with GTAW is an extremely efficient and simple way to make weld joints. The process is limited to a maximum thickness of $\frac{1}{16}$ in. (1.6 mm) of the sheet closest to the arc. Gas tungsten arc welding can be used for mild steels, low-alloy steels, stainless steels, and aluminum alloys.

A special welding gun, designed for arc spot welding, must be used. The nozzle of the gun is used to apply pressure to hold the parts in close contact. A gun for gas

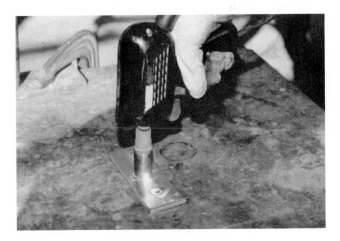

FIGURE 26–2 Making a GTAW spot weld.

tungsten arc spot welding is shown in Figure 26–2. The gun nozzle is made of copper or stainless steel and is normally water cooled since the arc is contained entirely within the nozzle. The nozzle design controls the distance between the tungsten electrode and the surface of the work; it should have ports for the shielding gas to escape. The inside diameter should be related to the size of the tungsten electrode being used. The 0.500-in. (12-mm) inside diameter is most common. The nozzles can also be designed to help locate the arc spot weld, especially with respect to corners or edges of the top sheet. Arc spot welding equipment can be used to make tack welds at inside or outside corner joints. The gas tungsten arc spot gun includes a trigger switch that will actuate the arc spot operation.

A timer controller is required. Some gas tungsten arc power sources include timers. However, a separate timer can be used, provided that the power source includes a contactor. The timer should have the capability of being adjusted from 0.5 second up to 5 seconds.

The tungsten electrode type would be the same as used for the material to be welded. The $\frac{1}{8}$-in. (3.2-mm) electrode diameter is recommended for all work. The point of the tungsten should be the standard ground point but then squared off or blunted on the end. This tends to improve the depth of penetration. An arc length of $\frac{1}{16}$ in. (1.6 mm) is recommended. If the arc length is too short, the weld area or spot weld size will be small. As the arc length increases, the size of this arc spot weld will be

larger. If the arc length is too long, an unstable arc will result and there will be a lack of uniformity.

The normal sequence of events is: The nozzle of the gun is placed on the joint and sufficient pressure is applied to bring the parts in intimate contact. The trigger is depressed, which starts the welding cycle. Gas flow is initiated to purge the area within the gun nozzle. If water cooling is employed, the cooling water will also start to flow. The arc will be initiated by the high-frequency current supplied by the power source. The arc will continue for the period of time established and will be extinguished as the contactor opens. The shielding gas will continue to flow for a predetermined post-flow time. This completes the cycle.

Normally, the thinnest metals joined are 24 gauge, which is 0.022 in. (0.56 mm). If both top and bottom sheets are of the same thickness, it is best to utilize a copper backup to prevent the weld from falling through the joint and having a depression on the top and excessive penetration on the bottom. A schedule for gas tungsten arc welding is shown in Figure 26–3. It is best to use high current and short time cycle rather than lower currents

FIGURE 26–3 Schedule of arc spot welds using GTAW.

Material Type	Metal Thickness (Top Sheet)			Welding Conditions (A)		Arc Spot Time (seconds)	Shielding Gas (Argon)	
	Gauge	in.	mm	DCEP + HF	AC-HF		ft³/hr	liters/min
Stainless steel	24	0.025	0.64	125	175	1.0	10	4.5
	24	0.025	0.64	110	175	1.25	10	4.5
	24	0.025	0.64	100	150	1.5	10	4.5
	22	0.031	0.79	125	175	1.5	10	4.5
	22	0.031	0.79	100	175	1.75	10	4.5
	18	0.050	1.27	140	200	1.5	12	5.6
	18	0.050	1.27	110	150	2.5	12	5.6
	16	0.062	1.57	170	250	3.0	12	5.6
	16	0.062	1.57	140	—	3.25	12	5.6
	16	0.062	1.57	115	—	5.25	12	5.6
		0.064	1.62	160	250	2.25	12	5.6
Mild steel	22	0.031	0.79	170	250	1.5	8	3.6
	22	0.031	0.79	140	200	2.0	8	3.6
	22	0.031	0.79	120	175	2.25	8	3.6
	18	0.050	1.27	170	250	1.75	10	4.5
	18	0.050	1.27	140	200	2.0	10	4.5
	18	0.050	1.27	135	200	2.5	10	4.5
	16	0.062	1.57	170	250	3.0	12	5.6
	16	0.062	1.57	155	225	3.5	12	5.6
Aluminum		0.022	0.56	—	170	1.1	8	3.6
		0.032	0.81	—	200	1.5	8	3.6
		0.048	1.21	—	220	1.7	8	3.6
		0.064	1.63	—	250	2.2	8	3.6

Notes: 1. The electrode is 2% thoriated tungsten, except for aluminum pure tungsten electrodes are used, $\frac{1}{8}$ in. in diameter.
2. High-frequency is used to start the arc when using dc and ac.
3. Arc-length electrode to work $\frac{1}{16}$ in.

with longer time period. The amount of current increases the size of the nuggets and the thickness of the materials that can be welded. Nugget diameter affects strength. The time factor tends to increase penetration but to a much lesser degree than the current. If too much current is used, it can cause splashing. This splashing of the molten metal from the top member may contaminate the tungsten and will result in an unsatisfactory weld. If the current is too low, the nugget may not penetrate completely through the top sheet to the bottom sheet.

The shielding gas will be either argon or helium with a flow rate of from 6 to 10 ft³/hr (2.5 to 4.5 liters/min). Helium will provide a smaller weld nugget with a greater depth of penetration. Argon produces a larger weld nugget with penetration not quite so deep.

Direct current should be used for all materials, except aluminum, with the electrode negative (straight polarity). Alternating current with continuous high frequency should be employed on aluminum. If aluminum is well cleaned, the electrode negative (straight polarity) can be used. Parts to be welded should be clean of oil, dirt, grease, scale, and so on. This is especially true at the interface and absolutely necessary when welding aluminum. The weld diameter is the basis for the shear strength of arc spot welds. The shear strength will be similar to resistance spot welds made in the same material.

Gas tungsten arc spot welding is widely used in the manufacture of automotive parts, appliances, precision metal parts, and parts for electronic components. It is normally applied as a semiautomatic process; however, it can be mechanized and used for high-volume production work.

Gas Metal and Flux Cored Arc Spot Welding

The gas metal arc and flux cored arc welding processes can both be used for making arc spot welds. The equipment used is identical. The difference is in the electrode wire employed.

The arc welding equipment used for normal semiautomatic welding can be used for gas metal and flux cored arc spot welding by the addition of a timer and a special torch nozzle. Time range for the timer should be the same as used for gas tungsten arc spot welding. The nozzle and torch should be sufficiently strong so that they can transmit the force to hold the parts together in intimate contact.

The sequence of events to produce an arc spot weld is almost the same as used for gas tungsten arc welding. The only difference is that when the arc starts, the wire feeder will feed electrode wire into the arc.

The gas metal and flux cored arc processes can weld greater thicknesses of metal than gas tungsten arc spot welding. With proper conditions, welds can be routinely made through the top plate of $\frac{1}{4}$ in. (6.4 mm) thickness. At the other end of the scale, welds can be made in 24-gauge material 0.022 in. (0.56 mm) thick. Welding is done primarily on the mild steels and low-alloy steels. Figure 26–4 shows an arc spot weld being made.

CO_2 shielding gas is normally used. Carbon dioxide is selected since it has the highest penetrating qualities of any shielding gas. If high penetration is not required, the 75% argon–25% CO_2 mixture can be used.

The size and type of electrode wire have a large effect on the depth of penetration and on the diameter of the weld nugget at the interface. For maximum strength it is desirable to have a large nugget at the interface. The electrode wire should be a deoxidized type; normally, the E705-4 or 5 electrode wire is used with solid wires or the E70T-1 electrode wire is used for the flux cored arc welding process.

The weld schedule for making GMAW and FCAW arc spot welds is shown in Figure 26–5. The electrode wire changes from a small diameter of 0.030-in. (0.8-mm) solid wire up through $\frac{1}{16}$-in. (1.6-mm) solid wire and to $\frac{3}{32}$-in. (2.4-mm)-diameter flux cored wire to accommodate all the weld thicknesses shown. The larger-diameter wires provide higher strength per arc spot weld, since they produce larger nuggets at the interface.

When welding thin materials, a backup bar should be used behind the second sheet. It is also possible to make arc spot welds through more than two sheets, provided that the combined thickness of the top two sheets is within the range of the welding schedule. Intimate contact is required. The same equipment with timers can be used for tack welds and for plug welds when the top sheet is too thick.

The advantages of arc spot welds over resistance spot welds are as follows:

1. Access is required to the top or front side of the joint.
2. The amount of pressure is not excessive, assuming that the parts are fitted properly.

FIGURE 26–4 Making a GMAW or FCAW spot weld.

FIGURE 26–5 Schedule of arc spot welds using FCAW or GMAW.

Gauge	Material Thickness		Electrode Wire Diameter		Electrode Type	Current (A dc)	Arc Voltage (EP V)	Arc Spot Time (seconds)	Wire Consumed per Spot		Typical Shear Strength per Spot	
	in.	mm	in.	mm					in.	mm	lb	kg
24	0.022	0.56	0.030	0.76	E60S-1	90	24	1.0	$4\frac{5}{8}$	115	625	283.60
22	0.032	0.81	0.030	0.76	E60S-1	120	27	1.2	5	125	730	331.13
20	0.037	0.94	0.030	0.76	E60S-1	120	27	1.2	$10\frac{1}{8}$	253	1337	606.46
22	0.032	0.81	0.035	0.89	E60S-1	140	26	1.0	6	150	800	362.88
20	0.037	0.94	0.035	0.89	E60S-1	140	26	1.0	6	150	1147	520.33
18	0.033	0.84	0.035	0.89	E60S-1	190	27	1.0	$8\frac{1}{2}$	212	1507	683.58
16	0.059	1.50	0.035	0.89	E60S-1	190	28	2.0	$17\frac{1}{4}$	431	1434	641.46
14	0.072	1.82	0.035	0.89	E60S-1	190	28	5.0	$40\frac{1}{4}$	1006	2600	1179.96
18	0.039	0.99	0.045	1.14	E60S-1	200	27	0.7	4	100	1414	641.39
16	0.059	1.50	0.045	1.14	E60S-1	260	29	1.0	6	150	2070	938.95
14	0.072	1.82	0.045	1.14	E60S-1	300	30	1.5	$12\frac{3}{4}$	319	3224	1462.41
12	0.110	2.79	0.045	1.14	E60S-1	300	30	3.5	$28\frac{1}{2}$	712	4300	1950.48
11	0.124	3.15	0.045	1.14	E60S-1	300	30	4.2	34	850	4114	1866.11
16	0.059	1.50	$\frac{1}{16}$	1.6	E60S-1	250	29	1.0	$2\frac{3}{4}$	69	1654	750.25
14	0.072	1.82	$\frac{1}{16}$	1.6	E60S-1	360	31	1.0	$5\frac{1}{2}$	137	3340	1515.02
12	0.110	2.79	$\frac{1}{16}$	1.6	E60S-1	440	32	1.0	$7\frac{1}{4}$	181	5000	2268.00
11	$\frac{1}{8}$	3.18	$\frac{1}{16}$	1.6	E60S-1	490	32	1.0	$8\frac{1}{2}$	212	5634	2556.58
	$\frac{5}{32}$	4.0	$\frac{1}{16}$	1.6	E60S-1	490	32	1.5	9	225	5447	2460.76
	$\frac{3}{16}$	4.76	$\frac{1}{16}$	1.6	E60S-1	490	32	2.0	$16\frac{3}{4}$	419	6834	3179.90
	$\frac{1}{4}$	6.4	$\frac{1}{16}$	1.6	E60S-1	490	34	3.5	$28\frac{1}{8}$	703	8667	4721.35
16	0.062	1.57	$\frac{3}{32}$	2.4	E70T-2	400	30	0.6	$1\frac{3}{8}$	34	2550	1156.68
11	$\frac{1}{8}$	3.18	$\frac{3}{32}$	2.4	E70T-2	500	34	0.8	3	75	3400	1442.24
	$\frac{3}{16}$	4.76	$\frac{3}{32}$	2.4	E70T-2	650	38	1.6	$8\frac{1}{4}$	206	7050	3197.88
	$\frac{1}{4}$	6.4	$\frac{3}{32}$	2.4	E70T-2	750	40	2.2	$15\frac{3}{4}$	394	10,300	4672.08

Notes: 1. Contact tip-to-work distance: $\frac{1}{4}$ to $\frac{3}{8}$ in. for fine wire, $\frac{7}{8}$ for large wire and flux cored wire, CO_2 $\frac{7}{8}$ in., gasless, $1\frac{1}{2}$ in.
2. CO_2 when used: 35 ft^3/hr (6.5 liters/min).

3. Operator skill is minimum, and a welder's helmet is not required.

4. The consistency and reproducibility of arc spot welds are excellent.

5. The amount of weld spatter, smoke, and flash is minimum, and metal finishing can be eliminated for many products.

6. Distortion is minimum.

7. Close cost control is obtained using arc spot welding.

8. It is readily adaptable to designs originally used for bolting or riveting.

Arc spot welding is being used in automobile body assemblies, the attachment of brackets to body assemblies, frame assemblies, the assembly of industrial products, the assembly of lattice structural beams, and for innumerable other applications. The strengths are in the same relationship as resistance spot welds on the same materials in the same thicknesses.

Variations of the Process

There is one variation that would be more accurately described as a plug weld. It is used to join dissimilar metals. It has been used for joining aluminum to copper and galvanized steel to aluminum. Aluminum filler wire is used with inert gas through a hole in the copper or galvanized part. Copper terminals can be joined to aluminum cables, and galvanized steel brackets can be joined to aluminum pans.[2] In plug welding it is important to establish the arc on the bottom piece for a good-quality weld.

Another variation of arc spot welding is done with the shielded metal arc welding process using covered electrodes. Special spot welding guns or holders are used. Small-diameter electrodes are used. The special holder causes the arc to strike and holds a short arc for several seconds without manual assistance. An arc shield surrounds the arc area, and a welding helmet is not required. This process variation will weld through 16-gauge steel in the flat, horizontal, and vertical positions. It is used in auto body repair shops.

26-2 SHEET METAL WELDING

Sheet metal is metal having a thickness of $\frac{1}{8}$ in. (3.2 mm) or less. This means that it has a gauge number of 11 and higher, the higher numbers indicating thinner thicknesses of metal. The thickness and gauge number relationship are shown in Figure 26–6. Welding can be performed on the thinnest metal produced, but special fixturing and automatic travel are required. Under normal conditions using a manual or semiautomatic process, sheet metal approximately 0.035 in. (0.9 mm) in thickness or roughly 20 gauge can be welded.

Thin sheets of stainless steels, aluminum alloys, and nickel alloys are also welded. The processes most commonly used are gas metal arc welding for thin sections and gas tungsten arc and plasma arc welding for the thinnest metals.

There are two major problems involved with welding sheet metal: (1) minimizing the distortion and (2) avoiding burn-through.

The problems of burn-through and distortion can be minimized by the use of tight fitup, clamping, fixtures, and backup bars. These are all recommended for production welding; however, for maintenance and repair welding accurate fitup, clamping, and the use of backup bars may not be possible.

The GMAW process, utilizing the short-circuiting arc transfer, is the most suitable. This process has replaced shielded metal arc welding for almost all sheet metal applications. The main reason is its ability to operate at a wide range of current levels. Its relatively high-speed travel, which balances the heat buildup problem, greatly reduces weld distortion. Brazing is used for galvanized steel applications and for joining copper alloys. The single carbon arc method and the gas torch method are both used but are losing popularity.

The problem of burn-through occurs with all of the processes, and steps to avoid burn-through include the use of close-tolerance cutting to provide tight, even fitup between parts. The preparation of sheet metal for welding is normally by shearing, which produces straight edges that can easily be aligned properly. A backup bar of copper and the ample use of clamps to hold the sheet metal in alignment and against the backing bar will aid in making good-quality weld joints. The travel speed should be as high as possible; this is a matter of welder skill and practice. The welder should try to travel at a uniform high speed but must be able to follow the joint accurately.

Fitup and distortion are closely related since distortion ahead of the arc will cause the fitup to vary and cause burn-through. A large number of small tack welds should be used. They should be relatively short but closely spaced. This will help maintain tight fitup and will reduce weld distortion. It is also helpful to use the push-travel angle. Another assist is to make the weld in a downhill position. If the work can be tilted so that welding can be done downhill, approximately 45°, a flatter bead will result, travel speed will be higher, and distortion will be reduced.

When using the gas metal arc welding process, the argon—CO_2 mixture (75% argon and 25% CO_2) helps improve the welding operation since it tends to reduce penetration into the joint, reduces the spatter level, and produces a smoother weld bead. The fine wire variation can be used at extremely low currents and should match the thickness of the metal. FCAW with very small electrode wire can be used on the heavier gauges.

FIGURE 26–6 Sheet metal gauges.

Gauge Number	Aluminum and Brass Brown and Sharp (in.)	Steel Sheets Mfrs. Std.[a] (in.)	Strip and Tubing and Copper Birmingham or Stubs (in.)	Nearest Metric Thickness[b]	Stress Wire Gauge[c]
6/0s	0.5800				0.4615
5/0s	0.5165		0.500		0.4305
4/0s	0.4600		0.454		0.3938
4/0s	0.4096		0.425		0.3625
3/0s	0.3648		0.380	10.0	0.3310
2/0s	0.3249		0.340	9.0	0.3065
1	0.2893		0.300	8.0	0.2830
2	0.2576		0.284	7.0	0.2625
3	0.2294	0.2391	0.259	6.0	0.2437
4	0.2043	0.2242	0.238	5.5	0.2253
5	0.1819	0.2092	0.220	5.0	0.2070
6	0.1620	0.1943	0.203	4.8	0.1920
7	0.1443	0.1793	0.180	4.5	0.1770
8	0.1285	0.1644	0.165	4.2	0.1620
9	0.1144	0.1495	0.148	3.8	0.1483
10	0.1019	0.1345	0.134	3.5	0.1350
11	0.0907	0.1196	0.120	3.0	0.1205
12	0.0808	0.1046	0.109	2.8	0.1055
13	0.0720	0.0897	0.095	2.2	0.0915
14	0.0641	0.0747	0.083	2.0	0.0800
15	0.0571	0.0673	0.072	1.8	0.0720
16	0.0508	0.0598	0.065	1.6	0.0625
17	0.0453	0.0538	0.058	1.4	0.0540
18	0.0403	0.0478	0.049	1.2	0.0475
19	0.0359	0.0418	0.042	1.1	0.0410
20	0.0320	0.0359	0.035	1.0	0.0348
21	0.0285	0.0329	0.032	0.090	0.0317
22	0.0253	0.0299	0.028	0.080	0.0286
23	0.0226	0.0269	0.025	0.070	0.0258
24	0.0201	0.0239	0.022	0.060	0.0230
25	0.0179	0.0209	0.020	0.055	0.0204
26	0.0159	0.0179	0.018	0.045	0.0181
27	0.0142	0.0164	0.016	0.040	0.0173
28	0.0126	0.0149	0.014	0.035	0.0162
29	0.0113	0.0135	0.013		0.0150
30	0.0100	0.0120	0.012		0.0140
31	0.0089	0.0105	0.010		0.0132
32	0.0080	0.0097	0.009		0.0128
33	0.0071	0.0090	0.008		0.0118
34	0.0063	0.0082	0.007		0.0104
35	0.0056	0.0075	0.005		0.0095
36	0.0050	0.0067	0.004		0.0090
37	0.0045	0.0064			0.0085
38	0.0040	0.0060			0.0080

[a]Replaces U.S. standard (revised) gauge.
[b]ANSI B32.3.
[c]Replaces Washburn and Moen gauge.

When using the shielded metal arc welding process, use the smallest size electrodes possible. This is the $\frac{3}{32}$-in. diameter size or, for heavier sheet metal, the $\frac{1}{8}$-in. diameter size. Either ac or dc can be used; however, this will dictate the electrode type that should be selected. If ac equipment is available, the E6013 electrodes should be used. Ac is preferred over the direct current for welding thinner sheet metal. If dc equipment is to be used, the selection would be E6012 electrodes. When using the electrode negative (straight polarity), penetration is reduced and the metal transfer is more of a spray type. A short arc length should be used, with the arc length equal to the core wire diameter. The pull travel angle is preferred, and the weld joint should be positioned at approximately 45° so that the welding can be done in the downhill direction.

When plasma arc welding or gas tungsten arc welding is employed, use high current and maximum travel speed. The gas tungsten arc process has the slowest travel speed of the arc processes, which tends to increase distortion. The plasma process can be used at a higher speed, which will reduce distortion.

When using the oxyfuel gas welding process, the forehand technique or forward pushing travel angle should be used. When using the single carbon arc process and a bronze rod, the arc is played on the rod and allowed to melt onto the sheet metal joint. This reduces heat into the joint, reduces distortion, and allows for rapid travel speed. The electrode or filler rod must be selected to match the base metal. Procedure information using these processes is given in Figure 26-7.

FIGURE 26–7   Sheet metal welding schedules.

Process	Sheet Metal Gauge[a]			Filler Rod or Electrode Diameter		Volt	A dc	Shielding Gas and Flow	Travel Speed	
	Gauge	in.	mm	in.	mm				in./min	mm/min
PAW	25	0.020	0.5	—	—	18	12	20	21	533
	20	0.030	0.8	—	—	18	34	20	17	432
	16	0.062	1.6	—	—	20	65	20	14	355
	$\frac{3}{32}$ in.	0.093	2.4	—	—	17	85	20	16	406
	$\frac{1}{8}$ in.	0.125	3.2	—	—	18	100	20	16	406
GTAW	20	0.032	0.8	$\frac{1}{16}$	1.6	11	75–100	10	13	330
	18	0.040	1.0	$\frac{1}{16}$	1.6	12	90–120	10	15	380
	16	0.063	1.6	$\frac{1}{16}$	1.6	11	95–135	10	15	380
	$\frac{3}{32}$ in.	0.094	2.4	$\frac{3}{32}$	2.4	12	135–175	10	14	355
	$\frac{1}{8}$ in.	0.125	3.2	$\frac{1}{8}$	3.2	12	145–205	12	11	280
GMAW	24	0.025	0.6	0.030	0.8	16	30–50	20	16	406
	22	0.031	0.8	0.030	0.8	16	40–60	20	19	482
	20	0.037	0.9	0.035	0.9	17	55–85	20	37	940
	18	0.050	1.3	0.035	0.9	18	70–100	20	37	940
	16	0.063	1.6	0.035	0.9	18	80–110	20	32	813
	$\frac{5}{64}$ in.	0.078	1.9	0.035	0.9	19	100–130	20	27	686
	$\frac{1}{8}$ in.	0.125	3.2	0.035	0.9	20	120–160	20	20	508
SMAW	24	0.025	0.1	$\frac{3}{32}$	2.4	25	40	—	20	508
	22	0.031	0.8	$\frac{3}{32}$	2.4	25	40	—	21	533
	20	0.038	0.9	$\frac{3}{32}$	2.4	25	50	—	28	711
	18	0.050	1.3	$\frac{3}{32}$	2.4	27	65	—	30	762
	16	0.063	1.6	$\frac{3}{32}$	2.4	27	75	—	30	762
	$\frac{5}{64}$ in.	0.078	1.9	$\frac{3}{32}$	2.4	28	100	—	35	889
	$\frac{1}{8}$ in.	0.125	3.2	$\frac{1}{8}$	3.2	26	120	—	40	1016
CAW[b]	24	0.025	0.1	$\frac{3}{16}$	4.8	18	25	—	8	203
	22	0.031	0.8	$\frac{3}{16}$	4.8	18	35	—	8	203
	20	0.038	0.9	$\frac{3}{16}$	4.8	18	45	—	10	255
	18	0.050	1.3	$\frac{3}{16}$	4.8	19	50	—	10	255
	16	0.062	1.6	$\frac{3}{16}$	4.8	20	50	—	12	305
	$\frac{5}{64}$ in.	0.078	1.9	$\frac{1}{4}$	6.4	20	80	—	13	330
	$\frac{1}{8}$ in.	0.125	3.2	$\frac{1}{4}$	6.4	19	95	—	15	380

[a]Parameters are suitable for square-groove butt or for fillet welds.
[b]Bronze cold wire [$\frac{1}{8}$ in. (3.2 mm)] used with carbon arc welding.

26-3 ONE-SIDE WELDING

The term *one-side welding* is not new, but it has been popularized and given a high degree of importance in the shipyards. One-side welding is the production of a butt weld from one side of the joint that achieves a 100% efficiency and produces a back side of the joint that is acceptable from an appearance and quality viewpoint.

All welds made on medium- and small-diameter pipe and tubes are one-side welds, since the back side of the weld is inaccessible. During the early days of submerged arc welding, techniques for backing the joint with flux were developed that produced welds made entirely from the top side of the joint.

In the shipbuilding industry, it is customary to handle very large welded subassemblies. Previously, automatic welding was done by a pass from each side of the joint. This required the assemblies to be turned over to complete the weld. Turning large assemblies involved extremely heavy capacity overhead cranes with sufficient height. Heavy-capacity cranes and high bays in buildings are extremely expensive. The elimination of the turning-over operation represented a substantial cost savings when building ships.

One-side welding can be divided into two different methods of operation. The first method uses a weld-backing apparatus. Flat plates are set on this apparatus and long straight joints are automatically welded in the flat position. The second method uses portable backing materials that are taken to and applied to the back side of the weld joint, wherever it is located.[3]

With one-side welding it is difficult to obtain a 100% joint efficiency over the entire length of the joint. Welds made from both sides usually have an overlap of penetration. The extent of this overlap is not important as long as it occurs. The problem of joining large assemblies is complicated by the material preparation tolerances. The alignment and flatness and warpage inherent in any welding operation create fitup problems of large subassemblies. The variations of fitup rapidly change the penetration of the weld and the reinforcement on the back side of the weld joint.

Another problem of welding large structures is the straightness or fairness of the parts at the joint. Misalignment is common and complicates the problem of welding. These problems can be eliminated with one-side welding.

One method of making one-side welds was developed during the early use of submerged arc welding. This method utilized submerged arc flux on the underside of the weld joint. The flux was brought into intimate contact with the back of the joint by means of air pressure in a hose. This system has been called the BF method, which indicates *flux backing* (Figure 26-8).

Another method used in seam welding machines is known as the *copper backing* method, indicated by the

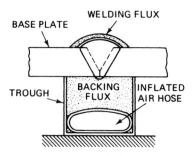

FIGURE 26–8 Backing flux (BF) method.

letters CB (Figure 26-9). The copper backing method utilized heavy copper bars brought into intimate contact with the back side of the weld joint. Often a recessed groove is placed in the copper bar immediately below the weld joint to allow for root penetration. For heavy-duty welding the copper bar would be water cooled. For high-quality work on nonferrous metals using gas metal arc welding, backing gas was introduced into the recessed groove.

Both of these methods were highly successful for thinner materials or for relatively short welds. For heavier plates or for long joints, neither of these methods was entirely satisfactory. A development combining the advantages of both while eliminating their disadvantages is known as *flux-copper backing* method, abbreviated FCB (Figure 26-10). This backing method uses a layer of granulated flux of consistent thickness in con-

FIGURE 26–9 Copper backing (CB) method.

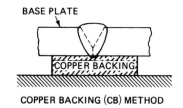

FIGURE 26–10 Flux-copper backing (FCB) method.

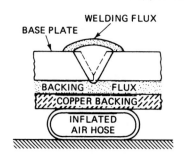

tact with the underside of both workpieces and in contact with the upper side of a copper backing bar. The copper backing bar helps control the uniform size of the reinforcement bead. The Japanese have a word, *uranami,* which describes the root bead viewed from the back side of the joint. The width of the flux layer is approximately 4 in. (100 mm), which will accommodate variation of the joint. The copper bar is approximately $\frac{1}{2}$ in. (12.5 mm) thick and 5 in. (125 mm) wide and as long as the joint. The thickness of the backing flux is about $\frac{1}{4}$ in. (6 mm). The backing flux is normally the same as the welding flux. Special backing flux may be used. The back pressure keeps the flux in intimate contact with the parts being welded.

There are variations to all three of these methods. For example, with the FB method, sometimes the flux is encased in a paper container, which is burned during the welding operation. This makes the fitup quicker. In the case of the CB method, different kinds of coatings are placed on the copper bar. Some users place fiberglass tape or other inorganic fibers on the copper bar to maintain pressure against the back side of the weld. In the FCB method, different systems are used to keep the backing against the work and different types of flux are used.

An improvement of the FB method utilizes special backing flux, called RF for *refractory flux.* This flux helps to form a uniform backing or uranami bead. The flux contains phenol resin binders, which undergo thermal hardening from the heat of the arc.

The second type of one-side welding utilizes portable backing material applied to the back side of the joint. The portable system uses short lengths of assemblies containing the backing materials. One of these, known as the KL method, is shown in Figure 26–11. This backing assembly is approximately 2 ft long. The top layer of flux is the beadforming type; underneath is a refractory material. Both are enclosed in a thin sheet metal trough. This steel trough is sufficiently flexible so that it can be formed to fit the changing contour of a ship hull.

Another portable method utilizes the FAB backing assembly (Figure 26–12). This assembly is composed of several layers of fiberglass tape; under this is a layer of insulating material, and below this is a corrugated cardboard pad. This entire assembly is enclosed in a cardboard

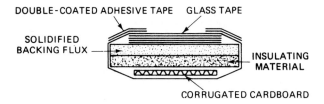

FIGURE 26–12 Portable flexible (FAB) backing method.

container. Double-coated adhesive tape is used to keep the backing assembly in intimate contact with the underside of the weld joint. This assembly is relatively flexible and can be formed to the contour of the part being welded.

There are several other configurations of portable backing devices. One of particular interest is known as a CRB method, for "coated rod backing." It is similar to a large covered electrode except that the flux covering on the rod does not melt. When the root opening is excessive, two rods may be used. Magnetic clamps are used to hold the backing assembly in place.

A popular portable backing system utilizes an adhesive tape that carries a layer of backing flux along with several layers of aluminum foil. This backing tape can be placed against the backside of the joint and will adhere to the parts. The flux will help form the backing bead. Tape is being used with the consumable electrode arc welding process. When using submerged arc welding, sufficient root face must be provided so that the welding arc does not come in contact with the backing flux in the tape. Figure 26–13 shows this tape with the flux. Figure 26–14 shows the tape in use joining cylindrical members. The tape is on the outside of the structure, which is the root side of the joint (Figure 26–15). The weld is made from inside with a mechanized submerged arc welding head.

This tape is very helpful for welding vertical joints using the gas metal arc process. The tape eliminates the draft of air blowing through the root opening, which reduces the efficiency of the shielding gas.

Another portable backing method utilizes short ceramic tiles (approximately 2 inch long) against the backside of the joint. This is known as ceramic weld backing (KATBAK). The tiles do not melt under the heat of the arc. The tiles are available with different cross-sectional profiles, with a groove allowing penetration of the root to provide a *uranami* bead. The tiles are attached to an adhesive-coated flexible aluminum tape, shown in Figure 26–16. The pressure-sensitive adhesive is heat resistant and allows the backing material to be attached to the joint in any position. Semiautomatic gas metal arc welding with solid or cored wires can be used.

FIGURE 26–11 Portable (KL) backing method.

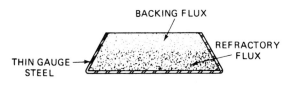

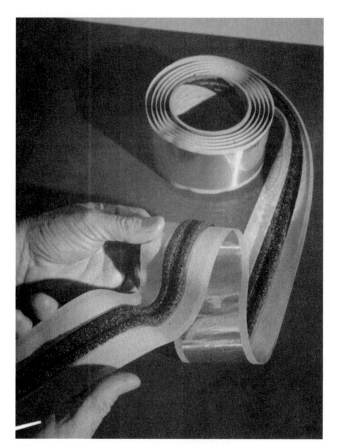

FIGURE 26–13 Backing tape with flux.

FIGURE 26–14 Backing tape in use, back side.

26-4 NARROW GAP WELDING

Narrow gap welding (NGW), officially called *narrow groove welding,* is a term applied to arc welds made in thick materials utilizing a square-groove weld joint or a V-groove weld joint with a groove angle of not over 10° and utilizing a root opening or gap between the parts $\frac{1}{4}$ in. (6.4 mm) to $\frac{3}{8}$ in. (9.5 mm) wide.[4] Narrow gap welding has been used on material from 2 in. (50 mm) thick up through 12 in. (300 mm) thick. A backing system or a U-groove design is required at the root.

For economic reasons, international attention is being directed toward narrow gap welding. Heavier plates are being used for pressure vessels, nuclear reactors, penstocks, ship decking, and so on. As the thickness increases, the amount of weld metal in the joint increases at a much greater rate. Figure 26–17 shows the cross-sectional area of narrow gap square-groove design versus the single-V-groove design with an included groove angle of 30°, 45°, or 60°. The narrow gap welding concept can also be used for welding heavy-wall pipe and tubing. This can be accomplished by rolling the pipe in the 1G position or orbiting the pipe in the 5G position. The cross-section view of a narrow gap weld is shown in Figure 26–18.

The economic advantage is obtained when welding plate thicknesses of $1\frac{1}{2}$ in. (38 mm) and above and is based on using less weld metal to produce the joint, which in turn decreases the labor cost.[5]

The advantages of narrow gap welding over conventional welding include the following:

1. High-quality welds, the result of low heat input and the multipass technique, which provides an extremely narrow heat-affected zone and fine-grained weld metal. The mechanical properties of the weld joint are excellent.
2. High productivity as a result of smaller cross section of the weld. This uses less weld metal and less labor for joint preparation and welding operations.
3. All-position capability, due to the small volume of the molten weld pool and low heat input.
4. Lower residual stresses, due to the smaller number of weld passes to produce the joint.

Narrow gap welding is not a welding process; it is a technique or procedure. It can utilize several arc welding processes. The most popular is gas metal arc welding us-

FIGURE 26–15 Making the submerged arc weld.

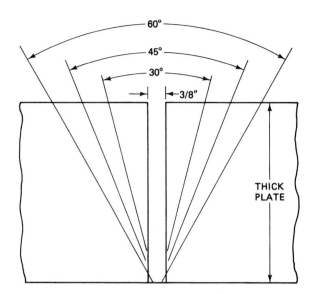

FIGURE 26–17 Comparison of cross-sectional area.

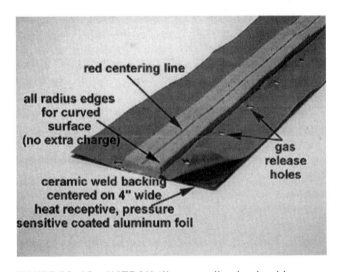

red centering line

all radius edges
for curved
surface
(no extra charge)

gas
release
holes

ceramic weld backing
centered on 4" wide
heat receptive, pressure
sensitive coated aluminum foil

FIGURE 26–16 KATBAK tiles on adhesive backing.

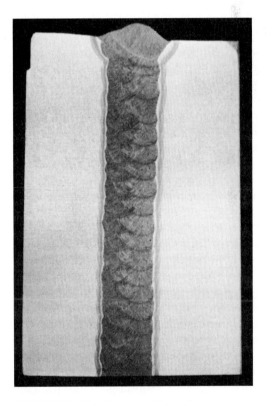

FIGURE 26–18 Cross section of narrow gap weld.

ing spray transfer. Narrow gap welds can also be made with flux cored arc welding using special electrodes. Submerged arc welding has also been used. Gas tungsten arc welding with the "hot wire" can also be used.

It is possible to consider electrogas and electroslag welding as narrow gap welding since they both employ square-groove welds with a relatively small root opening on heavy plate materials. However, these are not included in this section.

Narrow gap welding can be done on carbon steels and on low-alloy high-strength steels. The low heat input makes it particularly advantageous for welding quenched and tempered steels.

There are disadvantages to narrow gap welding, including:

1. Repair welding is difficult and must be done by conventional techniques.
2. The welding head and control are relatively complex and expensive.
3. Joint fitup must be accurate in order to have consistent results the entire length of the joint.
4. Magnetic arc blow can be a problem with gas metal arc welding.
5. Technology is more demanding, which requires better-trained operators.
6. Filler metals required are special and may be more expensive.

Most of the work on narrow groove welding has been done with gas metal arc welding. A variety of techniques has been developed.

The original narrow gap work involved two welding arcs, two electrode wires, and two contact tubes. One is directed toward one sidewall and the other toward the opposite sidewall. The wire feeders, contact tubes, and so on, are mounted on a special carriage with a fixed distance between them and the sidewalls. This ensures proper sidewall fusion, regardless of minor variations in the root opening. The shielding gas is introduced into the joint by special nozzles that extend to the bottom of the groove. A backing strip is used for making the first, and usually the second, pass. Subsequent passes are deposited on previous layers. Approximately 10 weld passes are required for each inch (25 mm) of joint thickness. Typical welding conditions would utilize an electrode diameter of either 0.035 in. (0.9 mm) or 0.045 in. (1.1 mm). The current would be in the neighborhood of 225 to 250 A dc electrode positive at 25 to 26 V. Constant-voltage power sources are used. The travel speed would be from 40 to 50 in./min (1000 to 1200 mm/min), and the shielding gas could be 75% argon plus 25% CO_2 or 95% argon and 5% oxygen. The head is arranged so that the contact tubes are approximately $\frac{1}{2}$ in. above the arc. Contact tubes are retracted as the weld is made. The major factor of this technique is the necessity to direct the electrode wire to the joint sidewall. This is done by introducing "cast" into the electrode wire immediately before it goes into the contact tube. This ensures that as the electrode wire leaves the contact tube it will travel to the side to which it is directed. In this way sidewall penetration is maintained and undercut is avoided. The entire head assembly must be accurately built, properly insulated, and adjustable for variations in gap and allowances made as the weld builds up. The control system is designed to provide automatic sequence to start and stop the electrodes at the same location in the joint. The major problem is directing the electrode wires into the sidewall to avoid undercutting and potential defects.

Another system utilizing gas metal arc welding is the *twist wire* technique. The electrode is actually two wires twisted together. This system uses a straight contact tube, and as the electrode melts, two arcs are generated from the tips of the two wires. They have a straight transfer mode into the sidewall and provide continuous rotational movement. The rate of arc rotation depends on the pitch of the twisted electrode wires and on the arc length. This technique achieves good sidewall penetration.

Another gas arc metal variation uses an oscillating or swinging torch or a swiveling torch. Another method uses a contact tip bent at the end. One of the most novel systems uses a rotating contact tip where the electrode is off center of the axis of rotation. This is known as the "rotating arc" technique. Other systems use preformed wire, bent wire, and so on.

Another slightly different system uses a wider root gap, $1\frac{1}{2}$ times larger, and uses a larger electrode wire, normally $\frac{1}{8}$ in. (3.2 mm) in diameter. The welding current ranges from 400 to 450 A, and the voltage ranges from 30 to 37 V with electrode negative. Shielding gas composed of one-third CO_2, one-third argon, and one-third helium is used. The contact tube is farther away from the arc and, with straight polarity, seems to help direct the electrode wire to the sidewalls for complete fusion. A backing strip is used for the first pass. Metal transfer is globular, but the spread of the arc is sufficient to make a pass as wide as the groove. A drag angle of the electrode wire is used. Heat input is greater with this method.

New variations of gas metal arc welding are still appearing for narrow groove welding. The latest system was developed in Canada. It utilizes two electrode wires feeding simultaneously but directly to opposite sides of the joint. Sidewall fusion is excellent and variations of the root opening, within broad limits, are accommodated. The special feature is the use of two power sources using pulsed current. The peak current pulse alternates from one electrode to the other and avoids the arc disturbance that normally occurs when two arcs are feeding the same weld pool. The control circuit automatically changes welding parameters to accommodate variations in the root opening.

One of the problems with gas metal arc welding and narrow gap welding is arc blow and the loss of arc stability. The Canadian system overcomes this problem.

Special flux cored or metal cored electrode wires have been developed for narrow gap welding. These are used in the same manner as solid wires with gas metal arc welding.

The submerged arc welding process can be used for narrow gap welding. Submerged arc welding employs ac current, which avoids the magnetic arc blow problem. Larger electrode wires are normally used. Higher heat input and greater deposition rates are also employed. Special submerged arc flux is used to avoid slag entrapment.

The gas tungsten arc welding process with a hot wire filler is used for narrow gap welding. This version provides good arc stability, good out-of-position capability, good side-wall penetration, and no spatter and slag. The deposition rates are lower, but multiple heads are used. A special head containing the tungsten electrode provides for arc oscillation and carries the filler wire to the arc.

Undoubtedly, additional variations of the arc welding processes will be developed to provide the economic and quality advantages of narrow groove welding.

26-5 UNDERWATER WELDING

Underwater welding began during World War I, when the British Navy used it to make temporary repairs on battleships. The repairs consisted of welding around leaking rivets on the ships' hulls. The introduction of covered electrodes made it possible to weld under water and to produce welds having approximately 80% of the strength and 40% of the ductility of welds made in air. Underwater welding was originally restricted to salvage operations and emergency repair work, and was limited to depths below the surface of not over 30 ft (10 m). Major advances have been made in underwater welding in recent years.

Underwater welding can be divided into two categories: *welding in the wet* environment and *welding in the dry* environment. Welding in the wet is used primarily for emergency repairs or salvage operations in relatively shallow water.

Welding in the Wet

The poor quality of welds made in the wet is due to the problem of heat transfer, welder visibility, and the presence of hydrogen in the arc atmosphere. When the base metal and the arc area are surrounded by water, there is no temperature or heat buildup of the base metal at the weld. This creates a high-temperature gradient or quench effect, which reduces the ductility of the weld metal. The arc area is composed of a high concentration of water vapor. The arc atmosphere of hydrogen and oxygen of the water vapor is absorbed in the molten weld metal and contributes to porosity and hydrogen cracking. In addition, welders working under water are restricted in their efforts to see and manipulate the welding arc. Under ideal conditions, the welds produced in the wet with covered electrodes are marginal. They may be used for short periods but should be replaced as quickly as possible. Improvements in underwater welding electrodes are helping in-the-wet weld quality.

Efforts have been made to produce a bubble of gas in which the weld can be made. This technique has not been able to ensure good-quality welds made with covered electrodes in-the-wet.

The general arrangements for underwater in-the-wet welding are shown in Figure 26–19. The power source for underwater welding should always be a dc machine rated at 300 or 400 A. Generator welding machines are often employed for underwater welding in-the-wet. The frame of the welding machine must be connected to the ship. The welding circuit must include a positive

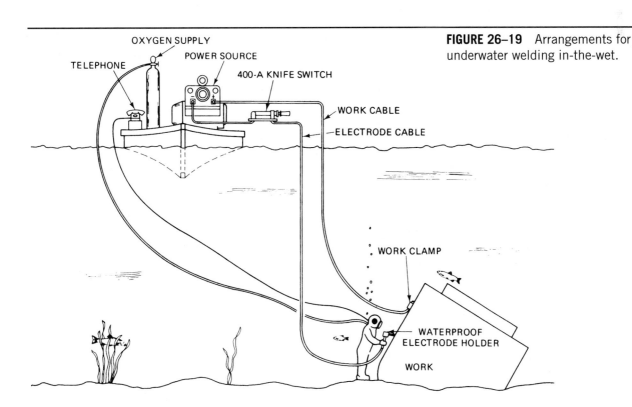

FIGURE 26–19 Arrangements for underwater welding in-the-wet.

switch, usually a knife switch that is operated on the surface upon the command of the welder-diver. The knife switch in the electrode circuit must be capable of breaking the full welding current. It is required for safety reasons. The welding power should be connected to the electrode holder only while the welder is welding. Direct current with electrode negative (straight polarity) is used. Special electrode holders with insulation against water are employed (Figure 26–20). The underwater welding electrode holder will accommodate two sizes of electrodes, normally $\frac{3}{16}$ in. (4.8 mm) and $\frac{5}{32}$ in. (4 mm). The electrode types normally used meet the E6013 classification, and must be waterproofed. This is done by wrapping them with waterproof tape or by dipping them in a sodium silicate mix or other waterproofing material. Electrodes for underwater welding are available commercially.

The welding lead and work lead should be at least 2/0 size, and the insulation must be perfect. If the total length of the lead exceeds 300 ft (100 m), they should be paralleled. With paralleled leads to the electrode holder the last 3 ft (1 m) should be a single cable. All connections must be thoroughly insulated so that the water cannot come in contact with the metal parts. If the insulation breaks, the seawater will contact the copper conductor and part of the current will leak away and will not be available at the arc. In addition, there will be rapid deterioration of the copper cable at the break. The workpiece lead should be connected to the work being welded within 3 ft (1 m) of the point of welding. Welding in-the-wet is shown in Figure 26–21.

A special underwater cutting torch that utilizes the oxygen arc cutting process with a tubular steel covered electrode is also shown in Figure 26–19. This torch is fully insulated and utilizes a twist-type collet for gripping the

FIGURE 26–21 Welding in-the-wet.

electrode. It includes an oxygen valve and connections for attaching the welding lead and an oxygen hose. It is equipped to handle up to $\frac{5}{16}$-in. (7.9-mm) tubular electrode. In this process the arc is struck in the normal fashion and oxygen is fed through the center hole of the electrode to provide the cutting action. The normal electrical connections are employed.

Complete information concerning underwater cutting and welding in-the-wet with covered electrodes is given in the U.S. Navy's "Underwater Cutting and Welding" technical manual.[6]

The need to produce high-quality welds under water has increased as oil and gas are found in deep water. Most offshore exploration, drilling, and production, until recently, was in water ranging from 30 to 50 ft (10 to 16 m) deep. When a pipeline needs to be repaired, it is raised to the surface, repaired, and lowered back to the ocean floor. Exploration, drilling, and production are moving into deeper water, up to the 1000 ft (305 m) depth. Modifications and work must be done on the ocean floor. More pipelines are damaged and there is a necessity for making tie-ins of pipelines on the ocean floor. The repairs and tie-ins must be high-quality welds to prohibit the possibility of leaks or oil spills. This type of work is now being done in depths of 200 to 600 ft (61 to 182 m).

Welding in-the-Dry

The development of welding in-the-dry or in a dry environment makes it possible to produce high-quality weld joints that meet x-ray code requirements. A number of welding processes are used for welding in-the-dry: the shielded metal arc, the gas tungsten arc, the plasma arc, the gas metal arc, and the flux cored arc welding process. The shielded metal arc welding process is rarely used for welding in-the-dry environment, because of the large amount of smoke and fumes produced. The gas tungsten arc welding process is being used to produce welds that

FIGURE 26–20 Underwater electrode holders.

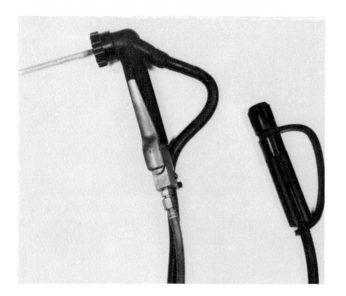

meet the quality requirements of API standard 1104. It is being used at depths of over 300 ft (91 m). The gas tungsten arc welding process is relatively slow, but is acceptable since the welding operation is a small part of the total repair operation.

Efforts are ongoing to develop plasma arc welding more fully for deep-water operations. Successful application of gas metal arc welding in-the-dry has been made to depths as great as 180 ft (51 m).

There are two basic types of in-the-dry underwater welding. One involves a welding chamber or habitat and is known as hyperbaric welding. The habitat or large welding chamber provides the welder-diver with all the necessary equipment for welding and related work in a dry environment. The weld chamber is made so that it can be sealed to the part to be welded. Since most work is on pipe, arrangements are made to seal the habitat to the pipe. The bottom of the chamber is exposed to open water and is covered by a grating. The pressure of the atmosphere inside the chamber is equal to the water pressure at the operating depth. Figure 26–22 shows a welding habitat.

Life-support equipment involves two-way telephone, video camera for continuous observation, life-sup-

FIGURE 26–22 Underwater pressurized habitat for pipe welding.

port atmosphere for the welder-diver (which may be different from the welding atmosphere), power for tools and for welding, and the gas supply for the welding atmosphere. Habitats usually include an atmosphere-conditioning system used to both cool and filter the atmosphere in the chamber. Filtering is important since metal vapor is released during welding. Air conditioning is employed since heat is generated by welding. For gas metal arc welding the welding power source is normally on the surface. Welding cables are lowered to the habitat to provide power for the arc. The electrode wire, wire feeder, and control unit are located in the habitat. The wire feeder must be protected from the high-pressure and high-humidity conditions in the chamber. The electrode wire must also be protected from the humidity.

The gas for breathing and for welding is designed for use at the high pressures that are involved. The pressure in the habitat increases 1 atmosphere or 14.7 lb/in.^{2} (1 kg/cm^2) for each 33 ft (10 m) of depth. The water pressure must be equalized by the pressure of the atmosphere within the habitat. This high pressure creates problems for the crew and for welding. With shielded metal arc welding, the problem involves the removal and filtering of the atmosphere in the habitat. With gas tungsten arc welding, very little smoke and fumes are generated, but the inert gas used for welding disrupts the breathing atmosphere. For saturated diving work the breathing atmosphere of the welder-diver is based on the working depth. Premixed gas is used and the oxygen content is based on the depth.[7]

Welding with the gas metal arc in a habitat presents specific problems.[8] In shallow depths the welding in the habitat is essentially the same as welding on the surface. When welding at depths of 125 ft (35 m), the pressure is about 4 atmospheres (approximately 50 psi gauge). The weld metal quality is essentially the same as surface-made welds; however, the welding voltage increases and the weld bead penetration increases. As the depth is increased, the atmospheric pressure increases and the arc becomes more constricted, and this leads to increased arc voltage and increased penetration and higher burnoff rates. This makes the weld pool difficult to handle. At depths beyond 125 ft (35 m) the weld pool becomes increasingly difficult to control and an increased amount of smoke is generated. Special welding power sources are required.

Wet-Dry Welding

Improvements have been made to provide more flexibility for gas metal arc welding when welding underwater.[9] The dry hyperbaric chambers or habitats are extremely expensive and must be designed for specific applications. It is now possible to take gas metal arc welding outside the habitat by the use of special nozzles or chambers that surround the torch. In using this apparatus the welder-diver is in-the-wet or in the water, but the nozzle of the welding gun and material to be welded is in the dry atmosphere (Figure 26-23). The gun and nozzle are in this small chamber, but the wire feeder and the electrode wire supply are in another watertight pressurized enclosure. The pressure of the shielding gas coming through the system is greater than the pressure of the water at the operating level. The gas flows through the wire feeder enclosure through the electrode conduit to the torch, where it provides the shielding atmosphere for welding. It also provides the pressure to evacuate the water to provide a dry atmosphere. The dry gas environment chambers are relatively inexpensive, small, and lightweight. They are provided with flexible seals to be used against the part being welded. They can be handheld or made with clamps for quick attachment to the part to be welded. The gun is hand manipulated inside the small chamber, in the same way as on the surface. The chambers are made of transparent material or have a sufficient number of windows so that the welder can see inside to properly manipulate and direct the welding arc. This technique can be utilized for welding with gas metal arc up to 125 ft (35 m) below the surface. High-quality welds that meet code requirements can be produced.

Special safety precautions must be followed when doing underwater welding. These include all precautions normally employed by divers, plus those required for welding. Welders-divers must be aware of the possibilities of entrapped gases in parts being welded or cut. These gases are usually rich in hydrogen and oxygen and may explode when ignited. Only experienced, well-trained personnel should do underwater welding.

FIGURE 26-23 Welder in-the-wet welding in-the-dry with small gas-filled enclosure.

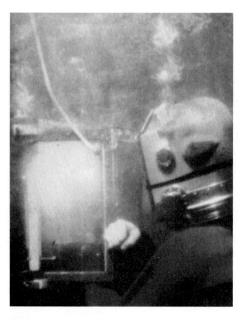

Underwater Cutting in-the-Wet

Underwater cutting may be more important than underwater welding. Cutting is required to prepare for repair welding of underwater structures. It is also used for salvage and demolition work of submersed structures.

There are two basic methods of underwater cutting: the oxyfuel gas methods and the oxyarc cutting methods, with many variations. These are similar to above-water applications, but with specialized equipment, and should only be performed by skilled welders-divers who have necessary training in the particular operation. This is due to the safety hazards and the precautions necessary to guard against injuries. The oxyfuel gas techniques all involve high-pressure gases, which can be dangerous. The oxyarc methods involve electrical circuits that require protection.

Oxyfuel gas underwater cutting was perfected by the U.S. Navy. They developed a special torch with an extra nozzle that fed compressed air into the cutting area, providing a bubble of air that displaced the water and enabled the cut to be made in an atmosphere of air. The underwater oxyfuel gas cutting process is essentially the same as in air wherein the metal is heated to its kindling temperature and then a jet of oxygen is introduced, which oxidizes (cuts) the steel. Excess oxygen is used to help blow away the molten metal. This extra shroud or nozzle is used with all oxyfuel gas cutting methods.

Oxyacetylene gas is used for cutting in shallow depths, usually to about 15 ft (3.5 m) in depth. This cutting depth limit is due to the fact that high-pressure acetylene is very dangerous. The gas pressures required increase rapidly as the operation is performed in deeper water. Propane, a liquefied petroleum gas, is used with oxygen for underwater cutting in depths down to 150 ft (50 m), but is related to water temperature. The most versatile fuel gas is hydrogen, which can be used for most underwater working depths. Gasoline is even used as a fuel for some underwater cutting applications.

It is very important that the proper type of torch is employed for the working depth and fuel gas involved. Torch ignition is somewhat involved. For very shallow depths, the torch is ignited before it is taken underwater and then adjusted for the particular depth involved. For deeper working depths a pilot flame, or "match," is involved. This is a tiny flame that burns continuously before the torch is ignited at the working depth. It is a part of the underwater torch mechanism.

Constant communication is required between the diver-cutter and the above-water surface handlers. Pressure must be adjusted for the fuel gas and the oxygen to maintain the proper operation. The oxyfuel gas systems will cut up to 2 in. (50 mm) of steel. Considerable skill is required for oxyfuel gas cutting, as well as good visibility. The quality of the cut depends on the skill of the diver-cutter. Mild and low-alloy steels can be cut underwater using the special air-modified cutting torch.

Oxyarc cutting was developed to reduce some of the safety hazards inherent with the oxyfuel gas systems. Much of the development work was done during World War II by the U.S. Navy. Initially, coated electrodes with a hollow core wire were used. These were similar to those used for cutting stainless and nonferrous metals. The outside of the electrode coating was covered with a waterproofing material, and a jet of pure oxygen ran down the hollow tubular electrode to the arc. The arc heated the metal to be cut to its kindling temperature, and the oxygen through the tubular electrode caused rapid oxidation (cutting) of the metal and helped carry the molten metal away from the cut. Special insulated waterproof electrode holders with oxygen valves were employed. This system utilized a regular welding power source.

Variations to this process involve different types of electrodes with different types of coatings. The basics are the same, that is, a tubular electrode with an oxygen stream through the center using a conventional power source. The arc provides the necessary heat to the metal to be cut to raise it to its kindling temperature, and the oxygen displaces the water in the cutting area and also promotes rapid oxidation of the metal being cut. The basic hollow electrode with protective coating is used. However, different types of materials are used for the core, including nonferrous metals, carbon, and ceramic materials with high electrical conductivity.

The latest development is the hollow steel rod known as exothermic reaction electrodes. This is also called thermal lance cutting (described in Section 9-2). These rods may or may not have a coating or a waterproof covering. They require an insulated waterproof electrode holder that includes an oxygen supply valve, shown in Figure 26-24. Most importantly, they do not require a welding machine. They can be started by a spark supplied by two 12-volt batteries and a special striker, shown in Figure 26-25. They generate very high temperatures,

---

FIGURE 26-24 Electrode holder for arc cutting.

FIGURE 26-25 Striker.

---

which will cut metals and nonmetals, including reinforced concrete. Cutting is terminated when the oxygen flow stops.

Several other oxyarc processes can be used, including GMAW with a water jet and plasma arc, with waterproof torches for underwater cutting.

26-6 WELDING IN SPACE

In the not-too-distant future, a space station like the artist's conception in Figure 26-26 will be orbiting the earth. It will be so large that it cannot be launched from the earth. In order to build it, a large number of subassemblies will be carried into space by the orbiter or similar vehicle, where they will be assembled. There are several possible methods for joining the subassemblies. The Russians expect to weld the subassemblies together, whereas the U.S. space agency prefers to use mechanical fasteners plus welding. It is estimated that there will be approximately 100 miles of pipe and tubing in the space station. Piping subassemblies will be welded on earth, but the method of joining them in space will probably be welding. The space station will have a design life of at least 30 years. Undoubtedly, the maintenance and repair of the structure and piping will involve welding. In addition to welding, there will be thermal cutting, brazing, and thermal spraying required in space.

There are many difficulties to welding in space. Conditions in space differ greatly from those on earth. This includes a high-vacuum atmosphere. Gravity would be minimal, and weightlessness is a problem. There will be high and low temperatures, ranging from almost absolute zero to high temperatures, resulting from solar radiation. There will be other radiation—electrical and magnetic fields—that may affect welding. In addition, it is not expected that astronauts and cosmonauts would possess welding skills. This would indicate that fully automatic mechanisms would be required for welding.

To weld in space successfully, a variety of scientific research and development projects need to be solved. The Paton Welding Institute in Kiev, Ukraine, in cooperation with the Russian Space Administration, has developed welding equipment for space research.[10] They

FIGURE 26–26 International space station now under construction.

have done work on earth utilizing vacuum chambers and low temperatures. They have experimented with several processes. Electron beam welding showed promise, as well as plasma welding, gas metal arc welding, and brazing. They also did experimental work in flying laboratories where weightlessness could be simulated for short periods of time.

The first record of space welding was done in October 1969 in the spacecraft *Soyuz 6.* The welding unit was named Vulkin, which weighed approximately 50 k (110 lb) and was battery powered. Cosmonauts G. S. Shonin and V. N. Kubaslov did manual electron beam welding, plasma welding, and gas metal arc welding.

In 1979 an improved power source known as Isparital was aboard orbital station *Salyut-6,* and additional experiments were performed, including electron beam cutting. In 1983 the power source was again improved, and additional experiments were performed on orbital station *Salyut-7.*

In 1984 an improved power source known as the versatile hand tool (VHT) was on board orbital station *Salyut-7.* During this flight, cosmonauts Svetlana Savtskeya and Vladimir Dzhanibekov made welds outside

the space capsule, shown in Figure 26–27. This included additional welding experiments plus a thermal spray experiment by cosmonauts who learned welding in special laboratories on the ground. One of the most significant space welding activities was performed in 1986 using the VHT machine when cosmonauts Kizin and Solovyer welded a 39-f truss-type girder in space. This was an experiment to determine the feasibility of using welding to construct a space vehicle. Figure 26–28 shows the latest VHT and power source with cosmonaut Vladimir Dzhanibekov.

In the United States, space experiments have progressed as well.[11,12] These include weightlessness experiments in an airplane and welding in a vacuum chamber; in June 1973 welding experiments took place on *Skylab 1.* Welds have been made with orbital tube heads in a space-simulated environment. Laser welding was considered but rejected, due to its low electrical efficiency. These experiments have proven the feasibility of welding in space. Additional experiments are ongoing, and there is every reason to believe that welding will be used to assemble the space station in the not-too-distant future.

FIGURE 26–27 Making welds outside the space capsule.

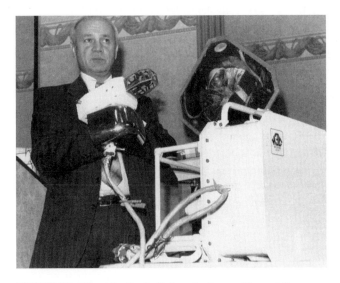

FIGURE 26–28 Vladimir Dzhanibekov with welding equipment.

26-7 MICROJOINING

Microjoining, or micro welding, is the joining of very small parts. From a practical point of view, it involves the welding of thin sheet metal up to 0.020 in. (0.5 mm) thick and wire or tubular components up to 0.040 in. (1 mm) in diameter. Microjoining is most widely used by the electronics, instrument, and packaging industries. Some of the common applications are attaching leads to microchips, attaching leads to electronic devices, packaging microchips, making medical instruments and devices such as heart pacers, and packaging these devices and other miniaturized welding applications.

Many of the welding and joining processes are used for microjoining. This includes gas tungsten arc, plasma arc, and stud welding. It also includes laser beam and electron beam welding and the resistance welding processes. The solid state processes are employed, including ultrasonic, pressure welding, and diffusion bonding. Soldering and brazing are both widely used. In addition, for certain applications adhesive bonding is employed.

The objective of microjoining is to produce a strong metallurgical joint between the parts being welded with good electrical conductivity.

The materials being microjoined include all the common metals plus exotic and precious metals utilized by the electronics industry. The most common metals welded are copper, aluminum, beryllium copper, stainless steel, titanium, gold, and silver. The nonmetals include ceramics and plastics.

Microjoining is normally done automatically or with automated equipment. For the larger parts, manual or semiautomatic welding may be used, usually under a magnifying glass with micrometer adjustments for location or movement of parts.

Microjoining requires miniaturized, specialized equipment. High-quality welds are demanded of microjoining since the weld is a part of life-support equipment, delicate computer equipment, and so on.

One of the major applications for microjoining is known as wire bonding. This is the joining of small wires in electro connections between chips and lead frames. Wires as small as the human hair are joined for this application. Components of this type are subject to vibration, shock loading, low temperatures, and high temperatures. Stresses due to vibrations, shock, and temperature changes of different materials require good-quality welds that are strong and have conductivity. Figure 26–29 shows the joining of small parts.

FIGURE 26–29 Micro welds.

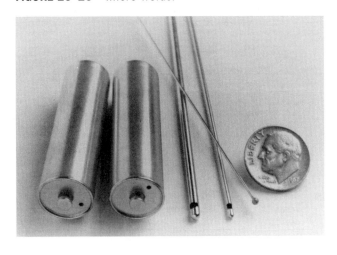

The arc welding processes mentioned previously use miniaturized equipment, sometimes called micro plasma or micro TIG. Welding currents range from $\frac{1}{2}$ to 10 A, and pulsed current is normally used. Precision motion and holding devices must be employed, as well as automatic control systems, which are driven by a microprocessor.

The high-energy beam processes, EB and laser beam welding, are commonly used with lower-powered equipment. Precision motion devices, focusing equipment, and computer controls are required. Special miniaturized resistance welding equipment is widely used. This includes spot welding, projection welding, roll seam and intermittent welding, and ultrasonic welding done with precision motion devices and accurate control systems.

Brazing and soldering utilize capillary attraction but have precision holding devices and precision methods for introducing fluxes and filler material. The method of heating is essentially the same as with large applications. Electric radiant lamps and the flame are often used. Some work is done in a vacuum or with protective gases. Iron soldering is not practiced.

For adhesive bonding, miniaturized adhesive applicators are required. Microjoining is a rapidly changing technology driven primarily by the electronics industry and medical devices.

 # QUESTIONS

26-1. Why is a timer used for arc spot welding?

26-2. Is it necessary to wear a welding helmet when arc spot welding?

26-3. Can arc spot welding be used to join aluminum?

26-4. Can flux cored arc welding be used for arc spot welding?

26-5. What process is recommended for steel sheet metal welding?

26-6. Why is good fixturing needed for thin sheet metal welding?

26-7. What is one-side welding?

26-8. Describe two different backing methods.

26-9. Why is one-side welding important in shipbuilding?

26-10. What is the primary advantage for narrow gap welding?

26-11. What design of weld joint is used for narrow gap welding?

26-12. What welding processes are used for narrow gap welding?

26-13. What is special about an underwater welding electrode?

26-14. Explain the difference between underwater wet welding and dry welding.

26-15. Can pipe welds be made under water to meet code requirements?

26-16. What is welding in a habitat?

26-17. What determines the pressure of the atmosphere in the habitat?

26-18. What industries use microjoining?

26-19. The Russians made the first weld in space in what year?

26-20. What welding processes were used in space?

REFERENCES

1. T. W. Shearer, "Arc Spot Welding," Bulletin 105, Welding Research Council, New York, May 1965.

2. R. A. Stoehr and F. R. Collins, "Gas Metal Arc Spot Welding Joins Aluminum to Other Metals," *Welding Journal* (April 1963).

3. J. T. Biskup, "One Side Welding Is Strong in Japan," *Canadian Machinery and Metal Working* (May 1970).

4. C. A. Butler, F. P. Meister, and M. D. Randall, "Narrow Gap Welding, A Process for All Positions," *Welding Journal* (February 1969).

5. V. Malin, "Monograph on Narrow-Gap Welding Technology," Bulletin 323, Welding Research Council, New York, May 1987.

6. "Underwater Cutting and Welding," Navships 0929-000-08010, U.S. Navy Supervisor of Welding, Naval Ship Systems Command, Washington, D.C.

7. P. T. DeLaune, Jr., "Offshore Structural Repair Using Specification for Underwater Welding, AWS D5.6," *Welding Journal* (February 1987).

8. "Specifications for Underwater Welding," ANSI/AWS D3.6, American Welding Society, Miami, Fla.

9. "Underwater Welding," Proceedings of International Conference, Trondham, Norway, June 27–28, 1983, International Institute of Welding, Pergamon Press, Elmsford, N.Y.

10. B. E. Paton, "Welding in Space," *Welding Engineer* (January 1972).

11. B. Irving, "Electron Beam Welding, Soviet Style: A Front Runner for Space," *Welding Journal* (July 1991).

12. B. Irving, "Joining Experts Meet to Exchange Ideas about Welding in Space," *Welding Journal* (January 1992).

APPENDIX

OUTLINE

A-1 GLOSSARY OF WELDING TERMS

Air*   Slang for oxygen, should not use.

All-Weld-Metal-Test-Specimen*   A test specimen with a reduced section composed wholly of weld metal.

Alternating Current or AC*   Electricity that reverses its direction periodically. For cycle current, the current goes in one direction and then in the other direction 60 times in the same second, so that the current changes its direction 120 times in 1 second.

Ammeter*   An instrument for measuring either direct or alternating electric current (depending on its construction). Its scale is usually graduated in amperes and milliamperes.

Anode*   The positive terminal of an electrical source.

Arc Blow   The deflection of an electric arc from its normal path because of magnetic forces.

Arc Length*   The distance from the end of the electrode to the point where the arc makes contact with the work surface.

Arc Spot Weld   A spot weld made by an arc welding process.

Arc Voltage   The voltage across the welding arc.

Arc Wander*   Wander or drifting of arc in various directions. (*See also* Arc Blow.)

*Asterisks indicate that the term is not the official AWS term.

As-Welded   The condition of weld metal, welded joints, and weldments after welding but prior to any subsequent thermal mechanical or chemical treatments.

Autogenous Weld   A fusion weld made with the addition of filler metal.

Backfire   The momentary recession of the flame into the welding tip or cutting tip, followed by immediate reappearance or complete extinction of the flame.

Backing   A material or device placed against the back side of the joint, or at both sides of a weld in electroslag and electrogas welding, to support and retain molten weld metal. The material may be partially fused or remain unfused during welding and may be either metal or nonmetal.

Backstep Sequence   A longitudinal sequence in which weld passes are made in the direction opposite to the progress of welding.

Base Material   The material to be welded, brazed, soldered, or cut. (*See also* Substrate.)

Bell Hole Welding*   A pipeline term whereby the pipe sections are welded together to the end of the transmission line in position.

Bevel   An angular type of edge preparation.

Blacksmith Welding   *See* the preferred term, Forge Welding.

Bottle*   *See* the preferred term, Cylinder.

Boxing   The continuation of a fillet weld around a corner of a member as an extension of the principal weld.

Braze   A weld produced by heating an assembly to suitable temperatures and by using a filler metal, having liquids above 450° C (842° F) and below the solidus of the base materials. The filler metal is distributed between the closely fitted surfaces of the joint by capillary attraction.

Buckling*   Distortion of sheet metal due to the forces of expansion and contraction caused by the application of heat.

Burner   *See* the preferred term, Oxygen Cutter.

Butt Joint   A joint between two members aligned approximately in the same place.

Butt Weld   An erroneous term for a weld in a butt joint. (*See also* Butt Joint.)

Button Weld*   *See* Arc Spot Weld.

Cap Pass*   A pipeline term that refers to the final or reinforcing pass of the weld joint.

Carbon Steel*   Carbon steel is a term applied to a broad range of material containing: carbon 1.7% max., manganese 1.65% max., silicon 0.60% max. Carbon steel is subdivided as follows:

Low-carbon steels	0.15% C max.
Mild-carbon steels	0.15–0.29% C
Medium-carbon steels	0.30–0.59% C
High-carbon steels	0.60–1.70% C

Cast iron*   A wide variety of iron-base materials containing carbon 1.7 to 4.5%; silicon 0.5 to 3%; phosphorus 0.8% max.; sulfur 0.2% max.; molybdenum, nickel, chromium, and copper can be added to produce alloyed cast irons.

Chamfer   *See* the preferred term, Bevel.

Covered electrode   A composite filler metal electrode consisting of a core of a bare electrode or metal cored electrode to which a covering sufficient to provide a slag layer on the weld metal has been applied. The converging may contain materials providing such functions as shielding from the atmosphere, deoxidation, and arc stabilization, and can serve as a source of metallic additions to the weld.

Crater   A depression at the termination of a weld bead.

Cup   *See* the preferred term, Nozzle.

Cylinder*   A portable container used for transportation and storage of a compressed gas.

Depth of Fusion   The distance that fusion extends into the base metal or previous pass from the surface melted during welding.

Direct Current or DC*   Electric current that flows in only one direction. It is measured by an ammeter.

Double Ending*   A pipeline term meaning welding two lengths of pipes together, usually roll welding in the flat position.

Downhand   *See* the preferred term, Flat Position.

Downhill Welding*   A pipe welding term indicating that the weld progresses from the top of the pipe to the bottom of the pipe. The pipe is not rotated.

Electrode: Tungsten Electrode   A nonfiller metal electrode used in arc welding or cutting, made principally of tungsten.

Elongation*   Extension produced between two gauge marks during a tensile test. Expressed as a percentage of the original gauge length.

FabCO Welding*   Trade name; *see* Flux Cored Arc Welding.

Face of Weld   The exposed surface of a weld on the side from which welding was done.

Filler Bead*   A pipeline term referring to the passes laid over the hot pass but not the next to last or final pass.

Fillet Weld   A weld of approximately triangular cross section joining two surfaces approximately at right angles to each other in a lap joint, T-joint, or corner joint.

Firing Line Welder*   A pipeline—a welder in the hot pass crew.

Flat Position   The welding position used to weld from the upper side of the joint; the face of the weld is approximately horizontal.

Flux   Material used to prevent, dissolve, or facilitate removal of oxides and other undesirable surface substances.

Flux Cored Arc Welding (FCAW)   An arc welding process that produces coalescence of metals by heating them with an arc between a continuous filler metal (consumable) electrode and the work. Shielding is provided by a flux contained within the tubular electrode. Additional shielding may or may not be obtained from an externally supplied gas or gas mixture. (*See also* Flux Cored Electrode.)

Flux Cored Electrode   A composite filler metal electrode consisting of a metal tube or other hollow configuration containing ingredients to provide such functions as shielding atmosphere, deoxidation, arc stabilization, and slag formation. Alloying materials may be included in the core. External shielding may or may not be used.

Forge Welding   A solid-state welding process that produces coalescence of metals by heating them in air in a forge and by applying pressure or blows sufficient to cause permanent deformation at the interface.

Friction Welding   A solid-state welding process that produces coalescence of materials by the heat obtained from a mechanically induced sliding motion between rubbing surfaces. The work parts are held together under pressure.

Furnace Brazing (FB)   A brazing process in which the workpieces are placed in a furnace and heated to the brazing temperature.

Gas Metal Arc Welding (GMAW)   An arc welding process that produces coalescence of metals by heating them with an arc between a continuous filler metal electrode and the workpieces. Shielding is obtained entirely from an externally supplied gas or gas mixture. Some methods of this process are called MIG or CO_2 welding.

Gas Pocket*   Pipeline term for porosity.

Gas Shielded Metal Arc Welding   A general term used to describe gas metal arc welding, gas tungsten arc welding, and flux cored arc welding when gas shielding is employed.

Gas Tungsten Arc Welding (GTAW)   An arc welding process that produces coalescence of metals by heating them with an arc between a tungsten (nonconsumable) electrode and the work. Shielding is obtained from a gas or gas mixture. Pressure may or may not be used, and filler metal may or may not be used. (This process has sometimes been called TIG welding.)

Groove Weld   A weld made in the groove between two members to be joined. The standard types of groove welds are as follows: double-bevel-groove weld, double-flare-bevel-groove weld, double-flare-V-groove weld, double-J-groove weld, double-U-groove weld, double-V-groove weld, single-bevel-groove weld, single-flare-bevel-groove weld, single-flare-V-groove weld, single-J-groove weld, single-U-groove weld, single-V-groove weld, and square-groove weld.

Ground Connection   An electrical connection of the welding machine frame to the earth for safety. (*See also* Workpiece Connection *and* Workpiece Lead.)

Ground Lead   *See* the preferred term, Workpiece Lead.

Heat-Affected Zone That portion of the base metal that has not been melted, but whose mechanical properties or microstructure have been altered by the heat of welding, brazing, soldering, or cutting.

Heliarc* Trade name; *See* Gas Tungsten Arc Welding.

Hollow Bead* Pipeline term for porosity in the root reinforcement.

Horizontal Position Fillet weld: the position in which welding is performed on the upper side of an approximately horizontal surface and against an approximately vertical surface. Groove weld: the position of welding in which the axis of the weld lies in an approximately horizontal plane and the face of the weld lies in an approximately vertical plane.

Hot Pass* A pipeline term that refers to the second pass or the pass over the stringer bead. The hot pass is usually made at high currents to practically remelt the entire stringer bead.

Impact Resistance* Energy absorbed during breakage by impact of specially prepared notched specimen, the result being commonly expressed in foot-pounds.

Induction Brazing A brazing process in which the heat required is obtained from the resistance of the work to induced electric current.

Inertia Friction Welding A variation of friction welding in which the energy required to make the weld is supplied primarily by the stored rotational kinetic energy of the welding machine.

Joint Penetration The depth of weld extends from its face into a joint, exclusive of reinforcement.

Joint Welding Procedure* The materials, detailed methods, and practices employed in the welding of a particular joint.

Kerf The width of the cut produced during a cutting process.

Land A nonstandard term for root face.

Lap Joint A joint between two overlapping members in parallel planes.

Lead Burning An erroneous term used to denote the welding of lead.

Low-Alloy Steel* Low-alloy steels are those containing low percentages of alloying elements.

Melting Rate The weight or length of electrode melted in a unit of time.

Micro-Wire Welding* Trade name; *see* Gas Metal Arc Welding.

MIG Welding* *See* the preferred terms, Gas Metal Arc Welding *and* Flux Cored Arc Welding.

Molten Weld Pool *See* the preferred term, Weld Pool.

Nose* *See* Root Face.

Nozzle A device that directs shielding media.

Off-Center Coating* When flux coating on a covered electrode is thicker on one side than the opposite side.

Open-Circuit Voltage The voltage between the output terminals of the welding machine when no current is flowing in the welding circuit.

Overfill* Excessive reinforcement.

Overhead Position The position in which welding is performed from the underside of the joint.

Overlap The protrusion of weld metal beyond the toe, face, or root of the weld.

Oxygen Cutter One who performs a manual oxygen cutting operation.

Parent Metal *See* the preferred terms, Base Material *and* Substrate.

Pass *See* Weld Pass.

Peening The mechanical working of metals using impact blows.

Penetration *See* the preferred terms, Joint Penetration *and* Root Penetration.

Performance Qualification* Methods, tests, and acceptable standards used to qualify a welding procedure.

Porosity Cavity-type discontinuities formed by gas entrapment during solidification.

Postheating *See* Postweld Heat Treatment.

Postweld Heat Treatment Any heat treatment after welding.

Preheating The application of heat to the base metal immediately before welding, brazing, soldering, or cutting.

Procedure Qualification The demonstration that welds made by a specific procedure can meet prescribed standards.

Procedure Specification* A very complete and formal welding procedure written in accordance with code.

Psi* Pounds per square inch.

Puddle *See* the preferred term, Weld Pool.

Quarter Weld* Pipe welding making the joint in four segments rotating the pipe 90° between each segment.

QWP* Qualified welding procedure.

Radiography* The use of radiant energy in the form of x-rays or gamma rays for the nondestructive examination of metals.

Reduction of Area* The difference between the original cross-sectional area and that of the smallest area at the point of rupture; usually states as a percentage of the original area.

Returning (Boxing)* The practice of continuing a weld around a corner as an extension of the principal weld.

Reverse Polarity A nonstandard term for direct current electrode positive.

Rheostat* A variable resistor that has one fixed terminal and a movable contact (often erroneously referred to as a "two-terminal potentiometer"). Potentiometers may be used as rheostats, but a rheostat cannot be used as a potentiometer, because connections cannot be made to both ends of the resistance element.

Root Face That portion of the groove face adjacent to the root of the joint.

Root Gap *See* the preferred term, Root Opening.

Root of Weld *See* Weld Root.

Root Opening The separation at the joint root between the workpieces.

Root Penetration The depth that a weld extends into the root of a joint measured on the centerline of the root cross section.

Seal Weld Any weld designed primarily to provide a specific degree of tightness against leakage.

Shielded Metal Arc Welding (SMAW) An arc welding process that produces coalescence of metals by heating them with an arc between a covered metal electrode and the work-piece. Shielding is obtained from decomposition of the electrode covering. Pressure is not used and filler metal is obtained from the electrode.

Shoulder   See the preferred term, Root Face.

Silver Soldering   Nonpreferred term used to denote brazing with a silver-base filler metal. (*See also* the preferred terms, Furnace Brazing, Induction Brazing, *and* Torch Brazing.)

Singer*   Electrode holder.

Size of Weld*   Groove weld: the joint penetration (depth of bevel plus the root penetration when specified). The size of a groove weld and its effective throat are one and the same. Fillet weld: for equal-leg fillet welds, the leg lengths of the largest isosceles right triangle that can be inscribed within the fillet weld cross section. For unequal-leg fillet welds, the leg lengths of the largest right triangle that can be inscribed within the fillet weld cross section. When one member makes an angle with the other member greater than 105°, the length (size) is of less significance than the effective throat that is the controlling factor for the strength of a weld.

Slag Inclusion   Nonmetallic solid material entrapped in weld metal or between weld metal and base metal.

Slugging   The act of adding a separate piece or pieces of material in a joint before or during welding that results in a welded joint, not complying with design, drawing, or specification requirements.

Spatter   The metal particles expelled during fusion welding and that do not form a part of the weld.

Squirt Welding*   Semiautomatic and submerged arc welding.

Stick Welding*   Welding using shielded metal arc welding.

Stove Pipe Welding*   A pipeline term whereby each length of pipe is joined to the transmission line in a progressive fashion with each joint made in position.

Straight Polarity   A standard term for direct current electrode negative.

Stress Relief Heat Treatment   Uniform heating of a structure or a portion thereof to a sufficient temperature to relieve the major portion of the residual stresses, followed by uniform cooling.

Stringer Bead   A type of weld bead made without appreciable weaving motion. (*See also* Weave Bead.)

Stripper*   A pipeline term referring to the pass that brings the weld groove flush with the surface of the pipe.

Substrate   Any base material to which a thermal sprayed coating or surfacing weld is applied.

Tack Weld   A weld made to hold parts of a weldment in proper alignment until the final welds are made.

Tensile Strength*   The maximum load per unit of original cross-sectional area obtained before rupture of a tensile specimen. Measured in pounds per square inch.

Theoretical Throat   The distance from the beginning of the root of the joint perpendicular to the hypotenuse of the largest right triangle that can be inscribed within the fillet weld cross section. Actual throat: the shortest distance from the root of a fillet weld to its face. Effective throat: the minimum distance minus any reinforcement from the root of a weld to its face.

TIG Welding*   See Gas Tungsten Arc Welding.

Toe of Weld   See Weld Toe.

Torch Brazing   A brazing process in which the heat required is furnished by a fuel gas flame.

Tungsten Electrode   See Electrode: Tungsten Electrode.

Ultimate Tensile Strength*   The maximum tensile stress that will cause a material to break (usually expressed in pounds per square inch).

Underbead Crack   A crack in the heat-affected zone generally not extending to the surface of the base metal.

Undercut   A groove melted into the base metal adjacent to the toe or root of a weld and left unfilled by weld metal.

Underfill   A depression on the face of the weld or root surface extended below the surface of the adjacent base metal.

Uphill Welding*   A pipe welding term indicating that the welds are made from the bottom of the pipe to the top of the pipe. The pipe is not rotated.

VAE*   Visually acceptable external—inspection of a weld.

Vertical Position   The position of welding in which the axis of the weld is approximately vertical.

Weave Bead   A type of weld bead made with transverse oscillation.

Weaving*   A technique of depositing weld metal in which the electrode is oscillated.

Weld   A localized coalescence of metals or nonmetals produced either by heating the materials to welding temperatures, with or without the application of pressure, or by the application of pressure alone, and with or without the use of filler material.

Weld Metal   That portion of a weld that has been melted during welding.

Weld Pass   A single progression of welding along a joint. The result of a pass is a weld bead or layer.

Weld Pool   The localized volume of molten metal in a weld prior to its solidification as weld metal.

Weld Puddle   A nonstandard term for weld pool.

Weld Root   The points, as shown in cross section, at which the back of the weld intersects the base metal surfaces.

Weld Toe   The junction of the weld face and the base metal.

Welder   One who performs a manual or semiautomatic welding operation. (Sometimes erroneously used to denote a welding machine.)

Welding Ground   See the preferred term, Workpiece connection.

Welding Procedure   The detailed methods and practices including all joint welding procedures involved in the production of a weldment. (*See also* Joint Welding Procedure.)

Welding Procedure Specification (WPS)   A document providing in detail the required variables for specific application to assure repeatability by properly trained welders and welding operators.

Welding Process   A joining process that produces coalescence of materials by heating them to the welding temperature, with or without the application of pressure or by the application of pressure alone, and with or without the use of filler metal. See also the master chart of welding and allied processes.

Welding Rod   A form of welding filler metal, normally packaged in straight lengths, that does not conduct electrical current.

Weldment An assembly whose component parts are joined by welding.

Welder One who performs manual or semiautomatic welding.

Whipping* A term applied to an inward and upward movement of the electrode that is employed in vertical welding to avoid undercut.

Wire Welding* *See* Gas Metal Arc Welding.

Workpiece Connection The connection of the work lead to the work.

Workpiece Lead The electrical conductor between the source of arc welding current and the work.

A-2 ORGANIZATIONS INVOLVED WITH WELDING

AA Aluminum Association, 900 19th St. N.W., Suite 300, Washington, DC 20006. An industry association of producers of aluminum. The association aim is to increase understanding of the aluminum industry, and to provide technical, statistical, and marketing information.

AAR Association of American Railroads, 1920 L Street, N.W., Washington, DC 20036. An industry association of railroads. Among other things, it publishes specifications for rolling stock and welding qualifications.

AASHTO American Association of State Highway Transportation Officials, Suite 341, National Press Building, Washington, DC 20045. An association of state transportation officials. AASHTO issues various standards and specifications.

ABET Accreditation Board for Engineering and Technology, 345 East 47th Street, New York, NY 10017. The official board that accreditates college and university engineering programs.

ABS American Bureau of Shipping, 45 Eisenhower Drive, Paramus, NJ 07653. (201) 368-9110. A nonprofit classification society. Classification is a service for shipowners to establish that the ship has been built to recognized standards. ABS provides rules for building ships and issues approvals for welding filler metals.

AFS American Foundrymen's Society, Golf and Wolf Roads, Des Plaines, IL 60016. A technical society devoted to the advancement of manufacture and utilization of castings through research, education, and dissemination of technology.

AIA Aerospace Industries Association of America, 1725 De Sales Street, N.W., Washington, DC 20036. A national industry association of companies engaged in the research, development, and manufacture of aerospace systems, missiles, and astronautical vehicles, and their propulsion of control units or associated equipment.

AISC American Institute of Steel Construction, No. 1, East Wacker Drive, Suite 3100, Chicago, IL 60601. (312) 670-2400. An industry association of fabricated structural steel producers. AISC provides design information and standards pertaining to structural steel.

AISI American Iron and Steel Institute, 1101 17 Street, N.W., Washington, DC 20036. www.steel.org. An industry association of the iron and steel producers. It provides statistics on steel production and use, and publishes the steel products manuals.

ANSI American National Standards Institute, 1430 Broadway, New York, NY 10018. http://www.ansi.org. ANSI is formerly the United States of America Standards Institute (USASI), formerly the American Standard Association (ASA). ANSI is the U.S. representative to ISO. It is a nonprofit corporation that publishes National Standards in cooperation with technical and engineering societies, trade associations, and government agencies.

API American Petroleum Institute, 1801 K Street, Washington, DC 20006. An association of the petroleum industry, it publishes various standards involved with welding including cross-country pipeline welding, storage tanks, and line pipe.

AREA American Railway Engineering Association, 59 East Van Buren Street, Chicago, IL 60605. An engineering society which, through committees, develops standards applicable to railroads. Several of these involve welding.

ASME American Society of Mechanical Engineers, 345 East 47th Street, New York, NY 10017. (212) 705-7740; http://www.asme.org. An engineering society which, among other things, publishes the boiler and pressure vessel code. Section IX is the welding qualification section of this code.

ASM International American Society for Metals, Materials Park, OH 44073. (216) 338-5151; http://www.asm-intl.org. A technical society that seeks to advance the knowledge of metals and materials, their engineering, design, processing, and fabricating through research, education, and dissemination of information.

ASNT American Society for Nondestructive Testing, 4153 Arlingate Plaza, Caller #28519, Columbus, OH 43228. (614) 274-6003. The purpose of this engineering society is scientific and educational, directed toward the advancement of theory and practice of nondestructive test methods for improved product quality and reliability.

ASQC American Society for Quality Control, Inc., 161 West Wisconsin Avenue, Milwaukee, WI 53203. An engineering society that seeks to create, promote, and stimulate interest in the advancement and diffusion of knowledge of the science of control and its application to the quality of industrial products.

ASTM American Society for Testing and Materials, 1916 Race Street, Philadelphia, PA 19103. http://www.astm.org. A scientific and technical organization for standards, materials, products, and systems. It is the world's largest source of voluntary consensus standards.

AWI American Welding Institute, 10628 Dutchtown Road, Knoxville, TN 37932. (615) 675-2150. A nonprofit development and technology transfer organization devoted to welding.

AWI Australian Welding Institute, Eagle House, 118 Alfred Street, Milson's Point, N.S.W., Australia 2061. The national welding society of Australia.

AWS American Welding Society, 550 N.W. LeJeune Road, P.O. Box 351040, Miami, FL 33126. (305) 443-9353; http://www.amweld.org. A nonprofit technical society organized and founded for the purpose of advancing the art and science of welding. The AWS publishes codes and standards concerning all phases of welding and the *Welding Journal*. www.aws.org

AWWA American Water Works Association, 6666 West Quincy Avenue, Denver, CO 80235. An industry association of water companies and companies serving the water supply industry. It publishes numerous standards, several in cooperation with AWS.

BSI British Standards Institution, 2 Park Street, London, England. A nonprofit concern. The principal object is to coordinate the efforts of producers and users for the improvement, standardization, and simplification of engineering and industrial material.

CDA Copper Development Association, 260 Madison Avenue, New York, NY 10016. (800) 232-3282. A trade association of copper producers. Publishes standards of commercial and copper mill products and standard designations for copper and copper alloys.

CGA Compressed Gas Association, 500 Fifth Avenue, New York, NY 10036. A nonprofit membership association and technical organization interested in both adequate data and sound utilization for gases.

CSA Canadian Standards Association, 178 Rexdale Boulevard, Rexdale, Ontario, Canada M9W 1R3. A National Association of Technical Committees to provide a national standardizing body for Canada. It publishes many standards involving welding.

CWB Canadian Welding Bureau, 254 Merton Street, Toronto, Ontario, Canada M4S 1A9. A division of the CSA, its purpose is to provide the necessary codes and standards covering all phases of welding, and the guidance of fabricators, designers, architects, consulting engineers, and governmental departments.

DVS Deutscher Verband für Schweisstechnik e.V., Aachener Strasse 172, Postfach 27 25, D-4000 Dusseldorf 1. The German Welding Society.

EWI Edison Welding Institute, 1250 Arthur E. Adams Dr., Columbus, OH 43221. (614) 688-5000; http://www.ewi.org. A nonprofit applied engineering center dedicated to welding and related joining technologies.

IEC International Electrotechnical Commission, 3 rue de Varembé (or PO Box 131), 1211 Geneva, 20, Switzerland. International organization that writes specifications for electrical machinery that are adopted as national standards by various countries.

IIW International Institute of Welding. Contact your local welding society. An international society of national associations to promote the development of welding and assist in the international standards for welding in collaboration with the ISO. Contact CSA (Canada) or AWS (US).

ILZRO International Lead Zinc Research Organization, 2525 Meridian Parkway, Durham, NC 27713. (919) 361-4647. International organization that promotes the use of lead and zinc through research and technical information.

IMA International Magnesium Association, 1303 Vincent Place, Suite 1, McLean, VA 22101-3615. (703) 442-8888. International organization of magnesium producers that promotes the use of magnesium. Provides technical information about magnesium.

ISO International Organization for Standardization, Paris, France, is a worldwide federation of national standards institutes. Contact CSA (Canada) or AWS (US).

ITA International Titanium Association, 1871 Folsom St, Suite 100, Boulder, CO 80302-5791. (303) 443-7515; Internet http://www.titanium.org. Formerly the Titanium Development Association. Promotes the use of titanium. Provides technical information concerning welding, and so on.

JIS Japanese Industrial Standards, I-24 Akasaka, 4-chome, Minatoku, Tokyo 107, Japan. The Japanese Standards Association publishes standards including metals, welding filler metals, and so on.

Lloyd's Register of Shipping Trust Corp., Inc. 71 Fenchurch Street, London EC3M 4BS, England. This society was established for the purpose of obtaining for the use of merchants, shipowners, and underwriters a faithful and accurate classification of mercantile shipping. The society approves design, surveys, apparatus, material, and so on.

MCAA Mechanical Contractors Association of America, 5530 Wisconsin Avenue, N.W., Washington, DC 20015. Formerly Heating, Piping and Air-Conditioning Contractors National Association. A trade association of contractors in the piping, heating, and air-conditioning business. It sponsors the National Certified Pipe Welding Bureau.

NBBPVI National Board of Boiler and Pressure Vessel Inspectors, 1055 Crupper Drive, Columbus, OH 43229. An organization of chief boiler inspectors of the states and cities in the U.S. and provinces of Canada. The national board enforces the various sections of the ASME boiler code.

NCPWB National Certified Pipe Welding Bureau, 5530 Wisconsin Avenue, Suite 750, Washington, DC 20015. A division of the Mechanical Contractors Association of America, Inc. Its purpose is to develop and test procedures and, through its local chapters, to establish pools of workers qualified to weld under these procedures.

NEMA National Electrical Manufacturers Association, 1300 N. 17th. St., Rosslyn, VA 22209. (703) 841-3200; http://www.nema.org. An industry association of manufacturers of electrical machinery. Publishes standards and industry statistics including welding.

NFPA National Fire Protection Association, 470 Atlantic Avenue, Boston, MA 02210. An organization dedicated to promoting the science and improving the methods of fire protection. NFPA publishes the *National Electrical Code.*® The code provides safety installation information for welding machines.

NiDI Nickel Development Institute, 214 King Street West, Suite 510, Toronto, M5H 356. Ont. Canada. (416) 591-7999. An organization of nickel producers that provides technical information concerning nickel and metals containing nickel corrosion resistance, welding, and so on.

NWSA National Welding Supply Association, 1900 Arch Street, Philadelphia, PA 19103. (215) 264-3484. An industry association of welding supply distributors.

PFI Pipe Fabrication Institute, 1326 Freeport Road, Pittsburgh, PA 15238. An association of the pipe-fabricating industry.

PLCA Pipe Line Contractors Association, 2800 Republic National Bank Building, Dallas, TX 75201. An industry association of contractors that builds underground pipelines, especially cross-country pipelines.

RWMA Resistance Welder Manufacturers Association, 1900 Arch Street, Philadelphia, PA 19103. www.rwma.org. An association of manufacturers of resistance welding equipment. Establishes standards for welding equipment and procedure information.

SAE Society of Automotive Engineers, Inc., 400 Commonwealth Drive, Warrendale, PA 15086. An engineering society with the objective of promoting the arts, sciences, standards, and engineering practices connected with the design, construction, and utilization of self-propelled mechanisms, prime movers, components thereof, and related equipment.

SFSA Steel Founders' Society of America, 20611 Center Ridge Road, Rocky River, OH 44116. The Steel Founders' Society is an association of companies engaged in the manufacture of steel castings. They publish technical bulletins, the *Steel Castings Handbook* and the *Journal of Steel Castings Research*.

SPFA Steel Plate Fabricators Association, 2400 South Downing Avenue, Westchester, IL 60154. (708) 562-8750. A nonprofit industry association of metal plate fabricators.

TWI The Welding Institute, Abington Hall, Cambridge, England CB1 6AL UK. A professional institute to further the exchange of technical knowledge through meetings, publications, a library and information service, and courses in its School of Welding Technology. Its research division provides research on welding.

UL Underwriters' Laboratories, Inc., 207 East Ohio Street, Chicago, IL 60611. The Underwriters' Laboratories, Inc., is a nonprofit organization that operates laboratories for the examination and testing of devices, systems, and materials. They publish standards for safety for oxyfuel gas torches, regulators, gauges, acetylene generators, transformer-type arc welding machines, and many other items.

WD&F Welding Design and Fabrication, Penton Publishing, 1100 Superior Avenue, Cleveland OH 44114-2543. (216) 696-7000. A monthly magazine devoted to welding and associated processes. The January issue each year contains a buyer's guide of welding equipment, materials, and supplies.

WRC Welding Research Council, 345 East 47th Street, New York, NY 10017. (212) 705-7080. A nonprofit association organized to provide a mechanism for conducting cooperative research work in the welding field.

A-3 COMPUTER SOFTWARE PROGRAMS

In order for a computer to perform any calculations or do any work, it must be programmed. To accomplish this a computer must be loaded with a software program. This is usually accomplished by loading a CD-ROM into the computer. CD stands for compact disc and ROM means read only memory. The disc programs the computer memory to produce the type of data requested. CD-ROMs can be very expensive and complex. Sometimes more than one CD may be required. The programs must be compatible with the hardware, with the computer, and with each other. Once the software is installed and operating properly, it is able to do the job for which the computer was designed. Software programs that provide a multitude of jobs are available from many different sources. Different software packages can be designed for a variety of tasks that relate to welding.

Some of the many different software programs of interest to the welding industry are:

Design Programs
Structural Design—related to AWS structural code
Pressure Vessel Design—related to ASME Section IX
Pressure Piping Design—related to ASME codes
Design to Transfer Loads Effectively
Weld Details for All Joints
Base Metal Specifications for All Pieces

Design Analysis Software Programs

There are many software programs that allow the designer to analyze the design. These are called Finite Element Analysis (FEA) programs. Some of them look at the design from all points of view, to find a better way of designing the part, eliminate pieces, reduce the number of thicknesses of material, and so on. Some of the programs impose theoretical stresses to reveal stress concentrations and overloaded joints. Other programs may apply hydraulic loads or fatigue or repeated loads. Many different programs can be applied to reveal any welding design weaknesses to obtain the optimum design.

Computer-Assisted Manufacturing (CAM)

There are many programs available that help provide input into different manufacturing tools. The program for nesting parts for automatic shape cutting is very popular. These programs go so far as to set conditions such as travel speeds and gas pressures based on type and thickness of material being cut. Some programs establish the welding parameters for automatic welding machines, and others will program welding robots.

Expert Systems

A variety of expert systems are available to help establish welding parameters for different situations. These are based on the needs for the particular weld joint. The information is stored to establish the parameters for making such a weld.

Calculations

There are numerous programs that provide calculations for determining the weld metal required for a particular joint or total weldment. With the input of cost information they can actually calculate the cost for each joint or weldment.

Welding Procedure Programs

These software packages provide methods for making different welds and also record the different welding procedures available as well as their requirements and their tests. Welding procedures can be completely documented and stored according to a type of metal and the particular code involved. These are then matched to individual welders who have taken qualification tests at different times. This eliminates the duplication of procedures and keeps track of welders and when they should update their test program.

Available Programs

There are many programs available from the different companies that produce welding software. Some of the more popular welding programs are as follows:

Arc Linc	Weldbest
Arc Works	Weldcost
Auto CAD	Welderqual
FF Weld	Weldgen
Filler Metal Data Manager	Welding Co-ordinator
NDT Spec	Welding Pro-writer
Turbo	Weldollar
Weld Spec	Weldpec Plus
Weld-it	

Undoubtedly there are other programs not mentioned here that could be extremely useful and time saving for the average welding operation.

Companies That Produce Welding Software in the U.S.

American Welding Society 550 N.W. LeJeune Road, Miami, FL 33126. (800) 443-9353; www.aws.org

Applied Production, Inc. 200 Technical Center Drive, Suite 202, Milford Center, OH 45150. (513) 831-8800; www.applied-production.com

Baysinger Engineered Software Technology 11660 N. Placita Marcela, Marana, AZ 85653. (520) 682-6383

Canadian Welding Bureau 7250 West Credit Avenue, Mississauga, ON L5N 5N1, Canada. (905) 542-1312; fax (905) 542-1318; www.cwbgroup.com

Codeware 11221 Richmond, Suite C-103, Houston, TX 77082. (713) 497-5705; www.codeware.com

Computer Engineering, Inc. P.O. Box 1657, Blue Springs, MO 64013. (800) 473-1976; fax (816) 228-0680; www.computereng.com

Computers Unlimited 2407 Montana Avenue, Billings, MT 59101-2336. (406) 255-9500; fax (406) 255-9595

C-spec (TWI) 1855 Gateway Boulevard, Suite 700, Concord, CA 94520. (925) 930-8223, fax (925) 930-8223; www.cspec.com

Deneb Robotics, Inc. 5500 New King Street, Troy, MI 48098. (248) 267-9696; www.deneb.com

Edison Welding Institute 1250 Arthur E. Adams Dr., Columbus, OH 43221. (614) 486-9400; fax (614) 486-9258; www.ewi.org

Engineering Systems Int'l. Corporation 570 Kirts Boulevard, Suite 231, Troy, MI 48084. (248) 362-4466; www.esi.fr

The Lincoln Electric Company, Cleveland, OH 44117-1199. (216) 481-8100; fax (216) 486-1751; www.lincolnelectric.com

Measurement Masters, Inc. 711 West 17th Street, Building E-11, Costa Mesa, CA 92627. (714) 631-6950.

Microcomputer Technology Consultants Ltd. P.O. Box 467, Bewley Building, Suite 342, Lockport, NY 14095-0467. (716) 433-7722; fax (716) 433-1554.

Miller Electric Manufacturing Co. 1535 West Spencer Street, Appleton, WI 45912-1079. (404) 735-4055; www.millerwelds.com

Penton Education Division 1100 Superior Avenue, Cleveland, OH 44114. (800) 321-7003; fax (216) 696-4369; www.penton.com

SDRC, 2000 Eastman Drive, Milford, OH 45150. (513) 576-2400; www.I-DEAS.com

Servo-Robot Inc. 1380 Graham Bell, Boucherville, PQ J4B 6H5, Canada. (514) 655-4223; fax (514) 655-4963

Weaver Engineering (Weldment Design Engineering) 1219 West Gate Avenue North, #210, Seattle, WA 98109. (206) 352-8027; www.weavereng.com

For more information on this subject contact Chris Pollack at AWS headquarters. He is the Secretary of the Committee on Computerization of Welding Information.

A-4 INFORMATION SOURCES USING A COMPUTER

Recent innovations have opened up a vast storehouse of information for the welding industry. Getting on the Internet is like tapping into the world's biggest library, the world's biggest catalog, and the world's largest collection of technical information. The Internet is a vast store of information available 24 hours a day anywhere in the world. To take advantage of this information it is necessary to have a computer that is plugged into a telephone line and to know how to use the Internet system. The Internet is a network of millions of computers linked by the world's telecommunication system. The following is a brief description of how it can be used to connect you to the worldwide source of welding information.

Those computers that are set up to supply information are called servers. Servers connect to each other to form networks. The "client" is a computer that allows connections to information services over the network. The World Wide Web (www) is a collection of protocols (rules) and standards used to access the information available. Each Web site has a unique address, such as http://www.AWS.org, which is the address of the American Welding Society. The specific address is termed Uniform Resource Locator (URL). It always starts out as http://www. plus the URL. For example http://www.welding.org is the Web site or address of the Hobart Institute of Welding Technology. This site offers extensive information regarding its welding skill training classes, technical programs, certification, and more.

An extension to the URL provides additional information. For example ".edu" indicates an educational institution, ".org" stands for a nonprofit organization, ".gov" stands for a government department, and ".com" is a commercial site. The www protocol covers most types of information, including multimedia, and the method of communication. To enable a user (you) to gain access to the Internet, an account is required with a service provider. When you log onto the provider's server, their software will allow you to browse the Internet, accessing and retrieving information from various servers at Web sites.

The amount of information on the Web is virtually limitless. Finding what you are seeking can be difficult and time consuming. Search engines create a Web site database of key words that is indexed and continually updated. There are many search engines. Most provide service free. Some popular ones are:

> www.yahoo.com
> www.lycos.com
> www.infoseek.com
> www.webmaster@aws.org
> www.altavista.digital.com
> www.excite.com
> www.hotbot.com
> www.google.com

Once in the search engine site, type in the keyword or name of information you are seeking. With a little experience you will soon be able to find much more information on welding than you require.

The following search categories are used in the welding industry:

1. Associations, societies, and institutes
2. Education and training
3. Welding publications
4. Materials-joining software
5. Materials-joining standards
6. Commercial companies that sell welding equipment and services

Under category 1, the following are some of those listed:

> AWS.org—American Welding Society
> ASTM.org—American Society for Testing and Materials
> ANSI.org—National Standards Institute

NWSA.com—National Welding Supply Association
EWI.org—Edison Welding Institute
iiw-iis.org—International Institute of Welding
TWI.co.uk—The Welding Institute (United Kingdom)

See Appendix A–2 for names of other organizations and their Internet addresses. You can also use a search engine such as www.ewi.org/resources, to locate organizations.

Under category 2, education and training, use a search engine to find the institution you are interested in.

Under category 3 are the following published in the U.S. (and there are others):

> *Welding Jounral*—www.aws.org
> *Welding Design and Fabrication*—www.penton.com
> *The American Welder*—www.aws.org
> *The Fabricator*—www.fmametalfab.org
> *Gases and Welding Distributor*—www.penton.com
> *Practical Welding Today*—www.fmametalfab.org

The list of materials-joining software, category 4, is similar to the companies shown in Appendix A–3. They should all be checked in order to make your best selection. Again, a search engine will be helpful.

Category 5 is a list of standards produced by different engineering and technical societies relating to welding. A handy search engine is available here: cssinfo.com. Documents can be ordered on-line in print format, and many are available for downloading.

Category 6 is a listing of companies with Web sites that provide a catalog of the products of that particular company. In many cases the URL is www., followed by the company name, followed by .com. These addresses are for some of the larger welding companies:

> www.esab.com
> www.hobartwelders.com
> www.jetline.com
> www.lincolnelectric.com
> www.millerwelds.com
> www.nationalstandard.com
> www.thermadyne.com
> www.webmaster@aws.org

Use one of the search engines and click on "general sites" to find welding machines, sites such as manufacturers.net and suppliersonline.com, or whatever you are looking for.

A-5 CONVERSION INFORMATION

Length, or Distance Speed Conversion
Approximate Conversion

Scales (paired Inch | mm etc.):

Thickness — Inch | mm
Wire Ø — Inch | mm
Travel — Inch/Min | mm/Min
Distance — Miles | km

Exact Conversion

For use with electronic calculators. First enter the conversion constant number. Press the ⊠ button. Enter the known quantity of the dimension. Press the ⊟. The desired value of the dimension will appear on the display.

25.40 ⊠ __ in   ⊟ __ mm
304.8 ⊠ __ ft   ⊟ __ mm
.0393 ⊠ __ mm   ⊟ __ in
.00328 ⊠ __ mm   ⊟ __ ft

.0621 ⊠ __ km   ⊟ __ mi
1.609 ⊠ __ mi   ⊟ __ km

.4233 ⊠ __ in/min ⊟ __ mm
2.362 ⊠ __ mm/s   ⊟ __ in/min

Flow Rate–Liquid Measure
Approximate Conversion

Gas Flow — Cu. ft/Hr. | Liters/Min
Liquid Measure — Gal. | Liters

Exact Conversion

For use with electronic calculators. First enter the conversion constant number. Press the ⊠ button. Enter the known quantity of the dimension. Press the ⊟. The desired value of the dimension will appear on the display.

0.4719 ⊠ __ cu ft/hr   ⊟ __ L/min
2.119 ⊠ __ L/min   ⊟ __ cu ft/hr

3.785 ⊠ __ gal/min   ⊟ __ L/min
0.264 ⊠ __ L/min   ⊟ __ gal/min

645.2 ⊠ __ in^2 ⊟ __ mm^2
0.00155 ⊠ __ mm^2 ⊟ __ in^2

Weight–Pressure–Load
Approximate Conversion

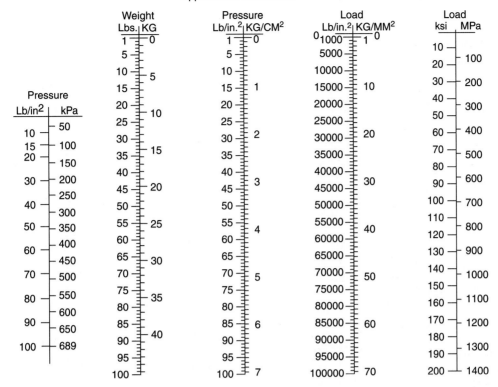

Pressure

Lb/in²	kPa
10	50
15	100
20	150
30	200
	250
40	300
50	350
	400
60	450
70	500
	550
80	600
90	650
100	689

Weight — Lbs. | KG
0–1 ... 100 / 0 ... 40

Pressure — Lb/in² | KG/CM²
0–1 ... 100 / 0 ... 7

Load — Lb/in² | KG/MM²
0 1000 ... 100000 / 1 ... 70

Load — ksi | MPa
10 ... 200 / 100 ... 1400

Exact Conversion

For use with electronic calculators. First enter the conversion constant number. Press the ⊠ button. Enter the known quantity of the dimension. Press the ⊟. The desired value of the dimension will appear on the display.

000703	⊠ __ Lb/in²	⊟ __	kg/mm²
6.8947	⊠ __ Lb/in²	⊟ __	Kilopascal (kPa)
0.006895	⊠ __ Lb/in²	⊟ __	Megapascal (MPa)
0.07030	⊠ __ Lb/in²	⊟ __	kg/cm²
14.2234	⊠ __ kg/cm²	⊟ __	Lb/in²
1422.34	⊠ __ kg/mm²	⊟ __	Lb/in² (PSI)
	⊠ __ kg/mm²	⊟ __	Pascal (Pa)
	⊠ __ kg/mm²	⊟ __	Kilopascal (kPa)
0.00145	⊠ __ kPa	⊟ __	Lb/in² (PSI)
0.145	⊠ __ kPa	⊟ __	Lb/in² (PSI)
	⊠ __ kPa	⊟ __	kg/mm²
	⊠ __ Pa	⊟ __	kg/mm²
4.448	⊠ __ lb	⊟ __	Newton (N)
9.807	⊠ __ kg	⊟ __	Newton (N)
.2248	⊠ __ N	⊟ __	Lb
.1009	⊠ __ N	⊟ __	kg
0.4536	⊠ __ Lb	⊟ __	kg
2.205	⊠ __ kg	⊟ __	Lb

Metric Prefixes

Exponential Expression	Multiplication Factor	Prefix	Symbol
10^{12}	1000000000000	tera	T
10^{9}	1000000000	giga	G
10^{6}	1000000	mega	M
10^{3}	1000	kilo	k
10^{2}	100	hecto*	h
10	10	deka*	da
10^{-1}	......................................0.1	deci*	d
10^{-2}	....................................0.01	centi*	c
10^{-3}	..................................0.001	milli	m
10^{-6}	.............................0.000001	micro	μ
10^{-9}	..........................0.000000001	nano	n
10^{-12}	....................0.000000000001	pico	p
10^{-15}	..............0.000000000000001	femto	f
10^{-18}	.........0.000000000000000001	atto	a

*Rarely Used

Temperature-Impact Values
Approximate Conversion

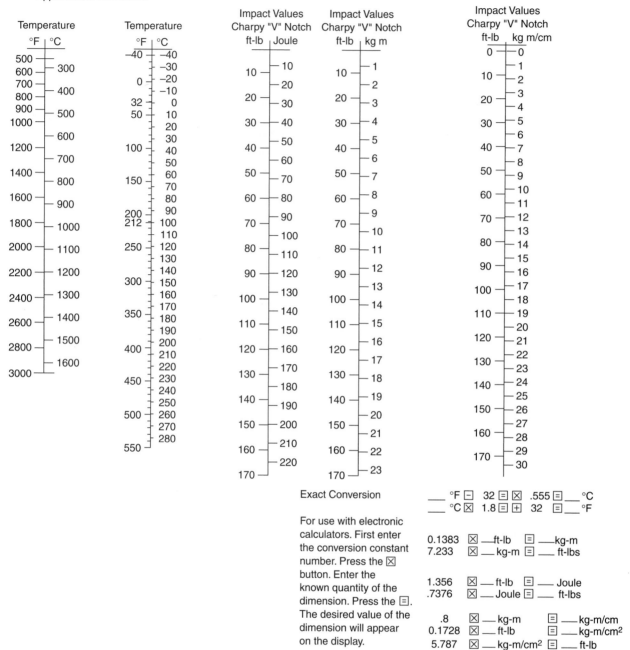

Exact Conversion

| | °F | − | 32 | = | × | .555 | = | ___ °C |
| | °C | × | 1.8 | = | + | 32 | = | ___ °F |

For use with electronic
calculators. First enter
the conversion constant
number. Press the ×
button. Enter the
known quantity of the
dimension. Press the =.
The desired value of the
dimension will appear
on the display.

0.1383 × ___ft-lb = ___kg-m
7.233 × ___ kg-m = ___ ft-lbs

1.356 × ___ ft-lb = ___ Joule
.7376 × ___ Joule = ___ ft-lbs

.8   × ___ kg-m   = ___ kg-m/cm
0.1728 × ___ ft-lb   = ___ kg-m/cm²
5.787 × ___ kg-m/cm² = ___ ft-lb

A-6 WEIGHTS AND MEASURES

Impact Values. Sometimes metric values are kg-m/cm^2; if so, multiply by 0.8 (area under notch).

Foot-Pounds	Kilogram-Meters	Joules	Foot-Pounds	Kilogram-Meters	Joules	Foot-Pounds	Kilogram-Meters	Joules
1	0.14	1.36	35	4.84	47.46	69	9.54	93.56
2	0.28	2.71	36	4.98	48.82	70	9.68	94.92
3	0.42	4.07	37	5.12	50.17	71	9.82	96.28
4	0.55	5.42	38	5.25	51.53	72	9.95	97.63
5	0.69	6.78	39	5.39	52.88	73	10.09	98.99
6	0.83	8.14	40	5.53	54.24	74	10.23	100.34
7	0.97	9.49	41	5.67	55.60	75	10.37	101.70
8	1.11	10.85	42	5.81	56.95	76	10.51	103.06
9	1.24	12.20	43	5.95	58.31	77	10.65	104.41
10	1.38	13.56	44	6.08	59.66	78	10.78	105.77
11	1.52	14.92	45	6.22	61.02	79	10.92	107.12
12	1.66	16.27	46	6.36	62.38	80	11.06	108.48
13	1.80	17.63	47	6.50	63.73	81	11.20	109.84
14	1.94	18.98	48	6.64	65.09	82	11.34	111.19
15	2.07	20.34	49	6.78	66.44	83	11.48	112.55
16	2.21	21.70	50	6.91	67.80	84	11.61	113.90
17	2.35	23.05	51	7.05	69.16	85	11.75	115.26
18	2.49	24.41	52	7.19	70.51	86	11.89	116.62
19	2.63	25.76	53	7.33	71.87	87	12.03	117.97
20	2.77	27.12	54	7.47	73.22	88	12.17	119.33
21	2.90	28.48	55	7.60	74.58	89	12.31	120.68
22	3.04	29.83	56	7.74	75.94	90	12.44	122.04
23	3.18	31.19	57	7.88	77.29	91	12.58	123.40
24	3.32	32.54	58	8.02	78.65	92	12.72	124.75
25	3.46	33.90	59	8.16	80.00	93	12.86	126.11
26	3.60	35.26	60	8.30	81.36	94	13.00	127.46
27	3.73	36.61	61	8.43	82.72	95	13.13	128.82
28	3.87	37.97	62	8.57	84.07	96	13.27	130.18
29	4.01	39.32	63	8.71	85.43	97	13.41	131.53
30	4.15	40.68	64	8.85	86.78	98	13.55	132.89
31	4.29	42.04	65	8.99	88.14	99	13.69	134.24
32	4.42	43.39	66	9.13	89.50	100	13.83	135.60
33	4.56	44.75	67	9.26	90.85			
34	4.70	46.10	68	9.40	92.21			

Wire Gauge Diameter

U.S. Steel Wire Gauge No.	Diameter		U.S. Steel Wire Gauge No.	Diameter		U.S. Steel Wire Gauge No.	Diameter	
	in.	mm		in.	mm		in.	mm
7/0s	0.4900	12.447	11	0.1205	3.0607	28	0.0162	0.4115
6/0s	0.4615	11.7221	12	0.1055	2.6797	29	0.0150	0.381
5/0s	0.4305	10.9347	13	0.0915	2.3241	30	0.0140	0.3556
4/0s	0.3938	10.0025	14	0.0800	2.032	31	0.0132	0.3353
3/0s	0.3625	9.2075	15	0.0720	1.8389	32	0.0128	0.3251
2/0s	0.3310	8.4074	16	0.0625	1.5875	33	0.0118	0.2997
0	0.3065	7.7851	17	0.0540	1.3716	34	0.0104	0.2642
1	0.2830	7.1882	18	0.0475	1.2065	35	0.0095	0.2413
2	0.2625	6.6675	19	0.0410	1.0414	36	0.0090	0.2286
3	0.2437	6.1899	20	0.0348	0.8839	37	0.0085	0.2159
4	0.2253	5.7226	21	0.0317	0.8052	38	0.0080	0.2032
5	0.2070	5.2578	22	0.0286	0.7264	39	0.0075	0.1905
6	0.1920	4.8768	23	0.0258	0.6553	40	0.0070	0.1778
7	0.1770	4.4958	24	0.0230	0.5842	41	0.0066	0.1678
8	0.1620	4.1148	25	0.0204	0.5182	42	0.0062	0.1575
9	0.1483	3.7668	26	0.0181	0.4597	43	0.0060	0.1524
10	0.1350	3.429	27	0.0173	0.4394	44	0.0058	0.1473

Metric Conversion

Inch: Fraction	Inch: Decimal	Millimeter	Inch: Fraction	Inch: Decimal	Millimeter	Inch: Fraction	Inch: Decimal	Millimeter
$\frac{1}{64}$	0.0158	0.3969	$\frac{11}{32}$	0.3437	8.7312	$\frac{43}{64}$	0.6719	17.0656
$\frac{1}{32}$	0.0312	0.7937	$\frac{23}{64}$	0.3594	9.1281	$\frac{11}{16}$	0.6875	17.4625
$\frac{3}{64}$	0.0469	1.1906	$\frac{3}{8}$	0.375	9.525	$\frac{45}{64}$	0.7031	17.8594
$\frac{1}{16}$	0.0625	1.5875	$\frac{25}{64}$	0.3906	9.9219	$\frac{23}{32}$	0.7187	18.2562
$\frac{5}{64}$	0.0781	1.9844	$\frac{13}{32}$	0.4062	10.3187	$\frac{47}{64}$	0.7344	18.6532
$\frac{3}{32}$	0.0937	2.3812	$\frac{27}{64}$	0.4219	10.7156	$\frac{3}{4}$	0.750	19.050
$\frac{7}{64}$	0.1094	2.7781	$\frac{7}{16}$	0.4375	11.1125	$\frac{49}{64}$	0.7656	19.4469
$\frac{1}{8}$	0.125	3.175	$\frac{29}{64}$	0.4531	11.5094	$\frac{25}{32}$	0.7812	19.8433
$\frac{9}{64}$	0.1406	3.5719	$\frac{15}{32}$	0.4687	11.9062	$\frac{51}{64}$	0.7969	20.2402
$\frac{5}{32}$	0.1562	3.9687	$\frac{31}{64}$	0.4844	12.3031	$\frac{13}{16}$	0.8125	20.6375
$\frac{11}{64}$	0.1719	4.3656	$\frac{1}{2}$	0.500	12.700	$\frac{53}{64}$	0.8281	21.0344
$\frac{3}{16}$	0.1875	4.7625	$\frac{33}{64}$	0.5156	13.0968	$\frac{27}{32}$	0.8437	21.4312
$\frac{13}{64}$	0.2031	5.1594	$\frac{17}{32}$	0.5312	13.4937	$\frac{55}{64}$	0.8594	21.8281
$\frac{7}{32}$	0.2187	5.5562	$\frac{35}{64}$	0.5469	13.8906	$\frac{7}{8}$	0.875	22.2250
$\frac{15}{64}$	0.2344	5.9531	$\frac{9}{16}$	0.5625	14.2875	$\frac{57}{64}$	0.8906	22.6219
$\frac{1}{4}$	0.25	6.35	$\frac{37}{64}$	0.5781	14.6844	$\frac{29}{32}$	0.9062	23.0187
$\frac{17}{64}$	0.2656	6.7469	$\frac{19}{32}$	0.5937	15.0812	$\frac{59}{64}$	0.9219	23.4156
$\frac{9}{32}$	0.2812	7.1437	$\frac{39}{64}$	0.6094	15.4781	$\frac{15}{16}$	0.9375	23.8125
$\frac{19}{64}$	0.2969	7.5406	$\frac{5}{8}$	0.625	15.875	$\frac{61}{64}$	0.9531	24.2094
$\frac{5}{16}$	0.3125	7.9375	$\frac{41}{64}$	0.6406	16.2719	$\frac{31}{32}$	0.9687	24.6062
$\frac{21}{64}$	0.3281	8.3344	$\frac{21}{32}$	0.6562	16.6687	$\frac{63}{64}$	0.9844	25.0031

Areas
    Parallelogram = base × altitude.
    Triangle = half base × altitude.
    Trapezoid = half the sum of the two parallel sides × the perpendicular distance between them.
    Regular polygon = half of perimeter × the perpendicular distance from the center to any one side.
    Circle = square of the diameter × 0.7854.
    Sector of circle = number of degrees in arc × square of radius × 0.008727.
    Segment of circle = area of sector with same arc minus area of triangle formed by radii of the arc and chord of the segment.
    Octagon = square of diameter of inscribed circle × 0.828.
    Hexagon = square of diameter of inscribed circle × 0.866.
    Sphere = area of its great circle × 4; or square of diameter × 3.1416 (π).

Volumes
    Prism = area of base × altitude.
    Wedge = length of edge plus twice length of base × one-sixth of the product of the height of the wedge and the breadth of its base.
    Cylinder = area of base × altitude.
    Cone = area of base × one-third of altitude.
    Sphere = cube of diameter × 0.5236.

Miscellaneous
    Diameter of circle = circumference × 0.31831.
    Circumference of circle = diameter × 3.1416 (π).

INDEX